HANDBOOK OF STOCHASTIC ANALYSIS AND APPLICATIONS

STATISTICS: Textbooks and Monographs

D. B. Owen, Founding Editor, 1972–1991

1. The Generalized Jackknife Statistic, *H. L. Gray and W. R. Schucany*
2. Multivariate Analysis, *Anant M. Kshirsagar*
3. Statistics and Society, *Walter T. Federer*
4. Multivariate Analysis: A Selected and Abstracted Bibliography, 1957–1972, *Kocherlakota Subrahmaniam and Kathleen Subrahmaniam*
5. Design of Experiments: A Realistic Approach, *Virgil L. Anderson and Robert A. McLean*
6. Statistical and Mathematical Aspects of Pollution Problems, *John W. Pratt*
7. Introduction to Probability and Statistics (in two parts), Part I: Probability; Part II: Statistics, *Narayan C. Giri*
8. Statistical Theory of the Analysis of Experimental Designs, *J. Ogawa*
9. Statistical Techniques in Simulation (in two parts), *Jack P. C. Kleijnen*
10. Data Quality Control and Editing, *Joseph I. Naus*
11. Cost of Living Index Numbers: Practice, Precision, and Theory, *Kali S. Banerjee*
12. Weighing Designs: For Chemistry, Medicine, Economics, Operations Research, Statistics, *Kali S. Banerjee*
13. The Search for Oil: Some Statistical Methods and Techniques, *edited by D. B. Owen*
14. Sample Size Choice: Charts for Experiments with Linear Models, *Robert E. Odeh and Martin Fox*
15. Statistical Methods for Engineers and Scientists, *Robert M. Bethea, Benjamin S. Duran, and Thomas L. Boullion*
16. Statistical Quality Control Methods, *Irving W. Burr*
17. On the History of Statistics and Probability, *edited by D. B. Owen*
18. Econometrics, *Peter Schmidt*
19. Sufficient Statistics: Selected Contributions, *Vasant S. Huzurbazar (edited by Anant M. Kshirsagar)*
20. Handbook of Statistical Distributions, *Jagdish K. Patel, C. H. Kapadia, and D. B. Owen*
21. Case Studies in Sample Design, *A. C. Rosander*
22. Pocket Book of Statistical Tables, *compiled by R. E. Odeh, D. B. Owen, Z. W. Birnbaum, and L. Fisher*
23. The Information in Contingency Tables, *D. V. Gokhale and Solomon Kullback*
24. Statistical Analysis of Reliability and Life-Testing Models: Theory and Methods, *Lee J. Bain*
25. Elementary Statistical Quality Control, *Irving W. Burr*
26. An Introduction to Probability and Statistics Using BASIC, *Richard A. Groeneveld*
27. Basic Applied Statistics, *B. L. Raktoe and J. J. Hubert*
28. A Primer in Probability, *Kathleen Subrahmaniam*
29. Random Processes: A First Look, *R. Syski*
30. Regression Methods: A Tool for Data Analysis, *Rudolf J. Freund and Paul D. Minton*
31. Randomization Tests, *Eugene S. Edgington*
32. Tables for Normal Tolerance Limits, Sampling Plans and Screening, *Robert E. Odeh and D. B. Owen*
33. Statistical Computing, *William J. Kennedy, Jr., and James E. Gentle*
34. Regression Analysis and Its Application: A Data-Oriented Approach, *Richard F. Gunst and Robert L. Mason*
35. Scientific Strategies to Save Your Life, *I. D. J. Bross*
36. Statistics in the Pharmaceutical Industry, *edited by C. Ralph Buncher and Jia-Yeong Tsay*
37. Sampling from a Finite Population, *J. Hajek*
38. Statistical Modeling Techniques, *S. S. Shapiro and A. J. Gross*
39. Statistical Theory and Inference in Research, *T. A. Bancroft and C.-P. Han*
40. Handbook of the Normal Distribution, *Jagdish K. Patel and Campbell B. Read*
41. Recent Advances in Regression Methods, *Hrishikesh D. Vinod and Aman Ullah*
42. Acceptance Sampling in Quality Control, *Edward G. Schilling*
43. The Randomized Clinical Trial and Therapeutic Decisions, *edited by Niels Tygstrup, John M Lachin, and Erik Juhl*
44. Regression Analysis of Survival Data in Cancer Chemotherapy, *Walter H. Carter, Jr., Galen L. Wampler, and Donald M. Stablein*
45. A Course in Linear Models, *Anant M. Kshirsagar*
46. Clinical Trials: Issues and Approaches, *edited by Stanley H. Shapiro and Thomas H. Louis*

47. Statistical Analysis of DNA Sequence Data, *edited by B. S. Weir*
48. Nonlinear Regression Modeling: A Unified Practical Approach, *David A. Ratkowsky*
49. Attribute Sampling Plans, Tables of Tests and Confidence Limits for Proportions, *Robert E. Odeh and D. B. Owen*
50. Experimental Design, Statistical Models, and Genetic Statistics, *edited by Klaus Hinkelmann*
51. Statistical Methods for Cancer Studies, *edited by Richard G. Cornell*
52. Practical Statistical Sampling for Auditors, *Arthur J. Wilburn*
53. Statistical Methods for Cancer Studies, *edited by Edward J. Wegman and James G. Smith*
54. Self-Organizing Methods in Modeling: GMDH Type Algorithms, *edited by Stanley J. Farlow*
55. Applied Factorial and Fractional Designs, *Robert A. McLean and Virgil L. Anderson*
56. Design of Experiments: Ranking and Selection, *edited by Thomas J. Santner and Ajit C. Tamhane*
57. Statistical Methods for Engineers and Scientists: Second Edition, Revised and Expanded, *Robert M. Bethea, Benjamin S. Duran, and Thomas L. Boullion*
58. Ensemble Modeling: Inference from Small-Scale Properties to Large-Scale Systems, *Alan E. Gelfand and Crayton C. Walker*
59. Computer Modeling for Business and Industry, *Bruce L. Bowerman and Richard T. O'Connell*
60. Bayesian Analysis of Linear Models, *Lyle D. Broemeling*
61. Methodological Issues for Health Care Surveys, *Brenda Cox and Steven Cohen*
62. Applied Regression Analysis and Experimental Design, *Richard J. Brook and Gregory C. Arnold*
63. Statpal: A Statistical Package for Microcomputers—PC-DOS Version for the IBM PC and Compatibles, *Bruce J. Chalmer and David G. Whitmore*
64. Statpal: A Statistical Package for Microcomputers—Apple Version for the II, II+, and IIe, *David G. Whitmore and Bruce J. Chalmer*
65. Nonparametric Statistical Inference: Second Edition, Revised and Expanded, *Jean Dickinson Gibbons*
66. Design and Analysis of Experiments, *Roger G. Petersen*
67. Statistical Methods for Pharmaceutical Research Planning, *Sten W. Bergman and John C. Gittins*
68. Goodness-of-Fit Techniques, *edited by Ralph B. D'Agostino and Michael A. Stephens*
69. Statistical Methods in Discrimination Litigation, *edited by D. H. Kaye and Mikel Aickin*
70. Truncated and Censored Samples from Normal Populations, *Helmut Schneider*
71. Robust Inference, *M. L. Tiku, W. Y. Tan, and N. Balakrishnan*
72. Statistical Image Processing and Graphics, *edited by Edward J. Wegman and Douglas J. DePriest*
73. Assignment Methods in Combinatorial Data Analysis, *Lawrence J. Hubert*
74. Econometrics and Structural Change, *Lyle D. Broemeling and Hiroki Tsurumi*
75. Multivariate Interpretation of Clinical Laboratory Data, *Adelin Albert and Eugene K. Harris*
76. Statistical Tools for Simulation Practitioners, *Jack P. C. Kleijnen*
77. Randomization Tests: Second Edition, *Eugene S. Edgington*
78. A Folio of Distributions: A Collection of Theoretical Quantile-Quantile Plots, *Edward B. Fowlkes*
79. Applied Categorical Data Analysis, *Daniel H. Freeman, Jr.*
80. Seemingly Unrelated Regression Equations Models: Estimation and Inference, *Virendra K. Srivastava and David E. A. Giles*
81. Response Surfaces: Designs and Analyses, *Andre I. Khuri and John A. Cornell*
82. Nonlinear Parameter Estimation: An Integrated System in BASIC, *John C. Nash and Mary Walker-Smith*
83. Cancer Modeling, *edited by James R. Thompson and Barry W. Brown*
84. Mixture Models: Inference and Applications to Clustering, *Geoffrey J. McLachlan and Kaye E. Basford*
85. Randomized Response: Theory and Techniques, *Arijit Chaudhuri and Rahul Mukerjee*
86. Biopharmaceutical Statistics for Drug Development, *edited by Karl E. Peace*
87. Parts per Million Values for Estimating Quality Levels, *Robert E. Odeh and D. B. Owen*
88. Lognormal Distributions: Theory and Applications, *edited by Edwin L. Crow and Kunio Shimizu*
89. Properties of Estimators for the Gamma Distribution, *K. O. Bowman and L. R. Shenton*
90. Spline Smoothing and Nonparametric Regression, *Randall L. Eubank*
91. Linear Least Squares Computations, *R. W. Farebrother*
92. Exploring Statistics, *Damaraju Raghavarao*
93. Applied Time Series Analysis for Business and Economic Forecasting, *Sufi M. Nazem*
94. Bayesian Analysis of Time Series and Dynamic Models, *edited by James C. Spall*
95. The Inverse Gaussian Distribution: Theory, Methodology, and Applications, *Raj S. Chhikara and J. Leroy Folks*
96. Parameter Estimation in Reliability and Life Span Models, *A. Clifford Cohen and Betty Jones Whitten*
97. Pooled Cross-Sectional and Time Series Data Analysis, *Terry E. Dielman*

98. Random Processes: A First Look, Second Edition, Revised and Expanded, *R. Syski*
99. Generalized Poisson Distributions: Properties and Applications, *P. C. Consul*
100. Nonlinear L_p-Norm Estimation, *Rene Gonin and Arthur H. Money*
101. Model Discrimination for Nonlinear Regression Models, *Dale S. Borowiak*
102. Applied Regression Analysis in Econometrics, *Howard E. Doran*
103. Continued Fractions in Statistical Applications, *K. O. Bowman and L. R. Shenton*
104. Statistical Methodology in the Pharmaceutical Sciences, *Donald A. Berry*
105. Experimental Design in Biotechnology, *Perry D. Haaland*
106. Statistical Issues in Drug Research and Development, *edited by Karl E. Peace*
107. Handbook of Nonlinear Regression Models, *David A. Ratkowsky*
108. Robust Regression: Analysis and Applications, *edited by Kenneth D. Lawrence and Jeffrey L. Arthur*
109. Statistical Design and Analysis of Industrial Experiments, *edited by Subir Ghosh*
110. *U*-Statistics: Theory and Practice, *A. J. Lee*
111. A Primer in Probability: Second Edition, Revised and Expanded, *Kathleen Subrahmaniam*
112. Data Quality Control: Theory and Pragmatics, *edited by Gunar E. Liepins and V. R. R. Uppuluri*
113. Engineering Quality by Design: Interpreting the Taguchi Approach, *Thomas B. Barker*
114. Survivorship Analysis for Clinical Studies, *Eugene K. Harris and Adelin Albert*
115. Statistical Analysis of Reliability and Life-Testing Models: Second Edition, *Lee J. Bain and Max Engelhardt*
116. Stochastic Models of Carcinogenesis, *Wai-Yuan Tan*
117. Statistics and Society: Data Collection and Interpretation, Second Edition, Revised and Expanded, *Walter T. Federer*
118. Handbook of Sequential Analysis, *B. K. Ghosh and P. K. Sen*
119. Truncated and Censored Samples: Theory and Applications, *A. Clifford Cohen*
120. Survey Sampling Principles, *E. K. Foreman*
121. Applied Engineering Statistics, *Robert M. Bethea and R. Russell Rhinehart*
122. Sample Size Choice: Charts for Experiments with Linear Models: Second Edition, *Robert E. Odeh and Martin Fox*
123. Handbook of the Logistic Distribution, *edited by N. Balakrishnan*
124. Fundamentals of Biostatistical Inference, *Chap T. Le*
125. Correspondence Analysis Handbook, *J.-P. Benzécri*
126. Quadratic Forms in Random Variables: Theory and Applications, *A. M. Mathai and Serge B. Provost*
127. Confidence Intervals on Variance Components, *Richard K. Burdick and Franklin A. Graybill*
128. Biopharmaceutical Sequential Statistical Applications, *edited by Karl E. Peace*
129. Item Response Theory: Parameter Estimation Techniques, *Frank B. Baker*
130. Survey Sampling: Theory and Methods, *Arijit Chaudhuri and Horst Stenger*
131. Nonparametric Statistical Inference: Third Edition, Revised and Expanded, *Jean Dickinson Gibbons and Subhabrata Chakraborti*
132. Bivariate Discrete Distribution, *Subrahmaniam Kocherlakota and Kathleen Kocherlakota*
133. Design and Analysis of Bioavailability and Bioequivalence Studies, *Shein-Chung Chow and Jen-pei Liu*
134. Multiple Comparisons, Selection, and Applications in Biometry, *edited by Fred M. Hoppe*
135. Cross-Over Experiments: Design, Analysis, and Application, *David A. Ratkowsky, Marc A. Evans, and J. Richard Alldredge*
136. Introduction to Probability and Statistics: Second Edition, Revised and Expanded, *Narayan C. Giri*
137. Applied Analysis of Variance in Behavioral Science, *edited by Lynne K. Edwards*
138. Drug Safety Assessment in Clinical Trials, *edited by Gene S. Gilbert*
139. Design of Experiments: A No-Name Approach, *Thomas J. Lorenzen and Virgil L. Anderson*
140. Statistics in the Pharmaceutical Industry: Second Edition, Revised and Expanded, *edited by C. Ralph Buncher and Jia-Yeong Tsay*
141. Advanced Linear Models: Theory and Applications, *Song-Gui Wang and Shein-Chung Chow*
142. Multistage Selection and Ranking Procedures: Second-Order Asymptotics, *Nitis Mukhopadhyay and Tumulesh K. S. Solanky*
143. Statistical Design and Analysis in Pharmaceutical Science: Validation, Process Controls, and Stability, *Shein-Chung Chow and Jen-pei Liu*
144. Statistical Methods for Engineers and Scientists: Third Edition, Revised and Expanded, *Robert M. Bethea, Benjamin S. Duran, and Thomas L. Boullion*
145. Growth Curves, *Anant M. Kshirsagar and William Boyce Smith*
146. Statistical Bases of Reference Values in Laboratory Medicine, *Eugene K. Harris and James C. Boyd*
147. Randomization Tests: Third Edition, Revised and Expanded, *Eugene S. Edgington*
148. Practical Sampling Techniques: Second Edition, Revised and Expanded, *Ranjan K. Som*

149. Multivariate Statistical Analysis, *Narayan C. Giri*
150. Handbook of the Normal Distribution: Second Edition, Revised and Expanded, *Jagdish K. Patel and Campbell B. Read*
151. Bayesian Biostatistics, *edited by Donald A. Berry and Dalene K. Stangl*
152. Response Surfaces: Designs and Analyses, Second Edition, Revised and Expanded, *André I. Khuri and John A. Cornell*
153. Statistics of Quality, *edited by Subir Ghosh, William R. Schucany, and William B. Smith*
154. Linear and Nonlinear Models for the Analysis of Repeated Measurements, *Edward F. Vonesh and Vernon M. Chinchilli*
155. Handbook of Applied Economic Statistics, *Aman Ullah and David E. A. Giles*
156. Improving Efficiency by Shrinkage: The James-Stein and Ridge Regression Estimators, *Marvin H. J. Gruber*
157. Nonparametric Regression and Spline Smoothing: Second Edition, *Randall L. Eubank*
158. Asymptotics, Nonparametrics, and Time Series, *edited by Subir Ghosh*
159. Multivariate Analysis, Design of Experiments, and Survey Sampling, *edited by Subir Ghosh*
160. Statistical Process Monitoring and Control, *edited by Sung H. Park and G. Geoffrey Vining*
161. Statistics for the 21st Century: Methodologies for Applications of the Future, *edited by C. R. Rao and Gábor J. Székely*
162. Probability and Statistical Inference, *Nitis Mukhopadhyay*
163. Handbook of Stochastic Analysis and Applications, *edited by D. Kannan and V. Lakshmikantham*
164. Testing for Normality, *Henry C. Thode, Jr.*

Additional Volumes in Preparation

Handbook of Applied Econometrics and Statistical Inference, *edited by Aman Ullah, Alan T. K. Wan, and Anoop Chaturvedi*

Visualizing Statistical Models and Concepts, *R. W. Farebrother and Michael Schyns*

HANDBOOK OF STOCHASTIC ANALYSIS AND APPLICATIONS

EDITED BY

D. KANNAN
University of Georgia
Athens, Georgia

V. LAKSHMIKANTHAM
Florida Institute of Technology
Melbourne, Florida

CRC Press
Taylor & Francis Group
Boca Raton London New York

CRC Press is an imprint of the
Taylor & Francis Group, an **informa** business

First published 2002 by Marcel Dekker, Inc.

Published 2020 by CRC Press
Taylor & Francis Group
6000 Broken Sound Parkway NW, Suite 300
Boca Raton, FL 33487-2742

First issued in paperback 2020

ISBN 13: 978-0-367-57873-2 (pbk)
ISBN 13: 978-0-8247-0660-9 (hbk)

Visit the Taylor & Francis Web site at
http://www.taylorandfrancis.com

and the CRC Press Web site at
http://www.crcpress.com

Preface

Various phenomena arising in physics, biology, finance, and other fields of study are intrinsically affected by random noise, (white or colored noise). One thus models any such phenomenon by an appropriate stochastic process or a stochastic equation. An analysis of the resulting process or equation falls in the realm of the so-called stochastic analysis. The applicatory value of stochastic analysis is therefore undeniable. In this handbook we present an overview of the analysis of some basic stochastic processes and stochastic equations along with some selective applications. The handbook is already voluminous even with this limited choice of topics, and therefore we hope that the reader will forgive us for omissions.

This handbook on stochastic analysis and applications contains 12 chapters. The first six chapters of the handbook may be considered the theoretical half (though they contain several illustrative applications) and the remaining six chapters the applied half. Markov processes and semimartingales are two predominant processes at the foundation of a stochastic analysis. The first two chapters present a clear exposition of these two basic processes. These chapters include material on Itô's stochastic calculus. To these we also add Chapter 3 presenting the important white noise theory of Hida. Stochastic differential equations (SDEs) are extensively used to model various phenomena that are subject to random perturbations. Chapter 4 details this topic. As in the case of deterministic equations, one needs numerical methods to analyze SDEs. The numerical analysis of SDEs is a fast-developing area that is not as rich in theory as its deterministic counterpart is. Chapter 5 presents an up-to-date account of the numerical analysis of SDEs. One can say without reservation that the study of large deviations is currently the most active area of research in probability, finding applications in a vast number of fields. Chapter 6 gives a thorough survey of this topic. The rest of the handbook is on applications. Stochastic control methods are needed or alluded to in some of these applications. We start the set of applied chapters with methods of control theory and the stabilization of control, Chapter 7. Game theoretic methods applied to economics helped at least one to earn a Nobel prize for economics. Chapter 8 presents a survey of stochastic game theory. We follow this with Chapter 9 on stochastic manufacturing systems where hierarchical control methods are used. Chapter 10 presents stochastic algorithms with several applications. Chapter 11 applies stochastic methods to optimization problems (as opposed to stochastic optimization methods). The final chapter is on stochastic optimization methods applied to (stochastic) financial mathematics. The introductory section of each chapter will provide details on the topics covered and the relevance of that chapter, so we refrain from summarizing them in detail here. Nevertheless, we will mention below a few simple facts just to introduce those chapters.

Markov chains and processes are, informally, randomized dynamical systems. These processes are used as models in a wide range of applications. Also, the theory of Markov processes is well developed. The handbook opens with an expository survey of some of the main topics in Markov process theory and applications. Professor Rabi Bhattacharya, who has published numerous research articles in this area and also has co-authored a popular first-year graduate level textbook on stochastic processes writes this chapter.

It would hardly be an exaggeration to say that semimartingale theory is central in any stochastic analysis. These processes form the most general integrators known in stochastic calculus. Chapter 2 presents an extensive survey of the theory of this important process. Professor Jia-an Yan, the author of Chapter 2, has co-authored an excellent book on this subject. Both Chapter 1 and Chapter 2 include several aspects of stochastic calculus that form a basis for understanding the remaining chapters.

Professor H.H. Kuo has researched extensively the white noise calculus of Hida, and also has written a substantial monograph on this subject. He authors Chapter 3.

Chapter 4 completes a cycle of stochastic calculus by presenting a well-rounded survey of the theory of stochastic differential equations (SDEs) and is written by Professor Bo Zhang, who specializes in the stability analysis of stochastic equations. This chapter reviews the theory of SDEs, which is fundamental in a vast number of applications in a variety of fields of study, and so forms a basis for what follows in the rest of the handbook (except for the chapter on large deviations).

The longest chapter (Chapter 5) in the handbook is on the numerical analysis of stochastic differential equations. The importance of the numerical analysis of deterministic systems is well known. Compared to the deterministic case, the study of the numerical methods for stochastic equations is still at a developing stage (and a fast one at that). This chapter is important due to its multidisciplinary character, the wide range of potential applications of stochastic differential equations, and the limitations of analytical methods for SDEs caused by their high complexity and partial intractability. Professor Henri Schurz, who wrote this chapter, has co-authored a textbook on the numerical analysis of SDEs and developed an accompanying program diskette. He presents an extensive list of references on this subject here.

One may say without much hesitation that the large deviation theory is currently the most active subject of research in probability. Professors Dembo and Zeitouni have not only done extensive research in this area but also co-authored a popular monograph on this topic. Chapter 6 is an up-to-date survey of this theory, which found applications in many areas including statistical physics, queuing systems, information theory, risk-sensitive control, stochastic algorithms, and communication networks. This chapter includes applications to hypothesis testing in statistics and the Gibbs conditioning principle in statistical mechanics.

The remaining half of the handbook is on applications; regrettably a lot of important applications are not included due to space constraints. Control theory and stabilization of controls is the subject matter of Chapter 7 written by Professor Pavel Pakshin. The dynamic programming and maximum principle methods are detailed in the chapter. The separation principle is used for the solution of the standard linear-quadratic Gaussian (LQG) control problem. Chapters 9 and 12 extensively use the control theory methods in applications to stochastic manufacturing systems and asset pricing, respectively.

Chapter 8, written by Professor K.M. Ramachandran, discusses stochastic game theory. Recently, three prominent researchers in game theory won the Nobel prize for economics. This vouches for the importance of game theory, both deterministic and stochastic. The chapter includes both the two-person zero-sum games and N-person non-cooperative games. Emphasis is placed on solution methods, old and new. Applications to defense, finances, economics, institutional investor speculation, etc, are presented.

Stochastic control theory enriched the analysis of manufacturing systems. Professor Qing Zhang who wrote Chapter 9 has also co-authored the first authoritative monograph on stochastic manufacturing systems. Chapter 9 includes the theory and applications developed since the appearance of that monograph. Manufacturing systems are usually large and complex, and are subject to various discrete events such as purchasing new equipment and machine failures and repairs. Due to the large size of these systems and the presence of these events, obtaining exact optimal feedback policies to run these systems is nearly impossible both theoretically and computationally. Only small-sized problems are addressed even in approximation of solutions. Therefore, these systems are managed in a hierarchical fashion. The reduction in complexity is achieved by decomposing the problem into problems of the smaller subsystems with a proper coordinating mechanism, aggregating products and subsequently disaggregating them, and replacing random processes with their averages. This chapter adopts the latter method.

Professor George Yin reviews stochastic approximations and their applications in Chapter 10. He presents various forms of stochastic approximation algorithms, projections and truncation procedures, algorithms with soft constraints, and global stochastic approximation algorithms, among other methods. The utility of stochastic approximation methods is demonstrated with applications to adaptive filtering, system identification, stopping time rules for least squares algorithm, adaptive step-size tracking algorithms, approximation of threshold control policies, GI/G/1 queues, distributed algorithms for supervised learning, etc. George Yin has co-authored a book on this topic and this chapter includes recent results.

Chapter 11, written by Professor Ron Shonkwiler, is on stochastic methods for global optimization. Until the stochastic methods came along, there were no good general methods addressing global optimization. Stochastic methods are simple to implement, versatile, and robust, and they parallelize effectively. These methods often mimic some natural process such as temperature-based annealing or biological recombination. The theory behind these methods is built on the theory of Markov chains and renewal theory, and it provides a framework for illuminating their strengths and weaknesses. Detailed descriptions of the basic algorithms are provided along with comparisons and contrasts.

Professor Thaleia Zariphopoulou wrote the final chapter (Chapter 12), which is on stochastic control methods in asset pricing, and she is an active researcher in this area. Most of the valuation models lead to stochastic optimization problems. This chapter presents an exposition of stochastic optimization methods used in financial mathematics along with a quick summary of results on the Hamilton-Jacobi-Bellman (HKB) equation. In addition to optimization models of expected utility in complete markets as well as markets with frictions, this chapter provides models of derivative pricing.

Acknowledgments

The editors express their deep sense of gratitude to all the authors who contributed to the *Handbook of Stochastic Analysis and Applications.* Obviously, this handbook would not have been possible without their help. Mrs. Sharon Southwick provided local computer help to D. Kannan. She developed the uniform code to compile all the articles in one file. The editors are very thankful for all her help. The editors are also thankful to the editorial staff of Marcel Dekker, Inc. in particular to Maria Allegra and Brian Black for their patience and cooperation during the long process of bringing out the handbook.

D. Kannan V. Lakshmikantham

Contents

Contributors

Rabi Bhattacharya Indiana University, Bloomington, Indiana

Amir Dembo Stanford University, Stanford, California

Hui-Hsiung Kuo Louisiana State University, Baton Rouge, Louisiana

Franklin Mendivil Georgia Institute of Technology, Atlanta, Georgia

P. V. Pakshin Nizhny Novgorod State Technical University at Arzamas, Arzamas, Russia

K. M. Ramachandran University of South Florida, Tampa, Florida

R. Shonkwiler Georgia Institute of Technology, Atlanta, Georgia

M. C. Spruill Georgia Institute of Technology, Atlanta, Georgia

H. Schurz University of Minnesota, Minneapolis, Minnesota

Jia-An Yan Chinese Academy of Sciences, Beijing, China

G. Yin Wayne State University, Detroit, Michigan

Thaleia Zariphopoulou The University of Texas at Austin, Austin, Texas

Ofer Zeitouni Technion, Haifa, Israel

Bo Zhang People's University of China, Beijing, China

Q. Zhang University of Georgia, Athens, Georgia

HANDBOOK OF STOCHASTIC ANALYSIS AND APPLICATIONS

Chapter 1

Markov Processes and Their Applications

RABI BHATTACHARYA
Department of Mathematics
Indiana University
Bloomington, Indiana

1.1 Introduction

For the most part in this chapter we will confine ourselves to time–homogeneous Markov processes. In discrete time, such a Markov process on a (measurable) *state space* $(S, \mathcal{S})$ is defined by a (one–step) *transition probability* $p(x, dy)$, $x \in S$, where (i) for each $x \in S$, $p(x, dy)$ is a probability measure on $(S, \mathcal{S})$ and (ii) for each $B \in \mathcal{S}, x \to p(x, B)$ is a measurable function on $(S, \mathcal{S})$ into $([0,1], \mathcal{B}([0,1])$. Here $\mathcal{B}(\mathcal{X})$ denotes the *Borel σ–field* on a topological space $\mathcal{X}$. Let $\Omega_0 = S^\infty$ be the space of all sequences $\mathbf{x} = (x_0, x_1, \cdots, x_n, \cdots)$ in S, Ω_0 being endowed with the product σ–field $\mathcal{F}_0 = \mathcal{S}^{\otimes\infty}$ generated by the class of all finite–dimensional measurable cylinders of the form $A = B \times S^\infty = \{\mathbf{x} \in S^\infty : x_j \in B_j, 0 \le j \le n\}$ with $B_j \in \mathcal{S}$ for $j = 0, 1, \cdots, n$ and n arbitrary. For any given probability measure μ on $(S, \mathcal{S})$ one can construct a unique probability measure P_μ on $(\Omega, \mathcal{F})$ by assigning to cylinder sets A of the above form the probability

$$P_\mu(A) = \int_{B_0}\int_{B_1}\cdots\int_{B_{n-2}}\int_{B_{n-1}} p(x_{n-1}, B_n)p(x_{n-2}, dx_{n-1})$$
$$p(x_{n-3}, dx_{n-2})\cdots p(x_0, dx_1)\mu(dx_0). \tag{1.1.1}$$

evaluated by iterated integration. In the case S is a *Polish space*, i.e., S is homeomorphic to a complete separable metric space, and $\mathcal{S} = \mathcal{B}(S)$, such a construction of a P_μ is provided by *Kolmogorov's Existence Theorem* (See Billingsley [1], pp. 486–490). For general state spaces $(S, \mathcal{S})$ this construction is due to Tulcea [2] (Also see Nevue [3], pp. 161–166).

The coordinate process $\{X_n : n = 0, 1, \cdots\}$ defined on $(S^\infty, \mathcal{S}^{\otimes\infty})$ by $X_n(\mathbf{x}) = x_n$ ($\mathbf{x} = (x_0, x_1, \cdots, x_n, \cdots)$) is a *Markov process* with transition probability $p(x, dy)$ and initial distribution μ. In other words, the conditional distribution of the process $X_n^+ := (X_n, X_{n+1}, \cdots)$ on $(S^\infty, \mathcal{S}^{\otimes\infty})$ given $\mathcal{F}_n := \sigma\{X_j : 0 \le j \le n\}$, namely the σ–field of past and present events up to time n, is P_{X_n}, where P_y is written for P_μ with $\mu = \delta_y$, i.e., $\mu(\{y\}) = 1$. Often one needs a larger probability space than this canonical model $(S^\infty, \mathcal{S}^{\otimes\infty}, P_\mu)$ e.g., to accommodate a family of random variables independent of the process $\{X_n : n = 0, 1, 2, \cdots\}$.

Hence we will consider a general probability space $(\Omega, \mathcal{F}, P)$ on which is defined a sequence $(X_0, X_1, \cdots, X_n, \cdots)$ whose distribution is P_μ given by (10.7.65).

Sometimes a Markov process, or its transition probability $p(x, dy)$, may admit an *invariant probability* $\pi(dy)$, i.e.,

$$\int_S p(x, B)\pi(dx) = \pi(B) \quad \forall B \in \mathcal{S}. \tag{1.1.2}$$

In this case if one takes $\mu = \pi$ as the initial distribution, then the Markov process $\{X_n : n = 0, 1, 2, \cdots\}$ is *stationary* in the sense that $X_n^+ \equiv (X_n, X_{n+1}, \cdots)$ has the same distribution as $(X_0, X_1, \cdots)$, namely P_π, for every $n > 0$. In particular, X_n has distribution π for all $n \geq 0$. We will often be concerned with the derivation of criteria for the existence of a *unique* invariant probability π, and then $\{X_n : n = 0, 1, 2, \cdots\}$ is *ergodic* in the sense of ergodic theory, i.e., the σ–field $\mathcal{F}_I$ of shift–invariant events is *trivial*; that is $P(B) = 0$ or 1 for $B \in \mathcal{F}_I$.

To describe an important strengthening of the Markov property described above, let $\{\mathcal{F}_n : n = 0, 1, 2, \cdots\}$ be an increasing sequence of sub–σ–fields of $\mathcal{F}$ such that X_n is $\mathcal{F}_n$–measurable for every n, and the conditional distribution of X_n^+, given $\mathcal{F}_n$, is $P_{X_n} (n \geq 0)$. Such a family $\{\mathcal{F}_n : n = 0, 1, 2, \cdots\}$ is called a *filtration*. For example, one may take $\mathcal{F}_n = \sigma\{X_j : 0 \leq j \leq n\} (n \geq 0)$, or $\mathcal{F}_n$ may be the σ–field generated by $\{X_j : 0 \leq j \leq n\}$ and a family of random variables independent of $\{X_j : j \geq 0\}$. A random variable $\tau : \Omega \to \{0, 1, 2, \cdots\} \cup \{\infty\}$ is said to be a $\{\mathcal{F}_n\}$*–stopping time* if $\{\tau \leq n\} \in \mathcal{F}_n$ for every n. Define the *pre–*τ σ–field $\mathcal{F}_\tau$ by $\mathcal{F}_\tau := \{A \in \mathcal{F} : A \cap \{\tau \leq n\} \in \mathcal{F}_n \forall n\}$. It is not difficult to check that if τ is a stopping time then the conditional distribution of $X_\tau^+ := (X_\tau, X_{\tau+1}, \cdots)$ given $\mathcal{F}_\tau$ is P_{X_τ} on the set $\{\tau < \infty\}$. This property is called the *strong Markov property* and it is extremely useful in deriving various distributions and expectations of random variables related to the Markov process.

We now turn to the case of continuous parameter Markov processes. Suppose one is given a family of transition probabilities $p(t; x, dy) (t > 0, x \in S)$ on a state space $(S, \mathcal{S})$, satisfying (i) $p(t; x, dy)$ is a probability measure on $(S, \mathcal{S})$ for all $t > 0, x \in S$, (ii) $x \to p(t; x, B)$ is measurable on $(S, \mathcal{S})$ for all $t > 0, B \in \mathcal{S}$, and (iii) the following *Chapman–Kolmogorov equation* holds

$$p(t + s; x, B) = \int_S p(s; z, B)p(t; x, dz) \ \ (t > 0, s > 0, x \in S, \mathcal{B} \in \mathcal{S}) \tag{1.1.3}$$

Given any initial distribution μ, one can then construct a probability measure P_μ on $(\Omega_0 = S^{[0,\infty)}, \mathcal{F}_0 = \mathcal{S}^{\otimes[0,\infty)})$ as follows. Note that $S^{[0,\infty)}$ is the set of all functions on $[0, \infty)$ into S, and $\mathcal{S}^{\otimes[0,\infty)}$ is the product σ–field on $S^{[0,\infty)}$ generated by the coordinate projections $X_t(\omega) = \omega(t), \omega \in \Omega_0$. Define P_μ on measurable cylinders of the form $A = \{\omega \in \Omega_0 : \omega_{t_i} \in B_i, i = 0, 1, \cdots, n\}$, $B_i \in \mathcal{S}$ $(0 \leq i \leq n), 0 < t_1 < t_2 < \cdots < t_n$, by

$$\begin{aligned} P_\mu(A) = \int_{B_0}\int_{B_1} \cdots \int_{B_{n-2}}\int_{B_{n-1}} & p(t_n - t_{n-1}; x_{n-1}, B_n)p(t_{n-1} - t_{n-2}; x_{n-2}, dx_{n-1}) \\ & p(t_{n-2} - t_{n-3}; x_{n-3}, dx_{n-2}) \cdots p(t_1; x_0, dx_1)\mu(dx_0) \end{aligned} \tag{1.1.4}$$

obtained by iterated integration. In the case of a metric space (S, ρ) it is generally advantageous to define such a Markov process to have some path regularity. The following results are due to Dynkin [4], pp. 91–97, and Snell [5]. First, on a metric space S define the transition probability $p(t; x, dy)$, or a Markov process with this transition probability, to have the *Feller property* if $x \to p(t; x, dy)$ is weakly continuous, i.e., for every $t > 0$

$$(T_t f)(x) := \int_S f(y)p(t; x, dy) \tag{1.1.5}$$

is a continuous function of x for every bounded continuous f.

Theorem 1.1.1 *(Dynkin–Snell) Let (S, ρ) be a metric space, and $p(t; x, dy)$ a transition probability on $(S, \mathcal{B}(S))$. (a) if*

$$\lim_{t\downarrow 0} \frac{1}{t}\{\sup_{x\in S} p(t; x, B_\varepsilon^c(x))\} = 0 \ \forall \varepsilon > 0 \ (B_\varepsilon(x) := \{y : \rho(y, x) < \varepsilon\}, \tag{1.1.6}$$

then, for every initial distribution μ, there exists a probability space $(\Omega, \mathcal{F}, P_\mu)$ on which is defined a Markov process $\{X_t : t \geq 0\}$ with continuous sample paths and transition probability $p(t; x, dy)$ so that (1.4) holds for $A = \{X_{t_i} \in B_i \text{ for } i = 0, 1, \cdots, n\}, 0 < t_1 < t_2 < \cdots < t_n, B_i \in \mathcal{S} = \mathcal{B}(S)(i = 0, 1, \cdots, n)$. (b) If, instead of (1.6), one has

$$\lim_{t\downarrow 0}\{\sup_{x\in S} p(t; x, B_\varepsilon^c(x))\} = 0 \quad \forall \varepsilon > 0, \tag{1.1.7}$$

then one may find a probability space $(\Omega, \mathcal{F}, P_\mu)$ on which is defined a Markov process $\{X_t : t \geq 0\}$ which has right–continuous sample paths with left limits, having the transition probability $p(t; x, dy)$ and initial distribution μ.

For right–continuous Markov processes with the Feller property, the *strong Markov property* holds. To be precise, let $\{\mathcal{F}_t : t \geq 0\}$ be a family of increasing sub–σ–fields of $\mathcal{F}$ such that X_t is $\mathcal{F}_t$–measurable, and the conditional distribution of $X_t^+ := \{X_{t+s} : s \geq 0\}$ given $\mathcal{F}_t$ is $P_{X_t}(t \geq 0)$. Let $\tau : \Omega \to [0, \infty]$ be a $\{\mathcal{F}_t\}$*–stopping time*, i.e., $\{\tau \leq t\} \in \mathcal{F}_t$ for every $t \geq 0$, and define the *pre–τ σ–field* $\mathcal{F}_\tau := \{A \in \mathcal{F} : A \cap \{\tau \leq t\} \in \mathcal{F}_t \ \forall t \geq 0\}$. Then the strong Markov property requires that the conditional distribution of $X_\tau^+ := \{X_{\tau+s} : s \geq 0\}$ given $\mathcal{F}_\tau$ is P_{X_τ}, on the set $\{\tau < \infty\}$.

It may be noted that, unlike the discrete parameter case, the transition probability $p(t; x, dy)$ needed to construct a continuous parameter Markov process must be given for *all* t at least in a small time interval $(0, \delta], \delta > 0$. One may then construct $p(t; x, dy)$ for all $t > 0$, by the Chapman–Kolmogorov equation (3.1.3). Thus, except in special cases such as for processes with independent increments, continuous parameter transition probabilities and corresponding Markov processes are constructed from infinitesimal characteristics. For jump Markov chains these characteristics are the infinitesimal transition rates $q_{ij} := \lim_{t\downarrow 0} \frac{1}{t} p(t; i, j)(i \neq j)$. More generally, one specifies the infinitesimal generator

$$(\mathbf{A}f)(x) := \lim_{t\downarrow 0} \frac{(T_t f)(x) - f(x)}{t} \tag{1.1.8}$$

for a suitable class of functions f. In the case of diffusion on $\mathbb{R}^k$, $\mathbf{A}$ is a second order elliptic operator of the form

$$(\mathbf{A}f)(x) = \frac{1}{2} \sum_{r,r'=1}^{k} a_{rr'}(x) \frac{\partial^2 f(x)}{\partial x^{(r)} \partial x^{(r')}} + \sum_{r=1}^{k} b_r(x) \frac{\partial f}{\partial x^{(r)}}, \tag{1.1.9}$$

where $b(x)$ is the so–called drift velocity, and $a(x)$ the diffusion matrix, of the process $\{X_t : t \geq 0\}$.

Finally, for a continuous parameter Markov process $\{X_t : t \geq 0\}$ an *invariant* (initial) *distribution* π, if it exists, satisfies

$$\int_S p(t; x, B)\pi(dx) = \pi(B), \quad \forall t > 0, B \in \mathcal{S}.$$

Under such an initial distribution π, the process $\{X_t : t \geq 0\}$ is *stationary*, i.e., the distribution of $X_t^+ := \{X_{t+s} : s \geq 0\}$ is the same as that of $\{X_s : s \geq 0\}$, namely P_π, for all $t > 0$.

1.2 Markov Chains

We will refer to a Markov process on a state space $(S, \mathcal{S})$ as a *Markov chain* if S is countable and $\mathcal{S}$ is the class of all subsets of S. Consider a time–homogeneous Markov chain $X_n (n = 0, 1, 2, \cdots)$ on a (countable) state space S. Its transition probabilities are specified by the matrix $p = ((p_{ij}))$ where $p_{ij} = p(i, \{j\}) \equiv P(X_{n+1} = j | X_n = i)$, for $i, j \in S$. Denote by $p^n = ((p_{ij}^{(n)}))$ the *n–step transition probability* matrix where $p_{ij}^{(n)} = P(X_{m+n} = j | X_m = i)$. Write $i \to j$ if $p_{ij}^{(n)} > 0$ for some $n \geq 1$. If $i \to j$ and $j \to i$ one says i and j *communicate.* A state i is *essential* if for every j such that $i \to j$ one has $j \to i$. All other states are *inessential.* On the class $\mathcal{E}$ of all essential states, the relation $\to$ is an *equivalence relation,* which therefore decomposes $\mathcal{E}$ into disjoint subsets of communicating states. For each $i \in \mathcal{E}$, define the *period* of i as the greatest common divisor d_i of the set $\{n \geq 1 : p_{ii}^{(n)} > 0\}$. It may be shown that $d = d_i$ depends only on the communicating class to which i belongs. For $d > 1$, each such class is divided into d subsets $C_0, C_1, \cdots, C_{d-1}$ such that the process $\{X_n\}$ moves cyclically among them: $C_0 \to C_1 \to C_2 \to \cdots \to C_{d-1} \to C_0$. In other words, if the chain is in C_r at time n, then it will move to $C_{r+1} (\text{mod } d)$ at time $n+1$. A Markov chain is said to be *irreducible* if it comprises a single equivalence class of essential states.

A state i is said to be *recurrent* if

$$P(X_n = i \text{ for infinitely many } n | X_0 = i) = 1, \tag{1.2.10}$$

and i is *transient* if

$$P(X_n = i \text{ for infinitely many } n | X_0 = i) = 0. \tag{1.2.11}$$

All inessential states are transient. However, an essential state may not be recurrent. Also, recurrence is a class property, i.e., if $i \in \mathcal{E}$ is recurrent then so are all states belonging to the equivalence class (of communicating states) to which i belongs. For further analysis of (12.2.1), (12.2.2), consider the *first passage time* to the state j, namely, $\tau_j := \inf\{n \geq 1 : X_n = j\}$, as well as the time for the rth passage to j defined for all r recursively by $\tau_j^{(r+1)} := \inf\{n > \tau_j^{(r)} : X_n = j\} (r = 0, 1, 2, \cdots)$. Here $\tau_j^{(0)} := 0$, and $\tau_j^{(1)} = \tau_j$. Let $\rho_{ij} = P(\tau_j < \infty | X_o = i) \equiv P(X_n = j \text{ for some } n \geq 1 | X_0 = i)$. It follows from the strong Markov property that $P(\tau_j^{(r+1)} < \infty | X_0 = i) = \rho_{ij} \rho_{jj}^r (r = 0, 1, \cdots)$. In particular, $P(\tau_i^{(r)} < \infty | X_0 = i) = \rho_{ii}^r (r = 1, 2, \cdots)$. Letting $r \to \infty$ one obtains the probability of the event in parenthesis in (2.1) or (2.2) as $\lim_{r \to \infty} \rho_{ii}^r$, which is 1 iff $\rho_{ii} = 1$, and 0 iff $\rho_{ii} < 1$. This criterion for transience and recurrence also establishes a dichotomy of S into recurrent and transient states.

Another useful criterion for transience and recurrence may be expressed in terms of the so–called Green's function $G(i, j)$, which is the expected number of visits to j by the process, starting at $X_0 = i$:

$$G(i, j) = E\Big(\sum_{n=0}^{\infty} \mathbf{1}_{\{X_n = j\}} | X_0 = i\Big) = \sum_{n=0}^{\infty} p_{ij}^{(n)}, \tag{1.2.12}$$

where $p_{ij}^{(0)} = 1$ or 0 according as $i = j$ or $i \neq j$. Denoting by $N(j)$ the number of visits to j, one has

$$G(i, j) = E(N(j) | X(0) = i) = \sum_{r=0}^{\infty} P(N(j) > r | X(0) = i)$$

$$\sum_{r=0}^{\infty} P(\tau_j^{(r+1)} < \infty | X(0) = i) = \sum_{r=0}^{\infty} \rho_{ij} \rho_{jj}^r = \frac{\rho_{ij}}{1 - \rho_{jj}}, \tag{1.2.13}$$

if $\rho_{jj} < 1$. If $\rho_{jj} = 1$, and $\rho_{ij} > 0$, then $G(i,j) = \sum_{r=0}^{\infty} \rho_{ij}\rho_{jj}^r = \infty$. In particular, $G(i,i) < \infty$ iff i is transient (and $G(i,i) = \infty$ iff i is recurrent).

A (recurrent) state is said to be *positive recurrent* if $E(\tau_i|X_0 = i) < \infty$. A recurrent state i is *null recurrent* if $E(\tau_i|X_0 = i) = \infty$. Positive recurrence is a class property, i.e., if i is positive recurrent then so is every state in the equivalence class to which i belongs.

It follows from the strong Markov property that if i is a recurrent state then the blocks $B_r := \{X_j : j \in [\tau_i^{(r)}, \tau_i^{(r+1)})\}$, $r = 1, 2, \cdots$, are independent and identically distributed (*i.i.d.*), no matter what the initial state X_0 is. Denoting $\tau_i^{(r)}$ by $\tau^{(r)}$, this means that the events $\{\tau^{(r+1)} - \tau^{(r)} = k, \ X_{\tau^{(r)}+1} = i_1, \cdots, X_{\tau^{(r)}+k-1} = i_{k-1}\}(r = 1, 2, \cdots)$ are independent and have the same probability, for every given $k \geq 1$ and every given k–triple $(i_0, i_1, \cdots, i_{k-1})$. Assume now that i is positive recurrent and let $\mathcal{E}_i$ denote the (equivalence) class of states communicating with i. Write $T_j^{(r)} = \#\{n \in [\tau_i^{(r)}, \tau_i^{(r+1)}) : X_n = j\}(r \geq 1)$. If $X_0 \in \mathcal{E}_i$, then for every $j \in \mathcal{E}_i$ the long–run proportion of times $X_n = j$ exists almost surely and is given by the strong law of large numbers (SLLN) as

$$\begin{aligned} \pi_j &= \lim_{r\to\infty} \frac{T_j^{(1)} + T_j^{(2)} + \cdots + T_j^{(r)}}{\tau_j^{(r+1)} - \tau_j^{(1)}} = \lim_{r\to\infty} \frac{T_j^{(1)} + \cdots + T_j^{(r)}}{r} \Big/ \left(\tau_j^{(r+1)} - \tau_j^{(1)}\right) \\ &= \frac{\theta_j}{m_j}, \quad \text{say,} \end{aligned} \tag{1.2.14}$$

where $\theta_j = ET_j^{(1)}, m_j = E(\tau_j^{(2)} - \tau_j^{(1)})$, which do not depend on the particular initial state X_0 in $\mathcal{E}_i$. It may now be checked that $\{\pi_j : j \in \mathcal{E}_i\}$ is an *invariant probability function* for the Markov chain. When the process is restricted to $\mathcal{E}_i$ then $\pi(A) := \sum_{j\in A} \pi_j$ defines a unique invariant probability for the chain with state space $\mathcal{E}_i$. If there is only one equivalence class (i.e., $\mathcal{E} = \mathcal{E}_i$ for $i \in \mathcal{E}$), then the Markov chain has the unique invariant probability π as described. If, on the other hand, there are N different positive recurrent equivalence classes, $\mathcal{E}^{(1)}, \mathcal{E}^{(2)}, \cdots, \mathcal{E}^{(N)} (N > 1)$, and $\pi^{(1)}, \pi^{(2)}, \cdots, \pi^{(N)}$ are the invariant probabilities on $\mathcal{E}^{(1)}, \mathcal{E}^{(2)}, \cdots, \mathcal{E}^{(N)}$, respectively, then any convex combination $\sum_u a_u \pi^{(u)}$ is an invariant probability.

1.2.1 Simple Random Walk

One may apply the criterion for recurrence described above in terms of $G(i,i)$ to simple symmetric random walks $X_n \equiv S_n = x + Y_1 + \cdots + Y_n (n \geq 1), S_0 = x$, where $Y_n (n = 1, 2, \cdots)$ are i.i.d. with values in the lattice $\mathbb{Z}^k : P(Y_n = \pm e_u) = 1/2k(u = 1, \cdots, k)$, where e_u has one in the uth coordinate and zeros elsewhere. In the case $k = 1, p_{00}^{(2n+1)} = 0 \ \forall n = 0, 1, 2, \cdots$, and $p_{00}^{(2n)} = P(S_{2n} = 0|S_0 = 0) = \binom{2n}{n}\frac{1}{2^{2n}} \sim \frac{1}{\sqrt{\pi n}}$, by Stirling's formula. Here the relation $\sim$ means that the ratio of its two sides goes to one as $n \to \infty$. Since $\sum_{n=1}^{\infty} 1/\sqrt{\pi n} = \infty$, it follows that $G(0,0) = \infty$ so that 0 is a recurrent state; since all states communicate with each other, all states are recurrent. Of course, one can apply the other criterion in terms of ρ_{00} also, for the case $k = 1$, showing directly that $\rho_{00} = 1$. For the simple symmetric random walk on $\mathbb{Z}^2$, one may similarly show that $p^{(2n)}(0,0) > c/n$ for some positive constant c. Hence 0 is a recurrent state and, therefore, all states are recurrent. For $k \geq 3$, one shows that $p_{00}^{(2n)} \leq c/n^{k/2}$, so that $G(0,0) < \infty$ if $k \geq 3$. Thus 0 is transient, as are all states in $\mathbb{Z}^k$. We have arrived at

Theorem 1.2.1 *(Polya). The simple symmetric random walk on $\mathbb{Z}^k$ is recurrent for $k =$ 1, 2 and transient for $k \geq 3$.*

If a simple random walk on $\mathbb{Z}^k$ is asymmetric, i.e., $P(Y_n = e_u) \neq 1/2k$ for some u $(u = 1, 2, \cdots, k)$, then by the strong law of large numbers S_n^u/n converges almost surely to a nonzero constant, where S_n^u denotes the uth coordinate of S_n. It follows that all states are transient. In the case $k = 1$, one may compute ρ_{ij}. Suppose $p \equiv P(Y_1 = +1) > \frac{1}{2}, q = 1 - p > 0$. Then

$$\rho_{ij} = \begin{cases} 1 & \text{if} \quad i < j, \\ 2q & \text{if} \quad i = j, \\ \left(\frac{q}{p}\right)^{i-j} & \text{if} \quad i > j \end{cases} \tag{1.2.15}$$

1.2.2 Birth–Death Chains and the Ehrenfest Model

Like a simple random walk, a *birth–death chain* moves one step at a time — either one unit to the right or one unit to the left. Unlike a simple random walk, the probabilities of moving to the right or to the left, say, p_i and q_i depend on the present state of i of the process. Let $S = \mathbb{Z}$ be the set of all integers, and let positive numbers $p_i, q_i (i \in \mathbb{Z})$ be given satisfying $p_i + q_i \leq 1$. Write $r_i = 1 - p_i - q_i$. Then a (birth–death) Markov chain $X_n (n \geq 0)$ with a given initial state X_0 is defined by transition probabilities

$$\begin{aligned} &p_{i,i+1} \equiv P(X_{n+1} = i+1 | X_n = i) = p_i, \ p_{i,i-1} \equiv P(X_{n+1} = i-1 | X_0 = i) = q_i, \\ &p_{i,i} \equiv P(X_{n+1} = i | X_n = i) = r_i, \ p_{i,j} \equiv P(X_{n+1} = j | X_0 = i) = 0 \text{ for } |j - i| > 1. \end{aligned} \tag{1.2.16}$$

Note that all states communicate with each other and $\mathcal{E} = \mathbb{Z}$ comprises a single equivalence class. If $r_i = 0$ for every i then the chain is periodic with period 2. An effective method of determining transience, recurrence, etc. is by means of the following recursive equations governing the probability $\psi(i) \equiv \psi_{c,d}(i) = P(\{X_n\}$ reaches c before $d | X_0 = i)$, where $c \leq i \leq d$ are integers:

$$\begin{aligned} &\psi(i) = r_i \psi(i) + p_i \psi(i+1) + q_i \psi(i-1), \ c < i < d, \\ &\psi(c) = 1, \ \psi(d) = 0. \end{aligned} \tag{1.2.17}$$

The first equation is arrived at by considering the three disjoint and exhaustive possibilities, $X_1 = i, X_1 = i+1, X_1 = i-1$, and conditioning on each of them. By casting this equation in the form $p_i(\psi(i+1) - \psi(i)) = q_i(\psi(i) - \psi(i-1))$ and proceeding recursively until one of the *boundaries* c or d is reached, and then using the boundary conditions $\psi(c) = 1, \psi(d) = 0$, one can prove that

$$\psi(i) = \frac{\sum_{x=i}^{d-1} \frac{q_x q_{x-1} \cdots q_{c+1}}{p_x p_{x-1} \cdots p_{c+1}}}{1 + \sum_{x=c+1}^{d-1} \frac{q_x q_{x-1} \cdots q_{c+1}}{p_x p_{x-1} \cdots p_{c+1}}}, \quad c + 1 \leq y \leq d - 1 \tag{1.2.18}$$

Letting $d \to \infty$ in (12.2.9) one obtains the probability ρ_{ic} that the process (ever) reaches c, starting at state i. In particular, for all $i > c$

$$\rho_{ic} = \begin{cases} 1 & \text{if} \quad \sum_{x=1}^{\infty} \frac{q_1 q_2 \cdots q_x}{p_1 p_2 \cdots p_x} = \infty, \\ < 1 & \text{if} \quad \sum_{x=1}^{\infty} \frac{q_1 q_2 \cdots q_x}{p_1 p_2 \cdots p_x} < \infty. \end{cases} \tag{1.2.19}$$

Similarly, one has for all $i < d$,

$$\rho_{id} = \begin{cases} 1 & \text{if} \quad \sum_{x=-\infty}^{0} \frac{p_x p_{x+1} \cdots p_0}{q_x q_{x+1} \cdots q_0} = \infty, \\ < 1 & \text{if} \quad \sum_{x=-\infty}^{0} \frac{p_x p_{x+1} \cdots p_0}{q_x q_{x+1} \cdots q_0} < \infty. \end{cases} \tag{1.2.20}$$

Hence all states are recurrent if both sums (in (12.2.10) and (12.2.11)) diverge, and they are all transient otherwise. The criterion $p \neq \frac{1}{2}$ for transience, and $p = \frac{1}{2}$ for recurence, for simple random walks on $\mathbb{Z}$ follows as a special case of this. Also, (12.2.6) may be derived from (12.2.9), after letting $d \uparrow \infty$, or $c \downarrow -\infty$.

One may also consider birth–death chains on a finite state space of $N+1$ consecutive integers, $S = \{a, a+1, \cdots, a+N\}$, with transition probabilities (12.2.7) if $a < i < N$, and with reflecting and/or absorbing boundary conditions at $i = a$ and $i = a+N$. *Reflecting boundary conditions* at a and $a+N$ are given by

$$p_{a,a+1} \equiv p_a > 0, \quad p_{a+N,a+N-1} \equiv q_{a+N} > 0, \tag{1.2.21}$$

while the *absorbing boundary conditions* are

$$p_{a,a} \equiv r_a = 1, \quad p_{a+N,a+N} \equiv r_{a+N} = 1. \tag{1.2.22}$$

If both boundaries are reflecting, as in (12.2.12), then it is not difficult to check that all states are positive recurrent, with the unique invariant probability function π_i given as the normalized solution of $\pi' p = \pi'$, where $\pi' = (\pi_a, \pi_{a+1}, \cdots, \pi_{a+N})$. This solution is obtained recursively as

$$\pi_i = \frac{p_a p_{a+1} \cdots p_{i-1}}{q_{a+1} q_{a+2} \cdots q_i} \pi_a (a < i \leq a+N), \pi_a = \left(1 + \sum_{i=a+1}^{N} \frac{p_a \cdots p_{i-1}}{q_{a+1} \cdots q_i}\right)^{-1}. \tag{1.2.23}$$

An important example of such a birth–death chain is provided by the *Ehrenfest model* for heat exchange, with $S = \{-d, -d+1, \cdots, 0, 1, 2, \cdots, d\}$, with $p_i = 1 - \frac{d+i}{2d}$ and $q_i = \frac{d+i}{2d}$ for $-d < i < d$, and with $p_{-d} = 1 = q_d$. From (12.2.14) it follows that the unique invariant probability function is given by the binomial

$$\pi_i = \binom{2d}{d+i} \frac{1}{2^{2d}} \quad (-d \leq i \leq d). \tag{1.2.24}$$

This model was used by the husband–wife team of physicists P. and T. Ehrenfest in 1907, and later by Smoluchowski in 1916, to resolve an apparent contradiction with well accepted laws of thermodynamics that threatened to wreck Boltzmann's kinetic theory at the turn of the century. One may think of $d+i$ as the temperature in body A when the state is i, and $d-i$ the corresponding temperature in an equal body B in contact with A. The mean temperature of each body under equilibrium is d (corresponding to $i = 0$). The thermodynamic equilibrium would be achieved when the temperatures of the two bodies are approximately, i.e., macroscopically, equal. By the second law of thermodynamics, the progression towards equilibrium is orderly and irreversible. On the other hand, heat exchange is a random process according to the kinetic theory, and it was pointed out by Poincaré and Zermelo that this meant that the process starting at a macroscopic equilibrium, i.e., from a state near $i = 0$, will sooner or later reach the state of $i = -d$ or $i = d$ thus reverting to a state of extreme disequilibrium. The Ehrenfests showed that the time to reach states $\pm d$ from 0 is so enormously large compared to the time to reach 0 from $\pm d$ as to be far beyond the reach of physical time.

1.2.3 Galton–Watson Branching Process

Particles such as neutrons or organisms such as bacteria can produce new particles or organisms of the same type. Suppose that the number of particles which a single particle can produce is a random variable with a probability mass function (p.m.f.) f. Assume also that if there are i particles in the nth generation then the numbers of offspring produced by them are i independent random variables each with p.m.f. f. Let X_n denote the size of the nth generation. Then $X_n (n = 0, 1, 2, \cdots)$ is a Markov chain with state space $S = \{0, 1, 2, \cdots\}$ and transition probabilities

$$p_{ij} = f^{*i}(j) \quad (i = 1, 2, \cdots; j = 0, 1, 2, \cdots)$$
$$p_{00} = 1. \tag{1.2.25}$$

Here f^{*i} is the *i–fold convolution* of f:

$$f^{*1} = f, f^{*(i+1)}(j) = \sum_{j'} f^{*i}(j') f^{*}(j - j'). \tag{1.2.26}$$

From (12.2.16) it follows that 0 is an absorbing state. Let ρ_i denote the *probability of extinction*, or eventual absorption at 0, when $X_0 = i$. Note that $\rho_i = \rho^i$ where $\rho := \rho_1$. Of course, if $f(0) = 0$ then $\rho_1 = 0 = \rho_i$ for all $i \geq 1$, and if $f(0) = 1$ then $\rho_1 = 1 = \rho_i$ for all i. We assume henceforth that

$$0 < f(0) < 1. \tag{1.2.27}$$

To compute ρ_1 let $\phi(z) := \sum_{j=0}^{\infty} f(j) z^j = f(0) + \sum_{j=1}^{\infty} f(j) z^j (|z| \leq 1)$ be the *generating function* of f. Then

$$\phi'(z) = \sum_{j=1}^{\infty} j f(j) z^{j-1}, \ \phi''(z) \sum_{j=2}^{\infty} j(j-2) z^{j-2} \ (|z| < 1). \tag{1.2.28}$$

If the *mean* $\mu = \sum_{j=1}^{\infty} j f(j)$ of the offspring distribution is finite, then

$$\mu = \phi'(1). \tag{1.2.29}$$

Since $\phi'(z) > 0$ for $0 < z < 1, \phi$ is strictly increasing on $[0, 1]$. Let us now assume, in addition to (1.2.27), that

$$f(0) + f(1) < 1. \tag{1.2.30}$$

Then $\phi''(z) > 0$ for $0 < z < 1$, so that ϕ' is strictly increasing. Thus ϕ is a strictly increasing and strictly convex function on $[0, 1]$, and $\max \phi'(z) = \mu = \phi'(1)$. Hence, if $\mu \leq 1$, then the graph of $y = \phi(z)$ lies strictly above the line $y = z$ on $[0, 1)$. Thus the only fixed point of ϕ on $[0, 1]$ in this case is $z = 1$, since $\phi(1) = 1$. If, on the other hand, $\mu > 1$, then the graph of $y = \phi(z)$ must intersect the line $y = z$ at another point $z_0 \in (0, 1)$, in addition to $z = 1$. Thus in this case ϕ has two fixed points Now note that $\rho = \rho_1$ is a fixed point of ρ. For, writing $\rho_0 = 1$,

$$\rho = \sum_{j=0}^{\infty} P(X_1 = j | X_0 = 1) \rho_j = \sum_{j=0}^{\infty} f(j) \rho^j = \phi(\rho). \tag{1.2.31}$$

It is not difficult to check that $\rho = z_0$ in case $\mu > 1$. Thus if $\mu > 1$ then there is a positive chance $1 - \rho \equiv 1 - z_0$ of survival if $X_0 = 1$; if $X_0 = i > 1$ then the chance of survival is $1 - \rho_i = 1 - \rho^i = 1 - z_0^i$.

1.2.4 Markov Chains in Continuous Time

Let $p_{ij}(t) := \text{Prob}(X_{s+t} = j | X_s = i)$ denote the transition probabilities of a time–homogeneous right–continuous Markov process $\{X_t : t \geq 0\}$ on a countable state space $(S, \mathcal{S})$. The Markov property leads to the *Chapman–Kolmogorov equation*

$$p_{ik}(t + s) = \sum_{j \in S} p_{ij}(s) p_{jk}(t) \quad (i, j \in S). \tag{1.2.32}$$

Denoting by q_{ij} the *transition rates* or *infinitesimal parameters*,

$$q_{ij} = p'_{ij}(0) \equiv \lim_{h\downarrow 0} \frac{p_{ij}(h) - p_{ij}(0)}{h} = \lim_{h\downarrow 0} \frac{p_{ij}(h) - \delta_{ij}}{h},$$
$$(\delta_{ij} = 1 \quad \text{if} \quad i = j; \delta_{ij} = 0 \quad \text{if} \quad i \neq j) \tag{1.2.33}$$

one arrives at *Kolmogorov's backward equations*

$$p'_{ik}(t) = \sum_j q_{ij} p_{jk}(t) \quad (i, k \in S). \tag{1.2.34}$$

by differentiating (1.2.32) with respect to s and setting $s = 0$. Similarly, differentiating (1.2.32) with respect to t and setting $t = 0$, one gets *Kolmogorov's forward equations*

$$p'_{ik}(s) = \sum_j p_{ij}(s) q_{jk} \quad (i, k \in S). \tag{1.2.35}$$

Note that $q_{ij} \geq 0$ for all $i \neq j, q_{ii} = -\lambda_i \leq 0$ for all i. Since $\sum_j p_{ij}(t) = 1$ for all i and for all $t > 0$, under appropriate conditions on q_{ij}'s (e.g., $\sup\{|q_{ii}| : i \in S\} < \infty$) one may differentiate $\sum_j p_{ij}(t)$ term–by–term to get

$$\sum_j q_{ij} = 0, \quad \text{or} \sum_{\{j:j\neq i\}} q_{ij} = \lambda_i \equiv -q_{ii} (i \in S).$$
$$(q_{ij} \geq 0 \ \forall i \neq j) \tag{1.2.36}$$

Conversely, given a matrix $((q_{ij}))_{i,j\in S}$ satisfying $q_{ij} \geq 0$ for all $i \neq j$, $q_{ii} \equiv -\lambda_i \leq 0$, one may solve Kolmogorov's equations (1.2.34), or (1.2.35), iteratively, to construct the transition probabilities $p_{ij}(t)$. Under suitable conditions on the q_{ij} (e.g., $\sup\{|q_{ii}| : i \in S\} < \infty$) the solution is unique, and one has unique transition probabilities $p_{ij}(t)$. Therefore, under such conditions, given any initial state (or initial distribution), one may construct a unique Markov process with right–continuous sample paths having the transition rates q_{ij}. If no growth conditions are attached to q_{ij} satisfying (1.2.36), then one may have more than one set of solutions to (1.2.34) (or, (1.2.35). One still has a unique *minimal solution* $p^{\circ}_{ij}(t)(i, j \in S)$ satisfying (i) $p^{\circ}_{ij}(t) \geq 0$ for all i, j, and for $t > 0$, (ii) Chapman–Kolmogorov equations (1.2.32), and (iii) $\sum_j p^{\circ}_{ij}(t) \leq 1$ for all $i \in S$. In the case $\sum_j p^{\circ}_{ij}(t) < 1$ for some i, one may introduce a new absorbing state Δ_∞, say, and define $p^{\circ}_{i\Delta_\infty}(t) = 1 - \sum_j p^{\circ}_{ij}(t)$, $p^{\circ}_{\Delta_\infty\Delta_\infty}(t) = 1$, for all $t > 0$, to have a transition probability of a Markov process on $S \cup \{\Delta_\infty\}$. Such a process, say $\{X^{(0)}_t : t \geq 0\}$ starting from a state $i \in S$, may again be constructed to have right–continuous paths up to a random time ξ, an *explosion time*, at which time the process is absorbed into Δ_∞. Apart from this *minimal process*, there are in general other Markov processes, with the given transition rates q_{ij}, which may be constructed by essentially specifying an appropriate behavior after explosion. For example, the process may jump back to S according to a specified jump distribution every time an explosion occurs, successive jumps back to S being independent of each other.

A convenient method of analyzing continuous parameter Markov chains $\{X_t : t \geq 0\}$ is to consider first the successive *holding times* $T_0, T_1, \cdots$, defined by

$$T_0 = \inf\{t > 0 : X_t \neq X_0\}, \ \tau_1 = T_0,$$
$$\tau_n = \inf\{t > \tau_{n-1} : X_1 \neq X_{\tau_{n-1}}\}, \ T_{n-1} = \tau_n - \tau_{n-1}$$
$$(n \geq 2), \ \tau_0 = 0. \tag{1.2.37}$$

By the strong Markov property, the process $Y_n := X_{\tau_n} (n = 0, 1, 2, \cdots)$ is a discrete parameter Markov process on $(S, \mathcal{S})$. The one–step transition probabilities of this process are given by

$$\left.\begin{array}{ll} k_{ij} & = \frac{q_{ij}}{-q_{ii}} \equiv \frac{q_{ij}}{\lambda_i} \quad i \neq j \\ k_{ii} & = 0 \end{array}\right\} \quad (q_{ii} \neq 0) \tag{1.2.38}$$

$$\left.\begin{array}{ll} k_{ij} & = 0 \quad i \neq j \\ k_{ii} & = 1 \end{array}\right\} \quad (q_{ii} = 0).$$

Also, conditionally given $\{Y_n = i_n (n = 0, 1, 2, \cdots)\}$, the holding times $T_0, T_1, \cdots$ are independent and exponentially distributed with means $-1/q_{i_n i_n}$ $(n = 0, 1, 2, \cdots)$, assuming $\lambda_n \equiv -q_{i_n i_n} > 0$ for all n. If $q_{i_n i_n} = 0$ for some $n = m$, say, then of course, by (2.2.3), the Markov chain is absorbed in state i_m after m transitions of $\{Y_n : n = 0, 1, 2, \cdots\}$, so that $T_m = \infty$ a.s.

A *Poisson process* $\{N_t : t \geq 0\}$ with mean parameter $\lambda > 0$ is the most familiar example of a continuous parameter chain, for which $p_{ij}(t) = e^{-\lambda t}(\lambda t)^{j-i}/(j - i)!$ if $j \geq i$, $p_{ij}(t) = 0$ if $j < i$ $(i, j \in S = \{0, 1, 2, \cdots\})$. In this case the transition rates are $q_{i,i+1} = \lambda, q_{ii} = -\lambda, q_{ij} = 0$ for all other pairs (i, j). This is an example of a *pure birth process* in which the embedded discrete time process $\{Y_n : n = 0, 1, 2, \cdots\}$ is deterministic, given Y_0, i.e., $Y_n = n + Y_0$ $(n \geq 0)$. Hence, given $Y_0 = i_0$, the holding times are independent exponential random variables with means $E(T_n|Y_0 = i_0) = 1/\lambda_{i_0+n} \equiv -1/q_{i_0+n,i_0+n} (n = 0, 1, 2, \cdots)$. A general pure birth process is specified by specifying $-q_{ii} \equiv \lambda_i > 0$ for all $i \in S = \{0, 1, 2, \cdots\}$. It is not difficult to show that for a pure birth process explosion occurs if and only if $\sum_0^\infty \lambda_n^{-1} \equiv \sum_0^\infty E(T_n|X_0 = 0) < \infty$.

For another example consider a model for chemical reaction kinetics. Suppose the total number of molecules of chemicals A and B together in a solution is N. Let X_t denote the number of molecules of A at time t. In the presence of a catalyst C elementary reactions $A \to B$, signifying a transformation of a molecule of A into a molecule of B, or vice versa $(B \to A)$ may take place. The transition rates are given by

$$q_{ij} = \begin{cases} (N - i)r_B & \text{for } j = i + 1,\ 0 \leq i \leq N - 1, \\ ir_A & \text{for } j = i - 1,\ 1 \leq i \leq N, \\ -\lambda_i = -\{ir_A + (N - i)r_B\} & \text{for } j = i,\ 0 \leq i \leq N, \\ 0 & \text{otherwise} \end{cases} \tag{1.2.39}$$

Here $r_A > 0, r_B > 0$. The unique equilibrium, or invariant probability $\pi = (\pi_0, \pi_1, \cdots, \pi_N)'$ satisfies $\pi_j = \sum_k \pi_k p_{kj}(t) (t > 0)$, so that on differentiation with respect to t (at $t = 0$) one gets the equation $\sum_k \pi_k q_{kj} = 0$ $j = 0, 1, \cdots, N$. That is,

$$\begin{aligned} &\pi_{j-1}(N - j + 1)r_B - \pi_j\{jr_A + (N - j)r_B\} + \pi_{j+1}(j + 1)r_A \\ &\quad = 0 \ (j = 1, 2, \cdots, N - 1) \\ &- \pi_0 N r_B + \pi_1 r_A = 0, \ \pi_{N-1} r_B - \pi_N N r_A = 0. \end{aligned} \tag{1.2.40}$$

The solution is given by the binomial distribution

$$\begin{aligned} \pi_i &= \binom{N}{i} p^i (1 - p)^{N-i} \ (i = 0, 1, \cdots, N), \\ p &:= \frac{r_B}{r_A + r_B}. \end{aligned} \tag{1.2.41}$$

1.2.5 References

For detailed proofs of results of this section and for more comprehensive accounts of random walks and Markov chains, one may refer to Bhattacharya and Waymire [6], Chung [7], Feller [8], Karlin and Taylor [9], Spitzer [10]. For branching processes, see Harris [11] and Athreya and Ney [12].

1.3 Discrete Parameter Markov Processes on General State Spaces

[Discrete Parameter Markov Process]

This section is devoted to time–homogeneous Markov processes on general state spaces $(S, \mathcal{S})$, and especially to their ergodic properties. A theory analogous to that for Markov chains (on countable state spaces) exists for φ–irreducible Harris processes introduced by Doeblin [13], [14], Harris [15] and Orey [16]. These processes are defined below. Subsection provides certain criteria for ergodicity of such processes due largely to Doeblin [13], Tweedie [17], Athreya and Ney [18] and Nummelin [19].

A Markov process is said to be *φ–irreducible* with respect to a nontrivial (i.e., nonzero) sigma finite measure φ on $(S, \mathcal{S})$ if for every A such that $\varphi(A) > 0$ one has

$$L(x, A) := \text{Prob}(\tau_A < \infty | X_0 = x) > 0 \ \forall x \in S. \tag{1.3.42}$$

Here τ_A is the *first return time to* A:

$$\tau_A = \inf \{n \geq 1 : X_n \in A\}, \quad (A \in \mathcal{S}). \tag{1.3.43}$$

Write

$$Q(x, A) = \text{Prob}(X_n \in A \text{ for infinitely many } n | X_0 = x), \ (x \in S, A \in \mathcal{S}). \tag{1.3.44}$$

A set A is said to be *inessential* if $Q(x, A) = 0$ for every $x \in S$, otherwise A is *essential.* For a φ–irreducible Markov process there exists sets $C_1, C_2, \cdots, C_d, D$, in $\mathcal{S}$ which form a partition of S such that

(i) $p(x, C_{i+1}) = 1 \ \forall x \in C_i \ (i = 1, 2, \cdots, d-1)$, $p(x, C_1) = 1 \ \forall x \in C_d$,

(ii) $\varphi(D) = 0$ and D is (at most) a countable union of inessential sets,

and (iii) if there is any other cycle of sets $C'_j (j = 1, 2, \cdots, d')$ satisfying (i) then d' divides d. This maximal d is called the *period* of the Markov process. If $d = 1$ then the Markov process is said to be *aperiodic.*

A Markov process is said to be *φ–recurrent*, or *Harris recurrent*, with respect to a (nonzero) sigma finite measure φ if $Q(x, A) = 1$ for $x \in S$ whenever $\varphi(A) > 0$. In particular, a φ–recurrent process is φ–irreducible.

A Markov process is said to be *ergodic* if it has a unique invariant probability, say, π.

1.3.1 Ergodicity of Harris Recurrent Processes

A basic result of Doeblin [13] sets the stage for the general results in this subsection. Let $p(x, dy)$ be the transition probability of a Markov process $\{X_n : n = 0, 1, 2, \cdots\}$ on $(S, \mathcal{S})$, and $p^{(n)}(x, dy)$ the n–step transition probability. The following condition is called the *Doeblin minorization: There exists $N \geq 1, \delta > 0$ and a probability measure ν on $(S, \mathcal{S})$ such that*

$$p^{(N)}(x, B) \geq \delta \nu(B) \quad \forall x \in S, B \in \mathcal{S}. \tag{1.3.45}$$

Theorem 1.3.1 *(Doeblin). Under the Doeblin minorization (12.3.20) there exists a unique invariant probability π, and one has*

$$|p^{(n)}(x,B) - \pi(B)| \leq (1-\delta)^{[\frac{n}{N}]} \ \forall x \in S, B \in \mathcal{S}, \tag{1.3.46}$$

where $[\frac{n}{N}]$ denotes the integer part of $\frac{n}{N}$.

To give an idea of the proof, let T^{*n} denote the linear operator on the space $\mathcal{P}(S)$ of all probability measures on $(S, \mathcal{S})$ defined by

$$T^{*n}\mu(B) = \int_S p^{(n)}(x,B)\mu(dx), \ B \in \mathcal{S}. \tag{1.3.47}$$

In other words, $T^{*n}\mu$ is the distribution of X_n when X_0 has distribution μ. Let $d_{TV}(\mu_1, \mu_2)$ denote the *total variation distance* on $\mathcal{P}(S)$ defined by

$$d_{TV}(\mu_1\mu_2) \equiv \|\mu - \nu\|_{TV} = \sup_{B \in \mathcal{S}} |\mu_1(B) - \mu_2(B)|, \ (\mu_1, \mu_2 \in \mathcal{P}(S)). \tag{1.3.48}$$

Condition (12.3.20) implies that T^{*N} is a strict contraction:

$$d_{TV}(T^{*N}\mu_1, T^{*N}\mu_2) \leq (1-\delta) d_{TV}(\mu_1, \mu_2). \tag{1.3.49}$$

Since $(\mathcal{P}(S), d_{TV})$ is a complete metric space, (1.3.49) implies the existence of a unique fixed point π of T^{*N}, and

$$d_{TV}(T^{*kN}\mu, \pi) \leq (1-\delta)^k d_{TV}(\mu, \pi) \leq (1-\delta)^k, \ (k = 1, 2, \cdots). \tag{1.3.50}$$

Since $d_{TV}(T^{*(kN+j)}\mu_1, \pi) = d_{TV}(T^{*kN}(T^{*j}\mu), \pi) \leq (1-\delta)^k$ for all $j = 0, 1, \cdots, N-1$, (12.3.21) follows.

It is known that Doeblin minorization is in fact necessary as well as sufficient for uniform (in x) exponential convergence in total variation distance to a unique invariant probability π (see Nummelin [20], Theorem 6.15).

We next consider a *local minorization condition* on a set $A_0 \in \mathcal{S}$ given by

$$p^{(N)}(x,B) \geq \delta\nu(B) \ \forall x \in A_0, B \in A_0 \cap \mathcal{S} \tag{1.3.51}$$

for some $N \geq 1, \delta > 0$, and a probability measure ν on $(S, \mathcal{S})$ such that $\nu(A_0) = 1$. A set A_0 satisfying (12.3.24) is sometimes called a $(\nu-)$ *small set*. If, in addition, A_0 is a *recurrent set*, i.e.,

$$L(x, A_0) = 1 \quad \forall x \in S, \tag{1.3.52}$$

and

$$\sup_{x \in A_0} E_x \tau_{A_0} < \infty, \tag{1.3.53}$$

where E_x denotes expectation when $X_0 \equiv x$, there exists a unique invariant probability. The following result is due to Athreya and Ney [18], Nummelin [19].

Theorem 1.3.2 *If the local minorization (12.3.24) holds on a recurrent set A_0, and (12.3.26) holds, then there exists a unique invariant probability π and, for all $x \in S$,*

$$\|\frac{1}{n}\sum_{j=1}^{n} p^{(j)}(x,\cdot) - \pi\|_{TV} \to 0 \text{ as } n \to \infty. \tag{1.3.54}$$

To understand the main ideas behind the proof, consider the Markov process $\{X_{\tau^{(n)}} : n = 0, 1, \cdots\}$ on $(A_0, A_0 \cap \mathcal{S})$ observed at successive return times $\tau^{(n)}$ to $A_0 : \tau^{(0)} = 0, \tau^{(1)} = \tau_{A_0}, \tau^{(n)} = \inf\{j > \tau^{(n-1)} : X_j \in A_0\}(n \geq 1)$. Its transition probability $p_{A_0}(x, dy)$ has the Doeblin minorization property (12.3.20) with $N = 1$; therefore, by Theorem 3.1, it has a unique invariant probability π_0 on $(A_0, A_0 \cap \mathcal{S})$. Given any $B \in \mathcal{S}$ the proportion of time spent in B during the time period $\{0, 1, \cdots, n\}$, namely, $n^{-1}\sum_{j=1}^{n} 1_{\{X_j \in B\}}$, can now be shown to converge (a.s. π_0) to

$$\pi(B) := \frac{1}{E_{\pi_0}\tau_{A_0}} E_{\pi_0} \sum_{1}^{\tau_{A_0}} \mathbf{1}_{\{X_j \in B\}} = \frac{1}{E_{\pi_0}\tau_{A_0}} \int_{A_0} p_{A_0}(x, B)\pi_0(dx), \tag{1.3.55}$$

where E_{π_0} denotes expectation under π_0 as the initial distribution (on A_0) and, for general $B \in \mathcal{S}, x \in A_0$,

$$p_{A_0}(x, B) := \sum_{n=1}^{\infty} \text{Prob}(X_n \in B, X_k \in A_0^c \text{ for } 1 \leq k < n). \tag{1.3.56}$$

Note that (12.3.29) is consistent with the notation $p_{A_0}(x, dy)$ as the transition probability of $\{X_{\tau^{(n)}} : n = 0, 1, \cdots\}$ on A_0. Viewed as a measure on $(S, \mathcal{S})$ (for each $x \in A_0$), the total mass of $p_{A_0}(x, \cdot)$ is $p_{A_0}(x, S) = E_x\tau_{A_0}$. The probability π in (12.3.28) is the unique invariant probability for $p(x, dy)$ on $(S, \mathcal{S})$.

It is known that if $\mathcal{S}$ is countably generated, then the local minorization condition (12.3.24) on a recurrent set A_0 is equivalent to Harris recurrence (i.e., ν–recurrence) of $\{X_n : n = 0, 1, \cdots\}$.

In order to apply Theorem 3.2 one needs to find a set A_0 satisfying (12.3.24), (12.3.25), (12.3.26). The following result provides an effective criterion for a set A_0 to satisfy (12.3.25), (12.3.26).

Theorem 1.3.3 *(Foster–Tweedie Drift Criterion). Suppose $A_0 \subset S$ is such that a local minorization condition (12.3.24) holds. Assume that, in addition, there exists a nonnegative measurable function V on S such that*

$$(i) \int_S V(y)p(x, dy) \leq V(x) - 1 \quad \forall x \in A_0^c,$$

$$(ii) \sup_{x \in A_0} \int_S V(y)p(x, dy) < \infty. \tag{1.3.57}$$

Then there exists a unique invariant probability π and (12.3.27) holds. If, in addition, the Markov process is strongly aperiodic in the sense that (12.3.24) holds with $N = 1$, then

$$\|p^{(n)}(x, \cdot) - \pi\|_{TV} \to 0 \text{ as } n \to \infty \quad (x \in S). \tag{1.3.58}$$

One proves this by showing that (12.3.30) implies (Meyn and Tweedie [21], p. 265)

$$E_x\tau_{A_0} \leq \begin{cases} V(x) & \forall x \in A_0^c \\ 1 + \int_S V(y)p(x, dy) & \forall x \in A_0, \end{cases} \tag{1.3.59}$$

so that (12.3.25) and (12.3.26) both hold under (12.3.30).

By strengthening (12.3.30)(i) to: *there exists* $\theta < 1$ *such that*

$$\int_S V(y)p(x,dy) \leq \theta V(x) \quad \forall x \in A_0^c, \tag{1.3.60}$$

One obtains *geometric ergodicity*, namely,

Theorem 1.3.4 *(Geometric Ergodicity). Suppose (12.3.24), (12.3.30)(ii) and (12.3.33) holds (for some* $\theta < 1$*) for a measurable function* V *having values, in* $[1,\infty)$*. Then there exists* $\rho \in (0,1)$ *and a function* $C(x)$ *with values in* $(0,\infty)$ *such that*

$$\|\frac{1}{n}\sum_{j=1}^{n} p^{(j)}(x,\cdot) - \pi\|_{TV} \leq C(x)\rho^n \quad \forall x \in S, n \geq 1. \tag{1.3.61}$$

If, in addition (12.3.24) holds with $N = 1$*, then one has*

$$\|p^{(n)}(x,\cdot) - \pi(\cdot)\|_{TV} \leq C(x)\rho^n \quad \forall x \in S, n \geq 1. \tag{1.3.62}$$

For a proof of this see Meyn and Tweedie [21], Chapter 15.

1.3.2 Iteration of I.I.D. Random Maps

Many Markov processes, if not a majority, that arise in applications are specified as stochastic, or randomly perturbed, dynamical systems. In continuous time there may be given by, e.g., stochastic differential equations, and in discrete time by stochastic difference equations or recursions. Among examples of the latter type are the autoregressive models. In general, these discrete time processes are represented as actions of random iteration of i.i.d. random maps on S.

Since such representations often arise from, and display, physical dynamical considerations, they also in many instances suggest special methods of analysis of large time behavior. Additionally, the present topic gains significance from the fact that most Markov processes may be represented as actions of iterated i.i.d. random maps, as the following proposition shows.

To be precise, let S be a Borel subset of a Polish space $\mathcal{X}$. Recall that a Polish space $\mathcal{X}$ is a topological space which is metrizable as a complete separable metric space. For example, S may be a Borel subset of a euclidean space. Let $\mathcal{S}$ be the Borel sigma field of S. For random maps $\alpha_n (n \geq 1)$ on S into itself we will write $\alpha_1 x := \alpha_1(x), \alpha_n\alpha_{n-1}\cdots\alpha_1 x := \alpha_n \circ \alpha_{n-1} \circ \cdots \circ \alpha_1 x$.

Proposition 1.3.5 *Let* $p(x,dy)$ *be a transition probability on* $(S,\mathcal{S})$*, where* S *is a Borel subset of a Polish Space and* $\mathcal{S}$ *is the Borel sigma field on* S*. There exists (i) a probability space* $(\Omega,\mathcal{F},P)$ *and (ii) a sequence of i.i.d. random maps* $\{\alpha_n : n = 1,2,\cdots\}$ *on* S *into itself such that* $\alpha_n x$ *has distribution* $p(x,dy)$*. In particular, the recursion*

$$X_0 \equiv x_0, \; X_n = \alpha_n X_{n-1} \quad (n \geq 1), \tag{1.3.63}$$

or, $X_n = \alpha_n\alpha_{n-1}\cdots\alpha_1 x_0 (n \geq 1), X_0 = x_0$*, defines a Markov process* $\{X_n : n \geq 0\}$ *with initial state* x_0 *and transition probability* p*.*

Conversely, given a sequence of i.i.d. random maps $\{\alpha_n : n \geq 1\}$ on any measurable state space $(S,\mathcal{S})$ (not necessarily a Borel subset of a Polish space), one may define the Markov process $\{X_n : n = 0,1,2,\cdots\}$ having the transition probability

$$p(x,B) := \text{Prob}(\alpha_1 x \in B), \quad x \in S, \; B \in \mathcal{S}. \tag{1.3.64}$$

Note that one requires the event $\{\omega \in \Omega : \alpha_1(\omega)x \in B\}$ to belong to the sigma field $\mathcal{F}$ (of the underlying probability spaces $(\Omega, \mathcal{F}, P)$), and also $x \to p(x, B)$ must be measurable on $(S, \mathcal{S})$. These two requirements are satisfied if $(\omega, x) \to \alpha_1(\omega)x$ is measurable on $(\Omega \times S, \mathcal{F} \otimes \mathcal{S})$ into $(S, \mathcal{S})$. A *random map* is defined to be a map satisfying the last measurability property.

Example 1 *(Random Walk)*. Here $S = \mathbb{Z}^k$ or $\mathbb{R}^k$, and

$$X_{n+1} = X_n + \varepsilon_{n+1} \; (n \geq 0), \; X_0 \equiv x_0, \tag{1.3.65}$$

where $\{\varepsilon_n\}$ are i.i.d. One may take $\alpha_n(\omega)x := x + \varepsilon_n(\omega) \; (x \in S), n \geq 1$.

Example 2 *(Linear Models)*. Here $S = \mathbb{R}^k$, and given a $k \times k$ matrix A and an i.i.d. sequence of mean zero random vectors $\{\varepsilon_n : n \geq 1\}$ one defines

$$X_{n+1} = AX_n + \varepsilon_{n+1} \; (n \geq 0), \; X_0 \equiv x_0. \tag{1.3.66}$$

Take $\alpha_n(\omega)$ to be the map $\alpha_n(\omega)x = Ax + \varepsilon_n(\omega)(n \geq 1)$.

Example 3 *(Autoregressive Models)*. Let $p \geq 1$, $\beta_0, \beta_1, \cdots, \beta_{p-1}$ real constants, $\{\eta_n : n \geq p\}$ an i.i.d. sequence of mean zero real–valued random variables, and let $Y_0, Y_1, \cdots, Y_{p-1}$ be independent of $\{\eta_n : n \geq p\}$. Define

$$Y_{n+p} = \sum_{i=0}^{p-1} \beta_i Y_{n+i} + \eta_{n+p} \quad (n \geq 0). \tag{1.3.67}$$

Then $\{Y_n : n \geq 0\}$ is said to be an autoregressive process of order p or, in brief, an $AR(p)$ process. Now let

$$\mathbf{X}_n = (Y_n, Y_{n+1}, \cdots, Y_{n+p-1})' \quad (n \geq 0). \tag{1.3.68}$$

Then one may write

$$\mathbf{X}_n = A\mathbf{X}_{n-1} + \boldsymbol{\varepsilon}_n \quad (n \geq 0), \tag{1.3.69}$$

where A is the $p \times p$ matrix

$$A = \begin{bmatrix} 0 & 1 & 0 & 0 & \cdot & \cdot & \cdot & \cdot & 0 & 0 \\ 0 & 0 & 1 & 0 & \cdot & \cdot & \cdot & \cdot & 0 & 0 \\ \cdot & \cdot & \cdot & \cdot & \cdot & \cdot & \cdot & \cdot & \cdot & \cdot \\ 0 & 0 & 0 & 0 & \cdot & \cdot & \cdot & \cdot & 0 & 1 \\ \beta_0 & \beta_1 & \beta_2 & \beta_3 & \cdot & \cdot & \cdot & \cdot & \beta_{p-2} & \beta_{p-1} \end{bmatrix}, \tag{1.3.70}$$

and

$$\boldsymbol{\varepsilon}_n = (0, 0, \cdots, 0, \eta_{n+p-1})' \quad (n \geq 1). \tag{1.3.71}$$

Thus one may treat this example as a special case of Example 1. If A is a *stable* matrix, i.e., the eigenvalues of A are all of modulus less than one (thus lying inside the unit circle in the complex plane), then the series

$$\sum_{n=0}^{\infty} A^n \boldsymbol{\varepsilon}_{n+1} \tag{1.3.72}$$

converges to a random vector $\mathbf{Z}$, say, and it follows that the Markov process $\{\mathbf{X}_n : n \geq 0\}$ has a unique invariant probability, say π, and $\mathbf{X}_n$ converges in distribution to π as $n \to \infty$, no matter what the initial distribution is. The eigenvalues of A are the (generally complex valued) solutions of the equation (in λ)

$$0 = \det(A - \lambda \mathbf{1}) = (-1)^{p+1}(\beta_0 + \beta_1\lambda + \cdots + \beta_{p-1}\lambda^{p-1} - \lambda^p). \tag{1.3.73}$$

The following result is now immediate.

Proposition 1.3.6 *If the roots of the polynomial equation (1.3.73) all lie inside the unit circle, namely in $\{z \in \mathbb{C} : |z| < 1\}$, then (a) the Markov process $\{\boldsymbol{X}_n : n \geq 0\}$ defined by (12.3.43) has a unique invariant probability π and $\boldsymbol{X}_n$ converges in distribution to π as $n \to \infty$, and (b) the AR(p) process $\{Y_n : n \geq 0\}$ is asymptotically stationary, with Y_n converging in distribution to the (marginal) distribution π_1 of Z_1 where $\boldsymbol{Z} = (Z_1, \cdots, Z_p)$ has distribution π.*

In the statement above the term asymptotic stationarity may be formally defined as follows. Let $Y_n (n \geq 1)$ be a sequence of random variables with values in a Polish spaces S with Borel sigma field $\mathcal{S}$. The sequence $\{Y_n : n \geq 1\}$ is said to be *asymptotically stationary* if the distribution Q_m of $\mathbf{Y}_m := (Y_m, Y_{m+1}, \cdots)$ on $(S^\infty, \mathcal{S}^{\otimes\infty})$ converges weakly to the distribution Q_∞, of a stationary process, say, $\mathbf{U} = (U_1, U_2, \cdots)$, as $m \to \infty$. It may be checked that weak convergence in the space $\mathcal{P}(S^\infty)$ of all probability measures on $(S^\infty, \mathcal{S}^{\otimes\infty})$ is equivalent to weak convergence of all finite–dimensional distributions.

Example 4 *(ARMA Models).* To define an *autoregressive moving–average process* of order (p, q), briefly *ARMA*(p, q), let $\beta_0, \beta_1, \cdots, \beta_{p-1}$ and $\theta_1, \theta_2, \cdots, \theta_q$ be $p+q$ real numbers (constants), $\{\eta_n : n \geq p\}$ an i.i.d. sequence of real–valued mean–zero random variables, and $(Y_0, Y_1, \cdots, Y_{p-1})$ a given p–tuple of real–valued random variables. The ARMA (p, q) process $\{Y_n : n \geq 0\}$ is then given, recursively, by

$$Y_{n+p} = \sum_{i=0}^{p-1} \beta_i Y_{n+i} + \sum_{j=1}^{q} \theta_j \eta_{n+p-j} + \eta_{n+p} \quad (n \geq 0). \tag{1.3.74}$$

As in Example 3, this admits a Markovian representation

$$\mathbf{X}_{n+1} = B\mathbf{x}_n + \boldsymbol{\varepsilon}_{n+1} \quad (n \geq 0), \tag{1.3.75}$$

where $\mathbf{X}_n, \boldsymbol{\varepsilon}_n$ are $(p+q)$–dimensional random vectors,

$$\begin{aligned} \mathbf{X}_n &:= (Y_n, Y_{n+1}, \cdots, Y_{n+p-1}, \eta_{n+p-q}, \cdots, \eta_{n+p-1})', \\ \boldsymbol{\varepsilon}_n &:= (0, 0, \cdots, 0, \eta_{n+p-1}, 0, 0, \cdots, 0, \eta_{n+p-1})', \quad (n \geq 0) \end{aligned} \tag{1.3.76}$$

with only the pth and $(p+q)$th coordinates of $\boldsymbol{\varepsilon}_n$ as η_{n+p-1} and the others are zero. The $(p+q) \times (p+q)$ matrix B in (3.34) is given by

$$\begin{bmatrix} A & \\ & C \\ 0 & \end{bmatrix} = B,$$

where A is the matrix (3.29) and

$$C = \begin{bmatrix} 0 & 0 & 0 & \cdot & \cdot & 0 & 0 \\ 0 & 0 & 0 & \cdot & \cdot & 0 & 0 \\ 0 & 0 & 0 & \cdot & \cdot & 0 & 0 \\ 0 & 0 & 0 & \cdot & \cdot & \cdot & 0 \\ \theta_q & \theta_{q-1} & \cdot & \cdot & & \theta_2 & \theta_1 \\ 0 & 1 & 0 & 0 & \cdot & 0 & 0 \\ 0 & 0 & 1 & 0 & \cdot & 0 & 0 \\ \cdot & \cdot & \cdot & \cdot & \cdot & \cdot & \cdot \\ 0 & 0 & 0 & 0 & \cdot & 1 & 0 \\ 0 & 0 & 0 & 0 & & 0 & 0 \end{bmatrix}. \tag{1.3.77}$$

The eigenvalues of B are the p eigenvalues of A and q zeros. Therefore, B is stable if and only if A is stable. Thus we arrive at

Proposition 1.3.7 *If the roots of (1.3.73) all lie inside the unit circle in the complex plane then (a) the Markov process* $\{\boldsymbol{X}_n : n \geq 0\}$ *defined by (12.3.47), (12.3.49), has a unique invariant probability* π*, and (b) the ARMA* (p, q) *process* $\{Y_n : n \geq 0\}$ *is asymptotically stationary with* Y_n *converging in distribution to the (marginal) distribution* π_1 *of* Z_1 *where* $\boldsymbol{Z} = (Z_1, \cdots, Z_{p+q})$ *has distribution* π.

Example 5 *(Nonlinear Autoregressive Models).* Let $p \geq 1$. Consider the real–valued process defined recursively by

$$Y_{n+p} = h(Y_n, Y_{n+1}, \cdots, Y_{n+p-1}) + \varepsilon_{n+p} \ (n \geq 0) \tag{1.3.78}$$

where (i) $\{\varepsilon_n : n \geq p\}$ is an i.i.d. sequence of mean–zero random variables with a common density which is positive on $\mathbb{R}$, (ii) h is a real–valued measurable function on $\mathbb{R}^p$ which is bounded on compacts, (iii) $(Y_0, Y_1, \cdots, Y_{p-1})$ is a given p–tuple of random variables independent of $\{\varepsilon_n : n \geq p\}$. By applying the Foster–Tweedie drift criterion for geometric ergodicity (Theorem 3.4) on may prove the following result.

Proposition 1.3.8 *In addition to assumptions (i)–(iii) above, assume that there exist* $a_i \geq 0, (i = 1, \cdots, p)$ *with* $\sum_1^p a_i < 1$*, and* $R > 0$ *such that*

$$|h(\boldsymbol{y})| \leq \sum_{i=1}^{p} a_i y_i \quad for \quad |\boldsymbol{y}| > R. \tag{1.3.79}$$

Then the Markov process

$$\boldsymbol{X}_n := (Y_n, Y_{n+1}, \cdots, Y_{n+p-1}), \quad (n \geq 0) \tag{1.3.80}$$

has a unique invariant probability π *and is geometrically ergodic. In particular,* $\{Y_n : n \geq 0\}$ *is asymptotically stationary and* Y_n *converges in distribution to a probability* π_1 *on* $\mathbb{R}$*, the convergence being exponentially fast in total variation distance.*

1.3.3 Ergodicity of Non–Harris Processes

The general criteria for ergodicity, or the existence of a unique invariant probability, presented in Section 1.3.1 apply only to processes which are Harris, i.e., φ–irreducible with respect to a non–trivial sigma finite measure φ. We now consider certain classes of Markov processes for which no such φ may exist. These often arise as actions of iterations of i.i.d. random maps on a state space $(S, \mathcal{S})$.

Theorem 1.3.9 below applies to Markov processes on a complete separable metric space (S, ρ), with a Borel sigma field $\mathcal{S}$, on which are defined an i.i.d. sequence of random Lipschitz maps $\{\alpha_n : n \geq 1\}$: $X_n := \alpha_n \alpha_{n-1} \cdots \alpha_1 X_0 (n \geq 1), X_0$ independent of $\{\alpha_n : n \geq 1\}$. *Lipschitz constant* of a function f, i.e., L_f is the smallest constant M such that $\rho(f(x), f(y)) \leq M\rho(x, y)$ for all x, y. The following result due to Diaconis and Freedman [22] says, roughly speaking, that if the (harmonic) average of L_{α_1} is less than one and if, for some $x_0 \in S$, $\alpha_1 x_0$ remains bounded on the average, then X_n has a unique invariant probability.

Theorem 1.3.9 *Let* (S, ρ) *be a complete separable metric space such that the i.i.d. maps* $\alpha_n (n \geq 1)$ *satisfy*

$$E \log L_{\alpha_1} < 0. \tag{1.3.81}$$

If, in addition, for some $x_0 \in S$ *one has*

$$E\rho(\alpha_1 x_0, x_0) < \infty, \tag{1.3.82}$$

then (a) there exists a unique invariant probability and the Markov process $\{X_n : n \geq 0\}$ *is asymptotically stationary and ergodic, no matter what the initial distribution is.*

To understand the main idea behind the proof, note that for the processes $X_n(x)$ and $X_n(y)$ corresponding to $X_0 \equiv x$ and $X_0 \equiv y$, respectively,

$$\rho(X_n(x), X_n(y)) \le L_{\alpha_n}\rho(X_{n-1}(x), X_{n-1}(y)) \le \cdots \le \Big(\prod_{j=1}^{n} L_{\alpha_j}\Big)\rho(x, y).$$

Taking logarithms, and using the strong law of large numbers and (12.3.55), one shows that $\sup_{x\in B}\rho(X_n(x), X_n(y)) \to 0$ a.s. as $n \to \infty$ for every bounded $B \subset S$. In particular, if there exists an invariant probability, say, $\pi, p^{(n)}(x, dz)$ converges weakly to it as $n \to \infty$, for every x. Therefore, there can not be more than one invariant probability. Now the condition (1.3.82), in conjunction with (12.3.55), implies tightness of $\{p^{(n)}(x_0, dz) : n = 1, 2, \cdots\}$, proving (a). Asymptotic stationarity follows from this (See the remark following the statement of Proposition 3.6).

It may be noted that for (3.40) one allows the possibility $E \log L_{\alpha_1} = -\infty$, and even that of $\mathrm{Prob}(L_{\alpha_1} = 0) > 0$.

One may also relax (12.3.55), (1.3.82) by requiring the inequalities to hold for the N–fold composition $\alpha_N \cdots \alpha_2\alpha_1$, for some $N \ge 1$. From this extension of Theorem 3.8 one may device the ergodicity of $\mathbf{X}_n$ in the $AR(p)$ model of Example 3, Section 1.3.1.

Among many applications we mention the construction of fractal images by iteration of i.i.d. $\alpha_n (n \ge 1)$, where α_n takes values in a given finite set of affine maps. The affine maps themselves are chosen using some features of the target image (Diaconis and Freedman [22], Diaconis and Shahshahami [23], Barnsley and Elton [24], Barnsley [25]).

The next result is a generalization of a theorem of Dubins and Freedman [26] on monotone maps on an interval. Following an earlier generalization by Bhattacharya and Lee [27] to closed subsets of $\mathbb{R}^k$, and open or semi open rectangles, the theorem below is derived in Bhattacharya and Majumdar [28].

Theorem 1.3.10 *Let $\{\alpha_n : n \ge 1\}$ be a sequence of i.i.d. maps on a measurable space $(S, \mathcal{S})$ with the following properties: There exists a class of sets $\mathcal{A} \subset \mathcal{S}$ such that (i) $\alpha_1^{-1}A \in \mathcal{A}$ a.s. $\forall A \in \mathcal{A}$, (ii) there exists $N \ge 1$ and $\delta > 0$ such that $Prob((\alpha_N \cdots \alpha_2\alpha_1)^{-1}A = S$ or $\phi) \ge \delta > \forall A \in \mathcal{A}$, and (iii) Under $d(\mu, \nu) := \sup\{|\mu(A) - \nu(A)| : A \in \mathcal{A}\}$, $(\mathcal{P}(S), d)$ is a complete metric space, where $\mathcal{P}(S)$ is the set of all probability measures on $(S, \mathcal{S})$. Then there exists a unique invariant probability π for the Markov process generated by the iterations of $\{\alpha_n : n \ge 1\}$, and one has*

$$\sup_{x\in S, A\in\mathcal{A}} |p^{(n)}(x, A) - \pi(A)| \le (1 - \delta)^{[\frac{n}{N}]}, n \ge 1, \tag{1.3.83}$$

where $[n/N]$ is the integer part of n/N.

For a proof note that for arbitrary $\mu, \nu \in \mathcal{P}(S)$,

$$\begin{aligned} d(T^{*N}\mu, T^{*N}\nu) &\equiv \sup_{A\in\mathcal{A}} |T^{*N}\mu(A) - T^{*N}\nu(A)| \\ &\le (1 - \delta)d(\mu, \nu), \end{aligned} \tag{1.3.84}$$

by assumptions (i) and (ii). Thus T^{*N} is a strict contraction on $(\mathcal{P}(S), d)$. By the assumption (iii) of completeness it now follows that T^{*N} has a unique fixed point π and that $d(T^{*kN}\mu\pi) \le (1 - \delta)^k d(\mu, \pi)$, for all $k = 1, 2, \cdots$. From this (3.42) follows on setting $\mu = p^{(n)}(x, dz)$.

As a first consequence of Theorem 1.3.9, one obtains the following result of Dubins and Freedman [26].

Corollary 1.3.11 *Let $\{\alpha_n : n \geq 1\}$ be i.i.d. monotone random maps on an interval $S = J$. If there exist $N \geq 1$, $x_0 \in J$ and $\delta > 0$ such that*

$$Prob(\alpha_N \cdots \alpha_1 x \leq x_0 \forall x \text{ or } \alpha_N \cdots \alpha_1 x \geq x_0 \forall x) \geq \delta, \tag{1.3.85}$$

then there exists a unique invariant probability π and one has

$$\sup_{x,c} |\text{Prob}(X_n(x) \leq c) - \pi(\{x \in J : x \leq c\})| \leq (1-\delta)^{[\frac{n}{N}]}. \tag{1.3.86}$$

To derive this from Theorem 1.3.9, one takes $\mathcal{A}$ to be the class of all intervals of the form $A = \{x \in J : x \leq c\}$, $A = \{x \in J : x < c\}$, and their complements. Assumption (i) of Theorem 1.3.9 is immediate due to monotonicity, while (ii) follows from (12.3.61). Finally, the distance d in this case is just the supremum distance between cumulative distribution functions, and therefore $(\mathcal{P}(S), d)$ is complete.

The above result for monotone maps has many applications. Among theoretical consequences, it provides criteria for ergodicity for processes generated by iterations of i.i.d. random quadratic maps (see Bhattacharya and Rao [29], Bhattacharya and Waymire [30]). For applications to economics see Bhattacharya and Waymire [6], pp. 178–180, and Bhattacharya and Majumdar [31].

It is possible to generalize (12.3.24) to i.i.d. monotone maps on arbitrary partially ordered spaces. See, e.g., Bhattacharya and Lee [27] and Hopenhayn and Prescott [32]. A particular application to the Ising model of mathematical physics due to Propp and Wilson [33] is derived in detail in Diaconis and Freedman [22].

1.3.4 References

Orey [16], Nummelin [20] and Meyn and Tweedie [21] are standard references for φ–irreducible and Harris recurrent processes. Doeblin's pioneering work on geometric ergodicity is contained in Doeblin [13], [14]. Theorem 3.2, obtained independently by Athreya and Ney [18] and Nummelin [19], has its roots in the notion of C–sets in Orey [16]. Theorems 3.3, 3.4 on drift criteria for ergodicity and geometric ergodicity were obtained by Tweedie [17], [34].

Proofs of Proposition 3.5 on representation by iteration of i.i.d. random maps may be found in Kifer [35], pp. 7–9, and Bhattacharya and Waymire [6], p. 228. The result is due to Blumenthal and Carson [36], at least under the condition of continuity. The treatment of $AR(p)$ and $ARMA(p,q)$ models given here follows Bhattacharya and Waymire [6], pp. 166–173.

Theorem 3.8 for ergodicity under (harmonic) average contraction is due to Diaconis and Freedman [22]. This generalizes earlier work by Dubins and Freedman [26], Bhattacharya and Lee [37], Barnsley and Elton [24] and others. The basic work for ergodicity under iteration of i.i.d. monotone maps is due to Dubins and Freedman [26], who obtained Corollary 3.10 for compact intervals and continuous maps. An extension to measurable maps on intervals was given by Yahav [38]. Bhattacharya and Lee [27] generalized the results to measurable monotone maps on closed subsets of $\mathbb{R}^k$ and other sets, and derive functional CLTs for appropriate classes of functions on the state space. Hopenhayn and Prescott [32] provide an extension to general partially ordered spaces. Theorem 3.10, which includes these results on ergodicity, is derived in Bhattacharya and Majumdar [28].

1.4 Continuous Time Markov Processes on General State Spaces

It is not difficult to extend the notions of ϕ–irreducibility, Harris recurrence, Doeblin minorization, etc. to continuous parameter Markov processes $\{X_t : t \geq 0\}$ on general state spaces $(S, \mathcal{S})$, and one can derive analogues of Theorems 3.1–3.4 in this case. Our aim in this section is, however, to briefly introduce jump processes and processes with independent increments.

1.4.1 Processes with Independent Increments

A process $\{X_t : t \geq 0\}$ with values in $\mathbb{R}$ is said to have *independent increments* if $X_{t_{i+1}} - X_{t_i}, i = 0, 1, \cdots, n-1$ are independent random variables for every finite set of time points $0 = t_0 < t_1 < \cdots < t_n$. This is equivalent to requiring that, for all $0 < s < t$, $X_t - X_s$ is independent of $\mathcal{G}_s := \sigma\{X_u : 0 \leq u \leq s\}$. Such a process is said to be *homogeneous* if the distribution of $X_t - X_s$ depends only on $t-s$ for all $0 \leq s < t$, i.e., $X_{t+h} - X_{s+h}$ has the same distribution as $X_t - X_s$ for all $0 \leq s < t$, and all $h > 0$. For processes with homogeneous independent increments, the distribution of $X_t - X_s$ is *infinitely divisible* (i.d.), since for each $n \geq 1, X_t - X_s$ may be expressed as the sum of n i.i.d. random variables. Conversely, given any infinitely divisible law γ, there exists a homogeneous process with independent increments $\{X_t : t \geq 0\}$ such that $X_{t+1} - X_t$ has distribution γ, for all $t \geq 0$. For a fixed initial value, say, $X_0 = 0$ (or $X_0 = x_0$), all finite–dimensional distributions of a process with homogeneous independent increments are determined by this i.d. law γ of $X_1 - X_0$. We will choose a version of such a process which has right–continuous sample paths with finite left limits. This is always possible (See Doob [39]).

Crucial to the sample path analysis of such processes is the following result of Paul Lévy [40]—[42].

Theorem 1.4.1 *Let $\{X_t : t \geq 0\}$ be a homogeneous process with independent increments. (a) If the process has continuous sample paths almost surely, then it is Gaussian, i.e., a Brownian motion. (b) If $t \to X_t$ is an increasing step function with jumps of size one only, almost surely, then the process is a Poisson process.*

This theorem may be proved by expressing $X_t - X_s$ as a sum of n i.i.d. random variables $(n = 2, 3, \cdots)$ and computing the limit of the characteristic function of the sum as $n \to \infty$. For a detailed proof see Itô [43], Section 1.4.

Consider now an arbitrary process $\{X_t : t \geq 0\}$ with homogeneous independent increments. Let $\mathcal{B}_\varepsilon$ denote the Borel sigma field of $\{x \in \mathbb{R} : |x| > \varepsilon\} \equiv \mathbf{R}_\varepsilon$. Let $\nu_t(B)$ denote the number of jumps of $\{X_u : 0 \leq u \leq t\}$ in $B \in \mathcal{B}_\varepsilon$. Then $\{\nu_t(B) : t \geq 0\}$ is a process with independent increments, whose sample paths are nondecreasing step functions with jumps of size one only. Hence, by Theorem 4.1(b), $\{\nu_t(B) : t \geq 0\}$ is a *Poisson process*, for each $B \in \mathcal{B}_\varepsilon$. We now use another remarkable result of Lévy, which says that two processes with independent increments are independent if they do not have any common points of discontinuity (a.s.). This implies that if $B_1, B_2, \cdots, B_n$ are disjoint sets in $\mathcal{B}_\varepsilon$, then the Poisson processes $\{\nu_t(B_i) : t \geq 0\}, i = 1, 2, \cdots, n$, are independent. Hence $\nu_t(dx)$ is a random measure on $(\mathbf{R}_\varepsilon, \mathcal{B}_\varepsilon)$, and defines a Poisson random field. The contribution to X_t of all jumps of size greater than ε is then given by

$$X_{t,\varepsilon} := \sum_{\{x \in \mathbf{R}_\varepsilon : X_s - X_{s-} = x \text{ for some } s \in [0,t]\}} x = \int_{\mathbf{R}_\varepsilon} x \nu_t(dx). \tag{1.4.87}$$

Write $\Psi_t(B) := E\nu_t(B), B \in \mathcal{B}_\varepsilon$. Then $\Psi_t = t\Psi_1$ is a measure on $(\mathbf{R}_\varepsilon, \mathcal{B}_\varepsilon)$. One may now think of letting $\varepsilon \downarrow 0$ in (4.1) to get $X_{t,0+}$, say, as the limit so that $X_t - X_{t,0+}$ is continuous. This limit, indeed, exists in probability provided

$$\lambda_t := \lim_{\varepsilon \downarrow 0} E\nu_t(\mathbf{R}_\varepsilon) < \infty. \tag{1.4.88}$$

In that case $\lambda_t = t\lambda$ for some $\lambda > 0$, in view of homogeneity, and the process $\{X_{t,0+} : t \geq 0\}$ has at most finitely many (jump) discontinuities in any finite interval of time. Such a process is called a *compound Poisson* process with the Poisson parameter λ and jump distribution $\mu(dx)$, where $\mu(B) = E(\nu_1(B))/\lambda$ for B a Borel subset of $\{x \in \mathbb{R} : x \neq 0\}$. Compound Poisson processes arise, for example as models of net receipts of insurance (i.e., premium receipts minus payoffs of claims). When the limit (12.4.76) is infinite, the limit (12.4.75) as $\varepsilon \downarrow 0$ may not exist. However, if $EX_t^2 < \infty$, one may center $X_{t,\varepsilon}$ to get a representation

$$X_t = X_t^{(0)} + \lim_{n\to\infty} \int_{\{x:\frac{1}{n}\leq|x|\leq 1\}} x\,[\nu_t(dx) - \Psi_t(dx)] + \int_{\{x:|x|>1\}} x\nu_t(dx), \tag{1.4.89}$$

where $\{X_t^{(0)} : t \geq 0\}$ is a process with continuous sample paths (a.s.) and, therefore, is a Brownian motion (possibly with a drift). Also, $\Psi_t(dx) = t\Psi_1(dx)$. More generally, without assuming $EX_t^2 < \infty$, one has the representation

$$X_t = X_t^{(0)} + \int_{\{x:|x|>0\}} \{x\nu_t(dx) - \frac{x}{1+x^2}\Psi_t(dx)\}. \tag{1.4.90}$$

these representations one can derive the *Lévy–Khinchin* representation of characteristic functions of infinitely divisible laws, recalling the characteristic functions of normal and Poisson random variables: *There exist constants* $m, \sigma^2 \geq 0$ *such that*

$$Ee^{i\xi(X_1 - X_0)} = \exp\{im\xi - \frac{\xi^2\sigma^2}{2} + \int_{\{x\neq 0\}} (e^{i\xi x} - 1 - \frac{i\xi x}{1+x^2})\Psi_1(dx)\}. \tag{1.4.91}$$

As a special case, suppose $\int_{\{x\neq 0\}} \frac{x}{1+x^2}\Psi_1(dx) = \theta$, say, exists. By centering the process so that $m+\theta = 0$ and assuming $\sigma^2 = 0$ (i.e., the continuous part does not exist), one arrives at a *pure jump process.* If, in addition, Ψ_1 is finite the process is a compound Poisson process.

The *stable process* of index $\alpha, 0 < \alpha \leq 2$, also arises as special cases ($\alpha = 2$ corresponds to $\Psi_1 = 0$).

1.4.2 Jump Processes

Jump processes may be thought of as generalizations of continuous parameter Markov chains considered in Section 1.2.4. As in the case of such chains, a jump process $\{X_t : t \geq 0\}$ may be represented by means of (a) an *embedded discrete parameter Markov process* $\{Y_n : n = 0, 1, 2, \cdots\}$ with an arbitrary (one–step) transition probability $\bar{p}(x, dy)$ on a general state space $(S, \mathcal{S})$, with the restriction $\bar{p}(x, \{x\}) = 0$ for all x, and (b) for each specification of $\{Y_n : n = 0, 1, 2, \cdots\}$ a sequence of *holding times* $T_0, T_1, \cdots$ whose conditional distribution given $\sigma\{Y_n : n \geq 0\}$ is that of an independent sequence of exponential random variables with parameters $\lambda(Y_0), \lambda(Y_1), \cdots$. Here $\lambda(\cdot)$ is a nonnegative measurable function on $(S, \mathcal{S})$. The process $X_t := Y_n$ for $\tau_n := T_0 + T_1 + \cdots + T_n \leq t < \tau_{n+1}, n \geq 0$, is then a jump Markov process. The infinitesimal transition rates are

$$\begin{aligned} q(x, B) &= \lambda(x)\bar{p}(x, B) \quad \text{for } x \notin B, \\ q(x, \{x\}) &= -\lambda(x), \end{aligned} \tag{1.4.92}$$

if $\lambda(x) > 0$. If $\lambda(x) = 0$, then the state x is absorbing, and one takes $q(x, B) = 0\ \forall B \in \mathcal{S}$.

1.4.3 References

Theorem 4.1 is due to Lévy [40], [41]. After the subject was introduced by de Finetti [44] and Kolmogorov [45], the general case was fully developed by Lévy [40] (Also see Lévy [42]). The elegant representation (4.4) of Lévy was made precise by Itô [46]. The so-called Lévy–Khinchin representation (4.5) of infinitely divisible laws is due essentially to Lévy [40], following the derivation of the case of finite variances by Kolmogorov [45]. Khinchin [47] was the first to give an analytic derivation of (4.5).

Jump processes, sometimes called the *pure jump processes*, have been studied by Pospisil [48], Feller [49], [50], Doeblin [51], and Doob [52]. For complete treatments see Doob [39] and Gikhman and Skorokhod [53].

1.5 Markov Processes and Semingroup Theory

Let $(S, \mathcal{S})$ be a measurable space and $p(t; x, dy)(t > 0, x \in S)$ a transition probability on it. That is, (i) for every $t > 0$ and every $x \in S, B \to p(t; x, B)$ is a probability measure on $(S, \mathcal{S})$, (ii) for every $t > 0$ and every $B \in \mathcal{S}, x \to p(t; x, B)$ is measurable on $(S, \mathcal{S})$, and (iii) the *Chapman–Kolmogorov equation* holds:

$$p(t + s; x, B) = \int_S p(t; z, B)p(s; x, dz). \tag{1.5.93}$$

Let $\mathbb{B}(S)$ denote the Banach space of all real–valued bounded measurable functions f on S, with the *sup norm* $\|f\| := \sup\{|f(x)| : x \in S\}$. Define the *transition operator* $T_t : \mathbb{B}(S) \to \mathbb{B}(S)$ by

$$(T_t f)(x) = \int_S f(y)p(t; x, dy), \quad (t > 0). \tag{1.5.94}$$

By the Chapman–Kolmogorov equation, $\{T_t : t > 0\}$ is a *one–parameter semigroup of* (commuting) *linear operators* on the Banach space $\mathbb{B}(S)$:

$$T_{t+s}f = T_t(T_s f) = T_s(T_t f). \tag{1.5.95}$$

Also, each T_t is a *contraction*, namely, $\|T_t\| := \sup\{\|T_t f\| : f \in \mathbb{B}(S), \|f\| \leq 1\} \leq 1$. For

$$\|T_t f\| \equiv \sup_x |(T_t f)(x)| \leq \sup_y |f(y)| \equiv \|f\|. \tag{1.5.96}$$

One–parameter semigroups arise in other contents also. For example, if $\mathcal{X}$ is an arbitrary Banach space and A is a bounded linear operator on $\mathcal{X}$, then

$$T_t h := e^{tA}h \equiv \sum_{n=0}^{\infty} \frac{t^n}{n!} A^n h \ (h \in \mathcal{X}; t > 0), \tag{1.5.97}$$

defines a one–parameter semigroup of operators on $\mathcal{X}$. One of our objectives in the present section is to identify semigroups on $\mathbb{B}(S)$ (or on supspaces of $\mathbb{B}(S)$) which yield transition probabilities and, therefore, Markov processes. One of the most useful applications of semigroup theory to Markov processes is Feller's construction (Feller [54]–[56]) of all one–dimensional diffusions. This is briefly presented in Section 1.5.2. Semigroup theory also provides a basic connection between second order linear parabolic equations and diffusions, discussed in Section 1.6.

1.5.1 The Hille–Yosida Theorem

We consider Banach spaces $\mathcal{X}$ which are either real or complex. We will denote the scalar field by F (i.e., $F = \mathbb{R}$ or $\mathbb{C}$), and the *norm* by $\|\cdot\|$. Let A be a linear operator defined on a linear supspace D_A into $\mathcal{X}$. A is *closed* if its graph $\{\langle f, Af\rangle : f \in D_A\}$ is a closed subset of $\mathcal{X} \times \mathcal{X}$ in the product topology. The *closed graph theorem* implies that if A is a closed operator and D_A is closed, then A is bounded (see Folland [57], p. 155).

Given a linear operator A defined on a linear subspace $D_A \subset \mathcal{X}$, the *resolvent set* of A is

$$\begin{aligned}\rho(A) := \{\lambda \in F : \lambda - A \equiv \lambda I - A : D_A \to \mathcal{X} \text{ is one–to–one and onto,}\\ (\lambda - A)^{-1} \text{ is bounded on } \mathcal{X}\}.\end{aligned} \tag{1.5.98}$$

Note that if $\rho(A) \neq \emptyset$ then A is closed, since $\lambda - A$ is closed for $\lambda \in \rho(A)$. For $\lambda \in \rho(A)$, $R_\lambda := (\lambda - A)^{-1}$ is called the *resolvent operator*.

Let $\{T_t : t > 0\}$ be a *one–parameter contraction semigroup* on $\mathcal{X}$, i.e., (i) T_t is a bounded linear operator on $\mathcal{X}$, for each $t > 0$, (ii) $T_{t+s} = T_tT_s\ \forall t > 0, s > 0$, and (iii) $\|T_tf\| \leq \|f\|\ \forall f \in \mathcal{X} (t > 0)$. Let $\mathcal{X}_0 := \{f \in \mathcal{X} : \|T_tf - f\| \to 0 \text{ as } t \downarrow 0\}$ be the *center* of the semigroup. Then $\mathcal{X}_0$ is a closed subset of $\mathcal{X}$ and is, therefore, a Banach space with the norm $\|\cdot\|$. In the case $\mathcal{X}_0 = \mathcal{X}$, the semigroup $\{T_t : t > 0\}$ is said to be *strongly continuous.*

The *infinitesimal generator* A of a one–parameter semigroup $\{T_t : t > 0\}$ is defined by

$$Af = \lim_{t\downarrow 0} \frac{T_t - f}{t} \tag{1.5.99}$$

for all $f \in \mathcal{X}$ for which the limit exists (in norm). The set of all such f is called the *domain of the infinitesimal generator* and denoted D_A. Clearly, $D_A \subset \mathcal{X}_0$. Some basic properties of $\{T_t : t > 0\}$ are as follows.

Proposition 1.5.1 *Let $\{T_t : t > 0\}$ be a strongly continuous one–parameter contraction semigroup on $\mathcal{X}$. Then*

(a) $t \to T_tf$ is continuous on $[0, \infty)$, with T_0 as the identity for all $f \in \mathcal{X}$;

(b) D_A is dense in $\mathcal{X}$;

(c) for every $f \in D_A$ and all $t > 0$, one has

$$\begin{aligned}&(i) && T_tf \in D_A,\\ &(ii) && AT_tf = T_tAf,\\ &(iii) && T_tf - f = \int_0^t T_sAf ds;\end{aligned} \tag{1.5.100}$$

(d) A is closed;

(e) if $Re\lambda > 0$, then $\lambda \in \rho(A)$, $\|R_\lambda\| \leq 1/Re\lambda$, and

$$R_\lambda f = \int_0^\infty e^{-\lambda t} T_tf dt,\ \forall f \in \mathcal{X}. \tag{1.5.101}$$

The proof of (a) follows from continuity at $t = 0$, and the semigroup property. The proof of (c) is also simple. For the rest one shows that, for $\lambda > 0$, the operator S_λ, defined on $\mathcal{X}$ by $S_\lambda f := \int_0^\infty e^{-\lambda t} T_t f dt$, satisfies (1) $\|S_\lambda\| \leq 1/\lambda$, (2) $(T_hS_\lambda f - S_\lambda f)/h \to \lambda S_\lambda f - f$ as $h \downarrow 0$. This implies $AS_\lambda f = \lambda S_\lambda f - f$, so that $(\lambda - A)S_\lambda f = f, \forall f \in \mathcal{X}$. Hence $S_\lambda = (\lambda - A)^{-1} \equiv R_\lambda$. In particular, $\lambda \in \rho(A)$. Relations (d), (e) follow from this. (b) follows from the fact that the distribution with the exponential p.d.f. $\lambda e^{-\lambda t}$ converges weakly to the Dirac measure δ_0 as $\lambda \to \infty$, so that $\lambda R_\lambda f \to f$ as $\lambda \to \infty, \forall f \in \mathcal{X}$. Note that the range of $\lambda R_\lambda \equiv \lambda(\lambda - A)^{-1}$ is the domain of A; hence, for any $f \in \mathcal{X}, \lambda R_\lambda f (\lambda > 0)$ are elements in D_A which converge to f as $\lambda \to \infty$.

The next result is the centerpiece of the theory of one–parameter contraction semigroups.

Theorem 1.5.2 *(Hille–Yosida). A linear operator A is the infinitesimal generator of a strongly continuous one–parameter contraction semigroup on $\mathcal{X}$ if and only if the following conditions hold: (1) A is densely defined, i.e., $\bar{D}_A = \mathcal{X}$, (2) $\forall \lambda > 0, \lambda \in \rho(A)$ and $\|R_\lambda\| \le 1/\lambda$.*

For a proof and for applications to Markov processes, see Dynkin [4], or Ethier and Kurtz [58].

Consider now the special case of the real Banach space $\mathcal{X} = \widehat{C}(S)$ of real–valued continuous functions on a locally compact separable metric space S which vanish at infinity. Endow $\widehat{C}(S)$ with the sup norm $\|f\| := \sup\{|f(x)| : x \in S\}$. Note that $\widehat{C}(S)$ satisfies the hypothesis of the Hille–Yosida theorem and, in addition, satisfies the "*maximum principle*"

$$Af(x_0) \le 0 \quad \text{if } f(x_0) \equiv \sup_{x \in S} f(x) \ge 0 \ (f \in D_A). \tag{1.5.102}$$

Then the contraction semigroup $\{T_t : t > 0\}$ generated by A are, by the Riesz Representation Theorem (see, e.g., Folland [57], p. 216), given by

$$(T_t f)(x) = \int_S f(y) p(t; x, dy) \quad f \in \widehat{C}(S), \tag{1.5.103}$$

where $p(t; x, dy)$ are nonnegative measures on $(S, \mathcal{S})$ (with $\mathcal{S}$ as the Borel sigma field) satisfying the Chapman–Kolmogorov equation (5.1). In general, however, one only knows that $p(t; x, S) \le 1 (t > 0, x \in S)$. Additional ("boundary") conditions may be required to ensure that $p(t; x, S) = 1 (t > 0, x \in S)$. But even in the defective case $p(t; x, S) < 1$ (for some $t > 0, x \in S$) one may introduce a state Δ_∞, the so–called *state at infinity*, and assign $p(t; x, \{\Delta_\infty\}) = 1 - p(t; x, S), p(t; \Delta_\infty, \{\Delta_\infty\}) = 1$ for all $t > 0$. Then, with this augmented $p(t; x, dy)$, one has a transition probability of a Markov process on $(S \cup \{\Delta_\infty\}, \bar{\mathcal{S}})$, where $\bar{\mathcal{S}} = \mathcal{S} \cup \{A \cup \{\Delta_\infty\} : A \in \mathcal{S}\}$. The state Δ_∞ is then an absorbing state. On the other hand, the transition operators $\bar{T}_t$, say, corresponding to this augmented p have an infinitesimal generator, say, $\hat{A}$ which satisfies, of course, the hypothesis of the Hille–Yosida theorem as well as the "maximum principle" (12.5.127). In general, in case of defective p, there are other extensions to Markov transition probabilities. In the next section we will consider these extensions for Feller's construction of all one–dimensional diffusions.

The simplest example of an infinitesimal generator for a Markov process is the matrix operator $Q = ((q_{ij}))$ on a finite state space S, as described in Section 1.2.4 (See (1.2.33)–(1.2.36)), satisfying $q_{ij} \ge 0 \ \forall i \ne j, \sum_{j \in S} q_{ij} = 0 \ \forall i \in S$. With the discrete topology, S is compact metric, and $\widehat{C}(S)$ is the set of all functions on S. The semigroup T_t is then given by

$$T_t = e^{tQ}, \ (T_t f)(x) = \sum_{n=0}^{\infty} \frac{t^n}{n!}(Q^n f)(x), \ p(t; i, j) = \sum_{n=0}^{\infty} \frac{t^n}{n!} q_{ij}^{(n)}, \tag{1.5.104}$$

where Q^n is the matrix product $Q \cdots Q$, and $Q^n \equiv ((q_{in}^{(n)}))_{i,j \in S}$. Such a construction also holds if S is denumerable, provided $\sup\{-q_{ii} \equiv \sum_{j \ne i} q_{ij} : i \in S\} < \infty$.

Before concluding this subsection let us point out that an abstract formulation of *Kolmogorov's backward equation* is given by the relation

$$\frac{d}{dt} T_t f = A T_t f \quad (f \in D_A), \tag{1.5.105}$$

which is obtained on differentiating the first and third expressions in (5.3) with respect to s and setting $s = 0$.

1.5.2 Semigroups and One-Dimensional Diffusions

In a series of classic papers Feller [54]–[56] adapted and extended the Hille–Yosida theory of semigroups to construct the class of all nonsingular one–dimensional Markov processes whose sample paths are continuous up to a killing time ζ or the time to reach a boundary, whichever is smaller. *Nonsingularity* in this context means that from any point in the interior of the state space the process can move to the left as well as to the right with positive probability. We will only consider the so–called Feller processes for which the transition probability $p(t;x,dy)$ is weakly continuous: that is $x \to p(t;x,dy)$ is weakly continuous. One also refers to this property as the *Feller property* of the transition probability.

Consider an interval (a,b) with $-\infty \leq a < b \leq \infty$. The compactification of a (or, b) is finite; this has the usual meaning of adding a (or, b) to the open interval. If a (or, b) is infinite, one still includes this infinite point $-\infty$ (or, ∞) and gives $[a,b]$ the topology under which $[a,b]$ is homeomorphic to a compact interval. Without loss of generality we assume $a < 0 < b$.

Assume for simplicity that there is no *killing*, i.e., the process does not terminate at a point in the interior (a,b). Such a process is characterized by (1) a strictly increasing and continuous *scale function* $s(x)$ on (a,b) and a strictly increasing and right–continuous function $m(x)$ on (a,b) called the *speed function*, and (3) *boundary conditions*, in case the boundary can be reached in finite time from the interior with positive probability. The scale function is determined up to an additive constant from the requirement that

$$\phi(x) := \frac{s(d)-s(x)}{s(d)-s(c)}, \quad c \leq x \leq d, \tag{1.5.106}$$

equals the probability that the process sarting at $x \in [c,d] \subset (a,b)$ reaches c before d. The speed function is determined up to the addition of an affine linear function of s by the requirement

$$\left(\frac{d}{dm(x)}\,\frac{d}{ds(x)}\right) M(x) = -1 \quad c < x < d, \ M(c) = 0 = M(d), \tag{1.5.107}$$

where $M(x)$ is the expected time to reach c or d starting at $x \in [c,d]$. Without loss of generality, assume $s(0) = 0, m(0) = 0$. Given $s(\cdot)$ and $m(\cdot)$ a unique Markov process is determined up to the time one of the boundary points a or b is reached. In the case the boundary point a is *inaccessible*, i.e.,

$$\int_a^0 m(x)ds(x) = -\infty, \tag{1.5.108}$$

the point a cannot be reached in finite time and no boundary condition needs to be prescribed at a. Similarly, if b is *inaccessible*, i.e.,

$$\int_0^b m(x)ds(x) = \infty, \tag{1.5.109}$$

the point b cannot be reached in finite time, and no boundary condition is necessary at b. In the case both boundary points are inaccessible, the operator $\frac{d}{dm(x)}\frac{d}{ds(x)} = A$, say, is the infinitesimal generator of a semigroup of transition operators $\{T_t : t > 0\}$ on the Banach space $C[a,b]$ of all real–valued continuous functions on $[a,b]$, with the "sup norm": $\|f\| = \max\{|f(x)| : x \in [a,b]\}$. The corresponding transition probability $p(t;x,dy)$ $(t > 0, x \in (a,b))$ then defines a conservative strong Markov process on the state space $S = (a,b)$, which may be constructed to have continuous sample paths.

If the boundary point, say b is *accessible*, i.e., the integral (5.17) is finite, but

$$\int_0^b s(x)dm(x) = \infty, \tag{1.5.110}$$

then b is called an *exit boundary*. If, on the other hand, b is inaccessible but the integral in (5.18) is finite, then b is called an *entrance boundary*. In the case of an exit boundary, say b, starting from $x \in (a,b)$ the process reaches b with positive probability; but it cannot reenter (a,b) in a continuous trajectory. If b is an entrance boundary, starting from $x \in (a,b)$ the process cannot reach b; but if it starts from b it reenters (a,b) instantaneously in a continuous trajectory. If the boundary point a is accessible, but

$$\int_a^0 s(x)dm(x) = -\infty, \tag{1.5.111}$$

then it is said to be an *exit boundary*. On the other hand, if a is inaccessible, but the integral in (5.19) is finite, then a is said to be an *entrance boundary*. If a is inaccessible and (5.19) holds then it is a *natural boundary*. Similarly, if b is inaccessible and (5.18) holds then b is a natural boundary.

Next suppose a boundary point, say b, is accessible and the integral (5.18) is finite, then b is a *regular* boundary. In this case the process starting at $x \in (a,b)$ reaches b with positive probability, and a boundary condition needs to be specified at b. The most general boundary condition at b is then given by

$$\theta f(b) + \int_{[a,b]} \frac{f(b)-f(x)}{s_1(b)-s_1(x)} q(dx) + \gamma(Af)(b) = 0, (A := (d/dm(x)(d/ds(x))), \tag{1.5.112}$$

for all f belonging to the domain D_{A_1} of the generator, say, A_1 of the diffusion. Here $\theta \geq 0, \gamma \geq 0, q$ is a finite measure on $[a,b]$, and s_1 is a continuous nondecreasing function on $[a,b]$ which equals the scale function s in a neighborhood of b. Also, one must have

$$\text{either} \quad \gamma > 0, \quad \text{or} \quad \int_{[a,b]} \frac{q(dx)}{s_1(b)-s_1(x)} = \infty. \tag{1.5.113}$$

The integrand in (5.20) equals $(df(x)/ds(x))_{x=b}$ at $x=b$. Thus one may write (5.20) as

$$\theta f(b) + \delta \left(\frac{df(x)}{ds(x)}\right)_{x=b} + \int_{[a,b)} (f(b)-f(x))q_1(dx) + \gamma(Af)(b) = 0, \tag{1.5.114}$$

where $\delta = q(\{b\})$ and $q_1(dx) = q(x)/(s_1(b)-s_1(x))$ on $[a,b)$. In particular, if q_1 is a finite measure then

$$\delta + \gamma > 0. \tag{1.5.115}$$

Similarly, if a is a regular boundary, then for $f \in D_{A_1}$

$$\theta' f(a) - \delta' \left(\frac{df(x)}{ds(x)}\right)_{x=a} + \int_{(a,b]} (f(a)-f(x))q_1'(dx) + \gamma'(A_1 f)(a) = 0, \tag{1.5.116}$$

for some nonnegative constants $\theta', \delta', \gamma'$, and a measure $q_1'(dx)$. If q_1' is a finite measure then

$$\delta' + \gamma' > 0. \tag{1.5.117}$$

We may now state Feller's characterization of all one–dimensional diffusions in Theorems 5.3, 5.4 below. We will say that a Markov process on a state space $(S, \mathcal{S})$ with a transition probability $p(t;x,dy)$ is *conservative* if $p(t;x,S) = 1$ for all $f > 0, x \in S$.

Theorem 1.5.3 *Let $A \equiv (d/dm(x))(d/ds(x))$ be defined on the Banach space $C[a,b]$. Let D_A denote the set of all $f \in C[a,b]$ such that $Af \in C[a,b]$. Then A with domain D_A is the infinitesimal generator of a Markov semigroup $\{T_t : t > 0\}$ if and only if a and b are both inaccessible, i.e., iff (5.16) and (5.17) hold. The corresponding Markov process is conservative and may be constructed to have continuous sample paths and the strong Markov property.*

The continuity of the sample paths may be ensured by verifying the Dynkin–Snell criterion (See Theorem 1.1). By virtue of the construction of the semigroup on $C[a,b], T_t f \in C[a,b]$ if $f \in C[a,b]$, i.e., $x \to p(t;x,dy)$ is weakly continuous. Therefore, the Markov process has the strong Markov property. For the processes characterized by the next theorem, the sample paths may be continuous only up to the time of reaching one of the boundaries; but in any case the sample paths can be taken to be right–continuous, and this together with the Feller property implies the strong Markov property.

We continue to use the notation of Theorem 5.3 to denote D_A the set of all $f \in C[a,b]$ such that $(d/dm(x))(d/ds(x))f(x) \in C[a,b]$. For a linear operator A_1 on $C[a,b]$, write $A_1 \subset A$ if $D_{A_1} \subset D_A$ and $A_1 f = Af$ for $f \in D_{A_1}$.

Theorem 1.5.4 *Let a be a regular boundary point, i.e., both integrals in (5.16) and (5.19) are finite. If $p(t;x,dy)$ is the Feller transition probability, possibly defective, corresponding to a semigroup $\{T_t : t > 0\}$ generated by an infinitesimal generator $A_1 \subset A$ then, for every $f \in D_{A_1}$, (5.24) holds along with (5.25) if q_1' is finite. If b is a regular boundary point, i.e., both integrals in (5.17) and (5.18) are finite, then for $A_1 \subset A$ to be an infinitesimal generator of a Markov semigroup on $C[a,b]$, (5.22) must hold, along with (5.23) if q_1 is finite for every $f \in D_{A_1}$.*

For self–contained proofs of Theorems 5.3, 5.4 see Mandl [59], or Itô and McKean [60].

The transition probability under the conditions of Theorem 5.4 is conservative if and only if θ' is zero (in (5.24)) in the case of a is a regular boundary, and θ is zero (in (5.22)) in the case of b is a regular boundary point.

Suppose b is a regular boundary, and $\theta = 0 = \gamma$, and q_1 is the zero measure, i.e., (5.22) reduces to

$$\left(\frac{df(x)}{ds(x)}\right)_{x=b} = 0, \quad f \in D_{A_1} \tag{1.5.118}$$

then the diffusion is said to have a (*pure*) *reflection* at b. In this case on reaching b the process instantaneously returns to (a,b) continuously. If all the terms in (5.22) vanish except the last term, i.e.,

$$(Af)(b) \equiv \left(\frac{d}{dm(x)}\,\frac{df(x)}{ds(x)}\right)_{x=b} = 0, \quad f \in D_{A_1} \tag{1.5.119}$$

then the diffusion is *absorbing* at b. In this case once the process reaches b it stays there ever after. If all the terms in (5.22) are zero except θ, i.e.,

$$f(b) = 0, \quad f \in D_{A_1}, \tag{1.5.120}$$

then the process is nonconservative and is not defined for times $t \geq \tau_b := \inf\{t > 0 : \omega(t) = b\}$, $\omega \in C([0,\infty) \to [a,b])$. If all terms in (5.22) vanish except for the integral term and γ, and q_1 is finite, then on reaching b the process remains there for an exponentially distributed *holding time* and then jumps into $[a,b)$ according to the (normalized) q_1 distribution, and continues in a continuous trajectory until a boundary is reached again, etc. If (5.22) reduces to

$$\delta\left(\frac{df(x)}{ds(x)}\right)_{x=b} + \gamma(Af)(b) = 0 \tag{1.5.121}$$

with $\delta > 0, \gamma > 0$, the process has continuous trajectories and on reaching b immediately enters (a, b) continuously, but its motion is slower than in the case (5.26). Such a boundary is sometimes called a *sticky boundary.* Entirely analogous descriptions apply, of course, to boundary behavior at a.

For examples, consider first the *standard one-dimensional Brownian motion* $\{B_t^x : t \geq 0\}$, starting at $x (x \in \mathbb{R})$. Its transition probability density $p(t; x, y)$ satisfies Kolmogorov's backward equation

$$\frac{\partial p(t; x, y)}{\partial t} = \frac{1}{2}\frac{\partial^2 p(t; x, y)}{\partial x^2} \quad (t > 0; -\infty < x < \infty, -\infty < y < \infty). \tag{1.5.122}$$

Here $a = -\infty, b = \infty$ are inaccessible boundaries. More generally, consider a diffusion on $(-\infty, \infty)$ with *drift velocity* $\mu(x)$ and *diffusion coefficient* $\sigma^2(x) > 0$, where $\mu(x)$ and $\sigma^2(x)$ are locally Lipschitzian. Its transition probability density satisfies the backward equation

$$\begin{aligned}\frac{\partial p(t; x, y)}{\partial t} &= \mu(x)\frac{\partial p(t; x, y)}{\partial x} + \frac{1}{2}\sigma^2(x)\frac{\partial^2 p(t; x, y)}{\partial x^2} (t > 0; x, y \in \mathbb{R}) \\ &= Ap(t; x, y)\end{aligned} \tag{1.5.123}$$

Here $a = -\infty, b = \infty$, and a, b are inaccessible boundaries. One may take

$$s(x) = \int_0^x e^{-I(0,z)} dz, \; m(x) = \int_0^x \frac{1}{2\sigma^2(z)} e^{I(0,z)} dz \tag{1.5.124}$$

with the usual convention $\int_c^d = -\int_d^c$, and with

$$I(0, z) := \int_0^z \frac{2\mu(y)}{\sigma^2(y)} dy. \tag{1.5.125}$$

When, as above, the state space is taken to be $(-\infty, \infty)$, a boundary point, say b, being accessible means that the process escapes to infinity from the state space in finite time with positive probability. This phenomenon is also known as *explosion.* In other words, one has explosion (i.e., accessibility of $-\infty$ and/or $+\infty$) if and only if (see (5.16), (5.17)) at least one of the following conditions holds:

$$\begin{aligned}&\int_{-\infty}^0 \left(\int_x^0 \frac{1}{2\sigma^2(z)} e^{I(0,z)} dz\right) e^{-I(0,x)} dx < \infty, \\ &\int_0^\infty \left(\int_0^x \frac{1}{2\sigma^2(z)} e^{I(0,z)} dz\right) e^{-I(0,x)} dx < \infty.\end{aligned} \tag{1.5.126}$$

Assume now that the diffusion governed by (5.31) is nonexplosive, i.e., conservative on $(-\infty, \infty)$, so that both integrals in (5.34) diverge. As in the case of Markov chains, one may define a diffusion $\{X_t : t \geq 0\}$ to be *recurrent* if

$$\text{Prob}(\tau_c < \infty | X(0) = x) = 1 \quad \forall x, c, \tag{1.5.127}$$

where

$$\tau_c := \inf\{t \geq 0 : X_t = c\}. \tag{1.5.128}$$

Sometimes (5.35) is referred to as *point recurrence*, which is special to one–dimensional diffusions. This probability of reaching a point c, starting from a point $x \neq c$, is zero for all nonsingular diffusions in multidimension.

If (5.35) does not hold, the diffusion on $(-\infty, \infty)$ (governed by (5.31)) is said to be *transient*. One may show that this is equivalent to

$$\text{Prob}(|X_t| \to \infty \quad \text{as} \quad t \to \infty | X_0 = x) = 1 \quad \forall x. \tag{1.5.129}$$

A (recurrent) diffusion is said to be *positive recurrent* if

$$E(\tau_c | X_0 = x) < \infty \quad \forall x, c. \tag{1.5.130}$$

A recurrent diffusion which is not positive recurrent is said to be *null recurrent*.

Theorem 1.5.5 *Consider the diffusion on $(-\infty, \infty)$ governed by (5.31) with $\mu(x), \sigma^2(x)$ locally Lipschitzian, $\sigma^2(x) > 0$ for all x. (a) The diffusion is recurrent if and only if both integrals in (5.34) diverge and*

$$\int_{-\infty}^{0} e^{-I(0,z)} dz \equiv \int_{-\infty}^{0} e^{I(z,0)} dz = \infty, \int_{0}^{\infty} e^{-I(0,z)} dz = \infty. \tag{1.5.131}$$

(b) The diffusion is positive recurrent if and only if both integrals in (5.34) diverge and

$$\int_{-\infty}^{\infty} \frac{1}{2\sigma^2(z)} e^{I(0,z)} dz < \infty. \tag{1.5.132}$$

If the diffusion is positive recurrent then it has a unique invariant probability whose density is given by

$$\pi(x) = \frac{c}{2\sigma^2(x)} e^{I(0,x)}, \tag{1.5.133}$$

c being the normalizing constant.

Part (a) may be proved by letting $c \downarrow -\infty$, or $d \uparrow \infty$, in (5.14) and showing that the limits converge to 1 if $s(x) \to -\infty$ as $x \to -\infty$, and $s(x) \to +\infty$ as $x \to +\infty$. Part (b) is similarly proved by solving (5.15) for $M(x)$ and letting $c \downarrow -\infty$, or $d \uparrow \infty$. The computation of the unique invariant probability may be checked by showing that $\pi(x)$ is the unique normalized solution of the adjoint equation $A^*\pi(x) = 0$, where

$$(A^*g)(x) = -\frac{d}{dx}(\mu(x)g(x)) + \frac{1}{2}\frac{d^2}{dx^2}(\sigma^2(x)g(x)). \tag{1.5.134}$$

Next we consider some examples, following Karlin and Taylor [61], of diffusions which model stochastic changes over time of gene frequencies in a large biological population. Consider a biological population of size N comprising two genetic types A and a.

The limiting form, as the size $N \to \infty$, of the proportion X_t of type A genes at time t in the absence of mutation and selection may be represented as a diffusion on $[0, 1]$ with $\mu(x) = 0$ and $\sigma^2(x) = \sigma^2 x(1-x), \sigma^2 > 0$. The boundaries 0 and 1 are easily seen to be regular, and one assigns absorbing boundary conditions at 0 and 1.

More generally, allowing mutations from type A to type a to occur at a rate γ_1 and that from type a to type A to occur at a rate γ_2, X_t is modeled as a diffusion on $[0, 1]$ with

$$\mu(x) = -\gamma_1 x + \gamma_2(1-x), \ \sigma^2(x) = \sigma^2 x(1-x). \tag{1.5.135}$$

Since 0 does not belong to the interior of the state space here, one may define the scale function by

$$\begin{aligned}
I(\tfrac{1}{2},x) &= \int_{\frac{1}{2}}^{x} \frac{2\mu(z)}{\sigma^2(z)}dz = -\frac{2\gamma_1}{\sigma^2}\int_{\frac{1}{2}}^{x}\frac{dz}{1-z} + \frac{2\gamma_2}{\sigma^2}\int_{\frac{1}{2}}^{x}\frac{dz}{z} \\
&= \frac{2\gamma_1}{\sigma^2}\log(1-x) + \frac{2\gamma_2}{\sigma^2}\log x - \left(\frac{2\gamma_1}{\sigma^2}+\frac{2\gamma_2}{\sigma^2}\right)\log\frac{1}{2}, \\
e^{I(\frac{1}{2},x)} &= c_1 x^{2\gamma_2/\sigma^2}(1-x)^{2\gamma_1/\sigma^2} \; (c_1 = 2^{2\gamma_1/\sigma^2+2\gamma_2/\sigma^2}), \\
e^{-I(\frac{1}{2},x)} &= (1/c_1)x^{-2\gamma_2/\sigma^2}(1-x)^{-2\gamma_1/\sigma^2}, \\
s(x) &= \int_{\frac{1}{2}}^{x} e^{-I(\frac{1}{2},y)}dy \sim (1/c_1)\int_{\frac{1}{2}}^{x} y^{-2\gamma_2/\sigma^2}dy \text{ (as } x \downarrow 0). \qquad (1.5.136)
\end{aligned}$$

Here $\sim$ indicates that the ratio of the two sides goes to one. Thus, as $x \downarrow 0$,

$$s(x) \sim \begin{cases} (1/c_1)(-2\gamma_2/\sigma^2+1)^{-1}[x^{-2\gamma_2/\sigma^2+1} - (\frac{1}{2})^{-2\gamma_2/\sigma^2+1}] \\ \to -\infty \quad \text{if} \quad \gamma_2/\sigma^2 > \frac{1}{2}, \\ \to \text{ a finite constant if } \gamma_2/\sigma^2 \le \frac{1}{2}. \end{cases} \qquad (1.5.137)$$

Similarly,

$$\begin{aligned}
m(x) &:= \int_{\frac{1}{2}}^{x} \frac{2}{\sigma^2(z)}e^{I(\frac{1}{2},z)}dz = \frac{2c_1}{\sigma^2}\int_{\frac{1}{2}}^{x} z^{2\gamma_2/\sigma^2-1}(1-z)^{2\gamma_2/\sigma^2-1}dz \\
&\to \text{ a finite constant as } x \downarrow 0 \quad \forall \gamma_2 \ge 0. \qquad (1.5.138)
\end{aligned}$$

It follows that there exists a constant $c' > 0$ such that

$$\int_{\frac{1}{2}}^{x} m(x)ds(x) \sim -c'\int_{\frac{1}{2}}^{x} z^{-2\gamma_2/\sigma^2}dz = \left(\frac{-c'}{-2\gamma_2/\sigma^2+1}\right)\left[x^{-2\gamma_2/\sigma^2+1} - \left(\frac{1}{2}\right)^{-2\gamma_2/\sigma^2+1}\right].$$

Hence the boundary 0 is inaccessible if $\gamma_2/\sigma^2 > \frac{1}{2}$, and accessible if $\gamma_2/\sigma^2 \le 1/2$. On the other hand, for some $c' > 0$

$$\int_{\frac{1}{2}}^{x} s(z)dm(z) \sim c_1'\int_{\frac{1}{2}}^{x} z^{-2\gamma_2/\sigma^2+1}z^{2\gamma_2/\sigma^2-1}dz$$

converges to a finite constant at $x \downarrow 0$, for all $\gamma_2 \ge 0$. Therefore, if $\gamma_2/\sigma^2 > \frac{1}{2}$ then 0 is an entrance boundary; and for $0 \le \gamma_2/\sigma^2 \le \frac{1}{2}$, 0 is a regular boundary point and in the latter case one prescribes an absorbing condition at 0. Similar conditions apply to boundary 1 to show that it is an entrance boundary if $\gamma_1/2\sigma^2 > 1/2$ and a regular boundary if $\gamma_1/2\sigma^2 \le \frac{1}{2}$.

As a final example, consider a diffusion on $[0,1]$ modeling the evolution of the fraction of type A genes without mutation, but with *selection* favoring A:

$$\mu(x) = \gamma x(1-x), \sigma^2(x) = \sigma^2 x(1-x) \; (\gamma > 0, \sigma^2 > 0). \qquad (1.5.139)$$

Here

$$\begin{aligned}
I(\tfrac{1}{2},x) &= \int_{\frac{1}{2}}^{x} \frac{2\gamma}{\sigma^2}dz = \frac{2\gamma}{\sigma^2}(x-\frac{1}{2}), \; e^{I(\frac{1}{2},x)} = e^{(2\gamma/\sigma^2)(x-\frac{1}{2})}, \\
s(x) &= \int_{\frac{1}{2}}^{x} e^{\gamma/\sigma^2}e^{-(2\gamma/\sigma^2)z}dz = \frac{1}{2\gamma/\sigma^2}\left(1-e^{-2\gamma/\sigma^2(x-\frac{1}{2})}\right) \\
m(x) &= \int_{\frac{1}{2}}^{x} \frac{2}{\sigma^2 z(1-z)}e^{-\gamma/\sigma^2}\cdot e^{(2\gamma/\sigma^2)z}dz. \qquad (1.5.140)
\end{aligned}$$

It is simple to check that 0 is inaccessible and is a natural boundary. On the other hand, 1 is accessible and is an exit boundary.

Finally, given a diffusion $\{X_t : t \geq 0\}$ one can introduce a *killing* at an exponential rate $k(\cdot)$ as follows. Conditionally, given a path $\{X_t(\omega) : t \geq 0\}$, the probability that the process is not killed up to time t is $\exp\{-\int_0^t k(X_s(\omega))ds\}; t \geq 0$. If $\{X_t : t \geq 0\}$ is governed by (5.31) with infinitesimal generator $A = \frac{1}{2}\sigma^2(x)d^2/dx^2 + \mu(x)d/dx$, then the new process $\{Y_t : t \geq 0\}$ observed up to the time of killing, has the generator $A_1 := A - k(\cdot)$, i.e.,

$$A_1 f(x) = Af(x) - k(x)f(x) \quad \forall f \in D_{A_1}. \tag{1.5.141}$$

For applications to genetics, see Karlin and Taylor [61], pp. 272–284.

1.5.3 References

Comprehensive treatments of connections between semigroup theory and Markov processes may be found in Dynkin [4] and Ethier and Kurtz [58]. Feller's derivation of all semigroups of transition operators of one–dimensional (Fellerian) Markov processes with continuous trajectories up to the boundary (and all possible boundary conditions for such processes) is contained in [54]–[56], with a correction due to Wentzell [62]. A detailed exposition of Feller's theory is given in Mandl [59].

Itô–McKean [60] contains a complete account of one–dimensional Fellerian diffusions, including singular cases, and gives a construction using Brownian local time. These local times and their many fascinating properties were found by Lévy [42]. A mathematically regorous derivation of the existence of a continuous Brownian local time is due to Trotter [63]. A very readable treatment of conservative one–dimensional diffusions and their construction by the use of local times may be found in Freedman [64].

1.6 Stochastic Differential Equations

In the last section Feller's construction of all one–dimensional diffusions by the use of Hille–Yosida theory of semigroups has been reviewed, and their properties discussed. There are two other general methods for constructing diffusions. One of them is to use the theory of partial differential equations (PDE) for second order linear parabolic equations to construct the transition probability densities $p(t; x, y)$ as fundamental solutions of the initial value problems for such equations. See, e.g., Dynkin [4] and Friedman [64] for this. Once these are constructed and their smoothness established, Markov processes with continuous trajectories, and having the strong Markov property, may be constructed by standard probabilistic methods. The PDE method is also the most effective in providing proofs of smoothness of the solutions of initial value problems (for parabolic equations) as well as boundary value problems (for parabolic and elliptic equations). In addition, in multidimension the theory of semigroups, without the intervention of PDE or probabilistic methods, has proved to be difficult to use to construct transition probabilities of diffusions.

The third method of constructing diffusions using the fascinating theory of stochastic differential equations (SDE), principally due to K. Itô [66], is the most attractive of all from the probabilistic point of view. Among its advantages are (1) the ease with which one can construct diffusions in multidimension with arbitrary locally Lipschitzian coefficients and without the requirement of ellipticity or nonsingularity, (2) analysis in function space $C([0, T] : \mathbb{R}^k)$ of such quantities as density of one diffusion with respect to another having the same nonsingular diffusion matrix but different drift velocities (Cameron–Martin–Girsanov Theorem), (3) relatively simple derivations of asymptotic properties, and (4) elegant extensions to manifolds. We will only briefly survey some basic facts in this section. For complete treatments one may look up a number of texts, e.g., McKean [67], Ikeda and Watanabe [68], Friedman [69], Karatzas and Shreve [70], and Rogers and Williams [71].

1.6.1 Stochastic Integrals, SDE, Itô's Lemma

Consider the stochastic differential equation

$$dX_t = b(X_t)dt + \sigma(X_t)dB_t \quad (t \geq 0), \tag{1.6.142}$$

where $\{X_t : t \geq 0\}$ is a k–dimensional process, $b(\cdot)$ a locally Lipschitzian vector field in $\mathbb{R}^k(b(x) = (b_1(x)b_2(x), \cdots, b_k(x)$ with each $b_i(x)$ real–valued), $\sigma(x)$ a locally Lipschitzian $(k \times k)$ matrix–valued function, and $\{B_t : t \geq 0\}$ a k–dimensional standard Brownian motion independent of the initial value X_0. The equation (6.1) is to be interpreted in its integral version

$$X_t = X_0 + \int_0^t b(X_s)ds + \int_0^t \sigma(X_s)dB_s, \quad (t \geq 0)$$

$$(X_t^{(i)} = X_0^{(i)} + \int_0^t b_i(X_s)ds + \int_0^t \sigma_{i\cdot}(X_s) \cdot dB_s \quad (i = 1, \cdots, k)) \tag{1.6.143}$$

Here $\sigma_{i\cdot}(x)$ denotes the ith row vector of $\sigma(x)$, and $\sigma_i(X_s) \cdot dB_s = \sum_{j=1}^k \sigma_{ij}$ $(X_s)dB^{(j)}(s)$, so that the stochastic integral $\int_0^t \sigma_{i\cdot}(X_s) \cdot dB_s$ is really a sum of k stochastic integrals, each with respect to a one–dimensional standard Brownian motion. Thus one needs to define the stochastic integral

$$\int_0^t f(s)d\bar{B}_s,$$

say, where $\{\bar{B}_s : s \geq 0\}$ is one of the standard one–dimensional Brownian motions $B^{(j)}(\cdot)$ and $\{f(s,\omega) : s \geq 0\}$ is a *nonanticipative* right–continuous real–valued square–integrable stochastic process. Nonanticipativity means $f(s,\cdot)$ is $\mathcal{F}_s$–measurable for every $s \geq 0$, where $\{\mathcal{F}_s : s \geq 0\}$ is an increasing family of sigma–fields on a probability space $(\Omega, \mathcal{F}, P), \mathcal{F}_s \subset \mathcal{F}$, such that B_s is $\mathcal{F}_s$–measurable and $\{B_t - B_s : t \geq s\}$ is independent of $\mathcal{F}_s$. First let $f(s,\omega)$ be a *nonanticipative step functional (n.a.s.f.)* on an interval $[0,T]$, i.e., there exist $0 = t_0 < t_1 < \cdots < t_m = T$ constants, and $\mathcal{F}_{t_i}$–measurable square integrable random variables $f_i(i = 0, 1, \cdots, m)$ such that $f(s) = f_i$ for $t_i \leq s < t_{i+1}(i = 0, 1, \cdots, m-1), f(T) = f_m$. Then define, for $s \in [t_{i_0}, t_{i_0+1}](i_0 = 0, 1, \cdots, m-1)$, the *stochastic integral*

$$\int_s^t f(u)d\bar{B}_u = \begin{cases} f_{i_0}(\bar{B}_t - \bar{B}_0) \text{ for } t_{i_0} \leq t \leq t_{i_0} + 1 \\ \sum_{i=i_0}^{j-1} f_i(\bar{B}_{t_{i+1}} - \bar{B}_{t_i}) + f_j(\bar{B}_t - \bar{B}_{t_j}) \text{ for } t \in (t_j, t_{j+1}], \\ (j = i_0 + 1, \cdots, m). \end{cases} \tag{1.6.144}$$

In view of the fact that $\bar{B}_t - \bar{B}_s$ is orthogonal to all square–integrable $\mathcal{F}_s$–measurable random variables $(0 \leq s < t)$, it is simple to see that $\int_0^t f(s)d\bar{B}_s(0 \leq t \leq T)$ is a continuous, additive, square–integrable $\mathcal{F}_t$–martingale $(0 \leq t \leq T)$:

$$\int_0^t f(u)d\bar{B}_u = \int_0^s f(u)d\bar{B}_u + \int_s^t f(u)d\bar{B}_u,$$

$$E\Big(\int_0^t f(u)d\bar{B}_u \Big| \mathcal{F}_s\Big) = \int_0^s f(u)d\bar{B}_u \ (0 \leq s < t \leq T). \tag{1.6.145}$$

and

$$E\Big[\Big(\int_0^t f(u)d\bar{B}_u\Big)^2\Big|\mathcal{F}_s\Big] = E\Big[\int_0^t f^2(u)du\Big|\mathcal{F}_s\Big],$$
$$E\Big[\int_0^t f(u)d\bar{B}_u\Big]^2 = E\int_0^t f^2(u)du = \int_0^t Ef^2(u)du, \Big(0 \le s \le T\Big). \tag{1.6.146}$$

Using (3.6.4), (3.6.5), one may now extend the definition of the stochastic integral to all square integrable right–continuous nonanticipative processes $f(s)(s \ge 0)$, and derive (3.6.4), (3.6.5) for the latter.

Having defined the stochastic integral, also called the *Itô integral*, for all square integrable nonanticipative stochastic processes, we may solve the *stochastic differential equations (SDE)*, or *Itô equations*, (3.6.2) by the method of successive approximations with the 0th approximation given by $X_t^{(0)} \equiv X_0$ for all $t \ge 0$. The successive approximation scheme is given by

$$X_t^{(n+1)} = X_0 + \int_0^t b(X_s^{(n)})ds + \int_0^t \sigma(X_s^{(n)})dB_s,\ t \ge 0, \tag{1.6.147}$$

Using Doob's maximal inequality for square integrable martingales (See, e.g., Bhattacharya and Waymire [6], p. 52), one shows that for Lipschitzian coefficients $b(\cdot), \sigma(\cdot)$, for every $T > 0, \{X_t^{(n)} : 0 \le t \le T\}$ converges (in probability in the sup–norm on $[0, T]$) to a continuous nonanticipative square integrable stochastic process $\{X_t : t \ge 0\}$ which satisfies (3.6.2). In the same manner, one shows that such a solution is unique up to a P–null set. This process $\{X_t : t \ge 0\}$ is the desired solution to the SDE (3.6.2). This is then shown to be a Markov process, which has the strong Markov property since it has continuous sample paths and the Feller property:

Theorem 1.6.1 *If $b(\cdot)$ and $\sigma(\cdot)$ are Lipschitzian then, for any given X_0 independent of the Brownian motion $\{B_t : t \ge 0\}$ there exists a unique (up to a P-null set) continuous nonanticipative solution to (6.1) having the strong Markov property.*

By allowing $b(\cdot), \sigma(\cdot)$ to be only *locally* Lipschitz, one may construct unique solutions to (3.6.1) up to stopping times $\tau_n : \{t \ge 0 : |X_t| = n\}$, starting from $X_0 = x$, with $n > |x|$, and piecing them together construct X_t up to an *explosion time* $\zeta = \lim_{n\uparrow\infty} \tau_n$. A sufficient conditon for nonexplosion (i.e., for $\zeta = \infty$ a.s. for all initial x) and one for explosion have been given by Khas'minskii [72] (Also see Bhattacharya and Waymire [6], pp. 615–616).

One of the most distinctive and important features of the Itô calculus is *Itô's Lemma*, which when applied to functions of X_t says the following:

Theorem 1.6.2 *(Itô's Lemma). Let $\phi(t, x)$ be a bounded real-valued function on $[0, \infty)\times\mathbb{R}^k$ such that $\phi_0(t, x) := \partial\phi(t, x)/\partial t$, $\phi_r(t, x) := \partial\phi(t, x)/\partial x^{(r)}$, $\phi_{rr'}(t, x) := \partial^2\phi(t, x)/\partial x^{(r)}\partial x^{(r')}$ are bounded and continuous, $1 \le r,\ r' \le k$. Then*

$$d\phi(t, X_t) = \{\phi_0(t, X_t) + L\phi(t, X_t)\}dt + \mathit{grad}\ \phi(t, X_t)\cdot\sigma(X_t)dB_t \tag{1.6.148}$$

where grad $\phi(t, x) = (\phi_1(t, x), \cdots, \phi_k(t, x))'$ and L is the generator of the diffusion $\{X_t : t \ge 0\}$, namely,

$$L\phi(t, x) = \frac{1}{2}\sum_{r,r'=1}^{k} a_{rr'}(x)\phi_{rr'}(t, x) + \sum_{r=1}^{k} b_r(x)\phi_r(t, x),$$
$$(a(x) := \sigma(x)\sigma'(x)). \tag{1.6.149}$$

In particular, it follows from Theorem 6.2 that

$$Z_t := \phi(t, X_t) - \int_0^t \{\phi_0(u, X_u) + L\phi(u, X_u)\}du \quad (t \geq 0) \tag{1.6.150}$$

is a martingale, so that $EZ_t = EZ_0$, i.e.,

$$E\phi(t, X_t) - E\int_0^t \{\phi_0(u, X_u) + L\phi(u, X_u)\}du = E\phi(0, X_0), \quad \forall t \geq 0. \tag{1.6.151}$$

The extra second order terms involving $\phi_{rr'}$ in (6.8) arise due to the fact that the Taylor expression has to be carried up to second order derivatives to take into account Brownian squared variation $(dB_t)^2 \simeq dt$.

Among diverse applications of Itô's Lemma let us mention the derivation of criteria for different types of asymptotic behavior of $\{X_t : t \geq 0\}$. First, let $\{X_t : t \geq 0\}$ be a one–dimensional diffusion with $\sigma^2(x) > 0$ for all x. Consider the solution $\phi(x)$ of the *two–point boundary value problem*

$$L\phi(x) = 0, \quad c < x < d, \quad \phi(c) = 1, \phi(d) = 0. \tag{1.6.152}$$

One can explicitly solve for ϕ and show that it can be extended to all of $\mathbb{R}$ so as to be bounded and have bounded and continuous first and second order derivatives. Applying the martingale property of (6.9) and the optional stopping rule one has (6.10) with t replaced by $\tau := \tau_c \wedge \tau_d$ where $\tau_y := \inf\{t \geq 0 : X_t = y\}$, so that

$$E\phi(X_\tau) = \phi(x), \quad c < x < d, \tag{1.6.153}$$

where $X_0 = x$. The left side is just the probability $P(\tau_c < \tau_d | X_0 = x)$. Letting $d \uparrow \infty$, one obtains $P(\tau_c < \infty | X_0 = x)$, for all $x > c$. Similarly, letting $c \downarrow -\infty$ in $\phi(x)$ one obtains $P(\tau_d < \infty | X_0 = x)$ for all $x < d$. If both these limiting probabilities are one, then the diffusion is *recurrent*, otherwise it is *transient*. In this way one obtains a simple derivation of the criteria of transience and recurrence of Section 1.5 for the case of diffusions with drift $b(\cdot)$ and diffusion coefficient $\sigma^2(x) > 0$ (See Theorem 5.5(a)). Next consider the two–point boundary value problem

$$L\psi(x) = -1, \quad c < x < d, \quad \psi(c) = 0 = \psi(d). \tag{1.6.154}$$

Then, by Itô's Lemma one gets (instead of (6.12)) the relation

$$0 = E\psi(X_\tau) + E(\tau | X_0 = x) = \psi(x), \quad c < x < d, \tag{1.6.155}$$

where $\tau = \tau_c \wedge \tau_d$ as before. Since one may explicitly solve (6.13) by successive integrations to obtain $\psi(x)$, on letting $d \uparrow \infty$ one gets $E(\tau_c | X_0 = x)$, for all $x > c$. Similarly one obtains $E(\tau_d | X_0 = x)$ for all $x < d$ by letting $c \downarrow -\infty$ in $\psi(x)$. If both these limits are finite then one has *positive recurrence*. Otherwise (assuming recurrence) one has *null recurrence*. This yields the criteria in Theorem 5.5(b). These considerations can be extended to multidimension. For example, let $\{X_t \equiv B_t : t \geq 0\}$ be a k–dimensional standard Brownian motion, $k \geq 2$. In this case $L = \frac{1}{2}\Delta$, where Δ is the *Laplacian*. The solution to the *Dirichlet problem*

$$\Delta\phi(x) = 0, \quad c < |x| < d, \ \phi(x) = 1 \text{ for } |x| = c, \ \phi(x) = 0 \text{ for } |x| = d, \tag{1.6.156}$$

can be computed explicitly, since in this ϕ must be *radial* $\phi(x) = g(|x|)$ for some g. Hence (6.15) reduces to a one–dimensional equation of the form (6.11) (but with $g(r)$ in place of

$\phi(x)$ in (6.11)). Indeed, the solution is

$$\phi(x) = \begin{cases} \frac{\log d - \log|x|}{\log d - \log c} & \text{for} \quad k = 2, \\ \frac{(c/|x|)^{k-2} - (c/d)^{k-2}}{1-(c/d)^{k-2}} & \text{for} \quad k > 2, \ c < |x| < d. \end{cases} \tag{1.6.157}$$

Using Itô's Lemma to $\phi(X_t)$ one shows, exactly as before, that $\phi(x) = P(X(\cdot)$ reaches the set $\{|y| = c\}$ before $\{|y| = d\}|X_0 = x)$. Once again letting $d \uparrow \infty$ one obtains the probability that Brownian motion reaches the set $\{|y| = c\}$ starting at a point x with $|x| > c$. If this limiting probability is one the process is *recurrent*, otherwise *transient*. In this manner it turns out from (6.16) that *two–dimensional Brownian motion* (as well as the one–dimensional B.M.) *is recurrent*, while *higher dimensional Brownian motions are transient*. One may apply this method to derive criteria for transience and recurrence for general multidimensional diffusions. Although one cannot in general solve explicitly the Dirichlet problem (1.6.156) for a general elliptic operator L (in place of the Laplacian Δ), one may derive appropriate inequalities by finding ϕ such that $L\phi(x) \leq 0$ for $c < |x| < d$. Similarly, one may obtain criteria for null and positive recurrence for multidimensional diffusions generated by elliptic operators L with nonsingular matrix–valued function $a(x) := \sigma(x)\sigma'(x)$, by solving $L\phi(x) = -1(c < |x| < d)$ with $\phi(x) = 1$ if $|x| = c$ and $\phi(x) = 0$ for $|x| = d$, or at least by finding a ϕ satisfying an inequality $L\phi(x) \leq -1$. For such criteria for transience, null recurrence and positive recurrence for multidimensional diffusions, see Khas'minskii [72], Friedman [69], pp. 196–201, Bhattacharya [73]. Positive recurrent diffusions X_t have unique invariant probabilities and approach the equilibrium in total variation distance as $t \uparrow \infty$.

Theorems 6.1, 6.2 both easily extend to the case of *nonhomogeneous diffusions* on $\mathbb{R}^k$ governed by the Itô equation

$$X_t = X_0 + \int_0^t b(u, X_u)du + \int_0^t \sigma(u, X_u)dB_u \quad (t \geq 0), \tag{1.6.158}$$

where there exists a constant $M > 0$ such that for all $s, t \in [0, \infty)$, and $x, y \in \mathbb{R}^k$,

$$\begin{aligned} |b(s,x) - b(t,y)| &\leq M(|t-s| + |x-y|), \\ \|\sigma(s,x) - \sigma(t,y)\| &\leq M(|t-s| + |x-y|), \end{aligned} \tag{1.6.159}$$

with $\|\cdot\|$ denoting the matrix norm. The solution of (6.17) is a Markov process; but it is nonhomogeneous in time with a transition probability $p(s, x; t, dy)$ denoting the conditional distribution of X_t, given $X_s = x$, which is not just a function of $t - s$ (and x), $0 \leq s < t$. The form of Itô's Lemma remains unchanged, with $b(t,x), a(t,x) := \sigma(t,x)\sigma'(t,x)$ in place of $b(x), a(x)$, respectively.

1.6.2 Cameron–Martin–Girsanov Theorem and the Martingale Problem

The distribution on a finite time interval $[0, t]$ of a Brownian motion with only a time–dependent drift, namely, $\{X_s := B_s - \int_0^t \gamma(u)du, \ 0 \leq s \leq t\}$ with a nonrandom function $\gamma(\cdot)$, was shown by Cameron and Martin [74] to be absolutely continuous with respect to the distribution of $\{B_s, 0 \leq s \leq t\}$. The density was represented as $\exp\{\int_0^t \gamma(u) \cdot dB_u - \frac{1}{2}\int_0^t |\gamma(u)|^2 du\}$. This was made use of in computing distributions of various functionals of the process X_t (See [75]). A far reaching generalization of this was given by Girsanov [76]. Crucial in this development is the fact that for a bounded nonanticipative functional $f(t), t \geq 0$, with value in $\mathbb{R}^k$,

$$M_t := \exp\left\{\int_0^t f(s) \cdot dB_s - \frac{1}{2}\int_0^t |f(s)|^2 ds\right\}, t \geq 0, \tag{1.6.160}$$

is a martingale. In particular, $EM_t = EM_0 = 1$ for all $t \geq 0$. This is relatively simple to establish for bounded nonanticipative step functionals (see (6.3)), and the general assertion follows by approximation. Note that for nonrandom $f(\cdot)$, as in the Cameron–Martin density, the result follows from (1) independence of Brownian increments and (2) $E[\exp\{(c \cdot B_s)^2 - \frac{1}{2}|c|^2 s\}] = 1$ for $c \in \mathbb{R}^k$.

Let $(\Omega, \mathcal{F}, P)$ denote the original probability space on which the standard Brownian motion $\{B_t : t \geq 0\}$ on $\mathbb{R}^k$ is defined, with respect to a filtration $\{\mathcal{F}_t : t \geq 0\}$ (i.e., B_t is $\mathcal{F}_t$–measurable and $\{B_t - B_s : t \geq s\}$ is independent of $\mathcal{F}_s$). Using the martingale property of (6.19) one may now define a new probability measure Q_T on $(\Omega, \mathcal{F}_T)$, $T > 0$ arbitrary finite, by

$$Q_T(A) = \int_A M_T dP \quad (A \in \mathcal{F}_T). \tag{1.6.161}$$

Note that if $A \in \mathcal{F}_t, t \leq T$, then by the martingale property of M_t $(t \geq 0)$, one has $Q_T(A) = Q_t(A)$. In the case of the Cameron–Martin nonrandom $\gamma(\cdot)$, one may show without much difficulty that under Q_T (i.e., on $(\Omega, \mathcal{F}_T, Q_T)$) the process $\{\widetilde{B}_t := B_t - \int_0^t \gamma(u)du : 0 \leq t \leq T\}$ is a standard k–dimensional Brownian motion on $[0, T]$. This last fact remains true for arbitrary bounded nonanticipative functionals $f(s), s \geq 0$, in place of $\gamma(\cdot)$. The essential tools for the proof of this are (1) Itô's Lemma and (2) a result of Lévy [42], which says that a process $\{Z_t = (Z_t^{(1)}, \cdots, Z_t^{(k)}) : t \geq 0\}$ is a k–dimensional standard Brownian motion with respect to a filtration $\{\mathcal{F}_t : t \geq 0\}$ if (a) $t \to Z_t$ is a.s. continuous, (b) $Z_t^{(i)} - Z_0^{(i)}, t \geq 0$, is a $\{\mathcal{F}_t\}$–martingale for each i, $1 \leq i \leq k$, and (c) $(Z_t^{(i)} - Z_0^{(i)})(Z_t^{(j)} - Z_0^{(j)}) = \delta_{ij}t, t \geq 0$ is a $\{\mathcal{F}_t\}$–martingale for every pair i, j. Here δ_{ij} is Kronecker's delta.

In the case $\gamma(\cdot)$ is nonrandom, it follows from the above that under $Q_T, \{B_t = \widetilde{B}_t + \int_0^t \gamma(u)du, 0 \leq t \leq T\}$ is a standard Brownian motion with a drift $\gamma(\cdot)$, since $\{\widetilde{B}_t : 0 \leq t \leq T\}$ is a standard Brownian motion on $[0, T]$. But under $P, \{B_t : 0 \leq t \leq T\}$ is a standard Brownian motion on $[0, T]$. The Cameron–Martin formula follows from this. Girsanov's generalization is given by

Theorem 1.6.3 *(Cameron–Martin–Girsanov Theorem). Let $b(t, x), \gamma(t, x)$ be Lipschitzian vector fields $(t \geq 0, x \in \mathbb{R}^k)$ and $\sigma(t, x)$ a nonsingular Lipschitzian matrix-valued function such that $\sigma^{-1}(t, x)$ is bounded on $[0, T] \times \mathbb{R}^k$ for every $T > 0$. Consider two diffusions defined on $(\Omega, \mathcal{F}, P)$ governed by*

$$\begin{aligned} X_t^x &= x + \int_0^t b(u, X_u^x)du + \int_0^t \sigma(u, X_u^x)dB_u, t \geq 0, \\ Y_t^x &= x + \int_0^t \{b(u, Y_u^x) + \gamma(u, Y_u^x)\}du + \int_0^t \sigma(u, Y_u^x)dB_u, t \geq 0, \end{aligned} \tag{1.6.162}$$

where $\{B_t : t \geq 0\}$ is a standard k–dimensional Brownian motion with respect to a filtration $\{\mathcal{F}_t : t \geq 0\}$. Then the distribution $P_{2,T}$, say, of $Y^{(T)} := \{Y_t^x : 0 \leq t \leq T\}$ is absolutely continuous with respect to the distribution $P_{1,T}$ of $X^{(T)} := \{X_t^x : 0 \leq t \leq T\}$, and for every real-valued bounded continuous function f on $C([0, T] \to \mathbb{R})$ one has

$$Ef \circ Y^{(T)} = E(f \circ X^{(T)})M_T, \tag{1.6.163}$$

where

$$M_t = \exp\left\{\int_0^t \sigma^{-1}(u, X_u^x)\gamma(u, X_u^x) \cdot dB_u - \frac{1}{2}\int_0^t |\sigma^{-1}(u, X_u^x)\gamma(u, X_u^x)|^2 du\right\}$$
$$(0 \leq t \leq T). \tag{1.6.164}$$

To prove this define Q_T as in (6.20) but with M_T defined by (6.23). Then $\{\widetilde{B}_t := B_t - \int_0^t \sigma^{-1}(u, X_u^x)\gamma(u, X_u^x)du : 0 \le t \le T\}$ is a standard k–dimensional Brownian motion on $[0, T]$ under Q_T. Now one may write

$$X_t^x x + \int_0^t \{b(u, X_u^x) + \gamma(u, X_u^x)\}du + \int_0^t \sigma(u, X_u^x)d\widetilde{B}_u, \tag{1.6.165}$$

so that the distribution of $X^{(T)}$ under Q_T is the same as the distribution of $Y^{(T)}$ under P. Therefore writing E for expectation under P, and E_{Q_T} that under Q_T,

$$E(f \circ Y^{(T)}) = E_{Q_T}(f \circ X^{(T)}) = E((f \circ X^{(T)})M_T). \tag{1.6.166}$$

In particular, writing $h(X^{(T)}) = E[M_T|X^{(T)}]$, one has

$$\frac{dP_{2,T}}{dP_{1,T}} = h(X^{(T)}). \tag{1.6.167}$$

It has been shown by Novikov [77] that the martingale property of $M_T(t \ge 0)$ in (6.19) holds if

$$E\exp\{\int_0^t |f(s)|^2 ds\} < \infty \quad \forall t > 0. \tag{1.6.168}$$

Therefore, one may define the probability measure Q_T on $\mathcal{F}_T$ for all $T > 0$ under the *Novikov condition* (6.27). This allows one to construct diffusions with nonsmooth coefficients, by extending the Cameron–Martin Girsanov theorem. But even broader classes of diffusions were constructed by Stroock and Varadhan [78], [79] by means of their *martingale problem* formulation. For simplicity let us only consider the case of time–homogeneous diffusions on $\mathbb{R}^k$. Note that for Lipschitzian coefficients $b(\cdot)$ and $\sigma(\cdot)$, for every twice continuously differentiable function ϕ with compact support, $Z_t := \phi(X_t) - \int_0^t L\phi(X_u)du, t \ge 0$, is a martingale (see (6.9)). Conversely, let $b(\cdot), \sigma(\cdot)$ be measurable and bounded, and consider the space $(\Omega = C([0, \infty) \to \mathbb{R}^k)$, $\mathcal{F} =$ Borel sigma field of $\Omega)$. If there exists for each $x \in \mathbb{R}^d$ a unique probability measure P_x such that on $(\Omega, \mathcal{F}, P_x)$, with $\{X_t : t \ge 0\}$ denoting the coordinate process $X_t(\omega) - \omega(t)$,

$$\phi(X_t) - \int_0^t L\phi(X_u)du, \quad t \ge 0, \tag{1.6.169}$$

is a martingale for every infinitely differentiable ϕ with compact support, then the margingale problem is said to be *well posed*. In this case the coordinate process $\{X_t : t \ge 0\}$ is a Markov process with the strong Markov property, for each initial state x. A sufficient condition for well posedness due to Stroock and Varadhan [79] is that (1) $b(\cdot)$ is bounded measurable and (2) $\sigma(\cdot)$ is nonsingular continuous, $\sigma(\cdot)$ and $\sigma^{-1}(\cdot)$ are bounded. The main results of the Itô calculus, such as Theorems 6.2, 6.3 extend in this case, although the pathwise unique solutions of the SDE (3.6.1), required for the validity of Theorem 6.1, may not exist. One may also relax the boundedness conditions on $b(\cdot)$ and $\sigma(\cdot)$ and the boundedness of $\sigma^{-1}(\cdot)$ by requiring them to hold on compacts. Then one may define a diffusion up to an *explosion time* ζ. Khas'minskii's test for nonexplosion may be extended in this case (See Stroock and Varadhan [79], Section 10.2, pp. 254–259). Similarly, the Khas'minskii criteria for transience, null recurrence and positive recurrence may also be extended (See Bhattacharya [73]).

Apart from extending the construction of diffusions to broader classes of coefficients, the martingale problem formulation leads to broad, simple and verifiable conditions for a sequence of discrete parameter Markov processes, with decreasing step sizes and increasing number of transitions per unit time, to converge in distribution to a diffusion. Similarly, broad conditions on convergence in distribution of a sequence of diffusions to a limiting diffusion may also be derived. For this see Stroock and Varadhan [79], Chapter 11.

1.6.3 Probabilistic Representation of Solutions to Elliptic and Parabolic Partial Differential Equations

It has been already pointed out in Section 1.6.1 that Itô's Lemma and the optional stopping theorem of martingale theory lead to the probabilistic representation of the solutions to certain boundary value problems. More generally, let L be an elliptic operator

$$L = \frac{1}{2}\sum_{r,r'=1}^{k} a_{rr'}(x)\frac{\partial^2}{\partial x^{(r)}\partial x^{(r')}} + \sum_{r=1}^{k} b_r(x)\frac{\partial}{\partial x^{(r)}} \tag{1.6.170}$$

where $a(x) := ((a_{rr'}(x)))$ is symmetric and positive definite, and $b(x)$ and $a(x)$ are locally Lipschitzian. In particular, $a(x)$ is *uniformly elliptic* in bounded domains G, i.e., the smallest eigenvalue of $a(x)$ is bounded away from zero for $x \in G$. An *elliptic boundary value problem*

$$\begin{aligned} L\phi(x) &= f(x), \quad x \in G, \\ \phi(x) &= g(x), \quad x \in \partial G, \end{aligned} \tag{1.6.171}$$

is said to be *well posed* if, for given bounded continuous f on $\bar{G}$ and continuous g on ∂G, there exists a unique ϕ satisfying (6.30) which is continuous on $\bar{G} = G \cup \partial G$. A well known sufficient condition for well posedness for a uniformly elliptic operator L in a bounded open set G is that every point $x \in \partial G$ be a *Poincaré point*, i.e., there exists a truncated cone C_x with vertex at x such that $C_x \backslash \{x\}$ is contained in the complement of $\bar{G}$.

Theorem 1.6.4 *Let G be a bounded open subset of $\mathbb{R}^k$ with all its boundary points as Poincaré points. Assume that L is uniformly elliptic in G and that $b(\cdot)$ and $a(\cdot)$ are Lipschitzian in $\bar{G}$. Then for every given continuous function f in $\bar{G}$ and every given continuous g on ∂G, the elliptic boundary value problem (6.30) has a unique solution ϕ, and ϕ has the representation*

$$\phi(x) = E(g(X_\tau^x)) + E\int_0^\tau f(X_s^x)ds, \quad x \in \bar{G}. \tag{1.6.172}$$

Here $\{X_t^x : t \geq 0\}$ is the diffusion on $\mathbb{R}^k$, starting at x, generated by L, and $\tau = \inf\{t \geq 0 : X_t^x \in \partial G\}$.

To derive the representation (1.6.172) one uses Itô's Lemma (Theorem 6.2) to ϕ (or to a smooth extension of ϕ to $\mathbb{R}^k$ having compact support), and optional stopping, to get (see (6.10) with τ in place of t, and $\phi_0 = 0$)

$$E\phi(X_\tau^x) - E\int_0^\tau L\phi(X_s^x)ds = \phi(x), \quad x \in \bar{G}. \tag{1.6.173}$$

Since $\phi(X_\tau^x) = g(X_\tau^x)$ and $L\phi(X_s^x) = f(X_s^x)$ for $s < \tau$, (6.32) is the same as (6.31).

We next turn to the *initial value problem*

$$\begin{aligned} \frac{\partial u}{\partial t}(t,x) &= Lu(t,x) \quad (t > 0, x \in \mathbb{R}^k), \\ \lim_{t \downarrow 0} u(t,x) &= f(x) \quad (x \in \mathbb{R}^k). \end{aligned} \tag{1.6.174}$$

where f is a given (initial) function. Under suitable conditions on L, the *fundamental solution* to the initial value (or Cauchy–) problem (6.33) is a function $p(t;x,y)(t>0;x,y\in\mathbb{R}^k)$ satisfying

$$\frac{\partial p(t;x,y)}{\partial t} = Lp(t;x,y) \quad (t>0, x,y\in\mathbb{R}^k),$$
$$\lim_{t\downarrow 0} p(t;x,y)dy = \delta_x(dy). \tag{1.6.175}$$

Here the limit is in the topology of weak convergence of probability measures, and δ_x is the Dirac measure at x. This fundamental solution is also the transition probability density of the diffusion $\{X_t : t\geq 0\}$. Thus (6.33) is just Kolmogorov's backward equation, and one has

$$u(t,x) = (T_t f)(x) \equiv Ef(X_t^x), \tag{1.6.176}$$

provided f is continuous and bounded. A more general result is the following. We will write $\sigma(x)$ for the positive square root of the matrix $a(x)$.

Theorem 1.6.5 *(Feynman–Kac Formula). Let $b(\cdot)$ and $\sigma(\cdot)$ be Lipschitzian, $\sigma(\cdot)$ non singular, f a bounded continuous function and V a continuous function which is bounded above. (a) Suppose $u(t,x)$ is a solution of*

$$\frac{\partial u(t,x)}{\partial t} = Lu(t,x) + V(x)u(t,x) \quad (t>0, x\in\mathbb{R}^k),$$
$$\lim_{t\downarrow 0} u(t,x) = f(x) \quad (x\in\mathbb{R}^k), \tag{1.6.177}$$

such that (i) $u(t,x)$ is bounded on $(0,c]\times\mathbb{R}^k$ for every $c>0$, (ii) $u(t,x)$ and $Lu(t,x)$ are continuous on $(0,\infty)\times\mathbb{R}^k$ and bounded on $[c,d]\times\mathbb{R}^k$ for all $0<c<d$. Then

$$u(t,x) = E\Big[f(X_t^x)\exp\Big\{\int_0^t V(X_s^x)ds\Big\}\Big] \quad (t\geq 0, x\in\mathbb{R}^k). \tag{1.6.178}$$

(b) In particular, the solution to (6.36) is unique, and (6.37) holds, if $b(\cdot),\sigma(\cdot)$, $\sigma^{-1}(\cdot)$ are all bounded and Lipschitzian.

The Feynman–Kac representation (6.37) follows from an application of Itô's Lemma to $\phi(s,Y^{(s)})$ for the function $\phi(s,y) := u(t-s,y^{(1)})\exp\{y^{(2)}\}$ for $0\leq s<t, y=(y^{(1)},y^{(2)})\in\mathbb{R}^k\times\mathbb{R}^1$, and with

$$Y^{(1)}(s) = X_s^x, \qquad Y^{(2)}(s) = \int_0^s V(X_u^x)du, \qquad Y(s) = (Y^{(1)}(s), Y^{(2)}(s)).$$

For the existence and uniqueness of the solution to (6.36), see Friedman [69].

The Feynman–Kac Formula is important in quantum mechanics. It is also a very useful result for the derivation of distributions of many important functionals of diffusions $\{X_t^x : t\geq 0\}$ (See Feynman [80] and Kac [81], [82]).

1.6.4 References

The Gaussian process with independent increments, now universally referrred to as Brownian motion, made its appearance as early as 1900 in an article by Bachelier [82] on financial mathematics. The name Brownian motion gained a permanent place in mathematics and science following Einstein's pioneering work [84] on the kinetic theory of the transport of a solute of dilute concentration through a liquid medium. In particular, this work provided an explanation of experimental observations by the English botanist Robert Brown on the movement of large colloidal molecules in a solution. The first rigorous construction of Brownian motion with continuous sample paths was given by Wiener [85].

An early occurrence of a stochastic differential equation may be found in Langevin [86]. The first rigorous introduction of stochastic integration with respect to Brownian increments seems to be due to Paley, Wiener and Zygmund [87] who considered only nonrandom integrands. The fundamental work on stochastic integration and stochastic differential equations outlined in this section is due to Itô [88], [89], [66]. Somewhat later, and independently of Itô, Gikhman [90], [91] derived many of the same results. A generalization to stochastic integration with respect to martingales was introduced by Doob [39], and this was extended much further to a complete theory of stochastic integration with respect to semi-martingales by the French school led by Meyer [92], [93].

For comprehensive modern treatments of the theory of stochastic differential equations we refer to the books by Ikeda and Watanabe [68], Karatzas and Shreve [70], Rogers and Williams [71], and Revuz and Yor [94]. Less comprehensive but readable accounts and applications may be found in Arnold [95], Friedman [65], McKean [67], Lipsler and Shiryaev [96], and Bhattacharya and Waymire [6].

Other Topics

Among the most notable omissions in the present survey is the theory of large deviations for Markov processes, developed largely by Donsker and Varadhan [97], [98], Varadhan [99], and Freidlin and Wentzell [100].

Another important topic omitted from our discussion is the precise estimation of the speed of convergence of the n-step transition probability $p^{(n)}(x, dy)$ of an ergodic Markov process to its unique equilibrium π. For Markov chains, including random walks on groups, with finite but large state spaces, the pioneering work is that of Diaconis [101], [102], who discovered the fascinating *cutoff phenomena* for certain important classes of chains. If n lies just a little to the left of the cutoff point then $\|p^{(n)}(x, dy) - \pi(dy)\|_{TV}$ is close to one, i.e., the approximation is almost as bad as it can be. But if n lies just a little to the right of the cutoff point then the above total variation distance is close to zero. For sharp bounds on the error of approximation for more general chains see Diaconis and Stroock [103], Diaconis and Saloff-Coste [104] and Fill [105]. For diffusions on compact Riemannian manifolds, Chen and Wang [106] have recently developed coupling methods to estimate the speed of convergence to equilibrium of the transition probability $p(t; x, dy)$, as $t \to \infty$, and used this to improve upon some of the best known estimates of the spectral gap of the Laplace–Beltrami operator which had been obtained by differential geometers and global analysts. Also see Holly, Kusuoka and Stroock [107] for a different method. The precise estimation of the speed of convergence is also important in the analysis of certain classes of multiscale phenomena arising in geosciences. See Bhattacharya and Götze [108] and Bhattacharya [109] for this analysis.

Bibliography

[1] P. Billingsley. Probability and Measure, 3rd. ed. New York: John Wiley, 1995.

[2] C. Ionescu Tulcea. Measures des les espaces produits. Atti Acad. Naz. Lincei Rend. **7**: 208–211, 1949.

[3] J. Nevue. Mathematical Foundations of the Calculus of Probability. San Francisco: Holden–Day, Inc. 1965.

[4] E. B. Dynkin. Markov Processes, vol. 1. Berlin: Springer–Verlag, 1965.

[5] J. L. Snell. Applications of martingale system theorems. Trans. Amer. Math. Soc. **73**: 293–312, 1952.

[6] R. N. Bhattacharya, E. C. Waymire. Stochastic Processes with Applications. New York: John Wiley, 1990.

[7] K. L. Chung. Markov Chains with Stationary Transition Probabilities. 2nd ed. Berlin: Springer–Verlas, 1967.

[8] W. Feller. An Introduction to Probability Theory and Its Applications. vol. 1, 3rd ed. New York: John Wiley, 1967.

[9] S. Karlin, M. Taylor. A First Course in Stochastic Processes. 2nd ed. New York: Academic Press, 1975.

[10] F. Spitzer. Principles of Random Walk. Princeton. Van Nostrand, 1964.

[11] T. E. Harris. The Theory of Branching Processes. New York: Springer–Verlag, 1963.

[12] K. B. Athreya, P. E. Ney. Branching Processes. Berlin: Springer–Verlag, 1972.

[13] W. Doeblin. Sur des proprietes asymptotiques de mouvement regis par certains types de chaines simples. Bull. Math. Soc. **39(1)**: 57–115, **39(2)**: 3–61, 1937.

[14] W. Doeblin. Elements d'une theorie generale des chaines simple constants de Markoff. Ann. Scient. Ec. Norm. Sup. III. **57**: 61–110, 1940.

[15] T. E. Harris. The existence of stationary measures for certain Markov processes. Third Berkeley Symposium on Math. Statist. and Probab. vol. II: 113–124, 1956.

[16] S. Orey. Limit Theorems for Markov Chain Transition Probabilities. New York: Van Nostrand, 1971.

[17] R. L. Tweedie. Sufficient conditions for ergodicity and recurrence of Markov chains on a general state space. *Stoch. Proc. Appl.* **3**: 385–403, 1975.

[18] K. B. Athreya, P. E. Ney. A new approach to the limit theory of recurrent Markov chains. Trans. Amer. Math. Soc. **245**: 493–501, 1978.

[19] E. Nummelin. A splitting technique for Harris recurrent chains. Z. Wahrscheinlichkeitstheorie und Verw. Geb. **43**: 309–318, 1978.

[20] E. Nummelin. General Irreducible Markov Chains and Nonnegative Operators. Cambridge: Cambridge University Press, 1984.

[21] S. P. Meyn, R. L. Tweedie. Markov Chains and Stochastic Stability. New York: Springer–Verlag, 1993.

[22] P. Diaconis, D. A. Freedman. Iterated Random Functions. SIAM Review **41(1)**: 45–76, 1999.

[23] P. Diaconis, M Shahshahani. Products of random matrices and computer image generation. Contemp. Math. **50**: 173–182, 1986.

[24] M. Barnsley, J. Elton. A new class of markov processes for image encoding. Adv. Appl. Probab. **20**: 14–32, 1988.

[25] M. Barnsley. Fractals Everywhere, 2nd. ed. New York: Academic Press, 1993.

[26] L. Dubins, D. A. Freedman. Invariant probabilities for certain Markov processes. Ann. Math. Statist. **37**: 837–844, 1966.

[27] R. N. Bhattacharya, O. Lee. Asymptotics of a class of Markov processes which are not in general irreducible. Ann. Probab. **16**: 1333–1347, 1988.

[28] R. N. Bhattacharya, M. Majumdar. On a theorem of Dubins and Freedman. J. Theor. Probab. **12**: 1165–1185, 1999.

[29] R. N. Bhattacharya, B. V. Rao. Random iteration of two quadratic maps. Stochastic Processes: A Festschrift in Honour of Gopinath Kallianpur. New York: Springer–Verlag, pp. 13–22, 1993.

[30] R. N. Bhattacharya, E. C. Waymire. An approach to the existence of unique invariant probabilities for Markov processes. Colloquium on Limit Theorems in Probability and Statistics, Janos Bolyai Math. Soc. (To appear), 2000.

[31] R. N. Bhattacharya, M. Majumdar. Convergence to equilibrium of random dynamical systems generated by i.i.d. monotone maps, with applications to economies. Asymptotics, Nonparametrics and Time Series: A Festschrift for M. L. Puri. S. Ghosh, editor. New York: Marcel Dekker, pp. 713–741, 1999.

[32] H. A. Hopenhayn, E. C. Prescott. Stochastic monotonicity and stationary distributions for dynamic economies. Econometrica **60**: 1387–1406, 1992.

[33] J. Propp, D. Wilson. Exact sampling and coupled Markov chains. Random Structures and Algorithms **9**: 223-252, 1996.

[34] R. L. Tweedie. Operator geometric stationary distributions for Markov chains with applications to queuing models. Adv. Appl. Probab. **14**: 368–391, 1981.

[35] Y. Kifer. Ergodic Theory of Random Transformations. Boston: Birkhauser, 1986.

[36] R. M. Blumenthal, H. K. Corson. On continuous collections of measures. Sixth Berkeley Symp. on Math. Statist. and Probab., vol. 2, 1972, pp. 33–40.

[37] R. N. Bhattacharya, O. Lee. Ergodicity and central limit theorems for a class of Markov processes. J. Multivariate Anal. **27**: 80–90, 1988.

[38] J. A. Yahav. On a fixed point theorem and its stochastic equivalent. J. Appl. Probab. **12**: 605–611, 1975.

[39] J. L. Doob. Stochastic Processes. New York: John Wiley, 1953.

[40] P. Lévy. Sur les intégrales dont les éléments sont des variables aléatoires indépendantes. Ann. Scuola Norm. Sup. Pisa (2) **3**: 337–366, 1934.

[41] P. Lévy. Theorie de l'addition des variables aléatiores. Paris: Gautier–Villars, 1937.

[42] P. Lévy. Processes Stochastiques et Mouvement Brownian. Paris: Gautier Villars, 1948.

[43] K. Itô. Lectures on Stochastic Processes. Bombay: Tata Institute of Fundamental Research, 1960.

[44] B. de Finetti. Sulle funzioni a increments aleatorio. Rend. Acad. Naz. Lincei. Cl. Sci. Fis. Mat. Nat. (6) **10**: 163–168, 1929.

[45] A. N. Kolmogorov. Sulla forma generale di una processo stocastico omogeneo. (Una problema di Bruno de Finetti.) Rend. Acad. Naz. Lincei. Cl. Sci. Fis. Mat. Nat. **15(6)**: 805–808, 1932.

[46] K. Itô. On stochastic processes (I) (Infinitely divisible laws of probability). Jap. J. Math. **18**: 261–301, 1942.

[47] A. Khinchin. Zur Theorie de ubeschränktteilbaren Verteilungsgesetze. Mat. Sb. **2**: 79–119, 1937.

[48] B. Pospisil. Sur un problème de M. M. S. Bernstein et A. Kolmogoroff. Casopis Pest. Mat. Fys. **65**: 64–76, 1935–36.

[49] W. Feller. Zur Theorie der stochastischen Prozesse (Existenz und Eindeutigkeitssätze). Math. Ann. **113**: 113–160, 1936.

[50] W. Feller. On the integro–differential equations of purely discontinuous Markoff processes. Trans. Ann. Math. Soc. **48**: 488–515, 1940. Errata. Ibid. **58**: 474, 1945.

[51] W. Doeblin. Sur certains mouvements aléatoires discontinus. Skand. Aktuarietidskr. **22**: 211–222, 1939.

[52] J. L. Doob. Markoff chains — denumerable case. Trans. Am. Math. Soc. **58**: 455–473, 1945.

[53] I. I. Gikhman, A. V. Skorokhod. Introduction to the Theory of Random Processes. Philadelphia: WB Saunders, 1969.

[54] W. Feller. The parabolic differential equation and the associated semigroups of transformations. Ann. Math. **55**: 468–519, 1952.

[55] W. Feller. Diffusion processes in one dimension. Trans. Amer. Math. Soc. **97**: 1–31, 1954.

[56] W. Feller. Generalized second order differential operators and their lateral conditions. IU J. Math. **1**: 459–504, 1957.

[57] G. B. Folland. Real Analysis. New York: John Wiley, 1984.

[58] S. N. Ethier, T. G. Kurtz. Markov Processes: Characterization and Convergence. New York: John Wiley, 1986.

[59] P. Mandl. Analytical Treatment of One–Dimensional Markov Processes. New York: Springer–Verlag, 1968.

[60] K. Itô, H. P. McKean. Diffusion Processes and Their Sample Paths. New York: Springer–Verlag, 1965.

[61] S. Karlin, H. M. Taylor. A Second Course in Stochastic Processes. New York: Academic Press, 1981.

[62] A. D. Wentzell. Semi–groups of operators associated with a generalized second order differential operator. Dokl. Akad. Nauk. SSSR **111**: 269–272, 1956.

[63] H. F. Trotter. A property of Brownian paths. Ill. J. Math. **2**: 425–433, 1958.

[64] D. Freedman. Brownian Motion and Diffusion. San Francisco: Holden–Day, 1971.

[65] A. Friedman. Partial Differential Equations of parabolic Type. Englewood Cliffs: Prentice Hall, 1964.

[66] K. Itô. On stochastic differential equations. Mem. Amer. Math. Soc. **4**: 1–51, 1951.

[67] H. P. McKean. Stochastic Integrals. New York: Academic Press, 1969.

[68] N. Ikeda, S. Watanabe. Stochastic Differential Equations and Diffusion Processes. 2nd ed. Amsterdam: North–Holland, 1989.

[69] A. Friedman. Stochastic Differential Equations and Applications. Vol. 1. New York: Academic Press, 1975.

[70] I. Karatzas, S. E. Shreve. Brownian Motion and Stochastic Calculus. 2nd ed. New York: Springer–Verlag, 1991.

[71] L. C. G. Rogers, D. Williams. Diffusions, Markov Processes, and Martingales. Vol. 2. New York: John Wiley, 1987.

[72] R. Z. Khas'minskii. Ergodic properties of recurrent diffusion processes and stabilization of the Cauchy problem for parabolic equations. Theor. Probab. Appl. **5**: 179–196, 1960.

[73] R. N. Bhattacharya. Criteria for recurrence and existence of invariant measures for multidimensional diffusions. Ann. Probab. **3**: 541–553, 1978. Correction, ibid. **8**: 1194–1195, 1980.

[74] R. H. Cameron, W. T. Martin. Transformation of Weiner integrals under translations. Ann. Math. **45**: 386–396, 1944.

[75] R. H. Cameron, W. T. Martin. Evaluations of various Weiner integrals by use of certain Sturm–Liouville differential equations. Bull. Amer. Math. Soc. **51**: 73–90, 1945.

[76] I. V. Girsanov. On transforming a certain class of stochastic processes by absolutely continuous substitution of measures. Theor. Probab. Appl. **5**: 285–301, 1960.

[77] A. A. Novikov. On an identity for stochastic integrals. Theor. Probab. Appl. **17**: 717–720, 1972.

[78] D. W. Stroock, S. R. S. Varadhan. Diffusion processes with continuous coefficients, I and II. Comm. Pure & Appl. Math. **22**: 345–400, 479–530, 1969.

[79] D. W. Stroock, S. R. S. Varadhan. Multidimensional Diffusion Processes. New York: Springer–Verlag, 1979.

[80] R. P. Feynman. Space–time approach to non–relativistic quantum mechanics. Rev. Mod. Physics. **20**: 367–387, 1948.

[81] M. Kac. On distributions of certain Weiner functionals. Trans. Amer. Math. Soc. **65**: 1–13, 1949.

[82] M. Kac. On some connections between probability theory and differential and integral equations. Proc. 2nd Berkeley Symp. on Math. Statist. & Probab., pp. 189–215, 1951.

[83] L. Bachelier. Theorie de las spéculation. Ann. Sci. École Norm. Sup. **17**: 27–86, 1900 (English translation in The Random Character of Stock Market Prices, P. H. Cootner, ed., Cambridge: MIT Press, 1964).

[84] A. Einstein. On the movement of small particles suspended in a stationary liquid demanded by the molecular–kinetic theory of heat. Ann. der Physik, **17** 1905; and On the theory of the Brownian movement. Ann. der Physik **19**, 1906 (English translation in the book Investigations on the Theory of the Brownian Movement. R. Fürth, ed., New York: Dover, 1954.

[85] N. Wiener. Differential space. J. Math. Phys. **2**: 131–174, 1923.

[86] P. Langevin. Sur la theorie du mouvement Brownian. C. R. Acad. Sci. Paris. **146**: 530–533, 1908.

[87] R.E.A.C. Paley, N. Wiener, A. Zygmund. Notes on random functions. Math. Zeit. **37**: 647–668, 1933.

[88] K. Itô. Differential equations determining Markov processes. Zenkoku Shijo Danwakai **1077**: 1352–1400, (in Japanese), 1942.

[89] K. Itô. Stochastic integral. Proc. Imperial Akad. Tokyo. **20**: 519–524, 1944.

[90] I.I. Gikhman. A method of constructing random processes. Akad. Nauk SSSR **58**: 961–964 (in Russian), 1947.

[91] I.I. Gikhman. On the theory of differential equations of random processes. Uskr. Math. Z. **2**: 37–63 (in Russian), 1950.

[92] P. A. Meyer. Intégrales stochastiques. Lecture Notes in Mathematics. **39**: 72–162. Berlin: Springer–Verlag, 1967.

[93] P. A. Meyer. Un cours sur les intégrales stochastiques. Lecture Notes in Mathematics. **511**: 245–398. Berlin: Springer–Verlag, 1976.

[94] D. Revuz, M. Yor. Continuous Martingales and Brownian Motion. New York: Springer–Verlag, 1994.

[95] L. Arnold. Stochastic Differential Equations: Theory and Applications. New York: John Wiley, 1973.

[96] R. S. Lipster, A. N. Shiryaev. Statistics of Random Processes. Vol. 1. New York: Springer–Verlag, 1977.

[97] M. D. Donsker, S.R.S. Varadhan. Asymptotic evaluation of certain Markov process expectations for large time, I and II. Comm. Pure & Appl. Math. **28**: 1–47, 279–301, 1975.

[98] M. D. Donsker, S.R.S. Varadhan. On the principal eigenvalue of second order elliptic differential operators. Comm. Pure & Appl. Math. **29**: 595–621, 1976.

[99] S.R.S. Varadhan. Large Deviations and Applications. Philadelphia: SIAM, 1984.

[100] M. Freidlin, A. D. Wentzell. Random Perturbations of Dynamical Systems. 2nd ed. New York: Springer–Verlag, 1998.

[101] P. Diaconis. Group Representations in Probability and Statistics. Hayward: IMS, 1986.

[102] P. Diaconis. The cutoff phenomenom in finite Markov chains. Proc. Natl. Acad. Sci. U.S.A., **93**: 1659–1664, 1996.

[103] P. Diaconis, D. W. Stroock. Geometric bounds for eigenvalues for Markov chains. Ann. Appl. Prob. **1**: 36–61, 1991.

[104] P. Diaconis, L. Saloff–Coste. Logarithmic Sobolev inequalities and finite Markov chains. Ann. Appl. Probab. **6**: 695–750, 1996.

[105] J. Fill. Eigenvalue bounds on convergence to stationarity for nonreversible Markov chains, with application to the exclusion process. Ann. Appl. Probab. **1**: 62–87.

[106] M. F. Chen, F. Y. Wang. Applications of coupling method to the first eigenvalue on a manifold. Sci. Sin. (A) **37**: 1–14, (English edition), 1994.

[107] R. A. Holley, S. Kusuoka, D. W. Stroock. Asymptotics of the spectral gap, with applications to simulated annealing. J. Funct. Anal. **83**: 333–347, 1989.

[108] R. N. Bhattacharya, F. Götze. Time–scales for Gaussian approximation and its break down under a hierarchy of periodic spatial heterogeneities. Bernoulli **1**: 81–123, 1995.

[109] R. N. Bhattacharya. Multiscale diffusion processes with periodic coefficients and an application to solute transport in porous media. Ann. Appl. Probab. **9**: 951–1020, 1999.

Chapter 2

Semimartingale Theory and Stochastic Calculus

JIA-AN YAN
Academy of Mathematics and System Sciences, Chinese Academy of Sciences
P.O. Box 2734, Beijing 100080, China

K. Itô invented his famous stochastic calculus on Brownian motion in 40's. In the same period, J.L. Doob developed a martingale theory and related stochastic processes to an increasing family of σ-algebras $(\mathcal{F}_t)$ of events, where $\mathcal{F}_t$ expresses the information avilable until time t. From 60's to 70's the "Strasbourg school", headed by P.A. Meyer, developed a modern theory of martingales, the general theory of stochastic processes, and stochastic caluculs on semimartingales. It turned out soon that semimartingales constitute the largest class of right-continuous adapted integrators with respect to which stochastic integrals of simple predictable integrands satisfy the theorem of dominated convergence in probability. Stochastic calculus on semimartingales not only became an important tool for modern probability theory and stochastic processes, but also has broad applications to many branches of mathematics (e.g. partial differential equations, differential geometry, stochastic control), physics, engineering, mathematical finance and all other domains in which random dynamic structures are involved.

This chapter offers a concise and detailed overview of semimartingale theory and stochastic calculus. In Section 1, we present main results about the martingale theory and the general theory of stochastic processes. In Section 2 we introduce systematically the stochastic integrals of real-valued and vector-valued local martingales and semimartingales, for both predictable and progressive integrands. We present Itô's formula, the Doléans exponential formula, Tanaka-Meyer's formula for local times of semimartingales, the Fisk-Stratonovich integral, and the Itô stochastic differential equation. A general result about the existence and uniqueness of solutions of a stochastic differential equation driven by a semimartingale is also presented in Section 2. Finally, in Section 3, we present main ingredients of stochastic calculus on semimartingales, which are: stochastic integration w.r.t. random measures, charateristics of semimartingales, calculus on Lévy processes, Girsanov's theorems, martingale representation theorems. The characterization theorem for semimartingales and some sufficient conditions for the uniform integrability of exponential martingales are also included in this section.

The author wishes to express his sincere thanks to Professor Kannan and Professor Lakshmikantham, the editors of the handbook, for inviting him to write this chapter on semimartingale theory and stochastic calculus. The financial support from the National Natural Science Foundation of China (grant 79790130) and the Ministry of Science and Technology (the 973 project on mathematics) is acknowledged by the author.

2.1 General Theory of Stochastic Processes and Martingale Theory

In this section we will introduce the general theory of stochastic processes and martingale theory. Both theories are not only important basis for semimartingale theory and stochastic calculus based on semimartingales, but also indispensable tools for studying Markov processes and random point processes. For the sake of completeness, we include a short review on the classical theory of martingales. Most of results presented in this section can be found in He et al. (1992) [Ref. 1]. For those results not included in Ref. 1, we will indicate their references.

2.1.1 Classical Theory of Martingales

Discrete Time Martingales

Let $(\Omega, \mathcal{F}, \mathbf{P})$ be a probability space and $(\mathcal{F}_n, n \geq 0)$ an increasing sequence of sub-σ-fields of $\mathcal{F}$. We call $(\mathcal{F}_n)$ a *filtration*. Put $\mathcal{F}_\infty \hat{=} \sigma(\bigcup_n \mathcal{F}_n)$ and $\mathcal{F}_{-1} = \mathcal{F}_0$. A sequence of r.v.'s $(X_n, n \geq 0)$ is said to be $(\mathcal{F}_n)$-*adapted* (resp.*predictable*), if each X_n is $\mathcal{F}_n$-(resp.$\mathcal{F}_{n-1}$)-measurable.

We denote $\overline{\mathbb{N}}_0 = \{0, 1, 2, \cdots, \infty\}$. Let T be an $\overline{\mathbb{N}}_0$-valued r.v.. If $\forall n \in \mathbb{N}_0$, $[T = n] \in \mathcal{F}_n$, T is called an $(\mathcal{F}_n)$-*stopping time*. For a stopping time T we put

$$\mathcal{F}_T = \{A \in \mathcal{F}_\infty \ : \ A \cap [T = n] \in \mathcal{F}_n \, , \forall n \geq 0\} \, ,$$

then $\mathcal{F}_T$ is a σ-field. Let (X_n) be an adapted sequence of r.v.'s and T a stopping time. Then $X_T I_{[T<\infty]}$ is $\mathcal{F}_T$-measurable.

Definition 2.1.1 *An $(\mathcal{F}_n)$-adapted sequence of r.v.'s $(X_n, n \geq 0)$ is called a martingale (supermartingale, submartingale) if each X_n is integrable and*

$$\mathbf{E}[X_{n+1} \mid \mathcal{F}_n] = X_n (\leq X_n, \geq X_n) \text{ a.s. } .$$

It is called a local martingale, if there exists an increasing sequence (T_n) of stopping times with $\lim_n T_n = \infty$ such that for each k $(X_{n \wedge T_k} I_{[T_k > 0]}, n \geq 0)$ is a martingale.

A martingales (or supermartingales) $(X_n, n \in \mathbb{N}_0)$ is called *right-closable*, if there exists an 0. $X_\infty \in \mathcal{F}_\infty$, such that for all $n \in \mathbb{N}_0$, $\mathbf{E}[X_\infty | \mathcal{F}_n] = X_n (\leq X_n)$ a.s.. In this case, $(X_n, n \in \overline{\mathbb{N}}_0)$ is called a *right-closed* martingale (or supermartingale).

It is obvious that if (X_n) and (Y_n) are (super)martingales, then $(X_n + Y_n)$ is a (super)martingale and $(X_n \wedge Y_n)$ is a supermartingale. If (X_n) is a (sub)martingale and $f : \mathbf{R} \longrightarrow \mathbf{R}$ is a (non-decreasing) convex function on $\mathbf{R}$, such that each $f(X_n)$ is integrable, then by Jensen's inequality, $\big(f(X_n)\big)$ is a submartingale.

The main results presented below about discrete martingales are due to J. L. Doob (1953) [Ref. 2].

Theorem 2.1.2 (maximal inequalities) *If $N \geq 1, (X_n)_{n \leq N}$ is a submartingale, then for $\lambda > 0$,*

$$\lambda \mathbf{P}(\sup_{n \leq N} X_n \geq \lambda) \leq \int_{[\sup_{n \leq N} X_n \geq \lambda]} X_N d\mathbf{P} \; ,$$

$$\lambda \mathbf{P}(\sup_{n \leq N} | X_n | \geq \lambda) \leq 2\mathbb{E}[X_N^+] - \mathbf{E}[X_0] .$$

Theorem 2.1.3 (Doob's inequalities) *Let $N \geq 1, (X_n)_{n \leq N}$ be a martingale or a non-negative submartingale. Put $X_N^* = \sup_{n \leq N} |X_n|$. Then for $\lambda > 0$, $p \geq 1$ and $q > 1$, we have*

$$\mathbf{P}(X_N^* \geq \lambda) \leq \lambda^{-p}\mathbf{E}[| X_N |^p] ,$$

$$(\mathbf{E}[(X_N^*)^q])^{1/q} \leq \frac{q}{q-1}(\mathbf{E}[|X_N|^q])^{1/q} ,$$

$$\mathbf{E}[X_N^*] \leq \frac{e}{e-1}\left(1 + \sup_{n \leq N} \mathbf{E}[|X_n| \log^+ |X_n|]\right) .$$

Let (X_n) be an $(\mathcal{F}_n)$-adapted sequence and $[a, b]$ a finite closed interval. Put

$$\begin{aligned} T_0 &= \inf\{n \geq 0 : X_n \leq a\} , \qquad T_1 = \inf\{n > T_0 : X_n \geq b\} , \\ T_{2j} &= \inf\{n > T_{2j-1} : X_n \leq a\} , \; T_{2j+1} = \inf\{n > T_{2j} : X_n \geq b\} . \end{aligned}$$

(T_n) is an increasing sequence of stopping times. We denote by $U_a^b[X, N]$ the number of upcrossing of $[a, b]$ by sequence $(X_0, \cdots, X_N)$. Then

$$\left[U_a^b[X, k] = j\right] = [T_{2j-1} \leq N < T_{2j+1}] \in \mathcal{F}_N ,$$

so that $U_a^b[X, N]$ is an $\mathcal{F}_N$-measurable r.v..

Theorem 2.1.4 (upcrossing inequality) *Let $N \geq 1, (X_n)_{n \leq N}$ be a supermartingale. Then*

$$\mathbf{E}U_a^b[X, N] \leq \frac{1}{b-a}\mathbf{E}[(X_N - a)^-] .$$

As an application of this inequality one obtains the following martingale convergence theorem.

Theorem 2.1.5 *Let (X_n) be a supermartingale (resp. martingale).*

1) If $\sup_n E[X_n^-] < \infty$ (or equivalently, $\sup_n E[|X_n|] < \infty$), then X_n a.s. converges to an integrable r.v. X_∞ as $n \to \infty$. If (X_n) is a non-negative supermartingale, then for each $n \geq 0$,

$$E[X_\infty \mid \mathcal{F}_n] \leq X_n \quad \text{a.s.}.$$

2) If (X_n) is uniformly integrable, then X_n a.s. and L^1-converges to an integrable r.v. X_∞, and $\forall n \geq 0$,

$$\mathbf{E}[X_\infty \mid \mathcal{F}_n] \leq X_n \,(\text{resp.} = X_n) \quad \text{a.s.}.$$

In particular, if $X_n = \mathbf{E}[\xi \mid \mathcal{F}_n]$ with ξ being an integrable r.v., then X_n a.s. and L^1-converges to $\mathbf{E}[\xi \mid \mathcal{F}_\infty]$.

As an application of Theorem 3.1.5, we obtain the following result which shows that martingales with bounded increments either converge or oscillate between $+\infty$ and $-\infty$.

Theorem 2.1.6 (Ref. 2) *Let (X_n) be a martingale with $X_0 = 0$ and $|X_{n+1} - X_n| \leq M < \infty$, where M is a constant. Put*

$$C = \{\lim_{n\to\infty} X_n \text{ exists and finite}\}, \quad D = \{\limsup_{n\to\infty} = +\infty, \text{and } \liminf_{n\to\infty} = -\infty\}.$$

Then $\mathbf{P}(C \cup D) = 1$.

Using Theorem 2.1.6 we can prove the following generalization of the Borel-Cantelli Lemma [Ref. 2]: Let $(\mathcal{F}_n)$ be a filtration with $\mathcal{F}_0 = \{\emptyset, \Omega\}$ and (A_n) a sequence of events with $A_n \in \mathcal{F}_n$. Then $\{A_n i.o.\} = \left\{\sum_{n=1}^{\infty} \mathbf{P}(A_n|\mathcal{F}_{n-1}) = \infty\right\}$. (Hints: Let $X_0 = 0, X_n = \sum_{m=1}^{n} I_{A_m} - \mathbf{P}(A_m|\mathcal{F}_{m-1}), n \geq 1$.)

We now turn to the convergence of "reverse" supermartingales with the index set $-\mathbf{N}_0 = \{\cdots, -2, -1, 0\}$. Let $(\mathcal{F}_n)_{n\in -\mathbf{N}_0}$ be a sequence of sub-σ-fields of $\mathcal{F}$ such that for all $n \in -\mathbf{N}_0$, $\mathcal{F}_{n-1} \subset \mathcal{F}_n$. An $(\mathcal{F}_n)$-adapted stochastic sequence $(X_n)_{n\in -\mathbf{N}_0}$ is called a *martingale* (*supermartingale*), if for each $n \in -\mathbf{N}_0$, X_n is integrable and

$$\mathbf{E}[X_n \mid \mathcal{F}_{n-1}] = X_{n-1} (\leq X_{n-1}) \text{ a.s. .}$$

Theorem 2.1.7 *Let $(X_n)_{n\in-\mathbf{N}_0}$ be a supermartingale. Then $\lim_{n\to-\infty} X_n$ exists a.s. If $\lim_{n\to-\infty} \mathbf{E}[X_n] < +\infty$, a.s., then (X_n) is uniformly integrable, X_n a.s. and L^1-converges to $X_{-\infty}$.*

Corollary 2.1.8 *Let ξ be an integrable r.v. and $(\mathcal{G}_n)_{n\in I\!N_0}$ be a decreasing sequence of sub-σ-fields of $\mathcal{F}$. Put $\xi_n = \mathbf{E}[\xi \mid \mathcal{G}_n]$, then ξ_n a.s. and L^1-converges to $\mathbf{E}[\xi \mid \bigcap_n \mathcal{G}_n]$.*

The following are *Doob's stopping theorems* for right-closed (super)martingales and general (super)martingales.

Theorem 2.1.9 *Let $(X_n, n \in \overline{I\!N}_0)$ be a martingale (resp. supermartingale), S and T two stopping times. Then X_S and X_T are integrable and*

$$\mathbf{E}[X_T \mid \mathcal{F}_S] = X_{S\wedge T} \text{ (resp. } \leq X_{S\wedge T}) \text{ a.s. .}$$

Theorem 2.1.10 *1) Let $(X_n, n \in I\!N_0)$ be a martingale, S and T two finite stopping times. If X_T is integrable, then*

$$\mathbf{E}[X_T \mid \mathcal{F}_S] = X_{S\wedge T}, \text{ a.s. ,} \tag{10.1}$$

if and only if

$$\lim_{n\to\infty} \mathbf{E}[X_n I_{[T\geq n]} \mid \mathcal{F}_S] = 0, \text{ a.s. .}$$

In particular, if $\liminf_{n\to\infty} \mathbf{E}[|X_n| I_{[T\geq n]}] = 0$, *(10.1) holds.*

2) Let $(X_n, n \in I\!N_0)$ be a supermartingale, S and T two finite stopping times. If X_T is integrable and

$$\limsup_{n\to\infty} \mathbf{E}[X_n I_{[T\geq n]} \mid \mathcal{F}_S] \geq 0, \text{ a.s. ,}$$

then we have

$$\mathbf{E}[X_T \mid \mathcal{F}_S] \leq X_{S\wedge T}, \text{ a.s. .} \tag{10.2}$$

In particular, if $\liminf_{n\to\infty} \mathbf{E}[X_n^- I_{[T\geq n]}] = 0$, *(10.2) holds.*

As a consequence of Theorem 2.1.10 we have

Corollary 2.1.11 *Let $(X_n, n \in I\!N_0)$ be a martingale (resp. supermartingale) and T a stopping time with $\mathbf{E}[T] < \infty$. If there exists a constant C such that for all n*

$$\mathbf{E}[|X_{n+1} - X_n| \mid \mathcal{F}_n] \leq C \text{ a.s. on } [\mathrm{T} \geq \mathrm{n}+1],$$

then $\mathbf{E}[|X_T|] < \infty$, and

$$\mathbf{E}[X_T] = \mathbf{E}[X_0] \text{ (resp. } \leq \mathbf{E}[X_0]).$$

Let $(M_n, n \geq 0)$ be a martingale and (H_n) a predictable sequence. We denote $\Delta M_n = M_n - M_{n-1}$ and put

$$X_0 = H_0 M_0, \; X_n = H_0 M_0 + \sum_{i=1}^{n} H_i \Delta M_i, \; n \geq 1.$$

The sequence (X_n) is called the *martingale transform* of M by H and denoted by $H.M$.

Theorem 2.1.12 (Ref. 3) *Let $X = (X_n, n \geq 0)$ be an adapted sequence. The following properties are equivalent:*

1) X is a local martingale;

2) For every n, X_{n+1} is σ-integrable w.r.t. $\mathcal{F}_n$ and $\mathbb{E}[X_{n+1} \mid \mathcal{F}_n] = X_n$, a.s.;

3) X is a martingale transform.

The following theorem solves an optimal stopping problem.

Theorem 2.1.13 (Ref. 3, 4) *(Snell envelope)] Let $(Z_n)_{0 \leq n \leq N}$ be an adapted sequence of integrable r.v. We define by backward induction a sequence (U_n) as follows: let $U_N = Z_N$, and*

$$U_n = \text{Max}(Z_n, \mathbf{E}[U_{n+1}|\mathcal{F}_n]), \quad n \leq N-1,$$

Then (U_n) is a supermartingale, and is the smallest supermartingale dominating (Z_n) (i.e. $U_n \geq Z_n$ for all n). Moreover, if we denote by $\mathcal{T}_{n,N}$ the set of stopping times taking values in $\{n, \cdots, N\}$ and let $T_n = \inf\{j \geq n : U_j = Z_j\}$, where $\inf \emptyset := N$, then each T_n is a stopping time, $(U_n^{T_0})$ is a martingale, and we have for all $0 \leq n \leq N$,

$$U_n = \mathbf{E}[Z_{T_n}|\mathcal{F}_n] = \text{ess sup}\{\mathbf{E}[Z_T|\mathcal{F}_T] : T \in \mathcal{T}_{n,N}\}.$$

Moreover, the maximum of expected values $\mathbf{E}[Z_T]$ on $\mathcal{T}_{n,N}$ is attended at T_n, and the optimal value is equal to $\mathbf{E}[U_n]$, namely,

$$\mathbf{E}[U_n] = \mathbf{E}[Z_{T_n}] = \sup\{\mathbf{E}[Z_T] : T \in \mathcal{T}_{n,N}\}.$$

We call (U_n) the Snell envelope of (Z_n).

Continuous Time Martingales

Let $(\Omega, \mathcal{F}, \mathbf{P})$ be a probability space and $\mathbf{F} = (\mathcal{F}_t)_{t \geq 0}$ an increasing sequence of sub-σ-fields of $\mathcal{F}$. Put $\mathcal{F}_\infty = \sigma(\bigcup_t \mathcal{F}_t)$. If for all $t \geq 0$, $\mathcal{F}_{t+} = \bigcap_{s>t} \mathcal{F}_s = \mathcal{F}_t$, $\mathbf{F}$ is said to be right-continuous.

An $\overline{\mathbf{R}}_+$-valued r.v. T is called an $\mathbf{F}$-stopping time, if for each $t \geq 0$, $[T \leq t] \in \mathcal{F}_t$. For any $\mathbf{F}$-stopping time T, put

$$\mathcal{F}_T = \{A \in \mathcal{F}_\infty : \forall t \in \mathbb{R}_+, A[T \leq t] \in \mathcal{F}_t\}.$$

$\mathcal{F}_T$ is a σ-field.

Let X, Y be two processes. If for each $t \in \mathbb{R}_+$, $X_t = Y_t$, a.s., then we call Y a *version* of X. If for almost all ω, two paths $X.(\omega)$ and $Y.(\omega)$ are the same, we say that X and Y are *indistinguishable* from each other. Here and hereafter we don't distinguish between two indistinguishable processes. In particular, by a right-continuous process we mean a process with almost all paths being right-continuous functions on $\mathbf{R}_+$.

A process X is called $\mathbf{F}$-*adapted*, if for each $t \geq 0$, X_t is $\mathcal{F}_t$-measurable. An $\mathbf{F}$-adapted process $X = (X_t)_{t \geq 0}$ is called an $\mathbf{F}$-*martingale (supermartingale, submartingale)*, if each X_t is integrable and for $0 \leq s < t$,

$$\mathbf{E}[X_t \mid \mathcal{F}_t] = X_s (\leq X_s, \geq X_s) \quad \text{a.s.}.$$

Theorem 2.1.14 *Assume* $\mathbf{F}$ *is right-continuous. An* $\mathbf{F}$*-supermartingale* (X_t) *has a right-continuous version if and only if* $t \mapsto \mathbb{E}[X_t]$ *is a right-continuous function on* $\mathbf{R}_+$*. In particular, any* $\mathbf{F}$*-martingale has a right-continuous version.*

The following five theorems are direct consequence of the corresponding results for discrete time case.

Theorem 2.1.15 (Doob's Inequalities) *Let* (X_t) *be a right-continuous martingale or nonnegative submartingale. Put* $X^* = \sup_{t\geq 0}|X_t|$. *For any* $\lambda > 0$ $p \geq 1$ *and* $q > 1$,

$$\mathbf{P}(X^* \geq \lambda) \leq \lambda^{-p} \sup_t \mathbb{E}\left[|X_t|^p\right].$$

$$(\mathbf{E}[(X^*)^q])^{1/q} \leq \frac{q}{q-1} \sup_{t\geq 0}(\mathbf{E}[|X_t|^q)^{1/q}.$$

$$\mathbf{E}[X^*] \leq \frac{e}{e-1}(1 + \sup_{t\geq 0} \mathbf{E}[|X_t| \log^+(|X_t|)]).$$

Theorem 2.1.16 *If* (X_t) *is a right-continuous supermartingale such that* $\sup_t \mathbb{E}\left[|X_t^-|\right] < \infty$ *(or equivalently,* $\sup_t \mathbb{E}\left[|X_t|\right] < \infty$*), then as* $t \to \infty$, X_t *a.s converges to an integrable r.v.* X_∞. *If* (X_t) *is nonnegative, then* $(X_t, t \in \overline{\mathbb{R}}_+)$ *is a (right-closed) supermartingale.*

Theorem 2.1.17 *If* (X_t) *is a uniformly integrable right-continuous martingale (supermartingale), then as* $t \to \infty$, X_t *a.s. and* L^1*-converges to an integrable r.v.* X_∞. *Moreover,* $(X_t, t \in \overline{\mathbb{R}}_+)$ *is a martingale (supermartingale).*

Theorem 2.1.18 *Let* $(X_t)_{t>0}$ *be a right-continuous supermartingale w.r.t.* $(\mathcal{F}_t)_{t>0}$. *Put* $\mathcal{F}_0 = \bigcap_{s>0}\mathcal{F}_s$. *If* $\sup_{t>0}\mathbf{E}[X_t] < \infty$, *then as* $t \downarrow 0$, X_t *a.s. and* L^1*-converges to an* $\mathcal{F}_0$*-measurable r.v.* X_0, *and* $(X_t)_{t\geq 0}$ *is a* $(\mathcal{F}_t)$*-supermartingale.*

Theorem 2.1.19 *Let* $(X_t, t \in \overline{\mathbb{R}}_+)$ *be a right-continuous martingale (supermartingale). If* S *and* T *are two stopping times with* $S \leq T$, *then* X_S, X_T *are integrable and*

$$\mathbf{E}[X_T \mid \mathcal{F}_S] = X_S (\leq X_S) \quad \text{a.s.}.$$

Theorem 2.1.20 *Let* X *be a non-negative right-continuous supermartingale. Put* $T = \inf\{t > 0 : X_t = 0 \text{ or } X_{t-} = 0\}$, *then for a.e.* ω *and all* $t \in [T(\omega), \infty)$, $X_t(\omega) = 0$.

Theorem 2.1.21 *Let* $X^1 \leq X^2 \leq \cdots$ *be a sequence of right-continuous supermartingales with* $\sup_n \mathbb{E}[X_0^n] < 0$. *Then* $X_t = \sup_n X_t^n$, $t \geq 0$, *is a right-continuous supermartingale.*

2.1.2 General Theory of Stochastic Processes

Let $(\Omega, \mathcal{F})$ be a measurable space. A process on $(\Omega, \mathcal{F})$ is simply a collection of measurable functions $\{X_t, t \in \Lambda\}$, defined on $(\Omega, \mathcal{F})$, where Λ is a time parameter set. If Λ is an interval of $\mathbf{R} = (-\infty, \infty)$, (X_t) is called a process in continuous time. If Λ is a subset of $\mathbf{N} = \{0, 1, 2, \cdots\}$, (X_t) is called a process in discrete time. For a fixed $\omega \in \Omega$, the function $t \mapsto X_t(\omega)$ defined on Λ is called a *sample path* of the process (X_t). A process in continuous time having continuous paths is called a continuous process.

In the sequel we assume the time parameter set Λ is $\mathbf{R}_+$. We call an increasing family $(\mathcal{F}_t)$ of sub-σ-algebras of $\mathcal{F}$ a *filtration*.

The general theory of stochastic processes contains four parts: 1) the measurable structure of stochastic processes; 2) the section theorem, which provides an approach of studying trajectory properties of a stochastic process through values of the process taken at stopping times; 3) the projection theory of measurable processes, which is a generalization of the conditional expectation in probability theory; 4) the dual projections of finite variation processes, which are defined via projections of random measures.

Optional and Predictable Processes

Let $(\Omega, \mathcal{F})$ be a measurable space equipped with a filtration $\mathbf{F} = (\mathcal{F}_t)_{t\geq 0}$. Set $\mathcal{F}_\infty = \vee_{t\geq 0}\mathcal{F}_t$ and

$$\begin{aligned} \mathcal{F}_{t+} &= \bigcap_{s>t} \mathcal{F}_s \ , \ t \geq 0 \, , \\ \mathcal{F}_{t-} &= \bigwedge_{s<t} \mathcal{F}_s = \sigma(\bigcup_{s<t} \mathcal{F}_s) \ , t > 0 \, . \end{aligned}$$

By convention, we put $\mathcal{F}_{0-} = \mathcal{F}_0$, $\mathcal{F}_{\infty-} = \mathcal{F}_\infty$. The filtration $\mathbf{F}$ is called right-continuous, if for each $t \geq 0, \mathcal{F}_t = \mathcal{F}_{t+}$. Obviously, $\mathbf{F}_+ = (\mathcal{F}_{t+})_{t\geq 0}$ is right-continuous.

For an $\mathbf{F}$-stopping time T, we put

$$\begin{aligned} \mathcal{F}_{T+} &= \{A \in \mathcal{F}_\infty : \forall t \in \mathbb{R}_+, A[T \leq t] \in \mathcal{F}_{t+}\} \ , \\ \mathcal{F}_T &= \{A \in \mathcal{F}_\infty : \forall t \in \mathbb{R}_+, A[T \leq t] \in \mathcal{F}_t\} \ , \\ \mathcal{F}_{T-} &= \mathcal{F}_0 \vee \sigma\{A[t < T] : \ A \in \mathcal{F}_t, \, t \in \mathbb{R}_+\}. \end{aligned}$$

Then $\mathcal{F}_{T+}, \mathcal{F}_T, \mathcal{F}_{T-}$ are all σ-fields and it holds that $\mathcal{F}_{T-} \subset \mathcal{F}_T \subset \mathcal{F}_{T+}$. For each natural number $n \geq 1$, put

$$T_n = \sum_{k=1}^{\infty} \frac{k}{2^n} I_{[\frac{k-1}{2^n} \leq T < \frac{k}{2^n}]} + (+\infty) I_{[T=+\infty]} \, .$$

Then T_n are stopping times and $T_n \downarrow T$.

Let $A \in \mathcal{F}_T$. Put

$$T_A = T I_A + (+\infty) I_{A^c} \ ,$$

then T_A is a stopping time. We call T_A the restriction of T on A.

Definition 2.1.22 *Let U, V be $\overline{\mathbf{R}}_+$-valued function on Ω with $U \leq V$. Put*

$$\llbracket U, V \rrbracket = \{(\omega, t) \in \Omega \times \mathbb{R}_+ : \ U(\omega) \leq t \leq V(\omega)\} \, ,$$

$$\llbracket U, V \llbracket = \{(\omega, t) \in \Omega \times \mathbb{R}_+ : \ U(\omega) \leq t < V(\omega)\} \, .$$

Similarly, we can define $\rrbracket U, V \rrbracket$ and $\rrbracket U, V \llbracket$. They are called random intervals. $\llbracket U, U \rrbracket$ will be denoted by $\llbracket U \rrbracket$ and called the graph of U.

A random set B is called a *thin set*, if it can be expressed as a countable union of graphs of stopping times.

Definition 2.1.23 *1) A process is called cadlag process, if its sample paths are right-continuous with left-limits. "cadlag" is an acronym from the French "continu à droit, limité à gauche." Similarly we can define "caglad" process.*

2) A process is called an increasing process, if it is a cadlag process with nonnegative initial values and its paths are increasing functions.

3) A process is called a finite variation (FV) process or process of finite variation, if it is cadlag and its paths are of finite variation on each compact interval of $\mathbf{R}_+$.

4) Let $X = (X_t)_{t\geq 0}$ be a stochastic process. If $X_t(\omega)$, as a function of (ω, t), is $\mathcal{F} \times \mathcal{B}(\mathbf{R}_+)$-measurable, X is said to be measurable; if for each $t \in \mathbb{R}_+$, the restriction of X on $\Omega \times [0, t]$ is $\mathcal{F}_t \times \mathcal{B}([0, t])$-measurable, X is said to be progressively measurable (or simply, progressive).

5) The smallest σ-field on $\Omega \times \mathbb{R}_+$ such that all cadlag (resp. left-continuous) adapted processes are measurable is called the optional (resp. predictable) σ-field and denoted by $\mathcal{O}$ (resp. $\mathcal{P}$). A random set or stochastic process is called an optional (resp. predictable) set or process if it is $\mathcal{O}$ (resp. $\mathcal{P}$)-measurable.

Theorem 2.1.24 *1) Every progressive process is measurable and adapted.*

2) Every right-continuous (or left-continuous) process is progressive.

3) If (X_t) is a progressive process and T is a stopping time, then $X_T I_{[T<\infty]}$ is $\mathcal{F}_T$-measurable.

4) Every optional process is progressive.

Theorem 2.1.25 *1) We denote by $\mathcal{T}$ the collection of all stopping times. Then*

$$\mathcal{O} = \sigma\{[\![S,\infty[\![: S \in \mathcal{T}\}.$$

2) Put

$$\begin{aligned} \mathcal{C}_1 &= \{A\times\{0\} : A\in\mathcal{F}_0\}\cup\Big\{A\times]s,t] : 0<s<t, s,t\in\mathcal{Q}_+, A\in\bigcup_{r<s}\mathcal{F}_r\Big\}, \\ \mathcal{C}_2 &= \{A\times\{0\} : A\in\mathcal{F}_0\}\cup\Big\{A\times[s,t[: 0<s<t, s,t\in\mathcal{Q}_+, A\in\bigcup_{r<s}\mathcal{F}_r\Big\}, \\ \mathcal{C}_3 &= \{A\times\{0\} : A\in\mathcal{F}_0\}\cup\{]\!]S,\infty[\![: S\in\mathcal{T}\}, \end{aligned}$$

where $\mathcal{Q}_+$ denotes the set of all positive rational numbers. Then $\sigma(\mathcal{C}_1)=\sigma(\mathcal{C}_2)=\sigma(\mathcal{C}_3)=\mathcal{P}$. In particular, $\mathcal{P}\subset\mathcal{O}$.

Corollary 2.1.26 *Let (X_t) be a predictable processes and T a stopping time. Then X^T is a predictable process and $X_T I_{[T<\infty]}\in\mathcal{F}_{T-}$. Here and henceforth, X^T stands for the process X stopped at T, namely,*

$$X_t^T(\omega) = X_{t\wedge T(\omega)}(\omega).$$

Theorem 2.1.27 *If $(X_t)_{t\geq 0}$ is an adapted cadlag process, then there exists a sequence of strictly positive stopping times (T_n) such that*

$$[\Delta X\neq 0] := \{(\omega,t) : 0<t<+\infty, X_t(\omega)\neq X_{t-}(\omega)\} = \bigcup_n [\![T_n]\!].$$

Definition 2.1.28 *An $\overline{\mathbf{R}}_+$-valued function T on Ω is called a predictable time, if $[\![T,\infty[\![$ is a predictable set. A stopping time T is called an accessible time, if there exists a sequence of predictable times (T_n) such that $[\![T]\!]\subset\bigcup_n[\![T_n]\!]$.*

The following theorem characterizes predictable processes within cadlag adapted processes.

Theorem 2.1.29 *Let $X=(X_t)$ be a cadlag adapted process. Then X is a predictable process if and only if X satisfies the following conditions:*

1) there exists a sequence of strictly positive predictable times (T_n) such that $[\Delta X\neq 0]\subset\bigcup_n[\![T_n]\!]$,

2) for each predictable time T, $X_T I_{[T<\infty]}\in\mathcal{F}_{T-}$.

Let $A=(A_t)_{t\geq 0}$ be an FV process. For each $\omega\in\Omega$, the function of finite variation $A_.(\omega)$ on $\mathbf{R}_+$ can be uniquely decomposed as $A_.(\omega)=A_.^c(\omega)+A_.^d(\omega)$, where $A_.^c(\omega)$ is a continuous function of finite variation, $A_.^d(\omega)$ is a purely discontinuous function of finite variation:

$$A_t^d(\omega) = \sum_{0<s\leq t}\Delta A_s(\omega).$$

We call A^c the continuous part of A and A^d the purely discontinuous part of A. An FV process A is said to be purely discontinuous if $A^c=0$.

The *Stieltjes integral* of a measurable process H w.r.t. an FV process A is defined path by path:

$$B_t(\omega) = \int_{[0,t]} H_s(\omega)dA_s(\omega) = H_0A_0 + \int_0^t H_s(\omega)dA_s(\omega) \; .$$

We denote (B_t) by $H_{\dot{s}}A$ or simply $H.A$. We denote by $L_S(A)$ the set of all measurable processes which are Stieltjes integrable w.r.t. A.

Theorem 2.1.30 *Let $A = (A_t)$ be an FV process and $H \in L_S(A)$.*
1) If H is progressive and A is adapted, then $H.A$ is adapted.
2) if H and A are predictable, then so is $H.A$.

Let (τ_t) be an increasing process. If for each $t \in \mathbb{R}_+$ τ_t is an $(\mathcal{F}_t)$-stopping time, we call (τ_t) a *random time-change.* Put $\mathcal{G}_t = \mathcal{F}_{\tau_t}$. We call $(\mathcal{G}_t)$ the filtration induced by τ.

Theorem 2.1.31 *Assume that $(\mathcal{F}_t)$ is right-continuous.*
1) Let $(\mathcal{G}_t)$ be the filtration induced by a random time-change (τ_t). Then $(\mathcal{G}_t)$ is right-continuous.
2) Let (A_t) be an adapted increasing process with $A_\infty = \infty$. Put

$$\tau_t = \inf\{s > 0 : A_s > t\}, \quad \mathcal{G}_t = \mathcal{F}_{\tau_t}.$$

Then (τ_t) is a random time-change, called the one associated to A. If (A_t) is continuous, then for any $(\mathcal{F}_t)$-stopping time σ, A_σ is a $(\mathcal{G}_t)$-stopping time, and we have $\mathcal{F}_\sigma \subset \mathcal{G}_{A_\sigma}$. If (A_t) is further strictly increasing, then $\mathcal{F}_\sigma = \mathcal{G}_{A_\sigma}$.

Section Theorem and Its Applications

Let $(\Omega, \mathcal{F}, \mathbf{P})$ be a probability space equipped with a filtration $\mathbf{F} = (\mathcal{F}_t)$. $\mathbf{F} = (\mathcal{F}_t)$ is said to be *complete,* if $(\Omega, \mathcal{F}, \mathbf{P})$ is complete and $\mathcal{F}_0$ contains all $\mathbf{P}$-null sets. If $\mathbf{F}$ is complete and right-continuous, we say that $\mathbf{F}$ satisfies the *usual conditions.*

A probability space $(\Omega, \mathcal{F}, \mathbf{P})$ equipped with a right-continuous filtration $\mathbf{F} = (\mathcal{F}_t)$ is called a *filtered probability space* or *stochastic basis* and denoted by $(\Omega, \mathcal{F}, \mathbf{F}, \mathbf{P})$. If $(\Omega, \mathcal{F}, \mathbf{P})$ is complete and $\mathbf{F}$ satisfies the usual conditions, we call $(\Omega, \mathcal{F}, \mathbf{F}, \mathbf{P})$ a *complete stochastic basis.* Any stochastic basis $(\Omega, \mathcal{F}, \mathbf{F}, \mathbf{P})$ can be completed as follows: First we complete the probability space $(\Omega, \mathcal{F}, \mathbf{F}, \mathbf{P})$ and then let $\mathcal{F}_t^P$ be the σ-field generated by $\mathcal{F}_t$ and all $\mathbf{P}$-null sets.

A subset Λ of $\Omega \times \mathbb{R}_+$ is called an *evanescent set* (w.r.t. $\mathbf{P}$), if the projection of Λ on Ω is a $\mathbf{P}$-null set. Two processes $X = (X_t)$ and $Y = (Y_t)$ are said to be *indistinguishable* (denoted by $X = Y$), if $\{(\omega, t) : X_t(\omega) \neq Y_t(\omega)\}$ is an evanescent set. If $\{(\omega, t) : X_t(\omega) > Y_t(\omega)\}$ is an evanescent set, we write $X \leq Y$.

The following theorem is called the *section theorem.* It is one of the most important results in the general theory of stochastic processes.

Theorem 2.1.32 *Let A be an optional (resp. accessible, predictable) set. Then for any $\epsilon > 0$ there exists a stopping time (resp. accessible time, predictable time) T such that*
1) $[\![T]\!] \subset A$;
2) $\mathbf{P}(T < \infty) \geq \mathbf{P}(\pi(A)) - \epsilon$.
Here $\pi(A) = \{\omega : \exists t \in \mathbf{R}_+ \text{ such that } (\omega, t) \in A\}$ is the projection of A on Ω.

We give below some applications of the section theorem.

Theorem 2.1.33 *Let $X = (X_t)$ and $Y = (Y_t)$ be two optional (resp. predictable) processes. If for each bounded stopping time (resp. predictable time) T we have $X_T \leq Y_T$ a.s., then $X \leq Y$. In particular, if for each bounded stopping time (resp. predictable time) T we have $X_T = Y_T$ a.s., then $X = Y$.*

Definition 2.1.34 *1) A stopping time T is said to be (a.s.) foretellable, if there exists a sequence of stopping times (T_n) such that on $[T > 0]$ we have $T_n < T$ (a.s.), for all n, and $\lim_n T_n = T$ (a.s.).*

2) A stopping time T is called a totally inaccessible time, if for each predictable time S we have $\mathbf{P}(T = S < \infty) = 0$.

Theorem 2.1.35 *For each stopping time there exists $A \subset [T < \infty], A \in \mathcal{F}_{T-}$ such that T_A is an accessible time and T_{A^c} is a totally inaccessible time.*

T_A and T_{A^c} are called the accessible and totally inaccessible part of T and denoted by T^a and T^i, respectively.

Theorem 2.1.36 *Let $X = (X_t)$ be a cadlag adapted process. Then there exists a sequence (T_n) of strictly positive stopping times satisfying the following conditions:*

i) $[\Delta X \neq 0] \subset \bigcup_n [\![T_n]\!]\}$,

ii) each T_n is predictable or totally inaccessible,

iii) $[\![T_n]\!] \cap [\![T_m]\!] = \emptyset$, for $n \neq m$.

The following theorem describes the structure of an adapted or predictable FV process.

Theorem 2.1.37 *If A is an adapted (resp. predictable) FV process, then so is A^d and there exists a sequence (S_n) of strictly positive stopping times (resp. predictable times) with disjoint graphs such that*

$$A_t^d = \sum_n \Delta A_{S_n} I_{[S_n \leq t]}.$$

Moreover, any adapted FV process A admits the following unique decomposition:

$$A = A^c + A^{da} + A^{di},$$

where A^c is a continuous adapted FV process, A^{da} and A^{di} are purely discontinuous adapted FV processes, A^{da} has only accessible jumps, A^{di} has only totally inaccessible jumps.

Definition 2.1.38 *Let $A \subset \Omega \times \mathbb{R}_+$. Put*

$$D_A(\omega) = \inf\{t \in \mathbb{R}_+ : (\omega, t) \in A\}, \ \omega \in \Omega,$$

D_A is called the debut of A. Here and henceforth, we follow the convention that $\inf \emptyset = +\infty$.

Theorem 2.1.39 *1) If $(\mathcal{F}_t)$ satisfies the usual conditions, the debut of any progressive set is a stopping time.*

2) All predictable times are a.s. foretellable. If $(\mathcal{F}_t)$ is complete, all predictable times are foretellable.

3) If $(\mathcal{F}_t)$ is complete, any evanescent measurable process is a predictable process and any right-continuous adapted process is an optional process.

4) If $(\mathcal{F}_t)$ is complete, any right-continuous supermartingale is indistinguishable to a cadlag process.

5) If $(\mathcal{F}_t)$ satisfies the usual conditions, any martingale has a cadlag version.

Theorem 2.1.40 *Assume that $(\mathcal{F}_t)$ is complete. If X is a cadlag adapted process, then X is predictable if and only if it satisfies the following conditions:*

i) For any totally inaccessible time S, on $[S<\infty]$ we have $X_S = X_{S-}$ a.s.,

ii) For any predictable time T, $X_T I_{[T<\infty]}$ is $\mathcal{F}_{T-}$-measurable.

The following theorem is the predictable form of Doob's stopping theorem. It is the basis for defining predictable projections of measurable processes.

Theorem 2.1.41 *Assume that $(\mathcal{F}_t)$ satisfies the usual conditions. If $(X_t, t \in \overline{\mathbf{R}}_+)$ is a cadlag supermartingale (resp. martingale), then for any predictable time T and stopping time U with $U \geq T$, X_U and X_{T-} are integrable and we have*

$$\mathbf{E}[X_U \mid \mathcal{F}_{T-}] \leq X_{T-} (\text{resp. } = X_{T-}) \text{ a.s.}.$$

In particular, if ξ is an integrable r.v. and S, T are two predictable times, then we have

$$\mathbf{E}[\mathbf{E}[\xi \mid \mathcal{F}_{S-}] \mid \mathcal{F}_{T-}] = \mathbf{E}[\xi \mid \mathcal{F}_{(S\wedge T)-}] \text{ a.s.}.$$

Corollary 2.1.42 *Assume that $(\mathcal{F}_t)$ satisfies the usual conditions.*

1) Any right-continuous predictable martingale is continuous.

2) Let T be a stopping time. Then T is a predictable time, if and only if for any bounded cadlag martingale M one has $\mathbf{E}[\Delta M_T] = 0$, where $\Delta M_0 = \Delta M_\infty = 0$ by convention.

Definition 2.1.43 *1) Let $\mathbf{F} = (\mathcal{F}_t)$ be a complete filtration. $\mathbf{F}$ is said to be quasi-left-continuous, if $\mathcal{F}_T = \mathcal{F}_{T-}$ for any predictable time T.*

2) An adapted cadlag process X is said to be quasi-left-continuous, if for each predictable time T we have $X_T = X_{T-}$, a.s. on $[T<\infty]$.

Theorem 2.1.44 *The following conditions are equivalent:*

1) $\mathbf{F}$ is quasi-left-continuous,

2) Every accessible time is a predictable time,

3) Every cadlag $\mathbf{F}$-martingale is quasi-left-continuous.

Projections of Measurable Processes

We assume that $(\Omega, \mathcal{F}, \mathbf{P})$ is a complete probability space and $\mathbf{F} = (\mathcal{F}_t)$ is a filtration satisfying the usual conditions. We shall define projections of processes via conditional expectations of random variables. For convenience we use the generalized conditional expectations.

Definition 2.1.45 *Let $(\Omega, \mathcal{F}, \mathbf{P})$ be a probability space and $\mathcal{G}$ a sub-σ-field of $\mathcal{F}$. A r.v. ξ is said to be σ-integrable w.r.t $\mathcal{G}$, if there exist $\Omega_n \in \mathcal{G}$, $\Omega_n \uparrow \Omega$ such that each ξI_{Ω_n} is integrable, or equivalently, there exists a $\mathcal{G}$-measurable real r.v. $\eta > 0$ such that $\xi\eta$ is integrable.*

Theorem 2.1.46 *Let ξ be a r.v., σ-integrable w.r.t. $\mathcal{G}$. Put*

$$\mathcal{C} = \{A \in \mathcal{G} : \ \mathbf{E}[\xi \mid I_A] < +\infty\}.$$

Then there exists uniquely a $\mathcal{G}$-measurable real r.v. η such that for all $A \in \mathcal{C}$ we have

$$\mathbf{E}[\xi I_A] = \mathbf{E}[\eta I_A].$$

We call η is the conditional expectation of ξ w.r.t. $\mathcal{G}$, and denote it by $\mathbf{E}[\xi | \mathcal{G}]$.

It is easy to prove that the above generalized conditional expectation posses all properties of the ordinary conditional expectation.

Theorem 2.1.47 *Let (X_t) be a measurable process such that for every stopping time T, $X_T I_{[T<\infty]}$ is σ-integrable w.r.t. $\mathcal{F}_T$. Then there exists a unique optional process, denoted by ${}^oX$, such that for every stopping time T we have*

$$\mathbf{E}[X_T I_{[T<\infty]} \mid \mathcal{F}_T] = {}^oX_T I_{[T<\infty]} \quad \text{a.s.}$$

In this case, we say that X has the optional projection ${}^oX$.

Obviously, every progressive process X has the optional projection and ${}^oX$ is an optional version of X.

Theorem 2.1.48 *Let $X = (X_t)$ be a measurable process such that for every predictable time T, $X_T I_{[T<\infty]}$ is σ-integrable w.r.t. $\mathcal{F}_{T-}$. Then there exists a unique predictable process, denoted by ${}^pX$, such that for every predictable time T we have*

$$\mathbf{E}[X_T I_{[T<\infty]} \mid \mathcal{F}_{T-}] = {}^pX_T I_{[T<\infty]} \quad \text{a.s.}.$$

In this case, we say that X has the predictable projection ${}^pX$.

Let X be a cadlag martingale. Then by Theorem 1.1.41, X_- is the predictable projection of X. Here by convention, $X_{0-} = X_0$.

The following theorem shows that the projection has a property, similar to the smoothing property of conditional expectation.

Theorem 2.1.49 *Let X be a measurable process and Y an optional (resp. predictable) process. If the optional (resp. predictable) projection of X exists, then so does XY and*

$${}^o(XY) = ({}^oX)Y \quad (\text{resp.}\, {}^p(XY) = ({}^pX)Y).$$

Dual Projections of FV Processes

First of all we define the measure on $\mathcal{F} \times \mathcal{B}(\mathbf{R}_+)$ generated by an increasing process.

Definition 2.1.50 *Let A be an increasing process. We define a set-function μ_A as follows:*

$$\mu_A(H) = \mathbf{E}\Big[\int_{[0,\infty[} I_H(\cdot, s) dA_s(\cdot)\Big], \; H \in \mathcal{F} \times \mathcal{B}(\mathbf{R}_+) \ .$$

Then μ_A is a measure on $\mathcal{F} \times \mathcal{B}(\mathbf{R}_+)$. We call it the measure generated by A.

Put

$$T_n(\omega) = \inf\{t \geq 0 : \; A_t(\omega) \geq n\} \ .$$

Then T_n is a r.v., $[\![0, T[\![\in \mathcal{F} \times \mathcal{B}(\mathbf{R}_+)$, $\bigcup_n [\![0, T_n[\![= \Omega \times I\!R_+$, and $\mu_A([\![0, T]\!]) \leq n$. Consequently, μ_A is a σ-finite measure on $\mathcal{F} \times \mathcal{B}(\mathbf{R}_+)$. Obviously, μ_A doesn't charge evanescent sets and for all $t \geq 0, F \in \mathcal{F}$, we have

$$\mu_A(F \times [0, t]) = \mathbf{E}[I_F A_t].$$

Theorem 2.1.51 *A measure μ on $\mathcal{F} \times \mathcal{B}(\mathbf{R}_+)$ is generated by certain increasing process if and only if for each $t \geq 0$ the set-function G_t on $(\Omega, \mathcal{F})$, defined by*

$$G_t(F) = \mu(F \times [0, t]), \; F \in \mathcal{F} \, ,$$

is a σ-finite measure and absolutely continuous w.r.t. $\mathbf{P}$. The increasing process generating μ is unique.

Definition 2.1.52 *Let μ be a measure on $\mathcal{F} \times \mathcal{B}(\mathbf{R}_+)$ not charging evanescent sets. μ is called an optional (resp. predictable) measure, if for any bounded measurable process X, we have*

$$\mu(X) = \mu^o(X) \quad (\text{resp.}\, \mu(X) = \mu({}^pX)),$$

where $\mu(X) = \int_{\Omega \times I\!R_+} X d\mu$.

Below we define the projections of measures. They are the basis for studying the dual predictable projection of an increasing process.

Theorem 2.1.53 *Let μ be a σ-finite measure on $\mathcal{F} \times \mathcal{B}(\mathbf{R}_+)$ not charging evanescent sets. For any positive bounded measurable process X, set*

$$\mu^o(X) = \mu({}^oX) \ , \quad \mu^p(X) = \mu({}^pX) \ .$$

Then μ^o (resp. μ^p) is an optional (resp. predictable) measure on $\mathcal{F} \times \mathcal{B}(\mathbf{R}_+)$ not charging evanescent sets. We call μ^o (resp. μ^p) the optional (resp. predictable) projections of μ.

Obviously, μ and μ^p coincide on the optional σ-field $\mathcal{O}$, μ and μ^p coincide on the predictable σ-field $\mathcal{P}$. Besides, in order for μ be an optional (resp. predictable) measure on $\mathcal{F} \times \mathcal{B}(\mathbf{R}_+)$ it is necessary and sufficient that $\mu = \mu^o$ (resp. $\mu = \mu^p$).

Theorem 2.1.54 *Let μ_A be the measure on $\mathcal{F} \times \mathcal{B}(\mathbf{R}_+)$ generated by an increasing process A. Then μ_A is optional (resp. predictable), if and only if A is adapted (resp. predictable).*

Theorem 2.1.55 *Let A and B be two adapted (resp. predictable) increasing processes. The following statements are equivalent:*

1) For almost all ω, $dB.(\omega) \ll dA.(\omega)$,

2) $\mu_B \ll \mu_A$ on $\mathcal{F} \times \mathcal{B}(\mathbf{R}_+)$,

3) $\mu_B \ll \mu_A$ on $\mathcal{O}$ (resp. $\mathcal{P}$,

4) There exists a non-negative optional (resp. predictable) process H, denoted by $\frac{dB}{dA}$, such that $B = H.A$, a.s..

Let A be an increasing process. If $A_\infty = \lim_{n\to\infty} A_n$ is integrable, A is called an *integrable increasing process.* If A_0 is σ-integrable w.r.t. $\mathcal{F}_0$ and there exist stopping times $T_n \uparrow \infty$ a.s. such that $A_{T_n} - A_0$ are integrable, A is said to be *locally integrable*. If there exist $T_n \uparrow \infty$ a.s. such that each $A_{T_n -} I_{[T_n > 0]}$ is integrable, A is said to be *prelocally integrable.* An FV process is called a *process of integrable variation,* if its total variation is integrable. Similarly, we can define processes of prelocally (resp. locally) integrable variation.

Obviously, any adapted FV process is of prelocally integrable variation, any predictable FV process is of locally integrable variation.

Theorem 2.1.56 *Let μ be a measure on $\mathcal{F} \times \mathcal{B}(\mathbf{R}_+)$ generated by an increasing process A, and μ^o (resp. μ^p) be the optional (resp. predictable) projection of μ. Then μ^o (resp. μ^p) is generated by an adapted (resp. predictable) increasing process if and only if A is prelocally (resp. locally) integrable.*

Theorem 1.1.56 hints us to give the following definition.

Definition 2.1.57 *Let A be a prelocally (resp. locally) integrable increasing process. We denote by A^o (resp. A^p) the adapted (predictable) increasing process generating the measure μ_A^o (resp. μ_A^p) and call A^o (resp. A^p) the dual optional (resp. predictable) projection of A. If A is adapted, we often use notation $\widetilde{A}$ to denote A^p and call $\widetilde{A}$ the compensator of A.*

The above definition can be extended naturally to processes of prelocally or locally integrable variation.

Theorem 2.1.58 *Let A be a process of prelocally (resp. locally) integrable variation and H be an optional (resp. predictable) process. If $H \in L_S(A)$ and $H.A$ is of prelocally (resp. locally) integrable variation, then $H \in L_S(A^o)$ (resp. $H \in L_S(A^p)$ and $(H.A)^o = H.A^o$ (resp. $(H.A)^p = H.A^p$). Moreover, for any stopping time T, we have*

$$\mathbf{E}\Big[\int_{[0,T]} | H_s || dA_s^o |\Big] \leq \mathbb{E}\Big[\int_{[0,T]} | H_s || dA_s |\Big] ,$$

and for any predictable time T, we have

$$\mathbf{E}\Big[\int_{[0,T]} | H_s || dA_s^p |\Big] \leq \mathbb{E}\Big[\int_{[0,T]} | H_s || dA_s |\Big] .$$

Theorem 2.1.59 *Let A be a process of prelocally (resp. locally) integrable variation and H be an optional (resp. predictable) process. If $H \in L_S(A) \cap L_S(A^o)$ (resp. $H \in L_S(A) \cap L_S(A^p)$) and $H.A^o$ (resp. $H.A^p$) is of prelocally (resp. locally) integrable variation, then $H.A$ itself is a process of prelocally (resp. locally) integrable variation.*

Theorem 2.1.60 *Let A be an adapted (resp. predictable) FV process and H be a measurable process having optional (resp. predictable) projection such that $H \in L_S(A)$ and $H.A$ is of prelocally (resp. locally) integrable variation. Then ${}^oH \in L_S(A)$ (resp. ${}^pH \in L_S(A)$) and $(H.A)^o = ({}^oH).A$ (resp. $(H.A)^p = ({}^pH).A$).*

The following theorem gives a martingale characterization of the dual predictable projection.

Theorem 2.1.61 *Let A be an adapted process of integrable variation and B a predictable process of integrable variation. Then B is the dual predictable projection of A if and only if $A - B$ is a uniformly integrable martingale with initial value zero. As a consequence, we have $\Delta(A^p) = {}^p(\Delta A)$.*

2.1.3 Modern Martingale Theory

We assume that $(\Omega, \mathcal{F}, \mathbf{P})$ is a complete probability space and $\mathbf{F} = (\mathcal{F}_t)$ is a filtration satisfying the usual conditions. All martingales we consider will be assumed to be cadlag. We use the following notations:

$\mathcal{A}$($\mathcal{A}_{\text{loc}}$)——the collection of all adapted processes of (locally) integrable variation.
$\mathcal{A}^+$ ($\mathcal{A}^+_{\text{loc}}$)——the collection of all adapted (locally) integrable increasing processes.
$\mathcal{V}$——the collection of all adapted FV processes.
$\mathcal{V}^+$——the collection of all adapted increasing processes.
$\mathcal{T}$——the collection of all stopping times.
$\mathcal{M}$——the collection of all uniformly integrable martingales.

Doob-Meyer's Decomposition

For any class $\mathcal{G}$ of processes we denote by $\mathcal{G}_0$ the sub-class of $\mathcal{G}$ consisting of all elements of $\mathcal{G}$ with null initial value. For an adapted process of integrable variation A we denote its predictable dual projection by $\widetilde{A}$ instead of A^p.

A measurable process X is said to be of class (D) if $\{X_T I_{[T<\infty]} : T \in \mathcal{T}\}$ is uniformly integrable. From Doob's stopping theorem we know that all uniformly integrable martingales and nonnegative right-closed submartingales are of class (D).

Let $Z = (Z_t)$ be a nonnegative supermartingale. If $\lim_{t\to\infty} \mathbf{E}[Z_t] = 0$, we call Z a *potential.*

Theorem 2.1.62 *Let $A = (A_t)$ be a predictable integrable increasing process with $A_0 = 0$ and $Z = (Z_t)$ be the optional projection of $(A_\infty - A_t)$. Then Z is a potential of class (D). We call Z the potential generated by A.*

Theorem 2.1.63 *Let Z be a potential of class (D). Then there exists a unique predictable integrable increasing process A with $A_0 = 0$ such that Z is generated by A.*

As a consequence of Theorem 2.1.63 we obtain the following Doob-Meyer's decomposition theorem for supermartingales of class (D), due to Meyer (1962) [Ref. 5].

Theorem 2.1.64 *Let X be a supermartingale of class (D). Then X can be decomposed uniquely as*

$$X = M - A,$$

where M is a uniformly integrable martingale, A is a predictable integrable increasing process with $A_0 = 0$. (1.6) is called the Doob-Meyer's decomposition of X.

Martingales with Integrable Variation and Uniformly Square Integrable Martingales

A martingale is called a *martingale with integrable variation,* if it is also an FV process of integrable variation. We denote by $\mathcal{W}$ the collection of all martingales with integrable variation.

Theorem 2.1.65 *If $M \in \mathcal{W}$, then for any bounded martingale N we have*

$$\mathbf{E}[M_\infty N_\infty] = \mathbf{E}\Big[M_0 N_0 + \sum_{s>0} \Delta M_s \Delta N_s\Big].$$

Moreover, $(L_t) = (M_t N_t - \sum_{s\le t} \Delta M_s \Delta N_s)$ is a uniformly integrable martingale.

The following theorem shows the special role of predictable processes in the theory of stochastic integration.

Theorem 2.1.66 *If $M \in \mathcal{W}$ and H is a predictable process such that*

$$\mathbf{E}\left[\int_{[0,\infty[} | H_s || dM_s |\right] < \infty,$$

then $H.M \in \mathcal{W}$.

A martingale M is called a *uniformly square integrable martingale,* if $\sup_t \mathbb{E}\,[M_t^2] < \infty$. We denote by $\mathcal{M}^2$ the collection of all uniformly square integrable martingales and denote by $\mathcal{M}^{2,c}$ the collection of all continuous uniformly square integrable martingales.

Let $M \in \mathcal{M}$. Then $M \in \mathcal{M}^2$ if and only if $\mathbf{E}[M_\infty^2] < \infty$. In fact we have

$$\mathbf{E}[M_\infty^2] = \sup_t \mathbb{E}\,[M_t^2].$$

Moreover, $\mathcal{M}^2$ is a Hilbert space with inner product given by $(M, N) = \mathbf{E}[M_\infty N_\infty]$, and it is isomorphic to $L^2(\Omega, \mathcal{F}, \mathbf{P})$ through the mapping $M \mapsto M_\infty$.

Theorem 2.1.67 *If $(M^n)_{n\ge1}$ converges to M in $\mathcal{M}^2$, then there exists a subsequence $(M^{n_k})_{k\ge1}$ such that for almost all ω, $M_t^{n_k}(\omega)$ converges to $M_t(\omega)$ uniformly in $t \in \mathbb{R}_+$. Consequently, $\mathcal{M}^{2,c}$ is a closed subspace of $\mathcal{M}^2$.*

Definition 2.1.68 *Let $\mathcal{M}^{2,d}$ denote the orthogonal complement of $\mathcal{M}^{2,c}$ in $\mathcal{M}^2$. We call elements of $\mathcal{M}^{2,d}$ purely discontinuous uniformly square integrable martingales.*

Let $M \in \mathcal{M}^{2,d}$. Obviously, we have $M_0 = 0$, a.s.. Let $M \in \mathcal{M}^2$. Then M admits the following unique decomposition:

$$M = M_0 + M^c + M^d ,$$

where $M^c \in \mathcal{M}_0^{2,c}$, $M^d \in \mathcal{M}^{2,d}$. We call M^c the *continuous martingale part* of M and M^d the *purely discontinuous martingale part* of M.

Let $M \in \mathcal{M}^2$ and T be a stopping time. Then

$$(M^T)^c = (M^c)^T, \quad (M^T)^d = (M^d)^T .$$

Theorem 2.1.69 *1) Let $M \in \mathcal{M}^2$. Then*

$$\mathbf{E}[M_0^2] + \mathbf{E}[\sum_{s>0}(\Delta M_s)^2] \leq I\!E\,[M_\infty^2],$$

and the equality holds if and only if $M - M_0 \in \mathcal{M}^{2,d}$.

2) If $M, N \in \mathcal{M}^2$, then

$$\mathbf{E}[M_0 N_0] + \mathbf{E}[\sum_{s>0} |\Delta M_s \Delta N_s|] \leq \sqrt{\mathbf{E}[M_\infty^2]}\sqrt{\mathbf{E}[N_\infty^2]}.$$

3) If $M \in \mathcal{M}^{2,d}$, then for any $N \in \mathcal{M}^2$ we have

$$\mathbf{E}[M_\infty N_\infty] = \mathbf{E}[\sum_s \Delta M_s \Delta N_s] .$$

In addition, $(L_t) = (M_t N_t - \sum_{s \leq t} \Delta M_s \Delta N_s)$ is a uniformly integrable martingale.

4) $\mathcal{M}_0^2 \cap \mathcal{W} \subset \mathcal{M}^{2,d}$.

Definition 2.1.70 *Let $M \in \mathcal{M}^2$. M^2 is a submartingale of class (D), since by Doob's inequality we have $M_\infty^* = \sup_t |M_t| \in L^2$. Thus according to Doob-Meyer's decomposition theorem there exists a unique predictable integrable increasing process, denoted by $\langle M \rangle$, such that $M^2 - \langle M \rangle \in \mathcal{M}_0$. $\langle M \rangle$ is called the predictable quadratic variation or the sharp bracket process of M. For $M, N \in \mathcal{M}^2$, put*

$$\langle M, N \rangle = \frac{1}{2}[\langle M + N \rangle - \langle M \rangle - \langle N \rangle].$$

$\langle M, N \rangle$ is called the predictable quadratic covariation or the sharp bracket process of M and N.

Definition 2.1.71 *For $M, N \in \mathcal{M}^2$, put*

$$[M, N]_t = M_0 N_0 + \langle M^c, N^c \rangle_t + \sum_{0<s\leq t} \Delta M_s \Delta N_s , \quad t \geq 0.$$

$[M, N]$ is an adapted process of integrable variation, called the quadratic covariation of M and N. The process $[M, M]$ (or simply, $[M]$) is an adapted integrable increasing process, called the quadratic variation or bracket process of M.

Theorem 2.1.72 *Let $M, N \in \mathcal{M}^2$.*

1) $[M, N]$ is the unique adapted process of integrable variation such that $MN - [M, N] \in \mathcal{M}_0$ and $\Delta[M, N] = \Delta M \Delta N$.

2) $\langle M, N\rangle$ is the dual predictable projection of $[M, N]$.

The following theorem is a basis for the definition of stochastic integrals.

Theorem 2.1.73 (Kunita-Watanabe inequality) *Let $M, N \in \mathcal{M}^2$, and H, K be two measurable processes. Then*

$$\int_{[0,\infty[} |H_s K_s||d\langle M, N\rangle_s| \leq \Big(\int_{[0,\infty[} H_s^2 d\langle M\rangle_s\Big)^{1/2} \Big(\int_{[0,\infty[} K_s^2 d\langle N\rangle_s\Big)^{1/2} \quad \text{a.s.},$$

$$\mathbf{E}\Big[\int_{[0,\infty[} |H_s K_s||d\langle M, N\rangle_s|\Big] \leq \Big\|\sqrt{\int_{[0,\infty[} H_s^2 d\langle M\rangle_s}\Big\|_p \Big\|\sqrt{\int_{[0,\infty[} K_s^2 d\langle N\rangle_s}\Big\|_q \quad \text{a.s.},$$

where p, q is a pair of conjugate indices, $\|\cdot\|_p$ is the L^p-norm. A similar result holds for $[M, N], [M]$ and $[N]$.

Local Martingales and Semimartingales

Definition 2.1.74 *Let M be a cadlag adapted process. If there exist stopping times $T_n \uparrow +\infty$ such that each $M^{T_n} - M_0$ is uniformly integrable martingale (resp. martingale of integrable variation). Then M is called a local martingale (resp. local martingale of locally integrable variation). We call (T_n) the localizing sequence for M.*

We denote by $\mathcal{M}_{\text{loc}}$ (resp. $\mathcal{W}_{\text{loc}}$) the collection of all local martingales (resp. local martingales of locally integrable variation). We set $\mathcal{M}_{\text{loc},0} = \{M \in \mathcal{M}_{\text{loc}} : M_0 = 0\}$.

Lemma 2.1.75 *Let M be a local martingale and $\epsilon > 0$. Put*

$$A = \sum_{s\leq \cdot} \Delta M_s I_{[|\Delta M_s|>\epsilon]},$$

then $A \in \mathcal{A}_{\text{loc}}$.

The following is the fundamental theorem for local martingales.

Theorem 2.1.76 *Let M be a local martingale. Then for any $\epsilon > 0$, M admits the following decomposition:*

$$M = M_0 + U + V,$$

where $U \in \mathcal{M}_{\text{loc},0}$ with $|\Delta U| \leq \epsilon$ and $V \in \mathcal{W}_{\text{loc},0}$.

Corollary 2.1.77 *1) If $M \in \mathcal{M}_{\text{loc}}$, then for all $t \geq 0$, $\sum_{s\leq t}(\Delta M_s)^2 < \infty$ a.s..*

2) $\mathcal{M}_{\text{loc}} \cap \mathcal{V} = \mathcal{W}_{\text{loc}}$.

3) Let $A \in \mathcal{V}$. Then $A \in \mathcal{A}_{\text{loc}}$ if and only if there exists a predictable FV process B such that $A - B$ is a local martingale.

If $M \in \mathcal{M}_{\mathrm{loc},0}$ has a decomposition $M = U + V$ with $U \in \mathcal{M}^{2,d}_{\mathrm{loc}}$ and $V \in \mathcal{W}_{\mathrm{loc}}$, we call M a *purely discontinuous local martingale.* We denote by $\mathcal{M}^c_{\mathrm{loc}}$ (resp.$\mathcal{M}^d_{\mathrm{loc}}$) the collection of all continuous (resp. purely discontinuous) local martingales.

We denote by $\mathcal{M}^{da}_{\mathrm{loc}}$ (resp. $\mathcal{M}^{di}_{\mathrm{loc}}$) the set of all purely discontinuous local martingales with accessible (resp. totally inaccessible) jumps.

Theorem 2.1.78 *Any local martingale M admits the following unique decomposition:*

$$M = M_0 + M^c + M^d = M_0 + M^c + M^{da} + M^{di},$$

where $M^c \in \mathcal{M}^c_{\mathrm{loc},0}$, $M^d \in \mathcal{M}^d_{\mathrm{loc}}$, $M^{da} \in \mathcal{M}^{da}_{\mathrm{loc}}$, and $M^{di} \in \mathcal{M}^{di}_{\mathrm{loc}}$. We call M^c the continuous martingale part of M and M^d the purely discontinuous martingale part of M.

Definition 2.1.79 *Let M and N be two local martingales. Put*

$$[M,N]_t = M_0N_0 + \langle M^c, N^c\rangle_t + \sum_{0<s\leq t} \Delta M_s \Delta N_s\,.$$

Then $[M,N]$ is an adapted FV process, called quadratic covariation of M and N. $[M,M]$ (or simply, $[M]$) is an adapted increasing process, called the quadratic variation or bracket process of M.

If $[M,N] \in \mathcal{A}_{\mathrm{loc}}$, we denote by $\langle M,N\rangle$ the dual predictable projection of $[M,N]$. If $M,N \in \mathcal{M}^2_{\mathrm{loc}}$, then $[M,N] \in \mathcal{A}_{\mathrm{loc}}$.

Theorem 2.1.80 *1) Let $M \in \mathcal{M}_{\mathrm{loc}}$. Then $M = 0$ iff $[M] = 0; M \in \mathcal{M}^c_{\mathrm{loc}}$ iff $[M]$ is continuous; $M \in \mathcal{M}^d_{\mathrm{loc}}$ iff $[M]$ is purely discontinuous.*

2) If $M \in \mathcal{M}_{\mathrm{loc}}$, then $\sqrt{[M]}$ is a locally integrable increasing process.

3) If $M,N \in \mathcal{M}_{\mathrm{loc}}$, then $[M,N]$ is the unique adapted FV process such that $MN - [M,N] \in \mathcal{M}_{\mathrm{loc},0}$ and $\Delta[M,N] = \Delta M \Delta N$.

The following theorem shows that martingale transforms can be considered as stochastic integrals of simple integrands w.r.t. a local martingale.

Theorem 2.1.81 *Let M be a local martingale, S and T two stopping times with $S \leq T$, and ξ an $\mathcal{F}_S$-measurable real r.v.. Put $H = \xi I_{]\!]S,T]\!]}$. Then $L = \xi(M^T - M^S)$ is a local martingale, and for any local martingale N we have*

$$[L,N] = \xi([M,N]^T - [M,N]^S) = H.[M,N]\,.$$

The following theorem gives a characterization for jump processes of local martingales. It plays an important role in the definition of stochastic integrals w.r.t. local martingales.

Theorem 2.1.82 *Let H be an optional process such that $[H \neq 0]$ is a thin set. Then H is a jump process of a local martingale, if and only if*

i) ${}^pH = 0$,

ii) $\sqrt{\Sigma_{s\leq \cdot} H_s^2} \in \mathcal{A}^+_{\mathrm{loc}}$.

Definition 2.1.83 *Let $X = (X_t)$ be a cadlag adapted process. If X can be expressed as the sum of a local martingale M and an adapted FV process A:*

$$X = M + A\,,$$

we call X a semimartingale. The continuous martingale part of M in the above decomposition is uniquely determined by X. We call it the continuous martingale part of X and denote it by X^c.

We denote by $\mathcal{S}$ the collection of all semimartingales.
Let X, Y be two semimartingales. Put

$$[X,Y]_t = X_0Y_0 + \langle X^c, Y^c\rangle_t + \sum_{s\le t} \Delta X_s \Delta Y_s \ , \ t \ge 0,$$

Then $[X,Y]$ is called the quadratic covariation of X and Y. $[X,X]$ (or simply, $[X]$) is an adapted increasing process, called the quadratic variation or bracket process of X. If $[X,Y] \in \mathcal{A}_{\text{loc}}$, we denote by $\langle X,Y\rangle$ the dual predictable projection of $[X,Y]$.

Definition 2.1.84 *Let $X \in \mathcal{S}$. If X can be expressed as $X = M + A$, where M is a local martingale and A is a process of locally integrable variation, we call X a special semimartingale.*
We denote by $\mathcal{S}_p$ the collection of all special semimartingales.

Theorem 2.1.85 *Let $X \in \mathcal{S}_p$. Then X admits the following unique decomposition:*

$$X = M + A \ ,$$

where M is a local martingale, A is a predictable FV process with $A_0 = 0$. We call this decomposition the canonical decomposition of X.

The following theorem gives some useful characterizations of special semimartingales.

Theorem 2.1.86 *Let X be a semimartingale. The following statements are equivalent:*
1) X is a special semimartingale,
2) $\sqrt{[X]}$ is a locally integrable increasing process,
3) $X^* = (X_t^*)$ is a locally integrable increasing process.

Definition 2.1.87 *Let X be an adapted cadlag process. If for each $t \in \mathbb{R}_+$, X_t is integrable, and*

$$\mathrm{Var}(X) = \sup_{\tau} \sum_{i=0}^{n} \mathbb{E}\left[|X_{t_i} - \mathbf{E}[X_{t_{i+1}} \mid \mathcal{F}_{t_i}]| \right] < +\infty,$$

where the supremum is taken over the set of all finite partitions τ of $[0,\infty]$ of the form $0 = t_0 < t_1 < \cdots < t_n < t_{n+1} = \infty$, and $X_\infty = 0$ by convention, then X is called a quasi-martingale.

Theorem 2.1.88 *Let X be an adapted cadlag process. Then X is a quasi-martingale if and only if X is the difference of two nonnegative cadlag supermartingales. In particular, any quasi-martingale is a special semi-martingale, and any special semimartingale is a local quasi-martingale. Moreover, if X is a quasi-martingale, then X can be uniquely decomposed as the difference of two nonnegative cadlag supermartingales V' and V'' such that Var(X) = $\mathbf{E}[V_0' + V_0'']$. This decomposition is called Rao's decomposition.*

From Theorem 1.1.88 it is easy to see that the quasi-martingale property is preserved under random time-changes or reductions of the filtration.

Martingale Spaces $\mathcal{H}^1, \mathcal{BMO}$ and $\mathcal{H}^p$

The contents of this subsection belong to the fine parts of modern martingale theory. The terminology $\mathcal{BMO}$, an acronym of bounded mean oscillation, is borrowed from modern analysis.

Definition 2.1.89 *We denote by $\mathcal{H}^1$ the set of all local martingales M such that*

$$\|M\|_{\mathcal{H}^1} := \mathbf{E}[\sqrt{[M]_\infty}] < \infty.$$

Each element of $\mathcal{H}^1$ is called an $\mathcal{H}^1$-martingale.

Obviously, $\mathcal{H}^1$ is a vector space. $\|\cdot\|_{\mathcal{H}^1}$ is a norm on $\mathcal{H}^1$.

Theorem 2.1.90 *1) $\mathcal{M}_{\text{loc}} = \mathcal{H}^1_{\text{loc}}$.*

2) If $M \in \mathcal{M}^2$, then $M \in \mathcal{H}^1$ and $\|M\|_{\mathcal{H}^\infty} \le \|M\|_{\mathcal{M}^2}$.

3) If $M \in \mathcal{W}$, then $M \in \mathcal{H}^1$ and $\|M\|_{\mathcal{H}^1} \le \|M\|_{\mathcal{A}} := \mathbf{E}[\int_{[0,\infty[} |dM_s|]$.

4) The collection of all bounded martingale (denoted by $\mathcal{M}^\infty$) is dense in $\mathcal{H}^1$. For $M \in \mathcal{M}^\infty$ we have

$$\|M\|_{\mathcal{BMO}} \le 2\|M_\infty\|_{L^\infty}.$$

Definition 2.1.91 *We denote by $\mathcal{BMO}$ the set of all uniformly square integrable martingales M such that*

$$\|M\|_{\mathcal{BMO}} := \sup_{T\in\mathcal{F}} \sqrt{\frac{\mathbf{E}(M_\infty - M_{T-}I_{[T>0]})^2}{\mathbf{P}(T<\infty)}} < \infty,$$

where $\mathcal{T}$ is the collection of all stopping times and $\frac{0}{0} = 0$ by convention. Each element of $\mathcal{BMO}$ is called a $\mathcal{BMO}$-martingale.

It is easy to check that $\mathcal{BMO}$ is a linear space, $\|\cdot\|_{\mathcal{BMO}}$ is a norm on $\mathcal{BMO}$.

Theorem 2.1.92 *Let M be a local martingale. The following statements are equivalent:*

1) $M \in \mathcal{BMO}$,

2) There exist constants $c_1, c_2 > 0$ such that $|M_0| \le c_1$ a.s., and for any stopping time T $|\Delta M_T| \le c_1$ a.s. and

$$\mathbf{E}([M]_\infty - [M]_T) \le c_2^2 \mathbf{P}(T < \infty),$$

3) There exists constants $c_1, c_2 > 0$ such that $|M_0| \le c_1$ a.s., $|\Delta M| \le c_1$ and for all $t \ge 0$,

$$\mathbf{E}[[M]_\infty|\mathcal{F}_t] - [M]_t \le c_2^2 \quad \text{a.s.}.$$

In particular, $\mathcal{BMO}$-martingales are locally bounded martingales.

The following theorem is a fundamental result about $\mathcal{H}^1$-and $\mathcal{BMO}$-martingales.

Theorem 2.1.93 (Fefferman's inequality) *Let M and N be two local martingales and U a progressive process. Then*

$$\mathbf{E}\Big[\int_{[0,\infty[} |U_s||d[M,N]_s|\Big] \le \sqrt{2}\mathbf{E}\Big[\Big(\int_{[0,\infty[} U_s^2 d[M]_s\Big)^{1/2}\Big]\|N\|_{\mathcal{BMO}}.$$

In particular, when $U = 1$, we have

$$\mathbf{E}\Big[\int_{[0,\infty[} |d[M,N]_s|\Big] \le \sqrt{2}\|M\|_{\mathcal{H}^1}\|N\|_{\mathcal{BMO}}.$$

Theorem 2.1.94 *Let $M \in \mathcal{H}^1$. Then M is a uniformly integrable martingale, and*

$$\|M_\infty\|_{L^1} \le 2\sqrt{2}\|M\|_{\mathcal{H}^1}.$$

The following theorem gives a useful characterization for $\mathcal{BMO}$-martingales.

Theorem 2.1.95 *Let $N \in \mathcal{M}^2$. Then $N \in \mathcal{BMO}$ if and only if there is a constant $c > 0$ such that for all $M \in \mathcal{M}^2$ (or equivalently, for all bounded martingale M),*

$$|\mathbf{E}[M,N]_\infty| \le c\|M\|_{\mathcal{H}^1}.$$

In this case, $\|N\|_{\mathcal{BMO}} \le \sqrt{5}c$.

The following theorem shows that $\mathcal{BMO}$ can be considered as the dual space of $\mathcal{H}^1$.

Theorem 2.1.96 *Let $(\mathcal{H}^1)^*$ be the Banach space formed by all bounded linear functionals on $\mathcal{H}^1$ (i.e., $(\mathcal{H}^1)^*$ is the dual space of $\mathcal{H}^1$). Let $N \in \mathcal{BMO}$. Put*

$$\varphi_N(M) = \mathbf{E}[M,N]_\infty, \quad M \in \mathcal{H}^1.$$

Then $N \mapsto \varphi_N$ is a one to one linear mapping from $\mathcal{BMO}$ onto $(\mathcal{H}^1)^*$ and

$$\frac{1}{\sqrt{2}}\|\varphi_N\| \le \|N\|_{\mathcal{BMO}} \le \sqrt{5}\|\varphi_N\|,$$

where $\|\varphi\|$ denotes the norm of bounded linear function φ. In particular, $\mathcal{BMO}$ with norm $\|\cdot\|_{\mathcal{BMO}}$ is a Banach space.

Theorem 2.1.97 (Davis' inequalities) *Let M be a local martingale. We have*

$$\mathbf{E}[M_\infty^*] \le 2\sqrt{6}\|M\|_{\mathcal{H}^1},$$

$$\|M\|_{\mathcal{H}^1} \le (7+4\sqrt{2})\mathbf{E}[M_\infty^*].$$

As an important consequence of Davis' inequalities, we have

Theorem 2.1.98 *Let M be a local martingale. Then $M \in \mathcal{H}^1$ if and only if $\mathbf{E}[M_\infty^*] < \infty$. Furthermore, $\|M\|_{\mathcal{H}^1}$ and $\|M_\infty^*\|_{L^1}$ are two equivalent norms on $\mathcal{H}^1$. In particular, $\mathcal{H}^1$ with norm $\|\cdot\|_{\mathcal{H}^1}$ is a Banach space.*

Definition 2.1.99 *Let $\Phi(t)$ be a nonnegative monotone increasing convex function $\mathbf{R}_+$ with $\Phi(0) = 0$. $\Phi(t)$ is called a moderate convex function if there is a constant $c > 0$ such that for all $t \ge 0$, $\Phi(2t) \le c\Phi(t)$.*

Let $\Phi(t)$ be a moderate convex function and φ be its right derivative. We define a constant ρ by:

$$\rho = \sup_{u \ge 0} \frac{u\varphi(u)}{\Phi(u)}.$$

The following is the celebrated Burkholder-Davis-Gundy inequality.

Theorem 2.1.100 (B-D-G inequality) *Let M be a local martingale and Φ be a moderate increasing convex function on $\mathbf{R}_+$ such that $\Phi(M_\infty^*)$ and $\Phi(\sqrt{[M]_\infty})$ are integrable. Then*

$$\rho^{-(\rho+1)}(7+4\sqrt{2})^{-\rho}\mathbf{E}[\Phi(\sqrt{[M]_\infty})]$$

$$\le \mathbf{E}[\Phi(M_\infty^*)] \le \rho^{\rho+1}(2\sqrt{6})^\rho \mathbf{E}[\Phi(\sqrt{[\sqrt{M}]_\infty})],$$

where ρ is defined in Definition 1.1.99.

Remark *Let $\Phi(t) = t^p (p > 1)$. The corresponding B-D-G inequality is called Burkholder's inequality.*

The next theorem gives a *John-Nirenberg type inequality* for $\mathcal{BMO}$ martingales.

Theorem 2.1.101 *Let $M \in \mathcal{BMO}$ and $\|M\|_{\mathcal{BMO}} = m$.*
1) If $\lambda < \frac{1}{8m}$, then

$$\mathbf{E}[e^{\lambda M^*_\infty}] < \frac{6}{1-8m\lambda}.$$

2) If $\lambda < \frac{1}{m^2}$, then for any stopping time T,

$$\mathbf{E}[\exp\{\lambda([M]_\infty - [M]_{T-}I_{[T>0]})\}|\mathcal{F}_T] \leq \frac{1}{1-\lambda m^2}.$$

Definition 2.1.102 *Let M be a local martingale, $1 < p < \infty$. Put*

$$\|M\|_{\mathcal{H}^p} = (\mathbf{E}[(\sqrt{[M]_\infty})^p])^{1/p} = \|\sqrt{[M]_\infty}\|_{L^p},$$
$$\mathcal{H}^p = \{M \in \mathcal{M}_{\text{loc}} : \|M\|_{\mathcal{H}^p} < \infty\}.$$

Each element of $\mathcal{H}^p$ is called an $\mathcal{H}^p$-martingale. Obviously, $\mathcal{H}^p$ is a linear space and $\|\cdot\|_{\mathcal{H}^p}$ is a norm on $\mathcal{H}^p$.

Theorem 2.1.103 *1) Let $1 < p < \infty$. Put*

$$\mathcal{M}^p = \{M \in \mathcal{M} : \|M_\infty\|_{L^p} < \infty\}.$$

Then $\mathcal{H}^p = \mathcal{M}^p$, $\|M\|_{\mathcal{H}^p}$, $\|M^*_\infty\|_{L^p}$ and $\|M_\infty\|_{L^p}$ are equivalent norms.
2) Let (p, q) be a pair of conjugate indexes. Then the dual space of $\mathcal{H}^p$ is $\mathcal{H}^q$. Moreover, if $M \in \mathcal{H}^p$ and $N \in \mathcal{H}^q$, then $K = MN - [M, N] \in \mathcal{H}^1$.

2.2 Stochastic Integrals

The stochastic integral is of the form $\int_{[0,t]} H_s dX_s$, where both the integrand (H_t) and the integrator (X_t) are stochastic processes. In 1944, K. Itô first defined the stochastic integrals of adapted measurable processes w.r.t. a Brownian motion (cf [Ref. 6, 7]). The key character of the stochastic integrals is that the resulting processes are martingales. In 1967, H. Kunita and S. Watanabe [Ref. 8] defined stochastic integrals of progressive processes w.r.t. square integrable martingales. In 1970, C. Doléans-Dade and P. A. Meyer [Ref. 9] defined the stochastic integrals of locally bounded predictable processes w.r.t. local martingales and semimartingales. In 1976, P. A. Meyer [Ref. 10] introduced the stochastic integrals of optional processes w.r.t. local martingales. In 1979, J. Jacod [Ref. 11] defined the stochastic integrals of unbounded predictable processes w.r.t. semimartingales (see also Ref. 12, 13). In this section we present the definition and properties of stochastic integrals, the change of variables formula (Itô's formula), Doléans-Dade exponential formula, the local times of semimartingales, and stochastic differential equations driven by semimartingales. As in Section 1, most of results in this section can be found in Ref. 1. We only indicate the references for those results which are not included in Ref. 1.

2.2.1 Stochastic Integrals w.r.t. Local Martingales

Predictable Integrands

We begin with the one-dimensional case. Let M be a real local martingale with the decomposition $M = M_0 + M^c + M^d$ and H be a predictable process. We want to define the "stochastic integral" of H w.r.t. M, denoted by $H.M$. If $H = \xi I_{]\!]S,T]\!]}$, where $S \leq T$ are two stopping times and ξ is $\mathcal{F}_S$-measurable, then $H.M$ should naturally defined as $H.M = \xi(M^T - M^S)$. Then by Theorem 1.1.81, for any local martingale N, we have $[H.M, N] = H.[M, N]$. This property characterizes uniquely an element $H.M$ of $\mathcal{M}_{\text{loc}}$. If we want that $H.M$ satisfies this property for general integrands H, then by Theorem 1.1.80, a necessary condition for H is that $H^2 \in L_S([M])$ and $\sqrt{H^2.[M]} \in \mathcal{A}^+_{\text{loc}}$. Fortunately, under this condition we can effectively define a local martingale $H.M$ to meet that property. First, by using the Kunita-Watanabe inequality (Theorem 1.1.73) we can define a continuous local martingale L' such that $[L', N] = H.[M^c, N]$ for any local martingale N. Second, by using the characterization for jump processes of local martingales (Theorem 1.1.82) we can define uniquely an $L'' \in \mathcal{M}^d_{\text{loc}}$ such that $\Delta L'' = H\Delta M$. Finally, we put $H.M = L' + L''$. Then for any local martingale N, we have

$$[H.M, N] = H.[M, N].$$

We call $H.M$ the *stochastic integral* of H w.r.t. M. Sometimes we denote also this integral by $H_{\dot{m}}M$ to insist that the obtained process is required to be a local martingale.

Let M be a local martingale. We denote by $L_m(M)$ the set of all predictable processes H such that $H^2 \in L_S([M])$ and $\sqrt{H^2.[M]} \in \mathcal{A}^+_{\text{loc}}$.

In the sequel, we also use the following notations to denote stochastic integrals: for $t \geq 0$

$$\int_{[0,t]} H_s dM_s = (H.M)_t,$$

$$\int_0^t H_s dM_s = \int_{(0,t]} H_s dM_s = ((HI_{]\!]0,\infty[\![}).M)_t.$$

The concept of stochastic integral will be generalized below, but we always use the same notations for stochastic integrals.

The following theorem characterizes the stochastic integrals.

Theorem 2.2.1 *Let M be a local martingale and $H \in L_m(M)$. Then $H.M$ is the unique local martingale such that $[H.M, N] = H.[M, N]$ holds for every local martingale N.*

The following theorem summarizes the fundamental properties of stochastic integrals.

Theorem 2.2.2 *Let M be a local martingale, $H, K \in L_m(M)$.*

1) $L_m(M) = L_m(M^c) \cap L_m(M^d)$, $(H.M)^c = H.M^c$, $(H.M)^d = H.M^d$.

2) $(H.M)_0 = H_0M_0, \Delta(H.M) = H\Delta M$.

3) $H + K \in L_m(M)$, and $(H + K).M = H.M + K.M$.

4) If H' is a predictable process, then $H' \in L_m(H.M)$ if and only if $HH' \in L_m(M)$. If it is the case, we have

$$H'.(H.M) = (H'H).M.$$

5) If T is a stopping time, then

$$(H.M)^T = H.M^T = (HI_{[\![0,T]\!]}).M.$$

Theorem 2.2.3 *Let M be a local martingale.*

1) If A is a predictable FV process, then $\Delta A \in L_m(M)$ and

$$(\Delta A).M = [M, A] - M_0A_0.$$

2) If $T > 0$ is a predictable time, then $I_{[\![T]\!]} \in L_m(M)$ and

$$I_{[\![T]\!]}.M = \Delta M_T I_{[\![T,\infty[\![}.$$

The following theorem shows that the stochastic integrals coincides with the Stieltjes integral when the integrator is a local martingale of finite variation and both integrals exist.

Theorem 2.2.4 *If $M \in \mathcal{W}_{\text{loc}}$ and $H \in L_m(M) \cap L_S(M)$, then $H_{\dot{m}}M = H_{\dot{s}}M$.*

Theorem 2.2.5 *Let $M \in \mathcal{W}_{\text{loc}}$.*
1) If $\sum_{s\leq \cdot} |H_s \Delta M_s| \in \mathcal{A}^+_{\text{loc}}$, then $H \in L_m(M) \cap L_S(M)$.
2) If $H \in L_m(M)$ and $\sum_{s\leq \cdot} |H_s \Delta M_s| \in \mathcal{V}^+$, then $H \in L_S(M)$.

Theorem 2.2.6 (Ref. 11 (Kunita-Watanabe Decomposition)) *If $M, N \in \mathcal{M}^2_{\text{loc}}$, then N has the following decomposition:*

$$N = N_0 + H.M + L,$$

where $H = \frac{d\langle M, N-N_0\rangle}{d\langle M\rangle}$, $H.M, L \in \mathcal{M}^2_{\text{loc}}$, and $L_0 = 0$, LM is a local martingale.

Now we turn to the vector stochastic integrals (cf. Ref. 12). Let $M = (M^i)_{i\leq n}$ be an $\mathbf{R}^n$-valued local martingale and $H = (H^i)_{i\leq n}$ an $\mathbf{R}^n$-valued predictable process. If for each i, $H^i \in L_m(M^i)$, then we define naturally the *componentwise stochastic integral* as

$$H.M = \sum_{i=1}^{n} H^i.M^i.$$

In order for the stochastic integral to have good properties, such as representing a real local martingale as a stochastic integral w.r.t. a vector local martingales, we need to consider a larger class of integrands. To this end, we take an adapted increasing process Γ (e.g., $\Gamma = \sum_{i=1}^n [M^i, M^i]$) such that $d[M^i, M^j] \ll d\Gamma, \forall i, j \leq n$, and let

$$\gamma^{ij} = \frac{d[M^i, M^j]}{d\Gamma}.$$

We denote by $L_m(M)$ the set of all $\mathbf{R}^n$-valued predictable processes H such that

$$\sqrt{(\sum_{i,j=1}^{n} H^i \gamma^{ij} H^j).\Gamma} \in \mathcal{A}^+_{\text{loc}}.$$

It is easy to see that the space $L_m(M)$ doesn't depend on the choice of Γ. Similar to the real local martingale case, for $H \in L_m(M)$ we can define uniquely a real local martingale, denoted by $H.M$, such that for any real local martingale N,

$$[H.M, N] = (\sum_{i=1}^{n} H^i \gamma^{iN}).\Gamma,$$

where $\gamma^{iN} = \frac{d[M^i,N]}{d\Gamma}$. We call $H.M$ the *(vector) stochastic integral* of H w.r.t. M. Sometimes we denote also this integral by $H_{\dot{m}}M$. If $H, K \in L_m(M)$, then

$$[H.M, K.M] = (\sum_{i,j=1}^{n} H^i \gamma^{ij} K^j).\Gamma.$$

The properties of vector stochastic integrals are similar to that of the scalar case.

Theorem 2.2.7 (Ref. 12) *Let $M = (M^i)_{i\leq n}$ be a vector local martingale. If $[M^i, M^j] = 0, \forall i \neq j$, then $L_m(M) = \{H = (H^i)_{i\leq n} : H^i \in L_m(M^i), \forall i \leq n\}$, and the vector stochastic integral coincides with the componentwise stochastic integral.*

Progressive and Adapted Integrands

Let M be a continuous local martingale and H a progressive process. Then there exists $L \in \mathcal{M}_{loc}$ such that $[L, N] = H.[M, N]$ holds for all $N \in \mathcal{M}_{loc}$ iff $H^2 \in L_S([M])$. In this case, there exists a predictable process $K \in L_m(M)$ such that $K.M = L$. We say that H is *integrable* w.r.t. M, and L is called the *stochastic integral* of H w.r.t. M, denoted by $H.M$.

Let M be a purely discontinuous local martingale and H a progressive process. If $H\Delta M$ has predictable projection and there exists a purely discontinuous local martingale L such that $\Delta L = H\Delta M - {}^p(H\Delta M)$, we call L the *compensated stochastic integral* of H w.r.t. M, and denote $L = H_{\dot{c}}M$.

The above observation leads to the following

Definition 2.2.8 *Let M be a local martingale and H a progressive process. If $H^2 \in L_S([M^c])$, ${}^p(H\Delta M)$ exists and*

$$\sqrt{\sum_{s\leq \cdot}(H_s\Delta M_s - {}^p(H\Delta M)_s)^2} \in \mathcal{A}^+_{\mathrm{loc}},$$

then we put

$$H_{\dot{c}}M = H_0M_0 + H.M^c + H_{\dot{c}}M^d.$$

$H_{\dot{c}}M$ is called the compensated stochastic integral of H w.r.t. M. We often write $H.M$ instead of $H_{\dot{c}}M$.

Example 2.2.9 *1) Let M be a purely discontinuous local martingale. Put $H = I_{[\Delta M \neq 0]}$. Then the compensated stochastic integral of H w.r.t. M exists and $H_{\dot{c}}M = M$.*

2) Let M be a local martingale and X a semimartingale. Then $\Delta X_{\dot{c}}M$ exists if and only if $[X, M] \in \mathcal{A}_{\mathrm{loc}}$. If it is the case then $\Delta X_{\dot{c}}M = [X, M] - \langle X, M\rangle$.

The compensated stochastic integral is a generalization of the predictable stochastic integral. However, the conditions for the existence of compensated stochastic integrals are hard to verify, and we have no characterization for compensated stochastic integrals. The following theorem gives a sufficient condition for the existence of compensated stochastic integrals, originally proposed by P. A. Meyer [Ref. 10].

Theorem 2.2.10 *Let M be a local martingale and H a progressive process. If $\sqrt{H^2.[M]} \in \mathcal{A}^+_{loc}$, then $H_{\dot{c}}M$ exists, and it is the unique local martingale L such that for any bounded martingale N, $[L, N] - H.[M, N] \in \mathcal{M}_{loc,0}$. Besides, if we assume already $H^2 \in L_S([M])$, then the condition $\sqrt{H^2.[M]} \in \mathcal{A}^+_{\mathrm{loc}}$ is also necessary for the existence of $H_{\dot{c}}M$.*

The following theorem generalizes Itô's stochastic integrals of adapted measurable processes w.r.t. a Brownian motion.

Theorem 2.2.11 *Let M be a continuous local martingale with $M_0 = 0$. Assume that there exists a deterministic continuous increasing function $a = (a_t)$ such that for almost all ω $d[M](\omega) \ll da$. Let H be an adapted measurable process. Then there exists $L \in \mathcal{M}_{loc}$ such that $[L, N] = H.[M, N]$ holds for all $N \in \mathcal{M}_{loc}$ if and only if $H^2 \in L_S([M])$. In this case, there exists a predictable process K such that $K \in L_m(M)$ and $K.M = L$. L is called the stochastic integral of H w.r.t. M, and denoted by $H.M$.*

2.2.2 Stochastic Integrals w.r.t. Semimartingales

Predictable Integrands

We begin with the real-valued semimartingale case.

Lemma 2.2.12 *Let X be a semimartingale and H a predictable process. Let $X = M + A$ and $X = N + B$ be two decompositions of X, where $M, N \in \mathcal{M}_{loc}$ and $A, B \in \mathcal{V}_0$. If $H \in L_m(M) \cap L_S(A)$ and $H \in L_m(N) \cap L_S(B)$, then*

$$H_{\dot{m}}M + H_{\dot{s}}A = H_{\dot{m}}N + H_{\dot{s}}B.$$

Based on Lemma 1.2.12 we propose the following definition.

Definition 2.2.13 *Let X be a semimartingale and H a predictable process. If there exists a decomposition $X = M + A$, where $M \in \mathcal{M}_{loc}$ and $A \in \mathcal{V}_0$, such that $H \in L_m(M) \cap L_S(A)$, we say that H is integrable w.r.t. X (or simply H is X-integrable), and call $X = M + A$ an H-decomposition of X. In this case we put*

$$H.X = H_{\dot{m}}M + H_{\dot{s}}A.$$

$H.X$ is independent of H-decompositions of X, and is called the stochastic integral of H w.r.t. X. We denote by $L(X)$ the collection of all predictable processes which are integrable w.r.t. semimartingale X.

Remark *1) Let X be a semimartingale and $X = M + A$ be a decomposition of X, where $M \in \mathcal{M}_{loc}$ and $A \in \mathcal{V}_0$. Then any locally bounded predictable process H is X-integrable, and $X = M + A$ is an H-decomposition of X.*

2) Let M be a local martingale. Then $L_m(M) \subset L(M)$ and for $H \in L_m(M)$ two definitions of stochastic integrals coincide. In general, $H \in L(M)$ does not imply that $H.M$ is a local martingale, unless we know $H.M$ is a special semimartingale (see below Corollary 1.2.16) or $H.M$ is bounded below by a constant (see below Theorem 1.2.20).

3) Let X be an adapted FV process. If $H \in L(X) \cap L_S(X)$, then $H.X = H_{\dot{s}}X$. In general, $H \in L(X)$ does not imply that $H \in L_S(X)$, unless $H.X \in \mathcal{V}$ (see below Theorem 1.2.17) or X is predictable (see below Theorem 1.2.33).

The next theorem summarizes the fundamental properties of stochastic integrals of predictable processes w.r.t. semimartingales.

Theorem 2.2.14 *Let X be a semimartingale, and $H \in L(X)$.*

1) $(H.X)^c = H.X^c$, $\Delta(H.X) = H\Delta X$, $(H.X)_0 = H_0X_0$.

2) For any stopping time T

$$(H.X)^T = H.X^T = (HI_{[\![0,T]\!]}).X, (H.X)^{T-} = H.X^{T-}.$$

3) For any semimartingale Y, $[H.X, Y] = H.[X, Y]$.

4) If Y is a semimartingale and $H \in L(Y)$, then $H \in L(X + Y)$ and $H.(X + Y) = H.X + H.Y$.

5) If K is a predictable process and $|K| \leq |H|$, then $K \in L(X)$.

Theorem 2.2.15 *Let X be a special semimartingale and $H \in L(X)$. Then $H.X$ is a special semimartingale if and only if the canonical decomposition of X is an H-decomposition of X.*

Corollary 2.2.16 *1) If M is a local martingale, $H \in L(M)$ and $H.M$ is a special semimartingale, then $H \in L_m(M)$ and $H.M$ is a local martingale. In particular, for any continuous local martingale M, we have $L_m(M) = L(M)$.*

2) If $X \in \mathcal{V}$ and $H \in L(X)$ with $H.X \in \mathcal{V}$, then $H \in L_S(X)$.

The next theorem is an important consequence of Theorem 1.2.16.

Theorem 2.2.17 *Let X be a semimartingale and $H \in L(X)$. Let U be an optional set such that $U \supset [|H\Delta X| > 1$ or $|\Delta X| > 1]$ and for almost all ω, for each $t > 0$, $\{s : (\omega, s) \in U\} \cap [0,t]$ contains at most a finite number of points. Put*

$$A_t = \sum_{s \leq t} \Delta X_s I_{\{(\cdot,s)\in U\}}, \quad Z_t = X_t - A_t, \ t \geq 0.$$

Then $H \in L(Z)$, and the canonical decomposition $Z = N + B$ of the special semimartingale Z is an H-decomposition of Z.

In Theorem 1.2.17, if we put $U = [|H\Delta X| > 1$ or $|\Delta X| > 1]$, then $X = N + (B + A)$ is an H-decomposition of X, where $N \in \mathcal{M}_{loc}$. Moreover, we have $|\Delta N| \leq 2$ (since $|\Delta B| \leq 1$), so N is a locally bounded martingale. Using this fact and Theorem 1.2.17 we can prove the following important properties of stochastic integrals.

Theorem 2.2.18 *Let X be a semimartingale.*

1) $H, K \in L(X) \Longrightarrow H + K \in L(X)$.

2) Let $H \in L(X)$ and K be a predictable process. Then $K \in L(H.X)$ if and only if $KH \in L(X)$. In this case, we have $K.(H.X) = (KH).X$.

3) Let H be a predictable process. If there exist stopping times $T_n \uparrow \infty$ such that $H \in L(X^{T_n})$ for each n, then $H \in L(X)$.

Let $\tau = (\tau_t)$ be a random time-change and X be an adapted cadlag process. We say that τ is X-continuous, if for any $t \in \mathbf{R}_+$, X is constant on $[\tau_{t-}, \tau_t]$, a.s., where $\tau_{t-} = 0$ by convention.

The following theorem shows how semimartingales, covariation processes and stochastic integrals are transformed by a random time-change.

Theorem 2.2.19 *Let X be an $\mathbf{F}$-local martingale (resp. semimartingale) and let $\tau = (\tau_t)$ be a random time-change with induced filtration $\mathbf{G} = (\mathcal{G}_t)$ such that τ is X-continuous. Then $X \circ \tau$ is a $\mathbf{G}$-local martingale (resp. semimartingale) and we have $[X \circ \tau] = [X] \circ \tau$, a.s.. Furthermore, if $H \in L_m(X)$ (resp. $\in L(X)$), then $H \circ \tau \in L_m(X \circ \tau)$ (resp. $\in L(X \circ \tau)$), and $(H \circ \tau).(X \circ \tau) = (H.X) \circ \tau$.*

Now we turn to the vector stochastic integrals of semimartingales. Let $X = (X^i)_{i \leq n}$ be an $\mathbf{R}^n$-valued semimartingale and $H = (H^i)_{i \leq n}$ an $\mathbf{R}^n$-valued predictable process. If for each i, $H^i \in L(X^i)$, then we define naturally the *componentwise stochastic integral* as

$$H.X = \sum_{i=1}^{n} H^i.X^i.$$

Like the martingale case, we can extend this componentwise integral to a vector integral allowing a larger class of integrands. To this end, we need to define the vector Stieltjes integral. Let $A = (A^i)_{i \leq n}$ be an $\mathbf{R}^n$-valued adapted FV process. We take an adapted increasing process Γ (e.g., $\Gamma_t = \sum_{i=1}^n \int_{[0,t]} |dA_s|$) such that $|dA^i| \ll d\Gamma, \forall i \leq n$, and let

$$\gamma^i = \frac{dA^i}{d\Gamma}.$$

We denote by $L_S(A)$ the set of all $\mathbf{R}^n$-valued measurable processes $H = (H^i)_{i\leq n}$ such that

$$\sum_{i=1}^{n} H^i \gamma^i \in L_S(\Gamma).$$

The space $L_S(A)$ doesn't depend on the choice of Γ. If $H = (H^i)_{i\leq n} \in L_S(A)$, put

$$H_{\dot{s}}A = (\sum_{i=1}^{n} H^i \gamma^i)_{\dot{s}}\Gamma.$$

We call $H_{\dot{s}}A$ the *vector Stieltjes integral* of H w.r.t. A.

For vector semimartingales we have a similar result as Lemma 1.2.12. So we can define the vector semimartingale integral in the same manner as in Definition 1.2.13. Its properties are similar to that in the one-dimensional case.

As pointed out before, the stochastic integral of a predictable process w.r.t. a local martingale is not necessarily a local martingale. However, we have the following two results: the first one is due to Emery (1980), the second one is due to Ansel-Stricker (1994).

Theorem 2.2.20 (Ref. 14, 15) *1) Let $M \in \mathcal{M}_{\text{loc}}$. If $H \in L(M)$ and $H\Delta M \geq 0$, then $H.M \in \mathcal{M}$. In particular, if $\Delta M \geq 0$, then $L_m(M) = L(M)$.*

2) Let $M = (M^1, \cdots, M^n)$ be a vector-valued local martingale and $H \in L(M)$. Then $H.M$ is a local martingale if and only if there exist a sequence of stopping tomes T_n tending to ∞ and a sequence of integrable r.v.'s θ_n taking negative values such that $H \cdot \Delta M^{T_n} \geq \theta_n$. In particular, if $H.M$ is bounded below by a constant, then $H.X$ is a local martingale.

The following is the so-called optional decomposition theorem for vector-valued semimartingales. This theorem has important applications in mathematical finance.

Theorem 2.2.21 (Ref. 16, 17) *Let S be an $\mathbf{R}^d$-valued semimartingale. We denote by $\mathcal{P}$ the set of all probability measures $\mathbf{Q}$ such that $\mathbf{Q}$ is equivalent to $\mathbf{P}$ and S is a local martingale under $\mathbf{Q}$. Assume that $\mathcal{P} \neq \emptyset$. If X is a local supermartingale under each $\mathbf{Q} \in \mathcal{P}$, then there exist an adapted increasing process with $C_0 = 0$ and an $\mathbf{R}^d$-valued predictable process H such that H is S-integrable under each $\mathbf{Q} \in \mathcal{P}$ and $X = X_0 + H.S - C$.*

Note that in contrast to the standard Doob-Meyer decomposition, the process C is in general not predictable and not uniquely determined.

The following result is a direct consequence of Theorem 1.2.20 and 1.2.21.

Theorem 2.2.22 (Ref. 16) *Let S be an $\mathbf{R}^d$-valued semimartingale and $\mathcal{P}$ be the set of all probability measures $\mathbf{Q}$ such that $\mathbf{Q}$ is equivalent to $\mathbf{P}$ and S is a local martingale under $\mathbf{Q}$. Assume that $\mathcal{P} \neq \emptyset$. If X is a local martingale under each $\mathbf{Q} \in \mathcal{P}$, then there exists an $\mathbf{R}^d$-valued predictable process H such that H is S-integrable under each $\mathbf{Q} \in \mathcal{P}$ and $X = X_0 + H.S$.*

Progressive Integrands

Now we extend stochastic integrals of predictable processes w.r.t. semimartingales to progressive integrands such that they include stochastic Stieltjies integrals (cf. Ref. 18).

We denote by $\mathcal{M}^q_{\text{loc}}$ (resp. $\mathcal{V}^q$) the set of all quasi-continuous local martingales (resp. adapted FV processes). We put

$$\mathcal{S}^q = \mathcal{M}^q_{\text{loc}} + \mathcal{V}^q, \quad \mathcal{S}^{da} = \mathcal{M}^{da}_{\text{loc}} + \mathcal{V}^{da}.$$

Then we have

$$\mathcal{S} = \mathcal{S}^{da} \oplus \mathcal{S}^{q} \text{ direct sum},$$

where $\mathcal{S}$ is the set of all semimartingales.

Let $X \in \mathcal{S}$. We denote by $X = X^{da} + X^{q}$ the decomposition of X following $\mathcal{S}^{da} \oplus \mathcal{S}^{q}$ direct sum. It is obvious that we have

$$L(X) = L(X^{da}) \cap L(X^{q}), \quad H.X = H.X^{da} + H.X^{q}, \quad \forall H \in L(X).$$

For a progressive process H we will define its integrals w.r.t. X^{da} and X^{q} separately and then make a summation.

Let $X \in \mathcal{S}^{da}$ and H be a predictable process. It is easy to prove that $H \in L(X)$ if only if there exists a (unique) $Y \in \mathcal{S}^{da}$ such that $\Delta Y = H\Delta X$. This suggests the following

Definition 2.2.23 (Ref. 18) *Let $X \in \mathcal{S}^{da}$. A progressive process H is said to be X-integrable, denoted by $H \in I(X)$, if there exists a (unique) $Y \in \mathcal{S}^{da}$ such that $\Delta Y = H\Delta X$. In this case we put $H.X = Y$ and call $H.X$ the stochastic integral of H w.r.t. X.*

Lemma 2.2.24 (Ref. 18) *Let $M \in \mathcal{M}^{q}_{\text{loc}}$. Let H be a progressive process and K be a predictable process such that $[{}^{o}H \neq K]$ is a thin set, where ${}^{o}H$ is the optional projection of H. Assume that $K \in L_m(M)$ and $\sum_{s\leq \cdot} |H_s - K_s||\Delta M_s| \in \mathcal{V}$. Then any predictable process K' such that $[{}^{o}H \neq K]$ is a thin set verifies the above condition. Moreover, we have $K.M = K'.M$ and*

$$\sum_{s\leq \cdot} |H_s - K_s||\Delta M_s| = \sum_{s\leq \cdot} |H_s - K'_s||\Delta M_s|.$$

In this case we say H is M-integrable in the local martingale sense and denote $H \in I_m(M)$. Its integral w.r.t. M is defined by

$$H.M = K.M + \sum_{s\leq \cdot} (H_s - K_s\Delta M_s).$$

Lemma 2.2.25 (Ref. 18) *Let $X \in \mathcal{S}^{q}$ and H be a progressive process. Assume there exists a so-called H-decomposition $X = M + A$ with $M \in \mathcal{M}^{q}_{\text{loc}}$ and $A \in \mathcal{V}^{q}$ such that $H \in I_m(M) \cap L_S(A)$. Then the sum $H.M + H.A$ doesn't depend on the choice of the H-decomposition. In this case, H is said to be X-integrable, denoted by $H \in I(X)$, and its integral w.r.t. M is defined by*

$$H.X = H.M + H.A.$$

Finally, we can give the following

Definition 2.2.26 (Ref. 18) *Let $X \in \mathcal{S}$. A progressive process H is said to be X-integrable, denoted by $H \in I(X)$, if H is separately X^{da}-integrable and X^{q}-integrable. If $H \in I(X)$, the integral of H w.r.t. X is defined by $H.X = H.X^{da} + H.X^{q}$.*

Remark *This integral extends that of predictable processes w.r.t. semimartingales and includes the stochastic Stieltjes integral of progressive processes w.r.t. adapted FV processes.*

2.2.3 Convergence Theorems for Stochastic Integrals

The following theorem, due to Lenglart [Ref. 19] is the key for the study of convergence of stochastic integrals.

Theorem 2.2.27 (Lenglart's Inequality) *Let X be an adapted cadlag process and A an adapted increasing process such that for any bounded (or, equivalently, finite) stopping time T,*

$$\mathbf{E}[|X_T|] \leq \mathbf{E}[A_T].$$

Then for any constants $c > 0$, $d > 0$, stopping time T and measurable set H, we have

$$\mathbf{P}(H \cap [X_T^* \geq c]) \leq \frac{1}{c}\mathbf{E}\big[A_T \wedge (d + \sup_{t \leq T} \Delta A_t)\big] + \mathbf{P}(H \cap [A_T \geq d])).$$

If furthermore A is predictable, we have

$$\mathbf{P}(H \cap [X_T^* \geq c]) \leq \frac{1}{c}\mathbf{E}[A_T \wedge d] + \mathbf{P}(H \cap [A_T \geq d])).$$

From Theorem 2.2.27 we can prove easily the following

Theorem 2.2.28 *Let $M \in \mathcal{M}^2_{\mathrm{loc}}$, T be a stopping time and B a measurable set. Assume H, $H^{(n)} \in L_m(M)$, $n \geq 1$, and $(H - H^{(n)}).M \in \mathcal{M}^2_{loc}$, $n \geq 1$. If*

$$I_B \int_{[0,T]} (H_s - H_s^{(n)})^2 d\langle M \rangle_s \xrightarrow{P} 0,$$

then

$$I_B \sup_{s \leq T} |(H.M)_s - (H^{(n)}.M)_s| \xrightarrow{P} 0.$$

The next theorem is a convergence theorem for stochastic integrals.

Theorem 2.2.29 *Let X be a semimartingale, T a finite stopping time, B a measurable set, and let $H, H^{(n)}, n \geq 1$, be locally bounded predictable processes. If for almost all $\omega \in B$ $(H^{(n)}_{\cdot}(\omega))_{n \geq 1}$ is uniformly bounded and convergent to $H_{\cdot}(\omega)$ on $[0, T(\omega)]$, then*

$$I_B \sup_{t \leq T} |(H^{(n)}.X)_t - (H.X)_t| \xrightarrow{P} 0.$$

Definition 2.2.30 *Let T be a finite stopping time and $(T_n)_{n \geq 0}$ an increasing sequence of stopping times with $T_0 = 0$ and $\sup_n T_n = T$. We say that $\tau : 0 = T_0 \leq T_1 \leq \cdots$ is a stochastic partition of interval $[0, T]$, if for almost all ω, the sequence $(T_n(\omega))$ is stationary (i.e., there exists a natural number $n(\omega)$ such that $T_n(\omega) = T(\omega)$ when $n \geq n(\omega)$); in other words, for almost all ω $(T_n(\omega))$ forms a finite partition of interval $[0, T(\omega)]$. Let*

$$\delta(\tau) = \sup_j |T_{j+1} - T_j|.$$

$\delta(\tau)$ is a finite r.v., and is called the mesh of partition τ.

The following theorem shows that the stochastic integrals of left-continuous processes w.r.t. semimartingales are of Riemann-Stieltjes type.

Theorem 2.2.31 *Let X be a semimartingale, H an adapted cadlag or left-continuous process, and T a finite stopping time. If*

$$\tau^{(n)} : 0 = T_0^{(n)} \leq T_1^{(n)} \leq \cdots, \ n \geq 1,$$

be a sequence of stochastic partitions of $[0, T]$ such that $\lim_n \delta(\tau^{(n)}) = 0$ a.s., then

$$\sup_{t \leq T} \left| \sum_i H_{T_i^{(n)}} (X_{T_{i+1}^{(n)} \wedge t} - X_{T_i^{(n)} \wedge t}) - \int_0^t H_{s-} dX_s \right| \xrightarrow{P} 0, \quad n \to \infty.$$

The following is the dominated convergence theorem for stochastic integrals.

Theorem 2.2.32 *Let X be a semimartingale, $H \in L(X)$, $K^{(n)}$ and K be predictable processes such that $|K^{(n)}| \le |H|$, $|K| \le |H|$. Let $B \in \mathcal{F}$ and T be a finite stopping time. If for almost all $\omega \in B$ we have $\lim_{n\to\infty} K_t^{(n)}(\omega) = K_t(\omega)$ for all $t \in [0, T(\omega)]$, then*

$$I_B \sup_{t\le T} \left|(K^{(n)}.X)_t - (K.X)_t\right| \xrightarrow{\mathbf{P}} 0, \ n \to \infty.$$

In particular, if we put $H^{(n)} = HI_{[|H|\le n]}$, then for all $t \ge 0$,

$$\sup_{s\le T} \left|(H^{(n)}.X)_s - (H.X)_s\right| \xrightarrow{\mathbf{P}} 0, \quad n \to \infty.$$

From Theorems 1.2.32 and Corollary 1.2.16 we can prove the following result.

Theorem 2.2.33 *If A is a predictable FV process and $H \in L(A)$, then $H \in L_S(A)$, and $H.A = H_{\dot{s}}A$.*

The following theorem is an easy consequence of Theorem 1.2.32.

Theorem 2.2.34 *Let X be a semimartingale and $H \in L(X)$ with $[H \neq 0]$ being a thin set. If for each $t \in \mathbb{R}$, $\sum_{s\le t} |H_s||\Delta X_s| < \infty$, a.s., then*

$$(H.X)_t = H_0X_0 + \sum_{s\le t} H_s\Delta X_s.$$

The following theorem justifies the terminologies "quadratic variation" and "quadratic covariation."

Theorem 2.2.35 *Let X and Y be two semimartingales. If T is a finite stopping time, and $\tau_n : 0 = T_0^n \le T_1^n \le \cdots$ is a sequence of stochastic partitions of $[0, T]$ with $\delta(\tau_n)$ tending to zero, then*

$$\sup_{t\le T} \left|\sum_i (X_{T_{i+1}^n\wedge t} - X_{T_i^n\wedge t})(Y_{T_{i+1}^n\wedge t} - Y_{T_i^n\wedge t}) + X_0Y_0 - [X, Y]_t\right| \xrightarrow{P} 0.$$

Lemma 2.2.36 *Let M be a local martingale, and H be a progressive process such that $H.\, M \in \mathcal{H}^1$ and*

$$\mathbf{E}\left[\left(\int_{[0,\infty[} H_s^2 d[M]_s\right)^{1/2}\right] < \infty,$$

Then for any $N \in \mathcal{BMO}$, $[H.\, M, N] - H.\, [M, N]$ is a martingale with integrable variation. In particular, $\mathbf{E}[H.\, M, N]_\infty = \mathbf{E}\left[\int_{[0,\infty[} H_s d[M, N]_s\right]$.

The following theorem is an extension of the first Davis inequality (see Theorem 1.1.97).

Theorem 2.2.37 *Let M be a local martingale, H be a progressive process such that $\sqrt{H^2.[M]}$ is locally integrable. Then for any stopping time T we have*

$$\mathbf{E}[(H.\, M)_T^*] \le 2\sqrt{6}\mathbf{E}\left[\left(\int_{[0,T]} H_s^2 d[M]_s\right)^{1/2}\right].$$

As an application of Theorem 1.2.37, we obtain the following convergence theorem for progressive stochastic integrals.

Theorem 2.2.38 *Let M be a local martingale. We denote by $L^0(M)$ the set of all progressive processes H such that $\sqrt{H^2.[M]} \in \mathcal{A}^+_{\text{loc}}$. Let $(H^{(n)}) \subset L^0(M), H \in L^0(M)$ and T be a stopping time.*

1) If $\mathbf{E}\Big[\Big(\int_{[0,T]}(H_s^{(n)} - H_s)^2 d[M]_s\Big)^{1/2}\Big] \to 0$, *then*

$$\mathbf{E}\Big[\sup_{t\leq T}\Big|(H^{(n)}.M)_t - (H.\ M)_t\Big|\Big] \to 0.$$

2) If $\sum_{n=1}^{\infty}\mathbf{E}\Big[\Big(\int_{[0,T]}(H_s^{(n)} - H_s)^2 d[M]_s)^{1/2}\Big] < \infty$, *then*

$$\sup_{t\leq T}\Big|(H^{(n)}.M)_t - (H.\ M)_t\Big| \to 0. \quad \text{a.s.}.$$

We end this section with a result about stochastic integrals of processes depending on a parameter.

Theorem 2.2.39 (Ref. 20) *Let $(S, \mathcal{S})$ be a measurable space and X be a continuous semimartingale. Let $(H_t(s))_{t\geq 0}, s \in \mathcal{S}$ be a family of processes which are progressive on $S \times \mathbf{R}_+$ in the sense that for every $t \geq 0$, the mapping $(s, t, \omega) \mapsto H_t(s, \omega)$ is $\mathcal{S} \times \mathcal{B}([0, t]) \times \mathcal{F}_t$-measurable. If for every $s \in S$, $H(s) \in L(X)$, then the family $Y_t(s) = (H(s).X)_t$ has a version that is progressive on $S \times \mathbf{R}_+$, and continuous for each $s \in S$.*

2.2.4 Itô's Formula and Doléans Exponential Formula

In this section we present the change of variables formula for semimartingales (Itô's formula), the most powerful tool in stochastic calculus.

To begin with, from Theorem 1.2.31 and 1.2.35 we can deduce the following

Theorem 2.2.40 *If X and Y are two semimartingales, then we have the following formula of integration by parts:*

$$X_tY_t = \int_0^t X_{s-}dY_s + \int_0^t Y_{s-}dX_s + [X, Y]_t, \quad t \geq 0.$$

From the formula of integration by parts one can prove easily the following

Theorem 2.2.41 (Itô's Formula) *Let $X^1, \cdots, X^d$ be semimartingales, and F be a C^2-function on $\mathbf{R}^d$ (i.e. F has continuous partial derivatives of the first and the second order). Put $X_t = (X_t^1, \cdots, X_t^d)$ ((X_t) is also called an n-dimensional semimartingale). Then*

$$F(X_t) - F(X_0) = \sum_{j=1}^d \int_0^t D_jF(X_{s-})dX_s^j + \sum_{0<s\leq t}\eta_s(F) + \frac{1}{2}A_t(F),$$

where

$$\eta_s(F) = F(X_s) - F(X_{s-}) - \sum_{j=1}^d D_jF(X_{s-})\Delta X_s^j,$$

$$A_t(F) = \sum_{i,j=1}^d \int_0^t D_{ij}F(X_{s-})d\langle (X^i)^c, (X^j)^c\rangle_s,$$

$D_jF = \frac{\partial F}{\partial x_j}$, $D_{ij}F = \frac{\partial^2 F}{\partial x_i \partial x_j}$, *and the series $\sum_{0<s\leq t}\eta_s(F)$ is absolutely convergent.*

Remark *1) We have the following refinement of Theorem 1.2.41. Let $d = n + m$, and $X^1, \cdots, X^n$ be semimartingales, and $X^{n+1}, \cdots, X^{n+m}$ be adapted FV processes. Let F be a continuous function on $\mathbf{R}^{n+m}$, of class C^2 w.r.t. the first n variables and of class C^1 w.r.t. the last m variables (it may be $n = 0$ or $m = 0$). Put $X_t = (X_t^1, \cdots, X_t^{n+m})$. Then*

$$F(X_t) - F(X_0) = \sum_{j=1}^{n+m} \int_0^t D_jF(X_{s-})dX_s^j + \sum_{0<s\le t} \eta_s(F)$$
$$+\frac{1}{2}\sum_{i,j=1}^{n} \int_0^t D_{ij}F(X_{s-})d\langle (X^i)^c, (X^j)^c\rangle_s.$$

2) Itô's formula can be applied to a function defined on an open domain of $\mathbf{R}^n$. For example, if X and Y are two semimartingales with $[Y = 0$ or $Y_- = 0]$ being evanescent, then by using Itô's formula we can prove that X/Y is a semimartingale.

3) One can apply Itô's formula to complex valued semimartingales. As an example, let X, Y be continuous semimartingales, and put $Z_t = X_t + iY_t$. Then for any analytic function f we have

$$f(Z_t) = f(Z_0) + \int_0^t f'(Z_s)dZ_s + \frac{1}{2}\int_0^t f''(Z_s)d[Z, Z]s.$$

As an application of Itô's formula, we obtain the *Lévy's characterization of Brownian motion.*

Theorem 2.2.42 *Let $B_t = (B_t^1, \cdots, B_t^d)$ be a d-dimensional $(\mathcal{F}_t)$-adapted continuous process. Then (B_t) is an $\mathcal{F}$-Brownian motion if and only if each (B_t^i) is an $(\mathcal{F}_t)$-local martingale and for $1 \le i, j \le d$, $B_t^iB_t^j - \delta_{ij}t$ is an $(\mathcal{F}_t)$-local martingale (i.e. $\langle B^i, B^j\rangle_t = \delta_{ij}t$).*

Lemma 2.2.43 *Let M be a continuous local martingale. Then for almost all ω $M.(\omega)$ and $\langle M\rangle.(\omega)$ have the same constancy intervals, i.e., for any $a < b$ if $M(\omega)$ is constant on $[a, b]$, so is $\langle M\rangle.(\omega)$ and vice versa.*

By Theorem 1.2.42 and Lemma 1.2.43 we obtain the following result, due to Knight (1971) [Ref. 21].

Theorem 2.2.44 *Let $M = (M^1, \cdots, M^d)$ be a d-dimensional continuous local martingale with $M_0 = 0$ such that $\langle M^i, M^j\rangle = 0$ for $i \ne j$ and $\langle M^i\rangle_\infty = \infty$ for each i. Put*

$$\tau_t^i = \inf\{s : \langle M^i\rangle_s > t\},\ B_t^i = M_{\tau_t}^i,\ \mathcal{G}_t = \mathcal{F}_{\tau_t},\ t \ge 0,\ 1 \le i \le d.$$

Then $B = (B^1, \cdots, B^d)$ is a standard d-dimensional Brownian motion.

Theorem 2.2.45 *Let X be a semimartingale. Put*

$$V_t = \prod_{0<s\le t} (1 + \Delta X_s)e^{-\Delta X_s} \quad (V_0 = 1).$$

Then for almost all ω the above infinite product is absolutely convergent for all $t > 0$, and $V = (V_t)$ is an adapted purely discontinuous process of finite variation.

Theorem 2.2.46 *Let*

$$Z_t = \exp\left\{X_t - X_0 - \frac{1}{2}\langle X^c\rangle_t\right\} \prod_{0<s\le t} (1 + \Delta X_s)e^{-\Delta X_s}. \tag{46.1}$$

Then $Z = (Z_t)$ is the unique semimartingale satisfying the stochastic integral equation

$$Z_t = 1 + \int_0^t Z_{s-}dX_s.$$

We call Z the Doléans (stochastic) exponential of X, and denote it by $\mathcal{E}(X)$. (46.1) is called the Doléans exponential formula, due to Doléans-Dade [Ref. 22].

By using the Doléans exponential formula, we obtain the following result on multiplicative decompositions of nonnegative submartingales.

Theorem 2.2.47 *Let X be a strictly positive submartingale with canonical decomposition $X = M + A$, where $M \in \mathcal{M}_{\text{loc}}, A \in \mathcal{A}^+$ with $A_0 = 0$ and A is predictable. Then X can be uniquely expressed as $X = BN$, where B is an increasing predictable process with $B_0 = 1$, and N is a martingale. Moreover, we have*

$$B = \mathcal{E}(X_-^{-1}.A), \quad N = B^{-1}.M.$$

More generally, we have

Theorem 2.2.48 (Ref. 11) *Let X be a strictly positive special semimartingale with $X_- > 0$ and $X_0 = 1$. Then X admits a unique multiplicative decomposition $X = MA$, where M is a positive local martingale and A is a positive predictable FV process with $A_0 = 1$. If furthermore X is a supermartingale, then A is decreasing.*

Theorem 2.2.49 *Let $X, Y \in \mathcal{S}$. Then $\mathcal{E}(X)\mathcal{E}(Y) = \mathcal{E}(X + Y + [X, Y])$.*

As an application of Theorem 1.2.49 we obtain a multiplicative decomposition of an exponential semimartingale.

Theorem 2.2.50 (Ref. 23) *Let X be a special semimartingale with the canonical decomposition $X = N + A$, where N is a local martingale and A is a predictable FV process. Assume that $X_0 = 0$ and $[\Delta A = -1]$ is evanescent. Then $\frac{1}{1+\Delta A}$ is locally bounded, and we have*

$$\mathcal{E}(X) = \mathcal{E}(M)\mathcal{E}(A),$$

where $M = \frac{1}{1+\Delta A}.N$.

In fact, by Theorem 1.2.49 we have $\mathcal{E}(M)\mathcal{E}(A) = \mathcal{E}(M+[M, A]+A)$. However, $M+[M, A]$ is a local martingale and has the same continuous martingale part and same jumps as N has, so we have $M + [M, A] = N$.

Theorem 2.2.51 (Ref. 24) *Let Z be a semimartingale with $[\Delta Z = -1]$ being evanescent and let H be an adapted cadlag process (not necessarily a semimartingale). Then the unique solution of the equation*

$$X_t = H_t + \int_0^t X_{s-} dZ_s, \quad t \geq 0$$

is given by

$$X_t = H_t + \mathcal{E}(Z)_t \Big\{ \int_0^t \mathcal{E}(Z)_{s-}^{-1} H_{s-} dZ_s - \int_0^t \mathcal{E}(Z)_s^{-1} H_{s-} d[Z]_s \Big\}.$$

If H is a semimartingale, X_t has another expression:

$$X_t = \mathcal{E}(Z)_t \Big\{ H_0 + \int_0^t \mathcal{E}(Z)_{s-}^{-1} dH_s - \int_0^t \mathcal{E}(Z)_s^{-1} d[H, Z]_s \Big\}.$$

2.2.5 Local Times of Semimartingales

Let X be a semimartingale, f be a continuous convex function on $\mathbf{R}$, and f' be its left derivative. Approximating f by C^∞-functions and using Itô's formula we can prove that $f(X)$ is a semimartingale and

$$f(X_t) = f(X_0) + \int_0^t f'(X_{s-})dX_s \tag{2.2.1}$$

$$+ \sum_{0<s\leq t} [f(X_s) - f(X_{s-}) - f'(X_{s-})\Delta X_s] + C_t, \tag{2.2.2}$$

where $C = (C_t)$ is a continuous adapted increasing process with $C_0 = 0$. In particular, if we take $f(x) = (x-a)^+$ or $f(x) = (x-a)^-$ we obtain

Theorem 2.2.52 *Let X be a semimartingale and $a \in \mathbf{R}$. Then*

$$\begin{aligned}(X_t - a)^+ = (X_0 - a)^+ &+ \int_0^t I_{[X_{s-}>a]}dX_s + \sum_{0<s\leq t} [I_{[X_{s-}>a]}(X_s - a)^- \\ &+ I_{[X_{s-}\leq a]}(X_s - a)^+] + \frac{1}{2}L_t^a(X),\end{aligned}$$

$$\begin{aligned}(X_t - a)^- = (X_0 - a)^- &- \int_0^t I_{[X_{s-}\leq a]}dX_s + \sum_{0<s\leq t} [I_{[X_{s-}>a]}(X_s - a)^- \\ &+ I_{[X_{s-}\leq a]}(X_s - a)^+] + \frac{1}{2}L_t^a(X),\end{aligned}$$

where $L_t^a(X)$ is a continuous adapted increasing process with $L_0^a(X) = 0$. For almost all ω the measure $dL^a(X)(\omega)$ does not charge the set $\{t : X_{t-}(\omega) \neq a\}$ and the interior of $\{t : X_{t-}(\omega) = a\}$.

$L_t^a(X)$ is called the local time of X at a. The above two equalities are called Tanaka-Meyer formulas.

Integrating $I_{[X_-=a]}$ and $I_{[X_-\leq a]}$ w.r.t. the two sides of the first Tanaka-Meyer formula, we obtain the following two formulas for local times.

Corollary 2.2.53 *Let X be a semimartingale and $a \in \mathbf{R}$. Then*

$$L_t^a(X) = 2\Big[\int_0^t I_{[X_{s-}=a]}d(X_s - a)^+ - \sum_{0<s\leq t} I_{[X_{s-}=a]}(X_s - a)^+\Big],$$

$$L_t^a(X) = 2\Big[\int_0^t I_{[X_{s-}\leq a]}d(X_s - a)^+ - \sum_{0<s\leq t} I_{[X_{s-}\leq a]}(X_s - a)^+\Big].$$

Expressing (C_t) in (2.2.1) by means of local times, we obtain a generalization of Itô's formula as follows.

Theorem 2.2.54 *Let X be a semimartingale and let f be a continuous convex function on $\mathbf{R}$ and f' its left derivative. Then*

$$\begin{aligned}f(X_t) &= f(X_0) + \int_0^t f'(X_{s-})dX_s \\ &+ \sum_{0<s\leq t} \Big[f(X_s) - f(X_{s-}) - f'(X_{s-})\Delta X_s\Big] + \frac{1}{2}\int_{-\infty}^{\infty} L_t^a(X)\rho(da),\end{aligned}$$

where ρ is the second order derivative of f in the sense of generalized functions (ρ is a Radon measure).

Corollary 2.2.55 *Let X be a semimartingale and g be a nonnegative or bounded Borel function. Then*

$$\int_0^t g(X_s)d\langle X^c\rangle_s = \int_{-\infty}^{\infty} L_t^a(X)g(a)da.$$

Theorem 2.2.56 *Let $X \in \mathcal{S}$ and f be the difference of two continuous convex functions on $\mathbf{R}$. For any $a \in \mathbf{R}$ we set $A(a) = \{x : f(x = a)\}$ and $B(a) = \{x : f(x) = a, |f_r'(x)| + |f_l'(x)| > 0\}$, where f_r' (resp. f_l') stands for the right (resp. left) derivative of f. Then $B(a)$ is at most countable and we have*

$$L_t^a(f(X)) = \sum_{x\in B(a)} [f_r'(x)^+ L_t^x(X) + f_l'(x)^- L_t^{-x}(-X)].$$

2.2.6 Fisk-Stratonovich Integrals

The content of this section is taken from Protter (1990) [Ref. 24]. Let X and Y be two continuous semimartingales. Let $t > 0$ and

$$\tau_n : 0 = t_0^n < t_1^n < \cdots < t_{m(n)}^n = t$$

be a sequence of finite partitions of $[0, t]$ with $\delta(\tau_n)$ tending to zero. According to Theorem 2.26, as $n \to \infty$,

$$\begin{aligned} I_n(t) &= \sum_i \frac{1}{2}(Y_{t_i^n} + Y_{t_{i+1}^n})(X_{t_{i+1}^n} - X_{t_i^n}) \\ &\xrightarrow{P} \int_0^t Y_s dX_s + \frac{1}{2}([X, Y]_t - X_0Y_0). \end{aligned}$$

We denote this limit by $\int_0^t Y_s \circ dX_s$. It is easy to verify that this integral obeys the rules of ordinary calculus. Namely, for any continuous semimartingale X in $\mathbf{R}^d$ and function $f \in C^3(\mathbf{R}^d)$, we have

$$f(X_t) = f(X_0) + \sum_{i=1}^d \int_0^t f_i'(X_s) \circ dX_s^i, \text{ a.s., } t \geq 0.$$

More generally, we pose the following

Definition 2.2.57 (Ref. 24) *Let X and Y be two semimartingales. We put*

$$\int_0^t Y_s \circ dX_s = \int_0^t Y_{s-} dX_s + \frac{1}{2}\langle X^c, Y^c\rangle.$$

We call this integral the Fisk-Stratonovich integral (F-S integral, for short) of Y w.r.t. X. In the literature, it is often called the Stratonovich integral.

Theorem 2.2.58 (Ref. 24) *Let $X = (X^1, \cdots, X^d)$ be an d-dimensional semimartingale, and F be a C^3-function on $\mathbf{R}^d$. Then*

$$\begin{aligned} F(X_t) - F(X_0) &= \sum_{j=1}^d \int_0^t D_j F(X_{s-}) \circ dX_s^j \\ &+ \sum_{0<s\leq t} [F(X_s) - F(X_{s-}) - \sum_{j=1}^d D_j F(X_{s-})\Delta X_s^j]. \end{aligned}$$

Now we extend the F-S integral to non-semimartingale integrands. To this end we need a general notion of quadratic covariation of stochastic processes.

Definition 2.2.59 (Ref. 24) *Let X and Y be adapted cadlag processes. The quadratic covariation of X and Y, if it exists, is defined to be an FV process, denoted by $[X,Y]$, such that*

$$\sup_{t\le T}\left|\sum_i (X_{T^n_{i+1}\wedge t}-X_{T^n_i\wedge t})(Y_{T^n_{i+1}\wedge t}-Y_{T^n_i\wedge t})+X_0Y_0-[X,Y]_t\right|\xrightarrow{P}0,\quad n\to\infty,$$

where $\tau_n : 0 = T^n_0 \le T^n_1 \le \cdots$ is any sequence of stochastic partitions of $[0,+\infty)$ with $\lim_{n\to\infty}\sup_m T^n_m=\infty$ and $\delta(\tau_n)$ tending to zero.

If $[X,X]$ exists, we say X has finite quadratic variation. If $[X,X]$ and $[Y,Y]$ exist then $[X+Y,X+Y]$ and $[X,Y]$ exist and the polarization identity holds:

$$[X,Y]=\frac{1}{2}([X+Y,X+Y]-[X,X]-[Y,Y]).$$

Lemma 2.2.60 (Ref. 24) *Let $X=(X^1,\cdots,X^d)$ be an d-dimensional semimartingale, and f be a C^1-function on $\mathbf{R}^d$. Then $f(X)$ has finite quadratic variation.*

Definition 2.2.61 (Ref. 24) *Let H be an adapted cadlag process and X a semimartingale. If $[H,X]$ exists, we put*

$$\int_0^t H_s\circ dX_s=\int_0^t H_s dX_s+\frac{1}{2}([X,Y]^c-X_0Y_0).$$

We call this integral the Fisk-Stratonovich integral (F-S integral, for short) of H w.r.t. X.

Theorem 2.2.62 (Ref. 24) *Let $X=(X^1,\cdots,X^d)$ be an d-dimensional semimartingale, and F be a C^2-function on $\mathbf{R}^d$. Then*

$$\begin{aligned}F(X_t)-F(X_0) &= \sum_{j=1}^d\int_0^t D_jF(X_{s-})\circ dX^j_s\\ &+\sum_{0<s\le t}[F(X_s)-F(X_{s-})-\sum_{j=1}^d D_jF(X_{s-})\Delta X^j_s].\end{aligned}$$

Theorem 2.2.63 (Ref. 24) *Let X be a semimartingale with $X_0=0$. Then the unique solution of the stochastic integral equation*

$$Z_t=Z_0+\int_0^t Z_{s-}\circ dX_s.$$

is given by

$$Z_t=Z_0\exp(X_t)\prod_{0<s\le t}(1+\Delta X_s)e^{-\Delta X_s},$$

and it is called the Fisk-Stratonovich exponential.

2.2.7 Stochastic Differential Equations

In this subsection we mainly present some basic results about the Itô stochastic differential equation (in short: SDE). We refer the reader to Karatzas and Shreve (1991) [Ref. 25]. A general result about the existence and uniqueness of solutions of stochastic differential equations driven by semimartingales is also mentioned.

Definition 2.2.64 *Let $(B_t)_{t\geq 0}$ be a d-dimensional $(\mathcal{F}_t)$-Brownian motion and $0 \leq t_0 \leq T$. Let $b : [t_0, T] \times I\!R^m \to I\!R^m$ and $\sigma : [t_0, T] \times I\!R^m \to M^{m,d}$ be Borel measurable maps, where $M^{m,d}$ is the set of all $m \times d$-matrices. An $I\!R^m$-valued continuous $(\mathcal{F}_t)$-adapted process X is said to be a solution of the following Itô stochastic differential equation*

$$dX_t = b(t, X_t)dt + \sigma(t, X_t)dB_t, \ t \in [t_0, T], \quad X_{t_0} = \xi, \tag{64.1}$$

with $\xi = (\xi^1, \cdots, \xi^m)$ being $\mathcal{F}_{t_0}$-measurable, if X satisfies the stochastic integral equation

$$X_t^i = \xi^i + \int_{t_0}^t b^i(s, X_s)ds + \sum_{j=1}^d \int_0^t \sigma_j^i(s, X_s)dB_s^j, \ 1 \leq i \leq m, \ t \in [t_0, T]. \tag{64.2}$$

Such a solution of (64.1) is called a strong solution meaning that it is based on the path of the underlying Brownian motion (B_t). In particular, a strong solution is adapted to the natural filtration of the Brownian motion (B_t). If such a strong solution doesn't exists, we have to find a Brownian motion (B_t) on a suitable stochastic basis and an adapted process (X_t) such that X_0 has the given distribution and (64.2) holds. Such a process (X_t) is known as a weak solution of (64.1).

In the sequel we denote

$$|x|^2 = \sum_{i=1}^m x_i^2, \quad |\gamma|^2 = \mathrm{tr}(\gamma\gamma^{\mathrm{T}}) = \sum_{\mathrm{j}=1}^{\mathrm{d}} \sum_{\mathrm{i}=1}^{\mathrm{m}} (\gamma^{\mathrm{ij}})^2$$

for $x \in \mathbb{R}^m$ and $\gamma \in M^{m,d}$. For notational simplicity, we take $t_0 = 0$.

Theorem 2.2.65 *If b and σ are Lipschitz in x:*

$$|b(t, x) - b(t, y)| + |\sigma(t, x) - \sigma(t, y)| \leq K|x - y| \tag{65.1}$$

and satisfy the linear growth condition in x:

$$|b(t, x)| + |\sigma(t, x)| \leq K(1 + |x|), \tag{65.2}$$

where K is a constant, then (64.1) has a unique solution X. Moreover, if on $[0, T]$ b and σ satisfies the polynomial growth condition in x:

$$\sup_{0\leq t\leq T} |b(t, x)| + |\sigma(t, x)| \leq C(1 + |x|^{2\mu}), \quad x \in I\!R^m, \tag{65.3}$$

for some constant $C > 0$, $\mu \geq 1$ and $I\!E\,[|\xi|^{2\mu}] < \infty$, then we have

$$I\!E\,[\sup_{0\leq t\leq T} |X_t|^{2\mu}] \leq K_1 + K_2 I\!E\,[|\xi|^{2\mu}].$$

Remark *If b and σ are only locally Lipschitz in the sense that for each positive constant L there is a constant K such that (65.1) is satisfied for x and y with $|x| \leq L, |y| \leq L$, then (64.1) still has a unique solution. If b and σ are continuous w.r.t. t, then one can prove that the unique solution to (64.1) is a diffusion process, usually called an Itô diffusion. Its drift vector is b and the diffusion matrix is $a = \sigma\sigma^\tau$.*

If in (64.1) b and σ are linear functions in x:

$$b(t,x) = G(t)x + g(t); \quad \sigma(t,x) = (H_1(t)x + h_1(t), \cdots, H_d(t)x + h_d(t)),$$

where G and $H_i(t)$ are $m \times m$ matrices, $g(t)$ and $h_i(t)$ are $\mathbb{R}^m$-valued functions, we call (64.1) a linear SDE.

The following theorem gives an explicit expression for the solution of a linear SDE.

Theorem 2.2.66 *Assume that G, g, H_i, h_i are measurable locally bounded functions. Then the unique solution of linear SDE (64.1) with $X_0 = c$ is given by*

$$X_t = \Phi_t\left(c + \int_0^t \Phi_s^{-1} dY_s\right),$$

where

$$dY_t = \left(g(t) - \sum_{i=1}^d H_i(t)h_i(t)\right)dt + \sum_{i=1}^d h_i(t)dB_t^i,$$

and Φ_t is the solution of the homogeneous SDE

$$d\Phi_t = G(t)\Phi(t)dt + \sum_{i=1}^d H_i(t)\Phi_t dB_t^i$$

with initial value $\Phi_0 = I$. In particular, if c is a constant or a normal r.v., the solution of a linear SDE is a Gaussian process.

Remark *If $G(t) = G$ and $H_i(t) = H_i, 1 \le i \le d$, do not depend on t and $G, H_1, \cdots, H_d$ commute:*

$$GH_i = H_iG, H_iH_j = H_jH_i, \forall i, j,$$

then

$$\Phi_i = \exp\left\{\left(G - \sum_{i=0}^d H_i^2/2\right)t + \sum_{i=1}^d H_i B_t^i\right\}.$$

Example 2.2.67 *Consider the following SDE:*

$$dX_t = -cX_t dt + \sigma dB_t, \ X_0 = \xi.$$

Its unique solution is

$$X_t = e^{-ct}\left(\xi + \sigma \int_0^t e^{cs)} dB_s\right).$$

It is called the Ornstein-Uhlenbeck process. The SDE is called the Langevin equation, because it was originally introduced by Langevin (1908) to model the velocity of a physical Brownian particle. If ξ is a constant or a normal r.v., then (X_t) is a Gaussian process.

For a one-dimensional SDE (i.e. $m = d = 1$), the following result due to Yamada and Watanabe (1971) [Ref. 26] relaxes considerably the conditions on the existence and uniqueness of the solutions to (64.1).

Theorem 2.2.68 (Ref. 26) *Assume $m = d = 1$. In order for (64.1) to have a unique solution it suffices that b is continuous and Lipschitz in x and σ is continuous with the property*

$$|\sigma(t,x) - \sigma(t,y)| \le \rho(|x-y|),$$

for all x and y and t, where $\rho : \mathbb{R}_+ \to \mathbb{R}_+$ is a strictly increasing function with $\rho(0) = 0$ and for any $\epsilon > 0$,

$$\int_{(0,\epsilon)} \rho^{-2}(x)dx = \infty.$$

The following theorem states the *Feynman-Kac formula*, which provides a probabilistic representation for the solution of a parabolic differential equation.

Theorem 2.2.69 *Let u be a continuous, real valued function on $[0,T] \times I\!R^d$, of class $C^{1,2}$ on $[0,T) \times I\!R^d$, which is the solution of the Cauchy problem*

$$-\frac{pu}{pt} + ku = \mathcal{A}_t u + g, \quad (t,x) \in [0,T) \times I\!R^d \tag{69.1}$$

subject to the terminal condition

$$u(T,x) = f(x), \quad x \in I\!R^d. \tag{69.2}$$

Here $f : I\!R^d \to I\!R, k : I\!R^d \to I\!R_+$, and $g : [0,T] \times I\!R^d$ are continuous functions. Assume that u, f and g satisfy the polynomial growth condition in x:

$$|f(x)| + |g(x)| + \sup_{0\le t\le T} |u(t,x)| \le C(1+|x|^{2\mu}), \quad x \in I\!R^d,$$

for some constant $C>0$, $\mu \ge 1$. Then u admits the representation

$$\begin{aligned} u(t,x) &= I\!E^{\,t,x}\Big[f(X_T)\exp\Big\{-\textstyle\int_t^T k(\theta,X_\theta)d\theta\Big\} \\ &\quad + \textstyle\int_t^T g(s,X_s)\exp\Big\{-\textstyle\int_t^s k(\theta,X_\theta)d\theta\Big\}ds\Big], \end{aligned}$$

where $\{\mathbf{P}^{t,x}, t\ge 0, x \in I\!R^d\}$ is the family of probability measures associated with the Markov process (X_t). In particular, such a solution to (69.1) and (69.2) is unique.

If k does not depend on t, then

$$\begin{aligned} u(t,x) &= I\!E^{\,0,x}\Big[f(X_t)\exp\Big\{-\textstyle\int_0^t k(X_\theta)d\theta\Big\} \\ &\quad + \textstyle\int_0^t g(t-s,X_s)\exp\Big\{-\textstyle\int_0^s k(X_\theta)d\theta\Big\}ds\Big] \end{aligned}$$

is the unique solution of the Cauchy problem

$$\frac{\partial u}{\partial t} + ku = \mathcal{A}_t u + g, \quad (t,x) \in [0,T) \times I\!R^d$$

subject to the initial condition

$$u(0,x) = f(x), \quad x \in I\!R^d.$$

Now we consider the following stochastic differential equation driven by an n-dimensional semimartingale:

$$X = H + \sum_{i=1}^n (F_i X).Z^i, \tag{$*$}$$

where $Z = (Z^1,\cdots,Z^n)$ is an n-dimensional semimartingale where $Z_0 = 0$, $H = (H^1,\cdots,H^m)$ is an m-dimensional cadlag adapted process (i.e., each component H^i is a cadlag adapted process) and $F_i, 1 \le i \le n$, are mappings from the set of all m-dimensional cadlag adapted processes to the set of all n-dimensional locally bounded predictable processes such that for each stopping time T, $F_i(X^{T-})$ coincides with $F_i X$ on $]\!]0,T]\!]$. $X = (X^{\cdot}\cdots,X^m)$ is the unknown process. For instance, let $f_i(\omega,s,x_1,\cdots,x_m)$ be an n-dimensional measurable function on $\Omega \times \mathbf{R}_+ \times \mathbf{R}^m$ such that 1) for fixed $x_1,\cdots x_m$ and $s, f_i(\cdot,s,x_1,\cdots,x_m)$ is $\mathcal{F}_s$-measurable; 2) for almost all ω and for fixed $x_1,\cdots,x_m$,

$f_i(\omega,\cdot,x_1,\cdots,x_m)$ is left-continuous with right limits; 3) for almost all ω and all s $f_i(\omega,s,\cdot)$ is continuous. Put $(F_iX)_t = f_i(\omega,t,X^1_{t-},\cdots,X^m_{t-})$. Then F_i meets the above requirements.

The equation $(*)$ was first introduced and studied independently by Doléans-Dade (1976) [Ref. 27] and Protter (1997) [Ref. 28]. For further studies see Emery (1978) [Ref. 29], Métivier (1982) [Ref. 30] and Protter (1990) [Ref. 24].

The following theorem gives a sufficient condition for the existence and uniqueness of the solution of equation $(*)$.

Theorem 2.2.70 *If each F_i satisfies the following Lipschitz condition*

$$\mathbf{E}[(F_iX - F_iY)^{*2}_\infty] \le C\mathbf{E}\Big[\sum_{i=1}^m (X^i - Y^i)^{*2}_\infty\Big],$$

where C is a constant, then equation $(*)$ has a unique solution.

2.3 Stochastic Calculus on Semimartingales

In this section we present main results about stochastic calculus on semimartingales, which are: stochastic integration w.r.t. random measures, characteristics of semimartingales, calculus on Lévy processes, Girsanov's theorems, martingale representation theorems. The profound characterization theorem for semimartingales and some sufficient conditions for the uniform integrability of exponential martingales are also included. As in the previous sections, for those results which can be found in He et al. (1992) [Ref. 1] we omit the citations of the reference.

2.3.1 Stochastic Integration w.r.t. Random Measures

Let $(\Omega,\mathcal{F},(\mathcal{F}_t),\mathbf{P})$ be a stochastic basis, $\mathcal{O}$ and $\mathcal{P}$ be optional and predictable σ-algebras on $\Omega\times\mathbf{R}_+$. Let $E = \mathbf{R}^d\setminus\{0\}$ and $\mathcal{B}(E)$ be its Borel σ-field. We put

$$(\tilde\Omega,\tilde{\mathcal{F}}) = (\Omega\times\mathbf{R}_+\times E,\ \mathcal{F}\times\mathcal{B}(\mathbf{R}_+)\times\mathcal{B}(E)),$$
$$\tilde{\mathcal{O}} = \mathcal{O}\times\mathcal{B}(E), \quad \tilde{\mathcal{P}} = \mathcal{P}\times\mathcal{B}(E).$$

$\tilde{\mathcal{O}}$ (resp. $\tilde{\mathcal{P}}$) is called *optional* (resp. *predictable*) *σ-field* in $\tilde\Omega$. An $\tilde{\mathcal{O}}$ (resp. $\tilde{\mathcal{P}}$)-measurable function defined on $\tilde\Omega$ is called an *optional* (resp. *predictable*) *function* on $\tilde\Omega$.

In the sequel, for a σ-field $\mathcal{G}$ on an abstract set G, we denote by $\mathcal{G}^+$ (resp. $\mathcal{G}^b$) the set of all nonnegative (resp. bounded) $\mathcal{G}$-measurable functions on G.

Definition 2.3.1 *An extended real function μ defined on $\Omega\times(\mathcal{B}(\mathbf{R}_+)\times\mathcal{B}(E))$ is called a random measure on $\mathbf{R}_+\times E$, if*

i) for each fixed $\omega\in\Omega$, $\mu(\omega,\cdot)$ is a σ-finite measure on $\mathcal{B}(\mathbf{R}_+)\times\mathcal{B}(E)$ with $\mu(\omega,\{0\}\times E) = 0$,

ii) for each $\hat B\in\mathcal{B}(\mathbf{R}_+)\times\mathcal{B}(E)$, $\mu(\cdot,\hat B)$ is a r.v. on $(\Omega,\mathcal{F})$.

For a random measure μ, we define

$$M_\mu(\tilde B) = \mathbf{E}\Big[\int_{\mathbf{R}_+\times E} I_{\tilde B}(\omega,t,x)\mu(\omega,dt,dx)\Big], \quad \tilde B\in\tilde{\mathcal{F}}.$$

Then M_μ is a measure on $(\tilde\Omega,\tilde{\mathcal{F}})$, called the measure generated by μ. A random measure μ is said to be integrable if M_μ is a finite measure: $M_\mu(\tilde\Omega)<\infty$. μ is said to be optionally (resp. predictably) σ-integrable, if the restriction of M_μ on $\tilde{\mathcal{O}}$ (resp. $\tilde{\mathcal{P}}$) is a σ-finite measure.

The concept of random measure is a generalization of the concept of measure generated by an increasing process. In fact, let $A = (A_t(\omega))$ be an increasing process. Take $E = \{x_0\}$, a set of one point, and $\mathcal{B}(E) = \{\emptyset, E\}$, then

$$\mu(\omega, dt, dx) = dA_t(\omega)\delta_{x_0}(dx)$$

is a random measure, and

$$M_\mu(F \times [0,t] \times E) = \mu_A(F \times [0,t]), \quad F \in \mathcal{F},$$

where δ_{x_0} denotes the Dirac measure at x_0 and μ_A is the measure on $\mathcal{F} \times \mathcal{B}(\mathbf{R}_+)$ generated by A (see Definition 1.1.50).

If $W \in \tilde{\mathcal{F}}^+$, then

$$\nu(\omega, \hat{B}) = \int_{\hat{B}} W(\omega, t, x)\mu(\omega, dt, dx), \quad \hat{B} \in \mathcal{B}(\mathbf{R}_+) \times \mathcal{B}(E),$$

is a random measure. We denote it by $\nu = W.\mu$ or $d\nu = W d\mu$. If $W \in \tilde{\mathcal{F}}$ is such that for every $t \geq 0$, $\int_{[0,t]\times E} |W| d\mu < \infty$, we define a FV process $W * \mu$ by

$$W * \mu_t = \int_{[0,t]\times E} W d\mu, \quad t \geq 0.$$

Definition 2.3.2 *A random measure μ is called optional (resp. predictable), if for any $W \in \tilde{\mathcal{O}}^+$ (resp. $\tilde{\mathcal{P}}^+$), $W * \mu$ is an optional (resp. predictable) process.*

Theorem 2.3.3 *1) If μ is a random measure such that for every $t \geq 0$, $1 * \mu_t < \infty$, then μ is optional (resp. predictable) if and only if for every $B \in \mathcal{B}(E), 1_B * \mu = (\mu([0,t] \times B))_{t\geq 0}$ is optional (resp. predictable).*

2) If μ is an optional (resp. predictable) random measure and $W \in \tilde{\mathcal{O}}^+$ (resp. $\tilde{\mathcal{P}}^+$), then so is $\nu = W.\mu$.

3) Let μ and ν be two optionally (resp. predictably) σ-integrable optional (resp. predictable) random measures. If the restrictions of M_μ and M_ν on $\tilde{\mathcal{O}}$ (resp. $\tilde{\mathcal{P}}$) are identical, then $\mu = \nu$.

Theorem 2.3.4 *Let m be a measure on $(\tilde{\Omega}, \tilde{\mathcal{F}})$ such that its restriction on $\tilde{\mathcal{O}}$ (resp. $\tilde{\mathcal{P}}$) is σ-finite. There exists an optional (resp. predictable) random measure μ such that $m = M_\mu$ if and only if*

i) for any evanescent set $N \subset \Omega \times \mathbf{R}_+$, $m(N \times E) = 0$.

ii) for any $\tilde{A} \in \tilde{\mathcal{O}}$ (resp. $\tilde{\mathcal{P}}$) with $m(\tilde{A}) < \infty$ and bounded measurable process X,

$$m(XI_{\tilde{A}}) = m({}^{o}XI_{\tilde{A}}) \quad (\text{resp. } m(XI_{\tilde{A}}) = m({}^{p}XI_{\tilde{A}})).$$

In this case, such a random measure μ is uniquely determined by m.

Corollary 2.3.5 *Let μ be a predictably σ-integrable random measure. Then there exists a unique predictable random measure ν such that the restrictions of M_μ and M_ν coincide on $\tilde{\mathcal{P}}$.*

We call ν the *predictable projection* or *compensator* of μ, and denote it by μ^p or $\tilde{\mu}$.

Theorem 2.3.6 *Let μ be a predictably σ-integrable random measure. If $W \in \tilde{\mathcal{F}}^+$ is such that $\nu = W.\mu$ is a predictably σ-integrable random measure, then*

$$\tilde{\nu} = U.\tilde{\mu},$$

where U is the Radon-Nikodym derivative of $\frac{d\nu}{d\mu}$ on $\tilde{\mathcal{P}}$. We denote $U = M_\mu[W|\tilde{\mathcal{P}}]$.

Corollary 2.3.7 *Let μ be a predictably σ-integrable random measure. If $W \in \tilde{\mathcal{F}}$ is such that $X = W * \mu$ is a process with locally integrable variation, then X has the dual predictable projection: $\tilde{X} = U * \tilde{\mu}$, where $U = M_\mu[W|\tilde{\mathcal{P}}]$.*

Theorem 2.3.8 *Let μ be a predictably σ-integrable random measure. If $W \in \tilde{\mathcal{P}}^+$ and T is a predictable time, then*

$$\int_E W(T,x)\tilde{\mu}(\{T\},dx)I_{[T<\infty]} = \mathbf{E}\Big[\int_E W(T,x)\mu(\{T\},dx)I_{[T<\infty]}\Big|\mathcal{F}_{T-}\Big] \quad a.s..$$

Definition 2.3.9 *A random measure μ is called an integer-valued random measure if μ takes values in $\{0,1,2,\cdots,+\infty\}$, for all $t \geq 0$ $\mu(\{t\}\times E) \leq 1$, and μ is optional and optionally σ-integrable.*

An integer-valued random measure μ on $\mathbf{R}_+ \times E$ is called an extended Poisson measure relative to the filtration $(\mathcal{F}_t)$, if

(i) the measure m defined by $m(A) = \mathbf{E}[\mu(A)]$ is σ-finite;

(ii) for every $s \in \mathbf{R}$ and every $A \in \mathcal{B}(\mathbf{R}_+)\times\mathcal{B}(E)$ such that $A \subset (s,\infty)\times E$ and that $m(A) < \infty$, the variable $\mu(\cdot, A)$ is independent of $\mathcal{F}_s$.

We call m the intensity measure of μ.

If m satisfies $m(\{t\}\times E) = 0$ for each $t \in I\!R_+$, then μ is called a Poisson measure. If m has the form $m(dt,dx) = dt \times F(dx)$, where F is a σ-finite measure on $(E,\mathcal{B}(E))$, then μ is called a homogeneous Poisson measure.

Theorem 2.3.10 *A random measure μ is an integer-valued random measure if and only if*

$$\mu(\omega,dt,dx) = \sum_s I_D(\omega,s)\delta_{(s,\beta_s(\omega))}(dt,dx),$$

where D is a thin set, $\beta = (\beta_t)$ is an optional process.

Definition 2.3.11 *Let $X = (X_t)$ be a d-dimensional adapted cadlag process. Put*

$$\mu(\omega,dt,dx) = \sum_{s>0} I_{[\Delta X_s(\omega)\neq 0]}(s)\delta_{(s,\Delta X_s(\omega))}(dt,dx).$$

Then μ is a predictably σ-integrable integer-valued random measure, called the jump measure of X.

We now turn to define the stochastic integral of a predictable function W w.r.t. the compensated random measure $\mu - \nu$, where μ is a predictably σ-integrable integer-valued random measure and ν is its compensator. If $W * \mu \in \mathcal{A}_{\text{loc}}$, then $W * \nu \in \mathcal{A}_{\text{loc}}$ and we can define the stochastic integral of W w.r.t. $\mu - \nu$ by

$$W * (\mu - \nu) = W * \mu - W * \nu,$$

which is a local martingale. If W satisfies $\int_E |W(t,x)|\hat{\nu}(\{t\},dx) < \infty$ for all $t \geq 0$, we put

$$\begin{aligned}\hat{W}_t &= \int_E W(t,x)\nu(\{t\},dx), \quad t \geq 0,\\ \widetilde{W}_t &= \int_E W(t,x)\mu(\{t\},dx) - \hat{W}_t, \qquad t \geq 0.\end{aligned}$$

Clearly, $\widetilde{W} = (\widetilde{W}_t)$ and $\hat{W} = (\hat{W}_t)$ are all thin processes, and $\hat{W}$ is predictable. By Theorem 1.3.8, we have ${}^p(\widetilde{W}) = 0$. Put

$$\mathcal{G}(\mu) = \Big\{W \in \tilde{\mathcal{P}} : \forall t \geq 0 \int_E |W(t,x)|\nu(\{t\},dx) < \infty \text{ and } \sqrt{\sum_s (\widetilde{W}_s)^2} \in \mathcal{A}^+_{\text{loc}}\Big\}.$$

Then by Theorem 1.1.82, for every $W \in \mathcal{G}(\mu)$ there exists a unique purely discontinuous local martingale M such that $\Delta M = \widetilde{W}$. We call M the *stochastic integral* of W w.r.t. $\mu - \nu$, and denoted by $W * (\mu - \nu)$, or symbolically,

$$M_t = \int_{[0,t]\times E} W(s,x)(\mu(ds,dx) - \nu(ds,dx)), \quad t \geq 0.$$

It is worth mentioning that the single integral $W * \mu$ or $W * \nu$ may be not defined.

Theorem 2.3.12 *Let $W \in \mathcal{G}(\mu)$, $M = W * (\mu - \nu)$, and H be a predictable process. Then H is integrable w.r.t M if and only if $HW \in \mathcal{G}(\mu)$. In this case, we have*

$$H.M = (HW) * (\mu - \nu).$$

2.3.2 Characteristics of a Semimartingale

In this subsection, for any semimartingale we give its canonical representation based on its jump measure and introduce its characteristics. The latter is an important tool for studying semimartingales.

Lemma 2.3.13 *Let X be a d-dimensional special semimartingale, and $X = X_0 + M + A$ be its canonical decomposition. Let μ be the jump measure of X and ν be its compensator. Then $W^i(\omega, t, x) = x^i$ belongs to $\mathcal{G}(\mu)$, and the purely discontinuous martingale part of M is given by*

$$M^d = x * (\mu - \nu).$$

Theorem 2.3.14 *Let X be a d-dimensional semimartingale, μ be its jump measure, and ν be the compensator of μ. Then*

$$X = X_0 + \alpha + X^c + (xI_{[|x|\leq 1]}) * (\mu - \nu) + (xI_{[|x|>1]})) * \mu. \tag{14.1}$$

where X^c is the continuous martingale part of X, α is a d-dimensional predictable FV process with $\alpha_0 = 0$. Moreover, we have

$$\nu(\{0\} \times E) = \nu(\mathbf{R}_+ \times \{0\}) = 0, \tag{14.2}$$

$$(|x|^2 \wedge 1) * \nu \in \mathcal{A}^+_{\text{loc}}, \tag{14.3}$$

$$\Delta\alpha_t = \int_{|x|\leq 1} x\nu(\{t\}, dx). \tag{14.4}$$

(14.1) is called the *canonical representation* of semimaringale X.

Let X be a d-dimensional semimartingale. Denote $\beta = (\beta_{ij})$, where

$$\beta_{ij} = \langle (X^i)^c, (X^j)^c \rangle.$$

The triple (α, β, ν) is called the *local characteristics* (or simply, *characteristics*) of semimartingale X, associated with the *truncation function* $h(x) = xI_{[|x|\leq 1]}$.

Corollary 2.3.15 *Let $X = (X^1, \cdots, X^d)$ be a d-dimensional semimartingale with jump measure μ and Lévy measure ν. Then X is a special semimartingale if and only if*

$$(|x|I_{[|x|>1]}) * \mu = \sum_{s\leq \cdot} (|\Delta X_s| I_{[|\Delta X_s|>1]}) \in \mathcal{A}_{\text{loc}}.$$

If X is a special semimartingale, its canonical decomposition is

$$X = (X_0 + X^c + x * (\mu - \nu)) + (\alpha + (xI_{[|x|>1]}) * \nu).$$

Theorem 2.3.16 *Let X be a d-dimensional semimartingale having canonical representation (14.1), and f be a bounded C^2-function on $\mathbf{R}^d$. Then the canonical decomposition of special semimartingale $f(X)$ is given by $f(X) = M + A$, where*

$$\begin{aligned} M &= f(X_0) + \text{grad} f(X_-).X^c + [f(X_- + x) - f(X_-)] * (\mu - \nu), \\ A &= \text{grad} f(X_-).\alpha + \frac{1}{2}\text{grad}^2 f(X_-).\beta \\ &\quad + [f(X_- + x) - f(X_-) - x^\tau \text{grad} f(X_-) I_{[|x|\le 1]}] * \nu. \end{aligned}$$

In particular, the special semimartingale $Y = e^{iu^\tau X}(u \in \mathbf{R}^d)$ has the following canonical decomposition:

$$Y = Y_0 + (Y_-).N + (Y_-).H,$$

where

$$N = iu^\tau X^c + (e^{iu^\tau x} - 1) * (\mu - \nu),$$

$$H = iu^\tau \alpha - \frac{|u|^2}{2}\beta + (e^{iu^\tau x} - 1 - iu^\tau x I_{[|x|\le 1]}) * \nu.$$

Theorem 2.3.17 *Let M be a real locally square integrable martingale with characteristics (α, β, ν) and $M_0 = 0$. Then*

$$\langle M \rangle = \beta + x^2 * \nu.$$

2.3.3 Processes with Independent Increments and Lévy Processes

In this subsection, we present some results about processes with independent increments in terms of characteristic of semimartingales. In particular, we collect main results about processes with independent increments which are also semimartingales. As an application, we obtain the classical Lévy-Itô decomposition of a Lévy process.

A d-dimensional stochastic process (X_t) is said to be *stochastically continuous (or continuous in probability)* if for all $t \ge 0$ and $\epsilon > 0$,

$$\lim_{s \to t} \mathbf{P}(|X_s - X_t| > \epsilon) = 0.$$

The following theorem characterizes stochastic continuous semimartingales in terms of their characteristics.

Theorem 2.3.18 *Let X be a d-dimensional semimartingale with characteristics (α, β, ν). Then X is stochastically continuous if and only if for every $t > 0$ $\nu(\{t\} \times E) = 0$, a.s.. In this case, α is also stochastically continuous.*

A d-dimensional *process with independent increments* (in short: PII) on a stochastic basis $(\Omega, \mathcal{F}, (\mathcal{F}_t), \mathbf{P})$ is an adapted cadalag $\mathbf{R}^d$-valued process X such that $X_0 = 0$ and for all $0 \le s \le t$ the variable $X_t - X_s$ is independent of $\mathcal{F}_s$. If the distribution of $X_t - X_s$ only depends on the difference $t - s$, the PII X is called a *process with stationary independent increments* (in short: PSII) or *Lévy process.* Remark that the stationarity of the increments excludes the possibility of fixed jumps. So every Lévy process is stochastically continuous.

A Poisson process and a Wiener process are Lévy processes.

A stochastically continuous PII has no fixed jumps (i.e. $X_t = X_{t-}$, a.s., for all t).

Theorem 2.3.19 *Let (X_t) be a d-dimensional PII. Then X is also a semimartingale if and only if for each $u \in \mathbf{R}^d$, the function $t \mapsto \mathbf{E}[e^{iu^\tau X_t}]$ has finite variation over finite intervals.*

Theorem 2.3.20 *Let (X_t) be a d-dimensional semimartingale with $X_0 = 0$. Then it is a PII if and only if its characteristic (m, β, ν) is deterministic. In this case, the set of all fixed times of discontinuity is $J = \{t : \nu(\{t\} \times \mathbf{R}^d) > 0\}$, and for all $s \le t, u \in \mathbf{R}^d$, we have:*

$$\begin{aligned}
&\mathbf{E}[e^{iu^\tau (X_t - X_s)}] \\
= \ & \exp\Big\{iu^\tau(m_t - m_s) - \frac{1}{2}u^\tau(\beta_t - \beta_s)u \\
+ \ & \int_{[s,t]\times E} (e^{iu^\tau x} - 1 - iu^\tau x I_{[|x|\le 1]}) I_{J^c}(r)\nu(dr, dx)\Big\} \\
\times \ & \prod_{s<r\le t, r\in J} \Big\{e^{-iu^\tau \Delta m_r}\Big[1 + \int_{\mathbf{R}^d} (e^{iu^\tau x} - 1)\nu(\{r\} \times dx)\Big]\Big\}.
\end{aligned}$$

Corollary 2.3.21 *A d-dimensional process X is a Lévy process if and only if it is a semimartingale whose characteristics has the form*

$$m_t = bt, \ \ \beta_t(\omega) = Ct, \ \ \nu(\omega; dt, dx) = dtF(dx),$$

where $b \in \mathbf{R}^d$, C is a symmetric nonnegative $d \times d$ matrix, F is a positive measure on $\mathbf{R}^d$ with $F(\{0\}) = 0$ and $\int(|x|^2 \wedge 1)F(dx) < \infty$. We call F the Lévy measure of X. In this case, for all $t \in \mathbf{R}_+, u \in \mathbb{R}^d$ we have

$$\mathbf{E}[e^{iu^\tau X_t}] = \exp\Big\{t\Big(iu^\tau b - \frac{1}{2}u^\tau Cu + \int_E (e^{iu^\tau x} - 1 - iu^\tau x I_{[|x|\le 1]})F(dx)\Big)\Big\}.$$

In particular, we have

$$\mathbf{E}[e^{iu^\tau X_t}] = \exp\{t\psi(u)\},$$

where

$$\psi(u) = \log \mathbb{E}\,[e^{iu^\tau X_1}].$$

Theorem 2.3.22 *Let (X_t) be a d-dimensional PII without fixed jumps.. Put*

$$\varphi_t(u) = \mathbf{E}[e^{iu^\tau X_t}], \quad u \in \mathbf{R}^d.$$

Then for each $u \in \mathbf{R}^d$, $\varphi_t(u)$ is continuous, and

$$Z_t(u) = \frac{e^{iu^\tau X_t}}{\varphi_t(u)}, \ t \ge 0$$

is a martingale.

Theorem 2.3.23 *Let X be a d-dimensional PII without fixed jumps. Then there exists an $\mathbf{R}^d$-valued continuous deterministic function f such that $X - f$ is a semimartingale. If X itself is a semimartingale, then for all $u \in \mathbf{R}^d$, $\varphi_t(u)$ is a function of finite variation. Conversely, if for some $u \neq 0$, $\varphi_t(u)$ is a function of finite variation, then X is a semimartingale.*

Theorem 2.3.24 *Let X be a d-dimensional semimartingale without fixed jumps. Then X is a PII if and only if its local characteristics (α, β, ν) is deterministic.*

The following theorem gives a description of PII without fixed jumps.

Theorem 2.3.25 *Let X be a d-dimensional PII without fixed jumps. Then*

$$X_t = m_t + G_t + \int_{[0,t]\times[|x|>1]} x d\mu + \int_{[0,t]\times[|x|\leq 1]} x d(\mu - \nu), \tag{25.1}$$

where

1) m_t is a deterministic continuous function in $\mathbf{R}^d$ with $m_0 = 0$, and G is a centered d-dimensional continuous Gaussian PII with $G_0 = 0$ (hence, G is a martingale);

2) μ is the jump measure of X which has the following properties:

i) For any $\hat{B} \in \mathcal{B}(\mathbf{R}_+) \times \mathcal{B}(E)$ with $\nu(\hat{B}) < \infty$, $\mu(\hat{B})$ obeys a Poisson law with parameter $\nu(\hat{B})$. If $\hat{B} \subset]s,\infty[\times E$ for some $s \geq 0$, then $\mu(\hat{B})$ is independent of $\mathcal{F}_s$,

ii) $\forall n \geq 1$ and disjoint sets $\hat{B}_1, \cdots, \hat{B}_n \in \mathcal{B}(\mathbf{R}_+) \times \mathcal{B}(E), \mu(\hat{B}_1), \cdots, \mu(\hat{B}_n)$ are independent;

3) ν, the compensator of μ, equals $\mathbf{E}[\mu]$ and is a σ-finite measure on $\mathcal{B}(\mathbf{R}_+) \times \mathcal{B}(E)$, and for each $t \geq 0$, $\nu(\mathbf{R}_+ \times \{0\}) = \nu(\{t\} \times E) = 0$, $\int_{[0,t]\times E} (x^2 \wedge 1) d\nu < \infty$;

4) X_0, G and μ are independent;

5) X is quasi-left continuous.

In addition, we have

$$\varphi_t(u) = \exp\Big\{iu^\tau m_t - \frac{1}{2}u^\tau \beta_t u + \int_{[0,t]\times E} (e^{iu^\tau x} - 1 - iu^\tau x I_{[|x|\leq 1]}) d\nu\Big\} \tag{25.2}$$

where β_t is the covariance matrix of G_t, which is equal to the $d \times d$-matrix $(\langle G^i, G^j \rangle_t)$.

(25.1) is the famous *Lévy-Itô decomposition* of a PII without fixed jumps. We also call (m, β, ν) in (25.2) the *characteristics* of X. The law of the process X is uniquely determined by its initial law and its characteristics.

Theorem 2.3.26 *Let $X^{(1)}, \cdots, X^{(n)}$ be PII-semimartingales without fixed jumps. Then $X^{(1)}, \cdots, X^{(n)}$ are mutually independent if and only if*

$$[X^{(j)}, X^{(k)}] = 0, \ \ j \neq k, j, k = 1, \cdots, n.$$

Theorem 2.3.27 *Let X be a d-dimensional PII without fixed jumps with $X_0 = 0$.*

1) If X is a semimartingale and $\mathbf{E}[|X_t|] < \infty$, $t > 0$, then X is a special semimartingale.

2) If X is a special semimartingale, then $\mathbf{E}[|X_t|] < \infty$, $t > 0$.

3) If X is a local martingale, then X is a martingale.

4) If ΔX is bounded, then for all $p > 0$ and $0 \leq s < t$, $\mathbf{E}[|X_t - X_s|^p] < \infty$.

Theorem 2.3.28 *Let X be a d-dimensional PII without fixed jumps. Then X is a Lévy process if and only if (25.1) holds with $m_t = bt, G_t = \sigma B_t$, and $d\nu = dt \times F(dx)$, where $b \in I\!R^d$, σ is a $d\times d$-matrix, F is a σ- finite measure on $E = \mathbf{R}^d \setminus \{0\}$ with $\int (|x|^2 \wedge 1) F(dx) < \infty$, and (B_t) is a d-dimensional Brownian motion independent with μ. In this case we have*

$$F(A) = \mathbf{E}[\nu([0,1] \times A), \ \ A \in \mathcal{B}(E).$$

Theorem 2.3.29 *Let X be a d-dimensional Lévy process with jump measure μ and Lévy measure F. Let g be a Borel function on $\mathbf{R}_+ \times E$.*

1) If $\int_{\mathbf{R}_+\times E} g^+ ds dF < \infty$, then

$$\mathbf{E}\left[\exp\left\{\int_{\mathbf{R}_+\times E} g d\mu\right\}\right] = \exp\left\{\int_{\mathbf{R}_+\times E} \Big(e^g - 1\Big) ds dF\right\}.$$

2) If $\int_{\mathbf{R}_+\times E} \frac{g^2}{1+|g|} ds dF < \infty$, *then*

$$\mathbf{E}\left[\exp\left\{\int_{[0,t]\times E} g(s,x)[\mu(ds,dx) - F(dx)ds]\right\}\right] = \exp\left\{\int_{[0,t]\times E} \left(e^g - 1 - g\right) ds dF\right\}.$$

The following theorem generalizes the Lévy theorem on the martingale characterization of Brownian motion.

Theorem 2.3.30 *Let* X *be a process with* $X_0 = 0$. *Then* X *is a Gaussian PII without fixed jumps if and only if the following conditions are satisfied:*

i) There is a continuous deterministic function f *such that* $Y = X - f$ *is a continuous local martingale,*

ii) The process $\langle Y \rangle$ *is deterministic.*

The following theorem gives a martingale characterization of Poisson process (due to S. Watanable).

Theorem 2.3.31 *Let* X *be an adapted point process, i.e.,*

$$X = \sum_{n=1}^{\infty} I_{[\![T_n,\infty[\![},$$

where (T_n) *is an increasing sequence of stopping times such that* $T_n \uparrow \infty$ *and for each* $n \geq 0$ $T_n < \infty \Rightarrow T_n < T_{n+1} (T_0 = 0)$. *Then the following two statements are equivalent:*

1) X *is a (inhomogeneous) Poisson process (i.e.* $\forall 0 \leq s < t$, $X_t - X_s$ *has a Poisson distribution).*

2) There is a continuous increasing function Λ_t *such that* $X - \Lambda$ *is a local martingale with initial value zero.*

As a corollary of Theorem 1.3.31 we obtain a counterpart of Theorem 1.2.44.

Theorem 2.3.32 *Let* X *be an adapted point process with* $X_0 = 0$ *and a predictable increasing process* A *be its compensator. Let* (τ_t) *be the random time-change associated with* A. *Then* (X_{τ_t}) *is a standard Poisson process.*

2.3.4 Absolutely Continuous Changes of Probability

In this subsection we present Girsanov's theorems which describe how to transform the compensator of a random measure and the canonical representation of a semimartingale under absolutely continuous changes of probability. Some basic results on the uniform integrability of Doléans exponential martingales are presented as well. Finally, we present the characterization theorem for semimartingales which shows that semimartingales constitute the largest class of integrators w.r.t. which stochastic integrals of predictable processes can be reasonably defined.

Girsanov's Theorems

Let $\mathbf{Q}$ be a probability measure on $(\Omega, \mathcal{F}_\infty)$. If $\forall t \in \mathbb{R}_+$ the restriction of $\mathbf{Q}$ on $\mathcal{F}_t$ is absolutely continuous w.r.t. $\mathbf{P}$, we denote $\mathbf{Q} \ll_{\text{loc}} \mathbf{P}$, and put $Z_t = \left.\frac{d\mathbf{Q}}{d\mathbf{P}}\right|_{\mathcal{F}_t}$ (If on $\mathcal{F}_\infty$ we have $\mathbf{Q} \ll \mathbf{P}$, then $Z_t = \mathbf{E}[\frac{d\mathbf{Q}}{d\mathbf{P}} \mid \mathcal{F}_t])$. We always take the cadlag version of (Z_t). If $\mathbf{Q} \ll_{\text{loc}} \mathbf{P}$, then under $\mathbf{Q}$ almost all trajectories of Z are strictly positive functions.

Let $\mathbf{Q} \ll_{\text{loc}} \mathbf{P}$. Then for any stopping time T, we have $d\mathbf{Q} = Z_T I_{[T<\infty]} d\mathbf{P}$ on $\mathcal{F}_T \cap [T < \infty]$. Moreover, for any bounded stopping time T, Z^T is a uniformly integrable martingale. Let (X_t) be an $(\mathcal{F}_t)$-adapted cadlag process and T a finite stopping time. Then by the Bayes rule for conditional expectation it is easy to prove that $(ZX)^T$ is a uniformly integrable martingale under $\mathbf{P}$ if and only if X^T is a uniformly integrable martingale under $\mathbf{Q}$.

In this subsection we always assume that $\mathbf{Q} \ll_{\text{loc}} \mathbf{P}$ and denote $Z_t = \left.\frac{d\mathbf{Q}}{d\mathbf{P}}\right|_{\mathcal{F}_t}$.

The following lemma is essential for studying the changes of probability.

Lemma 2.3.33 *Let (X_t) be an $(\mathcal{F}_t)$-adapted cadlag process. Then (X_t) is a $\mathbf{Q}$-local martingale if and only if there exist finite stopping times $T_n \uparrow +\infty$, $\mathbf{Q}$- a.s., such that $(ZX)^{T_n}$ is a $\mathbf{P}$-local martingale. In particular, if $\mathbf{Q}$ and $\mathbf{P}$ are equivalent on $\mathcal{F}_\infty$, then X is a $\mathbf{Q}$-local martingale if and only if ZX is a $\mathbf{P}$-local martingale.*

The following theorem shows that the semimartingale property is reserved under locally absolutely continuous changes of the probability.

Theorem 2.3.34 *If X is a P-semimartingale, then X is a $\mathbf{Q}$-semimartingale and the quadratic variation $[X](\mathbf{Q})$ of X under $\mathbf{Q}$ is equal to the quadratic variation $[X](\mathbf{P})$ of X under $\mathbf{P}$, $\mathbf{Q}$-a.s..*

The following theorem is a Girsanov's theorem for local martingales.

Theorem 2.3.35 *Let X be a P-local martingale. Put*

$$T(\omega) = \inf\{t : Z_t(\omega) = 0\}\,,\ U = \Delta X_T I_{[\![T,\infty[\![}.$$

Under $\mathbf{Q}$ we define

$$Y_t = X_t - \int_0^t Z_s^{-1} d[X,Z]_s + \widetilde{U}_t\,, \qquad t \geq 0\,,$$

where $\widetilde{U}$ is the dual predictable projection of U under $\mathbf{P}$. Then Y is a $\mathbf{Q}$-local martingale.

Corollary 2.3.36 *Let (X_t) be a continuous P-local martingale. If there exists a P-local martingale L such that $Z = \mathcal{E}(L)$, then $Y_t = X_t - [X,L]_t$ is a $\mathbf{Q}$-local martingale.*

The following corollary is the classic form of Girsanov's theorem. If (H_t) is a deterministic function, the corresponding result is the Cameron-Martin theorem.

Corollary 2.3.37 *Let $(B_t, 0 \leq t \leq T)$ be an $(\mathcal{F}_t)$-Brownian motion and (H_t) an adapted measurable process such that $\int_0^T H_s^2 ds < \infty$, a.s.. For $0 \leq t \leq T$, Put*

$$\begin{aligned} Z_t &= \exp\{\int_0^t H_s dB_s - \frac{1}{2}\int_0^t H_s^2 ds\}\,, \\ B'_t &= B_t - \int_0^t H_s ds. \end{aligned}$$

Assume $\mathbf{E}[Z_T] = 1$, i.e. $(Z_t, 0 \leq t \leq T)$ is a martingale. We define a new probability measure $\mathbf{Q}$ by $d\mathbf{Q} = Z_T d\mathbf{P}$. Then under $\mathbf{Q}$ process $(B'_t, 0 \leq t \leq T)$ is an $(\mathcal{F}_t)$-Brownian motion.

Remark *Let (H_t) be an adapted measurable process such that $\forall t \in \mathbf{R}_+$, $\int_0^t H_s^2 ds < \infty$ a.s., and that (Z_t) is a martingale. Put $d\mathbf{Q}_t = Z_t d\mathbf{P}$, then it can be proved that there exists a unique probability measure $\mathbf{Q}$ on $(\Omega, \mathcal{F}_\infty)$ such that $\mathbf{Q}|_{\mathcal{F}_t} = \mathbf{Q}_t$, $\forall t \in I\!R_+$. Thus under $\mathbf{Q}$, (B'_t) is an $(\mathcal{F}_t)$-Brownian motion.*

Definition 2.3.38 *Let A be an adapted FV process. If there exist stopping times $S_n \uparrow +\infty$ $\mathbf{Q}$-a.s. such that for each n, A^{S_n} is of locally P-integrable variation, we say that the dual predictable projection of A under $\mathbf{P}$ exists $\mathbf{Q}$-a.s.. We denote still by $\widetilde{A}$ the $\mathbf{Q}$-a.s. defined process such that $\widetilde{A}^{S_n} = \widetilde{(A^{S_n})}$.*

Theorem 2.3.39 *Let X be a P-local martingale. Then X is a special $\mathbf{Q}$-semimartingle, if and only if the dual predictable projection of $[X, Z]$ under $\mathbf{P}$ exists $\mathbf{Q}$-a.s., (denoted by $\langle X, Z\rangle$). If it is the case, then*

$$X' = X - \frac{1}{Z_-}.\langle X, Z\rangle$$

is a $\mathbf{Q}$-local martingale.

Corollary 2.3.40 *If X is a $\mathbf{P}$ semimartingale and X^c is the continuous martingale part of X under $\mathbf{P}$, then $(X^c)' = X^c - \frac{1}{Z_-}. < X^c, Z >$ is the continuous martingale part of X under $\mathbf{Q}$.*

Theorem 2.3.41 *Suppose $X \in \mathcal{M}_{loc,0}(\mathbf{P})$ and $[X, Z] \in \mathcal{A}_{loc}(\mathbf{P})$. Let H be a predictable process such that $H \in L_m(X)$ under $\mathbf{P}$ (i.e., $\sqrt{H^2.[X]} \in \mathcal{A}_{loc}(\mathbf{P})$) and $[H.X, Z] \in \mathcal{A}_{loc}(\mathbf{P})$. Set $X' = X - \frac{1}{Z_-}.\langle X, Z\rangle$. Then under $\mathbf{Q}$, $H \in L_m(X')$, and*

$$H.X' = H.X - \frac{1}{Z_-}.\langle H.X, Z\rangle\,.$$

Theorem 2.3.42 *If X is a $\mathbf{P}$-semimartingale and H is a predictable process such that under $\mathbf{P}$, $H.X$ exists (denoted by $H_{\dot{\mathbf{P}}}X$). Then under $\mathbf{Q}$, $H.X$ exists (denoted by $H_{\dot{\mathbf{Q}}}X$), and $H_{\dot{\mathbf{Q}}}X$ is $\mathbf{Q}$-indistinguishable from $H_{\dot{\mathbf{P}}}X$.*

As an application of Theorem 1.3.42, we obtain the following property of the stochastic integral.

Theorem 2.3.43 *Let X and Y be two semimartingales, H and K two predictable processes such that $H.X$ and $K.Y$ exist. If $A \in \mathcal{F}$ is such that on A, X and Y are indistinguishable, H and K are indistinguishable, then on A, $H.X$ and $K.Y$ are indistinguishable as well.*

The following theorem is the Girsanov's theorem for random measures.

Theorem 2.3.44 *Let μ be a predictably σ-integrable integer-valued random measure and ν its compensator. Let M'_μ (resp. M'_ν) be the measure generated by μ (resp. ν) on $\tilde{\mathcal{F}}$ under $\mathbf{Q}$. Then*

1) Under $\mathbf{Q}$, M'_μ and M'_ν are σ-finite on $\tilde{\mathcal{P}}$ and $M'_\mu \ll M'_\nu$ on $\tilde{\mathcal{P}}$.

2) We denote by Y the Radon-Nikodym derivative of M'_μ w.r.t. M'_ν on $\tilde{\mathcal{P}}$, then

$$Z_-Y = M_\mu[Z|\tilde{\mathcal{P}}],$$

and the predictable projection ν' of μ under $\mathbf{Q}$ is given by $Y.\nu$.

The following theorem is the Girsanov's theorem for semimartingales.

Theorem 2.3.45 *Let X be a $\mathbf{P}$-semimartingale and*

$$X = X_0 + \alpha + X^c + (xI_{[|x|>1]}) * \mu + (xI_{[|x|\leq 1]}) * (\mu - \nu)$$

be its canonical representation under **P**, *where* μ *is the jump measure of* X *and* ν *is its compensator under* **P**. *Then under* **Q** *the canonical representation of* X *is*

$$X = X_0 + \alpha' + (X^c)' + (xI_{[|x|>1]}) * \mu + (xI_{[|x|\leq 1]}) * (\mu - \nu'),$$

where

$$(X^c)' = X^c - \frac{1}{Z_-}.\langle X^c, Z\rangle,$$

$$\alpha' = \alpha + \frac{1}{Z_-}.\langle X^c, Z\rangle + ((Y-1)xI_{[|x|\leq 1]}) * \nu,$$

and $\nu' = Y.\nu$ *is the compensator of* μ *under* **Q**, Y *being the Radon-Nikodym derivative of* M'_μ *w.r.t.* M'_ν *on* $\tilde{P}$.

Uniform Integrability of Exponential Martingales

Let M be a local martingale null at zero such that $\Delta M \geq -1$. Then its Doléans stochastic exponential $\mathcal{E}(M)$ is a nonnegative local martingale. In applications of the Girsanov theorem, it is important to know when $\mathcal{E}(M)$ is a uniformly integrable martingale. For continuous martingale M, the following results are well known, due to Novikov (1972) [Ref. 31] and Kazamaki (1977) [Ref. 32], respectively.

Theorem 2.3.46 (Ref. 3) *Let* M *be a continuous local martingale with* $M_0 = 0$. *If*

$$\mathbf{E}\Big[\exp\{\frac{1}{2}\langle M, M\rangle_\infty\}\Big] < \infty$$

or

$$\sup_t \mathbb{E}\left[\exp\{\frac{1}{2}M_t\}\right] < \infty,$$

then $\mathcal{E}(M)$ *is a uniformly integrable martingale.*

Remark *We always have* $\sup_t \mathbb{E}\,[\exp\{\frac{1}{2}M_t\}] \leq (\mathbf{E}[\exp\{\frac{1}{2}\langle M, M\rangle_t\}])^{1/2}$, *so that the Kazamaki's condition is weaker than the Novikov's condition. If* (M) *is a uniformly integrable martingale, then the Kazamaki's condition becomes* $\mathbf{E}[\exp\{\frac{1}{2}M_\infty\}] < \infty$.

Using Theorem 1.2.50 it is easy to prove the following

Theorem 2.3.47 (Ref. 23) *1) If* M *is a square integrable martingale and its oblique variation process* $\langle M, M\rangle$ *is bounded, then* $\mathcal{E}(M)$ *is a square integrable martingale.*

2) If M *is a maringale of integrable variation and the compensator of the process* $\sum_{s\leq t}|\Delta M_s|$ *is bounded, then* $\mathcal{E}(M)$ *is a martingale of integrable variation.*

The following result, due to Lépingle and Mémin (1978), generalizes Theorem 1.3.46 to the noncontinuous martingale case.

Theorem 2.3.48 (Ref. 23) *Let* M *be a local martingale with* $M_0 = 0$ *and* $\Delta M \geq -1$. *Let* μ *be the jump measure and of* M *and* ν *its compensator. Put*

$$T = \inf\{t : \Delta M_t = -1\} = \inf\{t : \mathcal{E}(M)_t = 0\}.$$

1) If

$$\mathbf{E}\left[\exp\left\{\frac{1}{2}\langle M^c, M^c\rangle_\infty + ((1+x)\log(1+x) - x) * \nu_\infty\right\}\right] < \infty,$$

then $\mathcal{E}(M)$ is a uniformly integrable martingale and $[\mathcal{E}(M)_\infty > 0] = [T = \infty]$, a.s..

2) If $\Delta M > -1$ and

$$\mathbf{E}\left[\exp\{\frac{1}{2}\langle M^c, M^c\rangle_\infty\}\prod_t (1+\Delta M_t)\exp\Big\{-\frac{\Delta M_t}{1+\Delta M_t}\Big\}\right] < \infty,$$

then $\mathcal{E}(M)$ is a uniformly integrable martingale and $\mathcal{E}(M)_\infty > 0$, a.s..

3) If M is uniformly integrable and $\Delta M \geq -1+\delta$ with $0 < \delta \leq 1$, and if

$$\mathbf{E}\left[\exp\frac{1}{1+\delta}M_\infty\right] < \infty,$$

then $\mathcal{E}(M)$ is a uniformly integrable martingale.

For further extensions see Yan (1982) [Ref. 33].

The Characterization Theorem for Semimartingales

We denote by L^1 and L^∞ the spaces of all integrable r.v. and all bounded r.v., respectively. If G and H are subsets of L^1, we denote $G - H = \{x - y : x \in G, y \in H\}$, and denote by $\overline{G}$ the closure of G in L^1.

The following theorem is due to Yan (1980) [Ref. 34].

Theorem 2.3.49 *Let K be a convex set in L^1 and $0 \in K$. Then the following three statements are equivalent:*

1) For all $\eta \in (L^1)^+ \setminus \{0\}$, there exists $c > 0$ such that $c\eta \notin \overline{K - (L^\infty)^+}$,

2) For all $A \in \mathcal{F}$ with $\mathbf{P}(A) > 0$ there exists $c > 0$ such that $cI_A \notin \overline{K - (L^\infty)^+}$,

3) There exists a $\zeta \in L^\infty$ such that $\zeta > 0$ a.s. and $\sup_{\xi\in K}\mathbf{E}[\zeta\xi] < \infty$, where $(L^1)^+$ and $(L^\infty)^+$ are the sets of all nonnegative elements of L^1 and L^∞ respectively.

Denote by $\mathcal{H}$ the collection of all bounded predictable processes of the following form:

$$H = \sum_{i=0}^{n-1} \xi_i I_{]t_i,t_{i+1}]},$$

where $0 = t_0 < t_1 < \cdots < t_n < \infty$, $\xi_i \in \mathcal{F}_{t_i}$, $|\xi_i| \leq 1$, $i = 0, 1, \cdots, n-1$. Let X be a process. For every $H \in \mathcal{H}$ define a process $J(X, H)$ as follows:

$$J(X,H)_t = \sum_{i=0}^{n-1} \xi_i (X_{t_{i+1}\wedge t} - X_{t_i \wedge t}), \quad t \geq 0.$$

Obviously, for every t the mapping $(X, H) \longmapsto J(X, H)_t$ is bilinear. Moreover, if X is a semimartingale, then $J(X, H) = H.X$.

Based on Theorem 1.3.49 one can prove the characterization theorem for semimartingales.

Theorem 2.3.50 *Let X be an adapted cadlag process. In order for X to be a semimartingale it is necessary and sufficient that for every sequence $(H^{(n)}) \subset \mathcal{H}$ and all $t \geq 0$ $\frac{1}{n}J(X, H^{(n)})_t \xrightarrow{\mathbf{P}} 0$.*

Corollary 2.3.51 *Let $\mathbf{G} = (\mathcal{G}_t)$ be a filtration satisfying the usual conditions such that for all $t \geq 0$ $\mathcal{G}_t \subset \mathcal{F}_t$. Suppose X is an $\mathbf{F}$-semimartingale and $\mathbf{G}$-adapted. Then X is a $\mathbf{G}$-semimartingale, $[X](\mathbf{F})$ and $[X](\mathbf{G})$ are indistinguishable.*

Theorem 2.3.52 *Let $\mathbf{G} = (\mathcal{G}_t)$ be a filtration satisfying the usual conditions such that for all $t \geq 0$ $\mathcal{G}_t \subset \mathcal{F}_t$. Suppose X is an $\mathbf{F}$-semimartingale and $\mathbf{G}$-adapted. Let H be a $\mathbf{G}$-predictable process such that H is integrable w.r.t. X and $\mathbf{G}$ (the integral is denoted by $H_{\dot{\mathbf{G}}}X$), then $H_{\dot{\mathbf{F}}}X$ and $H_{\dot{\mathbf{G}}}X$ are indistinguishable.*

2.3.5 Martingale Representation Theorems

Let $(\Omega, \mathcal{F}, \mathbf{P}, \mathcal{F}_t)$ be a stochastic basis. Let M be a d-dimensional local martingale with $M_0 = 0$. We denote by μ the jump measure of M and by ν the compensator of μ. If every real local $(\mathcal{F}_t)$-martingale can be represented as a stochastic integral of an $\mathbf{R}^d$-valued predictable process w.r.t. M, we say that M has the *predictable representability.*

For a d-dimensional semimartingale we will define its predictable representability in the weak sense in Definition 1.3.66.

Predictable Representability for Local Martingales

Let M be a d-dimensional local martingale with $M_0 = 0$. Recall that $L_m(M)$ is the collection of all $\mathbf{R}^d$-valued predictable processes which are integrable w.r.t. M in the sense of local martingales. Put

$$\mathcal{L}(M) = \{H.\ M : H \in L_m(M)\}, \quad \mathcal{L}^1(M) = \mathcal{L}(M) \cap \mathcal{H}^1.$$

Saying that M has the predictable representability means that $\mathcal{L}(M) = \mathcal{M}_{\text{loc},0}$. It is easy to see that M has the predictable representability if and only if $\mathcal{M}^c_{\text{loc}} = \mathcal{L}(M^c), \mathcal{M}^d_{\text{loc}} = \mathcal{L}(M^d)$ and $\mathcal{L}(M) = \mathcal{L}(M^c) + \mathcal{L}(M^d)$.

Let X be a stochastic process on a complete probability space $(\Omega, \mathcal{F}, \mathbf{P})$. Put

$$\mathcal{F}^0_t = \sigma(X_s, s \leq t), \quad \mathcal{F}_t(X) = \bigcap_{s>t} \mathcal{F}^0_s.$$

$\mathcal{F}_t(X)$ is called the *natural filtration* of X. We denote by $\mathcal{F}^P_t = \mathcal{F}_t(X) \vee \mathcal{N}$, where $\mathcal{N}$ is the σ-field generated be all P-null sets. Then $(\mathcal{F}^P_t(X))$ satisfies the usual conditions. We call it the *completed natural filtration* of X.

Theorem 2.3.53 *A Brownian motion (B_t) has the predictable representability w.r.t. its completed natural σ-filtration $(\mathcal{F}^P_t(B))$. In particular, any $(\mathcal{F}^P_t(B))$-local martingale are continuous.*

The following theorem is essential for characterization of the predictable representability of local martingales.

Theorem 2.3.54 *Let M be a d-dimensional local martingale with $M_0 = 0$. Then the following statements are equivalent:*

1) $\mathcal{L}(M) = \mathcal{M}_{\text{loc},0}$,
2) $\mathcal{M}^\infty_0 \subset \mathcal{L}(M)$,
3) For all $L \in \mathcal{M}_{\text{loc},0}$, $LM \in \mathcal{M}_{\text{loc},0} \Rightarrow L = 0$,
4) For all $N \in \mathcal{M}^\infty_0$, $NM \in \mathcal{M}_{\text{loc},0} \Rightarrow N = 0$.
Here $\mathcal{M}^\infty$ is the collection of all bounded martingales.

Theorem 2.3.55 *Let M be a d-dimensional continuous local martingale with $M_0 = 0$. Then the following statements are equivalent:*

1) $\mathcal{L}(M) = \mathcal{M}^c_{\text{loc},0}$,
2) $\mathcal{M}^{\infty,c} \subset \mathcal{L}(M)$ ($\mathcal{M}^{\infty,c}$ is the space of all bounded continuous martingales),
3) For all $L \in \mathcal{M}^c_{\text{loc},0}$, $LM \in \mathcal{M}_{\text{loc},0} \Rightarrow L = 0$,
4) For all $N \in \mathcal{M}^{\infty,c}$, $NM \in \mathcal{M}_{\text{loc},0} \Rightarrow N = 0$.

Lemma 2.3.56 *Let M be a d-dimensional local martingale with $M_0 = 0$. If M has the predictable representability, then for any stopping time T, M^T has the predictable representability w.r.t. $(\mathcal{F}_{t\wedge T})_{t\geq 0}$.*

The following theorem is due to Jacod and Yor (1977) [Ref. 35] (see also Ref. 11).

Theorem 2.3.57 *Let M be a d-dimensional local martingale with $M_0 = 0$. Put*

$$\Gamma = \left\{ \mathbf{P}' : \begin{array}{l} \mathbf{P} \textit{is a probability measure on } \mathcal{F}, \\ \mathbf{P}' = \mathbf{P}|_{\mathcal{F}_0} \textit{ and } M \in \mathcal{M}_{\text{loc},0}(\mathbf{P}') \end{array} \right\}.$$

Then the following statements are equivalent:
1) M has the predictable representability,
2) $\mathbf{P}' \in \Gamma, \mathbf{P}' \ll_{\text{loc}} \mathbf{P} \Rightarrow \mathbf{P}' = \mathbf{P}$,
3) $\mathbf{P}' \in \Gamma, \mathbf{P}' \ll \mathbf{P} \Rightarrow \mathbf{P}' = \mathbf{P}$,
4) $\mathbf{P}' \in \Gamma, \mathbf{P}' \sim \mathbf{P} \Rightarrow \mathbf{P}' = \mathbf{P}$,
5) $\mathbf{P}' \in \Gamma, \mathbf{P}' \sim \mathbf{P}, \frac{d\mathbf{P}'}{d\mathbf{P}} \in L^\infty \Rightarrow \mathbf{P}' = \mathbf{P}$.
Let M be a d-dimensional local martingale with $M_0 = 0$. Put

$$\Gamma(M) = \{\mathbf{P}' : \mathbf{P}' \text{ is a probability measure on } \mathcal{F} \text{ and } M \in \mathcal{M}_{\text{loc}}(\mathbf{P}')\}.$$

Denote by $\Gamma_e(M)$ the set of extreme points of $\Gamma(M)$, i.e., $\mathbf{P}' \in \Gamma_e(M) \Longleftrightarrow \mathbf{P}' \in \Gamma(M)$ and if $\mathbf{P}' = a\mathbf{P}_1 + (1-a)\mathbf{P}_2, \mathbf{P}_1, \mathbf{P}_2 \in \Gamma(M), 0 < a < 1$, then $\mathbf{P}' = \mathbf{P}_1 = \mathbf{P}_2$. However, in general we do not know whether $\Gamma(M)$ is a convex set or not.

Theorem 2.3.58 *Let M be a d-dimensional local martingale with $M_0 = 0$. Then the following statements are equivalent:*

1) M has the predictable representability, and $\mathcal{F}_0$ is the trivial σ-field $\mathcal{N}$ (i.e., the σ-field generated by all $\mathbf{P}$-null sets),

2) $\mathbf{P} \in \Gamma_e(M)$.

Theorem 2.3.59 *Let M be a d-dimensional local $\mathbf{P}$-martingale with $M_0 = 0$. Assume $\mathbf{Q} \ll_{\text{loc}} \mathbf{P}$, $[M, Z] \in (\mathcal{A}_{\text{loc}}(\mathbf{P}))^B$ and under $\mathbf{P}$ M has the predictable representability. Then under $\mathbf{Q}$, $M' = M - \frac{1}{Z_-}.\langle M, Z\rangle \in \mathcal{M}_{\text{loc},0}(\mathbf{Q})$ has the predictable representability as well.*

Definition 2.3.60 *Let ν be a predictable random measure with $\nu(\{0\} \times E) = 0$. If*

$$\nu(\omega, dt, dx) = G(\omega, t, dx) dB_t(\omega), \tag{60.1}$$

where i) B is a predictable increasing process with $B_0 = 0$, ii) for fixed $(\omega, t), G(\omega, t, \cdot)$ is a measure on $(E, \mathcal{B}(E))$, iii) for fixed $K \in \mathcal{B}(E), G(\cdot, \cdot, E)$ is a predictable process, then (60.1) is called a predictable decomposition of ν. Moreover, if

$$I_A . B = 0, \quad A = \{(\omega, t) : G(\omega, t, E) = 0\}, \tag{60.2}$$

the predictable decomposition (60.1) is said to be canonical.

Lemma 2.3.61 *If μ is the jump measure of an adapted cadlag process, then its compensator ν has the canonical predictable decomposition.*

Theorem 2.3.62 *Let $M \in \mathcal{M}_{\text{loc},0}$ and (α, β, ν) be its characteristics. If the canonical predictable decomposition of ν is given by (61.1), then M has the predictable representability if and only if $\mathcal{M}^c_{\text{loc},0} = \mathcal{L}(M^c), \mathcal{M}^d_{\text{loc}} = \mathcal{L}(M^d)$ and $\mathbf{P}(d\beta_t \perp dB_t) = 1$.*

Corollary 2.3.63 *Assume $M \in \mathcal{M}^2_{\text{loc},0}$. Then M has the predictable representability if and only if*

$$\mathcal{M}^c_{\text{loc},0} = \mathcal{L}(M^c), \mathcal{M}^d_{\text{loc}} = \mathcal{L}(M^d) \text{ and } \mathbf{P}(d\langle M^c\rangle_t \perp d\langle M^d\rangle_t) = 1.$$

Theorem 2.3.64 *Let X be a step process and $\mathbf{F} = (\mathcal{F}_t)$ be the complete natural filtration of X. Assume X is quasi-left-continuous and $X \in \mathcal{A}_{\text{loc}}$. Then the following statements are equivalent:*

1) $M = X - \tilde{X}$ has the predictable representability,

2) For any stopping time T, we have $\mathcal{F}_T = \mathcal{F}_{T-}$,

3) The compensator ν of the jump measure of X has the form: $\nu(dt, dx) = \delta_{H_t}(dx)\Lambda(dt)$, where H is a predictable process and $\Lambda(dt) = \nu(dt, E)$.

Corollary 2.3.65 *Assume X is a point process and $\mathbf{F} = (\mathcal{F}_t)$ is the complete natural filtration of X. Then $M = X - \tilde{X}$ has the predictable representability.*

Predictable Representability for Semimartingales

Definition 2.3.66 *Let X be a d-dimensional semimartingale, μ, X^c and (α, β, ν) be its jump measure, continuous martingale part and local characteristics respectively. Write*

$$\mathcal{K}(\mu) = \{W * (\mu - \nu) : W \in \mathcal{G}(\mu)\}.$$

If $\mathcal{M}^c_{\text{loc},0} = \mathcal{L}(X^c)$ and $\mathcal{M}^d_{\text{loc}} = \mathcal{K}(\mu)$, or equivalently

$$\mathcal{M}_{\text{loc},0} = \mathcal{L}(X^c) + \mathcal{K}(\mu)$$

(the right side is the linear sum of two vector spaces), we say that X has the predictable representability in weak sense.

Theorem 2.3.67 *Assume $X \in \mathcal{M}_{\text{loc},0}$ and X has the predictable representability. Then X has the predictable representability in weak sense as well.*

Theorem 2.3.68 *Let X be a d-dimensional semimartingale. Then the following statements are equivalent:*

1) $\mathcal{M}^d_{\text{loc}} = \mathcal{K}(\mu)$,

2) $\mathcal{O} = \mathcal{P} \vee \sigma\{\Delta X\}$,

3) For all $M \in \mathcal{M}^d_{\text{loc}}$, $M_\mu[\Delta M|\tilde{\mathcal{P}}] = 0 \Rightarrow M = 0$,

4) For all $M \in \mathcal{M}^{\infty,d}$, $M_\mu[\Delta M|\tilde{\mathcal{P}}] = 0 \Rightarrow M = 0$,

5) For any totally inaccessible time T, $[\![T]\!] \subset [\Delta X \neq 0]$; For any stopping time T, $\mathcal{F}_T = \mathcal{F}_{T-} \vee \sigma\{\Delta X_T I_{[T<\infty]}\}$.

The following theorem is a consequence of Theorem 1.3.68 and 1.3.55.

Theorem 2.3.69 *Let X be a d-dimensional semimartingale. Then the following statements are equivalent:*

1) X has the predictable representability in weak sense,

2) For all $M \in \mathcal{M}_{\text{loc},0}$, $\langle M^c, X^c\rangle = 0$ and $M_\mu[\Delta M|\tilde{\mathcal{P}}] = 0 \Rightarrow M = 0$,

3) For all $N \in \mathcal{M}^\infty_0$, $\langle N^c, X^c\rangle = 0$ and $M_\mu[\Delta N|\tilde{\mathcal{P}}] = 0 \Rightarrow N = 0$.

Theorem 2.3.70 *Let X be a d-dimensional semimartingale and (α, β, ν) be its local characteristics. Put*

$$\Gamma = \left\{ \mathbf{P}' : \begin{array}{c} \mathbf{P}' \text{ is a probability measure on } \mathcal{F}, \mathbf{P}' = \mathbf{P}|_{\mathcal{F}_0}, \\ X \in \mathcal{S}(\mathbf{P}') \text{ and } (\alpha, \beta, \nu) \text{ is still the predictable} \\ \text{characteristics of } X \text{ under } \mathbf{P}' \end{array} \right\}.$$

Then the following statements are equivalent:
1) X has the predictable representability in weak sense,
2) $\mathbf{P}' \in \Gamma, \mathbf{P}' \ll_{\text{loc}} \mathbf{P} \Rightarrow \mathbf{P}' = \mathbf{P}$,
3) $\mathbf{P}' \in \Gamma, \mathbf{P} \ll \mathbf{P} \Rightarrow \mathbf{P}' = \mathbf{P}$,
4) $\mathbf{P}' \in \Gamma, \mathbf{P}' \sim \mathbf{P} \Rightarrow \mathbf{P}' = \mathbf{P}$,
5) $\mathbf{P}' \in \Gamma, \mathbf{P}' \sim \mathbf{P}, \frac{d\mathbf{P}'}{d\mathbf{P}} \in L^\infty \Rightarrow \mathbf{P}' = \mathbf{P}$.

From Theorem 1.3.70 we obtain immediately the next result about the predictable representability in weak sense for step processes.

Theorem 2.3.71 *Assume X is a step process and $\mathbf{F} = (\mathcal{F}_t)$ is the complete natural filtration of X: $\mathbf{F} = \mathbf{F}^P(X)$. Then X has the predictable representability in weak sense. In particular, each $\mathbf{F}$-local martingale is purely discontinuous.*

The following theorem, due to Jacod (1977) [Ref. 36], is a general result about the predictable representability in weak sense for semimartingales.

Theorem 2.3.72 *Let X be a d-dimensional semimartingale and (α, β, ν) be its local characteristics. Put*

$$\Gamma = \left\{ \mathbf{P}' : \begin{array}{c} \mathbf{P}' \text{ is a probability on } \mathcal{F},\ X \in \mathcal{S}(\mathbf{P}') \text{ and } (\alpha, \beta, \nu) \\ \text{is still the local characteristics of } X \text{ under } \mathbf{P}' \end{array} \right\}.$$

Then the following statements are equivalent:
1) X has the predictable representability in weak sense and $\mathcal{F}_0$ is the trivial σ-field.
2) $\mathbf{P}$ is an extreme point of Γ.

Theorem 2.3.73 *Assume $X \in \mathcal{S}$ has the predictable representability in weak sense. If $\mathbf{Q} \ll_{\text{loc}} \mathbf{P}$, then under $\mathbf{Q}$ X has the predictable representability in weak sense as well.*

Now we present some results on the predictable representabilty in weak sense for PII.

Theorem 2.3.74 *Let X be a d-dimensional PII-semimartingale. Let $\mathbf{F} = \mathbf{F}^P(X)$. Then X has the predictable representability in weak sense.*

The following result, due to Xue (1992)[Ref. 37], is more convenient for applications.

Theorem 2.3.75 (Ref. 37) *Let X be a d-dimensional PII-semimartingale. Let $\mathbf{F} = \mathbf{F}^P(X)$. Let X^c denote the continuous martingale part of X and X^d denote the purely discontinuous local martingale $x * (\mu - \nu)$, where μ is the jump measure of X and ν is the compensator of μ. If X has no fixed jumps, or equivalently, if X is quasi-left-continuous, then the $2d$-dimensional local martingale (X^c, X^d) has the predictable representability w.r.t. $\mathbf{F}$.*

Theorem 2.3.76 *Let X be a Lévy process and $\mathbf{F} = \mathbf{F}^P(X)$. Assume X is a martingale. Then X has the predictable representability w.r.t. $\mathbf{F}$ if and only if X is a standard Wiener process or a compensated Poisson process, up to a constant factor.*

Theorem 2.3.77 *Let μ be an integer-valued random measure on $\mathbf{R}_+ \times E$ such that for all $\omega \in \Omega, t \in \mathbb{R}_+, \mu(\omega, [0,t] \times E) < \infty$. Put*

$$\mathcal{F}_s^0 = \sigma(\mu([0,r] \times B) : r \le s, B \in \mathcal{B}(E)), \quad \mathcal{F}_t = \bigcap_{s>t} \mathcal{F}_s.$$

Then all $\mathbf{F}^P$-local martingales have the form

$$M_t = M_0 + W * \mu - W * \nu,$$

where ν is the compensator of μ and W is a $\widetilde{\mathcal{P}}$- measurable function such that $|W| * \mu$ is locally integrable.

References

1. SW He, JG Wang, JA Yan. Semimartingale theory and stochastic calculus. Beijing New York: Science Press and CRC Press Inc., 1992.

2. JL Doob. Stochastic Processes. New York: Wiley and Sons, 1953.

3. AN Shiryaev. Essentials of Stochastic Finance, Facts, Models, Theory. World Scientific, 1999.

4. JL Snell. Applications of martingale systems theorems, Trans. Amer. Math. Soc. 73: 293-312, 1952.

5. PA Meyer. A decomposition theorem for supermartingales, Illinois J. Math. 6:193-205, 1962.

6. K Itô. Stochastic integrals. Proc. Imp. Acad. Tokyo 20:519-524, 1944.

7. K Itô. On a formula concerning stochastic differentials. Nagoya Math. J. 3:55-65, 1951.

8. H Kunita, S Watanabe. On square integrable martingales. Nagoya Math. J. 39:209-245, 1967.

9. C Doléans-Dade, PA Meyer. Intégrales stochastiques par rapport aux martingales locales, Sém. Probab. IV. LN in Math. 124, Springer, 1970, pp 77-107.

10. PA Meyer. Un cours sur les intégrales stochastiques. Sém. Probab. X, LN in Math. 511, Springer, 1976, pp 246-400.

11. J Jacod. Calcul stochastique et problém de martingales. LN in Math. 714, Springer, 1979.

12. J Jacod. Intégrales stochastiques par rapport à une semimartingale vectorielle et changements de filtration. Sém. Probab. XIV, LN In Math. 784, Springer, 1980, pp 161-172.

13. JA Yan. Remarques sur l'intégrale stochastique de processus non bornés. Sém. Probab. XIV, LN in Math. 784, Springer, 1980, pp 128-139.

14. E Emery. Compensation de processus V.F. non localement intégrables, Sém. Prob. XIV, LN in Math. 784, Springer, 1980, pp 140-147.

15. JP Ansel, C Stricker. Couverture des actifs contingents et prix maximum. Ann Inst Henri Poincaré 30: 303-315, 1994.

16. DO Kramkov. Optional decomposition of supermartingales and hedging contingent claims in incomplete security markets. Probab. Theory and Related Fields 105(4):459-479, 1996.

17. H Föllmer, YuM Kabanov. Optional decomposition and Lagrang multipliers. Finance and Stochastics 2(1):69-81, 1998.

18. JA Yan. Some remarks on the theory of stochastic integration. Sém. Probab. XXIV, LN in Math. 1485, Springer, 1991, pp 95-107.

19. E Lenglart. Relation de domination entre deux processus. Ann. Inst. Henri Poincaré, Section B 13:171-179, 1977.

20. C Stricker, M Yor. Calcul stochastique dépendant d'un paramètre. Z.W. 45:109-133, 1978.

21. FB Knight. A reduction of continuous, square-integrable martingales to Brownian motion, Sém. Probab. V, LN in Math. 190, Springer, 1971, pp 19-31.

22. C Doléans-Dade. Quelques applications de la formule de changement de variables pour les semi-martingales. Z. W. 16 :181-194, 1970.

23. D Lépingle, J Mémin. Sur l'integrabilité uniforme des martingales exponentielles, Z. W. 42 :175-203, 1978.

24. PE Protter. Stochastic integration and differential equation: A new approach. Springer, 1990.

25. I Karatzas, SE Shreve. Brownian motion and stochastic calculus. 2nd ed., Springer, 1991.

26. T Yamada, S Watanabe. On the uniqueness of solutions of stochastic differential equations. J. of Math. of Kyoto Univ. **11**, 155-167, 1971.

27. C Doléans-Dade. Existence and unicity of solutions of stochastic differential equations, Z. W. **36**, 93-101, 1976.

28. PE Protter. On the existence, uniqueness, convergence, and explositions of solutions of systems of stochastic integral equations. Ann. Prob. 5:243-261, 1977.

29. E Emery. Stabilité des solutions des équations différentielles stochastiques, application aux intégrales multiplicatives stochastiques. Z. W. 41:241-262, 1978.

30. M Métivier. Semimartingales: A Course on Stochastic Processes, de Gruyter, Berlin New York , 1982.

31. AA Novikov. On an identity for stochastic integrals. Theory probab. Appl. 17:717-720, 1972.

32. N Kazamaki. On a problem of Girsanov, Tôhoko Math. J., 29:597-600, 1977.

33. JA Yan. A propos de l'intégrabilité uniforme des martingales exponentielles. Sém. Probab. XVI, LN in Math. 920, Springer, 1982, pp 338-347.

34. JA Yan. Caractérisation d'une classe d'ensembles convexes de L^1 on $\mathcal{H}^1$. Sém. Probab. XIV, LN in Math. 784, Springer, 1980, pp 220-222.

35. J Jacod, M Yor. Etude des solutions extrémales et representation integrable de solutions pour certains problèmes de martingales. Z.W. 38:83-125, 1977.

36. J Jacod. A general theorem of representation for martingales. Proc. Symp. Pure Math. 31:37–53, 1977.

37. XX Xue. Martingale representation for a class of processes with independent increments, and its applications, in: Applied Stochstic Analysis, I Karatzas and D Ocone (Eds.), LN in Control and Inform. Sciences 117, Springer, 1992, pp 279-311.

38. C Dellacherie, PA Meyer. Probabilities and potential. Amstrerdam New York: North-Holland, 1978.

39. C Dellacherie, PA Meyer. Probabilities and potential B. Amstrerdam New York: North-Holland, 1982.

40. RJ Elliott. Stochastic Calculus and Applications. New York: Springer, 1982.

41. E Emery. Une topologie sur l'espace des semimartingales. Sém. Prob. XIII, LN in Math. 721, Springer, 1979, pp 260-280.

42. IV Girsanov. On transforming a certain class of stochastic processes by absolutely continuous substitution of measures, Theory Probab. Appl. 5(3):285-301, 1962.

43. J Jacod, AN Shiryaev. Limit theorems for stochastic processes. Springer, 1987.

44. T Jeulin. Semi-martingales et grossissement d'une filtration. LN in Math. 873, Springer, 1980.

45. O Kallenberg. Foudation of Modern Probability. Springer, 1997.

46. P Lévy. Processus Stochastiques et Mouvement Brownien. Paris: Guthier-villars, 1948.

47. B Øksendal. Stochastic Differential Equations. 5th ed., Springer, 1998.

48. D Revuz, M Yor. Continuous martingale and Brownian motion. 2nd ed., Springer, 1994.

49. ICG Rogers, D Williams. Diffusions, Markov Processes, and Martingales, Vol. 2 *Ito Calculus*. Wiley & Sons, 1987.

Chapter 3

White Noise Theory

Hui-Hsiung Kuo

Department of Mathematics
Louisiana State University
Baton Rouge, LA 70803

3.1 Introduction

3.1.1 What is white noise?

White noise is a sound with equal intensity at all frequencies within a broad band. Rock music, the roar of a jet engine, and the noise at a stock market are just a few examples of white noise. We use the word "white" to describe this kind of noise because of its similarity to "white light" which is made up of all different colors (frequencies) of light combined together.

In applied science white noise is often taken as an idealization of phenomena involving sudden and extremely large fluctuations. Mathematically, one can think of white noise as a stochastic process $z(t)$ such that $z(t)$'s are independent and for each t, $z(t)$ has mean 0 and variance ∞ in the sense that

$$E\big(z(t)z(s)\big) = \delta(t-s), \tag{3.1.1}$$

where δ is the Dirac delta function. Thus it seems to be reasonable to claim that we can define an integral $\int_a^b f(t)z(t)\,dt$ such that

$$E\left(\int_a^b f(t)z(t)\,dt\right)^2 = \int_a^b \int_a^b f(t)f(s)E\big(z(t)z(s)\big)\,dtds = \int_a^b f(t)^2\,dt.$$

But what is the definition of the integral $\int_a^b f(t)z(t)\,dt$?

3.1.2 White noise as the derivative of a Brownian motion

White noise can be regarded as the derivative of a Brownian motion. But what is a Brownian motion? As is well-known, Robert Brown made microscopic observations in 1827 that small particles contained in the pollen of plants, when immersed in a liquid, exhibit highly irregular motions. This highly irregular motion is called a Brownian motion. Mathematically, a

Brownian motion is a continuous stochastic process $B(t)$ with independent increments and for $t < s$, $B(s) - B(t)$ is a Gaussian random variable with mean 0 and variance $s - t$. Thus $E\big(B(s) - B(t)\big)^2 = s - t$ and so it is plausible to say that

$$|B(s) - B(t)| \approx (s-t)^{1/2}, \qquad for\ small\ s - t.$$

But then this means that the derivative of $B(t)$, or the white noise $\dot{B}(t)$, does not exist. Hence the integral $\int_a^b f(t)\dot{B}(t)\,dt$ does not seem to be defined at all.

When $f(t)$ is a function of bounded variation, we can use the integration by parts formula to define the integral $\int_a^b f(t)\dot{B}(t)\,dt$ by

$$\int_a^b f(t)\dot{B}(t)\,dt = f(t)B(t)\Big]_a^b - \int_a^b B(t)\,df(t),$$

where the integral in the right-hand side is a Riemann-Stieltjes integral. However, we cannot use this definition for general $f \in L^2(a,b)$.

If we combine the white noise $\dot{B}(t)$ and dt together to get $\dot{B}(t)\,dt = dB(t)$ as an integrator, then the integral $\int_a^b f(t)\,dB(t)$ can be defined for all $f \in L^2(a,b)$. This integral, called a Wiener integral, is a Gaussian random variable with mean 0 and variance $\|f\|^2$.

But still the white noise, as the derivative of a Brownian motion, does not exist. Is it possible to give a mathematically sound definition of white noise $\dot{B}(t)$? Is it possible to define the integral $\int_a^b f(t)\dot{B}(t)\,dt$ directly without rewriting $\dot{B}(t)\,dt$ as $dB(t)$? Before we pursue these questions further we give a simple example in the next section to show how white noise can be used.

3.1.3 The use of white noise — a simple example

Consider the following second-order differential equation

$$\frac{d^2x}{dt^2} + x = F(t).$$

This differential equation describes the motion of an undamped harmonic oscillator with external force $F(t)$. It has fundamental solutions $\sin t$ and $\cos t$. A particular solution is informally given by

$$x_p(t) = -\cos t \int_0^t F(s)\sin s\,ds + \sin t \int_0^t F(s)\cos s\,ds. \tag{3.1.2}$$

What are the integrals in the right-hand side of this equation? The answer depends on what the function $F(t)$ is. Let us consider some special cases.

1. $F(t) = \dot{B}(t)$.

We can use the integration by parts formula to derive

$$\begin{aligned} \int_0^t \dot{B}(s)\sin s\,ds &= B(t)\sin t - \textstyle\int_0^t B(s)\cos s\,ds \\ \int_0^t \dot{B}(s)\cos s\,ds &= B(t)\cos t + \textstyle\int_0^t B(s)\sin s\,ds. \end{aligned}$$

Thus the particular solution in Equation (3.1.2) for $F(t) = \dot{B}(t)$ is given by

$$x_p(t) = \int_0^t B(s)\cos(t-s)\,ds.$$

One can easily check that $x_p(t)$ is a Gaussian random variable with mean 0 and variance $2^{-1}t - 4^{-1}\sin(2t)$.

2. $F(t) = \ddot{B}(t)$ (the second derivative of $B(t)$).

Again we can informally apply the integration by parts formula to Equation (3.1.2) with $F(t) = \ddot{B}(t)$ to get

$$x_p(t) = B(t) - \dot{B}(0)\sin t + \int_0^t B(s)\sin(s-t)\,ds.$$

This time $x_p(t)$ contains a bad term $\dot{B}(0)\sin t$. Fortunately, we can drop it because $\sin t$ is a fundamental solution. Hence a particular solution is given by

$$x_p(t) = B(t) + \int_0^t B(s)\sin(s-t)\,ds.$$

But is the term $\dot{B}(0)\sin t$ really that bad? Can we give a mathematically sound meaning for $\dot{B}(0)$?

3. $F(t) =$ a positive colored noise.

Consider a positive colored noise $C(t)$ such that $C(t)$'s are independent and for each t, $C(t)$ is positive and has infinite fluctuations. One may think that $|\dot{B}(t)|$ is such a noise. As it turns out $|\dot{B}(t)|$ has no renormalization. However, the following renormalization of $e^{\dot{B}(t)}$ is such a noise:

$$C(t) = \mathcal{N}e^{\dot{B}(t)} \overset{def}{=\!=} \frac{e^{\dot{B}(t)}}{Ee^{\dot{B}(t)}},$$

where $Ee^{\dot{B}(t)}$ is the informal expectation and $\mathcal{N}$ denotes a renormalization. We will explain what $\mathcal{N}e^{\dot{B}(t)}$ is in 3.1.5 and 12.2.3.

If we take the external force $F(t) = \mathcal{N}e^{\dot{B}(t)}$, then $x_p(t)$ is given by

$$x_p(t) = -\cos t\int_0^t \mathcal{N}e^{\dot{B}(t)}\sin s\,ds + \sin t\int_0^t \mathcal{N}e^{\dot{B}(t)}\cos s\,ds.$$

3.1.4 White noise as a generalized stochastic process

We now have an informal understanding of white noise and its use in a simple example. But then, what is a mathematically sound definition of white noise? In order to motivate the concept, we make a comparison between functions and stochastic processes. An (ordinary) function on $\mathbf{R}$ is a function $f(t)$ for $t \in \mathbf{R}$. A generalized function is a function $f(\xi)$ depending linearly on test functions ξ. For example, the Dirac delta function δ is the generalized function such that

$$\delta(\xi) = \xi(0), \qquad \xi\text{: } test function.$$

On the other hand, an (ordinary) stochastic process is a function $X(t)$ such that for each t, $X(t)$ is a random variable. Therefore, by a generalized stochastic process, we mean a function $X(\xi)$ depending linearly on test functions ξ such that for each ξ, $X(\xi)$ is a random variable.

So here is a mathematically sound definition of white noise, namely, a white noise is a generalized stochastic process $X(\xi)$ such that for each test function ξ, the random variable $X(\xi)$ is Gaussian with mean 0 and variance $\int_{\mathbf{R}} \xi(t)^2\,dt$.

What is the relationship between this definition of white noise and the informal one in 10.7.65? Note that $X(\xi+\eta) = X(\xi) + X(\eta)$ for any test functions ξ and η. Square both sides of the equality, take expectation, and then simplify to get

$$E\big(X(\xi)X(\eta)\big) = \int_{\mathbf{R}} \xi(t)\eta(t)\,dt. \tag{3.1.3}$$

Suppose $X(\xi) = \int_{\mathbf{R}} \xi(t)z(t)\,dt$. Then it follows informally from Equation (3.1.3) that

$$\int_{\mathbf{R}^2} \xi(t)\eta(s)E\big(z(t)z(s)\big)\,dtds = \int_{\mathbf{R}} \xi(t)\eta(t)\,dt, \qquad \forall \xi, \eta.$$

Hence we must have $E\big(z(t)z(s)\big) = \delta(t-s)$, which is exactly Equation (3.1.1).

Does white noise as such a generalized stochastic process $X(\xi)$ exist? Take two independent Brownian motions $B_1(t)$ and $B_2(t)$ for $t \geq 0$ and define

$$B(t) = \begin{cases} B_1(t), & \text{if } t \geq 0; \\ B_2(-t), & \text{if } t < 0. \end{cases}$$

Then the Wiener integral $X(\xi) = \int_{\mathbf{R}} \xi(t)\,dB(t)$ defines a white noise $\dot{B}(t)$ as a generalized process. How about the second derivative $\ddot{B}(t)$ of $B(t)$? We can regard it as a generalized stochastic process defined by

$$Y(\xi) = -\int_{\mathbf{R}} \xi'(t)\,dB(t).$$

How about the colored noise $\mathcal{N}e^{\dot{B}(t)}$ in 3.1.3? It is much more complicated to define $\mathcal{N}e^{\dot{B}(t)}$ as a generalized stochastic process.

3.1.5 White noise as an infinite dimensional generalized function

In the previous section we defined the white noise $\dot{B}(t)$ and its derivative $\ddot{B}(t)$ are generalized stochastic processes. In order to see how to define the colored noise $\mathcal{N}e^{\dot{B}(t)}$ as a generalized stochastic process, let us consider the product $\dot{B}(t)\dot{B}(s)$. It is a generalized stochastic process defined by

$$X(\xi,\eta) = \int_{\mathbf{R}^2} \xi(t)\eta(s)\,dB(t)dB(s) + \int_{\mathbf{R}} \xi(t)\eta(t)\,dt,$$

where the first integral in the right-hand side is a Wiener integral of order 2. Thus X acts on test functions of two variables. It is plausible that this is also the case for the renormalization of $\dot{B}(t)^2$. Similarly, the renormalization of $\dot{B}(t)^n$ is some kind of generalized stochastic process $X(\varphi)$ acting on test functions φ of n variables. Therefore, it is reasonable to say that $\mathcal{N}e^{\dot{B}(t)}$ is some kind of generalized stochastic process $X(\varphi)$ acting on test functions φ of infinitely many variables!

Thus in order to study the white noise $\dot{B}(t)$ and functions of $\dot{B}(t)$ such as $\mathcal{N}e^{\dot{B}(t)}$, it is necessary to define test functions of infinitely many variables. This is the motivation for T. Hida to introduce the theory of white noise in 1975 [18].

When white noise is defined as a generalized stochastic process, it is regarded as a whole $\dot{B}$ and for each t, the quantity $\dot{B}(t)$ still has no meaning. However, in Hida's theory of white noise, each $\dot{B}(t)$ is meaningful as a generalized function on an infinite dimensional space.

The functions $\mathcal{N}e^{\dot{B}(t)}$ and the renormalization of $\dot{B}(t)^n$ are also generalized functions on the same space.

The collection $\{\dot{B}(t);\ t \in \mathbf{R}\}$ of generalized functions is taken as a continuum coordinate system. With this system, we can take time propagation explicitly into account in many applications.

Nowadays Hida's theory of white noise is regarded as an infinite dimensional distribution theory. In this chapter we will give a brief survey of this theory and describe some applications. For details and more information, see [40]. Other excellent sources of the white noise theory and applications can be found in [21] [22] [27] [48].

3.2 White noise as a distribution theory

3.2.1 Finite dimensional Schwartz distribution theory

A complex-valued function ξ on $\mathbf{R}$ is called *rapidly decreasing* if it is a smooth function and for any nonnegative integers j and k,

$$\lim_{|x|\to\infty} \left|x^j \xi^{(k)}(x)\right| = 0.$$

Let A be the operator $A = -d^2/dx^2 + x^2 + 1$. Obviously, if ξ is rapidly decreasing, then $A\xi$ is also rapidly decreasing.

Let $\mathcal{S}(\mathbf{R})$ denote the space of all rapidly decreasing functions on $\mathbf{R}$. It is easy to see that $\mathcal{S}(\mathbf{R}) \subset L^2(\mathbf{R})$. For each integer $p \geq 0$, define an inner product norm by

$$|\xi|_p = |A^p \xi|_0, \qquad \xi \in \mathcal{S}(\mathbf{R}),$$

where $|\cdot|_0$ is the $L^2(\mathbf{R})$-norm. Then we have a sequence $\{|\cdot|_p\}_{p=0}^{\infty}$ of norms on $\mathcal{S}(\mathbf{R})$. This sequence of norms generates a topology on $\mathcal{S}(\mathbf{R})$. Thus $\mathcal{S}(\mathbf{R})$ is a topological vector space. It is called the *Schwartz space of test functions* on $\mathbf{R}$. Its dual space $\mathcal{E}'(\mathbf{R})$ is called the space of *generalized functions* (or *tempered distributions*) on $\mathbf{R}$.

The one dimensional Schwartz distribution theory is the study of test functions in $\mathcal{S}(\mathbf{R})$ and generalized functions in $\mathcal{S}'(\mathbf{R})$ and continuous linear operators acting on these spaces.

On the space $\mathcal{S}(\mathbf{R})$ of test functions, there are continuous linear operators such as differential operator, Laplacian operator, translation operator, scaling operator, multiplication, convolution, Fourier transform. These operators can be extended by continuity to continuous operators on the space $\mathcal{S}'(\mathbf{R})$ of tempered distributions. The extensions use the translation invariance of the Lebesgue measure.

In general, a complex-valued function ξ on $\mathbf{R}^n$ is called *rapidly decreasing* if it is smooth and for any multi-indices $\mathbf{j} = (j_1, j_2, \cdots, j_n)$ and $\mathbf{k} = (k_1, k_2, \cdots, k_n)$,

$$\lim_{|x|\to\infty} \left|x^{\mathbf{j}} D^{\mathbf{k}} \xi(x)\right| = 0,$$

where $x^{\mathbf{j}} = x_1^{j_1} x_2^{j_2} \cdots x_n^{j_n}$ and $D^{\mathbf{k}} = \partial^{k_1+k_2+\cdots+k_n}/\partial x_1^{k_1} \partial x_2^{k_2} \cdots \partial x_n^{k_n}$.

Let $\mathcal{S}(\mathbf{R}^n)$ denote the space of all rapidly decreasing functions on $\mathbf{R}^n$. Then $\mathcal{S}(\mathbf{R}^n) \subset L^2(\mathbf{R}^n)$. For each integer $p \geq 0$, define an inner product norm by

$$|\xi|_p = |(A^p)^{\otimes n} \xi|_0, \qquad \xi \in \mathcal{S}(\mathbf{R}^n),$$

where $|\cdot|_0$ is the $L^2(\mathbf{R}^n)$-norm. This sequence of norms generates a topology on $\mathcal{S}(\mathbf{R}^n)$. The resulting topological vector space $\mathcal{S}(\mathbf{R}^n)$ is called the *Schwartz space of test functions* on $\mathbf{R}^n$.

Its dual space $\mathcal{E}'(\mathbf{R}^n)$ is called the space of *generalized functions* (or *tempered distributions*) on $\mathbf{R}^n$.

We also have those continuous linear operators mentioned above on the space $\mathcal{S}(\mathbf{R}^n)$ and their extensions to the space $\mathcal{S}'(\mathbf{R}^n)$. Again the extensions use the translation invariance of the Lebesgue measure.

White noise distribution theory is a generalization of the Schwartz distribution theory to infinite dimensional spaces. It is well-known that the Lebesgue measure does not exist in infinite dimensional spaces. A natural measure to use for infinite dimensional analysis is the standard Gaussian measure.

3.2.2 White noise space

When T. Hida introduced the theory of white noise in 1975, he used the infinite dimensional space $\mathcal{S}'(\mathbf{R})$ of tempered distributions as a base space. As is well-known that finite dimensional theory is built with the Lebesgue measure. But the Lebesgue measure does not exist in infinite dimensional spaces. Therefore, we need to look for another measure on $\mathcal{S}'(\mathbf{R})$.

In the Schwartz space $\mathcal{S}(\mathbf{R})$ we have a sequence of norms $\{|\cdot|_p\}_{p=0}^{\infty}$. Let $\mathcal{S}_p(\mathbf{R})$ be the completion of $\mathcal{S}(\mathbf{R})$ with respect to the norm $|\cdot|_p$. The topological vector space $\mathcal{S}(\mathbf{R})$ is a nuclear space, which means that for any $q \geq 0$ there exists some $p > q$ such that the inclusion mapping $\mathcal{S}_p(\mathbf{R}) \hookrightarrow \mathcal{S}_q(\mathbf{R})$ is a Hilbert-Schmidt operator. This assertion follows from the fact that the operator A has eigenvalues $2n, n = 1, 2, \ldots,$ (see page 17 in [40]).

Since $\mathcal{S}(\mathbf{R})$ is a nuclear space, we can apply the Minlos theorem to obtain a probability measure μ on its dual space $\mathcal{S}'(\mathbf{R})$ such that

$$\int_{\mathcal{S}'(\mathbf{R})} e^{i\langle x,\xi\rangle}\, d\mu(x) = e^{-\frac{1}{2}|\xi|_0^2}, \qquad \xi \in \mathcal{S}(\mathbf{R}).$$

The probability space $(\mathcal{S}'(\mathbf{R}), \mu)$ is called a *white noise space*. The measure μ is called the *white noise measure* on $\mathcal{S}'(\mathbf{R})$. It is also called the standard Gaussian measure on $\mathcal{S}'(\mathbf{R})$.

The reason that $\mathcal{S}'(\mathbf{R})$ is called a white noise space is because elements in $\mathcal{S}'(\mathbf{R})$ can be regarded as the "sample paths" of white noise. To see this, define

$$X(\xi)(x) = \langle x, \xi\rangle, \qquad \xi \in \mathcal{S}(\mathbf{R}),\ x \in \mathcal{S}'(\mathbf{R}).$$

Then $X(\cdot)$ is a generalized stochastic process and for each $\xi \in \mathcal{S}(\mathbf{R})$, the random variable $X(\xi)$ defined on $\mathcal{S}'(\mathbf{R})$ is Gaussian with mean 0 and variance $|\xi|_0^2$. Hence $X(\cdot)$ defines a white noise. In informal notation, we have

$$X(\xi)(x) = \int x(t)\xi(t)\, dt.$$

On the other hand, since $\dot{B}(t)$ is regarded as white noise, we also have

$$X(\xi)(\cdot) = \int \xi(t)\dot{B}(t)\, dt.$$

It follows from the last two equations above that $x = \dot{B}$ and so elements in $\mathcal{S}'(\mathbf{R})$ can be regarded as $\dot{B}$, i.e., sample paths of white noise.

3.2.3 Hida's original idea

Recall from 12.2.1 that we have the following triple for the Schwartz distribution theory on $\mathbf{R}^r$

$$\mathcal{S}(\mathbf{R}^r) \subset L^2(\mathbf{R}^r) \subset \mathcal{S}'(\mathbf{R}^r).$$

We can follow the same idea to extend Schwartz distribution theory to infinite dimensional spaces. The space $\mathbf{R}^r$ is replaced by $\mathcal{S}'(\mathbf{R})$ and the Lebesgue measure is replaced by the white noise measure μ. Then the space $L^2(\mathbf{R}^r)$ is replaced by the space $L^2(\mathcal{S}'(\mathbf{R}))$, denoted by (L^2) for simplicity. Thus we need to find a nuclear space V such that

$$V \subset (L^2) \subset V^*,$$

where the inclusion mappings are continuous and V is dense in (L^2). The space V is a space of test functions and its dual space V^* is a space of generalized functions. Such a triple is often called a *Gel'fand triple.*

Note that for each $\xi \in \mathcal{S}(\mathbf{R})$, $\langle \cdot, \xi \rangle$ is defined everywhere on $\mathcal{S}'(\mathbf{R})$ and has the Gaussian distribution with mean 0 and variance $|\xi|_0^2$. If $h \in L^2(\mathbf{R})$, then $\langle \cdot, h \rangle$ is defined almost everywhere on $\mathcal{S}'(\mathbf{R})$ and has the Gaussian distribution with mean 0 and variance $|h|_0^2$.

For $a > b$, we let $1_{[a,b)} = -1_{[b,a)}$ by convention. For each $t \in \mathbf{R}$, define

$$B(t) = \langle \cdot, 1_{[0,t)} \rangle.$$

Then $B(t), t \geq 0$ is a Brownian motion. Moreover, we can define multiple Wiener integrals $I_n(f)$ with respect to $B(t)$ for any $f \in L^2(\mathbf{R}^n)$.

One way to construct V and V^* is to utilize the Wiener-Itô decomposition theorem for the space (L^2). The theorem says that every $\varphi \in (L^2)$ can be represented by

$$\varphi = \sum_{n=0}^{\infty} I_n(f_n), \qquad f_n \in \widehat{L}^2(\mathbf{R}^n), \tag{3.2.4}$$

where $\widehat{L}^2$ denotes the symmetric L^2-functions. The (L^2)-norm of φ is given by

$$\|\varphi\|_0 = \left(\sum_{n=0}^{\infty} n! |f_n|_0^2 \right)^{1/2}.$$

Here we have used the same notation $|\cdot|_0$ to denote the norm on $L^2(\mathbf{R}^n)$ for any n.

Question: What is $\dot{B}(t)$ for each $t \in \mathbf{R}$?

Whatever the definition $\dot{B}(t)$ is, we must have

$$\dot{B}(t) = \lim_{\Delta \to 0} \frac{B(t+\Delta) - B(t)}{\Delta} = \lim_{\Delta \to 0} \frac{1}{\Delta} \langle \cdot, 1_{[t,t+\Delta)} \rangle = \lim_{\Delta \to 0} \langle \cdot, \Delta^{-1} 1_{[t,t+\Delta)} \rangle.$$

Let $\varphi_\Delta = \langle \cdot, \Delta^{-1} 1_{[t,t+\Delta)} \rangle$. Observe that $\Delta^{-1} 1_{[t,t+\Delta)}$ does not converge in $L^2(\mathbf{R})$ as $\Delta \to 0$. Hence φ_Δ does not converge in (L^2) as $\Delta \to 0$. Thus in order to answer the above question, we need to find a weaker norm on (L^2) so that φ_Δ would converge with respect to this weaker norm. What weaker norm should we take? Note that $\Delta^{-1} 1_{[t,t+\Delta)}$ converges to the Dirac delta function δ_t at t in the distribution sense as $\Delta \to 0$. Actually, the convergence can be shown to be in the dual space $\mathcal{S}'_p(\mathbf{R})$ of $\mathcal{S}_p(\mathbf{R})$ for any $p > 5/12$ (see page 21 of [40].) This shows that the norms $|\cdot|_{-p}$ on $\mathcal{S}'_p(\mathbf{R})$ for $p \geq 0$ should be used to generate weaker norms on the space (L^2) in order to get a space V^* of generalized functions on the white noise space $\mathcal{S}'(\mathbf{R})$. By this choice of topology, $\dot{B}(t)$ is a generalized function for each $t \in \mathbf{R}$. Symbolically,

$$\dot{B}(t) = I_1(\delta_t).$$

Question: What is $\dot{B}(t)^2$ for each $t \in \mathbf{R}$?

We would define $\dot{B}(t)^2 = \lim_{\Delta\to 0}\varphi_\Delta^2$. Note that $E\varphi_\Delta^2 = \Delta^{-1} \to +\infty$ as $\Delta \to 0^+$. Thus even if we consider a weaker norm on (L^2), we could not expect φ_Δ^2 to converge with respect to this weaker norm. We need a *renormalization.* Note that

$$\varphi_\Delta^2 - \frac{1}{\Delta} = I_2\big((\Delta^{-1}1_{[t,t+\Delta)}) \otimes (\Delta^{-1}1_{[t,t+\Delta)})\big)$$

Observe that $(\Delta^{-1}1_{[t,t+\Delta)}) \otimes (\Delta^{-1}1_{[t,t+\Delta)})$ does not converge in $L^2(\mathbf{R}^2)$. However, it converges to $\delta_t \otimes \delta_t$ in $\mathcal{S}_p'(\mathbf{R}) \otimes \mathcal{S}_p'(\mathbf{R})$ for any $p > 5/12$. Thus if we use the norms $|\cdot|_{-p}$ to generate a space of generalized functions, then the renormalization of $\dot{B}(t)^2$, denoted by $:\dot{B}(t)^2:$, is a generalized function for each $t \in \mathbf{R}$. Symbolically,

$$:\dot{B}(t)^2: = I_2(\delta_t \otimes \delta_t).$$

Question: What is $e^{\dot{B}(t)}$ for each $t \in \mathbf{R}$?

Consider e^{φ_Δ}. It can be easily checked that $Ee^{\varphi_\Delta} = e^{1/(2\Delta)}$. Hence

$$\frac{e^{\varphi_\Delta}}{Ee^{\varphi_\Delta}} = e^{\varphi_\Delta - \frac{1}{2\Delta}} = \sum_{n=0}^{\infty} \frac{1}{n!} I_n\big((\Delta^{-1}1_{[t,t+\Delta)})^{\otimes n}\big).$$

By the same idea as above for $\dot{B}(t)$ and $\dot{B}(t)^2$, we can use the norms $|\cdot|_{-p}$ to generate a space of generalized functions. Then as $\Delta \to 0$, $e^{\varphi_\Delta}/Ee^{\varphi_\Delta}$ converges to a generalized function, denoted by $\mathcal{N}e^{\dot{B}(t)}$. Symbolically,

$$\mathcal{N}e^{\dot{B}(t)} = \sum_{n=0}^{\infty} \frac{1}{n!} I_n\big(\delta_t^{\otimes n}\big).$$

3.2.4 Spaces of test and generalized functions

In the last section we gave an intuitive motivation as how to define generalized functions on the white noise space $\mathcal{S}'(\mathbf{R})$. We will be more precise in this section to define the space of generalized functions. In fact, we will first define the space of test functions (just as in the finite dimensional Schwartz distribution theory) and then define its dual space as the space of generalized functions.

Let $\varphi \in (L^2)$. By the Wiener-Itô decomposition theorem, φ can be uniquely represented by Equation (3.2.4). For each integer $p \geq 0$, define

$$\|\varphi\|_p = \left(\sum_{n=0}^{\infty} n!|(A^p)^{\otimes n} f_n|_0^2\right)^{1/2}. \tag{3.2.5}$$

Let $(\mathcal{S}_p) = \{\varphi \in (L^2);\ \|\varphi\|_p < \infty\}$. Then $(\mathcal{S}_p)$ is a Hilbert space with norm $\|\cdot\|_p$. Let $(\mathcal{S})$ be the projective limit of the family $\{(\mathcal{S}_p);\ p \geq 0\}$. This projective limit can be thought of as $\cap_{p\geq 0}(\mathcal{S}_p)$ with the topology given by the sequence of norms $\{\|\cdot\|_p\}_{p=0}^\infty$.

Elements in $(\mathcal{S})$ are called *test functions* on the white noise space $\mathcal{S}'(\mathbf{R})$. Let $(\mathcal{S})^*$ be the dual space of $(\mathcal{S})$. Elements in $(\mathcal{S})^*$ are called *generalized functions* on $\mathcal{S}'(\mathbf{R})$. Thus we have the following Gel'fand triple

$$(\mathcal{S}) \subset (L^2) \subset (\mathcal{S})^*.$$

This Gel'fand triple is one example of the triple $V \subset (L^2) \subset V^*$ we mentioned in the beginning of the previous section. It is an infinite dimensional analogue of the Gel'fand triple $\mathcal{S}(\mathbf{R}^r) \subset L^2(\mathbf{R}^r) \subset \mathcal{S}'(\mathbf{R}^r)$ for the finite dimensional Schwartz distribution theory.

As it turns out $(\mathcal{S})^* = \cup_{p\geq 0}(\mathcal{S}_p)^*$ and for each $p \geq 0$, $(\mathcal{S}_p)^*$ is the completion of (L^2) with respect to the following norm $\|\cdot\|_{-p}$:

$$\|\varphi\|_{-p} = \left(\sum_{n=0}^{\infty} n!|(A^{-p})^{\otimes n} f_n|_0^2\right)^{1/2}.$$

The topology on $(\mathcal{S})^*$ is the inductive limit topology, namely, the finest locally convex topology such that for each p the inclusion mapping from $(\mathcal{S}_p)^*$ into $(\mathcal{S})^*$ is continuous. A sequence Φ_n converges in $(\mathcal{S})^*$ if and only if there exists some $p \geq 0$ such that $\Phi_n \in (\mathcal{S}_p)^*$ for all n and Φ_n converges in $(\mathcal{S}_p)^*$.

Note that for any $q \geq p$, we have the following continuous inclusion mappings

$$(\mathcal{S}) \hookrightarrow (\mathcal{S}_q) \hookrightarrow (\mathcal{S}_p) \hookrightarrow (L^2) \hookrightarrow (\mathcal{S}_p)^* \hookrightarrow (\mathcal{S}_q)^* \hookrightarrow (\mathcal{S})^*.$$

Each element Φ in $(\mathcal{S}_p)^*$ can be represented by

$$\Phi = \sum_{n=0}^{\infty} I_n(F_n), \qquad F_n \in \widehat{\mathcal{S}'_p}(\mathbf{R}^n),$$

where $I_n(F_n)$ can be regarded as a generalized multiple Wiener integral. Moreover,

$$\|\Phi\|_{-p} = \left(\sum_{n=0}^{\infty} n!|(A^{-p})^{\otimes n} F_n|_0^2\right)^{1/2}. \tag{3.2.6}$$

For any φ represented by Equation (3.2.4), we have

$$\langle\langle \Phi, \varphi \rangle\rangle = \sum_{n=0}^{\infty} n!\langle F_n, f_n \rangle,$$

where $\langle\langle \cdot, \cdot \rangle\rangle$ is the bilinear pairing of $(\mathcal{S})^*$ and $(\mathcal{S})$, and $\langle \cdot, \cdot \rangle$ denotes the bilinear pairing of $\mathcal{S}'(\mathbf{R}^n)$ and $\mathcal{S}(\mathbf{R}^n)$ for any n.

3.2.5 Examples of test and generalized functions

In this section we give some simple examples of test and generalized functions. More example of generalized functions will be given later in 12.5.119.

Example 3.2.1 *The white noise $\dot{B}(t)$ is a generalized function in $(\mathcal{S})^*$ for each $t \in \mathbf{R}$. In fact, $\dot{B}(t) = I_1(\delta_t)$ and*

$$\|\dot{B}(t)\|_{-p} = |\delta_t|_{-p} < \infty \qquad \textit{for any } p > \tfrac{5}{12}.$$

Thus $\dot{B}(t) \in (\mathcal{S}_p)^*$ for any $p > \frac{5}{12}$.

Example 3.2.2 *The kth derivative $B^{(k)}(t)$ is a generalized Brownian motion in $(\mathcal{S})^*$ for each $t \in \mathbf{R}$. It is given by*

$$B^{(k)}(t) = I_1\Big((-1)^{k-1}\delta_t^{(k-1)}\Big),$$

where the derivative of δ_t is in the distribution sense. We have

$$\big\|\dot{B}^{(k)}(t)\big\|_{-p} = \big|\delta_t^{(k-1)}\big|_{-p} < \infty \qquad \textit{for any } p > \tfrac{5}{12} + \tfrac{k-1}{2}.$$

Hence $B^{(k)}(t) \in (\mathcal{S}_p)^*$ for any $p > \frac{5}{12} + \frac{k-1}{2}$.

Example 3.2.3 *The renormalization* $:\dot{B}(t)^n:$ *is a generalized function in* $(\mathcal{S})^*$ *given by* $:\dot{B}(t)^n:= I_n(\delta_t^{\otimes n})$ *and*

$$\|:\dot{B}(t)^n:\|_{-p} = \sqrt{n!}\,|\delta_t|_{-p}^n.$$

Hence the generalized function $:\dot{B}(t)^n:$ *belongs to the space* $(\mathcal{S}_p)^*$ *for any* $p > \frac{5}{12}$.

Example 3.2.4 *The renormalization* $\mathcal{N}e^{\dot{B}(t)}$ *is given by*

$$\mathcal{N}e^{\dot{B}(t)} = \sum_{n=0}^{\infty} \frac{1}{n!} I_n(\delta_t^{\otimes n}).$$

Hence by Equation (3.2.6), we have

$$\|\mathcal{N}e^{\dot{B}(t)}\|_{-p} = \left(\sum_{n=0}^{\infty} n! \frac{1}{(n!)^2} |\delta_t|_{-p}^{2n}\right)^{1/2} = \exp\left[\frac{1}{2}|\delta_t|_{-p}^2\right].$$

Therefore, $\mathcal{N}e^{\dot{B}(t)}$ *is a generalized function in the space* $(\mathcal{S}_p)^*$ *for any* $p > \frac{5}{12}$.

Example 3.2.5 *(Donsker's delta function) The Dirac delta function* δ_a *at* a *has the following expansion in the distribution sense (page 357 in [40])*

$$\delta_a(x) = \frac{1}{\sqrt{2\pi}\,\sigma} e^{-\frac{a^2}{2\sigma^2}} \sum_{n=0}^{\infty} \frac{1}{n!\sigma^{2n}} :a^n:_{\sigma^2} :x^n:_{\sigma^2},$$

where $:x^n:_{\sigma^2}$ *is the Hermite polynomial of order* n *with parameter defined by*

$$:x^n:_{\sigma^2} = (-\sigma^2)^n e^{x^2/2\sigma^2} D_x^n e^{-x^2/2\sigma^2}$$

For more information on $:x^n:_{\sigma^2}$, *see page 354 in [40]. Put* $x = B(t)$ *and* $\sigma^2 = t$ *to get Donsker's delta function* $\delta_a(B(t))$ *represented by*

$$\begin{aligned}\delta_a(B(t)) &= \tfrac{1}{\sqrt{2\pi t}} e^{-\frac{a^2}{2t}} \textstyle\sum_{n=0}^{\infty} \frac{1}{n!t^n} :a^n:_t :B(t)^n:_t \\ &= \tfrac{1}{\sqrt{2\pi t}} e^{-\frac{a^2}{2t}} \textstyle\sum_{n=0}^{\infty} \frac{1}{n!t^n} :a^n:_t\, I_n\Big(1_{[0,t)}^{\otimes n}\Big).\end{aligned}$$

See pages 64 and 357 in [40]. Thus Donsker's delta function is a generalized function in $(\mathcal{S})^*$. *Actually, we can use the following facts to show that* $\delta_a(B(t)) \in (\mathcal{S}_p)^*$ *for any* $p > \frac{1+t}{2}$:

(1) *(page 353 in [40]) Let* e_n *be the Hermite function of order* $n \geq 0$ *defined by*

$$e_n(x) = \frac{1}{\sqrt{\sqrt{\pi}\,2^n n!}} H_n(x) e^{-x^2/2},$$

where $H_n(x) = (-1)^n e^{x^2} D_x^n e^{-x^2}$. *Then the set* $\{e_n;\, n \geq 0\}$ *is an orthonormal basis for* $L^2(\mathbf{R})$.

(2) *(page 354 in [40])* $Ae_n = (2n+2)e_n$, $n \geq 0$. $(A = -d^2/dx^2 + x^2 + 1)$.

(3) *(page 355 in [40])* $\displaystyle\sup_{\sigma>0, x\in\mathbf{R}} \left|\frac{1}{\sqrt{n!}\,\sigma^n} :x^n:_{\sigma^2} e^{-x^2/4\sigma^2}\right| = O(n^{-1/12})$.

Example 3.2.6 *Let $\xi \in \mathcal{S}(\mathbf{R})$. The renormalized exponential function* $:e^{\langle \cdot, \xi \rangle}:$ *is defined by*

$$:e^{\langle \cdot, \xi \rangle}: = \sum_{n=0}^{\infty} \frac{1}{n!} I_n(\xi^{\otimes n}).$$

For any integer $p \geq 0$, by Equation (3.2.5),

$$\| :e^{\langle \cdot, \xi \rangle}: \|_p^2 = \sum_{n=0}^{\infty} n! \frac{1}{(n!)^2} |\xi|_p^{2n} = \exp\left[|\xi|_p^2 \right].$$

Since $\xi \in \mathcal{S}(\mathbf{R})$, $|\xi|_p < \infty$ for all $p \geq 0$. Hence $\| :e^{\langle \cdot, \xi \rangle}: \|_p < \infty$ for all $p \geq 0$. Therefore, $:e^{\langle \cdot, \xi \rangle}:$ is a test function in $(\mathcal{S})$ for any $\xi \in \mathcal{S}(\mathbf{R})$.

3.3 General spaces of test and generalized functions

3.3.1 Abstract white noise space

As we mentioned In 12.2.2, T. Hida used the white noise space $(\mathcal{S}'(\mathbf{R}), \mu)$ when he introduced the theory of white noise in 1975. Later on in 1980, I. Kubo and S. Takenaka [33] constructed the spaces of test and generalized functions defined on a general base space. We now describe this base space.

Let $\mathcal{E}$ be a real topological vector space with topology generated by a sequence of inner product norms $\{|\cdot|_p\}_{p=0}^{\infty}$. Assume that $\mathcal{E}$ is complete with respect to the metric defined by

$$d(\xi, \eta) = \sum_{p=0}^{\infty} 2^{-p} \frac{|\xi - \eta|_p}{1 + |\xi - \eta|_p}.$$

Let $\mathcal{E}_p$ be the completion of $\mathcal{E}$ with respect to the norm $|\cdot|_p$. Then $\mathcal{E}_p$ is a Hilbert space with norm $|\cdot|_p$.

We impose the following conditions on the sequence of norms $\{|\cdot|_p\}_{p=0}^{\infty}$:

(a) There exists a constant $0 < \rho < 1$ such that for any $p \geq 0$,

$$|\cdot|_0 \leq \rho |\cdot|_1 \leq \rho^2 |\cdot|_2 \leq \cdots \leq \rho^p |\cdot|_p \leq \cdots$$

(b) For any $p \geq 0$, there exists some $q \geq p$ such that the inclusion mapping $i_{q,p} : \mathcal{E}_q \hookrightarrow \mathcal{E}_p$ is a Hilbert-Schmidt operator.

Conditions (a) and (b) imply that $\lim_{q \to \infty} \|i_{q,p}\|_{HS} = 0$. Condition (b) says that $\mathcal{E}$ is a nuclear space. By identifying $\mathcal{E}_0$ with its dual space we get a Gel'fand triple

$$\mathcal{E} \subset \mathcal{E}_0 \subset \mathcal{E}',$$

where $\mathcal{E}'$ is the dual space of $\mathcal{E}$.

By the Minlos theorem there exists a unique probability measure μ on $\mathcal{E}'$ such that

$$\int_{\mathcal{E}'} e^{i\langle x, \xi \rangle} \, d\mu(x) = e^{-\frac{1}{2}|\xi|_0^2}, \qquad \xi \in \mathcal{E}.$$

The probability space $(\mathcal{E}', \mu)$ is called an *abstract white noise space.*

Example 3.3.1 *Take $\mathcal{E} = \mathcal{S}(\mathbf{R})$ and $|\xi|_p = |A^p\xi|_0$. Here $A = -d^2/dx^2 + x^2 + 1$ and $|\cdot|_0$ is the $L^2(\mathbf{R})$-norm. Then $\mathcal{E}$ satisfies the above conditions (a) and (b) with $\rho = 1/2$. The white noise space $\mathcal{E}'$ is $\mathcal{S}'(\mathbf{R})$ given in 12.2.2.*

Example 3.3.2 *Take a real Hilbert space H with norm $|\cdot|_0$. Let $\{e_n; n \geq 1\}$ be an orthonormal basis for H. Define a linear operator T on H by $Ae_n = \lambda_n e_n$ with eigenvalues satisfying the conditions:*

(1) $1 < \lambda_1 \leq \lambda_2 \leq \cdots \leq \lambda_n \leq \cdots$

(2) $\sum_{n=1}^{\infty} \lambda_n^{-a} < \infty$ *for some positive constant a.*

For an integer $p \geq 0$, let $\mathcal{H}_p$ be the domain of the operator T. Then $\mathcal{H}_p$ is a Hilbert space with norm $|u|_p = |Tu|_0$. Let $\mathcal{H} = \cap_{p \geq 0} \mathcal{H}_p$ with topology generated by the sequence of norms $\{|\cdot|_p\}_{p=0}^{\infty}$. Then $\mathcal{H}$ satisfies the above conditions (a) and (b) with $\rho = 1/\lambda_1$. The resulting white noise space $\mathcal{H}'$ is used in [40].

3.3.2 Wick tensors

Let $(\mathcal{E}', \mu)$ be an abstract white noise space and $\mathcal{E} \subset \mathcal{E}_0 \subset \mathcal{E}'$ the associated Gel'fand triple. We will use the subindex c to denote the complexification of a real vector space. The same notation $\langle x, \xi \rangle$ will be used to denote the bilinear pairing of $x \in \mathcal{E}'$ and $\xi \in \mathcal{E}_c$.

The *Wick tensor* $:x^{\otimes n}:$ of an element x in $\mathcal{E}'$ is defined by

$$:x^{\otimes n}: = \sum_{k=0}^{[n/2]} \binom{n}{2k} (2k-1)!!(-1)^k x^{\otimes(n-2k)} \widehat{\otimes} \tau^{\otimes k},$$

where τ is the trace operator, i.e., $\langle \tau, \xi \otimes \eta \rangle = \langle \xi, \eta \rangle$ for $\xi, \eta \in \mathcal{E}_c$.

The definition of Wick tensor is motivated by the following well-known formula for Hermite polynomials with parameter σ^2:

$$:x^n:_{\sigma^2} = \sum_{k=0}^{[n/2]} \binom{n}{2k} (2k-1)!!(-\sigma^2)^k x^{n-2k}.$$

Let $f \in \mathcal{E}_{0,c}^{\widehat{\otimes} n}$, ($\mathcal{E}_{0,c}$ denotes the complexification of $\mathcal{E}_0$.) The bilinear pairing $\langle :x^{\otimes n}:, f \rangle$ is defined for μ-a.e. on $\mathcal{E}'$ and the equality $\langle :\cdot^{\otimes n}:, f \rangle = I_n(f)$ holds (see Theorem 5.4 in [40].)

For simplicity, we will use (L^2) to denote the space $L^2(\mathcal{E}', \mu)$. Let $\varphi \in (L^2)$. The Wiener-Itô decomposition of φ can be written in terms of Wick tensors as

$$\varphi(x) = \sum_{n=0}^{\infty} \langle :x^{\otimes n}:, f_n \rangle, \qquad f_n \in \mathcal{E}_{0,c}^{\widehat{\otimes} n}. \tag{3.3.7}$$

Moreover, the (L^2)-norm of φ is given by

$$\|\varphi\|_0 = \left(\sum_{n=0}^{\infty} n! |f_n|_0^2 \right)^{1/2}.$$

The use of Wick tensors (rather than multiple Wiener integrals) in Equation (3.3.7) has some advantages. The dependence of $x \in \mathcal{E}'$ in the expansion is very precise. The calculation involving the expansion can be easily manipulated.

Let $\xi \in \mathcal{E}_{0,c}$. We define the *renormalized exponential function* $:e^{\langle x,\xi\rangle}:$ by

$$:e^{\langle x,\xi\rangle}: = e^{\langle x,\xi\rangle - \frac{1}{2}\langle \xi,\xi\rangle}.$$

To find the Wiener-Itô decomposition of $:e^{\langle x,\xi\rangle}:$, note that the generating function of the Hermite polynomials is given by

$$\exp\left[tu - \frac{1}{2}\sigma^2 t^2\right] = \sum_{n=0}^{\infty} \frac{t^n}{n!} :u^n:_{\sigma^2} .$$

(See page 354 in [40].) Put $t = 1, u = \langle x, h\rangle, \sigma^2 = |h|_0^2$ $(h \in \mathcal{E}_0)$ to get

$$\exp\left[\langle x, h\rangle - \frac{1}{2}|h|_0^2\right] = \sum_{n=0}^{\infty} \frac{1}{n!} :\langle x, h\rangle^n:_{|h|_0^2} .$$

But $:\langle x, h\rangle^n:_{|h|_0^2} = \langle :x^{\otimes n}:, h^{\otimes n}\rangle$ (see Theorem 5.4 in [40].) Hence

$$\exp\left[\langle x, h\rangle - \frac{1}{2}|h|_0^2\right] = \sum_{n=0}^{\infty} \frac{1}{n!} \langle :x^{\otimes n}:, h^{\otimes n}\rangle.$$

Now, we can replace h by $\xi \in \mathcal{E}_{0,c}$ and $|h|_0^2$ by $\langle \xi, \xi\rangle$ in this equation to get the Wiener-Itô decomposition of $:e^{\langle x,\xi\rangle}:$,

$$:e^{\langle x,\xi\rangle}: = \sum_{n=0}^{\infty} \frac{1}{n!} \langle :x^{\otimes n}:, \xi^{\otimes n}\rangle. \tag{3.3.8}$$

From this equality we can easily find the (L^2)-norm of $:e^{\langle \cdot,\xi\rangle}:$

$$\| :e^{\langle \cdot,\xi\rangle}: \|_0 = \exp\left[\frac{1}{2}|\xi|_0^2\right].$$

3.3.3 Hida-Kubo-Takenaka space

Let $\varphi \in (L^2)$ be represented by Equation (3.3.7). For each integer $p \geq 0$, define

$$\|\varphi\|_p = \left(\sum_{n=0}^{\infty} n!|f_n|_p^2\right)^{1/2}. \tag{3.3.9}$$

Let $(\mathcal{E}_p) = \{\varphi \in (L^2);\ \|\varphi\|_p < \infty\}$. Then $(\mathcal{E}_p)$ is a Hilbert space with norm $\|\cdot\|_p$. Let $(\mathcal{E})$ be the projective limit of the family $\{(\mathcal{E}_p);\ p \geq 0\}$. Note that $(\mathcal{E}_0) = (L^2)$. Let $(\mathcal{E})^*$ be the dual space of $(\mathcal{E})$. By identifying (L^2) with its dual space, we get the following Gel'fand triple

$$(\mathcal{E}) \subset (L^2) \subset (\mathcal{E})^*.$$

This Gel'fand triple is often referred to as the *Hida-Kubo-Takenaka space.*

Let $(\mathcal{E}_p)^*$ be the dual space of $(\mathcal{E}_p)$. Then we have continuous inclusion mappings for any $q \geq p$,

$$(\mathcal{E}) \hookrightarrow (\mathcal{E}_q) \hookrightarrow (\mathcal{E}_p) \hookrightarrow (L^2) \hookrightarrow (\mathcal{E}_p)^* \hookrightarrow (\mathcal{E}_q)^* \hookrightarrow (\mathcal{E})^*.$$

Note that $(\mathcal{E})^* = \cup_{p\geq 0}(\mathcal{E}_p)^*$ and for each $p \geq 0$, $(\mathcal{E}_p)^*$ is the completion of (L^2) with respect to the norm $\|\cdot\|_{-p}$

$$\|\varphi\|_{-p} = \left(\sum_{n=0}^{\infty} n!|f_n|_{-p}^2\right)^{1/2}, \tag{3.3.10}$$

where we use the same notation $|\cdot|_{-p}$ to denote the norm on $\mathcal{E}'_{p,c}$ and its nth symmetric tensor product space for any n.

Example 3.3.3 *Let $\xi \in \mathcal{E}_c$ and consider the renormalized exponential function*

$$:e^{\langle\cdot,\xi\rangle}: = e^{\langle\cdot,\xi\rangle - \frac{1}{2}\langle\xi,\xi\rangle} = \sum_{n=0}^{\infty}\frac{1}{n!}\langle :\cdot^{\otimes n}:, \xi^{\otimes n}\rangle.$$

We can use Equation (3.3.9) to check that for any $p \geq 0$,

$$\|:e^{\langle\cdot,\xi\rangle}:\|_p = \exp\left[\frac{1}{2}|\xi|_p^2\right]. \tag{3.3.11}$$

It follows that $:e^{\langle\cdot,\xi\rangle}: \in (\mathcal{E})$ for any $\xi \in \mathcal{E}_c$.

Example 3.3.4 *Let $y \in \mathcal{E}'_c$. Being motivated by the equality in Equation (3.3.8), we define the renormalized exponential function $:e^{\langle\cdot,y\rangle}:$ by*

$$:e^{\langle\cdot,y\rangle}: = \sum_{n=0}^{\infty}\frac{1}{n!}\langle :\cdot^{\otimes n}:, y^{\otimes n}\rangle.$$

Since $\mathcal{E}'_c = \cup_{p\geq 0}\mathcal{E}'_{p,c}$, there exists some $p \geq 0$ such that $y \in \mathcal{E}'_{p,c}$ and so $|y|_{-p} < \infty$. We can use Equation (3.3.10) to find that

$$\|:e^{\langle\cdot,y\rangle}:\|_{-p} = \exp\left[\frac{1}{2}|y|_{-p}^2\right]. \tag{3.3.12}$$

This shows that $:e^{\langle\cdot,y\rangle}: \in (\mathcal{E})^*$ for any $y \in \mathcal{E}'_c$.

3.3.4 Kondratiev-Streit space

Let $0 \leq \beta < 1$ be a fixed number. For $\varphi \in (L^2)$ being represented by Equation (3.3.7) and an integer $p \geq 0$, define

$$\|\varphi\|_{p,\beta} = \left(\sum_{n=0}^{\infty}(n!)^{1+\beta}|f_n|_p^2\right)^{1/2}. \tag{3.3.13}$$

Let $(\mathcal{E}_p)_\beta = \{\varphi \in (L^2);\ \|\varphi\|_{p,\beta} < \infty\}$. Then $(\mathcal{E}_p)_\beta$ is a Hilbert space with norm $\|\cdot\|_{p,\beta}$. Let $(\mathcal{E})_\beta$ be the projective limit of the family $\{(\mathcal{E}_p)_\beta;\ p \geq 0\}$. Note that $(\mathcal{E}_0)_\beta \neq (L^2)$ unless $\beta = 0$. But we do have $(\mathcal{E}_0)_\beta \subset (L^2)$. Let $(\mathcal{E})_\beta^*$ be the dual space of $(\mathcal{E})_\beta$. Then we have the following Gel'fand triple

$$(\mathcal{E})_\beta \subset (L^2) \subset (\mathcal{E})_\beta^*.$$

This triple was introduced in [30] [31] and is called the *Kondratiev-Streit space.*

Let $(\mathcal{E}_p)^*_\beta$ be the dual space of $(\mathcal{E}_p)_\beta$. Then we have continuous inclusion mappings for any $q \geq p$,

$$(\mathcal{E})_\beta \hookrightarrow (\mathcal{E}_q)_\beta \hookrightarrow (\mathcal{E}_p)_\beta \hookrightarrow (L^2) \hookrightarrow (\mathcal{E}_p)^*_\beta \hookrightarrow (\mathcal{E}_q)^*_\beta \hookrightarrow (\mathcal{E})^*_\beta.$$

When $\beta = 0$, we have $(\mathcal{E})_0 = (\mathcal{E})$. Moreover, for any $0 \leq \beta < 1$,

$$(\mathcal{E})_\beta \subset (\mathcal{E}) \subset (L^2) \subset (\mathcal{E})^* \subset (\mathcal{E})^*_\beta.$$

Note that $(\mathcal{E})^*_\beta = \cup_{p \geq 0}(\mathcal{E}_p)^*_\beta$ and for each $p \geq 0$, $(\mathcal{E}_p)^*_\beta$ is the completion of (L^2) with respect to the norm $\|\cdot\|_{-p,-\beta}$

$$\|\varphi\|_{-p,-\beta} = \left(\sum_{n=0}^{\infty} (n!)^{1-\beta} |f_n|_{-p}^2\right)^{1/2}. \tag{3.3.14}$$

Example 3.3.5 *The renormalized exponential function* $:e^{\langle \cdot, \xi\rangle}:$ *is a test function in* $(\mathcal{E})_\beta$ *for any* $\xi \in \mathcal{E}_c$. *By Equation (3.3.13), we have*

$$\|:e^{\langle \cdot, \xi\rangle}:\|_{p,\beta} = \left(\sum_{n=0}^{\infty} \frac{1}{(n!)^{1-\beta}} |\xi|_p^{2n}\right)^{1/2}. \tag{3.3.15}$$

Example 3.3.6 *The function* $:e^{\langle \cdot, y\rangle}:$ *is a generalized function in* $(\mathcal{E})^*_\beta$ *for any* $y \in \mathcal{E}'_c$. *By Equation (3.3.14) we have*

$$\|:e^{\langle \cdot, y\rangle}:\|_{-p,-\beta} = \left(\sum_{n=0}^{\infty} \frac{1}{(n!)^{1+\beta}} |y|_{-p}^{2n}\right)^{1/2}. \tag{3.3.16}$$

The next example shows that $(\mathcal{E})^*$ is a proper subspace of $(\mathcal{E})^*_\beta$ for $0 < \beta < 1$. Later on we will give a more interesting example to show the need to study the Kondratiev-Streit space.

Example 3.3.7 *Let* $0 < \beta < 1$. *Take a nonzero* $x \in \mathcal{E}'$ *and define the function*

$$\Phi = \sum_{n=0}^{\infty} (n!)^{\frac{\beta-2}{4}} \langle :\cdot^{\otimes n}:, x^{\otimes n}\rangle.$$

It is easy to check that $\|\Phi\|_{-p} = \infty$ *for all* $p \geq 0$. *Hence* $\Phi \notin (\mathcal{E})^*$. *On the other hand,* $\Phi \in (\mathcal{E})^*_\beta$. *In fact,* $\Phi \in (\mathcal{E}_p)^*_\beta$ *if* $x \in \mathcal{E}'_p$.

3.3.5 Cochran-Kuo-Sengupta space

Let $\{\alpha(n)\}_{n=0}^\infty$ be a sequence of real numbers satisfying the conditions:

(A1) $\alpha(0) = 1$ and $\inf_{n \geq 0} \alpha(n)\sigma^n > 0$ for some $\sigma \geq 1$.

(A2) $\lim_{n \to \infty} \left(\frac{\alpha(n)}{n!}\right)^{1/n} = 0$.

In [14] a stronger condition $\inf_{n \geq 0} \alpha(n) > 0$ is assumed. But the weaker condition $\inf_{n \geq 0} \alpha(n)\sigma^n > 0$ for some $\sigma \geq 1$ is good enough to get a Gel'fand triple. This condition was introduced in [7].

Let $\varphi \in (L^2)$ be represented by Equation (3.3.7) and $p \geq 0$ an integer, define

$$\|\varphi\|_{p,\alpha} = \left(\sum_{n=0}^{\infty} n!\alpha(n)|f_n|_p^2\right)^{1/2}. \tag{3.3.17}$$

Let $[\mathcal{E}_p]_\alpha = \{\varphi \in (L^2);\ \|\varphi\|_{p,\alpha} < \infty\}$. Then $[\mathcal{E}_p]_\alpha$ is a Hilbert space with norm $\|\cdot\|_{p,\alpha}$. Let $[\mathcal{E}]_\alpha$ be the projective limit of the family $\{[\mathcal{E}_p]_\alpha;\ p \geq 0\}$.

Note that by condition (a) in 12.3.17 and condition (A1),

$$\begin{aligned}\sum_{n=0}^{\infty} n!\alpha(n)|f_n|_p^2 &\geq \textstyle\sum_{n=0}^{\infty} n!\alpha(n)\rho^{-2pn}|f_n|_0^2\\ &= \textstyle\sum_{n=0}^{\infty} n!\big(\alpha(n)\sigma^n\big)\sigma^{-n}\rho^{-2pn}|f_n|_0^2\\ &\geq \inf_{n\geq 0}\big(\alpha(n)\sigma^n\big)\textstyle\sum_{n=0}^{\infty} n!\sigma^{-n}\rho^{-2pn}|f_n|_0^2.\end{aligned}$$

Choose large p so that $p \geq (-2\log\rho)^{-1}\log\sigma$. Then we have $\sigma^{-1}\rho^{-2p} \geq 1$ and so

$$\sum_{n=0}^{\infty} n!\alpha(n)|f_n|_p^2 \geq \inf_{n\geq 0}\big(\alpha(n)\sigma^n\big)\sum_{n=0}^{\infty} n!|f_n|_0^2.$$

This implies that $[\mathcal{E}_p]_\alpha \subset (L^2)$ for all $p \geq (-2\log\rho)^{-1}\log\sigma$. Hence $[\mathcal{E}]_\alpha \subset (L^2)$. Let $[\mathcal{E}]_\alpha^*$ be the dual space of $[\mathcal{E}]_\alpha$. Then we have a Gel'fand triple

$$[\mathcal{E}]_\alpha \subset (L^2) \subset [\mathcal{E}]_\alpha^*.$$

This triple was introduced in [14] and is called the *Cochran-Kuo-Sengupta space.*

Let $[\mathcal{E}_p]_\alpha^*$ be the dual space of $[\mathcal{E}_p]_\alpha$. For $p \geq (-2\log\rho)^{-1}\log\sigma$, $[\mathcal{E}_p]_\alpha^*$ is the completion of (L^2) with respect to the norm $\|\cdot\|_{-p,1/\alpha}$ defined by

$$\|\varphi\|_{-p,1/\alpha} = \left(\sum_{n=0}^{\infty} \frac{n!}{\alpha(n)}|f_n|_{-p}^2\right)^{1/2}. \tag{3.3.18}$$

When $\alpha(n) = 1$ for all n, the associated triple is the Hida-Kubo-Takenaka space. When $\alpha(n) = (n!)^\beta$, the associated triple is the Kondratiev-Streit space. Moreover, we have

$$[\mathcal{E}]_\alpha \subset (\mathcal{E}) \subset (L^2) \subset (\mathcal{E})^* \subset [\mathcal{E}]_\alpha^*. \tag{3.3.19}$$

Example 3.3.8 *The renormalized exponential function* $:e^{\langle\cdot,\xi\rangle}:$ *is a test function in* $[\mathcal{E}]_\alpha$ *for any* $\xi \in \mathcal{E}_c$. *By Equation (3.3.17), we have*

$$\|:e^{\langle\cdot,\xi\rangle}:\|_{p,\alpha} = \left(\sum_{n=0}^{\infty} \frac{\alpha(n)}{n!}|\xi|_p^{2n}\right)^{1/2}. \tag{3.3.20}$$

Note that condition (A2) implies that the series converges.

Example 3.3.9 *For any* $y \in \mathcal{E}_c'$, *the renormalized exponential function* $:e^{\langle\cdot,y\rangle}:$ *is a generalized function in* $[\mathcal{E}]_\alpha^*$. *By Equation (3.3.18) we have*

$$\|:e^{\langle\cdot,y\rangle}:\|_{-p,1/\alpha} = \left(\sum_{n=0}^{\infty} \frac{1}{n!\alpha(n)}|y|_{-p}^{2n}\right)^{1/2}. \tag{3.3.21}$$

It can be easily checked from condition (A1) that the series converges.

Example 3.3.10 *Consider the following sequence*

$$\alpha(n) = \frac{n!}{\big(\log(n+2)\big)^n}, \qquad n \geq 0. \tag{3.3.22}$$

Conditions (A1) and (A2) can be easily checked. Take a nonzero $x \in \mathcal{E}'$ *and define the function*

$$\Phi = \sum_{n=0}^{\infty} e^{-n \log\log(n+2)} \langle :\cdot^{\otimes n}:, x^{\otimes n} \rangle.$$

This function defines a generalized function in $[\mathcal{E}]_\alpha^*$ *for the sequence in Equation (3.3.22). However, it does not belong to any of the Kondratiev-Streit spaces* $(\mathcal{E})_\beta^*$.

Example 3.3.11 *(***Bell numbers and Bell number spaces***)*

Let $\exp_k$ *be the kth iteration of the exponential function, i.e.,*

$$\exp_k(x) = \underbrace{\exp\big(\exp \cdots \big(\exp(x)\big)\big)}_{k-times}.$$

This function has the Taylor series expansion

$$\exp_k(x) = \sum_{n=0}^{\infty} \frac{B_k(n)}{n!} x^n.$$

The Bell numbers of order k *are the numbers defined by*

$$b_k(n) = \frac{B_k(n)}{\exp_k(0)}, \qquad n \geq 0. \tag{3.3.23}$$

The Bell numbers $\{b_2(n)\}_{n=0}^{\infty}$ *of order* 2 *are usually called the Bell numbers. The first few of them are* $1, 1, 2, 5, 15, 52, 203$.

The Bell numbers of any order obviously satisfy conditions (A1) and (A2). The associated Gel'fand triple

$$[\mathcal{E}]_{b_k} \subset (L^2) \subset [\mathcal{E}]_{b_k}^*$$

is called the Bell number space of order k. *It can be easily checked that for any* $k \geq 2$ *and* $0 \leq \beta < 1$,

$$[\mathcal{E}]_{b_k} \subset (\mathcal{E})_\beta \subset (L^2) \subset (\mathcal{E})_\beta^* \subset [\mathcal{E}]_{b_k}^*.$$

3.4 Continuous versions and analytic extensions

3.4.1 Continuous versions

In this section we study test functions in the Hida-Kubo-Takenaka space

$$(\mathcal{E}) \subset (L^2) \subset (\mathcal{E})^*.$$

Since $(\mathcal{E})$ is defined as the projective limit of $\{(\mathcal{E}_p);\, p \geq 0\}$, a test function in $(\mathcal{E})$ is defined only μ-a.e.

A fundamental fact due to Kubo and Yokoi [37] says that every test function in $(\mathcal{E})$ has a unique continuous version. We give an intuitive explanation of this fact. For the complete proof, see the book [40]. Let $\varphi \in (\mathcal{E})$. Then $\varphi \in (L^2)$ and by Equation (3.3.7)

$$\varphi = \sum_{n=0}^{\infty} \langle :\cdot^{\otimes n}:, f_n \rangle, \qquad \mu\text{-}a.e.$$

Note that $\|\varphi\|_p < \infty$ for all $p \geq 0$. Hence for each n, $|f_n|_p < \infty$ for all $p \geq 0$ and so $f_n \in \mathcal{E}_c^{\widehat{\otimes} n}$. Therefore, the pairing $\langle :x^{\otimes n}:, f_n\rangle$ is defined for all $x \in \mathcal{E}$. For each $x \in \mathcal{E}'$, define

$$\widetilde{\varphi}(x) = \sum_{n=0}^{\infty} \langle :x^{\otimes n}:, f_n \rangle.$$

By Proposition 6.1 in [40] this series converges absolutely for each $x \in \mathcal{E}'$. Moreover, by Theorem 6.4 in [40], $\widetilde{\varphi}$ is a continuous function on $\mathcal{E}'$. The function $\widetilde{\varphi}$ is a version of φ. Thus we have the next theorem.

Theorem 3.4.1 *Every test function in $(\mathcal{E})$ has a unique continuous version.*

From now on a test function in $(\mathcal{E})$ is understood to be its continuous version. Hence it can be represented pointwise for $x \in \mathcal{E}'$ as

$$\varphi(x) = \sum_{n=0}^{\infty} \langle :x^{\otimes n}:, f_n \rangle, \qquad f_n \in \mathcal{E}_c^{\widehat{\otimes} n}. \tag{3.4.24}$$

Now, let $x \in \mathcal{E}'$ be fixed. Define a linear functional T on $(\mathcal{E})$ by

$$T(\varphi) = \varphi(x), \qquad \varphi \in (\mathcal{E}). \tag{3.4.25}$$

It follows from Equation (3.4.24) that

$$|T(\varphi)| \leq \sum_{n=0}^{\infty} |f_n|_p \, |:x^{\otimes n}:|_{-p}.$$

Write $|f_n|_p\,|:x^{\otimes n}:|_{-p}$ as $\left(\sqrt{n!}\,|f_n|_p\right)\left(|:x^{\otimes n}:|_{-p}/\sqrt{n!}\right)$ and then apply the Schwarz inequality to get

$$\begin{aligned} |T(\varphi)| &\leq \left(\textstyle\sum_{n=0}^{\infty} n!|f_n|_p^2\right)^{1/2} \left(\textstyle\sum_{n=0}^{\infty} \frac{1}{n!}|:x^{\otimes n}:|_{-p}^2\right)^{1/2}. \\ &= \|\varphi\|_p \left(\textstyle\sum_{n=0}^{\infty} \frac{1}{n!}|:x^{\otimes n}:|_{-p}^2\right)^{1/2}. \end{aligned}$$

By Lemma 7.10 in [40], we have the inequality

$$|:x^{\otimes n}:|_{-p} \leq \sqrt{n!}\left(|x|_{-p} + |\tau|_{-p}^{1/2}\right)^n,$$

where τ is the trace operator (see 12.3.18.) Therefore,

$$|T(\varphi)| \leq \|\varphi\|_p \left(\sum_{n=0}^{\infty} \left(|x|_{-p} + |\tau|_{-p}^{1/2}\right)^{2n} \right)^{1/2}. \tag{3.4.26}$$

Note that $\lim_{p\to\infty} |x|_{-p} = 0$ and $\lim_{p\to\infty} |\tau|_{-p} = 0$. Hence we can choose large p such that $|x|_{-p} + |\tau|_{-p}^{1/2} < 1$. Then the series in Equation (3.4.26) is convergent. Thus T is a continuous linear functional on $(\mathcal{E})$.

The continuous linear functional T defined by Equation (3.4.25) is called *Kubo-Yokoi delta functional* at x. This functional, denoted by $\widetilde{\delta}_x$, is a generalized function in the space $(\mathcal{E})^*$.

Suppose the Wiener-Itô decomposition of $\widetilde{\delta}_x$ is given by

$$\widetilde{\delta}_x = \sum_{n=0}^{\infty} \langle :\cdot^{\otimes n}:, F_n \rangle. \tag{3.4.27}$$

From Equations (3.4.24) and (3.4.27),

$$\varphi(x) = \langle\langle \widetilde{\delta}_x, \varphi \rangle\rangle = \sum_{n=0}^{\infty} n! \langle F_n, f_n \rangle. \tag{3.4.28}$$

Upon comparing Equations (3.4.24) and (3.4.28) we see that F_n is given by

$$F_n = \frac{1}{n!} :x^{\otimes n}:, \qquad n \geq 0.$$

Thus we have proved the following theorem.

Theorem 3.4.2 *The Kubo-Yokoi delta function $\widetilde{\delta}_x$ at $x \in \mathcal{E}'$ is a generalized function in $(\mathcal{E})^*$. It has the Wiener-Itô decomposition*

$$\widetilde{\delta}_x = \sum_{n=0}^{\infty} \frac{1}{n!} \langle :\cdot^{\otimes n}:, :x^{\otimes n}: \rangle.$$

By Theorem 7.9 in [40] there is some constant K_p independent of x such that

$$\|\widetilde{\delta}_x\|_{-p} \leq K_p e^{\frac{1}{2}|x|_{-p}^2}.$$

Moreover, by Theorem 7.18 in [40] the function

$$x \longmapsto \widetilde{\delta}_x$$

is continuous from $\mathcal{E}'$ into $(\mathcal{E})^*$ with the inductive limit topology for both spaces.

We have a similar result for Donsker's delta function $\delta_a(B(t))$ in Example 3.2.5. By Theorem 7.15 in [40] the function

$$a \longmapsto \delta_a(B(t))$$

is continuous from $\mathbf{R}$ into $(\mathcal{S})^*$ with the inductive limit topology for $(\mathcal{S})^*$.

3.4.2 Analytic extensions

Define a linear operator Θ from $(\mathcal{E})$ into itself such that

$$\Theta: \ e^{\langle \cdot, \xi \rangle} \longmapsto :e^{\langle \cdot, \xi \rangle}:, \qquad \forall \xi \in \mathcal{E}_c.$$

Since the linear span of the set $\{e^{\langle \cdot, \xi \rangle};\ \xi \in \mathcal{E}_c\}$ is dense in $(\mathcal{E})$, a linear operator on $(\mathcal{E})$ is uniquely determined by its action on this set. By Theorem 6.2 in [40] the linear operator Θ is continuous from $(\mathcal{E})$ into itself. Hence its adjoint Θ^* is a continuous linear operator from $(\mathcal{E})^*$ into itself.

Let $x \in \mathcal{E}'$ and $\xi \in \mathcal{E}_c$ be fixed. Note that

$$:e^{\langle \cdot, x\rangle}: = \sum_{n=0}^{\infty} \frac{1}{n!} \langle :\cdot^{\otimes n}:, x^{\otimes n}\rangle, \quad :e^{\langle \cdot, \xi\rangle}: = \sum_{n=0}^{\infty} \frac{1}{n!} \langle :\cdot^{\otimes n}:, \xi^{\otimes n}\rangle.$$

Therefore,

$$\langle\langle :e^{\langle \cdot, x\rangle}:, :e^{\langle \cdot, \xi\rangle}: \rangle\rangle = \sum_{n=0}^{\infty} n! \left\langle \frac{1}{n!} x^{\otimes n}, \frac{1}{n!} \xi^{\otimes n} \right\rangle = e^{\langle x, \xi\rangle}. \tag{3.4.29}$$

Now, let $x \in \mathcal{E}'$ be fixed. Then by the definition of Θ and Equation (3.4.29),

$$\langle\langle \Theta^*(:e^{\langle \cdot, x\rangle}:), e^{\langle \cdot, \xi\rangle} \rangle\rangle = \langle\langle :e^{\langle \cdot, x\rangle}:, \Theta(e^{\langle \cdot, \xi\rangle}) \rangle\rangle = \langle\langle :e^{\langle \cdot, x\rangle}:, :e^{\langle \cdot, \xi\rangle}: \rangle\rangle = e^{\langle x, \xi\rangle}.$$

On the other hand, we have $\langle\langle \widetilde{\delta}_x, e^{\langle \cdot, \xi\rangle} \rangle\rangle = e^{\langle x, \xi\rangle}$. Hence we have shown that

$$\langle\langle \Theta^*(:e^{\langle \cdot, x\rangle}:), e^{\langle \cdot, \xi\rangle} \rangle\rangle = \langle\langle \widetilde{\delta}_x, e^{\langle \cdot, \xi\rangle} \rangle\rangle, \qquad \forall \xi \in \mathcal{E}_c.$$

This implies that for any $x \in \mathcal{E}'$,

$$\Theta^*(:e^{\langle \cdot, x\rangle}:) = \widetilde{\delta}_x.$$

By using this equality we see that

$$\varphi(x) = \langle\langle \widetilde{\delta}_x, \varphi \rangle\rangle = \langle\langle \Theta^*(:e^{\langle \cdot, x\rangle}:), \varphi \rangle\rangle = \langle\langle :e^{\langle \cdot, x\rangle}:, \Theta\varphi \rangle\rangle.$$

Hence we conclude that for any test function φ in $(\mathcal{E})$,

$$\varphi(x) = \langle\langle :e^{\langle \cdot, x\rangle}:, \Theta\varphi \rangle\rangle, \qquad x \in \mathcal{E}'. \tag{3.4.30}$$

This representation of test functions is very useful. Observe that the variable x in φ goes over to the renormalized exponential function $:e^{\langle \cdot, x\rangle}:$. Obviously, the function $:e^{\langle \cdot, x\rangle}:$ has nice regularity and growth properties, which can automatically be transferred to test functions.

Recall from Example 3.3.4 that the renormalized exponential function $:e^{\langle \cdot, y\rangle}:$ is defined for any $y \in \mathcal{E}_c'$. Therefore, Equation (3.4.30) shows that a test function φ (defined on $\mathcal{E}'$) can be extended to a function defined on $\mathcal{E}_c'$. We use the same notation φ to denote this extension, i.e.,

$$\varphi(y) = \langle\langle :e^{\langle \cdot, y\rangle}:, \Theta\varphi \rangle\rangle, \qquad y \in \mathcal{E}_c'.$$

A complex-valued function defined on a complex Hilbert space is called *analytic* if it is locally bounded and Fréchet differentiable. The next theorem (see Theorem 6.13 in [40]) says that every test function has a unique analytic extension.

Theorem 3.4.3 *Every test function φ in $(\mathcal{E})$ has a unique extension to a function $\varphi(y)$, $y \in \mathcal{E}_c'$, such that φ is analytic on $\mathcal{E}_{p,c}'$ for any $p \geq 0$.*

3.4.3 Integrable functions

An interesting consequence of the representation of test functions in Equation (3.4.30) is the following inclusion:

$$(\mathcal{E}) \subset \bigcap_{1 \leq r < \infty} L^r(\mu),$$

where μ is the white noise measure on $\mathcal{E}'$. This fact is due to Obata [48] (see also Section 8.5 of [40].) Below we give a different and very simple proof.

First note that by condition (b) in 12.3.17 there exists some $p \geq 0$ such that the inclusion mapping $i_{p,0} : \mathcal{E}_p \hookrightarrow \mathcal{E}_0$ is a Hilbert-Schmidt operator. This implies that $(\mathcal{E}_0, \mathcal{E}_{-p})$ is an abstract Wiener space. Hence the white noise measure μ is supported on $\mathcal{E}_{-p}$.

Next we state a theorem which can be easily proved by direct calculations.

Theorem 3.4.4 *Suppose the inclusion mapping $i_{p,0} : \mathcal{E}_p \hookrightarrow \mathcal{E}_0$ is a Hilbert-Schmidt operator. Then for any $r \leq \|i_{p,0}\|_{HS}^{-2}$,*

$$\int_{\mathcal{E}'} e^{\frac{r}{2}|x|_{-p}^2}\, d\mu(x) \leq e^{\frac{r}{2}\|i_{p,0}\|_{HS}^2}.$$

Now, let $\varphi \in (\mathcal{E})$ and $1 \leq r < \infty$ be fixed. By Equations (3.4.30) and (3.3.12) we have the inequality for any $p \geq 0$,

$$|\varphi(x)| \leq \|\Theta\varphi\|_p\, e^{\frac{1}{2}|x|_{-p}^2}, \qquad x \in \mathcal{E}'.$$

Recall from 12.3.17 that $\lim_{p\to\infty} \|i_{p,0}\|_{HS} = 0$. Hence, for the given fixed number r, we can choose p such that $\|i_{p,0}\|_{HS} \leq 1/\sqrt{r}$. Then apply Theorem 3.4.4 to get

$$\int_{\mathcal{E}'} |\varphi(x)|^r\, d\mu(x) \leq \|\Theta\varphi\|_p^r \int_{\mathcal{E}'} e^{\frac{r}{2}|x|_{-p}^2}\, d\mu(x) \leq \|\Theta\varphi\|_p^r\, e^{\frac{r}{2}\|i_{p,0}\|_{HS}^2}.$$

Hence for any $r \geq 0$ and p such that $\|i_{p,0}\|_{HS} \leq 1/\sqrt{r}$, we have

$$\|\varphi\|_{L^r(\mu)} \leq \|\Theta\varphi\|_p\, e^{\frac{1}{2}\|i_{p,0}\|_{HS}^2}, \qquad \forall \varphi \in (\mathcal{E}). \tag{3.4.31}$$

This inequality proves the next theorem.

Theorem 3.4.5 *The inclusion $(\mathcal{E}) \subset \cap_{1\leq r<\infty} L^r(\mu)$ holds and for each $1 \leq r < \infty$, the inclusion mapping $(\mathcal{E}) \hookrightarrow L^r(\mu)$ is continuous.*

Let $f \in L^s(\mu)$, $1 < s \leq \infty$. Define a linear functional Φ_f on $(\mathcal{E})$ by

$$\Phi_f(\varphi) = \int_{\mathcal{E}'} \varphi(x) f(x)\, d\mu(x). \tag{3.4.32}$$

Let r be given by $r^{-1} + s^{-1} = 1$. Then $1 \leq r < \infty$. Choose $p \geq 0$ such that $\|i_{p,0}\|_{HS} \leq 1/\sqrt{r}$. Then by Equation (3.4.31),

$$|\Phi_f| \leq \|f\|_{L^s(\mu)}\, \|\varphi\|_{L^r(\mu)} \leq \|f\|_{L^s(\mu)}\, \|\Theta\varphi\|_p\, e^{\frac{1}{2}\|i_{p,0}\|_{HS}^2}.$$

This shows that Φ_f is continuous. Hence it induces a generalized function in $(\mathcal{E})^*$. We state this fact as the next theorem.

Theorem 3.4.6 *The inclusion $\cup_{1<s\leq\infty} L^s(\mu) \subset (\mathcal{E})^*$ holds and for each $1 < s \leq \infty$, the inclusion mapping $L^s(\mu) \hookrightarrow (\mathcal{E})^*$ is continuous.*

3.4.4 Generalized functions induced by measures

In the previous section we see that functions in $L^s(\mu)$ with $1 < s \leq \infty$ induce generalized functions in the space $(\mathcal{E})^*$. Let Φ_f be the linear functional given by f as in Equation (3.4.32). Observe that if φ is a nonnegative test function, then $\Phi_f(\varphi) \geq 0$. This leads to the concept of positive generalized functions.

A generalized function Φ is called *positive* if $\langle\langle \Phi, \varphi \rangle\rangle \geq 0$ for all nonnegative test functions φ.

The following theorem is due independently to Kondratiev [28] and Yokoi [58]. For the proof, see Theorem 15.3 in [40].

Theorem 3.4.7 *A generalized function $\Phi \in (\mathcal{E})^*$ is positive if and only if there exists a finite measure ν on $\mathcal{E}'$ such that $(\mathcal{E}) \subset L^1(\nu)$ and*

$$\langle\langle \Phi, \varphi \rangle\rangle = \int_{\mathcal{E}'} \varphi(x)\, d\nu(x), \qquad \forall \varphi \in (\mathcal{E}).$$

Being motivated by this theorem, we say that a measure ν on $\mathcal{E}'$ is a *Hida measure* if $(\mathcal{E}) \subset L^1(\nu)$ and the linear functional

$$\varphi \longmapsto \int_{\mathcal{E}'} \varphi(x)\, d\nu(x), \qquad \varphi \in (\mathcal{E}),$$

is continuous. Thus ν induces a generalized function $\widetilde{\nu} \in (\mathcal{E})^*$ such that

$$\langle\langle \widetilde{\nu}, \varphi \rangle\rangle = \int_{\mathcal{E}'} \varphi(x)\, d\nu(x), \qquad \varphi \in (\mathcal{E}).$$

We can replace $(\mathcal{E})$ by $(\mathcal{E})_\beta$ and $[\mathcal{E}]_\alpha$. In that case, ν induces a generalized function in $(\mathcal{E})_\beta^*$ and $[\mathcal{E}]_\alpha^*$, respectively.

Note that the Kubo-Yokoi delta function at $x \in \mathcal{E}'$ (see 12.4.75) is the generalized function induced by the Dirac measure δ_x at x on $\mathcal{E}'$.

The next theorem gives a characterization of Hida measures. For the proof, see Theorem 15.17 in [40]. The case $\beta = 0$ is due to Lee [43].

Theorem 3.4.8 *A measure ν on $\mathcal{E}'$ is a Hida measure with $\widetilde{\nu} \in (\mathcal{E})_\beta^*$ if and only if ν is supported in $\mathcal{E}_p'$ for some $p \geq 0$ such that*

$$\int_{\mathcal{E}_p'} \exp\left[\frac{1}{2}(1+\beta)|x|_{-p}^{\frac{2}{1+\beta}}\right] d\nu(x) < \infty.$$

Recently, Asai et al. [8] have extended this characterization theorem to Hida measures which induce positive generalized functions in the Cochran-Kuo-Sengupta space. Here we briefly describe their result.

Let $C_{+,1/2}$ denote the set of positive continuous functions u on $[0, \infty)$ satisfying the condition

$$\lim_{r\to\infty} \frac{\log u(r)}{\sqrt{r}} = \infty.$$

We assume that $u \in C_{+,1/2}$ satisfies the following conditions:

(U1) u is increasing and $u(0) = 1$.

(U2) $\lim_{r\to\infty} r^{-1} \log u(r) < \infty$.

(U3) $\log u(x^2)$ is a convex function on $[0, \infty)$.

Define the Legendre transform of u by

$$\ell_u(t) = \inf_{r>0} \frac{u(r)}{r^t}, \qquad t \geq 0.$$

For more information about the Legendre transform, see [6]. With the function u, we associate a sequence of real numbers defined by $\alpha(n) = (\ell_u(n)n!)^{-1}$, $n \geq 0$.

Now, let $[\mathcal{E}]_u \subset (L^2) \subset [\mathcal{E}]_u^*$ denote the CKS-space given by the sequence

$$\alpha(n) = \frac{1}{\ell_u(n)n!}, \qquad n \geq 0. \tag{3.4.33}$$

For more information about this Gel'fand triple, see [7] [8].

Theorem 3.4.9 *Let $u \in C_{+,1/2}$ satisfy conditions (U1) (U2) (U3). Then a measure ν on $\mathcal{E}'$ is a Hida measure with $\widetilde{\nu} \in [\mathcal{E}]_u^*$ if and only if ν is supported in $\mathcal{E}_p'$ for some $p \geq 0$ such that*

$$\int_{\mathcal{E}_p'} u\big(|x|_{-p}^2\big)^{1/2} \, d\nu(x) < \infty.$$

Note that Theorem 3.4.8 is a special case of Theorem 3.4.9 with the function

$$u(r) = \exp\left[(1+\beta)\, r^{\frac{1}{1+\beta}}\right].$$

An important class of Hida measures is given by the distribution laws of the solution of an $\mathcal{E}'$-valued stochastic integral equation

$$X(t) = x + \int_0^t F(s, X(s))\, ds + \int_0^t G(s, X(s))\, dW(s).$$

Under certain assumptions on F and G, it is proved in Theorem 3.1 in [42] that the distribution laws of $X(t)$ are Hida measures inducing generalized functions in the space $(\mathcal{E})^*$.

3.4.5 Generalized Radon-Nikodym derivative

Suppose a measure ν on $\mathcal{E}'$ is absolutely continuous with respect to the white noise measure μ and its Radon-Nikodym derivative $d\nu/d\mu$ belongs to $L^s(\mu)$ for some $1 < s \leq \infty$. Then by Theorem 3.4.6, the measure ν induces a generalized function $\widetilde{\nu}$ in $(\mathcal{E})^*$ such that

$$\langle\!\langle \widetilde{\nu}, \varphi \rangle\!\rangle = \int_{\mathcal{E}'} \frac{d\nu}{d\mu}(x) \varphi(x)\, d\mu(x).$$

Thus ν is a Hida measure and we can interpret $\widetilde{\nu}$ as the Radon-Nikodym derivative $d\nu/d\mu$.

On the other hand, suppose ν is a Hida measure. Then it induces a generalized function $\widetilde{\nu}$ such that

$$\langle\!\langle \widetilde{\nu}, \varphi \rangle\!\rangle = \int_{\mathcal{E}'} \varphi(x)\, d\nu(x).$$

Symbolically, this equation can be rewritten as

$$\int_{\mathcal{E}'} \widetilde{\nu}(x) \varphi(x)\, d\mu(x) = \int_{\mathcal{E}'} \varphi(x)\, d\nu(x).$$

Thus we can interpret $\widetilde{\nu}$ as the *generalized Radon-Nikodym derivative* $d\nu/d\mu$. If $\widetilde{\nu}$ is given by a function in $L^s(\mu)$ for some $1 < s \leq \infty$, then ν is absolutely continuous with respect to μ and $\widetilde{\nu}$ is the ordinary Radon-Nikodym derivative $d\nu/d\mu$.

Next we examine Gaussian measures on $\mathcal{E}'$ to explore the idea of generalized Radon-Nikodym derivative a little bit further. For $t > 0$ and $y \in \mathcal{E}'$, let $\mu_{y,t}$ be the Gaussian measure defined by $\mu_{y,t}(C) = \mu(t^{-1/2}(C-y))$, $C \in \mathcal{B}(\mathcal{E}')$. The well-known dichotomy theorem (e.g., see [38]) says that $\mu_{y,t}$ is either equivalent or singular to μ, and they are equivalent if and only if $t = 1$ and $y \in \mathcal{E}_0$. Moreover, for $h \in \mathcal{E}_0$, the Radon-Nikodym derivative of $\mu_{h,1}$ with respect to μ is given by

$$\frac{d\mu_{h,1}}{d\mu}(x) = \exp\left[\langle x, h\rangle - \frac{1}{2}|h|_0^2\right], \qquad x \in \mathcal{E}'. \tag{3.4.34}$$

First consider the measure $\mu_y(\cdot) = \mu(\cdot - y)$ with $y \in \mathcal{E}'$. Recall from Example 3.3.3 that $:e^{\langle \cdot, \xi\rangle}: \in (\mathcal{E})$ for any $\xi \in \mathcal{E}_c$. On the one hand, it is easy to check that

$$\int_{\mathcal{E}'} :e^{\langle x,\xi\rangle}: \, d\mu_y(x) = e^{\langle y,\xi\rangle}, \qquad \forall \xi \in \mathcal{E}_c. \tag{3.4.35}$$

On the other hand, $:e^{\langle \cdot, y\rangle}: \in (\mathcal{E})^*$ and by Equation (3.4.29) we have

$$\langle\!\langle :e^{\langle \cdot,y\rangle}:, :e^{\langle \cdot,\xi\rangle}: \rangle\!\rangle = e^{\langle y,\xi\rangle}, \qquad \forall \xi \in \mathcal{E}_c. \tag{3.4.36}$$

Since the linear span of the set $\{: e^{\langle \cdot,\xi\rangle} :; \xi \in \mathcal{E}_c\}$ is dense in $(\mathcal{E})$, we can conclude from Equations (3.4.35) and (3.4.36) that μ_y is a Hida measure and its generalized Radon-Nikodym derivative with respect to μ is given by

$$\frac{d\mu_y}{d\mu} = \widetilde{\mu}_y = :e^{\langle \cdot,y\rangle}: .$$

Observe that if $y = h \in \mathcal{E}_0$, then $: e^{\langle \cdot,h\rangle} : = \exp\left[\langle \cdot, h\rangle - \frac{1}{2}|h|_0^2\right]$ and so the above formula becomes the one in Equation (3.4.34).

Now, let $t > 0$ and consider the measure $\mu^{(t)}(\cdot) = \mu(t^{-1/2}(\cdot))$. We can easily check that

$$\int_{\mathcal{E}'} :e^{\langle x,\xi\rangle}: \, d\mu^{(t)}(x) = \exp\left[\frac{1}{2}(t-1)\langle \xi, \xi\rangle\right], \qquad \forall \xi \in \mathcal{E}_c. \tag{3.4.37}$$

Define a function Φ_t by

$$\Phi_t = \sum_{n=0}^{\infty} \frac{(t-1)^n}{n!2^n} \langle :\cdot^{\otimes 2n}:, \tau^{\otimes n}\rangle, \tag{3.4.38}$$

where τ is the trace operator. This function defines a generalized function in $(\mathcal{E})^*$ and for all $\xi \in \mathcal{E}_c$,

$$\begin{aligned} \langle\!\langle \Phi_t, :e^{\langle x,\xi\rangle}: \rangle\!\rangle &= \textstyle\sum_{n=0}^{\infty}(2n)! \left\langle \frac{(t-1)^n}{n!2^n}\tau^{\otimes n}, \frac{1}{(2n)!}\xi^{\otimes 2n}\right\rangle \\ &= \textstyle\sum_{n=0}^{\infty} \frac{(t-1)^n}{n!2^n} \langle \xi, \xi\rangle^n \\ &= \exp\left[\frac{1}{2}(t-1)\langle \xi, \xi\rangle\right]. \end{aligned} \tag{3.4.39}$$

By comparing Equations (3.4.37) and (3.4.39) we conclude right away that $\mu^{(t)}$ is a Hida measure and its generalized Radon-Nikodym derivative with respect to μ is given by Equation (3.4.38), i.e.,

$$\frac{d\mu^{(t)}}{d\mu} = \widetilde{\mu}^{(t)} = \sum_{n=0}^{\infty} \frac{(t-1)^n}{n!2^n} \langle :\cdot^{\otimes 2n}:, \tau^{\otimes n}\rangle. \tag{3.4.40}$$

In fact, for any $y \in \mathcal{E}'$ and $t > 0$, the measure $\mu_{y,t}(\cdot) = \mu(t^{-1/2}(\cdot - y))$ is a Hida measure and

$$\langle\langle \widetilde{\mu}_{y,t}(\cdot), :e^{\langle x,\xi\rangle}: \rangle\rangle = \exp\left[\langle y, \xi\rangle + \frac{1}{2}(t-1)\langle \xi, \xi\rangle\right].$$

The expression for the Wiener-Itô decomposition of $\widetilde{\mu}_{y,t}(\cdot)$ is very complicated. But, without knowing this decomposition, how can we tell that $\mu_{y,t}$ is really a Hida measure? One way is to apply Theorem 3.4.8. The other way is to use the S-transform which we will discuss in 12.5.118 and 12.5.119.

3.5 Characterization theorems

3.5.1 The S-transform

In 12.4.6 we saw that a generalized function can be identified by its action on test functions $: e^{\langle \cdot, \xi\rangle} :$ for $\xi \in \mathcal{E}_c$. This way of identifying a generalized function Φ is quite useful, in particular, when it is very hard or impossible to find the explicit form for the Wiener-Itô decomposition of Φ.

Recall that for any $\xi \in \mathcal{E}_c$, the function $: e^{\langle \cdot, \xi\rangle} :$ is a test function in all of the spaces $(\mathcal{E})$ (ref3.3), $(\mathcal{E})_\beta$ with $0 \leq$ ¡1 (12.3.20), and $[\mathcal{E}]_\alpha$ (12.3.21).

The *S-transform* of a generalized function Φ is defined to be the function

$$S\Phi(\xi) = \langle\langle \Phi, :e^{\langle \cdot, \xi\rangle}: \rangle\rangle, \qquad \xi \in \mathcal{E}_c.$$

This concept of S-transform is due to Kubo and Takenaka [33]. When Hida introduced the theory of white noise in 1975 [18], he used the $\mathcal{T}$-transform

$$\mathcal{T}\Phi(\xi) = \langle\langle \Phi, e^{i\langle \cdot, \xi\rangle} \rangle\rangle, \qquad \xi \in \mathcal{E}.$$

Obviously, we can also regard $(\mathcal{T}\Phi)(\xi)$ as defined for $\xi \in \mathcal{E}_c$. The relationship between S- and $\mathcal{T}$-transforms is given by

$$\mathcal{T}\Phi(\xi) = e^{-\frac{1}{2}\langle \xi, \xi\rangle} S\Phi(i\xi), \quad S\Phi(\xi) = e^{-\frac{1}{2}\langle \xi, \xi\rangle} \mathcal{T}\Phi(-i\xi).$$

The restriction of the S-transform to the Hilbert space (L^2) is known as the Segal-Bargmann transform [9].

Since the linear span of the set $\{: e^{\langle \cdot, \xi\rangle} :;\ \xi \in \mathcal{E}_c\}$ is dense in each of the three spaces of test functions, a generalized function is uniquely determined by its S-transform. Of course, the linear span of the set $\{: e^{\langle \cdot, \xi\rangle} :;\ \xi \in \mathcal{E}\}$ is also dense and we could have defined $S\Phi(\xi)$ for $\xi \in \mathcal{E}$. However, we use the space $\mathcal{E}_c$ instead of $\mathcal{E}$ in the definition of the S-transform because of its convenience for the characterization theorems in 12.5.119 and 12.5.121.

Suppose a generalized function Φ is represented by

$$\Phi = \sum_{n=0}^{\infty} \langle :\cdot^{\otimes n}:, F_n\rangle.$$

Then its S-transform is given by

$$S\Phi(\xi) = \sum_{n=0}^{\infty} \langle F_n, \xi^{\otimes n}\rangle, \qquad \xi \in \mathcal{E}_c.$$

Now, observe that if φ is a test function, then its S-transform can be written as

$$S\varphi(\xi) = \langle\!\langle :e^{\langle\cdot,\xi\rangle}:, \varphi\rangle\!\rangle, \qquad \xi \in \mathcal{E}_c.$$

Note that $:e^{\langle\cdot,x\rangle}:$ is a generalized function for any $x \in calE'$. Thus we can restrict the S-transform to the space of test functions and define

$$\breve{S}\varphi(x) = \langle\!\langle :e^{\langle\cdot,x\rangle}:, \varphi\rangle\!\rangle, \qquad x \in \mathcal{E}'.$$

Then in view of Equation (3.4.30) we have the $\breve{S}\theta\varphi = \varphi$ for all $\varphi \in (\mathcal{E})$. Thus $\breve{S} = \Theta^{-1}$. Hence $\breve{S}$ is the continuous linear operator from $(\mathcal{E})$ into itself such that

$$\breve{S}: \,:e^{\langle\cdot,\xi\rangle}: \longmapsto e^{\langle\cdot,\xi\rangle}, \qquad \forall \xi \in \mathcal{E}_c.$$

3.5.2 Characterization of generalized functions

The S-transform of a generalized function is a function on $\mathcal{E}_c$. In order to specify a generalized function by its S-transform, we must have a precise description of those functions on $\mathcal{E}_c$ which are S-transforms of generalized functions. This precise description is known as the characterization of generalized functions.

Let Φ be a generalized function and $F = S\Phi$. For any fixed $\xi, \eta \in \mathcal{E}_c$, we have

$$F(\xi + z\eta) = e^{-\frac{1}{2}(\langle\xi,\xi\rangle + 2z\langle\xi,\eta\rangle + z^2\langle\eta,\eta\rangle)} \langle\!\langle \Phi, e^{\langle\cdot,\xi\rangle + z\langle\cdot,\eta\rangle}\rangle\!\rangle.$$

It is almost obvious that the function $F(\xi + z\eta)$ is an entire function of $z \in \mathbf{C}$. For a proof, see Lemma 8.1 in [40]. This analyticity condition does not depend on what kind of generalized function Φ is.

The other condition which F must satisfy is the growth condition. The growth condition plays the most crucial role in the characterization. It depends on the spaces of generalized functions, namely, $(\mathcal{E})^*$, $(\mathcal{E})^*_\beta$, $[\mathcal{E}]^*_\alpha$, $[\mathcal{E}]^*_u$.

A. Hida-Kubo-Takenaka space (generalized functions)

Let $\Phi \in (\mathcal{E})^*$ and $F = S\Phi$. Then there exists some $p \geq 0$ such that $\Phi \in (\mathcal{E}_p)^*$. Hence by Equation (3.3.11),

$$|F(\xi)| \leq \|\Phi\|_{-p} \, \|:e^{\langle\cdot,\xi\rangle}:\|_p = \|\Phi\|_{-p} \exp\left[\frac{1}{2}|\xi|_p^2\right].$$

This is a growth condition for $F = S\Phi$ with $\Phi \in (\mathcal{E})^*$.

The next theorem is due to Potthoff and Streit [50]. For the proof see [22] or [40]. Actually we have proved the trivial part, i.e., necessity part, of this theorem.

Theorem 3.5.1 *A function $F\colon \mathcal{E}_c \to \mathbf{C}$ is the S-transform of a generalized function in $(\mathcal{E})^*$ if and only if it satisfies the conditions:*

(1) For any $\xi, \eta \in \mathcal{E}_c$, the function $F(z\xi + \eta)$ is an entire function of $z \in \mathbf{C}$.

(2) There exist constants $K, a, p \geq 0$ such that

$$|F(\xi)| \leq K \exp\left[a|\xi|_p^2\right], \qquad \xi \in \mathcal{E}_c.$$

The growth condition (2) is equivalent to the condition: there exist constants $K, p \geq 0$ such that

$$|F(\xi)| \leq K \exp\left[|\xi|_p^2\right], \qquad \xi \in \mathcal{E}_c.$$

The equivalence can be checked by using the inequality $|\xi|_p \leq \rho^{q-p}|\xi|_q$ for any $q \geq p$, which follows from condition (a) in 12.3.17. Having the constant a in the inequality is just for convenience to check the growth condition. This remark also applies to the growth conditions for other spaces.

Example 3.5.2 *In Example 3.2.5 we defined Donsker's delta function $\delta(B(t) - a)$. Here we give another definition. In the distribution sense we have the equality*

$$\delta_a(x) = \frac{1}{2\pi} \int_{\mathbf{R}} e^{iu(x-a)}\, du.$$

Put $x = B(t)$ to get

$$\delta_a(B(t)) = \frac{1}{2\pi} \int_{\mathbf{R}} e^{iu(B(t)-a)}\, du.$$

Apply the S-transform and interchange it with the integral to derive the equality

$$S(\delta_a(B(t)))(\xi) = \frac{1}{\sqrt{2\pi t}} \exp\left[-\frac{1}{2t}\left(a - \int_0^t \xi(u)\, du\right)^2\right].$$

Obviously, this function satisfies conditions (1) and (2) of Theorem 3.5.1. Hence Donsker's delta function is a generalized function in the space $(\mathcal{S})^*$ (see 12.2.4).

Example 3.5.3 *Consider the function $F(\xi) = \sin\langle\xi, \xi\rangle$, $\xi \in \mathcal{E}_c$. We can easily check that this function satisfies conditions (1) and (2) in Theorem 3.5.1. Hence it is the S-transform of a generalized function in $(\mathcal{E})^*$. Similarly, the functions $\cos\langle\xi, \xi\rangle$, $\sinh\langle\xi, \xi\rangle$, $\cosh\langle\xi, \xi\rangle$, are all S-transforms of generalized functions in $(\mathcal{E})^*$.*

B. Kondratiev-Streit space (generalized functions)

Let $\Phi \in (\mathcal{E})_\beta^*$ and $F = S\Phi$. Then there exists some $p \geq 0$ such that $\Phi \in (\mathcal{E}_p)_\beta^*$. Hence by Equation (3.3.15),

$$|F(\xi)| \leq \|\Phi\|_{-p,-\beta} \|{:}e^{\langle\cdot,\xi\rangle}{:}\|_{p,\beta} = \|\Phi\|_{-p,-\beta}\, G^{(\beta)}\left(|\xi|_p^2\right)^{1/2},$$

where the function $G^{(\beta)}$ is defined by

$$G^{(\beta)}(r) = \sum_{n=0}^{\infty} \frac{1}{(n!)^{1-\beta}} r^n.$$

We can use the function $G^{(\beta)}$ as a growth condition. However, this is not so good because the series for $G^{(\beta)}$ cannot be summed up in a closed form unless $\beta = 0$ (the Hida-Kubo-Takenaka case.) Thus the growth condition using $G^{(\beta)}$ as a growth function is almost impossible to check when $0 < \beta < 1$.

Fortunately we have the inequalities from page 358 in [40] and Lemma 7.1 in [14]:

$$\exp\left[(1-\beta)r^{\frac{1}{1-\beta}}\right] \leq G^{(\beta)}(r) \leq 2^\beta \exp\left[(1-\beta)2^{\frac{\beta}{1-\beta}} r^{\frac{1}{1-\beta}}\right], \qquad r \geq 0.$$

Hence we can replace $G^{(\beta)}(r)$ by $\exp\left[r^{\frac{1}{1-\beta}}\right]$ as a growth function.

The next theorem is due to Kondratiev and Streit [30] [31]. For the proof, see the book [40].

Theorem 3.5.4 *A function $F : \mathcal{E}_c \to \mathbf{C}$ is the S-transform of a generalized function in $(\mathcal{E})^*_\beta$ if and only if it satisfies the conditions:*

(1) For any $\xi, \eta \in \mathcal{E}_c$, the function $F(z\xi + \eta)$ is an entire function of $z \in \mathbf{C}$.

(2) There exist constants $K, a, p \geq 0$ such that

$$|F(\xi)| \leq K \exp\left[a|\xi|_p^{\frac{2}{1-\beta}}\right], \qquad \xi \in \mathcal{E}_c.$$

Example 3.5.5 *The grey noise measure was introduced in [73] (see also [40].) It is the measure ν_λ, $0 < \lambda \leq 1$, on $\mathcal{E}'$ with characteristic function given by*

$$\int_{\mathcal{E}'} e^{i\langle x,\xi\rangle}\, d\nu_\lambda(x) = L_\lambda\big(|\xi|_0^2\big), \qquad \xi \in \mathcal{E},$$

where $L_\lambda(t)$ is the Mittag-Leffler function with parameter λ, i.e.,

$$L_\lambda(t) = \sum_{n=0}^\infty \frac{(-t)^n}{\Gamma(1+\lambda n)}.$$

Here Γ is the gamma function. It is shown in Example 8.5 in [40] that ν_λ is a Hida measure. The generalized function $\widetilde{\nu}_\lambda$ induced by ν_λ has S-transform given by

$$S\widetilde{\nu}_\lambda(\xi) = e^{-\frac{1}{2}\langle\xi,\xi\rangle} L_\lambda\big(-\langle\xi,\xi\rangle\big), \qquad \xi \in \mathcal{E}_c.$$

Therefore, $S\widetilde{\nu}_\lambda$ satisfies the inequality

$$|S\widetilde{\nu}_\lambda(\xi)| \leq C_\lambda \exp\left[2^{-1}|\xi|_0^2 + \lambda|\xi|_0^{2/\lambda}\right], \qquad \xi \in \mathcal{E}_c.$$

where C_λ is a constant depending only on λ. Hence by the above theorem, $\widetilde{\nu}_\lambda$ is a generalized function in the space $(\mathcal{E})^*_{1-\lambda}$.

C. Cochran-Kuo-Sengupta space (generalized functions)

Let $\Phi \in [\mathcal{E}]^*_\alpha$ and $F = S\Phi$. Then there exists some $p \geq 0$ such that $\Phi \in [\mathcal{E}_p]^*_\alpha$. Hence by Equation (3.3.20),

$$|F(\xi)| \leq \|\Phi\|_{-p,1/\alpha} \|:e^{\langle\cdot,\xi\rangle}:\|_{p,\alpha} = \|\Phi\|_{-p,1/\alpha}\, G_\alpha\big(|\xi|_p^2\big)^{1/2},$$

where G_α is the exponential generating function of the sequence $\{\alpha(n)\}_{n=0}^\infty$, i.e.,

$$G_\alpha(r) = \sum_{n=0}^\infty \frac{\alpha(n)}{n!} r^n.$$

We state two conditions on the sequence $\{\alpha(n)\}_{n=0}^\infty$:

- (B1) $\displaystyle\limsup_{n\to\infty} \left(\frac{n!}{\alpha(n)} \inf_{r>0} \frac{G_\alpha(r)}{r^n}\right)^{1/n} < \infty.$

(B2) The sequence $\gamma(n) = \alpha(n)/n!$, $n \geq 0$, is log-concave, i.e.,

$$\gamma(n)\gamma(n+2) \leq \gamma(n+1)^2, \qquad \forall n \geq 0.$$

It follows from Theorem 4.3 in [14] that condition (B2) implies condition (B1). Obviously, the sequence $\alpha(n) = 1$ for all n (for the Hida-Kubo-Takenaka space) satisfies conditions (B1) and (B2). The sequence $\alpha(n) = (n!)^\beta$ (for the Kondratiev-Streit space) satisfies condition (B2), hence also (B1). In [14] the Bell numbers (see Example 3.3.11) are shown to satisfy condition (B1). But it is proved in [4] that the Bell numbers actually satisfy condition (B2).

The next theorem is due to Cochran et al. [14].

Theorem 3.5.6 *If F is the S-transform of $\Phi \in [\mathcal{E}]_\alpha^*$, then F satisfies the conditions:*

(1) For any $\xi, \eta \in \mathcal{E}_c$, the function $F(z\xi + \eta)$ is an entire function of $z \in \mathbf{C}$.

(2) There exist constants $K, a, p \geq 0$ such that

$$|F(\xi)| \leq K G_\alpha\big(a|\xi|_p^2\big)^{1/2}, \qquad \xi \in \mathcal{E}_c.$$

Conversely, suppose condition (B1) holds and let $F : \mathcal{E}_c \to \mathbf{C}$ be a function satisfying conditions (1) and (2). Then F is the S-transform of a generalized function in $[\mathcal{E}]_\alpha^*$.

Observe that under condition (B1) or the stronger condition (B2), a complex-valued function F on $\mathcal{E}_c$ is the S-transform of a generalized function in $[\mathcal{E}]_\alpha^*$ if and only if it satisfies the above conditions (1) and (2).

Example 3.5.7 *The Poisson noise measure on $\mathcal{S}'(\mathbf{R})$ is the measure $\wp$ having the characteristic function*

$$\int_{\mathcal{S}'(\mathbf{R})} e^{i\langle x,\xi\rangle}\, d\wp(x) = \exp\left[\int_{\mathbf{R}} \big(e^{i\xi(t)} - 1\big)\, dt\right], \qquad \xi \in \mathcal{S}(\mathbf{R}).$$

It follows from this equality that

$$\int_{\mathcal{S}'(\mathbf{R})} :e^{\langle x,\xi\rangle}:\, d\wp(x) = \exp\left[\int_{\mathbf{R}} \Big(e^{\xi(t)} - 1 - \frac{1}{2}\xi(t)^2\Big)\, dt\right], \qquad \xi \in \mathcal{S}_c(\mathbf{R}).$$

Therefore,

$$\left|\int_{\mathcal{S}'(\mathbf{R})} :e^{\langle x,\xi\rangle}:\, d\wp(x)\right| \leq \exp\left[\frac{1}{2}|\xi|_0^2 + \int_{\mathbf{R}} \big|e^{\xi(t)} - 1\big|\, dt\right], \qquad \xi \in \mathcal{E}_c.$$

From this inequality we can check by elementary calculations that there exist constants $K, a, p \geq 0$ such that

$$\left|\int_{\mathcal{S}'(\mathbf{R})} :e^{\langle x,\xi\rangle}:\, d\wp(x)\right| \leq K G_{b_2}\big(a|\xi|_p^2\big)^{1/2}, \qquad \xi \in \mathcal{E}_c,$$

where G_{b_2} is the exponential generating function of the Bell numbers $\{b_2(n)\}_{n=0}^\infty$ of order 2, i.e., from Equation (3.3.23),

$$G_{b_2}(r) = \sum_{n=0}^\infty \frac{b_2(n)}{n!} r^n = \frac{1}{\exp_2(0)} \sum_{n=0}^\infty \frac{B_2(n)}{n!} r^n = \exp\left[e^r - 1\right].$$

Hence by the above theorem, the Poisson noise measure $\wp$ induces a generalized function $\widetilde{\wp}$ in the Bell number space $[\mathcal{S}]_{b_2}^*$ (see Example 3.3.11.)

D. CKS-space associated with a growth function (generalized functions)

Let u be a growth function in $C_{+,1/2}$ satisfying the conditions (U1) (U2) (U3) in 3.4.4. Define the *dual Legendre transform* of u by

$$u^*(r) = \sup_{s\geq 0} \frac{e^{2\sqrt{rs}}}{u(s)}, \qquad r \in [0,\infty).$$

Let $[\mathcal{E}]_u \subset (L^2) \subset [\mathcal{E}]_u^*$ be the CKS-space associated with u as defined in 3.4.4. The following theorem is due to Asai et al. [7] [8].

Theorem 3.5.8 *Let $u \in C_{+,1/2}$ satisfy conditions (U1) (U2) (U3). Then a function $F : \mathcal{E}_c \to \mathbf{C}$ is the S-transform of a generalized function in $[\mathcal{E}]_u^*$ if and only if it satisfies the conditions:*

(1) For any $\xi, \eta \in \mathcal{E}_c$, the function $F(z\xi + \eta)$ is an entire function of $z \in \mathbf{C}$.

(2) There exist constants $K, a, p \geq 0$ such that

$$|F(\xi)| \leq K u^*\big(a|\xi|_p^2\big)^{1/2}, \qquad \xi \in \mathcal{E}_c.$$

3.5.3 Convergence of generalized functions

We have several spaces of test functions, namely, $(\mathcal{E})$, $(\mathcal{E})_\beta$, $[\mathcal{E}]_\alpha$, $[\mathcal{E}]_u$. They are all nuclear spaces. Hence the strong topology and the inductive limit topology on each of the dual spaces are the same.

Since a generalized function can be understood by its S-transform as shown in the previous section, we need to express the convergence of generalized functions in terms of their S-transforms. The next theorem is due to Potthoff and Streit [50] for the case $\beta = 0$. See [40] for the proof.

Theorem 3.5.9 *Let $\Phi_n \in (\mathcal{E})_\beta^*$ and $F_n = S\Phi_n$. Then Φ_n converges strongly in $(\mathcal{E})_\beta^*$ if and only if the following conditions are satisfied:*

(1) $\lim_{n\to\infty} F_n(\xi)$ exists for each $\xi \in \mathcal{E}_c$.

(2) There exist constants $K, a, p \geq 0$, independent of n, such that

$$|F_n(\xi)| \leq K \exp\Big[a|\xi|_p^{\frac{2}{1-\beta}}\Big], \qquad \forall n \geq 1,\ \xi \in \mathcal{E}_c.$$

This theorem can be extended to the space $[\mathcal{E}]_\alpha^*$ by replacing condition (2) with the condition: There exist constants $K, a, p \geq 0$, independent of n, such that

$$|F_n(\xi)| \leq K G_\alpha\big(a|\xi|_p^2\big)^{1/2}, \qquad \forall n \geq 1,\ \xi \in \mathcal{E}_c.$$

Similarly, for the space $[\mathcal{E}]_u^*$ associated with a growth function u satisfying the conditions (U1) (U2) (U3), we simply replace the above condition (2) with the condition: There exist constants $K, a, p \geq 0$, independent of n, such that

$$|F_n(\xi)| \leq K u^*\big(a|\xi|_p^2\big)^{1/2}, \qquad \forall n \geq 1,\ \xi \in \mathcal{E}_c.$$

3.5.4 Characterization of test functions

Suppose φ is a test function. Obviously, the S-transform $S\varphi$ of φ must satisfy the same analyticity condition for generalized functions. However, the growth condition is different.

A. Hida-Kubo-Takenaka space (test functions)

Let $\varphi \in (\mathcal{E})$ and $F = S\varphi$. In order to find the growth condition, write F as

$$F(\xi) = \langle\langle :e^{\langle \cdot, \xi \rangle}:, \varphi \rangle\rangle.$$

For any $q \geq p \geq 0$, use Equation (3.3.12) and condition (a) in 12.3.17 to get

$$|F(\xi)| \leq \| :e^{\langle \cdot, \xi \rangle}: \|_{-q} \|\varphi\|_q \leq \|\varphi\|_q \exp\left[\frac{1}{2}|\xi|_{-q}^2\right] \leq \|\varphi\|_q \exp\left[\frac{1}{2}\rho^{2(q-p)}|\xi|_{-p}^2\right].$$

Hence for any $a, p \geq 0$, we can choose $q > p$ such that $2^{-1}\rho^{2(q-p)} < a$. Then

$$|F(\xi)| \leq \|\varphi\|_q \exp\left[a|\xi|_{-p}^2\right].$$

This gives the growth condition for the S-transform of a test function in $(\mathcal{E})$. The next theorem is due to Kuo et al. [41]. For the proof see [22] or [40].

Theorem 3.5.10 *A function $F\colon \mathcal{E}_c \to \mathbf{C}$ is the S-transform of a test function in $(\mathcal{E})$ if and only if it satisfies the conditions:*

(1) For any $\xi, \eta \in \mathcal{E}_c$, the function $F(z\xi + \eta)$ is an entire function of z1$\mathbf{C}$.

(2) For any $a, p \geq 0$, there exists a constant $K \geq 0$ such that

$$|F(\xi)| \leq K \exp\left[a|\xi|_{-p}^2\right], \qquad \xi \in \mathcal{E}_c.$$

B. Kondratiev-Streit space (test functions)

Let $\varphi \in (\mathcal{E})_\beta$ and $F = S\varphi$. For any $q \geq p \geq 0$, use Equation (3.3.16) and condition (a) in 12.3.17 to get

$$\begin{aligned} |F(\xi)| &\leq \| :e^{\langle \cdot, \xi \rangle}: \|_{-q,-\beta} \|\varphi\|_{q,\beta} \\ &= \|\varphi\|_{q,\beta} G^{(-\beta)}\left(|\xi|_{-q}^2\right)^{1/2} \\ &\leq \|\varphi\|_{q,\beta} G^{(-\beta)}\left(\rho^{2(q-p)}|\xi|_{-p}^2\right)^{1/2}, \end{aligned}$$

where the function $G^{(-\beta)}$ is defined by

$$G^{(-\beta)}(r) = \sum_{n=0}^{\infty} \frac{1}{(n!)^{1+\beta}} r^n.$$

It is not practical to use the growth function $G^{(-\beta)}$ as for the case of generalized functions in $(\mathcal{E})_\beta^*$. But we have the inequalities for $r \geq 0$

$$2^{-\beta} \exp\left[(1+\beta)2^{-\frac{\beta}{1+\beta}} r^{\frac{1}{1+\beta}}\right] \leq G^{(-\beta)}(r) \leq \exp\left[(1+\beta) r^{\frac{1}{1+\beta}}\right]. \tag{3.5.41}$$

Therefore, we get

$$|F(\xi)| \leq \|\varphi\|_{q,\beta} \exp\left[2^{-1}(1+\beta)\rho^{\frac{2(q-p)}{1+\beta}} |\xi|_{-p}^{\frac{2}{1+\beta}}\right].$$

Hence for any $a, p \geq 0$, we can choose $q > p$ such that $2^{-1}(1+\beta)\rho^{\frac{2(q-p)}{1+\beta}} < a$. Then

$$|F(\xi)| \leq \|\varphi\|_{q,\beta} \exp\left[a|\xi|_{-p}^{\frac{2}{1+\beta}}\right].$$

This is the growth condition for the S-transform of a test function in $(\mathcal{E})_\beta$.

The next theorem is due to Kondratiev and Streit [30] [31]. For the proof, see the book [40].

Theorem 3.5.11 *A function $F\colon \mathcal{E}_c \to \mathbf{C}$ is the S-transform of a test function in $(\mathcal{E})_\beta$ if and only if it satisfies the conditions:*

(1) For any $\xi, \eta \in \mathcal{E}_c$, the function $F(z\xi + \eta)$ is an entire function of $z \in \mathbf{C}$.

(2) For any $a, p \geq 0$, there exists a constant $K \geq 0$ such that

$$|F(\xi)| \leq K \exp\left[a|\xi|_{-p}^{\frac{2}{1+\beta}}\right], \qquad \xi \in \mathcal{E}_c.$$

C. Cochran-Kuo-Sengupta space (test functions)

Let $\varphi \in (\mathcal{E})_\alpha$ and $F = S\varphi$. For any $q \geq p \geq 0$, use Equation (3.3.21) and condition (a) in 12.3.17 to get

$$\begin{aligned} |F(\xi)| \quad &\leq \|{:}e^{\langle\cdot,\xi\rangle}{:}\|_{-q,1/\alpha} \|\varphi\|_{q,\alpha} \\ &= \|\varphi\|_{q,\alpha} G_{1/\alpha}\big(|\xi|_{-q}^2\big)^{1/2} \\ &\leq \|\varphi\|_{q,\alpha} G_{1/\alpha}\big(\rho^{2(q-p)}|\xi|_{-p}^2\big)^{1/2}, \end{aligned}$$

where $G_{1/\alpha}$ is the exponential generating function of $\{1/\alpha(n)\}_{n=0}^\infty$, i.e.,

$$G_{1/\alpha}(r) = \sum_{n=0}^\infty \frac{1}{n!\alpha(n)} r^n.$$

Hence for any $a, p \geq 0$, we can choose $q > p$ such that $\rho^{2(q-p)} < a$. Then

$$|F(\xi)| \leq \|\varphi\|_{q,\alpha} G_{1/\alpha}\big(a|\xi|_{-p}^2\big)^{1/2}.$$

This is the growth condition for the S-transform of a test function in $(\mathcal{E})_\alpha$.

We state two conditions on the sequence $\{1/\alpha(n)\}_{n=0}^\infty$:

- ($\widetilde{B}1$) $\displaystyle\limsup_{n\to\infty}\left(n!\alpha(n) \inf_{r>0} \frac{G_{1/\alpha}(r)}{r^n}\right)^{1/n} < \infty.$

  ($\widetilde{B}2$) The sequence $\left\{\dfrac{1}{n!\alpha(n)}\right\}$ is log-concave.

Similar to the conditions (B1) and (B2), condition ($\widetilde{B}2$) implies condition ($\widetilde{B}1$). Moreover, it is shown in [4] that the Bell numbers satisfy condition ($\widetilde{B}2$).

The next theorem is due to Asai et al. [5].

Theorem 3.5.12 *If F is the S-transform of a test function in $[\mathcal{E}]_\alpha$, then F satisfies the conditions:*

(1) *For any $\xi, \eta \in \mathcal{E}_c$, the function $F(z\xi + \eta)$ is an entire function of $z \in \mathbf{C}$.*

(2) *For any $a, p \geq 0$, there exists a constant $K \geq 0$ such that*

$$|F(\xi)| \leq K G_{1/\alpha}\big(a|\xi|_{-p}^2\big)^{1/2}, \qquad \xi \in \mathcal{E}_c.$$

Conversely, suppose condition $(\widetilde{B}1)$ holds and let $F : \mathcal{E}_c \to \mathbf{C}$ be a function satisfying conditions (1) and (2). Then F is the S-transform of a test function in $[\mathcal{E}]_\alpha$.

D. CKS-space associated with a growth function (test functions)

Recall from 3.4.4 that we have a Gel'fand triple $[\mathcal{E}]_u \subset (L^2) \subset [\mathcal{E}]_u^*$ associated with a growth function $u \in C_{+,1/2}$. The next theorem is due to Asai et al. [7] [8].

Theorem 3.5.13 *Let $u \in C_{+,1/2}$ satisfy conditions (U1) (U2) (U3) in 3.4.4. Then a function $F : \mathcal{E}_c \to \mathbf{C}$ is the S-transform of a test function in $[\mathcal{E}]_u$ if and only if it satisfies the conditions:*

(1) *For any $\xi, \eta \in \mathcal{E}_c$, the function $F(z\xi + \eta)$ is an entire function of $z \in \mathbf{C}$.*

(2) *For any $a, p \geq 0$, there exists a constant $K \geq 0$ such that*

$$|F(\xi)| \leq K u\big(a|\xi|_{-p}^2\big)^{1/2}, \qquad \xi \in \mathcal{E}_c.$$

3.5.5 Intrinsic topology for the space of test functions

In the finite dimensional Schwartz distribution theory, a test function is infinitely differentiable and rapidly decreasing. In the white noise distribution theory, this property is replaced by the analyticity and growth conditions. This idea is due to Y.-J. Lee [43] for the test functions in $(\mathcal{E})$. The extension to the space $(\mathcal{E})_\beta$ involves only delicate computations. However, the extension to the space $[\mathcal{E}]_\alpha$ and $[\mathcal{E}]_u$ requires new concepts and techniques.

Recall from Example 3.3.4 that $:e^{\langle \cdot, x\rangle}:$ is a generalized function for any $x \in \mathcal{E}_c'$. Thus Equation (3.4.30) for a test function $\varphi \in (\mathcal{E})$ can be extended to $x \in \mathcal{E}_c'$,

$$\varphi(x) = \langle\!\langle :e^{\langle \cdot, x\rangle}:, \Theta\varphi \rangle\!\rangle, \qquad x \in \mathcal{E}_c',$$

where Θ is a continuous linear operator from $(\mathcal{E})$ into itself defined in 12.4.76. From the above equality, we get

$$|\varphi(x)| \leq \| :e^{\langle \cdot, x\rangle}: \|_{-p} \, \|\Theta\varphi\|_p = \|\Theta\varphi\|_p \exp\left[\frac{1}{2}|\xi|_{-p}^2\right], \qquad x \in \mathcal{E}_{p,c}'.$$

Being motivated by this inequality, we define a norm $\|\cdot\|_{\mathcal{A}_p}$ on the space $(\mathcal{E})$ by

$$\|\varphi\|_{\mathcal{A}_p} = \sup_{x \in \mathcal{E}_{p,c}'} |\varphi(x)| \exp\left[-\frac{1}{2}|x|_{-p}^2\right].$$

The next theorem is due to Y.-J. Lee [43]. See Theorem 4.60 in [22].

Theorem 3.5.14 *The topology on $(\mathcal{E})$ generated by $\{\|\cdot\|_{\mathcal{A}_p};\, p \geq 0\}$ is the same as the one generated by $\{\|\cdot\|_p;\, p \geq 0\}$.*

Next, consider test functions in $(\mathcal{E})_\beta$. We can derive from Equations (3.3.16) and (3.5.41) the following inequality

$$|\varphi(x)| \leq \|\Theta\varphi\|_{p,\beta} \exp\left[\frac{1}{2}(1+\beta)|x|_{-p}^{\frac{2}{1+\beta}}\right], \qquad x \in \mathcal{E}'_{p,c}.$$

where $\Theta\colon (\mathcal{E})_\beta \to (\mathcal{E})_\beta$ is continuous by Theorem 6.2 in [40]. In view of this inequality we define the following norm for each $p \geq 0$,

$$\|\varphi\|_{\mathcal{A}_{p,\beta}} = \sup_{x \in \mathcal{E}'_{p,c}} |\varphi(x)| \exp\left[-\frac{1}{2}(1+\beta)|x|_{-p}^{\frac{2}{1+\beta}}\right].$$

The next theorem is from Theorem 15.14 in [40].

Theorem 3.5.15 *The topology on $(\mathcal{E})_\beta$ generated by $\{\|\cdot\|_{\mathcal{A}_{p,\beta}};\, p \geq 0\}$ is the same as the one generated by $\{\|\cdot\|_{p,\beta};\, p \geq 0\}$.*

Now, we consider the test functions in the space $[\mathcal{E}]_u$. Let $u \in C_{+,1/2}$ be a growth function satisfying conditions (U1) (U2) (U3) in 3.4.4. Recall that the space $[\mathcal{E}]_u$ of test functions is $[\mathcal{E}]_\alpha$ associated with the sequence

$$\alpha(n) = \frac{1}{\ell_u(n)n!}, \qquad n \geq 0,$$

where ℓ_u is the Legendre transform of u. Thus the norms on $[\mathcal{E}]_u$ are given by

$$\|\varphi\|_{p,u} = \left(\sum_{n=0}^{\infty} \frac{1}{\ell_u(n)} |f_n|_p^2\right)^{1/2}, \qquad p \geq 0.$$

Being motivated by the growth condition in Theorem 3.5.13, we define another family of norms $\{\|\cdot\|_{\mathcal{A}_{p,u}}\}$ on $[\mathcal{E}]_u$ by

$$\|\varphi\|_{\mathcal{A}_{p,u}} = \sup_{x \in \mathcal{E}'_{p,c}} |\varphi(x)|\, u\big(|x|_{-p}^2\big)^{-1/2}.$$

The next theorem is due to Asai et al. [6] [8].

Theorem 3.5.16 *Let $u \in C_{+,1/2}$ satisfy conditions (U1) (U2) (U3) in 3.4.4. The topology on $[\mathcal{E}]_u$ generated by $\{\|\cdot\|_{\mathcal{A}_{p,u}};\, p \geq 0\}$ is the same as the one generated by $\{\|\cdot\|_{p,u};\, p \geq 0\}$.*

3.6 Continuous operators and adjoints

In 3.6.1 to 3.6.5 we will discuss various continuous linear operators acting on the Kondratiev-Streit space $(\mathcal{E})_\beta \subset (L^2) \subset (\mathcal{E})_\beta^*$. In 1.6.147 we will extend these results to the CKS-spaces $[\mathcal{E}]_\alpha \subset (L^2) \subset [\mathcal{E}]_\alpha^*$ and $[\mathcal{E}]_u \subset (L^2) \subset [\mathcal{E}]_u^*$.

3.6.1 Differential operators

Let $\varphi \in (\mathcal{E})$ and $y \in \mathcal{E}'$. The directional derivative of φ in y is defined by

$$D_y\varphi(x) = \lim_{\epsilon \to 0} \frac{\varphi(x+\epsilon y) - \varphi(x)}{\epsilon}.$$

Let φ be represented by $\varphi(x) = \sum_{n=0}^{\infty}\langle :\cdot^{\otimes n}:, f_n\rangle$. Then $D_y\varphi$ is represented by

$$D_y\varphi = \sum_{n=1}^{\infty} n\langle :\cdot^{\otimes(n-1)}:, \langle y, f_n\rangle\rangle,$$

where $\langle y, f_n\rangle$ is the bilinear pairing of y and one variable of f_n. Since f_n is assumed to be symmetric, this is well-defined. Note that after the pairing of y and f_n, $\langle y, f_n\rangle$ is a function of $n-1$ variables.

For the proof of the next theorem, see Theorem 9.1 in [40]

Theorem 3.6.1 *For any $y \in \mathcal{E}'$, the differential operator D_y is continuous from $(\mathcal{E})_\beta$ into itself.*

Thus for any $y \in \mathcal{E}'$, the adjoint operator D_y^* is continuous from $(\mathcal{E})_\beta^*$ into itself. If $\Phi \in (\mathcal{E})_\beta^*$ is represented by

$$\Phi = \sum_{n=0}^{\infty}\langle :\cdot^{\otimes n}:, F_n\rangle, \qquad F_n \in (\mathcal{E}_c')^{\widehat{\otimes} n},$$

then $\partial_y^*\Phi$ is represented by

$$\partial_y^*\Phi = \sum_{n=0}^{\infty}\langle :\cdot^{\otimes(n+1)}:, y\widehat{\otimes}F_n\rangle,$$

where $y\widehat{\otimes}F_n$ denotes the symmetric tensor product of y and F_n.

We have the following properties for the operators D_y and D_y^*:

(1) For any fixed $\varphi \in (\mathcal{E})_\beta$, the linear mapping $y \mapsto D_y\varphi$ is continuous from $\mathcal{E}'$ into $(\mathcal{E})_\beta$ (Theorem 9.3 in [40]).

(2) For any fixed $\varphi \in (\mathcal{E})_\beta$ and $x \in \mathcal{E}'$, the linear functional $y \mapsto D_y\varphi(x)$ is continuous on $\mathcal{E}'$ (Corollary 9.4 in [40]).

(3) For any fixed $\Phi \in (\mathcal{E})_\beta^*$, the linear mapping $y \mapsto D_y^*\Phi$ is continuous from $\mathcal{E}'$ into $(\mathcal{E})_\beta^*$ (Theorem 9.12(b) in [40]).

(4) Let $\eta \in \mathcal{E}$. The differential operator D_η from $(\mathcal{E})_\beta$ into itself has a unique extension by continuity to a continuous linear operator $\widetilde{D}_\eta$ from $(\mathcal{E})_\beta^*$ into itself (Theorem 9.10 in [40]).

(5) For any $\varphi \in (\mathcal{E})_\beta$ and $y \in \mathcal{E}'$, the S-transform of $D_y\varphi$ is given by

$$SD_y\varphi(\xi) = \frac{d}{d\lambda}\breve{S}\varphi(\xi+\lambda y)\bigg|_{\lambda=0}, \qquad \xi \in \mathcal{E}_c.$$

(Theorem 9.7 in [40])

(6) For any $\Phi \in (\mathcal{E})_\beta^*$ and $y \in \mathcal{E}'$, the S-transform of $D_y^*\Phi$ is given by

$$SD_y^*\Phi(\xi) = \langle y, \xi\rangle S\Phi(\xi), \qquad \xi \in \mathcal{E}_c.$$

(Theorem 9.13 in [40])

The operators D_y and D_y^* are also called *annihilation* and *creation operators*, respectively. We have the following commutation identities for these operators from Theorem 9.15 in [40]. The commutator $[A, B]$ is defined by $[A, B] = AB - BA$.

(1) $[D_x, D_y] = 0$ on $(\mathcal{E})_\beta$ for all $x, y \in \mathcal{E}'$.

(2) $[D_x^*, D_y^*] = 0$ on $(\mathcal{E})_\beta^*$ for all $x, y \in \mathcal{E}'$.

(3) $[\widetilde{D}_\xi, \widetilde{D}_\eta] = 0$ on $(\mathcal{E})_\beta^*$ for all $\xi, \eta \in \mathcal{E}$.

(4) $[\widetilde{D}_\eta, D_y^*] = \langle y, \eta\rangle I$ on $(\mathcal{E})_\beta^*$ for all $\eta \in \mathcal{E}$ and $y \in \mathcal{E}'$.

(5) $[D_y, D_\eta^*] = \langle y, \eta\rangle I$ on $(\mathcal{E})_\beta$ for all $y \in \mathcal{E}'$ and $\eta \in \mathcal{E}$.

Now, let $\mathcal{E}$ be the Schwartz space $\mathcal{S}(\mathbf{R})$. The differential operator D_{δ_t} with $y = \delta_t$ is denoted by ∂_t. The operator ∂_t is often referred to as the *white noise differential operator* (or *Hida differential operator*.) It is a continuous linear operator from $(\mathcal{S})_\beta$ into itself. The adjoint ∂_t^* is a continuous linear operator from $(\mathcal{S})_\beta^*$ into itself. We have the following facts.

(1) If φ is represented by $\varphi = \sum_{n=0}^\infty \langle :\cdot^{\otimes n}:, f_n\rangle$, then $\partial_t\varphi$ is represented by

$$\partial_t\varphi = \sum_{n=1}^\infty n\langle :\cdot^{\otimes(n-1)}:, f_n(t,\cdot)\rangle.$$

(2) If Φ is represented by $\Phi = \sum_{n=0}^\infty \langle :\cdot^{\otimes n}:, F_n\rangle$, then $\partial_t^*\Phi$ is represented by

$$\partial_t^*\Phi = \sum_{n=0}^\infty \langle :\cdot^{\otimes(n+1)}:, \delta_t \widehat{\otimes} F_n\rangle.$$

(3) For any $\varphi \in (\mathcal{S})_\beta$, the function $t \mapsto \partial_t\varphi$ is continuous from $\mathbf{R}$ into $(\mathcal{S})_\beta$.

(4) For any $\Phi \in (\mathcal{S})_\beta^*$, the function $t \mapsto \partial_t^*\Phi$ is continuous from $\mathbf{R}$ into $(\mathcal{S})_\beta^*$.

(5) $[\partial_s, \partial_t] = 0$, $[\partial_s^*, \partial_t^*] = 0$, $[\partial_s, \partial_t^*] = \delta_s(t)I$.

Now, let (H, B) be an abstract Wiener space [38]. The *Gross Laplacian* $\Delta_G\varphi$ of a twice H-differentiable function φ on B is defined by

$$\Delta_G\varphi(x) = trace_H D^2\varphi(x).$$

As pointed out in the beginning of 12.4.77 that there exists some $p \geq 0$ such that $(\mathcal{E}_0, \mathcal{E}_{-p})$ is an abstract Wiener space. Hence we can define the Gross Laplacian for functions on $\mathcal{E}'$ by

$$\Delta_G\varphi(x) = trace_{\mathcal{E}_0} D^2\varphi(x).$$

For the proof of the next theorem, see Theorems 10.11 and 10.12 in [40].

Theorem 3.6.2 *The Gross Laplacian Δ_G is a continuous linear operator from $(\mathcal{E})_\beta$ into itself. If $\varphi \in (\mathcal{E})_\beta$ is represented by $\varphi = \sum_{n=0}^\infty \langle :\cdot^{\otimes n}:, f_n\rangle$, then*

$$\Delta_G\varphi = \sum_{n=0}^\infty (n+2)(n+1)\langle :\cdot^{\otimes n}:, \langle \tau, f_{n+2}\rangle\,\rangle.$$

Another infinite dimensional Laplacian is the *number operator* N. Let $\varphi \in (\mathcal{E})_\beta$ be represented by $\varphi = \sum_{n=0}^\infty \langle :\cdot^{\otimes n}:, f_n\rangle$. Then we define $N\varphi$ by

$$N\varphi = \sum_{n=1}^\infty n\langle :\cdot^{\otimes n}:, f_n\rangle.$$

For the proof of the next theorem, see Theorems 9.23 and 9.25 in [40]

Theorem 3.6.3 *The number operator N is a continuous linear operator from $(\mathcal{E})_\beta$ into itself and also from $(\mathcal{E})_\beta^*$ into itself. Moreover, for any $\Phi \in (\mathcal{E})_\beta^*$ and $\varphi \in (\mathcal{E})_\beta$, we have*

$$\langle\langle N\Phi, \varphi\rangle\rangle = \langle\langle \Phi, N\varphi\rangle\rangle.$$

The Gross Laplacian and number operator are related by the *lambda operator* Λ. For $\varphi \in (\mathcal{E})_\beta$ being represented by $\varphi = \sum_{n=0}^\infty \langle :\cdot^{\otimes n}:, f_n\rangle$, $\Lambda\varphi$ is defined by

$$\Lambda\varphi(x) = \sum_{n=1}^\infty n\langle x\widehat{\otimes} :x^{\otimes(n-1)}:, f_n\rangle.$$

By Theorem 10.18 in [40], the lambda operator Λ is continuous from $(\mathcal{E})_\beta$ into itself and $\Lambda = \Delta_G + N$.

When $\mathcal{E}$ is the Schwartz space $\mathcal{S}(\mathbf{R})$, we can express the Gross Laplacian Δ_G, its adjoint Δ_G^*, and the number operator N in terms of the operators ∂_t and ∂_t^* by

$$\Delta_G = \int_{\mathbf{R}} \partial_t^2\, dt, \quad \Delta_G^* = \int_{\mathbf{R}} (\partial_t^*)^2\, dt, \quad N = \int_{\mathbf{R}} \partial_t^*\partial_t\, dt.$$

3.6.2 Translation and scaling operators

Let $\varphi \in (\mathcal{E})_\beta$ and $y \in \mathcal{E}'$. The *translation* $T_y\varphi$ of φ by y is defined by

$$T_y\varphi(x) = \varphi(x+y), \qquad x \in \mathcal{E}'.$$

Note that the Wick tensor $:x^{\otimes n}:$ defined in 12.3.18 satisfies the identity

$$:(x+y)^{\otimes n}: = \sum_{k=0}^n \binom{n}{k} :x^{\otimes(n-k)}:\widehat{\otimes} y^{\otimes k}. \tag{3.6.42}$$

See Lemma 7.16 in [40] for the proof.

The identity in Equation (3.6.42) can be used to prove the next theorem. For details, see Theorem 10.21 in [40].

Theorem 3.6.4 *Let $y \in \mathcal{E}'$. The translation operator T_y by y is continuous from $(\mathcal{E})_\beta$ into itself.*

The adjoint T_y^* is a continuous linear operator from $(\mathcal{E})_\beta^*$ into itself. For any $\Phi \in (\mathcal{E})_\beta^*$, the S-transform of $T_y^*\Phi$ is given by

$$S(T_y^*\Phi)(\xi) = e^{\langle y,\xi\rangle} S\Phi(\xi), \qquad \xi \in \mathcal{E}_c.$$

Moreover, we have the following two facts from Theorems 10.22 and 10.26 in [40]

(1) Let $\eta \in \mathcal{E}$. The translation operator T_η extends by continuity to a continuous linear operator $\widetilde{T}_\eta$ from $(\mathcal{E})^*_\beta$ into itself.

(2) For any $y \in \mathcal{E}'$, the equality holds

$$T_y^* T_y \varphi = :e^{\langle \cdot, y \rangle}: \varphi, \qquad \forall \varphi \in (\mathcal{E})_\beta.$$

Next we discuss the scaling operator. Let $\varphi \in (\mathcal{E})_\beta$. The *scaling* of φ by a complex number λ is defined by

$$S_\lambda \varphi(x) = \varphi(\lambda x), \qquad x \in \mathcal{E}'.$$

The Wick tensor $:x^{\otimes n}:$ defined in 12.3.18 satisfies the identity

$$:(\lambda x)^{\otimes n}: = \sum_{k=0}^{[n/2]} \binom{n}{2k} (2k-1)!! \lambda^{n-2k} (\lambda^2 - 1)^k :x^{\otimes (n-2k)}: \widehat{\otimes} \tau^{\otimes k}. \tag{3.6.43}$$

See Lemma 11.17 in [40] for the proof.

The identity in Equation (3.6.43) can be used to prove the next theorem. For details, see Theorem 11.18 in [40].

Theorem 3.6.5 *Let $\lambda \in \mathbf{C}$. The scaling operator S_λ by λ is continuous from $(\mathcal{E})_\beta$ into itself.*

The adjoint S_λ^* is a continuous linear operator from $(\mathcal{E})^*_\beta$ into itself. Moreover, we can easily check the following facts:

(1) For any $\varphi \in (\mathcal{E})_\beta$ and $\lambda \neq 0$, the S-transform of $S_\lambda \varphi$ is given by

$$S(S_\lambda \varphi)(\xi) = e^{\frac{1}{2}\left(\frac{1}{\lambda^2} - 1\right)\langle \xi, \xi \rangle} S\big(\varphi \widetilde{\mu}^{(\lambda^2)}\big)\left(\frac{\xi}{\lambda}\right),$$

where for a complex number λ, the function $\widetilde{\mu}^{(\lambda^2)}$ is the generalized function defined by Equation (3.4.40).

(2) For any $\Phi \in (\mathcal{E})^*_\beta$ and $\lambda \in \mathbf{C}$, the S-transform of $S_\lambda^* \Phi$ is given by

$$S(S_\lambda^* \Phi)(\xi) = e^{\frac{1}{2}(\lambda^2 - 1)\langle \xi, \xi \rangle} S\Phi(\lambda \xi).$$

(3) For any $\varphi \in (\mathcal{E})_\beta$ and $\lambda \in \mathbf{C}$, the equality holds

$$S_\lambda^* S_\lambda \varphi = \varphi \widetilde{\mu}^{(\lambda^2)}, \qquad \forall \varphi \in (\mathcal{E})_\beta.$$

3.6.3 Multiplication and Wick product

An important property of the space $(\mathcal{E})_\beta$ of test functions is the fact that $(\mathcal{E})_\beta$ is an algebra, i.e., $\varphi\psi \in (\mathcal{E})_\beta$ for any $\varphi, \psi \in (\mathcal{E})_\beta$. From Theorem 8.18 in [40] we have the next theorem.

Theorem 3.6.6 *The pointwise product $\varphi\psi$ of two test functions φ and ψ in $(\mathcal{E})_\beta$ is also a test function in $(\mathcal{E})_\beta$. Moreover, the mapping $(\varphi, \psi) \mapsto \varphi\psi$ is continuous from $(\mathcal{E})_\beta \times (\mathcal{E})_\beta$ into $(\mathcal{E})_\beta$.*

Let M_ψ denote the multiplication operator by $\psi \in (\mathcal{E})_\beta$, i.e.,

$$M_\psi \varphi = \psi\varphi, \qquad \varphi \in (\mathcal{E})_\beta.$$

We have the following facts about the multiplication operator M_ψ:

(1) For any $\psi \in (\mathcal{E})_\beta$, the operator M_ψ is continuous from $(\mathcal{E})_\beta$ into itself.

(2) For any $\psi \in (\mathcal{E})_\beta$, the operator M_ψ extends by continuity to a continuous linear operator $\widetilde{M_\psi}$ from $(\mathcal{E})_\beta^*$ into itself. Moreover, we have $\widetilde{M_\psi} = M_\psi^*$.

In particular, for $\eta \in \mathcal{E}$, let Q_η denote the multiplication by $\langle \cdot, \eta\rangle$, i.e., $Q_\eta = M_{\langle \cdot, \eta\rangle}$. We have the following properties:

(1) For any $\eta \in \mathcal{E}$, Q_η is a continuous linear operator from $(\mathcal{E})_\beta$ into itself.

(2) For any $\eta \in \mathcal{E}$, Q_η has a unique extension by continuity to a continuous linear operator $\widetilde{Q}_\eta$ from $(\mathcal{E})_\beta^*$ into itself and $\widetilde{Q}_\eta = Q_\eta^*$.

(3) For any $\eta \in \mathcal{E}$, $Q_\eta = D_\eta + D_\eta^*$ as continuous operators from $(\mathcal{E})_\beta$ into itself.

(4) For any $\eta \in \mathcal{E}$, $\widetilde{Q}_\eta = \widetilde{D}_\eta + D_\eta^*$ as continuous operators from $(\mathcal{E})_{\beta^*}$ into itself.

(5) For any $y \in \mathcal{E}'$, $Q_y = D_y + D_y^*$ as continuous linear operators from $(\mathcal{E})_\beta$ into $(\mathcal{E})_\beta^*$.

For the proofs of (3) (4) (5), see Theorems 9.18 and 9.20 in [40].

For a special case of the above property (5), take $\mathcal{E} = \mathcal{S}(\mathbf{R})$ and $y = \delta_t$. Recall that elements in the white noise space $\mathcal{S}'(\mathbf{R})$ can be denoted by $\dot{B}$. Thus $\langle \dot{B}, \delta_t\rangle = \dot{B}(t)$ and so Q_{δ_t} is the multiplication by $\dot{B}(t)$. The operator Q_{δ_t} is called *white noise multiplication* and is simply denoted by $\dot{B}(t)$. Hence we have

$$\dot{B}(t) = \partial_t + \partial_t^* \tag{3.6.44}$$

as continuous linear operators from $(\mathcal{E})_\beta$ into $(\mathcal{E})_\beta^*$.

Next, we discuss the Wick product of two generalized functions. Let $\Phi, \Psi \in (\mathcal{E})_\beta^*$. By Theorem 3.5.4 the product $(S\Phi)(S\Psi)$ is the S-transform of a unique generalized function in $(\mathcal{E})_\beta^*$. This unique generalized function is defined to be the *Wick product* $\Phi \diamond \Psi$ of Φ and Ψ. Hence we have

$$S(\Phi \diamond \Psi)(\xi) = \big(S\Phi(\xi)\big)\big(S\Psi(\xi)\big), \qquad \xi \in \mathcal{E}_c.$$

Note that the pointwise product of two generalized functions cannot be defined. But the Wick product is always defined for any two generalized functions.

Example 3.6.7 *Note that $S\langle :\cdot^{\otimes n}:, f_n\rangle = \langle f_n, \xi^{\otimes n}\rangle$. Hence we have*

$$\langle :\cdot^{\otimes m}:, f_m\rangle \diamond \langle :\cdot^{\otimes n}:, f_n\rangle = \langle :\cdot^{\otimes m+n}:, f_m \widehat{\otimes} f_n\rangle.$$

In fact, we can use this equality to define the Wick product of two generalized functions in terms of their Wiener-Itô expansions.

Example 3.6.8 *Note that $S(:e^{\langle \cdot, x\rangle}:)(\xi) = e^{\langle x, \xi\rangle}$. Hence for any $x, y \in \mathcal{E}_c'$,*

$$:e^{\langle \cdot, x\rangle}: \diamond :e^{\langle \cdot, y\rangle}: = :e^{\langle \cdot, x+y\rangle}:.$$

From Theorem 8.12 and its remarks in [40] we have the next theorem.

Theorem 3.6.9 *(1) The mapping* $(\Phi, \Psi) \mapsto \Phi \diamond \Psi$ *is continuous from* $(\mathcal{E})^*_\beta \times (\mathcal{E})^*_\beta$ *into* $(\mathcal{E})^*_\beta$. *(2) We have* $\varphi \diamond \psi \in (\mathcal{E})_\beta$ *for any* $\varphi, \psi \in (\mathcal{E})_\beta$ *and the mapping* $(\varphi, \psi) \mapsto \varphi \diamond \psi$ *is continuous from* $(\mathcal{E})_\beta \times (\mathcal{E})_\beta$ *into* $(\mathcal{E})_\beta$.

There is a relationship between the pointwise and Wick products from Theorem 8.17 in [40]. For any $\varphi, \psi \in (\mathcal{E})_\beta$,

$$\varphi\psi = \breve{S}\big((\Theta\varphi) \diamond (\Theta\psi)\big), \quad \varphi \diamond \psi = \Theta\big((\breve{S}\varphi)(\breve{S}\psi)\big), \tag{3.6.45}$$

where $\breve{S}$ and Θ are defined in 12.5.118 and 12.4.76, respectively.

Even though there is no Lebesgue measure on the space $\mathcal{E}'$, we can define the convolution of two generalized functions. Let $\Phi, \Psi \in (\mathcal{E})^*_\beta$. Define the *convolution* of Φ and Ψ by

$$\Phi * \Psi = \Phi \diamond \Psi \diamond g_{-2},$$

where g_{-2} is the generalized function with S-transform given by

$$Sg_{-2}(\xi) = \exp\left[\frac{1}{2}\langle \xi, \xi \rangle\right], \qquad \xi \in \mathcal{E}_c.$$

Obviously, the mapping $(\Phi, \Psi) \mapsto \Phi * \Psi$ is continuous from $(\mathcal{E})^*_\beta \times (\mathcal{E})^*_\beta$ into $(\mathcal{E})^*_\beta$.

The *convolution* of two finite measures ν_1 and ν_2 is defined by

$$\nu_1 * \nu_2(\cdot) = \int_{\mathcal{E}'} \nu_1(\cdot - x)\, d\nu_2(x).$$

The convolution $\nu_1 * \nu_2$ of two Hida measures (see 3.4.4) ν_1 and ν_2 is also a Hida measure. Moreover, we have

$$(\nu_1 * \nu_2)^{\widetilde{}} = \widetilde{\nu}_1 * \widetilde{\nu}_2.$$

3.6.4 Fourier-Gauss transform

In this section we briefly discuss the Fourier-Gauss transform from Chapter 11 of [40].

There are several infinite dimensional Fourier type transforms. In 1947 Cameron and Martin introduced the Fourier-Wiener transform acting on the L^2-space of the Wiener measure. In 1956 Segal [54] introduced the Fourier-Wiener transform for the normal distribution on a Hilbert space. In 1961 Bargman [9] defined a transform which is nowadays called the Segal-Bargman transform. In 1967 Gross [17] used the μ-convolution on an abstract Wiener space. In 1975 Hida [18] used the $\mathcal{T}$-transform to develop the white noise theory. Later in 1980 Kubo and Takenaka [33] introduced the S-transform to study white noise functionals. In 1982 Kuo [39] defined Fourier transform on the space of generalized functions. In 1987 Lee [43] introduced the Fourier-Gauss transform which includes all the previous Fourier type transforms.

Let $a, b \in \mathbf{C}$. The *Fourier-Gauss transform* $\mathcal{G}_{a,b}\varphi$ of a function φ is defined by

$$\mathcal{G}_{a,b}\varphi(y) = \int_{\mathcal{E}'} \varphi(ax + by)\, d\mu(x).$$

We have several special cases:

(1) $\mathcal{G}_{1,1}$ is the operator $\breve{S}$ in 12.5.118.

(2) $\mathcal{G}_{i,1}$ is the operator Θ in 12.4.76.

(3) $\mathcal{G}_{\sqrt{2},i}$ is the Fourier-Wiener transform.

(4) $\mathcal{G}_{\sqrt{t},1}$ is the convolution with the measure $\mu^{(t)}$ in 12.4.6.

(5) $\mathcal{G}_{\sqrt{2},-i}$ is the second quantization $\Gamma(-iI)$ of the operator $-iI$.

(6) $\mathcal{G}^*_{1,-i}$ is the Fourier transform acting on the space of generalized functions.

For the proof of the next theorem, see Theorem 11.28 in [40]

Theorem 3.6.10 *Let $\varphi \in (\mathcal{E})_\beta$ be represented by $\varphi = \sum_{n=0}^\infty \langle :\cdot^{\otimes n}:, f_n\rangle$. Then for any $a, b \in \mathbf{C}$, $\mathcal{G}_{a,b}\varphi$ belongs to $(\mathcal{E})_\beta$ and*

$$\mathcal{G}_{a,b}\varphi = \sum_{n=0}^{\infty} \langle :\cdot^{\otimes n}:, h_n\rangle,$$

where h_n is given by

$$h_n = b^n \sum_{k=0}^{\infty} \binom{n+2k}{2k} (2k-1)!!(a^2+b^2-1)^k \langle \tau^{\otimes k}, f_{n+2k}\rangle.$$

Here are some properties of the Fourier-Gauss transform:

(1) For any $a, b \in \mathbf{C}$, the operator $\mathcal{G}_{a,b}$ is continuous from $(\mathcal{E})_\beta$ into itself. (See Theorem 11.29 in [40].)

(2) $\mathcal{G}_{0,1} = I$.

(3) $\mathcal{G}_{s,t} \circ \mathcal{G}_{a,b} = \mathcal{G}_{\pm\sqrt{a^2+b^2s^2},bt}$. (See Theorem 11.30 in [40].)

(4) If $b \neq 0$, then $\mathcal{G}_{a,b}$ is invertible and $\mathcal{G}_{a,b}^{-1} = \mathcal{G}_{\pm ia/b,1/b}$. (This follows from properties (2) and (3).)

(5) If $a^2 + b^2 = 1$ and $|b| = 1$, then $\mathcal{G}_{a,b}$ is a unitary operator of $(\mathcal{E}_p)_\beta$ for any $p \geq 0$. Conversely, if $\mathcal{G}_{a,b}$ is a unitary operator of $(\mathcal{E}_p)_\beta$ for some $p \geq 0$, then $a^2 + b^2 = 1$ and $|b| = 1$. (See Theorem 11.34 in [40].)

The adjoint $\mathcal{G}^*_{a,b}$ is a continuous linear operator from $(\mathcal{E})^*_\beta$ into itself. Moreover, if $b \neq 0$, then $\mathcal{G}^*_{a,b}$ is invertible.

Now, we consider a special Fourier-Gauss transform given by $a = 1$ and $b = -i$. For convenience, let $\mathcal{G}$ denote $\mathcal{G}_{1,-i}$ and let $\mathcal{F} = \mathcal{G}^*$. The operators $\mathcal{G}$ and $\mathcal{F}$ are called the *Gauss* and *Fourier transforms*, respectively.

The next theorem on the Gauss transform $\mathcal{G}$ is from Theorem 11.33 in [40].

Theorem 3.6.11 *The operator $\mathcal{G}\colon (\mathcal{E})_\beta \to (\mathcal{E})_\beta$ satisfies the following equalities:*

$$\begin{aligned}
\mathcal{G}^4 &= I, \\
\mathcal{G}Q_\eta &= -iD^*_\eta \mathcal{G}, \quad \eta \in \mathcal{E}, \\
\mathcal{G}D^*_\eta &= -iQ_\eta \mathcal{G}, \quad \eta \in \mathcal{E}, \\
\mathcal{G}D_x &= iD_x \mathcal{G}, \quad x \in]'.
\end{aligned}$$

On the other hand, we have a corresponding theorem for the Fourier transform $\mathcal{F}$ from Theorems 11.7 and 11.11 in [40].

Theorem 3.6.12 *The operator $\mathcal{F}\colon (\mathcal{E})^*_\beta \to (\mathcal{E})^*_\beta$ satisfies the following equalities:*

$$\begin{aligned}
\mathcal{F}^4 &= I, \\
\mathcal{F}\widetilde{D}_\eta &= i\widetilde{Q}_\eta \mathcal{F}, \quad \eta \in \mathcal{E}, \\
\mathcal{F}\widetilde{Q}_\eta &= i\widetilde{D}_\eta \mathcal{F}, \quad \eta \in \mathcal{E}, \\
\mathcal{F}D^*_x &= -iD^*_x \mathcal{F}, \quad x \in \mathcal{E}'.
\end{aligned}$$

The next two theorems are from Theorems 11.36 and 11.38 in [40]. They are the characterization theorems for the Gauss transform $\mathcal{G}$ and Fourier transform $\mathcal{F}$ in terms of differential and multiplication operators.

Theorem 3.6.13 *The Gauss transform $\mathcal{G} = \mathcal{G}_{1,-i}$ is the unique (up to a constant) continuous linear operator T from $(\mathcal{E})_\beta$ into itself satisfying the equalities:*

$$TQ_\xi = -iD^*_\xi T, \quad TD^*_\xi = -iQ_\xi T, \quad \forall \xi \in \mathcal{E}.$$

Theorem 3.6.14 *The Fourier transform $\mathcal{F} = \mathcal{G}^*_{1,-i}$ is the unique (up to a constant) continuous linear operator T from $(\mathcal{E})^*_\beta$ into itself satisfying the equalities:*

$$T\widetilde{D}_\xi = i\widetilde{Q}_\xi T, \quad T\widetilde{Q}_\xi = i\widetilde{D}_\xi T, \quad \forall \xi \in \mathcal{E}.$$

Finally, we consider an important special case of the Fourier-Gauss transform. For a real number θ, let

$$a = \pm(1 - e^{i\theta}\cos\theta)^{1/2}, \quad b = e^{i\theta}. \tag{3.6.46}$$

We use $\mathcal{G}_\theta$ to denote the Fourier-Gauss transform $\mathcal{G}_{a,b}$, i.e., $\mathcal{G}_\theta = \mathcal{G}_{a,b}$ with a and b given by Equation (3.6.46). Note that $\mathcal{G}_\theta$ does not depend on the choice of plus and minus signs for a.

The adjoint $\mathcal{F}_\theta = \mathcal{G}^*_\theta$ is called the *Fourier-Mehler transform.* The next two theorems are from Theorems 11.39 and 11.40 in [40].

Theorem 3.6.15 *The family $\{\mathcal{G}_\theta;\, \theta \in \mathbf{R}\}$ is a strongly continuous one-parameter group acting on $(\mathcal{E})_\beta$ with infinitesimal generator $iN + \frac{i}{2}\Delta_G$.*

Theorem 3.6.16 *The family $\{\mathcal{F}_\theta;\, \theta \in \mathbf{R}\}$ is a strongly continuous one-parameter group acting on $(\mathcal{E})^*_\beta$ with infinitesimal generator $iN + \frac{i}{2}\Delta^*_G$.*

The transform $\mathcal{G}_\theta$ and Fourier-Mehler transform $\mathcal{F}_\theta$ can also be characterized in similar ways as in Theorems 3.6.13 and 3.6.14, respectively. For the proofs and further information on the Fourier-Mehler transform, see [40].

3.6.5 Extensions to CKS-spaces

In this section we will study continuous linear operators and their adjoints on a CKS-space $[\mathcal{E}]_\alpha \subset (L^2) \subset [\mathcal{E}]^*_\alpha$ associated with a sequence $\{\alpha(n)\}_{n=0}^\infty$ of positive numbers satisfying conditions (A1) and (A2) in 12.3.21. For this purpose we need to impose the following conditions on the sequence $\{\alpha(n)\}_{n=0}^\infty$:

- (C1) There exists a constant c_1 such that for all $n \le m$,

$$\alpha(n) \le c_1^m \alpha(m).$$

(C2) There exists a constant c_2 such that for all n and m,

$$\alpha(n+m) \le c_2^{n+m}\alpha(n)\alpha(m).$$

(C3) There exists a constant c_3 such that for all n and m,

$$\alpha(n)\alpha(m) \le c_3^{n+m}\alpha(n+m).$$

These conditions were introduced by Kubo et al. in [60]. Note that condition (C1) is satisfied if the sequence $\{\alpha(n)\}$ is increasing. It can be easily checked that condition (C3) implies condition (C1). Moreover, the Bell numbers satisfy conditions (C2) and (C3) (for the proof, see [60].)

1. Differential operators

For the proof of the next theorem, see Theorem 3.1 in [60].

Theorem 3.6.17 *Assume that $\alpha(n) \le c_1^{n+1}\alpha(n+1)$ for all $n \ge 0$. In particular, let condition (C1) be satisfied. Then for any $y \in \mathcal{E}'$, the differential operator D_y is continuous from $[\mathcal{E}]_\alpha$ into itself.*

In fact, the condition $\alpha(n) \le c_1^{n+1}\alpha(n+1)$ for all $n \ge 0$ is also necessary for the continuity of a differential operator D_y with $y \ne 0$.

Those properties concerning the operators D_y and D_y^* in 3.6.1 are all true under the condition $\alpha(n) \le c_1^{n+1}\alpha(n+1)$ for all $n \ge 0$, in particular, under condition (C1). This is also the case for the Gross Laplacian, i.e., under this condition, the Gross Laplacian Δ_G is continuous from $[\mathcal{E}]_\alpha$ into itself. Hence its adjoint Δ_G^* is continuous from $[\mathcal{E}]_\alpha^*$ into itself.

However, for the number operator N, we do not need to assume any C-condition. For any sequence $\{\alpha(n)\}$ satisfying conditions (A1) and (A2), the number operator N is continuous from $[\mathcal{E}]_\alpha$ into itself and also from $[\mathcal{E}]_\alpha^*$ into itself.

2. Translation and scaling operators

From Equation (3.3.19) we have $[\mathcal{E}]_\alpha \subset (\mathcal{E})$ and so every $\varphi \in [\mathcal{E}]_\alpha$ has a unique continuous version and has a unique analytic extension (see 12.4.75 and 12.4.76.) Thus we can define translation and scaling operators acting on the space $[\mathcal{E}]_\alpha$ as in 3.6.2.

The next theorem is from section 3.2 in [60]

Theorem 3.6.18 *Assume that condition (C1) is satisfied. Then for any $y \in \mathcal{E}'$ and $\lambda \in \mathbf{C}$, the translation operator T_y and scaling operator S_λ are continuous from $[\mathcal{E}]_\alpha$ into itself.*

We can easily see that those properties and identities in 3.6.2 concerning the adjoints T_y^* and S_λ, and the extension $\widetilde{T}_\eta$ are all valid for the CKS-space under condition (C1).

3. Multiplication and Wick product

From Theorems 3.4 and 3.5 in [60] we have the next two theorems concerning the Wick product.

Theorem 3.6.19 *Assume that condition (C2) is satisfied. Then $[\mathcal{E}]_\alpha$ is closed under the Wick multiplication and the mapping $(\varphi, \psi) \mapsto \varphi \diamond \psi$ is continuous from $[\mathcal{E}]_\alpha \times [\mathcal{E}]_\alpha$ into $[\mathcal{E}]_\alpha$.*

Theorem 3.6.20 *Assume that condition (C3) is satisfied. Then $[\mathcal{E}]_\alpha^*$ is closed under the Wick multiplication and the mapping $(\Phi, \Psi) \mapsto \Phi \diamond \Psi$ is continuous from $[\mathcal{E}]_\alpha^* \times [\mathcal{E}]_\alpha^*$ into $[\mathcal{E}]_\alpha^*$.*

As for the pointwise multiplication of two test functions, recall the first identity from Equation (3.6.45)

$$\varphi\psi = \breve{S}\big((\Theta\varphi) \diamond (\Theta\psi)\big).$$

We can check that under condition (C1) the operators Θ and $\breve{S}$ are continuous from $[\mathcal{E}]_\alpha$ into itself (this fact also follows from the continuity of the Fourier-Gauss transform below.) Hence we have the next theorem.

Theorem 3.6.21 *Assume that conditions (C1) and (C2) are satisfied. Then $[\mathcal{E}]_\alpha$ is closed under pointwise multiplication and the mapping $(\varphi, \psi) \mapsto \varphi\psi$ is continuous from $[\mathcal{E}]_\alpha \times [\mathcal{E}]_\alpha$ into $[\mathcal{E}]_\alpha$.*

4. Fourier-Gauss transform

The next theorem is from Theorem 3.3 in [60].

Theorem 3.6.22 *Assume that condition (C1) is satisfied. Then for any $a, b \in \mathbf{C}$, the Fourier-Gauss transform $\mathcal{G}_{a,b}$ is continuous from $[\mathcal{E}]_\alpha$ into itself.*

In particular, the operators $\breve{S}$, Θ, $\mathcal{G}$, and the Fourier-Wiener transform are all continuous from $[\mathcal{E}]_\alpha$ into itself.

The properties of $\mathcal{G}_{a,b}$ and $\mathcal{G}^*_{a,b}$ and the characterization theorems in 3.6.4 are all valid under the condition (C1).

As for the CKS-space $[\mathcal{E}]_u \subset (L^2) \subset [\mathcal{E}]^*_u$ given by a growth function u, we need to assume that u satisfies conditions (U1) (U2) (U3) in 3.4.4. It is shown in [7] that for such a function u, the associated sequence $\{\alpha(n)\}_{n=0}^\infty$ in Equation (3.4.33) satisfies condition (C2) and (C3) (hence also (C1) since (C3) implies (C1).) Thus the results for the CKS-space $[\mathcal{E}]_\alpha \subset (L^2) \subset [\mathcal{E}]^*_\alpha$ can be automatically carried over to the CKS-space $[\mathcal{E}]_u \subset (L^2) \subset [\mathcal{E}]^*_u$.

3.7 Comments on other topics and applications

At the end of this survey article, we mention some other topics and applications of white noise theory.

1. *Lévy and Volterra Laplacians*

In 3.6.1 we discussed the Gross Laplacian and number operator. There are two more Laplacian operators: the Lévy and Volterra Laplacians.

Let f be a function defined on a Hilbert space H. The Lévy Laplacian $\Delta_L f$ of f, as originally proposed by P. Lévy, is defined to be the function

$$\Delta_L f(x) = \lim_{n\to\infty} \frac{1}{n} \sum_{k=1}^{n} f''(x)(e_k, e_k),$$

where $\{e_n\}$ is an orthonormal basis for H. Obviously, if $f''(x)$ is a trace class operator of H, then $\Delta_L f(x) = 0$. Thus if the Gross Laplacian $\Delta_G f$ exists, then $\Delta_L f = 0$. On the other hand, when $\Delta_G f$ does not exist, the Lévy Laplacian $\Delta_L f$ may be defined. For example, consider the function $f(x) = |x|^2$ ($|\cdot|$ is the norm on H.) We have $f''(x) = I$ and so $\Delta_G f$ does not exist. However, $\Delta_L f(x) = 1$.

One of the original motivations for Hida to introduce white noise theory was to understand the Lévy Laplacian from the white noise viewpoint. The above function f, when written in white noise language, is the function $F(\xi) = \int_{\mathbf{R}} \xi(t)^2\, dt$. This function is the S-transform of the generalized function $\Phi = \int_{\mathbf{R}} :\dot{B}(t)^2:\, dt$ in $(\mathcal{S})^*$. The Gross Laplacian $\Delta_G \Phi$

is not defined. However, $\Delta_L \Phi = 1$. In general, the Gross Laplacian acts on ordinary functions, while the Lévy Laplacian acts on generalized functions. Thus $\Delta_L \Phi$ is defined through the S-transform $F = S\Phi$ of Φ and expressed in terms of the second functional derivative $F''(\xi)$ of F. However, the Lévy Laplacian picks up only the singular part of $F''(\xi)$. The regular part of $F''(\xi)$ gives another Laplacian, called the Volterra Laplacian $\Delta_V \Phi$ of Φ. In the absence of the singular part, the regular part would give the Gross Laplacian.

For the precise definitions and a comprehensive discussion of the Lévy and Volterra Laplacian in terms of the S-transform, see [40] and the references therein. For the recent development about the Lévy Laplacian, see [51] [52].

We remark that Accardi [1] has discovered a very important relationship between the Lévy Laplacian and Yang-Mills equations.

2. *Integral kernel operators and quantum probability*

Integral kernel operators were introduced by Hida et al. in [23]. They are operators of the form

$$\Xi_{j,k}(\theta) = \int_{T^{j+k}} \theta(s_1, \ldots, s_j, t_1, \ldots, t_k)\partial_{s_1}^* \cdots \partial_{s_j}^* \partial_{t_1} \cdots \partial_{t_k}\, ds_1 \cdots ds_j dt_1 \cdots dt_k,$$

where $T \subset \mathbf{R}$ is an interval and $\theta \in (\mathcal{S}')^{\otimes(j+k)}$. The integral kernel operator $\Xi_{j,k}(\theta)$ is a continuous linear operator from $(\mathcal{S})_\beta$ into $(\mathcal{S})_\beta^*$. In fact, it is also continuous from $[\mathcal{S}]_\alpha$ into $[\mathcal{S}]_\alpha^*$ if the sequence $\{\alpha(n)\}$ satisfies condition (C1) in 3.6.5. Integral kernel operators with $\theta \in L^2(\mathbf{R}^{j+k})$ had already been studied in quantum probability before Hida et al. [23]. The case with θ being a tempered distribution is a significant progress in quantum probability.

For more information on integral kernel operators, see [40] (chapter 10) and [48]. For more recent development related to integral kernel operators and the application to quantum probability, see [12] [13] [49].

One of the most important applications of white noise theory is to quantum probability. The noncommutative Itô lemma can be formulated in a very natural way by using the white noise theory. For an excellent account of the white noise approach to quantum stochastic calculus, see [2] which is part of a forthcoming book by Accardi et al.

3. *Stochastic integration*

An Itô integral is an integral of the form $\int_a^b f(t)\, dB(t)$ with the integrand f being nonanticipating and almost all sample paths of f are square integrable. Thus the integral $\int_0^1 B(1)\, dB(t)$ is not an Itô integral, even though intuitively we would have $\int_0^1 B(1)\, dB(t) = B(1)\int_0^1 dB(t) = B(1)^2$. This simple example served as a motivation for Itô himself in 1976 to extend Itô's integral for integrands which may be anticipating.

In fact, being motivated by the problem to extend Itô's lemma for functions of the form $g(B(t), B(1))$, $0 \le t \le 1$, Hitsuda already defined stochastic integrals for anticipating integrands in 1972 during the Japan-USSR joint probability conference. Skorokhod, influenced by Hitsuda's lecture in the conference, published a paper in 1975 to extend the Itô integral without assuming the nonanticipating property for the integrand. Both Hitsuda and Skorokhod used the Wiener-Itô decomposition of the integrand to define this new stochastic integral, which is nowadays called the Hitsuda-Skorokhod integral.

From the white noise viewpoint we can write the Itô integral $\int_a^b f(t)\, dB(t)$ as $\int_a^b f(t)\dot{B}(t)\, dt$ (cf. 10.7.66.) But if we regard $f(t)\dot{B}(t)$ as multiplication by $\dot{B}(t)$, then the class of functions f that we can integrate is small. A better idea is to write the integral as $\int_a^b \dot{B}(t)f(t)\, dt$ and regard $\dot{B}(t)$ as a multiplication operator $\dot{B}(t) = \partial_t + \partial_t^*$ (see Equation (3.6.44).) This leads to the integral $\int_a^b(\partial_t + \partial_t^*)f(t)\, dt$. On the one hand, for nonanticipating f, we have

$\int_a^b \partial_t^* f(t)\,dt = \int_a^b f(t)\,dB(t)$ as pointed out by Kubo and Takenaka in [35]. Thus the integral $\int_a^b \partial_t^* f(t)\,dt$ is an extension of the Itô integral. It turns out that this integral is the Hitsuda-Skorokhod integral. On the other hand, the integral $\int_a^b \partial_t f(t)\,dt$ is not well-defined and gives rise to the integrals $\int_a^b \partial_{t+} f(t)\,dt$ and $\int_a^b \partial_{t-} f(t)\,dt$.

See chapter 13 of [40] for more information on the above discussion and applications such as intersection local times, Donsker's delta function, and Tanaka formula, among other things. Recently, de Faria et al. have proved the Clark-Ocone formula for certain generalized functions in [15].

4. *Infinite dimensional harmonic analysis*

The finite dimensional Fourier transform, depending on which properties one wants to keep, has several infinite dimensional analogues. In 3.6.4 we gave some of these analogues in white noise theory: the S-transform, Fourier-Wiener transform, second quantization operator, and Fourier transform. Recently, Lee and Stan [73] have used the second quantization operator to obtain a white noise generalization of the Heisenberg uncertainty principle. On the other hand, the S-transform is used by Stan [56] to generalize the Paley-Wiener theorem to a white noise space.

An important motivation for Hida to introduce white noise theory is to study infinite dimensional rotation groups. Let $\mathcal{E} \subset \mathcal{E}_0 \subset \mathcal{E}'$ be an abstract white noise space in 12.3.17 with $\mathcal{E}$ understood to be infinite dimensional. The set $\mathcal{O}(\mathcal{E};\mathcal{E}_0)$ of linear homeomorphisms g from $\mathcal{E}$ onto itself such that $|g(\xi)|_0 = |\xi|_0$ is referred to as an infinite dimensional rotation group.

The infinite dimensional rotation group $\mathcal{O}(\mathcal{E};\mathcal{E}_0)$ contains many subgroups. The trivial ones are rotations on any fixed finite dimensional subspace of $\mathcal{E}$ preserving the $\mathcal{E}_0$-norm. An important subgroup is the Lévy group. Let $\mathcal{P}_L$ denote the set of permutations σ of $\mathbf{N}$ (natural numbers) such that

$$\lim_{N\to\infty} \frac{1}{N} \Big|\{n \le N, \sigma(n) > N\}\Big| = 0.$$

Let $\{e_n\}_{n=1}^\infty \subset \mathcal{E}$ be an orthonormal basis for $\mathcal{E}_0$. For each $\sigma \in \mathcal{P}_L$, let g_σ be the linear operator on $\mathcal{E}$ defined by

$$g_\sigma(\xi) = \sum_{n=1}^\infty \langle \xi, e_n\rangle\, e_{\sigma(n)}, \qquad \xi \in \mathcal{E}.$$

The set $\mathcal{G}_L = \{g_\sigma; \sigma \in \mathcal{P}_L\}$ is called the *Lévy group*. It is a subgroup of $\mathcal{O}(\mathcal{E};\mathcal{E}_0)$. Obviously, no g_σ in $\mathcal{G}_L$ can be approximated by finite dimensional rotations. The Lévy group is closely related to the Lévy Laplacian (see the papers by Hida [20] and Obata [46] [47].)

In the special case when $\mathcal{E}_0 = L^2(\mathbf{R}^d)$, i.e., $\mathcal{E}$ is a nuclear subspace of $L^2(\mathbf{R}^d)$, we can use the structure of $L^2(\mathbf{R}^d)$ to get fascinating subgroups of $\mathcal{O}(\mathcal{E};L^2(\mathbf{R}^d))$. These subgroups are one-parameter groups called *whiskers* in [18]. Some examples of whiskers are shifts, isotropic dilations, special conformal transformations, and special orthogonal groups.

For a full account of infinite dimensional rotation groups, see the forthcoming book by Hida [21].

5. *Mathematical physics*

White noise theory provides a very natural approach to define and study Feynman integrals as initiated by Hida and Streit [72]. Consider the Schrödinger equation

$$i\hbar \frac{\partial\psi}{\partial t} = -\frac{\hbar^2}{2m}\Delta\psi + V(x)\psi, \qquad \psi(0,x) = \delta_0(x),$$

where $\hbar$ is the Planck constant and δ_0 is the Dirac delta function at 0. It is shown in Section 14.2 of [40] that the white noise formulation of the solution is given by

$$\psi(t,x) = \int_{\mathcal{S}'(\mathbf{R})} \left(\mathcal{N}\exp\left[\tfrac{1}{2}\left(\tfrac{im}{\hbar}+1\right)\int_0^t \dot{B}(u)^2\,du\right]\right) \times \exp\left[-\tfrac{i}{\hbar}\int_0^t V\big(x-B(u)\big)\,du\right]\delta_x(B(t))\,d\mu(\dot{B}), \tag{3.7.47}$$

where $\delta_x(B(t))$ is Donsker's delta function in Example 3.2.5 and $\mathcal{N}\exp[\cdots]$ denotes the renormalization of $\exp[\cdots]$. What we need to show is, for a given potential function V, the integrand $\Phi_{t,x}$ in Equation (3.7.47) is a generalized function in some space of generalized functions. Then the Feynman integral, given as the expectation of $\Phi_{t,x}$ in Equation (3.7.47), can be defined by $\psi(t,x) = \langle\langle \Phi_{t,x}, 1\rangle\rangle$.

In chapter 14 of [40], the integrand $\Phi_{t,x}$ is shown to be a generalized function in the space $(\mathcal{S})^*$ for the cases $V(x) = 0$ (free particle), $-ax$ (constant external force), $\frac{1}{2}gx^2$ (harmonic oscillator). When V is the Fourier transform of a finite measure m (given by Albeverio and Høegh-Krohn in [3]), it is wrongly stated in [40] (page 316) that the corresponding integrand $\Phi_{t,x}$ is a generalized function in $(\mathcal{S})^*$. Actually, the integrand $\Phi_{t,x}$ should be a generalized function in a Bell number space $[\mathcal{S}]^*_{b_k}$ (see Example 3.3.11 in 12.3.21) of some order k, which depends on the growth order of the finite measure m.

There are other important applications of white noise theory to mathematical physics such as Dirichlet forms and quantum field theory. See Chapters 10 and 11 of [22] and the references therein.

6. *Random fields and stochastic variational calculus*

A rich area for applications of white noise theory is stochastic variational calculus for random fields. Let $\mathcal{C}$ be a class of smooth manifolds diffeomorphic to the sphere S^{d-1} in $\mathbf{R}^d$. For each $C \in \mathcal{C}$, let $X(C)$ be a generalized function in the space $(\mathcal{E})^*$ (in general, it can be $(\mathcal{S})^*_\beta$, $[\mathcal{S}]^*_\alpha$, or $[\mathcal{E}]^*_u$.) In [20] Hida describes a stochastic variational equation for a random field $\{X(C); C \in \mathcal{C}\}$ as

$$\delta X(C) = \Phi\big(X(C'),\, (C') \subset (C),\, Y(s),\, s \in C,\, \delta C, C\big), \tag{3.7.48}$$

where (C) denotes the domain enclosed by C, $Y(s)$ is the innovation for $X(C)$. The formulation in Equation (3.7.48) is motivated by an attempt to understand the Tomonaga equation in quantum mechanics.

Important results for special cases of Equation (3.7.48) have been obtained by Hida and Si Si [24] and Si Si [55]. For further information see the forthcoming book by Hida [21].

7. *Stochastic partial differential equations*

Earlier applications of white noise theory to study stochastic partial differential equation was done by Chow [11] and Lindstrøm et al. [45]. Later Øksendal and his colleagues developed the techniques much further. In particular, we mention the Burgers equation driven by a non-Gaussian noise

$$u_t + \lambda u \cdot u_x = \nu u_{xx} + F(t,x,\omega). \tag{3.7.49}$$

In [25] [26], Holden et al. regarded this equation as an equation taking values in a space of generalized functions. In order to do so, they replaced the multiplication by the Wick product. Thus Equation (3.7.49) is replace by the following equation

$$\Phi_t + \lambda\Phi \diamond \Phi_x = \nu\Phi_{xx} + F(t,x,\omega).$$

The S-transform provides a very useful tool to study this kind of stochastic partial differential equations. For a comprehensive account of the progress in this area, see [27] and the reference therein.

Recently Kondratiev et al. [10] have also made a rather significant progress about Burgers equations with random noises such as the Poisson and gamma noises [29].

Bibliography

[1] Accardi, L.: Yang-Mills equations and Lévy Laplacian; in: *Dirichlet Forms and Stochastic Processes*, Z. M. Ma et al. (eds.) (1995) 1–24, Walter de Gruyter, Berlin.

[2] Accardi, L., Lu, Y. G., and Volovich, I. V.: White noise approach to classical and quantum stochastic calculi; *Centro V. Volterra, Universita di Roma "Tor Vergata" Preprint* #375 (1999).

[3] Albeverio, S. and Høegh-Krohn, R.: *Mathematical Theory of Feynman Path Integrals.* Lecture Notes in Math. **523**, Springer-Verlag, 1976.

[4] Asai, N., Kubo, I., and Kuo, H.-H.: Bell numbers, log-concavity, and log-convexity; in: *Classical and Quantum White Noise*, L. Accardi et al. (eds.) Kluwer Academic Publishers (1999).

[5] Asai, N., Kubo, I., and Kuo, H.-H.: Characterization of test functions in CKS-space; in: *Proc. International Conference on Mathematical Physics and Stochastic Processes*, A. Albeverio et al. (eds.) World Scientific (1999).

[6] Asai, N., Kubo, I., and Kuo, H.-H.: Log-concavity, log-convexity, and growth order in white noise analysis; *Preprint* (1999).

[7] Asai, N., Kubo, I., and Kuo, H.-H.: CKS-space in terms of growth functions; *Preprint* (1999).

[8] Asai, N., Kubo, I., and Kuo, H.-H.: General characterization theorems and intrinsic topologies in white noise analysis; *Preprint* (1999).

[9] Bargmann, V.: On a Hilbert space of analytic functions and an associated integral transform, I; *Comm. Pure Appl. Math.* **14** (1961) 187–214.

[10] Benth, F. E. and Streit, L.: The Burgers equation with a non-Gaussian random force; *Preprint* (1995).

[11] Chow, P. L.: Generalized solution of some parabolic equations with a random drift; *J. Appl. Math. Optim.* **20** (1989) 81–96.

[12] Chung, D. M. and Ji, U. C.: Transformation groups on white noise functionals and their applications; *J. Appl. Math. Optim.* **37** (1998) 205–223.

[13] Chung, D. M., Ji, U. C., and Obata, N.: Transformations on white noise functions associated with second order differential operators of diagonal type; *Nagoya Math. J.* **149** (1998) 173–192.

[14] Cochran, W. G., Kuo, H.-H., and Sengupta, A.: A new class of white noise generalized functions; *Infinite Dimensional Analysis, Quantum Probability and Related Topics* **1** (1998) 43–67.

[15] de Faria, M., Oliveira, M. J., and Streit, L.: A generalized Clark-Ocone formula; *Preprint* (1998).

[16] Gannoun, R., Hachaichi, R., Ouerdiane, H., and Rezgui, A.: Un Thórème de Dualite Entré Espaces de Fonctions Holomorphes à Croissance Exponentiele; *J. Funct. Anal.* (to appear).

[17] Gross, L.: Potential theory on Hilbert space; *J. Funct. Anal.* **1** (1967) 123–181.

[18] Hida, T.: *Analysis of Brownian Functionals.* Carleton Mathematical Lecture Notes **13**, 1975.

[19] Hida, T.: Infinite-dimensional rotation group and unitary group; *Lecture Notes in Math.* **1379** (1989) 125–134, Springer-Verlag.

[20] Hida, T.: White noise analysis: An overview and some future directions; *IIAS Reports* 1995-001 (1995).

[21] Hida, T.: *White Noise and Functional Analysis.* (to appear).

[22] Hida, T., Kuo, H.-H., Potthoff, J., and Streit, L.: *White Noise: An Infinite Dimensional Calculus.* Kluwer Academic Publishers, 1993.

[23] Hida, T., Obata, N., and Saitô: Infinite dimensional rotations and Laplacians in terms of white noise calculus; *Nagoya Math. J.* **128** (1992) 65–93.

[24] Hida, T. and Si Si: Innovations for random fields; *Infinite Dimensional Analysis, Quantum Probability and Related Topics* **1** (1998) 499–509.

[25] Holden, H., Lindstrøm, T., Økendal, B., Ubøe, J., and Zhang, T. S.: The Burgers equation with a noisy force and the stochastic heat equation; Comm PDE **19** (1994).

[26] Holden, H., Lindstrøm, T., Økendal, B., Ubøe, J., and Zhang, T. S.: The stochastic Wick-type Burgers equation; *Preprint* (1994).

[27] Holden, H., Økendal, B., Ubøe, J., and Zhang, T. S.: *Stochastic Partial Differential Equations.* Birkhäuser, 1996.

[28] Kondratiev, Yu. G.: Nuclear spaces of entire functions in problems of infinite-dimensional analysis; *Soviet Math. Dokl.* **22** (1980) 588–592.

[29] Kondratiev, Yu. G., da Silva, J. L., Streit, L., and Us, G. F.: Analysis on Poisson and gamma spaces; *Infinite Dimensional Analysis, Quantum Probability and Related Topics* **1** (1998) 91–117.

[30] Kondratiev, Yu. G. and Streit, L.: A remark about a norm estimate for white noise distributions; *Ukrainian Math. J.* **44** (1992) 832–835.

[31] Kondratiev, Yu. G. and Streit, L.: Spaces of white noise distributions: Constructions, Descriptions, Applications. I; *Reports on Math. Phys.* **33** (1993) 341–366.

[32] Kubo, I., Kuo, H.-H., and Sengupta, A.: White noise analysis on a new space of Hida distributions; *Infinite Dimensional Analysis, Quantum Probability and Related Topics* (in press).

[33] Kubo, I. and Takenaka, S.: Calculus on Gaussian white noise I; *Proc. Japan Academy* **56A** (1980) 376–380.

[34] Kubo, I. and Takenaka, S.: Calculus on Gaussian white noise II; *Proc. Japan Academy* **56A** (1980) 411–416.

[35] Kubo, I. and Takenaka, S.: Calculus on Gaussian white noise III; *Proc. Japan Academy* **57A** (1981) 433–437.

[36] Kubo, I. and Takenaka, S.: Calculus on Gaussian white noise IV; *Proc. Japan Academy* **58A** (1982) 186–189.

[37] Kubo, I. and Yokoi, Y.: A remark on the space of testing random variables in the white noise calculus; *Nagoya Math. J.* **115** (1989) 139–149.

[38] Kuo, H.-H.: *Gaussian Measures in Banach Spaces*. Lecture Notes in Math. **463**, Springer-Verlag, 1975.

[39] Kuo, H.-H.: On Fourier transform of generalized Brownian functionals; *J. Multivariate Analysis* **12** (1982) 415–431.

[40] Kuo, H.-H.: *White Noise Distribution Theory*. CRC Press, Boca Raton, 1996.

[41] Kuo, H.-H., Potthoff, J., and Streit, L.: A characterization of white noise test functionals; *Nagoya Math. J.* **121** (1991) 185–194.

[42] Kuo, H.-H. and Xiong, J.: Stochastic differential equations in white noise space; *Infinite Dimensional Analysis, Quantum Probability, and Related Topics* **1** (1998) 611–632.

[43] Lee, Y.-J.: Analytic version of test functionals, Fourier transform and a characterization of measures in white noise calculus; *J. Funct. Anal.* **100** (1991) 359–380.

[44] Lee, Y.-J. and Stan, A.: An infinite-dimensional Heisenberg uncertainty principle; *Taiwanese J. Math.* (1999) (to appear).

[45] Lindstrøm, T., Øksendal, B., and Ubøe, J.: Stochastic differential equations involving positive noise; *Stochastic Analysis*, M. Barlow and N. Bingham (eds.) (1991) 261–303, Cambridge University Press.

[46] Obata, N.: Analysis of the Lévy Laplacian; *Soochow J. Math.* **14** (1988) 105–109.

[47] Obata, N.: A characterization of the Lévy Laplacian in terms of infinite dimensional rotation groups; *Nagoya Math. J.* **118** (1990) 111–132.

[48] Obata, N.: *White Noise Calculus and Fock Space*. Lecture Notes in Math. **1577**, Springer-Verlag, 1994.

[49] Obata, N.: Wick product of white noise operators and quantum stochastic differential equations; *J. Math. Soc. Japan* **51** (1999) 613–641.

[50] Potthoff, J. and Streit, L.: A characterization of Hida distributions; *J. Funct. Anal.* **101** (1991) 212–229.

[51] Saitô, K.: A (C_0)-group generated by the Lévy Laplacian; *J. Stochastic Analysis and Appl.* **16** (1998) 567–584.

[52] Saitô, K.: A (C_0)-group generated by the Lévy Laplacian II; *Infinite Dimensional Analysis, Quantum Probability and Related Topics* **1** (1998) 425–437.

[53] Schneider, W. R.: Grey noise; *Stochastic Processes, Physics and Geometry*, S. Albeverio et al. (eds.) (1990) 676–681, World Scientific.

[54] Segal, I. E.: Tensor algebras over Hilbert spaces, I; *Trans. Amer. Math. Soc.* **81** (1956) 106–134.

[55] Si Si: A variational formula for some random fields; an analogue of Ito's formula; *Infinite Dimensional Analysis, Quantum Probability and Related Topics* **2** (1999) 305–313.

[56] Stan, A.: Paley-Wiener theorem for white noise analysis; *J. Funct. Anal.* (to appear).

[57] Streit, L. and Hida T.: Generalized Brownian functionals and the Feynman integral; *Stochastic Processes and Their Applications* **16** (1983) 55–69.

[58] Yokoi, Y.: Positive generalized white noise functionals; *Hiroshima Math. J.* **20** (1990) 137–157.

Chapter 4

Stochastic Differential Equations and Their Applications

Bo Zhang
Department of Statistics
People's University of China, Beijing, China

Introduction

The concept of stochastic differential equations was introduced in 1902 for the first time by Gibbs [21] in which the integral of Hamilton-Jacobi differential equations for conservation systems in statistical mechanics with random initial states was studied. However stochastic differential equations was not rigorously described in terms of mathematical language until 1951, when the famous article — on stochastic differential equations — was published by Itò [40]. Since then, stochastic differential equations have been well known and are widely used outside of mathematics. There are many fruitful connections to other mathematical disciplines, such as measure theory, partial differential equation, differential geometry and potential theory. The subject has also rapidly developed its own life as a fascinating research field with many interesting unanswered questions.

General speaking, the basic theoretical problems concerned with stochastic differential equations are the same as those in the case of deterministic differential equatioins, namely: existence and uniqueness of a solution, analytical properties of the solutions, and dependence on the solutions on the initial values.

In the first part of this chapter, we will deal with Itò type stochastic differential equations with respect to the Brownian motion process, and its applications in which stochastic differential equations on manifold and backward stochastic differential equations and application are also discussed. The second part discusses some generalizations, which include stochastic differential equation with respect to Poisson point processes, stochastic differential equation governed by C-valued Levy process, stochastic differential equations with respect to semimartingale and stochastic differential equations with respect to nonlinear integrators. Thirdly, we will discuss functional stochastic differential equations. At last, we will give a short review of stochastic differential equation in abstract spaces.

4.1 SDEs with respect to Brownian motion

4.1.1 Ito type SDEs

Let us begin this part with some definitions. Let R^d be the d-dimentional Euclidean space and let $W^d = C([0,\infty) \to R^d)$ be the space of all continuous functions w defined on $[0,\infty)$ with values in R^d. For $w_1, w_2 \in W^d$, let

$$\rho(w_1, w_2) = \sum_{k=1}^{\infty} (\max_{0 \le t \le k} |w_1(t) - w_2(t)| \wedge 1)$$

where $|\cdot|$ denotes the Euclidean metric in R^d. W^d is a complete separable metric space under metric ρ. Let $\mathcal{B}(W^d)$ be the topological σ field on W^d and $\mathcal{B}_t(W^d)$ be the sub-σ-field of $\mathcal{B}(W^d)$ generated by $w(s)$, $0 \le s \le t$. In other words, $\mathcal{B}_t(W^d)$ is the inverse σ-field $\rho_t^{-1}[\mathcal{B}(W^d)]$ of $\mathcal{B}(W^d)$ under the mapping $\rho_t : W^d \to W^d$ defined by $(\rho_t w)(s) = w(t \wedge s)$. We define $R^d \otimes R^r$ as the set of all real $d \times r$ matrices; $\mathcal{B}(R^d \otimes R^r)$ is the topological σ field on $R^d \otimes R^r$ obtained by identifying $R^d \otimes R^r$ with dr dimensional Euclidean space.

We denote by $\mathcal{A}^{d,r}$ the set of all functions $a(t,w) : [0,\infty) \times W^d \to R^d \otimes R^r$ such that
(i) it is $\mathcal{B}([0,\infty)) \times \mathcal{B}(W^d)/\mathcal{B}(R^d \otimes R^r)$-measurable, and
(ii) for each $t \in [0,\infty)$, $W^d \ni w \longmapsto (t,w) \in R^d \otimes R^r$ is $\mathcal{B}(W^d)/\mathcal{B}(R^d \otimes R^r)$-measurable.

Let $(\Omega, \mathcal{F}, \{\mathcal{F}_t\}_{t\ge 0}, P)$ be a complete probability space which satisfies the usual conditions, i.e. $\{\mathcal{F}_t\}_{t\ge0}$ is an increasing and right continuous family of sub-σ-algebras of $\mathcal{F}$ and $\mathcal{F}_0$ containing all P-null sets.

A stochastic process $X = (X_t(\omega))_{t\ge0}$ is said to be *continuous* (or *left continuous* or *right continuous*) if for almost all $\omega \in \Omega$, the function $X_t(\omega)$ is continuous on $t \in R_+$ (or left continuous or right continuous). It is said to be *cadlag* (right continuous and left limit) if it is right continuous and for almost all $\omega \in \Omega$ the left limit $X_{t-}(\omega) = \lim_{s\uparrow t} X_s(\omega)$ exists and is finite for all $t > 0$. It is said to be *adapted* (to $\{\mathcal{F}_t\}$) if for every t, $X_t(\omega)$ is $\mathcal{F}_t$-measurable in ω.

Consider the following d-dimensional stochastic differential equation:

$$dX(t) = b(t,X)dt + a(t,X)dB(t) \tag{4.1.1}$$

where $b(t,x) = (b_1(t,x), \cdots, b_d(t,x))^T$ is Borel measurable function $(t,x) \in [0,\infty) \times R^d \to R^d$, and $a(t,x) = (a_{ij}(t,x))_{d\times r}$ is a $d \times r$-matrix Borel measurable function $(t,x) \in [0,\infty) \times R^d \to R^d \otimes R^r$, and $B(t)$ is an r-dimensional standard Browian motion process.

Definition 4.1.1 *Let $a = (a_{ij}(t,w)) \in \mathcal{A}^{d,r}$ and $b = (b_i(t,w)) \in \mathcal{A}^{d,1}$ be given. By a solution of the equation (4.1.1), we mean a d-dimensional continuous stochastic process $X = (X(t))_{t\ge0}$ defined on a probability space $(\Omega, \mathcal{F}, P)$ with a reference family $(\mathcal{F}_t)_{t\ge0}$ such that*

- *there exists an r-dimensional $(\mathcal{F}_t)$-Brownian motion process $B(t) = (B(t))_{t\ge0}$ with $B(0) = 0$ a.s.;*
- *$X = (X(t))$ is a d-dimensional continuous process adapted to $(\mathcal{F}_t)_{t\ge0}$, i.e. for each $t \in [0,\infty)$, X is a mapping: $\omega \longmapsto X(t,\omega) \in W^d$ which is $\mathcal{F}_t/\mathcal{B}_t(W^d)$-measurable;*
- *the family of adapted processes $a_{ij}(t, X(t,\omega))$ and $b(t, X(t,\omega))$ belong to the spaces $\mathcal{L}_2^{loc}$ and $\mathcal{L}_1^{loc}$ respectively, where $\mathcal{L}_1^{loc} = \{\Psi = (\Psi(t))_{t\ge0} | \Psi$ is measurable $(\mathcal{F}_t)$-adapted process and $\forall t \ge 0, \int_0^t |\Psi(s,\omega)|ds < \infty, a.s.\}$, and $\mathcal{L}_2^{loc} = \{\Psi = (\Psi(t))_{t\ge0} | \Psi$ is measurable $(\mathcal{F}_t)$-adapted process and $\forall t \ge 0, \int_0^t \Psi^2(s,\omega)ds < \infty, a.s.\}$;*

- *$X = (X_1(t), \cdots, X_r(t))$ and $B = (B_1(t), \cdots, B_r(t))$ satisfy*

$$X_i(t) - X_i(0) = \int_0^t b_i(s, X(s))ds + \sum_{j=0}^{r} \int_0^t a_{ij}(s, X(s))dB_j(s), i = 1, 2, \cdots, d, \quad (4.1.2)$$

with probability one, where the integral by $dB_j(t)$ is Itò integral.

The stochastic differential equations which are most important and which are mainly studied are of the following type.

Definition 4.1.2 *Let $a(t, x)$ be a Borel measurable function $(t, x) \in [0, \infty) \times R^d \to R^d \otimes R^r$ and $b(t, x)$ be a Borel measurable function $(t, x) \in [0, \infty) \times R^d \to R^d$. Then $a(t, w)$ and $b(t, w)$ defined by $a(t, w) = a(t, w(t))$ and $b(t, w) = b(t, w(t))$ clearly satisfy $a \in \mathcal{A}^{d,r}$, and $b \in \mathcal{A}^{d,1}$. In such a case, the stochastic differential equation (4.1.1) is said to be of the Markovian type. The equation then has the following form:*

$$dX(t) = b(t, X(t))dt + a(t, X(t))dB(t) \quad (4.1.3)$$

Furthermore, if a and b do not depend on t and are functions of $x \in R^d$ alone, then the equation (4.1.1) is said to be of the time-independent (or time homogeneous) Markovian type.

Note that an equation of Markovian type reduces to a system of ordinary differential equations (a dynamical system) $\dot{X} = b(t, X_t)$ when $a \equiv 0$. Thus a stochastic differential equation generalizes the notion of an ordinary differential equation by adding the effect of random fluctuation.

Suppose that at least one solution of (4.1.1) exists. We will present several definitions concerning the uniqueness of solutions.

Definition 4.1.3 *We say that the uniqueness of solutions for (4.1.1) holds if whenever X and X' are two solutions whose initial laws on R^d coincide, then the laws of the processes X and X' on the space W^d coincide.*

This is so-called "the uniequeness in the sense of probability law." On the other hand if we consider stochastic differential equations as a tool for defining sample paths of a random process as functionals of Brownian paths, then the following definition might be more natural.

Definition 4.1.4 *We say that the pathwise uniqueness of solution for (4.1.1) holds if whenever X and X' are any two solutions defined on the same probability space $(\Omega, \mathcal{F}, P)$ with the same reference family $(\mathcal{F}_t)$ and the same r-dimensional $(\mathcal{F}_t)$-Brownian motion such that $X(0) = X'(0) a.s.$, then $X(t) = X'(t)$ for all $t \geq 0$ a.s.*

Definition 4.1.5 (strong solution) *A solution $X = (X(t))$ of (4.1.1) is called a strong solution if there exists a function $F(x, w) : R^d \times W_0^r \to W^d$ which is $\hat{C}(R^d \times W_0^r)$-measurable, that means for any Borel probability measure μ on R^d there exists a function $\tilde{F} : R^d \times W_0^R \to W^d$ which is $\overline{(R^d \times W_0^r)}^{\mu \times P^W} / \mathcal{B}(W^d)$-meaurable and for almost all $x(\mu)$ it holds $F(x, w) = \tilde{F}_\mu(x, w), a.s. w(P^W)$, here P^W is the Wiener measure on W_0^r. For each $x \in R^d, w \longmapsto F(x, w)$ is $\overline{\mathcal{B}}_t(W_0^r)^{P^W} / \mathcal{B}_t(W^d)$-measurable for every $t \geq 0$ and it holds*

$$X = F(X(0), B) a.s. \quad (4.1.4)$$

Definition 4.1.6 *We say that the equation (4.1.1) has a unique strong solution if there exists a function $F(x, w) : R^d \times W_0^r \to W^d$ with the same properties as in Definition 1.2 such that the following is ture*

(i) for any r-dimensional $(\mathcal{F}_t)$-Brownian motion process $B = (B(t))$ $(B(0) = 0)$ on a probability space with a reference family $(\mathcal{F}_t)$ and any R^d-valued random variable ξ which is $\mathcal{F}_0$-measurable, the continuous process $X = F(\xi, B)$ is a solution of (4.1.1) on this space with $X(0) = \xi$ a.s.;

(ii) for any solution (X, B) of (4.1.1), $X = F(X(0), B)$ holds a.s.

Theorem 4.1.7 *Given $a \in \mathcal{A}^{d,r}$ and $b \in \mathcal{A}^{d,1}$, the equation (4.1.1) has a unique strong solution if and only if for any Borel probability measure μ on R^d, a solution X of (4.1.1) exists such that the law of initial value $X(0)$ coincides with μ and the pathwise uniqueness of solutioins holds.*

Theorem 4.1.8 (Existence) *Suppose that $a \in \mathcal{A}^{d,r}$ and $b \in \mathcal{A}^{d,1}$ are bounded and continuous. Then, for any given probability μ on $(R^d, \mathcal{B}(R^d))$ with compact support, there exists a solution (X, B) of the equation (4.1.1) such that the law of $X(0)$ coincides with μ,i.e., $P\{X(0) \in A\} = \mu(A)$ for any $A \in \mathcal{B}(R^d)$.*

Remark *The boundedness assumption on a and b can be weakened, but some kind of restriction on the growth order of a and b is necessary in order to guarantee the existence of a global solution. (see e.g. [94]) The condition that μ has compact support can be removed. (see. [39]) If we remove this condition of boundedness, then a solution does exist locally but, in general, explodes in finite time. Let $\hat{R}^d = R^d \cup \{\Delta\}$ be the one-point compactification of R^d and $\hat{W}^d = \{w; [0, \infty) \ni t \mapsto w(t) \in C(\hat{R}^d), w(t') = \Delta \forall t' \geq t, \;\; if w(t) = \Delta\}$. Let $\mathcal{B}(\hat{W}^d)$ be the σ-field generated by Borel cylinder sets. For $w \in \hat{W}^d$, we say $e(w) = \inf\{t; w(t) = \Delta\}$ the explosion time of the trajectory w. We now can modify the notion of a solution as follows.*

Let $a(x) = (a_{ij}(x)) : R^d \to R^d \otimes R^r$ and $b(x) = (b_i(x)) : R^d \to R^d$ be continuous. Consider the following stochastic differential equation

$$dX(t) = b(X(t))dt + a(X(t))dB(t) \tag{4.1.5}$$

Definition 4.1.9 *We say that a $(\hat{W}^d, \mathcal{B}(\hat{W}^d))$-valued continuous stochastic process $X = (X(t))_{t\geq 0}$ defined on a probability space $(\Omega, \mathcal{F}, P)$ with a reference family $(\mathcal{F}_t)_{t\geq 0}$ is a solution of (4.1.5) if*

(i) there exists an r-dimensional $(\mathcal{F}_t)$-Brownian motion process $B(t) = (B(t))_{t\geq 0}$ with $B(0) = 0$ a.s.;

(ii) $X = (X(t))$ is adapted to $(\mathcal{F}_t)_{t\geq 0}$, i.e. for each $t, \omega \longmapsto X(t, \omega) \in \hat{R}^d$ is $\mathcal{F}_t$-measurable and

(iii) if $e(w) = e(X(\omega))$ is the explosion time of $X(\omega) \in \hat{W}^d$, then for almost all ω,

$$X_i(t) - X_i(0) = \int_0^t b_i(X(s))ds + \sum_{j=0}^{r} \int_0^t a_{ij}(X(s))dB_j(s), i = 1, 2, \cdots, d, \tag{4.1.6}$$

for all $t \in [0, e(w))$.

Theorem 4.1.10 *(Existence) Suppose $a(t, x)$ and $b(t, x)$ are locally Lipschitz continuous uniformly, i.e., for every $N > 0$ there exists a constant $K_N > 0$ such that*

$$||a(t, x) - a(t, y)||^2 + |b(t, x) - b(t, y)|^2 \leq K_N |x - y|^2 \forall x, y \in B_N, t \geq 0, \tag{4.1.7}$$

where $||a(t, x)||^2 = trace(aa^T), B_N = \{x \in R^d; |x| \leq N\}$. Then the pathwise uniqueness of solutions of (4.1.1) holds and hence it has a unique strong solution.

If we consider the case of equation of the Markovian type and $d = 1$, the condition (4.1.7) for the pathwise uniqueness of solutions of the equation:

$$dX(t) = b(x)dt + a(x)dB(t) \tag{4.1.8}$$

can be weakened in the following theorem.

Theorem 4.1.11 *Let $d = r = 1$ and suppose that $b(x)$ and $a(x)$ are bounded. Assume further that the following conditions are satisfied*

(i) there exists an strictly increasing function $\rho(t)$ on $[0, \infty)$ such that $\rho(0) = 0$, $\int_{0+} \rho^{-2}(t)dt = \infty$ and $|a(x) - a(y)| \le \rho(|x - y|)$ for all $x, y \in R^1$;

(ii) there exists an increasing and concave function $\kappa(t)$ on $[0, \infty)$ such that $\kappa(0) = 0$, $\int_{0+} \kappa^{-1}(t)dt = \infty$ and $|b(x) - b(y)| \le \kappa(|x - y|)$ for all $x, y \in R^1$.

Then the pathwise uniqueness of solution holds for the equation (4.1.8) and hence it has the unique strong solution.

Besides pathwise uniqueness, we have the beautiful and important result on uniqueness of solutions in the sense of probability law in the following theorem. For the general case, refer to [88] and [49].

Theorem 4.1.12 *Consider the equation of the time homogeneous Markovian case (4.1.8). If $a(x)a(x)$ is uniformly positive definite, bounded and continuous and $b(x)$ is bounded and Borel measurable, then the uniqueness of solutions holds.*

4.1.2 Properties of solutions

First let us summarize the basic properties of the solution below. Before this we would like to consider the Ito equations (4.1.1) in the following form which the initial condition $(t, y) \in [0, T) \times L^2(\Omega, \mathcal{F}_t, P; R^d)$

$$X_t = y + \int_s^t b(r, X_r) + \int_s^t a(r, X_r)dB_r. \tag{4.1.9}$$

We assume that the coefficients are Lipschitz continuous with respect to $x \in R^d$ uniformly in $t \in [0, T]$. The solution of (4.1.9) will be dnoted as $X_{s,t}(y), X_{s,t}(y, \omega)$ or $X_{s,t}$ simply. Then we have the following theorem. For the proof, see [51] and related papers.

Theorem 4.1.13 *We can choose a modification of the solution in the following way. For almost all ω,*

(i) $X_{s,t}(y, \omega)$ is continuous in (s, t, y);

(ii) for each $s \in [0, T)$, and $\forall y, y' \in L^2(\Omega, \mathcal{F}_t, P; R^d)$,

$$E[\sup_{t\in[s,T]} |X_{s,t}(y) - X_{s,t}(y')|^2] \le C|y - y'|^2,$$

and for each $p \ge 2$,

$$E[sup_{t\in[s,T]}|X_{s,t}(y)|^p] \le C_p|y|^p, \forall y \in L^p(\Omega, \mathcal{F}_t, P; R^d)$$

where C and C_p are constants which depend on the coefficients of (4.1.9) and Lipschitz constant, also C_p depends on p.

(iii) $X_{t_1,t_3}(y, \omega) = X_{t_2,t_3}(X_{t_1,t_2}(y, \omega), \omega)$ holds for any $t_1 \le t_2 \le t_3$ and y.

(iv)the map $X_{s,t}(\cdot, \omega) : R^d \to R^d$ is homeomorphism for any $s \le t$;

Futhermore if the coefficients of (4.1.9) are of c^k -class in x. Then

(v)$X_{s,t}(\cdot, \omega) : R^d \to R^d$ is a c^{k-1}-diffeomorphism for any $s \le t$ almost surely.

We know that it is a Markov diffusion solution $X(t)$ of (4.1.3) if the coefficients are continuous with respect to (t, x). We can ask: when is it a stationary process? The following theorem holds, for the proof refer to Khasminskii's book [43].

Theorem 4.1.14 *Assume that coefficients of (4.1.5) do not depend on time and satisfy the Lipschitz condition (4.1.7), growth condition:*

$$|b(x)| + ||a(x)|| \leq K(1 + |x|) \tag{4.1.10}$$

for some constant $K > 0$ in domain $U_M = \{y : |y| < M\}$ for each $M > 0$, and $X(0) = X_0$ is a random variable independent of $B(t)$; let, additionally, exist a positive definite function $V(y) \in C^2(R^d)$ such that:

$$\sup_{|y|>M} LV(y) = -A_M \to -\infty, asM \to \infty$$

where

$$LV(y) = \sum_{i=1}^{d} b_i(y)\frac{\partial V(y)}{\partial y_i} + \frac{1}{2}\sum_{i,j=1}^{d} \{a(y)a(y)^T\}_{ij}\frac{\partial^2 V}{\partial y_i \partial y_j}; \tag{4.1.11}$$

then, there exists a solution of (4.1.5) being a stationary Markov process.

In the following, we mention some asymptotic behaviour of solutions. One of asymptotic property of solutions is associated with the existence of an ergodic distribution for the process $X(t)$. The next theorem holds [23].

Theorem 4.1.15 *Assume that the coefficients of equation (4.1.8) fulfill the conditions*
(i) $a(x), b(x)$ and $a'(x)$ the derivative of $a(x)$ satisfy the Lipschitz condition;
(ii)$a(x > 0$ and $\lim_{|x|\to\infty} a(x) = \frac{1}{s_0} > 0$ exists;
(iii) $\int_{-\infty}^{\infty} \frac{b(x)}{a(x)^2}dx = 0$.
Then

$$\lim_{t\to\infty} P\{\frac{X(t)}{\sqrt{t}} < \frac{x}{s_0}\} = \frac{1}{\sqrt{2\pi}}\int_{-\infty}^{x} \exp\{-\frac{u^2}{2}\}du$$

Another important class of asymptotic problem in the theory of stochastic differential equations is associated with stability of stochastic dynamic systems. The stability of a dynamic system is usually understood as the insensitivity of the state of the system within an unbounded time interval $[0, \infty)$ to small changes in the initial state or in the parameter of the system. In contrast with the deterministic case, in the stochastic case the number of different stability notions is greater due to the larger variety of concepts of stochastic convergence. The concepts of stochastic stability which are studied most often are the stability in probability, stability with probability one, and stability of moments.

In the past decades the problem of stability of stochastic systems has generated a great deal of interest. Both the stability in the Lyapunov sense and the non-Lyapunov sense have been well studied. See [43], [57],[65],[66], [42], [47], [58], and [97] in which stabilities of various kinds are discussed in detail. Let us end this section with some notions of stability in the Lypunov sense in the following definition. Let $X_t(x_0)$ be a global solution on $[0, \infty)$ of (4.1.1) with initial value x_0. Without loss of generality the trivial solution $X_t \equiv 0$ can be studied.

Definition 4.1.16 *A trivial solution of (4.1.1) is said to be:*

1. *stable in probability, if for every* $\epsilon > 0$,

$$\lim_{x_0 \to 0} P[\sup_{t_0 \le t < \infty} |X_t(x_0)| \ge \epsilon] = 0;$$

2. *asymptotically stable in probability, if it is stable in probability and*

$$lim_{x_0 \to 0} P[lim_{t \to \infty} X_t(x_0) = 0] = 1;$$

3. *asymptotically stable in large in probability, if it is stable in probability and*

$$P[lim_{t \to \infty} X_t(x_0) = 0] = 1, \ for \ all \ x_0 \in R^d;$$

4. *p-stable, if for every* $\epsilon > 0$, *there exists a* $\delta > 0$ *such that*

$$\sup_{t_0 \le t < \infty} E|X_t(x_0)|^p \le \epsilon, \ for |x_0| \le \delta;$$

5. *asymptotically p-stable,if it is p-stable and if*

$$lim_{t \to \infty} E|X_t(x_0)|^p = 0, \ for \ all x_0 \ in \ a \ neighborhood \ of \ 0;$$

6. *exponentially p-stable, if there exist positive constants* c_1 *and* c_2 *such that, for all sufficiently small* $\delta > 0$

$$E|X_t(x_0)|^p \le c_1|x_0|^p \exp\{-c_2(t - t_0)\}, for|x_0| \le \delta.$$

7. *weak exponentially stable in mean, if there exists a wedge function* $\lambda(s)$ *and positive constants* $c_1 > 0, c_2 > 0$ *such that for* $\delta > 0$ *small enough, and* $x_0 \in U_\delta = \{x \in R^d : |x| \le \delta\}$, *implies*

$$E[\lambda(|X_t(x_0)|)] \le c_1 \lambda(|x_0|) \exp\{-c_2 t\} \qquad t > 0$$

where a wedge function means that a continuous function $\lambda(s)$ *defined on* $[0, h)$ *satisfying* $\lambda(0) = 0$ *and* $\lambda(s) > 0$ *for* $s > 0$.

4.1.3 Equations depending on a parameter

Let $A(t,x), F(t,x)$ be a random d-vector and $n \times n$ matrix respectively, defined for $(t,x) \in [s, \infty) \times R^d$ for some $s \ge 0$

(i) $A(t,x), F(t,x)$ are continuous in (t,x), for each $\omega \in \Omega$;

(ii) $A(t,x), F(t,x)$ are measurable in (t,x,ω);

(iii) $A(t,x), F(t,x)$ are $\mathcal{F}_t$ meaurable for each (t,x), where $\mathcal{F}_t$ is an increasing family of σ-fields such that $B(t)$ is $\mathcal{F}_t$ measurable and $\sigma(B(t+\lambda) - B(t), \lambda \ge 0)$ is independent of $\backslash calF_t$ for all $t \ge 0$;

(iv) there is a constant K such that

$$|A(t,x)| \le K(1 + |x|), |F(t,x)| \le (1 + |x|) \ \ a.s., \tag{4.1.12}$$

and

$$|A(t,x) - A(t,x')| \leq K(|x-x'|), |F((t,x) - F(t,x')| \leq K|x-x'| \quad a.s. \tag{4.1.13}$$

Denote by $M^p_\omega[\alpha,\beta](1 \leq p \leq \infty)$ the class of all nonanticipative functions $f(t)$ satisfying

$$E[\int_\alpha^\beta |f(t)|^p dt] < \infty, (E[ess \sup_{\alpha \leq t \leq \beta} |f(t)|] < \infty, if \quad p = \infty).$$

Let $\phi(t)$ be a functioin in $M^\infty_\omega[s,T]$. Consider the equation

$$X(t) = \phi(t) + \int_s^t A(u, X(u))du + \int_s^t F(u, X(u))dB(u). \tag{4.1.14}$$

This equation is so-called stochastic differential equations with random coefficients.

Theorem 4.1.17 *If (i)-(iv) hold and $\phi \in M^\infty_\omega[s,T]$, then there exists a unique solution $X(t)$ in $M^2_\omega[s,T]$; further, $X(t)$ belongs to $M^\infty_\omega[s,T]$.*

Theorem 4.1.18 *Let $A_\alpha(t,x), F_\alpha(t,x), \phi_\alpha(t)$ satisfy the assumptions of Theorem 4.1.17. for any $0 \leq \alpha \leq 1$, with the constant K (in (4.1.12),(4.1.13)) independent of α, and with*

$$\sup_{s \leq t \leq T} E|\phi_\alpha(t)|^2 \leq C,$$

where c is a constant independent of α. Suppose that for any $N > 0, t \in [s,T], \epsilon > 0$,

$$\lim_{\alpha \downarrow 0} P\{\sup |x| \leq N|A_\alpha(t,x) - A_0(t,x)| > \epsilon\} = 0$$

$$\lim_{\alpha \downarrow 0} P\{\sup |x| \leq N|F_\alpha(t,x) - F_0(t,x)| > \epsilon\} = 0$$

Suppose also that

$$\lim_{\alpha \downarrow 0} \sup |x| \leq NE|\phi_\alpha(t) - \phi_0(t)|^2 = 0$$

Consider the solutions $X_\alpha(t)$ of the equations

$$X_\alpha(t) = \phi_\alpha(t) + \int_s^t A_\alpha(u, X_\alpha(u))du + \int_s^t F_\alpha(u, X_\alpha(u))dB(u)$$

Then,

$$\sup_{s \leq t \leq T} E|X_\alpha(t) - X_0(t)|^2 \to 0 \ \backslash mbox\{if\} \ \alpha \downarrow 0.$$

Now we can study the behavior of the solution $X_{x,s}(t)$ in the parameters s, x via Theorem 4.1.17 and 4.1.18. Recall that

$$X_{x,s}(t) = x + \int_s^t b(u, X_{x,s}(u))du + \int_s^t \sigma(u, X_{x,s}(u))dB(u). \tag{4.1.15}$$

We need the following condition: $\frac{\partial b}{\partial x_i}, \frac{\partial \sigma}{\partial x_i}$ exist and are continous $(1 \leq i \leq d)$ in the following sense.

Definition 4.1.19 *Let $g(x) = g(x_1, x_2, \cdots, x_d), f(x) = f(x_1, x_2, \cdots, x_d)$ be random functions for x in some open set. If*

$$\int |\frac{1}{h}[g(x_1, x_2, \cdots, x_i + h, x_{i+1}, \cdots, x_d) - g(x_1, \cdots, x_d)] - f(x_1, \cdots, x_d)|^2 dp \to 0$$

as $h \to 0$, then we say that $g(x)$ has a derivative with respect to x_i in the $L^2(\Omega)$ sense, and the derivative is equal to $f(x)$. We write $(\partial/\partial x_i)g(x) = f(x)$. Similarly one defines the derivative $D^\alpha g(x)$ in the $L^2(\Omega)$ sense, for any $\alpha = (\alpha_1, \cdots, \alpha_d)$.

Theorem 4.1.20 *If the coefficients of (4.1.15) are Lispchitz in the sense that they are continous, of linear growth, and $\frac{\partial b}{\partial x_i}, \frac{\partial \sigma}{\partial x_i}$ exist continously $(1 \le i \le d)$, then the derivatives $\partial X_{x,s}(t)/\partial x_i$ exist in the $L^2(\Omega)$ sense and the functions $\zeta_i(t) = \partial X_{x,s}(t)/\partial x_i$ satisfy the stochastic differential equation with random coefficients*

$$\zeta_i(t) = e_i + \int_s^t \zeta_i(u) \cdot b_x(u, X_{x,s}(u))du + \int_s^t \zeta_i(u) \cdot \sigma_x(u, X_{x,s}(u))dB(u)$$

where e_i is the vector with components δ_{ij}.

Theorem 4.1.21 *If the conditions in Theorem 1.3.3 hold and assume that $D_x^\alpha b(t,x), D_x^\alpha \sigma(t,x)$ exist and are continuous if $|\alpha| \le 2$, and*

$$|D_x^\alpha b(t,x)| + |D_x^\alpha \sigma(t,x)| \le K_0(1 + |x|^\beta) \quad (|x| \le 2)$$

where K_0, β are positive constants. Then the second derivatives $\frac{\partial^2 X_{x,s}(t)}{\partial x_i \partial x_j}$ exist in the $L^2(\Omega)$ sense, and they satisfy the stochastic differential equations with random coefficients obtained by applying formally $\partial^2/\partial x_k \partial x_j$ to (4.1.15).

4.1.4 Stratonovich Stochastic Differential Equations

We shall consider SDEs written with Stratonovich integrals

$$dX_t = X_0 + b(s, X_t)dt + a(t, X_t) \circ dB(t). \tag{4.1.16}$$

where $b(t,x) = (b_1(t,x), \cdots, b_d(t,x))^T$ is Borel measurable function $(t,x) \in [0,\infty) \times R^d \to R^d$, and $a(t,x) = (a_{ij}(t,x))_{d\times r}$ is a $d \times r$-matrix Borel measurable function $(t,x) \in [0,\infty) \times R^d \to R^d \otimes R^r$, and $B(t)$ is an r-dimensional Brownian motion process, and $\circ dB(t)$ denotes the Stratonovich integral. Further we assume that $b_i(t,x)(i = 1, 2, \cdots, d)$ are continuous in (t,x) , continuosly derivatiable in t, twice continuously differentiable in x and their first dervatives in x are bounded. Then (4.1.16) can be writen as a Itò type SDEs:

$$dX_t = \tilde{b}(t, X_t) + a(t, X_t)dB(t), \tag{4.1.17}$$

where

$$\tilde{b}_i = b_i(t,x) + \frac{1}{2}\sum_{j=1}^{r}\sum_{k=1}^{d} \frac{\partial a_{ij}}{\partial x_k} a_{kj}; i = 1, 2, \cdots, d.$$

Hence the existence and uniqueness of the Stratonovich equation (4.1.16) can be proceed via Itô equation (4.1.17). For detail, please refer to [89].

4.1.5 Stochastic Differential Equations on Manifolds

Let M be a d-dimensional C^∞-manifold i.e., M is a Hausdorff topological space with an open covering $\{U_\alpha\}_{\alpha\in\Lambda}$ of M, each U_α provided with a homeomorphism ϕ_α with an open subset $\phi_\alpha(U_\alpha)$ of R^d such that, if $U_\alpha \cap U_\beta \not= \phi$ the function $\phi_\beta \circ \phi_\alpha$ from $\phi_\alpha(U_\alpha \cap U_\beta)$ in to $\phi_\beta(U_\alpha \cap U_\beta)$ is a C^∞ -function. U_α is called a coordinate nerghborhood and for $x \in U_\alpha, \phi_\alpha(x) = (x^1, \cdots, x^d) \in R^d$ is called a local coordinate of x. For the sake of simplicity, we assume that M is connected and σ-compact throughout this section. It is then well known that M is paracompact and has a countable open base.

A function $f(x)$ defined on an open subset D of M is called C^∞ if it is C^∞ as a function of the local coordinate, i.e.,$f\circ\phi_\alpha$ is C^∞ on $\phi_\alpha(U_\alpha \cap D)$ for every α. Let $F(M)$ be the totality of all real valued C^∞-functions on M and $F_0(M)$ be the subclass of $F(M)$ consisting of all functions in $F(M)$ with compact support. $F(M)$ and $F_0(M)$ are algebras over the field of real numbers R with the usual rules of $f+g, fg$ and $\lambda f(f,g \in F(M))$ or $F_0(M), \lambda \in R)$. Let $x \in M$. By a tangent vector at x we mean a linear mapping V of $F(M)$ into R such that

$$V(fg) = V(f)g(x) + f(x)V(g).$$

Denote by $T_x(M)$, the set of all tangent vectors at x which is a linear space, it is called the tangent space at x, with the rules

$$(V+V')(f) = V(f) + V'(f), and (\lambda V)(f) = \lambda V(f).$$

Let $(x^1, \cdots, X^d)$ be a local coordinate in a coordinate neighborhood U of x. Every $f \in F(M)$ is expressed on U as a C^∞-function $f(x^1, \cdots, x^d)$. Then $f \mapsto (\frac{\partial f}{\partial x^i})(x)$ is a tangent vector at x for every $i = 1, 2, \cdots, d$. This is denoted by $(\frac{\partial}{\partial x^i})_x$. It is easy to see that $\{(\frac{\partial}{\partial x^i})_x\}_{i=1,2,\cdots,d}$ forms a base for $T_x(M)$. By a vector field we mean a mapping $V : x \ni M \mapsto V(x) \in T_x(M)$. V is called a C^∞-vector field if for every $f \in F(M), (Vf)(x) := V(x)f$ is a C^∞-function. Thus V is a C^∞-vector field if and only if V is a linear mappping of $F(M)$ into $F(M)$(or $F_0(M)$ into $F_0(M)$) such that $V(fg) = V(f)g + fV(g)$. In the following we only consider a C^∞-vetor field, $\mathcal{X}(M)$ denoting the totality of C^∞-vector fields. Let $A_0, A_1, \cdots, A_r \in \mathcal{X}(M)$. Consider the following stochastic differential equation

$$dX(t) = A_0(X(t))dt + \sum_{k=1}^{r} A_k(X(t))dB^k(t). \tag{4.1.18}$$

Let $\hat{M} = M$ or $M \cup \{\Delta\}$ (= the one-point compactification of M) accordingly as M is compact or noncompact. Let $\hat{W}(M)$ be the path space defined by

$$\begin{aligned} \hat{W}(M) &= \{w; w \text{is a continuous mapping } [0,\infty) \to \hat{M} \text{ such that} \\ &\quad w(0) \in M \text{ and if } w(t) = \Delta \text{ then} W(t') = \Delta \forall t' \geq t\} \end{aligned}$$

and let $\mathcal{B}(\hat{W}(M))$ be the σ-field generated by the Borel cylinder sets. The explosion time $e(w)$ is defined by $e(w) = \inf\{t; w(t) = \Delta\}$.

Definition 4.1.22 *A solution $X = (X(t))$ of (4.1.18) is any $(\mathcal{F}_t)$-adapted $\hat{W}(M)$-valued random variable (i.e., a continuous process on $\hat{M}$ with Δ as a trap) defined on a probability* *space with a reference family* $(\mathcal{F}_t)$ *and an r-dimensional* $(\mathcal{F}_t)$*-Brownian motion* $B = (B(t))$ *with* $B(0) = 0$ *such that the following is satisfied: for every* $f \in F_0(M)$,

$$f(X(t)) - f(X(0)) = \int_0^t (A_0 f)(X(s))ds + \int_0^t \sum_{k=1}^r (A_k f)(X(s)) \circ dB^k(s), \qquad (4.1.19)$$

where the second term on the right-hand side is understood in the sense of Fisk-Stratonovich integral (cf. [39] Chapter III). We have the following theorem.

Theorem 4.1.23 *There exists a function* $F : M \times W_0^r \to \hat{W}(M)$ *which is*

$$\cap_\mu \overline{\mathcal{B}(M) \times \mathcal{B}_t(W_0^r)}^{\mu \times P^W} / \mathcal{B}_t(\hat{W}(M))$$

measurable for every $t \geq 0$ *, here* μ *runs over all probabilities on* $(M, \mathcal{B}(M))$*, such that*
(i) for every solution $X = (X(t))$ *with respect to the Brownian motion* $B = (B(t))$*, it holds that*

$$X = F(X(0), B) \quad a.s.,$$

and
(ii) for every r-dimensional $(\mathcal{F}_t)$*-Brownian motion* $B(B(t))$ *with* $B(0) = 0$ *defined on a probability space with a reference family* $(\mathcal{F}_t)$ *and an M-valued* $(\mathcal{F}_0)$*-measurable random variable* $\xi, X = F(\xi, B)$ *is a solution of (4.1.18) with* $X(0) = \xi$ *a.s.*

Given vector fields $A_\alpha \in \mathcal{X}(M), \alpha = 0, 1, 2, \cdots, r$, we construct a mapping $X = (X(t,x,w)) : M \times W_0^r \ni (x,w) \mapsto X(\cdot, x, w) \in \hat{W}(M)$. This may also be regarded as a mapping: $[0,\infty) \times M \times W_0^r \ni (t, xw) \mapsto X(t,x,w) \in \hat{M}$. Smilary to the flat space, we can show that the mapping $M \ni x \mapsto X(t,x,w) \in \hat{M}$ is a local diffeomorphism of M for each fixed $t \geq 0$ and for almost all w such that $X(t,x,w) \in M$. We have the following theorem:

Theorem 4.1.24 *Assume that* M *is a compact manifold.* $X(t,x,w)$ *has a modification, which is denoted by* $X(t,x,w)$ *again, such that the mapping* $X_t(w) : x \mapsto X(t,x,w)$ *is* C^∞ *in the sense that* $x \mapsto f(X(t,x,w))$ *is* C^∞ *for every* $f \in F(M)$ *and all fixed* $t \in [0,\infty)$*,a.s. Furthermore, for each* $x \in M$ *and* $t \in [0,\infty)$*, the differential* $X(t,x,w)$ *of the mapping* $x \mapsto X(t,x,w)_*$

$$X(t,x,w)_* : T_x(M) \mapsto T_{X(t,x,w)}(M)$$

is an isomorphism a.s. on the set $\{w; X(t,x,w) \in M\}$.

4.2 Applications

4.2.1 Diffusions

In a stochastic differential equation of the form

$$dX_t = b(t, X_t)dt + a(t, X_t)dB_t, t \geq s, X_s = x \qquad (4.2.20)$$

where $X_t \in R^d, b(t,x) \in R^d, a(t,x) \in R^{d \times r}$ and B_t is a r -dimensional Brownian motion, we will call b the *drift coefficient* and a or sometimes $\frac{1}{2}aa^T$ the *diffusion coefficient*. We assume the the coefficients of (4.2.20) satisfy the Lipschtz condition and hence there exists a unique solution. Denote by $X_t = X_t^{s,x}$ the unique solution of (4.2.20) and $X_t = X_t^x$ simply if $s = 0$. The solution of a SDE may be considered as the mathematical description of the motion of a small particle in a moving fluid. Therefore such stochastic processes are called Ito diffusions. Further if we assume that the coefficients do not depend on t but on x only, the resulting process $X_t(\omega)$ will have the property of being *time-homogeneous*, in the sense that $\{X_{s+h}^{s,x}\}_{h\geq 0}$ and $\{X_h^{0,x}\}_{h\geq 0}$ have the same P^0-distributions, i.e.$\{X_t\}_{t\geq 0}$ is time-homogeneous. We inroduce the probability laws Q^x of $\{X_t\}_{t\geq 0}$, for $x \in R^n$. Let $\mathcal{M} = \sigma(\omega \to X_t = X_t^y, t \geq 0, y \in R^n)$ and define Q^x on the members of $\mathcal{M}$ by

$$Q^x[X_{t_1} \in E_1, \cdots, X_{t_k} \in E_k] = P^0[X_{t_1}^x \in E_1, \cdots, X_{t_k}^x \in E_k] \tag{4.2.21}$$

where $E_i \subset R^n$ are Borel sets; $1 \leq i \leq k$. If denote $\mathcal{M}_t = \sigma(X_r; r \leq t)$, then $\mathcal{M}_t \subset \mathcal{F}_t$ due to X_t is measuable with respect to $\mathcal{F}_t$. Denote by $\mathcal{F}_\tau = \sigma(X_{s\wedge\tau}; s \geq 0)$, here τ is a stopping time.

Markov property

We can give the following theorem on the important Markov property and strong Markov property :

Theorem 4.2.1 *Let f be a bounded Borel function from R^n to R. Then, for $t, h > 0$*

$$E^x[f(X_{t+h})|\mathcal{F}_t](\omega) = E^{X_t(\omega)}[f(X_h)], \tag{4.2.22}$$

where E^x denotes the expectation w.r.t the probability measure Q^x. Thus $E^y[f(X_h)]$ means $E[f(X_h^y)]$, where E denotes the expectation w.r.t.P^0. The right hand side is the function $E^y[f(X_h)]$ evaluated at $y = X_t(\omega)$.

Remark *It is easy to see that*

$$E^x[f(X_{t+h})|\mathcal{M}_t] = E^{X_t}[f(X_h)]$$

due to $\mathcal{M}_t \subset \mathcal{F}_t$.

Theorem 4.2.2 *(The strong Markov property for Ito diffusion) Let f be a bounded Borel function on R^n, τ a stopping time w.r.t.$\mathcal{F}_t$, $\tau < \infty$ a.s. Then*

$$E^x[f(X_{\tau+h})|\mathcal{F}_\tau] = E^{X_\tau}[f(X_h)], \forall h \geq 0. \tag{4.2.23}$$

The generator of an Ito diffusion

It is fundamental for many applications that we can associate a second order partial differential operator A to an Ito diffusion X_t. The basic connection between A and X_t is that A is the generator of the process X_t:

Definition 4.2.3 *Let $\{X_t\}$ be a time homogeneous Ito diffusion in R^n. The infinitesimal generator A of X_t is defined by*

$$AF(s) = \lim_{t\downarrow 0} \frac{E^x[f(X_t)] - f(x)}{t}; x \in R^n$$

The set of functions $f : R^5 \to R$ such that the limit exists for all $x \in R^n$ is denoted by $\mathcal{D}_A$.

In the following, we will find out the relation between A and the coefficients b, a in the stochastic differential equation:

$$dX_t = b(X_t)dt + a(X_t)dB_t, t \geq s, X_s = x. \tag{4.2.24}$$

Theorem 4.2.4 *Let X_t be the Ito diffusion (4.2.24), if $f \in C_0^2(R^n)$, i.e. f is twice differentiable and has compact support, then $F \in \mathcal{D}_A$ and*

$$Af(x) = \sum_i b_i(x)\frac{\partial f}{\partial x_i} + \frac{1}{2}\sum_{i,j}(aa^T)_{i,j}(x)\frac{\partial^2 f}{\partial x_i \partial x_j}$$

Example 4.2.5 *The n-dimensional Brownian motion is the solution of the stochastic differential equation*

$$dX_t = dB_t,$$

i.e. we have $b = 0$ and $a = I_n$ the n-dimensional identity matrix. So the generator of B_t is

$$Af = \frac{1}{2}\sum \frac{\partial^2 f}{\partial x_i^2}; f = f(x_1, \cdots, x_n) \in C_0^2(R^n)$$

i.e. $A = \frac{1}{2}\Delta$, where Δ is the Laplace operator.

By using the generator A, we now have the Dynkin formula:

Theorem 4.2.6 *Let $f \in C_0^2(R^n)$, τ be a stopping time, $E^x[\tau] < \infty$. Then*

$$E^x[f(X_\tau)] = f(x) + E^x[\int_0^\tau Af(X_s)ds] \tag{4.2.25}$$

We now introcuce an operator which is related to the generator A and is used in the solution of the Dirichlet problem.

Definition 4.2.7 *Let $\{X_t\}$ be an Ito diffusion. The characteristic operator $\mathcal{A} = \mathcal{A}_X$ of $\{X_t\}$ is defined by*

$$\mathcal{A}f(x) = \lim_{U \downarrow x} \frac{E^x[f(X_{\tau_U})] - f(x)}{E^x[\tau_U]} \tag{4.2.26}$$

where the U's are open sets U_k decreasing to the point x, in the sense that $U_{k+1} \subset U_k$ and $\bigcap_k U_k = \{x\}$, and $\tau_U = \inf\{t > 0; X_t \notin U\}$ is the first exit time from U for X_t.

Kolmogorov's backward equation

If we choose $f \in C_0^2(R^n)$ and $\tau = t$ in the Dynkin's formula (4.2.25), we know that

$$u(t, x) = E^x[f(X_t)] \tag{4.2.27}$$

is differentiable with respect to t and

$$\frac{\partial u}{\partial t} = E^x[Af(X_t)].$$

We can get the following Kolmogorov's backward equation:

Theorem 4.2.8 *Let $f \in C_0^2(R^n)$. Then $u(t, \cdot) \in \mathcal{D}_A$ for each t and*

$$\frac{\partial u}{\partial t} = Au, t > 0, x \in R^n \tag{4.2.28}$$

$$u(0, x) = f(x); x \in R^n. \tag{4.2.29}$$

Moreover, if $w(t, x) \in C^{1,2}(R \times R^n)$ is bounded function satisfying (4.2.28) and (4.2.29) then $w(t, x) = u(t, x)$, given by (4.2.27).

We can obtain the following useful generalization of Kolmogorov's backward equation:

Theorem 4.2.9 *(The Feynman-Kac formula) Let $f \in C_0^2(R^n)$ and $q \in C(R^n)$. Assume that q is lower bounded.*

(i) Put

$$v(t,x) = E^x[\exp(-\int_0^t q(X_s)ds)f(X_t)]. \tag{4.2.30}$$

Then

$$v(t,x) = Av - qv; t > 0, x \in R^n \tag{4.2.31}$$

$$v(0,x) = f(x); x \in R^n. \tag{4.2.32}$$

Moreover, if $w(t,x) \in C^{1,2}(R \times R^n)$ is bounded on $K \times R^n$ for each compact $K \subset R$ and w solves (4.2.31), (4.2.32), then $w(t,x) = v(t,x)$ given by (4.2.30).

The Girsanov theorem

Before we give the Girsanov theorem, we introduce a definition.

Definition 4.2.10 *Let $\mathcal{V} = \mathcal{V}[S,T]$ be the class of functions $f(t,\omega) : [0,\infty) \times \Omega \to R$ such that*

(i)$(t,\omega) \to f(t,\omega)$ is $\mathcal{B} \times \mathcal{F}$ -measurable, where $\mathcal{B}$ denotes the Borel σ-algebra on $[0,\infty)$.

(ii) $f(t,\omega)$ is $\mathcal{F}_t$-adapted.

(iii)$E[\int_S^T f(t,\omega)^2 dt] < \infty$.

Theorem 4.2.11 *(The Girsanov theorem I)*

Let $Y(t) \in R^n$ be an Ito process of the form

$$dY(t) = a(t,\omega)dt + dB(t); t \leq T, Y_0 = 0,$$

where $T \leq \infty$ is a given constant and $B(t)$ is n-dimensional Brownian motion. Put

$$M_t = \exp(-\int_0^t a(s,\omega)dB_s - \frac{1}{2}\int_0^t a^2(s,\omega)ds); t \leq T. \tag{4.2.33}$$

Assume that $a(s,\omega)$ satisfies Novikov's condition

$$E[\exp(\frac{1}{2}\int_0^T a^2(s,\omega_d s)] < \infty \tag{4.2.34}$$

where $E = E^{P^0}$ is the expectation w.r.t.P^0. Define the measure Q on $(\Omega, \mathcal{F}_T)$ by

$$dQ(\omega) = M_T(\omega)dP^0(\omega) \tag{4.2.35}$$

Then $Y(t)$ is an n-dimesional Brownian motion w.r.t. the probability law Q, for $t \leq T$.

Theorem 4.2.12 (The Girsanov theorem II) *Let $Y(t) \in R^n$ be an Ito process of the form*

$$dY(t) = \beta(t,\omega)dt + \theta(t,\omega)dB(t); t \leq T,$$

where $\beta(t,\omega) \in R^n, \theta(t,\omega) \in R^{n\times m}$ and $B(t) \in R^m$. Suppose there exist $\mathcal{V}[0,T]$-process $u(t,\omega) \in R^m$ and $\alpha(t,\omega) \in R^n$ such that

$$\theta(t,\omega)u(t,\omega) = \beta(t,\omega) - \alpha(t,\omega)$$

and assume that $u(t,\omega)$ satisfies Novikov's condition

$$E[\exp(\frac{1}{2}\int_0^T u^2(s,\omega_d s)] < \infty \tag{4.2.36}$$

Put

$$M_t = \exp(-\int_0^t u(s,\omega)dB_s - \frac{1}{2}\int_0^t u^2(s,\omega)ds); t \leq T. \tag{4.2.37}$$

and

$$dQ(\omega) = M_T(\omega)dP^0(\omega) \quad on \;\; \mathcal{F}_T. \tag{4.2.38}$$

Then

$$\hat{B}(t) := \int_0^t u(s,\omega)ds + B(t); t \leq T \tag{4.2.39}$$

is a Brownian motion w.r.t. the probability law Q and in terms of $\hat{B}(t)$ the process $Y(t)$ has the stochastic integral repesentation

$$dY(t) = \alpha(t,\omega)dt + \theta(t,\omega)d\hat{B}(t).$$

Theorem 4.2.13 (The Girsanov theorem III) *Let $X(t) = X^x(t) \in R^n$ and $Y(t) \in R^n$ be an Ito diffusion and an Ito process, respectively, of the forms*

$$\begin{aligned} dX(t) &= b(X(t))dt + a(X(t))dB(t); t \leq T, X(0) = x \qquad (4.2.40)\\ dY(t) &= [\gamma(t,\omega) + b(Y(t))]dt + a(Y(t))dB(t); t \leq T, Y(0) = x \end{aligned}$$

where the functions $b : R^n \to R^n$ and $a : R^n \to R^{n\times m}$ satisfy the Lipschitz condition and linear growth condition and $\gamma(t,\omega) \in \mathcal{V}[0,T], x \in R^n$. Suppose there exists a $\mathcal{V}[0,T]$-process $u(t,\omega)$ satisfying Novikov's condition (4.2.36). Define M_t, Q and $\hat{B}(t)$ as in (4.2.37) (4.2.38) and (4.2.39). Then

$$dY(t) = b(Y(t))dt + a(Y(t))d\hat{B}(t).$$

Therefore the Q-law of $Y(t)$ is the same as the P^0-law of $X^x(t)$; $t \leq T$.

Remark *The Girsanov theorem says that if we change the drift coefficient of a given Ito process, then the law of process will not change dramatically. In fact, the law of the new process will be absolutely continuous w.r.t. the law of the original process and we can compute explicitly the Randon-Nikodym derivative.*

4.2.2 Boundary value problem

The Dirichlet problem

We now use diffusion type SDE to solve the following generalization of the Dirichlet problem: Given a domain D in R^n and a continuous function ϕ on ∂D the boundary of D. Find a function $\tilde{\phi}$ continuous on the closure $\bar{D}$ of D such that

$$\begin{aligned} L\tilde{\phi} &= 0 \;\; in \;\; D, \qquad (4.2.41)\\ \lim_{x\to y, x\in D}\tilde{\phi}(x) &= \phi(y), \;\; for \;\; all \;\; (regular) \;\; y \in \partial D. \end{aligned}$$

where L is a semi-elliptic partial differential operator on $C^2(R^n)$ of the form

$$L = \sum_{i,j=1}^{n} a_{ij}(x)\frac{\partial^2}{\partial x_i \partial x_j} + \sum_{i=1}^{n} b_i(x)\frac{\partial}{\partial x_i} \tag{4.2.42}$$

L is called semi-elliptic (elliptic) when all the eigenvalues of the symmetric matrix (a_{ij}) are nonnegative (positive), for all x. A point $y \in \partial D$ is called regular for D (w.r.t. X_t) if

$$Q^y[\tau_D = 0] = 1.$$

Otherwise the point y is called irregular.

The idea to solve this problem is simple. First we find an Ito diffusion (X_t) whose generator A coincides with L on $C_0^2(R^n)$. Let X_t be the solution of

$$dX_t = b(X_t) + \sigma(X_t)dB_t, \tag{4.2.43}$$

where B_t is an n-dimiensional Brownian motion, $\frac{1}{2}\sigma\sigma^T = (a_{ij})$. Then the candidate for the solution $\tilde{\phi}$ is

$$\tilde{\phi}(x) = E^x[\phi(X_{\tau_D})].$$

Unfortunately, some examples show that this problem is not solvable in general (we refer to [74]). However, it is possible to formulate a weak, stochastic version of the problem. This stochastic version will always have a solution, which coincides with the solution of the original problem in case such a solution exists.

Definition 4.2.14 *Let f be a locally bounded, measurable function on D. Then f is called X-harmonic in D if*

$$f(x) = E^x[f(X_{\tau_U})]$$

for all $x \in D$ and all bounded open sets U with $\bar{U} \subset D$.

We now can give the stochastic version of Dirichlet problem as following: Given a bounded measurable function ϕ on ∂D, find a function $\tilde{\phi}$ on D such that

$$(i)_s \qquad \tilde{\phi} \text{ is X-harmonic} \tag{4.2.44}$$

$$(ii)_s \qquad \lim_{t\uparrow\tau_D} \tilde{\phi}(X_t) = \phi(X_{\tau_D}) \text{ a.s.} Q^x, x \in D. \tag{4.2.45}$$

We have the following theorem:

Theorem 4.2.15 *Let ϕ be a bounded measurable function on ∂D. Define*

$$\tilde{\phi}(x) = E^x[\phi(X_{\tau_D})], \tag{4.2.46}$$

then $\tilde{\phi}$ solves the stochastic Dirichlet problem (4.2.44). On the other hand suppose g is a bounded function on D such that

(1)g is X-harmonic,

(2)$\lim_{t\uparrow\tau_D} g(X_t) = \phi(X_{\tau_D}) a.s. Q^x, x \in D$,

then $g(x) = E^x[\phi(X_{\tau_D})], x \in D$.

We now can ask that under what conditions the solution $\tilde{\phi}$ of the stochastic Dirichlet problem (4.2.44) will also be a solution of the original Dirichlet problem (4.2.41). The following theorem answers this problem partially.

Theorem 4.2.16 *Suppose L is uniformly elliptic in D, i.e. the eigenvalues of (a_{ij}) are bounded away from 0 in D, and the coefficients b and σ satisfy the Lipschitz condition and the linear growth condition. Let ϕ be a bounded continuous function on ∂D. Put*

$$\tilde{\phi}(x) = E^x[\phi(X_{\tau_D})].$$

Then $\tilde{\phi} \in C^{2+\alpha}(D)$ for all $\alpha < 1$ and $\tilde{\phi}$ solves the Dirichlet problem (4.2.41).

The Poisson problem

Consider the Poisson problem: Given a continuous function g on D find a C^2 function f in D such that

$$\begin{aligned} Lf &= -g \;\; in \;\; D, \\ \lim_{x\to y, x\in D} f(x) &= 0, \text{ for all (regular)} y \in \partial D. \end{aligned} \tag{4.2.47}$$

where L is a semi-elliptic partial differential operator on a domain $D \subset R^n$ as before. Let X_t be an associated Ito diffusion described by (4.2.43). Similar to the discussion in the Dirichlet problem we have the following theorems.

Theorem 4.2.17 *(Solution of the stochastic Poisson problem) Assume that*

$$E^x[\int_0^{\tau_D} |g(X_s)|ds] < \infty, \forall x \in D. \tag{4.2.48}$$

Define

$$\hat{g}(x) = E^x[\int_0^{\tau_D} g(X_s)ds]. \tag{4.2.49}$$

Then

$$\mathcal{A}\hat{g} = -g \;\; in \;\; D,$$

and

$$\lim_{t\uparrow\tau_D} \hat{g}(X_t) = 0, a.s.Q^x, \forall x \in D.$$

Theorem 4.2.18 *(Solution of the combined stochastic Dirichlet and Poisson equation) Let $\phi \in C(\partial D)$ be bounded and let $g \in C(D)$ satisfy (4.2.48), define*

$$h(x) = E^x[\int_0^{\tau_D} g(X_s)ds] + E^x[\phi(X_{\tau_D})], x \in D. \tag{4.2.50}$$

a) Then

$$\mathcal{A}h = -g \;\; in \;\; D \tag{4.2.51}$$

and

$$\lim_{t\uparrow\tau_D} \phi(X_t) = \phi(X_{\tau_D})], a.s.Q^x, \forall x \in D. \tag{4.2.52}$$

Moreover, it there exists a function $h_1 \in C^2(D)$ and a constant C such that

$$|h_1(x)| < C(1+), x \in D, E^x[\int_0^{\tau_D} |g(X_s)|ds]$$

and h_1 satisfies (4.2.51) and (4.2.52), then $h_1 = h$.

Remark *We have the similar result that if L is unifomly elliptic in D and $g \in C^\alpha(D)$ (for some $\alpha > 0$) is bounded, then the function h given by (4.2.50) solves the Dirichlet-Poisson problem,i.e.*

(i) $Lh = -g$ in D,

(ii)$\lim_{D \ni x \to y} h(x) = \phi(y)$ for all regular $y \in \partial D$.

4.2.3 Optimal stopping

The time homogeneous case

Let us consider the optimal stopping problem in the section. Let X_t be an Ito diffusion on R^n and let g (the reward function) be a given function on R^n, satisfying

$$\begin{array}{ll} a) & g(\xi) \geq 0, \quad \forall \xi \in R^n; \\ b) & g \text{ is continuous.} \end{array}$$

The problem is to find a stopping time $\tau^* = \tau^*(x, \omega)$ for (X_t) such that

$$E^x[g(X_{\tau^*})] = \sup_\tau E^x[g(X_\tau)], \forall x \in R^n, \tag{4.2.53}$$

the supremum being taken over all stopping times τ for all (X_t), E^x denotes the expectation with respect to the probability law Q^x of the process $(X_t)_{t \geq 0}$. We also want to find the corresponding optimal expected reward

$$g^* = E^x[g(X_{\tau^*})]. \tag{4.2.54}$$

We can regard X_t as the state of a game at time t, each ω corresponds to one sample of the game. For each time t we have to take an option to either stop the game, thereby obtaining the reward $g(X_t)$, or continue the game in the hope that stopping it at a later time will give a bigger reward. The problem of course is that we do not know what state the game is in at future times, except the probability distribution of the future. Hence it is really a stopping time problem. So, among all possible stopping times, we are seeking for the optimal one, τ^* which gives the best result, i.e. the biggest expected reward in the sense of (4.2.53) .

We now can discuss the problem. We need the following concepts.

Definition 4.2.19 *A measurable function $f : R^n \to [0, \infty]$ is called supermeanvalued (w.r.t. X_t) if*

$$f(x) \geq E^x[f(X_\tau)] \tag{4.2.55}$$

for all stopping times τ and all $x \in R^n$; it is called superharmonic (w.r.t. X_t) if, in addition, it is also lower semicontinuous.

Definition 4.2.20 *Let h be a measurable function on R^n. If f is a superharmonic (supermeanvalued) function and $f \geq h$ we say that f is a superharmonic (supermeanvalued) majorant of h (w.r.t. X_t). The function*

$$\bar{h}(x) = \inf_f f(x); x \in R^n, \tag{4.2.56}$$

the infimum being taken over all supermeanvalued majorant f of h. It is easy to show that it is supermeanvalued and therefore $\bar{h}$ is the least supermeanvalued majorant of h. Similarly, if function $\hat{h}$ is a superharmonic majorant of h and for any other superharmonic majorant f of h we have $\hat{h} \leq f$. Then $\hat{h}$ is called the least superharmonic majorant of h.

We are now ready for the existence and uniqueness result on the optimal stopping problem.

Theorem 4.2.21 *(Existence theorem for optimal stopping)*
Let g^* denote the optimal reward and $\hat{g}$ the least superharmonic majorant of a continuous reward function $g \geq 0$.

a) Then

$$g^*(x) = \hat{g}(x). \tag{4.2.57}$$

b) For $\epsilon > 0$ let

$$D_\epsilon = \{x; g(x) < \hat{g}(x) - \epsilon\}. \tag{4.2.58}$$

suppose g is bounded. Then stopping at the first time τ_ϵ of exit from D_ϵ is close to being optimal, in the sense that

$$|g^*(x) - E^x[g(X_{\tau_\epsilon})] \leq 2\epsilon \tag{4.2.59}$$

for all x.

c) For arbitrary continuous $g \geq 0$, let

$$D = \{x; g(x) < g^*(x)\} \textit{(the continuation region)}. \tag{4.2.60}$$

For $N = 1, 2, \cdots$ define $g_N = g \wedge N, D_N = \{x; g_N(x) < (g_N)^\wedge(x)\}$ and $\sigma_N = \tau_{D_N}$. Then $D_N \subset D_{N+1}, D_N \subset D \bigcap g^{-1}([0, N)), D = \bigcup_N D_N$. If $\sigma_N < \infty$ a.s. Q^x for all N then.

$$g^*(x) = \lim_{N \to \infty} E^x[g(X_{\sigma_N})]. \tag{4.2.61}$$

d) In particular, if $\tau_D < \infty$ a.s. Q^x and the family $\{g(X_{\tau_N})\}_N$ is uniformly integrable w.r.t. Q^x, then

$$g^*(x) = E^x[g(X_{\tau_D})]. \tag{4.2.62}$$

and $\tau^* = \tau_D$ is an optimal stopping time.

Remark *This theorem gives a sufficient condition for the existence of an optimal stopping time τ^*. Unfortunately, τ^* need not exist in general. For example, if $X_t = t$, for $t \geq 0$ (deterministic) and $g(\xi) = \frac{\xi^2}{1+\xi^2}; \xi \in R$, then $g^*(x) = 1$, but there is no stopping time τ such that $E^x[g(X_\tau)] = 1$. However, we can prove that if an optimal stopping time τ^* exists, then the stopping time given in the last theorem is optimal:*

Theorem 4.2.22 *(Uniqueness theorem for optimal stopping)*
Define as before

$$D = \{x; g(x) < g^*(x)\} \subset R^n.$$

Suppose there exists an optimal stopping time $\tau^* = \tau^*(x, \omega)$ for the problem (4.2.53) for all x. Then

$$\tau^* \geq \tau_D \quad \forall x \in D \tag{4.2.63}$$

and

$$g^*(x) = E^x[g(X_{\tau_D})], \quad \forall x \in R^n. \tag{4.2.64}$$

Hence τ_D is an optimal stopping time for the problem (4.2.53).

Remark *Let $\mathcal{A}$ be the characteristic operator of X. Assume $g \in C^2(R^n)$. Define*

$$U = \{x; \mathcal{A}g(x) > 0\}.$$

Then $U \subset D$. Consequently, from (4.2.63) we conclude that it is never optimal to stop the process before it exits from U. But there may be cases when $U \neq D$, so that it is optimal to proceed beyond U before stopping.

The time inhomogeneous case

Let ust now consider the case when the reward function g depends on both time and space, i.e.

$$g = g(t,x) : R \times R^n \to [0,\infty), g \text{ is continuous.} \tag{4.2.65}$$

Then the problem is to find $g_0(x)$ and τ^* such that

$$g_0(x) = \sup_\tau E^x[g(\tau, X_\tau)] = E^x[g(\tau^*, X_{\tau^*})]. \tag{4.2.66}$$

To reduce this case to the time homogeneous case, we proceed as follows:
Suppose the Ito diffusion $X_t = X_t^x$ has the form

$$dX_t = b(X_t)dt + \sigma(X_t)dB_t; t \geq 0, X_0 = x$$

where $b : R^n \to R^n$ and $\sigma : R^n \to R^{n\times m}$ are given functions satisfying the Lipschtz condition and linear growth condition, B_t is m -dimensional Brownian motion. Define the Ito diffusion $Y_t = Y^{(s,x)}$ in R^{n+1} by

$$Y(t) = \begin{bmatrix} s+t \\ X_t^x \end{bmatrix}; t \geq 0$$

Then

$$dY_t = \begin{bmatrix} 1 \\ b(X_t) \end{bmatrix} dt + \begin{bmatrix} 0 \\ \sigma(X_t) \end{bmatrix} dB_t = \hat{b}(Y_t)dt + \hat{\sigma}(Y_t)dB_t$$

where

$$\hat{b}(\eta)(t) = \hat{b}(t,\xi) = \begin{bmatrix} 1 \\ b(\xi) \end{bmatrix} \in R^{n+1}, \sigma(\eta)(t) = \hat{\sigma}(t,\xi) = \begin{bmatrix} 0 \\ \sigma(\xi) \end{bmatrix} \in R^{(n+1)\times m},$$

with $\eta = (t,\xi) \in R \times R^n$. So Y_t is an Ito diffusion starting at $y = (s,x)$. Let $Q^y = Q^{(s,x)}$ denote the probability law of $\{Y_t\}$ and let $E^y = E^{(s,x)}$ denote the expectation w.r.t. Q^y. In terms of Y_t the problem (4.2.66) can be writen

$$g_0(x) = g^*(0,x) = \sup_\tau E^{(0,x)}[g(Y_\tau)] = E^{(0,x)}[g(Y_{\tau^*})] \tag{4.2.67}$$

which is a special case of the problem

$$g^*(s,x) = \sup_\tau E^{(s,x)}[g(Y_\tau)] = E^{(s,x)}[g(Y_{\tau^*})] \tag{4.2.68}$$

which is of the form (4.2.53) and (4.2.54) with X_t replaced by Y_t.

Remark *The characteristic operator $\hat{\mathcal{A}}$ of Y_t is given by*

$$\hat{\mathcal{A}}g(s,x) = \frac{\partial g}{\partial s}(s,x) + \mathcal{A}g(s,x); g \in C^2(R \times R^n)$$

where $\mathcal{A}$ is the characteristic operator of X_t.

Example 4.2.23 *(When is the right time to sell the stocks?)*
Suppose the price X_t at time t of a person's assets varies according to a stochastic differential equation of the form:

$$dX_t = rX_t dt + \alpha X_t dB_t, X_0 = x > 0,$$

where B_t is 1-dimensional Brownian motion and r, α are known constants. Suppose that connected to the sale of the assets there is a fixed fee/tax or transaction cost $a > 0$. Then if the person decides to sell at time t, the discounted net of the sale is

$$e^{-\rho t}(X_t - a),$$

where $\rho > 0$ is the given discounting factor. The problem is to find a stopping time τ that maximizes

$$E^{(s,x)}[e^{-\rho\tau}(X_t - a)] = E^{(s,x)}[g(\tau, X_\tau],$$

where

$$g(t,\xi) = e^{-\rho t}(\xi - a).$$

The characteristic operator $\hat{\mathcal{A}}$ of the process $Y_t = (s+t, X_t)$ is given by

$$\hat{\mathcal{A}}f(s,x) = \frac{\partial f}{\partial s} + rx\frac{\partial f}{\partial x} + \frac{1}{2}\alpha^2 x^2 \frac{\partial^2 f}{\partial x^2}; f \in C^2(R^2)$$

Hence $\hat{\mathcal{A}}g(s,x) = -\rho e^{-\rho s}(x-a) + rxe^{-\rho s} = e^{-\rho s}((r-\rho)x + \rho a)$. So

$$U := \{(s,x); \hat{\mathcal{A}}g(s,x) > 0\} = \begin{cases} R \times R_+ & : \ if \ \ r \geq \rho \\ \{(s,x); x < \frac{a\rho}{\rho - r}\} & : \ if \ \ r < \rho \end{cases}$$

Therefore if $r \geq \rho$ we have $U = D = R \times R_+$, hence τ^* does not exist. If $r > \rho$ then $g^* = \infty$, while if $r = \rho$, then

$$g^*(s,x) = xe^{-\rho s}.$$

For the case $r < \rho$. We can conclude that D is invariant w.r.t.t in the sense that

$$D + (t_0, 0) = D, \forall t_0.$$

And the D has only the connected component which contains U with the from $D(x_0) = \{(t,x); 0 < x < x_0\}$ for some $x \geq \frac{a\rho}{\rho - r}$. Put $\tau(x_0) = \tau_{D(x_0)}$ and let

$$\tilde{g}(s,x) = \tilde{g}_{x_0}(s,x) = E^{(s,x)}[g(Y_{\tau(x_0)})].$$

We know that $f = \tilde{g}$ is the solution of the boundary value problem

$$\begin{aligned} &\frac{\partial f}{\partial s} + rx\frac{\partial f}{\partial x} + \frac{1}{2}\alpha^2 x^2 \frac{\partial^2 f}{\partial x^2} = 0 \quad 0 < x < x_0 \\ &f(s, x_0) = e^{-\rho s}(x_0 - a). \end{aligned} \tag{4.2.69}$$

If we try a solution of (4.2.69) of the form

$$f(s,x) = e^{-\rho s}\phi(x)$$

we have the following 1-dimensional problem

$$\begin{aligned} -\rho\phi + rx\phi'(x) + \frac{1}{2}\alpha^2 x^2 \phi''(x) &= 0 \quad for \quad s < x < x_0 \\ \phi(x_0) &= x_0 - a \end{aligned} \qquad (4.2.70)$$

The general solution ϕ of (4.2.70) is

$$\phi(x) = C_1 x^{\gamma_1} + C_2 x^{\gamma_2},$$

where C_1, C_2 are arbitrary constants and

$$\gamma_i = \alpha^{-2}[\frac{1}{2}\alpha^2 - r \pm \sqrt{(r - \frac{1}{2}\alpha^2)^2 + 2\rho\alpha^2}], (i = 1, 2), \gamma_2 < 0 < \gamma_1.$$

since $\phi(x)$ is bounded as $x \to 0$ we have $C_2 = 0$ and the boundary requirement $\phi(x_0) = x_0 - a$ gives $C_1 = x^{-\gamma_1}(x_0 - a)$. So the bounded solution f of (4.2.69) is

$$\tilde{g}_{x_0}(s,x) = f(s,x) = e^{-\rho s}(x_0 - a)(\frac{x}{x_0})^{\gamma_1}.$$

If we fix (s,x) then the value of x_0 which maximizes $\tilde{g}(s,x)$ is easily seen to be given by

$$x_0 = x_{\max} = \frac{a\gamma_1}{\gamma_1 - 1}$$

Hence we have that

$$\begin{aligned} g^*(s,x) &= \sup_{\tau} E^{(s,x)}[g(\tau, X_\tau)] = \sup_{x_0} E^{(s,x)}[g(\tau(x_0), X_{\tau(x_0)})] \\ &= \sup_{x_0} \tilde{g}_{x_0}(s,x) = \tilde{g}_{x_{max}}(s,x). \end{aligned}$$

The conclusion is therefore that one should sell the assets the first time the price of them reaches the value $x_{\max} = \frac{a\gamma_1}{\gamma_1 - 1}$. The expected discounted profit obtained from this strategy is

$$g^*(s,x) = e^{-\rho s}(\frac{\gamma_1 - 1}{a})^{\gamma_1 - 1}(\frac{x}{\gamma_1})^{\gamma_1}.$$

4.2.4 Stochastic control

We consider stochastic controlled system of the type:

$$dX_t = dX_t^u = b(t, X_t, u_t)dt + \sigma(t, X_t, u_t)dB_t, \qquad (4.2.71)$$

where $X_t \in R^n, b : R \times R^n \times U \to R^n, \sigma : R \times R^n \times U \to R^{n \times m}$ and B_t is m-dimensional Brownian motion. Here $u_t \in U \subset R^k$ is a parameter whose value we can choose in the given Borel set U at any instant t in order to control the process X_t. Thus $u_t = u(t,\omega)$ is a stochastic process. Since our decision at time t must be based upon what has happened up to time t, the function $\omega \to u(t,\omega)$ must be measurable w.r.t. $\mathcal{F}$, i.e. the process u_t must be $\mathcal{F}$-adapted. Thus the right hand side of (4.2.71) is well defined as a stochastic integral, under suitable assumptions on the functions b and σ.

Let $\{X_h^{s,x}\}_{h\geq s}$ be the solution of (4.2.71) such that $X_s^{s,x} = x$, i.e.

$$X_h^{s,x} = x + \int_s^h b(r, X_r^{s,x}, u_r)dr + \int_s^h \sigma(r, X_r^{s,x}, u_r)dB_r; h \geq s$$

and let the probability law of X_t starting at x for $t = s$ be denoted by $Q^{s,x}$, so that

$$Q^{s,x}[X_{t_1} \in F_1, \cdots, X_{t_k} \in F_k] = P^0[X_{t_1}^{s,x} \in F_1, \cdots, X_{t_k}^{s,x} \in F_k]$$

for $s \leq t_i, F_i \subset R^n; 1 \leq i \leq k, k = 1, 2, \cdots$.

Let $F : R \times R^n \times U \to R$ (the "utility rate" function) and $K : R \times R^n \to R$ (the "bequest" function) be ginven continuous functions, let G be a given domain in $R \times R^n$ and let $\hat{T}$ be the first exit time after s from G for the process $\{X_r^{s,x}\}_{r\geq s}$, i.e.

$$\hat{T} = \hat{T}^{s,x}(\omega) = \inf\{r > s; (r, X_r^{s,x}(\omega)) \notin G\}.$$

Suppose

$$E^{s,x}[\int_s^{\hat{T}} |F^{u_r}(r, X_r)|dr + |K(\hat{T}, X_{\hat{T}})|\chi_{\{\hat{T}<\infty\}}] < \infty, \forall s, x, u$$

where $F^u(r, z) = F(r, z, u)$. Then we define the perfomance function $J^u(s, x)$ by

$$J^u(s, x) = E^{s,x}[\int_s^{\hat{T}} F^{u_r}(r, X_r)dr + K(\hat{T}, X_{\hat{T}})\chi_{\{\hat{T}<\infty\}}].$$

Let

$$Y_t = (s + t, X_{s+t}^{s,x}), \quad for \quad t \geq 0, Y_0 = (s, x),$$

and substitute this in (4.2.71), we have

$$dY_t = dY_t^u = b(Y_t, u_t)dt + \sigma(Y_t, u_t)dB_t. \tag{4.2.72}$$

The probability law of Y_t starting at $y = (s, x)$ for $t = 0$ is also denoted by $Q^{s,x} = Q^y$. Let

$$T := \inf\{t > 0; Y_t \notin G\} = \hat{T} - s,$$

and

$$K(\hat{T}, X_{\hat{T}}) = K(Y_{\hat{T}-s} = K(Y_T),$$

then the performance function may be written in terms of Y as follows, wtih $y = (s, x)$,

$$J^u(y) = E^y[\int_0^T F^{u_t}(Y_t)dt + K(Y_T)\chi_{\{T<\infty\}}].$$

So the problem is — for each $y \in G$ — to find the number $\Phi(y)$ and a control $u^* = u^*(t, \omega) = u^*(y, t, \omega)$ such that

$$\Phi(y) := \sup_{u(t,\omega)} J^u(y) = J^{u^*}(y)$$

where the supremum is taken over all $\mathcal{F}_t$-adapted processes $\{u_t\}$ with values in U. Such a control U^* — if it exists — is called an optimal control and Φ is called optimal perfomance or the valued function. For the sake of simplicity, we consider the control functions $u(t,\omega)$ of the form $u(t,\omega) = u_0(t, X_t(\omega))$ for some function $u_0 : R^{n+1} \to U \subset R^k$. In this case we assume that u does not depend on the starting point $y = (s,x)$, i.e., the value we choose at time t only depends on the state of the system at this time. These are called Markov controls, because with such u the corresponding process X_t becomes an Ito diffusion, in particular a Markov process. In the following we will not distinguish between u and u_0. Thus we will identify a function $u : R^{n+1} \to U$ with the Markov contol $u(Y) = u(t, X_t)$ and simply call such functions Markov controls. In such a case the system (4.2.72) becomes

$$dY_t = b(Y_t, u(Y_t))dt + \sigma(Y_t, u(Y_t))dB_t. \tag{4.2.73}$$

For $v \in U$ and $f \in C_0^2(R \times R^n)$ define

$$(L^v f)(y) = \frac{\partial f}{\partial s}(y) + \sum_{i=1}^{n} b_i(y,v)\frac{\partial f}{\partial x_i} + \sum_{i,j=1}^{n} a_{ij}(y,v)\frac{\partial^2 f}{\partial x_i \partial x_j} \tag{4.2.74}$$

where $a_{ij} = \frac{1}{2}(\sigma\sigma^T)_{ij}, y = (s,x)$ and $x = (x_1, \cdots, x_n)$. Then for each choice of the function h the solution $Y_t = Y_t^u$ is an Ito diffusion with generator A given by:

$$(Af)(y) = (L^{u(y)} f)(y)$$

for $f \in C_0^2(R \times R^n)$. For $v \in U$ define

$$F^v(y) = F(y,v).$$

Then we have the following Hamilton-Jacobi-Bellman (HJB) equation.

Theorem 4.2.24 *Define*

$$\Phi(y) = \sup\{J^u(y); u = u(Y) \ \ Markov \ \ control\}.$$

Suppose that $\Phi \in C^2(G) \cap C(\bar{G})$ is bounded, $T < \infty$ a.s. Q^y for all $y \in G$ and that an optimal Markov control u^* exists. Suppose ∂G is regular for $Y_t^{u^*}$. Then

$$\sup_{v \in U}\{F^v(y) + (L^v\Phi)(y)\} = 0, \forall y \in G \tag{4.2.75}$$

and

$$\Phi(y) = K(y), \forall y \in \partial G. \tag{4.2.76}$$

The supremum in (4.2.75) is obtained if $v = U^*(y)$ where u^* is optimal. In other words,

$$F(y, u^*(y)) + (L^{u^*}\Phi)(y) = 0, \forall y \in G. \tag{4.2.77}$$

Remark *This theorem states that if an optimal control u^* exists then we know that its value v at the point y is a point v where the function*

$$v \to F^v(y) + (L^v\Phi)(y); v \in U$$

attains its maximum. This is a necessary condition for the optimal control. The next theorem states that if at each point y we have found $v = u_0(y)$ such that $F^v(y) + (L^v\Phi)(y)$ is maximal, then $u_0(y)$ be an optimal control.

Theorem 4.2.25 *Let ϕ be a function in $C^2(G) \cap C(\bar{G})$ such that, for all $v \in U$*

$$F^v(y) + (L^v\phi)(y) \leq 0; y \in G$$

with boundary values

$$\lim_{t \to T} \phi(Y_t) = K(Y_T)\chi_{\{T<\infty\}}, a.s.Q^y$$

and such that $\{\phi(Y_\tau)\}_{\tau \leq T}$ is uniformly Q^y -integrable for all Markov control u and all $y \in G$. Then $\phi(y) \geq J^u(y)$ for all Markov controls u and all $y \in G$. Moreover, if for each $y \in G$ we have found $u_0(y)$ such that

$$F^{u_0(y)}(y) + (L^{u_0(y)}\phi)(y) = 0$$

then $u_0 = u_0(y)$ is a Markov control such that

$$\phi(y) = J^{u_0(y)}(y)$$

and hence u_0 must be an optimal control and $\phi(y) = \Phi(y)$.

Remark *These two theorems provide a very nice solution to the stochastic control problem in the case where only Markov controls are considered. It seems that considering only Markov controls is too restrictive, but fortunately one can always obtain as good a performance with a Markov control as with an arbitrary $\mathcal{F}_t$-adapted control (under some conditions). We have the following theorem.*

Theorem 4.2.26 *Let*

$$\Phi_M(y) = \sup\{J^u(y); u = u(Y) \text{ Markov control}\}$$

and

$$\Phi_\alpha(y) = \sup\{J^u(y); u = u(t,\omega)\mathcal{F}_t - \text{adapted control}\}$$

Suppose there exists an optimal Markov control $u_0 = u_0(Y)$ for the Markov control problem such that all the boundary points of G are regular w.r.t. $Y_t^{u_0}$ and that Φ_M is a bounded function in $C^2(G) \cap C(\bar{G})$. Then

$$\Phi_M(y) = \Phi_\alpha(y), \forall y \in G.$$

Example 4.2.27 *(An optimal portfolio selection problem)*
Let X_t denote the wealth of a person at time t. Suppose that the person has the choice of two different investments. The price $p_1(t)$ at time t of one of the assets is assumed to satisfy the equation

$$\frac{dp_1}{dt} = p_1(a + \alpha W_t) \tag{4.2.78}$$

where W_t denotes white noise and $a, \alpha > 0$ are constants measuring the average relative rate of change of p and the size of the noise respectively. As we have known we can interpret (4.2.78) as the Ito stochastic differential equation

$$dp_1 = p_1 a dt + p_1 \alpha dB_t. \tag{4.2.79}$$

This investment is called risky, since $\alpha > 0$. We assume that the price p_2 of the other asset satisfies a similar equation, but without noise:

$$dp_2 = p_2 b dt. \tag{4.2.80}$$

This investment is called safe. So it is natural to assume $b < a$. At each instant the person can choose how big of a fraction of u of his wealth he will invest in the risky asset, thereby investing the fraction $1-u$ in the safe one. This gives the following stochastic differential equation for the wealth $X_t = X_t^u$:

$$\begin{aligned} dX_t &= uX_t a dt + uX_t \alpha dB_t + (1-u)X_t b dt \\ &= X_t(au + b(1-u))dt + \alpha u X_t dB_t. \end{aligned} \tag{4.2.81}$$

Suppose that, starting with the wealth $X_t = x > 0$ at time t, the person wants to maximize the expected utility of the wealth at some future time $t_0 > t$. If we allow no borrowing (i.e. require $X \geq 0$) and are given a utility function $N : [0, \infty) \to [0, \infty), N(0) = 0$ (usually assumed to be increasing and concave) the problem is to find $\Phi(s, x)$ and a Markov control $u^* = u^*(t, X_t), 0 \leq u^* \leq 1$ such that

$$\Phi(s, x) = \sup\{J^u(s, x); u \text{ Markov control, } 0 \leq u \leq 1\} = J^{u^*}(s, x),$$

where $J^u(s, x) = E^{s,x}[N(X_T^u)]$ and T is the first exit time from the region $G = \{(r, z); r < t_0, z > 0\}$. This is a performance criterion of the form (4.2.72)/(4.2.73) with $F = 0$ and $K = N$. The differential operator L^v has the form

$$(L^v f)(t, x) = \frac{\partial f}{\partial t} + x(av + b(t-v))\frac{\partial f}{\partial x} + \frac{1}{2}\alpha^2 v^2 x^2 \frac{\partial^2 f}{\partial x^2}$$

The HJB equation becomes

$$\sup_v\{(L^v\Phi)(t, x)\} = 0, \quad for \quad (t, x) \in G;$$

and

$$\Phi(t, x) = N(x), \ fort = t_0, \Phi(t, 0) = N(0), \ fort < t_0.$$

Therefore, for each (t, x) we try to find the value $v = u(t, x)$ which maximizes the function

$$\eta(v) = L^v\Phi = \frac{\partial \Phi}{\partial t} + x(b + (a-b)v) \tag{4.2.82}$$

$$\frac{\partial \Phi}{\partial x} + \frac{1}{2}\alpha^2 v^2 x^2 \frac{\partial^2 \Phi}{\partial x^2}. \tag{4.2.83}$$

If $\Phi_x := \frac{\partial \Phi}{\partial x} > 0$ and $\Phi_{xx} := \frac{\partial^2 \Phi}{\partial x^2} < 0$, the solution is

$$v = u(t, x) = -\frac{(a-b)\Phi_x}{x\alpha^2\Phi_{xx}}. \tag{4.2.84}$$

If we substitute this into the HJB equation (4.2.82) we obtain the following nonlinear boundary value problem for Φ:

$$\begin{aligned} \Phi_t + bx\Phi_x - \frac{(a-b)^2\Phi_x^2}{x\alpha^2\Phi_{xx}} &= 0, \ fort < t_0, x > 0 \\ \Phi(t, x) &= N(x), \ for\ t = t_0, \ orx = 0. \end{aligned} \tag{4.2.85}$$

The problem (4.2.85) is hard to solve for general N. Important examples of increasing and concave functions are the power functions

$$N(x) = x^r \text{ where } 0 < r < 1.$$

If we choose such a utility function N, we try to find a solution of (4.2.84) of the form

$$\phi(t, x) = f(t)x^r.$$

Substituting we obtain

$$\phi(t, x) = e^{\lambda(t_0 - t)} x^r,$$

where $\lambda = br + \frac{(a-b)^2 r}{2\alpha^2(1-r)}$. Using (4.2.84) we obtain the optimal control

$$u^*(t, x) = \frac{a - b}{\alpha^2(1 - r)}.$$

If $\frac{a-b}{\alpha^2(1-r)} \in (0, 1)$ this is the solution to the problem.

4.2.5 Backward SDE and applications

The adapted solution for a linear backward stochastic differential equations was first investigated by Bismut(1973) and in 1978 [5], then by Bensoussan (1982), and others. The first result for the existence of an adapted solution to a continuous nonlinear BSDE with Lipschitzian coefficient was obtained by Pardoux and Peng(1990). Later Peng and Pardoux developed the theory and applications of such BSDEs in a series of papers (1991, 1992, 1993, 1994). We would like to introduce some basic resulst on BSDEs in Peng's survey papers.

Backwards Stochastic Differential Equations

First, let us recall backward intergrals. Let $[0, T]$ be a fixed time intval, $B_t, t \in [0, T]$ be a standard Brownian motion defined on a complete probability space $(\Omega, \mathcal{F}, P)$. Let $0 \le s < t \le T$. We denote by $\mathcal{F}_s^t$ the least complete σ-field for which all random variables $B_u - B_v : s \le v \le u \le t$ are measurable. Denote $\mathcal{F}_0^t = \mathcal{F}_t$. Let t be a fixed time in $[0, T]$, let $f(r), r \in [0, t]$ be a continuous stochastic process which is $\mathcal{F}_r^t$ -measurable for each r. The stochastic integrals can be defined for the backward direction. The Itô backward integral is defined as

$$\int_s^t f(r) dB_r \equiv \lim_{|\Delta| \to 0} \sum_{k=0}^{n-1} f(t_{k+1})(B_{t_{k+1}} - B_{t_k}). \tag{4.2.86}$$

where Δ denotes the partition $\{s = t_0 < t_1 < \cdots < t + n = t\}$ and $|\Delta| = \max_k |t_{k+1} - t_k|$. Precisely speaking, the limit of the right hand side exists in probability and it does not depend on the choice of a sequence of partitions. It has these properties:

$$E[\int_s^t f(r) dB_r] = 0$$

$$E[|\int_s^t f(r) dB_r|^2] = E[\int_s^t f(r)^2 dr].$$

Consider backward stochastic differential equation

$$Y_t = \xi + \int_t^T g(s, Y_s, Z_s)ds - \int_t^T Z_s dB_s, \tag{4.2.87}$$

where

$$g(\omega, t, y, z) : \Omega \times [0, T] \times R^m \times R^{m\times d} \to R^m$$

is such that $g(\cdot, y, z)$ is a R^m-valued $\mathcal{F}_t$-adapted process for each $(y, z) \in R^m \times R^{m\times d}$, satisfying

$$\int_0^T |g(\cdot, 0, 0)ds \in L^2(\Omega, \mathcal{F}_T, P; R). \tag{4.2.88}$$

and

$$|g(t, y, z) - g(t, y', z')| \le C(|y - y'| + |z - z'|). \tag{4.2.89}$$

The problem is to find out a pair of processes $(Y_t, Z_t) \in \mathcal{M}(0, T; R^m \times R^{m\times d})$ which are $\mathcal{F}_t$-adapted satisfying equation (4.2.87), where $\mathcal{M}(0, \tau; R^m)$ is the space of a all $(\mathcal{F}_t)$-adapted R^m-valued processes that satisfies

$$E\int_0^T |v_t|^2 dt < \infty.$$

Remark *Here the uniqueness of a pair of processes means uniqueness in the space* $\mathcal{M}(0, T; R^m \times R^{m\times d})$. *That means if there are two processes* (Y^1, Z^1) *and* (Y^2, Z^2) *satisfying (4.2.87), then we have*

$$E[\int_0^T |Y_t^1 - Y_t^2|^2] = 0, \quad and \quad E[\int_0^T |Z_t^1 - Z_t^2|^2] = 0.$$

We have the following existence and uniqueness theorem.

Theorem 4.2.28 *Let* $g(\omega, t, y, z) : \Omega \times [0, T] \times R^m \times R^{m\times d} \to R^m$ *be given as above and satisfying conditions (4.2.88) and (4.2.89). Then for any given* $\xi \in L^2(\Omega, \mathcal{F}_T, P; R^m)$, *the stochastic differential equation (4.2.87) has a unique solution, i.e. there exists a unique* $\mathcal{F}_t$ *-adapted process* $(Y, Z) \in \mathcal{M}(0, T; R^m \times R^{m\times m})$ *satisfying (4.2.87).*

Also we have the continuous dependence of initial parameters. In fact we have the following theorem.

Theorem 4.2.29 *Let conditions in Theorem 4.2.28 hold and assume that* $\xi^1, \xi^2 \in L^2(\Omega, \mathcal{F}_T, P; R^m$ $\phi^1, \phi^2 \in \mathcal{M}(0, T; R^m)$. *Then the solutions* (Y^1, Z^1), *and* (Y^2, Z^2) *of*

$$Y_t^1 = \xi^1 + \int_t^T [g(s, Y_s^1, Z_s^1) + \phi_s^1]ds - \int_t^T Z_s^1 dB(s)$$

and

$$Y_t^2 = \xi^2 + \int_t^T [g(s, Y_s^2, Z_s^2) + \phi_s^2]ds - \int_t^T Z_s^2 dB(s)$$

satisfy the following estimation

$$|Y_t^1 - Y_t^2|^2 + \frac{1}{2}E^{\mathcal{F}_t}\int_t^T [|Y_s^1 - Y_s^2|^2 + |Z_s^1 - Z_s^2|^2]e^{\beta(s-t)}ds$$
$$\le \; E^{\mathcal{F}_t}|\xi^1 - \xi^2|^2 e^{\beta(T-t)} + E^{\mathcal{F}_t}\int_t^T |\phi_s^1 - \phi_s^2|e^{\beta(s-t)}ds,$$

where $\beta = 16(1 + C^2)$.

For one dimensional case, i.e., $m = 1$. We have the following compression theorem which will be used in the later discussion.

Theorem 4.2.30 *If* $g(\omega, t, y, z)$ *and a cadlag process* $V_t \in \mathcal{M}(0, T; R)$ *satisfy in addition to (4.2.88) and (4.2.89) the condition:*

$$\sup_{t\le T} E|V|^2 < \infty, \tag{4.2.90}$$

$(\bar{Y}, \bar{Z})$ *be the solution of BSDE:*

$$\bar{Y} = \bar{\xi} + \int_t^T \bar{g}_s + V_T - V_t - \int_t^T \bar{Z}_s dB_s, \tag{4.2.91}$$

where $(\bar{g}_t)(\bar{V}_t) \in L^2_{\mathcal{F}}(0, T; R)$ *and* $\bar{\xi} \in L^2(\Omega, \mathcal{F}, P; R)$ *are given and satisfy*

$$\xi \ge \bar{\xi}, g(t, \bar{Y}_t, \bar{Z}_t) \ge \bar{g}_t, a.s., a.e.,$$

and such that $\bar{V} - V$ *is an increasing process. Then*

$$Y_t \ge \bar{Y}_t, a.e., a.s.$$

Hence we now have

$$Y_0 = \bar{Y}_0 \iff \xi = \bar{\xi}, g(s, \bar{Y}_s, \bar{Z}_s) \equiv \bar{g}_s, V_s \equiv \bar{V}_s.$$

Example 4.2.31 *We often meet the case of comparing BSDEs*

$$Y_t^1 = \xi^1 + \int_t^T [g(s, Y_s^1, Z_s^1) + c_s^1]ds - \int_t^T Z_s^1 dB_s, \tag{4.2.92}$$

and

$$Y_t^2 = \xi^2 + \int_t^T [g(s, Y_s^2, Z_s^2) + c_s^2]ds - \int_t^T Z_s^2 dB_s, \tag{4.2.93}$$

where $c^1(\cdot), c^2(\cdot) \in \mathcal{M}(0, T, R)$*. If we assume that* $C_s^1 \ge c_s^2, a.e., a.s.$*, and* $\xi \ge \bar{\xi} a.s.$*, then by Theorem 2.4.2, we have* $Y_t \ge \bar{Y}_t, a.s., a.e.$*. In financial markets (say in the Merton model),* $c(\cdot)$ *denotes the rate of consumption of an investor,* $Y(t)$ *denotes one's wealth at time* t*, while* $Z(t)$ *denote one's potfolio selection strategy. In this case, we can explain the compression theorem as follows: if an investor wants to get a higher financial return at a time in the future, then either one put more money in the financial market or reduce one's consumption before time* T*.*

Example 4.2.32 *We consider a special case of (4.2.92) when $g(s,0,0) \equiv 0$. It is easy to see that if $c_s^2 \equiv 0$ and $\xi^2 = 0$, then (4.2.93) has a unique solution $(Y_s^2, Z_s^2) \equiv 0$. However if both ξ and $c^1(\cdot)$ are nonnegtive, then the solution of (4.2.92) Y^1 is also nonnegative. Moreover we have:*

$$Y_0^1 = 0 \iff c_x^1 \equiv 0, \xi^1 = 0.$$

We can explain this result in finance. Such a financial market is non arbitrage: If an investor wants to get an oportunity riskless in the future time T (i.e. $\xi^1 \geq 0$ and $E\xi^1 > 0$), then one's investment $y_0^1 > 0$ at the moment $t = 0$.

We have seen that the solution of BSDE is always discribed by a pair of processes (Y, Z), however, the main part is the first term Y. We will see that the process Y satisfies a backwards semigroup property. Now given $t_1 \leq T$ and $\eta \in L^2(\Omega, \mathcal{F}_{t_1}, P; R)$, consider the following BSDE on $[0, t_1]$:

$$y_r = \eta + \int_r^{t_1} g(s, y_s, z_s)ds - \int_r^{t_1} z_s dB_s, r \in [0, t_1]. \tag{4.2.94}$$

Define

$$G_{r,t_1}[\eta] := y_r : L^2(\Omega, \mathcal{F}_{t_1}, P; R) \to L^2(\Omega, \mathcal{F}_r, P; R). \tag{4.2.95}$$

From the uniqueness of solution of BSDE, we know that for $t < r \leq t_1$,

$$G_{t,t_1}[\eta] = G_{t,r}[y_r] = G_{t,r}[G_{r,t_1}[\eta]].$$

Furthermore, we have the following properties.
(i) $G_{t_1,t}[\eta] = G_{t_1,t_2}[G_{t_2,t}[\eta]], \forall 0 \leq t_1 \leq t_2 \leq t$;
(ii) $\lim_{r \uparrow T} G_{r,t}[\eta] = \eta, \forall \eta \in L^2(\Omega, \mathcal{F}, P; R)$;
(iii) $\lim_{i \to \infty} E|G_{r,t}[\eta_i] - G_{r,t}[\eta]|^2$, if $E|\eta_i - \eta|^2 \to 0$;
(iv) $\eta_1 \geq \eta_2$, a.s.$\Rightarrow G_{r,t}[\eta_1] \geq G_{r,t}[\eta_2]$,a.s.

A generalized dynamic programming principle

In this subsection, we formulate a stochastic optimal contol problem where the cost function is determined by a backward stochastic differential equation of the form (4.2.87). We get the dynamic programming principle, known as Bellman's principle, in this situation.

Suppose that given $\mathcal{M}(0, T; R^k)$ and a Borel set U in R^k, we denote by $\mathcal{U}$ the class of admissible controls, i.e. all processes for which are valued in U. For simplicity, we assume that U is compact set.

For a given admissible control $\alpha(t), t \geq 0$ valued in U , and a given initial data $x \in R^n$, consider the following stochastic control problem

$$\begin{aligned} dy(s) &= b(y(s), \alpha(s))ds + \sigma(y(s), \alpha(s))dB(s), s \in [t, T], \\ y(0) &= x. \end{aligned} \tag{4.2.96}$$

where $b(x, \alpha), \sigma(x, \alpha)$ are R^n-valued and $\mathcal{L}(R^d, R^n)$ valued functions defined on $R^n \times R^k$. Further we asumme that

Assumption 4.2.97 *b and σ are continuous in (x, α), and continuously differentiable in x, their derivatives b_x, σ_x are bounded.*

Obviously, the corresponding solution $y(\cdot) = y^{x,\alpha}(\cdot)$ *is well defined and*

$$E|y^{x,\alpha}(s)|^2 \leq C|x|^2,$$

where C *is independent of* x, α, *and* s.

We now introduce the following backward stochastic differential equation: let $f(p, q, x\alpha)$ *be a real function defined on* $R \times R^d \times R^n \times R^d$. *For any given continuous function* $g(x) : R^n \to R$ *satisfying*

$$|g(x)| \leq C(1 + |x|),$$

$$p(s) = g(y(t)) + \int_s^t f(p(r), q(r), y(r), \alpha(r))dr - \int_s^t q(r)dB(r), s \in [0, t]. \quad (4.2.98)$$

We assume that

Assumption 4.2.99 f *is continous in* (p, q, x, α) *and continuously differentiable in* (p, q, x), *the derivatives* f_p, f_q, f_x *are bounded.*

It is easy to see that $p(s)$ *is* $\mathcal{F}_s$*-adapted and* $p(0) = Ep(0)$. *We can introduce the following generalized cost function*

$$J(x, t; g(\cdot), \alpha(\cdot)) = p(0)(= Ep(0)), \forall t \in (0, T).$$

Since for given $g(\cdot), J(x, t; g(\cdot), \alpha(\cdot))$ *is uniformly bounded in* $\mathcal{A}$. *Thus we can define the value function*

$$V((x, t; g(\cdot)) = \inf_{\alpha(\cdot) \in \mathcal{A}} J(x, t; g(\cdot), \alpha(\cdot)).$$

If we assume that

Assumption 4.2.100 $g(x)$ *is a uniform Lipschitz function.*

Then we have the following generalized dynamic programming principle.

Theorem 4.2.33 *Let Assumptions 4.2.97, 4.2.99 and 4.2.100 hold. Then we have*

$$V(x, t + h; g(\cdot)) = V(x, t; V(\cdot, h, g)), \forall x, \forall t + h \leq T. \quad (4.2.101)$$

Example 4.2.34 *A trivial situation of the above optimal control problem is when* f *depends only on* (p, q, α) *and* $g = 0$:

$$\inf_{\alpha(\cdot) \in \mathcal{A}} p(0) = \inf_{\alpha(\cdot) \in \mathcal{A}} \{\int_0^t f(p(r), q(r), \alpha(r))dr - \int_0^t q(r)dB(r)\},$$

where, for given $\alpha(\cdot) \in \mathcal{A}, (p(\cdot), q(\cdot))$ *solves*

$$p(s) = \int_s^t f(f(p(r), q(r), \alpha(r))dr - \int_s^t q(r)dB(r), 0 \leq s \leq t.$$

In this case it is easily seen that

$$\begin{aligned} \inf_{\alpha(\cdot) \in \mathcal{A}} p(0) &= \inf_{\alpha(\cdot) \in \mathcal{A}} E \int_0^t f(p(r), q(r), \alpha(r))dr \\ &= \inf_{\alpha(\cdot) \in \mathcal{A}} E \int_0^t f(p_1(r), 0, \alpha(r))dr, \end{aligned}$$

where $\mathcal{A}_0 = \{\alpha(\cdot) \in L^2(0,T); \alpha(s) \in U, a.e.\}$ *and* $p_1(s), 0 \le s \le t$, *solves*

$$p_1(s) = \int_s^t f(f(p(r), 0, \alpha(r))dr, \alpha(\cdot) \in \mathcal{A}_0.$$

From the dynamic programming principle, we can derive Hamilton-Jacobi-Bellman equation.

For a fixed $g(x)$, define value function:

$$u(x,t) = V(x, T-t, g(\cdot)), (x,t) \in R^n \times [0,T].$$

The dynamic programming principle (4.2.101) now can be written in the form

$$u(x,t) = \inf_{\alpha(\cdot)\in\mathcal{A}} \{\int_0^h f(p(r), q(r), y(r), \alpha(r))dr - \int_0^h q(r)dB(r) + u(y(h), t+h)\}, \tag{4.2.102}$$

where $y(\cdot)$ is the trajectory corresponding to $\alpha(\cdot)$ with initial data $y(0) = x$, $(p(\cdot), q(\cdot))$ solves the following backward equation

$$p(s) = u(y(h), t+h) + \int_s^h f(p(r), q(r), y(r), \alpha(r))dr - \int_s^h q(r)dB(r), 0 \le s \le h.$$

Similar to the classical optimal control, function u can be solved by a nonlinear partial differential equation: this is the following generalized Hamilton-Jacobi-Bellman equation.

$$\begin{aligned} \partial_t u + H(D^2u, Du, u, x, t) &= 0, \\ u(x,T) &= g(x), \end{aligned} \tag{4.2.103}$$

where Du and D^2u denote respectively the gradient and the Hessian of u and

$$\begin{aligned} H(D^2u, Du, ux) &= \inf_{\alpha\in\mathcal{A}}\{\mathcal{L}(x,\alpha)u + f(u, \sigma^T(x,\alpha)Du, x, \alpha)\}, \\ \mathcal{L}(x,\alpha)u &= \frac{1}{2}trace(\sigma(x,\alpha)\sigma^T(x,\alpha)D^2u) + (Du, b(x,\alpha)). \end{aligned}$$

Definition 4.2.35 *Let u be a continuous function on $R \times (0,T)$; u is said to be a viscosity subsolution (resp. supersolution) of (4.2.103), if for all $\phi \in C^{2,1}(R^n \times [0,T])$ the following inequality holds, at each minimum (resp. maximum) point (x,t) of $\phi - u$*

$$\partial_t\phi(x,t) + H(D^2\phi(x,t), D\phi(x,t), \phi(x,t), x) \ge 0,$$

$$(resp.\ \ \partial_t\phi(x,t) + H(D^2\phi(x,t), D\phi(x,t), \phi(x,t), x) \le 0);$$

u is said to be a viscosity solution of (4.2.103) if u is both a viscosity subsolution and a viscosity supersolution of (4.2.103).

We now end this section with the folowing theorem.

Theorem 4.2.36 *If assumptions 2.5.1, 2.5.2, and 2.5.3 hold, then the value function $u(x,t)$ is the viscosity solution of (4.2.103).*

4.3 Some generalizations of SDEs

So far we have only considered stochastic differential equations with respect to Brownian motion. For such equations, the solutions are always continuous processes. Now in this section, we discuss more general stochastic differential equations which include Poisson point processes as well as Brwonian motions; stochastic differential equation with respect to semimartaingale; stochastic differential equation with respect to nonlinear integrators.

4.3.1 SDEs of the jump type

Forward SDE with jumps

Let $\{U, \mathcal{B}_U\}$ be a measurable space and $n(du)$ be a σ-finite measure on it. Let U_0 be a set in $\mathcal{B}_U$ such that $n(U\backslash U_0) < \infty$. Let $b(s,x)$ be a Borel measurable function $[0,\infty) \times R^d \to R^d$, $a(s,x)$ be a Borel measurable function $[0,\infty) \times R^d \to R^d \otimes R^r$, and $f(t,x,u)$ be a $\mathcal{B}(R^+) \times \mathcal{B}(R^d) \times \mathcal{B}_U$ measurable function $[0,\infty) \times R^d \times U \to R^d$ such that for some positive constant K,

$$||a(t,x)||^2 + |b(t,x)|^2 + \int_{U_0} ||f(t,x,u)||^2 n(du) \leq K(1+|x|^2, x \in R^d, t \leq 0; \qquad (4.3.104)$$

and

$$\begin{aligned} &||a(t,x) - a(t,y)||^2 + |b(t,x) - b(t,y)|^2 \qquad (4.3.105) \\ + &\int_{U_0} ||f(t,x,u) - f(t,y,u)||^2 n(du) \leq K|x-y|^2, \quad t \geq 0, x, y \in R^d. \end{aligned}$$

Consider the SDEs

$$\begin{aligned} X(t) &= X(0) + \int_0^t b(s,X(s))ds + \int_0^t a(s,X(s))dB(s) \\ &+ \int_0^{t+} \int_U f(s-,X(s-),u) 1_{U_0}(u) \tilde{N}_p(dsdu) \qquad (4.3.106) \\ &+ \int_0^{t+} \int_U f(s-,X(s-),u) 1_{U\backslash U_0}(u) N_p(dsdu) \end{aligned}$$

where $B(t)$ is an r-dimensional standard Brownian motion process, $p(\cdot)$ is a stationary Poisson point process taking values in a measurable space $(U, \mathcal{B}(U))$, with characteristic measure $n(\cdot)$ and $\overline{N}_k(ds,dz)$ is the Poisson counting measure defined by $p(\cdot)$ with compensator $n(du)ds$, $\tilde{N}_k(ds,dz)$ is the martingale measure such that

$$\tilde{N}_k(ds,du) = N_k(ds,du) - n(du)ds.$$

By a solution of the equation (4.3.106), we mean a right continuous process $X = (X(t))$ with left hand limits on R^d defined on a probability space $(\Omega, \mathcal{F}, P)$ with a reference family $(\mathcal{F}_t)$ such that X is $\mathcal{F}_t$-adapted and there exists an r-dimensional $\mathcal{F}_t$-Borwnian motion $B(t)$ and an $(\mathcal{F}_t)$-stationary Poisson point process p on U with characteristic measure n such that the equation (4.3.106) holds a.s. We have the following existence theorem; for the proof, the reader may refer to [39].

Theorem 4.3.1 *If $b(s,x), a(t,x)$ and $f(t,x,u)$ satisfy (4.3.104) and (4.3.105), then for any given R-dimensionial $(\mathcal{F}_t)$-Brownian motion $B = (B(t))$, any $(\mathcal{F}_t)$-stationary Poisson point process p with characteristic measure n and any R^d-valued $\mathcal{F}_0$ -measurable random variable Ξ defined on a probability space with a reference family $(\mathcal{F}_t)$, there exists a unique d-dimensional $(\mathcal{F}_t)$-adapted right-continuous process $X(t)$ with left-hand limits which satisfies equation (4.3.106) and such that $X(0) = \xi$ a.s.*

Backward SDE with jump

The adapted solution for a backward stochastic differential equation with respect to Brownian motion have been discussed in last section. Tang and Li (1994) applied Peng's idea to get the first result on the existence of an adapted solution to a BSDE with Poisson jumps for a fixed terminal time and with Lipschitzian coefficients. We state here a new result by Situ Rong (1997) on the existence and uniqueness of an adapted solution of BSDE with jumps and with non-Lipschitzian coefficient.

Consider a BSDE in R^d

$$\begin{aligned} X(t) &= X_0 + \int_{t\wedge\tau}^{\tau} b(s, X(s), g(s), h(s), \omega)ds \\ &- \int_{t\wedge\tau}^{\tau} g(s)dB(s) - \int_{t\wedge\tau}^{\tau}\int_U h(s,u)(\tilde{N}_p(ds,du), t \geq 0, \end{aligned} \tag{4.3.107}$$

where $B(t)$ is an r-dimensional standard Brownian motion process, $p(\cdot)$ is a Poisson point process taking values in a measurable space $(U, \mathcal{B}(U)), \overline{N}_p(ds,du)$ is the Poisson counting measure defined by $p(\cdot)$ with compensator $n(dz)ds, \tilde{N}(ds,du)$ is the martingale measure such that

$$\tilde{N}_p(ds,du) = N_p(ds,du) - n(du)ds,$$

$n(\cdot)$ is a σ-finite measure on $\mathcal{B}(U)$, τ is a bounded $\mathcal{T}_t$-stopping time, and X_0 is a $\mathcal{T}_t$-measurable and R^d-valued random variable, where $\mathcal{T}_t$ is the σ-algebra generated (and completed) by all $B(s), s \leq t$, and $N_p((0,s], s \leq t, U \in \mathcal{B}(U)$. We also use the following notation $\mathcal{P}^2_{k,(\mathcal{G}_t)}([0,\tau] : R^d)$ is the set of R^d-valued $\mathcal{T}_t$-predictable processes $f(t,u,\omega)$ such that

$$E\int_0^{\tau}\int_U |f(t,u,\omega)|^2 n(du)dt < \infty;$$

$L^2_{(\mathcal{T}_t)}([0,\tau] : R^d)$ is the set of $f(t,\omega)$, which is $\mathcal{T}_t$-adapted, jointly measurable and R^d-valued such that

$$E\int_0^{\tau}\int_U |f(t,\omega)|^2 dt < \infty;$$

and $L^2_{(\mathcal{T}_t)}([0,\tau] : R^{d\otimes r})$ is defined similarly. Denote by $L^2_{n(\cdot)}(R^d)$ the set of R^d-valued functions $f(z), u \in U$, which is $\mathcal{B}(U)$ measurable such that $|||f||| = (\int_U |f(u)|^2 n(ds))^{1/2} < \infty$. Denote by $< a, b >= a \cdot b$ the inner product of $a, b \in R^d$; $||g||$ the norm of the matrix $g \in R^{d\otimes r}$.

Definition 4.3.2 *$(X(t), g(t), h(t))$ is said to be a solution of (4.3.107), iff it satisfies (4.3.107) and $(X(t), g(t), h(t)) \in L^2_{(\mathcal{T}_t)}([0,\tau] : R^d) \times L^2_{(\mathcal{T}_t)}([0,\tau] : R^{d\otimes r}) \times \mathcal{P}^2_{k,(\mathcal{T}_t)}([0,\tau] : R^d)$.*

Theorem 4.3.3 *Assume that $\tau \leq T$ and*

(i). $b = b_1 + b_2, b_i = b_i(t,x,g,h,\omega) : [0,T] \times R^d \times R^{d\otimes r} \times L^2_{n(\cdot)}(R^d) \times \Omega \to R^d, (i = 1,2)$ are $\mathcal{T}_t$-adapted and measurable processes such that P-a.s.

$$|b_1(t,x,g,h,\omega)| \leq c(t),$$

$$|b_1(t,x,g,h,\omega)| \leq c(t)(1 + |X| + ||g|| + |||h|||),$$

where $c(t) \geq 0$ is real and nonrandom such that

$$\int_0^T c(t)^2 dt < \infty;$$

(ii). for all $t \in [0,T]; x, x_i \in R^d; g, g_i \in R^{d\times r}; p_i \in L^2_{\Pi(\cdot)}(R^d), i = 1,2,$

$$\begin{aligned}
& (x_1 - x_2, b(t, x_1, g_1, h_1) - b_1(t, x_2, g_2, h_2, \omega) \\
\leq\ & c(t)(\rho(|x_1 - x_2|^2) + |x_1 - x_2|(||g_1 - g_2|| + |||h_1 - h_2|||)), \\
& b_1(t, x, g, h_1, \omega) - b_1(t, x, g, h_2, \omega)| \\
\leq\ & c(t)|||h_1 - h_2|||, \\
& |b_2(t, x_1, g_1, h_1, \omega) - b_2(t, x_2, g_2, h_2, \omega)| \\
\leq\ & c(t)(|x_1 - x_2| + ||g_1 - g_2|| + |||h_1 - h_2|||),
\end{aligned}$$

where $c(t)$ has the same property in (i), and $\rho(u)$ is a real function which is increasing, concave and continous such that $\rho(0) = 0$ and $\rho(u) > 0$ if $u > 0$ and

$$\int_{0+} du/\rho(u) = +\infty,$$

(iii). $b(t, x, g, h, \omega)$ is continuous in $(x, g, h) \in R^d \times R^{d\otimes r} \times L^2_{n(\cdot)}(R^d)$;
(iv). X_0 is $\mathcal{T}_\tau$-measurable, and $E|X_0|^2 < \infty$.

Then (4.3.107) has a unique solution $(X(t), g(t), h(t))$.

Remark *Here the "uniqueness" means that if $(X_i(t), g_i(t), h_i(t)), i = 1,2$ are two solutions of (4.3.107), then $E\int_0^T |X_1(t) - X_2(t)^2 dt = 0$, $E\int_0^T ||g_1(t) - g_2(t)||^2 dt = 0$, and $E\int_0^T |||h_1(t) - h_2(t)|||^2 dt = 0$.*

Here is an example to show that under conditions of Theorem 4.3.3 coefficient b can be non-Lipschitzian in x.

Example 4.3.4 *Let $b = b_1 + b_2$, where b_2 satisfies conditions of Theorem 4.3.13 and we assume that $0 \leq v_i(t,\omega), i = 1,2,3$, are $\mathcal{T}_t$-progressive processes on $[0,T]$ and uniformly bounded. Let*

$$b_1 = \sum_{i=1}^3 b_{1i}$$

where

$$b_{11}(t, x, g, h, \omega) = -|x|^{r_0 - 2} x \cdot v_1(t,\omega) 1_{\{x\neq 0\}},$$

$$b_{12}(t, x, g, h, \omega) = -|x|^{r_1 - 2} x \cdot v_2(t,\omega) 1_{\{x\neq 0\}}(||g|| 1_{\{||g|| \leq k_0\}} + k_0 1_{\{||g|| > k_0\}}),$$

$$b_{13}(t, x, g, h, \omega) = -|x|^{r_2 - 2} x \cdot v_3(t,\omega) 1_{\{x\neq 0\}}(||g|| 1_{\{|||h||| \leq k_1\}} + k_1 1_{\{|||h||| > k_1\}}),$$

and $k_i \geq 0, i = 0,1,; r_j \in (1,2), j = 0,1,2$; which are all constants. Then b_1 satisfies all conditions in (i), (ii) and (iii), but it is not Lipschitzian continuous.

There are still some existence theorems, examples and some convergence theorems and applications in Situ's paper, with various type conditions. For the detail, we refer to his paper [86].

SDEs Governed by C-valued Lévy process

Let $\mathcal{C} = \mathcal{C}(R^d, R^d)$ be the Fréchet space of all continuous maps from R^d into R^d equipped the compact uniform topology determined by the metric

$$\rho(f,g) = \sum_{N=1}^{\infty} \frac{1}{2^N} \frac{\sup_{|x|\leq N} |f(x) - g(x)|}{1 + \sup_{|x|\leq N} |f(x) - g(x)|}.$$

A $\mathcal{C}$-valued stochastic process $Y_t = Y_t(\omega), t \geq 0$, is called a Lévy process if it is continuous in probability, is right continuous with finite limit on the left in ρ-topology, and has independent increments. In particular, if almost all paths of X_t are continuous in t, then X_t is called a Brownian motion. A $\mathcal{C}$-valued Lévy process Y_t is said to be stationary if the law of $Y_t - Y_s$ depends only on $t - s$, and that $Y_0 = 0$.

Given a $\mathcal{C}$-valued stationary Lévy process η_t, we define a point process associated with it. Define

$$\eta_{s-} := \lim_{r \uparrow s} \eta_r, \quad \Delta\eta_s := \eta - \eta_{s-}$$

and

$$D_p := \{s > t_0 : \Delta\eta_s \not= 0\}.$$

Let p_t be a $\mathcal{C}$-valued point process defined by $p_t := \Delta\eta_t$, and let $N_p((0,t], A)$ be the counting measure of p_t, that is

$$N_p((0,t], A) := \#\{s \in D_p \bigcap (0,t] : p_s \in A\},$$

where $A \in \mathcal{B}(\mathcal{C})$, the Borel σ-algebra of $\mathcal{C}$, and $\#\{B\}$ denotes the cardinality of the set B. It is a stationary Poisson random measure. The intensity measure defined by

$$\hat{\nu}((t_0,t] \times A) := E[N_p((t_0,t], A)]$$

is of the form $\nu(A)t$. The measure ν satisfies the following property.

Condition I

$$\nu\{f : f = 0\} = 0.$$

We assume the existence of an open neighborhood U of $0 \in \mathcal{C}$ such that $\nu(U^c) < \infty$ and $\int_U |f(x)|^2 \nu(df) < \infty$ holds for any x.

Let $X_t(x)$ denote the restriction of the C-valued Levy process X_t at the point $x \in R^d$. Then for any $x_1, x_2, \cdots, x_n \in R^d$, the n-point process $(X_t(x_1), \cdots, X_t(x_n))$ is an nd-dimensional Levy process. Hence the characteristic function admits the Levey-Khinchin's formula:

$$\begin{aligned} E[e^{it\sum_{k=1}^n (\alpha_k, X_t(x_k))}] &= \exp t\{i\sum_{k=1}^{n}(\alpha_k, b(x_k)) - \frac{1}{2}\sum_{k,l}\alpha_k a(x_k, x_l)\alpha_l \\ &\quad + \int_U (e^{i\sum_k^n(\alpha_k, f(x_k))} - 1 - i\sum_k(\alpha_k, f(x_k)))\nu(df) \\ &\quad + \int_{U^c} (e^{i\sum_k(\alpha_k, f(x_k))} - 1)\nu(df)\} \end{aligned} \quad (4.3.108)$$

where,

Condition II $b(x)$ is an R^d-valued function,

Condition III $a(x,y)$ is a $d \times d$-matrix valued function such that $a^{k,l}(x,y) = a^{l,k}(y,x)$ for and $k,l = 1,2,\cdots,n$ and $x,y \in R^d$, and $\sum_{i,j} a_i a(x_i,x_j)a_j \geq 0$ for any $x_i, a_i \in R^d, i = 1,2,\cdots,n$.

Note that the law of $X_t(\omega)$ is uniquely determined by the system (a,b,ν,U) which is called the characteristics of X_t. We need the following conditions on it.

(**A.1**) $a(x,y)$ is bi-Lipschitz continuous in the sense that

$$||a(x,x) - 2a(x,y) + a(y,y)|| \leq L|x-y|^2, \quad \forall x,y \in R^d,$$

where $||a|| = \sum_i a_{ii}$; and

(**A.2**) $b(x)$ is a Lipschitz continuous, i.e.,

$$|b(x) - b(y)| \leq L_1|x-y|, \qquad \forall x,y \in R^d.$$

(**A.3**) There is a positive constant L such that

$$\int_U |f(x) - f(y)|^{p'} \nu(df) \leq L_1|x-y|^{p'}, \quad \forall x,y \in R^d,$$

and

$$\int_U |f(x)|^{p'} \nu(df) \leq L_1(1+|x|)^{p'}, \quad \forall x \in R^d,$$

holds for any $p' \in [2,p]$, where $p > d$.

Under these assumptions as above, we have the following theorem on the C-valued Levy process.

Theorem 4.3.5 *Let (a,b,ν,U) be a system satisfying Conditions I,II,III, and (A.1),(A.2) and (A.3) for some $p > d$. Then there is a C-valued Levy process with the characteristics (a,b,ν,U).*

Let $X_t(x), x \in R^d, t \in [0,T]$ be a C-valued Levy process with characteristics (a,b,ν,U) saisfying (A.1),(A.2) and (A.3) for some $p > d$. Let $s < t$ and $\mathcal{F}_{s,t}$ be the least sub σ-field of $\mathcal{F}$ for which $X_u - X_v; s \leq u \leq v \leq t$ are measurable. Then for each s and $x, X_t(x) - X_s(x), t \in [s,T]$ is an $\mathcal{F}_{s,t}$-adapted semimartingale. $X_t(x)$ is decomposed to the sum of the process of bounded varation

$$V_t(x) \equiv b(x)t + \int_{U^c} f(x)N_p((0,t],df)$$

and an L^2-martingale $Y_t(x) \equiv X_t(x) - V_t(x)$. Let $< Y^i(x), Y^i(y) >$ be the continuous process of bounded variation such that

$$(Y_t^i(x) - Y_s^i(x))(Y_t^j(x) - Y_t^j(y)) - \{< Y^i(x), Y^j(y) >_t - < Y^i(x), Y^j(y) >_s\}$$

is an $\mathcal{F}_{s,t}$-martingale. Then it holds $< Y^i(x), Y^j(y) >_t = A^{ij}(x,y)$, where

$$A^{ij}(x,y) = a^{ij}(x,y) + \int_U f^i(x)f^j(y)\nu(df).$$

Let $s > 0$ be a fixed number and let $\phi_t(\omega)$ be an $\mathcal{F}_{s,t}$-adapted R^d-valued process, right continuous with the left limits. The Ito integral of ϕ_t by dY_t is defined by

$$\int_s^t dY_r^i(\phi_{r-}) = \lim_{\delta\to 0}\sum_{k=0}^{n-1}\{Y_{t_{k+1}}^i(\phi_{t_{k+1}\wedge t}) - Y_{t_k\wedge t}^i)\}$$

where δ are partitions $\{s = t_0 < t_1 < \cdots < t_n = T\}$. The limit exists in probability and is a local martingale. Let $\psi_t(\omega)$ be an $\mathcal{F}_{s,t}$adapted process having the same property as ϕ_t. Then it holds

$$\left\langle \int_s^t dY_r^i(\phi_{r-}), \int_s^t dY_r^i(\psi_{r-})\right\rangle = \int_s^t A^{ij}(\phi_{r-},\psi_{r-})dr$$

Now the stochastic integral by C-valued Lévy process X_t is defined by

$$\int_s^t dX_r(\phi_{r-}) + \int_s^t b(\phi_{r-})dr + \int_t^t \int_{U^c} f(\phi_{r-})N_p(dr, df).$$

Now we can consider the following stochastic differential equation defined by the C-valued Lévy process X_t:

$$d\xi_t = dX_t(\xi_{t-}) \tag{4.3.109}$$

Definition 4.3.6 *Given a time s and a state x, an R^d -valued $\mathcal{F}_{s,t}$-adapted process ξ_t right continuous with the left limit is called a solution of the equatioin (4.3.109) if it satisfies*

$$\xi_t = x + \int_s^t dX_r(\xi_{r-}). \tag{4.3.110}$$

Firstly, we have the following existence and uniqueness theorem and continuity theorems.

Theorem 4.3.7 *For each s, x, the equation (4.3.110) has a unique solution.*

Theorem 4.3.8 *Let $X_t(x)$ be a C-valued Lévy process with characteristics (a, b, ν, U) satisfying (A.1),(A.2) and (A.3) for some $p > d$. Then the solution of equation (4.3.110) has a modification $\xi_{s,t}$ with the following properties.*

(i) For each $s, \xi_{s,t}, t \in [s, T]$ is a right continuous C -valued process with the left limits.

(ii) For any $0 \le t_0 < t_1 < \cdots < t_n, \xi_{t_i,t_{i+1}}, i = 0, 1, \cdots n-1$ are mutually independent.

(iii) For each s, it holds that $\xi_{s,u} = \xi_{t,u}\circ\xi_{s,t}$ a.s., for any $s < t < u$.

Secondly, if we make the following assumptions, we can obtain the regularity of the solution with respect to the initial data.

(B.1) $a(x, y) = (a^{ij}(x, y))$ are m-times continuously differentiable in both x and y. Further, $D_x^k D_y^k a(x, y)$ is bi-Lipschitz continuous for any k with $|k| \le m$.

(B.2)$b(x) = (b^i(x))$ is a C^m-function and $D^k b(x)$ is Lipschitz continuous for any k with $|k|leqm$.

(B.3) The measure ν is supported by C^m. There is a positive constant L such that

$$\int_U |D^k f(x) - D^k f(y)|^{p'}\nu(df) \le L|x-y|^{p'}, \forall x, y \in R^d$$

and

$$\int_U |D^k f(x)|^{p'}\nu(df) \le L, \forall x \in R^d$$

hold for any k with $1 \leq |k| \leq m$ and

$$\int_U |f(x)|^{p'} \nu(df) \leq L_1(1+|x|)^{p'}, \quad \forall x \in R^d,$$

for $p' \in [2, p]$.

Let us define the product of two elements f, g of $C(R^d; R^d)$ by the composition $f \circ g$ of the maps. Then $C(R^d, R^d)$ becomes a topological semigroup by the topology ρ. We denote the semigroup by G_+. From Theorem 3.1.5 we know the solution $\xi_{s,t}(x)$ of equation (4.3.110) defines a Levy process in the semigroup G_+. The associated C-valued Levy process X_t is called the infinitesimal generator of $\xi_{s,t}$ and $\xi_{s,t}$ is said to be generated by the C-valued Lévy process X_t. Denote by G_+^m the sub-semigroup of G_+ consisting of C^m-maps. It is a topological semigroup by the metric

$$\rho_m(f, g) = \sum_{|k| \leq m} \rho(D^k f, D^k g).$$

The Lévy process with values in G_+^m is defined similarly as that with values in G_+.

Theorem 4.3.9 *Suppose the characteristics of a C-valued Levy process satisfy (B.1),(B.2) and (B.3) for some $p > (m+1)^2 d$. Then the solution $\xi_{s,t}(x)$ of (4.3.110) has a modification such that it is a C_+^m-valued Levy process. Futhermore, in case $U = C$, there is a constant M such that*

$$E[\sup_{s \leq r \leq t} |D^k \xi_{s,r}(x) - D^k \xi_{s,r}(y)|^{p'}] \leq M(t-s)|x-y|^{p'}, \forall x, y \in R^d,$$

and

$$E[\sup_{s \leq r \leq t} |D^k (\xi_{s,r}(x) - x)|^{p'}] \leq M(t-s), \forall x \in R^d$$

hold for any k with $1 \leq |k| \leq m$ and $p' \in [2, p/(m+1)^2]$.

Finally, we discuss the homeomorphic property of the solution. Denote by G the totality of homeoporphisms of R^d. It is a subgroup of G_+, and is a topological group by the metric

$$d(f, g) = \rho(f, g) + \rho(f^{-1}) + \rho(g^{-1}).$$

However, we use the metic ρ instead of this d. The definition of the G-valued Levy process is similar to that of G_+-valued Levy process. For the case that the intesity measure ν of the Poisson point process is finite measure, we have

Theorem 4.3.10 *Let $X_t(x)$ be a C-valued Levy process satisfying (A.1),(A.2) and (A.3) for some $p > d$. Suppose the following*

*(**A.4**) The intensity measure ν is finite and is supported by f such that $\phi_f \equiv f + id \in G$. Then the solution of equation (4.3.110) defines a G-valued Levy process.*

For the case that the intensity measure ν is σ-finite, we need the following assumptions.

(**A.5**) $\phi_f \equiv f + id$ are homeomorphisms a.s. ν. ν satisfies

$$\int_C \frac{||f||^2}{1 + ||f||^2} \nu(df) < \infty,$$

where

$$||f|| = \sup_x \frac{|f(x)|}{1 + |x|} + \sup_{x \neq y} \frac{|f(x) - f(y)|}{|x - y|}.$$

Theorem 4.3.11 *Assume (A.1),(A.2) and (A.5) hold. The solution of equation (4.3.110) defines a G-valued Levy process.*

4.3.2 SDE with respect to semimartingale

Ito's stochastic differential equation

Let us begin with introducing some notation. Let D be a domain in R^d and R^e be another Euclidean space. Let m be a nonnegative integer. Denote by $C^m(D, R^e)$ or C^m the set of all maps $f : D \to R^e$ which are m-times continuously differentiable. In case $m = 0$, it is often denoted by $C(D, R^e)$. For multi-index of nonnegative integers $\alpha = (\alpha_1, \cdots, \alpha_d)$, we define the differential operator

$$D_x^\alpha = \frac{\partial^{|\alpha|}}{(\partial x^1)^{\alpha_1} \cdots (\partial x^d)^{\alpha_d}},$$

where $|\alpha| = \sum \alpha_i$. Let K be a subset of D. We set

$$||F||_{m:K} = \sup_{x \in K} \frac{|f(x)|}{(1+|x|)} + \sum_{1 \le |\alpha| \le m} \sup_{x \in K} |D^\alpha f(x)|.$$

Then $C(D, R^e)$ is a Frechet space under the family of seminorms $\{|| \ ||_{m:K} : K$ are compacts in $D\}$. When $K = D$, we write $|| \ ||_{m:K}$ as $|| \ ||_m$. Denote by $C_b^m(D, R^e)$ or C_b^m the set $\{f \in C^m : ||f||_m < \infty\}$. Then it is a Banach space with the norm $|| \ ||_m$. Now let δ be a positive number less than or equal to 1. Denote by $C^{m,\delta}(D, R^e)$ or simply by $C^{m,\delta}$ the set of all f of C^m such that $D^\alpha f, |\alpha| = m$ are δ-Holder continuous. By the seminorms

$$||f||_{m+\delta:K} = ||f||_{m:K} + \sum_{|\alpha|=m} \sup_{x,y \in K \& x \neq y} \frac{|D^\alpha f(x) - D^\alpha f(y)|}{|x-y|^\delta},$$

it is a Frechet space. When $K = D$ we write $|| \ ||_{m+\delta:m}$ as $|| \ ||_{m+\delta}$. Denote by $C_b^{m,\delta}(D, R^e)$ or $C_b^{m,\delta}$ the set of $\{f \in C^{m,\delta} : ||f||_{m+\delta} < \infty\}$.

A continuous function $f(x,t), x \in D, t \in [0,T]$ is said to belong to the class $C^{m,\delta}$ if for every $t, f(t) \equiv f(\cdot, t)$ belong to $C^{m,\delta}$ and $||f(t)||_{m+\delta:K}$ is integrable on $[0,T]$ with respect to t for any compact subset K. If the set K is replaced by D, f is said to belong to the class $C_b^{m,\delta}$. We define the set $\tilde{C}_m$ of all R^e-valued functions $g(x,y), x, y \in D$ which are m-times continuously differentiable with respect to each variable x and y . For $g \in \tilde{C}^m$, define

$$||g||_{m:K} = sup_{x,y \in K} \frac{|g(x,y)|}{(1+|x|)(1+|y|)} + \sum_{1 \le |\alpha| \le m} \sup_{x,y \in K} |D_x^\alpha D_y^\alpha g(x,y)|,$$

and for $0 < \delta \le 1$,

$$||g||_{m+\delta:K} = ||g||_{m:K} + \sum_{|\alpha|=m} ||D_x^\alpha D_y^\alpha g||_{\delta:K},$$

where

$$||g||_{\delta:K} = \sup_{x \neq x', y \neq y' \in K} \frac{|g(x,y) - g(x',y) - g(x,y') + g(x',y')|}{|x-x'|^\delta |y-y'|^\delta}.$$

The function g is said to belong to the space $\tilde{C}^{m,\delta}$ if $||g||_{m+\delta:K} < \infty$ for any compact set K in D. We denote $||\ ||_{m:D}$ and $||\ ||_{m+\delta:D}$ by $||\ ||_m$ and $||\ ||_{m+\delta}$, respectively. We set $\tilde{C}_b^m = \{g : ||g||_m < \infty\}$ and $\tilde{C}_b^{m,\delta} = \{g : ||g||_{m+\delta} < \infty\}$.

A continuous function $g(x, y, t), x, y \in D, t \in [0, T]$ is said to belong to the class $\tilde{C}^{m,\delta}$ if for every $t, g(t) \equiv g(\cdot, \cdot, t)$ belongs to the space $\tilde{C}^{m,\delta}$ and $||g(t)||_{\delta:K}$ is integrable on $[0, T]$ with respect to t for any compact subset K. The classes $\tilde{C}_b^{m,\delta}$ is defined similarly.

Let $F(x, t) = (F^1(x, t), \cdots, F^d(x, t)), x \in R^d$ be a continuous semimartingale with values in $C = C(R^d, R^d)$. We will discuss the following stochastic differential equation

$$d\phi_t = F(\phi_t, dt). \tag{4.3.111}$$

We shall first introduce assumptions for a continuous semimartingale F so that equation (4.3.111) is well defined. Let $F^i(x, t) = M^i(x, t) + B^i(x, t)$ be the decomposition such that $M^i(x, t)$ is a contimuous local martingale and $B^i(x, t)$is a continuous process of bounded varation. Set $A^{ij}(x, y, t) =< M^i(x, t), M^j(y, t) >$. Let $(a(x, y, t), b(x, t), A_t)$ be the local characteristic of the semimartingale F, i.e. A_t is a continuous strictly increasing process such that both $A^{ij}(x, y, t)$ and $B^i(x, t)$ are absolutely continuous with respect to A_t a.s. for any $x, y \in R^d$. Hence there exist predictable processes $a^{ij}(x, y, t)$ and $b^i(x, t)$ with parameters x, y such that

$$A^{ij}(x, y, t) = \int_0^t a^{ij}(x, y, s)dA_s, \quad B^i(x, t) = \int_0^t b^i(x, s)dA_s.$$

Let $b(x, t) = (b^1(x, t), \cdots, b^d(x, t))$ and $a(x, y, t) = (a^{ij}(x, y, t)), i, j = 1, \cdots, d$. Then $a(x, y, t)$ is a $d \times d$-matrix valued function with the following properties.

(a) **symetric:** $a^{ij}(x, y, t) = a^{ji}(y, x, t)$ holds a.e. μ for and x, y and i, j.

(b) **nonnegative definite:** $\sum_{i,j,p,q} a^{ij}(x_p, x_q, t)\xi_p^i\xi_q^j \geq 0$ holds a.e.μ for any x_p, $(\xi_p^1, \cdots, \xi_p^d)$, $p = 1, \cdots, n$.

Now let us classify the family of semimartingales $F(x, t), x \in D$ according to the regularity of its local characteristic. The local characteristic (a, A) is said to belong to the class $B^{m,\delta}$ if $a(x, y, t)$ is a predictable process with values in $\tilde{C}^{m,\delta}(D, R)$ and for any compact subset K of D, $||a(t)||_{m+\delta:K} \in L^1(A)$. In particular if $||a(t)||_{m+\delta} \subset L^1(A)$ holds, (a, A_t) is said to belong to the class $B_b^{m,\delta}$. The local characteristic (b, A_t) is said to belong to the class $B^{m,\delta}$ if $b(x, t)$ is a predictable process with values in $C^{m,\delta}(D, R)$ and for any compact subset K of D $||b(t)||_{m+\delta:K} \in L^1(A)$. In particular if $||b(t)||_{m+\delta} \in L^1(A)$ holds, (b, A_t) is said to belong to the class $B^{m,\delta}$. The triple (a, b, A_t) is then said to be the class $(B^{m,\delta}, B^{m',\delta'})$ if (a, A_t) belong to the class $B^{m,\delta}$ and (b, A_t) belong to the class $B^{m',\delta'}$. When $m = m'$ and $\delta = \delta'$, the triple is simply said to belong to the class $B^{m,\delta}$.

Now if F is a contniuous C-semimartingale with the local characteristic belonging to the class $B^{0,\delta}$; that is for every i, the local characteristic of $F^i = M^i(x, t) + B^i(x, t)$ belonging to the class $B^{0,\delta}$, then Itô's stochastic integral $\int_0^t F(\phi_s, ds)$ is well defined for any continuous R^d-valued predictable process ϕ_t . We will give the definition of the solution of the stochastic differential equation.

Definition 4.3.12 *Let $t_0 \in [0, T]$ and $x_0 \in R^d$. A continuous R^d-valued process $\phi_t, t_0 \leq t \leq T$ adapted to $\mathcal{F}_t$ is called a solution of Ito's stochastic differential equation based on $F(x, t)$ starting at x_0 at time t_0 if it satisfies*

$$\phi_t = x_0 + \int_{t_0}^t F(\phi_s, ds). \tag{4.3.112}$$

Also ϕ_t is said to be governed by Ito's stochastic differetial equation based on $F(x,t)$.

We have the existence and uniqueness theorem of the solution of the above equation.

Theorem 4.3.13 *Let $F(x,t)$ be a continuous semimartingale with values in $C(R^d, R^d)$ with local characteristic belonging to the class $B_b^{0,1}$. Then for each t_0 and x_0, equation (4.3.112) has a unique solution.*

For proof of this theorem and the following theorems in this section we refer to Kunita's book [54]. This is the development of Itô's stochastic differential equation to general processes. This is not without difficulties because of the presence of the jumps. In any case, the integrator is to be viewed as a process in the usual sense indexed by an extra parameter which will be eventually replaced by the integrand. The stochastic intgral in the right hand side of (4.3.112) is defined in Fujiwara-Kunita as a limit of Itô-Riemann sums.

If we do not assume the uniform Lipschitz condition for the local characteristic, then, the explosion may occur at a finite time. So we shall define a local solution of a stochastic differential equation and give the existence theorem.

Let $\phi_t, t \in [t_0, \sigma_\infty)$ be a continuous local process with values in R^d adapted to $(\mathcal{F}_t)$. It is called a local solution of equation (4.3.112) if

$$\phi_{t\wedge\sigma_N} = x_0 + \int_{t_0}^{t\wedge\sigma_N} F(\phi_{s\wedge\sigma_N}, ds)$$

is satisfied for any N where $\{\sigma_N\}$ is a sequence of stopping times such that $\sigma_N < \sigma_\infty$ and $\sigma_N \uparrow \sigma_\infty$. Furthermore if $\lim_{t\uparrow\sigma_\infty} \phi_t = \infty$ is satisfied when $\sigma_\infty < T$, it is called a *maximal solution* and σ_∞ is called the *explosion time.* If the explosion time is equal to T a.s., the solution $\phi_t, t \in [t_0, T)$ is called a *global solution.* Further if equation (4.3.112) has a global solution for any initial condition, equation (4.3.112) or the corresponding C-semimartingale F is called *complete (to the forward).*

Theorem 4.3.14 *Let $F(x,t)$ be a continuous semimartingale with values in $C(R^d, R^d)$ with local characteristic belonging to the class $B^{0,1}$. Then for each t_0 and x_0 the stochastic differential equation (4.3.112) has a unique maximal solution.*

Up to the global solution, we have the following theorem.

Theorem 4.3.15 *Assume that the local characteristic (a, b, A_t) of a continuous C -semimartingale F belongs to the class $B^{0,1}$ and is of linear growth, i.e. there exists a positive predictable process K_t with $\int_0^T K_t dA_t < \infty$ such that*

$$||a(x,x,t)|| \leq K_t(1+|x|)^2, \tag{4.3.113}$$

$$|b(x,t)| \leq K_t(t+|x|). \tag{4.3.114}$$

Then for each t_0 and x_0 equation (4.3.112) has a unique global solution. Furthermore, if the process K_t satisfies

$$E[\exp\{\lambda \int_0^T K_u dA_u\}] < \infty, \forall\lambda > 0 \tag{4.3.115}$$

the global solution has finite monents of any order.

Next is a theorem about the homeomorphic property of solutions of SDE. That means that a system of solutions of a stochastic differential equation defines a stochastic flow of homeomorphisms provided that the local characteristic of $F(x,t)$ governing the stochastic differential equation belongs to $B_b^{0,1}$.

Let $F(x,t) = (F^1(x,t), \cdots, F^d(x,t)), x \in R^d$ be a continuous $C(R^d, R^d)$-valued semimartingale with local characteristic belonging to $B_b^{0,1}$. Consider an Itô stochastic differential equation

$$\phi_t = x + \int_s^t F(\phi_r, dr). \tag{4.3.116}$$

We have seen in Theorem 3.2.1 that equation (4.3.116) has a unique global solution for any s, x. We denote its solution by $\phi_{s,t}(x), t \geq s$. Then we have the folowing theorem about the homoeomorphic property of solution of SDE.

Theorem 4.3.16 *(i) Assume that the local characteristic of $F(x,t)$ belongs to the class $B_b^{0,1}$. Then there exists a modification of the system of solutions denoted by $\phi_{s,t}, 0 \leq s \leq t \leq T$ such that it is a forward stochastic flow of homeomorphisms. Further for every $s, \phi_{s,t}, t \in [s,T]$ is a $C^{0,\gamma}$-semimartingale flow for any $\gamma < 1$.*

(ii) Assume that $F(x,t)$ is a Brownian motion with values in $C^{0,\gamma}$ with mean vector $\int_0^t b(x,r)dr$ and covariance $\int_0^t a(x,y,r)dr$ where a belongs to the class $\tilde{C}_{ub}^{0,1}$ Then the associated flow is a Brownian flow with infinitesimal mean b and infinitesimal covariance a.

Now let's see the diffeomorphic property of solutions of SDE.

Let $G(\lambda,\tau,t), (\lambda,\tau) \in R^e \times [0,T]$ be a family of continuous R^d-semimartingales with parameter (λ,τ) with local characteristic $(a(\lambda,\tau,\lambda',\tau',t), b(\lambda,\tau,t),t)$. Let $0 \leq \delta \leq 1, 0 < \gamma \leq \frac{1}{2}$ and $p \geq 1$. We assume both a and b are continuous random fields and continuously differentiable with respect to λ and λ'. Let $a' = D_\lambda^\alpha D_{\lambda'}^\alpha a, b' = D_\lambda^\alpha b$ for $|\alpha| \leq 1$. Set

$$L_1^{\alpha,\delta,\gamma}(\lambda,\tau,\lambda',\tau',t) = \frac{||a'(\lambda,\tau,\lambda,\tau,t) - 2a'(\lambda,\tau,\lambda',\tau',t) + a'(\lambda',\tau',\lambda',\tau',t)||}{|\lambda-\lambda'|^{2\delta} + |\tau-\tau'|^{2\gamma}}$$

$$L_2^{\alpha,\delta,\gamma}(\lambda,\tau,\lambda',\tau',t) = \frac{|b'(\lambda,\tau,t) - b'(\lambda',\tau',t)|}{\lambda-\lambda'|^{\delta} + |\tau-\tau'|^{\gamma}}.$$

These are called L^p-bounded if $E[|L_i^{\alpha,\delta,\gamma}|^p]$ are bounded with respect to $(\lambda,\tau,\lambda',\tau',t)$. Now the local characteristic (a,b) is said to belong to the class $B_p^{1,\delta,\gamma}$ if $a, b, L_i^{0,1,\gamma}$ and $L_i^{\alpha,\delta,\gamma}$ for $|\alpha| = 1$ are all L^p-bounded.

Now let $G_1(\lambda,\tau,t), (\lambda\tau) \in D \times [0,T]$ be a family of continuous R^d-valued semimartingales with parameter $(\lambda\tau)$ and let $G_2(\lambda,\tau,t)$ be a family of continuous $R^d \otimes R^d$-valued semimartingales with parameter (λ,τ). Let $G_3(y,t)$ be a continuous $C(R^d : R^d)$-valued semimartingale. We need the following assumption:

Condition 4.3.1 *(i) The local characteristic (a_1, b_1, t) of G_1 belongs to the class $B_p^{1,\delta,\gamma}$ for any $p \geq 1$.*
(ii) The local characteristic (a_2, b_2, t) of G_2 belongs to the class $B_p^{1,\delta,\gamma}$ for any $p \geq 1$. Further, a_2, b_2 are uniformly bounded.
(iii) The local characteristic (a_3, b_3, t) of G_3 belongs to the class $B_{ub}^{1,\delta}$.

Define

$$G(y,\lambda,\tau,t) = G_1(\lambda,\tau,t) + G_2(\lambda,\tau,t) + G_3(y,t).$$

It is a family of continuous $C(R^d : R^d)$-valued semimartingales with parameter (λ, τ). Consider stochastic differential equation with parameter (λ, τ):

$$\eta_t = y + \int_s^t G(\eta_u, \lambda, \tau, du). \tag{4.3.117}$$

For each y, λ, τ and s it has a unique solution denoted by by $\eta_{s,t}(y, \lambda, \tau)$. Given a C^∞-function $q(\lambda)$ with values in R^d, we set

$$\eta_{s,t}(\lambda) \equiv \eta_{s,t}(q(\lambda), \lambda, s).$$

We will study its continuity with respect to (s, t, λ) and its differentiability with respect to λ. We have the following theorems.

Theorem 4.3.17 *Assume that $G(y, \lambda, \tau, t)$ satisfies Condition 4.3.1 for some $\delta, \gamma > 0$. Let $\eta_{s,t}(y, \lambda, \tau)$ be the solution of equation (4.3.117). Set $\eta_{s,t}(\lambda) \equiv \eta_{s,t}(q(\lambda), \lambda, s)$. for $q(\lambda)$ a smooth function. Then $\eta_{s,t}(\lambda)$ has modification which is continuous in (s, t, λ). Any continuous modification is differentiable with respect to λ for any s, t and the dervatives are continuous in (s, t, λ) a.s. Futher if q and its first derivatives are bounded and the latter is uniformly δ-Holder continuous, then for every $p > 1$ there exists a positive constant c such that the modification $\eta_{s,t}(\lambda)$ satisfies*

$$E\left[|\frac{\partial \eta_{s,t}}{\partial \lambda^i}(\lambda) - \frac{\partial \eta_{s,t}}{\partial \lambda^i}(\lambda')|^{2p}\right] \leq c\{|\lambda - \lambda'|^{2p\delta} + |s - s'|^{2p\gamma} + |t - t'|^p\},$$

and

$$E\left[|\frac{\partial \eta_{s,t}}{\partial \lambda^i}(\lambda)|^{2p}\right] \leq c$$

for any $s, t, s', t', \lambda, \lambda'$. Furthermore, for every s it is a contimuous $C^{1,\epsilon}$-semimartingale for any $\epsilon < \delta$.

Theorem 4.3.18 *Assume that the local characteristic of the contnuous C-semimartingale $F(x, t)$ belongs to the class $B_b^{k,\delta}$ for some $k \geq 1$ and $\delta > 0$. Then the solution of stochastic differential equation based on F has a modification $\phi_{s,t}, 0 \leq s \leq t \leq T$ such that it is a forward stochastic flow of C^k-diffeomorphisms. Further it is a $C^{k,\epsilon}$-semimartingale for any $\epsilon < \delta$.*

Stratonovich's stochastic differential equation

Next we shall consider stochastic differential equations described in terms of Stratonovich integrals. As we will soon see, Stratonovich's stochastic differential equation can be rewritten as an Ito's equation. Hence most problems involving a Stratonovich's stochastic differential equation can be reduced to a problem involving an Ito's equation.

Let $\tilde{F}(x, t)$ be a continuous C-semimartingale with local characteristic belonging to the class $(B^{2,\delta}, B^{1,0})$ for some $\delta > 0$. A continuous R^d-valued local semimartingale $\phi_t, t \in [t_0, \sigma_\infty)$ is called a local solution of Stratonovich's stochastic differential equation based on $\tilde{F}(x, t)$ starting at x_0 at time t_0 if it satisfies

$$\phi_{t \wedge \sigma_N} = x_0 + \int_{t_0}^{t \wedge \sigma_N} \tilde{F}(\phi_{s \wedge \sigma_N}, \circ ds) \tag{4.3.118}$$

for any N where $\{\sigma_N\}$ is a sequence of stopping times such that $\sigma_N < \sigma_\infty$ and $\sigma_N \uparrow \sigma_\infty$. Also ϕ_t is said to be governed by the Stratonovich's stochastic differential equation based on $\tilde{F}(x,t)$. Furthermore if $\lim_{t\uparrow\sigma_\infty} \phi_t = \infty$ is satisfied when $\sigma_\infty < T$, it is called a *maximal solution* and σ_∞ is called the *explosion time*. We have the similar theorem with previous section.

Theorem 4.3.19 *Let $\tilde{F}(x,t)$ be a continuous C^1-semimartingale with local characteristic (a, b, A_t) belonging to the class $(B^{2,\delta}, B^{1,0})$ for some $\delta > 0$. Then for each t_0 and x_0, the Stratonovich's equation (4.3.118) has a unique maximal solution. Further the solution satisfies Itô's equation based on $\tilde{F}(x,t) + C(x,t)$ where*

$$C(x,t) = \frac{1}{2}\int_0^t \{\sum_{j=1}^d \frac{\partial a^{\cdot j}}{\partial x^j}(x,y,s)|_{y=x}\} dA_s \tag{4.3.119}$$

Conversely let $F(x,t), x \in R^d$ be a continuous C^1-semimartingale with local characteristic (a, b, A_t) belonging to the class. $(B^{2,\delta}, B^{1,0})$ for some $\delta > 0$. Then the solution of the Itô equation based on F satisfies the Stratonovich's equation based on $F(x,t) - C(x,t)$. The term $C(x,t)$ here is often called the correction term of the semimartingale F or $\tilde{F}$.

Backward equation

In this section, we shall give the definition of the solution of the backward stochastic differential equations. The arguments are completely parallel to those of (forward) stochastic differential equations. The only difference is that these are defined to the backward direction. We would like to start with the definition on backward stochastic integral with respect to semimartingale.

Let $\{\mathcal{F}_{s,t} : 0 \le s \le t \le T\}$ be a family of sub-σ-field of $\mathcal{F}$ which contain all null sets and satisfy $\mathcal{F}_{s,t} \subset \mathcal{F}_{s',t'}$ if $s' \le s \le t \le t'$, $\cap_\epsilon \mathcal{F}_{s,t+\epsilon} = \mathcal{F}_{s,t}$ and $\cap_\epsilon \mathcal{F}_{s-\epsilon,t} = \mathcal{F}_{s,t}$. It is called a filtration (with two parameters) of sub σ-field of $\mathcal{F}$. A continuous process $\hat{M}_t$ is called a backward martingale adapted to $(\mathcal{F}_{s,t})$ if it is integrable, $\hat{M}_t - \hat{M}_s$ is $\mathcal{F}_{s,t}$-measurable and satisfies $E[\hat{M}_r - \hat{M}_t|\mathcal{F}_{s,t}] = \hat{M}_s - \hat{M}_t$ for any $r \le s \le t$. A backward localmartingale is defined similarly to the (forward) localmartingale. Let $\hat{X}_t$ be a continuous process such that $\hat{X}_t - \hat{X}_s$ is $(\mathcal{F}_{s,t})$-adapted. It is called a backward semimartingale if it can be written as the sum of a continuous backward localmartingale and a process of bounded variation.

Now let $F(\cdot,t)$ be a continuous backward semimartingale with values in C with local characteristic belonging to the class $B^{0,0}$. Fix the time t for the moment. Let $F_s, s \in [0,t]$ be a continuous $(\mathcal{F}_{s,t})$-adapted process. Then the backward Itô integral of f_s based on $F(\cdot,t)$ is defined by

$$\int_s^t F(f_r, \hat{d}r) = \lim_{|\Delta|\to 0} \sum_{k=0}^{n-1} \{F(f_{t_{k+1}\vee s}, t_{k+1} \cup s) - F(f_{t_{k+1}\vee s}, t_k \vee s)\}, \tag{4.3.120}$$

where $\Delta = \{0 = t_0 < t_1 < \cdots < t_n = t\}, t \vee s = \max\{t,s\}$ if the right hand side converges in probability. It is a continuous backward semimartingale with respect to s.

Suppose that $F(\cdot,t)$ is a continuous backward C-semimartingale with local characteristic belonging to the class $(B^{2,\delta}, B^{1,0})$ for some $\delta > 0$, and f_s is a continuous backward semimartingale. The backward Stratonovich integral is well defined.

$$\begin{aligned}\int_s^t F(f_r, \circ\hat{dr}) &= \lim_{|\Delta|\to 0} \sum_{k=0}^{n-1} \frac{1}{2}\{F(f_{t_{k+1}\vee s}, t_{k+1} \vee s) \\ &\quad + F(f_{t_k \vee s}, t_{k+1} \vee s) - F(f_{t_{k+1}\vee s}, t_k \vee s)\} \\ &\quad - F(f_{t_k\vee s}, t_k \vee s), \end{aligned} \tag{4.3.121}$$

since the right hand side converges in probability. These two integrals are related by

$$\int_s^t F(f_r, \circ\hat{dr}) = \int_s^t F(f_r, \hat{dr}) - \frac{1}{2}\sum_j < \int_s^t \frac{\partial F}{\partial x^j}(f_r, \hat{dr}), f_t^j >, \tag{4.3.122}$$

where $<,>$ denote the joint quadratic variation.

Now a continuous $(\mathcal{F}_{s,t_0})$-adapted process $\phi_s, s \in [0, t_0]$ with values in R^d is called the solution of the backward Itô stochastic differential equation based on $F(x, t)$ starting at x_0 at time t_0 if it satisfies

$$\phi_s = x_0 - \int_s^{t_0} F(\phi_r, \hat{dr}). \tag{4.3.123}$$

We can define the solution of the backward Stratonovich's SDE similarly.

4.3.3 SDE driven by nonlinear integrator

Introduction

We have discussed the stochastic differtial equations driven by semimartingale in the previous section. In this section, we will present the stochastic differtial equations driven by nonlinear integrators. Since this is a totally different way from which we familiar with, we would like to begin with the definition of nonlinear integrator and stochastic calculs, and then discuss stochastic differential equation driven by nonlinear integrator. The reader is refered to the book of Carmona and Nualart for the details on the concepts and results.

Nonlinear Integrators and Integrals

Let $\mathcal{X}$ denote a separable Banach space endowed with its Borel σ-field $\mathcal{B}_{\mathcal{X}}$ and $\mathcal{D}(\mathcal{X})$ (resp. $\mathcal{L}(\mathcal{X})$ denote the space of cadlag i.e. right contniuous with left limits, (resp. cadlag, i.e. left continuous with right limits,) functions from $[0, \infty)$ into $\mathcal{X}$. We endow these spaces with the topology of the uniform convergence in probability on compact sets (UCP topology for short) given by the distance:

$$d_{UCP}(X, Y) = E\{d_{\infty,loc}(X(\cdot), Y(\cdot))\} \tag{4.3.124}$$

where

$$d_{\infty,loc}(X(\cdot), Y(\cdot)) = \sum_{n=1}^{\infty} \frac{1}{2^n} \frac{\sup_{0\le t\le n} ||X(t) - Y(t)||}{1 + \sup_{0\le t\le n} ||X(t) - Y(t)||}.$$

For simplicity, the discussion is most of the time limited to these processes which are bounded. So for each $X \in \mathcal{D}(\mathcal{X})$ we set

$$||X||_{\mathcal{S}^{(p)}} = ||X^*||_p, \tag{4.3.125}$$

where $X^* = \sup_{t\geq 0} ||X(t)||$ and we denote by $\mathcal{S}^{(p)}$ the space of càdlàg adapted processes X for which the quantity (4.3.125) is finite. We will also use the standard notation:

$$X^t(s) = X(s \wedge t), s, t \geq 0$$

$$X^* = \sup_{t\geq 0} ||X(t)|| \quad \textit{and} \quad X_t^* = (X^t)^* = \sup_{0\leq s\leq t} ||X(s)||$$

whenever X is an $\mathcal{X}$-valued function defined on $[0, \infty)$.

The notion of simple predictable process is crucial to the stochastic integral. We will need the follwoing definition.

Definition 4.3.20 *An $\mathcal{X}$-valued process X is said to be simple predictable if it has representation of the form:*

$$X(t) = X_{-1}1_{\{0\}}(t) + \sum_{j=0}^{n} X_j 1_{(\tau_j,\tau_{j+1}]}(t), \tag{4.3.126}$$

where $0 = \tau_0 \leq \tau_1 \leq \cdots \leq \tau_{n+1} < \tau_{n+2} = +\infty$ is finite sequence of stopping times and where X_{-1} is bounded and $\mathcal{F}_l$-measurable and where X_j is for each $j = 0, 1, \cdots, n$ a bounded $\mathcal{X}$-valued $\mathcal{F}_{\tau_j}$ measurable random variable (denote by $X_j \in \mathcal{F}_{\tau_j}$ from now on). We also assume that the stopping times τ_j's and the random variables X_j's take only finitely many values. Denote by $\mathcal{S}(\mathcal{X})$ (or $\mathcal{S}$ if no confusion is possible) the collection of $\mathcal{X}$-valued simple predictable processes.

Let us assume that, for each $h \in \mathcal{L}(\mathcal{X}), \{Z_t(h); t \geq 0\}$ is a R^d-valued cadlag adapted process such that $Z_0(h) = Z_0(h^0)$ and such that, for all $t > 0$ and $h, h' \in \mathcal{L}(\mathcal{X})$ we have :

$$Z_t(h) = Z_t(h'), \tag{4.3.127}$$

outside a P-null set (possibly depending upon h, h' and t) whenever:

$$h(s) = h'(s), \forall s \leq t, \text{and h } h(t+) = h'(t+).$$

In the sequel, we will say that a family of random variables

$$\{Z_t(h); t \geq 0, h \in \mathcal{L}(X)\}$$

satisfies condition (4.3.127) if the above holds.

We define mapping I_Z from $\mathcal{S}$ into $L^0(\Omega, R^d)$ by:

$$I_Z(X) = Z_0(X^0) + \sum_{j=0}^{n+1} (Z_{\tau_{j+1}}(X^{\tau_j+}) - Z_{\tau_j}(X^{\tau_j+})) \tag{4.3.128}$$

whenever X is a simple predictable process admitting the representation (4.3.126) and where we used $Y^{\tau+}$ to denote the process Y stopped after time τ, i.e. the process defined by:

$$Y^{\tau+}(t) = \begin{cases} Y(t), & : \text{ if } t \leq \tau; \\ Y(\tau_+), & : \text{ if } t > \tau. \end{cases}$$

It is easy to see that for any simple predictable process X possessing the representation (4.3.126), we have

$$X^{\tau_j}(\tau_j) = X^{\tau_j}(\tau_j+) = X_{j-1}, X^{\tau_j+}(\tau_j) = X_{j-1}, \text{ and } X^{\tau_j+}(\tau_j+) = X_j. \tag{4.3.129}$$

Definition 4.3.21 *Z is said to be a (nonlinear) integrator if, for each $t \geq 0$ and each stopping time τ taking finitely many values less than t, the mapping I_{Z^τ} defined by (4.3.128) for Z^Z is locally uniformly continuous when $\mathcal{S}$ is equipped with the topology of uniform convergence in (t, ω). This continuity is also uniform in τ restricted to the set $\mathcal{T}_t$ of stopping times taking finitely many values bounded by t. This means that we have :*

$$\lim_{\epsilon \to 0} \sup_{\tau \in \mathcal{T}_t, ||X^*||_\infty \leq K, ||Y^*||_\infty \leq K, ||(X-Y)^*||_\infty \leq \epsilon} ||I_{Z^\tau}(X) - I_{Z^\tau}(Y)||_0 = 0, \tag{4.3.130}$$

for any $K \geq 0$ and $t \geq 0$ where $||f - g||_0$ denotes the distance between f and g in L^0 (the space of all equivalence classes of random variables endowed with the topology of the convergence in probability).

A nonlinear integrator in the sense of the above definition will be called an L^0-integrator because of the use of the cconvergence in probability. In fact the above definition means that the mapping I_{Z^τ} has uniformly continuous extension from the space of bounded processes which are uniform limits of simple predictable processes to the space L^0. More generally, we will say that Z is a (nonlinear) L^p-integrator when the map I_{Z^τ} has a uniformly continuous extension into the space L^p. In other words, Z is a (nonlinear) L^P-integrator if (4.3.130) is satisfied with $|| \cdot ||_p$ instead of $|| \cdot ||_0$.

Before further discussion of the notion of the nonlinear integrator, we will give an example to show that this definition is appropriate.

Example 4.3.22 *Let $\mathbf{z} = \{z(t); t \geq 0\}$ be a real valued adapted cadlag process such that $z(0) = 0$ and for each $h \in L(\mathcal{X})$ we set:*

$$Z_t(h) = h(t+)z(t). \tag{4.3.131}$$

If X is a simple predictable process with the decomposition (4.3.126) we have

$$\begin{aligned} I_Z(X) &= \sum_{j=0}^{n+1} Z_{\tau_{j+1}}(X^{\tau_j}+) - Z_{\tau_j}(X^{\tau_j}+) \\ &= \sum_{j=0}^{n+1} X^{\tau_j}+(\tau_{j+1}+)z(\tau_{j+1}) - X^{\tau_j}+(\tau_j+)z(\tau_j) \\ &= \sum_{j=0}^{n} X_j(z(\tau_{j+1}) - z(\tau_j)), \end{aligned} \tag{4.3.132}$$

where we used the definition of $I_Z(X)$ (4.3.128), the Definition (4.3.131) of Z and relation (4.3.129).

Next we list some simple properties of integrators in the sense of Definition 4.3.21.

Properties:

(i) The set of integrators is a vector space;

(ii) An integrator remains so after an absolutely continuous change of probability;

(iii) If $\{P_k; k \geq 1\}$ is a sequence of probabilities such that Z is a P_k-integrator for each $k \geq 1$, then Z is also a P-integrator where P is defined by$P + \sum_{k=1}^{\infty} \lambda_k P_k$ for some sequence $\{\lambda_k; k \geq 1\}$ of nonnegative numbers such that $\sum_{k=1}^{\infty} = 1$.

(iv) Let Z be an integrator for the filtration $\{\mathcal{F}_t; t \geq 0\}$ and let $\{\mathcal{G}_t; t \geq 0\}$ be a subfiltration such that $Z(h)$ is still adapted to $\{\mathcal{G}_t; t \geq 0\}$ for all $h \in \mathcal{L}(\mathcal{X})$. Then, Z is also an integrator for the filtration $\{\mathcal{G}_t; t \geq 0\}$.

(v) For any given Z, if there exists a sequence $\{\tau_n; n \geq 1\}$ of nonnegative finite random variables increasing to $+\infty$ and a sequence $\{Z_n; n \geq 1\}$ of integrators such that $Z^{\tau_n -} = (Z_n)^{\tau_n -}$, then Z is an integrator.

(vi) If $Z(h)$ is adapted and cadlag for every $h \in \mathcal{L}(\mathcal{X})$ and if there exists a sequence $\{\tau_n; n \geq 1\}$ of stopping times increasing to $+\infty$ such that Z^{τ_n} is an integrator for each $n \geq 1$, then Z is also an integrator.

Let us give the most common example of a nonlinear integrator. It corresponds to the integration of processes that do not depend on the entire past but merely on the present. Assume that, for each $x \in \mathcal{X}, \{Z_t(x); t \geq 0\}$ is an R^d-valued càdlàg adapted process and we will assume (temporarily) that each process $Z_.(x)$ is defined at ∞. Also we assume that $\{Z_t(x); x \in \mathcal{X}\}$ is measurable for each $t \geq 0$, and we define for each $h \in L(\mathcal{X})$ the R^d valued stochastic process $\{\hat{Z}_t(h); t \geq 0\}$ by $\hat{Z}_0(h) = Z_0(h(0)), \hat{Z}_\infty(h) = Z_\infty(h(\infty-))$ and

$$\hat{Z}_t(h) = Z_t(h(t_+)). \tag{4.3.133}$$

for finite $t > 0$.

Definition 4.3.23 *We will say (simply) that Z is a (nonlinear) integrator whenever $\hat{Z}$ defined by (4.3.133) is an integrator in the sense of Definition 4.3.21*

In such a case we use the same notation for the integral of simple predictable processes. In particular, if a process X in $\mathcal{S}$ has a decomposition of the form (4.3.128) we have

$$I_z(X) = Z_0(X_0) + \sum_{i=0}^{n+1}(Z_{\tau_{i+1}}(X_i) - Z_{\tau_i}(X_i)).$$

As in the classical theory, we will consider stochastic integrals as processes rather than random variables or random vectors. We proceed in the usual way. If $\{Z_t(h); t \geq 0, h \in \mathcal{L}(\mathcal{X})\}$ is a nonlinear integrator then the formula:

$$I_Z(X)_t = I_{Z^t}(X) \tag{4.3.134}$$

can be used to define the stochastic integral of X with respect to the integrator Z as a stochastic process. This process is cadlag, and consequently, formula (4.3.134) defines a mapping from $\mathcal{S}(\mathcal{X})$ into $\mathcal{D}(R^d)$. Note that

$$I_Z(X)_t = Z_0(X_{-1}) + \sum_{j=0}^{n}[Z_{\tau_{j+1}\wedge t}(X^{\tau_j}+) - Z_{\tau_j \wedge t}(X^{\tau_j}+)], \tag{4.3.135}$$

for all the processes $X = \{X(t); t \geq 0\}$ having a decomposition of the form (4.3.126). Finally, we also note that the above integrals use the integrators Z^t which is always defined at ∞. In other words, we will not require the definition of Z_∞. The notion $X \cdot Z$ is standard for the integral $I_Z(X)$, which it is convenient to use.

Now we have the following result which is the strict analog of the corresponding one in the classical case:

Proposition 4.3.24 *If Z is an integrator, the mapping $I_Z : \mathcal{S} \to \mathcal{D}(R^d)$ is locally uniformly continuouos for the UCP topology, and consequently, it can be extended by uniform continuity to the subspace $\mathcal{L}_b(\mathcal{X})$ of $\mathcal{L}(\mathcal{X})$ formed by the bounded processes.*

We list some simple properties of the nonlinear integrals. In the following X is always an element of $\mathcal{L}$, and Z is an integrator in the sense of Definition 4.3.21 or Definition 4.3.23.

Property (1) If τ is any stopping time, then $I_Z(X)^\tau = I_{Z^\tau}(X)$.

Remark *This peroperty makes possible the definition of the nonlinear integral $I_Z(X)$ not only for bounded left continuous processes but also for all left coninuous processes by a simple localization argument.*

Property (2) If X is simple predictable, then the jump process $\Delta I_Z(X)$ is indistingushable from the process $\{Z_t(X^t) - Z_{t-}(X^t); t \geq 0\}$ (or $\{Z_t(X(t)) - Z_{t-}X(t)); t \geq 0\}$ when Z is an integrator in the sense of Definition 3.3.2').

Property (3) Let P and Q be any given probability measures and let us assume that Z is an integrator for both P and Q. Then, there exists a stochastic process $I_Z(X)$ which is a version of both $I_Z^P(X)$ and $I_Z^Q(X)$.

Property (4) Let $\{\mathcal{G}_t; t \geq 0\}$ be another filtration and let us assume that X is also $\{\mathcal{G}_t; t \geq 0\}$-adapted and that Z is also a $\{\mathcal{G}_t; t \geq 0\}$-nonlinear integrator. Then $I_Z^{\mathcal{F}}(X) = I_Z^{\mathcal{G}}(X)$.

Definition 4.3.25 *A family $\{Z_t(h); t \geq 0, h \in L(\mathcal{X})\}$ (resp. $\{Z_t(x); t \geq 0, x \in \mathcal{X}\}$) is said to be a strong (nonlinear) integrator if the corresponding family $\{\tilde{Z}_t((y,h)); t \geq 0, (y,h) \in L(R \times \mathcal{X})\}$ (resp. $\{\tilde{Z}_t((y,x)); t \geq 0, (y,x) \in R \times \mathcal{X}\}$) is a (nonlinear) integrator in the sense of Definitioin 4.3.21(resp. 4.3.23)*

Property (5) If $Z_.(h)$ is a martingale for each $h \in \mathcal{L}(\mathcal{X})$ then so is $I_Z(X)$ for each $X \in \mathcal{S}(\mathcal{X})$.

Remark *$Z_t(h)$ is a strong nonlinear integrator if for each $t \geq 0$ and any $K > 0$ one has*

$$\lim_{\epsilon \longrightarrow 0} \sup ||I_{\tilde{Z}^t}(Y, X) - I_{\tilde{Z}^t}(Y', X')||_0 = 0$$

where the supremum is taken over all the X, X', Y, Y' satisfying $||X^*||_\infty \leq K, ||Y^*||_\infty \leq K, ||(X')^*||_\infty \leq K, ||(Y')^*||_\infty \leq K$ and such that $||(X - X')^*||_\infty \leq \epsilon$ and $||(Y - Y')^*||_\infty \leq \epsilon$. The reason why we do not have to consider stopping times τ taking finitely many values and bounded by t is that for such a stopping time τ one has

$$I_{\tilde{Z}^\tau}(Y, X) = I_{\tilde{Z}^t}(Y \cdot 1_{[}0, \tau], X).$$

Next, let us look at the change of variables formula in the spirit of the famous Itô's formula.

Theorem 4.3.26 *Let $Z = \{Z_t(x); t \geq 0, x \in R^d\}$ be a R-valued strong nonlinear L^1-integrator satisfying the following properties.*

(i) $Z = \{Z_t; t \geq 0\}$ is a $C^{(2)}(R^2; R)$-valued cadlag process.

(ii) The partial derivatives $\partial Z_t(x)/\partial x_i$ are also strong nonlinear L^1-integrators. Then for every R^d-valued continuous semimartingale $X = \{X_t; t \geq 0\}$ the following formula holds:

$$\begin{aligned} Z_t(X_t) &= Z_0(X_0) + \sum_{i=1}^d \int_0^t \frac{\partial Z_s}{\partial X_i}(X_s) dX_s^i \\ &+ \frac{1}{2} \sum_{i,j=1}^d \int_0^t \frac{\partial^2 Z_s}{\partial x_i \partial x_j}(X_s) d < X^i, X^j >_s \\ &+ \sum_{i=1}^d I_{<X^i, \partial Z^c/\partial x_i>}(X)_t + I_Z(X)_t. \end{aligned} \tag{4.3.136}$$

Notice that the fact that X^i is a continuous semimartingale implies that we have identity of the two types of brackets, i.e.

$$< X^i, \frac{\partial Z^c}{\partial x_i}(y) >= [X^i, \frac{\partial Z}{\partial x_i}(y)],$$

and that this process is a strong nonlinear integrator. This theorem can be extended to the case that Z is R^d-valued.

Stochastic Differential Equations

We have seen that Lipschitz hypothesis is crucial to the classical stochastic differential equations driven by semimartingale. No surprisingly, that the Lipschitz hypothesis will be used in the discussion of existence and the uniqueness of the solutions of SDEs driven by nonlinear integrator. However since the notions of integrand and integrator are confounded in the present (nonlinear) theory, the Lipschitz assumptions have to be reformulated appropriately. We will begin with the following definition, then discuss the theory of SDEs, some existence and uniquedess of solution, dependency of initial condition, differentiability and homeomorphic property of the solution are given without proof. The reader may refer to the text of Carmona and Nualart. For the sake of simplicity, assume that $\mathcal{X} = R^d$ from now on. Let us recall that a cadlag adapted process $\{X(t); t \geq 0\}$ is said to be a semimartingale if it possesses a decomposition of the form

$$X(t) = X(0) + M(t) + V(t), t \geq 0, \tag{4.3.137}$$

where the process $V = \{V(t); t \geq 0\}$ is adapted and cadlag and has sample paths locally of bounded variation which is called the Stieljes process, and where the process $M = \{M(t); t \geq 0\}$ is a local martingale such that $V(0) = M(0) = 0$. If in the decomposition (4.3.137), Stieljes process V is predictable, then the semimartingale is called special semimartingale. In this case, the decomposition is called the canonical decomposition of special semimartingale, the decomposition is unique.

Definition 4.3.27 *A family $\{Z_t(h); t \geq 0, h \in L(R^d)\}$ of R^d-valued special semimartingales is said to have a canonical decomposition uniformly controlled by the nondecreasing right-continuous process $\{A_t; t \geq 0\}$ (with $A_0 = 0$) if for each $h \in L(R^d)$ and $i = 1, \cdots, d$, the canonical decomposition*

$$Z_t^i(h) = Z_0^i(h) + M_t^i(h) + V_t^i(h), \tag{4.3.138}$$

of the ith coordinate process $\{Z_t^i(h); t \geq 0\}$ is that:
i) The processes $M_\cdot^i(h)$ and $V(t)$ satisfy condition (4.3.127) $Z_0(h) = Z_0(h^0)$, and :

$$|Z_0(h) - Z_0(h')| \leq B|h(0) - h'(0)| \tag{4.3.139}$$

for some positive random variable B and all h and h' in $L(R^d)$.
ii) For each $h \in L(R^d), \{M_t^i(h); t \geq 0\}$ is a locally square integrable martingale such that $M_0^i(h) = 0$ and for all h and h' in $L(R^d)$ one has:

$$< M_\cdot^i(h) - M_\cdot^i(h') > ((s,t]) \leq \int_{(s,t]} |h_u - h'_u|^2 dA_u \tag{4.3.140}$$

for all $0 \leq s \leq t < \infty$ and all $h, h' \in L(R^d)$.

iii) For each $h \in L(R^d)$, $\{V_t^i(h); t \geq 0\}$ is a predictable Stieltjes process such that $V_0^i(h) = 0$ and:

$$V_\cdot^i(h) - V_\cdot^i(h')|_{var}((s,t]) \leq \int_{(s,t]} |h_u - h'_u| dA_u \tag{4.3.141}$$

for all $0 \leq s \leq t < \infty$ and all $h, h' \in L(R^d)$.

We can now give the existence and uniqueness theorem of the SDE driven by nonlinear integrator.

First let us recall the stochastic measure associated with the jumps of the $C^{(0)}$-valued adapted cadlag process $Z = \{Z_t; t \geq 0\}$, namely:

$$\mu_Z(df, dt) = \sum_{s>0} 1_{\{\Delta Z_s \neq 0\}} \delta_{(\Delta Z(s), s)}(df, dt).$$

The dual predictable projection of μ_Z is the unique predictable random measure ν satisfying:

$$E\left\{\int_{[0,t]\times C^{(0)}} W(f,s)\nu(df,ds)\right\} = E\left\{\sum_{s>0} 1_{\{\Delta Z_s \neq 0\}} W(\Delta Z_s, s)\right\},$$

for all $t > 0$ and for all the nonnegative $\mathcal{B}_{C^{(0)}} \times \mathcal{P}$-measurable functions $(f, s, \omega) \to W(f, s, \omega)$.

Theorem 4.3.28 *Let $\{Z_t(h); t \geq 0, h \in L(R^d)\}$ be a family of R^d valued special semimartingales with canonical decompositioin uniformly controlled by the nondecreasing predictable process $\{A_t; t \geq 0\}$ and let $J = \{J(t); t \geq 0\}$ be a R^d valued cadlag process. Then, there exists a unique (up to indistinguishable) solution $\{X(t); t \geq 0\}$ of the system of equatioins:*

$$X^i(t) = J^i(t) + \int_0^t dZ_s^i(X_-) i = 1, \cdots, d \tag{4.3.142}$$

and the solution is a semimartingale whenever J is, where

$$X_- = \{X(t-); t \geq 0\}.$$

The following theorems are devoted to the investigation of the properties of the solution of the stochastic differential equation

$$X_t = x + \int_0^t dZ_s(X_{s-}) \tag{4.3.143}$$

whose existence and uniqueness was given in Theorem 4.3.28 The next theorem shows the dependence of the solution upon the initial conditions of $X_0 = x$. Since the equation (4.3.143) is parametrized by the space $\mathcal{X} = R^d$, the notion of control of an integrator by an increasing process needs to be reformulated as following definition.

Definition 4.3.29 *A family $\{X_t(x); t \geq 0. x \in R^d\}$ of R^d-valued special semimartingales is said to have a canonical decomposition controlled by the nondecreasing predictable process $\{A_t; t \geq 0\}$ (with $A_0 = 0$) if for each $x \in R^d$ and each $i \in \{i, \cdots, d\}$, the ith coordinate process $Z^i(x) = \{Z_t^i(x); t \geq 0\}$ satisfies $Z_0^i(x) = 0$ and has a canonical decomposition $Z_t^i(x) = M_t^i(x) + V_t^i(x)$ such that the predictable Stieltjes process $Z_\cdot^i(x) + \{V_t^i(x); t \geq 0\}$ satisfies*

$$|V_\cdot^i(x) - V_\cdot^i(y)((s,t])| \leq |x - y|(A_t - A_s),$$

and the martingale part $M^i_\cdot(x) = \{M^i_t(x); t \geq 0\}$ is a locally square integrable martingale satisfying

$$< M^i_\cdot(x) - M^i_\cdot(y) > ((s,t]) \leq |x-y|^2(A_t - A_s)$$

for all nonnegative numbers s and t satisfying $s \leq t$ and all x and y in R^d.

Definition 4.3.30 *If $A = \{A_t; t \geq 0\}$ is a nondecreasing predictable process, we denote by $\mathcal{S}_{spe}(A)$ the set of families $\{Z_t(x); t \geq 0, x \in R^d\}$ of special semimartingales such that*
(i) $\{Z_t(\cdot); t \geq)\}$ is a $C^{(0)}(R^d, R^d)$-valued cadlag adapted process.
(ii) its canonical decomposition is controlled by A.

Theorem 4.3.31 *Let $\{Z_t(x); t \geq 0, x \in R^d\}$ be a family in $\mathcal{S}_{spe}(A)$ which satisfies:*

$$\int_{R^d} |\xi|^p \nu_{Z_\cdot(x)-Z_\cdot(y)}(d\xi, (s,t]) \leq |x-y|^p(A_t - A - s) \tag{4.3.144}$$

for some $p \geq d \wedge 2$, all $s \leq t$ and all $x, y \in R^d$. Then, for each $x \in R^d$ one can choose a version $\{X_t(x); t \geq 0\}$ of the solution of the stochastic differential equation (4.3.143) in such a way that the mapping $x \to X_\cdot(x)$ is continuous from R^d into $D(R^d)$.

We used the symbol ν_Z for the dual predictable projection of the jump measure $\mu_Z = \sum_{s\geq 0} 1_{\{\Delta_s \neq 0\}} \delta_{(\Delta_s, s)}$ mentioned at the begining this section. In order to give the differentiability of the solution of (4.3.143), we need the following definition and hypotheses

Definition 4.3.32 *A symmetric function $a : R^d \times R^d \to R^k$ is said to be bi-Lipschitz continuous if there exists a positive constant L such that*

$$|a(x,x) - 2a(x,y) + a(y,y)| \leq |x-y|^2, \tag{4.3.145}$$

for all $x, y \in R^d$.

Hypotheses:

(Hy1) For each fixed $t \geq 0$, the functions $x \to V^i_t(x), 1 \leq i \leq d$, are continuously differentiable, and their differentials satisfy

$$|D^K_x V^i_t(x) - D^K_y V^i_t(y) - D^K_x V^i_s(x) + D^K_y V^i_s(y)| \leq (A - t - A - s)|x-y|, \tag{4.3.146}$$

where $D^K_x = \frac{\partial}{\partial x_k}$ denotes the first-order dervative with respect to the kth coordinate of x,

(Hy2) For each fixed $t \geq 0$ and for each $1 \leq i \leq d$ the function $(x,y) \to < M^i(x), M^i(y) >_t$ is twice continuously differentiable and for each $k \in \{1, \cdots, d\}$ and $0 \leq s \leq t$ the function

$$(x,y) \to D^k_x D^k_y \{< M^i(x), M^i(y) > ((s,t])\}$$

is bi-Lipschitz function with constant $A_t - A_s$.

(Hy3) The measure ν when viewed as a measure on function space is concentrated on the space of continuously differentiable functions $C^{(1)} = C^{(1)}(R^d; R^d)$ and satisfies

$$\int_{C^{(1)}} |\nabla(f(x) - f(y))|^p \nu_Z(df, (s,t]) \leq (A_t - A_s)|x-y|^p, \tag{4.3.147}$$

for some $p > d+1$ and all $x, y \in R^d$ and $0 \leq s \leq t$.

We now can state the theorem.

Theorem 4.3.33 *Suppose that the hypotheses (Hy1), (Hy2) and (Hy3) above hold for a family* $\{Z_.(x); x \in R^d\}$ *of special semimartingales for some number* $p > 2(d+1)$ *and also for* $p/2$. *Then there exists a version* $X_t(x)$ *of the solution of equation (4.3.143) such that* $\{X_t; t \geq 0\}$ *is a cadlag* $C^{(1)}$*-valued process whose derivative is the unique solution of the linear stochastic differential equation:*

$$D^k X_t^i(x) = \delta_k^i + \sum_{l=1}^{d} [D^k X_-(x)^l \cdot I_{D^l Z^i}(X_-(x))]_t \tag{4.3.148}$$

where δ_k^i *denotes the usual Kronecker delta which is 1 when* $i = k$ *and 0 otherwise.*

Under suitable conditions, one can obtain the homeomorphic property of the continuous mapping $x \to X_t(x)$ in Theorem 4.3.31.

Theorem 4.3.34 *Let us assume that* $\{Z_t(x); t \geq 0, x \in R^d\}$ *is an element of* $\mathcal{S}_{spe}(A)$ *such that condition (4.3.144) is satisfied for some* $p > 6d$. *Further we assume that the mappings* $x \to \phi_t(x) = x + Z(x)_t - Z(x)_{t-}$ *are one-to-one for any fixed* $(t, \omega) \in [0, \infty) \times \Omega$ *and also that either one of the following two condition holds:*

$$\int_{R^d} |\xi|^2 |x - y + \xi|^{-(p+2)} \nu_{Z_.(x) - Z_.(y)}(d\xi, (s,t]) \leq (A_t - A_s)|x - y|^{-p}$$

or

$$|Z_t(x) - Z_{t-}(x) - Z_t(y) + Z_{t-}(y)| \leq |x - y| \Delta B_t$$

for some increasing and adapted process $\{B_t; t \geq 0\}$ *satisfying* $B_0 = 0$. *Then, there exists a version* $X_t(x)$ *for the solution of the stochastic differential equation (4.3.143) such that, for* P*-almost all* $\omega \in \Omega$ *the mapping* $x \to X_t(x, \omega)$ *is one-to-one for every* $t > 0$, *and* $x \to X_.(x)$ *is continuous from* R^d *to* $D(R^d)$.

4.4 Stochastic Functional Differential Equations

4.4.1 Existence and Uniqueness of Solution

Let $r > 0$, $J = [-r, 0]$ and $C(J, R^n)$ be the Banach space of all continuous paths $\gamma : J \to R^n$ with the sup-norm $|\gamma|_C = \sup_{s \in J} |\gamma(s)|$ where $|\cdot|$ denotes the Euclidean norm on $R^n (n \geq 1)$. As a metric space we associate with $C(J, R^n)$ its Borel σ-field Borel $(C(J, R^n))$.

Denote by $\mathcal{L}^2(\Omega, C(J, R^n))$ the space of all $\mathcal{F}$-measurable stochastic processes: $\theta : \Omega \to C(J, R^n)$ such that the function $\Omega \ni \omega \mapsto ||\theta(\omega)||_C \in R$ is of class $\mathcal{L}^2$, i.e. $\int_\Omega ||\theta(\omega)||_C^2 dP(\omega) < \infty$. Then $\mathcal{L}^2(\Omega, C)$ is complete when endowed with the semi-norm

$$||\theta||_{\mathcal{L}^2(\Omega, C)} = \left[\int_\Omega ||\theta(\omega)||_C^2 dP(\omega)\right]^{1/2}$$

For any $a > 0$, denote by $C([0,a], \mathcal{L}^2(\Omega, C(J, R^n)))$ the space of all $\mathcal{L}^2$-continuous $C(J, R^n)$-valued processes $y : [0, a] \to \mathcal{L}^2(\Omega, C(J, R^n))$; again this is complete under the semi-norm:

$$||y||_{C([0,a], \mathcal{L}^2(\Omega, C)} = \sup_{t \in [0,a]} ||y(t)||_{\mathcal{L}^2(\Omega, C)}$$

Denote by $C_A([0,a], \mathcal{L}^2(\Omega, C(J, R^n)))$ the set of all processes $y \in C([0,a], C(J, R^n)))$ which are adapted to $(\mathcal{F})_{0 \leq t \leq a}$. It is easy to prove that $C_A([0,a], \mathcal{L}^2(\Omega, C(J, R^n)))$ is closed in $C([0,a], \mathcal{L}^2(\Omega, C(J, R^n)))$.

We consider stochastic functional differential equations

$$\begin{aligned} dX(t) &= g(t, X_t)dz(t) \\ X_0 &= \theta(\omega)(t), \end{aligned} \tag{4.4.149}$$

in the sense of the following stochastic functional integral equation

$$X(\omega)(t) = \begin{cases} \theta(\omega)(0) + \int_0^t g(u, X_u)dz(\cdot)(u) & , \quad 0 \le t \le a \\ \theta(\omega)(t) & , \quad t \in J, a.a.\omega \in \Omega, \end{cases}$$

where $g : [0,a] \times \mathcal{L}^2(\Omega, C(J, R^n)) \to \mathcal{L}^2(\Omega, L(R^m, R^n))$, $\theta \in \mathcal{L}^2(\Omega, C(J, R^n))$ is a given initial process, and the "noise process" $z : \Omega \to C([0,a], R^m)$ which is m-dimensional and has continuous sample paths. The stochastic integral in (4.4.149) is a McShange belated integral [70]; and for each $u \in [0,a], X_u \in \mathcal{L}^2(\Omega, C(J, R^n))$ is defined by $X_u(\omega)(s) = X(\omega)(u + s), a.a.\omega \in \Omega$, for all $s \in J$. $\mathcal{L}^2(\Omega, C(J, R^n))$ will be our basic configuration space of initial processes for the stochastic FDE (4.4.149). Obviously this space entails that the initial data as well as the solution process of the SFDE will necessarily have almost all their sample paths continuous. The trajectory $[0,a] \ni t \mapsto X_t \in \mathcal{L}^2(\Omega, C(J, R^n))$ will be sought in the space $C_A([0,a], \mathcal{L}^2(\Omega, C(J, R^n)))$. In order to solve the SFDE (4.4.149), we impose the following conditions of existence (cf. Gihman and Skorohod [23], McShane [70]):

Conditions (E)

(i) The noise process $z : \Omega \to C([0,a], R^m)$ is expressible in the form

$$z(\omega)(t) = \lambda(t) + z_m(\omega)(t) \quad \forall t \in [0,a], a.a.\omega \in \Omega \tag{4.4.150}$$

where $\lambda : [0,a] \to R^m$ is a Lipschitz function and $z_m : \Omega \to C([0,a], R^m)$ is a martingale adapted to $(\mathcal{F})_{0 \le t \le a}$ and is such that there is a $K > 0$ with

$$|E(z_m(\cdot)(t_2) - z_m(\cdot)(t_1)|\mathcal{F}_{t_1})| \le K(t_2 - t_1) \tag{4.4.151}$$

$$E(|z_m(\cdot)(t_2) - z_m(\cdot)(t_1)|^2 \mathcal{F}_{t_1}) \le K(t_2 - t_1) \tag{4.4.152}$$

a.s. whenever $t_1, t_2 \in [0,a]$ and $t_1 \le t_2$.

(ii) The coefficient process g is continuous and is also uniformly Lipschitz in the second varibale with respect to the first i.e. there exists $L > 0$ such that

$$||g(t, \Psi_1) - g(t, \Psi_2)||_{\mathcal{L}^2} \le L||\Psi_1 - \Psi_2||_{\mathcal{L}^2(\Omega, C)} \tag{4.4.153}$$

for all $t \in [0,a]$, and all $\Phi_1, \Psi_2 \in \mathcal{L}^2(\Omega, C(J, R^n))$.

(iii) For each process $y \in C_A([0,a], \mathcal{L}^2(\Omega, C(J, R^n))$ the process

$$[0,a] \ni t \mapsto g(t, y(t)) \in \mathcal{L}^2(\Omega, L(R^m, R^n))$$

is also adapted to $(\mathcal{F})_{0 \le t \le a}$.

Here is the existence and uniqueness theorem for solutions of the stochastic functional differential equations (4.4.149):

Theorem 4.4.1 *Suppose Conditions (E) are satisfied, and let $\theta \in \mathcal{L}^2(\Omega, C(J, R^n))$ be $\mathcal{F}_0$-measurable. Then the stochastic functional differential equations (4.4.149) has a solution $X \in \mathcal{L}^2(\Omega, C([-r,a], R^n))$ adapted to $(\mathcal{F})_{0 \le t \le a}$ and with initial process θ. Furthermore,*
(i) X is unique up to equivalence (of stochastic processes) among all solutions of (4.4.149) belonging to $\mathcal{L}^2(\Omega, C([-r,a], R^n))$ and adapted to $(\mathcal{F})_{0 \le t \le a}$, i.e. if $\tilde{X} \in \mathcal{L}^2(\Omega, C([-r,a], R^n))$ is a solution of (4.4.149) adapted to $(\mathcal{F})_{0 \le t \le a}$ and with initial process θ, then

$$X(\cdot)(t) = \tilde{X}(\cdot)(t) \;\; \mathcal{F}_t \;\; a.s., \forall t \in [0,a];$$

(ii) the trajectory $[0,a] \ni t \mapsto X_t \in \mathcal{L}^2(\Omega, C(J,R^n))$ *is a* $C(J,R^n)$*-valued process adapted to* $(\mathcal{F})_{0\le t\le a}$ *with a.a. sample paths continuous (It belongs to* $C_A([0,a],\mathcal{L}^2(\Omega,C(J,R^n)))$*).*

Remark *Let* $0 \le t_1 \le t \le a$*. Then one can solve the following stochastic FDE for any process* $\Psi \in \mathcal{L}^2(\Omega, C(J,R^n); \mathcal{F}_{t_1})$ *at time* t_1*:*

$$X(\omega)(t) = \begin{cases} \Psi(\omega)(0) + (\omega)\int_{t_1}^t g(u, X_u)dz(\cdot)(u) & , \quad t_1 \le t \le a \\ \Psi(\omega)(t-t_1) & , \quad t_1 - r \le t \le t_1. \end{cases}$$

where the (unique) solution $X \in \mathcal{L}^2(\Omega, C([t_1 - r, t_1], R^n))$*. This gives a family of maps*

$$T_t^{t_1} : \mathcal{L}^2(\Omega, C(J,R^n); \mathcal{F}_{t_1}) \to \mathcal{L}^2(\Omega, C(J,R^n); \mathcal{F}_t), t \ge t_1$$

with $\Psi \longmapsto X_t$*. When* $t_1 = 0$*, we define* $T_t, t \ge 0$*, to be*

$$T_t = T_t^0 : \mathcal{L}^2(\Omega, C(J,R^n); \mathcal{F}_0) \to \mathcal{L}^2(\Omega, C(J,R^n); \mathcal{F}_t).$$

The following theorem on continuation of trajectories of a stochastic FDE is consequence of Theorem 4.4.1.

Theorem 4.4.2 *Assume Conditions (E) are satisfied. If* $0 \le t_1 \le t_2 \le a$*, then*

$$T_{t_2} = T_{t_2}^{t_1} \circ T_{t_1}.$$

We can prove the following theorems of solutions dependence on the initial process.

Theorem 4.4.3 *Suppose the conditions of Theorem 4.4.1 are satisfied by* g *and* z *in stochastic FDE (4.4.149). Then each map*

$$T_t = T_t^0 : \mathcal{L}^2(\Omega, C(J,R^n); \mathcal{F}_0) \to \mathcal{L}^2(\Omega, C(J,R^n); \mathcal{F}_t), \quad t \in [0,a],$$

is Lipschitz; indeed for all $t \in [0,a], \theta_1, \theta_2 \in \mathcal{L}^2(\Omega, C(J,R^n); \mathcal{F}_0)$

$$||T_t(\theta_1) - T_t(\theta_2)||_{\mathcal{L}^2(\Omega,C)} \le \sqrt{2}||\theta_1 - \theta_2||_{\mathcal{L}^2(\Omega,C)} \exp\{ML^2t\}$$

where L *is the Lipschitz constant of* g *and* M *is a constant which doesn't depend on the coefficient process* g *but only on the noise* z*.*

With suitable Frechet differentiability hypotheses on the coefficient process g, one can prove that T_t is C^1.

Condition (D):

The coefficient process g has continuous partial derivatives with respect to the second variable i.e. the mapping

$$[0,a] \times \mathcal{L}^2(\Omega, C(J,R^n)) \to L(^2(\Omega, C(J,R^n)), L^2(\Omega, L(R^m, R^n)))$$

$$(t, \Psi) \longmapsto D_{(2)}g(t, \Psi)$$

is continuous, where $D_{(2)}g(t,\Psi)$ is the partial derivative of g in the second variable at (t,Ψ).

Theorem 4.4.4 *Suppose the stochastic FDE (4.4.149) satisfies conditions (E) and (D). Then for each* $t \in [0,a]$*,*

$$T_t : \mathcal{L}^2(\Omega, C(J,R^n); \mathcal{F}_0) \to \mathcal{L}^2(\Omega, C(J,R^n); \mathcal{F}_t) \hookrightarrow L^2(\Omega, C(J,R^n); \mathcal{F}_t)$$

is C^1*.*

4.4.2 Markov property

In this subsection we consider the following stochastic functional differential equations

$$\begin{aligned} dX(t) &= H(t, X_t)dt + G(t, X_t)dW(t), t > t_1 \geq 0 \\ X_{t_1} &= \Psi \in C(J, R^n) \end{aligned} \tag{4.4.154}$$

which is the symbolization by differential notation of the following equations

$$X(\omega)(t) = \begin{cases} \Psi(\omega)(0) + \int_0^t H(u, X_u(\omega))du + & \\ (\omega)) \int_{t_1}^t G(u, X_u(\cdot))dW(u), & t_1 \leq t \\ \Psi(\omega)(t - t_1), & t_1 - r \leq t \leq t_1 .a.a.\omega \in \Omega. \end{cases}$$

Compared with SFDE(4.4.149) in the last subsection, here the coefficient process g factors through a drift $H : [0, a] \times C(J, R^n) \to R^n$ and a diffusion $G : [0, a] \times C(J, R^n) \to L(R^m, R^n)$, while the noise process takes the form $\{t + W(t) : t \in [0, a]\}$ with W an m-dimensional Brownian motion on a filtered probability space $(\Omega, \mathcal{F}, (\mathcal{F}_t)_{0 \leq t \leq a}, D)$. In order to establish Markov behavior for the trajectory $\{X_t : t \geq 0\}$ of the Stochastic FDE (4.4.154). We need the following conditions.

Condition (M):

(i) Let $(\mathcal{F}_t)_{0 \leq t \leq a}$ be the Brownian filtration $\mathcal{F}_t = \sigma\{W(u) : 0 \leq u \leq t\}, \mathcal{F}_a = \mathcal{F}$. Assume without loss of any generality that each $\mathcal{F}_t$ contains all P-null sets in $\mathcal{F}$.

(ii) The coefficients H, G are globally Lipschitz, i.e. there is an $L > 0$ such that

$$||G(t, \eta^1) - G(t, \eta^2)|| \leq L||\eta^1 - \eta^2||_C, |H(t, \eta^1) - H(t, \eta^2)| \leq L||\eta^1 - \eta^2||_C,$$

for all $t \in [0, a]$ and all $\eta^1, \eta^2 \in C(J, R^n)$.

(iii) H and G are continuous.

Since the above Condition (M) imply the conditions of existence (E) in last subsection, we know that the stochastic FDE (4.4.154) will have a unique solution. We can obtain the Lipschitz maps

$$T_t^{t_1} : \mathcal{L}^2(\Omega, C(J, R^n); \mathcal{F}_{t_1}) \to \mathcal{L}^2(\Omega, C(J, R^n); \mathcal{F}_t), t \geq t_1,$$

$$\Psi \mapsto X_t$$

When $t_1 = 0$,

$$T_t^0 : \mathcal{L}^2(\Omega, C(J, R^n); \mathcal{F}_0) \to \mathcal{L}^2(\Omega, C(J, R^n); \mathcal{F}_t).$$

Hence the trajectory of the stochastic FDE (4.4.154) can be viewed as a Markov process

$$[0, a] \times \Omega \longrightarrow C(J, R^n)$$

$$(t, \omega) \longmapsto T_t(\Psi)(\omega)$$

taking values in $C(J, R^n)$. We have the theorem on Markov property.

Theorem 4.4.5 *(The Markov Property) Suppose Condition (M) is satisfied by the stochastic FDE (4.4.154). Then its trajectories*

$$\{X_t^\eta : t \in [0, a], \eta \in C(J, R^n)\}$$

describe a Markov process on $C(J,R^n)$ *with transition probabilities* $p(t_1,\eta,t_2,\cdot)$ *given by*

$$p(t_1,\eta,t_2,B) = P\{\omega : \omega \in \Omega, T_{t_2}^{t_1}(\eta)(\omega) \in B\} \tag{4.4.155}$$

for $0 \le t_1 \le t_2 \le a, \eta \in C(J,R^n)$ *and* $B \in Borel - C(J,R^n)$. *Indeed for any* $\theta \in \mathcal{L}^2(\Omega, C(J,R^n); \mathcal{F}_0)$ *the Markov property*

$$P(T_{t_2}(\theta) \in B|\mathcal{F}_{t_1}) = p(t_1, T_{t_1}(\theta)(\cdot), t_2, B) = P(T_{t_2}(\theta) \in B|T_{t_1}(\theta))$$

holds a.s. on Ω.

If we consider the time-homogenious case, i.e., if the coefficients $H : C(J,R^n) \to R^n, G : C(J,R^n) \to L(R^m,R^n)$. In this case, the SFDE(4.4.154) becomes

$$\begin{aligned} dX(t) &= H(X_t)dt + G(X_t)dW(t), t > 0 \\ X_0 &= \theta \in C(J,R^n) \end{aligned} \tag{4.4.156}$$

And the condition (M) becomes

Conditions (A):

(i) For each $t \in [0,a]$, $(\mathcal{F}_t)$ is the σ-algebra generated by $\{W(\cdot)(s) : 0 \le s \le t\}$, together with all sets of P-measure zero in $\mathcal{F} = \mathcal{F}_a$.

(ii) There is an $L > 0$ such that

$$||G(\eta^1) - G(\eta^2)|| \le L||\eta^1 - \eta^2||_C, |H(\eta^1) - H(\eta^2)| \le L||\eta^1 - \eta^2||_C,$$

for all $\eta^1, \eta^2 \in C(J,R^n)$.

We have the following therorem.

Theorem 4.4.6 *(Time-homogeneity) Suppose that the autonomous stochastic FDE (4.4.156) satisfies Conditions (A). For* $0 \le t_1 \le t_2 \le a, \eta \in C(J,R^n)$ *let* $p(t_1,\eta,t_2,\cdot)$ *be the associated transition probabilites of trajectories of (4.4.156). Then the trajectory Markov process* $\{X_t^\eta : 0 \le t \le a, \eta \in C(J,R^n)\}$ *is time-homogeneous, i.e.*

$$p(t_1,\eta,t_2,\cdot) = p(0,\eta,t_2 - t_1,\cdot), 0 \le t_1 \le t_2 \le a, \eta \in C(J,R^n).$$

Denote by C_b *the Banach space of all bounded uniformly continuous functions* $\phi : C(J,R^n) \to R$ *furnished with the supremum norm*

$$||\phi||_{C_b} = \sup\{|\phi(\eta)| : \eta \in C(J,R^n)\}.$$

Let $0 \le t_1 \le t_2 \le a, \eta \in C(J,R^n)$ *and define* $T_{t_2}^{t_1}(\eta) \in \mathcal{L}^2(\Omega, C(J,R^n); \mathcal{F}_{t_2})$ *as above by*

$$\eta \mapsto X_{t_2}$$

the trajectory of the stochastic FDE(4.4.154). For each $\phi \in C_b$ *define the function* $P_{t_2}^{t_1}(\phi) : C(J,R^n) \to R$ *by setting*

$$P_{t_2}^{t_1}(\phi) = E(\phi \circ T_{t_2}^{t_1}(\eta)) = \int_C \phi(\xi)p(t_1,\eta,t_2,d\xi)$$

where $p(t_1,\eta,t_2,\cdot)$ *are the transition probabilities of (4.4.154). Since* ϕ *is bounded, it is clear that* $P_{t_2}^{t_1}(\phi)$ *is also bounded. Furthermore, each* $P_{t_2}^{t_1}(\phi)$ *is uniformly continuous on* $C(J,R^n)$. *We end the subsection with the following theorem.*

Theorem 4.4.7 *For the stochastic functional differential equation (4.4.154) suppose Conditions (M) are satisfied. Then the family* $\{P_{t_2}^{t_1} : 0 \leq t_1 \leq t_2 \leq a\}$ *is a contraction semigroup on* C_b *i.e.*
(i) Each $P_{t_2}^{t_1} : C_b \to C_b$ *is a continous linear operator*

$$\phi \mapsto P_{t_2}^{t_1}(\phi)$$

with $||P_{t_2}^{t_1}|| \leq 1$ *for all* $0 \leq t_1 \leq t_2 \leq a$*;*
(ii) $P_{t_2}^{t_1} \circ P_{t_3}^{t_2} = P_{t_3}^{t_1}, 0 \leq t_1 \leq t_2 \leq t_3 \leq a$.

In particular, for the autonomous stochastic FDE (4.4.156), the family $\{P_t \equiv P_t^0 : 0 \leq t \leq a\}$ *is a one-parameter contraction semigroup on* C_b *i.e.*

$$P_{t_1} \circ P_{t_2} = P_{t_1+t_2} \quad if \quad t_1, t_2, t_1 + t_2 \in [0, a].$$

For topological properties of such semigroups, the reader may refer to [71].

4.4.3 Regularity of the trajectory field

This subsection concerns various regularity properties of the trajectory field $\{X_t^\eta : t \in [0, a], \eta \in C(J, R^n)\}$ generated by the autonomous stochastic FDE:

$$\begin{aligned} dX(t) &= H(X_t)dt + G(X_t)dW(t), 0 < t < a \\ X_0 &= \eta \in C \equiv C(J, R^n) \end{aligned} \tag{4.4.157}$$

with coefficients $H : C \to R^n, G : C \to L(R^m, R^n)$ and m-dimensional Brownian motion W.

The following condition on the diffusion coefficient g will be required in the discussion of the regularity property of the trajectory field.

The Frobenius Condition (F): A map $g : R^n \to L(R^m, R^n)$ is said to be satisfy the Frobenius condition if it is C^1 with $Dg : R^n \to L(R^m, R^n))$ globally bounded, locally Lipschitz and such that

$$\{Dg(x)[g(x)(v_1)]\}(v_2) = \{Dg(x)[g(x)(v_2)]\}(v_1), \forall x \in R^n, v_1, v_2 \in R^m.$$

Consider the stochastic FDE with ordinary diffusion coefficient:

$$\begin{aligned} dX(t) &= H(X_t)dt + g(X(t))dW(t), 0 < t < a \\ X_0 &= \eta \in C \equiv C(J, R^n), J = [-r, 0]. \end{aligned} \tag{4.4.158}$$

The coefficients $H : C \to R^n, g : C \to L(R^m, R^n)$ are Lipschitz maps and W is m-dimensional Brownian motion adapted to the filtered probability space $(\Omega, \mathcal{F}(\mathcal{F}_t)_{0 \leq t \leq a}, P)$. For such a stochastic FDE, we have the following theorem.

Theorem 4.4.8 *In the stochastic FDE (4.4.158), suppose* H *is Lipschitz and* g *is a* C^2 *map satisfying the Frobenius condition. Then the trajectory field* $\{X_t^\eta : t \in [0, a], \eta \in C(J, R^n)\}$ *has a version* $X : \Omega \times [0, a] \times C(J, R^n) \to C(J, R^n)$ *having the following properties. For any* $0 < \alpha < \frac{1}{2}$*, there is a set* $\Omega_\alpha \subset \Omega$ *of full* P*-measure such that, for every* $\omega \in \Omega_\alpha$*,*
(i) the map $X(\omega, \cdot, \cdot) : [0, a] \times C(J, R^n) \to C(J, R^n)$ *is continuous;*
(ii) for every $t \in [r, a]$ *and* $\eta \in C(J, R^n), X(\omega, t, \eta) \in C^\alpha(J, R^n)$*; where* $C^\alpha(J, R^n)$ *is the Banach space of all* α*-Holder continuous paths* $\eta : J \to R^n$ *with the* α*-Holder norm:*

$$||\eta||_{C^\alpha} = ||\eta||_C + \sup\{\frac{|\eta(s_1) - \eta(s_2)|}{|s_1 - s_2|^\alpha} : s_1, s_2 \in J, s \neq s_2\}.$$

(iii) $X(\omega, \cdot, \cdot) : [r, a] \times C(J, R^n) \to C^\alpha(J, R^n)$ *is continuous;*
(iv) for each $t \in [r, a], X(\omega, t, \cdot) : C(J, R^n) \to C^\alpha(J, R^n)$ *is Lipschitz on every bounded set in* $C(J, R^n)$*, with a Lipschitz constant independet of* $t \in [r, a]$*. In particular each* $X(\omega, t, \cdot) : C(J, R^n) \to C(J, R^n)$ *is a compact map.*

But if the diffusion coefficient G in (4.4.157) depends on the past, we have an example in which all versions of the trajectory field are almost surely highly irregular. For the detail, we refer to Mohammed's book [71]. Now we can investigate regularity in probability of the trajectory field for the SFDE(4.4.157).

Theorem 4.4.9 *Let E be a real Banach space and* $y : \Omega \times [0, a] \to E$ *an (*$\mathcal{F} \otimes$ *Borel* $[0, a]$*, Borel E) measurable process with almost all sample paths continuous. Suppose that for each* $t \in [0, a], y(\cdot, t) \in \mathcal{L}^{2k}(\Omega, E; \mathcal{F})$ *in the Bochner sense and there is a number* $c = c(a, k) > 0$ *such that*

$$E|y(\cdot, t_1) - y(\cdot, t_2)|_E^{2k} \leq c|t_1 - t_2|^k$$

for all $t_1, t_2 \in [0, a]$*. Then for every* $0 < \alpha < \frac{1}{2}(1 - \frac{1}{k})$ *and any real* $N > 0$*, one has*

$$\begin{aligned} & P\left(\sup_{t_1 \neq t_2 \in [0,a]} \frac{|y(\cdot, t_1) - y(\cdot, t_2)|_E}{|t_1 - t_2|} \geq \frac{8(k\alpha + 1)}{\alpha} 4^{\frac{1}{2k}} N \right) \\ \leq & \frac{2ca^{k(1-2\alpha)}}{k(1 - 2\alpha)[k(1 - 2\alpha) - 1]} \frac{1}{N^{2k}}. \end{aligned}$$

Theorem 4.4.10 *Suppose* $H : \Omega \times C \to R^n, G : \Omega \times C \to L(R^m, R^n)$ *satisfy the conditions:*
(i) For each $\eta \in C, H(\cdot, \eta)$ *and* $G(\cdot, \eta)$ *are* $\mathcal{F}_0$*-measurable;*
(ii) There is a constant $K > 0$ *such that*

$$|H(\omega, \eta)| leq K(1 + ||\eta||_C), ||G(\omega, \eta)|| \leq K(1 + ||\eta||_C)$$

for a.a. $\omega \in \Omega$*, all* $\eta \in C$*;*
(iii) For each $N > 0$*, there exists* $L_N > 0$ *such that*

$$|H(\omega, \eta_1) - H(\omega, \eta_2)| \leq L_N ||\eta_1 - \eta_2||,$$

and

$$||G(\omega, \eta_1) - G(\omega, \eta_2)|| \leq L_N ||\eta_1 - \eta_2||$$

for a.a. $\omega \in \Omega$*, all* $\eta_1, \eta_2 \in C$ *with* $||\eta_1||_C \leq N, ||\eta_2||_C \leq N$*.*

Then for each $\eta \in C$ *the stochastic FDE*

$$\begin{aligned} dX(t) &= H(\cdot, X_t) + G(\cdot, X_t)dW(t), 0 < t < a \\ X_0 = \eta \in C & \end{aligned} \tag{4.4.159}$$

has a (pathwise) unique solution X^η *in the sense of Theorem ??. Furthermore,* $X^\eta \in \mathcal{L}^{2k}(\Omega, C([-r, a], R^n)), X_t^\eta \in \mathcal{L}^{2k}(\Omega, C)$ *for every integer* $k > 0$*, and each* $t \in [0, a]$*. Indeed there is a constant* $c_k^1 > 0$ *depending only on* K, k, a *such that*

$$E||X_t^\eta||_C^{2k} \leq c_k^1(1 + ||\eta||_C^{2k})$$

for all $\eta \in C, t \in [0, a], k = 1, 2, \cdots$*.*

Theorem 4.4.11 *Let $\eta \in C$ and $0 < \alpha < \frac{1}{2}$. Then the solution X^η of (4.4.157) satisfies*

$$P\{\omega : \omega \in \Omega, X^\eta(\omega)|[0,a] \in C^\alpha([0,a], R^n)\} = 1$$

and

$$P\{\omega : \omega \in \Omega, X_t^\eta(\omega) \in C^\alpha \quad for \quad all \quad r \leq t \leq a\} = 1$$

Corollary If $\eta \in C$, then for any $0 < \alpha < \frac{1}{2}$ the trajectory $\{X_t^\eta : t \in [r,a]\}$ is a process $\Omega \times [r,a] \to C$ with almost all sample paths being α-Holder continuous.

For further results on stochastic functional differential equations, we refer to Mohammed book [71] and Gikhman and Skorohod's book [25].

4.5 Stochastic Differential Equations in Abstract Spaces

In this section we summarize some basic results on stochastic evolution equation on infinite dimensional spaces. For the existence of a regular solution to a class of evolution equations with Lipschitz or locally Lipschitz drift and diffusion coefficients, the dissipative systems and the regular dependence of solutions on initial data are given. For the details of proof and discussion, we refer to Da Prato and Zabczyk [13] and [14].

4.5.1 Stochastic evolution equations

Consider stochastic differential equations of the form:

$$\begin{aligned} dX &= (AX + F(X))dt + B(X)dW(t), \\ X(0) &= \xi, \end{aligned} \tag{4.5.160}$$

where ξ is a random variable on a given pobability space $(\Omega, \mathcal{F}, P)$. $W(t), t \geq 0$ is a cylindrical Weiner process on a Hilbert space U. F and B are nonlinear transformations and A the infinitesimal generator of a strongly continuous semigroup $S(t), t \geq 0$.

We assume that U and H are separable Hilbert spaces, A is the infinitesimal generator of a C_0-semigroup $S(t), t \geq 0$, on H and B is a bounded linear operator from U into H. Let $W(t), t \geq 0$, be a cylindrical Wiener process on U, given by a formal expansion

$$W(t) = \sum_{n=1}^{\infty} \beta_n(t)e_n, t \geq 0$$

where $e_n, n \in N$, is an orthonormal basis on U. ξ is an H-valued $\mathcal{F}_0$-measurable random variable. Denote by $||R||_{HS}$ or $||R||_2$ the Hilbert-Schmidt norm of the opreator $R \in L(U,H)$. The space of all Hilbert-Schmidt operators from U into H (endowed with the Hilbert-Schmidt norm) will be denoted by $L_2(U,H)$. This is again a separable Hilbert space.

Definition 4.5.1 *An H-valued $\mathcal{F}_t$-adapted stochastic process $Z(t), t \geq 0$, is said to be a weak solution to the equation*

$$DZ = AZdt + BdW, Z(0) = \xi, \tag{4.5.161}$$

if for arbitrary $h \in D(A^*)$ and all $t \geq 0, P$-a.s.

$$< h, Z(t) >=< h, \xi > + \int_0^t < A^*h, Z(s) > ds+ < B^*h, W(t) > . \tag{4.5.162}$$

One can show that there exists a solution to (4.5.161) if and only if the operators

$$Q_t = \int_0^t S(t)BB^*S^*(r)dr, \quad t \geq 0$$

are of trace class. In this case the solution is given by the formula

$$Z(t) = S(t)\xi + \int_0^t S(t-s)BdW(s), \quad t \geq 0.$$

Example 4.5.2 *Let* $H = L^2(0,1) = U, B = I$,

$$A = \frac{d^2}{d\xi^2}, D(A) = H^2(0,1) \cap H_0^1(0,1).$$

Then the stochastic convolution

$$Z(t) = W_A(t) = \int_0^t S(t-s)IDW(s), t \geq 0$$

is a weak solution of the equation

$$dZ(t) = AZ(t)dt + dW(t)$$

In order to investigate the existence and uniqueness results on the stochastic evolutioin equations (4.5.160) on a separable Hilbert space H. *We need to make some assumptions.*

Hypothesis 5.1
(i) A is the infinitesimal generator of a strongly continuous semigroup $S(t), t \geq 0$ on H.
(ii) F is a mapping from H into H and there exists a constant $c_0 > 0$ such that

$$|F(x)| \leq c_)(1 + |x|), \quad x \in H, |F(x) - F(y)| \leq c_0|x - y|, \quad x, y \in H.$$

(iii) B is a strongly continuous mapping from H into $L(U; H)$ such that for any $t > 0$ and $x \in H, S(t)B(x)$ belong to $L_2(U; H)$, and there exists a locally square integrable mapping

$$K : [0, +\infty) \to [0, +\infty), \quad t \mapsto K(t)$$

such that

$$||S(t)B(x)||_{HS} \leq K(t)(1 + |x|), t > 0, x \in H,$$

$$||S(t)B(x) - S(t)B(y)||_{HS} \leq K(t)|x - y|, t > 0, x, y \in H.$$

Definition 4.5.3 *An* $\mathcal{F}_t$*-adapted process* $X(t), t \geq 0$, *is said to be a mild solution of (4.5.160) if it satisfies the folowing integral equation,*

$$\begin{aligned} X(t) &= S(t)\xi + \int_0^t S(t-s)F(X(s))ds \\ &+ \int_0^t S(t-s)B(X(s))dW(s), t \in [0, T]. \end{aligned} \tag{4.5.163}$$

Denote by $\mathcal{H}_{p,T}$ the Banach space of all equivalence classes of predictable H-valued processes $Y(t), t \geq 0$, such that

$$||Y||_{p,T} = \sup_{t \in [0,T]} (E|Y(t)|^p)^{1/p} < +\infty.$$

We have the following theorem.

Theorem 4.5.4 *Assume Hypothesis 5.1 and let $p \geq 2$. Then for an arbitrary $\mathcal{F}_0$-measurable initial condition ξ such that $E|\xi|^p < \infty$ there exists a unique mild solution X of (4.5.160) in $\mathcal{H}_{p,T}$ and there exists a constant C_T, independent of ξ, such that*

$$\sup_{t\in[0,T]} E|X(t)|^p \leq C_T(1 + E|\xi|^p).$$

Finally, if there exists $\alpha \in (0, 1/2)$ such that

$$\int_0^1 s^{-2\alpha} K^2(s)ds > +\infty,$$

where K is the function from Hypothesis 5.1-(iii), then the solution $X(\cdot)$ (Denoted by $X(\cdot, \xi)$) is continuous P-a.s.

If we assume that the coefficients F and B of equation (4.5.160) are smooth then the solution to the equation are also smooth in a proper sense. We have the following result.

Theorem 4.5.5 *Assume that the mappings A, F and B satisfy Hypothesis 5.1.*

(i) If F and B have first Frechet derivatives bounded and continuous, then the solution $X(\cdot, x)$ to problem (4.5.160) is continuously differentiable in x as a mapping from H into $\mathcal{H}_{2,T}$. Moreover, for any $h \in H$, the process $\zeta^h(t) = X_x(t, x)h, t \in [0, T]$, is a mild solution of the following equation,

$$d\zeta^h = (A\zeta^h + F_x(X)\cdot\zeta^h)dt + B_x(X)\cdot\zeta^h dW(t), \tag{4.5.164}$$

$$\zeta^h(0) = h. \tag{4.5.165}$$

In additioin there exists a constant $C_{1,T}$, independent of h, such that

$$\sup_{t\in[0,T]} E|X_x(t, x)h|^2 \leq C_{1,T}|h|^2.$$

(ii) Assume in addition that F and B have bounded and continuous Frechet derivatives and that for any $t > 0, x, y, z \in H, S(t)B_{xx}(x)(y, z)$ belongs to $L_2(U; H)$ and there exists a locally square integrable mapping

$$K_1 : [0, +\infty) \to [0, +\infty), t \mapsto K_1(t),$$

such that

$$||S(t)B_{xx}(x)(y, z)u||_{HS} \leq K_1(t)|y||z||u|, \forall x, y, z \in H, u \in U.$$

Then the solution $X(\cdot, x)$ to problem (4.5.160) is twice continuously differentiable, and for any $h, g \in H$, the process $\eta^{h,g}(t) = X_{xx}(t, x)(h, g)$, $t \in [0, T]$, is a mild solution of the following equation,

$$\begin{aligned} d\eta^{h,g} &= (A\eta^{h,g} + F_x(X)\cdot\eta^{h,g})dt + B_x(X)\cdot\eta^{h,g}dW(t) \\ &\quad + F_{xx}(X)\cdot(\zeta^h, \zeta^g))dt + B_{xx}(X)\cdot(\zeta^h, \zeta^g)dW(t), \qquad (4.5.166) \\ \eta^{h,g}(0) &= 0. \end{aligned}$$

Corollary 4.5.6 *Assume that all conditions in Theorem 4.5.5 are satisfied, then for any $x \in H, X(t, x), t \geq 0$, is a Markov process. The corresponding transition semigroup $P_t, t \geq 0$ is defined by*

$$P_t\phi(x) = E[\phi(X(t, x))], x \in H, \phi \in B_b(H).$$

From above theorem we can also get an important result about the Kolmogorov backward equation associated to (4.5.160) with $\xi = x$:

$$\begin{aligned} \frac{\partial}{\partial t}v(t,x) &= \frac{1}{2}Tr[B^*(x)v_{xx}(t,x)B(x)] \\ &\quad + < Ax + F(x), v_x(t,x) >, t > 0, x \in D(A), \\ v(0,x) &= \phi(x); x \in H. \end{aligned} \tag{4.5.167}$$

Definition 4.5.7 *A strict solution of problem (4.5.167) is a continuous function $v : [0, +\infty) \times H \to R$ having continuous first and second partial derivatives with respect to x, such that $v(\cdot, x)$ is continuously differentiable in t for all $x \in D(A)$, and fulfilling equation (4.5.167) for all $x \in D(A)$ and $t \geq 0$.*

We need the following stronger conditions (than Hypothesis 5.1):
Hypothesis 5.2 (i) Hypothesis 5.1 -(i)-(ii) holds.
(ii) B is a mapping from H into $L_2(U; H)$, and there exists a constant $c_1 > 0$ such that

$$||B(x)||_{HS} \leq c_1(1 + |x|), x \in H,$$

and

$$||B(x) - B(y)||_{HS} \leq c_1|x - y|, x, y \in H.$$

Theorem 4.5.8 *Assume that the mappings F and B satisfy Hypothesis 5.2. If in addition the first and the second derivatives of F and B are bounded and continuous and $\phi \in C_b^2(H)$ then equation (4.5.167) has a unique strict solution v and it is given by the formula*

$$v(t,x) = E(\phi(X(t,x))) = P_t\phi(x), t \geq 0, x \in H.$$

4.5.2 Dissipative stochastic systems

In this section we present some methods which imply existence and uniqueness of stochastic equations in Hilbert spaces and Banach spaces.

First let us recall some properties of the subdifferential of the norm on a Banach space E. The subdifferential $\partial||x||$ of $||\cdot||$ at x is defined as follows,

$$\partial||x|| = \{x^* \in E^* : ||x + y|| - ||x|| \geq < y, x^* >, \forall y \in E\},$$

where E^* is the dual of E. One can show that the set $\partial||x||$ is convex, closed, nonempty and given by

$$\partial||x|| = \begin{cases} \{x^* \in E^* : < x, x^* > = ||x||, ||x^*|| = 1\} &, \quad \text{if } x \neq 0, \\ \{x^* \in E^* : ||x^*|| \leq 1\} &, \quad \text{if } x = 0. \end{cases}$$

A mapping $f : D(f) \subset E \to E$ is said to be dissipative if and only if for any $x, y \in D(f)$ there exists $z^* \in \partial||x - y||$ such that

$$< f(x) - f(y), z^* > \leq 0$$

A dissipative mapping f is called m-dissipative if the range of $\lambda I - f$ is the whole space E for some $\lambda > 0$ (and then for any $\lambda > 0$).

Now we can discuss the following problem

$$\begin{aligned} dX &= (AX + F(X))dt + BdW(t), \\ X(0) &= x, \end{aligned} \tag{4.5.168}$$

where A, F satisfy some dissipativity assumptions on appropriate spaces and B is a bounded operator. Let H be a Hilbert space and let K be a reflexive Banach space included in H. We assume that K is a dense Borel subset of H and such that the embedding of K in H is continuous. We need the following conditions on A and F.

Hypothesis 5.3 (i) There exists $\eta \in R$ such that the operator $A - \eta$ and $F - \eta$ are m-dissipative on H.

(ii) The parts on K of $A - \eta$ and $F - \eta$ are m-dissipative on K.

(iii) $D(F) \supset K$ and F maps bounded sets in K into bounded sets of H.

Denote by A_K and F_K the parts of A and F respectively, that is

$$D(A_K) = \{x \in D(A) \cap K : A_K x \in K\}, A_K x = Ax, x \in D(A_K),$$

and

$$D(F_K) = \{x \in D(F) \cap K : F_K x \in K\}, F_K(x) = F(x), x \in D(F_K),$$

$S(t), t \geq 0$, is the semigroup generated by A in H.

Hypothesis 5.4 The process $W_A(t), t \geq 0$, is continuous on H, takes values in the domain $D(F_K)$ of the part of F in K, and for any $T > 0$ we have

$$\sup_{t\in[0,T]} (||W_A(t)||_K + ||F(W_A(t))||_K) < +\infty, P - a.s.$$

Where

$$W_A(t) = \int_0^t S(t-s)BdW(s), t \geq 0$$

is the solution to the linear equation

$$dZ = AZdt + BdW(t), Z(0) = 0.$$

Definition 4.5.9 *An H-continuous, adapted process $X(t), t \geq 0$, is said to be a mild solution to (4.5.168) if it satisfies P-a.s. the integral equation*

$$X(t) = S(t)x + \int_0^t S(t-s)F(X(s))ds + W_A(t), t \geq 0.$$

If, for an H-valued process X, there exists a sequence $\{X_n\}$ of mild solutions of (4.5.168) such that P-a.s., $X_n(\cdot) \to X(\cdot)$ uniformly on any interval $[0, T]$, then X is said to be a generalized solution to (4.5.168).

Theorem 4.5.10 *Assume that Hypothesis 5.3 and 5.4 are fulfilled. Then for arbitrary $x \in K$ there exists a unique mild solution of (4.5.168) and for arbitrary $x \in H$ there exists a unique generalized solution of (4.5.168).*

Remark *The generalized solutions $X(t, x), t \in [0, T], x \in H$, of (4.5.168) are all Markov processes in H with a Feller transition semigroup $P_t, t \geq 0$, given by*

$$P_t\phi(x) = E[\phi(X(t, x))], x \in H, \phi \in B_b(H),$$

where $B_b(H)$ is the set of all bounded Borel functions on H.

Next we consider the problem

$$dX = (AX + F(X))dt + dW(t), \quad (4.5.169)$$
$$X(0) = x \in E,$$

on a Banach space $E \subset H$. We assume

Hypothesis 5.5 (i) $A : D(A) \subset E \to E$ generates a semigroup $S(t), t \geq 0$, on E that is strongly continous in $(0, +\infty)$.

(ii) There exists $\omega \in R$ such that

$$||S(t)|| \leq e^{\omega t}, t \geq 0.$$

(iii) $F : E \to E$ is continuous.

(iv) There exists $\eta \in R$ such that $A + F - \eta$ is dissipative.

(v) $W(\cdot)$ is a cylindrical Wiener process on H such that the stochastic convolution $W_A(t), t \geq 0$, belongs to $C([0, T]; E)$ for arbitrary $T > 0$.

If we define the mild solution of (4.5.169) $X \in C([0, T]; E)$ by

$$X(t) = S(t)x + \int_0^t S(t-s)F(X(s))ds + W_A(t).$$

We have the following theorem.

Theorem 4.5.11 *Assume that Hypothesis 5.5 holds. Then for any $x \in E$ problem (4.5.169) has a unique solution.*

4.6 Anticipating Stochastic Differential Equation

Stochastic calculus have been developed to allow non-adapted, or anticipating integrands, which makes it possible to study various classes of equations where the coefficients and solutions are nonadapted processes. The simplest such equation is the following

$$X_t = X_0 + \int_0^t f(X_s)ds + \int_0^t g(x_s)dB_s$$

where the given initial condition X_0 at time zero is not independent of the driving Brownian motion process $\{B_t\}$. The second type of equation of interest is a stochastic equation with a "boundary condition" of the type $h(X_0, X_1) = \bar{h}$, instead of an initial condition at time zero. The third type of stochastic differential equation with anticipating coefficients is given by a stochastic Voltrrra equation where the coefficents anticipate the driving Brownian motion process. In this section, we will view some basic results on stochastic differential equations with anticipating initial condition and coefficients. For more details and the "boundary condition" problem please refer to Huang [33], [73] and [77].

4.6.1 Volterra equations with anticipating kernel

Let $\Omega = C(R_+; R^k)$, equipped with the topology of uniform convergence on compact subsets of R_+, $\mathcal{F}$ be the Borel σ-field over Ω, P is standard Wiener measure,

$$W_t(\omega) = (W_t^1(\omega), \cdots, W_t^K(\omega))' = \omega(t)$$

If $h \in L^2(R_+)$, we denote by $\delta_j(h)$ the Wiener integral:

$$\delta_i(h) = \int_0^\infty h(t)dW_t^i$$

Let $\mathcal{S}$ denote the dense subset of $L^2(\Omega, \mathcal{F}, P)$ consisting of those random variables F which take the form

$$F = f(\delta_{i_1}(h_1), \cdots, \delta_{i_n}(h_n)) \tag{4.6.170}$$

where $n \in \mathbf{N}, f \in C_b^\infty(R^n), h_1, \cdots, h_n \in L^2(R_+), i_1, \cdots, i_n \in \{1, \cdots, k\}$. If F has the form (4.6.170), we define its derivative in the direction i as the process $\{D_t^i F, t \geq 0\}$ defined by

$$D_t^i F = \sum_{\{l:i_l=i\}} \frac{\partial f}{\partial x_l}(\delta_{i_1}(h_1), \cdots, \delta_{i_n}(h_n))h_l(t)$$

More generally, we define the pth order derivative of F

$$D_{t_1 \cdots t_p}^{i_1 \cdots i_p} F = D_{t_p}^{i_p} \cdots D_{t_1}^{i_1} F$$

DF will stand for the k-dimensional process

$$\{D_t F = (D_t^1 F, \cdots, D_t^k F)' < t \geq 0\}$$

We know that for $i = 1, \cdots, k, D^i$ is an unbounded closable operator from $L^2(\Omega)$ into $L^2(\Omega \times R_+)$. We identify D^i with its closed extension, and denote by $\mathcal{D}_i^{1,2}$ its domain. D^i is a local operator in the sense that if $F \in \mathcal{D}_i^{1,2}$, then $D_t^i F = 0, dP \times dt a.e.$ on $\{F = 0\} \times R_+$.

$\mathcal{D}^{1,2} = \cap_{i=1}^k \mathcal{D}_i^{1,2}$ is the domain of the closed unbounded operator D from $L^2(\Omega)$ into $L^2(\Omega \times R_+; R^k)$. More generally the spaces $\mathcal{D}_i^{1,p}$ and $\mathcal{D}^{1,p} = \cap_{i=1}^k \mathcal{D}_i^{1,p}$ for $p \geq 2$. $\mathcal{D}_i^{1,p}$ is the closure of $\mathcal{S}$ with respect to the norm:

$$||F||_{i,1,p} = ||F||_p + ||||D^i F||_{L^2(R_+)}||_p$$

where $|| \cdot ||_p$ denotes the norm in $L^p(\Omega)$.

Furthermore let $\mathcal{D}_j^{2,p}$ and $\mathcal{D}^{2,p}$ are the closures of $\mathcal{S}$ with respect to respectively the norms:

$$||F||_{j,2,p} = ||F||_p + ||||D^j D^j F||_{L^2(R_+^2)}||_p$$

and

$$||F||_{2,p} = ||F||_p + ||\sum_{i,j=1}^k ||D^i D^j F||_{L^2(R_+^2)}||_p.$$

Define

$$\mathcal{L}_j^{l,p} = L^p(R_+, dt; \mathcal{D}_j^{l,p}), j = 1, \cdots, k; l = 1 \text{ or } 2$$

and

$$\mathcal{L}^{l,p} = L^p(R_+, dt; \mathcal{D}^{l,p}), l = 1 \text{ or } 2.$$

$\mathcal{L}_{i,C}^{1,p}$ will denote the set of those elements u of $\mathcal{L}_i^{1,p}$ which satisfy

(i) For any $T > 0$, the set of functions $\{s \to D_t^i u_s; s \in [0,T]-\{t\}\}_{t\in[0,T]}$ is equicontinuous with values in $L^p(\Omega)$.

(ii) $ess\sup_{(s,t)\in[0,T]^2} E(|D_s^i u_t|^p) < \infty, \forall T > 0$.

Moreover, $\mathcal{L}_C^{1,p} = \cap_{i=1}^k \mathcal{L}_{i,C}^{1,p}$ and $\mathcal{L}_C^{2,p} = \mathcal{L}_C^{1,p} \cap \mathcal{L}^{2,p}$. If $u \in \mathcal{L}_{i,C}^{1,p}$, we define

$$(D_+^i u)_t = L^p(\Omega) - \lim_{\epsilon\downarrow 0} D_t u_{t+\epsilon}$$

$$(D_-^i u)_t = L^p(\Omega) - \lim_{\epsilon\downarrow 0} D_t u_{t-\epsilon}$$

$$(\nabla^i u)_t = (D_+^i u)_t + (D_+^i u)_t$$

$(\nabla u)_t$ will denote the k-dimensional vector $((\nabla^1 u)_t, \cdots, (\nabla^k u)_t)^T$.

Denote by $\mathcal{D}_{loc}^{1,p}$ the set of all random variables F which are such that there exists a sequence $\{(\Omega_n, F_n), n \in \mathbf{N}\} \subset \mathcal{F} \times \mathcal{D}^{1,p}$ with the following two properties $\Omega_n \uparrow \Omega$ a.s., as $n \to \infty$ and for each n, $F = F_n$ a.s. on Ω_n. We then say that the sequence $\{F_n\}$ localizes F in $\mathcal{D}^{1,p}$, and $D_t F$ is defined without ambiguity by

$$D_t F = D_t F_n \ on\ \Omega_n \times R^+, \quad n \in \mathbf{N}$$

$\mathcal{D}_{i,loc}^{1,p}$ is defined analogously. We define $\mathcal{L}_{loc}^{1,p}$ as the set of measurable processes u which are such that for any $T > 0$, there exists a sequence $\{(\Omega_n^T, u_n^T); n \in \mathbf{N}\} \subset \mathcal{F} \times \mathcal{L}^{1,p}$ such that $\Omega_n^T \uparrow \Omega$ a.s. and $u = u_n^T dP \times dt$ a.e. on $\Omega_n^T \times [o,T], n \in \mathbf{N}$. In that case , $\{u_n^T, n \in \mathbf{N}\}$ will be said to localize u in $\mathcal{L}^{1,p}$ on the time interval $[0,T]$. $\mathcal{L}_{i,loc}^{1,p}, \mathcal{L}_{C,loc}^{1,p}$ and $\mathcal{L}_{i,C,loc}^{1,p}$ are defined similarly.

Denote by $\mathcal{L}^{1,loc}$ the set of all measurable processes u such that for any $T > 0$ there exists a sequence $\{\beta_n^T, n \in \mathbf{N}\} \subset \cap_{p\geq 2}\mathcal{D}^{1,p}$ satisfying

(i) $\{\beta_n^T = 1\} \uparrow \Omega$ a.s.,

(ii) $\gamma_T \beta_n^T u \in \cap_{p\geq 2}\mathcal{L}^{1,p}$ for every n,

(iii) $\beta_n^T D.u. \in \cap_{p\geq 2} L^p(\Omega; L^2([0,T]^2))$ for every n,

where $\gamma_T(t) = 1_{[0,T]}(t)$.

$\mathcal{L}_C^{1,loc}$ is defined similarly with $\cap_{p\geq 2}\mathcal{L}^{1,p}$ in (ii) replaced by $\cap_{p\geq 2}\mathcal{L}_C^{1,p}$. The set of sequences $\{\beta_n^T\}_T$ will be called localizer. Note that

$$\mathcal{L}^{1,loc} \subset \mathcal{L}_{loc}^{1,p} \ and\ \mathcal{L}_C^{1,loc} \subset \mathcal{L}_{C,loc}^{1,p}, \forall p \geq 2.$$

Consider the stochastic differential equation in R^d

$$X_t = X_0 + \int_0^t F(t,s,X_s)ds + \sum_{i=1}^k \int_0^t G_i(t,s,X_s)dW_s^i \tag{4.6.171}$$

where the coefficients $F, G_1, \cdots, G_k$ are random functions of (t,s,x). and are $\mathcal{F}_t$ measurable for each t. Unfortunately, we cannot treat such a situation in general. Rather, we shall assume that G_i is of the form

$$G_i(t,s,x) = G_i(H_t,t,s,x)$$

where $G_i(h,t,s,x)$ is $\mathcal{F}_s$ measurable, and $\{H_t\}$ is $\mathcal{F}_t$-progressively measurable. For the sake of simplify the notation, we shall assume from now on that F and G_i do not depend on (t,x,ω), and we consider the Volterra equation of the type

$$X_t = X_0 + \int_0^t F(X_s)ds + \sum_{i=1}^k \int_0^t G_i(H_t,X_s)dW_s^i \tag{4.6.172}$$

where X_t takes values in R^d, and $\{H_t\}$ is a given p-dimensional progressively measurable process. We shall assume that $G_i \in C^{1.0}, 1 \leq i \leq k$, and first postulate the following set of hypotheses.

There exists $q > p$, bounded set $B \subset R^p$ and $K > 0$ s.t.

$$X_0 \in L^q(\Omega, \mathcal{F}_0, P; R^d)$$
$$H_t \in B \; a.s., \forall t \geq 0$$
$$H \in (\mathcal{L}^{1,2})^p; |D_s H_t| \leq K a.s., 0 \leq s \leq t$$
$$F(x) + \sum_{i=1}^{k} |G - i(h,x)| + \sum_{i=1}^{k} |\frac{\partial G_i}{\partial h}(h,x)| \leq K(1 + |x|)$$
$$|F(x) - F(y)| + \sum_{i=1}^{k} |G_i(h,x) - G_i(h,y)|$$
$$+ \sum_{i=1}^{k} |\frac{\partial G_i}{\partial h}(h,x) - \frac{\partial G_i}{\partial h}(h,y)| leq K|x - y|$$

for $0 \leq s \leq t, h \in B, x, y \in R^d$. We have the following theorem.

Theorem 4.6.1 *Under the above conditions, there exists a unique element of* $\cap_{t>0} L^q_{prog}(\Omega \times (0,t))$ *which solves equation (4.6.172).*

Further if we assume that

(i) X_0 is $\mathcal{F}_0$ measurable;

(ii) $H \in (\mathcal{L}^{1,2}_{loc})^p$, is progressively measurable and can be localized in $(\mathcal{L}^{1,2})^p$ by progressively measurable processes;

(iii) $|H_t| + \sum_{i=1}^{k} |D^i_s H_t| \leq U_t a.s., 0 \leq s \leq t$ where U_t is increasing.

(iv) Same growth and Lipschitz conditions as above on F, G_i, and $\frac{\partial G_i}{\partial h}$, but with K replaced by increasing processes $\{V^N_t, t \geq 0\}$, the inequality with V^N_t being satisfied $\forall h \in R^p$ with $|h| \leq N$.

Then equation (4.6.172) has a unique solution which is progressively measurable and belonging a.s. to $\cap_{t>0} L^q(0,t)$.

Moreover if,

(v) $t \to D^i_s H_t$ is a.s. continuous on $[s, +\infty)$;

(vi) $h \to \frac{\partial G_i}{\partial h}(h,x)$ is continuous, $\forall x$;

then the solution $\{X_t\}$ of equation (4.6.172) is a.s. continuous.

Finally, if H_t is a semi-martingale with appropriate properties, and $h \to G_i(h,x)$ is of class $C^2, \forall x$, the second derivative being Lipschitz in h, then the solution $\{X_t\}$ is a semi-martingale.

4.6.2 SDEs with anticipating drift and initial condition

We consider the stochastic differential equation

$$X_t = X_0 + \int_0^t b(X_s)ds + \sum_{i=1}^{k} \int_0^t \sigma^i(X_s) \circ dW^i_s \tag{4.6.173}$$

where

(i) $X_0 \in \cap_{p \geq 2} \mathcal{D}^{1,p}_{loc}$, with $1_{\{|X^i_0| \leq n\}} \sup_{s \leq T} |D_s X^i_0| \in \cap_{p \geq 2} L^p(\Omega)$, $\forall T > 0, n \in \mathbf{N}$ and $1 \leq i \leq d$.

(ii) $b : \Omega \times R^d \to R^d$ is a measurable mapping s.t. $b \in C^2(R^d; R^d)$; $b, b'_{x_1}, \cdots, b_{x_d} \in \mathcal{L}^{1,2}(L^2(R^d;\mu)^d)$ where $\mu = N(0, I)$, and $D_t b, D_t b'_{x_1}, \cdots, D_t b'_{x_d} \in C(R^d; R^d)$, (t, ω) a.e. and moreover

$$\forall \epsilon > 0, \exists C_\epsilon s.t. |b(x)| \leq C_\epsilon(1 + |x|^{1-\epsilon})$$

and $\exists p, C_{p,T}$ such that

$$|D_t b(x)| + |b'_x(x)| + |b''_{xx}(x)| + D_t b'_x(x)| \leq C_{p,T}(1 + |x|^p, \forall t \in [0, T], x \in R^d$$

(iii)$\sigma^i \in C^\infty(R^d, R^d); 1 \leq i \leq k$, all its derivatives are bounded, as well as those of

$$m(x) = \frac{1}{2}\sum_{i=1}^k \frac{\partial \sigma^i}{\partial x}(x)\sigma^i(x)$$

we associate to (4.6.173) with the equation

$$\frac{dY_t}{dt} = (\phi_t^{*-1} b)(Y_t); Y_0 = X_0 \tag{4.6.174}$$

where $\phi_t^{*-1} b(x) = [\frac{\partial \phi_t}{\partial x}(x)]^{-1} b(\phi_t(x))$ and $\{\phi_t(x); t \geq 0\}$ is the flow associated to equation (4.6.173) with $b \equiv 0$, i.e.:

$$\begin{aligned} \phi_t(x) &= x + \sum_{i=1}^k \int_0^t \sigma^i(\phi_s(x)) \circ dW_s^i \\ &= x + \int_0^t m(\phi_s(x))ds + \sum_{i=1}^k \int_0^t \sigma^i(\phi_s(x))dW_s^i \end{aligned}$$

then we have the follwoing theorem.

Theorem 4.6.2 *Under the above assumptions, equation (4.6.174) possesses a unique non exploding solution $\{Y_t, t \geq 0\}$. If $X_t = \phi_t(Y_t), t \geq 0$, then X is the unique a.s. continuous element of $\mathcal{L}_C^{1,p}$ which solves (4.6.173).*

Bibliography

[1] Arnold L., *Stochastic Differential Equations : Theory and Applications*, Krieger Publishing Company, Malabar, Florida 1992.

[2] Bainov D. and Simeonov P., *Integral Inequalities and Applications*, Kluwer Academic Publishers, Dordrecht, Boston, London, 1992.

[3] Bassan B., Some results about stochastic flows with and without jumps, *Translation: Lithuanian Mathematical Journal* 30(1990), No.3, 208-215 (1991).

[4] Bertram L.E. and Sarachik P.E., Stability of circuits with randomly time-varying parameters. *IRE. Trans. Circuit Theory CT-6. Special supplement* 260-270 (1959).

[5] Bismut J.M. Anintroductory approach to duality, in:Optimal Stochastic Control, SIAM Rev., 20(1978), 62—78.

[6] Buckdahn R., Linear Skorohod stochastic differential equations, *Probability Theory and Related Fields,* 90, 223-240 (1991).

[7] Buckdahn R., Skorohod stochastic differential equations of diffusion type, *Probability Theory and Related Fields*, 93,297-323 (1992).

[8] Buckdahn R. and Nualart D., Linear stochastic differential equations and Wick products, *Proba. Theory Relat. Fields* 99, 501-526 (1994).

[9] Carmona R. A. and Nualart D., *Nonlinear Stochastic Integrators, Equations and Flows*, Gordon and Breach Science Publishers (1990).

[10] Chow P.L., Stability of nonlinear stochastic evolution equations, *J. Math. Anal. Appl.* 89(1982), 400-409.

[11] Cox J. Ingersoll J. and Ross S., An extmporal general equilibrium model of asset prices. *Econometrica* 53 353-384 (1985).

[12] Curtain R.F. and Pritchard A.J.,*Infinite Dimensional Linear System Theory*, Lecture Notes in Control and Information Sciences No.8, Springer-Verlag, New York Berlin, 1978.

[13] Da Prato G. and Zabczyk J. *Stochastic Equations in Infinite Dimensions*, Encyclopedia of Mathematics and its Application, Cambridge University Press, 1992.

[14] Da Prato G. and Zabczyk J. *Ergodicity for Infinite Dimensional Systems* London Mathematical Society Lecture Note Series 229, Cambridge University Press, 1996.

[15] Duffie D. and Huang C., Stochastic production-exchange equilibria. *Reseach paper* 974, Graduate School of Business, Stanford University (1989).

[16] Feng Z. S., Liu Y.Q. and Guo F.W., Criteria for practical stability in the p-th mean of nonlinear stochastic systems, *Appl. Math. Comput.* 49(1992), No.2-3, 251-260

[17] Flandoli F. and Schaumloffel K.U., Stochastic parabolic equations in bounded domains: random evolution operator and Lyapunov exponents, *Stochastics and Stochastic Reports.* Vol. 29, (1990) 461-485.

[18] Flandoli F.,*Regularity Theory and Stochastic Flows for Parabolic SPDEs*, Gordon and Breach Science Publishers (1995).

[19] Fujiwara T. and Kunita H., Stochastic differential equations of jump type and Levy processed in diffeomorphisms group,*J. Math. Kyoto Univ. (JMKYAZ)***25-1**, 71-106 (1985).

[20] Gatarek D. and Sobczyk K., On the existence of optimal controls of Hilbert space-valued diffusions. *SIAM J. Control and Optimization.* Vol.32 (1994) No.1, 170-175.

[21] Gibbs J.W., *Elementary principles in statistical mechanics*, New Haven Yale University Press (1902).

[22] Gikhman I.I. and Skorohod A. V., *Introduction to The Theory of Random Processes*, Saunders Company, Philadelphia, London, Toronto, 1969 (translation from Russian).

[23] Gikhman I.I. and Skorohod A. V., *Stochastic Differential Equations.* Springer-Verlag, Berlin, 1972.

[24] Gikhman I.I. and Skorohod A. V., *Controlled Stochastic Processes.* Springer-Verlag, 1979.

[25] Gikhman I.I. and Skorohod A. V., *Stochastic Differential Equations and their Applications* , Kijev, Naukovaja Dumka, 1982

[26] Haussman U.G., Asymptotic stability of the linear Ito equation in infinite dimensions, *J. Math. Anal. Appl.*, 65(1978), 219-135.

[27] Haussman U.G. and Lepeltier J.P., On the existence of optimal controls,*SIAM J. Control and Optimization*, Vol. 28, 4(1990), 851-902.

[28] Hida T. Kuo H.H. Potthoff J. and Streit L., *White Noise,* Kluwer Academic Publishers, (1993) Dodrecht, Boston, London.

[29] Hida T. and Potthoff J., White noise analysis - An overview, in: *White Noise Analysis - Mathematics and Applications*, T. Hida, H.H. Kuo, J. Potthoff and L. Streit (eds.) (1990) 140-165, World Scientific, Singapore.

[30] Hille E. and Phillips, R.S.,*Functional Analysis and Semigroups*, Vol.31, American Mathematical Socity, Colloquium Publication, Providence, R.I., 1957.

[31] Hu Xuanda, *Stability Theory of Stochastic Differential Equations,* Nanjing University Publishing Company, 1986.

[32] Huang F. Characteristic conditions for exponential stability of linear dynamic systems , *Ann.of Differential Equations*, 1(1985), 1:43-56.

[33] Huang Z.Y. On the generalized solutions of stochastic boundary value problems, *Stochastics 11*, 237-248, 1984.

[34] Ichikawa A., Dynamic programming approach to stochastic evolutions, *SIAM J. Control and Optimization*, 17(1979), 152-174

[35] Ichikawa A., Stability of semilinear stochastic evolution equations, *J. Math. Anal. Appl.* 90(1982), 12-44.

[36] Ichikawa A., Equivalence of L_p-stability and exponential stability for a class of nonlinear aemigroups, *Nonlinear Analysis*, 8, 7(1984), 805-817.

[37] Ichikawa A., Semilinear stochastic evolution equations: boundedness, stability and invariant measures, *Stochastics*, 12(1984), 1-39.

[38] Ichikawa A., Filtering and control of stochastic differential equations with unbounded coefficients, *Stochastic Anal. Appl.* 4(1986), 187-212.

[39] Ikeda N. and Watanabe S., *Stochastic Differential Equations and Diffusion Processes*,(Second Edition) North-Holland Publishing Company, Amsterdam. Oxford. New York (1989).

[40] Itô K., *On Stochastic Differential Equations*, Memoirs, Amer. Math. Soc. No.4(1951).

[41] Itô K., Stochastic differential of continuous local quasi-martingales, Lecture Notes in Mathematics 294, *Stability of Stochastic Dynamical Systems,* Springer-Verlag 1972, 1-7.

[42] Kannan D., Tsoi A.H. and Zhang B. Practical stability in pth mean and controlability of Lévy flow, *Communications of Applied Analysis*, 1(1998), 65-80.

[43] Khas'minskii R. Z., *Stochastic Stability of Differential Equations* , Sijthoff and Noordhoff, Alphen aan Rijn, Holland, 1980.

[44] Khas'minskii R. Z. and Mandrekar V., On the stability of stochastic evolution equations , *The Dynkin Frestschrift Markov Processes and their Applications*, Birkhauser Boston Basel Berlin 1994, 185-198.

[45] Kozin F., On almost sure asymptotic sample properties of diffusion processes defined by stochastic differential equations, *J. of Math. of Kyoto Univ.* 4(1965), 515-528.

[46] Kozin F., Stability of the linear stochastic system, Lecture Notes in Mathematics 294, *Stability of Stochastic Dynamical Systems,* Springer-Verlag 1972, 186-229.

[47] Kozin F., Some results on stability of stochastic dynamical systems, Probabilistic Eng. Mech., 1, No.1, March 1986.

[48] Krylov N.V., and Zvonkin A.K., On strong solutions of stochastic differential equations. Sel. Math. Sov. I,19—61.

[49] Krylov N.V., *Introduction to the Theory of Diffusion Processes,* Translations of Mathematical Monographs Vol.142 Amer. Math. Soc. 1995.

[50] Kunita H., On the decomposition of solutions of stochastic differential equations, Proceedings, LMS Durham Symposium, 1980, Lecture Notes in Math 851, 213—255 (1981).

[51] Kunita H., Stochastic partial differential equations connected with non-linear filtering, *Nonlinear filtering and stochastic control (proceedings, Cortona 1982),* Lecture Notes in Math 972 100—169.

[52] Kunita H., *Lectures on Stochastic Flows and Applications,* Springer-Verlag, Berlin Heidelberg New York Tokyo (1986).

[53] Kunita H., Convergerce of stochastic flows with jumps and Lévy processes in diffeomorphisms group, *Ann. Inst. Henri Poincaré,* Vol.22, No.3,1986, 287-321.

[54] Kunita H., *Stochastic Flows and Stochastic differential equations,* Cambridge University Press (1990).

[55] Kuo H.H., Lectures on white noise analysis, *Soochow J. Math.* 18, 229-300.

[56] Kuo H.H. and Potthoff J., Anticipating stochastic integrals and stochastic functional equations; in: *White Noise Analysis - Mathematics and Applications,* T. Hida, H.H. Kuo, J. Potthoff and L. Streit (eds.) (1990) 256-273, World Scientific, Singapore.

[57] Kushner, H.J., *Stochastic Stability and Control,* New York, Academic Press, 1967.

[58] Kushner H.J., Stochastic stability, Lecture Notes in Mathematics 294,*Stability of Stochastic Dynamical Systems,* Springer-Verlag 1972, 97-124

[59] Ladde G.S. and Lakshmikantham V., *Random Differential Inequalities,* Academic Press, New York,1980.

[60] Lakshmikantham V., Leela S., and Martynyuk A.A., *Pactical Stability of Nonlinear Systems,* World Scientific, Singapore,1990.

[61] Lakshmikantham V., Practical stabilization of control systems, *Mathematical Theory of Control, Lecture Notes in Pure and Appl. Math.* 142, Marcel Dekker, Inc. New York, Basel, Hong Kong 1993.

[62] LaSalle J.P. and Lefschetz S., *Stability by Lyapunov's Derict Method with Applications* , Academic Press, New York,1961.

[63] Loeve M. *Probability Theory II*, Springer-Verlag,New York Berlin Hong Kong 1977.

[64] Lyapunov A.M., Probléme général de la stabilité du mouvement, *Annals of Math. Studies* 17, Princeton Univ. Press, Princeton, New Jersey, 1949.

[65] Mao X., *Stability of Stochastic Differential equations With Respect to Semimartingales*, Longman Scientific & Technical, 1991.

[66] Mao X., *Exponential Stability of Stochastic Differential Equations.* Longman Scientific & Technical, 1994.

[67] Maria Jolis and Marta Sanz-solé, Integrator properties of the Skorohold integral, *Stochastics and Stochastics Reports,* Vol.41, 163-176.

[68] Martynyuk A.A., *Practical Stability of Motion (in Russian)* , Naukava Dumka, Kiev, 1983.

[69] Martynyuk A.A. , Methods and Problems of Practical Stability of Motion theory, *Nonlinear Vibr.Problems* 22:1, (1984), 9-46.

[70] McShane,H.p.*Stochastic calculus and stochastic models,* Academic Press, London-New York (1980).

[71] Mohammed,S-E A. *Stochastic functional differential equations,* Pitman Publishing limited, Boston-London-Melbourne (1984).

[72] Nualart D. and Pardoux E., Stochastic calculus with anticipating integrands, *Prob. Theory and Rel. Fields* 78(1988), 535-581.

[73] Ocone D. and Pardoux E., A generalized Itô-Ventzell formula. Application to a class of anticipating stochastic differential equations, *Ann. Inst. Henri Poicaré,* Vol.25, No.1 39-71 (1989).

[74] Oksendal B. *Stochastic Differential Equatioins* An introduction with applications. Springer-Verlag 1995.

[75] Pardoux E., Stochastic partial differential equations and filtering of diffusion processes,*Stochastics,* 3(1979), 127-167.

[76] Pardoux E. and Protter P., Stochastic Volterra equations with anticipating coefficients. *Ann. Proba.* 18(1990), No.4, 1635-1655.

[77] Pardoux E., Applications of anticipating stochastic calculus to stochastic differential equations, *Stochastic Analysis and Related Topics II, Lecture Notes in Math. 1444*, Springer-Verlag 1990, 63-105.

[78] Pazy A., On the applicability of Lyapunov's theorem in Hilbert space, *SIAM J. math. Analysis* 3(1972), 291-294.

[79] Pazy A., *Semigroup of Linear Operators and Application to Partial Differential Equations.* Springer-Verlag, New York, Berlin, Heidelberg, Tokyo, 1983.

[80] Prato G. D. and Ichikawa A., Quadratic control for linear time varying systems, *SIAM J. Control and Optimization Vol* 28 No.2 (1990), 359-381.

[81] Prato G. D. and Zabczyk J., *Stochastic Equations in Infinite Dimensions*, Cambridge University Press, 1992.

[82] Prato G. D. and Tubero L. , *Stochastic Partial Differential Equations and Applications*, Longman Scientific and Technical (1992)

[83] Pritchard A.J. and Zabczyk J., Stability of stabilizability of infinite dimensional systems, *SIAM Review,* 23(1981) 1:25-52.

[84] Rozovskii B.L., *Stochastic Evolution Systems,* Kuwer Academic Publishers, Boston, 1990.

[85] Skorohod A.S., On a generalization of a stochastic integral, *Theory Probab. Appl.* 20(1975), 219-233

[86] Situ R. On solutions of backward stochastic differential equations with jumps and applications, *Stochastic processes and their applications,* 66(1997), No.1, 209—236.

[87] Sobczyk K., *Stochastic Differential Equations*, Kluwer Academic Publishers , Dordrecht, Boston, London.

[88] Stroock D.W. and Varadhan S.R.R., Diffusion processes with continuous coefficients, I, II, Comm. Pure. Appl. Math., 22(1969), 345—400, 479—530.

[89] Stratonovich R.L. Anew representation for stochastic integrals and equations, J. Siam Control 4, 362—371.

[90] Tsoi A. H. and Zhang B., Weak exponential stability of stochastic differential equations. *Stochastic Analysis and Applications*, 4(1997), 643-649.

[91] Tsoi A. H. and Zhang B., Lyapunov functions in weak exponential stability and controlled stochastic systems,*Journal of Ramanujan Math. Society*, 2(1996), 85-102.

[92] Tsoi A. H. and Zhang B., Practical stabilities of Itó type nonlinear stochastic differential systems and related control problems, *Dynamic Systems and Applications*, 1(1997), 107-124.

[93] Yan J.A,Peng S.G.,Fang and Wu, *Selection topics in stochastic analysis*, Ke xue chu ban she, 1997 (in chinese).

[94] Yershov M.P., Localization of conditions on the coefficients of diffusion type equations in existence theorems, Proc. Intern.Symp.SDE Kyoto 1976(ed. by K Ito), 493—507, Kinokuniya, Tokyo, 1978.

[95] Yosida K., *Functional analysis*, Springer-Verlag Berlin, Gottingen, Heidelberg, 1965. Academic Press, New York, 1980.

[96] Zabczyk J., Remarks on the control of discrete-time distributed parameter systems, *SIAM J. Control Optim.* 12 (1974), 721-735.

[97] Zhang B. Stability theory and stochastic control systems, *Bulltin of Hong Kong Math.Soc., 1(1997), 197—202.*

Chapter 5

Numerical Analysis of Stochastic Differential Equations Without Tears

H. SCHURZ[1]
School of Mathematics
Institute of Technology
University of Minnesota
127 Vincent Hall
Minneapolis, Minnesota, MN 55455

5.1 Introduction

Noise plays a significant role in many physical situations, in particular when the corresponding dynamical system (differential or difference equations) undergoes bifurcations (i.e. changes in its qualitative behavior). How the noise occurring in the observed dynamics should best be modeled remains a challenging problem. However, by mathematical tools like rescaling and limit theorems, we know that the Gaussian white noise case plays a central role. It is desirable to study the qualitative behavior of the arising systems of stochastic differential equations (SDEs) as approximations of real natural phenomena. Unfortunately, most of their explicit solutions are not known. Thus, one has to resort to numerical methods. The challenge consists of constructing sequences of approximations by difference equations which replicate the qualitative behavior of the original dynamics of stochastic differential equations. This is where modern numerical analysis starts and where the topic of our survey is placed.

The survey is organized as follows. Section 2 describes the main setting for ordinary stochastic differential equations. Thereafter, in Section 3 we develop the idea of Taylor expansions of their solutions. By truncation of these expansions one systematically arrives at schemes for numerical methods. We present a comprehensive toolbox of numerical schemes by Section 4. In Section 5 the basic concepts of the following presentation are combined by the main principle of numerical analysis: namely consistency (i.e. local approximation), stability (control on global growth behavior of solution), contractivity (control on global error propagation) under geometric invariances (like positivity or $\mathbb{D}$-invariance). How these

[1]Research partially supported by the University of Minnesota, School of Mathematics and IMA in Minneapolis and Weierstrass Institute in Berlin. Current version is from December 15, 1999.

four requirements imply global convergence is shown there. The importance of that principle is manifested by the following sections. Section 6 summarizes the most general convergence results which form, together with the stochastic Taylor expansions, the backbone of any theoretical convergence analysis. We will exhibit pth mean, strong pth mean, double L^p and weak convergence concepts for numerical approximations. In the Sections 7 and 8 we exhibit the issues of numerical stability, stationarity, boundedness and contractivity. The family of stochastic Theta methods is examined in this respect in a fairly thorough presentation. Section 9 discusses some problems related to implementation, simulation, variable step size algorithms, random number generation and illustrative examples. Finally, Section 10 concludes this survey by some final comments, further developments and outlook.

All in all, the results are presented in a general, but not the most general, form. As it is natural for surveys, we shall concentrate on the main ideas rather than on all details, all facets or all cross relations.

5.2 The Standard Setting For (O)SDEs

Assume that the physical process is described by an ordinary stochastic differential system ((O)SDE) with Gaussian white noise, integrated in the sense of Itô (without loss of generality) on a given, fixed, deterministic time-interval $[0, T]$. A system of such stochastic differential equations (SDEs) can be written in terms of differentials as

$$dX_t = a(t, X_t)\,dt + \sum_{j=1}^{m} b^j(t, X_t)\,dW_t^j \tag{5.2.1}$$

where $a, b^j : [0, T] \times \mathbb{R}^d \longrightarrow \mathbb{R}^d$ are the drift and diffusion parts, and $\{W_t^j : 0 \leq t \leq T\}$ represent m mutually independent Wiener processes on the complete probability space $(\Omega, \mathcal{F}, (\mathcal{F}_t)_{t\in[0,T]}, \mathbb{P})$. To ensure the meaningfulness of systems (5.2.1), throughout the survey we impose the uniform Lipschitz-continuity of coefficients a, b^j, i.e.

$$(ULC) \qquad \exists K_L \in \mathbb{R}\; \forall x, y \in \mathbb{R}^d\; \forall t \in [0, T] \quad \|f(t, x) - f(t, y)\| \leq K_L\|x - y\|$$

and the linear-polynomial boundedness on a, b^j, i.e. we have

$$(UBC) \qquad \forall x \in \mathbb{R}^d\; \forall t \in [0, T] \quad \|f(t, x)\| \leq K_B(1 + \|x\|)$$

where function $f : [0, T] \times \mathbb{R}^d \longrightarrow \mathbb{R}^d$ is either a or b^j, and K_L, K_B are appropriate real constants. These requirements together with

$$(IMC) \qquad \mathbb{E}\,\|X_0\|^p < +\infty$$

(for a suitable $p \in \mathbb{R}$, $p \geq 1$), where $\|\cdot\|$ denotes the Euclidean vector norm in $\mathbb{R}^d$, guarantee the existence and uniqueness of (strong) solutions of system (5.2.1) with finite and uniformly bounded absolute moments $\mathbb{E}\,\|X_t\|^p$ for all admissible times t. In fact, one may relax the conditions for existence and uniqueness of solutions to (5.2.1) to (uniform) **one sided Lipschitz continuity** of the coefficient system (a, b^j), i.e.

$$(OLC) \qquad \exists K_{OL} \in \mathbb{R}\; \forall x, y \in \mathbb{R}^d\; \forall t \in [0, T]$$

$$< a(t, x) - a(t, y), x - y >_d + \frac{p-1}{2}\sum_{j=1}^{m}\|b^j(t, x) - b^j(t, y)\|^2 \leq K_{OL}\|x - y\|^2$$

and to (uniform) **one sided linear-polynomial boundedness** of coefficient system (a, b^j), i.e. we have

$$(OBC) \qquad \exists K_{OB}, K_{OB}^I, K_{OB}^H \in \mathbb{R} \; \forall x \in \mathbb{R}^d \; \forall t \in [0,T]$$

$$< a(t,x), x >_d + \frac{p-1}{2} \sum_{j=1}^{m} \|b^j(t,x)\|^2 \; \leq \; K_{OB}^I + K_{OB}^H \|x\|^2 \; \leq \; K_{OB}(1 + \|x\|^2)$$

where $< x, y >_d = \sum_{i=1}^d x_i y_i$ is identified with the d-dimensional Euclidean scalar product of $\mathbb{R}^d$, and $K_{OL}, K_{OB}, K_{OB}^I, K_{OB}^H$ are appropriate real constants throughout this survey. For sake of simplicity of this work, we shall carry out our studies here only for the case of non-time-dependent constants K in the conditions above. However, one may generalize all the presented ideas to the case $K = K(t)$ where $K(t)$ is Lebesgue-integrable (i.e. where $K \in L^1([0,T], \mathcal{B}([0,T]), \mu)$ with σ-algebra $\mathcal{B}([0,T])$ of Borel-measurable subsets of $[0,T]$ and Lebesgue-measure μ).

Throughout our survey let $X_{s,x}(t)$ denote the solution of system (5.2.1), started at value $x \in \mathbb{R}^d$ at time s with $0 \leq s \leq t \leq T$. (Therefore one may identify $X_{0,x}(t) = X_{s,X(s)}(t)$ for all $0 \leq s \leq t \leq T$.). Moreover, any stochastic process X, Y occurring here will be considered and viewed in that form $Y(t) = Y_{s,Y(s)}(t)$ to indicate the functional dependence which will be exploited at several places.

The following lemma performs an important part of the "analytical backbone" for the course of numerical analysis of (O)SDEs, stating under which main conditions and properties of exact solutions of (O)SDEs (5.2.1) all numerical analysis is carried out here.

Lemma 5.2.1 *(Schurz (1996)): Assume $p \geq 2$ and $K_{OB}^I, K_{OB} \geq 0$. Let $X = (X_t)_{0 \leq t \leq T}$ satisfy (5.2.1) on $[0,T]$ with $\mathbb{E}\,\|X_0\|^p < +\infty$.*
If (OLC) is valid, then for all $\mathcal{F}_s$-measurable $X(s), Y(s) \in \mathbb{R}^d$ with the property $\mathbb{E}\,\|X(s)\|^p < +\infty$, $\mathbb{E}\,\|X(s) - Y(s)\|^p < +\infty$ we have

$$\mathbb{E}\,\left\|X_{s,X(s)}(t) - X_{s,Y(s)}(t)\right\|^p \; \leq \; \mathbb{E}\,\|X(s) - Y(s)\|^p \exp\left(pK_{OL}(t-s)\right) \tag{5.2.2}$$

where x, y are independent of all $\mathcal{F}_T^j$, for all $0 \leq s \leq t \leq T$.
If (OBC) is satisfied then $\forall x \in \mathbb{R}^d$

$$\begin{aligned} \mathbb{E}\,\left\|X_{s,X(s)}(t)\right\|^p & \qquad\qquad\qquad\qquad\qquad\qquad\qquad\qquad (5.2.3) \\ \leq \; & \left[\mathbb{E}\,\|X(s)\|^p + 2K_{OB}^I \frac{1 - \exp\left(-[(p-2)K_{OB}^I + pK_{OB}^H](t-s)\right)}{(p-2)K_{OB}^I + pK_{OB}^H}\right] \cdot \\ & \cdot \exp\left([(p-2)K_{OB}^I + pK_{OB}^H](t-s)\right) \\ \leq \; & \left[\mathbb{E}\,\|X(s)\|^p + \frac{1 - \exp(-2(p-1)K_{OB}(t-s))}{p-1}\right] \exp\left(2(p-1)K_{OB}(t-s)\right), \end{aligned}$$

for all $0 \leq s \leq t \leq T$ (If $K_{OB}^I < 0$ or $K_{OB} < 0$, then similar estimates also hold, but only up to the stopping time t^* when $v(t^*) = 0$. From then onwards only the inhomogeneous part of the inequalities contributes to the estimates considered, and hence $v(t) = 0$ for all t with $t^* \leq t \leq T$.).

The proof is an immediate consequence of the constants-of-variation inequality from Schurz (1997). It is worth noting that these estimates are sharp (e.g. take stationary Ornstein Uhlenbeck process and Geometric Brownian Motion). To the author's current knowledge the assumptions (OBC) and (OLC) are the most general ones under which one has carried out rigorous mathematical analysis of numerical methods applied to general classes of $(O)SDEs$ with respect to convergence, stability and perturbation concepts. A similar statement as in Lemma 2.1 can be formulated for the case $1 \leq p < 2$. Due to the sharpness of the estimates obtained in Lemma 2.1, one could not expect more general statements within a consistent numerical L^p approach for (O)SDEs. See forthcoming papers of Schurz (1999). It is interesting as well that under the stronger assumptions (ULC) and (UBC) the estimates of Lemma 5.2.1 simplify to those with constants $K_{OB} = pK_B$ and $K_{OL} = pK_L$. When assumptions (OBC) are not met, explosions in the solutions may occur (numerically confirmed by exploding numerical approximations), whereas when (OLC) is not met, nonuniqueness can lead to serious branching effects of different numerical approximations, since they might follow different solution paths then. Condition (OBC) can be relaxed by Lyapunov-type techniques which we will not follow in this presentation due to lack of space. Of course, the uniformity in estimates above could be relaxed as well towards L^1-integrable kernels $K_{OB}(t), K_{OL}(t)$. These generalizations will not be touched by this survey because of the character imposed naturally on a lesser technical presentation. Roughly speaking, condition (OBC) ensures the control on stability, and (OLC) the control on the propagation of initial errors. Thus, these conditions are very crucial for adequate numerical analysis, cf. the main principle of numerics below.

We assume enough smoothness (e.g. $b^j \in C^{0,1}_{Lip}([0,T] \times \mathbb{R}^d)$). Thus, the restriction to Itô equations (5.2.1) is not so essential at this point, since one may use a well-known transformation formula between different stochastic calculi to convert the results in an equivalent way under some mild smoothness assumptions ($b^j \in C^{0,1}_{Lip}([0,T] \times \mathbb{R}^d)$, see Arnold (1974). However, for practical reasons, such as modeling issues and the implementation of numerical algorithm, it could be important. For the important special case of Stratonovich calculus, see Stratonovich (1966). Note also in the Stratonovich case the additional assumption that the coefficient system $(a + \frac{1}{2}c, b^1, ..., b^m)$, with

$$c(t,x) := \sum_{j=1}^{m} \sum_{k=1}^{d} b^j_k(t,x) \frac{\partial b^j(t,x)}{\partial x_k}$$

satisfies conditions (OBC), and (OLC) is generally needed in order to ensure existence and uniqueness, unless one can apply a stochastic Lyapunov-type technique under local Lipschitz continuities of a, c, b^j. More details can be found in Dynkin (1965), Gikhman and Skorochod (1971), Arnold (1974), Khas'minskiĭ (1980), Gard (1988), Protter (1990), Karatzas and Shreve (1991), Krylov (1995), Øksendal (1998) among many others. We also suppose that X_0 is independent of all natural filtrations $\mathcal{F}^j_t = \{W^j_s : 0 \leq s \leq t\}$. For example, one often assumes that X_0 is deterministic. A system of the form

$$dX_t = a(t, X_t)\, dt + \sum_{j=1}^{m} b^j(t)\, dW^j_t \qquad (5.2.4)$$

is said to be a system with *additive noise*, otherwise one with *multiplicative noise*. It is worth stressing that the stochastic calculi coincide for systems with additive noise. If systems (5.2.1) or (5.2.4) have coefficients a, b^j which do not depend on time t then they are called *autonomous*, otherwise *nonautonomous*. In passing we note that systems (5.2.1) also arise as finite-dimensional approximations of stochastic partial differential equations (SPDEs) in engineering, e.g. after application of method of lines to parabolic SPDEs, or as diffusion limits of stochastic interacting particle systems in Mathematical Biology. Throughout this exposition we presume that the readership is familiar with basic facts on probability theory, stochastic processes and deterministic differential equations.

Now, it is natural to ask for the construction of accurate approximations of systems of (O)SDEs (5.2.1) and their justification. As in deterministic analysis, the main tool for providing them and their local convergence analysis is given by Taylor-type expansions.

5.3 Stochastic Taylor Expansions

Let us sketch the main idea of Taylor expansions. For this purpose, we recite the famous Itô formula in abbreviated operator form, originating from Itô (1951). Define

$$\nabla_x = \left(\frac{\partial}{\partial x_1}, \frac{\partial}{\partial x_2}, ..., \frac{\partial}{\partial x_d} \right)^T$$

as the d-dimensional gradient in the x-direction.

5.3.1 The Itô Formula (Itô's Lemma)

Define linear partial differential operators

$$\mathcal{L}^0 = \frac{\partial}{\partial t} + < a(t,x), \nabla_x >_d + \frac{1}{2} \sum_{j=1}^{m} \sum_{i,k=1}^{d} b_i^j(t,x) b_k^j(t,x) \frac{\partial^2}{\partial x_k \partial x_i}$$

and $\mathcal{L}^j =< b^j(t,x), \nabla_x >_d$ where $j = 1, 2, ..., m$. Then, thanks to the fundamental contribution of Itô (1951), we have the following lemma.

Lemma 5.3.1 *(Stopped Itô Formula in Integral Form). Assume that the given deterministic mapping $V \in C^{1,2}([0,T] \times I\!R^d, I\!R^k)$. Let τ with $0 \le t \le \tau \le T$ be a finite $\mathcal{F}_t$-adapted stopping time.*
Then, we have

$$V(\tau, X_\tau) = V(t, X_t) + \sum_{j=0}^{m} \int_t^\tau \mathcal{L}^j V(s, X_s)\, dW_s^j . \tag{5.3.1}$$

5.3.2 The main idea of stochastic Itô-Taylor expansions

By iterative application of Itô formula we gain the family of stochastic Taylor expansions. This idea is due to Wagner and Platen (1978). Suppose we have enough smoothness of V and of coefficients a, b^j of the Itô SDE. Remember, thanks to Itô's formula, for $t \ge t_0$

$$V(t, X_t) = V(t_0, X_{t_0}) + \int_{t_0}^t \mathcal{L}^0 V(s, X_s) ds + \sum_{j=1}^{m} \int_{t_0}^t \mathcal{L}^j V(s, X_s)\, dW_s^j.$$

Now, take $V(t,x) = x$ at the first step, and set $b^0(t,x) \equiv a(t,x), W_t^0 \equiv t$. Then one derives

$$\begin{aligned} X_t &= X_{t_0} + \int_{t_0}^t a(s, X_s) ds + \sum_{j=1}^{m} \int_{t_0}^t b^j(s, X_s)\, dW_s^j \\ V &\equiv b^j \\ &= X_{t_0} + \int_{t_0}^t \left[a(t_0, X_{t_0}) + \sum_{k=0}^{m} \int_{t_0}^s \mathcal{L}^k a(u, X_u) dW_u^k \right] ds \\ &\quad + \sum_{j=1}^{m} \int_{t_0}^t \left[b^j(t_0, X_{t_0}) + \sum_{k=0}^{m} \int_{t_0}^s \mathcal{L}^k a(u, X_u) dW_u^k \right] dW_s^j \end{aligned}$$

$$b^0 \equiv a$$

$$= \underbrace{X_{t_0} + \sum_{j=0}^{m} b^j(t_0, X_{t_0}) \int_{t_0}^{t} dW_s^j}_{\textbf{Euler Increment}} + \underbrace{\sum_{j,k=0}^{m} \int_{t_0}^{t} \int_{t_0}^{s} \mathcal{L}^k b^j(u, X_u)\, dW_u^k dW_s^j}_{\textbf{Remainder Term R}_E}$$

$$V \equiv \mathcal{L}^k b^j$$

$$= \underbrace{X_{t_0} + \sum_{j=0}^{m} b^j(t_0, X_{t_0}) \int_{t_0}^{t} dW_s^j + \sum_{j,k=1}^{m} \mathcal{L}^k b^j(t_0, X_{t_0}) \int_{t_0}^{t} \int_{t_0}^{s} dW_u^k dW_s^j}_{\textbf{Increment of Mil'shtein Method}}$$

$$\underbrace{+\sum_{j=1}^{m} \int_{t_0}^{t} \int_{t_0}^{s} \mathcal{L}^0 b^j(u, X_u)\, du\, dW_s^j + \sum_{k=1}^{m} \int_{t_0}^{t} \int_{t_0}^{s} \mathcal{L}^k a(u, X_u)\, dW_u^k ds + \sum_{j,k=1,l=0}^{m} \int_{t_0}^{t} \int_{t_0}^{s} \int_{t_0}^{u} \mathcal{L}^l \mathcal{L}^k b^j(z, X_z)\, dW_z^l dW_u^k dW_s^j}_{\textbf{Remainder Term R}_M}$$

$$V \equiv \mathcal{L}^k b^j$$

$$= \underbrace{X_{t_0} + \sum_{j=0}^{m} b^j(t_0, X_{t_0}) \int_{t_0}^{t} dW_s^j + \sum_{j,k=0}^{m} \mathcal{L}^k b^j(t_0, X_{t_0}) \int_{t_0}^{t} \int_{t_0}^{s} dW_u^k dW_s^j}_{\textbf{Increment of 2nd order Taylor Method}}$$

$$+ \underbrace{\sum_{j,k,l=0}^{m} \int_{t_0}^{t} \int_{t_0}^{s} \int_{t_0}^{u} \mathcal{L}^l \mathcal{L}^k b^j(z, X_z)\, dW_z^l dW_u^k dW_s^j}_{\textbf{Remainder Term R}_{TM2}}$$

$$V \equiv \mathcal{L}^r \mathcal{L}^k b^j$$

$$= \underbrace{X_{t_0} + \sum_{j=0}^{m} b^j(t_0, X_{t_0}) \int_{t_0}^{t} dW_s^j + \sum_{j,k=0}^{m} \mathcal{L}^k b^j(t_0, X_{t_0}) \int_{t_0}^{t} \int_{t_0}^{s} dW_u^k dW_s^j}_{\textbf{Increment of 3rd order Taylor Method}}$$

$$\underbrace{+ \sum_{j,k,r=0}^{m} \mathcal{L}^r \mathcal{L}^k b^j(t_0, X_{t_0}) \int_{t_0}^{t} \int_{t_0}^{s} \int_{t_0}^{u} dW_v^r dW_u^k dW_s^j}_{\textbf{Increment of 3rd order Taylor Method}}$$

$$+ \underbrace{\sum_{j,k,r,l=0}^{m} \int_{t_0}^{t} \int_{t_0}^{s} \int_{t_0}^{u} \int_{t_0}^{v} \mathcal{L}^l \mathcal{L}^r \mathcal{L}^k b^j(z, X_z)\, dW_z^l dW_v^r dW_u^k dW_s^j}_{\textbf{Remainder Term R}_{TM3}}$$

....

This process can be continued under appropriate assumptions of smoothness and boundedness of the involved expressions. Thus, this is the place from which all numerical methods systematically originate, and where the main tool for consistency analysis is coming from. One has to expand the functionals in a hierarchical way, otherwise one would lose important order terms, and the implementation would be inefficient. Of course, for qualitative, smoothness and efficiency reasons we do not have to expand all terms in the Taylor expansions at the same time (e.g. cf. Mil'shtein increment versus 2nd order Taylor increments).

The *Taylor method* can be read down straightforwardly by truncation of stochastic Taylor expansion. *Explicit* and *implicit methods*, *Runge-Kutta methods*, *linear-implicit* or *partially implicit* methods are considered as modifications of Taylor methods by substitution of derivatives by corresponding difference quotients, explicit expressions by implicit ones, respectively. However, it necessitates finding a more efficient form for representing stochastic Taylor expansions and hence Taylor methods.

5.3.3 Hierarchical sets, coefficient functions, multiple integrals

Kloeden and Platen (1991) based on Wagner and Platen (1978) introduced a more compact and hence a very efficient formulation of stochastic Taylor expansions. For its statement, we have to formulate what is meant by multiple indices, hierarchical sets, remainder sets, coefficient functions and multiple integrals in the Itô sense.

Definition 5.3.2 *A multiple index has the form* $\alpha = (\alpha_1, \alpha_2, ..., \alpha_{l(\alpha)})$ *where* $l(\alpha) \in I\!N$ *is called the length of the multiple index* α, *and* $n(\alpha)$ *is the total number of zero entries of* α. *The symbol* ν *denotes the empty multiple index with* $l(\nu) = 0$. *The operations* $\alpha- = (\alpha_1, ..., \alpha_{l(\alpha)-1})$ *and* $-\alpha = (\alpha_2, ..., \alpha_{l(\alpha)})$ *are called right- and left-subtraction, respectively (in particular,* $(\alpha_1)- = -(\alpha_1) = \nu$*). The set of all multiple indices is defined to be*

$$\mathcal{M}_{k,m} = \left\{ \alpha = (\alpha_1, \alpha_2, ..., \alpha_{l(\alpha)}) : \alpha_i \in \{k, k+1, ..., m\}, i = 1, 2, ..., l(\alpha), l(\alpha) \in I\!N \right\}.$$

A hierarchical set $Q \subset \mathcal{M}_{0,m}$ *is any set of multiple indices* $\alpha \in \mathcal{M}_{0,m}$ *such that* $\nu \in Q$ *and* $\alpha \in Q$ *implies* $-\alpha \in Q$. *The hierarchical set* Q_k *denotes the set of all multiple indices* $\alpha \in \mathcal{M}_{0,m}$ *with length smaller than* $k \in I\!N$, *i.e.*

$$Q_k = \left\{ \alpha \in \mathcal{M}_{0,m} : l(\alpha) \leq k \right\}.$$

The set

$$R(Q) = \left\{ \alpha \in \mathcal{M}_{0,m} \setminus Q : \alpha- \in Q \right\}$$

is called the remainder set $R(Q)$ *of the hierarchical set* Q. *A multiple (Itô) integral* $I_{\alpha,s,t}[V(.,.)]$ *is defined to be*

$$I_{\alpha,s,t}[V(.,.)] = \begin{cases} \int_s^t I_{-\alpha,s,u}[V(.,.)]\, dW_u^{\alpha_1} & \text{if} \quad l(\alpha) > 1 \\ \int_s^t V(u, X_u)\, dW_u^{\alpha_{l(\alpha)}} & \text{otherwise} \end{cases}$$

for a given process $V(t, X_t)$ *where* $V \in C^{0,0}$ *and fixed* $\alpha \in \mathcal{M}_{0,m} \setminus \{\nu\}$. *A multiple (Itô) coefficient function* $V_\alpha \in C^{0,0}$ *for a given mapping* $V = V(t,x) \in C^{l(\alpha),2l(\alpha)}$ *is defined to be*

$$V_\alpha(t,x) = \begin{cases} \mathcal{L}^{l(\alpha)} V_{\alpha-}(t,x) & \text{if} \quad l(\alpha) > 0 \\ V(t,x) & \text{otherwise} \end{cases}.$$

Similar notions can be introduced with respect to Stratonovich calculus (in fact, in general with respect to any stochastic calculus), see Kloeden and Platen (1991) for Itô and Stratonovich calculus.

5.3.4 A more compact formulation

Now we are able to state a general form of Itô-Taylor expansions. Stochastic Taylor expansions for Itô diffusion processes have been introduced and studied by Wagner and Platen

(1978), Sussmann (1988), Arous (1989), and Hu (1992). Stratonovich Taylor expansions can be found in Kloeden and Platen (1991). We will follow the main idea of Wagner and Platen (1978).

An Itô-Taylor expansion for an Itô SDE (5.2.1) is of the form

$$V(t, X_t) = \sum_{\alpha \in Q} V_\alpha(s, X_s) I_{\alpha,s,t} + \sum_{\alpha \in R(Q)} I_{\alpha,s,t}[V_\alpha(.,.)] \qquad (5.3.2)$$

for a given mapping $V = V(t,x) : [0,T] \times \mathbb{R}^d \longrightarrow \mathbb{R}^k$ which is smooth enough, where $I_{\alpha,s,t}$ without the argument $[\cdot]$ is understood to be $I_{\alpha,s,t} = I_{\alpha,s,t}[1]$. Sometimes this formula is also referred to as *Wagner-Platen expansion.* Now, for completeness, let us restate the Theorem 5.1 of Kloeden and Platen (1991).

Theorem 5.3.3 *(Wagner-Platen Expansion).*
Let ρ and τ be two $\mathcal{F}_t$-adapted stopping times with $t_0 \le \rho \le \tau \le T < +\infty$ (a.s.). Assume $V : [0,T] \times \mathbb{R}^d \longrightarrow \mathbb{R}^k$. Take any hierarchical set $Q \in \mathcal{M}_{0,m}$.
Then, each Itô SDE with coefficients a, b^j possesses a Itô-Taylor expansion (5.3.2) with respect to the hierarchical set Q, provided that all derivatives of V, a, b^j (related to Q) exist.

The proof is carried out in Kloeden and Platen (1991) using the Itô formula and induction on the maximum length $\sup_{\alpha \in Q} l(\alpha) \in \mathbb{N}$. A similar expansion holds for Stratonovich SDEs.

5.3.5 The example of Geometric Brownian Motion

Consider the well-known equation of *Geometric Brownian motion* which is sometimes also called the *lognormal model* in $\mathbb{R}^1$. It is governed by

$$dX_t = a\, X_t\, dt + \sigma\, X_t\, dW_t$$

where a, σ are real constants. Now, let us apply deterministic (since we know an explicit solution expression) and stochastic Taylor expansions (see above) to this equation. This leads to

$$\begin{aligned} X_t &= X_{t_0} \sum_{i,j=0}^{+\infty} \frac{[a - \frac{\sigma^2}{2}(t-t_0)]^i}{i!} \frac{[\sigma(W_t - W_{t_0})]^j}{j!} \\ &= X_{t_0} \exp\left((a - \frac{\sigma^2}{2})(t-t_0) + \sigma(W_t - W_{t_0}) \right) \\ &= X_{t_0} \left(1 + \sum_{\alpha \in \mathcal{M}_{0,1} \setminus \{\nu\}} a^{n(\alpha)} \sigma^{l(\alpha)-n(\alpha)} I_{\alpha,t_0,t} \right) \end{aligned}$$

$$= X_{t_0} \left(1 + a(t-t_0) + \sigma(W_t - W_{t_0}) + \sum_{\alpha \in \mathcal{M}_1 \setminus \{\nu,(0),(1)\}} a^{n(\alpha)} \sigma^{l(\alpha)-n(\alpha)} I_{\alpha,t_0,t} \right)$$

where the coefficient functions are

$$V(t,x) = x, V_\alpha(t,x) = a^{n(\alpha)} \sigma^{l(\alpha)-n(\alpha)} x$$

with $n(\alpha)$ as the total number of zeros of $\alpha \in \mathcal{M}_{0,1}$, and ν as the empty index. Consequently, the stochastic Taylor expansion can generate a kind of *Geometric Wiener Chaos expansion.*

5.3.6 Key relations between multiple integrals

The following lemma connects different multiple integrals. In particular, its formula can be used to express multiple integrals by other ones and to reduce the computational effort of their generation. The following lemma is a slightly generalized version of an auxiliary lemma taken from Kloeden and Platen (1995), see proposition 5.2.3, p. 170.

Lemma 5.3.4 *Let $\alpha = (j_1, j_2, ..., j_{l(\alpha)}) \in \mathcal{M}_{0,m} \setminus \{\nu\}$ with $l(\alpha) \in I\!N$. Then, $\forall k \in \{0, 1, ..., m\}$ $\forall t, s : 0 \leq s \leq t \leq T$ we have*

$$(W_t^k - W_s^k) I_{\alpha,s,t} = \sum_{i=0}^{l(\alpha)} I_{(j_1,j_2,...,j_i,k,j_{i+1},...,j_{l(\alpha)}),s,t} + \sum_{i=0}^{l(\alpha)} \chi_{\{j_i=k\neq 0\}} I_{(j_1,j_2,...,j_{i-1},0,j_{i+1},...,j_{l(\alpha)}),s,t} \tag{5.3.3}$$

$$= I_{(k,j_1,j_2,...,j_{l(\alpha)}),s,t} + I_{(j_1,k,j_2,...,j_{l(\alpha)}),s,t} + I_{(j_1,j_2,k,j_3,...,j_{l(\alpha)}),s,t} + ... + I_{(j_1,j_2,j_3,...,j_{l(\alpha)},k),s,t} + \sum_{i=0}^{l(\alpha)} \chi_{\{j_i=k\neq 0\}} I_{(j_1,j_2,...,j_{i-1},0,j_{i+1},...,j_{l(\alpha)}),s,t}$$

where $\chi_{\{.\}}$ denotes the characteristic function of the subscribed set. Hence, it obviously suffices to generate basis sets of multiple integrals. See also Gaines and Lyons (1994) in respect to minimal sets of multiple Stratonovich integrals which need to be generated. In order to get a more complete picture on the structure of multiple integrals, we note the following assertion.

Lemma 5.3.5 *(Hermite Polynomial Recursion Formula). Suppose that the multiple index $\alpha = (j_1, j_2, ..., j_{l(\alpha)}) \in \mathcal{M}_{0,m}$ with $j_1 = j_2 = ... = j_{l(\alpha)} = j \in 0, 1, ..., m$ and $l(\alpha) \geq 2$. Then, for all t with $t \geq s \geq 0$ we have*

$$I_{\alpha,s,t} = \begin{cases} \dfrac{(t-s)^{l(\alpha)}}{l(\alpha)!}, & j = 0 \\ \dfrac{(W_t^j - W_s^j) I_{\alpha-,s,t} - (t-s) I_{(\alpha-)-,s,t}}{l(\alpha)!}, & j \geq 1 \end{cases} \tag{5.3.4}$$

This lemma corresponds to a slightly generalized version of Corollary 5.2.4 (p.171) of Kloeden and Platen (1995). It is also interesting to note that this recursion formula for multiple Itô integrals coincides with the recursion formula for hermite polynomials. Let us conclude with a list of relations between multiple integrals which exhibit some consequences of Lemmas 3.2 and 3.3. For more details, see Kloeden and Platen (1995). Take $j, k \in \{0, 1, ..., m\}$

and $0 \le s \le t \le T$.

$$
\begin{aligned}
I_{(j),s,t} &= W_t^j - W_s^j \\
I_{(j,j),s,t} &= \frac{1}{2!}\left(I^2_{(j),s,t} - (t-s)\right) \\
I_{(j,j,j),s,t} &= \frac{1}{3!}\left(I^3_{(j),s,t} - 3(t-s)I_{(j),s,t}\right) \\
I_{(j,j,j,j),s,t} &= \frac{1}{4!}\left(I^4_{(j),s,t} - 6(t-s)I^2_{(j),s,t} + 3(t-s)^2\right) \\
I_{(j,j,j,j,j),s,t} &= \frac{1}{5!}\left(I^5_{(j),s,t} - 10(t-s)I^3_{(j),s,t} + 15(t-s)^2 I_{(j),s,t}\right) \\
& \quad \ldots\ldots\ldots \\
(t-s)I_{(j),s,t} &= I_{(j,0),s,t} + I_{(0,j),s,t} \\
(t-s)I_{(j,k),s,t} &= I_{(j,k,0),s,t} + I_{(j,0,k),s,t} + I_{(0,j,k),s,t} \\
I_{(j),s,t}I_{(0,j),s,t} &= 2I_{(0,j,j),s,t} + I_{(j,0,j),s,t} + I_{(j,j,0),s,t} \\
I_{(j),s,t}I_{(j,0),s,t} &= I_{(0,j,j),s,t} + I_{(j,0,j),s,t} + 2I_{(j,j,0),s,t} \\
& \quad \ldots\ldots\ldots\ldots
\end{aligned}
$$

The efficient approximation of multiple stochastic integrals still remains a challenge to be tackled. First approaches in this respect are found in Kloeden, Platen and Wright (1992) using the Karhunen-Loeve expansion (Fourier series of the Wiener process) for Itô and Stratonovich integrals, and Lyons and Gaines (1994) using Box counting methods to treat Stratonovich integrals by looking at Levy areas. In particular, Gaines (1994, 1995) has analyzed the algebra of iterated integrals and could establish some basis sets of intergals which need to be generated to approximate the entire set of multiple integrals.

5.4 A Toolbox of Numerical Methods

By truncation of Taylor expansions and locally implicit or explicit substitutions of the results of differential operators for the coefficient functions, one arrives at an infinite set of possibilities to form stochastic approximation techniques. We will exhibit just a few of them. In the following, and later, let $(Y_n)_{n\in\mathbb{N}}$ denote the sequence of approximation values for the solution at time t_n along the *time-discretization*

$$0 = t_0 \le t_1 \le t_2 \le \ldots \le t_{n_T} = T$$

(for simplicity, we suppose that $t_0 = 0$ and $t_{n_T} = T$). The time-discretization is said to be *equidistant* if there is a number $\Delta \in \mathbb{R}_+$ (called the *step size*) such that $\Delta = t_{i+1} - t_i$ for all $i = 0, 1, \ldots, n_T - 1$. In general, we define

$$\Delta = \max_{i=0,1,\ldots,n_T-1} |t_{i+1} - t_i|$$

as the *step size*, and $\Delta_i = t_{i+1} - t_i$ as the *local step size*. Consider $\Delta W_n^j = W^j_{t_{n+1}} - W^j_{t_n}$ as the current increment of the Wiener process component W^j.

5.4.1 The explicit and fully drift-implicit Euler method

The most well-known numerical method is given by the explicit *Euler method*. It was firstly studied by Maruyama (1955). That is why it is sometimes called *Euler-Maruyama method*.

The scheme of the explicit Euler method is defined by

$$Y_{n+1} = Y_n + a(t_n, Y_n)\,\Delta_n + \sum_{j=1}^{m} b^j(t_n, Y_n)\,\Delta W_n^j\,. \tag{5.4.1}$$

Its convergence has been proved by Gikhman and Skorochod (1971). It represents the most-studied, best-understood and simplest-implementable numerical method. Nowadays, it is even used to understand existence and uniqueness proofs of solutions of SDEs, see Krylov (1990, 1995). A drawback of method (5.4.1) can be seen in the lack of numerical stability (in fact "substable" behavior), the low convergence order, incorrect stationary laws and some problems with the geometrical invariance properties (e.g. nonsimplectic integrator). Despite these facts it is a very popular and very easily implemented, hence practical, method. It is natural to ask for a counterpart to the deterministic *implicit Euler method.* Its *drift-implicit* scheme is given by

$$Y_{n+1} = Y_n + a(t_{n+1}, Y_{n+1})\,\Delta_n + \sum_{j=1}^{m} b^j(t_n, Y_n)\,\Delta W_n^j\,. \tag{5.4.2}$$

The use of the drift-implicit Euler method can be seen to control numerical stability of certain moments, boundary value replication and to reduce variance effects. However, there are the drawbacks of superstability, asymptotic nonexactness of stationary laws to be replicated, and more computational effort due to additional practical implementation of resolution algorithms of nonlinear algebraic equations.

5.4.2 The family of stochastic Theta methods

A first natural generalization of explicit and implicit Euler methods is presented by *stochastic Theta methods.* They are convex linear combinations of explicit and implicit Euler imcrement functions of the drift part, whereas the diffusion part is explicitly treated due to the problem of adequate integration within one and the same stochastic calculus. The scheme of a stochastic Theta method is written as

$$Y_{n+1} = Y_n + (\Theta_n a(t_{n+1}, Y_{n+1}) + (I - \Theta_n) a(t_n, Y_n))\,\Delta_n + \sum_{j=1}^{m} b^j(t_n, Y_n)\,\Delta W_n^j, \tag{5.4.3}$$

where I represents the $d \times d$ real unit matrix, and Θ_n is a uniformly bounded parameter matrix in $\mathbb{R}^{d\times d}$, which is also called the *matrix of implicitness parameters.* This family has been introduced by Ryashko and Schurz (1997) as a generalization of deterministic Theta methods. If $\theta = 0$ then its scheme reduces to classical (forward) *Euler method,* if $\theta = 1$ to the *backward Euler* or often called *implicit Euler method,* and if $\theta = 0.5$ to the *implicit trapezoidal Method.* Originally they were invented by Talay (1982), who proposed $\Theta_n = \theta_0 I$ with scalar $\theta_0 \in [0, 1]$. A study of the qualitative behavior of these methods can be found in Stewart and Peplow (1991) in deterministics, and in Schurz (1997) in stochastics. Another generalization is given by the *drift explicit-implicit Euler method* following

$$Y_{n+1} = Y_n + a\,(t_n + \theta_n \Delta_n, \Theta_n Y_{n+1} + (I - \Theta_n) Y_n)\,\Delta_n + \sum_{j=1}^{m} b^j(t_n, Y_n)\,\Delta W_n^j \tag{5.4.4}$$

where $\theta_n \in \mathbb{R}$, $\Theta_n \in \mathbb{R}^{d\times d}$ such that local algebraic resolution can be guaranteed.

5.4.3 Trapezoidal and midpoint methods

For the integration of conservation laws and Hamiltonian systems, it is recommended to take derivates of the *implicit midpoint method*

$$Y_{n+1} = Y_n + a\left(\frac{t_{n+1}+t_n}{2}, \frac{Y_{n+1}+Y_n}{2}\right) \Delta_n + \sum_{j=1}^{m} b^j(t_n, Y_n)\, \Delta W_n^j. \tag{5.4.5}$$

This method seems to be very promising for the control of numerical stability, area-preservation and boundary laws in stochastics as well. The drawback can be the local resolution of nonlinear algebraic equations, which can be circumvented by predictor-corrector methods (PCMs), see below. A natural extension of trapezoidal integration techniques is represented by the *implicit trapezoidal method* governed by

$$Y_{n+1} = Y_n + \frac{1}{2}\left(a(t_{n+1}, Y_{n+1}) + a(t_n, Y_n)\right) \Delta_n + \sum_{j=1}^{m} b^j(t_n, Y_n)\, \Delta W_n^j. \tag{5.4.6}$$

Both the trapezoidal and midpoint method have an improved local mean consistency behavior (they are of mean convergence order 2, locally considered of mean order 3, under enough smoothness of $a \in C_b^{3,3}([0,T] \times \mathbb{R}^d)$), compared to the explicit and implicit Euler methods. The trapezoidal method has problems when one integrates high-dimensional systems with boundary conditions, as reported by numerous deterministic numerical analysts. However, it is the only numerical method from the class of Theta methods with $\Theta_n = \theta I, \theta \in \mathbb{R}^1$ which asymptotically integrates linear stochastic systems without bias (i.e. asymptotically exact method with respect to stationary laws), see below or Schurz (1996, 1997, 1999).

5.4.4 Rosenbrock methods (RTMs)

In the methods before it appears that one needs algebraic resolution of implicit equations at each integration step. This can be circumvented by the use of *linear-implicit* methods. A specific form of a linear-implicit method which exploits the information of the Jacobian matrices is given by *stochastic Rosenbrock methods.* The idea of linear implicitness traces back to Rosenbrock's fundamental work in deterministic numerical analysis and the idea to treat at least linear systems asymptotically more adequately. In stochastic analysis these methods have been studied in particular by the school of Artemiev, see Averina and Artemiev (1997) for a more detailed overview. An *r-stage Rosenbrock method* (RTM) can be written as

$$Y_{n+1} = Y_n + \sum_{i=1}^{r}\sum_{j=0}^{m} p_i^j k_i^j \tag{5.4.7}$$

$$k_i^j = \Delta W_n^j \left(I - \Delta_n c_j \frac{\partial a}{\partial y}(t_n, Y_n)\right)^{-1} b^j \left(t_n + \nu_i^j \Delta_n, Y_n + \sum_{l=1}^{i-1} \beta_{i,l}^j k_l^j\right)$$

$$\sum_{i=1}^{r} p_i^j = 1, \nu_i^j \in [-1,1], \beta_{1,0}^j = 0, j = 0,1,...,m, i = 1,2,...,r$$

where c_j, $\beta_{i,l}^j$ are appropriate real constants, described by $m+1$ Butcher tableaus. If

$$\max\left(\|b^j(t,y)\|, \sup_{0 \le t \le T, y \in \mathbb{R}^d} \left\| \frac{\partial a}{\partial x}(t,y) \right\|_{d \times d} \right) \le K_J < +\infty$$

with the Euclidean vector norm and a compatible matrix norm $\|.\|_{d\times d}$, the Jacobian is uniform Lipschitz continuous and some natural step size restriction

$$\Delta_n \le \min\{\frac{1}{2c_j K_J} : j = 0, 1, ..., m\}$$

is satisfied, one can show mean square convergence, depending on the choice of k_i^j and β_i^j. For example, a converging two stage RTM (i.e. $r = 2$) is given by

$$2c_j p_1^j + 2(c_j + \beta_{2,1}^j)p_2^j = 1.$$

The big advantage of these methods can be seen in the significant improvement of the linear stability behavior and better integration of linear systems of SDEs. They are also quite useful in certain nonlinear situations, when the linear part controls the behavior of underlying nonlinear dynamic. These methods are preferable when one is integrating in the moment sense, and where the deterministic part plays the most significant role in the course of dynamics. Their drawback is apparent with the additional computation of Jacobian matrices (sometimes even at each step) and algebraic resolution of high-dimensional systems. These methods do not incorporate the stochastic pathwise influence of random integration (not appropriate for the computation of almost sure characteristics like almost sure Lyapunov exponents!).

5.4.5 Balanced implicit methods (BIMs)

For the control on almost sure path-behavior, for the incremental growth and error propagation, Mil'shtein, Platen and Schurz (1998) have introduced the class of *Balanced implicit methods* determined by

$$Y_{n+1} = Y_n + a(t_n, Y_n)\,\Delta_n + \sum_{j=1}^{m} b^j(t_n, Y_n)\,\Delta W_n^j + \sum_{j=0}^{m} c^j(t_n, Y_n)(Y_{n+1} - Y_n)|\Delta W_n^j| \tag{5.4.8}$$

with appropriate weight matrices $C^j(t, x)$ such that the inverse of $d \times d$ matrix.

$$M(t, x) = I + \sum_{j=0}^{m} \theta_j C^j(t, x)$$

exists and is uniformly bounded for all values $\theta_j \in \mathbb{R}_+^1$, $0 \le \theta_0 \le \hat{\theta}_0 < +\infty$ and $(t, x) \in [0, T] \times \mathbb{R}^d$. This class has been studied in Schurz (1996, 1997) and Fischer and Platen (1999). It represents a linear-implicit integration technique, and hence local resolution can be guaranteed and made very simple as well. However, the choice of the matrix weights $C^j(t, x)$ is still a challenge for future research and exhibits a very problematic and practically oriented question (basically C^j has to be chosen according to the desired qualitative properties of discussed discretization, and thanks to Schurz (1996, 1997, 1999) it is proved that the coefficients C^j with $j = 1, 2, ..., m$ are not really needed to have asymptotically exact control on the first and second moments of approximation Y. However, these coefficients are needed in context of pathwise control, see Schurz (1996, 1997, 1999)).

5.4.6 Predictor-corrector methods (PCMs)

A simple, but computationally efficient, idea to circumvent the computational problem of implicit algebraic equations is provided by the predictor-corrector techniques. The predictor scheme is used to forecast the future solution value and plugged into the corrector scheme to have the final approximation values. This procedure leads to an improvement of the numerical stability behavior almost to that of fully implicit schemes, but without the trouble of solving implicit equations at each integration step. Let us manifest this by the example of explicit midpoint and explicit trapezoidal methods introduced and tested by Peterson (1994) in stochastic numerics. The *explicit midpoint method* satisfies

$$Y_{n+1} = Y_n + a\left(\frac{t_{n+1}+t_n}{2}, \frac{Y_{n+1}^E + Y_n}{2}\right)\Delta_n + \sum_{j=1}^{m} b^j(t_n, Y_n)\,\Delta W_n^j \tag{5.4.9}$$

using the explicit Euler method Y_{n+1}^E as its predictor. The *explicit trapezoidal method* is governed by the scheme

$$Y_{n+1} = Y_n + \frac{1}{2}\left(a(t_{n+1}, Y_{n+1}^E) + a(t_n, Y_n)\right)\Delta_n + \sum_{j=1}^{m} b^j(t_n, Y_n)\,\Delta W_n^j \tag{5.4.10}$$

using the explicit Euler method Y_{n+1}^E as its predictor scheme. More generally, one could think of *explicit Theta methods* following

$$Y_{n+1} = Y_n + \left(\Theta_n a(t_{n+1}, Y_{n+1}^E) + (I - \Theta_n) a(t_n, Y_n)\right)\Delta_n + \sum_{j=1}^{m} b^j(t_n, Y_n)\,\Delta W_n^j, \tag{5.4.11}$$

where the parameter matrix Θ_n is as in the Theta methods before. Of course, more complicated predictor-corrector methods can be constructed from Taylor or Runge-Kutta methods through the substitution of implicit expressions by predicting values of other schemes as well. However, care needs to be taken to do it in an efficient way (maximum convergence order should be kept along with substantial improvements of qualitative properties). As this procedure would sprinkle our brief survey goal, it is left to the taste of the readership. The art of appropriate combinations heavily depends on the qualitative goal what one wants to achieve by these new "hybrid" methods.

5.4.7 Explicit Runge-Kutta methods (RKMs)

Stochastic Runge-Kutta methods have been studied by many authors, for example Rümelin (1982), Burrage and Platen (1994), Averina and Artemiev (1997), or more recently by Burrage and Burrage (1996, 1997, 1998). Let us follow the presentation of Burrage and Burrage. They devote their studies to Stratonovich equations

$$dX_t = a(X_t)\,dt + \sum_{j=1}^{m} b^j(X_t) \circ dW_t^j \tag{5.4.12}$$

since the nature of Stratonovich integration appears to be closer to the deterministic case, and in fact the Stratonovich-Taylor expansions exhibit slightly simpler structures. An r-stage Runge-Kutta method is given by

$$\begin{aligned}
Z^j &= J_{(j),t_n,t_{n+1}} C^j,\ z^j = J_{(j),t_n,t_{n+1}} \gamma^j, j = 0, 1, ..., m \\
k_i &= Y_n + \sum_{j=0}^{m}\sum_{l=1}^{i-1} Z_{i,l}^j b^j(k_{i-1}), i = 1, 2, ..., r \\
Y_{n+1} &= Y_n + \sum_{j=0}^{m}\sum_{l=1}^{r} z_l^j b^j(Y_l)
\end{aligned} \tag{5.4.13}$$

where $J_{(j),t_n,t_{n+1}} = \Delta W_n^j$ and C^j represent suitable $r \times r$ real matrices and γ^j appropriate r-dimensional real vectors. Rümelin (1982) has shown the order restriction of these methods. The maximum attainable strong and mean square convergence orders are 0.5 for the entire class of multidimensional (i.e. $d > 1$) Stratonovich SDEs (5.4.12) with noncommutative noise even under C^∞ smoothness of a, b^j, c. The situation dramatically changes when commutative noise is met. Then one may attain any order of convergence under C^∞ conditions, as it is with the case $d = 1$. For systems with noncommutative noise, the meaningfulness of methods (5.4.13) is still questioned, since we may obtain the same order of convergence by much simpler numerical methods. All in all, by Burrage and Burrage (1998), it is clear that classical deterministic Runge-Kutta techniques using only multiple products of $J_{(j)}q$ with different components j do not really help to increase the order of convergence. From Kloeden, Platen and Wright (1992) it becomes apparent that new random variables are needed to increase the order of convergence – a fact originating from the series expansions of stochastic processes. In Burrage, Burrage and Belward (1997) it is pointed out that, if one incorporates all multiple Stratonovich integrals up to order $p \in N$, then the order of strong convergence cannot exceed $\min\{\frac{p+1}{2}, \frac{r-1}{2}\}$ when $p \geq 2$, $r \geq 3$ (and 1 when $p = 1$) for an r-stage stochastic Runge-Kutta method. For more details on maximum attainable order bounds, see Clark and Cameron (1980), Rümelin (1982), Burrage and Burrage (1997, 1998), Schurz (1999) and Roman (2000).

5.4.8 Newton's method

A very important task consists of minimizing the leading error coefficients of numerical approximations. For this purpose, N. Newton (1986, 1991) has introduced the concept of asymptotically efficient, $\mathcal{F}_T^N$-measurable numerical methods. Consider

$$\mathcal{F}_T^N = \sigma\Big\{W_{t_n}^j : n = 0, 1, ..., N; j = 1, 2, ..., m\Big\} \tag{5.4.14}$$

along a discretization $0 = t_0 \leq t_1 \leq ... \leq t_N = T$ for a fixed time interval $[0, T]$.

Definition 5.4.1 *An $\mathcal{F}_T^N$-measurable numerical method $(Y_n)_{n\in\{0,1,...,N\}}$ is said to be pth mean asymptotically efficient iff either $\mathbb{E}\ [\|X_T - Y_N\|^p|\mathcal{F}_T^N] = 0$ or, for any other $\mathcal{F}_T^N$-measurable numerical approximation $(\hat{Y}_n)_{n\in\{0,1,...,N\}}$, we have*

$$\liminf_{N\to+\infty} \frac{\mathbb{E}\left[\|X_T - \hat{Y}_N\|^p \Big| \mathcal{F}_T^N\right]}{\mathbb{E}\left[\|X_T - Y_N\|^p \Big| \mathcal{F}_T^N\right]} \geq 1\,. \tag{5.4.15}$$

It is clear, in the case $p = 2$ – the mean square case, the "best" approximation is achieved by the conditional expectation $\mathbb{E}\ [X_T|\mathcal{F}_T^N]$ which has the minimum mean square error distance to the exact solution X. However, it is very hard to compute that expression analytically. Newton (1991) has given a partial answer of how to construct asymptotically efficient $\mathcal{F}_T^N$-measurable methods for both autonomous Itô and autnomous Stratonovich SDEs with one-dimensional Wiener process $(W_t)_{0\leq t\leq T}$ as driving noise (i.e. when $m = 1$). Set $X_0 = Y_0$. The *efficient Euler-Runge-Kutta method* for Itô SDEs with $m = 1$ follows the scheme

$$
\begin{aligned}
k_0^0 &= a(Y_n),\ k_0^1 = b(Y_n) \\
k_1^0 &= a(Y_n + k_0^0\Delta_n + k_0^1\Delta W_n),\ k_1^1 = b(Y_n - \frac{2}{3}k_0^0(\Delta W_n + \sqrt{3\Delta_n})) \\
k_2^1 &= b(Y_n + \frac{2}{9}k_0^0(3\Delta W_n + \sqrt{3\Delta_n})) \\
k_3^1 &= b(Y_n - \frac{20}{27}k_0^0\Delta_n + \frac{10}{27}(k_1^1 - k_0^1)\Delta W_n - \frac{10}{27}k_1^1\sqrt{3\Delta_n}) \qquad (5.4.16) \\
Y_{n+1} &= Y_n + \frac{1}{2}(k_1^0 + k_0^0)\Delta_n + \frac{1}{40}(37k_0^1 + 30k_2^1 - 27k_3^1)\Delta W_n \\
&\quad + \frac{1}{16}(8k_0^1 + k_1^1 - 9k_2^1)\sqrt{3\Delta_n}
\end{aligned}
$$

for all $n = 0, 1, ..., n_T - 1$. This scheme possesses a mean square convergence order $\gamma = 1.0$, hence it already represents a method of higher order in case $m = 1$. A similar method for Stratonovich equations with $m = 1$ is found in Newton (1991), as well as the proofs of their asymptotic efficiency under the main conditions $a, b \in C_b^1(\mathbb{R}^d) \cap C_p^3(\mathbb{R}^d)$ and $b \in C_p^4(\mathbb{R}^d)$. In principle, that concept of efficiency can be extended to enlarged discretized filtrations $\hat{\mathcal{F}}_T^N$ where more information of higher order multiple integrals is incorporated (however, the computations for asymptotically efficient approximations turns out to be very laborious and hardly feasible by hand).

5.4.9 The explicit and implicit Mil'shtein methods

The simplest higher order method is due to Mil'shtein (1974). It has the scheme

$$
\begin{aligned}
Y_{n+1} &= Y_n + a(t_n, Y_n)\,\Delta_n + \sum_{j=1}^{m} b^j(t_n, Y_n)\,\Delta W_n^j \qquad (5.4.17) \\
&\quad + \sum_{j,k=1}^{m} \mathcal{L}^k b^j(t_n, Y_n) \int_{t_n}^{t_{n+1}} \int_{t_n}^{s} dW_u^k dW_s^j.
\end{aligned}
$$

This method has limited use when numerical stability is an important issue and multidimensional Wiener processes ($m > 1$) drive the dynamics (except for certain noise commutativity conditions). The generation of multiple integrals $I_{(i,j)} = \int\int dW^k dW^j$ is described in Kloeden, Platen and Wright (1992) by using Karhunen-Loeve expansion. There is an idea to make the Mil'shtein method implicit (see Kloeden and Platen). Then the *family of implicit Mil'shtein methods* follows the scheme

$$
\begin{aligned}
Y_{n+1} &= Y_n + (\theta a(t_{n+1}, Y_{n+1}) + (1-\theta)a(t_n, Y_n))\,\Delta_n + \sum_{j=1}^{m} b^j(t_n, Y_n)\,\Delta W_n^j \qquad (5.4.18) \\
&\quad + \sum_{j,k=1}^{m} \mathcal{L}^k b^j(t_n, Y_n) \int_{t_n}^{t_{n+1}} \int_{t_n}^{s} dW_u^k dW_s^j
\end{aligned}
$$

where $\theta \in [0, 1]$ is an implicitness parameter to be chosen. The convergence orders are as that of explicit Mil'shtein method. However, the numerical stability behavior cannot be improved compared to corresponding Theta methods with $\Theta = \theta I$. For more details in this respect, see below or Schurz (1996, 1997). Thus, the balance between convergence and stability requirements is already a problem here with growing order of convergence. More generally, one might think of the usage of *implicit Theta–Mil'shtein methods* governed by

$$Y_{n+1} = Y_n + (\Theta_n a(t_{n+1}, Y_{n+1}) + (I - \Theta_n) a(t_n, Y_n))\,\Delta_n + \sum_{j=1}^{m} b^j(t_n, Y_n)\,\Delta W_n^j \qquad (5.4.19)$$

$$+ \sum_{j,k=1}^{m} \mathcal{L}^k b^j(t_n, Y_n) \int_{t_n}^{t_{n+1}} \int_{t_n}^{s} dW_u^k dW_s^j$$

where $\Theta_n \in \mathbb{R}^{d\times d}$ is a certain matrix of implicitness parameters, and the usage of *drift explicit-implicit Mil'shtein methods*

$$Y_{n+1} = Y_n + a\,(\Theta_n Y_{n+1} + (I - \Theta_n) Y_n)\,\Delta_n + \sum_{j=1}^{m} b^j(t_n, Y_n)\,\Delta W_n^j \qquad (5.4.20)$$

$$+ \sum_{j,k=1}^{m} \mathcal{L}^k b^j(t_n, Y_n) \int_{t_n}^{t_{n+1}} \int_{t_n}^{s} dW_u^k dW_s^j$$

where Θ_n, as before, is small enough that the local resolution of implicit algebraic equations can be guaranteed. But the meaningfulness of the last two methods (5.4.19) and (5.4.20) is still in question.

5.4.10 Gaines's representation of Mil'shtein method

By algebraic rearrangement of multiple integrals and using the fundamental relations between them in the explicit Mil'shtein method one gains the representation of Gaines, which clearly exhibits the role of efficient generation of stochastic area integrals (in particular, of Levy integrals).

$$\begin{aligned} Y_{n+1} &= Y_n + a(t_n, Y_n)\Delta_n + \sum_{j=1}^{m} b^j(t_n, Y_n)\Delta W_n^j \\ &\quad + \frac{1}{2}\sum_{j=1}^{m} \mathcal{L}^j b^j(t_n, Y_n)\Big((\Delta W_n^j)^2 - \Delta_n\Big) \\ &\quad + \frac{1}{2}\sum_{j,k=1, j<k}^{m} \left(\mathcal{L}^k b^j(t_n, Y_n) + \mathcal{L}^j b^k(t_n, Y_n)\right) \Delta W_n^j \Delta W_n^k \\ &\quad + \frac{1}{2}\sum_{j,k=1, j<k}^{m} \left(\mathcal{L}^k b^j(t_n, Y_n) - \mathcal{L}^j b^k(t_n, Y_n)\right) A_n^{j,k} \end{aligned} \qquad (5.4.21)$$

where $A_n^{j,k} = I_{(j,k),t_n,t_{n+1}} - I_{(k,j),t_n,t_{n+1}}$ represent the Levy areas. The advantage of this representation may be seen in the significant simplification under noise commutativity of $\mathcal{L}^k b^j$ (i.e. when $\mathcal{L}^k b^j = \mathcal{L}^j b^k$, which is obviously fulfilled in the case $d = m = 1$) and which results in a more efficient implementation of Mil'shtein methods. On the other hand, this representation clearly shows that the art of applying Mil'shtein methods consists of the efficient generation of Levy area integrals and hence of multiple integrals. The highlighted influence of commutativity conditions is made visible by Gaines representation. The efficient generation of Levy integrals $A_n^{i,j}$ still seems to be a problem. The problem of generation of Levy areas and multiple integrals is described and studied in Gaines (1994, 1995). In order to approximate these Levy areas there is a kind of box-counting algorithm, see Gaines and

Lyons (1994), that is an alternative to truncation of Fourier series approximating stochastic multiple integrals. The Gaines representation can also be exploited to receive more efficient implementations of the implicit Theta–Mil'shtein methods and drift explicit-implicit Mil'shtein methods. Also the development of balanced implicit Mil'shtein methods using Gaines representation could be useful.

5.4.11 Generalized Theta–Platen methods

The natural substitution of the differential quotients arising from the differential operators in the Mil'shtein's methods lead to the *generalized Theta–Platen's method* governed by

$$Y_{n+1} = Y_n + \Big(\Theta_n a(t_{n+1}, Y_{n+1}) + (I - \Theta_n) a(t_n, Y_n)\Big)\Delta_n + \sum_{j=1}^{m} b^j(t_n, Y_n)\,\Delta W_n^j + \tag{5.4.22}$$

$$+\frac{1}{\sqrt{\Delta_n}} \sum_{j,k=1}^{m} \Big[b^j\Big(t_n, Y_n + a(t_n, Y_n)\Delta_n + \sqrt{\Delta_n}\, b^k(t_n, Y_n)\Big) - b^j(t_n, Y_n)\Big] I_{(j,k),t_n,t_{n+1}}$$

where we remember that $I_{(j,k),t_n,t_{n+1}} = \int_{t_n}^{t_{n+1}} \int_{t_n}^{s} dW_u^k dW_s^j$ and $\Theta_n \in \mathbb{R}^{d\times d}$ is a certain matrix of implicitness parameters. Platen (1987) suggested a similar variant to this method in the case $m = 1$ and $\Theta_n = 0$. A practical advantage becomes clear, since it is a derivative free method belonging to the class of implicit Runge-Kutta methods with strong order 1.0. However, to our knowledge, a qualitative study of this method has not been carried out so far, except for convergence statements. Of course, one could immediately apply this idea to arrive at *drift explicit-implicit Runge-Kutta methods* of strong order 1.0 following

$$Y_{n+1} = Y_n + a\Big(t_n + \theta_n\Delta_n, \Theta_n Y_{n+1} + (I - \Theta_n)Y_n\Big)\Delta_n + \sum_{j=1}^{m} b^j(t_n, Y_n)\,\Delta W_n^j + \tag{5.4.23}$$

$$+\frac{1}{\sqrt{\Delta_n}} \sum_{j,k=1}^{m} \Big[b^j(t_n, Y_n + \sqrt{\Delta_n}\, b^k(t_n, Y_n)) - b^j(t_n, Y_n)\Big] \int_{t_n}^{t_{n+1}} \int_{t_n}^{s} dW_u^k dW_s^j$$

where $\Theta_n \in \mathbb{R}^{d\times d}$ is a certain matrix of implicitness parameters and $\theta_n \in [-1, 1]$ such that the local algebraic resolution can be ensured. Furthermore, the idea of making the variants of explicit and implicit Mil'shtein–type methods into derivative free ones can be applied in conjunction with the Gaines's representation. Predictor-corrector implementations of Theta–Platen's techniques are useful to arrive at easily implementable numerical procedures as well. However, one does not get rid of the problem of efficient generation of Levy areas or stochastic multiple integrals by none of these algorithms (unless complete commutativity of (a, b^j) holds!).

5.4.12 Talay-Tubaro extrapolation technique and linear PDEs

A very efficient method for the computation of characteristics of probability distributions is presented by the Talay-Tubaro extrapolation method based on the well-known Euler methods and deterministic extrapolation idea. More precisely speaking, it is when one wants to compute $\mathbb{E}\,[f(X_T)|X_0 = x]$ for a given deterministic function $f : \mathbb{R}^d \longrightarrow R$ (f smooth enough or $a.b^j \in C^\infty$) and fixed terminal times T (nonvarying deterministic terminal times) using equidistant approximations exclusively. Define $\Delta := \frac{T-t_0}{N}$ as the equidistant step size of numerical approximation to be constructed.

$$Y_{n+1}^{\Delta} = Y_n^{\Delta} + a(t_n, Y_n^{\Delta})h + \sum_{j=1}^{m} b^j(t_n, Y_n^h)\Delta W_n^j \tag{5.4.24}$$

$$Y_{k+1}^{2\Delta} = Y_k + a(\hat{t}_k, Y_k^{2\Delta})2h + \sum_{j=1}^{m} b^j(\hat{t}_k, Y_k^{2\Delta})\widehat{\Delta W_k^j} \tag{5.4.25}$$

where ΔW_n^j and $\widehat{\Delta W_k^j}$ can be chosen as independent Wiener process increments or more efficiently taken as $\widehat{\Delta W_k^j} = \Delta W_{2k}^j + \Delta W_{2k-1}^j$, with independent random variables ΔW_k^j substituting the Wiener process increments by some discrete random variables satisfying certain moment relations (see Talay (1995) or Mil'shtein (1995)). Now, set

$$v_n^{\Delta} = 2\mathbb{E}\left[g(Y_{2n}^{\Delta})\right] - \mathbb{E}\left[g(Y_n^{2\Delta})\right]. \tag{5.4.26}$$

Then, based on error expansions by Talay-Tubaro (1990) in analogy of deterministic numerical analysis, it has been shown that $v_{n_T}^{\Delta}$ approximates

$$u(T,x) = \mathbb{E}\left[g(X_T)|X_0 = x\right]$$

where $u(t,x)$ solves the initial value problem (IVP) of the linear PDE (Drift-Diffusion equation)

$$\frac{\partial u}{\partial t}(t,x) = \frac{1}{2}\sum_{j=1}^{m}\sum_{i,k}^{d} b_i^j(t,x)b_k^j(t,x)\frac{\partial^2 u}{\partial x_i \partial x_k}(t,x) + \sum_{k=1}^{d} a_k(t,x)\frac{\partial u}{\partial x_k}(t,x) \tag{5.4.27}$$

where $u : [0,T] \times \mathbb{D} \longrightarrow \mathbb{R}^l$ and $u(0,x) = g(x), 0 \le t \le T$. The striking advantage is the increased order of weak convergence of approximations $v_{n_T}^{\Delta}$ to $\mathbb{E}\ u(0, X_T)$. Moreover, this approach seems to be very appropriate within approximation of Feynman-Kac representations of solutions of deterministic linear PDEs. Possible simplifications of random number generation can be applied to approximations aiming at weak convergence, see Section 9 dealing with implementation issues and the original works of Mil'shtein and Talay. There is a general opinion among Monte Carlo specialists that the approximation of systems of deterministic PDEs with very "complex" domains or whenever one needs only approximations at very specific points of the underlying domain is the field of potential applications where the Monte Carlo techniques as exhibited by Talay-Tubaro extrapolations are superior compared to standard deterministic techniques. Anyway, we should not forget that the new problem of reliable statistical estimation of mean values occurs in the stochastic approach now (which causes new errors). The drawback which is currently seen is that these extrapolation techniques have been suggested only for equidistant approximations with fixed deterministic terminal times T. It is also not quite clear how more complicated boundary conditions on $\partial\mathbb{D}$ can be incorporated in the stochastic approach. One should not forget that many smoothness assumptions on system ingredients must be made as well. An open problem arises with the applicability towards pth mean and pathwise integration. The idea of Talay and Tubaro (1990) has been continued to the case of Taylor approximations as basis methods by Kloeden, Platen and Hofmann (1995).

5.4.13 Denk-Hersch method for highly oscillating systems

To integrate highly oscillating systems, like that of electronic circuits, it is advisable to use a method due to Denk (1993) using an idea of Hersch (1958). The principal d-dimensional equation describing the behavior of electronic circuits is given by

$$\dot{x} + Ax = \hat{a}(t,x) \tag{5.4.28}$$

where A is a $d \times d$ matrix, and $\hat{a}$ is a highly oscillating function which might be noisy due to thermal noise. Then it is advisable to apply Adams-type methods based on the *principle of coherence* due to Hersch (1958). This has been worked out by Denk (1993) in the deterministic setting, using step size depending coefficients in the corresponding Adams methods. The principle of coherence roughly says that the numerical results in "two successive approximation steps should not contradict each other." Let us illustrate this principle in the linear case (i.e. linear coherence principle). Starting from the homogeneous problem $\dot{z} + Az = 0$ related to system (5.4.28), identifying $x_{n+1} = \Phi(\Delta_n)x_n$ as the description of related numerical method applied to the linear homogeneous IVP problem with $\Delta = \frac{t_1 - t_0}{2}$, we get

$$\begin{aligned} \text{using step size}\,\Delta: \quad z(t_1) &= \Phi(\Delta)z(t_0+\Delta) = \Phi(\Delta)\Phi(\Delta)z(t_0) = \Phi^2(\Delta)z(t_0) \\ \text{using step size}\,\Delta: \quad z(t_1) &= \Phi(2\Delta)z(t_0)\,. \end{aligned}$$

Thus, for a coherent numerical method, it must hold that $\Phi^2(\Delta) = \Phi(2\Delta)$. Of course, this would naturally be satisfied for the matrix exponential of the continuous time linear homogeneous system for z. However, only coherent numerical methods preserve the same semigroup property under discretization. Therefore, a coherent integration scheme for x must satisfy the condition $\Phi(h) = \exp(-hA)$ for all $h > 0$. Denk (1993) has combined Hersch's idea of coherence with the standard multistep approach applied to the fully inhomogeneous equation (5.4.28). This gives the *Denk-Hersch method*

$$Y_{n+k} - \exp(-\Delta A)Y_{n+k-1} = \Delta \sum_{l=0}^{k} B^l \hat{a}(t_{n+l-1}, Y_{n+l-1}) \tag{5.4.29}$$

with certain matrix coefficients $B^l = B^l(\Delta) \in \mathbb{R}^{d \times d}$. For example, the Denk-Hersch method with $k = 1$ is established with

$$B^1 = -\Big(I - [I - \exp(-\Delta A)](\Delta A)^{-1}\Big)(\Delta A)^{-1},$$

$$B^0 = [I - \exp(-\Delta A)](\Delta A)^{-1} - B^1\,.$$

It turns out that this scheme is consistent with order k, $A(0)$-stable and therefore convergent (see Lax-Richtmeyer equivalence theorem in deterministics). Note that these facts do not contradict the famous Dahlquist's order barriers for linear multistep methods since the coefficients B^l always depend on the step size Δ. Practical implementations are realized by predictor-corrector techniques. Even the problem of phase lags can be circumvented by the use of this numerical method due to Denk. This method has been further developed and applied to SDEs (5.2.3) with additive noise in circuit modeling and simulation (in fact it leads to the numerical treatment of (Ordinary) Stochastic Differential-Algebraic Equations ((O)SDAEs by stochastic Adams techniques). For more details, see Denk and Schäffler (1997) and Denk, Penski and Schäffler (1998), by using the concept of weak coherence (i.e. coherence for the nonnoisy system equation is guaranteed). Unfortunately, during the writing this survey, the author did not have access to any work which extends this idea to the fully multiplicative noise case (moment or almost sure stochastic coherence should be of interest) leading to coherence with stochastic fundamental matrix solution Φ. For example, in the almost sure sense, when complete commutativity $[B^j, B^k] = B^j B^k - B^k B^j = 0$ with

$B^0 = -A$ for all $j, k \in \{0, 1, ..., m\}$ is met, then

$$\Phi(\Delta_n, \theta(\omega)|_n) = \exp\left(-[A + \frac{1}{2}\sum_{j=1}^{m}(B^j)^2]\Delta_n + \sum_{j=1}^{m} B^j \Delta W_n^j(\theta(\omega)|_n \right)$$

for autonomous SDEs, where $\theta(\omega)_n$ denotes the random shift operator on sample space Ω to render the random dynamics to be a stochastic flow. This implies our idea of the new *fully stochastic Denk-Hersch method (DHS)* following the scheme

$$Y_{n+k} - \Phi(\Delta_{n+k},\ \theta(\omega)|_{n+k})\, Y_{n+k-1} = \sum_{l=0}^{k} C^l(\Delta_l)\, \hat{a}(t_{n+l-1}, Y_{n+l-1})\, \Delta_l + \sum_{l=0}^{k} c^j(t_{n+l-l})\, \Delta W_{n+l-1} \tag{5.4.30}$$

for splitted Itô SDEs

$$dX_t = (-A\, X_t + \hat{a}(t, X_t))\; dt + \sum_{j=1}^{m} \left(B^j X_t + b^j(t)\right) dW_t^j \tag{5.4.31}$$

with additive noise coefficients $c^j(t)$ and multiplicative noise coefficients $b^j(t, x) = B^j(t)x$, where $C^l(\Delta)$ are suitable matrix-valued Adams coefficients. Moreover, in the noncommutative situation one has to incorporate Lie brackets, as presented by the stochastic Magnus formula due to Magnus (1954). But the resulting procedure is fairly complex, and this should be of future interest. Basically, one could generally think of a generalization to a construction of a numerically exact integrator at given time instants at least for linear systems as indicated above (cf. approach of Mickens (1994) to numerically exact integrators in deterministic numerical analysis, and consult its standard references).

5.4.14 Stochastic Adams-type methods

In Denk and Schäffler (1997) and Brugnano, Burrage and Burrage (1999) stochastic analogs to well-known Adams-type methods which belong to the class of linear multi-step methods were developed. For example, following Brugnano et al (1999), the simplest *Adams method* applied to Stratonovich systems (5.4.12) is given by

$$Y_{n+k} = Y_{n+k-1} + \Delta_n \sum_{i=0}^{k} \beta_i g_{n+i}^0 + \frac{1}{2}\sum_{j=1}^{m} J_{(j),t_{n+k-1},t_{n+k}}(g_{n+k}^j + g_{n+k-1}^j) \tag{5.4.32}$$

where $g_{j,l} = b^j(Y_l), j = 0, 1, ..., m, l = 0, 1, ..., k$. The coefficients β_i are those of the deterministic Adams-Moulton method of order $k + 1$. This scheme can be rewritten as

$$Y_{n+1} = Y_n + \Delta_n \sum_{i=0}^{k} \beta_{k-i} g_{n+1-i}^0 + \frac{1}{2}\sum_{j=1}^{m} J_{(j),t_n,t_{n+1}}(g_{n+1}^j + g_n^j) \tag{5.4.33}$$

where $J_{(j),t_n,t_{n+1}} = \Delta W_n^j$. Method (5.4.32) can be combined with predictor-corrector implementations as well, which we omit here.

5.4.15 The two step Mil'shtein method of Horváth-Bokor

Horváth-Bokor (1997) has suggested the following equidistant *two step Mil'shtein scheme* applied to Itô SDEs (5.2.1) with $m = 1$ and componentwise governed by

$$\begin{aligned} Y_{n+1}^k &= (1-\gamma_k)Y_n^k + \gamma_k Y_{n-1}^k + a_k(t_n, Y_n)\Delta + V_n^k \\ &\quad + \gamma_k \Big[\Big((1-\alpha_k)a_k(t_n, Y_n) + \alpha_k a_k(t_{n-1}, Y_{n-1})\Big)\Delta + V_{n-1}^k\Big] \end{aligned} \tag{5.4.34}$$

with

$$V_n^k = b_k(t_n, Y_n)\Delta W_n + \frac{1}{2}\mathcal{L}^1 b_k(t_n, Y_n)\Big((\Delta W_n)^2 - \Delta\Big),$$

where $Y_0 = X_0$ and Y_1 is chosen by one explicit Mil'shtein step ($k = 1, 2, ..., d$), and parameters $\alpha_k, \gamma_k \in [0, 1]$, based on equidistant discretizations of time-intervals $[0, T]$. The strong convergence order 1.0 is also proven in that paper. In addition she proves the same convergence order for a *new multistep method*

$$\begin{aligned} Y_{n+1}^k &= (1-\gamma_k)Y_n^k + \gamma_k Y_{n-1}^k + a_k(t_n, Y_n)\Delta + b_k(t_n, Y_n)\Delta W_n \\ &\quad + \frac{1}{2}L^1 b_k(t_n, Y_n)\Delta W_n + \gamma_k \Big[\Big((1-\alpha_k)a_k(t_n, Y_n) + \alpha_k a_k(t_{n-1}, Y_{n-1})\Big)\Delta \\ &\quad + \frac{1}{2}\Big(b_k(t_n, Y_n) + b_k(t_{n-1}, Y_{n-1})\Big)\Delta W_{n-1} - \frac{1}{2}L^1 b_k(t_{n-1}, Y_{n-1})\Delta\Big] \end{aligned} \tag{5.4.35}$$

where $Y_0, Y_1, \gamma_k, \alpha_k$ are as in scheme (5.4.34) above. Finally, she reports about some numerical evidence that this new scheme "behaves better" than Mil'shtein methods.

5.4.16 Higher order Taylor methods

After substituting in the Taylor expansions X_t by Y_n or Y_{n+1}, respectively, and neglecting the remainder parts, one arrives at the following explicit and implicit Taylor method.

$$Y_{n+1} = Y_n + \sum_{\alpha \in Q_\gamma} x_\alpha(t_n, Y_n) I_{\alpha, t_{n+1}, t_n} \qquad \begin{aligned} Q_\gamma &= \left\{ \alpha \in \mathcal{M}_m : \begin{array}{l} l(\alpha) + n(\alpha) \le 2\gamma \text{ or} \\ l(\alpha) = n(\alpha) = \gamma + \frac{1}{2} \end{array} \right\} \\ Q_\beta &= \{\alpha \in \mathcal{M}_m : l(\alpha) \le \beta\}; 2\gamma, \beta \in \mathbb{N} \end{aligned}$$

where x_α is the Itô coefficient function which one gets by applying to $V(t, x) \equiv x$. The advantages are the obtained higher order of convergence (i.e. larger step sizes could be used) and approximating dynamics can have better geometric properties in accordance with those of underlying continuous time dynamics (e.g. during visualization of stochastic flows, see Kloeden, Platen and Schurz (1991) or, in filtering, see Kloeden, Platen and Schurz (1993)). On the other hand, there are serious problems with numerical instabilities, a large complexity for practical implementation, many smoothness assumptions on a, b^j, and in particular the problem of efficient generation of stochastic multiple integrals which has to be clarified. The efficient use of Taylor methods is limited in general, because of their growing complexity caused by very complex random number generation for involved multiple integrals, and accompanied loss of stability properties. However, for specific dynamics, the situation may change dramatically, and hence one should always check whether there are considerable simplifications (like those with commutative noise or one-dimensional situations) and if one meets limitations on the use of very small step sizes.

5.4.17 Splitting methods of Petersen-Schurz

Petersen (1998) and Schurz (1996, 1997, 1999) use the idea of splitting of the drift and diffusion parts, and treat the obtained split parts by different numerical procedures/techniques.

There are two basic cases of splittings: *additive and multiplicative.* An example for additively split dynamics is provided by the stochastic Duffing oscillator, see Schurz (1996, 1997) and Yannios and Kloeden (1996). For example, *additive splitting* is given when

$$b^j(t,x) = b^{j,1}(t,x) + b^{j,2}(t,x).$$

Then it is tempting to apply different numerical techniques to the separated parts since one may or one has to control only one part of the dynamics. The same is true for the more general *multiplicative splitting* when

$$b^j(t,x) = \hat{b}^j(t,x,x).$$

An example for multiplicatively splitted dynamics is provided by the *modified Van der Pol oscillator* with drift $a(t,x,y,z) = -\omega^2 x + \gamma^2(1 - c^2x^2 - d^2y^2)z$ with $z = y$. Under the usual smoothness and boundedness assumptions Petersen (1998) and Schurz (1996, 1997, 1998) have proven corresponding convergence orders, i.e. Petersen in the weak convergence sense and Schurz in pth mean sense. For more details, see their papers. The simplest example would be the *linear-implicit Theta-Euler methods*

$$Y_{n+1} = Y_n + \Big(\Theta_n A(t_{n+1})Y_{n+1} + (I - \Theta_n)A(t_n)Y_n + \hat{a}(t_n, Y_n)\Big)\Delta_n \qquad (5.4.36)$$
$$+ \sum_{j=1}^{m} b^j(t_n, Y_n)\Delta W_n^j$$

or

$$Y_{n+1} = Y_n + \Big[A(t_n + \theta_n^0\Delta_n)\Big(\Theta_n Y_{n+1} + (I - \Theta_n)Y_n\Big) + \hat{a}(t_n + \hat{\theta}_n^0, Y_n)\Big]\Delta_n \qquad (5.4.37)$$
$$+ \sum_{j=1}^{m} b^j(t_n + \theta_n^j\Delta_n, Y_n)\Delta W_n^j$$

with appropriate implicitness matrices $\Theta_n \in \mathbb{R}^{d\times d}$ and scalars $\theta_n^j, \hat{\theta}_n^0 \in [-1, 1]$, suggested by Schurz (1996). So far the linear part has been implicitly treated. Sometimes it is even more advantageous to treat the nonlinear part of coefficients b^j by implicit numerical techniques, since only that part controls the qualitative behavior. This idea was originally suggested by Öttinger (1996) for problems of polymeric fluids (however, without any mathematical proof). A *nonlinear-implicitly splitted Theta-Euler method* follows

$$Y_{n+1} = Y_n + \Big(\Theta_n a(t_{n+1}, Y_{n+1}) + (I - \Theta_n)a(t_n, Y_n) + \hat{a}(t_n, Y_n)\Big)\Delta_n \qquad (5.4.38)$$
$$+ \sum_{j=1}^{m} b^j(t_n, Y_n)\Delta W_n^j,$$

and in particular the *nonlinear-implicitly splitted trapezoidal method* is governed by

$$Y_{n+1} = Y_n + \Big(\frac{1}{2}[a(t_{n+1}, Y_{n+1}) + a(t_n, Y_n)] + \hat{a}(t_n, Y_n)\Big)\Delta_n \qquad (5.4.39)$$
$$+ \sum_{j=1}^{m} b^j(t_n, Y_n)\Delta W_n^j,$$

and *nonlinear-implicitly splitted midpoint method* by

$$Y_{n+1} = Y_n + \Big(a(\frac{t_{n+1} + t_n}{2}, \frac{1}{2}[Y_{n+1} + Y_n]) + \hat{a}(t_n, Y_n)\Big)\Delta_n \qquad (5.4.40)$$

$$+\sum_{j=1}^{m} b^j(t_n, Y_n)\Delta W_n^j.$$

We have noticed that an introduction of implicitness in the diffusion parts $b^j, j = 1, 2, ..., m$ has not been observed in the splitting methods within the Itô calculus so far. This fact is due to the a priori convention of Itô integration to take only the left-hand side values in the course of Itô-Riemann sum approximations. This can be done partially in the Stratonovich calculus (however, with a lot of care, since explosions may occur in certain dynamics). For this purpose, one can exploit the technique of suitable truncation of random variables (i.e. with the care of not destroying convergence orders) or using the sign of random variables (such as by the balanced implicit methods case (see BIMs above)). The practical value of random variable substitutions can be seen best in the case of weak approximation techniques, see later or Mil'shtein (1995) and Talay (1995).

Thus, Petersen (1998) has introduced the following 2nd order weakly converging *2nd order drift-splitted explicit-implicit method*

$$\begin{aligned} Y_{n+1} &= Y_n + \frac{1}{2}\Big(a(Y_{n+1}) + \hat{a}(Y_n^E) + a(Y_n) + \hat{a}(Y_n)\Big)\Delta_n + \qquad (5.4.41)\\ &+\frac{1}{2}\sum_{j=1}^{m}\Big[b^j\Big(Y_n + \frac{\Delta_n}{2}(a(Y_n)+\hat{a}(Y_n)) + \frac{\sqrt{2}}{2}\sum_{k=1}^{m} b^k(Y_n)\xi_n^{1,j}\Big) + \\ &+b^j\Big(Y_n + \frac{\Delta_n}{2}(a(Y_n)+\hat{a}(Y_n)) - \frac{\sqrt{2}}{2}\sum_{k=1}^{m} b^k(Y_n)\xi_n^{1,j}\Big)\Big]\xi_n^{2,j} + \sum_{j,k=1}^{m}\mathcal{L}^k b^j(Y_n)\hat{I}_n^{j,k} \end{aligned}$$

for autonomous SDEs (5.2.1) with drift $b^0(x) = a(x) + \hat{a}(x)$ and diffusion coefficients $b^j(x)$, where

$$Y_n^E = Y_n + (a(Y_n) + \hat{a}(Y_n))\Delta_n + \sum_{j=1}^{m} b^j(Y_n)\xi_n^{0,j}$$

and $\xi_n^{0,j}, \xi_n^{1,j}, \xi_n^{2,j}, I_n^{j,k}$ are appropriate random variables satisfying certain moment relations (see Petersen (1998) for more details). This method can be rendered to a derivative free one - an approach which leads to the Öttinger-Petersen method (see Öttinger (1996) and Petersen (1998)).

The question of optimal splitting represents a quite complex problem. There is only one rough rule in general: the part of the dynamics of b^j which is responsible for the stabilizing branch in the continuous time system should be treated implicitly and the other instable branch should be treated explicitly. In general, one has to deal with *partial-implicit methods*, i.e. one splits the dynamics of $b^j(t,x) = \hat{b}^j(t,t,x,x)$ by the partial treatment $\hat{b}^j(t_n, t_{n+1}, Y_{n+1}^I, Y_n^E)$ with implicit methods Y^I and explicit methods Y^E. Only some care is needed to keep the finiteness, boundedness, desired order of convergence and some other qualitative properties of discrete time dynamics. For more details, see Schurz (1998). All in all, the adequate introduction of implicitness and splitting forms turns out to be a very case-sensitive problem. Thus, there are no fully generalizable conclusions, except for the additive splitting case when there is one study with respect to asymptotic mean square behavior and numerical characteristic exponents already available, see Schurz (1999) and/or Sections 6 – 8 below.

5.4.18 The ODE method with commutative noise

It is promising to use deterministic algorithms under certain circumstances. For example, under commutative noise one can exploit the Doss representation of diffusions, see Doss

(1977). The resulting splitting idea was pursued by Talay (1983), Bensoussan et al. (1989), and picked up by Castell and Gaines (1994), Roman (2000), see also Schurz (1999). The idea goes as follows. The entire ODE approach is based on the key assumption of **commutative noise** stated by $\forall k, j \in \{1, ..., m\}$ $\forall (t,x) \in [0,T] \times \mathbb{R}^d$

$$< b^j, \nabla_x > b^k = \sum_{i=1}^{d} b_i^j(t,x) \frac{\partial b^k(t,x)}{\partial x_i} = \sum_{i=1}^{d} b_i^k(t,x) \frac{\partial b^j(t,x)}{\partial x_i} =< b^k, \nabla_x > b^j.$$

Then Doss (1977) has given an explicit representation $X_t = h(D(t), W_t)$ of the solution of SDEs. For example, in case $d = m = 1, b = b(x)$ the deterministic function h satisfies the PDE

$$\frac{\partial h(u,v)}{\partial v} = b(h(u,v))$$

with initial condition $h(u,0) = u$, provided that $b \in C^2(\mathbb{R}^d)$ fulfills the Lipschitz condition (ULC), and where $D(t)$ satisfies the random initial value problem for randomized differential equation

$$D'(t) = \exp\left(- \int^{W_t} b'(h(D(t),v))dv \right) b(h(D(t),W_t)$$

started at $D_0 = X_0$. Thanks to a conjecture of E. Pardoux, Talay (1983) has made use of this idea in the fully multidimensional case

$$X_t = h(D_1(t), D_2(t), ..., D_d(t), W_t^1, W_t^2, ..., W_t^m).$$

With these contributions in hand, one obtains the procedure to approximate the composed solution $X = h(D(t), W_t)$ by solving the deterministic PDE for h first (e.g. by deterministic analytical or numerical methods), and then one may numerically integrate the related ODE for $D(t)$ for each random path of the underlying Wiener process W_t by deterministic methods started at $D_0 = X_0$. Under appropriate commutativity conditions for all a, b^j one may even show that the increments of Taylor methods for certain functionals $V(t, X_t)$ can only locally depend on the Wiener process increments $\Delta_{s,t} W^j = W_t^j - W_s^j$ and time-increments $\Delta = t - s$, see Schurz (1999). This is useful to approximate certain conditional expectations $\mathbb{E}\,[V(t,X_t)|\mathcal{F}_s]$ and the algebra of iterated multiple integrals, where $V(t, X_t)$ is an appropriate $\mathcal{F}_t$-measurable functional of X and $\mathcal{F}_s$ represents the σ-field of underlying natural filtration at time $s \le t$. Of course, one can now successfully apply deterministic numerical methods like that of higher order Runge-Kutta methods to approximate the involved random differential equations with high accuracy and sophisticated knowledge on deterministic numerical analysis based on deterministic Taylor expansions. Thus, "pathwise" approximations of SDEs are possible, exploiting the full knowledge of deterministic numerical analysis. However, the user may be warned that this cannot be done in general (i.e. when commutativity does not hold)! For further reading, we recommend to consult the paper of Talay (1983) where he proves the convergence of numerical methods using the Doss representation, but without calculating D, h explicitly. In Roman (2000) one finds a discussion on convergence orders in conjunction with Runge-Kutta techniques applied to Stratonovich SDEs without drift part under the additional condition of noise commutativity (i.e. then even the calculation of D is not needed to establish approximations using the Doss approach, since then $D(t) = D(0) = X_0$). Under appropriate smoothness conditions of a, b^j, V with some very restrictive commutativity conditions (i.e. V-commutativity of order 2γ) involving a, b^j, V, an alternative is given by Schurz (1999) without using the Doss approach, but resulting into the achievement of any desired convergence order for Taylor approximations of $V(t, X_t)$ (i.e. even infinite Taylor series can be obtained, which of course need to be truncated for the practical implementation).

5.4.19 Random local linearization methods (LLMs)

In Mechanical Engineering a localization technique has widely been used for a long time. This technique basically says one should linearize the dynamics locally at each step, and then approximate the original nonlinear dynamics by the linearized dynamics. For example, see Iyengar (1988). This method can be applied with some care to (O)SDEs as well. To our current knowledge, as one of the first, Ozaki (1985) recognized the power of this technique in stochastic hydrology. The mathematics of stochastic linearization was later treated independently by Jimenez et al (1996), Roy and Schurz (1996), Ozaki and Shoji (1998), Shoji (1998) among others. Let us follow somehow their ideas in conjunction with the method of removing multiplicative noise terms in the original dynamics. The key assumptions (connected to the more general and challenging problem of when a stochastic dynamics is qualitatively represented by its corresponding linearization in an adequate manner) are that the drift $a(t,x)$ and diffusion parts $b^j(t,x)$ are smooth enough (e.g. continuously differentiable with respect to time, at least twice continuously differentiable with respect to the space coordinate and sufficiently smooth such that the corresponding function f from below is in $C^{1,3}([0,T]\times \mathbb{D})$), the dynamics for the stochastic process X lives on a compact bounded set $\mathbb{D}$ of R^d (a.s.), the diffusion part $\inf_{(t,x)\in[0,T]\times\mathbb{D}} ||b^j(t,x)|| > 0$ is uniformly bounded away from zero (such diffusions are called **nondegenerate**) and the information on the Wiener process is given at all discrete time-instants. For simplicity of illustration, we shall confine ourselves to one-dimensional Itô SDEs

$$dX_t = a(t,X_t)dt + b(t,X_t)dW_t.$$

It is convenient to transform this equation to an SDE with additive noise

$$dZ_t = \left(a(t,X_t)\left.\frac{d\phi(z)}{dz}\right|_{z=Z_t} + \frac{[b(t,X_t)]^2}{2}\left.\frac{d^2\phi(z)}{dz^2}\right|_{z=Z_t}\right) dt + \sigma(t)dW_t$$

provided that an invertible $\phi(x)$ as solution of

$$b(t,x)\frac{d\phi(x)}{dx} = \sigma(t)$$

exists. This equation for Z is an equation with random drift coefficients

$$g = g(t,X_t,Z_t) = a(t,X_t)\left.\frac{d\phi(z)}{dz}\right|_{z=Z_t} + \frac{[b(t,X_t)]^2}{2}\left.\frac{d^2\phi(z)}{dz^2}\right|_{z=Z_t}$$

$$= \sigma(t)\left(a(t,\phi^{-1}(Z_t))\left.\frac{d\phi(z)}{dz}\right|_{z=Z_t} + \frac{[b(t,\phi^{-1}(Z_t))]^2}{2}\left.\frac{d^2\phi(z)}{dz^2}\right|_{z=Z_t}\right) = f(t,Z_t).$$

Since $g(t,\phi^{-1}(Z_t),Z_t) = f(t,Z_t)$, this yields the equivalent Itô SDE

$$dZ_t = f(t,Z_t)dt + \sigma(t)dW_t$$

with additive noise and deterministic drift function $f(t,z) = g(t,\phi^{-1}(z),z)$. Now we may apply the Itô formula to f in order to linearize the dynamics. Therefore it locally follows that

$$f(t,Z_t) \approx f(s,Z_s) + \left.\left(\frac{\sigma^2(t)}{2}\frac{\partial^2 f(t,z)}{\partial z^2} + \frac{\partial f(t,z)}{\partial t}\right)\right|_{z=Z_s}(t-s) + \left.\frac{\partial f(t,z)}{\partial z}\right|_{z=Z_s}(Z_t - Z_s),$$

i.e.

$$f(t,Z_t) \approx l_s Z_t + m_s t + n_s$$

where

$$\begin{aligned} l_s &= \frac{\partial f(s,Z_s)}{\partial z}, \; m_s = \frac{\sigma^2(s)}{2}\frac{\partial^2 f(s,Z_s)}{\partial z^2} + \frac{\partial f(s,Z_s)}{\partial t} \\ n_s &= f(s,Z_s) - \frac{\partial f(s,Z_s)}{\partial z} Z_s - \left(\frac{\sigma^2(s)}{2}\frac{\partial^2 f(s,Z_s)}{\partial z^2} + \frac{\partial f(s,Z_s)}{\partial t} \right) s. \end{aligned}$$

Consequently, on each subinterval $[s, s+h]$ we have to solve the linear SDE

$$dZ_t = (l_s Z_t + m_s t + n_s)dt + \sigma(t)\, dW_t.$$

This SDE has the explicit solution

$$Z_{s+h} = Z_s + \int_s^{s+h} (m_s u + n_s) \exp(-l_s u)\, du + \int_s^{s+h} \exp(-l_s u)\sigma(u)\, dW_u$$

which one can obtain by local application of Itô's formula to $\exp(-l_s t)Z_t$ on $[s, s+h]$. Now, this equation can be solved by

$$\begin{aligned} Z_{s+h} &= Z_s + \frac{f(s,Z_s)}{l_s}\Big(\exp(l_s h - 1\Big) + \frac{m_s}{l_s^2}\Big(\exp(l_s h) - 1 - l_s h\Big) + \\ &\quad + \int_s^{s+h} \exp(l_s(s+h-u))\sigma(u)\, dW_u \qquad (5.4.42) \\ &= Z_s + \frac{f(s,Z_s)}{l_s}\Big(\exp(l_s h - 1\Big) + \frac{m_s}{l_s^2}\Big(\exp(l_s h) - 1 - l_s h\Big) + \qquad (5.4.43) \\ &\quad - \int_s^{s+h} \exp(l_s(s+h-u))(ls(s+h-u)\sigma(u) + \sigma\prime(u))W_u du + \\ &\quad + \sigma(s+h)W_{s+h} - \sigma(s)W_s. \end{aligned}$$

The generation of the random integral in (5.4.42) is easy to manipulate since it follows a Gaussian distribution with mean zero and variance

$$\int_s^{s+h} \exp(2l_s(s+h-u))\sigma^2(u)\, du$$

provided that it is square-integrable with respect to Lebesgue measure. For example, when $\sigma = \sigma_0$ is constant, then the local variance is equal to

$$\sigma_0^2 \frac{\exp(2l_s h) - 1}{2l_s h}.$$

As an alternative, we may exploit the second identity (5.4.43) gained by the formula the of partial integration for Brownian motions to generate the local increments $Z_t - Z_s$ by optimal quadrature formulas for the integral expressions (see e.g. Egorov et al). Consequently, we have a local random linearization technique available for approximation of solutions of SDEs through composition of the local increment formulas we have received before. Shoji (1998) provides some experimental results which show the slightly improved error behavior of LLMs compared to numerical results using the classical Euler methods for generation of both the exact solution and numerical approximation. He also gives a proof of global pth mean convergence rates $\gamma_g = 1.0$ of the obtained LLMs for SDEs with additive noise on compact real domains based on the almost sure continuity of the related diffusion process with additive noise for $p \geq 2$ (thus the same convergence order as classical Euler methods applied to additive noise dynamics). It remains a question how efficient the presented approach really is in the fully multidimensional framework.

5.4.20 Simultaneous time and chance discretizations

Gelbrich (1995) has found a new approach to "weak approximations" of SDEs. He presents a combination of time-discretization methods of Euler and Mil'shtein methods with a "chance-discretization" based on well-known invariance principles. The grid is constructed to tune the discretization. The convergence of the approximate solutions is shown using $(\mathbb{E}\, ||.\cdot||^p_{C[t_0,T]})^{1/p}$ norms for $p \in [2, +\infty)$. The obtained convergence rates can be interpreted as rates for the L^p Wasserstein metrics ($p \in [1, +\infty)$) between the distributions of exact and approximate solutions.

5.4.21 Stochastic waveform relaxation methods

Schneider and Schurz (1998) have recently developed a stochastic version of deterministically well-known waveform iteration methods on pth mean Banach spaces of solutions of SDEs. These methods are designed particularly for high-dimensional systems of SDEs as obtained after discretizing stochastic partial differential equations (SPDEs) by the common space discretization. Stochastic waveform relaxation algorithms (Jacobi, Gauss-Seidel, SOR, etc.) are easily parallelizable iteration methods for SDEs with no functional delay effects (i.e. for Markov processes), hence their efficiency is seen in application to high-dimensional systems of SDEs. The construction and proof of pth mean convergence is heavily based on the fixed point principles and the efficient estimates of related contractivity constants, and depends on finding appropriate splittings of the original system into subsystems to introduce windowing techniques for local iterations with global exchange of data for the global iteration. For more details, see Schneider and Schurz (1998).

5.4.22 Comments on numerical analysis of SPDEs

Stochastic partial differential equations (SPDEs) have been studied for a fairly long time. For example, Benssousan and Temam (1972, 1973), Krylov and Rozovskiĭ (1977 - 1986), Pardoux (1979), Gyöngy and Krylov (1980, 1982, 1996), Schmalfuss (1986), Rozovskiĭ (1990), Da Prato and Zabczyk (1992, 1996), Flandoli and Crauel (1994, 1998), Greksch and Tudor (1995), Krylov (1996), Kuo (1996), Crauel, Debussche and Flandoli (1997), Holden et al (1997) and Krylov and Lototsky (1999). Stochastic Navier-Stokes equations are treated in Bensoussan and Temam (1971, 1972), Greksch and Schmalfuss (1996). Stochastic evolution equations are studied by Rosovski (1990) and Greksch and Tudor (1995). Da Prato and Zabczyk (1992, 1996) follow the classical deterministic semigroup approach to treat linear SPDEs which leads to many direct computations. A systematic L^p-theory has been developed by Krylov (1995, 1996). Holden, Øksendahl, Ubøe and Zhang (1996) report on a systematic approach to SPDEs based on Wick-type white noise calculus. There are already a few papers on numerical analysis of stochastic partial differential equations (SPDEs) available. As one of the first, Gyöngy (1989) and Gaines (1995) outlined the role of stochastic numerical methods for the solution of SPDEs. Gyöngy (1991, 1998) introduces stochastic lattice methods, Gyöngy and Nualart (1995, 1997) provide with an implicit numerical scheme, and Gaines (1995) basically makes use of a stochastic generalization of well-known method of lines, leading to finite-dimensional approximations of SPDEs by (O)SDEs. Grecksch and Kloeden (1996), Grecksch and Wadewitz (1996) study stochastic Galerkin approximation and derive space - time step size convergence orders for evolutionary systems. Convergence proofs are also given in Gyöngy (1998), Davie and Gaines (1999). Hoo (1998, 1999) has recently carried out a work where he exploits techniques of discrete Sobolev spaces and the analytical L^p-theory due to Krylov (1996). All in all, we can confirm that this area is rapidly growing and has a very promising future. Stochastic finite element techniques must be further developed (first approaches, mainly motivated by Mechanical Engineering

(Crack Growth), are found in Dey (1979), Contreras (1980), Wong (1984), Skurt (1986), Faravelli (1988), Germani and Piccioni (1988), Ghanem and Spanos (1990, 1991, 1997), Hien and Kleiber (1990), Skurt and Michel (1990, 1992), Kleiber and Hien (1992), Araujo and Awruch (1994), Elishakoff, Ren and Shinozuka (1995), Papadrakakis and V. Papadopoulos (1996), Ren, Elishakoff and Shinozuka (1997), Allen, Novosel and Zhang (1998), Benth and Gjerde (1998), Peng (1998), Ghanem (1999), Matthies and Bucher (1999)). It would be advantageous to know when a difference method can be preferred to a finite element one. However, one does need a very profound knowledge of basics of numerical analysis for systems of (O)SDEs to understand the numerics of SPDEs.

5.4.23 General concluding comment on numerical methods

Although it is very daring to make any statement about the preferences of numerical methods, the author's current opinion is as follows. In general, splitting techniques together with ODE techniques (Doss splittings), BIMs, RTMs, PCMs, Newtons method, Talay-Tubaro extrapolations, Denk's method, the Burrage-Butcher school of stochastic Runge-Kutta methods and the local linearization approach to approximate probability densities and to phase plane analysis represent the most advanced and efficient numerical methods which are currently available in the market of academic literature for general systems of (O)SDEs, as of 1999. However, in specific situations the classical Taylor methods do perform very well, see Kloeden, Platen and Schurz (1991, 1993) with respect to qualitative dynamical pattern behavior, in filtering and Schurz (1999) under V-commutativity. Also in general it is advisable to form a test set of different numerical procedures, to apply to one and the same continuous time dynamics, and then, if the results qualitatively coincide one should accept the received joint approximation as the approximation result (similar to the general philosophy of statistical estimations). There is still a lack of knowledge on efficient numerical integration of high-dimensional systems of SDEs, how to perform very reliable variable step size and order techniques and how to control dynamics with non-Lipschitz continuous coefficients. The entire analysis can consist only of a careful study of both the qualitative behavior of continuous and discrete time dynamics, exploiting the specific structure of underlying systems and taking into account the following main principles of (stochastic) numerical analysis.

5.5 On the Main Principles of Numerics

The key to understanding the analysis and mathematically justified construction of numerical methods (and above all their behavior more profoundly) in the pth mean sense, inspired by Schurz (1999), can be illustrated as follows. Fix $p \geq 1$. Let $X_{t,x}(t+h), Y_{t,x}(t+h)$ denote the one-step integral representations of exact and approximate process started at x at time t and evaluated at time $t+h$.

5.5.1 $\mathbb{D}$-invariance

An important fact which is neglected by many authors is that, for a fair comparison between exact solution and numerical approximation, we need to find a common (random) normed space where one could and should do numerical analysis. Since this problem seems to be very difficult on bounded domains in stochastic numerical analysis, most of the authors in stochastics circumvent it by treating the numerical approximation procedures on the whole vector space like that of $\mathbb{R}^d$. This embedding is always possible, but surely not always necessary and not the most desirable procedure. For example, one may consider the simple logistic equation or the innovation diffusion due to Schurz (1996,1997) where a closed manifold is a geometrically invariant region for both the exact and approximate dynamics embedded in the entire vector space for the exact solution. For simplicity, let $\mathbb{D}$ be an open or a closed subdomain of $\mathbb{R}^d \setminus \{-\infty, +\infty\}$. Thus we ask for

Definition 5.5.1 *The numerical sequence* $Y = (Y_n)_{n \in \mathbb{N}}$ **leaves the domain $\mathbb{D}$ invariant** *(a.s.) (or in short Y is* **$\mathbb{D}$-invariant***) iff*

$$\mathbb{P}\,\{Y_n \in \mathbb{D} | Y_0 \in \mathbb{D}\} \;=\; 1.$$

The construction of such sequences can be a very tough task in stochastic analysis. In this respect the class of BIMs (as shown in Sections 6,7 and Schurz (1997)) is very promising. Another problem which arises is how to study and guarantee stochastic boundary conditions. The latter question is not touched here, unfortunately, due to its complexity. For the special case of a.s. nonnegativity, see Section 7.

5.5.2 Numerical pth mean consistency

Next, we want to have at least locally accurate behavior of our approximations to be constructed, representing an obvious requirement. Therefore we ask for

Definition 5.2. The numerical sequence $Y = (Y_n)_{n \in \mathbb{N}}$ is said to be **pth mean consistent with order** $\gamma \in \mathbb{R}_+$ with respect to X solving SDE (5.2.1) on $[0,T]$ iff there is a real constant $K_p^C \geq 0$ such that

$$\mathbb{E}\;\|X_{t,x}(t+h) - Y_{t,x}(t+h)\|^p \;\leq\; (K_p^C)^p(1 + \|x\|^p)h^{p\gamma_p}$$

for all sufficiently small $h \leq \min\{1, T-t\}$ and all $t \in [0, T-h]$. Y is said to be **mean consistent with order** $\gamma_0 \in \mathbb{R}_+$ if there is a real constant K_0 such that

$$\|\mathbb{E}\,[X_{t,x}(t+h)] - \mathbb{E}\,[Y_{t,x}(t+h)]\| \;\leq\; K_0^C(1 + \|x\|)h^{\gamma_0}$$

for all sufficiently small $h \leq \min\{1, T-t\}$ and all $t \in [0, T-h]$.

Consistency always says how good a numerical method locally approximates the underlying exact dynamics (i.e. consistency = local approximation of corresponding vector fields $(a, b^1, ..., b^m)$). The consistency behavior and order of a method can be found by comparison with Taylor expansion on the same local subinterval. For example, the Euler method has mean square consistency order 1.0, and the Mil'shtein method has mean square consistency order 1.5 under enough smoothness of the SDE coefficients. The Euler method possesses a mean consistency rate 2.0, the same as that for the Mil'shtein method, provided there is enough smoothness in the system (5.2.1) to guarantee a comparison of this kind. Unfortunately, it is not well worked out for all methods in the literature (i.e. there is still some demand to do it very carefully in the future). We will see that the interplay of mean and pth mean consistency rates will be essential for the global convergence rate on $[0,T]$, see the following main theorems with respect to stochastic L^p-numerics.

5.5.3 Numerical pth mean stability

The next very important requirement is the control on the evolution of the state process Y_n of the numerical methods. To guarantee nonexploding behavior, and in analogy to that of the continuous time solution, one naturally asks for

Definition 5.5.2 *The numerical sequence* $Y = (Y_n)_{n\in\mathbb{N}}$ *is said to be* **numerically** p**th mean stable on** $[0,T]$ *for a stochastic process* $X = (X_t)_{0\le t\le T}$ *governed by SDE (5.2.1) iff it satisfies the estimates*

$$\mathbb{E}\left[\|Y_{t,x}(t+h)\|^p | Y_{t,x}(t) = x\right] \le (K^Y_{S1})^p \exp(pK^Y_{S2}h)(1+\|x\|^p)$$

for all $x \in \mathbb{D}$, *all* $0 \le t \le t+h \le T$, *with appropriate real constants* $K^Y_{S1} \ge 0, K^Y_{S2}$. *The numerical sequence* $Y = (Y_n)_{n\in\mathbb{N}}$ *is said to be* **asymptotically numerically** p**th mean stable on** $[0,T]$ *for a stochastic process* $X = (X_t)_{0\le t\le T}$ *governed by SDE (5.2.1) iff it is numerically pth mean stable with constant* $K^Y_{S2} < 0$.

This definition is slightly different from the classical definitions of stability of dynamical systems. At a first glance, we need rather the property of uniform boundedness of pth moments and linear boundedness with respect to the initial moments. The more strict requirements of classical stability notions are a little bit too restrictive for a fairly general qualitative convergence analysis. They are "better" covered by the notion of contractivity in our opinion, see below. In short, we consider here only exponential-type stability. One might also think of polynomial stability.

5.5.4 Numerical pth mean contractivity

It is always desirable to have a control on the error growth behavior (propagation of initial errors) as integration time advances. The optimal situation is when small initial errors produce no significant effect on the total accuracy of numerical approximations. Sometimes this property is also called *perturbation stability*, but here it is referred to as *contractivity*, originating from the well-known concept of B-stability in deterministic numerical analysis. Then we ask for

Definition 5.5.3 *The numerical sequence* $Y = (Y_n)_{n\in\mathbb{N}}$ *is said to be* **numerically** p**th mean contractive on** $[0,T]$ *for a stochastic process* $X = (X_t)_{0\le t\le T}$ *governed by SDE (5.2.1) iff it satisfies the estimates*

$$\mathbb{E}\left[\|Y_{t,x}(t+h) - Y_{t,y}(t)\|^p | Y_{t,x}(t) = x, Y_{t,y}(t) = y\right] \le \exp(pK^Y_C h)\|x-y\|^p$$

for all $x \in \mathbb{D}$, *all* $0 \le t \le t+h \le T$, *with appropriate real constant* K^Y_C. *The numerical sequence* $Y = (Y_n)_{n\in\mathbb{N}}$ *is said to be* **asymptotically numerically** p**th mean contractive on** $[0,T]$ *for a stochastic process* $X = (X_t)_{0\le t\le T}$ *governed by SDE (5.2.1) iff it is numerically pth mean contractive with constant* $K^Y_C < 0$.

Of course, the growth of perturbations can also be controlled somehow by numerical stability (use Minkowski's inequality to realize that fact), however only up to a certain extent. In fact, there are many more systems which have asymptotically contractive, but not asymptotically stable, behavior (take e.g. pth mean dissipative systems with additive noise, since we switch off the influence of inhomogeneities by the requirement of contractivity.)

5.5.5 Numerical pth mean convergence

Last but not least, we need to talk about pth mean convergence of numerical approximations. As we always assume, the processes X and Y are constructed on one and the same probability space $(\Omega, \mathcal{F}, (\mathcal{F}_t)_{0\le t\le T}, \mathbb{P})$.

Definition 5.5.4 *Fix $p \geq 1$ and time-interval $[0,T]$. Assume that*

$$\mathbb{E}\,\|X(0)\|^p + \mathbb{E}\,\|Y(0)\|^p < +\infty\,.$$

A numerical sequence $Y = (Y_n)_{n\in\mathbb{N}}$ (method, scheme, etc.) constructed along time-discretizations $0 = t_0 \leq t_1 \leq ... \leq t_n \leq ... \leq t_{n_T} = T$ with maximum step size $\Delta > 0$ is said to be **numerically pth mean converging** *to a stochastic process $X = (X_t)_{0\leq t\leq T}$ iff*

$$\lim_{\Delta\to 0}\ \sup_{n=0,1,\ldots,n_T}\ \mathbb{E}\,\|X_{t_n} - Y_n\|^p \ = \ 0.$$

A numerically pth mean converging sequence $Y = (Y_n)_{n\in\mathbb{N}}$ (method, scheme, etc.)

$$0 = t_0 \leq t_1 \leq ... \leq t_n \leq ... \leq t_{n_T} = T$$

with maximum step size $\Delta > 0$ is said to be **numerically pth mean converging with order $\gamma_g \in \mathbb{R}_+$** *to a stochastic process $X = (X_t)_{0\leq t\leq T}$ iff there is a deterministic constant $K = K(p,T,\mathbb{E}\,\|Y_0|^p, \mathbb{E}\,\|X_0\|^p) > 0$ such that*

$$\bar{\varepsilon}_{n_T} \ := \ \sup_{n=0,1,\ldots,n_T} \left(\mathbb{E}\,\|X_{t_n} - Y_n\|^p\right)^{\frac{1}{p}} \ \leq \ K(p,T,\mathbb{E}\,\|Y_0\|^p, \mathbb{E}\,\|X_0\|^p)\cdot\Delta^{\gamma_g}$$

for all sufficiently small step sizes Δ.

There are many interesting cross relations between the concepts of mean, pth mean consistency, stability, contractivity and pth mean convergence. For some more details, see below. Roughly speaking, consistency refers to the property of local approximation of corresponding vector fields (a, b^j) and its accuracy, whereas convergence relates to the property of global approximation of the entire dynamics on fixed time intervals $[0,T]$. Contractivity describes how initial perturbations grow in the course of dynamics, and stability controls that no undesired explosions occur. This leads to the following main principles of (numerical) approximation theory.

5.5.6 The main principle: combining all concepts from 5.1-5.5

Finally we are able to combine the main four concepts we have presented under the a.s. invariance of domain $\mathbb{D} \subset \mathbb{R}^d$ for both the exact solution and numerical methods. Let $p \geq 1$ and $q \geq 1$ be conjugate exponents, i.e. $\frac{1}{p} + \frac{1}{q} = 1$. We find

Proposition 5.5.5 *Assume that SDE (5.2.1) satisfies (OLC), (OBC) and (IMC), and we have a locally* **mean consistent** *with order $\gamma_0 \in \mathbb{R}_+$, and* **$p$th mean consistent** *with order $\gamma_p \in \mathbb{R}_+$, numerical approximation $Y = (Y_n)_{n\in\mathbb{N}}$ for the diffusion process $X = (X_t)_{t\in[0,T]}$ satisfying SDE (5.2.1). Let*

$$\gamma_g := \max\left(\frac{\gamma_0}{p} + \frac{\gamma_p}{q}, \frac{\gamma_0}{q} + \frac{\gamma_p}{p}\right) - 1 \geq 0.$$

Then the following **main principle of stochastic-numerical analysis for SDEs** *holds, namely*

[1] **consistency of Y + contractivity of X + stability of Y $\Longrightarrow$ pth mean convergence with worst case order $\gamma \geq \gamma_g$** *and / or*

[2] **consistency of Y + stability of X + contractivity of Y $\Longrightarrow$ pth mean convergence with worst case order $\gamma \geq \gamma_g$.**

under the **$\mathbb{D}$-invariance** *of X, Y, and more precisely, the order of pth mean convergence is at least γ_g. Moreover, if the assumptions (OLC) and (OBC) on the given SDE and the consistency requirements are uniformly satisfied with respect to all finite time-intervals $[0,T]$ with finite uniform constants K_{OL}, K_{OB}, K_0^C and K_p^C, either $K_{OL} < 0$ and $Y = (Y_n)_{n\in\mathbb{N}}$ is asymptotically pth mean stable or $K_{OB} < 0$ and $Y = (Y_n)_{n\in\mathbb{N}}$ is asymptotically pth mean contractive, then the pth mean error tends to zero as the terminal time T tends to $+\infty$ as well. The* **numerical pth mean error process** *$(\varepsilon_n)_{n\in\mathbb{N}}$ on $[0,T]$ satisfies*

$$\varepsilon_n := \left(\mathbb{E}\,\|X_{t_n} - Y_n\|^p\right)^{\frac{1}{p}} \leq \sup_{n\in\mathbb{N}} \left(\mathbb{E}\,\|X_{t_n} - Y_n\|^p\right)^{\frac{1}{p}}$$
$$\leq \exp\Big([K_{OL}]_+(T-t_0)\Big)\varepsilon_0 + \hat{K}_1\Delta^{\gamma_g}\frac{1-\exp\Big(-K_{OL}(T-t_0)\Big)}{K_{OL}} \tag{5.5.1}$$

with

$$\hat{K}_1 = \max(K_1^C, K_0^C)\left[1 + K_{S1}^Y \exp\Big([K_{S2}^Y]_+(T-t_0)\Big)(1+\|y\|)\right]$$

if Y is pth mean stable with stability constants K_{S1}^Y, K_{S2}^Y on $[0,T]$, and

$$\sup_{n\in\mathbb{N}} \varepsilon_n \leq \exp\Big([K_C^Y]_+(T-t_0)\Big)\varepsilon_0 + \hat{K}_1\Delta^{\gamma_g}\frac{1-\exp\Big(-K_{OL}(T-t_0)\Big)}{K_{OL}} \tag{5.5.2}$$

with

$$\hat{K}_1 = \max(K_1^C, K_0^C)\left[1 + \exp\Big([K_S^X]_+(T-t_0)\Big)(1+\|y\|)\right]$$

if Y is pth mean contractive with contractivity constant K_C^Y on $[0,T]$, where $p \geq 2$ and $K_S^X = \frac{2(p-1)}{p}K_{OB}$.

Remark *One can even show the convergence orders $\gamma_g = \gamma_2 - \frac{1}{2}$ with consistency orders $\gamma_p = \gamma_2 - \frac{1}{p}$ and $\gamma_0 \geq \gamma_p + \frac{1}{p}$ for $p \geq 2$, using the almost sure sample continuity of stochastic process X governed by SDE (5.2.1).*

Further Comments on Main Principles. Thus, with some care, we can exchange contractivity and stability assumptions between the exact solution X and the numerical approximation Y as it is more convenient to deduce some convergence statements or as it is more apparent to verify the corresponding properties by X, Y. This general principle has been proved for stochastic processes on randomized Banach spaces by Schurz (1999). Moreover, it can be shown that contractivity of X, contractivity of Y and consistency of Y may already imply stability of Y due to stability of X, and also stability of X and consistency of Y may imply convergence of Y by help of well-posedness of the SDE (5.2.1) (see theorems below for the case $p = 2$). These latter statements are not so trivial, since one can construct counterexamples where these implications between contractivity and stability can not be concluded for all stochastic dynamical systems (see Schurz (1999), in asymptotic sense as time T tends to infinity.) They turn out to be true implications on fixed, finite intervals under the assumptions of Proposition 5.1 for SDEs (5.2.1). Another interesting observation is the interplay between mean and pth mean consistency. This really becomes apparent when $p \geq 2$. Then we do need to ask for the additional assumption of higher order of mean consistency with order $\gamma_g + 1$ for very efficient error estimates (this comes from the supplement with the conjugate exponent q belonging to $p > 1$ by the conjugacy requirement $\frac{1}{p} + \frac{1}{q} = 1$ during application of the Hölder inequality to squeeze out the suitable local order of convergence for the total error estimation; we do not have the space here to explain it in more detail, see Schurz (1999) for a more general explanation of numerical principles on Banach spaces). This proposition is a stochastic counterpart of the forward direction of the Lax-Richtmeyer equivalence principle in deterministic numerical analysis, supported by a conjecture of P. Lax (1956). In fact, we believe that this idea originates from a more general construct of L.V. Kantorovich (1948). It can be split into the two directions mentioned by Proposition 12.5.2, depending on whether we have the property of numerical stability or numerical contractivity available during the error estimation process (see splitting below in the estimation process). The interesting interplay between mean and pth mean consistency of numerical approximations in achieving a suitable order of convergence, which originates from the main principles, can be illustrated, for the improvement of general understanding, as follows. Define the **pth mean global error**

$$\varepsilon(t) \;=\; (\mathbb{E}\,[\|X_{0,x}(t) - Y_{0,y}(t)\|^p | X_{0,x}(0) = x, Y_{0,y} = y])^{\frac{1}{p}}$$

along the time-discretization $0 = t_0 \leq t_1 \leq ... \leq t_{n_T} = T$. Under the commonly met assumptions on smoothness and linear-polynomial boundedness of coefficients a, b^j it suffices to control this error at instants t_{n+1} only. Identify $\varepsilon_n = \varepsilon(t_n)$ for $n = 0, 1, ..., n_T$ and fixed p. For simplicity (to avoid further technical and laborious computations), take $p = 1$. Define

$$Z_n^1 = X_{t_n, X_{0,x}(t_n)}(t_{n+1}) - X_{t_n, Y_{0,y}(t_n)}(t_{n+1})$$

$$Z_n^2 = X_{t_n, Y_{0,y}(t_n)}(t_{n+1}) - Y_{n, Y_{0,y}(t_n)}(t_{n+1}).$$

Now we have reached a point where the global error estimation process is split into two directions depending whether we will make use of numerical stability or numerical contractivity of approximation Y (depending which knowledge is available on Y, but note that one property out of contractivity of Y, the stability of Y that has to be fulfilled to have control on error propagation). Let us assume numerical stability of the approximation Y. Then one arrives at

$$\begin{aligned}
\varepsilon_{n+1} \;&=\; \mathbb{E}\,\|X_{t_n, X_{0,x}(t_n)}(t_{n+1}) - Y_{n, Y_{0,y}(t_n)}(t_{n+1})\| \\
&\leq\; \mathbb{E}\,\|Z_n^1 + Z_n^2\| \qquad\qquad (5.5.3)\\
&\leq\; \underbrace{\mathbb{E}\,\|X_{t_n, X_{0,x}(t_n)}(t_{n+1}) - X_{t_n, Y_{0,y}(t_n)}(t_{n+1})\|}_{\textbf{controlled by contractivity of X and } \varepsilon_n} \\
&\quad +\; \underbrace{\mathbb{E}\,\|X_{t_n, Y_{0,y}(t_n)}(t_{n+1}) - Y_{n, Y_{0,y}(t_n)}(t_{n+1})\|}_{\textbf{controlled by consistency of Y / stability of Y}} \\
&\leq\; \exp(K_{OL}\Delta_n)\underbrace{\mathbb{E}\,\|X_{0,x}(t_n) - Y_{0,y}(t_n)\|}_{=\varepsilon_n} \\
&\quad + \max(K_1^C, K_0^C)(1 + \mathbb{E}\,\|Y_{0,y}(t_n)\|)(\Delta_n)^{\gamma_g + 1} \\
&\leq\; \exp(K_{OL}\Delta_n)\varepsilon_n \\
&\quad + \max(K_1^C, K_0^C)\left[1 + K_{S1}^Y \exp\left(K_{S2}^Y(t_n - t_0)\right)(1 + \|y\|)\right](\Delta_n)^{\gamma_g + 1} \\
&\leq\; \exp(K_{OL}\Delta_n)\varepsilon_n \\
&\quad + \max(K_1^C, K_0^C)\left[1 + K_{S1}^Y \exp\left([K_{S2}^Y]_+(T - t_0)\right)(1 + \|y\|)\right]\Delta^{\gamma_g}\Delta_n
\end{aligned}$$

where $[.]_+$ denotes the nonnegative part of the inscribed expression (i.e. $z = [z]_+ + [-z]_+ = [z]_+ - [z]_-$). This estimation can only be done if X, Y leave the same domain $\mathbb{D}$ invariant! (i.e. the need of $\mathbb{D}$-invariance, which is not a big issue for approximations with efficient estimates and constructions on the entire space $\mathbb{R}^d$.) Using the following elementary nonautonomous discrete time version of Bellman-Gronwall inequality (linear variation-of-constants inequality in Schurz (1996, 1997), proof by induction)

Lemma 5.5.6 *(Schurz (1996)). Assume the sequence $v = (v_n)_{n\in\mathbb{N}}$ satisfies*

$$0 \le v_{n+1} \ \le \ v_n(1 + c_H(n)) + c_I(n)$$

or

$$0 \le v_{n+1} \ \le \ v_n \exp(c_H(n)) + c_I(n)$$

with appropriate finite, real constants c_H, c_I for all $n \in \mathbb{N}$.
Then v must satisfy the linear **discrete time constants-of-variation inequality,** *i.e.*

$$v_n \ \le \ v_0 \exp\Big(\sum_{i=0}^{n} c_H(i)\Big) + \sum_{i=0}^{n} c_I(i) \exp\Big(\sum_{l=i+1}^{n} c_H(l)\Big)$$

for all $n \in \mathbb{N}$.

Now

$$\begin{aligned}
\varepsilon_n \ &\le \ \exp\Big(K_{OL}(t_n - t_0)\Big)\varepsilon_0 + \sum_{i=0}^{n} K_I(i) \exp\Big(-K_{OL}(t_i - t_0)\Big)\Delta_i^{\gamma_g+1} \\
&\le \ \exp\Big(K_{OL}(t_n - t_0)\Big)\varepsilon_0 + \hat{K}_1 \Delta^{\gamma_g} \sum_{i=0}^{n} \exp\Big(-K_{OL}(t_i - t_0)\Big)\Delta_i \\
&\le \ \exp\Big(K_{OL}(t_n - t_0)\Big)\varepsilon_0 + \hat{K}_1 \Delta^{\gamma_g} \frac{1 - \exp\Big(-K_{OL}(t_n - t_0)\Big)}{K_{OL}} \\
&\le \ \exp\Big([K_{OL}]_+(T - t_0)\Big)\varepsilon_0 + \hat{K}_1 \Delta^{\gamma_g} \frac{1 - \exp\Big(-K_{OL}(T - t_0)\Big)}{K_{OL}}
\end{aligned}$$

uniformly for $n = 0, 1, ..., n_T - 1$, using the elementary fact that $\frac{1-\exp(-Kx)}{K}$ is a positive, monotonically increasing function at x, where

$$K_I(i) = \max(K_1^C, K_0^C)\Big[1 + K_{S1}^Y \exp\Big(K_{S2}^Y(t_i - t_0)\Big)(1 + \|y\|)\Big]$$

and

$$\hat{K}_1 = \max(K_1^C, K_0^C)\Big[1 + K_{S1}^Y \exp\Big([K_{S2}^Y]_+(T - t_0)\Big)(1 + \|y\|)\Big]$$

with appropriate finite, real constants c_H, c_I for all $n \in \mathbb{N}$.

Thus, if the initial error ε is controlled by

$$\varepsilon_0 \le K_{init}\Delta^{\gamma_g}$$

with some appropriate real constant $K_{init} \ge 0$, one finds

$$\varepsilon_n \ \le \ K_g \Delta^{\gamma_g}$$

where

$$K_g \le \max\{K_{init} \exp\Big([K_{OL}]_+(T - t_0)\Big), \hat{K}_1 \frac{1 - \exp\Big(-K_{OL}(T - t_0)\Big)}{K_{OL}}\}.$$

Consequently, the total error is uniformly bounded in terms of Δ^{γ_g} – a fact which justifies speaking of a global convergence rate γ_g of the related numerical method on the interval $[0,T]$. Moreover, the total error continuously depends on the initial values x, y of exact solution X and approximation Y, respectively, and also on the numerical consistency constants K_0^C, K_1^C of Y, the numerical stability constants K_{S1}^Y, K_{S2}^Y of Y, and on the length of the integration interval $T - t_0$.

Now let us return to the splitting (5.5.3) and assume that numerical contractivity of Y with contractivity constant K_C^Y is available instead of stability estimates for Y. Define

$$Z_n^1 = X_{t_n,X_{0,x}(t_n)}(t_{n+1}) - Y_{t_n,X_{0,x}(t_n)}(t_{n+1})$$

$$Z_n^2 = Y_{t_n,X_{0,x}(t_n)}(t_{n+1}) - Y_{t_n,Y_{0,y}(t_n)}(t_{n+1}).$$

$$\begin{aligned}
\varepsilon_{n+1} &= \mathbb{E}\,\|X_{t_n,X_{0,x}(t_n)}(t_{n+1}) - Y_{t_n,Y_{0,y}(t_n)}(t_{n+1})\| \\
&\leq \mathbb{E}\,\|Z_n^1 + Z_n^2\| \qquad (5.5.4)\\
&\leq \underbrace{\mathbb{E}\,\|X_{t_n,X_{0,x}(t_n)}(t_{n+1}) - Y_{t_n,X_{0,x}(t_n)}(t_{n+1})\|}_{\textbf{controlled by consistency of Y / stability of X}} \\
&\quad + \underbrace{\mathbb{E}\,\|Y_{t_n,X_{0,x}(t_n)}(t_{n+1}) - Y_{t_n,Y_{0,y}(t_n)}(t_{n+1})\|}_{\textbf{controlled by contractivity of Y and } \varepsilon_n} \\
&\leq \max(K_1^C, K_0^C)(1 + \mathbb{E}\,\|X_{0,x}(t_n)\|)(\Delta_n)^{\gamma_g+1} \\
&\quad + \exp(K_C^Y\Delta_n)\,\underbrace{\mathbb{E}\,\|X_{0,x}(t_n) - Y_{0,y}(t_n)\|}_{=\varepsilon_n} \\
&\leq \exp\Big(K_C^Y\Delta_n\Big)\varepsilon_n \\
&\quad + \max(K_1^C, K_0^C)\Big[1 + \exp\Big([K_S^X]_+(T-t_0)\Big)(1 + \|x\|^p)^{\frac{1}{p}}\Big]\Delta^{\gamma}\Delta_n \\
&\leq \exp\Big([K_C^Y]_+(T-t_0)\Big)\varepsilon_0 + \hat{K}_1\Delta^{\gamma_g}\frac{1 - \exp\Big(-K_C^Y(T-t_0)\Big)}{K_C^Y} \qquad (5.5.5)
\end{aligned}$$

for $n = 0, 1, ..., n_T - 1$, using Lemma 5.2 as before, where

$$\hat{K}_1 = \max(K_1^C, K_0^C)\Big[1 + \exp\Big([K_S^X]_+(T-t_0)\Big)(1 + \|y\|)\Big]$$

with appropriate stability constants K_S^X, which can be extracted from statements such as Lemma 2.1. Thus we get a similar uniform estimate for the global error ε_n as above. An analogous estimation process, but more technical and laborious with the use of Hölder's and Minkowski's inequalities, can be carried out for general $p > 1$. In particular, for the case $p = 2$, see also the general convergence theorem of Mil'shtein-Schurz.

A general warning is sent out to all who are tempted to neglect the interplay of key concepts in this basic principle combining the concepts of $\mathbb{D}$-invariance, pth mean consistency, and stability or contractivity to achieve global uniform error estimates for the class of SDEs satisfying (OBC), (OLC), (IMC) with Caratheodory drift a and drift b^j functions. The proofs can even be made to show some sharp estimates for the subclass of mentioned SDEs (5.2.1). There are also plenty of deterministic examples which might illustrate undesirable effects in numerical approximations compared to those of underlying continuous time dynamics (for example, take the logistic equation or other chaotic systems) to manifest the danger of its neglect. It can be argued that a consistent approach to numerical analysis and mathematically meaningful maximum of step size Δ for well-posed equations (5.2.1) should be selected according to criterion

$$\Delta \leq 1, \ \max(K_0^C, K_p^C)\Delta^{\gamma_l - 1} \leq 1, \ \max([K_C^X]_+, [K_C^Y]_+)\Delta \leq 1$$

as argued by theorems below with **local pth mean convergence order**

$$\gamma_l := \max\left(\frac{\gamma_0}{p} + \frac{\gamma_p}{q}, \frac{\gamma_0}{q} + \frac{\gamma_p}{p}\right) > 1.$$

It is not surprising that there is a corresponding relation for the minimum step size as well. However, these estimates would go beyond the goal of this survey. For more details, see the forthcoming papers of the author.

5.5.7 On fundamental crossrelations

The above mentioned main principle may be simplified in case of SDEs with (OBC) and (IMC) under some circumstances. We have already seen that (OBC) and (IMC) imply the stability of X, thanks to Lemma 5.2.1. Furthermore, the stability of Y can be concluded by the consistency of Y with local convergence order

$$\gamma_l := \max\left(\frac{\gamma_0}{p} + \frac{\gamma_p}{q}, \frac{\gamma_0}{q} + \frac{\gamma_p}{p}\right) > 1$$

and the stability of X with stability constants $K_{OB}^I = 0, K_S^X = 2(p-1)K_{OB}^H$, using the well-known Minkowski's inequality for L^p-spaces. Assume $\mathbb{D}$-invariance of both X and Y with respect to one and the same domain $\mathbb{D} \subset \mathbb{R}^d$, and sufficiently small discretization meshes such that $0 \leq \max(K_p^C, K_0^C)(\Delta)^{\gamma_l - 1} \leq 1$. Consider the estimate

$$\begin{aligned}
v(t) \ &:= \ \left(\mathbb{E}\, \|Y_{0,Y(0)}(t)\|^p\right)^{\frac{1}{p}} \ = \ \left(\mathbb{E}\, \|Y_{s,Y(s)}(t)\|^p\right)^{\frac{1}{p}} \\
&= \ \left(\mathbb{E}\, \|Y_{s,Y(s)}(t) - X_{s,Y(s)}(t) + X_{s,Y(s)}(t)\|^p\right)^{\frac{1}{p}} \\
&\leq \ \left(\mathbb{E}\, \|Y_{s,Y(s)}(t) - X_{s,Y(s)}(t)\|^p\right)^{\frac{1}{p}} + \left(\mathbb{E}\, \|X_{s,Y(s)}(t)\|^p\right)^{\frac{1}{p}} \\
&\leq \ \max(K_p^C, K_0^C)(1 + \mathbb{E}\, \|Y(s)\|^p)^{\frac{1}{p}}(t-s)^{\gamma_l} + \exp(K_S^X(t-s))(\mathbb{E}\, \|Y(s)\|^p)^{\frac{1}{p}} \\
&\leq \ v(s) + \max(K_p^C, K_0^C)(t-s)^{\gamma_l} + \exp(\max(K_S^X, K_p^C, K_0^C)(t-s)]v(s) \\
&\leq \ v(s) + \max(K_p^C, K_0^C)(t-s) + \exp(\max(K_S^X, K_p^C, K_0^C)(t-s)]v(s).
\end{aligned}$$

By application of linear variation-of-constants inequality from Lemma 5.2 due to Schurz (1996), we gain

$$\begin{aligned}
v(t) \ &\leq \ v(s)\exp\Big(\max(K_S^X, K_p^C, K_0^C)(t-s)\Big) + \\
&\quad + \max(K_p^C, K_0^C)\frac{1 - \exp\Big(-\max(K_S^X, K_p^C, K_0^C)(t-s)\Big)}{\max(K_S^X, K_p^C, K_0^C)},
\end{aligned}$$

therefore $\sup_{0 \leq t \leq T} v(t) \ < \ +\infty$ if $\mathbb{E}\, \|Y_0\|^p < +\infty$. In other words, we know that Y is pth mean numerically stable with suitable constants

$$K_{S1}^Y = 1, K_{S2}^Y = \max(K_S^X, K_p^C, K_0^C)$$

hence uniformly bounded on the fixed interval $[0, T]$.

Theorem 5.5.7 *Let $p \geq 1$. Assume that $Y = (Y_n)_{n \in \mathbb{N}}$ with maximum step size Δ governed by*

$$\max(K_p^C, K_0^C)(\Delta)^{\gamma_l - 1} \leq 1$$

represents a pth mean consistent numerical method with order $\gamma_p \geq 1$ for stochastic process $X = (X_t)_{0 \leq t \leq T}$ satisfying SDE (5.2.1) with (IMC), (OBC) and (OLC). Let its stability constants satisfy $K_{OB}^I = 0, K_S^X = 2(p-1)K_{OB}^H$. Assume X and Y are $\mathbb{D}$-invariant with respect to the same domain $\mathbb{D} \subset \mathbb{R}^d$.
Then the stochastic process Y is numerically pth mean stable, and hence has uniformly bounded pth moments, provided that $\mathbb{E}\,\|Y_0\|^p < +\infty$ and $Y_0 \in \mathbb{D}$.

Let us now describe the relation between convergence, consistency, stability and contractivity once more. For this purpose, we have to say a few words on contractivity. Contractivity is in general a weaker requirement than stability. This could be seen in Schurz (1996, 1997, 1999), since the concept of contractivity does not take into account any influence which might originate from the inhomogeneous parts of the dynamics (i.e. loosely speaking, the concept of contractivity represents the concept of stability of the homogeneous part of underlying dynamics, and in a certain sense it can be viewed as the stability property of the associated linearized nonautonomous flow.). Moreover, with the help of stability properties of underlying exact solution X, one can conclude stability of Y by contractivity of Y using Minkowski's inequality for L^p-spaces. Assume $\mathbb{D}$-invariance of both X and Y with respect to one and the same domain $\mathbb{D} \subset \mathbb{R}^d$, and sufficiently small discretization meshes such that $K_S^X \Delta \leq p, \Delta \leq 1$ with $K_S^X = (p-2)K_{OB}^I + pK_{OB}^H$. Consider the estimate

$$\begin{aligned}
v(t) &:= \left(\mathbb{E}\,\|Y_{0,Y(0)}(t)\|^p\right)^{\frac{1}{p}} = \left(\mathbb{E}\,\|Y_{s,Y(s)}(t)\|^p\right)^{\frac{1}{p}} \\
&= \left(\mathbb{E}\,\|Y_{s,Y(s)}(t) - Y_{s,X(s)}(t) + Y_{s,X(s)}(t) - X_{s,X(s)}(t) + X_{s,X(s)}(t)\|^p\right)^{\frac{1}{p}} \\
&\leq \underbrace{\left(\mathbb{E}\,\|Y_{s,Y(s)}(t) - Y_{s,X(s)}(t)\|^p\right)^{\frac{1}{p}}}_{\textbf{controlled by contractivity of Y}} + \underbrace{\left(\mathbb{E}\,\|X_{s,X(s)}(t) - Y_{s,X(s)}(t)\|^p\right)^{\frac{1}{p}}}_{\textbf{controlled by consistency of Y}} \\
&+ \underbrace{\left(\mathbb{E}\,\|X_{s,Y(s)}(t)\|^p\right)^{\frac{1}{p}}}_{\textbf{controlled by stability of X}}
\end{aligned}$$

$$\begin{aligned}
&\leq \exp(K_C^Y(t-s)) \underbrace{\left(\mathbb{E}\,\|Y_{0,Y(0)}(s) - X_{0,X(0)}(s)\|^p\right)^{\frac{1}{p}}}_{\textbf{controlled by convergence}} \\
&\quad + \max(K_p^C, K_0^C)\mathbb{E}\,(1 + \|X_{0,X(0)}(s)\|^p)^{\frac{1}{p}}\Delta^{\gamma_l} \\
&+ \exp(\frac{2(p-1)K_{OB}}{p}(t-s))\Big((\mathbb{E}\,\|Y(s)\|^p)^{\frac{1}{p}} + \frac{1 - \exp(-2(p-1)K_{OB}(t-s)}{p-1}\Big) \\
&\leq \exp(K_C^Y \Delta)K(p, T, \mathbb{E}\,\|Y(s)\|^p, \mathbb{E}\,\|X(s)\|^p)\Delta^{\gamma_g} \\
&+ \max(K_p^C, K_0^C) \sup_{0 \leq t \leq T} \mathbb{E}\,(1 + \|X_{0,X(0)}(s)\|^p)^{\frac{1}{p}}\Delta^{\gamma_l} + \exp(\frac{2(p-1)[K_{OB}]_+}{p}\Delta)v(s) \\
&+ \frac{\exp(2(p-1)K_{OB}\Delta) - 1}{p-1}\exp(\frac{2(p-1)[K_{OB}]_+}{p}\Delta)
\end{aligned}$$

where we set $t = t_{n+1}$ and $s = t_n$. Using the linear variation-of-constants inequality (see Schurz (1996,1997)), we easily see that there is a real constant $K > 0$ such that $\sup_{0\le t\le T} v(t) \le K < +\infty$. Thus, the Theorem 5.4 is established.

Theorem 5.5.8 *Let $p \ge 1$. Assume that $Y = (Y_n)_{n\in\mathbb{N}}$ with maximum step size Δ governed by*

$$\max(K_p^C, K_0^C)(\Delta)^{\gamma_l - 1} \le 1$$

represents a pth mean contractive numerical method with contractivity constant $K_C^Y = K_C^Y(p)$ and pth mean converging with order $\gamma_g \ge 0$ to stochastic process $X = (X_t)_{0\le t\le T}$ satisfying SDE (5.2.1) with (IMC), (OBC) and (OLC). Assume X and Y are $\mathbb{D}$-invariant with respect to the same domain $\mathbb{D} \subset \mathbb{R}^d$.
Then the stochastic process Y is numerically pth mean stable, and hence has uniformly bounded pth moments, provided that $\mathbb{E}\,\|Y_0\|^p < +\infty$ and $Y_0 \in \mathbb{D}$.

Furthermore, consistency and contractivity may already imply convergence in the pth mean sense. In a similar way as before we conclude this assertion.

Theorem 5.5.9 *Let $p \ge 1$. Assume that the numerical method $Y = (Y_n)_{n\in\mathbb{N}}$ with maximum step size $[K_C^Y]_+\Delta \le 1$ is pth mean contractive with contractivity constant $K_C^Y = K_C^Y(p)$, mean consistent with order $\gamma_0 = \gamma_g + 1$ and pth mean consistent with order $\gamma_p = \gamma_g + \frac{1}{p} \ge 0$ to stochastic process $X = (X_t)_{0\le t\le T}$ satisfying SDE (5.2.1) with (IMC), (OBC) and (OLC). Let X and Y be $\mathbb{D}$-invariant with respect to the same domain $\mathbb{D} \subset \mathbb{R}^d$.*
Then the stochastic process Y is numerically pth mean converging with order $\gamma_g \in \mathbb{R}_+$ on the fixed time-interval $[0,T]$, provided that $\mathbb{E}\,\|Y_0\|^p < +\infty$, $Y_0 \in \mathbb{D}$ and $\mathbb{E}\,\|X_0 - Y_0\|^p \le K_{init}\Delta^{p\gamma_g}$. Moreover, the error process $(\varepsilon_n)_{n\in\mathbb{N}}$ satisfies the estimate

$$\sup_{n\in\mathbb{N}} \varepsilon_n \le \exp([K_C^Y]_+T)\Big(\varepsilon_0 + \max(K_p^C, K_0^C)(1 + \sup_{0\le t\le T}(\mathbb{E}\,\|X_{0,X(0)}(t)\|^p)^{\frac{1}{p}})T\Delta^{\gamma_g}\Big).$$

As a consequence of presented analysis, we arrive at a **stochastic Kantorovich-Lax-Richtmeyer equivalence principle** for (O)SDEs. The proof is just a fancy, but trivial, combination of our previous results.

Proposition 5.5.10 *Fix $p \ge 1$. Assume the numerical sequence $Y = (Y_n)_{n\in\mathbb{N}}$ with $\mathbb{E}\,\|Y_0\|^p < +\infty$ and maximum step size Δ restricted by $[K_C^Y]_+\Delta \le 1$ and*

$$\max(K_p^C, K_0^C)(\Delta)^{\gamma_l - 1} \le 1$$

is $\mathbb{D}$-invariant, pth mean contractive with contractivity constant $K_C^Y = K_C^Y(p)$ and mean consistent with order γ_0 for the $\mathbb{D}$-invariant stochastic process $X = (X_t)_{0\le t\le T}$ satisfying SDE (5.2.1) with (IMC), (OBC) and (OLC).
Then it holds

Y is numerically pth mean stable and pth mean consistent with some order γ_p such that $\gamma_l > 1$	**iff**	*Y is numerically pth mean converging.*

with local pth mean convergence order $\gamma_l = \max\left(\frac{\gamma_0}{p} + \frac{\gamma_p}{q}, \frac{\gamma_0}{q} + \frac{\gamma_p}{p}\right)$.

One can also show that, from pth mean convergence with order $\gamma_g > 0$, it follows the property of local contractivity of Y on any bounded domain $\mathbb{D}$ which is left invariant by X and Y. However, this would sprinkle the scope of our survey. The noticed convergence orders are only "worst case estimates." There are some refinements for special cases $p = 2$ in the literature (almost all only for equidistant approximations), see Mil'shtein (1995) or also in Section 6 below. For a variable step size selective algorithm, one still has to take care of "too small step size"; thus a ratio between maximum and minimum step size is reasonable, see also Schurz (1996, 1997, 1999). Also one should never apply step sizes larger than 1 as seen before in our argument (unless one treats dissipative dynamics by appropriate implicit techniques), and, in particular, the maximum admissible step size should be restricted by $[K_C^Y]_+\Delta \leq 1$. For more general principles for numerical approximations of stochastic processes with values on randomized Banach spaces, see Schurz (1999).

Let us summarize the main principles of numerical analysis for stochastic differential equations by the following more generally valid **Diagram**:

Approximative Well-posedness: Stability of X Contractivity of Y	Consistency of (X, Y) $\Longleftrightarrow$	Approximative Well-posedness: Stability of Y Contractivity of X
	$\Updownarrow$	
	Convergence of (X, Y)	

which describes the main crossrelations and the fundamental equivalence principle in the context of stochastic approximations as well, which is the point where we arrived at the heart of the sophisticated numerical approximation theory for stochastic processes. Our remaining goal is just to make it come alive in conjunction with SDEs (5.2.1) and their numerical analysis in a concise course.

5.6 Results on Convergence Analysis

There is a variety of possible different convergence notions. We shall only collect the most frequent ones. Recall the numerical convergence notions from Section 5.

5.6.1 Continuous time convergence concepts

One of the weakest notions one could think of is that of weak convergence. One of the essential contributions of Mil'shtein relies on the following concepts of weak and mean square approximations, generalized by stepping down from pth mean to weak convergence. We shall pursue convergence analysis up to the strongest notion which is given by that of strong pth mean convergence. In the statements below, let $||\cdot||$ be a vector norm of $\mathbb{R}^d$ and $K_0, K_p(p \in [1, +\infty])$ be deterministic, real constants which may depend on smoothness and boundedness parameters of the explicit solution, as well as initial values, the length of time interval $[0, T]$, the dimensions d, m and some parameter of the corresponding numerical method. Remember

$$\Delta = \sup\{|t_{n+1} - t_n| : n = 0, 1, 2, ..., n_T - 1\}.$$

Fix the finite deterministic start instant $t_0 \in [0, T]$ with fixed terminal time $T > t_0$ where $T \in \mathbb{R}^1$. Let $Y = (Y_t)_{0 \leq t \leq T}$ denote a right-continuous time approximation of process $X = (X_t)_{0 \leq t \leq T}$.

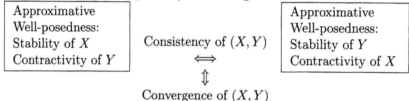

Definition 5.6.1 *A stochastic process* $Y = (Y_t^\Delta)_{0 \le t \le T}$ *(method, scheme, etc.) is called a* p**th mean approximation of** $X = (X_t)_{t \in [t_0,T]}$ **with order (rate)** $\gamma \ge 0$ *if*

$$\sup_{0 \le t \le T} \left(\mathbb{E} \, ||X_t - Y_t^\Delta \,||^p \right)^{1/p} \le K_p \cdot \Delta^\gamma, \tag{5.6.1}$$

a **mean square approximation of** $X = (X_t)_{t \in [t_0,T]}$ **with order (rate)** $\gamma \ge 0$ *if*

$$\sup_{0 \le t \le T} \left(\mathbb{E} \, ||X_t - Y_t^\Delta \,||^2 \right)^{1/2} \le K_2 \cdot \Delta^\gamma, \tag{5.6.2}$$

a **strong approximation of** $X = (X_t)_{t \in [t_0,T]}$ **with order (rate)** $\gamma \ge 0$ *if*

$$\sup_{0 \le t \le T} \mathbb{E} \, ||X_t - Y_t^\Delta \,|| \le K_1 \cdot \Delta^\gamma \tag{5.6.3}$$

a **strong mean square approximation of** $X = (X_t)_{t \in [t_0,T]}$ **with order (rate)** $\gamma \ge 0$ *if*

$$\left(\mathbb{E} \sup_{0 \le t \le T} ||X_t - Y_t^\Delta \,||^2 \right)^{1/2} \le K_2 \cdot \Delta^\gamma, \tag{5.6.4}$$

a **strong** p**th mean approximation of** $X = (X_t)_{t \in [t_0,T]}$ **with order (rate)** $\gamma_p \ge 0$ *if*

$$\left(\mathbb{E} \sup_{0 \le t \le T} ||X_t - Y_t^\Delta \,||^p \right)^{1/p} \le K_p \cdot \Delta^\gamma, \tag{5.6.5}$$

a **double** L^p**-approximation of** $(X_t)_{t \in [t_0,T]}$ **with order (rate)** $\gamma \ge 0$ *if*

$$\left(\mathbb{E} \int_0^T K(t) ||X_t - Y_t^\Delta \,||^p \, \mu(dt) \right)^{1/p} \le K_p \cdot \Delta^\gamma, \tag{5.6.6}$$

with a positive, μ*-integrable kernel* $K(t)$ *where* μ *is an appropriate positive, finite measure on* $([0,T], \mathcal{B}([0,T]))$ *(*$\mathcal{B}([0,T])$ *denotes the* σ*-field of Borel sets of* $[0,T]$*), a* **weak approximation of** $X = (X_t)_{t \in [t_0,T]}$ **with order (rate)** $\beta \ge 0$ *if*

$$\sup_{g \in F} \sup_{0 \le t \le T} ||\mathbb{E} \; g\,(X_t) - \mathbb{E} \; g(Y_t^\Delta) \,|| \le K_0 \cdot \Delta^\beta \tag{5.6.7}$$

and a **weak** τ**-convergent approximation of** $X = (X_t)_{t \in [t_0,T]}$ **with order (rate)** $\beta \ge 0$ *if*

$$\sup_{g \in F} \sup_{0 \le \tau \le T} ||\mathbb{E} \; g\,(X_\tau) - \mathbb{E} \; g(Y_\tau^\Delta) \,|| \le K_0 \cdot \Delta^\beta \tag{5.6.8}$$

for all time-discretizations of $[t_0, T]$ *with* $\Delta < \delta_0 < +\infty$*, where the supremum is taken over all finite stopping times* τ *and* F *is an appropriate class of real-valued functions.*

Remark *One also speaks of* p**th mean, mean square, strong, strong** p**th mean, double** L^p**, and weak orders (rates)** $\gamma, \beta \in \mathbb{R}_+$ *of convergence. The function class is frequently chosen to be*

$$F_r = \left\{ f : \mathbb{R}^d \longrightarrow \mathbb{R}^k, f \in C^\infty(\mathbb{D}), \exists K \, \forall x \in \mathbb{D} \, \|f(x)\| \le K(1 + \|x\|^r) \right\}$$

where $r \in \mathbb{R}, r \ge 1$*, and* $d, k \in \mathbb{N}$ *are fixed, but there are also attempts to relax conditions in* F *to certain classes of Lebesgue-measurable functions (see e.g. Bally and Talay (1996)* *under further conditions on the differential dynamics for X). The weak τ-convergence is introduced for the delicate problem of convergence and convergence rates for functionals involving random stopping times instead of deterministic terminal times, something of great use and very reasonable in optimal stochastic control problems related to diffusions X. Note also that, for pth mean convergence, it suffices to evaluate the error expressions at discretization points t_n under the commonly met assumptions on SDE coefficients and on approximating integrands arising by the related numerical method. This becomes clear from looking at the continuous time behavior of remainder terms of stochastic Taylor expansions. pth mean convergence analysis has enormous importance for estimation of noncontinuously differentiable or path-dependent functionals of SDEs. The* **main tools for stochastic-numerical analysis** *are the* **Itô Formula, Dynkin Formula, Wagner-Platen Expansion, Variation-of-Constants Inequalities, Burkholder-Davis-Gundy Inequalities, Semimartingale Decompositions and Stochastic Integration Theory, Stochastic Equivalence Principles** *like* **Stochastic Kantorovich-Lax-Richtmeyer Theorems** *(see for some variants, the main principle of numerics before) in conjunction with the fundamental convergence theorems presented below. These tools explain the construction and behavior of* **one-step approximations, local convergence (consistency), error propagation control (contractivity, stability) and global convergence***, and other qualitative features at which one might look.*

5.6.2 On key relations between convergence concepts

As a consequence of the Lyapunov inequality and fast L^p-convergence (Borel-Cantelli Theorem) we may notice

Proposition 5.6.2 *Assume that $F = C^1_{Lip}(\mathbb{R}^d, \mathbb{R}^k)$, $\sup_{0\leq t\leq T} \mathbb{E}\, ||X_t|| < +\infty$, and fix $p \geq 1$.*
Then the following implications hold

Strong pth mean	$\Longrightarrow$	*pth mean*	$\Longrightarrow$	*strong conv.*	$\Longrightarrow$	*weak conv.*
Strong pth mean	$\Longrightarrow$	*double L^p*				
Strong pth mean	$\Longrightarrow$	*a. s. conv.*				
Weak τ-convergence	$\Longrightarrow$	*weak conv.*				

where the related convergence orders are carried over one to one (at least when $p \geq 2$).

How the convergence rates for noncontinuously differentiable functions f are transferred in this diagram is a fairly complex and partially open question. For a partial answer, compare with Subsection 6.4. If F is the class of Hölder continuous functions with exponent $\alpha_H \in (0,1)$, then the orders are reduced by α_H (i.e. $\beta = \alpha_H\gamma_p$ are the related weak convergence orders, cf. Theorem 6.12). The weak τ-convergence orders are transferred to weak orders one to one (but not necessarily vice versa in all cases). In the nonsmooth situation of class F we also suggest to take the standard mollifying procedures and then to apply a favorite numerical method to the mollified problem (however, also here it has to be clarified how the convergence rates are carried over).

5.6.3 Fundamental theorems of mean square convergence

A refinement of the main principle of numerical analysis restricted to the concept of mean square convergence could be found by Mil'shtein (1988, 1995) who exclusively proved the statement for equidistant discretizations under usual conditions (ULC), (UBC) and (IMC)

at first. Schurz (1996) generalized that theorem to the case of variable step sizes under one-sided Lipschitz continuity and one-sided boundedness conditions as stated below, which considerably relax the original conditions and proof steps of Mil'shtein in a maximally possible way within mean square convergence framework. A corresponding variant for the general pth mean convergence case is in progress, see Schurz (1999). The following theorem can be considered as a fundamental theorem on the relation of mean square convergence rates and a very good starting point to understand pth mean convergence analysis in stochastic settings.

Theorem 5.6.3 *(Mil'shtein 1995, Schurz 1996): Assume a, b^j are Caratheodory functions, and*

(o) $X_{0,x_0}(0) = x_0 \in \mathbb{D}$ *independent of* $\mathcal{F}^j = \sigma(W_s^j, s \geq 0)$

(i) $\mathbb{E}\,\|x_0\|^2 < +\infty$

(ii) (one sided) mean square boundedness condition: $\exists K_0\ \forall t \in [0,T]\ \forall x \in \mathbb{D}$

$$< a(t,x), x > + \frac{1}{2}\sum_{j=1}^{m} \|b^j(t,x)\|^2 \;\leq\; K_0(1 + \|x\|^2)$$

(iii) (one sided) mean square Lipschitz condition: $\exists K_c\ \forall t \in [0,T]\ \forall x, y \in \mathbb{D}$

$$< a(t,x) - a(t,y), x - y > + \frac{1}{2}\sum_{j=1}^{m} \|b^j(t,x) - b^j(t,y)\|^2 \;\leq\; K_c\|x - y\|^2$$

(iv) $X_{0,x_0}(t)$, $Y_{0,x_0}(t)$ *regular on domain* $\mathbb{D} \subset \mathbb{R}^d$

(v) **one-step mean accuracy**: $\exists K_1\ \forall t \in [0,T]\ \forall h : 0 < h \leq \Delta\ \forall z \in \mathbb{D}$

$$\|\mathbb{E}\,[X_{t,z}(t+h) - Y_{t,z}(t+h)]\,\| \;\leq\; K_1(1 + \|z\|)h^{\gamma_0}$$

(vi) **one-step mean square accuracy**: $\exists K_2\ \forall t \in [0,T]\ \forall h : 0 < h \leq \Delta\ \forall z \in \mathbb{D}$

$$\mathbb{E}\,\|X_{t,z}(t+h) - Y_{t,z}(t+h)\|^2 \;\leq\; (K_2)^2(1 + \|z\|^2)h^{2\gamma_2}$$

(vii) $\gamma_0 \geq \gamma_2 + \frac{1}{2}$, $\gamma_2 \geq \frac{1}{2}$

Then

Fundamental Mean Square Convergence Relation

$$\varepsilon_2(T) = \sup_{0 \leq t \leq T} \left(\mathbb{E}\,\|X_{0,x_0}(t) - Y_{0,x_0}(t)\|^2\right)^{1/2} \;\leq\; K_3(1 + \|x_0\|^2)^{1/2}\Delta^{\gamma_2 - \frac{1}{2}}$$

where $K_0, ..., K_3, K_c$ are real constants, maximum step size $\Delta \leq 1$, and $\gamma_g(2) = \gamma_2 - \frac{1}{2}$.

The constants K_i can be determined very precisely by means of the same analysis as in Section 5. Under $(UBC), (ULC), (IMC)$ (which are the most reasonable conditions under strong pth mean convergence analysis) and with $p = 2$ Mil'shtein (1995) has sharpened the convergence assertion on mean square rates to those of strong mean square convergence with equidistant step sizes.

Theorem 5.6.4 *(Mil'shtein, 1995). Assume that SDE (5.2.1) satisfies the conditions* (IMC), (UBC), (ULC), $\mathbb{E}\,\|X_0\|^4 < +\infty$, *and all conditions [i] - [vii] from Theorem 6.2 are fulfilled. Furthermore, assume that*

(viii) **one-step 4th mean accuracy**: $\exists K_4\, \forall t \in [0,T]\, \forall h : 0 < h \leq \Delta\, \forall z \in \mathbb{D}$

$$\mathbb{E}\,\|X_{t,z}(t+h) - Y_{t,z}(t+h)\|^4 \;\leq\; (K_4)^4(1+\|z\|^4)h^{4\gamma_2 - 1}$$

with $\gamma_2 \geq \frac{3}{4}$.
Then

Fundamental 4th Mean Convergence Relation

$$\varepsilon_4(T) = \sup_{0\leq t\leq T} \left(\mathbb{E}\,\|X_{0,x_0}(t) - Y_{0,x_0}(t)\|^4\right)^{1/4} \;\leq\; K_5(1+\|x_0\|^4)^{1/4}\Delta^{\gamma_2 - \frac{1}{2}}$$

where $K_0, ..., K_5, K_c$ *are real constants, and* $\Delta \leq 1$, *i.e.* $\gamma_g(4) = \gamma_2 - \frac{1}{2}$.

This theorem can be generalized to the case of pth mean convergence when $p > 2$. As we know from Section 5, the maximum step size should be restricted to a sufficiently small one (at least smaller than 1), depending on contractivity, stability and consistency constants of (X, Y). As an application, one easily verifies the pth mean convergence of the Euler methods towards the explicit solution of SDEs with Hölder-(0.5) time-continuous and Lipschitz space-continuous coefficient functions a, b^j with order $\gamma_g = 0.5$ for all $p \geq 2$. Corresponding proofs can be worked out for other numerical methods.

5.6.4 Strong mean square convergence theorem

Mil'shtein (1988,1995) proved the "strengthened convergence theorem" concerning numerical strong mean square convergence. This is generalized by the author to the following continuous time variant (trivially covering the numerical convergence issues as originally defined by Mil'shtein (1988)).

Theorem 5.6.5 *Assume that the conditions of Theorem 6.3 are satisfied with* $\gamma_2 \geq \frac{3}{4}$ *and* $\gamma_0 \geq \gamma_2 + \frac{1}{2}$.
Then

Fundamental Strong Mean Square Convergence Relation

$$\varepsilon_2^s(T) = \left(\mathbb{E}\,\sup_{0\leq t\leq T} \|X_{0,x_0}(t) - Y_{0,x_0}(t)\|^2\right)^{1/2} \;\leq\; K_6(1+\|x_0\|^4)^{1/4}\Delta^{\gamma_2 - \frac{1}{2}}$$

where $K_0, ..., K_4, K_6, K_c$ *are real constants, and* $\Delta \leq 1$, *i.e.* $\gamma_g^s(2) = \gamma_2 - \frac{1}{2}$.

5.6.5 The Clark-Cameron mean square order bound in $\mathbb{R}^1$

Clark and Cameron (1980) could prove the following very remarkable result on maximum order bounds of partition $\mathcal{F}_T^N$-measurable approximations.

Definition 5.6.6 *The stochastic process* $Y = (Y_t)_{0\le t\le T}$ *is called* **partition $\mathcal{F}_T^N$-measurable** *iff all values* Y_{t_n} $(t_n \in [0,T], 0 \le n \le N)$ *are* $\mathcal{F}_{t_n}^N$*-measurable with*

$$\mathcal{F}_{t_n}^N = \sigma\left\{W_{t_k}^j : k = 0,1,...,n; j = 1,2,...,m\right\}$$

for all $n = 0,1,...,N$*, along a given* $\mathcal{F}_{t_n}^N$*-measurable discretization* $0 = t_0 \le t_1 \le ... \le t_N = T$ *for the fixed deterministic time-interval* $[0,T]$.

Remark *The conditional expectations* $\mathbb{E}\,[X_{t_{n+1}}|\mathcal{F}_{t_n}^N]$ *provide the partition* $\mathcal{F}_T^N$*-measurable stochastic approximations with the minimal mean square error due to their inherent projection property in Hilbert spaces* $L^2(\Omega,\mathcal{F},\mathbb{P})$*. Thus it is natural to study their error and practical implementation at first.*

Theorem 5.6.7 *Suppose* $X = (X_t)_{0\le t\le T}$ *satisfies a one-dimensional autonomous SDE*

$$dX_t = a(X_t)\,dt + dW_t \tag{5.6.9}$$

with $a \in C^3(\mathbb{R})$ *and all derivatives of* a *are uniformly bounded. Then*

$$\mathbb{E}\,[(X_T - \mathbb{E}\,[X_T|\mathcal{F}_T^N])^2] = \frac{cT}{N^2} + o(\frac{1}{N^2})$$

where

$$c = \frac{T^3}{12}\int_0^T \mathbb{E}\left[\exp\left(2\int_s^T a'(X_u)du\right)[a'(X_s)]^2\right] ds.$$

Thus, for systems with additive noise, we obtain the general mean square order bound 1.0 for numerical approximations using only the increments of underlying Wiener process. A similar result holds also for diffusions with variable diffusion coefficients $b(x)$ when

$$c(x) := a(x) - \frac{1}{2}b(x)b'(x) \not\equiv Kb(x)$$

for any real constant K, see Clark and Cameron (1980). They also provide a constructive example with multiplicative noise. Consider the two-dimensional SDE

$$\begin{aligned} dX_t^{(1)} &= dW_t^1 \\ dX_t^{(2)} &= X_t^{(1)}\,dW_t^2 \end{aligned}$$

driven by two independent scalar Wiener processes W^1, W^2. This system obviously has the solutions $X_t^{(1)} = W_t^1$ and

$$X_t^{(2)} = \int_0^t W_s^1 dW_s^2 = \int_0^t dW_t^1\,dW_t^2$$

(in fact it is a one-dimensional example with multidimensional "Wiener process differentials" (i.e. $m = 2$)). Then they compute the slow best convergence rate $\gamma_2 = 0.5$ (in mean square sense) for partition $\mathcal{F}_T^N$-measurable approximations using any set of N equidistant,

$\mathcal{F}_{t_n}^N$-measurable time-instants $t_n = n\frac{T}{N}$, and the mean square minimally attainable approximation error

$$\left(\mathbb{E}\left[\left|X_T^{(2)} - \mathbb{E}\left[X_T^{(2)}|\mathcal{F}_T^N\right]\right|^2\right]\right)^{\frac{1}{2}} = \frac{T}{2}\left[\frac{1}{N}\right]^{\frac{1}{2}}.$$

It is worth noting that $X^{(2)}$ represents the simplest nontrivial multiple integral with length $l(\alpha) > 1$. Liske (1982) has studied its joint distribution with (W_t^1, W_t^2). In this case the error order bound for $\mathcal{F}_T^N$-measurable approximations of X_T^2 is already attained with 0.5, since X^2 cannot be expanded in a linear combination of W^1, W^2. This system also exhibits an interesting test equation for the qualitative behavior of numerical methods (e.g. compare the numerically estimated distribution with that of the exact solution derived by Liske (1982)). Since in the L^2 sense one cannot provide better partition $\mathcal{F}_T^N$-measurable approximations than that of the projection done by conditional expectations, there are natural (convergence) order restrictions for $\mathcal{F}_T^N$-measurable approximations. Thus we cannot exceed the order 1 in L^2-sense for $\mathcal{F}_T^N$-measurable approximations. On the other hand, if one wants higher order of convergence in general, one has to enlarge the condition σ-field substantially (actually done by higher order multiple integrals and Levy areas). Note also this is not always necessary for approximations of functionals $V(t, X_t)$ of diffusion processes X with V-commutativity, see Schurz (1999). In fact, for example for pure one-dimensional diffusions X (i.e. when drift a is zero), the x-commutativity condition (i.e. $V(x) = x$), is then identical with the condition of *commutative noise* (in short: noise-commutativity) under the absence of drift terms

$$b^j(x)\frac{db^k(x)}{dx} = b^k(x)\frac{db^j(x)}{dx}$$

for all $j, k = 0, 1, 2, ..., m$. This requirement, together with $b^j \in C^\infty(\mathbb{R})$, effects that $b^j(x) = K_{j,k}b^k(x)$ with some deterministic real constants $K_{j,k}$. In this trivial case one could even obtain any order of pth mean convergence ($p \leq 1$). (This is no surprise after one has carefully analyzed the observation of Clark and Cameron which implies the approximation error 0 by the projection operator of conditional expectation under $a'(x) \equiv 0$ and the noise-commutativity assumption in the situation $d = 1$). Unfortunately, the situation in view of convergence order bounds is much more complicated in the fully multidimensional framework and needs more care in the near future.

5.6.6 Exact mean square order bounds of Cambanis and Hu

Cambanis and Hu (1996) noticed the following result concerning exact mean square convergence error bounds (i.e. for the asymptotic behavior of leading error coefficients of numerical schemes with respect to mean square convergence criteria). For the statement, we introduce the following definition of partition density.

Definition 5.6.8 *A strictly positive, differentiable function $h \in C^0([0,T]^2, \mathbb{R}_+)$ with uniformly bounded derivatives is said to be a* **regular partition density** *of the time-interval $[0,T]$ iff*

$$\int_{t_n^{N(t)}}^{t_{n+1}^{N(t)}} h(t,s)ds = \frac{1}{N(t)}$$

for $n = 0, 1, ..., N(t) - 1$, $t_0 = 0$, where $N = N(t)$ denotes the number of subintervals $[t_n^{N(t)}, t_{n+1}^{N(t)}]$ for a total time interval $[0,t]$ with terminal times $t \leq T$.

Regular partition densities possess the property that

$$\lim_{N(t)\to+\infty, t_n\to s} N(t)(t_{n+1}^{N(t)} - t_n^{N(t)}) = \frac{1}{h(t,s)}.$$

Therefore they describe the distribution of time-instants in discretizations of intervals $[0,T]$ in a fancy manner. Since the conditional approximation provides the mean square $\mathcal{F}^N$-measurable approximation (with $N = N(t)$) with minimal mean square error, one arrives at

Theorem 5.6.9 *Assume that X satisfies a one-dimensional SDE (5.2.1) with coefficients $a, b \in C^3(\mathbb{R}, \mathbb{R})$ possessing bounded derivatives up to third order, $\mathbb{E}\,|X_0|^2 < +\infty$, and all time-discretizations are exclusively done along a given regular partition density h on $[0,T]^2$. Then, there exists a Gaussian process $\eta = (\eta_t)_{0\le t\le T}$ on $(\Omega, \mathcal{F}, (\mathcal{F}_t)_{0\le t\le T}, \mathbb{P})$ such that*

$$\lim_{N(t)\to+\infty} N(t)\Big(X_t - \mathbb{E}\,[X_t|\mathcal{F}_t^{N(t)}]\Big) = \eta_t$$

with mean 0 and covariance matrix $C(t) =$

$$\int_0^t \frac{[(\mathcal{L}^1 a - \mathcal{L}^0 b)(X_s)]^2}{6[h(s)]^2} \exp\left(\int_s^t (2a\prime(X_u) - [b\prime(X_u)]^2)du + 2\int_s^t b\prime(X_u)dW_u\right) ds$$

which is the unique solution of

$$dC(t) = \left((2a\prime(X_t) + [b\prime(X_t)]^2)C(t) + \frac{[(\mathcal{L}^1 a - \mathcal{L}^0 b)(X_t)]^2}{12}\right)dt + 2b\prime(X_t)C(t)\,dW_t$$

with $\eta_0 = 0$ and has the property

$$\lim_{N(t)\to+\infty} N(t)\mathbb{E}\left[X_t - \mathbb{E}\,[X_t|\mathcal{F}_t^{N(t)}]\right]^2 = \mathbb{E}\,\eta_t = \int_0^t \frac{H(t,s)}{h(t,s)}ds$$

where $H(t,s) =$

$$\frac{1}{6}\mathbb{E}\left[[(\mathcal{L}^1 a - \mathcal{L}^0 b)(X_s)]^2 \exp\left(\int_s^t (2a\prime(X_u) - [b\prime(X_u)]^2)du + 2\int_s^t b\prime(X_u)dW_u\right)\right].$$

The optimal double mean square approximation error satisfies a similar relation. For more details, see Cambanis and Hu (1996). Also their results can be generalized to multidimensional diffusions with some care. This result is fundamental with respect to asymptotically optimal mean square discretizations. This fact can be seen from the fact that the function $h^* \in C^0([0,T], \mathbb{R}_+)$ established by

$$h^*(t,s) = \frac{[H(t,s)]^{1/3}}{\int_0^t [H(t,s)]^{1/3} ds}$$

minimizes the functional $\int_s^t \frac{H(t,s)}{[h(t,s)]^2} ds$ where $H(t,s) \ge 0$ among all regular partition densities h with $h(t,s) > 0$ and $\int_0^t h(t,s)ds = 1$. Therefore, any asymptotically mean square optimal approximation has to use a discretization following that optimal partition law. However, the practical value is still in doubt, since it will be hard to evaluate those expressions in the fully multidimensional framework or has any reader another suggestion in the case $m, d > 1$?

5.6.7 A theorem on double L^2-convergence with adaptive Δ_n

Consider nondegenerate SDEs (5.2.3) with additive noise coefficient $b = b(t)$ on a fixed time interval $[0, T]$. Define the **double L^2-approximation error** as

$$e_2 = e_2(Y^\Delta, a, b, X_0, T) = \left(\mathbb{E}\, \|X - Y^\Delta\|_{L^2}^2\right)^{1/2} = \left(\mathbb{E} \int_0^T \|X_t - Y_t^\Delta\|^2 \mu(dt)\right)^{1/2}$$

with respect to Lebesgue measure μ. Introduce **adaptive step size strategy**

$$\Delta_n = \min\left(\frac{h}{\|b(t_n)\|}, T - t_n, 1\right) \tag{5.6.10}$$

for a basic step size $h > 0$, tending later to zero. Let $N = N(h, b, T)$ denote the total number of steps necessary to integrate, i.e. $N = N(h, b, T) = \sup\{n : t_n \leq T\}$. Let $C_{UL}^{1,2} = C_{UL}^{1,2}([0, T], \mathbb{R}^d)$ be defined by

$$C_{UL}^{1,2} := \left\{ f \in C^{1,2}([0,T] \times \mathbb{R}^d) : \begin{array}{l} \exists \text{ constants } K_1, K_2, K_2 \forall x \in \mathbb{R}^d, \forall s, t[0,T] \\ \|\frac{\partial f}{\partial x}(t,x)\| \leq K_1 < +\infty \\ \|\frac{\partial^2 f}{\partial x^2}(t,x)\| \leq K_2 < +\infty \\ \|f(t,x) - f(s,x)\| \leq K_3(1 + \|x\|)|t - s| \end{array} \right\}.$$

Theorem 5.6.10 *Assume that X satisfies SDE (5.2.3) with drift $a \in C_{UL}^{1,2}$, additive noise with $b = b(t) \in L^2([0,T], \mathcal{B}([0,T]), \mu)$, $\mathbb{E}\,\|X_0\|^2 < +\infty$ and*

$$\inf_{0 \leq t \leq T} \|b(t)\| > 0.$$

Then the Euler method (5.4.1) applied to (5.2.3) with constant step size $\Delta = \frac{T}{N}$ generates double L^2-approximation errors with

$$\lim_{\Delta \to 0} \sqrt{N}\, e_2(Y^\Delta, a, b, X_0, T) = K_2 \|b\|_{L^2}$$

with K_2 an appropriate constant (e.g. $K_2 = \frac{1}{\sqrt{6}}$ if $d = 1, T = 1$), whereas the Euler method (5.4.1) applied to (5.2.3) with adaptive step size strategy (5.6.10) and basic step size h yields

$$\lim_{h \to 0} \sqrt{N(h, b, T)}\, e_2(Y^\Delta, a, b, X_0, T) = \hat{K}_2 \|b\|_{L^1}$$

with a suitable positive real constant $\hat{K}_2$ (e.g. $\hat{K}_2 = \frac{1}{\sqrt{6}}$ if $d = 1, T = 1$).

Hofmann, Müller-Gronbach and Ritter (1999) have noticed a similar result in one dimension (i.e. $d = m = 1$), for continuously differentiable b and $T = 1$. Under their conditions they prove that the estimates in Theorem 6.7 are the lower bounds for all $\mathcal{F}_{t_n}^N$-measurable approximations Y^Δ for SDEs in $\mathbb{R}^1$ with additive noise, i.e.

$$\lim_{\Delta \to 0} \sqrt{N} \inf_{Y^\Delta} e_2(Y^\Delta, a, b, X_0, T) = K_2 \|b\|_{L^1}$$

with $K_2 = \frac{1}{\sqrt{6}}$, $T = 1$, hence the Euler method with the mentioned adaptive strategy of step size selection (5.6.10) already produces asymptotically mean square optimal $\mathcal{F}_{t_n}^N$-measurable numerical approximations. However, one can carry it over to d-dimensional

SDEs with additive noise and L^p-integrable b as well (i.e. $p \geq 1$), as indicated by Theorem 6.7. It is worth noting that that step size selection suggested originally by Hofmann, Müller-Gronbach and Ritter (1999) is only designed to control large diffusion fluctuations, and it seems not to be very appropriate as one takes the limit as b goes to zero (i.e. incomplete adaptability is obtained in the presence of significant drift parts - an approach which leads to inconsistent results in view of deterministic limit equations, however which might be appropriate for pure diffusions with large diffusion coefficients $b(t) > 1$). We stress again by our main principles of numerics that the step size selection should be adapted rather to the consistency, contractivity and stability constants of the considered SDEs and according to the goal of achieving the requirements of $\mathbb{D}$-invariances in view of the behavior of dynamics of SDE to be discretized. However, all in all, it is clear that the asymptotics of the leading error coefficients of the related numerical method, which one wishes to squeeze out by those limiting procedures, heavily depends on the choice of possible step sizes. Thus, one should further study the (asymptotic) behavior of leading error coefficients (e.g. as done above with $K_2||b||_{L^p}$).

5.6.8 The fundamental theorem of weak convergence

The key contribution in this direction starts with fundamental works of Mil'shtein (1978), Platen (1980) and Talay (1982). Compare also with Kushner and Dupuis (1992) who give an alternative by Markov chain constructions. In Mil'shtein (1995) one can find the most general theorem on weak convergence. For this purpose, define

$$\mathcal{P} = \left\{ f = f(t,x) \text{ Lebesgue-measurable} : \begin{array}{l} \exists K, \kappa \in \mathbb{R}_+ \, \forall (t,x) \in [0,T] \times \mathbb{R}^d \, s.t. \\ ||f(t,x)||_d \leq K(1 + ||x||^\kappa) \end{array} \right\}$$

and a **one-step representation** of approximation Y by

$$Y_{t,x}(t+h) = x + \sum_{j=0}^{m} \phi^j(t,h,x,\xi^j).$$

Furthermore, set

$$\delta_{t,x}^X(t+h) := X_{t,x}(t+h) - x, \; \delta_{t,y}^Y(t+h) := Y_{t,y}(t+h) - y,$$

where $X_{t,x}(t+h)$ denotes the solution of SDE (5.2.1) started at x at time t, evaluated at time $t+h$.

Theorem 5.6.11 *(Mil'shtein, 1995). Assume that X satisfies SDE (5.2.1) with drift and diffusion vector functions*

$$a = a(t,x), b^j = b^j(t,x) \in C^{p+1,2p+2}([0,T] \times \mathbb{R}^d)$$

under the conditions (IMC), (UBC), (ULC), $\mathbb{E}\, ||X_0||^{2p_I} < +\infty$ for sufficiently large $p_I \geq 2$. Furthermore, let

(i) $a(t,x), b^j(t,x)$ together with all their partial derivatives belong to class $\mathcal{P}$,

(ii) $f = f(x)$ together with all their partial derivatives up to order $2(p_f + 1)$ belong to class $\mathcal{P}$,

(iii) Y have uniformly bounded moments

$$\sup_{k=0,1,\ldots,n_T} \mathbb{E}\, ||Y_{0,X_0}||^{2p_I} < +\infty,$$

(iv) Y fulfills the moment consistency conditions with a real $K(x) \in \mathcal{P}$ such that for all $r = 1, 2, ..., 2p+1$:

$$\left\| \mathbb{E}\left(\prod_{j=1}^{r}[\delta_{t,x}^{X}]^{i_j} - \prod_{j=1}^{r}[\delta_{t,x}^{Y}]^{i_j}\right)\right\| \leq K(x)h^{p+1} \tag{5.6.11}$$

$$\mathbb{E}\prod_{j=1}^{2p+2} \|\delta_{t,x}^{Y}\|^{i_j} < K(x)h^{p+1}. \tag{5.6.12}$$

Then Y is weakly converging towards X on the time interval $[0,T]$ with order p, i.e. there is a constant $K_w = K_w(T, a, b^j, p, d, p_I, p_f, X_0)$ such that

$$\sup_{f\in\mathcal{P}(r,K)}\ \sup_{0\leq t_k\leq T} \left\| \mathbb{E}\, f(X_{0,X_0}(t_k)) - \mathbb{E}\, f(Y_{0,X_0}(t_k))\right\| \leq K_w\Delta^p. \tag{5.6.13}$$

Of course, these are worst case estimates as well. For some specific classes of SDEs (5.2.1) the considered numerical methods may perform even better. For more details on weak convergence, it is recommended to consult Talay (1995) for a report on original results related to equidistant discretizations.

5.6.9 Approximation of some functionals

An interesting question is how the pth mean convergence orders can be carried over to the weak convergence order during the approximation of functionals of SDE solutions. This question was answered for the case of nonsmooth and path-dependent functionals by Schurz (1995). One important aim is to approximate

$$F(t,X) = f(t, X_t, \inf_{s\leq t}\|X_s\|, \sup_{s\leq t}\|X_s\|) \tag{5.6.14}$$

where $f = f(t,x,y,z)$ is Lebesgue-measurable at t, x, y, z. At first consider

$$F_0(t,X) = \mathbb{E}\, f(T,X_t) = \mathbb{E}\, f_T(X_t) \quad (t \in [0,T], T \text{ fixed}) \tag{5.6.15}$$

where $f : [0,T] \times \mathbb{D} \longrightarrow \mathbb{R}$ is convex at x with its second space derivative μ_t. Let Y_{n_t} be a right-continuous approximation as step function, $\mathcal{F}_t$–adapted numerical approximation of X_t, based on a numerical method generating random values Y_n and $n_t = \sup\{n : t_n \leq t\}$. The expression $p_X = p_X(t,x)$ denotes the probability density of process $X = (X_t)_{0\leq t\leq T}$ at point $x \in \mathbb{D}$ at time t, with support $supp(p_X(t,x))$. Let $\tau^\Delta([0,T])$ denote the collection of $\mathcal{F}_t$-adapted time instants belonging to time discretization of $[0,T]$ with maximum step size Δ.

Theorem 5.6.12 *(Schurz (1995)). Let $\mathcal{I} = [0,T]$ or $\mathcal{I} = \tau^\Delta([0,T])$. Assume that*

(0) *$\mathbb{D}$ is an open, deterministic subset of $\mathbb{R}^1$*

(i) *$f = f(t,x)$ is convex at $x \in \mathbb{D}$ with second (weak) derivative $\mu_T = f''$*

(ii) $\forall t \in \mathcal{I} \quad \int_{supp(p_X(t,x))\cap\mathbb{D}} |a|\mu_T(da) < +\infty$

(iii) $\exists p \geq 1 (p \in \mathbb{R})\ \forall t \in \mathcal{I} \quad \left(\mathbb{E}\,|X_t|^p\right)^{1/p} + \left(\mathbb{E}\,|Y_{n_t}|^p\right)^{1/p} \leq K_0 < +\infty$

(iv) $\mathbb{P}\,\{\omega \in \Omega : \forall t \in \mathcal{I}\, X_t(\omega) \in \mathbb{D}\} = \mathbb{P}\,\{\omega \in \Omega : \forall t \in \mathcal{I}\, Y_{n_t}(\omega) \in \mathbb{D}\} = 1$

(v) $\exists K_p = K_p(T) > 0\, \exists \gamma \geq 0 \quad \sup_{t\in\mathcal{I}} \left(\mathbb{E}\,|X_t - Y_{n_t}|^p\right)^{1/p} \leq K_p \cdot \Delta^\gamma$
$\tau^\Delta([0,T])$

(vi) *$supp(p_X = p_X(t,x)) \cap \mathbb{D}$ is compact*

Then there is a real constant $K = K(p,T) > 0$ such that

$$\epsilon := \sup_{t \in \mathcal{I}} \left| I\!E\, f(T, X_t) - I\!E\, f(T, Y_{n_t}) \right| \leq K \cdot \Delta^{\gamma} \tag{5.6.16}$$

Remark *This result is not so surprising since convex functions are quasi-linearizable and, on compact sets, even Lipschitz-continuous. However, it possesses an interesting proof. For any Lipschitz-continuous function f the pth mean convergence rates γ_g carry over one to one to weak convergence rates $\beta = \gamma_g$. With this result in hand, one can justify using numerical approximation with the highest possible accuracy, depending on regularity of price process X, to estimate European call and put options.*

Corollary 5.6.13 *(Schurz (1995)). Assume conditions $(0) - (v)$ of Theorem 6.9, that*

$$supp(p_{\|X-c\|} = p_{\|X-c\|}(t, z)) \cap I\!D \text{ is compact}$$

and consider functionals of the form

$$F_1(t) = f(t, \|X_t - c\|), c = const, t \in \mathcal{I} \tag{5.6.17}$$

where $f(t, z)$ is convex with respect to the space coordinate $z \in R^1$.
Then, there is a real constant $K = K(p,T)$ such that for all $t \in [0,T]$

$$\varepsilon(t) = |I\!E\, f(t, \|X_t - c\|) - I\!E\, f(t, \|Y_{n_t} - c\|)| \leq K \cdot \Delta^{\gamma}. \tag{5.6.18}$$

Remark *For* **concave functionals**, *similar results hold. The latter result can be verified for some path-dependent functionals as well.*

Corollary 5.6.14 *(Schurz (1995)). Assume conditions $(0) - (v)$ of Theorem 6.9, that*

$$supp(p_{\sup \|X\|} = p_{\sup_{0 \leq s \leq t} \|X_s\|}(t, z)) \cap I\!D \text{ is compact}$$

and $X_t - Y_{n_t}$ be a right-continuous submartingale with respect to the natural filtration $\mathcal{F}_t = \sigma\{W_s^j : 0 \leq s \leq t, j = 1, 2, ..., m\}$. Consider functionals of the form

$$F_2(T, t) = f_T(\sup_{0 \leq s \leq t} \|X_s\|) \tag{5.6.19}$$

$$\text{or} \quad F_3(T, t) = f_T(\sup_{0 \leq s \leq t} X_s^i)(i \in 1, 2, ..., d \text{ fixed}) \tag{5.6.20}$$

where $f_T(z)$ is convex with respect to the space coordinate $z \in R^1$.
Then, error estimate (5.6.16) is also valid for F_2, F_3 (with a constant $K > 0$ which may differ from that constant above, see (5.6.16)).

Thus, for path-dependent convex functionals and problems of optimal stochastic control, clarification of the problem of practical construction of approximations with a $\mathcal{F}_t$-submartingale error process remains to be done. The latter problem seems to be solvable for the class of X-subharmonic functionals f (but in general it is an open question).

Remark for Application to Mathematical Finance. Asset- and option price processes X for **Randomly Exercised Exotic Options** (**American Lookback Call Option**) may cause the following payoff functionals

$$F_i(\tau, X) = \exp\left(- \int_\tau^T r(s)ds \right) \left(\sup_{0 \le t \le \tau} |X_t^{(i)}| - K_i(T,\tau) \right)_+$$

for calls of the ith component of true observable price-process X (for puts respectively), where $K_i(T,\tau)$ represents the strike price at randomly stopped moment τ which is $\mathcal{F}_\tau$-adapted. Now, for example, there is the task of finding the optimal stopping strategy $0 \le \tau \le T < +\infty$ (i.e. random exercise time τ of the call option with bounded deterministic maximal terminal time) such that the expected discounted loss caused by the payoff at time τ is minimal under the amount of information $\mathcal{F}_t$ at current time t and discounted by $\mathcal{F}_s$-adapted random short interest rate $r(s)$, i.e. one wishes to approximate the optimal solution of the stochastic control problem

$$c(\tau^*) := \inf_{0 \le \tau \le T} \mathbb{E} \left[\exp\left(- \int_\tau^T r(s)ds \right) \left(\sup_{0 \le t \le \tau} |X_t^{(i)}| - K_i(T,\tau) \right)_+ \Big| \mathcal{F}_\tau \right],$$

where $c = c(\tau^*) = \mathbb{E}\,[F_i(\tau^*, X)|\mathcal{F}_{\tau^*}]$. This represents a composition of convex functionals, and with our results before we have to construct a pth mean converging numerical approximation which is right-continuous and which has a $\mathcal{F}_t$-submartingale as its error process $X_t - Y_{n_t}$. Then the convergence rate will be $\beta = \gamma_g$, and numerical approaches reported in the Mathematical Finance literature can be justified by our convergence approach, even for convex, path-dependent functionals of X which can be noncontinuously differentiable at some countable points. The practical construction is still a problem since the construction procedure which guarantees the submartingale error process may strongly depend on the structure of the price process X governed by some SDE.

For Hölder-continuous functionals one encounters the following result. Let $\mathbb{D}$ denote an open, deterministic domain of $\mathbb{R}^d$. Fix $d, k \in \mathbb{N}_+$. Define

$$C^0_{H(K_H,\alpha)} := \left\{ f : \mathbb{D} \subseteq \mathbb{R}^d \longrightarrow \mathbb{R}^k : \|f(x) - f(y)\|_k \le K_H \|x-y\|_d^\alpha \right\}$$

with Hölder constant K_H and Hölder exponent $\alpha \in [0,1]$. One arrives at

$$\|\mathbb{E}\, f(X_t) - \mathbb{E}\, f(Y^\Delta_{n_t})\|_k \le \mathbb{E}\, \|f(X_t) - f(Y^\Delta_{n_t})\|_k \le K_H \mathbb{E}\, \|X_t - Y^\Delta_{n_t}\|_d^\alpha$$

$$\le K_H (\mathbb{E}\, \|X_t - Y_t^\Delta\|_d^p)^{\alpha/p} \le K_H \cdot [K(p,T)]^\alpha \Delta^{\alpha\gamma}.$$

Taking the supremum leads to the following uniform convergence order estimation determined by the Hölder exponent α, uniformly with respect to the class of Hölder-continuous mappings, exhibiting a natural loss of convergence speed with decreasing Hölder exponent α. Now, fix real constants $\alpha \in [0,1]$ and $K_H \ge 0$.

Theorem 5.6.15 *(Schurz (1995)). Assume $f \in C^0_{H(K_H,\alpha)}$, $X = (X_t)_{0 \le t \le T}$ and $Y = (Y^\Delta_{n_t})_{0 \le t \le T}$ are two $\mathbb{D}$-invariant, $\mathcal{F}_t$-adapted stochastic processes with respect to the same stochastic basis $(\Omega, \mathcal{F}, (\mathcal{F}_t)_{0 \le t \le T}, \mathbb{P}\,)$ satisfying*

$$\sup_{t \in \mathcal{I}} \left(\mathbb{E}\, \|X_t - Y^\Delta_{n_t}\|_d^p \right)^{1/p} \le K(p,T)\Delta^\gamma.$$

with some $\gamma \in \mathbb{R}_+$.
Then, we have

$$\sup_{f \in C^0_{H(K,\alpha)}} \sup_{t \in \mathcal{I}} \| \mathbb{E}\, f(X_t) - \mathbb{E}\, f(Y^{\Delta}_{n_t}) \|_k \leq K_w(p,T,K_H,\alpha)\Delta^{\alpha\gamma}$$

with appropriate deterministic constant $K_w(p,T,K_H,\alpha) = K_H \cdot [K(p,T)]^{\alpha}$.

5.6.10 The pathwise error process for explicit Euler methods

Jacod and Protter (1998), motivated from Rootzén (1980) and Kurtz and Protter (1991), have statistically analyzed the pathwise error process of discrete time and continuous time explicit Euler methods using equidistant time-discretizations and applied to stochastic differential equations driven by more general semimartingales than assumed by SDE (5.2.1). For the statement of the fundamental result (Theorem 3.1, p. 275, in Jacod and Protter (1998)), let us define

$$\varepsilon^N_t := X_t - Y^N_t,\ t \in [0,T]$$

and only state the application of their result to the case of SDEs (5.2.1).

Theorem 5.6.16 *Assume that the SDE (5.2.1) has locally Lipschitz continuous coefficients* a, b^j *with at most linear growth.*
Then the continuous time error processes $\varepsilon^N_t, \varepsilon^N_{[\frac{Nt}{N}]}$ *tends to* 0 *in probability as* N *goes to* $+\infty$.

They also establish rates of stable convergence. In fact, they arrive at a certain stochastic differential equation for the limit of related normalized error processes $U^N_t = \sqrt{N}\varepsilon^N_t$ and $\hat{U}^N_t = \sqrt{N}\varepsilon^N_{\frac{[Nt]}{N}}$ as N tends to 0. See their paper for more details. In principle, this procedure can be continued for other and higher order methods under corresponding assumptions.

5.6.11 Almost sure convergence

It is clear from L^p-convergence that there exists a subsequence of $(Y_n)_{n\in\mathbb{N}}$ which almost surely converges to the exact solution X_t. The only works (to our knowledge) available at the time of writing this survey are that of O. Faure (1992), which is not accessible to the author at the moment, that of Talay (1983), and Pardoux and Talay (1985), who use the Doss representation (cf. the ODE method above) to obtain almost surely converging approximations. However, it is an open problem as to how in general an almost sure convergence order is transmitted when commutativity conditions are not met (and Doss representation could not be used to verify convergence orders so far. Remember that Talay (1983) makes use of commutative noise conditions for Doss representation in the fully multidimensional situation). Also, what happens when we have variable step sizes? This is an open problem to be left to the future. As a supplement, let us start with a trial of a definition of numerical almost sure convergence.

Definition 5.6.17 *Let* X, Y^{Δ} *be two* $\mathcal{F}_t$*-adapted stochastic processes with respect to one and the same stochastic basis* $(\Omega, \mathcal{F}, (\mathcal{F}_t)_{0\leq t\leq T}, \mathbb{P})$. *Assume* Y^{Δ} *uses a sufficiently small maximum step size* $\Delta \leq 1$ *for all time-discretizations of fixed time-interval* $[0,T]$.

Then the stochastic process Y^Δ is called **numerically a.s. converging** *to process X on $[0,T]$ iff*

$$\lim_{\Delta\to 0}\sup_{0\le t_n\le T}||X_{t_n}-Y^\Delta_{t_n}||=0\ (a.s.),$$

continuous time a.s. converging *to process X on $[0,T]$ iff*

$$\lim_{\Delta\to 0}\sup_{0\le t\le T}||X_t-Y^\Delta_t||=0\ (a.s.),$$

numerically a.s. converging with order $\gamma\in I\!R_+$ *to process X on $[0,T]$ iff*

$$\lim_{\Delta\to 0}\frac{1}{\Delta^{\gamma-\varepsilon}}\sup_{0\le t_n\le T}||X_{t_n}-Y^\Delta_{t_n}||=0\ (a.s.),$$

for all ε with $0<\varepsilon\le\gamma$, and **continuous time a.s. converging with order** $\gamma\in I\!R_+$ *to process X on $[0,T]$ iff*

$$\lim_{\Delta\to 0}\frac{1}{\Delta^{\gamma-\varepsilon}}\sup_{0\le t\le T}||X_t-Y^\Delta_t||=0\ (a.s.),$$

for all ε with $0<\varepsilon\le\gamma$, with respect to a class of admissible time-discretizations of $[0,T]$ with sufficiently small maximum step sizes Δ.

This definition is based on the concept of an admissible time-discretization.

Definition 5.6.18 *A time-discretization $(t_n)_{n\in N}\in[0,T]$ of a fixed time-interval $[0,T]$ is called* **uniformly admissible** *iff all t_n are $\mathcal{F}_{t_n}$-adapted and there exist a* **minimum step size** Δ_{min} *and* **maximum step size** Δ_{max} *with*

$$0<\Delta_{min}=\inf_{n\in I\!N}|t_{n+1}-t_n|\le\Delta_n\le\sup_{n\in I\!N}|t_{n+1}-t_n|=\Delta_{max}<+\infty.$$

Remark *This latter definition corresponds very well to the experience of practical numerical computations, where mostly the variable step size implementations possess an upper and lower bound for minimum and maximum step sizes. A corresponding work by the author is in progress in order to explain the (optimal) discretization problem in more details in the case of converging stochastic approximations and step size selection.*

For equidistant approximations one finds the following one-dimensional results in the literature. We shall extract the versions from Talay's review paper (1995), p. 66.

Theorem 5.6.19 *(Faure (1992)). Assume that $a,b^j\in C^0_{Lip}(I\!R^1)$.*

(1). If for some positive even integer $p=2k$ the initial condition (IMC) is satisfied, then the interpolated Euler method $Y^\Delta(t)$ applied to autonomous SDE (5.2.1) with equidistant step size $\Delta=\frac{T}{N}$ continuously time a.s. converges to the exact solution of (5.2.1) as the number N of equidistant subintervals tends to $+\infty$.

(2). If all initial moments exist, then the order of its continuous time a.s. convergence is $\gamma=0.5$.

Theorem 5.6.20 *(Talay (1983)). Assume that $a, b^j \in C^3_{Lip}(\mathbb{R}^1)$ with bounded derivatives up to third order, and the deterministic real-valued function $u = u(t)$ approximates the given trajectory of the underlying Wiener process $W = (W_t)_{0 \le t \le T}$ in the sense of uniform convergence topology on the space $C^0([0,T])$ of continuous functions on fixed time interval $[0,T]$. Let u have a zero 3-variation on $[0,T]$, i.e.*

$$\lim_{\Delta = \frac{T}{N} \to 0} \sum_{n=0}^{N} |u(t_n) - u(t_{n-1})|^3 = 0$$

for any partitions of $[0,T]$.
Then the right-continuous, piecewise constant approximation generated by the Euler method applied to autonomous SDE (5.2.1) continuous time a.s. converges to the exact solution of (5.2.1) as the number N of equidistant subintervals tends to $+\infty$, provided that the following **noise commutativity** *condition holds:*

$$\forall j,k \in \{1,2,...,m\} \quad b^k(x)\frac{\partial b^j(x)}{\partial x} = b^j(x)\frac{\partial b^k(x)}{\partial x}.$$

5.7 Numerical Stability, Stationarity, Boundedness and Invariance

After treating the concept of convergence (convergence on fixed, deterministic, finite time intervals T), we now devote our attention to the other important column of the main principle of numerical analysis: namely, that of numerical stability. The more one is interested in a control on nonexploding state processes and also nonexploding error propagation, the more necessary this is, whereas it is a must for adequate numerical integration on infinite time intervals (i.e. when one takes the limit as terminal times T tend to $+\infty$).

For simplicity of consideration, we start with the treatment of linear systems (It is not possible here to discuss the full extent of the problem of stochastic test equations in a mathematically rigorous way.). In view of stochastic terms, it is necessary to distinguish between the three main classes: linear systems with additive noise, linear systems with multiplicative noise, and of fully nonlinear systems. The case of multiplicative noise is the closest to the deterministic situation, since we could use the deterministic trivial equilibrium $X \equiv 0$ as the unique reference solution. Therefore it represents the best understood case from the three main cases. The case of additive noise really needs a stochastic approach to tackle the problem of numerical stability. See also Section 8 for an alternative by contractivity concept.

We shall also examine the problem of almost sure boundedness which is obviously connected with stability and invariance issues. For motivation, remember the main principles of stochastic numerical analysis in Section 5.

For the sake of simplicity, at first let us confine ourselves to the concept of mean square stability.

5.7.1 Stability of linear systems with multiplicative noise

Consider (for simplicity, autonomous) linear system of Itô SDEs

$$dX_t = A X_t \, dt + \sum_{j=1}^{m} B^j X_t \, dW_t^j \tag{5.7.1}$$

Assume that a unique stationary solution $X_\infty = 0$ of (5.7.1) exists. Then the necessary condition

$$\forall i = 1, 2, ..., d : \quad Re(\lambda_i(A)) < 0 \tag{5.7.2}$$

with $\lambda_i(A)$ as the ith eigenvalue of matrix A must be satisfied, at least for any moment stability with $p \geq 1$. (Note that condition (5.7.2) implies the stability of first moments $\mathbb{E}\ X_t$ since we obtain a kind of direct projection to the deterministic case, which can be easily seen in the linear systems case. However, in the nonlinear case we would observe the problem of closure of moment equations.) For the sake of simple illustration, let us confine ourselves to the case of mean square stability (i.e. $p = 2$ in moment stability).

Definition 5.7.1 *Assume $X \equiv 0$ is an equilibrium for (5.2.1). The equilibrium solution $X \equiv 0$ is called* **globally (asymptotically) mean square stable** *for the stochastic process $X = (X_t)_{t \geq 0}$ if*

$$\forall X_0 : \mathbb{E}\ \|X_0\|^2 < +\infty \implies \lim_{t \to +\infty} \mathbb{E}\ \|X_t\|^2 = 0. \tag{5.7.3}$$

Assume that $Y \equiv 0$ is an equilibrium for the numerical approximation $Y = (Y_n)_{n \in \mathbb{N}}$ for system (5.2.1). The equilibrium solution $Y \equiv 0$ is called **globally (asymptotically) numerically mean square stable** *for the numerical approximation $Y = (Y_n)_{n \in \mathbb{N}}$ if*

$$\forall X_0 : \mathbb{E}\ \|Y_0\|^2 < +\infty \implies \lim_{n \to +\infty} \mathbb{E}\ \|Y_n\|^2 = 0. \tag{5.7.4}$$

As a first illustrative result, consider the **family of drift-implicit Euler methods** (see Kloeden, Platen and Schurz (1994)), of the form

$$Y_{n+1} = Y_n + (\alpha A Y_{n+1} + (1 - \alpha) A Y_n) \Delta_n + \sum_{j=1}^{m} B^j Y_n \Delta W_n^j, \tag{5.7.5}$$

applied to equation (5.7.1), where $\alpha \in \mathbb{R}^1$ is the implicitness parameter, and the step size is sufficiently small such that the local algebraic resolution of (5.7.5) can be guaranteed (the latter requirement would be irrelevant when $\alpha \geq 0$ under condition (5.7.2)).

Theorem 5.7.2 *(Schurz (1996)). For all equidistant approximations $Y^\alpha = (Y_n^\alpha)_{n \in \mathbb{N}}$ generated by method (5.7.5) with fixed step size $\Delta > 0$, it holds that*

$$\begin{aligned} X \equiv 0 \text{ mean square stable } &\iff Y^\alpha \equiv 0 \text{ mean square stable with } \alpha = 0.5, \\ X \equiv 0 \text{ mean square stable and } \alpha \geq 0.5 &\implies Y^\alpha \equiv 0 \text{ mean square stable}, \\ Y^{\alpha_1} \equiv 0 \text{ mean square stable with } \alpha_1 \leq \alpha_2 &\implies Y^{\alpha_2} \equiv 0 \text{ mean square stable}. \end{aligned}$$

The proof can be seen in Schurz (1996, 1997) using the study of a stochastic version of Lyapunov equation

$$\boxed{AM + MA^T + \sum_{j=1}^{m} B^j M {B^j}^T = -C}$$

and basic facts from spectral theory of monotone operators. In fact, Schurz (1996, 1997) has developed the concept of mean square operators which describe the mean square evolution and stability behavior of related numerical method on a systems level. The **family of mean square operators** related to approximation sequence $Y = (Y_n)_{n \in \mathbb{N}}$ is defined by the sequence of $\mathcal{F}_{t_n}$-adapted operators $(\mathcal{L}_n)_{n \in \mathbb{N}}$ mapping from the set $\mathcal{S}^+_{d \times d}$ of positive semi-definite $d \times d$ matrices into itself by

$$\mathbb{E}\,[Y_{n+1} Y_{n+1}^T] = \mathbb{E}\,\mathcal{L}_n(Y_n Y_n^T) = \mathbb{E}\,\mathcal{L}_n \mathcal{L}_{n-1} ... \mathcal{L}_0(Y_0 Y_0^T).$$

Thus the asymptotic behavior of the related numerical method is connected to the study of the limit

$$\lim_{n \to +\infty} \mathbb{E}\left[\left(\prod_{i=0}^{n} \mathcal{L}_i\right)(Y_0 Y_0^T)\right]$$

along the mentioned operator family on the space of positive semi-definite matrices. In the equidistant case this study can be carried out by standard fixed point analysis and the tool of the spectral radius of related operator $\mathcal{L}$. However the concept of mean square operators even works for nonautonomous systems (5.7.1) and variable step size implementations using monotonicity argumentations. Thanks to that representation, Schurz (1996, 1997) could establish a systematic stability analysis of systems of discrete random mappings, the concept of stochastic A-stability on a systems level, the principle of monotonic nesting of stability domains for monotone systems. More generally, it is possible to develop a corresponding theory of pth mean stability operators for nonlinear stochastic dissipative systems, see Schurz (1997, 1999). This has been suggested and constructed with the family of drift-implicit Euler methods (5.7.5) therein. The study of that operator family still needs to be continued for other numerical methods. An interesting, illustrative and simple complex-valued test equation is given by the **stochastic Kubo oscillator** perturbed by multiplicative white noise in the Stratonovich sense

$$dX_t \;=\; i\,X_t\,dt \;+\; i\,\rho\,X_t \circ dW_t$$

where $\rho \in \mathbb{R}^1, i^2 = -1$. This equation describes rotations on the circle with radius $||X_0||$. Schurz (1994, 1996) has studied this example and shown that the corresponding discretization of implicit Mil'shtein methods explodes for any step size selection, whereas the lower order trapezoidal method or appropriately balanced implicit methods (BIMs) could stay close to the circle of the exact solution even for large integration times! This is a test equation which manifests that stochastically coherent (i.e. asymptotically exact) numerical methods are needed and the search for efficient higher order convergent methods is somehow restricted even under linear boundedness and infinitely smooth assumptions on drift and diffusion coefficients.

The illustrative example of one-dimensional complex-valued test SDE. Many authors (e.g. Mitsui and Saito (1996), Schurz (1996)) have studied the SDE

$$dX_t \;=\; \lambda X_t\,dt + \gamma X_t\,dW_t, \tag{5.7.6}$$

with $X_0 = x_0 \in \mathbb{C}^1$, representing a test equation for the class of completely commutative systems of SDEs with multiplicative white noise. This stochastic process has the unique exact solution $X_t \;=\; X_0 \cdot \exp((\lambda - \gamma^2/2)t + \gamma W_t)$ with second moment

$$\begin{aligned} \epsilon X_t X_t^* &= \epsilon |X_t|^2 \;=\; \epsilon \exp(2(\lambda - \gamma^2/2)_r t + 2\gamma_r W_t) \cdot |x_0|^2 \\ &= |x_0|^2 \cdot \exp(2(\lambda_r - \gamma_r^2/2 + \gamma_i^2/2)t + 2\gamma_r^2 t) \;=\; |x_0|^2 \cdot \exp((2\lambda_r + |\gamma|^2)t) \end{aligned}$$

where $x_0 \in \mathbb{C}^1$ is nonrandom (z_r is the real part, z_i the imaginary part of $z \in \mathbb{C}^1$) and $*$ denotes the complex conjugate value. The trivial solution $X \equiv 0$ of (5.7.6) is mean square stable for the process $\{X_t : t \geq 0\}$ iff $2\lambda_r + |\gamma|^2 < 0$. Now, let us compare the numerical approximations of families of (drift-)implicit Theta and Mil'shtein methods. Applied to equation (5.7.6), the drift-implicit Mil'shtein (5.4.18) and drift-implicit Theta methods (5.4.3) are given by

$$Y_{n+1}^{(M)} = \frac{1+(1-\theta)\lambda\Delta + \gamma\xi_n\sqrt{\Delta} + \gamma^2(\xi_n^2-1)\Delta/2}{1-\theta\lambda\Delta} \cdot Y_n^{(M)} \tag{5.7.7}$$

and

$$Y_{n+1}^{(E)} = \frac{1+(1-\theta)\lambda\Delta + \gamma\xi_n\sqrt{\Delta}}{1-\theta\lambda\Delta} \cdot Y_n^{(E)}\,, \tag{5.7.8}$$

respectively. Their second moments $P_n^{(E/M)} = \epsilon Y_n^{(E/M)} {Y_n^{(E/M)}}^*$ satisfy

$$\begin{aligned} P_{n+1}^{(M)} &= \left(\epsilon \left|\frac{1+(1-\theta)\lambda\Delta + \gamma\xi_n\sqrt{\Delta}}{1-\theta\lambda\Delta}\right|^2 + \epsilon\left|\frac{\gamma^2(\xi_n^2-1)}{1-\theta\lambda\Delta}\right|^2 \cdot \Delta^2/4 \right) \cdot P_n^{(M)} \\ &= P_0^{(M)} \cdot \left(\frac{|1+(1-\theta)\lambda\Delta|^2 + |\gamma|^2\Delta + |\gamma|^4\Delta^2/2}{|1-\theta\lambda\Delta|^2} \right)^{n+1} \\ &> P_0^{(E)} \cdot \left(\frac{|1+(1-\theta)\lambda\Delta|^2 + |\gamma|^2\Delta}{|1-\theta\lambda\Delta|^2} \right)^{n+1} = P_{n+1}^{(E)} \qquad (n = 0, 1, 2, ...), \end{aligned}$$

provided that $P_0^{(M)} = \epsilon Y_0^{(M)} {Y_0^{(M)}}^* \geq \epsilon Y_0^{(E)} {Y_0^{(E)}}^* = P_0^{(E)}$, and

$$\begin{aligned} P_{n+1}^{(M)} &= P_{n+1}^{(E)} \cdot \left(\frac{|1+(1-\theta)\lambda\Delta|^2 + |\gamma|^2\Delta + |\gamma|^4\Delta^2/2}{|1+(1-\theta)\lambda\Delta|^2 + |\gamma|^2\Delta} \right)^{n+1} \\ &= P_{n+1}^{(E)} \cdot \left(1 + \frac{|\gamma|^4\Delta^2/2}{|1+(1-\theta)\lambda\Delta|^2 + |\gamma|^2\Delta} \right)^{n+1}. \end{aligned}$$

while assuming identical initial values $P_0^{(M)} = P_0^{(E)}$. Hence, if the drift-implicit Mil'shtein method (5.7.7) possesses a mean square stable null solution then the corresponding drift-implicit Theta method (5.7.8) possesses it too. The mean square stability domain of (5.7.7) is smaller than that of (5.7.8) for any implicitness $\theta \in [0, 1]$. Besides, the drift-implicit Theta method (5.7.8) has a mean square stable null solution if $\theta \geq \frac{1}{2}$ and $2\lambda_r + |\gamma|^2 < 0$. The latter condition coincides with the necessary and sufficient condition for the mean square stability of the null solution of SDE (5.7.6). Thus, the drift-implicit Theta method (5.7.8) with implicitness $\theta = 0.5$ is useful to indicate mean square stability of the equilibrium solution of (5.7.6). More general theorems concerning the latter observations can be found in Schurz (1996, 1997).

5.7.2 Stationarity of linear systems with additive noise

Consider (for brevity, autonomous) linear system of SDEs

$$dX_t = A X_t \, dt + \sum_{j=1}^{m} b^j \, dW_t^j \tag{5.7.9}$$

Assume that there is a stationary solution X_∞ of (5.7.9). Then, for stationarity of autonomous systems (5.7.9) with additive white noise, it is a necessary and sufficient requirement that (5.7.2) is fulfilled.

Definition 5.7.3 *The random sequence* $(Y_n)_{n\in I\!N}$ *is said to be* **asymptotically** p**th mean preserving** *if*

$$\lim_{n\to+\infty} I\!E\,\|Y_n\|^p = I\!E\,\|X_\infty\|^p,$$

(asymptotically) mean preserving *if*

$$\lim_{n\to+\infty} I\!E\,Y_n = I\!E\,X_\infty,$$

(asymptotically) equilibrium preserving *if*

$$\mathcal{L}aw(Y_\infty) = \mathcal{L}aw(X_\infty)$$

with respect to systems (5.7.9).

This definition has been originally introduced by Schurz (1994). For an extension to systems (5.2.1), see the concept of asymptotically exact methods below.

Now, consider the **family of drift-implicit Euler methods** (see Kloeden, Platen and Schurz (1994)) with implicitness parameter $\alpha \in [0,1] \subset \mathbb{R}^1$, governed by

$$Y_{n+1} \;=\; Y_n + (\alpha A\,Y_{n+1} + (1-\alpha)A\,Y_n)\,\Delta_n + \sum_{j=1}^{m} b^j\,\Delta W_n^j \tag{5.7.10}$$

with independently Gaussian distributed increments $\Delta W_n^j = W_{t_{n+1}}^j - W_{t_n}^j$.

Theorem 5.7.4 *(Schurz (1997,1999)). Assume that*

(i) $\forall\lambda(\mathit{eigenvalue}(A))\; Re(\lambda(A)) < 0$

(ii) (X_0, Y_0) *independent of* $\mathcal{F}_\infty^j = \sigma\{W_s^j : 0 \le s < +\infty\}$

(iii) $I\!E\,\|X_0\|^p + I\!E\,\|Y_0\|^p < +\infty$ *for* $p \ge 2$

(iv) $A \in I\!R^{d\times d}, b^j \in I\!R^d$ *deterministic*

Then, the **trapezoidal method** *(i.e. (5.7.5) with* $\alpha = 0.5$*) applied to system (5.7.9) with any equidistant step size* $\Delta = \Delta_n$ *is asymptotically mean, pth mean and equilibrium preserving. Moreover, it is* **the only method from the entire family of implicit Euler methods** *with that behavior (i.e.,* $\Longrightarrow$ **asymptotic equivalence** *for systems with additive noise).*

Under diagonalizability of drift matrix A (real eigenvalues for simplicity) and condition (5.7.2) the conclusion of Theorem 7.2 can be seen very easily. First, the limit distribution of Y_n exists (for all implicitness parameters $\alpha > 0.5$). Second, the limit is Gaussian for all $\alpha \in [0.5, +\infty)$. Third, $\mathbb{E}\;Y_n \longrightarrow 0$ as n tends to $+\infty$ (as in deterministics if $\alpha \ge 0.5$). Fourth, due to uniqueness of Gaussian laws, it remains for one to look at second moments for all constant step sizes $\Delta > 0$. We notice

$$\lim_{n\to+\infty} \mathbb{E}\;\left([Y_n Y_n^T]\right)_{i,k} = -\sum_{j=1}^{m} \frac{b_i^j b_k^j}{\lambda_i + \lambda_k + (1-2\alpha)\Delta\lambda_i\lambda_k}$$

and

$$\mathbb{E}\left([X_\infty X_\infty^T]\right)_{i,k} = -\sum_{j=1}^{m} \frac{b_i^j b_k^j}{\lambda_i + \lambda_k}$$

Then

$$\lim_{n\to+\infty} \mathbb{E}\left([Y_n Y_n^T]\right)_{i,k} = \mathbb{E}\left([X_\infty X_\infty^T]\right)_{i,k} \quad \Longleftrightarrow \quad \alpha = 0.5.$$

Thus the stationarity with exact stationary Gaussian probability law is obvious. More general argumentations use fixed point principles. For more details, see Schurz (1996, 1997, 1999).

5.7.3 Asymptotically exact methods for linear systems

There do exist numerical methods which possess the same asymptotic probability law as the exact solution, for example those we have seen before. Schurz (1996, 1997, 1999) could constructively prove that fact for general linear systems of SDEs with additive and multiplicative noise. There the trapezoidal and midpoint method which coincide for linear autonomous systems of SDEs provide an asymptotically exact numerical method. Let $(t_n)_{n\in\mathbb{N}} \in [0, +\infty)$ be a sequence of $\mathcal{F}_{t_n}$-adapted stopping times with $\lim_{n\to+\infty} t_n = +\infty$ (a.s.).

Definition 5.7.5 *Suppose that* $\lim_{t\to+\infty} \mathbb{E}\,\|X_t\|^p = \mathbb{E}\,\|X_\infty\|^p < +\infty$ *for the stochastic process* $X = (X_t)_{t\geq 0}$ *on stochastic basis* $(\Omega, \mathcal{F}, (\mathcal{F}_t)_{t\geq 0}, \mathbb{P})$. *Then the random sequence* $Y = (Y_n)_{n\in\mathbb{N}}$ *on the same stochastic basis* $(\Omega, \mathcal{F}, (\mathcal{F}_{t_n})_{t_n\geq 0}, \mathbb{P})$ *is said to be* **asymptotically pth mean exact** *with respect to X iff*

$$\lim_{n\to+\infty} \mathbb{E}\,\|Y_n\|^p = \mathbb{E}\,\|X_\infty\|^p$$

and, in particular, if $p = 2$ *then Y is called* **asymptocially mean square exact**.

Theorem 5.7.6 *Assume that the stochastic process* $X = (X_t)_{t\geq 0}$ *on the stochastic basis* $(\Omega, \mathcal{F}, (\mathcal{F}_t)_{t\geq 0}, \mathbb{P})$ *satisfies SDE (5.7.1) discretized by the trapezoidal method (5.7.3) (i.e.* $\alpha = 0.5$*) or SDE (5.7.9) discretized by the trapezoidal method (5.7.10) (i.e.* $\alpha = 0.5$*) on the same* $(\Omega, \mathcal{F}, (\mathcal{F}_t)_{t\geq 0}, \mathbb{P})$. *Suppose* $\mathbb{E}\,||X_0||^p = \mathbb{E}\,||Y_0||^p$ *for all* $p > 0$.
Then the random sequence $Y = (Y_n)_{n\in\mathbb{N}}$ *with equidistant step sizes is asymptotically pth mean exact for all* $p > 0$.

Theorem 5.7.6, in its full extent, exhibits an unproved conjecture in the case of SDEs with multiplicative noise. For additive noise, it is an immediate consequence of results due to Schurz (1996, 1997, 1999). The proof for $p \geq 2$ can be carried out more easily. See forthcoming works of the author. It remains an open question whether one can construct other numerical methods which possess the properties of exactness and asymptotic exactness. In its full extent, this is a really challenging task for mathematics in the 21st century. A partial answer can be given for systems of linear SDEs with Gaussian white noise. Since we have no bias in the moments and due to linearity of the problem, it is clear that the trapezoidal method must approximate the conditional expectation asymptotically exactly. Since the conditional expectation is almost surely unique, we have the striking result that there is only one numerical method for well-posed linear systems which integrates linear systems of autonomous SDEs with Gaussian white noises asymptotically exactly, out of the class of all numerical methods (5.4.1) - (5.4.6) with $\Theta_n = \alpha I \in \mathbb{R}^{d\times d}, \theta_n = \alpha(1, 1, ..., 1)^T \in \mathbb{R}^d$

($\alpha = [0, 1]$) using any form of $\mathcal{F}_{t_n}$-adapted discretizations with lower order pth mean convergence. That method must be connected to the trapezoidal and midpoint methods in asymptotic sense.

5.7.4 Almost sure nonnegativity of numerical methods

A general problem of interest is the a.s. preservation of natural boundary conditions by discretization methods under the presence of random noise. The simplest form of an algebraic side condition which might arise in practice is the (a.s.) nonnegativity of numerical approximations. To give some illustration and a first solution consider the autonomous Itô SDEs

$$dX_t = a(X_t)dt + \sum_{j=1}^{m} b^j(X_t)\, dW_t^j \tag{5.7.11}$$

where a and b^j are such that a (strong) solution X_t on $\mathbb{R}_+^d$ exists (define $b^0(x) \equiv a(x)$). Now, consider the **family of Balanced Implicit Methods (BIMs)**, see Mil'shtein, Platen and Schurz (1992, 1994, 1998), governed by

$$\begin{aligned} Y_{n+1}^B &= Y_n^B + a(Y_n^B)\,\Delta_n + \sum_{j=1}^{m} b^j(Y_n^B)\,\Delta W_n^j \\ &\quad + \Big(\Delta_n C_0 + \sum_{j=1}^{m} |\Delta W_n^j| C_j\Big)\,(Y_n^B - Y_{n+1}^B) \end{aligned} \tag{5.7.12}$$

where C_0, C_1 are bounded matrices depending on Y_n^B such that

$$(I + \Delta_n C_0 + \sum_{j=1}^{m} |\Delta W_n^j| C_j)^{-1}$$

always exists and is uniformly bounded.

Theorem 5.7.7 *(Schurz (1996)). Assume that there are bounded, real $d \times d$ matrices $C_0, ..., C_m$ with nonnegative entries and positive constants K_3 and K_4 such that for all real-valued vectors x with nonnegative components*

(i) $[a(x) + C_0(x)x]_i \geq 0$ for all $i = 1, 2, ..., d$,

(ii) $[C_j(x)x]_i \geq |[b^j(x)]_i|$ for all $i = 1, 2, ..., d; j = 1, 2, .., m$,

and such that for all real-valued vectors $x \in \mathbb{R}^d$

(iii) $\sum_{j=0}^{m} \|C_j(x)b^j(x)\|^2 \leq K_3^2(1 + \|x\|^2)$

(iv) $\forall(\alpha_j \geq 0)_{j=0,1,...,m}, \alpha_0 \leq \hat{\alpha}$
$\exists M^{-1} = M^{-1}(x)$ with $M(x) = I + \sum_{j=0}^{m} \alpha_j C_j(x)$ and $\|M^{-1}(x)\| \leq K_4$ and

(v) $M^{-1} = M^{-1}(x)$ has only nonnegative entries for nonnegative vectors x.

Then, for any step sizes $(\Delta_n > 0)_{n \in \mathbb{N}}$, the BIMs (5.7.12) with weight matrices C_j are **positively invariant** *on $\mathbb{R}_+^d$, and provide* **strongly** *and* **mean square converging** *numerical approximations towards the exact solution of (5.7.11) with order $\gamma = 0.5$.*

This result can be verified constructively. For this purpose, consider BIMs generated by the scheme (5.7.12) with weight–matrices $C_j(.)$. Suppose these matrices satisfy the conditions $(i)-(v)$. The numerical approximations provided by BIMs converge strongly towards the SDE (5.7.11), with order $\gamma = 0.5$. This can be immediately concluded from the exposition Mil'shtein, Platen and Schurz (1992, 1994, 1998). Under condition (iv) of Theorem 5.7.7 the scheme of BIM (5.7.12) is rewritten as

$$Y_{n+1} = M_n^{-1}(Y_n)\left(Y_n + \sum_{j=0}^{m}\left(b^j(Y_n)\Delta W_n^j + C_j(Y_n)Y_n|\Delta W_n^j|\right)\right) \tag{5.7.13}$$

where $\Delta W_n^0 = \Delta_n$ and $M_n(x) = I + \sum_{j=0}^{m} C_j(x)|\Delta W_n^j|$. Suppose that $[Y_n]_i \geq 0$. We notice that the weight matrix M_n^{-1} preserves nonnegativity because of requirement (v). Thereby, we have to check only whether the random vector–valued function $\phi(x)$ with

$$\phi(x) = x + \sum_{j=0}^{m}\left(b^j(x)\Delta W_n^j + C_j(x)x|\Delta W_n^j|\right) \tag{5.7.14}$$

takes nonnegative values for nonnegative vectors $x \in \mathbb{R}^d$. Now, one arrives at the componentwise estimate

$$\begin{aligned}\phi_i(x) &\geq \sum_{j=0}^{m}\left([b^j(x)]_i\Delta W_n^j + [C_j(x)x]_i|\Delta W_n^j|\right)\\ &\geq \left([a(x) + C_0(x)x]_i\right)\Delta_n + \sum_{j=1}^{m}\left([C_j(x)x]_i - |[b^j(x)]_i|\right)|\Delta W_n^j| \geq 0\,.\end{aligned}$$

Each component of this random sum is nonnegative under assumptions $(i)-(ii)$. Hence, function ϕ takes nonnegative values for any random input ΔW_n^j. Therefore, the new vector Y_{n+1} only possesses nonnegative components. Consequently, Theorem 5.7.7 can be understood easily.

Further Remarks. Positive semi-definite C_j or nonnegative diagonal matrices C_j trivially fulfill (iv). Condition (5) is more restrictive, but it is satisfied by nonnegative diagonal matrices. The following conclusion for one-step approximations of some multidimensional SDEs is verified.

Theorem 5.7.8 *(Schurz (1996)). Assume that there are nonnegative constants K_j^0 and K_j^1 $(j = 0, 1, ..., m)$ such that*

$$|[b^j(x)]_i| \leq (K_j^0 + K_j^1|x_i|) \qquad \forall i = 1, 2, ..., d\,.$$

Then there exist numerical approximations $(Y_n)_{n\in\mathbb{N}}$ generated by BIMs (5.7.12) which strongly converge with order $\gamma = 0.5$ and maximize the one-step ε-probabilities of positivity, i.e. $\mathbb{P}\{Y_{n+1} > 0 | [Y_n]_i \geq \varepsilon, i = 1, 2, ..., d\} = 1$ for fixed, small values $\varepsilon > 0$.

In a constructive way one realizes the verification of Theorem 5.7.8. For this purpose, take the BIMs with diagonal matrices $C_j(x) = (c_j^{i,i})$ and elements

$$c_j^{i,i}(x) = \frac{2K_j^0}{\varepsilon + |x_i|} + K_j^1\,, \quad x = (x_1, ..., x_d)^T;\ j = 0, 1, ..., m;\ i = 1, 2, ..., d\,.$$

Thus, these functions are bounded and satisfy the conditions for strong convergence as stated in Mil'shtein, Platen and Schurz (1992, 1994, 1998). Therefore strong and mean square convergence of BIMs with order 0.5 is established. Nonnegativity of the one-step approximation (a.s.) is recognized as above as well.

Remark *A local one-step control in a reasonable distance to boundaries is possible without space discretization and with deterministic step sizes for random problems. However, in the vicinity of boundaries, one possibly needs to switch to careful random step size selection. pth mean convergence of BIMs can also be proved.*

5.7.5 Numerical invariance of intervals $[0, M]$

A problem of practical interest (e.g. in population dynamics, genetics and polymer physics) is that of getting numerically reasonable values in a given deterministic interval $[0, M]$ (a.s.). Since convergence statements are more of an asymptotic nature as step sizes are infinitesimally small, this question is not coverd by most of the authors. However, the main principle of numerical analysis in Section 5 has already shown the importance of the incorporation of geometric invariances (otherwise proofs have to be embedded in enlarged, nonnatural spaces, which can be a very laborious task to do or even infeasible if one is not aware of these geometric invariance properties). Schurz (1996) presents a way in context of innovation diffusions governed by one-dimensional SDEs

$$dX_t = [(p + \frac{q}{M}X_t)(M - X_t)]\,dt + \sigma X_t^{\alpha}(M - X_t)^{\beta}\,dW_t \tag{5.7.15}$$

driven by a given standard Wiener process W_t, started at $X_0 \in \mathbb{D} = [0, M] \subset \mathbb{R}^1$, where p, q, M, σ are positive and $\alpha, \beta \geq 0.5$ are real parameters. Here p can be understood as the coefficient of innovation, q as the coefficient of imitation and M as the total adoption size. However, in view of marketing issues, model (5.7.15) only makes sense within deterministic algebraic constraints. This fact generally leads to **S**tochastic **D**ifferential **A**lgebraic **E**quations (**SDAEs**) with nonanticipating boundary conditions. One can prove the $[0, M]$-invariance of SDE (5.7.15) whenever $\alpha, \beta \geq 0.5$ and $p, q, M > 0$. The natural question arises as to what happens then with the standard numerical approximation. Unfortunately, the classical (most-known and most-used) approximations such as explicit Euler and Mil'shtein method fail to preserve that $[0, M]$ invariance property with positive probability - a fact which can easily be seen by numerical experiments. However, some appropriate BIMs do have the $[0, M]$ property (a.s.). Consider the BIMs generated by

$$\begin{aligned} Y_{n+1} \;=\; & Y_n + (p + \frac{q}{M}Y_n)(M - Y_n)\Delta_n + \sigma Y_n^{\alpha}(M - Y_n)^{\beta}\Delta W_n + \\ & + \sigma K_d Y_n^{\alpha-1}(M - Y_n)^{\beta-1}|\Delta W_n|(Y_n - Y_{n+1}), \end{aligned} \tag{5.7.16}$$

where $K = K(M)$ is an appropriate positive constant and $Y_0 \in \mathbb{D} = [0, M]$ (a.s.). Then it holds the following theorem.

Theorem 5.7.9 *(Schurz (1995, 1996)). The numerical approximation $(Y_n)_{n \in \mathbb{N}}$ governed by (5.7.16) for SDE (5.7.15) is $\mathbb{D}$-invariant (a.s.) with $\mathbb{D} = [0, M]$, strongly and mean square convergent with order $\gamma = 0.5$ on any interval $[0, T]$ if*

$$Y_0 \in [0, M](a.s.),\; K_d = K_d(M) \geq M > 0,\; \alpha \geq 1,\; \beta \geq 1,\; 0 < \Delta_n \leq \frac{1}{p+q}\,(n \in \mathbb{N}).$$

More recently, we extended this idea to approximate interacting particle dynamics standardized on the interval $[0,1]$. Again the BIMs with adapted weighted coefficients, which take into account the current distance of approximations to the boundary, provide promising results without using projection methods and with keeping the same convergence order as Euler methods would have on the entire space $\mathbb{R}^d$. For more details, see the forthcoming paper of Schurz (1999). The simplest example is provided by one-dimensional Itô diffusion

$$dX_t \;=\; (\mu_1 + \mu_2 X_t)(1 - X_t)\,dt + \sum_{j=1}^{m} \sigma_j X_t^{\alpha_j}(1 - X_t)^{\beta_j}\,dW_t^j \tag{5.7.17}$$

where $\mu_1 \geq 0, \mu_2 \in R, \alpha_j \geq 0.5, \beta_j \geq 0.5, \sigma_j \in R$. Then one can show the almost sure $\mathbb{D}$-invariance property of SDE (5.7.17) with respect to domain $\mathbb{D} = [0,1]$ for any finite terminal time $T > 0$. This fact also seems to be a very natural requirement when one discusses genetic compositions and their asymptotic behavior in Mathematical Biology or stochastic innovation diffusions in Marketing Sciences, as seen before. In contrast to that property, numerical experiments easily show that the classical Euler and Mil'shtein methods may exit the domain $[0,1]$ at finite random times τ with positive probability. The problem of appropriate stopping rules for numerical approximations which do not destroy the order of convergence arises here, compared to the orders obtained for unstopped problems. In such cases we prefer a method from the family of BIMs once again instead of the simpler Euler method applied to SDE (5.7.17). For example, take $[0,1]$-boundary-adapted weights c^j with

$$\begin{aligned} C_0(x) &= [\mu_2]_-(1 - x) \\ C_j(x) &= |\sigma_j| x^{\alpha_j - 1}(1 - x)^{\beta^j - 1},\ x \in [0,1]\ if\ \alpha_j, \beta_j \geq 1 \end{aligned} \tag{5.7.18}$$

where $[.]_-$ represents the negative part of the inscribed expression (thus $\mu = [\mu]_+ - [\mu]_-$), and in the case of $0.5 \leq \alpha_j < 1$ or $0.5 \leq \beta_j < 1$ take

$$C_j(x) = \begin{cases} 0 \quad if\ x = 0\ or\ x = 1 \\ |\sigma_j| \cdot \left(\max(\frac{\alpha_j}{\alpha_j+\beta_j-1}, \frac{n+1}{n+2})\right)^{\alpha_j} \cdot \left(1 - \min(\frac{\alpha_j}{\alpha_j+\beta_j-1}, \frac{n}{n+1})\right)^{\beta_j - 1} \\ \qquad if\ \beta_j \geq 1 \text{ and } x \in \left(\frac{n}{n+1}, \frac{n+1}{n+2}\right] \\ |\sigma_j| \cdot \left(\max(\frac{\alpha_j}{\alpha_j+\beta_j-1}, \frac{n+1}{n+2})\right)^{\alpha_j} \cdot \left(1 - \max(\frac{\alpha_j}{\alpha_j+\beta_j-1}, \frac{n}{n+1})\right)^{\beta_j - 1} \\ \qquad if\ \alpha_j > 0.5 \text{ and } 0.5 \leq \beta_j < 1, x \in \left(\frac{n}{n+1}, \frac{n+1}{n+2}\right] \\ |\sigma_j| \cdot \sqrt{n+1} \quad if\ \alpha_j = \beta_j = 0.5, x \in \left(\frac{n}{n+1}, \frac{n+1}{n+2}\right] \end{cases} . \tag{5.7.19}$$

Then the following result can be concluded.

Theorem 5.7.10 *(Schurz (1999)). The BIMs $(Y_n^B)_{n \in \mathbb{N}}$ applied to (5.7.17) with scheme (5.7.12) using weights C_j specified by (5.7.18), (5.7.19) and with maximum step size Δ satisfying $(\mu_1 + [\mu_2]_+)\Delta \leq 1$ possess the invariance property with respect to interval $\mathbb{D} =$ $[0,1]$, i.e.*

$$\mathbb{P}\,\{Y_n^B \in [0,1] : \forall n \in N | Y_0^B \in [0,1]\}$$

for all $\mathcal{F}_0$-measurable $Y_0^B \in \mathbb{D} = [0,1]$. Therefore, they provide strongly converging, pth mean and strongly pth mean converging, double L^p-converging approximations on any finite time interval $[0,T]$ with order $\gamma = 0.5$ to the exact solution of SDEs (5.7.17) for $p \geq 2.0$.

Remark *More precisely, the result of [0,1]-invariance holds for all their paths by our specific deterministically boundary-adapted construction.*

5.7.6 Preservation of boundaries for Brownian Bridges

For simple illustration, consider **Brownian Bridges** (pinned Brownian motion). They can be generated by the one-dimensional SDE

$$dX_t = \frac{b - X_t}{T - t} dt + dW_t \tag{5.7.20}$$

started at $X_0 = a$, pinned to $X_T = b$ and defined on $t \in [0, T]$, where a and b are some fixed real numbers. According to the Corollary 6.10 of Karatzas and Shreve (1991), the process

$$X_t = \begin{cases} a(1 - \frac{t}{T}) + b\frac{t}{T} + (T - t)\int_0^t \frac{dW_s}{T - s} & \text{if } 0 \le t < T \\ b & \text{if } t = T \end{cases} \tag{5.7.21}$$

is the pathwise unique solution of (5.7.20) with the properties of having Gaussian distribution, continuous paths (a.s.) and expectation function

$$m(t) = \mathbb{E}\, X_t = a(1 - \frac{t}{T}) + b\frac{t}{T} \quad \text{on } [0, T] \tag{5.7.22}$$

Here problems are caused by unboundedness of drift

$$a(t, x) = \frac{b - x}{T - t}.$$

What happens now with approximations when we are taking the limit toward terminal time T? Can we achieve a preservation of the boundary condition $X_T = b$ in approximations Y under nonboundedness of the drift part of the underlying SDE at all?
A partial answer is given as follows. Consider the behavior of numerical solutions by the **family of implicit Euler methods**

$$Y_{n+1} = Y_n + \left[\alpha \frac{b - Y_{n+1}}{T - t_{n+1}} + (1 - \alpha)\frac{b - Y_n}{T - t_n}\right] \Delta_n + \Delta W_n \tag{5.7.23}$$

where $\alpha \in \mathbb{R}^+ = [0, +\infty), Y_0 = a$ and $n = 0, 1, ..., n_T - 1$. Obviously, in the case $\alpha = 0$, it holds that

$$Y^0(T) := Y_{n_T} = \lim_{n \to n_T} Y_n = b + \Delta W_{n_T - 1}\,. \tag{5.7.24}$$

Thus, the *explicit Euler method ends in random terminal values*, which is a contradiction to the behavior of exact solution (5.7.21)! Otherwise, in the case $\alpha > 0$, rewrite (5.7.23) as $Y_{n+1} =$

$$\frac{T - t_{n+1}}{T - t_{n+1} + \alpha\Delta_n} Y_n - \frac{(1 - \alpha)(T - t_{n+1})\Delta_n}{(T - t_n)(T - t_{n+1} + \alpha\Delta_n)} Y_n + \frac{T - t_{n+1}}{T - t_{n+1} + \alpha\Delta_n} \Delta W_n$$

$$+ \frac{(1 - \alpha)(T - t_{n+1})\Delta_n}{(T - t_n)(T - t_{n+1} + \alpha\Delta_n)} b + \frac{\alpha\Delta_n}{T - t_{n+1} + \alpha\Delta_n} b \tag{5.7.25}$$

which implies

$$Y^\alpha(T) := Y_{n_T} = \lim_{n \to n_T} Y_n = b\,. \tag{5.7.26}$$

Thus, the *implicit Euler methods can preserve (a.s.) the right terminal conditions*!

Theorem 5.7.11 *(Schurz (1996)). For any choice of step sizes $\Delta_n > 0, n = 0, 1, ..., n_T - 1$, it holds that*

$$\begin{array}{llcll} [1]. & \mathbb{E}\, Y_{n_T} & = & b & \text{if} \quad \alpha \geq 0 \\ [2]. & \mathbb{E}\,(Y_{n_T} - b)^2 & = & \Delta_{n_T - 1} & \text{if} \quad \alpha = 0 \\ [3]. & \mathbb{P}\,(Y_{n_T} = b) & = & 0 & \text{if} \quad \alpha = 0 \\ [4]. & \mathbb{P}\,(Y_{n_T} = b) & = & 1 & \text{if} \quad \alpha > 0 \end{array}$$

where the random sequence $(Y_n)_{n=0,1,\ldots,n_T}$ is generated by implicit Euler method (5.7.23) with step size $\Delta W_n \in \mathcal{N}(0, \Delta_n)$ where $\mathcal{N}(0, \Delta_n)$ denotes the Gaussian distribution with mean 0 and variance Δ_n (supposing deterministic step size).

Remark *Discontinuities in drift part may destroy convergence orders. A guarantee of algebraic constraints through implicit stochastic numerical methods can be observed. The example of Brownian Bridges supports the preference of implicit techniques, not only in so-called stiff problems as often cited.*

5.7.7 Nonlinear stability of implicit Euler methods

On nonlinear stability of stochastic numerical methods we could not find any treatments, except for that of Schurz (1996, 1997, 1999). In general one might think of nonlinear asymptotic pth mean stability. Let $p \in (0, +\infty)$.

Definition 5.7.12 *Assume $X \equiv 0$ is an equilibrium for system (5.2.1). The equilibrium solution $X \equiv 0$ is called* **globally (asymptotically) pth mean stable** *for the stochastic process $X = (X_t)_{t \geq 0}$ satisfying SDE (5.2.1) if*

$$\forall X_0 : \mathbb{E}\,\|X_0\|^p < +\infty \implies \lim_{t \to +\infty} \mathbb{E}\,\|X_t\|^2 = 0. \tag{5.7.27}$$

The equilibrium solution $X \equiv 0$ is called **(globally) exponentially pth mean stable** *for the stochastic process $X = (X_t)_{t \geq 0}$ satisfying SDE (5.2.1) if*

$$\exists K_0, K_1 > 0\, \forall t_1 \geq t_0\, \forall X_{t_0} :\ \mathbb{E}\,\|X_{t_0}\|^p < +\infty \implies$$

$$\mathbb{E}\,\|X_{t_1}\|^p \leq K_0 \exp(-K_1(t_1 - t_0))\mathbb{E}\,\|X_{t_0}\|^p. \tag{5.7.28}$$

Assume that $Y \equiv 0$ is an equilibrium for the numerical approximation $Y = (Y_n)_{n \in \mathbb{N}}$ applied to SDE (5.2.1). The equilibrium solution $Y \equiv 0$ is called **globally (asymptotically) numerically pth mean stable** *for the numerical approximation $Y = (Y_n)_{n \in \mathbb{N}}$ if*

$$\forall Y_0 : \mathbb{E}\,\|Y_0\|^p < +\infty \implies \lim_{n \to +\infty} \mathbb{E}\,\|Y_n\|^p = 0. \tag{5.7.29}$$

The equilibrium solution $Y \equiv 0$ is called **(globally) exponentially numerically pth mean stable** *for the numerical approximation $Y = (Y_n)_{n \in \mathbb{N}}$ if*

$$\exists K_0, K_1 > 0\, \forall n_1 \geq n_0\, \forall Y_{n_0} :\ \mathbb{E}\,\|Y_{n_0}\|^p < +\infty \implies$$

$$\mathbb{E}\,\|Y_{n_1}\|^p \leq K_0 \exp(-K_1(t_{n_1} - t_{n_0}))\mathbb{E}\,\|Y_{n_0}\|^p. \tag{5.7.30}$$

This definition leads to the following first result. Unfortunately, other results concerning nonlinear stability of stochastic numerical methods for SDEs are not known to the author at this writing.

Theorem 7.9. (Schurz (1996, 1997, 1999)). Assume that the SDE (5.2.1) has an exponentially mean square stable equilibrium solution $X \equiv 0$ with some constants $K_{OB}^{I} \leq 0, K_{OB}^{H} < 0$ (i.e. $p = 2$).
Then the drift-implicit Euler method applied to that SDE (5.2.1) possesses an exponentially mean square stable equilibrium solution $Y \equiv 0$ provided that

$$0 < \Delta_n \leq \sup_{k \in \mathbb{N}} \Delta_k < +\infty.$$

5.7.8 Linear and nonlinear A-stability

A-stability is one of the most desired properties of numerical algorithms. We should distinguish between the linear A-stability and nonlinear A-stability concepts, depending on the corresponding linear and nonlinear test classes of dissipative SDEs. However, one may find a unified treatment of the classical A-stability concept. Following Schurz (1996, 1997, 1999) we have these definitions, motivated by the fundamental works of Dahlquist in deterministic numerical analysis. Fix $p \in [1, +\infty)$.

Definition 5.7.13 *The numerical sequence* $Y = (Y_n)_{n \in \mathbb{N}}$ *(method, approximation, etc.) is called* **pth mean A-stable** *if it has an asymptotically numerically pth mean stable equilibrium solution* $Y \equiv 0$ *for all autonomous SDEs (5.2.1) having an asymptotically pth mean stable equilibrium solution* $X \equiv 0$ *with any constants* $K_{OB}^{I} \leq 0, K_{OB}^{H} < 0$*, using any admissible step size sequence* Δ_n *with* $\sup_{n \in \mathbb{N}} \Delta_n < +\infty$*. The numerical sequence* $Y = (Y_n)_{n \in \mathbb{N}}$ *(method, approximation, etc.) is said to be* **pth mean AN-stable** *if it has an asymptotically numerically pth mean stable equilibrium solution* $Y \equiv 0$ *for all SDEs (5.2.1) having an asymptotically pth mean stable equilibrium solution* $X \equiv 0$ *with some constants* $K_{OB}^{I} \leq 0$, $K_{OB}^{H} < 0$*, using any admissible step size sequence* Δ_n *with* $\sup_{n \in \mathbb{N}} \Delta_n < +\infty$.

Definition 5.7.14 *(Schurz (1996, 1997, 1999)). The drift-implicit Euler method applied to SDEs (5.2.1) provides mean square A- and AN-stable numerical approximations (i.e. when* $p = 2$*).*

Therefore, the implicit Euler methods are on the "sure numerically stable" side. However, we must notice that they provide "superstable" numerical approximations - a property which may lead to undesired stabilization effects of numerical dynamics, and then it would be better to make use of asymptotically exact numerical methods. In passing we note that linear *A*-stability of stochastic algorithms has been discussed by Artemiev (1994), Mitsui and Saito (1996) and Schurz (1993 - 1999), where Mitsui and Saito (1996) have only discussed the case of one-dimensional SDEs using the traditional stability function approach from deterministic numerical analysis, whereas Artemiev (1994) and Schurz (1993, 1996, 1997) have already treated the fully multidimensional setting. Nonlinear A-stability investigations could only be found in Schurz (1996, 1997, 1999) so far, according to our current knowledge. There is also an approach using the concept of **weak A-stability**, i.e. the A-stability of related deterministic numerical dynamics discretizing linear SDEs. However, this concept does not lead to new insights into the effects of stochasticity with respect to stability. For such attempts leading to recitation of known facts from deterministic numerical analysis, see Mil'shtein (1988, 1995), Kloeden and Platen (1992, 1995) or Platen (1999).

5.7.9 Stability exponents of explicit-implicit methods

The art of stability-adequate methods consists of construction of appropriate explicit-implicit methods which replicate some reasonable estimates for exponential growth rates or which even show evident coincidence with the corresponding growth rates of underlying continuous time dynamics. For this purpose, we introduce the following d-dimensional **explicit-implicit splitting methods**

$$\begin{aligned} X_{n+1} &= X_n + \Phi_0^I(X_i : i \leq n+1)\Delta_n + \Phi_0^E(X_i : i \leq n)\Delta_n \\ &+ \sum_{j=1}^{m} \Phi_j(X_i : i \leq n)\xi_n^j \sqrt{\Delta_n} \end{aligned} \tag{5.7.31}$$

where $\Delta_n = t_{n+1} - t_n$ is interpreted as a sequence of step sizes with monotonically increasing time-instants $(t_i)_{i \in \mathbb{N}}$ and $\lim_{i \to +\infty} t_i = +\infty$; $\Phi_0^I, \Phi_0^E, \Phi_j$ where $j = 1, 2, ..., m$ represent deterministic mappings from all currently generated values into $\mathbb{R}^d$ (They may admit past–path–dependence in general!), and ξ_n^j are real-valued, independent random variables on $(\Omega, \mathcal{F}, \mathbb{P})$ with moments

$$\mathbb{E}\, \xi_n^j = 0 \quad \text{and} \quad \mathbb{E}\, |\xi_n^j|^2 = (\sigma_n^j)^2 < +\infty .$$

Let $\tau^{\Delta}([0,T]) = \{t_i \in [0,T] : t_i \leq t_{i+1}, i \in \mathbb{N}\}$. Then we want to classify these additive splitting techniques by their exponential growth or decay exponents.

Definition 5.7.15 *Let $\mathcal{I} = [0, +\infty)$ or $\mathcal{I} = \tau^{\Delta}([0, +\infty))$. Then the* **upper (forward pth moment) stability exponent** *of a given stochastic process $(X(t))_{(t \in \mathcal{I})}$ in domain $\mathbb{D}$ is defined to be*

$$\overline{\lambda}_p := \limsup_{t \to +\infty} \frac{1}{t} \ln \mathbb{E}\, \|X(t)\|^p \tag{5.7.32}$$

for $X(t_0) \in \mathbb{D}$ (a.s.), provided that this limit exists. The **lower (forward pth moment) stability exponent** *of a given stochastic process $(X(t))_{(t \in \mathcal{I})}$ in domain $\mathbb{D}$ is defined to be*

$$\underline{\lambda}_p := \liminf_{t \to +\infty} \frac{1}{t} \ln \mathbb{E}\, \|X(t)\|^p \tag{5.7.33}$$

for $X(t_0) \in \mathbb{D}$ (a.s.), provided that this limit exists.

To save space we have stated this definition for the case of discrete and continuous time stochastic processes using time scales $\mathcal{I}$; then one only has to substitute the related continuous and discrete time scales, where the discrete elements $t_n \in \mathcal{I} = \tau^{\Delta}([0, +\infty)$ are also identified by integers $n \in \mathbb{N}$.

Theorem 5.7.16 *(Schurz (1999)). Let process $(X_n)_{n\in\mathbb{N}}$ satisfy the stochastic difference equation (5.7.31) under the above mentioned conditions for all $n \in \mathbb{N}$, whereas all ξ_n^j are independent of X_0 as well. Assume that $\forall n \in \mathbb{N}\ \forall x^{(l)} \in \mathbb{R}^d (l = 0, 1, ..., n+1)\ \forall j = 1, 2, ..., m$:*

$$\begin{array}{rcl} < \Phi_0^I(x^{(l)} : l \le n+1), x^{(n+1)} > & \le & \overline{k}_I(n)\|x^{(n+1)}\|^2 \\ < \Phi_0^E(x^{(l)} : l \le n), x^{(n)} > & \le & \overline{k}_E(n)\|x^{(n)}\|^2 \\ \|\Phi_0^I(x^{(l)} : l \le n+1)\|^2 & \ge & \overline{k}_0^I(n)\|x^{(n+1)}\|^2 \\ \|\Phi_0^E(x^{(l)} : l \le n)\|^2 & \le & \overline{k}_0^E(n)\|x^{(n)}\|^2 \\ \|\Phi_j(x^{(l)} : l \le n)\|^2 & \le & \overline{k}_j(n)\|x^{(n)}\|^2 \\ 2\overline{k}_I(n)\Delta_n & < & 1 + \overline{k}_0^I(n)\Delta_n^2 \end{array} \tag{5.7.34}$$

where $\overline{k}_I(n), \overline{k}_E(n), \overline{k}_0^I(n), \overline{k}_0^E(n), \overline{k}_j(n)$ are finite, deterministic, real numbers. Then

$$\overline{\lambda}_2 \le \limsup_{n\to+\infty} \frac{\displaystyle\sum_{i=0}^{n} \left(\frac{2\overline{k}_E(i) + 2\overline{k}_I(i) + (\overline{k}_0^E(i) - \overline{k}_0^I(i))\Delta_i + \sum_{j=1}^{m} (\sigma_i^j)^2 \overline{k}_j(i)}{1 - 2\overline{k}_I(i)\Delta_i + \overline{k}_0^I(i)\Delta_i^2} \right) \Delta_i}{t_{n+1}}. \tag{5.7.35}$$

Furthermore, if $\forall n \in \mathbb{N}\, \forall x^{(l)} \in \mathbb{R}^d (l = 0, 1, ..., n+1)\ \forall j = 1, 2, ..., m$:

$$\begin{array}{rcl} < \Phi_0^I(x^{(l)} : l \le n+1), x^{(n+1)} > & \ge & \underline{k}_I(n)\|x^{(n+1)}\|^2 \\ < \Phi_0^E(x^{(l)} : l \le n), x^{(n)} > & \ge & \underline{k}_E(n)\|x^{(n)}\|^2 \\ \|\Phi_0^I(x^{(l)} : l \le n+1)\|^2 & \le & \underline{k}_0^I(n)\|x^{(n+1)}\|^2 \\ \|\Phi_0^E(x^{(l)} : l \le n)\|^2 & \ge & \underline{k}_0^E(n)\|x^{(n)}\|^2 \\ \|\Phi_j(x^{(l)} : l \le n)\|^2 & \ge & \underline{k}_j(n)\|x^{(n)}\|^2 \\ -2\underline{k}_E(n)\Delta_n & < & 1 + \underline{k}_0^E(n)\Delta_n^2 + \sum_{j=1}^{m} (\sigma_n^j)^2 \underline{k}_j(n)\Delta_n \end{array} \tag{5.7.36}$$

where $\underline{k}_I(n), \underline{k}_E(n), \underline{k}_0^I(n), \underline{k}_0^E(n), \underline{k}_j(n)$ are finite, deterministic, real numbers, then

$$\underline{\lambda}_2 \ge \liminf_{n\to+\infty} \frac{\displaystyle\sum_{i=0}^{n} \left(\frac{2\underline{k}_E(i) + 2\underline{k}_I(i) + (\underline{k}_0^E(i) - \underline{k}_0^I(i))\Delta_i + \sum_{j=1}^{m} (\sigma_i^j)^2 \underline{k}_j(i)}{1 + 2\underline{k}_E(i)\Delta_i + \underline{k}_0^E(i)\Delta_i^2 + \sum_{j=1}^{m} (\sigma_i^j)^2 \underline{k}_j(i)\Delta_i} \right) \Delta_i}{t_{n+1}}. \tag{5.7.37}$$

Remark *This theorem provides a uniform estimate of the "spectrum" of (forward) mean square stability exponents for the class of stochastic difference equations satisfying (5.7.34) and (5.7.36). Of course these estimates are "worst case estimates," but they are sharp ones (see linear systems where equality is satisfied). Since the analysis of nonlinear, nonautonomous, discrete time stochastic mappings turns out to be very difficult, we restrict our attention only to the feasible case of mean square calculus. All in all, the art consists in* *finding the right splittings to guarantee the conditions of this theorem. Loosely speaking, as the main result, one has to apply explicit techniques to follow the unstable branches of the underlying continuous time stochastic dynamics, and one should apply implicit techniques to follow the stable branches of underlying continuous time stochastic dynamics of SDEs through numerical methods. Sometimes one even needs multiplicative splitting techniques to follow that rule of thumb precisely. The critical case of conservative systems (like during integration of stochastic Hamiltonian systems) is the most interesting case. Then numerical approximations should only be realized by exact (coherent) and asymptotically exact numerical techniques (e.g. by implicit midpoint rules). In this respect there is still plenty of work to do - a challenge for the new millennium. Some initial illustrative examples can be found in Schurz (1999).*

5.7.10 Hofmann-Platen's M-stability concept in $\mathbb{C}^1$

Here we refer to a specific test equation and a stability concept introduced originally by Hofmann and Platen (1994, 1996), whereas its meaningfulness still has to be discussed. Consider the one-dimensional complex-valued Stratonovich SDE

$$dX_t \;=\; (1-\rho)\lambda X_t\,dt \,+\, \sigma\sqrt{\rho}\,X_t \circ dW_t \tag{5.7.38}$$

which is equivalent to the complex-valued Itô SDE

$$dX_t \;=\; (1-\frac{1}{2}\rho)\lambda X_t\,dt \,+\, \sqrt{\rho\lambda}\,X_t\,dW_t \tag{5.7.39}$$

with $\lambda = Re(\lambda) + iIm(\lambda), \sigma = Re(\sigma) + iIm(\sigma) \in \mathbb{C}$ where $\sigma^2 = \lambda$, $i^2 = -1$, where $W = (W_t)_{t\geq 0}$ represents a one-dimensional, real-valued Wiener process. The real-valued parameter $\rho \in [0,2]$ describes the degree of stochasticity in test equation (5.7.38). For $\rho = 0$ one has a purely deterministic equation, for $\rho = 1$ a pure Stratonovich SDE with no drift, while for $\rho = 2$ we have an Itô SDE with no drift term. Numerical methods applied to test equation (5.7.38) can be written as

$$Y_{n+1} \;=\; G(\lambda\Delta_n,\rho)Y_n \;=\; \left(\prod_{k=0}^{n} G(\lambda\Delta_k,\rho)\right) Y_0 \tag{5.7.40}$$

in recursive form with complex-valued **stability transfer function** G related to the corresponding numerical method applied to SDE (5.7.38) such that $G : \mathbb{C} \times [0,2] \longrightarrow \mathbb{C}$ does not depend on the sequence $Y = (Y_n)_{n\in\mathbb{N}}$.

Definition 5.7.17 *The* **Hofmann-Platen M-stability set** *of a numerical method* $Y = (Y_n)_{n\in\mathbb{N}}$ *applied to test SDE (5.7.38) is defined to be* $\Gamma = \{\Gamma_\rho : 0 \leq \rho \leq 2\}$ *with* **HPM-stability regions**

$$\Gamma_\rho = \{\lambda\Delta \in \mathbb{C} : Re(\lambda) < 0, ess_\omega \sup |G(\lambda\Delta,\rho)| < 1\}$$

where $ess_\omega \sup$ *denotes the essential supremum with respect to all* $\omega \in \Omega$.

Thus the concept of HPM-stability refers to the worst case scenario which might arise by numerical dynamics. Regions of HPM-stability of some numerical methods, like that of the explicit Euler method and drift-implicit Euler method; are depicted in Hofmann and Platen (1994, 1995) and Platen (1999). For example, when $\rho = 0$, then the HPM-stability region is presented by the common circle region Γ_ρ of linear deterministic A-stability. With increasing $\rho \in [0,2]$ and $Re(\lambda)$ the HPM-stability region may monotonically shrink for the explicit Euler method (as it happens for linear, real-valued test SDEs (5.7.38) anyway) – a fact which does not surprise us much due to the specifically inherent structure of test equation (5.7.38) and simultaneously growing noise intensities (For example, compare with the qualitative behavior of equivalent Itô dynamics (5.7.39), where the increase of real parameter $\rho \in [0,2]$ yields destabilization effects on the moments under the condition $Re(\lambda) < 0$ and $p \geq 2!$). The drift-implicit Euler method has a larger HPM-stability region than the explicit Euler method, however the HPM-stability regions for both methods do not contain the entire ρ-axis due to the test equation (5.7.38). This fact implies step size restrictions leading to the natural choice of maximum and minimum step sizes - an experience which is incorporated in any sophisticated deterministic variable step size algorithm anyway. The concept of HPM-stability seems to be designed especially for dealing with stability issues of weakly converging numerical methods using equidistant step sizes, for which more degrees of freedom while simulating involved random variables are observed in general. For strong, pth mean and almost sure converging numerical methods, the concept of HPM-stability does not seem to be very appropriate. It even is too impractical, due to the very erratic behavior of random noise terms. However, the choice of test equation (5.7.38) together with the concept of HPM-stability represent one of the strongest criteria of numerical stability one might ask for and exhibit an interesting combination of effects of different stochastic calculi on the qualitative asymptotic behavior of numerical dynamics.

Our alternative suggestion: Take the complex-valued one-dimensional test equation

$$dX_t \;=\; \lambda X_t\, dt \;+\; \sigma_1 X_t \underset{(\nu_1)}{*} dW_t^1 \;+\; \sigma_2 X_t \underset{(\nu_2)}{*} dW_t^2 \tag{5.7.41}$$

interpreted in the stochastic (ν_1,ν_2)-calculus sense with deterministic parameters $\nu_1,\nu_2 \in [0,1]$ (i.e. $\nu_k = 0$ corresponds to Itô calculus, $\nu_k = 0.5$ to Stratonovich calculus), which is equivalent to the complex-valued Itô SDE

$$dX_t \;=\; (\lambda + \nu_1\sigma_1^2 + \nu_2\sigma_2^2)\, X_t\, dt \;+\; \sigma_1 X_t\, dW_t^1 \;+\; \sigma_2 X_t\, dW_t^2 \tag{5.7.42}$$

with $\lambda = Re(\lambda) + iIm(\lambda)$, $\sigma_k = Re(\sigma_k) + iIm(\sigma_k) \in \mathbb{C}$, where $W^k = (W_t^k)_{t\geq 0}$ represent two independent, one-dimensional, real-valued Wiener processes. The real-valued parameters $\nu_k \in [0,1]$ describe the influence of changes of stochastic calculus in the test equation (5.7.41). Such a test equation would be representative for at least the class of SDEs with varying stochastic integration calculus and with some commutativity (between drift and diffusion terms, which is trivially fulfilled in the one-dimensional real-valued situation under the absence of drift terms). Moreover, the **essential stability region** for equidistant numerical approximations with **stability transfer function** G_{ν_1,ν_2} should rather be defined by

$$\Gamma_{\nu_1,\nu_2} = \left\{ (\lambda\Delta, \sigma_1\sqrt{\Delta}, \sigma_2\sqrt{\Delta}) \in \mathbb{C}^3 : \begin{array}{l} Re(\lambda + \nu_1\sigma_1^2 + \nu_2\sigma_2^2) < 0 \\ ess_\omega \sup |G_{\nu_1,\nu_2}(\lambda\Delta, \sigma_1\sqrt{\Delta}, \sigma_2\sqrt{\Delta})| < 1 \end{array} \right\}.$$

for numerical methods with $Y_{n+1} = G_{\nu_1,\nu_2}(\lambda\Delta_n, \sigma_1\sqrt{\Delta}, \sigma_2\sqrt{\Delta})\, Y_n$ to test the influence of stochastic integration calculus. Certainly, for stability investigations of variable step size algorithms or with multidimensional test equations, more care is needed, and the theory of monotone operators could be exploited. Note that, for linear, one-dimensional test equations, one does not need any numerical method at all to generate its solution since one knows the explicit solution expression due to the naturally induced commutativity property in one dimension (Note the immense complexity of the stochastic test equation problem which continues to be a worthwhile and open discussion.).

5.7.11 Asymptotic stability with probability one

A very subile question is represented by the problem of asymptotic stability of numerical approximations with probability one. This question can be studied for equidistant approximations applied to linear SDEs as follows. For an approach, let us recall the concept of numerical asymptotic stability with probability one.

Definition 5.7.18 *The numerical sequence* $Y = (Y_n)_{n \in I\!N}$ *is called* **numerically asymptotically stable with probability one** *if*

$$\lim_{n \to +\infty} \|Y_n\| = 0$$

with probability one.

An application of the well-known strong Law of Large Numbers (SLLN) and the law of iterated logarithm provides the following crucial tool to verify asymptotic stability with probability one.

Theorem 5.7.19 *Assume that a discrete time stochastic process* $Z = (Z_n)_{n \in I\!N}$ *with nonnegative real values (a.s.) has a independently and identically distributed positive stability transfer function* $G(k)$ *satisfying*

$$Z_{n+1} = \left(\prod_{k=0}^{n} G(k) \right) Z_0$$

and $I\!E\left[ln(G(k))\right]^2 < +\infty$ *for all* $k \in I\!N$.
Then $Z = (Z_n)_{n \in I\!N}$ *converges to zero with probability one iff* $I\!E\left[ln(G(k))\right] < 0$.

To establish asymptotic stability with probability one for an originally given numerical method $Y = (Y_n)_{n \in \mathbb{N}}$ one may consider the pathwise quadratic evolution $Z_n = \|Y_n\|^2$ by taking the squared Euclidean norm of Y_n. Then one can identify the nonnegative random variables $G(n)$ such that $Z_{n+1} = G(n) Z_n$ for linear or linearly dominated problems, and it remains to check the equivalence criterion stated by Theorem 5.7.19. For example, the drift-implicit Theta methods applied to bilinear SDE with equidistant step sizes may fail to produce asymptotically stable approximations with probability one, even though they possess mean square A-stable numerical approximations for $\theta \geq 0.5$. However, the fully implicit weakly converging Euler methods (see Kloeden and Platen (1995), p. 337)) and the balanced implicit methods (BIMs) with any equidistant step size produce asymptotically stable approximations with probability one. For example, consider the one-dimensional Itô diffusion equation

$$dX_t = \sigma X_t \, dW_t,$$

as suggested by Mil'shtein, Platen and Schurz (1992, 1994, 1998). Take BIMs with scalar weights $c^0 = 0$ and $c^1 = |\sigma|$. Then asymptotic stability with probability one is established by application of Theorem 5.7.19 with

$$I\!E \left[ln \left| \frac{1 + |\sigma \Delta W_n| + \sigma \Delta W_n}{1 + |\sigma \Delta W_n|} \right| \right] < 0$$

with independently identically Gaussian distributed increments $\Delta W_n \in \mathcal{N}(0, \Delta)$, provided that one has the nontrivial situation $|\sigma| > 0$.

5.8 Numerical Contractivity

To study the numerical stability behavior which corresponds to the often–cited property of control on error propagation, one has to introduce the concept of numerical contractivity. This concept replicates the needs of a numerical approximation algorithm better than that of numerical stability in respect to control of error propagation in the course of numerical integration. Unfortunately, a lot is not known about this. The only contribution in this respect in stochastic analysis could be found in Schurz (1997). In that monograph one basically exploits the monotonicity of coefficient systems (a, b^j) of the related test class of SDEs. It is also worth noting that, for linear systems with multiplicative noise, the concepts of stability and contractivity coincide. However, for general nonlinear systems or systems with additive noise they do not. For systems with additive noise, the concept of contractivity is apparently much more appropriate than that of stability in describing the initial error propagation in numerical algorithms and in stochastic processes for controlling their convergence.

5.8.1 Contractivity of SDEs with monotone coefficients

Since a lot is not known about contractivity of continuous time SDE according to our knowledge until 1999 (please, feel free to check the literature), we feel a necessity to report about one specific result in the case of SDEs with monotone coefficients. This is taken from Schurz (1996, 1997). Suppose that $[t_1, t_2] \subset [0, +\infty)$ with $\mathcal{F}_{t_1}$-adapted instants t_1, t_2 and $t_1 < t_2$ (e.g. t_1, t_2 deterministic) only contains $\mathcal{F}_{t_1}$-adapted times s, t. Let x, y be deterministic or any $\mathcal{F}_s$-adapted values in the statement of the following definition.

Definition 5.8.1 *A stochastic process* $X = (X_t)_{t\geq 0}$ *with basis* $(\Omega, \mathcal{F}, (\mathcal{F}_t)_{0\leq t\leq T}, \mathbb{P})$ *is said to be* **strictly uniformly** p**th mean contractive on** $[t_1, t_2]$ *with respect to domain* $\mathbb{D}$ *iff* $\exists K_C^X \in \mathbb{R}\ \forall t, s \in [t_1, t_2]\ \forall x, y \in \mathbb{D}$

$$\mathbb{E}\left[||X_{s,x}(t) - X_{s,y}(t)||^p \Big| \mathcal{F}_s\right] = \exp\left(pK_C^X|t - s|\right)||x - y||^p \tag{5.8.1}$$

with **strictly uniform** p**th mean contractivity constant** K_C^X.

In general K_C^X, $\mathbb{D}$ could be random, but then some necessary extra assumptions on $X, K_C^X, \mathbb{D}$ must be made to ensure the meaningfulness of the introduced concept. In particular, *local* and *global contractivity* can be discussed within the same definition as well. One can also discuss concepts with *forward* and *backward contractivity*, but this is omitted here due to lack of space. Now, we confine ourselves to SDEs.

Definition 5.8.2 *A SDE (5.2.1) is said to have a* **strictly uniformly** p**th monotone coefficient system** (a, b^j) **on** $[t_1, t_2]$ *with respect to open domain* $\mathbb{D}$ *iff* $\exists K_{UC} \in \mathbb{R}\ \forall t[t_1, t_2]\ \forall x, y \in \mathbb{D}$

$$< a(t,x) - a(t,y), x - y >_d + \frac{1}{2}\sum_{j=1}^{m} ||b^j(t,x) - b^j(t,y)||^2$$
$$+ \frac{p-2}{2}\sum_{j=1}^{m} \frac{< b^j(t,x) - b^j(t,y), x - y >_d^2}{||x - y||^2} \leq K_{UC}\,||x - y||^2. \tag{5.8.2}$$

Theorem 5.8.3 *(Schurz (1996, 1997)). X satisfies SDE (5.2.1) with pth mean monotone coefficient system (a, b^j).*
Then X is pth mean contractive for all $p \geq 2$. The "worst case" pth mean contractivity constant K_C^X can be estimated by

$$K_C^X \leq K_{UC} \leq K_{OL}.$$

Thus the propagation of initial perturbations is under control in the case of SDEs with pth mean contractive coefficient systems (a, b^j). For nonautonoumous variants, see Schurz (1996, 1997, 1999).

5.8.2 Contractivity of implicit Euler methods

The only class of numerical methods which is known so far and provides pth mean contractive approximations for SDEs with monotone coefficient systems (a, b^j) is the drift-implicit Euler method.

Theorem 5.8.4 *(Schurz (1996, 1997)). Assume the SDE (5.2.1) has a mean square monotone coefficient system (a, b^j).*
Then the drift-implicit Euler method applied to (5.2.1) performs a mean square contractive numerical approximation for all uniformly admissible step sizes $(\Delta_n)_{n \in I\!N}$.

For an elementary proof, see Schurz (1996, 1997, 1999).

5.8.3 pth mean B- and BN-stability

It is natural to ask to transfer the deterministic concept of B-stability to the stochastic case. This can be done in the pth mean moment sense fairly straightforwardly, and it has been studied by Schurz (1996, 1997, 1999) in the case of SDEs at first. From those references we recall the definition

Definition 5.8.5 *A numerical sequence $Y = (Y_n)_{n \in I\!N}$ (method, scheme, approximation, etc.) is called* **pth mean B-stable** *if it is pth mean contractive for all autonomous SDEs (5.2.1) with pth mean monotone coefficient systems (a, b^j) and for all admissible step sizes. It is said to be* **pth mean BN-stable** *if it is pth mean contractive for all nonautonomous SDEs (5.2.1) with pth mean monotone coefficient systems (a, b^j) for all admissible step sizes.*

Theorem 5.8.6 *(Schurz (1996, 1997)). The drift-implicit Euler method applied to Itô SDEs (5.2.1) performs a mean square BN-stable and B-stable numerical approximation.*

The proof is an immediate consequence of the proof of Theorem 8.2. See Schurz (1996, 1997, 1999) for more details.

5.8.4 Contractivity exponents of explicit-implicit methods

In general, one is aiming at implementation of methods which have the controlled error propagation and stabilized numerical evolutions toward some invariant manifolds. In particular, the exponential growth behavior of errors in discretized dynamics is of special interest. More generally, consider **stochastic dynamical systems** $X(t,z)$ on $(\Omega, \mathcal{F}, (\mathcal{F}_t)_{t\in\mathbb{R}}, \mathbb{P})$ started at $X(0,z) = z \in \mathbb{D} \subseteq \mathbb{R}^d$ at time $t = s$. Again, for brevity, we shall state the following very general definition, referring to continuous and discrete time scales simultaneously.

Definition 8.4. The **upper (forward pth moment) contractivity exponent** of a stochastic dynamical system $X(t,z)$ on $\mathbb{D}$ is defined to be

$$\overline{\kappa}_p := \limsup_{t\to+\infty} \frac{1}{t} \, ln \, (\mathbb{E} \, \|X(t,x) - X(t,y)\|^p) \tag{5.8.3}$$

for $X(t_0,x) = x, X(t_0,y) = y \in \mathbb{D}$ (a.s.). The **lower (forward pth moment) contractivity exponent** of $X(t,z)$ on $\mathbb{D}$ is defined to be

$$\underline{\kappa}_p := \liminf_{t\to+\infty} \frac{1}{t} \, ln \, (\mathbb{E} \, \|X(t,x) - X(t,y)\|^p) \tag{5.8.4}$$

for $X(t_0,x) = x, X(t_0,y) = y \in \mathbb{D}$.

Let us now look at uniform estimates of those contractivity exponents in the case of a class of nonlinear stochastic difference equations with monotone coefficients. Fix a deterministic domain $\mathbb{D} \subset \mathbb{R}^d$. Now consider again the d-dimensional iterative mappings

$$\begin{aligned} X_{n+1}(z) &= X_n(z) + \Phi_0^I(X_i(z) : i \leq n+1)\Delta_n + \Phi_0^E(X_i(z) : i \leq n)\Delta_n \\ &\quad + \sum_{j=1}^m \Phi_j(X_i(z) : i \leq n)\xi_n^j \sqrt{\Delta_n} \end{aligned} \tag{5.8.5}$$

on the fixed deterministic domain $\mathbb{D}$ (a.s.), started at any $z \in \mathbb{D}$, where deterministic $\Delta_n = t_{n+1} - t_n$ is a sequence of step sizes with monotonically increasing time-instants $(t_i)_{i\in\mathbb{N}}$ and $\lim_{i\to+\infty} t_i = +\infty$, and ξ_n^j are real-valued, independent random variables on probability space $(\Omega, \mathcal{F}, \mathbb{P})$ with moments

$$\mathbb{E} \, \xi_n^j = 0 \quad \text{and} \quad \mathbb{E} \, |\xi_n^j|^2 = (\sigma_n^j)^2 < +\infty .$$

For convenience of statement, define $\delta_n(x,y) := x^{(n)} - y^{(n)}$. Since the analysis of nonlinear, nonautonomous, discrete time stochastic mappings turns out to be very difficult, we restrict our attention to the case of mean square calculus, as before.

Theorem 5.8.7 *(Schurz (1999)). Let process $(X_n(z))_{n\in\mathbb{N}}$ satisfy the stochastic difference equation (5.8.5) started at value $z \in \mathbb{D}$ under the above-mentioned conditions for all $n \in \mathbb{N}$, where all ξ_n^j are independent of $X_0(z)$ as well. Assume that $\forall n \in \mathbb{N}\ \forall x^{(l)}, y^{(l)} \in \mathbb{R}^d (l = 0, 1, ..., n+1)\ \forall j = 1, 2, ..., m$: $< \Phi_0^I(x^{(l)} : l \leq n+1) - \Phi_0^I(y^{(l)} : l \leq n+1), \delta_{n+1}(x,y) >$ $\leq \overline{c}_I(n)\|\delta_{n+1}(x,y)\|^2$*

$$\begin{aligned} < \Phi_0^E(x^{(l)} : l \leq n) - \Phi_0^E(y^{(l)} : l \leq n), \delta_n(x,y) > &\leq \overline{c}_E(n)\|\delta_n(x,y)\|^2 \\ \|\Phi_0^I(x^{(l)} : l \leq n+1) - \Phi_0^I(y^{(l)} : l \leq n+1)\|^2 &\geq \overline{c}_0^I(n)\|\delta_{n+1}(x,y)\|^2 \\ \|\Phi_0^E(x^{(l)} : l \leq n) - \Phi_0^E(y^{(l)} : l \leq n)\|^2 &\leq \overline{c}_0^E(n)\|\delta_n(x,y)\|^2 \\ \|\Phi_j(x^{(l)} : l \leq n) - \Phi_j(y^{(l)} : l \leq n)\|^2 &\leq \overline{c}_j(n)\|\delta_n(x,y)\|^2 \\ 2\overline{c}_I(n)\Delta_n &< 1 + \overline{c}_0^I(n)\Delta_n^2 \end{aligned} \tag{5.8.6}$$

where $\overline{c}_I(n), \overline{c}_E(n), \overline{c}_0^I(n), \overline{c}_0^E(n), \overline{c}_j(n)$ *are finite, deterministic, real numbers. Then*

$$\overline{\kappa}_2 \leq \limsup_{n\to+\infty} \frac{\sum_{i=0}^{n}\left(\frac{2\overline{c}_E(i)+2\overline{c}_I(i)+(\overline{c}_0^E(i)-\overline{c}_0^I(i))\Delta_i+\sum_{j=1}^{m}(\sigma_i^j)^2\overline{c}_j(i)}{1-2\overline{c}_I(i)\Delta_i+\overline{c}_0^I(i)\Delta_i^2}\right)\Delta_i}{t_{n+1}}. \quad (5.8.7)$$

Furthermore, if $\forall n \in I\!N\, \forall x^{(l)}, y^{(l)} \in I\!R^d (l = 0,1,...,n+1)\, \forall j = 1,2,...,m$: $< \Phi_0^I(x^{(l)} : l \leq n+1) - \Phi_0^I(y^{(l)} : l \leq n+1), \delta_{n+1}(x,y) > \geq \underline{c}_I(n)\|\delta_{n+1}(x,y)\|^2$

$$\begin{aligned}
< \Phi_0^E(x^{(l)} : l \leq n) - \Phi_0^E(y^{(l)} : l \leq n), \delta_n(x,y) > &\geq \underline{c}_E(n)\|\delta_n(x,y)\|^2 \\
\|\Phi_0^I(x^{(l)} : l \leq n+1) - \Phi_0^I(y^{(l)} : l \leq n+1)\|^2 &\leq \underline{c}_0^I(n)\|\delta_{n+1}(x,y)\|^2 \quad (5.8.8)\\
\|\Phi_0^E(x^{(l)} : l \leq n) - \Phi_0^E(y^{(l)} : l \leq n)\|^2 &\geq \underline{c}_0^E(n)\|\delta_n(x,y)\|^2 \\
\|\Phi_j(x^{(l)} : l \leq n) - \Phi_j(y^{(l)} : l \leq n)\|^2 &\geq \underline{c}_j(n)\|\delta_n(x,y)\|^2 \\
1+\underline{c}_0^E(n)\Delta_n^2+\sum_{j=1}^{m}(\sigma_n^j)^2\underline{c}_j(n)\Delta_n &> -2\underline{c}_E(n)\Delta_n
\end{aligned}$$

where $\underline{c}_I(n), \underline{c}_E(n), \underline{c}_0^I(n), \underline{c}_0^E(n), \underline{c}_j(n)$ *are finite, deterministic, real numbers, then*

$$\underline{\kappa}_2 \geq \liminf_{n\to+\infty} \frac{\sum_{i=0}^{n}\left(\frac{2\underline{c}_E(i)+2\underline{c}_I(i)+(\underline{c}_0^E(i)-\underline{c}_0^I(i))\Delta_i+\sum_{j=1}^{m}(\sigma_i^j)^2\underline{c}_j(i)}{1+2\underline{c}_E(i)\Delta_i+\underline{c}_0^E(i)\Delta_i^2+\sum_{j=1}^{m}(\sigma_i^j)^2\underline{c}_j(i)\Delta_i}\right)\Delta_i}{t_{n+1}}. \quad (5.8.9)$$

Remark *This theorem provides a uniform estimate of the "spectrum" of (forward) pth moment contractivity exponents for the class of stochastic difference equations satisfying monotonicity conditions (5.8.6) and (5.8.8). Of course, these estimates are again "worst case estimates" (but sharp ones, consider linear equations). The obtained result is useful in controlling the propagation of initial errors by explicit–implicit numerical methods. The splitting into an explicit part* Φ_0^E *and implicit part* Φ_0^I *should be realized such that Theorem 8.4 can be applied, and uniform boundedness of contractivity exponents of discrete dynamic from below and above can be achieved in accordance with the estimates of contractivity exponents of the underlying continuous time system. This estimation procedure can be used to prove convergence of nonlinear numerical methods as in deterministic analysis, built upon the role of contractivity in the interplay of main principles of numerical analysis.*

5.8.5 General V-asymptotics of discrete time iterations

Now we are interested to estimate the exponential growth behavior of discrete time stochastic processes along certain functionals rather than for the process itself. More generally, we may consider **stochastic dynamical systems** $X(t,z)$ on $(\Omega, \mathcal{F}, (\mathcal{F}_t)_{t\in\mathcal{I}}, I\!P)$ started at $X(0,z) = z \in I\!D \subseteq I\!R^d$ at time $t = s$. The time scale $t \in \mathcal{I} \subseteq I\!R$ could be discrete or continous. Again, for brevity, we shall state the following very general definition, referring to continuous and discrete time scales simultaneously.

Definition 5.8.8 *The* **upper (forward moment)** V**-exponent** *of a stochastic dynamical system* $X(t,z)$ *on* $\mathbb{D}$ *is defined to be*

$$\overline{\lambda}_V := \limsup_{t\to+\infty} \frac{1}{t} \, ln \, (\mathbb{E}\, V(t, X(t,z))) \tag{5.8.10}$$

for $X(t_0,z) = z \in \mathbb{D}$ *(a.s.). The* **lower (forward moment)** V**-exponent** *of* $X(t,z)$ *on* $\mathbb{D}$ *is defined to be*

$$\underline{\lambda}_V := \liminf_{t\to+\infty} \frac{1}{t} \, ln \, (\mathbb{E}\, V(t, X(t,z))) \tag{5.8.11}$$

for $X(t_0,z) = z \in \mathbb{D}$.

This definition and related concept of V-exponents have been introduced by Schurz (1999). By enlargement of dimension one may relate to both properties: contractivity and stability along functionals $V(X)$ of the dynamics of X. We are particularly interested in estimation of these V-exponents belonging to stochastic numerical methods. Then, in analogy to deterministic analysis, the following discrete time inequality turns out to be very useful. Take $\Delta_n = t_{n+1} - t_n$ as the current step size. Let $(t_n)_{n\in\mathbb{N}}$ be a monotonically nondecreasing sequence of deterministic time-instants with t_n diverging to $+\infty$ as n tends to $+\infty$, and define

$$\Delta\mathbb{E}\, V_n := \mathbb{E}\, V(n+1, X_{n+1}) - \mathbb{E}\, V(n, X_n)$$

for a discrete time $\mathbb{D}$-valued stochastic process $X = (X_n)_{n\in\mathbb{N}}$ on the probability space $(\Omega, \mathcal{F}, (\mathcal{F}_n)_{n\in\mathbb{N}}, \mathbb{P})$. Suppose that $\Delta\mathbb{E}\, V_n \le \overline{k}_n \mathbb{E}\, V(n, X_n)$ (for all $n \in \mathbb{N}$). Making use of elementary splitting

$$z(n+1) = z(n) + z(n+1) - z(n)$$

with $z(n+1) := \mathbb{E}\, V(n+1, X_{n+1})$, one concludes

$$z(n+1) \le z(n)(1+\overline{k}_n) \le z(0) \prod_{i=0}^{n} (1+\overline{k}_i)_+ \le z(0) \exp\left(\sum_{i=0}^{n} \overline{k}_i\right).$$

On the other hand, when $\Delta\mathbb{E}\, V_n \le \underline{k}_n \mathbb{E}\, V(n, X_n)$ and $1 + \underline{k}_n > 0$ (for all $n \in \mathbb{N}$), one recognizes the validity of

$$z(n) \le \frac{z(n+1)}{1+\underline{k}_n} \le z(n+1) \exp\left(\frac{-\underline{k}_n}{1+\underline{k}_n}\right)$$

which implies

$$z(n+1) \ge z(n) \exp\left(\frac{\underline{k}_n}{1+\underline{k}_n}\right) \ge z(0) \exp\left(\sum_{i=0}^{n} \frac{\underline{k}_i}{1+\underline{k}_i}\right),$$

using elementary inequality

$$\frac{1}{1+x} \le \exp(-\frac{x}{1+x}).$$

Taking the exponential logarithm and limit when integration time t_n advances implies the following fairly general assertion.

Theorem 5.8.9 *(Schurz (1999)). Assume that* $\mathbb{E}\, V(0, X_0) < +\infty$ *for a function* $V : \mathbb{N} \times \mathbb{D} \longrightarrow \mathbb{R}^1_+$ *with*

$$\underline{k}_n \mathbb{E}\, V(n, X_n) \leq \Delta \mathbb{E}\, V_n \leq \overline{k}_n \mathbb{E}\, V(n, X_n)$$

for all $n \in \mathbb{N}$*, where* $\underline{k}_i, \overline{k}_i$ *are deterministic, real constants along the dynamics of process* $(X_n)_{n \in \mathbb{N}}$*, and for all* $n \in \mathbb{N}$

$$1 + \underline{k}_n > 0\,.$$

Then, for all $n \in \mathbb{N}$*, we have*

$$\exp\left(\sum_{i=0}^{n} \frac{\underline{k}_i}{1+\underline{k}_i}\right) \mathbb{E}\, V(0, X_0) \leq \mathbb{E}\, V(n+1, X_{n+1}) \leq \exp\left(\sum_{i=0}^{n} \overline{k}_i\right) \mathbb{E}\, V(0, X_0)$$

and, if the limits exist, then

$$\liminf_{n\to+\infty} \frac{\sum_{i=0}^{n-1} \frac{\underline{k}_i}{1+\underline{k}_i}}{t_n} \leq \underline{\lambda}_V \leq \overline{\lambda}_V \leq \limsup_{n\to+\infty} \frac{\sum_{i=0}^{n-1} \overline{k}_i}{t_n}\,.$$

Remark *Theorem 5.8.9 can be used to prove some useful results concerning the estimation of moment stability and contractivity exponents of discrete time random iterations with* $V(x)$ *where* V *is an appropriate nonnegative function or functional for random iterations as they occur while applying numerical methods to SDEs. An example is given by* $V(x) = \|x\|^2$ *with Euclidean vector norm* $\|\cdot\|$*, as used for the mean square criterion (both stability and contractivity). But, often other functionals are more appropriate. For an example in this respect, see the next subsection. Under the existence of Riemann-integrals* $\int_{t_0}^{t} K(s)ds$ *with* $k_i = K(t_i)\Delta_i$ *one can also derive corresponding continuous time versions by taking the limit of arising Riemann sums in corresponding discrete time inequalities. It is always possible to find* $\underline{k}_i$ *such that* $1 + \underline{k}_i \geq 0$*. If only one* $\underline{k}_{i*}$ *with* $\underline{k}_{i*} = -1$ *exists, then our estimate of sequence* $z(n)$ *from below reduces to the trivial one, i.e.* $z(n) \geq 0$ *at least for all* $n \geq i*$*. Thus, this latter case would not be very meaningful in the estimation process anyway. The expectation operation in the Theorem 8.5 can be dropped even, and the result would still be valid.*

5.8.6 An example for discrete time V-asymptotics

For the sake of simplicity and illustration, we shall consider the stochastic oscillator with multiplicative white noise

$$\ddot{x} + 2\zeta\omega\dot{x} + \omega^2 x = \sigma\dot{x}\,\xi_t \tag{5.8.12}$$

where $\zeta, \omega > 0$ and the stochastic integration is understood in the sense of Itô. Due to linearity with multiplicative noise, the stability and contractivity issues coincide with each other for the system (5.8.12). Then the corresponding deterministic equation has an asymptotically stable zero solution if $0 < \zeta < 1$, and does not exponentially grow if $0 \leq \zeta \leq 1$. Thanks to Theorem 8.5, we know about the stochastic version that the **upper** V**-exponent** characterizing the maximally attainable exponential growth of trajectories of the stochastic dynamics along $V(x, y) = y^2 + \omega^2 x^2$ is not larger than zero if $0 \leq \sigma^2 \leq 4\zeta\omega$. Let us now look at the discretization of such a equation by numerical methods. Define

$$V(n+1, x, y) := \omega^2 x^2 + (1 + 2\zeta\omega\Delta_n)y^2$$

where $\Delta_n = t_{n+1} - t_n$ is current step size, and $v_{n+1} := \mathbb{E}\ V(n+1, X_{n+1}, Y_{n+1})$. For illustration purposes, the stochastic oscillator (5.8.12) is discretized by the **fully drift-implicit Euler method** given by

$$\begin{aligned} X_{n+1} &= X_n + Y_{n+1}\Delta_n \\ Y_{n+1} &= Y_n - (2\zeta\omega Y_{n+1} + \omega^2 X_{n+1})\Delta_n + \sigma Y_n \Delta W_n \end{aligned} \tag{5.8.13}$$

where $\Delta W_n = W(t_{n+1}) - W(t_n)$ along a time-discretization $(t_n)_{n\in\mathbb{N}}$, and

$$\mathbb{E}\,[X_0^2 + Y_0^2] < +\infty.$$

Now, let us look at the growth behavior of discretized oscillator (5.8.13). First, we equivalently rearrange the scheme (5.8.13) to an explicit one. Thus, one arrives at

$$\begin{aligned} X_{n+1} &= \frac{1+2\zeta\omega\Delta_n}{1+2\zeta\omega\Delta_n+\omega^2\Delta_n^2}X_n + \frac{(1+\Delta W_n)\Delta_n}{1+2\zeta\omega\Delta_n+\omega^2\Delta_n^2}Y_n \\ Y_{n+1} &= -\frac{\omega^2\Delta_n}{1+2\zeta\omega\Delta_n+\omega^2\Delta_n^2}X_n + \frac{(1+\Delta W_n)}{1+2\zeta\omega\Delta_n+\omega^2\Delta_n^2}Y_n\,. \end{aligned} \tag{5.8.14}$$

After some calculations this relation implies

$$v_{n+1} = \omega^2\mathbb{E}\left[\frac{1+2\zeta\omega\Delta_n}{1+2\zeta\omega\Delta_n+\omega^2\Delta_n^2}X_n^2\right] + \mathbb{E}\left[\frac{1+\sigma^2\Delta_n}{1+2\zeta\omega\Delta_n+\omega^2\Delta_n^2}Y_n^2\right],$$

hence

$$-\frac{\omega^2\Delta_a\Delta_n}{1+2\zeta\omega\Delta_b+\omega^2\Delta_b^2}v_n - \mathbb{E}\left[\frac{(\sigma^2-2\zeta\omega)\Delta_n - 2\zeta\omega\Delta_{n-1}(1+2\zeta\omega\Delta_n)}{1+2\zeta\omega\Delta_n+\omega^2\Delta_n^2}Y_n^2\right]_- \leq$$

$$\begin{aligned} \Delta\mathbb{E}\,V_n &= -\mathbb{E}\left[\frac{\omega^2\Delta_n^2}{1+2\zeta\omega\Delta_n+\omega^2\Delta_n^2}\omega^2X_n^2\right] \\ &\quad +\mathbb{E}\left[\frac{(\sigma^2-2\zeta\omega)\Delta_n - \omega^2\Delta_n^2 - 2\zeta\omega\Delta_{n-1}(1+2\zeta\omega\Delta_n+\omega^2\Delta_n^2)}{1+2\zeta\omega\Delta_n+\omega^2\Delta_n^2}Y_n^2\right] \\ &= -\frac{\omega^2\Delta_n^2}{1+2\zeta\omega\Delta_n+\omega^2\Delta_n^2}v_n + \mathbb{E}\left[\frac{(\sigma^2-2\zeta\omega)\Delta_n - 2\zeta\omega\Delta_{n-1}(1+2\zeta\omega\Delta_n)}{1+2\zeta\omega\Delta_n+\omega^2\Delta_n^2}Y_n^2\right] \\ &\leq -\frac{\omega^2\Delta_b\Delta_n}{1+2\zeta\omega\Delta_a+\omega^2\Delta_a^2}v_n + \mathbb{E}\left[\frac{(\sigma^2-2\zeta\omega)\Delta_n - 2\zeta\omega\Delta_{n-1}(1+2\zeta\omega\Delta_n)}{1+2\zeta\omega\Delta_n+\omega^2\Delta_n^2}Y_n^2\right]_+ \end{aligned}$$

Now, we may choose $\overline{k}_n, \underline{k}_n$ as indicated below in Theorem 5.8.10, and apply Theorem 5.8.9 to our situation with those $\overline{k}_n, \underline{k}_n$. Thus, the assertions of Theorem 5.8.10 follow straight forward by elementary analysis of the obtained exponential expressions.

Theorem 5.8.10 *(Schurz (1999). Assume that the stochastic oscillator (5.8.12) is discretized by the fully drift-implicit Euler method (5.8.13) along a time discretization $(t_n)_{n\in\mathbb{N}}$, and*

$$\mathbb{E}\,[X_0^2 + Y_0^2] < +\infty.$$

Then, for all $n \in \mathbb{N}$, all $l \in \mathbb{N}$ with $1 \leq l < n$, we have

$$v_l \exp\left(\sum_{i=l}^{n} \frac{\underline{k}_i}{1+\underline{k}_i}\right) \leq v_{n+1} = \mathbb{E}\,V(n+1, X_{n+1}, Y_{n+1}) \leq v_l \exp\left(\sum_{i=l}^{n} \overline{k}_i\right)$$

for the fully drift-implicit Euler method (5.8.13), where

$$\overline{k}_i = \frac{-\omega^2\Delta_i^2(1+2\zeta\omega\Delta_{i-1}) + [(\sigma^2-2\zeta\omega)\Delta_i - 2\zeta\omega\Delta_{i-1}(1+2\zeta\omega\Delta_i)]_+}{(1+2\zeta\omega\Delta_{i-1})(1+2\zeta\omega\Delta_i+\omega^2\Delta_i^2)}$$

and

$$\underline{k}_i = \frac{-\omega^2\Delta_i^2(1+2\zeta\omega\Delta_{i-1}) - [(\sigma^2-2\zeta\omega)\Delta_i - 2\zeta\omega\Delta_{i-1}(1+2\zeta\omega\Delta_i)]_-}{(1+2\zeta\omega\Delta_{i-1})(1+2\zeta\omega\Delta_i+\omega^2\Delta_i^2)}.$$

Furthermore, if $(\Delta_n)_{n\in\mathbb{N}}$ is a deterministic sequence then the V-exponents of numerical method (5.8.13) can be estimated by

$$\liminf_{n\to+\infty}\frac{1}{t_n}\sum_{i=1}^{n-1}\frac{\underline{k}_i}{1+\underline{k}_i} \le \underline{\lambda}_V \le \overline{\lambda}_V \le \limsup_{n\to+\infty}\frac{1}{t_n}\sum_{i=1}^{n-1}\overline{k}_i\,.$$

Additionally, in the following assume that

$$\exists\Delta_a,\Delta_b\in\mathbb{R}_+ : \forall n\in\mathbb{N}\quad 0<\Delta_b\le\Delta_n\le\Delta_a<+\infty\,. \qquad (5.8.15)$$

If

$$(\sigma^2-2\zeta\omega)\Delta_n - 2\zeta\omega\Delta_{n-1}(1+2\zeta\omega\Delta_n) \le 0 \qquad (5.8.16)$$

for all $n\in\mathbb{N}$ then

$$\overline{\lambda}_V \le -\frac{\omega^2\Delta_b}{1+2\zeta\omega\Delta_a+\omega^2\Delta_a^2}\,.$$

If

$$(\sigma^2-2\zeta\omega)\Delta_n - 2\zeta\omega\Delta_{n-1}(1+2\zeta\omega\Delta_n) \ge 0 \qquad (5.8.17)$$

for all $n\in\mathbb{N}$ then

$$\underline{\lambda}_V \ge -\frac{\omega^2\Delta_a}{1+2\zeta\omega\Delta_b}\,.$$

Remark *Most of the clever variable step size algorithms have implemented conditions on the step size selection like that of (5.8.15). We can conclude from our assertion that the fully drift-implicit Euler method (5.8.13) applied to stochastic oscillator (5.8.12) produces overdamped approximations compared to the asymptotics of the exact solution. This can be seen particularly in the critical case (the energy-conservative case) when $\sigma^2 = 4\zeta\omega$ under the condition (5.8.15). However, the observed effect of numerical stabilization also explains that the requirement (5.8.15) is meaningful in variable step size algorithms in order to achieve asymptotically stable approximations (i.e. with "sure side argumentation"). Asymptotically considered, when maximum step size Δ_a tends to zero, the V-exponents of the continuous time dynamics (5.8.12) are correctly replicated by the discretization method (5.8.13), which is what we would naturally expect, and with a convergence order in terms of Δ_a. Unfortunately, at this writing, the author does not know any other stochastic numerical method which has been examined with respect to "nonstandard" stability and contractivity behavior, as exhibited by the concept of V-exponents and applied to SDEs here.*

5.8.7 Asymptotic contractivity with probability one

Adapting Theorem 5.7.19 to the case of stochastic numerical contractivity, we may verify the contractivity of numerical approximations with probability one, applied to SDEs with Lipschitz continuous coefficient systems (a, b^j). Since, for linear systems with multiplicative noise, the concepts of contractivity and stability coincide, the major interest lies only in application to the case of nonlinear SDEs. Thus one can prove the asymptotic contractivity of balanced implicit methods (BIMs) with probability one, applied to nonlinear SDEs with strictly dissipative drift a and Lipschitz continuous coefficient systems (a, b^j) using any equidistant step size Δ and appropriate uniform estimates of one sided Lipschitz constant $K^a_{OLC} < 0$ and Lipschitz constants $K^{b^j}_L \geq 0$ of the linearly dominated coefficients (a, b^j) as their scalar weights $C_0 = -K^a_{OLC} I$, $C_j = K^{b^j}_L I$. Other with probability one asymptotically contractive numerical methods are not known to the author during at this writing, and the interesting question arises as to whether one can construct probability one asymptotically contractive numerical methods other than certain classes of BIMs?

5.9 On Practical Implementation

Although the theory of numerical algorithms is understood fairly well nowadays, there are plenty of interesting questions left to be discussed for the efficient implementation of stochastic numerical methods, such as the questions of parallelization, efficient generation of multiple integrals, variance reduction, preservation of algebraic boundary conditions, approximation of stopping times and nonsmooth, path-dependent functionals, optimal control, the role of random and quasi-random number generation, and the influence of statistical and roundoff errors for nonidentically distributed random variables.

5.9.1 Implementation issues: some challenging examples

In general it is advisable to study the underlying continuous time dynamics as much as one can before implementing numerical routines. Often, by this procedure the computational effort can be reduced significantly, as manifested in the following.
Stochastic Duffing Van der Pol Oscillator with White Noised Velocity. Oscillations of a magnetic pendulum can be described by **Duffing Van der Pol oscillators** to some extent. If the velocity component $X^{(2)}_t := \dot{x}(t)$ is only multiplicatively perturbed by white noise, then one arrives at the SDE

$$\begin{aligned} dX^{(1)}_t &= X^{(2)}_t\, dt \\ dX^{(2)}_t &= -\Big([(X^{(1)}_t)^2 - \alpha]X^{(1)}_t + X^{(2)}_t\Big)dt + \sigma X^{(1)}_t\, dW_t \end{aligned} \tag{5.9.1}$$

where W is a standard Wiener process, and $\alpha > 0$ and σ are real parameters determining the displacement $X^{(1)}_t := x(t)$, velocity $X^{(2)}_t = \dot{x}(t)$, location of asymptotically stable rest points $(-\sqrt{\alpha}, 0)$, $(\sqrt{\alpha}, 0)$ and noise intensity, respectively. First, we note that Stratonovich and Itô interpretations of arising stochastic integrals coincide here. This results in Euler methods being identical with Mil'shtein methods, Taylor methods with strong order 1.5 being identical with Taylor methods with strong order 2.0, and so on. For example, the explicit Mil'shtein method applied to (5.9.1) is implemented by linear-implicit, explicit-implicit Euler or explicit Euler method

$$\begin{aligned} Y^{(1)}_{n+1} &= Y^{(1)}_n + Y^{(2)}_n \Delta_n \\ Y^{(2)}_{n+1} &= Y^{(2)}_n - \Big([(Y^{(1)}_n)^2 - \alpha]Y^{(1)}_n + Y^{(2)}_n\Big)\Delta_n + \sigma Y^{(1)}_n \Delta W_n, \end{aligned} \tag{5.9.2}$$

thus no higher order multiple integrals need to be generated. However, the explicit Taylor 1.5 (= Taylor 2.0 here) method needs the generation of $I_{(0,1),t_n,t_{n+1}}$ and $I_{(1,0,1),t_n,t_{n+1}}$ which can be done by truncation of Karhunen-Loeve expansions up to some desired accuracy, see Kloeden and Platen (1992). This example shows that higher order methods are implementable with lesser computational effort than theoretically predicted, and these methods preserve the stochastic flow properties (e.g. "neighbors stay neighbors") a longer time than lower order methods, caused by the specific smooth structure of SDE system (5.9.1). For simulation results, see Kloeden, Platen and Schurz (1991).

A Stochastic Flow on the Unit Circle. Carverhill, Chappel and Elworthy (1986) discussed the **gradient stochastic flow** generated by the SDE

$$dX_t(x) = sin(X_t(x)) \circ dW_t^1 + cos(X_t(x)) \circ dW_t^2 \tag{5.9.3}$$

driven by two scalar, independent, real-valued Wiener processes W^1, W^2, and interpreted modulo 2π, started at initial angle $x \in [0, 2\pi)$. We also note that the Stratonovich and Itô versions coincide here. The flow belonging to equation (5.9.3) can be implemented by the Mil'shtein scheme

$$\begin{aligned} Y_{n+1} &= Y_n + sin(Y_n)\Delta W_n^1 + cos(Y_n)\Delta W_n^2 + \frac{1}{4} sin(2Y_n)[(\Delta W_n^1)^2 - (\Delta W_n^2)^2] + \\ &\quad + [cos(Y_n)]^2 \Delta W_n^1 \Delta W_n^2 - I_{(2,1),t_n,t_{n+1}} \end{aligned} \tag{5.9.4}$$

where $Y_n = Y_n(x) \bmod 2\pi$, exploiting the elementary relation that

$$I_{(1,2),t_n,t_{n+1}} + I_{(2,1),t_n,t_{n+1}} = \Delta W_n^1 \Delta W_n^2$$

which can be concluded from Lemma 3.2. This is an example of noncommutative noise. We need only to generate the multiple integral $I_{(2,1)}$ (or alternatively $I_{(1,2)}$) to achieve a pth mean convergent numerical approximation of order 1.0. For first simulation results, see Kloeden, Platen and Schurz (1991)

A Stochastic Flow on the Torus $S^2 = [0, 2\pi) \times [0, 2\pi)$. Baxendale (1986) has dealt with the calculation of Lyapunov exponents (i.e. characteristic numbers to describe the exponential growth or decrease of trajectories in the phase plane) of the two-dimensional **angular stochastic flow** generated by

$$dX_t(\alpha, x) = \sum_{j=1}^{m} b^j(X_t(\alpha, x)) \circ dW_t^j \tag{5.9.5}$$

driven by four scalar, independent, real-valued Wiener processes W^j with diffusion coefficients

$$b^1(x_1, x_2) = \begin{pmatrix} cos(\alpha) \\ sin(\alpha) \end{pmatrix} sin(x_1), \; b^2(x_1, x_2) = \begin{pmatrix} cos(\alpha) \\ sin(\alpha) \end{pmatrix} cos(x_1), \tag{5.9.6}$$

$$b^3(x_1, x_2) = \begin{pmatrix} -sin(\alpha) \\ cos(\alpha) \end{pmatrix} sin(x_2), \; b^4(x_1, x_2) = \begin{pmatrix} -sin(\alpha) \\ cos(\alpha) \end{pmatrix} cos(x_2), \tag{5.9.7}$$

started at initial angle $x = (x_0^1, x_0^2)$. Here α represents a bifurcation parameter, and Baxendale (1986) has calculated the bifurcation point $0.8 < \alpha_* < 0.9$ when the system (5.9.5) switches from the asymptotically stable mode (i.e. a Brownian motion remains as stable mode) to an asymptotically instable one (i.e. no strict contraction can be observed to a stable mode). Contractivity of that flow in the wide and strict senses can be observed in terms of clusters of its trajectories in the phase plane. This flow can be generated by

Mil'shtein methods as well. However, the full Taylor expansion is needed, and the generation of multiple integrals is more laborious, but desirable due to geometric properties of the stochastic flow to be visualized. Numerical evidence of the bifurcation point α_* can be obtained by higher order numerical methods too. For simulations, see Kloeden, Platen and Schurz (1991).

A Stochastic Planar Brusselators. For modeling unforced periodic oscillations in certain chemical reactions it is common to use the Brusselator equations. After neglecting spatial diffusion and centering at an equilibrium point, the following **planar Brusselator** occurs:

$$\begin{aligned} \frac{dx_1}{dt} &= (a-1)x_1 + ax_1^2 + (1+x_1)^2 x_2 \\ \frac{dx_2}{dt} &= -ax_1(1+x_1) - (1+x_1)^2 x_2 \end{aligned} \tag{5.9.8}$$

where positive parameter $a \in \mathbb{R}_+$ represents a Hopf bifurcation point for that system. A stochastic version of planar Brusselator given by the Stratonovich interpreted system (the model is motivated by Ehrhardt (1983) who investigated the existence of related invariant measures.)

$$\begin{aligned} dX_t^{(1)} &= \Big((a-1)X_t^{(1)} + a[X^{(1)}]_t^2 + (1+X_t^{(1)})^2 X_t^{(2)}\Big)dt + \sigma X_t^{(1)}(1+X_t^{(1)}) \circ dW_t \\ dX_t^{(2)} &= (1+X_t^{(1)})\Big((-aX_t^{(1)} - (1+X_t^{(1)})X_t^{(2)})dt - \sigma X_t^{(1)} \circ dW_t\Big). \end{aligned} \tag{5.9.9}$$

has been studied numerically by linearization in Schurz (1994, 1996, 1997). We recommend using the class of balanced implicit methods with appropriate weights, since multiplicative noise is essential for the modeling process here. However, care must be taken while choosing adequate weights in order to not to destroy the property of linear asymptotic exactness.

A Generalized Stochastic KPP Equation. Elworthy, Truman, Zhao and Gaines (1994) have studied approximate traveling waves for the **generalized stochastic KPP equation.** The related SPDE in $\mathbb{R}^1$ is given by

$$du_\mu(t,x) = [\frac{1}{2}\mu^2\Delta_x u_\mu(t,x) + \frac{1}{\mu^2}(1-u_\mu(t,x))u_\mu(t,x)]\,dt + \sigma u_\mu(t,x)\,dW_t \tag{5.9.10}$$

where Δ_x is the Laplace operator and $\sigma = \begin{cases} \frac{k}{\mu} & \text{mild noise} \\ \frac{k}{\mu^2} \quad \text{if} & \text{strong noise} \\ k & \text{weak noise} \end{cases}$

with $x, k, \mu \in \mathbb{R}^1$. They have been particularly interested in studying the behavior of $u_\mu(t,x)$ as μ tends to zero - a situation which represents a real challenge for adequate numerical integration. For simplicity, consider the mild noise case only. As commonly practiced for parabolic PDEs, they discretize the space variable x on the subinterval $[x_1, x_d]$ with space step size $h = \frac{x_d - x_1}{d}$ at first and arrive at the d-dimensional system of SDEs

$$dX_t = (A\,X_t + \frac{1}{\mu^2}a(X_t))\,dt + \frac{k}{\mu}X_t\,dW_t \tag{5.9.11}$$

with multiplicative (diagonal) white noise, tridiagonal drift matrix

$$A = \frac{\mu^2}{2h^2}\begin{pmatrix} -1 & 1 & 0 & 0 & . & . & 0 & 0 \\ 1 & -2 & 1 & 0 & . & . & 0 & 0 \\ . & . & . & . & . & . & . & . \\ 0 & . & . & . & 0 & 1 & -2 & 1 \\ 0 & 0 & . & . & . & 0 & 1 & -1 \end{pmatrix} \tag{5.9.12}$$

and drift vector components $a_i(x_1, x_2, ..., x_d) = (1 - x_i)x_i$. They suggested using the drift-implicit Mil'shtein method (5.4.18) with $\theta = 0.5$ for numerical integration of system (5.9.11). Since system (5.9.11) exhibits diagonal noise (hence commutative noise), the **linearly drift-implicit Mil'shtein method** simplifies to resolution of

$$(I - \frac{\Delta_n}{2}A)Y_{n+1} = (I + \frac{\Delta_n}{2}A)Y_n + \frac{1}{\mu^2}a(Y_n)\Delta_n + \frac{k}{\mu}Y_n\Delta W_n + \frac{k^2}{2\mu^2}Y_n[(\Delta W_n)^2 - \Delta_n] \tag{5.9.13}$$

where $Y_n = (Y_{n,i})_{i=1,2,...,d}$ denotes the **linearly drift-implicit Crank-Nicholson-Mil'shtein approximation** of $u_\mu(t,x)$ at position (t_n, x_i) for fixed parameters k, μ. Note, in case of an explicit numerical method one would have to require the Courant-Friedrichs-Levy-type condition $2\mu\Delta < h^2$. Then, using sufficiently small space step sizes h and initial conditions as step functions or point mass (e.g. $\phi(x) = 1$ if $x = 0$ and 0 otherwise), the numerical approximations (5.9.13) replicate the correct speed of wave propagation (even with k=0), as Elworthy et al (1994) report. Note the theoretically predicted speed of propagation, when starting with a δ-function, is proved to be $\sqrt{2 - k^2}$ due to Elworthy and Zhao. They also used that numerical method to visualize the related stochastic flow. We might also use balanced implicit Mil'shtein methods with diagonal implicit weights to control the asymptotic stability with probability one or stochastic waveform relaxation methods as introduced in Schneider and Schurz (1998) to exploit the computational efficiency of parallel computers. However, future research is still needed to understand those complex numerical dynamics better.

A Stochastic Heat Equation with Space-Time White Noise. A version of the **stochastic heat equation** driven by space-time white noise $\dot{W}(t,x)$ is given by

$$\frac{\partial u(t,x)}{\partial t} = \mu^2 \frac{\partial^2 u(t,x)}{\partial x^2} + \sigma[u(t,x)]^\kappa \dot{W}(t,x) \tag{5.9.14}$$

with initial conditions $u(0,x) = u_0(x)$, boundary conditions $u(t,0) = u(t,1)$ on the domain $[0,1]$, where μ, σ are certain real parameters. The nonlinearity parameter $\kappa \in [0.5, 2]$ in diffusion controls the long time behavior, invariant measures and possible blowups for the stochastic heat equation (5.9.14). Mueller (1991, 1993), Mueller and Perkins (1992), Mueller and Sowers (1993) have studied the qualitative behavior of one-dimensional SPDE (5.9.14). The space-time discretization of this quasilinear parabolic SPDE is not so trivial as it was with the KPP equation (due to presence of nonlinearities and space-time white noise). Luckily, the resulting SDE

$$dX_t = A X_t \, dt + \frac{\sigma}{\sqrt{h}} \sum_{j=1}^{d} b^j(X_t) \, dW_t^j \tag{5.9.15}$$

with $b^j(x) = (\delta_{i,j} x_i^\kappa)_{i=1,2,...,d}$ (where $\delta_{i,j}$ represents the Kronecker symbol), space step size $h = \frac{1}{d}$, and tridiagonal drift matrix A fulfills the commutative noise condition

$$A = \frac{\mu^2}{h^2} \begin{pmatrix} -2 & 1 & 0 & 0 & . & . & 0 & 1 \\ 1 & -2 & 1 & 0 & . & . & 0 & 0 \\ . & . & . & . & . & . & . & . \\ 0 & . & . & . & 0 & 1 & -2 & 1 \\ 1 & 0 & . & . & . & 0 & 1 & -2 \end{pmatrix} \tag{5.9.16}$$

(due to diagonal noise structure) and again one might apply the linearly drift-implicit Mil'shtein methods as before, but now with multidimensional white noise. Under commutative noise the generation of multiple integrals simplifies to trivial products of noise and time increments, cf. Gaines representation of Mil'shtein methods in Section 4. However, for pathwise simulation, as needed for investigating the flow structure, one has to take care of appropriately adding discretized Wiener paths (one may appreciate using Levy's construction of Brownian paths). See Gaines (1995) for some details. We recommend to apply the technique of balanced implicit methods to (5.9.15) to control convergence, boundedness and stability.

A "So Simple Looking" Nonlinear Test Equation From Quantum Mechanics. Several authors report serious problems such as suddenly–occurring, unnatural spikes during simulation of the complex-valued intensity of the cavity mode to describe the photon number while using positive *p*representation in Quantum Mechanics. For example, see Smith and Gardiner (1989) and Gilchrist, Gardiner and Drummond (1993) for details. The simplest model of a **cavity mode oscillator damped by one and two photon absorption** is governed by Itô SDE

$$dN_t \;=\; -(\frac{1}{2}\lambda + N_t)\, N_t\, dt \,+\, i\, N_t\, dW_t \tag{5.9.17}$$

where $N_t \in \mathbb{C}$ describes the intensity of cavity mode with parameter $\lambda \in \mathbb{C}$. We are still searching for a stable numerical method to apply to SDE (5.9.17). Who does know an asymptotically exact and stable numerical method or the solution of that puzzle for all meaningful parameters of complex system (5.9.17)?

Nonlinear Test Equations From Polymer Physics. Öttinger (1996) has investigated the polymerization process of polymeric fluids. In particular, motivated by the model of **Hookean dumbbells**, one may arrive at models similar to the system of SDEs

$$dX_t \;=\; \frac{b - \mathbb{E}\, X_t}{1 - \left[\frac{||X_t||}{||b||}\right]^2}\, dt \,+\, \Big(\sigma + B(b - X_t)\Big)\, dW_t \tag{5.9.18}$$

describing the length $X_t \in \mathbb{R}^d_+$ of polymer chains in a polymeric fluid, where $b \in \mathbb{R}^d_+$, $\sigma \in \mathbb{R}^d$ and $B \in \mathbb{R}^{d\times d}$ are appropriate parameters. This model can serve as an excellent nonlinear test system for the qualitative behavior of stochastic numerical algorithms, which offers plenty of challenging features concerning variance reduction, boundedness and stability issues due to its inherent nonlinearity in the drift part with some mean field interaction. Certain partial–implicit methods seem to perform best from all methods known up to now. See Öttinger (1996) for a first approach and some numerical experiments for similar models.

Numerous further applications of stochastic numerical methods can be found in the Physics and Chemistry literature, see Section 9.8. All in all, the role of practical implementation should not be underestimated, and we could only indicate a little bit of what may specifically happen and which issues seem to be important ones.

5.9.2 Generation of pseudorandom numbers

In order to implement the presented numerical techniques one needs to talk about how to generate the resulting random variables (Wiener process increments, in general increments of multiple integrals). To date, the most commonly accepted way is that of replacing randomness by pseudorandomness of those variables. Of course this is done with care and with the knowledge of introducing new errors whose propagation can be controlled by the concepts of numerical stability and contractivity as presented in sections before. Note that the resulting errors must be consistent with the convergence order to be achieved (i.e. only errors which are locally of higher order of convergence). How to replace random by pseudorandom variables is an entire industry nowadays. For a recent survey on pseudorandomness, see Goldreich (1999). We will suppose for our survey that the reader is familiar with the concepts of pseudorandomness, Kolmorgorov's complexity approach, Shannon's information theory and computational indistinguishability (in fact we are already pleased to be able to generate random variables by pseudorandom ones with some desired error order which does not destroy the order of numerical convergence).

Let us restate the most–used random number generators based on sequences of uniformly $U[0,1]$-distributed, independent pseudorandom numbers (U_n, V_n).

The Inverse Transform Method. An invertible distribution $F_X = F_X(x)$ of random variable X can be generated from uniformly distributed random numbers U by taking

$$x(U) = \inf\{x : U \leq F_X(x)\},$$

so $x(U) = F_X^{-1}(U)$ if F^{-1} exists.

Box-Muller Method. A more efficient method to generate independent, standard Gaussian distributed pseudorandom numbers is given by the Box-Muller method. This method takes the transform

$$G_n^1 = \sqrt{-2\ln(U_n)}\,\cos(2\pi V_n), G_n^2 = \sqrt{-2\ln(U_n)}\,\sin(2\pi V_n)$$

to obtain two independent Gaussian distributed numbers (G_n^1, G_n^2). Correlated random variables can be generated from those independent pairs by algebraic multiplication with corresponding matrices arising from Cholesky factorization of given correlation matrix, e.g. the factorization of correlation matrix

$$CC^T = \begin{pmatrix} \sqrt{\Delta} & 0 \\ \dfrac{\Delta^{3/2}}{2} & \dfrac{\Delta^{3/2}}{2\sqrt{3}} \end{pmatrix} \begin{pmatrix} \sqrt{\Delta} & 0 \\ \dfrac{\Delta^{3/2}}{2} & \dfrac{\Delta^{3/2}}{2\sqrt{3}} \end{pmatrix}^T = \begin{pmatrix} \Delta & \dfrac{\Delta^2}{2} \\ \dfrac{\Delta^2}{2} & \dfrac{\Delta^3}{3} \end{pmatrix}$$

to generate the pair of multiple integrals

$$(I_{(j),t_n,t_{n+1}}, I_{(j,0),t_n,t_{n+1}})^T = C(G_n^1, G_n^2)^T.$$

Polar Marsaglia Method. The Polar Marsaglia method also generates independent, standard Gaussian distributed pseudorandom numbers, which exhibits a slightly more computationally efficient generator than that of Box-Muller. This method avoids the time-consuming generation of trigonometric functions by the following procedure. At first, transform $\hat{U}_n = 2U_n - 1$, $\hat{V}_n = 2V_n - 1$ in order to achieve uniformly $[-1,1]$-distributed random numbers. Next, check whether

$$W_n := \hat{U}_n^2 + \hat{V}_n^2 \leq 1$$

or repeat until acceptance of pair $(\hat{U}_n, \hat{V}_n)$. Then, using the transform $W_n \leq 1$, we get

$$G_n^1 = \hat{U}_n\sqrt{-2\ln(W_n)/W_n},\ G_n^2 = \hat{V}_n\sqrt{-2\ln(W_n)/W_n}$$

which represent a pair of independent Gaussian distributed pseudorandom numbers (since $\cos(\arctan(U_n/V_n)) = U_n/\sqrt{W_n}$ and $\sin(\arctan(U_n/V_n)) = V_n/\sqrt{W_n}$, this follows from the Box-Muller method by noting that $\arctan(U_n/V_n)$ is uniformly $(0, 2\pi)$-distributed.). The probability of acceptance of the numbers $(\hat{U}_n, \hat{V}_n)$ is calculated to be $\pi/4 \approx 0.7864816$. Despite the possible nonacceptance, the Polar Marsaglia method is more efficient when generating large quantities as needed for statistical estimations related to the stochastic numerical algorithms (like multiple integrals at each integration step).

There are the commonly used methods of linear and nonlinear congruential generators (see Niederreiter (1988, 1992, 1995), Eichenhauer and Lehn (1986)) and the Fibonacci generator (for practical usage on supercomputers, see Petersen (1994)) to produce pseudorandom numbers (U_n, V_n) needed for the Box-Muller and Polar Marsaglia methods. See the citations for more details. We believe that it is important to be aware of the properties of the pseudorandom number generator which one uses for the simulation procedures during implementation of stochastic algorithms on computers. In particular, the measure of departing from independence of the used pseudo- or quasi random sequence might affect the quality of simulation results. Unbiased long range "random" number generators are needed for reliable simulations. In this respect, the Fibonacci generators seem to be very promising.

5.9.3 Substitutions of randomness under weak convergence

A substitution of random variables representing the algebra of multiple integrals is possible with some care. Mil'shtein (1988) and Talay (1995) suggest some "simplifications" of random variables by multipoint distributed ones instead of Gaussian increments of the Wiener processes. For example, the resulting *simplified Euler method* (5.4.1) uses independent, two-point distributed variables $\widehat{\Delta W}_n^j$ satisfying

$$\mathbb{P}\,\{\widehat{\Delta W}_n^j = \pm\sqrt{\Delta_n}\} = \frac{1}{2}.$$

In fact they conclude general moment conditions for the random number substitution without destroying the original convergence order. A *simplified weak order 2.0 convergent Taylor method* would use any variables $\widehat{\Delta W}_n^j$ with

$$|\mathbb{E}\,\widehat{\Delta W}_n^j| + |\mathbb{E}\,[\widehat{\Delta W}_n^j]^3| + |\mathbb{E}\,[\widehat{\Delta W}_n^j]^5| + |\mathbb{E}\,[\widehat{\Delta W}_n^j]^2 - \Delta_n| + +|\mathbb{E}\,[\widehat{\Delta W}_n^j]^4 - 3\Delta_n^2| \\ \leq K_m \Delta_n$$

where K_m is some real constant. For example, the three-point distribution with

$$\mathbb{P}\,\{\widehat{\Delta W}_n^j = \pm\sqrt{3\Delta_n}\} = \frac{1}{6}, \mathbb{P}\,\{\widehat{\Delta W}_n^j = 0\} = \frac{2}{3}$$

satisfies this relation. In passing we note that these substitutions are justified when one constructs and investigates weak approximations of smooth, nonpath–dependent functionals of solutions of SDEs. Practical experience shows that substitutions by continuous distributions (like appropriate sawtooth distributions or uniform distributions fulfilling the mentioned requirements of moments) perform better than the multipoint distributed random variables in the numerical simulation of weak approximations. For strong and pth mean approximations one might also think of possible random number substitutions, but certainly much more care should be taken in order to keep the convergence orders. For contractive numerical dynamics, the influence of errors caused by "approximate random numbers" instead of perfectly distributed ones is controlled by the magnitudes of the local convergence errors. Roughly speaking, then pth mean errors in the probability distribution of the random numbers should not exceed the magnitudes of local pth mean convergence errors controlled by the consistency property related to the used contractive numerical method. To date, we do not know what happens in the case of noncontractive numerical dynamics. From all our model assumptions, we believe that the property of being independently distributed is the most essential one, since we have dealt with approximations of stochastic processes with independent increments. Thus, the role of deviation from independence and Markovian character should be studied in the near future (one needs appropriate measures describing the deviation from independence of random number sequences). It is interesting to note that truly multistep schemes for ordinary SDEs with deterministic nonpathdependent coefficients (a, b^j) may already generate discrete time stochastic processes with dependent increments (however, in the limit as maximum step size tends to zero, they approximate processes with independent increments).

5.9.4 Are quasi random numbers useful for (O)SDEs?

First of all, we can not completely answer this question at present. Certainly, within the framework of weak convergence one has to discuss their use to approximate

$$u(T,x) = \mathbb{E}\left[f(X_T)\right] = \int_{\mathbb{R}^d} f(x)p_X(T,x)\,dx = \int_0^1 f(F_{X_T}^{-1}(z)\,dz,$$

$$\text{by sums} \qquad \frac{1}{M}\sum_{i=1}^{M} f\Big(F_{X_T}^{-1}(Z_i)\Big)$$

with appropriate random numbers Z_i according to Monte Carlo theory, where F_X denotes the distribution function of X_T. These random numbers can be replaced by members of low-discrepancy sequences (i.e. quasi random number sequences), see Niederreiter (1992), in view of numerical approximation of the integrals by (quasi) Monte Carlo methods. The main notion here is the notion of **discrepancy** of a point set in some r-dimensional unit cube, where r represents a positive real number. For any $a \in (0,1]^r$ with $a = (a_1, a_2, ..., a_r)$, define the cube

$$[0,a) := \{x = (x_1, x_2, ..., x_r) \in [0,1]^r : x_i < a_i, i = 1,2,...,r\}.$$

Then the **$*$-discrepancy** $D_M^*(Z_i)_{1\le i\le M}$ of a given sequence $(Z_i)_{1\le i\le M}$ with values $Z_i \in [0,1)^r$ is defined to be the quantity

$$D_{M,r}^*(Z_i)_{1\le i\le M} = \sup_{a\in(0,1]^r}\left|\frac{\#\{Z_i : Z_i \in [0,a), 1 \le i \le M\}}{M} - \prod_{i=1}^{r} a_i\right|$$

as a measure of uniformity of generated empirical r-dimensional distribution belonging to sequence $(Z_i)_{1\le i\le M}$ and depending on the sampling size $M \in \mathbb{N}_+$. An advantage of using quasi random numbers is that sampling errors controlled by $*$-discrepancy are proportional to $1/M$ compared to $1/\sqrt{M}$ achieved by standard Monte Carlo methods, where M denotes the used sampling size. However, care should be taken while using quasi random numbers. It is not quite clear to which advantages this leads in the case of approximating functionals of diffusion processes, although the quasi-random numbers exhibit a smaller deviation from uniformity compared to the uniformity of so-called uniformly distributed pseudorandom numbers. The reason of a rather negative answer to the usage of quasi-random numbers for numerical integration of SDEs is that we have to generate independent random numbers. Exactly this independence property of increments of involved Markov processes works against the property of having the lowest possible discrepancy, as it is generally aimed with quasi-random numbers; more precisely, independence and low discrepancy are contradictory requirements. The central questions are whether the use of quasi-random numbers leads to faster convergence of related approximations, to really more efficient integration techniques and to which class of SDEs we observe an advantage compared to the pseudo random number generators which are implemented in most modern computers. A first approach to the numerical treatment of a one-dimensional SDE with additive noise by the use of quasi-random numbers is found in the paper of Hofmann and Mathé (1997). They make use of the Koksma-Hlawka inequality for any function of bounded variation on $[0,1]^r$ as integrand - an inequality which provides an error estimate for the quadrature formulas in terms of the discrepancy of the (Z_i) and of the q-variation of f in the sense of Vitali $(q = 0,1,...,r-1)$. By this fundamental tool they prove that low discrepancy sequences of quasi-random numbers must not be used for simulation of one-dimensional Langevin equations (i.e. linear test SDE with additive noise). Low discrepancy sequences can destroy the (mean square) consistency property of the constructed approximation for the Langevin equation. This is demonstrated by the quasi random sequences of Kronecker-Weyl and van der Corput by Hofmann and Mathé (1997). However, restricting to sequences of completely uniformly distributed numbers yields sequences which may serve as quasi-random numbers, since these sequences have discrepancy bounded from below as necessary for (mean square) consistency. For more details, see their paper.

5.9.5 Variable step size algorithms

As one of the first implementations, stochastic variable step size algorithms could be found by the school of Artemiev since 1985. For example, Averina, Artemiev and Schurz (1994) have suggested adapting the deterministic procedures to construct variable step size algorithms. A variable step size technique based on the comparison of deterministically 2nd order and 3rd order embedded Runge-Kutta-Fehlberg methods applied to Itô SDEs on finite, fixed time-intervals $[0, T]$ works as follows (For Stratonovich systems one has a similar procedure.).
[1]. At first, fix a tolerance level $\varepsilon > 0$ for the local error and choose the initial step size Δ_0 with $0 < \Delta_0 \le \min(1, T - t_0)$.
[2]. Second, evaluate the schemes

$$Y^*_{n+1} = Y_n + \frac{1}{2}(k_1 + k_2) + \sum_{j=1}^{m} b^j(t_n, Y_n)\,\Delta W^j_n \tag{5.9.19}$$

$$Y_{n+1} = Y_n + \frac{1}{6}(k_1 + k_2 + k_3) + \sum_{j=1}^{m} b^j(t_n, Y_n)\,\Delta W^j_n \tag{5.9.20}$$

where

$$k_1 = \Delta_n a(t_n, Y_n), k_2 = \Delta_n a(t_n + \Delta_n, Y_n + k_1)$$

and

$$k_3 = \Delta_n a(t_n + \frac{\Delta_n}{2}, Y_n + \frac{1}{4}(k_1 + k_2)).$$

[3]. Third, calculate the locally scaled error prediction

$$\delta_n := \sqrt{\frac{1}{d}\sum_{i=1}^{d}\left(\frac{Y^*_{n+1,i} - Y_{n+1,i}}{\max(1, |Y_{n+1,i}|, |Y_{n,i}|)}\right)^2}.$$

[4]. Fourth, accept the step size Δ_n if $\delta_n \le 5\varepsilon$, and otherwise choose the new step size

$$\Delta_n^{new} = \frac{\Delta_n}{\max(fac_1, \min(fac_2, (\frac{\delta_n}{\varepsilon})^{1/3}/fac))} \tag{5.9.21}$$

with $fac = 0.9$ as suitable adjustment factor and repeat this procedure with the second step. The factors $fac_1 = 0.1$ and $fac_2 = 5$ control the ratio between maximum and minimum acceptable step size, i.e. fac_1 is understood as the coefficient for maximum increasing step size, and fac_2 as the coefficient for minimum decreasing step size.
Obviously, this algorithm circumvents the time-consuming statistical estimation for pathwise step size control. However, this technique seems to be appropriate especially for systems with "small noises," since one suppresses the influence of noise terms and large noise intensities in statistical decision making. This adaptive variable step size technique has been tested by Averina, Artemiev and Schurz (1994) with great success. This algorithm can also be realized with other numerical methods as basis methods (5.9.19) and (5.9.20) (e.g. Mil'shtein methods for treatment of the diffusion part, explicit-implicit or midpoint-trapezoidal methods for the treatment of the drift part).

Other contributions to step size and order selection for numerical approximations have been carried out by the dissertations of Hofmann (1995) for weak approximations (using extrapolation ideas) and recently by Mauthner (1999). The concept of variable and adaptive step size and order selective numerical algorithm needs to be studied further, due to their widely practical importance.

5.9.6 Variance reduction techniques

An important practical problem is that of reduction of occurring variances in the computational estimation process. Significant contributions in this respect have been made by Wagner (1987-1989) and Newton (1994). They develop the standard methods of **importance sampling** and **control variates**, see Hammersley and Handscomb (1964) for a general description. In both cases the **Clark-Funke-Shevlyakov-Haussmann integral representation theorem** for functionals of Itô diffusion processes provides the perfect variate in the sense that it is unbiased and has zero variance, in order to reduce the variance of functionals of simulated diffusions. However, a balance between variance reducing effects and computational efficiency has to be taken into account during practical implementation, due to resulting **computational complexity** of stochastic algorithms.

Recall that the criterion of weak convergence involves the problem of approximating the quantities $\mathbb{E}\, f(X_T)$. Two errors arise during approximation of these quantities, namely the discretization error and the error of statistical estimation of expectations motivated by the trivial observation

$$
\begin{aligned}
&\left\| \mathbb{E}\, f(X_T) - \frac{1}{M}\sum_{i=1}^{M} f(Y_{n_T,i}) \right\| \\
\leq\ & \underbrace{\left\| \mathbb{E}\, f(X_T) - \mathbb{E}\, f(Y_{n_T}) \right\|}_{\text{controlled by the discretization error}} + \underbrace{\left\| \mathbb{E}\, f(Y_{n_T}) - \frac{1}{M}\sum_{i=1}^{M} f(Y_{n_T,i}) \right\|}_{\text{controlled by statistical error}} \\
\leq\ & K_w(T,a,b^j,X_0,Y_0)\Delta^{\beta} + K_{stat}\frac{1}{\sqrt{M}}
\end{aligned}
$$

with appropriate constants $K_w(T,a,b^j,X_0,Y_0)$ and K_{stat}, maximum step size $\Delta > 0$, weak convergence rate $\beta \in \mathbb{R}_+$ and sample size $M \in \mathbb{N}$. Thus, the main problem for weakly converging approximations is the balanced control on the discretization and statistical errors, and these errors should not be considered separately to achieve a desired accuracy in weak approximation procedures. Moreover, the statistical error is increasing with growing variance

$$V_M(f) = \mathbb{E}\, ||\mathbb{E}\, f(Y_{n_T}) - \mathbb{E}\,\widehat{f(Y_{n_T})}||^2$$

where $\mathbb{E}\,\widehat{f(Y_{n_T})}$ is the substitution of $\mathbb{E}\, f(Y_{n_T})$ by statistical sampling procedures, e.g. like $\mathbb{E}\,\widehat{f(Y_{n_T})} = \frac{1}{M}\sum_{i=1}^{M} f(Y_{n_T,i})$. Now, it is natural to ask for methods to reduce that statistical error by variance reducing techniques. The following basic techniques originating from Monte Carlo integration theory are suggested.

Method of Control Variates. Roughly speaking, a control variate is a secondary variate which is simulated along the primary variate of the Monte Carlo method for $f(X)$. This secondary (control) variate has known mean, it should be a square-integrable random variable, and it is positively correlated with the primary variate. The control variate ξ can be constructed by the Clark-Funke-Shevlyakov-Haussmann integral representation theorem, involving certain Fréchet derivatives of $f(X)$ and the linearized dynamics of the underlying SDE. Newton (1994) then suggests then to use projection methods on certain Hilbert spaces of random variables to calculate control variates. By subtraction of the secondary from the primary variate $f(X)$ one obtaines a lower variance than $f(X)$, and whose mean differs from that of $f(X)$ by a certain known amount. For more details, see Newton (1994). As a simple example of the method of control variates, an unbiased estimate would be given by

$$f(X) - \rho(\xi - \mathbb{E}\,\xi)$$

where the parameter

$$\rho = \frac{Cov(f(X))}{Var(\xi)}$$

is chosen such that the variance

$$Var(f(X)) + \rho^2 Var(\xi) - 2\rho\, Cov(f(X), \xi)$$

is minimized. The latter procedure could be done for both variables X and Y_{n_T}.

Method of Importance Sampling. Roughly speaking, the technique of importance sampling involves the transformation of the underlying probability measure according to the Radon-Nikodym Theorem before averaging. Thus one has

$$\mathbb{E}\, f(X) = \int_{\mathbb{R}^d} f(x)\, d\mathbb{P}\,(x) = \int_{\mathbb{R}^d} f(x)\, \frac{d\,\mathbb{P}}{d\,\hat{\mathbb{P}}}(x)\, d\,\hat{\mathbb{P}}\,(x)$$

where $\hat{\mathbb{P}}$ is the new probability measure. If X is drawn according to that new measure $\hat{\mathbb{P}}$, then $f(X)\, d\mathbb{P}\,/d\,\hat{\mathbb{P}}\,(X)$ is an unbiased estimator for $\mathbb{E}\, f(X)$. The theoretical way to construct such a new measure is given by the Girsanov transformation under the validity of Novikov criterion. Then X must be chosen from

$$dX_t = \left(a(t, X_t) - \sum_{j-1}^{m} b^j(t, X_t) u_t^j\right) dt + \sum_{j=1}^{m} b^j(t, X_t)\, d\hat{W}_t^j$$

by discretization with Wiener process $\hat{W}_t^j = W_t^j + \int_0^t u_s^j\, ds$ such that

$$var(\mu_T f(X)) := \widetilde{\mathbb{E}}\,(\mu_T f(X))^2 - (\widetilde{\mathbb{E}}\,\mu_T f(X))^2 = \widetilde{\mathbb{E}}\,(\mu_T f(X))^2 - (\mathbb{E}\, f(X))^2$$

is "small" - as an optimal control problem with Radon-Nikodym derivative

$$\mu_T = \mu_T(u) = \exp\left(\sum_{j=1}^{m} \int_0^T u_t^{j'}\, d\hat{W}_t^j + \frac{1}{2} \sum_{j=1}^{m} \int_0^T |u_t^j|^2\, dt\right)$$

originating from the Girsanov transformation. The optimal $u = (u^j)$ is given by the Clark-Funke-Shevlyakov-Haussmann integral representation theorem.

Method of Antithetic Variates. The simplest version of this very general method uses symmetries of already generated random variables compensating heavy contributions with more variance for the estimator. For example, centered Gaussian distributed random pairs (G_1, G_2) can be multiplied by the factor -1, and one would save computational time and get more symmetry in the random number generation - a technique which may lead to smaller variances of simulated estimators In the spirit of this method is also the idea to take the average $(U+V)/2$ of two already simulated random numbers (U, V) as a further realization.

Method of Variance Reduction by Conditioning. It can also be convenient to use the conditional expectation $\mathbb{E}\,[f(X)|\widetilde{\mathcal{F}}]$ with some appropriate σ-field $\widetilde{\mathcal{F}}$ as variance reducing estimator. The variance is reduced according to inequality

$$Var(\mathbb{E}\,[f(X)|\widetilde{\mathcal{F}}]) \leq Var(f(X)).$$

It could be a problem, however, to find that σ-field. It is interesting to note that some implicit numerical methods like trapezoidal, midpoint or some balanced implicit methods (and asymptotically exact integrators) reduce the occurring variances through their inherent property of preconditioning in an almost optimal way.

All in all, the variance reduction problem exhibits a very challenging problem from the practical point of view. This problem arises in particular when very small quantities $f(X)$ must be estimated, as often is met in reliability investigation of structures in Mechanical Engineering, and efficiently practically implementable and mathematically justified new methods are urgently needed (cf. problems of reliability analysis in Earthquake Engineering).

5.9.7 How to estimate pth mean errors

An important practical question is how to estimate the resulting errors by statistical methods - often pointed out by potential applicants. This question can be answered under the existence of corresponding moments $\mathbb{E}\,\|.\|^{2p}, p \in (0, +\infty)$. For example,

$$\mathbb{E}\,\|X_i - Y_i\|^p \approx \frac{1}{M}\sum_{k=1}^{M} \|X_i^{(k)} - Y_i^{(l)}\|^p$$

where $X_i^{(k)}, Y_i^{(k)}$ denotes the kth sample of stochastic process values X, Y at time $t_i \in [0,T]$. This procedure is justified by the Laws of Large Numbers (LLN). More precisely, it can be proved that there is a finite real constant $K_{stat} > 0$ such that

$$\mathbb{P}\left\{\left|\mathbb{E}\,\|X_i - Y_i\|^p - \frac{1}{M}\sum_{k=1}^{M} \|X_i^{(k)} - Y_i^{(k)}\|^p\right| \geq \varepsilon\right\} \leq \frac{K_{stat}}{\varepsilon\sqrt{M}}$$

for all $\varepsilon > 0$, thanks to Chebyshev inequality. Moreover, there is a Gaussian distributed random variable C such that

$$\left|\mathbb{E}\,\|X_i - Y_i\|^p - \frac{1}{M}\sum_{k=1}^{M} \|X_i^{(k)} - Y_i^{(k)}\|^p\right| \approx \frac{|C|}{\sqrt{M}},$$

thanks to the Central Limit Theorem (CLT). Corresponding confidence intervals are constructed by standard statistical procedures. It is worth noting that the rapidity of convergence in the Central Limit Theorem (CLT) is usually estimated by the Berry-Esseen Theorem which provides estimates of the convergence of probability distribution of given the estimator in the form

$$\sup_x \left|\mathbb{P}\left\{\frac{\sum_{k=1}^{M} \eta_k}{\sqrt{M Var(\eta_1)}} \leq x\right\} - \Phi(x)\right| \leq K_{BE}\frac{\mathbb{E}\,|\eta_1|^3}{\sqrt{M[Var(\eta_1)]^3}}$$

with appropriate real constant $K_{BE} > 0$ satisfying $(2\pi)^{-1/2} \leq K_{BE} < 0.8$, where $\Phi = \Phi(x)$ represents the standard Gaussian probability distribution function, and provided that

$$\eta_k = \|X_i^{(k)} - Y_i^{(k)}\|^p - \mathbb{E}\,\|X_i - Y_i\|^p$$

are i.i.d. random variables with $\mathbb{E}\,|\eta_1|^3 < +\infty$ for fixed index i. (Recall that $Var(.)$ denotes the variance of inscribed random variable.). Besides, for reliable statistical estimation when the moments $\mathbb{E}\,\|.\|^{2p}$ do not exist, we advise consulting sophisticated literature on mathematical statistics. Mostly, one does not know the exact solution X. Then, for heuristic comparison studies, one could substitute the values of X with the values of another very accurate approximation process Z, e.g. with using "very small" step sizes compared to those of Y, in order to get some rough picture about the error process at least. Of course, the error process might also be depicted by the simulation of the corresponding error differential equation. For stochastic error process equations in case of Euler method, see Kurtz and Protter (1991), and Jacod and Protter (1998).

5.9.8 On software and programmed packages

To our current knowledge, there are the following programmed systems mentioned in literature or known to us:

(i) Fortran programs built in PRESTO by D. Talay (1994)

(ii) Fortran programs built in DYNAMICS & CONTROL by S. Artemiev et al.

(iii) TURBO-PASCAL programs on Diskette of Kloeden, Platen, Schurz (1994)

(iv) C programs built in GNANS on UNIX platforms by B. Martensen

(v) OSCIL - a C simulation code on UNIX platforms for our private use

Furthermore, there are MAPLE codes written by Cyganovski (1995, 1996), and MAPLE scripts by Kloeden and Scott available (1993). The latter codes are important in the sense that the messy differential operator products resulting from the stochastic Taylor expansions can be evaluated by symbolic manipulation routines as MAPLE fairly easily, compared to classical handworks. Thus, using symbolic manipulation higher order Taylor methods could be implemented much more easily than in the early days of stochastic numerics (remember the problem of efficient generation of multiple integrals remains a problem, at least up to the time of writing this paper at the end of 1999).

All in all, it seems to be still recommendable to develop corresponding software for stochastic numerical analysis and simulations. Which package should be preferred (like MATHEMATICA, REDUCE, MAPLE, MATLAB, etc.) is an fairly open question, too. Personally, we recommend to write your own specific codes, since an optimal implementation surely depends on the specific nature of a given problem, after you have gained some experience with an available standard package (e.g. as mentioned above). However, there is no hope of finding a universal, platform-independent toolbox for stochastic numerical methods, since the field itself seems to be too complex and too rapidly expanding into new directions.

5.9.9 Comments on applications of numerics for (O)SDEs

There is no need to emphasize the huge potential range of applications of stochastic algorithms and their numerical analysis. To name a few applications which are already treated in literature, see in Catchment Modeling by Unny (1984), in Stochastic Water Storage Models by Ozaki (1985), in Random Vibrations by Iyengar (1988), Quantum Physics by Smith and Gardiner (1989), for the approximation of Lyapunov exponents by Talay (1991), Grorud and Talay (1996), in Stochastic Hydrology by Karmeshu and Schurz (1993), in Markov Chain Filtering by Kloeden, Platen and Schurz (1993), in Seismology by Karmeshu and Schurz (1995), in Polymer Chemistry by Öttinger (1996), in Mechanical Engineering by Roy and Schurz (1996), in Stochastic Marketing by Schurz (1996), in Mathematical Finance by Rogers and Talay (1997), in Nonlinear Filtering by Kannan and Zhang (1998), to Schrödinger equations by Schurz (1999), among many others. We personally see that the most challenging field is in the adequate numerical treatment of stochastic infinite dimensional systems, such as stochastic partial differential equations (SPDEs).

5.10 Comments, Outlook, Further Developments

By no means can we claim any completeness of this survey. It should be understood only as a tentative, first course introduction to the theory and applications of numerical analysis of (ordinary) stochastic differential equations - nothing more. However, we hope that we have given some more insight into the theory and related problems of stochastic numerical analysis as well. There are a few of recommendable survey papers in the literature which all readers are cordially invited to look at and compare. Just to mention few of them, see Mil'shtein (1988, 1995), Kloeden and Platen (1989), Kloeden, Platen and Schurz (1991), Talay (1995), Newton (1996), Platen (1999).

5.10.1 Recent and further developments

The recent research is currently concentrated on numerical analysis for jump diffusions (e.g. see Liu and Li), Levy processes (e.g. see Protter and Talay), stochastic delay (functional) equations (e.g. see Tudor), reflected diffusions (Lépingle, Slominski), forward-backward equations (e.g. Ma, Protter and Yong), stochastic particle approximations (e.g. see Kurtz and Xiong, Bossy and Talay), stochastic partial differential equations (SPDEs), the latter area as its own field of development (e.g. see Hoo, Wong, Grecksch, Gyöengy, Davie and Gaines, Allen, Novosel and Zhang, Matthies and Bucher, etc.) to name a few of those "hot topics." Most of these contributions try to exploit purely deterministic ideas in this rapidly growing field of research (such as Galerkin approximation, the method of lines, finite elements techniques, discrete Sobolev space techniques and/or spectral methods for PDEs).

Fairly new fields of research are given by the numerical treatment of stochastic functional-differential equations, stochastic singularly perturbed systems, stochastic differential-algebraic equations, stochastic integro-differential equations or stochastic difference-differential equations and their combinations. Promising results in those fields require an immense preknowledge of several mathematical disciplines, and hence they represent a real mathematical challenge for the 21st century. For example, the field of systems of nonautonomous stochastic difference equations should be studied to understand the adequate construction of numerical methods with variable step sizes and error control better, or, last but not least, the convergence rates of approximations for stopping times rather than deterministic, fixed terminal times.

5.10.2 General comments

The attached reference list is comprehensive, but not complete. We have only concentrated on citing the key references, and we are sure that more ideas can be read from the physics literature (e.g. from Gardiner (1997) or Öttinger (1996)).

All in all, it only remains to warn everybody not to go deeper into new fields of numerics without studying the analytical theory before hand. Otherwise, they will one day have to recognize that their numerical algorithms do not replicate the behavior of natural phenomena. We also recommend understanding the so called "simpler case" of numerical analysis of systems of (O)SDE at first. Explosions or "strange numerical behavior" are mostly due to ill-posedness, a lack of understanding, or too fast approaches to generalizing or putting the learned things into practice. One should return to the theoretical studies and check the presuming conditions of mathematical statements very carefully. In this respect the study of qualitative behavior of related stochastic dynamical systems will gain more and more importance in the challenging interface of deterministic and stochastic analysis.

5.10.3 Acknowledgements

The author expresses his deepest thanks to the support and understanding given by my family (not being able to share lots of time with me over the last three years of part time absence) and also to my first academic teacher Prof. Dr. P.H. Müller at Technical University of Dresden (Germany) who has taught me with patience. We are also thankful to the University of Minnesota, Minneapolis, which provided me with a very academically inspiring atmosphere. Grant D. Erdmann, as the first reader in December 1999, deserves my sincere thanks for correction of numerous misprints and pointing out poor English phrases, which naturally occur when the author is exhausted from intensive work and when he is not a native English speaker.

5.10.4 New trends - 10 challenging problem areas

- Randomized fractal calculus, stochastic-fractal Taylor and integral expansions
- Stochastic weak derivatives, numerics for stochastic distributions (on stochastic Schwarz-spaces, stochastic Sobolev spaces)
- Numerics for p-variation stochastic integration calculus
- SPDEs, Stochastic Functional-Difference-Differential-Equations (SFDDEs)
- Stochastic Lyapunov-type numerical techniques, Numerical orbital stability
- Efficient statistical methods for all of that areas, Numerical computational complexity
- Numerics for optimal random stopping time problems, stochastic control, stochastic resonance, stochastically coherent (adequate) methods
- Numerics for interacting particle systems in Mathematical Biology
- Efficient generation of random variables and (fractal) multiple integrals
- Numerics for the Schrödinger equation and **serious real-world applications**

Bibliography

[1] M.I. Abukhaled and E.J. Allen: A recursive integration method for approximate solution of stochastic differential equations, Int. J. Comput. Math. **66** (1998), No. 1-2, p. 53-66.

[2] M.I. Abukhaled and E.J. Allen: A class of second-order Runge-Kutta methods for numerical solution of stochastic differential equations. Stochastic Anal. Appl. **16** (1998), No. 6, p. 977-991.

[3] M.F. Allain: Sur quelques types d'approximation des solution d'equations différentielles stochastiques, PhD thesis, Univ. Rennes, 1974.

[4] E.J. Allen, S.J. Novosel and Z. Zhang: Finite element and difference approximation of some linear stochastic partial differential equations, Stochastics Stochastic Rep. **64** (1998), No. 1-2, p. 117-142.

[5] E.J. Allen and C.J. Nunn: Difference methods for numerical solution of stochastic two-point boundary-value problems, In: Elaydi, Saber N. (ed.) et al., Proceedings of the first international conference on difference equations, Trinity University, San Antonio, TX, USA, May 25-28, 1994. London: Gordon and Breach. p. 17-27, 1995.

[6] S.L. Anderson: Random number generators on vector supercomputers and other advanced structures, SIAM Review **32** (1990), p. 221-251.

[7] V.V. Anh, W. Grecksch and A.A. Wadewitz: A splitting method for a stochastic Goursat problem, Stochastic Anal. Appl. **17** (1999), No. 3, p. 315-326.

[8] M.V. Antipov: Congruence operator of the pseudo-random numbers generator and a modification of Euclidean decomposition, Monte Carlo Methods Aplic. **1** (1995), p. 203-219.

[9] M.V. Antipov: Sequences of numbers for Monte Carlo methods, Monte Carlo Methods Applic. **2** (1996), p. 219-235.

[10] J.M. Araujo and A.M. Awruch: On stochastic finite elements for structural analysis, Comput. Struct. **52** (1994), No. 3, p. 461-469.

[11] L. Arnold: Stochastic Differential Equations: Theory and Applications, Krieger Publishing Company, Malabar, 1992 (reprint of the original, John Wiley and Sons, Inc. from 1974, German original, Oldenbourg Verlag from 1973).

[12] L. Arnold: Random Dynamical Systems, Springer, Berlin, 1998.

[13] G.B. Arous: Flots et series de Taylor stochastiques, Probab. Theory and Rel. Fields **81** (1989), p. 29-77.

[14] S.S. Artemiev: A variable step size algorithm for the numerical solution of stochastic differential equations (in Russian), Numer. Meth. Cont. Mech. **16** (1985), p. 14-23.

[15] S.S. Artemiev: Certain aspects of application of numerical methods for solving SDE systems, Bull. Novosibirsk. Comp. Center, Numer. Anal. **1** (1993), p. 1-16.

[16] S.S. Artemiev: Stability of numerical methods for solving stochastic differential equations (in Russian), Sib. Matemat. Journal **35** (1994), No. 6, p. 1210-1214.

[17] S.S. Artemiev: The mean square stability of numerical methods for solving stochastic differential equations, Russian J. Numer. Anal. Math. Modeling **9** (1994), No. 5, p. 405-416.

[18] S.S. Artemiev and T.A. Averina: Numerical analysis of systems of ordinary and stochastic differential equations, VSP, Utrecht, 1997.

[19] S.S. Artemiev and H. Schurz: Stiff systems of stochastic differential equations with small noise and their numerical solution (in Russian), Prepr. Vychisl. Tsentr Ross. Akad. Nauk Sib. Otd. **1995** (1995), No. 1039, p. 1-24.

[20] S.S. Artemiev and I.O. Shkurko: An algorithm of variable order and variable step-size based on Rosenbrock-type methods, U.S.S.R. Comput. Math. Math. Phys. **26** (1986), No. 4, p. 193-195 (Translation of "A variable step size order algorithm based on Rosenbrock-type methods" (in Russian), Zh. Vychisl. Math. Math. Fiz. **26** (1986), No. 8, p. 1256-1257).

[21] S.S. Artemiev and I.O. Shkurko: A variable step algorithm for numerical integration of stiff systems of ordinary differential equations with oscillating solutions (in Russian), Model. Mekh. **2** (1988), No. 5, p. 17-25.

[22] S.S. Artemiev and I.O. Shkurko: Numerical analysis of dynamics of oscillatory stochastic systems, Sov. J. Numer. Anal. Math. Modeling **6** (1991), No. 4, p. 277-298.

[23] S. Asmussen, P. Glynn and J. Pitman: Discretization error in simulation of onedimensional reflecting Brownian motion, Ann. Appl. Probab. **5** (1995), p. 875-896.

[24] M.A. Atalla: Finite-difference approximations for stochastic differential equations, in Probabilistic methods for the Investigation of systems with an Infinite Number of Degrees of freedom (in Russian), Collection of Scientific Works, Kiev, p. 11-16.

[25] T.A. Averina and S.S. Artemiev: A new family of numerical methods for solving stochastic differential equations, Sov. Math. Dokl. **33** (1986), No. 3, p. 736-738.

[26] T.A. Averina and S.S. Artemiev: Numerical solution of stochastic differential equations, Sov. J. Numer. Anal. Math. Modeling **3** (1988), No. 4, p. 267-285.

[27] T.A. Averina, S.S. Artemiev and H. Schurz: Simulation of stochastic auto-oscillating systems through variable step size algorithms with small noise, Preprint No. **116**, WIAS, Berlin, 1994, Numerical analysis of stochastic auto-oscillating systems, Bull. Novosib. Comput. Cent., Ser. Numer. Anal. **1995** (1995), No. 6, p. 9-27 (Translation of "Numerical analysis of stochastic auto-oscillating systems" (in Russian), Prepr. Vychisl. Tsentr Ross. Akad. Nauk Sib. Otd. **1995**, No. 1028, p. 1-28, 1995).

[28] E.O. Ayoola: On numerical procedures for solving Lipschitzian quantum SDEs, Ph.D. Thesis, University of Ibadan, Nigeria, 1998.

[29] R. Azencott: Stochastic Taylor formula and asymptotic expansion of Feynman integrals, in Séminaire de probabilités XVI, Supplement, Springer Lect. Notes Math. **921** (1982), p. 237-285.

[30] V. Bally: Approximation for the solutions of stochastic differential equations I: L^p-convergence, Stochastics Stochastic Rep. **28** (1989), p. 209-246.

[31] V. Bally: Approximation for the solutions of stochastic differential equations II: Strong convergence, Stochastics Stochastic Rep. **28** (1989), p. 357-385.

[32] V. Bally: Approximation for the solutions of stochastic differential equations III: Jointly weak convergence, Stochastics Stochastic Rep. **30** (1990), p. 171-191.

[33] V. Bally, P. Protter and D. Talay: The law of the Euler scheme for stochastic differential equations, Z. Angew. Math. Mech. **76** (1996), Suppl. 3, p. 207-210.

[34] V. Bally and D. Talay: The Euler scheme for stochastic differential equations: Error analysis with Malliavin calculus, Math. Comput. Simul. **38** (1995), No. 1-3, p. 35-41.

[35] V. Bally and D. Talay: The law of the Euler scheme for stochastic differential equations I. Convergence rate of the distribution function, Probab. Theory Relat. Fields **104** (1996), p. 43-60.

[36] V. Bally and D. Talay: The law of the Euler scheme for stochastic differential equations. II: Convergence rate of the density, Monte Carlo Methods Appl. **2** (1996), No.2, p. 93-128.

[37] N. Bellomo and F. Flandoli: Stochastic partial differential equations in continuum physics - on the foundations of the stochastic interpolation methods for Itô type equations, Math. Comp. Simul. **31** (1989), p. 3-17.

[38] S. Benachour, B. Roynette, D. Talay and P. Vallois: Nonlinear self-stabilizing processes. I. Existence, invariant probability, propagation of chaos, Stochastic Process. Appl. **75** (1998), No. 2, p. 173-201.

[39] J.F. Bennaton: Discrete time Galerkin approximations to the nonlinear filtering solution: J. Math. Anal. Appl. **110** (1985), p. 364-383.

[40] A. Bensoussan, R. Glowinski and A. Rascanu: Approximation of the Zakai equation by the splitting up method, SIAM J. Control Optimiz. **28** (1990), p. 1420-1431.

[41] A. Bensoussan, R. Glowinski and A. Rascanu: Approximation of some stochastic differential equations by the splitting up method, Appl. Math. Optimiz. **25** (1990), p. 81-106

[42] A. Bensoussan and R. Temam: Equations aux dèrivées partielles stochastiques (I), Israel J. Math. **11** (1972), p. 95-129.

[43] A. Bensoussan and R. Temam: Equations stochastiques du type Navier-Stokes, J. Funct. Analysis **13** (1973), p. 195-222.

[44] F.E. Benth and J. Gjerde: Convergence rates for finite element approximations of stochastic partial differential equations, Stochastics Stochastics Rep. **63** (1998), No. 3-4, p. 313-326.

[45] P. Bernard, D. Talay and L. Tubaro: Vitesse de convergence d'une methode particulaire stochastique pour des equations de convection-diffusion-reaction [Convergence rate of a stochastic particle method for convection-reaction-diffusion equations (in French)], C. R. Acad. Sci., Paris, Ser. **I 317** (1993), No. 4, p. 381-384.

[46] P. Bernard, D. Talay and L. Tubaro: Rate of convergence of a stochastic particle method for the Kolmogorov equation with variable coefficients, Math. Comput. **63** (1994), No. 208, p. 555-587.

[47] R. Biscay, J.C. Jimenez, J.J. Riera, P.A. Valdes: Local linearization method for the numerical solution of stochastic differential equations, Ann. Inst. Statist. Math. **48** (1996), No. 4, p. 631-644.

[48] M. Bossy and D. Talay: Vitesse de convergence d'un algorithme particulaire stochastique pour l'equation de Burgers [Convergence rate of a stochastic particles method for the Burgers equation (in French)], C. R. Acad. Sci., Paris, Ser. **I 320** (1995), No. 9, p. 1129-1134.

[49] M. Bossy and D. Talay: Convergence rate for the approximation of the limit law of weakly interacting particles: Application to the Burgers equation, Ann. Appl. Probab. **6** (1996), No. 3, p. 818-861.

[50] M. Bossy and D. Talay: A stochastic particle method for the McKean-Vlasov and the Burgers equation, Math. Comput. **66** (1997), No. 217, p. 157-192.

[51] N. Bouleau: On effective computation of expectations in large or infinite dimension: Random numbers and simulation, J. Comput. Appl. Math. **31** (1990), p. 23-34.

[52] N. Bouleau and D. Lépingle: Numerical Methods for Stochastic Processes, Wiley, New York, 1993.

[53] N. Bouleau and D. Talay (eds.): Probabilites numeriques, Collection Didactique **10**, INRIA, Rocquencourt, 205 p., 1992.

[54] G. Box and M. Muller: A note on the generation of random normal variables, Ann. Math. Statist. **29** (1958), p. 610-611.

[55] W.E. Boyce: Approximate solutioin of random ordinary differential equations, Adv. Appl. Probab. **10** (1978), p. 172-184.

[56] P.P. Boyle: A Monte Carlo approach, J. Financial Economics **4** (1977), p. 323-338.

[57] H. Brezis: Analyse Fonctionelle: Théorie et Applications (in French), 2nd edition, Masson A., Paris, 1987.

[58] L. Brugnano, K. Burrage and P.M. Burrage: Adams-type methods for the numerical solution of stochastic ordinary differential equations, Manuscript, University of Queensland, Brisbane, 1999.

[59] K. Burrage: Parallel and sequential methods for ordinary differential equations, Clarendon Press, Oxford University Press, Oxford, 1995.

[60] K. Burrage and P.M. Burrage: High strong order explicit Runge-Kutta methods for stochastic ordinary differential equations, Appl. Numer. Math. **22** (1996), p. 81-101.

[61] K. Burrage, P.M. Burrage and J.A. Belward: A bound on the maximum strong order of stochastic Runge-Kutta methods for stochastic ordinary differential equations, BIT **37** (1997) No. 4, p. 771-780.

[62] K. Burrage and P.M. Burrage: General order conditions for stochastic Runge-Kutta methods for both commuting and non-commuting stochastic ordinary differential equation systems, Appl. Numer. Math. **28** (1998), No. 2-4, p. 161-177.

[63] K. Burrage and P.M. Burrage: High strong order methods for non-commutative stochastic ordinary differential equation systems and the Magnus formula, Manuscript, University of Queensland, Brisbane, 1999 (to appear in Physica D, special issue on Quantifying Uncertainty).

[64] P.M. Burrage: Runge-Kutta methods for stochastic differential equations, Ph.D. Thesis, University of Queensland, Brisbane, 1999.

[65] K. Burrage and E. Platen: Runge-Kutta methods for stochastic differential equations, Ann. Numer. Math. **1** (1994), p. 63-78.

[66] K. Burrage and T. Tian: The composite Euler method for stiff stochastic differential equations, Manuscript, University of Queensland, Brisbane, 1999.

[67] K. Burrage and T. Tian: A note on the stability properties of the Euler methods for solving stochastic differential equations, Manuscript, University of Queensland, Brisbane, 1999.

[68] J.C. Butcher: The numerical analysis of ordinary differential equations: Runge-Kutta and general linear methods, Wiley, Chichester, 1987.

[69] S. Cambanis and Y.Z. Hu: Exact convergence rate of the Euler-Maruyama scheme with application to sampling design, Stochastics Stochastic Rep. **59** (1996), No. 3-4, p. 211-240.

[70] L.L. Casasus: On the numerical solution of stochastic differential equations and applications (in Spanish), in Proceedings of the 9th Spanish-Portuguese Conference on Mathematics, Acta Salmanticensia Ciencias **46**, Universidad de Salamanca, Salamanca, p. 811-814, 1982.

[71] L.L. Casasus: On the convergence of numerical methods for stochastic differential equations, in Proceedings of the 5th Congress on Differential Equations and Applications, Informes **14**, Universidad de la Laguna, Puerto de la Cruz, p. 493-501, 1984.

[72] F. Castell and J. Gaines: An efficient approximation method for stochastic differential equations by means of the exponential Lie series, Math. Comput. Simul. **38** (1995), No. 1-3, p. 13-19.

[73] F. Castell and J. Gaines: The ordinary differential equation approach to asymptotically efficient schemes for solution of stochastic differential equations, Ann. Inst. Henri Poincaré **32** (1996), No. 2, p. 231-250.

[74] K.S. Chan and O. Stramer: Weak consistency of the Euler method for numerically solving stochastic differential equations with discontinuous coefficients, Stoch. Proc. Applic. **76** (1998), p. 33-44.

[75] C.C. Chang: Numerical solution of stochastic differential equations with constant diffusion coefficients, Math. Comp. **49** (1987), No. 180, p. 523-542.

[76] D. Chevance: Numerical methods for backward stochastic differential equations, in L.C.G. Rogers and D. Talay (eds.) Numerical Methods in Finance, Cambridge University Press, Cambridge, p. 232-244, 1997.

[77] P.L. Chow, J.L. Jiang, J.L. Menaldi: Pathwise convergence of approximate solutions to Zakai's equation in a bounded domain. Stochastic partial differential equations and applications (Trento, 1990), Longman Sci. Tech., Harlow, Pitman Res. Notes Math. Ser. **268** (1992), p. 111-123.

[78] J.M.C. Clark: The design of robust approximations to the stochastic differential equations of nonlinear filtering, in J.K. Skwirzynski (ed.) Communication Systems and Random Processes Theory, NATO ASI Series E: Applied Sciences **25**, Sijthoff and Noordhoff, Alphen aan den Rijn, p. 721-734, 1978.

[79] J.M.C. Clark: An efficient approximation scheme for a class of stochastic differential equations, in Advances in Filtering and Optimal Stochastic Control, Springer Lect. Notes in Contr. Inf. Sci. **42** (1982), p. 69-78.

[80] J.M.C. Clark: A nice discretization for stochastic line integrals, in B. Grigelionis (ed.) Stochastic Differential Systems, Springer Lect. Notes in Contr. Inf. Sci. **69** (1982), p. 131-142.

[81] J.M.C. Clark and R.J. Cameron: The maximum rate of convergence of discrete approximations for stochastic differential equations, in Stochastic Differential Systems, ed. B. Grigelionis, Springer Lect. Notes Contr. Inform. Sys. **25** (1980), p. 162-171.

[82] D.J. Clements and B.D.O. Anderson: Well behaved Itô equations with simulations that always misbehave, IEEE Trans. Automat. Control **18** (1973), p. 676-677.

[83] H. Contreras: The stochastic finite-element method, Comput. Struct. **12** (1980), p. 341-348.

[84] H. Crauel and F. Flandoli: Attractors for random dynamical systems. Probab. Theory Related Fields **100** (1994), No. 3, p. 365-393.

[85] H. Crauel and F. Flandoli: Hausdorff dimension of invariant sets for random dynamical systems, J. Dynam. Differential Equations **10** (1998), No. 3, p. 449-474.

[86] H. Crauel, A. Debussche and F. Flandoli: Random attractors, J. Dynam. Differential Equations **9** (1997), No. 2, p. 307-341.

[87] H. Crauel and M. Gundlach (ed.): Stochastic dynamics. Conference on random dynamical systems, Bremen, Germany, April 28 - May 2, 1997. Dedicated to Ludwig Arnold on the ocassion of his 60th birthday, Springer, New York, 440 p., 1999.

[88] D. Crisan, J. Gaines and T. Lyons: Convergence of a branching particle method to the solution of the Zakai equation, SIAM J. Appl. Math. **58** (1998), No. 5, p. 1568–1590.

[89] S.O. Cyganowski: A Maple package for stochastic differential equations, A.K. Easton and R.L. May (eds.) Computational Techniques and Applications: CTAC95, World Scientific, Singapore, 1995.

[90] S.O. Cyganowski: Solving stochastic differential equations with Maple, Maple Tech. **3** (1996), p. 38.

[91] G. Da Prato and J. Zabczyk: Stochastic Equations in Infinite Dimensions, Encyclopedia of Mathematics and its Applications **44**. Cambridge University Press, Cambridge, 1992.

[92] G. Da Prato and J. Zabczyk: Ergodicity for Infinite-dimensional Systems, London Mathematical Society Lecture Note Series **229**, Cambridge University Press, Cambridge, 1996.

[93] M.I. Dashevski and R.S. Liptser: Simulation of stochastic differential equations connected with the disorder problem by means of analog computer (in Russian), Automat. Remote Control **27** (1966), p. 665-673.

[94] A.M. Davie and J.G. Gaines: Convergence of implicit schemes for numerical solutions of parabolic stochastic partial differential equations, Manuscript, University of Edinburgh, Edinburgh, 1999.

[95] G. Deelstra and F. Delbaen: Long-term returns in stochastic interest rate models: different convergence results, Appl. Stochastic Models Data Anal. **13** (1997), No. 3-4, p. 401-407.

[96] G. Deelstra and F. Delbaen: Convergence of discretized stochastic (interest rate) processes with stochastic drift term, Appl. Stochastic Models Data Anal. **14** (1998), No. 1, p. 77-84.

[97] G. Denk: A new numerical method for the integration of highly oscillatory second-order ordinary differential equations, Sixth Conference on the Numerical Treatment of Differential Equations (Halle, 1992), Appl. Numer. Math. **13** (1993), No. 1-3, p. 57–67.

[98] G. Denk, C. Penski and S. Schäffler: Noise analysis in circuit simulation with stochastic differential equations, Z. Angew. Math. Mech. **78** (1998), Suppl. 3, S887-S890.

[99] G. Denk and S. Schäffer: Adam's methods for the efficient solution of stochastic differential equations with additive noise, Computing **59** (1997), No. 2, p. 153-161.

[100] S.S. Dey: Finite element method for random response of structures due to stochastic excitation, Comput. Methods Appl. Mech. Eng. **20** (1979), p. 173-194.

[101] P. Donnelly and T.G. Kurtz: Particle representations for measure-valued population models, Ann. Probab. **27** (1999), No. 1, p. 166–205.

[102] H. Doss: Liens entre équations différentielles stochastiques et ordinaires, Ann. Inst. Henri Poincaré **XIII** (1977), Section B, No. 2, p. 99-125.

[103] J. Douglas, J. Ma and P. Protter: Numerical methods for forward-backward stochastic differential equations, Ann. Appl. Probab. **6** (1996), No. 3, p. 940-968.

[104] I.T. Drummond, A. Hoch and R.R. Hogan: The stochastic method for numerical simulations: Higher order corrections, Nuc. Phys. **B220 FS8** (1983), p. 119-136.

[105] I.T. Drummond, A. Hoch and R.R. Hogan: Numerical integration of stochastic differential equations with variable diffusivity, J. Phys. A: Math. Gen. **19** (1986), p. 3871-3881.

[106] P.D. Drummond and I.K. Mortimer: Computer simulation of multiplicative stochastic differential equations, J. Comput. Phys. **93** (1991), No. 1, p. 144-170.

[107] A.A. Dsagnidse and R.J. Tschitashvili: Approximate integration of stochastic differential equations (in Russian), Tiblisi State University, Inst. Appl. Math. Trudy **IV** (1975), p. 267-279.

[108] E.B. Dynkin: Markov processes I, II, Springer, New York, 1965.

[109] J. Eichenhauer and J. Lehn: A non-linear congruential pseudo random number generator, Statist. Paper **27** (1986), p. 315-326.

[110] A. Einstein: Zur Theorie der Brownschen Bewegung, Ann. Phys. IV **19** (1906), p. 371.

[111] M. Ehrhardt: Invariant probabilities for systems in a random environment - with applications to the Brusselator, Bull. Math. Biol. **45** (1983), p. 579-590.

[112] I. Elishakoff, Y.J. Ren and M. Shinozuka: Improved finite element method for stochastic problems, Chaos Solitons Fractals **5** (1995), No. 5, p. 833-846.

[113] R. Elliott and R. Glowinski: Approximations to solutions of the Zakai filtering equation, Stochastic Anal. Appl. **7** (1989), p. 145-168.

[114] K.D. Elworthy, A. Truman, H.Z. Zhao and J.G. Gaines: Approximate traveling waves for generalized KPP equations and classical mechanics, Proc. Roy. Soc. London Ser. **A 446** (1994), No. 1928, p. 529-554.

[115] K. Entacher, A. Uhl and S. Wegenkittl: Linear congruential generators for parallel Monte Carlo: the leap-frog case, Monte Carlo Methods Appl. **4** (1998), No. 1, p. 1-16.

[116] S.M. Ermakov: Die Monte Carlo Methode und verwandte Fragen (in German), VEB Deutscher Verlag der Wissenschaften, Berlin, 1975.

[117] S.M. Ermakov and G.A. Mikhajlov: Statistical simulation. Textbook (in Russian: Statisticheskoe modelirovanie. Uchebnoe posobie), 2nd edition, Ministerstvo Vysshego i Srednego Spetsial'nogo Obrazovaniya SSSR, "Nauka" Glavnaya Redaktsiya Fiziko-Matematicheskoj Literatury, Moskva, 1982. statisticheskogo

[118] L. Fahrmeier: Schwache Konvergenz gegen Diffusionprozesse, Z. Angew. Math. Mech. **54** (1974), p. 245.

[119] L. Fahrmeier: Approximation von stochastischen Differentialgleichungen auf Digital- und Hybridrechnern, Computing **16** (1976), p. 359-371.

[120] L. Faravelli: Response variables correlation in stochastic finite element analysis, Meccanica **23** (1988), No. 2, p. 102-106.

[121] O. Faure: Simulation de mouvement brownien et des diffusions, Thèse ENPC, Paris, 1992.

[122] O. Faure and J.G. Gaines: Simulation trajectorielle des diffusions, in Probabilités Numériques, N. Bouleau and D. Talay (eds.), INRIA, Rocquencourt, 1992, p. 186-192.

[123] J.F. Feng: Numerical solution of stochastic differentialM equations, Chinese J. Num. Appl. **12** (1990), p. 28-41.

[124] J.F. Feng, G.Y. Lei and M.P. Qian: Second order methods for solving stochastic differential equations, J. Comput. Math. **10** (1992), p. 376-387.

[125] P. Fischer and E. Platen: Applications of the balanced method to stochastic differential equations in filtering, Monte Carlo Methods Appl. **5** (1999), No. 1, p. 19-38.

[126] G.S. Fishman: Monte carlo: Concepts, Algorithms and Applications, Series in Operations Research, Springer, New York, 1992.

[127] E. Fournie, J. Lebuchoux and N. Touzi: Small noise expansion and importance sampling, Asymptot. Anal. **14** (1997), p. 331-376.

[128] R.F. Fox: Second-order algorithm for the numerical integration of colored-noise problems, Phys. Rev. **A 43** (1991), p. 2649-2654.

[129] A. Friedman: Stochastic Differential Equations and Applications, Vol. **I**, **II**, Academic Press, Boston, 1975.

[130] J.N. Franklin: Difference methods for stochastic ordinary differential equations, Math. Comput. **19** (1965), p. 552-561.

[131] R. Funke and A.Yu. Shevlyakov: On a generalization of a formula of Clark (in Russian), Theory Random Processes **7** (1977), p. 93-96.

[132] J.G. Gaines: The algebra of iterated stochastic integrals, Stochastics Stoch. Reports **49** (1994), p. 169-179.

[133] J.G. Gaines: A basis for iterated stochastic integrals, Math. Comput. Simulation **38** (1995), No. 1-3, p. 7-11.

[134] J.G. Gaines: Numerical experiments with S(P)DE's, In Stochastic Partial Differential Equations, A.M. Etheridge (ed.), London Math. Soc. Lect. Note Series **216**, Cambridge Univ. Press, Cambridge, 1995, p. 55-71.

[135] J.G. Gaines and T.J. Lyons: Random generation of stochastic area integrals, SIAM J. Appl. Math. **54** (1994), No. 4, p. 1132-1146.

[136] J.G. Gaines, T.J. Lyons: Variable step size control in the numerical solution of stochastic differential equations, SIAM J. Appl. Math. **57** (1997), No. 5, p. 1455-1484.

[137] T.C. Gard: Introduction to stochastic differential equations, Marcel Dekker, Basel, 1988.

[138] C.W. Gardiner: Handbook of Stochastic Methods for Physics, Chemistry and Natural Sciences (2nd editioj), Springer Series in Synergetics **13**. Springer, Berlin, 1997.

[139] C.W. Gardiner, A. Gilchrist and P.D. Drummond: Using the positive P-representation, Manuscript, 1993.

[140] M. Gelbrich: Simultaneous time and chance discretization for stochastic differential equations, J. Comput. Appl. Math. **58** (1995), No. 3, p. 255-289.

[141] M. Gelbrich and S.T. Rachev: Discretization for stochastic differential equations, L^p Wasserstein metrics, and econometrical models, in Distributions with Fixed Marginales and Related Topics, IMS Lecture Notes Monogr. Ser. **28**, Inst. Math. Statist. Hayward, CA, p. 97-119, 1996.

[142] J.E. Gentle: random number generation and Monte Carlo methods, Series in Statistics and Computing, Springer, New York, 1998.

[143] A. Gerardi, F. Marchetti and A.M. Rosa: Simulation of diffusions with boundary conditions, Systems Control Lett. **4** (1984), No. 5, p. 253-261.

[144] A. Germani and M. Piccioni: Semi-discretization of stochastic partial differential equations on $\mathbb{R}^d$ by a finite-element technique, Stochastics **23** (1988), p. 131-148.

[145] R.G. Ghanem: Ingredients for a general purpose stochastic finite elements implementation, Comput. Methods Appl. Mech. Engrg. **168** (1999), No. 1-4, p. 19-34.

[146] R.G. Ghanem and P.D. Spanos: Polynomial chaos in stochastic finite elements, J. Appl. Mech. **57** (1990), No. 1, p. 197-202.

[147] R.G. Ghanem and P.D. Spanos: Stochastic finite elements: a spectral approach, Springer, New York, 1991.

[148] R.G. Ghanem and P.D. Spanos: A spectral formulation of stochastic finite elements, in Guedes Soares, C. (ed.) Probabilistic Methods for Structural Design, Kluwer Academic Publishers, Dordrecht, Solid Mech. Appl. **56** (1997), p. 289-312.

[149] R.G. Ghanem and P.D. Spanos: Spectral techniques for stochastic finite elements, Arch. Comput. Methods Engrg. **4** (1997), No. 1, p. 63-100.

[150] I.I. Gikhman and A.V. Skorochod: Stochastische Differentialgleichungen, Akademie-Verlag, Berlin, 1971.

[151] S.A. Gladyshev and G.N. Mil'shtein: The Runge-Kutta method for calculation of wiener integrals of functionals of exponential type (in Russian), Zh. Vychisl. Mat. Mat. Fiz. **24** (1984), p. 1136-1149.

[152] P.Y. Glorennec: Estimation a priori des erreurs dans la résolution numérique d'équations différentielles stochastiques, Seminaire de probabilités, Universite de Rennes **1**, p. 57-93, 1977.

[153] P.W. Glynn and O.L. Iglehart: Importance sampling for stochastic simulations, Management Science **35** (1989), p. 1367-1392.

[154] O. Goldreich: Pseudorandomness, Notices of AMS **46** (1999), No. 10, p. 1209-1216.

[155] J. Golec: Stochastic averaging principle for systems with pathwise uniqueness, Stochastic Anal. Appl. **13**, p. 307-322.

[156] J. Golec: Averaging Euler-type difference schemes, Stoch. Anal. Appl. **15** (1997), p. 751-758.

[157] J. Golec and G.S. Ladde: Euler–type approximation for systems of stochastic differential equations, J. Appl. Math. Simul. **28** (1989), p. 357-385.

[158] J. Golec and G.S. Ladde: On an approximation method for a class of stochastic singularly perturbed systems, Dynam. Systems Appl. **2** (1993), No. 1, p. 11-20.

[159] S.T. Goodlett and E.J. Allen: A variance reduction technique for use with the extrapolated Euler method for numerical solution of stochastic differential equations, Stochastic Anal. Appl. **12** (1994), No. 1, p. 131-140.

[160] L.G. Gorostiza: Rate of convergence of an approximate solution of stochastic differential equations, Stochastics **3** (1980), p. 267-276, Erratum in Stochastics **4** (1981), p. 85.

[161] H.S. Greenside and E. Helfand: Numerical integration of stochastic differential equations II, Bell System Techn. J. **60** (1981), p. 1927-1940.

[162] W. Grecksch and V.V. Anh: Approximation of stochastic differential equations with modified fractional Brownian motion, Z. Anal. Anwendungen **17** (1998), No. 3, p. 715-727.

[163] W. Grecksch and V.V. Anh: A parabolic stochastic differential equation with fractional Brownian motion input, Statist. Probab. Lett. **41** (1999), No. 4, p. 337-346.

[164] W. Greksch and P.E. Kloeden: Time-discretized Galerkin approximations of parabolic stochastic PDEs, Bull. Austral. Math. Soc. **54** (1996), No. 1, p. 79-85.

[165] W. Greksch and B. Schmalfuss: Approximation of the stochastic Navier-Stokes equation, Mat. Apl. Comput. **15** (1996), No. 3, p. 227-239.

[166] W. Grecksch and C. Tudor: Stochastic Evolution Equations: A Hilbert Space Approach, Mathematical Research **85**. Akademie Verlag, Berlin, 1995.

[167] W. Greksch and A. Wadewitz: Approximation of solutions of stochastic differential equations by discontinuous Galerkin methods, J. Anal. Appl. **15** (1996), p. 901-916.

[168] A. Greiner, W. Strittmatter and J. Honerkamp: Numerical integration of stochastic differential equations, J. Statist. Phys. **51** (1988), No. 1-2, p. 95-108.

[169] A. Grorud and D. Talay: Approximation of Lyapunov exponents of stochastic differential systems on compact manifolds, in analysis and optimization of systems, Proc. 9th Int. Conf., Antibes/Fr. 1990, Springer Lect. Notes Control Inf. Sci. **144** (1990), p. 704-713.

[170] A. Grorud and D. Talay: Approximation of Lyapunov exponents of nonlinear stochastic differential equations, SIAM J. Appl. Math. **56** (1996), No. 2, p. 627-650.

[171] S.J. Guo: On the mollifier approximation for solutions of stochastic differential equations, J. Math. Kyoto Univ. **22** (1982), p. 243-254.

[172] S.J. Guo: Approximation theorems based on random partitions for stochastic differential equations and applications, Chinese Ann. Math. **5** (1984), p. 169-183.

[173] I. Gyöngy: On stochastic equations with respect to semimartingales III, Stochastics **7** (1982), p. 231-254.

[174] I. Gyöngy: On approximation of Itô stochastic equations, Math. SSR Sbornik **70** (1991), p. 165-173.

[175] I. Gyöngy: A note on Euler's approximations, Potential Anal. **8** (1998), No. 3, p. 205-216.

[176] I. Gyöngy: Lattice approximations for stochastic quasi-linear parabolic partial differential equations driven by space-time white noise I, Potential Anal. **9**, No. 1, p. 1-25 (1998).

[177] I. Gyöngy and N.V. Krylov: On stochastic equations with respect to semimartingales I, Stochastics **4** (1980), p. 1-21.

[178] I. Gyöngy and N.V. Krylov: On stochastic equations with respect to semimartingales II. Itô formula in Banach spaces, Stochastics **6** (1982), p. 153-173.

[179] I. Gyöngy and N.V. Krylov: Existence of strong solutions for Ito's stochastic equations via approximations, Probab. Theory Relat. Fields **105** (1996), No. 2, p. 143-158.

[180] I. Gyöngy and D. Nualart: Implicit scheme for quasi-linear parabolic partial differential equations perturbed by space-time white noise, Stochastic Processes Appl. **58** (1995), No. 1, p. 57-72.

[181] I. Gyöngy and D. Nualart: Implicit scheme for stochastic partial differential equations driven by space-time white noise, Potential Analysis **7** (1997), p. 725-757.

[182] J.H. Halton: On the efficiency of certain quasi-random sequences of points in evaluating multi-dimensional integrals, Numer. Math. **2** (1960), p. 84-90.

[183] J.M. Hammersley and D.C. Handscomb: Monte Carlo Methods, Wiley, New York, 1964.

[184] C.J. Harris: Simulation of nonlinear stochastic equations with applications in modeling water pollution, in C.A. Brebbi (ed.) Mathematical Models for Environmental Problems, Pentech Press, London, p. 169-282, 1976.

[185] C.J. Harris and Y. Maghsoodi: Approximate integration of a class of stochastic differential equations. in Control Theory, Proc. 4th IMA Conf., Cambridge/Engl. 1984, p. 159-168, 1985.

[186] E. Hausenblas: A MonteCarlo method with inherited parallelism for solving partial differential equations with boundary conditions numerically, Manuscript, University of Salzburg, Salzburg, 1999.

[187] E. Hausenblas: A numerical scheme using excursion theory for simulating stochastic differential equations with reflection and local time at a boundary, Manuscript, University of Salzburg, Salzburg, 1999.

[188] E. Hausenblas: A numerical scheme using Itô excursions for simulating local time resp. stochastic differential equations with reflection, Osaka J. Math. **36** (1999), No. 1, p. 105-137.

[189] U.G. Haussmann: On the integral representation of fucntionals of Itô processes, Stochastics **3** (1979), p. 17-28.

[190] D.C. Haworth and S.B. Pope: A second–order Monte Carlo method for the solution of the Itô stochastic differential equation, Stochastic Anal. Appl. **4** (1986), p. 151-186.

[191] D. Heath and E. Platen: Valuation of FX barrier options under stochastic volatility, Financial Engineering and the Japanese Markets **3** (1996), p. 195-215.

[192] E. Helfand: Numerical integration of stochastic differential equations, Bell System Techn. J. **58** (1979), 2289-2299.

[193] D.B. Hernandez and R. Spigler: A–stability of implicit Runge–Kutta methods for systems with additive noise, BIT **32** (1992), p. 620-633.

[194] D.B. Hernandez and R. Spigler: Convergence and stability of implicit Runge–Kutta methods for systems with multiplicative noise, BIT **33** (1993), p. 654-669.

[195] J. Hersch: Contribution à la méthode des équations aux différences, ZAMP IXa (1958), No. 2, p. 129-180.

[196] T.D. Hien and M. Kleiber: Finite element analysis based on stochastic Hamilton variational principle, Comput. Struct. **37** (1990), No. 6, p. 893-902.

[197] D.J. Higham: Mean-square and asymptotic stability of numerical methods for stochastic differential equations, Strathclyde Mathematics Research Report **39**, University of Strathclyde, Glasgow, 1999.

[198] N. Hofmann: Beiträge zur schwachen Approximation stochastischer Differentialgleichungen (in German), Dissertation, Humboldt University Berlin, Berlin, 1995.

[199] N. Hofmann: Stability of weak numerical schemes for stochastic differential equations, Math. Comput. Simulation **38** (1995), No. 1-3, p. 63-68.

[200] N. Hofmann and P. Mathé: On quasi-Monte Carlo simulation of stochastic differential equations, Math. Comp. **66** (1997), No. 218, p. 573-590.

[201] N. Hofmann, T. Müller-Gronbach and K. Ritter: Optimal approximation of stochastic differential equations by adaptive step-size control, Math. Comp. (1999), to appear.

[202] N. Hofmann and E. Platen: Stability of weak numerical schemes for stochastic differential equations, Computers Math. Appl. **28** (1994), No. 10-12, p. 45-57.

[203] N. Hofmann and E. Platen: Stability of superimplicit numerical methods for stochastic differential equations, in Nonlinear Dynamics and Stochastic Mechanics (Waterloo, ON, 1993), p. 93-104, Fields Inst. Commun. **9**, Amer. Math. Soc., Providence, RI, 1996.

[204] H. Holden, B. Øksendal, J. Ubøe and T. Zhang: Stochastic Partial Differential Equations. A Modeling, White Noise Functional Approach, Probability and its Applications, Birkhäuser Boston, Inc., Boston (MA), 1996.

[205] R. Horváth-Bokor: On two-step methods for stochastic differential equations, Acta Cybernet. **13** (1997), No. 2, p. 197-207.

[206] R. Horváth-Bokor: On the stability of two-step methods for SDE, in Proceedings of the 7th International Conference on Operational Research, KOI'98, Rovinj, Croatia, September 30 - October 2, 1998.

[207] R. Horváth-Bokor: A theorem on the order of mean square convergence of multistep approximations of solutions of stochastic ordinary differential equations, submitted to Acta Hung. Math. (1999).

[208] Y.Z. Hu: Series de Taylor stochastique et formule de Campbell-Hausdorff d'apres Ben Arous, Séminaire de Probabilit es XXVI, Springer, New York, Lecture Notes in Math. **1626** (1992), p. 587-594.

[209] Y.Z. Hu: Strong and weak order of time discretization schemes of stochastic differential equations, In Azema, J. (ed.) et al., Seminaire de probabilites XXX, Springer Lect. Notes Math. **1626** (1996), p. 218-227.

[210] Y.Z. Hu: Semi-implicit Euler-Maruyama scheme for stiff stochastic equations, In: Koerezlioglu, H. (ed.) et al., Stochastic analysis and related topics V: The Silivri Workshop, held in Silivri, Norway, July 18-29, 1994, Proceedings, Boston, MA: Birkhäuser. Prog. Probab. **38** (1996), p. 183-302.

[211] Y.Z. Hu: Itô-Wiener chaos expansion with exact residual and correlation, variance inequalities, J. Theor. Probab. **10** (1997), No. 4, p. 835-848.

[212] Y.Z. Hu and H. Long: Symmetric integral and the approximation theorem of stochastic integral in the plane, Acta Math. Sci. **13** (1993), No. 2, p. 153-166.

[213] Y.Z. Hu and P.A. Meyer: On the approximation of multiple Stratonovich integrals, In Cambanis, S. (ed.) et al., Stochastic Processes: A Festschrift in Honour of Gopinath Kallianpur, Springer, New York, p. 141-147, 1993.

[214] Y.Z. Hu and S. Watanabe: Donsker's delta functions and approximation of heat kernels by time discretization methods, J. Math. Kyoto Univ. **36** (1996), No. 3, p. 499-518.

[215] J.C. Hull: Options, Futures, And Other Derivatives, (3rd ed.), Prentice Hall, Upper Saddle River (NJ), 1997.

[216] J. Hull and A. White: The use of control variate techniques in option pricing, J. Financial and Quantative Analysis **23** (1988), p. 237-251.

[217] N. Ikeda and S. Watanabe: Stochastic Differential Equations and Diffusions Processes (2nd ed.), North-Holland, Amsterdam, 1989.

[218] K. Itô: Stochastic integral, Proc. Imp. Acad. Tokyo **20** (1944), p. 519-524.

[219] K. Itô: On a formula concerning stochastic differential equations, Nagoya Math. J. **3** (1951), p. 55-65.

[220] R.N. Iyengar: Higher order linearization in nonlinear random vibration, Internat. J. Non-Linear Mech. **23** (1988), No. 5-6, p. 385-391.

[221] R.N. Iyengar and D. Roy: Extensions of the phase space linearization (PSL) technique for non-linear oscillators, J. Sound Vibration **211** (1998), No. 5, p. 877-906.

[222] J. Jacod and P. Protter: Asymptotic error distributions for the Euler method for stochastic differential equations, Ann. Probab. **26** (1998), No. 1, p. 267-307.

[223] J. Jacod and A.N. Shiryaev: Limit Theorems for Stochastic Processes, Springer, New York, 1987.

[224] A. Janicki: Numerical and Statistical Approximation of Stochastic Differential Equations with Non-Gaussian Measures, H. Steinhaus Center for Stochastic Methods in Science and Technology, Wroclaw, 1996.

[225] A. Janicki and A. Weron: Simulation of Chaotic Behavior of α-stable Stochastic processes, Monographs and Textbooks in Pure and Applied Mathematics, Marcel Dekker, New York, 1994.

[226] A. Janicki, Z. Michna and A. Weron: Approximation of stochastic differential equations driven by α-stable Lévy motion, Applicationes Mathematicae **24** (1996), p. 149-168.

[227] R. Janssen: Difference-methods for stochastic differential equations with discontinuous drift, Stochastics **13** (1994), p. 199-212.

[228] R. Janssen: Discretization of the Wiener process in difference methods for stochastic differential equations, Stochastic Process. Appl. **18** (1994), p. 361-369.

[229] J.C. Jimenez, I. Shoji and T. Ozaki: Simulation of stochastic differential equations through the local linearization method. A comparative study, J. Statist. Phys. **94** (1999), No. 3-4, p. 587-602.

[230] J.C. Jimenez, P.A. Valdes, L.M. Rodriguez, J.J. Riera and R. Biscay: Computing the noise covariance matrix of the local linearization scheme for the numerical solution of stochastic differential equations, Appl. Math. Lett. **11** (1998), No. 1, p. 19-23.

[231] C. Joy, P.P. Boyle and K.S. Tan: Quasi Monte Carlo methods in numerical finance, Management Science **42** (1996), p. 926-938.

[232] M.H. Kalos and P.A. Whitlock: Monte Carlo Methods, Wiley-Interscience, New York, 1986.

[233] S. Kanagawa: On the rate of convergence for Maruyama's approximation solutions of stochastic differential equations, Yokohama Math. J. **36** (1988), No. 1, p. 79–86.

[234] S. Kanagawa: The rate of convergence for approximate solutions of stochastic differential equations, Tokyo J. Math. **12** (1989), p. 33-48.

[235] S. Kanagawa: Estimates of convergence rates for approximate solutions of stochastic differential equations, in Various Problems in Stochastic Numerical Analysis, II (Japanese) (Kyoto, 1995), Sūrikaisekikenkyūsho Kōkyūroku **932** (1995), p. 125-134.

[236] S. Kanagawa: Error estimations for the Euler-Maruyama approximate solutions of stochastic differential equations. Monte Carlo Methods Appl. **1** (1995), No. 3, p. 165-171.

[237] S. Kanagawa: Convergence rates for the Euler-Maruyama type approximate solutions of stochastic differential equations, in Probability Theory and Mathematical Statistics (Tokyo, 1995), p. 183-192, World Sci. Publishing, River Edge, NJ, 1996.

[238] S. Kanagawa: Confidence intervals of discretized Euler-Maruyama approximate solutions of SDE's, in Proceedings of the Second World Congress of Nonlinear Analysts, Part 7 (Athens, 1996), Nonlinear Anal. **30** (1997), No. 7, p. 4101-4104.

[239] T. Kaneko and S. Nakao: A note on approximations for stochastic differential equations, in Séminaire de probabilités XXII, Springer lecture Notes in Math. **1321** (1988), p. 155-162.

[240] D. Kannan and De Ting Wu: A numerical study of the additive functionals of solutions of stochastic differential equations, Dynam. Systems Appl. **2** (1993), No. 3, p. 291-310.

[241] D. Kannan and Q. Zhang: Nonlinear filtering of an interactive multiple model with small observation noise: Numerical methods, Stochastic Anal. Appl. **16** (1998), No. 4, p. 631-659.

[242] L.V. Kantorovič: Functional analysis and applied mathematics (in Russian), Uspehi Matem. Nauk (N.S.) **3** (1948), No. 6(28), p. 89-185.

[243] L.V. Kantorovič: Functional analysis and applied mathematics (in Russian), Vestnik Leningrad. Univ. **3** (1948), No. 6, p. 3-18.

[244] I. Karatzas and S. Shreve: Brownian Motion and Stochastic Calculus, Springer, New York, 1988.

[245] Karmeshu and H. Schurz: Moment evolution of the outflow-rate from nonlinear conceptual reservoirs, in V.P. Singh and B. Kumar (eds.) Proc. International Conference on Hydrology and Water Resources, New Delhi, December 1993, Surface Water-Hydrology **1**, Kluwer Academic Publishers, Dordrecht, p. 403-413, 1996.

[246] Karmeshu and H. Schurz: Effects of distributed delays on the stability of structures under seismic excitation and multiplicative noise, Sādhanā **20** (1995), No. 2-4, p. 451-474

[247] Karmeshu and H. Schurz: Stochastic stability of structures under active control with distributed time delays, in M. Lemaire, J.-L. Favre, A. Mebarki (eds.) Applications of Statistics and Probability: Civil Engineering Reliability and Risk Analysis, Proc. ICASP 7 (Paris, July 1995), A.A. BALKEMA Publishers, Rotterdam, p. 1111-1119, 1995.

[248] W.S. Kendall: Doing stochastic calculus with Mathematica, in Economic and Financial Modeling with Mathematica, TELOS, Sanata Clara (CA), p. 214-238, 1993.

[249] R.Z. Khas'minskiĭ: Stochastic stability of differential equations, Sijthoff Noordhoff, Alphen aan den Rijn, 1980.

[250] J.R. Klauder and W.P. Petersen: Numerical integration of multiplicative–noise stochastic differential equations, SIAM J. Numer. Anal. **22** (1985), p. 1153-1166.

[251] M. Kleiber and T.D. Hien: The Stochastic Finite Element Method. Basic Perturbation Technique and Computer Implementation. Incl. 1 disc, Wiley, Chichester, 1992

[252] W. Kliemann and N. Sri Namachchivaya (eds.): Nonlinear Dynamics and Stochastic Mechanics, CRC Mathem. Modeling Series **5**, CRC Press, Boca Raton, 1995.

[253] P.E. Kloeden and R.A. Pearson: The numerical solution of stochastic differential equations, J. Austral. Math. Soc. **20** (1977), Series B, p. 8-12.

[254] P.E. Kloeden and E. Platen: A survey of numerical methods for stochastic differential equations, J. Stoch. Hydrol. Hydraul. **3** (1989), p. 155-178.

[255] P.E. Kloeden and E. Platen: Stratonovich and Itô Taylor expansions, Math. Nachr. **151** (1991), p. 33-50.

[256] P.E. Kloeden and E. Platen: Relations between multiple Itô and Stratonovich integrals, Stochastic Anal. Appl. **9** (1991), p. 86-96.

[257] P.E. Kloeden and E. Platen: Higher–order implicit strong numerical schemes for stochastic differential equations, J. Statist. Phys. **66** (1992), p. 283-314.

[258] P.E. Kloeden and E. Platen: Numerical solution of stochastic differential equations (2nd edition), Springer, Berlin, 1995.

[259] P.E. Kloeden and E. Platen: Numerical methods for stochastic differential equations, in W. Kliemann and N. Sri Namachchivaya (eds.) Nonlinear Dynamics and Stochastic Mechanics, CRC Math. Model. Series, CRC Press, Boca Raton, p. 437-461, 1995.

[260] P.E. Kloeden and L. Grüne: Pathwise approximation of random ordinary differential equations, Preprint **26/99**, Johann-Wolfgang-Goethe University, Frankfurt am Main, 1999.

[261] P.E. Kloeden, H. Keller and B. Schmalfuß: Towards a theory of random numerical dynamics, in M. Gundlach (ed.) Stochastic Dynamics (Bremen, 1997), Springer, New York, p. 259-282, 1999.

[262] P.E. Kloeden, E. Platen and N. Hofmann: Stochastic differential equations: Applications and numerical methods, in Proceedings of 6th IAHR International Symposium on Stochastic Hydraulics, National Taiwan University, Taipeh, p. 75-81, 1992.

[263] P.E. Kloeden, E. Platen and N. Hofmann: Extrapolation methods for the weak approximation of Itô diffusions, SIAM J. Numer. Anal. **32** (1995), No. 5, p. 1519-1534.

[264] P.E. Kloeden, E. Platen and H. Schurz: The numerical solution of nonlinear stochastic dynamical systems: a brief introduction, Int. J. Bifur. Chaos Appl. Sci. Eng. **1** (1991), No. 2, p. 277-286.

[265] P.E. Kloeden, E. Platen and H. Schurz: Effective simulation of optimal trajectories in stochastic control, Optimization **1** (1992), p. 633-644.

[266] P.E. Kloeden, E. Platen and H. Schurz: Higher order approximate Markov chain filters, in S. Cambanis et al. Stochastic Processes: A Festschrift in Honor of Gopinath Kallianpur, Springer, New York, p. 181-190, 1993.

[267] P.E. Kloeden, E. Platen, H. Schurz: Numerical solution of SDEs through computer experiments (1st edition), Springer, Berlin, 1994 (2nd edition, 1997).

[268] P.E. Kloeden, E. Platen, H. Schurz and M. Sørensen: On effects of discretization on estimators of drift parameters for diffusion processes, J. Appl. Probab. **33** (1996), No. 4, p. 1061-1076.

[269] P.E. Kloeden, E. Platen and I. Wright: The approximation of multiple stochastic integrals, Stochastic Anal. Appl. **10** (1992), No. 4, p. 431-441.

[270] P.E. Kloeden and W.D. Scott: Construction of stochastic numerical schemes through Maple, Maple Technical Newspaper **10** (1993), p. 60-65.

[271] A. Kohatsu-Higa: High order Itô-Taylor approximations to heat kernels, J. Math. Kyoto Univ. **37** (1997), p. 129-150.

[272] A. Kohatsu-Higa and S. Ogawa: Weak rate of convergence for an Euler scheme of nonlinear SDE's, Monte Carlo Methods Appl. **3** (1997), No. 4, p. 327-345.

[273] A. Kohatsu-Higa and P. Protter: The Euler scheme for SDE's driven by semimartingales, in H. Kunita and H.H. Kuo (eds.) Stochastic Analysis on Infinite-dimensional Spaces (Baton Rouge, LA, 1994), p. 141-151, Pitman Res. Notes Math. Ser. **310**, Longman Sci. Tech., Harlow, 1994.

[274] W.E. Kohler and W.E. Boyce: A numerical analysis of some first order stochastic initial value problems, SIAM J. Appl. Math. **27** (1974), p. 167-179.

[275] A.N. Kolmogorov: Grundbegriffe der Wahrscheinlichkeitsrechnung (in German), Springer, Berlin, 1933 (Reprint, 1973); Foundations of the Theory of Probability, Chelsea, New York, 1956.

[276] Y. Komori and T. Mitsui: Stable ROW-type weak scheme for stochastic differential equations, Monte Carlo Methods Appl. **1** (1995), No. 4, p. 279-300.

[277] Y. Komori and T. Mitsui: Stable ROW-type weak scheme for stochastic differential equations, in Various Problems in Stochastic Numerical Analysis, II (Japanese) (Kyoto, 1995), Sūrikaisekikenkyūsho Kōkyūroku **932**, (1995), p. 29-45.

[278] Y. Komori, T. Mitsui and H. Sugiura: Rooted tree analysis of the order conditions of row-type scheme for stochastic differential equations, BIT **37** (1997) (1), 43-66.

[279] Y. Komori, Y. Saito and T. Mitsui: Some issues in discrete approximate solution for stochastic differential equations, Workshop on Stochastic Numerics (Japanese) (Kyoto, 1993), Sūrikaisekikenkyūsho Kōkyūroku **850** (1993), p. 1-13.

[280] Y. Komori, Y. Saito and T. Mitsui: Some issues in discrete approximate solution for stochastic differential equations, in Recent Trends and Applications in the Numerical Solution of Ordinary Differential Equations, Comput. Math. Appl. **28** (1994), No. 10-12, p. 269-278.

[281] A. Korzeniowski: On computer simulation of Feynman–Kac path-integrals, J. Comp. Appl. Math. **66** (1996), p. 333-336.

[282] A. Korzeniowski and D.L. Hawkins: On simulating Wiener integrals and their expectations, Probab. Engng. Inform. Sci. **5** (1991), p. 101-112.

[283] R.I. Kozlov and M.G. Petryakov: The construction of comparison systems for stochastic differential equations and numerical methods (in Russian), Nauka Sibirsk Otdel. Novosibirsk, p. 45-52, 1986.

[284] T. Kurtz and P. Protter: Wong–Zakai corrections, random evolutions and numerical schemes for SDE's, in E.M.E. Meyer-Wolf and A. Schwartz (ed.) Stochastic Analysis: Liber Amicorum for Moshe Zakai, Academic Press, Boston, p. 331-346, 1991.

[285] T. Kurtz and P. Protter: Weak limit theorems for stochastic integrals and stochastic differential equations, Ann. Probab. **19** (1991), No. 3, p. 1035-1070.

[286] T.G. Kurtz and Jie Xiong: Particle representations for a class of nonlinear SPDEs, Stochastic Process. Appl. **83** (1999), No. 1, p. 103-126.

[287] H.J. Kushner: On the weak convergence of interpolated Markov chains to a diffusion, Ann. Probab. **2** (1974), p. 40-50.

[288] H.J. Kushner and P.G. Dupuis: Numerical Methods for Stochastic Control Problems in Continuous Time, Appl. of Math. **24**, Springer, New York, 1992.

[289] D.F. Kuznetzov: Some questions in the theory of numerical solution of Itô stochastic differential equations (in Russian), State Technical University Publisher, St. Petersburg, 1998.

[290] N.V. Krylov: A simple proof of the existence of s solution to the Itô equation with monotone coefficients, Theory Probab. Appl. **35** (1990), No. 3, p. 576-580.

[291] N.V. Krylov: Introduction to the theory of diffusion processes, Translations of Mathematical Monographs **142**, AMS, Providence, 1995.

[292] N.V. Krylov: Lectures on elliptic and parabolic equations in Hölder spaces, Graduate Studies in Mathematics **12**, American Math. Soc., Providence (RI), 1996.

[293] N.V. Krylov: On L_p-theory of stochastic partial differential equations in the whole space, SIAM J. Math. Anal. **27** (1996), No. 2, p. 313-340.

[294] N.V. Krylov and S.V. Lototsky: A Sobolev space theory of SPDE with constant coefficients on a half line, SIAM J. Math. Anal. **30** (1999), No. 2, p. 298-325.

[295] N.V. Krylov and B.L. Rozovskii: On the Cauchy problem for linear stochastic partial differential equations, Math. USSR, Izv. **11** (1977), p. 1267-1284.

[296] N.V. Krylov and B.L. Rozovskii: Stochastic partial differential equations and diffusion processes, Russ. Math. Surv. **37** (1982), No. 6, p. 81-105.

[297] H.H. Kuo: White Noise Distribution Theory, Probability and Stochastics Series, CRC Press, Boca Raton (FL), 1996.

[298] A.M. Law and W.D. Kelton: Simulation Modeling and Analysis (2nd edition), McGraw-Hill, New York, 1991.

[299] F. LeGland: Splitting-up approximation for SPDEs and SDEs with application to nonlinear filtering, in Stochastic Partial Differential Equations and their Applications, Springer Lect. Notes in Contr. Inform. Sci. **176** (1992), p. 177-187.

[300] D. Lépingle: An Euler scheme for stochastic differential equations with reflecting boundary conditions, Computes Rendus Acad. Sci. Paris, Séries I Math. **316** (1993), p. 601-605.

[301] D. Lépingle: Euler scheme for reflected stochastic differential equations, Math. Comput. Simul. **38** (1995), No. 1-3, p. 119-126.

[302] D. Lépingle and A. Ould Eida: Approximating systems of differential equations with random inputs or boundary conditions, Stochastic Anal. Appl. **16** (1998), No. 2, p. 313-324.

[303] D. Lépingle and B. Ribémont: Un schema multipas d'approximation de l'equation de Langevin (in French: A multistep approximation method for the Langevin equation), Stochastic Processes Appl. **37** (1991), No. 1, p. 61-69.

[304] C.W. Li and X.Q. Liu: Algebraic structure of multiple stochastic integrals with respect to Brownian motions and Poisson processes, Stochastics Stochastic reports **61** (1997), p. 107-120.

[305] C.W. Li and X.Q. Liu: Approximation of multiple stochastic integrals and its application to stochastic differential equations, Nonlinear Anal. Theory Methods Appl. **30** (1997), No. 2, p. 697-708.

[306] H. Liske: On the distribution of some functional of the Wiener process (in Russian), Theory of Random Processes **10** (1982), Naukova Dumka, Kiew, p. 50-54.

[307] H. Liske: Solution of an initial-boundary value problem for a stochastic equation of parabolic type by the semi-discretization method (in Russian), Theory of Random Processes **113** (1985), p. 51-56.

[308] H. Liske and E. Platen: Simulation studies on time discrete diffusion approximations, Math. Comput. Simul. **29** (1987), p. 253-260.

[309] X.Q. Liu and C.W. Li: Discretization of stochastic differential equations by the product expansion for the Chen series, Stochastics Stochastic Reports **60** (1997), No. 1-2, p. 23-40.

[310] X.Q. Liu and C.W. Li: Weak approximation and extrapolations of stochastic differential equations with jumps, submitted to SIAM J. Numer. Anal. (1999).

[311] S.V. Lototsky: Problems in statistics of stochastic differential equations, Thesis, University of Southern California, Los Angeles, 1996.

[312] J. Ma, P. Protter and J.M. Yong: Solving forward-backward stochastic differential equations explicitly – a four step scheme, Probab. Theory Related Fields **98** (1994), No. 3, p. 339-359.

[313] V. Mackevičius: On Ikeda-Nakao-Yamato type approximations, Litovsk. Mat. Sb. **30** (1990), No. 4, p. 752-757 (translation in Lithuanian Math. J. **30** (1991), No. 4, p. 350-354).

[314] V. Mackevičius: On approximation of stochastic differential equations with coefficients depending on the past, Liet. Mat. Rink. **32** (1992), No. 2, p. 285-298 (translation in Lithuanian Math. J. **32** (1993), No. 2, p. 227-237).

[315] Mackevičius: Second order weak approximations for Stratonovich stochastic differential equations, Liet. Mat. Rink. **34** (1994), No. 2, p. 226-247 (translation in Lithuanian Math. J. **34** (1995), No. 2, p. 183-200).

[316] V. Mackevičius: Extrapolation of approximations of solutions of stochastic differential equations, in Probability Theory and Mathematical Statistics (Tokyo, 1995), p. 276-297, World Sci. Publishing, River Edge, NJ, 1996.

[317] V. Mackevičius: Convergence rate of Euler scheme for stochastic differential equations: functionals of solutions, Math. Comput. Simulation **44** (1997), No. 2, 109-121.

[318] Y. Maghsoodi: Mean square efficient numerical solution of jump-diffusion stochastic differential equations, Sankhya, Ser. **A 58** (1996), No. 1, p. 25-47.

[319] Y. Maghsoodi: Exact solutions and doubly efficient approximations of jump-diffusion Itô equations, Stochastic Anal. Appl. **16** (1998), No. 6, p. 1049-1072.

[320] Y. Maghsoodi and C.J. Harris: In-probability approximation and simulation of nonlinear jump-diffusion stochastic differential equations, IMA J. Math. Control Inf. **4** (1987), p. 65-92.

[321] W. Magnus: On the exponential solution of differential equations for a linear operator, Comm. Pure Appl. Math. **7** (1954), p. 649-673.

[322] A. Makroglou: Numerical treatment of stochastic Volterra integro-differential equations, J. Comput. Appl. Math. **II** (1991), p. 307-313. Dublin/Irel. 1991,

[323] A. Makroglou: Collocation methods for stochastic Volterra integro-differential equations with random forcing function. Collected papers on stochastic systems modeling, Math. Comput. Simulation **34** (1992), No. 5, p. 459-466.

[324] F.H. Maltz and D.L. Hitzl: Variance reduction in MonteCarlo computations using multidimensional Hermite polynomials, J. Comput. Phys. **32** (1979), p. 345-376.

[325] R. Manella and V. Palleschi: Fast and precise algorithm for computer simulation of stochastic differential equations, Phys. Rev. **A 40** (1989), p. 3381-3386.

[326] S.I. Marcus: Modeling and approximation of stochastic differential equations driven by semimartingales, Stochastics **4** (1981), p. 223-245.

[327] G. Marsaglia and T.A. Bray: A convenient method for generating normal variables, SIAM Review **6** (1964), p. 260-264.

[328] G. Marsaglia, B. Narasimham and A. Zaman: A random number generator for PC's, Comput. Phys. Commun. **60** (1990), No. 3, p. 345-349.

[329] G. Maruyama: Continuous Markov processes and stochastic equations, Rend. Circ. Mat. Palermo **4** (1955), p. 48-90.

[330] H.G. Matthies and C. Bucher: Finite elements for stochastic media problems, Comput. Methods Appl. Mech. Engrg. **168** (1999), No. 1-4, p. 3-17.

[331] S. Mauthner: Step size control in the numerical solution of stochastic differential equations, J. Comput. Appl. Math. **100** (1998), No. 1, p. 93-109.

[332] S. Mauthner: Step size Schrittweitensteuerung bei der numerischen Loesung stochastischer Differentialgleichungen, Ph.D. Thesis, TH Darmstadt, Fortschritt-Berichte VDI. Reihe **10**, Informatik/Kommunikationstechnik. 578, VDI Verlag, Duesseldorf, p. 114, 1999.

[333] R.E. Mickens: Nonstandard Finite Difference Models of Differential Equations, World Scientific, Singapore, 1994.

[334] R. Mikulevicius and E. Platen: Time discrete Taylor approximations for Itô processes with jump component, Math. Nachr. **138** (1988), p. 93-104.

[335] R. Mikulevicius and E. Platen: Rate of convergence of the Euler approximation for diffusion processes, Math. Nachr. **151** (1991), p. 233-239.

[336] G.N. Mil'shtein: Approximate integration of stochastic differential equations, Theor. Probab. Applic. **19** (1974), p. 557-562.

[337] G.N. Mil'shtein: A method of second order accuracy integration of stochastic differential equations, Theor. Probab. Applic. **23** (1978), p. 396-401.

[338] G.N. Mil'shtein: Weak approximation of solutions of systems of stochastic differential equations, Theor. Probab. Applic. **30** (1985), p. 750-766.

[339] G.N. Mil'shtein: A theorem on the order of convergence of mean square approximations of solutions of systems of stochastic differential equations, Theor. Probab. Applic. **32** (1988), p. 738-741.

[340] G.N. Mil'shtein: Numerical integration of stochastic differential equations, Kluwer, Dordrecht, 1995 (translation of Russian original, Uralski University Press, Sverdlovsk, 1988).

[341] G.N. Mil'shtein: The solving of boundary value problems by numerical integration of stochastic equations, Math. Comput. Simul. **38** (1995), p. 77-85.

[342] G.N. Mil'shtein: Solving the first boundary value problem of parabolic type by numerical integration of stochastic differential equations, Theor. Probab. Applic. **40** (1995), p. 657-665.

[343] G.N. Mil'shtein: Application of numerical integration of stochastic equations for solving boundary value problems with Neumann boundary condition, Theor. Probab. Applic. **41** (1996), p. 210-218.

[344] G.N. Mil'shtein: Weak approximation of a diffusion process in a bounded domain, Stochastics Stoch. Reports **62** (1997), p. 147-200.

[345] G.N. Mil'shtein and E. Platen: The integration of stiff stochastic differential equations with stable second moments, Technical Report SRR **014-94**, ANU, Canberra, 1994.

[346] G.N. Mil'shtein, E. Platen and H. Schurz: Balanced implicit methods for stiff stochastic systems, SIAM J. Numer. Anal. **35** (1998), No. 3, p. 1010-1019 (Preprint No. **33**, WIAS, Berlin, 1992).

[347] G.N. Mil'shtein and M.V. Tret'yakov: Numerical solution of differential equations with colored noise, J. Statist. Phys. **77** (1994), p. 691-715.

[348] G.N. Mil'shtein and M.V. Tret'yakov: Numerical methods in the weak sense for stochastic differential equations with small noise, SIAM J. Numer. Anal. **34** (1997), p. 2142-2167.

[349] G.N. Mil'shtein and M.V. Tret'yakov: Mean square numerical methods for stochastic differential equations with small noises, SIAM J. Sci. Comput. **18** (1997), No. 4, p. 1067-1087.

[350] G.N. Mil'shtein and M.V. Tret'yakov: Numerical algorithms for semilinear parabolic equations with small parameter based on approximation of stochastic equations, Math. Comp. **69** (2000), No. 229, p. 237-267.

[351] B.J. Morgan: Elements of Simulation, Chapmann & Hall, London, 1984.

[352] M. Mori: Low discrepancy sequences generated by piecewise linear maps, Monte Carlo Methods Appl. **4** (1998), p. 141-162.

[353] C. Mueller: Long-time existence for the heat equation with a noise term, Probab. Theory Rel. Fields **90** (1991), p. 505-517.

[354] C. Mueller: Coupling and invariant measures for the heat equation with noise, Ann. Probab. **21** (1993), p. 2189-2199.

[355] C. Mueller and E.A. Perkins: The compact support property for solutions of the heat equation with noise, Probab. Theory Rel. Fields **93** (1992), p. 287-320.

[356] C. Mueller and R. Sowers: Blowup for the heat equation with a noise term, Probab. Theory Rel. Fields **97** (1993), p. 287-320.

[357] T. Müller-Gronbach: Optimal design for approximating the path of a stochastic process, J. Statist. Planning Inf. **49** (1996), No. 3, p. 371-385.

[358] T. Müller-Gronbach: Optimal designs for approximating a stochastic process with respect to a minimax criterion, Statistics **27** (1996), No. 3-4, p. 279-296.

[359] T. Müller-Gronbach: Asymptotically optimal designs for approximating the path of a stochastic process with respect to the L^∞-norm, in J. Anděl (ed.) ProbaStat '94 (Smolenice Castle, 1994), Tatra Mt. Math. Publ. **7** (1996), p. 87-95.

[360] T. Müller-Gronbach: Hyperbolic cross designs for approximation of random fields, J. Statist. Plann. Inference **66** (1998), No. 2, p. 321-344.

[361] T. Müller-Gronbach and K. Ritter: Uniform reconstruction of Gaussian processes, Stochastic Process. Appl. **69** (1997), No. 1, p. 55-70.

[362] T. Müller-Gronbach and K. Ritter: Spatial adaption for predicting random functions, Ann. Statist. **26** (1998), No. 6, p. 2264-2288.

[363] T. Müller-Gronbach and R. Schwabe: On optimal allocations for estimating the surface of a random field, Metrika **44** (1996), No. 3, p. 239-258.

[364] H. Nakazawa: Numerical procedures for sample structures on stochastic differential equations, J. Math. Phys. **31** (1990), p. 1978-1990.

[365] N.J. Newton: An asymptotically efficient difference formula for solving stochastic differential equations, Stochastics **19** (1986), No. 3, p. 175-206.

[366] N.J. Newton: Asymptotically optimal discrete approximations for stochatic differential equations, in Theory and Applications of Nonlinear Control Systems, p. 555-567, North-Holland, Amsterdam, 1986.

[367] N.J. Newton: An efficient approximation for stochastic differential equations on the partition of symmetrical first passage times, Stochastics **29** (1990), No. 2, p. 227-258.

[368] N.J. Newton: Asymptotically efficient Runge–Kutta methods for a class of Itô and Stratonovich equations, SIAM J. Appl. Math. **51** (1991), No. 2, p. 542-567.

[369] N.J. Newton: Variance reduction for simulated diffusion, SIAM J. Appl. Math. **54** (1994), No. 6, p. 1780-1805.

[370] N.J. Newton: Numerical methods for stochastic differential equations, Z. Angew. Math. Mech. **76** (1996), Suppl. 3, I–XVI, p. 211-214.

[371] N.J. Newton: Continuous-time Monte Carlo methods and variance reduction, Numerical Methods in Finance, p. 22-42, Publ. Newton Inst., Cambridge Univ. Press, Cambridge, 1997.

[372] H.J. Niederreiter: Remarks on nonlinear pseudo random numbers, Metrika **35** (1988), p. 321-328.

[373] H.J. Niederreiter: Random Number Generation and Quasi-Monte-Carlo Methods, SIAM, Philadelphia (PA), 1992.

[374] H.J. Niederreiter and P.J. Shine: Monte Carlo and Quasi-Monte Carlo Methods in Scientific Computing, Lecture Notes in Statistics **106**, Springer, New York, 1995.

[375] N.N. Nikitin, S.V. Pervachev, V.D. Razevig: On computer solution of servomechanisms (in Russian), Avtomatik. i Telemekhanik. **36** (1975), No. 4, p. 133-137.

[376] N.N. Nikitin, V.D. Razevig: Methods of numerical modeling of stochastic differential equations and estimates of their error (in Russian), Zh. Vychisl. Mat. i Mat. Fiz. **18** (1978), No. 1, p. 106-117.

[377] A.A. Novikov: On an identity for stochastic integrals, Theory Probab. Appl. **17** (1972), p. 717-720.

[378] D. Ocone: Malliavin's calculus and stochastic integral representations of functionals of diffusion processes, Stochastics **12** (1984), p. 161-185.

[379] S. Ogawa: A partial differential equation with the white noise as a coefficient, Z. Wahrscheinlichkeitstheorie und Verw. Gebiete **28** (1973/74), p. 53-71.

[380] S. Ogawa: Processus de Markov en interaction et systéme semi-linéaire d'équations d'évolution (in French), Ann. Inst. H. Poincaré Sect. B (N.S) **10** (1974), p. 279-299.

[381] S. Ogawa: Le bruit blanc et calcul stochastique (in French), Proc. Japan Acad. **51** (1975), p. 384-388.

[382] S. Ogawa: Équation de Schrödinger et équation de particule brownienne, J. Math. Kyoto Univ. **16** (1976), No. 1, p. 185-200.

[383] S. Ogawa: Sur la question d'existence de solutions d'une équation différentielle stochastique du type noncausal (in French) [On the existence of solutions of a stochastic differential equation of noncausal type], J. Math. Kyoto Univ. **24** (1984), No. 4, p. 699-704.

[384] S. Ogawa: Quelques propriétés de l'intégrale stochastique du type noncausal (in French) [Some properties of the stochastic integral of noncausal type], Japan J. Appl. Math. **1** (1984), No. 2, 405-416.

[385] S. Ogawa: Correction: "Remark on approximating a stochastic integral of noncausal type by a sequence of Stieltjes integrals" (in French), Tôhoku Math. J. (2) **36** (1984), No. 3, p. 483.

[386] S. Ogawa: Une remarque sur l'approximation de l'intégrale stochastique du type noncausal par une suite des intégrales de Stieltjes (in French: Remark on approximating a stochastic integral of noncausal type by a sequence of Stieltjes integrals), Tôhoku Math. J. (2) **36** (1984), No. 1, p. 41-48.

[387] S. Ogawa: The stochastic integral of noncausal type as an extension of the symmetric integrals, Japan J. Appl. Math. **2** (1985), No. 1, p. 229-240.

[388] S. Ogawa: Topics in the theory of noncausal stochastic integral equations, in Diffusion Processes and Related Problems in Analysis **I** (Evanston, IL, 1989), p. 411-420, Progr. Probab. **22**, Birkhäuser Boston, Boston, MA, 1990.

[389] S. Ogawa: Monte Carlo simulation of nonlinear diffusion processes, Japan J. Industrial and Appl. Math. **9** (1992), No. 1, p. 22-33.

[390] S. Ogawa: Monte Carlo simulation of nonlinear diffusionproc esses II, Japan J. Industrial and Appl. Math. **2** (1994), No. 1, p. 31-45.

[391] S. Ogawa: Some problems in the simulation of nonlinear diffusion processes, Math. Comput. Simul. **38** (1995), p. 217-223.

[392] S. Ogawa: On a robustness of the random particle method, Monte Carlo Methods Appl. **2** (1996), No. 3, p. 175-189.

[393] S. Ogawa: On a robustness of the random particle method. Pseudorandom numbers and chaos (in Japanese), Sūrikaisekikenkyūsho Kōkyūroku **1011** (1997), p. 28-41.

[394] S. Ogawa: Erratum to the article: "On a robustness of the random particle method" [Monte Carlo Methods Appl. **2** (1996), No. 3, p. 175-189], Monte Carlo Methods Appl. **3** (1997), No. 1, p. 83.

[395] S. Ogawa: Recent topics concerning numerical solution methods for nonlinear SDEs (in Japanese), in Problems in Stochastic Numerical Analysis **III** (Kyoto, 1997), Sūrikaisekikenkyūsho Kōkyūroku **1032**, (1998), p. 46-61.

[396] S. Ogawa and T. Sekiguchi: On the Itô formula of noncausal type, Proc. Japan Acad. Ser. A Math. Sci. **60** (1984), No. 7, p. 249-251.

[397] V.A. Ogorodnikov and S.M. Prigarin: Numerical Modeling of Random Processes and Fields: Algorithms and Applications, VSP, Utrecht, 1996.

[398] B. Øksendahl: Stochastic Differential Equations: An Introduction with Applications (5th edition), Springer, New York, 1998.

[399] H.C. Öttinger: Stochastic Processes in Polymeric Fluids, Springer, Berlin, 1996.

[400] T. Ozaki: A local linearization of nonlinear dynamical systems and time series models, (in Japanese), Proc. Inst. Statist. Math. **32** (1984), No. 2, p. 129-139.

[401] T. Ozaki: Statistical identification of storage models with application to stochastic hydrology, Water Resources Bulletin **21** (1985), p. 663-675.

[402] T. Ozaki: A bridge between nonlinear time series models and nonlinear stochastic dynamical systems: a local linearization approach, Statist. Sinica **2** (1992), No. 1, p. 113-135.

[403] E. Pardoux: Stochastic partial differential equations and filtering of diffusion processes, Stochastics **3** (1979), p. 127-167.

[404] E. Pardoux and D. Talay: Discretization and simulation of stochastic differential equations, Acta Applicandae Math. **3** (1985), p. 23-47.

[405] E. Pardoux and D. Talay: Stability of linear differential systems with parametric excitation, in Nonlinear Stochastic Dynamic Engineering Systems, Proc. IUTAM Symp., Innsbruck/Igls/Austria 1987, p. 153-168, 1989.

[406] M. Papadrakakis and V. Papadopoulos: Robust and efficient methods for stochastic finite element analysis using Monte Carlo simulation, Comput. Methods Appl. Mech. Engrg. **134** (1996), No. 3-4, p. 325-340.

[407] S. Paskov and J. Traub: Faster valuation of financial derivatives, J. Portfolio Manag. (1995), p. 113-120.

[408] X.Q. Peng, G. Liu, L. Wu, G.R. Liu and K.Y. Lam: A stochastic finite element method for fatigue reliability analysis of gear teeth subjected to bending, Comput. Mech. **21** (1998), No. 3, p. 253-261.

[409] W.P. Petersen: Numerical simulation of Itô stochastic differential equations on supercomputers, in Random Media (Minneapolis, Minn., 1985), p. 215-228, IMA Vol. Math. Appl. **7**, Springer, New York, 1987.

[410] W.P. Petersen: Some vectorized random number generators for uniform, normal and Poisson distributions for CRAY X-MP, J. Supercomputing **1** (1988), p. 318-335.

[411] W.P. Petersen: Lagged Fibonacci series random number generators for the NEC SX-3, Intern. J. High. Speed Computing **6** (1994), p. 387-398.

[412] W.P. Petersen: Some experiments on numerical simulations of stochastic differential equations and a new algorithm, J. Comput. Phys. **113** (1994), No. 1, p. 75-81.

[413] W.P. Petersen: A general implicit splitting for stabilizing numerical simulations of Itô stochastic differential equations, SIAM J. Numer. Anal. **35** (1998), No. 4, p. 1439-1451.

[414] R. Petterson: Approximations for stochastic differential equations with reflecting convex boundaries, Stochastic Processes Appl. **59** (1995), p. 295-308.

[415] E. Platen: Weak convergence of approximations of Itô integral equations, Z. Angew. Math. Mech. **60** (1980), No. 11, p. 609-614.

[416] E. Platen: Approximation of Itô integral equations, in Stochastic Differential Systems, Lecture Notes in Contr. Inform. Sci. **25** (1980), p. 172-176.

[417] E. Platen: An approximation method for a class of Itô processes, Litovsk. Mat. Sb. **21** (1981), No. 1, p. 121-133.

[418] E. Platen: A generalized Taylor formula for solutions of stochastic differential equations, Sankhya **44A** (1982), No. 2, p. 163-172.

[419] E. Platen: An approximation method for a class of Itô processes with jump component, Litovsk. Mat. Sb. **22** (1982), No. 2, p. 124-136.

[420] E. Platen: Approximation of first exit times of diffusions and approximate solutions of parabolic equations, Math. Nachrichten **111** (1983), p. 127-146.

[421] E. Platen: Zur zeitdiskreten Approximation von Ito Prozessen (in German), Dissertation B, IMATH, Berlin, 1984.

[422] E. Platen: On first exit times of diffusions, Stochastic Differential Systems (Marseille-Luminy, 1984), p. 192-195, Lecture Notes in Control and Information Sci. **69**, Springer, Berlin-New York, 1985.

[423] E. Platen: Derivative free numerical methods for stochastic differential equations, in Stochastic Differential Systems, Proc. IFIP-WG 7/1 Work. Conf. (Eisenach/GDR 1986), Lect. Notes Control Inform. Sci. **96** (1987), p. 187-193.

[424] E. Platen: Derivative free numerical methods for stochastic differential equations, in Stochastic Differential Systems, Proc. IFIP-WG 7/1 Work (Conf., Eisenach/GDR 1986), Lect. Notes Control Inf. Sci. **96** (1987), p. 187-193.

[425] E. Platen: On weak implicit and predictor-corrector methods, in Probabilités Numériques (Paris, 1992), Math. Comput. Simulation **38** (1995), No. 1-3, p. 69-76.

[426] E. Platen: An introduction to numerical methods for stochastic differential equations, Acta Numerica **8** (1999), p. 195-244.

[427] E. Platen and R. Rebolledo: Weak convergence of semimartingales and discretization methods, Stoch. Process. Appl. **20** (1985), p. 41-58.

[428] E. Platen and W. Wagner: On a Taylor formula for a class of Itô processes, Prob. Math. Statist. **3** (1982), No. 1, p. 37-51.

[429] P. Protter: On the existence, uniqueness, convergence and explosions of solutions of systems of stochastic integral equations, Ann. Probab. **5** (1977), 243-261. stochastic differential equations.

[430] P. Protter: Approximations of solutions of stochastic differential equations driven by semimartingales, Ann. Probab. **13** (1985), p. 716-743.

[431] P. Protter: Stochastic integration and differential equations, Springer, New York, 1990.

[432] P. Protter and D. Talay: The Euler scheme for Levy driven stochastic differential equations, Ann. Probab. **25** (1997), No. 1, p. 393-423.

[433] I. Radovic, I.M. Sobol and R.F. Tichy: Quasi-Monte Carlo methods for numerical integration: Comparison of different low discrepancy sequences, Monte Carlo Methods Appl. **2** (1996), p. 1-14.

[434] M.M. Rao: Stochastic Processes and Integration, Sijthoff & Noordhoff, Alphen aan den Rijn, 1979.

[435] N.J. Rao, J.D. Borwankar and D. Ramkrishna: Numerical solution of Itô integral equations, SIAM J. Control **12** (1974), No. 1, p. 124-139.

[436] V.D. Razevig: Digital modeling of multi-dimensional dynamics under random perturbations (in Russian), Automat. Remote Control **4** (1980), p. 177-186.

[437] Y.J. Ren, I. Elishakoff and M. Shinozuka: Finite element method for stochastic beams based on variational principles, J. Appl. Mech. **64** (1997), No. 3, p. 664-669.

[438] B.D. Ripley: Stochastic Simulation, Wiley, New York, 1983.

[439] B.D. Ripley: Computer generation of random variables: A tutorial letter, Inter. Statist. Rev. **45** (1993), p. 301-319.

[440] L.C.G. Rogers and D. Talay (eds.): Numerical Methods in Finance. Session at the Isaac Newton Institute, Cambridge, GB, 1995, Cambridge Univ. Press, Cambridge, 1997. 2000 (expected).

[441] L. Roman: A Runge-Kutta type scheme to solve $dX_t = \sigma(X_t) \circ dW_t$ under commutative noise, IMA Report **1658**, Minneapolis, December 1999 (Ph.D. Thesis, University of Minnesota, Minneapolis, expected 2000).

[442] W. Römisch and A. Wakolbinger: On the convergence rates of approximate solutions of stochastic equations, Lect. Notes Contr. Inform. Sci. **96** (1987), p. 204-212.

[443] H. Rootzén: Limit distributions for the error in approximations of stochastic integrals, Ann. Probab. **8** (1980), No. 2, p. 241-251.

[444] S.M. Ross: A Course in Simulation, MacMillan, New York, 1991.

[445] A. Răşcanu and C. Tudor: Approximation of stochastic equations by the splitting up method, in Qualitative Problems for Differential Equations and Control Theory, p. 277-287, World Sci. Publishing, River Edge, NJ, 1995.

[446] D. Roy and H. Schurz: A semi–analytical pathwise method for numerical solution of nonlinear oscillators, Manuscript, University of Innsbruck, Innsbruck, 1996.

[447] R.Y. Rubinstein: Simulation and the Monte Carlo Method, Wiley, New York, 1991.

[448] W. Rümelin: Numerical treatment of stochastic differential equation, SIAM J. Numer. Anal. **19** (1982), p. 604-613.

[449] L.B. Ryashko and H. Schurz: Mean square stability analysis of some linear stochastic systems, Dynam. Systems Appl. **6** (1997), No. 2, p. 165-190.

[450] B.L. Rozovskiĭ: Stochastic Evolution Systems, Kluwer, Dordrecht, 1990.

[451] K.K. Sabelfeld: On the approximate computation of Wiener integrals by Monte Carlo method (in Russian), Zh. Vychisl. Mat. Mat. Fiz. **19** (1979), p. 29-43.

[452] Y. Saito: T-stability analysis of numerical schemes for stochastic differential equations (in Japanese), Various Problems in Stochastic Numerical Analysis **II** (Kyoto, 1995), Sūrikaisekikenkyūsho Kōkyūroku **932** (1995), p. 15-28.

[453] Y. Saito and T. Mitsui: Simulation of stochastic differential equations, Ann. Inst. Statist. Math. **45** (1993), No. 3, p. 419-432.

[454] Y. Saito and T. Mitsui: T-stability of numerical scheme for stochastic differential equations, Contributions in numerical mathematics, p. 333-344, World Sci. Ser. Appl. Anal. **2**, World Sci. Publishing, River Edge, NJ, 1993.

[455] Y. Saito and T. Mitsui: Stability of numerical schemes for stochastic differential equations (in Japanese), in Workshop on Stochastic Numerics (Kyoto, 1993), Sūrikaisekikenkyūsho Kōkyūroku **850** (1993), p. 124-138.

[456] Y. Saito and T. Mitsui: Statistical error analysis in numerical simulation for stochastic integral processes, in Numerical Analysis of Ordinary Differential Equations and its Applications (Kyoto, 1994), p. 219-228, World Sci. Publishing, River Edge, NJ, 1995.

[457] Y. Saito and T. Mitsui: Stability analysis of numerical schemes for stochastic differential equations, SIAM J. Numer. Anal. **33** (1996), No. 6, p. 2254-2267. Nagoya, 1992.

[458] Y. Saito, K. Shingu and T. Mitsui: A numerical solution method for Langevin diffusion equations (equations of KPZ types) (in Japanese) Problems in Stochastic Numerical Analysis **III** (Kyoto, 1997), Sūrikaisekikenkyūsho Kōkyūroku **1032** (1998), p. 86-100.

[459] O. Schein and G. Denk: Numerical solution of stochastic differential-algebraic equations with applications to transient noise simulation of microelectronic circuits, J. Comput. Appl. Math. **100** (1998), No. 1, p. 77-92.

[460] B. Schmalfuss: Zur Approximation der der stochastischen Navier-Stokesschen Gleichungen (in German), Z. Tech. Hochsch. Leuna-Merseburg **27** (1985), No. 5, p. 605-612.

[461] B. Schmalfuss: Endlichdimensionale Approximation der Lösung der stochastischen Navier-Stokes-Gleichung (in German), Statistics **21** (1990), No. 1, p. 149-157.

[462] K.R. Schneider and H. Schurz: Stochastic waveform iteration methods for SDEs, Manuscript, WIAS, Berlin, 1999 (to appear as Report at WIAS Berlin and IMA Minneapolis, 1999).

[463] H. Schurz: Asymptotical stability of numerical solutions for multiplicative noise, Preprint **47**, IAAS, Berlin, 1993.

[464] H. Schurz: Mean square stability for discrete linear stochastic systems, Preprint **72**, IAAS, Berlin, 1993.

[465] H. Schurz: Approximation of some nonsmooth and path–dependent functionals of SDEs, Unpublished Manuscript, WIAS, Berlin, 1995.

[466] H. Schurz: Asymptotical mean square stability of an equilibrium point of some linear numerical solutions with multiplicative noise, Stochastic Anal. Appl. **14** (1996), No. 3, p. 313-354.

[467] H. Schurz: Numerical regularization for SDEs: Construction of nonnegative solutions, Dynam. Systems Appl. **5** (1996), p. 323-352.

[468] H. Schurz: Modeling and analysis of stochastic innovation diffusion, Z. Angew. Math. Mech. **76** (1996), Suppl. 3, I–XV, p. 366-369.

[469] H. Schurz: Lecture notes on Analytical and Numerical Numerical Methods for SDEs, Humboldt University Berlin, 1996.

[470] H. Schurz: Stability, stationarity, and boundedness of some implicit numerical methods for stochastic differential equations and applications (original: Report No. **11**, WIAS, Berlin, 1996), Logos-Verlag, Berlin, 1997.

[471] H. Schurz: Linear- and partial-implicit numerical methods for nonlinear SDEs, Unpublished Manuscript, Universidad de Los Andes, Bogotá, 1998.

[472] H. Schurz: Preservation of asymptotical laws through Euler methods for Ornstein-Uhlenbeck process, Stochastic Anal. Appl. **17** (1999), No. 3, p. 463-486.

[473] H. Schurz: The invariance of asymptotic laws of linear stochastic systems under discretization, Z. Angew. Math. Mech. **79** (1999), No. 6, p. 375-382.

[474] H. Schurz: On moment-dissipative stochastic dynamical systems, Technical Report No. **214**, University of Kaiserslautern, Kaiserslautern, 1999 (submitted).

[475] H. Schurz: Moment contractivity and stability exponents of nonlinear stochastic dynamical systems, Technical Report No. **215**, University of Kaiserslautern, Kaiserslautern, Report **1656**, IMA, University of Minnesota, Minneapolis, 1999 (submitted).

[476] H. Schurz: On Taylor series expansions and conditional expectations for Stratonovich SDEs with complete V-commutativity, Report **1671**, IMA, University of Minnesota, Minneapolis, December 1999 (submitted).

[477] H. Schurz: General principles for numerical approximation of stochastic processes on some stochastically weak Banach spaces, Report **1669**, IMA, University of Minnesota, Minneapolis, December 1999 (submitted).

[478] H. Schurz: Qualitative properties of balanced implicit methods (BIMs), Manuscript, Fields Institute, Toronto, 1999.

[479] H. Schurz: Introduction to Numerical and Analytical Methods of Stochastic Differential Equations, University of Minnesota, Minneapolis, 1999 (2 volumes in progress).

[480] Z. Schuss: Theory and Application of Stochastic Differential Equations, Wiley, New York, 1980.

[481] A. Shimizu and T. Kawachi: Approximate solutions of stochastic differential equations, Bull. Nagoya Inst. Tech. **36** (1984), p. 105-108.

[482] M. Shinozuka: Simulation of multivariate and multidimensional random differential processes, J. Acoust. Soc. Amer. **49** (1971), p. 357-367.

[483] M. Shinozuka: Monte Carlo solution of structural dynamics, J. Comp. Struct. **2** (1972), p. 855-874.

[484] I.O. Shkurko: Numerical solution of linear systems of stochastic differential equations (in Russian), in Numerical Methods for Statistics and Modeling, Novosibirsk, p. 101-109, Collected Scientific Works, 1987.

[485] I.O. Shkurko: On the order of convergence of some approximations of solutions of linear systems of stochastic differential equations (in Russian) Numerical Mathematics and Modeling in Physics (Russian), p. 45-55, Akad. Nauk SSSR Sibirsk. Otdel., Vychisl. Tsentr, Novosibirsk, 1989.

[486] I.O. Shkurko: Numerical treatment of SDEs with oscillatory solutions, Monte Carlo Methods and Parallel Algorithms (Primorsko, 1989), p. 71-74, World Sci. Publishing, Teaneck, NJ, 1991.

[487] I. Shoji: A note on asymptotic properties of the estimator derived from the Euler method for diffusion processes at discrete times, Statist. Probab. Lett. **36** (1997), No. 2, p. 153-159.

[488] I. Shoji: Approximation of continuous time stochastic processes by a local linearization method, Math. Comp. **67** (1998), No. 221, p. 287-298.

[489] I. Shoji: A comparative study of maximum likelihood estimators for nonlinear dynamical system models, Internat. J. Control **71** (1998), No. 3, 391-404.

[490] I. Shoji and T. Ozaki: Comparative study of estimation methods for continuous time stochastic processes, J. Time Ser. Anal. **18** (1997), No. 5, p. 485-506.

[491] I. Shoji and T. Ozaki: Estimation for nonlinear stochastic differential equations by a local linearization method, Stochastic Anal. Appl. **16** (1998), No. 4, p. 733-752.

[492] I. Shoji and T. Ozaki: A statistical method of estimation and simulation for systems of stochastic differential equations, Biometrika **85** (1998), No. 1, p. 240-243.

[493] L. Skurt: Anwendung einer stochastischen Finite-Elemente-Methode in der Bruchmechanik (in German), FMC-Ser., Akad. Wiss. DDR, Inst. Mech. **19** (1986), p. 45-54.

[494] L. Skurt and B. Michel: Stochastische Finite-Elemente-Methoden (in German), FMC-Ser., Akad. Wiss. DDR, Inst. Mech. **50** (1990), p. 95-104.

[495] L. Skurt and B. Michel: Stochastic finite element method for solid mechanic problems with uncertain values, in H. Bandemer (ed.) Modeling Uncertain Data, Akademie Verlag, Berlin, Math. Res. **68**, p. 28-33, 1992.

[496] I.H. Sloan and H. Wozniakowski: When are quasi-Monte-Carlo algorithms efficient for high dimensional integrals?, J. Complexity **14** (1998), p. 1-33.

[497] L. Slominski: On approximation of solutions of multidimensional SDEs with reflecting boundary conditions, Stochastic Process. Appl. **50** (1994), p. 197-219.

[498] A.M. Smith and C.W. Gardiner: Simulation of nonlinear quantum damping using the positive P representation, Phys. Rev. **39** (1989), p. 3511-3524.

[499] I.M. Sobol: The distribution of points in a cube and the approximate evaluation of integrals, USSR Comput. Math. Math. Phys. **19** (1967), p. 86-112.

[500] J.M. Steele and R.A. Stine: Mathematica and diffusions, in Economic and Financial Modeling with Mathematica, TELOS, Santa Clara (CA), p. 192-213, 1993.

[501] R.L. Stratonovich: A new representation for stochastic integrals and equations, SIAM J. Control **4** (1966), p. 362-371.

[502] D.W. Stroock and S.R.S. Varadhan: Multidimensional Diffusion Processes, Springer, New York, 1982.

[503] Y. Su and S. Cambanis: Sampling designs for estimation of a random process, Stochastic Process. Appl. **46** (1993), No. 1, p. 47-89.

[504] H. Sugita: Pseudo-random number generator by means of irrational rotation, Monte Carlo Methods Appl. **1** (1995), p. 35-57.

[505] M. Sun and R. Glowinski: Pathwise approximation and simulation for the Zakai filtering equation through operator splitting, Calcolo **30** (1994), p. 219-239.

[506] H.J. Sussmann: Product expansions of exponential Lie series and the discretization of stochastic differential equations, in W. Fleming and J. Lions (eds.) Stochastic Differential Systems, Stochastic Control Theory, and Applications, Springer IMA Series, Vol. **10** (1988), p. 563-582.

[507] D. Talay: Analyse Numérique des Equations Différentielles Stochastiques, Thése 3éme Cycle, Université de Provence, Centre Saint Charles, 1982.

[508] D. Talay: Convergence, pour chaque trajectoire, d'un schema d'approximation des E.D.S. (in French), C. R. Acad. Sci., Paris, Ser. **I 295** (1982), p. 249-252.

[509] D. Talay: How to discretize stochastic differential equations, in Nonlinear filtering and stochastic control, Proc. 3rd 1981 Sess. C.I.M.E., Cortona/Italy 1981, Lect. Notes Math. **972** (1982), p. 276-292.

[510] D. Talay: Resolution trajectorielle et analyse numerique des equations differentielles stochastiques (in French) Stochastics **9** (1983), p. 275-306.

[511] D. Talay: Efficient numerical schemes for the approximation of expectations of functionals of the solution of an SDE and applications, Springer Lect. Notes Contr. Inf. Sci. **61** (1984), p. 294-313.

[512] D. Talay: Discrétisation d'une équation différentielle stochastique et calcul approché d'espérance de fonctionelles de la solution, Math. Modél. Numér. Anal. **20** (1986), No. 1, p. 141-179.

[513] D. Talay: Classification of discreterization schemes of diffusions according to an ergodic criterium, in Stochastic Modeling and Filtering, Proc. IFIP-WG 7/1 Work. Conf., Rome/Italy 1984, Springer Lect. Notes Control Inf. Sci. **91** (1987), p. 207-218.

[514] D. Talay: Second-order discretization schemes of stochastic differential systems for the computation of the invariant law, Stochastics **29** (1990), p. 13-36.

[515] D. Talay: Approximation of upper Lyapunov exponents of bilinear stochastic differential equations, SIAM J. Numer. Anal. **28** (1991), p. 1141-1164.

[516] D. Talay: Presto: a software package for the simulation of diffusion processes, Statistics and Computing Journal **4** (1994), No. 4.

[517] D. Talay: Simulation of stochastic differential systems, in Probabilistic Methods in Applied Physics, ed. P. Krée and W. Wedig, Springer Lecture Notes in Physics **451** (1995), p. 54-96. Springer, Berlin, 1995.

[518] D. Talay: Probabilistic numerical methods for partial differential equations: elements of analysis, in Probabilistic Models for Nonlinear Partial Differential Equations (Montecatini Terme, 1995), p. 148-196, Lecture Notes in Math. **1627**, Springer, Berlin, 1996.

[519] D. Talay: The Lyapunov exponent of the Euler scheme for stochastic differential equations, in H. Crauel and M. Gundlach (eds.) Stochastic Dynamics, p. 241-258, Springer, New York, 1999.

[520] D. Talay and L. Tubaro: Expansion of the global error for numerical schemes solving stochastic differential equations, Stochastic Anal. Appl. **8** (1990), p. 483-509.

[521] S. Tanaka and S. Kanagawa: The accuracy of testing methods for pseudorandom numbers and approximation methods for SDEs (in Japanese), in Problems in Stochastic Numerical Analysis, III (Japanese) (Kyoto, 1997), Sūrikaisekikenkyūsho Kōkyūroku **1032** (1998), p. 21–45.

[522] U. Tetzlaff and H.U. Zschiesche: Näherungslösungen für Itô–differentialgleichungen mittels Taylorentwicklungen für Halbgruppen von Operatoren (in German), Wiss. Z. Techn. Hochschule Leuna-Merseburg **2** (1984), p. 332-339.

[523] S. Tezuka: Polynomial arithmetic analogue of Halton sequences, ACM Trans. Model. Comput. Simul. **3** (1993), p. 99-107.

[524] J. Timmer: Parameter estimation in nonlinear stochastic differential equations, Manuscript, University of Freiburg, Freiburg i.B., 1999.

[525] C. Törok: Numerical solution of linear stochastic differential equations, Comput. Math. Appl. **27** (1994), p. 1-10.

[526] J.F. Traub, G.W. Wasilkowski and H. Wozniakowski: Information-Based Complexity, Academic Press, New York, 1988.

[527] C. Tudor: Successive approximation of solutions of two-parameter Itô equations, (Romanian) Stud. Cerc. Mat. **36** (1984), No. 1, 50-61.

[528] C. Tudor: On the successive approximation of solutions of delay stochastic evolution equations, An. Univ. Bucureşti Mat. **34** (1985), 70-86. I, Rennes,

[529] C. Tudor: Approximation of delay stochastic equations with constant retardation by usual Itô equations, Rev. Roumaine Math. Pures Appl. **34** (1989), No. 1, p. 55-64.

[530] C. Tudor: Minimal and maximal solutions for stochastic equations driven by continuous semimartingales, An. Univ. Bucureşti Mat. **38** (1989), No. 1, 71-76.

[531] C. Tudor: A variation of constants formula for delay stochastic equations in Hilbert spaces, Stud. Cerc. Mat. **41** (1989), No. 2, 135-142.

[532] C. Tudor: Procesos Estocásticos (in Spanish). Mathematical Contributions: Texts **2**, Sociedad Matemática Mexicana, México City, 1994. Mexicana, Mexico, 1996 381–392.

[533] C. Tudor and M. Tudor: On approximation in quadratic mean for the solutions of two parameter stochastic differential equations in Hilbert spaces, An. Univ. Bucureşti Mat. **32** (1983), p. 73-88.

[534] C. Tudor and M. Tudor: On approximation of solutions for stochastic delay equations, Stud. Cerc. Mat. **39** (1987), No. 3, p. 265-274.

[535] C. Tudor and M. Tudor: Approximation of linear stochastic functional equations, Rev. Roumaine Math. Pures Appl. **35** (1990), No. 1, p. 81-99.

[536] C. Tudor, Constantin and M. Tudor: Approximation schemes for Iô-Volterra stochastic equations. Bol. Soc. Mat. Mexicana **3** (1995), No. 1, p. 73-85.

[537] C. Tudor and M. Tudor: Approximate solutions for multiple stochastic equations with respect to semimartingales, Z. Anal. Anwendungen **16** (1997), No. 3, p. 761-768.

[538] M. Tudor: Some second-order approximation schemes for stochastic equations with hereditary argument (in Romanian), Stud. Cerc. Mat. **44** (1992), No. 2, p. 147-158.

[539] M. Tudor: Approximation schemes for two-parameter stochastic equations, Probab. Math. Statist. **13** (1992), No. 2, p. 177-189. 1993

[540] M. Tudor: Difference approximations for linear stochastic functional equations, Stud. Cerc. Mat. **45** (1993), No. 4, p. 351-362.

[541] M. Tudor: Approximate solutions for integrodifferential and Volterra stochastic equations, Stud. Cerc. Mat. **48** (1996), No. 3-4, p. 285-292

[542] M. Tudor: Note on the Chaplygin method for planar stochastic differential equations, Stud. Cerc. Mat. **48** (1996), No. 1-2, p. 109-114.

[543] M. Tudor: Newton's method for stochastic integrodifferential equations, Stud. Cerc. Mat. **49** (1997), No. 1-2, p. 137-142.

[544] B. Tuffin: On the use of low discrepancy sequences in Monte Carlo methods, Monte Carlo Methods Appl. **2** (1996), p. 295-320.

[545] B. Tuffin: Comments on "On the use of low discrepancy sequences in Monte Carlo methods", Monte Carlo Methods Appl. **4**, p. 87-90.

[546] T.E. Unny: Numerical integration of stochastic differential equations in catchment modeling, Water Res. **20** (1984), p. 360-368.

[547] E. Valkeila: Computer algebra and stochastic analysis, CWI Quarterly **4** (1991), No. 3, p. 229-238.

[548] W. Wagner: Unbiased Monte Carlo evaluation of certain functional integrals, J. Comput. Phys. **71** (1987), p. 21-33.

[549] W. Wagner: Monte Carlo evaluation of functionals of solutions of stochastic differential equations. Variance reduction and numerical examples, Stochastic Anal. Appl. **6** (1988), p. 447-468.

[550] W. Wagner: Unbiased multi-step estimators for the Monte-Carlo evaluation of certain functionals, J. Comput. Phys. **79** (1988), p. 336-352.

[551] W. Wagner: Stochastische numerische Verfahren zur Berechnung von Funktionalintegralen (in German), Habilitation, Report **02/89**, IMATH, Berlin, 1989.

[552] W. Wagner: Unbiased Monte–Carlo estimators for functionals of weak solutions of stochastic differential equations, Stochastics Stoch. Reports **28** (1989), p. 1-20.

[553] W. Wagner and E. Platen: Approximation of Itô integral equations, February Report at ZIMM of Academy of Sciences of GDR, Berlin, 1978.

[554] A.D. Wentzell: A Course in the Theory of Random Processes (in Russian), Nauka, Moscow, 1975.

[555] A.D. Wentzell, S.A. Gladyshev and G.N. Mil'shtein: Piecewise constant approximation for the Monte-Carlo calculation of Wiener integrals, Theory Anal. Appl. **6** (1985), p. 745-752.

[556] M.J. Werner and P.D. Drummond: Robust algorithms for solving stochastic partial differential equations, J. Comput. Phys. **132** (1997), p. 312-326.

[557] N. Wiener: Differential space, J. Math. Phys. **2** (1923), p. 131-174.

[558] F.S. Wong: Stochastic finite element analysis of a vibrating string, J. Sound Vib. **96** (1984), p. 447-459.

[559] E. Wong and M. Zakai: On the convergence of ordinary integrals to stochastic integrals, Ann. Math. Statist. **36** (1965), p. 1560-1564.

[560] H. Wozniakowski: Average case complexity of multivariate integration, Bull. Amer. Math. Soc. **24** (1991), p. 185-194.

[561] D.J. Wright: The digital simulation of stochastic differential equations, IEEE Trans. Automat. Control **19** (1974), p. 75-76.

[562] D.J. Wright: Digital simulation of Poisson stochastic differential equations, Intern. J. Systems Sci. **11** (1980), p. 781-785.

[563] K. Xu: Stochastic pitchfork bifurcation: numerical simulations and symbolic calculations using Maple, Math. Comput. Simul. **38** (1995), No. 1-3, p. 199-209.

[564] S.J. Yakowitz: Computational Probability and Simulation, Addison Wesley, Reading (MA), 1977.

[565] T. Yamada: Sur l'approximation des solutions d'équations différentielles stochastiques, Z. Wahrsch. Verw. Gebiete **36** (1976), p. 153-164.

[566] N. Yannios and P.E. Kloeden: Time-discretization solution of stochastic differential equations, in R.L. May and A.K. Easton (eds.) Computational Techniques and Applications (Proc. CTAC 95), p. 823-830, World Scientific, Singapore, 1996.

[567] Y.Y. Yen: A stochastic Taylor formula for functionals of two-parameter semimartingales, Acta Vietnamica **13** (1988), p. 45-54.

[568] H. Yoo: An analytic approach to stochastic partial differential equations and its applications, Thesis, University of Minnesota, Minneapolis, 1998.

[569] H. Yoo: Semi-discretization of stochastic partial differential equations on $\mathbb{R}^1$ by a finite-difference method, Math. Comp. (1999), to appear.

[570] H. Yoo: On L_2-theory of discrete stochastic evolution equations and its application to finite difference approximations of stochastic PDEs, Probab. Theory Rel. Fields (1999), submitted.

A. Einstein said: Only a few are capable of free own thinking!
... let us go to C.F. Gauss ...

Chapter 6

Large Deviations and Applications

AMIR DEMBO
Department of Mathematics and Department of Statistics
Stanford University
Stanford, California
and
OFER ZEITOUNI
Department of Electrical Engineering, Technion
Haifa, Israel

6.1 Introduction

This chapter of the handbook is intended to give a review of the theory of large deviations and its applications. Here, "large deviations" are understood as the evaluation, for a family of probability measures parameterized by a real valued variable, of the probabilities of events which decay exponentially in the parameter. Except when stated otherwise, the proof of statements in the text can be found in the book [DeZ98], and we will not repeat this fact throughout the chapter.

We followed here largely the logical structure of [DeZ98]. That is, Section 6.2 describes the definition of the large deviation principle (LDP) and some of its equivalent formulations and basic properties. Section 6.3 provides an overview of large deviation theorems in $\mathbb{R}^d$. Moving to a more abstract setup where the underlying variables take values in a topological space, Section 6.4 presents, after a short discussion on properties of the LDP, a collection of methods aimed at establishing the LDP. These methods include transformations of the LDP (i.e., how the LDP behaves under maps between spaces), relations between the LDP and Laplace's method for the evaluation for exponential integrals, properties of the LDP in topological vector spaces, and the behavior of the LDP under projective limits. Section 6.5 deals with LDPs for the sample paths of certain stochastic processes and the application of such LDPto the problem of the exit of randomly perturbed solutions of differential equations from the domain of attraction of stable equilibria. Section 6.6 deals with LDPs for the empirical measure of (discrete time) random processes: Sanov's theorem for the empirical measure of an i.i.d. sample and its extensions to Markov processes and mixing sequences are discussed. The section ends with two particular applications of the LDP: one to hypothesis testing problems in statistics, the other to the Gibbs conditioning principle in statistical mechanics.

We have not made an attempt here to give proper credit to all theorems and statements in the text. The historical notes in the book [DeZ98] should be consulted for the history of the subject and of particular theorems. In what follows, we describe only the major steps in the development of the theory up to the mid 80s, referring the reader again to [DeZ98] for details and extensive references. We conclude by mentioning topics which are not covered in this chapter and references to them.

While much of the credit for the modern theory of large deviations and its various applications goes to Donsker and Varadhan (in the West) and to Freidlin and Wentzell (in the East), the topic is much older and may be traced back to the early 1900s and in particular to the work of statisticians like Cramér, Chernoff, and Khinchine, culminating in the work of Bahadur [Bah71] on the power of statistical tests. In a slightly different direction, Sanov [San57] obtained his theorem in the mid-fifties, for real valued random variables.

The abstract framework for the LDP was proposed by Varadhan [Var66]. At that time, the only "modern" large deviation principles available were the theorems of Schilder and Sanov. At the same time sample path results began to be available in Russia through the work of Borovkov [Bor67], and a few years later, through the seminal work of Freidlin and Wentzell [VF70], [VF72], who introduced also an abstract foundation to the LDP.

The next crucial step forward was achieved through a series of papers of Donsker and Varadhan [DV75a], [DV75b], [DV76], [DV83], starting in the mid-seventies, where they developed systematically the large deviations theory for empirical measures in the i.i.d. and Markov cases, and later showed its relevance to problems arising in statistical mechanics. Related ideas were also introduced by Gärtner.

Essential tools in the theory of large deviations also emerged around that time: subadditivity, which was used by Ruelle [Rue67] and Lanford [Lan73] in the context of thermodynamics, was introduced into large deviations theory proper by Bahadur and Zabell [BaZ79]. Contraction principles which were introduced by Varadhan in his seminal paper [Var66], were greatly expanded by Azencott [Aze80], who systematized the use of exponential approximations. The use of convexity considerations was greatly advanced through the work of Gärtner [Gär77] and later refined by Ellis [Ell84]. The systematic use of projective limits was introduced by Dawson and Gärtner in [DaG87].

Since the mid-eighties, there has been an exponential explosion in the quantity of work devoted to large deviations theory and its applications. We refer the reader to [DeZ98] for an overview of this work. Other treatments in book form, of Large Deviations Theory may be found in [Var84], [FW84] (with emphasis on sample path results and the problem of exit from a domain), [Ell85] (with special emphasis on statistical mechanics), [DeuS89], [Buc90] (with special emphasis on engineering applications), [SW95] (with special emphasis on queuing problems), and [DuE97].

We conclude this introduction by noting topics which were completely left out from this chapter: we barely discuss refinements of large deviation principles (in the form of precise asymptotics occurring mainly in statistics and statistical mechanics), or subexponential probabilities of large deviations (see [Nag79] for an account of the latter). In our discussion of concentration inequalities via martingale differences, we do not discuss the beautiful recent work of Talagrand [Tal95], [Tal96]. We have not discussed the intimate relation between large deviations and equilibrium statistical mechanics, referring instead the reader to [Ell85]. Similarly, we have omitted a discussion of the relation between large deviations estimates for Markov chains and analytic properties of their generators, referring the reader to [DeuS89], [Sal97] and [Mar98]. We have completely omitted the discussion of hydrodynamic limits, an updated account of which can be found in [KL99]. When dealing with empirical measures, we do not consider at all continuous time processes, referring instead the reader to [DeuS89] for the required modifications. We do not cover at all the important topics of large deviations in Banach spaces (see [DeuS89] for an account), large deviations for abstract gaussian processes (see [BeLd93] for a particularly transparent derivation of sample path results and [DV87] for the empirical process results), the relations between large deviations and information theory and engineering (see [CsK81]), or the application of large deviations and refinements to the study of heat-kernels (see [As81] for early results and [BeLa91], [KuS91], [KuS94] for more recent work). Our treatment of large deviations for the empirical measure of Markov chains does not cover the beautiful approach via regenerations, developed by Ney and Nummelin [INN85], [NN87a], [NN87b]. Finally, we have not discussed large deviations in the context of dynamical systems, and refer instead the reader to [Kif90], [Kif92].

6.2 The Large Deviation Principle

The *large deviation principle* (LDP) characterizes the limiting behavior, as $\epsilon \to 0$, of a family of probability measures $\{\mu_\epsilon\}$ on $(\mathcal{X}, \mathcal{B})$ in terms of a *rate function*. This characterization is via asymptotic upper and lower exponential bounds on the values that μ_ϵ assigns to measurable subsets of $\mathcal{X}$. Throughout, $\mathcal{X}$ is a topological space so that open and closed subsets of $\mathcal{X}$ are well-defined, and the simplest situation is when elements of $\mathcal{B}_\mathcal{X}$, the Borel σ-field on $\mathcal{X}$, are of interest. To reduce possible measurability questions, all probability spaces are assumed to have been completed, and, with some abuse of notations, $\mathcal{B}_\mathcal{X}$ always denotes the thus completed Borel σ-field.

Definitions *A rate function I is a lower semicontinuous mapping $I : \mathcal{X} \to [0, \infty]$ (such that for all $\alpha \in [0, \infty)$, the level set $\Psi_I(\alpha) \triangleq \{x : I(x) \leq \alpha\}$ is a closed subset of $\mathcal{X}$). A good rate function is a rate function for which all the level sets $\Psi_I(\alpha)$ are compact subsets of $\mathcal{X}$. The effective domain of I, denoted $\mathcal{D}_I$, is the set of points in $\mathcal{X}$ of finite rate, namely, $\mathcal{D}_I \triangleq \{x : I(x) < \infty\}$. When no confusion occurs, we refer to $\mathcal{D}_I$ as the domain of I.*

Note that if $\mathcal{X}$ is a metric space, the lower semicontinuity property may be checked on sequences, i.e., I is lower semicontinuous if and only if $\liminf_{x_n \to x} I(x_n) \geq I(x)$ for all $x \in \mathcal{X}$. A consequence of a rate function being good is that its infimum is achieved over closed sets.

The following standard notation is used throughout. For any set Γ, $\overline{\Gamma}$ denotes the closure of Γ, Γ^o the interior of Γ, and Γ^c the complement of Γ. The infimum of a function over an empty set is interpreted as ∞.

Definition $\{\mu_\epsilon\}$ *satisfies the large deviation principle with a rate function I if, for all $\Gamma \in \mathcal{B}$,*

$$-\inf_{x \in \Gamma^o} I(x) \leq \liminf_{\epsilon \to 0} \epsilon \log \mu_\epsilon(\Gamma) \leq \limsup_{\epsilon \to 0} \epsilon \log \mu_\epsilon(\Gamma) \leq -\inf_{x \in \overline{\Gamma}} I(x) . \tag{6.2.1}$$

The right- and left-hand sides of (6.2.1) are referred to as the upper and lower bounds, respectively.

Remark: Note that in (6.2.1), $\mathcal{B}$ need not necessarily be the Borel σ-field. Thus, there can be a separation between the sets on which probability may be assigned and the values of the bounds. In particular, (6.2.1) makes sense even if some open sets are not measurable. Except for this section, we always assume that $\mathcal{B}_\mathcal{X} \subseteq \mathcal{B}$ unless explicitly stated otherwise. The sentence "μ_ϵ satisfies the LDP" is used as shorthand for "$\{\mu_\epsilon\}$ satisfies the large deviation principle with rate function I." It is obvious that if μ_ϵ satisfies the LDP and $\Gamma \in \mathcal{B}$ is such that

$$\inf_{x \in \Gamma^o} I(x) = \inf_{x \in \overline{\Gamma}} I(x) \triangleq I_\Gamma , \tag{6.2.2}$$

then

$$\lim_{\epsilon \to 0} \epsilon \log \mu_\epsilon(\Gamma) = -I_\Gamma \,. \tag{6.2.3}$$

A set Γ that satisfies (6.2.2) is called an *I continuity set.* In general, the LDP implies a precise limit in (6.2.3) only for I continuity sets. Finer results may well be derived on a case-by-case basis for specific families of measures $\{\mu_\epsilon\}$ and particular sets. While such results do not fall within our definition of the LDP, a few illustrative examples are included. (See Sections 6.3.1 and 6.3.5.)

Some remarks on the definition now seem in order. Note first that in any situation involving nonatomic measures, $\mu_\epsilon(\{x\}) = 0$ for every x in $\mathcal{X}$. Thus, if the lower bound of (6.2.1) was to hold with the infimum over Γ instead of Γ^o, it would have to be concluded that $I(x) \equiv \infty$, contradicting the upper bound of (6.2.1) because $\mu_\epsilon(\mathcal{X}) = 1$ for all ϵ. Therefore, some topological restrictions are necessary, and the definition of the LDP codifies a particularly convenient way of stating asymptotic results that, on the one hand, are accurate enough to be useful and, on the other hand, are loose enough to be correct.

Since $\mu_\epsilon(\mathcal{X}) = 1$ for all ϵ, it is necessary that $\inf_{x \in \mathcal{X}} I(x) = 0$ for the upper bound to hold. When I is a good rate function, this means that there exists at least one point x for which $I(x) = 0$. Next, the upper bound trivially holds whenever $\inf_{x \in \overline{\Gamma}} I(x) = 0$, while the lower bound trivially holds whenever $\inf_{x \in \Gamma^o} I(x) = \infty$. This leads to an alternative formulation of the LDP which is actually more useful when proving it. Suppose I is a rate function and $\Psi_I(\alpha)$ its level set. Then (6.2.1) is equivalent to the following bounds.
(a) (Upper bound) For every $\alpha < \infty$ and every measurable set Γ with $\overline{\Gamma} \subset \Psi_I(\alpha)^c$,

$$\limsup_{\epsilon \to 0} \epsilon \log \mu_\epsilon(\Gamma) \leq -\alpha \,. \tag{6.2.4}$$

(b) (Lower bound) For any $x \in \mathcal{D}_I$ and any measurable Γ with $x \in \Gamma^o$,

$$\liminf_{\epsilon \to 0} \epsilon \log \mu_\epsilon(\Gamma) \geq -I(x) \,. \tag{6.2.5}$$

Inequality (6.2.5) emphasizes the local nature of the lower bound.

When $\mathcal{B}_\mathcal{X} \subseteq \mathcal{B}$, the LDP is also equivalent to the following bounds:
(a) (Upper bound) For any closed set $F \subseteq \mathcal{X}$,

$$\limsup_{\epsilon \to 0} \epsilon \log \mu_\epsilon(F) \leq -\inf_{x \in F} I(x). \tag{6.2.6}$$

(b) (Lower bound) For any open set $G \subseteq \mathcal{X}$,

$$\liminf_{\epsilon \to 0} \epsilon \log \mu_\epsilon(G) \geq -\inf_{x \in G} I(x). \tag{6.2.7}$$

In many cases, a countable family of measures μ_n is considered (for example, when μ_n is the law governing the empirical mean of n random variables). Then the LDP corresponds to the statement

$$\begin{aligned} -\inf_{x \in \Gamma^o} I(x) \;\; &\leq \;\; \liminf_{n \to \infty} a_n \log \mu_n(\Gamma) \leq \limsup_{n \to \infty} a_n \log \mu_n(\Gamma) \\ &\leq -\inf_{x \in \overline{\Gamma}} I(x) \end{aligned} \tag{6.2.8}$$

for some sequence $a_n \to 0$. Note that here a_n replaces ϵ of (6.2.1) and similarly, the statements (6.2.4)–(6.2.7) are appropriately modified. For consistency, the convention $a_n = 1/n$ is used throughout and μ_n is renamed accordingly to mean $\mu_{a^{-1}(1/n)}$, where a^{-1} denotes the inverse of $n \mapsto a_n$.

Often, a natural approach to proving the large deviations upper bound is to prove it first for compact sets. This motivates the following, where in the sequel all topological spaces are assumed to be Hausdorff.

Definition *Suppose that all the compact subsets of $\mathcal{X}$ belong to $\mathcal{B}$. A family of probability measures $\{\mu_\epsilon\}$ is said to satisfy the weak LDP with the rate function I if the upper bound (6.2.4) holds for every $\alpha < \infty$ and all compact subsets of $\Psi_I(\alpha)^c$, and the lower bound (6.2.5) holds for all measurable sets.*

It is important to realize that there are families of probability measures that satisfy the weak LDP with a good rate function but do not satisfy the full LDP. For example, let μ_ϵ be the probability measures degenerate at $1/\epsilon$. This family satisfies the weak LDP in $\mathbb{R}$ with the good rate function $I(x) = \infty$. On the other hand, μ_ϵ can not satisfy the LDP with this or any other rate function.

In view of the preceding example, strengthening the weak LDP to a full LDP requires a way of showing that most of the probability mass (at least on an exponential scale) is concentrated on compact sets. The tool for doing that is the following:

Definition *Suppose that all the compact subsets of $\mathcal{X}$ belong to $\mathcal{B}$. A family of probability measures $\{\mu_\epsilon\}$ on $\mathcal{X}$ is exponentially tight if for every $\alpha < \infty$, there exists a compact set $K_\alpha \subset \mathcal{X}$ such that*

$$\limsup_{\epsilon \to 0} \epsilon \log \mu_\epsilon(K_\alpha^c) < -\alpha. \tag{6.2.9}$$

Remarks:

(a) Beware of the logical mistake that consists of identifying exponential tightness and the goodness of the rate function: The measures $\{\mu_\epsilon\}$ need not be exponentially tight in order to satisfy a LDP with a good rate function. In some situations, however, and in particular whenever $\mathcal{X}$ is locally compact or, alternatively, Polish, exponential tightness is implied by the goodness of the rate function. For details, *c.f.* Lemma 6.4.5.

(b) Whenever it is stated that μ_ϵ satisfies the weak LDP or μ_ϵ is exponentially tight, it will be implicitly assumed that all the compact subsets of $\mathcal{X}$ belong to $\mathcal{B}$.

(c) Obviously, for $\{\mu_\epsilon\}$ to be exponentially tight, it suffices to have pre-compact K_α for which (6.2.9) holds.

In the following lemma, exponential tightness is applied to strengthen a weak LDP.

Lemma 6.2.1 *Let $\{\mu_\epsilon\}$ be an exponentially tight family.*

(a) If the upper bound (6.2.4) holds for some $\alpha < \infty$ and all compact subsets of $\Psi_I(\alpha)^c$, then it also holds for all measurable sets Γ with $\overline{\Gamma} \subset \Psi_I(\alpha)^c$. In particular, if $\mathcal{B}_\mathcal{X} \subseteq \mathcal{B}$ and the upper bound (6.2.6) holds for all compact sets, then it also holds for all closed sets.

(b) If the lower bound (6.2.5) (the lower bound (6.2.7) in case $\mathcal{B}_\mathcal{X} \subseteq \mathcal{B}$) holds for all measurable sets (all open sets), then $I(\cdot)$ is a good rate function.

Thus, when an exponentially tight family of probability measures satisfies the weak LDP with a rate function $I(\cdot)$, then I is a good rate function and the LDP holds.

6.3 Large Deviation Principles for Finite Dimensional Spaces

This section is devoted to the study of the LDP in finite dimensional spaces. We start with the empirical measure of i.i.d. random variables taking values in a finite set, moving on to the empirical mean of i.i.d. $\mathbb{R}^d$-valued variables, then relaxing the independence assumption. We conclude with a brief introduction to concentration inequalities and various refinements of the LDP . The material in this section is taken from Sections 2.1.1, 2.2, 2.3, 2.4.1 and 3.7 of [DeZ98], and the reader is referred there for more details, historical notes, and proofs.

6.3.1 The Method of Types

Throughout Section 6.3.1, all random variables assume values in a finite set $\Sigma = \{a_1, a_2, \dots, a_N\}$; Σ, which is also called the *underlying alphabet*, satisfies $|\Sigma| = N$, where for any set A, $|A|$ denotes its cardinality, or size. Combinatorial methods are then applicable for deriving LDPs for the empirical measures of Σ-valued processes and for the corresponding empirical means. While the scope of these methods is limited to finite alphabets, they illustrate the results one can hope to obtain for more abstract alphabets. Unlike other approaches, this method for deriving the LDP is based on point estimates and thus yields more information than the LDP statement. Throughout, $M_1(\Sigma)$ denotes the space of all probability measures on the alphabet Σ. Here $M_1(\Sigma)$ is identified with the standard probability simplex in $\mathbb{R}^{|\Sigma|}$, i.e., the set of all $|\Sigma|$-dimensional real vectors with nonnegative components that sum to 1. Open sets in $M_1(\Sigma)$ are obviously induced by the open sets in $\mathbb{R}^{|\Sigma|}$.

Let $Y_1, Y_2, \dots, Y_n$ be a sequence of random variables that are independent and identically distributed according to the law $\mu \in M_1(\Sigma)$. Let Σ_μ denote the support of the law μ, i.e., $\Sigma_\mu = \{a_i : \mu(a_i) > 0\}$. In general, Σ_μ could be a strict subset of Σ; When considering a single measure μ, it may be assumed, without loss of generality, that $\Sigma_\mu = \Sigma$ by ignoring those symbols that appear with zero probability.

Definition 6.3.1 *The type $L_n^{\mathbf{y}}$ of a finite sequence $\mathbf{y} = (y_1, \dots, y_n) \in \Sigma^n$ is the empirical measure (law) induced by this sequence. Explicitly, $L_n^{\mathbf{y}} = (L_n^{\mathbf{y}}(a_1), \dots, L_n^{\mathbf{y}}(a_{|\Sigma|}))$ is the element of $M_1(\Sigma)$ where*

$$L_n^{\mathbf{y}}(a_i) = \frac{1}{n} \sum_{j=1}^{n} 1_{a_i}(y_j) \,, \quad i = 1, \dots, |\Sigma| \,,$$

i.e., $L_n^{\mathbf{y}}(a_i)$ is the fraction of occurrences of a_i in the sequence $y_1, \dots, y_n$.

Let $\mathcal{L}_n$ denote the set of all possible types of sequences of length n. Thus, $\mathcal{L}_n \triangleq \{\nu : \nu = L_n^{\mathbf{y}} \text{ for some } \mathbf{y}\} \subset \mathbb{R}^{|\Sigma|}$, and the empirical measure $L_n^{\mathbf{Y}}$ associated with the sequence $\mathbf{Y} \triangleq (Y_1, \dots, Y_n)$ is a *random* element of the set $\mathcal{L}_n$. These concepts are useful for finite alphabets because of the following volume and approximation distance estimates.

Lemma 6.3.2 *(a) $|\mathcal{L}_n| \leq (n+1)^{|\Sigma|}$.*
(b) For any probability vector $\nu \in M_1(\Sigma)$,

$$d_V(\nu, \mathcal{L}_n) \triangleq \inf_{\nu' \in \mathcal{L}_n} d_V(\nu, \nu') \leq \frac{|\Sigma|}{2n} \,, \tag{6.3.10}$$

where $d_V(\nu, \nu') \triangleq \sup_{A \subset \Sigma} [\nu(A) - \nu'(A)]$ is the variational distance between the measures ν and ν'.

Proof. Note that every component of the vector $L_n^{\mathbf{y}}$ belongs to the set $\{\frac{0}{n}, \frac{1}{n}, \dots, \frac{n}{n}\}$, whose cardinality is $(n+1)$. Part (a) of the lemma follows, since the vector $L_n^{\mathbf{y}}$ is specified by at most $|\Sigma|$ such quantities.

To prove part (b), observe that $\mathcal{L}_n$ contains all probability vectors composed of $|\Sigma|$ coordinates from the set $\{\frac{0}{n}, \frac{1}{n}, \dots, \frac{n}{n}\}$. Thus, for any $\nu \in M_1(\Sigma)$, there exists a $\nu' \in \mathcal{L}_n$ with $|\nu(a_i) - \nu'(a_i)| \leq 1/n$ for $i = 1, \dots, |\Sigma|$. The bound of (6.3.10) now follows, since for finite Σ,

$$d_V(\nu, \nu') = \frac{1}{2} \sum_{i=1}^{|\Sigma|} |\nu(a_i) - \nu'(a_i)| \,. \qquad \square$$

Definition 6.3.3 *The type class $T_n(\nu)$ of a probability law $\nu \in \mathcal{L}_n$ is the set $T_n(\nu) = \{\mathbf{y} \in \Sigma^n : L_n^{\mathbf{y}} = \nu\}$.*

Note that a type class consists of all permutations of a given vector in this set. In the definitions to follow, $0 \log 0 \triangleq 0$ and $0 \log(0/0) \triangleq 0$.

Definition 6.3.4 *(a) The entropy of a probability vector ν is*

$$H(\nu) \triangleq -\sum_{i=1}^{|\Sigma|} \nu(a_i) \log \nu(a_i)\,.$$

(b) The relative entropy of a probability vector ν with respect to another probability vector μ is

$$H(\nu|\mu) \triangleq \sum_{i=1}^{|\Sigma|} \nu(a_i) \log \frac{\nu(a_i)}{\mu(a_i)}\,.$$

Remark: By applying Jensen's inequality to the convex function $x \log x$, it follows that the function $H(\cdot|\mu)$ is nonnegative. Note that $H(\cdot|\mu)$ is finite and continuous on the compact set $\{\nu \in M_1(\Sigma) : \Sigma_\nu \subseteq \Sigma_\mu\}$, because $x \log x$ is continuous for $0 \le x \le 1$. Moreover, $H(\cdot|\mu) = \infty$ outside this set, and hence $H(\cdot|\mu)$ is a good rate function.

The probabilities of the events $\{L_n^{\mathbf{Y}} = \nu\}$, $\nu \in \mathcal{L}_n$, are estimated in the following two lemmas. First, it is shown that outcomes belonging to the same type class are equally likely, and then the exponential growth rate of each type class is estimated.

Let Prob_μ denote the probability law $\mu^{\mathbb{Z}_+}$ associated with an infinite sequence of i.i.d. random variables $\{Y_j\}$ distributed following $\mu \in M_1(\Sigma)$.

Lemma 6.3.5 *If $\mathbf{y} \in T_n(\nu)$ for $\nu \in \mathcal{L}_n$, then*

$$Prob_\mu((Y_1, \dots, Y_n) = \mathbf{y}) = e^{-n[H(\nu)+H(\nu|\mu)]}\,.$$

Proof. The random empirical measure $L_n^{\mathbf{Y}}$ concentrates on types $\nu \in \mathcal{L}_n$ for which $\Sigma_\nu \subseteq \Sigma_\mu$ i.e., $H(\nu|\mu) < \infty$. Therefore, assume without loss of generality that $L_n^{\mathbf{y}} = \nu$ and $\Sigma_\nu \subseteq \Sigma_\mu$. Then

$$\text{Prob}_\mu((Y_1, \dots, Y_n) = \mathbf{y}) = \prod_{i=1}^{|\Sigma|} \mu(a_i)^{n\nu(a_i)} = e^{-n[H(\nu)+H(\nu|\mu)]}\,,$$

where the last equality follows by the identity

$$H(\nu) + H(\nu|\mu) = -\sum_{i=1}^{|\Sigma|} \nu(a_i) \log \mu(a_i)\,. \qquad \square$$

In particular, since $H(\mu|\mu) = 0$, it follows that for all $\mu \in \mathcal{L}_n$ and $\mathbf{y} \in T_n(\mu)$,

$$\text{Prob}_\mu((Y_1, \dots, Y_n) = \mathbf{y}) = e^{-nH(\mu)}\,. \tag{6.3.11}$$

Elementary combinatorics also yield that for every $\nu \in \mathcal{L}_n$,

$$(n+1)^{-|\Sigma|} e^{nH(\nu)} \le |T_n(\nu)| \le e^{nH(\nu)}\,. \tag{6.3.12}$$

By Lemma 6.3.5,

$$\begin{aligned}\text{Prob}_\mu(L_n^{\mathbf{Y}} = \nu) &= |T_n(\nu)|\, \text{Prob}_\mu((Y_1, \dots, Y_n) = \mathbf{y}\,,\ L_n^{\mathbf{y}} = \nu)\\ &= |T_n(\nu)|\, e^{-n[H(\nu)+H(\nu|\mu)]}\,.\end{aligned}$$

Hence, by (6.3.12), we have that

Lemma 6.3.6 (Large deviations probabilities) *For any $\nu \in \mathcal{L}_n$,*

$$(n+1)^{-|\Sigma|}\, e^{-nH(\nu|\mu)} \le Prob_\mu(L_n^{\mathbf{Y}} = \nu) \le e^{-nH(\nu|\mu)}\,.$$

Combining Lemmas 6.3.2 and 6.3.6, one obtains Sanov's theorem in the finite alphabet context. See Section 6.6.2 for the general case.

Theorem 6.3.7 (Sanov) *For every set Γ of probability vectors in $M_1(\Sigma)$,*

$$\begin{aligned} -\inf_{\nu\in\Gamma^o} H(\nu|\mu) &\le \liminf_{n\to\infty} \frac{1}{n}\log Prob_\mu(L_n^{\mathbf{Y}} \in \Gamma) \\ &\le \limsup_{n\to\infty} \frac{1}{n}\log Prob_\mu(L_n^{\mathbf{Y}} \in \Gamma) \le -\inf_{\nu\in\Gamma} H(\nu|\mu)\,, \end{aligned} \tag{6.3.13}$$

where Γ^o is the interior of Γ considered as a subset of $\mathbb{R}^{|\Sigma|}$.

6.3.2 Cramér's Theorem in $\mathbb{R}^d$

Consider the empirical means $\hat{S}_n \stackrel{\triangle}{=} \frac{1}{n}\sum_{j=1}^n X_j$, for i.i.d., d-dimensional random vectors $X_1,\ldots,X_n,\ldots$, with X_1 distributed according to the probability law $\mu \in M_1(\mathbb{R}^d)$. The *logarithmic moment generating function* associated with the law μ is defined as

$$\Lambda(\lambda) \stackrel{\triangle}{=} \log M(\lambda) \stackrel{\triangle}{=} \log E[e^{\langle\lambda, X_1\rangle}]\,, \tag{6.3.14}$$

where $\langle\lambda, x\rangle \stackrel{\triangle}{=} \sum_{j=1}^d \lambda^j x^j$ is the usual scalar product in $\mathbb{R}^d$, and x^j the jth coordinate of x. Another common name for $\Lambda(\cdot)$ is the *cumulant generating function.* In what follows, $|x| \stackrel{\triangle}{=} \sqrt{\langle x, x\rangle}$, is the usual Euclidean norm. Note that $\Lambda(0) = 0$, and while $\Lambda(\lambda) > -\infty$ for all λ, it is possible to have $\Lambda(\lambda) = \infty$. Let μ_n denote the law of $\hat{S}_n$ and $\overline{x} \stackrel{\triangle}{=} E[X_1]$. When $\overline{x}$ exists and is finite, and $E[|X_1 - \overline{x}|^2] < \infty$, then $\hat{S}_n \xrightarrow[n\to\infty]{\text{Prob}} \overline{x}$, by an application of Markov's inequality. Hence, in this situation, $\mu_n(F) \underset{n\to\infty}{\longrightarrow} 0$ for any closed set F such that $\overline{x} \notin F$. Cramér's theorem characterizes the logarithmic rate of this convergence by the following (rate) function.

Definition 6.3.8 *The Fenchel–Legendre transform of $\Lambda(\lambda)$ is*

$$\Lambda^*(x) \stackrel{\triangle}{=} \sup_{\lambda\in\mathbb{R}^d} \{\langle\lambda, x\rangle - \Lambda(\lambda)\}\,.$$

Theorem 6.3.9 (Cramér) *The sequence of measures $\{\mu_n\}$ satisfies the weak LDP on $\mathbb{R}^d$ with the convex rate function $\Lambda^*(\cdot)$; Moreover, for every open convex $A \subset \mathbb{R}^d$,*

$$\lim_{n\to\infty} \frac{1}{n} \log \mu_n(A) = -\inf_{x\in A} \Lambda^*(x)\,.$$

If $d = 1$ the full LDP holds, and for any $d < \infty$ the assumption that $\Lambda(\lambda) < \infty$ for all $|\lambda|$ small enough implies the full LDP and that $\Lambda^*(\cdot)$ is a good rate function.

Remarks:

(a) The definition of the Fenchel–Legendre transform for (topological) vector spaces and some of its properties are presented in Section 6.4.4. It is also shown there that the Fenchel–Legendre transform is a natural candidate for the rate function, since the LDP upper bound holds for compact sets in a general setup.

(b) When $d = 1$, for all n, and any closed set $F \subset \mathbb{R}$, one has the nonasymptotic upper bound

$$\mu_n(F) \leq 2e^{-n \inf_{x \in F} \Lambda^*(x)} .$$

We close this section by indicating the basic steps in the proof of Cramér's theorem. The upper bound is deduced from the case of a half-space, that is, an interval $[x, \infty)$ for $d = 1$. The latter is a rewrite of Chebycheff's inequality: for $\lambda \geq 0$, and $x \geq \overline{x}$,

$$\mu_n\Big([x,\infty)\Big) \leq E\Big(e^{n\lambda(\hat{S}_n - x)}\Big) = e^{-n[\lambda x - \Lambda(\lambda)]} ,$$

where optimizing $\lambda x - \Lambda(\lambda)$ over $\lambda > 0$ yields for $x \geq \overline{x}$ the value of $\Lambda^*(x)$, hence the stated upper bound.

The lower bound requires a more sophisticated idea, based on an "exponential change of measure." We present the sketch for the case $d = 1$:

Define the measure

$$\frac{d\tilde{\mu}}{d\mu}(x) = e^{\eta x - \Lambda(\eta)}$$

where η is such that $E_{\tilde{\mu}}(X_1) = x$ (we assume that such an η exists, otherwise one needs to approximate). Then, $\eta x - \Lambda(\eta) = \Lambda^*(x)$, and by the law of large numbers, $\hat{S}_n \to x$ in probability under the law $\tilde{\mu}^n$. Now,

$$\mu_n\Big([x-\delta, x+\delta]\Big) \geq \tilde{\mu}_n\Big([x-\delta, x+\delta]\Big) e^{n[\Lambda(\eta) - \eta x - \delta|\eta|]}$$

and the lower bound follows by considering first $n \to \infty$ and then $\delta \to 0$.

6.3.3 The Gärtner–Ellis Theorem

Consider a sequence of random vectors $Z_n \in \mathbb{R}^d$, where Z_n possesses the law μ_n and logarithmic moment generating function

$$\Lambda_n(\lambda) \triangleq \log E\left[e^{\langle \lambda, Z_n \rangle}\right] . \tag{6.3.15}$$

The existence of a limit of properly scaled logarithmic moment generating functions indicates that μ_n may satisfy the LDP. Specifically, the following assumption is imposed throughout Section 6.3.3.

Assumption 6.3.16 *For each $\lambda \in \mathbb{R}^d$, the logarithmic moment generating function, defined as the limit*

$$> \Lambda(\lambda) \triangleq \lim_{n \to \infty} \frac{1}{n} \Lambda_n(n\lambda)$$

exists as an extended real number. Further, the origin belongs to the interior of $\mathcal{D}_\Lambda \triangleq \{\lambda \in \mathbb{R}^d : \Lambda(\lambda) < \infty\}$.

In particular, if μ_n is the law governing the empirical mean $\hat{S}_n$ of i.i.d. random vectors $X_i \in \mathbb{R}^d$, then for every $n \in \mathbb{Z}_+$,

$$\frac{1}{n} \Lambda_n(n\lambda) = \Lambda(\lambda) \triangleq \log E[e^{\langle \lambda, X_1 \rangle}] ,$$

and Assumption 6.3.16 holds whenever $0 \in \mathcal{D}_\Lambda^o$.

Let $\Lambda^*(\cdot)$ be the Fenchel–Legendre transform of $\Lambda(\cdot)$, with $\mathcal{D}_{\Lambda^*} = \{x \in \mathbb{R}^d : \Lambda^*(x) < \infty\}$. Motivated by Theorem 6.3.9, it is our goal to state conditions under which the sequence μ_n satisfies the LDP with the rate function Λ^*.

Definition 6.3.10 *$y \in \mathbb{R}^d$ is an exposed point of Λ^* if for some $\lambda \in \mathbb{R}^d$ and all $x \neq y$,*

$$\langle \lambda, y \rangle - \Lambda^*(y) > \langle \lambda, x \rangle - \Lambda^*(x) . \tag{6.3.17}$$

λ in (6.3.17) is called an exposing hyperplane.

Definition 6.3.11 *A convex function $\Lambda : \mathbb{R}^d \to (-\infty, \infty]$ is essentially smooth if:*
(a) $\mathcal{D}_\Lambda^o$ is non-empty.
(b) $\Lambda(\cdot)$ is differentiable throughout $\mathcal{D}_\Lambda^o$.
(c) $\Lambda(\cdot)$ is steep, namely, $\lim_{n\to\infty} |\nabla\Lambda(\lambda_n)| = \infty$ whenever $\{\lambda_n\}$ is a sequence in $\mathcal{D}_\Lambda^o$ converging to a boundary point of $\mathcal{D}_\Lambda^o$.

The following theorem is the main result of Section 6.3.3.

Theorem 6.3.12 (Gärtner–Ellis) *Let Assumption 6.3.16 hold.*
(a) For any closed set F,

$$\limsup_{n\to\infty} \frac{1}{n} \log \mu_n(F) \leq - \inf_{x\in F} \Lambda^*(x) . \tag{6.3.18}$$

(b) For any open set G,

$$\liminf_{n\to\infty} \frac{1}{n} \log \mu_n(G) \geq - \inf_{x\in G\cap\mathcal{F}} \Lambda^*(x) , \tag{6.3.19}$$

where $\mathcal{F}$ is the set of exposed points of Λ^* whose exposing hyperplane belongs to $\mathcal{D}_\Lambda^o$.
(c) If Λ is an essentially smooth, lower semicontinuous function, then the LDP holds with the good rate function $\Lambda^*(\cdot)$.

Remarks:
(a) Theorem 6.3.12 is valid, as in the statement (6.2.8) of the LDP, when $1/n$ is replaced by a sequence of constants $a_n \to 0$, or even when a continuous parameter family $\{\mu_\epsilon\}$ is considered, with Assumption 6.3.16 properly modified.
(b) Although the Gärtner–Ellis theorem is quite general in its scope, it does not cover all $\mathbb{R}^d$ cases in which an LDP exists. As an illustrative example, consider $Z_n \sim$ Exponential(n). Assumption 6.3.16 then holds with $\Lambda(\lambda) = 0$ for $\lambda < 1$ and $\Lambda(\lambda) = \infty$ otherwise. Moreover, the law of Z_n possesses the density $ne^{-nz}1_{[0,\infty)}(z)$, and consequently the LDP holds with the good rate function $I(x) = x$ for $x \geq 0$ and $I(x) = \infty$ otherwise. A direct computation reveals that $I(\cdot) = \Lambda^*(\cdot)$. Hence, $\mathcal{F} = \{0\}$ while $\mathcal{D}_{\Lambda^*} = [0, \infty)$, and therefore the Gärtner–Ellis theorem yields a trivial lower bound for sets that do not contain the origin.
(c) Assumption 6.3.16 implies that $\Lambda^*(x) \leq \liminf_{n\to\infty} \Lambda_n^*(x)$ for $\Lambda_n^*(x) = \sup_\lambda\{\langle\lambda, x\rangle - n^{-1}\Lambda_n(n\lambda)\}$. However, pointwise convergence of $\Lambda_n^*(x)$ to $\Lambda^*(x)$ is not guaranteed. For example, when $P(Z_n = n^{-1}) = 1$, we have $\Lambda_n(\lambda) = \lambda/n \to 0 = \Lambda(\lambda)$, while $\Lambda_n^*(0) = \infty$ and $\Lambda^*(0) = 0$. This phenomenon is relevant when trying to go beyond the Gärtner–Ellis theorem, as for example in [Zab92, DeZ95].

Two auxiliary lemmas which play a crucial role in the proof are next stated. Lemma 6.3.13 presents the elementary properties of Λ and Λ^*, which are needed for proving parts (a) and (b) of the theorem, and moreover highlights the relation between exposed points and differentiability properties.

Lemma 6.3.13 *Let Assumption 6.3.16 hold.*
(a) $\Lambda(\lambda)$ is a convex function, $\Lambda(\lambda) > -\infty$ everywhere, and $\Lambda^*(x)$ is a convex good rate function.
(b) Suppose that $y = \nabla\Lambda(\eta)$ for some $\eta \in \mathcal{D}_\Lambda^o$. Then

$$\Lambda^*(y) = \langle \eta, y \rangle - \Lambda(\eta) . \tag{6.3.20}$$

Moreover $y \in \mathcal{F}$, with η being the exposing hyperplane for y.

The essential ingredients for the proof of parts (a) and (b) of the Gärtner–Ellis theorem are those presented in the course of proving Cramér's theorem in $\mathbb{R}^d$; namely, Chebycheff's inequality is applied for deriving the upper bound and an exponential change of measure is used for deriving the lower bound. However, since the law of large numbers is no longer available *a priori*, the large deviations *upper* bound for exponentially tilted measures is used in order to prove the lower bound.

The proof of part (c) of the Gärtner–Ellis theorem depends on rather intricate convex analysis considerations that are summarized in the following lemma. Here, $ri\mathcal{D}_{\Lambda^*}$ is the relative interior of the set $\{x : \Lambda^*(x) < \infty\}$. For the case of $\mathcal{D}_\Lambda = \mathbb{R}^d$, one may instead use a regularization of the random variables Z_n by adding asymptotically negligible Normal random variables.

Lemma 6.3.14 (Rockafellar) *If $\Lambda : \mathbb{R}^d \to (-\infty, \infty]$ is an essentially smooth, lower semi-continuous, convex function, then $ri\,\mathcal{D}_{\Lambda^*} \subseteq \mathcal{F}$.*

6.3.4 Inequalities for Bounded Martingale Differences

The precise large deviations estimates presented so far are all related to rather simple functionals of an independent sequence of random variables, namely to empirical means of such a sequence. We digress here from this theme by, while still keeping the independence structure, allowing for more complicated functionals. In such a situation, it is often hopeless to have a LDP, and one is content with the rough concentration properties of the random variables under investigation.

We next present concentration inequalities for discrete time martingales of bounded differences and show how these may apply for certain functionals of independent variables.

Our starting point is a bound on the moment generating function of a random variable in terms of its maximal possible value and first two moments.

Lemma 6.3.15 (Bennett) *Suppose $X \le b$ is a real-valued random variable with $\overline{x} = E(X)$ and $E[(X-\overline{x})^2] \le \sigma^2$ for some $\sigma > 0$. Then, for any $\lambda \ge 0$,*

$$E(e^{\lambda X}) \le e^{\lambda \overline{x}} \left\{ \frac{(b-\overline{x})^2}{(b-\overline{x})^2+\sigma^2} e^{-\frac{\lambda\sigma^2}{b-\overline{x}}} + \frac{\sigma^2}{(b-\overline{x})^2+\sigma^2} e^{\lambda(b-\overline{x})} \right\}. \tag{6.3.21}$$

Corollary 6.3.16 *Fix $a < b$. Suppose that $a \le X \le b$ is a real-valued random variable with $\overline{x} = E(X)$. Then, for any $\lambda \in \mathbb{R}$,*

$$E(e^{\lambda X}) \le \frac{\overline{x}-a}{b-a} e^{\lambda b} + \frac{b-\overline{x}}{b-a} e^{\lambda a}. \tag{6.3.22}$$

Once uniform bounds on the log moment generating function are available, one may apply Chebycheff's upper bound to deduce concentration inequalities. One uses successive conditioning and the martingale property to control the mean of the random variables involved, with boundedness of the increments allowing to use Lemma 6.3.15 or Corollary 6.3.16.

Corollary 6.3.17 *Suppose $v > 0$ and the real valued random variables $\{Y_n : n = 1, 2, \dots\}$ are such that both $Y_n \le 1$ almost surely, and $E[Y_n|S_{n-1}] = 0$, $E[Y_n^2|S_{n-1}] \le v$ for $S_n \triangleq \sum_{j=1}^n Y_j$, $S_0 = 0$. Then, for any $\lambda \ge 0$,*

$$E\left[e^{\lambda S_n}\right] \le \left(\frac{e^{-v\lambda} + ve^{\lambda}}{1+v}\right)^n. \tag{6.3.23}$$

Moreover, for all $x \geq 0$,

$$Prob(n^{-1}S_n \geq x) \leq \exp\Big(-nH\big(\frac{x+v}{1+v}\Big|\frac{v}{1+v}\big)\Big), \tag{6.3.24}$$

where $H(p|p_0) \triangleq p\log(p/p_0) + (1-p)\log((1-p)/(1-p_0))$ *for* $p \in [0,1]$ *and* $H(p|p_0) = \infty$ *otherwise. Finally, for all* $y \geq 0$,

$$Prob(n^{-1/2}S_n \geq y) \leq e^{-2y^2/(1+v)^2}. \tag{6.3.25}$$

A typical application of Corollary 6.3.17 is as follows, where in order not to be distracted by measurability concerns, assume that Σ is a Polish space, that is, a complete separable metric space. In applications, Σ is often either a finite set or a subset of $\mathbb{R}$.

Corollary 6.3.18 *Let* $Z_n = g_n(X_1, \dots, X_n)$ *for independent* Σ*-valued random variables* $\{X_i\}$ *and real-valued, measurable* $g_n(\cdot)$. *Let* $\{\hat{X}_i\}$ *be an independent copy of* $\{X_i\}$. *Suppose that for* $k = 1, \dots, n$,

$$|g_n(X_1, \dots, X_n) - g_n(X_1, \dots, X_{k-1}, \hat{X}_k, X_{k+1}, \dots, X_n)| \leq 1, \tag{6.3.26}$$

almost surely. Then, for all $x \geq 0$,

$$Prob(n^{-1}(Z_n - EZ_n) \geq x) \leq \exp\Big(-nH\big(\frac{x+1}{2}\Big|\frac{1}{2}\big)\Big), \tag{6.3.27}$$

and for all $y \geq 0$,

$$Prob(n^{-1/2}(Z_n - EZ_n) \geq y) \leq e^{-y^2/2}. \tag{6.3.28}$$

6.3.5 Moderate Deviations and Exact Asymptotics

Cramér's theorem deals with the tails of the empirical mean $\hat{S}_n$ of i.i.d. random variables. On a finer scale, the random variables $\sqrt{n}\hat{S}_n$ possess a limiting Normal distribution by the central limit theorem. In this situation, for $\beta \in (0, 1/2)$, the renormalized empirical mean $n^\beta \hat{S}_n$ satisfies an LDP but always with a quadratic (Normal-like) rate function. This statement is made precise in the following theorem. (Choose $a_n = n^{(2\beta-1)}$ in the theorem to obtain $Z_n = n^\beta \hat{S}_n$.)

Theorem 6.3.19 (Moderate Deviations) *Let* $X_1, \dots, X_n$ *be a sequence of* $\mathbb{R}^d$*-valued i.i.d. random vectors such that* $\Lambda_X(\lambda) \triangleq \log E[e^{\langle\lambda, X_i\rangle}] < \infty$ *in some ball around the origin,* $E(X_i) = 0$, *and* $\mathbf{C}$, *the covariance matrix of* X_1, *is invertible. Fix* $a_n \to 0$ *such that* $na_n \to \infty$ *as* $n \to \infty$, *and let* $Z_n \triangleq \sqrt{a_n/n}\sum_{i=1}^n X_i = \sqrt{na_n}\hat{S}_n$. *Then, for every measurable set* Γ,

$$\begin{aligned} -\frac{1}{2}\inf_{x\in\Gamma^o}\langle x, \mathbf{C}^{-1}x\rangle &\leq \liminf_{n\to\infty} a_n \log P(Z_n \in \Gamma) \\ &\leq \limsup_{n\to\infty} a_n \log P(Z_n \in \Gamma) \\ &\leq -\frac{1}{2}\inf_{x\in\overline{\Gamma}}\langle x, \mathbf{C}^{-1}x\rangle. \end{aligned} \tag{6.3.29}$$

The proof combines an application of the Gärtner-Ellis theorem with a Taylor expansion of logarithmic moment generation functions around $\lambda = 0$.

Remarks:

(a) A similar result may be obtained in the context of Markov additive processes.

(b) Theorem 6.3.19 is representative of the so-called *Moderate Deviation Principle* (MDP), in which for some $\gamma(\cdot)$ and a whole range of $a_n \to 0$, the sequences $\{\gamma(a_n)Y_n\}$ satisfy the LDP with the same rate function. Here, $Y_n = \sqrt{n}\hat{S}_n$ and $\gamma(a) = a^{1/2}$ (as in other situations in which Y_n obeys the central limit theorem).

Another refinement of Cramér's theorem involves a more accurate estimate of the law μ_n of $\hat{S}_n$. Specifically, for a "nice" set A, one seeks an estimate J_n^{-1} of $\mu_n(A)$ in the sense that $\lim_{n\to\infty} J_n \mu_n(A) = 1$. Such an estimate is an improvement over the normalized logarithmic limit implied by the LDP. The following theorem, a representative of the so-called *exact asymptotics*, deals with the estimate J_n for certain half intervals $A = [q, \infty) \subset \mathbb{R}$.

Theorem 6.3.20 (Bahadur and Rao) *Let μ_n denote the law of $\hat{S}_n = \frac{1}{n}\sum_{i=1}^n X_i$, where X_i are i.i.d. real valued random variables with logarithmic moment generating function $\Lambda(\lambda) = \log E[e^{\lambda X_1}]$. Consider the set $A = [q, \infty)$, where $q = \Lambda'(\eta)$ for some positive $\eta \in \mathcal{D}_\Lambda^o$.*
(a) If the law of X_1 is nonlattice, then

$$\lim_{n\to\infty} J_n \mu_n(A) = 1\,, \tag{6.3.30}$$

where $J_n = \eta\sqrt{\Lambda''(\eta)\, 2\pi n}\; e^{n\Lambda^*(q)}$.
(b) Suppose X_1 has a lattice law, i.e., for some x_0, d, the random variable $d^{-1}(X_1 - x_0)$ is (a.s.) an integer number, and d is the largest number with this property. Assume further that $1 > Prob\,(X_1 = q) > 0$. (In particular, this implies that $d^{-1}(q - x_0)$ is an integer and that $\Lambda''(\eta) > 0$.) Then

$$\lim_{n\to\infty} J_n \mu_n(A) = \frac{\eta d}{1 - e^{-\eta d}}\,. \tag{6.3.31}$$

Remarks:
(a) It can be shown that $\Lambda^*(q) = \eta q - \Lambda(\eta)$, $\Lambda(\cdot)$ is C^∞ in some open neighborhood of η, $\eta = \Lambda^{*\prime}(q)$ and $\Lambda^{*\prime\prime}(q) = 1/\Lambda''(\eta)$. Hence, $J_n = \Lambda^{*\prime}(q)\sqrt{2\pi n/\Lambda^{*\prime\prime}(q)}e^{n\Lambda^*(q)}$.
(b) Theorem 6.3.20 holds even when A is a small interval of size of order $\mathbf{O}(\log n/n)$.

The proof of Theorem 6.3.20 is based on an exponential translation of a local CLT. This approach is applicable for the dependent case of Section 6.3.3 and to a certain extent applies also in $\mathbb{R}^d$, $d > 1$.

6.4 General Properties

We focus our attention now on the abstract statement of the LDP as presented in Section 6.2 and give conditions for the existence of such a principle and various approaches for the identification of the resulting rate function. Section 6.4.1 explores the relations between the topological structure of the space, the existence of certain limits, and the existence and uniqueness of the LDP. Section 6.4.2 describes how to move around the LDP from one space to another. Thus, under appropriate conditions, the LDP can be proved in a simple situation and then effortlessly transferred to a more complex one. Section 6.4.3 is about the relation between the LDP and the computation of exponential integrals. Although in some applications the computation of the exponential integrals is a goal in itself, it is more often the case that such computations are an intermediate step in deriving the LDP. Section 6.4.4 exploits convexity, in the case of topological vector spaces, either to derive the LDP or to identify its rate function. Section 6.4.5 shows that the LDP is preserved under projective limits. This approach is quite general and may lead from finite dimensional computations to the LDP in abstract spaces.

The material for this section is taken from Chapter 4 of [DeZ98] to which the reader is referred for additional details, bibliography, and proofs.

6.4.1 Existence of an LDP and Related Properties

If a set $\mathcal{X}$ is given the coarse topology $\{\emptyset, \mathcal{X}\}$, the only information implied by the LDP is that $\inf_{x\in\mathcal{X}} I(x) = 0$, and many rate functions satisfy this requirement. To avoid such trivialities, we must put some constraint on the topology of the set $\mathcal{X}$. Recall that a topological space is Hausdorff if, for every pair of distinct points x and y, there exist disjoint neighborhoods of x and y. The natural condition that prevails throughout this chapter is that, in addition to being Hausdorff, $\mathcal{X}$ is a *regular space* as defined next.

Definition 6.4.1 *A Hausdorff topological space $\mathcal{X}$ is regular if, for any closed set $F \subset \mathcal{X}$ and any point $x \notin F$, there exist disjoint open subsets G_1 and G_2 such that $F \subset G_1$ and $x \in G_2$.*

In the rest of the chapter, the term *regular* will mean Hausdorff and regular. We recall that every metric space is regular. Moreover, if a real topological vector space is Hausdorff, then it is regular. All examples of an LDP considered in this chapter are either for metric spaces, or for Hausdorff real topological vector spaces.

We collect below some simple consequences of the definition of the LDP and our topological assumptions. The first desirable consequence of the assumption that $\mathcal{X}$ is a regular topological space is the uniqueness of the rate function associated with the LDP.

Lemma 6.4.2 *A family of probability measures $\{\mu_\epsilon\}$ on a regular topological space can have at most one rate function associated with its LDP.*

Remarks:
(a) If $\mathcal{X}$ is a locally compact space, or a Polish space, the rate function is unique as soon as a weak LDP holds.
(b) The uniqueness of the rate function does not depend on the Hausdorff part of the definition of regular spaces. However, the rate function assigns the same value to any two points of $\mathcal{X}$ that are not separated. Thus, in terms of the LDP, such points are indistinguishable.

As shown in the next lemma, the LDP is preserved under suitable inclusions. Hence, in applications, one may first prove an LDP in a space that possesses additional structure (for example, a topological vector space), and then use this lemma to deduce the LDP in the subspace of interest. It is then often convenient that Lemma 6.4.3 holds even when $\mathcal{B}_\mathcal{X} \subsetneq \mathcal{B}$.

Lemma 6.4.3 *Let $\mathcal{E}$ be a measurable subset of $\mathcal{X}$ such that $\mu_\epsilon(\mathcal{E}) = 1$ for all $\epsilon > 0$. Suppose that $\mathcal{E}$ is equipped with the topology induced by $\mathcal{X}$.*
(a) If $\mathcal{E}$ is a closed subset of $\mathcal{X}$ and $\{\mu_\epsilon\}$ satisfies the LDP in $\mathcal{E}$ with rate function I, then $\{\mu_\epsilon\}$ satisfies the LDP in $\mathcal{X}$ with rate function I' such that $I' = I$ on $\mathcal{E}$ and $I' = \infty$ on $\mathcal{E}^c$.
(b) If $\{\mu_\epsilon\}$ satisfies the LDP in $\mathcal{X}$ with rate function I and $\mathcal{D}_I \subset \mathcal{E}$, then the same LDP holds in $\mathcal{E}$. In particular, if $\mathcal{E}$ is a closed subset of $\mathcal{X}$, then $\mathcal{D}_I \subset \mathcal{E}$ and hence the LDP holds in $\mathcal{E}$.

Lemma 6.4.3 also holds for the weak LDP, since compact subsets of $\mathcal{E}$ are just the compact subsets of $\mathcal{X}$ contained in $\mathcal{E}$. Similarly, under the assumptions of the lemma, I is a good rate function on $\mathcal{X}$ iff it is a good rate function when restricted to $\mathcal{E}$.

The following is an important property of good rate functions.

Lemma 6.4.4 *Let I be a good rate function.*
(a) Let $\{F_\delta\}_{\delta>0}$ be a nested family of closed sets, i.e., $F_\delta \subseteq F_{\delta'}$ if $\delta < \delta'$. Define $F_0 = \cap_{\delta>0} F_\delta$. Then

$$\inf_{y\in F_0} I(y) = \lim_{\delta\to 0} \inf_{y\in F_\delta} I(y)\,.$$

(b) Suppose $(\mathcal{X}, d)$ *is a metric space. Then, for any set* A,

$$\inf_{y\in\overline{A}} I(y) = \lim_{\delta\to 0} \inf_{y\in A^\delta} I(y) \,, \tag{6.4.32}$$

where

$$A^\delta \stackrel{\triangle}{=} \{y : \ d(y, A) = \inf_{z\in A} d(y, z) \le \delta\} \tag{6.4.33}$$

denotes the closed blowup of A.

The next lemma is a partial converse of Lemma 6.2.1.

Lemma 6.4.5 *Let* $\{\mu_n\}$ *be a sequence of probability measures on a Polish space* $\mathcal{X}$ *that satisfies the large deviations upper bound with a good rate function. Then* $\{\mu_n\}$ *is exponentially tight.*

When a non-countable family of measures $\{\mu_\epsilon, \epsilon > 0\}$ satisfies the large deviations upper bound in a Polish space with a good rate function, Lemma 6.4.5 yields the exponential tightness of every sequence $\{\mu_{\epsilon_n}\}$, where $\epsilon_n \to 0$ as $n \to \infty$. As far as large deviations results are concerned, this is indistinguishable from exponential tightness of the whole family.

The following theorem introduces a general, indirect approach for establishing the *existence* of a *weak LDP*.

Theorem 6.4.6 *Let* $\mathcal{A}$ *be a base of the topology of* $\mathcal{X}$*. For every* $A \in \mathcal{A}$*, define*

$$\mathcal{L}_A \stackrel{\triangle}{=} - \liminf_{\epsilon\to 0} \ \epsilon \log \mu_\epsilon(A) \tag{6.4.34}$$

and

$$I(x) \stackrel{\triangle}{=} \sup_{\{A\in\mathcal{A}:\, x\in A\}} \mathcal{L}_A \ . \tag{6.4.35}$$

Suppose that for all $x \in \mathcal{X}$,

$$I(x) = \sup_{\{A\in\mathcal{A}:\, x\in A\}} \left[- \limsup_{\epsilon\to 0} \ \epsilon \log \mu_\epsilon(A) \right] . \tag{6.4.36}$$

Then μ_ϵ *satisfies the weak LDP with the rate function* $I(x)$.

Remarks:
(a) Observe that condition (6.4.36) holds when the limits $\lim_{\epsilon\to 0} \epsilon \log \mu_\epsilon(A)$ exist for all $A \in \mathcal{A}$ (with $-\infty$ as a possible value).
(b) When $\mathcal{X}$ is a locally convex, Hausdorff topological vector space, the base $\mathcal{A}$ is often chosen to be the collection of open, convex sets. This is done for example when proving Cramér's Theorem 6.6.1.
(c) It is easy to extend Theorem 6.4.6 to the context of a family of probability measures $\{\mu_{\epsilon,\sigma}\}$ that is indexed by an additional parameter σ. For example, σ may be the initial state of a Markov chain.

It is aesthetically pleasing to know that the following partial converse of Theorem 6.4.6 holds.

Theorem 6.4.7 *Suppose that* $\{\mu_\epsilon\}$ *satisfies the LDP in a regular topological space* $\mathcal{X}$ *with rate function* I*. Then, for any base* $\mathcal{A}$ *of the topology of* $\mathcal{X}$*, and for any* $x \in \mathcal{X}$,

$$\begin{aligned} I(x) &= \sup_{\{A\in\mathcal{A}:\, x\in A\}} \left\{ - \liminf_{\epsilon\to 0} \ \epsilon \log \mu_\epsilon(A) \right\} \\ &= \sup_{\{A\in\mathcal{A}:\, x\in A\}} \left\{ - \limsup_{\epsilon\to 0} \ \epsilon \log \mu_\epsilon(A) \right\} . \end{aligned} \tag{6.4.37}$$

Remark: For a Polish space $\mathcal{X}$ suffices to assume in Theorem 6.4.7 that $\{\mu_\epsilon\}$ satisfies the weak LDP. Consequently, by Theorem 6.4.6, in this context (6.4.37) is *equivalent* to the weak LDP.

The characterization of the rate function in Theorem 6.4.6 involves the supremum over a large collection of sets. Hence, it does not yield a convenient explicit formula. As shown in Section 6.4.4, if $\mathcal{X}$ is a Hausdorff topological vector space, this rate function can sometimes be identified with the Fenchel–Legendre transform of a limiting logarithmic moment generating function. This approach requires an *a priori* proof that the rate function is *convex*. The following lemma improves on Theorem 6.4.6 by giving a sufficient condition for the convexity of the rate function. Throughout, for any sets $A_1, A_2 \in \mathcal{X}$,

$$\frac{A_1 + A_2}{2} \triangleq \{x : \, x = (x_1 + x_2)/2, \, x_1 \in A_1, x_2 \in A_2\} \, .$$

Lemma 6.4.8 *Let $\mathcal{A}$ be a base for a Hausdorff topological vector space $\mathcal{X}$, such that in addition to condition (6.4.36), for every $A_1, A_2 \in \mathcal{A}$,*

$$\limsup_{\epsilon \to 0} \epsilon \log \mu_\epsilon \left(\frac{A_1 + A_2}{2} \right) \geq -\frac{1}{2} \left(\mathcal{L}_{A_1} + \mathcal{L}_{A_2} \right) . \tag{6.4.38}$$

Then the rate function I of (6.4.35), which governs the weak LDP associated with $\{\mu_\epsilon\}$, is convex.

When combined with exponential tightness, Theorem 6.4.6 implies the following large deviations analog of Prohorov's theorem.

Lemma 6.4.9 *Suppose the topological space $\mathcal{X}$ has a countable base. For any family of probability measures $\{\mu_\epsilon\}$, there exists a sequence $\epsilon_k \to 0$ such that $\{\mu_{\epsilon_k}\}$ satisfies the weak LDP in $\mathcal{X}$. If $\{\mu_\epsilon\}$ is an exponentially tight family of probability measures, then $\{\mu_{\epsilon_k}\}$ also satisfies the LDP with a good rate function.*

The next lemma applies for tight Borel probability measures μ_ϵ on metric spaces. In this context, it allows replacement of the assumed LDP in either Lemma 6.4.2 or Theorem 6.4.7 by a weak LDP.

Lemma 6.4.10 *Suppose $\{\mu_\epsilon\}$ is a family of tight (Borel) probability measures on a metric space $(\mathcal{X}, d)$, such that the upper bound (6.2.6) holds for all compact sets and some rate function $I(\cdot)$. Then, for any base $\mathcal{A}$ of the topology of $\mathcal{X}$, and for any $x \in \mathcal{X}$,*

$$I(x) \leq \sup_{\{A \in \mathcal{A}:\, x \in A\}} \left\{ -\limsup_{\epsilon \to 0} \epsilon \log \mu_\epsilon(A) \right\} . \tag{6.4.39}$$

6.4.2 Contraction Principles and Exponential Approximation

Section 6.4.2 is devoted to transformations that preserve the LDP, although, possibly, changing the rate function. Once the LDP with a good rate function is established for μ_ϵ, the basic *contraction principle* yields the LDP for $\mu_\epsilon \circ f^{-1}$, where f is any continuous map. The *inverse contraction principle* deals with f which is the inverse of a continuous bijection, and this is a useful tool for strengthening the topology under which the LDP holds. The remainder of the section is devoted to exponentially good approximations and their implications; for example, it is shown that when two families of measures defined on the same probability space are exponentially equivalent, then one can infer the LDP for one family from the other. A direct consequence is Theorem 6.4.19, which extends the contraction principle to "approximately continuous" maps.

The LDP is preserved under continuous mappings, as the following elementary theorem shows.

Theorem 6.4.11 (Contraction principle) *Let $\mathcal{X}$ and $\mathcal{Y}$ be Hausdorff topological spaces and $f : \mathcal{X} \to \mathcal{Y}$ a continuous function. Consider a good rate function $I : \mathcal{X} \to [0, \infty]$.*
(a) For each $y \in \mathcal{Y}$, define

$$I'(y) \stackrel{\triangle}{=} \inf\{I(x) : x \in \mathcal{X}, \quad y = f(x)\} . \tag{6.4.40}$$

Then I' is a good rate function on $\mathcal{Y}$, where as usual the infimum over the empty set is taken as ∞.
(b) If I controls the LDP associated with a family of probability measures $\{\mu_\epsilon\}$ on $\mathcal{X}$, then I' controls the LDP associated with the family of probability measures $\{\mu_\epsilon \circ f^{-1}\}$ on $\mathcal{Y}$.

Proof. (a) Clearly, I' is nonnegative. Since I is a good rate function, for all $y \in f(\mathcal{X})$ the infimum in the definition of I' is obtained at some point of $\mathcal{X}$. Thus, the level sets of I', $\Psi_{I'}(\alpha) \stackrel{\triangle}{=} \{y : I'(y) \le \alpha\}$, are

$$\Psi_{I'}(\alpha) = \{f(x) : I(x) \le \alpha\} = f(\Psi_I(\alpha)) ,$$

where $\Psi_I(\alpha)$ are the corresponding level sets of I. As $\Psi_I(\alpha) \subset \mathcal{X}$ are compact, so are the sets $\Psi_{I'}(\alpha) \subset \mathcal{Y}$.
(b) The definition of I' implies that for any $A \subset \mathcal{Y}$,

$$\inf_{y \in A} I'(y) = \inf_{x \in f^{-1}(A)} I(x) . \tag{6.4.41}$$

Since f is continuous, the set $f^{-1}(A)$ is an open (closed) subset of $\mathcal{X}$ for any open (closed) $A \subset \mathcal{Y}$. Therefore, the LDP for $\mu_\epsilon \circ f^{-1}$ follows as a consequence of the LDP for μ_ϵ and (6.4.41). □

Remarks:
(a) This theorem holds even when $\mathcal{B}_{\mathcal{X}} \subsetneq \mathcal{B}$, since for any (measurable) set $A \subset \mathcal{Y}$, both $f^{-1}(A) \subset f^{-1}(\overline{A})$ and $f^{-1}(A^o) \subset (f^{-1}(A))^o$.
(b) Note that the upper and lower bounds implied by part (b) of Theorem 6.4.11 hold even when I is not a good rate function. However, if I is *not* a good rate function, it may happen that I' is not a rate function, as the example $\mathcal{X} = \mathcal{Y} = \mathbb{R}$, $I(x) = 0$, and $f(x) = e^x$ demonstrates.
(c) Theorem 6.4.11 holds as long as f is continuous at every $x \in \mathcal{D}_I$; namely, for every $x \in \mathcal{D}_I$ and every neighborhood G of $f(x) \in \mathcal{Y}$, there exists a neighborhood A of x such that $A \subseteq f^{-1}(G)$. This suggests that the contraction principle may be further extended to cover a certain class of "approximately continuous" maps. Such an extension is pursued in Theorem 6.4.19.

We remind the reader that in what follows, it is always assumed that $\mathcal{B}_{\mathcal{X}} \subseteq \mathcal{B}$, and therefore open sets are always measurable. The following theorem shows that in the presence of exponential tightness, the contraction principle can be made to work in the reverse direction. This property is extremely useful for strengthening large deviations results from a coarse topology to a finer one, as in Corollary 6.4.13.

Theorem 6.4.12 (Inverse contraction principle) *Let $\mathcal{X}$ and $\mathcal{Y}$ be Hausdorff topological spaces. Suppose that $g : \mathcal{Y} \to \mathcal{X}$ is a continuous injection, and that $\{\nu_\epsilon\}$ is an exponentially tight family of probability measures on $\mathcal{Y}$. If $\{\nu_\epsilon \circ g^{-1}\}$ satisfies the LDP with the rate function $I : \mathcal{X} \to [0, \infty]$, then $\{\nu_\epsilon\}$ satisfies the LDP with the good rate function $I'(\cdot) \stackrel{\triangle}{=} I(g(\cdot))$.*

Corollary 6.4.13 *Let $\{\mu_\epsilon\}$ be an exponentially tight family of probability measures on $\mathcal{X}$ equipped with the topology τ_1. If $\{\mu_\epsilon\}$ satisfies an LDP with respect to a Hausdorff topology τ_2 on $\mathcal{X}$ that is coarser than τ_1, then the same LDP holds with respect to the topology τ_1.*

In order to extend the contraction principle beyond the continuous case, it is obvious that one should consider approximations by continuous functions. It is beneficial to consider a somewhat wider question, namely, when the LDP for a family of laws $\{\tilde{\mu}_\epsilon\}$ can be deduced from the LDP for a family $\{\mu_\epsilon\}$. The application to approximate contractions follows from these general results.

Definition 6.4.14 *Let $(\mathcal{Y}, d)$ be a metric space. The probability measures $\{\mu_\epsilon\}$ and $\{\tilde{\mu}_\epsilon\}$ on $\mathcal{Y}$ are called exponentially equivalent if there exist probability spaces $\{(\Omega, \mathcal{B}_\epsilon, P_\epsilon)\}$ and two families of $\mathcal{Y}$-valued random variables $\{Z_\epsilon\}$ and $\{\tilde{Z}_\epsilon\}$ with joint laws $\{P_\epsilon\}$ and marginals $\{\mu_\epsilon\}$ and $\{\tilde{\mu}_\epsilon\}$, respectively, such that the following condition is satisfied:*
For each $\delta > 0$, the set $\{\omega : (\tilde{Z}_\epsilon, Z_\epsilon) \in \Gamma_\delta\}$ is $\mathcal{B}_\epsilon$ measurable, and

$$\limsup_{\epsilon \to 0} \epsilon \log P_\epsilon(\Gamma_\delta) = -\infty \,, \tag{6.4.42}$$

where

$$\Gamma_\delta \stackrel{\triangle}{=} \{(\tilde{y}, y) : d(\tilde{y}, y) > \delta\} \subset \mathcal{Y} \times \mathcal{Y} \,. \tag{6.4.43}$$

Remarks:
(a) The random variables $\{Z_\epsilon\}$ and $\{\tilde{Z}_\epsilon\}$ in Definition 6.4.14 are called *exponentially equivalent.*
(b) The measurability requirement is satisfied whenever $\mathcal{Y}$ is a separable space, or whenever the laws $\{P_\epsilon\}$ are induced by separable real-valued stochastic processes and d is the supremum norm.

As far as the LDP is concerned, exponentially equivalent measures are indistinguishable, as the following theorem attests.

Theorem 6.4.15 *If an LDP with a good rate function $I(\cdot)$ holds for the probability measures $\{\mu_\epsilon\}$, which are exponentially equivalent to $\{\tilde{\mu}_\epsilon\}$, then the same LDP holds for $\{\tilde{\mu}_\epsilon\}$.*

As pointed out in the beginning of this section, an important goal in considering exponential equivalence is the treatment of approximations. To this end, the notion of exponential equivalence is replaced by the notion of exponential approximation, as follows.

Definition 6.4.16 *Let $\mathcal{Y}$ and Γ_δ be as in Definition 6.4.14. For each $\epsilon > 0$ and all $m \in \mathbb{Z}_+$, let $(\Omega, \mathcal{B}_\epsilon, P_{\epsilon,m})$ be a probability space, and let the $\mathcal{Y}$-valued random variables $\tilde{Z}_\epsilon$ and $Z_{\epsilon,m}$ be distributed according to the joint law $P_{\epsilon,m}$, with marginals $\tilde{\mu}_\epsilon$ and $\mu_{\epsilon,m}$, respectively. $\{Z_{\epsilon,m}\}$ are called exponentially good approximations of $\{\tilde{Z}_\epsilon\}$ if, for every $\delta > 0$, the set $\{\omega : (\tilde{Z}_\epsilon, Z_{\epsilon,m}) \in \Gamma_\delta\}$ is $\mathcal{B}_\epsilon$ measurable and*

$$\lim_{m \to \infty} \limsup_{\epsilon \to 0} \epsilon \log P_{\epsilon,m}(\Gamma_\delta) = -\infty \,. \tag{6.4.44}$$

Similarly, the measures $\{\mu_{\epsilon,m}\}$ are exponentially good approximations of $\{\tilde{\mu}_\epsilon\}$ if one can construct probability spaces $\{(\Omega, \mathcal{B}_\epsilon, P_{\epsilon,m})\}$ as above.

It should be obvious that Definition 6.4.16 reduces to Definition 6.4.14 if the laws $P_{\epsilon,m}$ do not depend on m. It can be shown that when $(\mathcal{Y}, d)$ is a Polish space, $\{\mu_{\epsilon,m}\}$ are exponentially good approximations of $\{\tilde{\mu}_\epsilon\}$ if and only if for any $\delta > 0$

$$\lim_{m \to \infty} \limsup_{\epsilon \to 0} \epsilon \log \sup \Big\{ \mu_{\epsilon,m}(A) - \tilde{\mu}_\epsilon(A^\delta) : \; A \in \mathcal{B}_\mathcal{Y} \Big\} = -\infty \,.$$

The main but somewhat technical consequence of Definition 6.4.16 is the following relation between the LDPs of exponentially good approximations.

Theorem 6.4.17 *Suppose that for every m, the family of measures $\{\mu_{\epsilon,m}\}$ satisfies the LDP with rate function $I_m(\cdot)$ and that $\{\mu_{\epsilon,m}\}$ are exponentially good approximations of $\{\tilde{\mu}_\epsilon\}$. Then*
(a) $\{\tilde{\mu}_\epsilon\}$ satisfies a weak LDP with the rate function

$$I(y) \triangleq \sup_{\delta>0} \liminf_{m\to\infty} \inf_{z\in B_{y,\delta}} I_m(z) , \qquad (6.4.45)$$

where $B_{y,\delta}$ denotes the ball $\{z : d(y,z) < \delta\}$.
(b) If $I(\cdot)$ is a good rate function and for every closed set F,

$$\inf_{y\in F} I(y) \leq \limsup_{m\to\infty} \inf_{y\in F} I_m(y) , \qquad (6.4.46)$$

then the full LDP holds for $\{\tilde{\mu}_\epsilon\}$ with rate function I.

Remarks:
(a) The sets Γ_δ may be replaced by sets $\tilde{\Gamma}_{\delta,m}$ such that the sets $\{\omega : (\tilde{Z}_\epsilon, Z_{\epsilon,m}) \in \tilde{\Gamma}_{\delta,m}\}$ differ from $\mathcal{B}_\epsilon$ measurable sets by $P_{\epsilon,m}$ null sets, and $\tilde{\Gamma}_{\delta,m}$ satisfy both (6.4.44) and $\Gamma_\delta \subset \tilde{\Gamma}_{\delta,m}$.
(b) If the rate functions $I_m(\cdot)$ are independent of m, and are good rate functions, then by Theorem 6.4.17, $\{\tilde{\mu}_\epsilon\}$ satisfies the LDP with $I(\cdot) = I_m(\cdot)$. In particular, Theorem 6.4.15 is a direct consequence of Theorem 6.4.17.
(c) In the context of part (a) of Theorem 6.4.17, if $(\mathcal{Y}, d)$ is a Polish space and $I_m(\cdot)$ are good rate functions, then $\{\tilde{\mu}_\epsilon\}$ satisfies the full LDP with the good rate function $I(\cdot)$ of (6.4.45). However, for general $(\mathcal{Y}, d)$ one cannot dispense with condition (6.4.46) in Theorem 6.4.17.

It should be obvious that the results on exponential approximations imply results on approximate contractions. We now present two such results. The first is related to Theorem 6.4.15 and considers approximations that are ϵ dependent. The second allows one to consider approximations that depend on an auxiliary parameter.

Corollary 6.4.18 *Suppose $f : \mathcal{X} \to \mathcal{Y}$ is a continuous map from a Hausdorff topological space $\mathcal{X}$ to the metric space $(\mathcal{Y}, d)$ and that $\{\mu_\epsilon\}$ satisfy the LDP with the good rate function $I : \mathcal{X} \to [0,\infty]$. Suppose further that for all $\epsilon > 0$, $f_\epsilon : \mathcal{X} \to \mathcal{Y}$ are measurable maps such that for all $\delta > 0$, the set $\Gamma_{\epsilon,\delta} \triangleq \{x \in \mathcal{X} : d(f(x), f_\epsilon(x)) > \delta\}$ is measurable, and*

$$\limsup_{\epsilon\to 0} \epsilon \log \mu_\epsilon(\Gamma_{\epsilon,\delta}) = -\infty . \qquad (6.4.47)$$

Then the LDP with the good rate function $I'(\cdot)$ of (6.4.40) holds for the measures $\mu_\epsilon \circ f_\epsilon^{-1}$ on $\mathcal{Y}$.

Proof. The contraction principle (Theorem 6.4.11) yields the desired LDP for $\{\mu_\epsilon \circ f^{-1}\}$. By (6.4.47), these measures are exponentially equivalent to $\{\mu_\epsilon \circ f_\epsilon^{-1}\}$, and the corollary follows from Theorem 6.4.15. □

A special case of Theorem 6.4.17 is the following extension of the contraction principle to maps that are not continuous, but that can be approximated well by continuous maps.

Theorem 6.4.19 *Let $\{\mu_\epsilon\}$ be a family of probability measures that satisfies the LDP with a good rate function I on a Hausdorff topological space $\mathcal{X}$, and for $m = 1, 2, \ldots$, let $f_m : \mathcal{X} \to \mathcal{Y}$ be continuous functions, with $(\mathcal{Y}, d)$ a metric space. Assume there exists a measurable map $f : \mathcal{X} \to \mathcal{Y}$ such that for every $\alpha < \infty$,*

$$\limsup_{m\to\infty} \sup_{\{x:I(x)\leq\alpha\}} d(f_m(x), f(x)) = 0 . \qquad (6.4.48)$$

Then any family of probability measures $\{\tilde{\mu}_\epsilon\}$ for which $\{\mu_\epsilon \circ f_m^{-1}\}$ are exponentially good approximations satisfies the LDP in $\mathcal{Y}$ with the good rate function $I'(y) = \inf\{I(x) : y = f(x)\}$.

The condition (6.4.48) implies that for every $\alpha < \infty$, the function f is continuous on the level set $\Psi_I(\alpha) = \{x : I(x) \leq \alpha\}$. Suppose that in addition,

$$\lim_{m\to\infty} \inf_{x\in\overline{\Psi_I(m)^c}} I(x) = \infty . \tag{6.4.49}$$

Then the LDP for $\mu_\epsilon \circ f^{-1}$ follows as a direct consequence of Theorem 6.4.19 by considering a sequence f_m of continuous extensions of f from $\Psi_I(m)$ to $\mathcal{X}$. (Such a sequence exists whenever $\mathcal{X}$ is a completely regular space.) That (6.4.49) need not hold true, even when $\mathcal{X} = \mathbb{R}$, may be seen by considering the following example. It is easy to check that $\mu_\epsilon = (\delta_{\{0\}} + \delta_{\{\epsilon\}})/2$ satisfies the LDP on $\mathbb{R}$ with the good rate function $I(0) = 0$ and $I(x) = \infty, x \neq 0$. On the other hand, the closure of the complement of any level set is the whole real line. If one now considers the function $f : \mathbb{R} \to \mathbb{R}$ such that $f(0) = 0$ and $f(x) = 1, x \neq 0$, then $\mu_\epsilon \circ f^{-1}$ does not satisfy the LDP with the rate function $I'(y) = \inf\{I(x) : x \in \mathbb{R},\ y = f(x)\}$, i.e., $I'(0) = 0$ and $I'(y) = \infty, y \neq 0$.

6.4.3 Varadhan's Lemma and its Converse

Throughout Section 6.4.3, $\{Z_\epsilon\}$ is a family of random variables taking values in the regular topological space $\mathcal{X}$, and $\{\mu_\epsilon\}$ denotes the probability measures associated with $\{Z_\epsilon\}$. The next theorem could actually be used as a starting point for developing the large deviations paradigm. It is a very useful tool in many applications of large deviations. For example, the asymptotics of the partition function in statistical mechanics can be derived using this theorem.

Theorem 6.4.20 (Varadhan) *Suppose that $\{\mu_\epsilon\}$ satisfies the LDP with a good rate function $I : \mathcal{X} \to [0,\infty]$, and let $\phi : \mathcal{X} \to \mathbb{R}$ be any continuous function. Assume further either the tail condition*

$$\lim_{M\to\infty} \limsup_{\epsilon\to 0} \epsilon \log E\left[e^{\phi(Z_\epsilon)/\epsilon}\, 1_{\{\phi(Z_\epsilon)\geq M\}}\right] = -\infty , \tag{6.4.50}$$

or the following moment condition for some $\gamma > 1$,

$$\limsup_{\epsilon\to 0} \epsilon \log E\left[e^{\gamma\phi(Z_\epsilon)/\epsilon}\right] < \infty . \tag{6.4.51}$$

Then

$$\lim_{\epsilon\to 0} \epsilon \log E\left[e^{\phi(Z_\epsilon)/\epsilon}\right] = \sup_{x\in\mathcal{X}} \{\phi(x) - I(x)\} .$$

Theorem 6.4.20, often referred to as "Varadhan's lemma" in the literature, is a direct consequence of the following three lemmas. For bounded $\phi(\cdot)$ the main Lemma 6.4.22 is proved by covering the compact level sets of $I(\cdot)$ by small neighborhoods using the lower semicontinuity of $I(\cdot)$ and the upper semicontinuity of $\phi(\cdot)$.

Lemma 6.4.21 *If $\phi : \mathcal{X} \to \mathbb{R}$ is lower semicontinuous and the large deviations lower bound holds with $I : \mathcal{X} \to [0,\infty]$, then*

$$\liminf_{\epsilon\to 0} \epsilon \log E\left[e^{\phi(Z_\epsilon)/\epsilon}\right] \geq \sup_{x\in\mathcal{X}} \{\phi(x) - I(x)\} . \tag{6.4.52}$$

Lemma 6.4.22 *If $\phi : \mathcal{X} \to \mathbb{R}$ is an upper semicontinuous function for which the tail condition (6.4.50) holds, and the large deviations upper bound holds with the good rate function $I : \mathcal{X} \to [0,\infty]$, then*

$$\limsup_{\epsilon\to 0} \epsilon \log E\left[e^{\phi(Z_\epsilon)/\epsilon}\right] \leq \sup_{x\in\mathcal{X}} \{\phi(x) - I(x)\} . \tag{6.4.53}$$

Lemma 6.4.23 *Condition (6.4.51) implies the tail condition (6.4.50).*

We next state a partial converse to Varadhan's lemma, due to Bryc [Bry90]. For each Borel measurable function $f : \mathcal{X} \to \mathbb{R}$, define

$$\Lambda_f \stackrel{\triangle}{=} \lim_{\epsilon \to 0} \epsilon \log \int_{\mathcal{X}} e^{f(x)/\epsilon} \mu_\epsilon(dx) , \qquad (6.4.54)$$

provided the limit exists. The main result of this section is that the LDP is a consequence of exponential tightness and the existence of the limits (6.4.54) for every $f \in \mathcal{G}$, for appropriate families of functions $\mathcal{G}$.

To this end it is assumed in the rest of the section that $\mathcal{X}$ is a completely regular topological space, i.e., $\mathcal{X}$ is Hausdorff, and for any closed set $F \subset \mathcal{X}$ and any point $x \notin F$, there exists a continuous function $f : \mathcal{X} \to [0, 1]$ such that $f(x) = 1$ and $f(y) = 0$ for all $y \in F$. Recall that Hausdorff topological vector spaces are completely regular.

The class of all bounded, real valued continuous functions on $\mathcal{X}$ is denoted throughout by $C_b(\mathcal{X})$.

Theorem 6.4.24 (Bryc) *Suppose that the family $\{\mu_\epsilon\}$ is exponentially tight and that the limit Λ_f in (6.4.54) exists for every $f \in C_b(\mathcal{X})$. Then $\{\mu_\epsilon\}$ satisfies the LDP with the good rate function*

$$I(x) = \sup_{f \in C_b(\mathcal{X})} \{ f(x) - \Lambda_f \} . \qquad (6.4.55)$$

Furthermore, for every $f \in C_b(\mathcal{X})$,

$$\Lambda_f = \sup_{x \in \mathcal{X}} \{ f(x) - I(x) \} . \qquad (6.4.56)$$

Remark: In the case where $\mathcal{X}$ is a topological vector space, it is tempting to compare (6.4.55) and (6.4.56) with the Fenchel–Legendre transform pair $\Lambda(\cdot)$ and $\Lambda^*(\cdot)$ of Section 6.4.4. Note, however, that here the rate function $I(x)$ need not be convex.

Sketch of Proof: Since $\Lambda_0 = 0$, it follows that $I(\cdot) \geq 0$. Moreover, $I(x)$ is lower semicontinuous, since it is the supremum of continuous functions. Due to the exponential tightness of $\{\mu_\epsilon\}$, the LDP asserted follows once the weak LDP (with rate function $I(\cdot)$) is proved. Moreover, by an application of Varadhan's lemma (Theorem 6.4.20), the identity (6.4.56) then holds. It remains, therefore, only to prove the weak LDP, which is a consequence of the following two lemmas.

Lemma 6.4.25 (Upper bound) *If Λ_f exists for each $f \in C_b(\mathcal{X})$, then, for every compact $\Gamma \subset \mathcal{X}$,*

$$\limsup_{\epsilon \to 0} \epsilon \log \mu_\epsilon(\Gamma) \leq - \inf_{x \in \Gamma} I(x) .$$

Lemma 6.4.26 (Lower bound) *If Λ_f exists for each $f \in C_b(\mathcal{X})$, then, for every open $G \subset \mathcal{X}$ and each $x \in G$,*

$$\liminf_{\epsilon \to 0} \epsilon \log \mu_\epsilon(G) \geq -I(x) .$$

This proof works because indicators on open sets are approximated well enough by bounded continuous functions. It is clear, however, that not all of $C_b(\mathcal{X})$ is needed for that purpose. The following definition is the tool for relaxing the assumptions of Theorem 6.4.24.

Definition 6.4.27 *A class $\mathcal{G}$ of continuous, real valued functions on a topological space $\mathcal{X}$ is said to be well-separating if:*
(1) $\mathcal{G}$ contains the constant functions.
(2) $\mathcal{G}$ is closed under finite pointwise minima, i.e., $g_1, g_2 \in \mathcal{G} \Rightarrow g_1 \wedge g_2 \in \mathcal{G}$.
(3) $\mathcal{G}$ separates points of $\mathcal{X}$, i.e., given two points $x, y \in \mathcal{X}$ with $x \neq y$, and $a, b \in \mathbb{R}$, there exists a function $g \in \mathcal{G}$ such that $g(x) = a$ and $g(y) = b$.

Remark: It is easy to check that if $\mathcal{G}$ is well-separating, so is $\mathcal{G}^+$, the class of all bounded above functions in $\mathcal{G}$.

When $\mathcal{X}$ is a vector space, a particularly useful class of well-separating functions exists.

Lemma 6.4.28 *Let $\mathcal{X}$ be a locally convex, Hausdorff topological vector space. Then the class $\mathcal{G}$ of all continuous, bounded above, concave functions on $\mathcal{X}$ is well-separating.*

The following lemma, states the specific approximation property of well-separating classes of functions that allows their use instead of $C_b(\mathcal{X})$. It is the key to the proof of Theorem 6.4.30.

Lemma 6.4.29 *Let $\mathcal{G}$ be a well-separating class of functions on $\mathcal{X}$. Then for any compact set $\Gamma \subset \mathcal{X}$, any $f \in C_b(\Gamma)$, and any $\delta > 0$, there exists an integer $d < \infty$ and functions $g_1, \cdots, g_d \in \mathcal{G}$ such that*

$$\sup_{x \in \Gamma} |f(x) - \max_{i=1}^{d} g_i(x)| \leq \delta$$

and

$$\sup_{x \in \mathcal{X}} g_i(x) \leq \sup_{x \in \Gamma} f(x) < \infty .$$

Theorem 6.4.30 (Bryc) *Let $\{\mu_\epsilon\}$ be an exponentially tight family of probability measures on a completely regular topological space $\mathcal{X}$, and suppose $\mathcal{G}$ is a well-separating class of functions on $\mathcal{X}$. If Λ_g exists for each $g \in \mathcal{G}$, then Λ_f exists for each $f \in C_b(\mathcal{X})$. Consequently, all the conclusions of Theorem 6.4.24 hold.*

The following variant of Theorem 6.4.24 dispenses with the exponential tightness of $\{\mu_\epsilon\}$, assuming instead that (6.4.56) holds for some good rate function $I(\cdot)$.

Theorem 6.4.31 *Let $I(\cdot)$ be a good rate function. A family of probability measures $\{\mu_\epsilon\}$ satisfies the LDP in $\mathcal{X}$ with the rate function $I(\cdot)$ if and only if the limit Λ_f in (6.4.54) exists for every $f \in C_b(\mathcal{X})$ and satisfies (6.4.56).*

6.4.4 Convexity Considerations

In Section 6.3.3, it was shown that when a limiting logarithmic moment generating function exists for a family of $\mathbb{R}^d$-valued random variables, then its Fenchel–Legendre transform is the natural candidate rate function for the LDP associated with these variables. The goal of Section 6.4.4 is to extend this result to topological vector spaces. As will be seen, convexity plays a major role as soon as the linear structure is introduced. For this reason, after the upper bound is established for all compact sets, some generalities involving the convex duality of Λ and Λ^* are presented. These convexity considerations play an essential role in applications. Finally, Theorem 6.4.36 is a (weak) version of the Gärtner–Ellis theorem in an abstract setup.

Throughout Section 6.4.4 $\mathcal{X}$ is a Hausdorff (real) topological *vector* space. Recall that such spaces are completely regular, so the results of Sections 6.4.1 and 6.4.3 apply. The dual space of $\mathcal{X}$, namely, the space of all *continuous linear functionals* on $\mathcal{X}$, is denoted throughout by $\mathcal{X}^*$. Let Z_ϵ be a family of random variables taking values in $\mathcal{X}$, and let $\mu_\epsilon \in M_1(\mathcal{X})$ denote the probability measure associated with Z_ϵ. By analogy with the $\mathbb{R}^d$ case presented in Section 6.3.3, the logarithmic moment generating function $\Lambda_{\mu_\epsilon} : \mathcal{X}^* \to (-\infty, \infty]$ is defined to be

$$\Lambda_{\mu_\epsilon}(\lambda) = \log E\left[e^{\langle \lambda, Z_\epsilon \rangle}\right] = \log \int_{\mathcal{X}} e^{\lambda(x)} \mu_\epsilon(dx)\,, \quad \lambda \in \mathcal{X}^*\,,$$

where for $x \in \mathcal{X}$ and $\lambda \in \mathcal{X}^*$, $\langle \lambda, x \rangle$ denotes the value of $\lambda(x) \in \mathbb{R}$.

Let

$$\bar{\Lambda}(\lambda) \stackrel{\triangle}{=} \limsup_{\epsilon \to 0} \epsilon \Lambda_{\mu_\epsilon}\left(\frac{\lambda}{\epsilon}\right)\,, \tag{6.4.57}$$

using the notation $\Lambda(\lambda)$ whenever the *limit exists*. In many cases, when $\epsilon \Lambda_{\mu_\epsilon}(\cdot/\epsilon)$ converges pointwise to $\Lambda(\cdot)$ for $\mathcal{X} = \mathbb{R}^d$ and an LDP holds for $\{\mu_\epsilon\}$, the rate function associated with this LDP is the Fenchel–Legendre transform of $\Lambda(\cdot)$. In the current setup, the Fenchel–Legendre transform of a function $f : \mathcal{X}^* \to [-\infty, \infty]$ is defined as

$$f^*(x) \stackrel{\triangle}{=} \sup_{\lambda \in \mathcal{X}^*} \{\langle \lambda, x \rangle - f(\lambda)\}\,, \quad x \in \mathcal{X}\,. \tag{6.4.58}$$

Thus, $\bar{\Lambda}^*$ denotes the Fenchel–Legendre transform of $\bar{\Lambda}$, and Λ^* denotes that of Λ when the latter exists for all $\lambda \in \mathcal{X}^*$.

The following upper bound is a consequence of Chebycheff's inequality and the covering of the compact set Γ by an appropriate half-space.

Theorem 6.4.32
(a) $\bar{\Lambda}(\cdot)$ of (6.4.57) is convex on $\mathcal{X}^*$ and $\bar{\Lambda}^*(\cdot)$ is a convex rate function.
(b) For any compact set $\Gamma \subset \mathcal{X}$,

$$\limsup_{\epsilon \to 0} \epsilon \log \mu_\epsilon(\Gamma) \leq - \inf_{x \in \Gamma} \bar{\Lambda}^*(x)\,. \tag{6.4.59}$$

Remarks:
(a) In Theorem 6.3.12, which corresponds to $\mathcal{X} = \mathbb{R}^d$, it was assumed, for the purpose of establishing exponential tightness, that $0 \in \mathcal{D}^o_\Lambda$. In the abstract setup considered here, the exponential tightness does not follow from this assumption, and therefore must be handled on a case-by-case basis.
(b) Note that any bound of the form $\bar{\Lambda}(\lambda) \leq K(\lambda)$ for all $\lambda \in \mathcal{X}^*$ implies that the Fenchel–Legendre transform $K^*(\cdot)$ may be substituted for $\bar{\Lambda}^*(\cdot)$ in (6.4.59). This is useful in situations in which $\bar{\Lambda}(\lambda)$ is easy to bound but hard to compute.
(c) The inequality (6.4.59) may serve as the upper bound related to a weak LDP. Thus, when $\{\mu_\epsilon\}$ is an exponentially tight family of measures, (6.4.59) extends to all closed sets. If in addition, the large deviations lower bound is also satisfied with $\bar{\Lambda}^*(\cdot)$, then this is a good rate function that controls the large deviations of the family $\{\mu_\epsilon\}$.

The implications of the existence of an LDP with a convex rate function to the structure of Λ and Λ^* are next explored. Building on Varadhan's lemma and Theorem 6.4.32, it follows that when the quantities $\epsilon \Lambda_{\mu_\epsilon}(\lambda/\epsilon)$ are uniformly bounded (in ϵ) and an LDP holds with a good convex rate function, then $\epsilon \Lambda_{\mu_\epsilon}(\cdot/\epsilon)$ converges pointwise to $\Lambda(\cdot)$ and the rate function equals $\Lambda^*(\cdot)$. Consequently, the assumptions of Lemma 6.4.8 together with the exponential tightness of $\{\mu_\epsilon\}$ and the uniform boundedness mentioned earlier, suffice to establish the LDP with rate function $\Lambda^*(\cdot)$.

Before proceeding with the identification of the rate function of the LDP as $\Lambda^*(\cdot)$, note that while $\Lambda^*(\cdot)$ is always convex by Theorem 6.4.32, the rate function may well be nonconvex. For example, such a situation may occur when contractions using nonconvex functions are considered. However, it may be expected that $I(\cdot)$ is identical to $\Lambda^*(\cdot)$ when $I(\cdot)$ is convex.

An instrumental tool in the identification of I as Λ^* is the following duality property of the Fenchel–Legendre transform, which is a consequence of the Hahn-Banach theorem.

Lemma 6.4.33 (Duality lemma) *Let $\mathcal{X}$ be a locally convex Hausdorff topological vector space. Let $f : \mathcal{X} \to (-\infty, \infty]$ be a lower semicontinuous, convex function, and define*

$$g(\lambda) = \sup_{x \in \mathcal{X}} \{\langle \lambda, x\rangle - f(x)\} .$$

Then $f(\cdot)$ is the Fenchel–Legendre transform of $g(\cdot)$, i.e.,

$$f(x) = \sup_{\lambda \in \mathcal{X}^*} \{\langle \lambda, x\rangle - g(\lambda)\} . \qquad (6.4.60)$$

This lemma has the following geometric interpretation. For every hyperplane defined by λ, $g(\lambda)$ is the largest amount one may push up the tangent before it hits $f(\cdot)$ and becomes a tangent hyperplane. The duality lemma states the "obvious result" that to reconstruct $f(\cdot)$, one only needs to find the tangent at x and "push it down" by $g(\lambda)$.

The first application of the duality lemma is in the following theorem, where convex rate functions are identified as $\Lambda^*(\cdot)$.

Theorem 6.4.34 *Let $\mathcal{X}$ be a locally convex Hausdorff topological vector space. Assume that μ_ϵ satisfies the LDP with a good rate function I. Suppose in addition that*

$$\bar{\Lambda}(\lambda) \stackrel{\triangle}{=} \limsup_{\epsilon \to 0} \epsilon \Lambda_{\mu_\epsilon}(\lambda/\epsilon) < \infty, \quad \forall \lambda \in \mathcal{X}^* . \qquad (6.4.61)$$

(a) For each $\lambda \in \mathcal{X}^*$, the limit $\Lambda(\lambda) = \lim_{\epsilon \to 0} \epsilon \Lambda_{\mu_\epsilon}(\lambda/\epsilon)$ exists, is finite, and satisfies

$$\Lambda(\lambda) = \sup_{x \in \mathcal{X}} \{\langle \lambda, x\rangle - I(x)\} . \qquad (6.4.62)$$

(b) If I is convex, then it is the Fenchel–Legendre transform of Λ, namely,

$$I(x) = \Lambda^*(x) \stackrel{\triangle}{=} \sup_{\lambda \in \mathcal{X}^*} \{\langle \lambda, x\rangle - \Lambda(\lambda)\} .$$

(c) If I is not convex, then Λ^* is the affine regularization of I, i.e., $\Lambda^*(\cdot) \le I(\cdot)$, and for any convex rate function f, $f(\cdot) \le I(\cdot)$ implies $f(\cdot) \le \Lambda^*(\cdot)$.

Remark: The weak* topology on $\mathcal{X}^*$ makes the functions $\langle \lambda, x\rangle - I(x)$ continuous in λ for all $x \in \mathcal{X}$. By part (a), $\Lambda(\cdot)$ is lower semicontinuous with respect to this topology, which explains why lower semicontinuity of $\Lambda(\cdot)$ is necessary in Rockafellar's lemma (Lemma 6.3.14).

Corollary 6.4.35 *Suppose that both condition (6.4.61) and the assumptions of Lemma 6.4.8 hold for the family $\{\mu_\epsilon\}$, which is exponentially tight. Then $\{\mu_\epsilon\}$ satisfies in $\mathcal{X}$ the LDP with the convex, good rate function Λ^*.*

Theorem 6.4.34 is not applicable when $\Lambda(\cdot)$ exists but is infinite at some $\lambda \in \mathcal{X}^*$, and moreover, it requires the *full* LDP with a convex, *good* rate function. As seen in the case of Cramér's theorem in $\mathbb{R}$, these conditions are not necessary. Of course, there is a price to pay: The resulting Λ^* may not be a good rate function and only the weak LDP is proved.

Having seen a general upper bound in Theorem 6.4.32 we turn next to sufficient conditions for the existence of a complementary lower bound. To this end, recall that a point $x \in \mathcal{X}$ is called an *exposed* point of $\bar{\Lambda}^*$ if there exists an *exposing hyperplane* $\lambda \in \mathcal{X}^*$ such that

$$\langle \lambda, x \rangle - \bar{\Lambda}^*(x) > \langle \lambda, z \rangle - \bar{\Lambda}^*(z) \,, \quad \forall z \neq x \,.$$

An exposed point of $\bar{\Lambda}^*$ is, in convex analysis parlance, an exposed point of the epigraph of $\bar{\Lambda}^*$. The following is an infinite-dimensional extension of the Gärtner-Ellis theorem. Note however that its assumption (6.4.63) is stronger, while part (c) is weaker than the finite dimensional counterpart because there is no explicit criterion for checking (6.4.64).

Theorem 6.4.36 (Baldi) *Suppose that $\{\mu_\epsilon\}$ are exponentially tight probability measures on $\mathcal{X}$.*
(a) For every closed set $F \subset \mathcal{X}$,

$$\limsup_{\epsilon \to 0} \epsilon \log \, \mu_\epsilon(F) \leq - \inf_{x \in F} \bar{\Lambda}^*(x) \,.$$

(b) Let $\mathcal{F}$ be the set of exposed points of $\bar{\Lambda}^*$ with an exposing hyperplane λ for which

$$\Lambda(\lambda) = \lim_{\epsilon \to 0} \epsilon \Lambda_{\mu_\epsilon}\left(\frac{\lambda}{\epsilon}\right) \text{ exists and } \bar{\Lambda}(\gamma\lambda) < \infty \quad \text{for some} \quad \gamma > 1 \,. \tag{6.4.63}$$

Then, for every open set $G \subset \mathcal{X}$,

$$\liminf_{\epsilon \to 0} \epsilon \log \, \mu_\epsilon(G) \geq - \inf_{x \in G \cap \mathcal{F}} \bar{\Lambda}^*(x) \,.$$

(c) If for every open set G,

$$\inf_{x \in G \cap \mathcal{F}} \bar{\Lambda}^*(x) = \inf_{x \in G} \bar{\Lambda}^*(x) \,, \tag{6.4.64}$$

then $\{\mu_\epsilon\}$ satisfies the LDP with the good rate function $\bar{\Lambda}^*$.

6.4.5 Large Deviations for Projective Limits

In Section 6.4.5, we develop a method of lifting a collection of LDPs in "small" spaces into the LDP in the "large" space $\mathcal{X}$, which is their projective limit. (See definition below.) The motivation for such an approach is as follows. Suppose we are interested in proving the LDP associated with a sequence of random variables $X_1, X_2, \ldots$ in some abstract space $\mathcal{X}$. The identification of $\mathcal{X}^*$ (if $\mathcal{X}$ is a topological vector space) and the computation of the Fenchel–Legendre transform of the moment generating function may involve the solution of variational problems in an infinite dimensional setting. Moreover, proving exponential tightness in $\mathcal{X}$, the main tool of getting at the upper bound, may be a difficult task. On the other hand, the evaluation of the limiting logarithmic moment generating function involves probabilistic computations at the level of real-valued random variables, albeit an infinite number of such computations. It is often relatively easy to derive the LDP for every finite collection of these real-valued random variables. Hence, it is reasonable to inquire if this implies that the laws of the original, $\mathcal{X}$-valued random variables satisfy the LDP.

An affirmative result is presented shortly in a somewhat abstract setting. The idea is to identify $\mathcal{X}$ with the projective limit of a family of spaces $\{\mathcal{Y}_j\}_{j\in J}$ with the hope that the LDP for any given family $\{\mu_\epsilon\}$ of probability measures on $\mathcal{X}$ follows as the consequence of the fact that the LDP holds for any of the projections of μ_ϵ to $\{\mathcal{Y}_j\}_{j\in J}$.

To make the program described precise, we first review a few standard topological definitions. Let $(J,\leq)$ be a partially ordered, right-filtering set. (The latter notion means that for any i,j in J, there exists $k\in J$ such that both $i\leq k$ and $j\leq k$.) Note that J need not be countable. A projective system $(\mathcal{Y}_j,p_{ij})_{i\leq j\in J}$ consists of Hausdorff topological spaces $\{\mathcal{Y}_j\}_{j\in J}$ and continuous maps $p_{ij}:\mathcal{Y}_j\to\mathcal{Y}_i$ such that $p_{ik}=p_{ij}\circ p_{jk}$ whenever $i\leq j\leq k$ ($\{p_{jj}\}_{j\in J}$ are the appropriate identity maps). The *projective limit* of this system, denoted by $\mathcal{X}=\varprojlim\mathcal{Y}_j$, is the subset of the topological product space $\mathcal{Y}=\prod_{j\in J}\mathcal{Y}_j$, consisting of all the elements $\mathbf{x}=(y_j)_{j\in J}$ for which $y_i=p_{ij}(y_j)$ whenever $i\leq j$, equipped with the topology induced by $\mathcal{Y}$. Projective limits of closed subsets $F_j\subseteq\mathcal{Y}_j$ are defined analogously and denoted $F=\varprojlim F_j$. The canonical projections of $\mathcal{X}$, which are the restrictions $p_j:\mathcal{X}\to\mathcal{Y}_j$ of the coordinate maps from $\mathcal{Y}$ to $\mathcal{Y}_j$, are continuous.

The following theorem yields the LDP in $\mathcal{X}$ as a consequence of the LDPs associated with $\{\mu_\epsilon\circ p_j^{-1},\epsilon>0\}$. In order to have a specific example in mind, think of $\mathcal{X}$ as the space of all maps $f:[0,1]\to\mathbb{R}$ such that $f(0)=0$, equipped with the topology of pointwise convergence. Then $p_j:\mathcal{X}\to\mathbb{R}^d$ is the projection of functions onto their values at the time instances $0\leq t_1<t_2<\cdots<t_d\leq 1$, with the partial ordering induced on the set $J=\cup_{d=1}^{\infty}\{(t_1,\ldots,t_d):0\leq t_1<t_2<\cdots<t_d\leq 1\}$ by inclusions. For details of this construction, see Section 6.5.1.

Theorem 6.4.37 (Dawson–Gärtner) *Let $\{\mu_\epsilon\}$ be a family of probability measures on $\mathcal{X}$, such that for any $j\in J$ the Borel probability measures $\mu_\epsilon\circ p_j^{-1}$ on $\mathcal{Y}_j$ satisfy the LDP with the good rate function $I_j(\cdot)$. Then $\{\mu_\epsilon\}$ satisfies the LDP with the good rate function*

$$I(\mathbf{x})=\sup_{j\in J}\{\,I_j(p_j(\mathbf{x}))\,\}\,,\qquad \mathbf{x}\in\mathcal{X}\,. \tag{6.4.65}$$

Remark: Throughout Section 6.4.5, we drop the blanket assumption that $\mathcal{B}_{\mathcal{X}}\subseteq\mathcal{B}$. This is natural in view of the fact that the set J need not be countable. It is worthwhile to note that $\mathcal{B}$ is required to contain all sets $p_j^{-1}(B_j)$, where $B_j\in\mathcal{B}_{\mathcal{Y}_j}$.

The following lemma is often useful for simplifying the formula (6.4.65) of the Dawson–Gärtner rate function.

Lemma 6.4.38 *If $I(\cdot)$ is a good rate function on $\mathcal{X}$ such that*

$$I_j(y)=\inf\{I(\mathbf{x}):\ \mathbf{x}\in\mathcal{X},\quad y=p_j(\mathbf{x})\}\,, \tag{6.4.66}$$

for any $y\in\mathcal{Y}_j$, $j\in J$, then the identity (6.4.65) holds.

The preceding theorem is particularly suitable for situations involving topological vector spaces that satisfy the following assumptions.

Assumption 6.4.67 *Let $\mathcal{W}$ be an infinite dimensional real vector space, and $\mathcal{W}'$ its algebraic dual, i.e., the space of all linear functionals $\lambda\mapsto\langle\lambda,x\rangle:\mathcal{W}\to\mathbb{R}$. The topological (vector) space $\mathcal{X}$ consists of $\mathcal{W}'$ equipped with the $\mathcal{W}$-topology, i.e., the weakest topology such that for each $\lambda\in\mathcal{W}$, the linear functional $x\mapsto\langle\lambda,x\rangle:\mathcal{X}\to\mathbb{R}$ is continuous.*

Remark: The $\mathcal{W}$-topology of $\mathcal{W}'$ makes $\mathcal{W}$ into the topological dual of $\mathcal{X}$, i.e., $\mathcal{W}=\mathcal{X}^*$.

For any $d\in\mathbb{Z}_+$ and $\lambda_1,\ldots,\lambda_d\in\mathcal{W}$, define the projection $p_{\lambda_1,\ldots,\lambda_d}:\mathcal{X}\to\mathbb{R}^d$ by $p_{\lambda_1,\ldots,\lambda_d}(x)=(\langle\lambda_1,x\rangle,\langle\lambda_2,x\rangle,\ldots,\langle\lambda_d,x\rangle)$.

Assumption 6.4.68 *Let $(\mathcal{X}, \mathcal{B}, \mu_\epsilon)$ be probability spaces such that:*
(a) $\mathcal{X}$ satisfies Assumption 6.4.67.
(b) For any $\lambda \in \mathcal{W}$ and any Borel set B in $\mathbb{R}$, $p_\lambda^{-1}(B) \in \mathcal{B}$.

Remark: Note that if $\{\mu_\epsilon\}$ are Borel measures, then Assumption 6.4.68 reduces to Assumption 6.4.67.

Theorem 6.4.39 *Let Assumption 6.4.68 hold. Further assume that for every $d \in \mathbb{Z}_+$ and every $\lambda_1, \ldots, \lambda_d \in \mathcal{W}$, the measures $\{\mu_\epsilon \circ p_{\lambda_1,\ldots,\lambda_d}^{-1}, \epsilon > 0\}$ satisfy the LDP with the good rate function $I_{\lambda_1,\ldots,\lambda_d}(\cdot)$. Then $\{\mu_\epsilon\}$ satisfies the LDP in $\mathcal{X}$, with the good rate function*

$$I(x) = \sup_{d \in \mathbb{Z}_+} \sup_{\lambda_1,\ldots,\lambda_d \in \mathcal{W}} I_{\lambda_1,\ldots,\lambda_d}((\langle \lambda_1, x\rangle, \langle \lambda_2, x\rangle, \ldots, \langle \lambda_d, x\rangle)) . \tag{6.4.69}$$

Remark: In most applications, one is interested in obtaining an LDP on $\mathcal{E}$ that is a non-closed subset of $\mathcal{X}$. Hence, the relatively effortless projective limit approach is then followed by an application specific check that $\mathcal{D}_I \subset \mathcal{E}$, as needed for Lemma 6.4.3. For example, in the study of empirical measures on a Polish space Σ, it is known *a priori* that $\mu_\epsilon(M_1(\Sigma)) = 1$ for all $\epsilon > 0$, where $M_1(\Sigma)$ is the space of Borel probability measures on Σ, equipped with the $B(\Sigma)$-topology, and $B(\Sigma) = \{f : \Sigma \to \mathbb{R}, f \text{ bounded, Borel measurable}\}$. Identifying each $\nu \in M_1(\Sigma)$ with the linear functional $f \mapsto \int_\Sigma f d\nu$, $\forall f \in B(\Sigma)$, it follows that $M_1(\Sigma)$ is homeomorphic to $\mathcal{E} \subset \mathcal{X}$, where here $\mathcal{X}$ denotes the algebraic dual of $B(\Sigma)$ equipped with the $B(\Sigma)$-topology. Thus, $\mathcal{X}$ satisfies Assumption 6.4.67, and $\mathcal{E}$ is not a closed subset of $\mathcal{X}$. It is worthwhile to note that in this setup, μ_ϵ is not necessarily a Borel probability measure.

When using Theorem 6.4.39, either the convexity of $I_{\lambda_1,\ldots,\lambda_d}(\cdot)$ or the existence and smoothness of the limiting logarithmic moment generating function $\Lambda(\cdot)$ are relied upon in order to identify the good rate function of (6.4.69) with $\Lambda^*(\cdot)$, in a manner similar to that encountered in Theorem 6.4.34. This is spelled out in the following corollary.

Corollary 6.4.40 *Let Assumption 6.4.68 hold.*
(a) Suppose that for each $\lambda \in \mathcal{W}$, the limit

$$\Lambda(\lambda) = \lim_{\epsilon \to 0} \epsilon \log \int_{\mathcal{X}} e^{\epsilon^{-1}\langle \lambda, x\rangle} \mu_\epsilon(dx) \tag{6.4.70}$$

exists as an extended real number, and moreover that for any $d \in \mathbb{Z}_+$ and any $\lambda_1, \ldots, \lambda_d \in \mathcal{W}$, the function

$$g((t_1, \ldots, t_d)) \triangleq \Lambda(\sum_{i=1}^{d} t_i \lambda_i) : \mathbb{R}^d \to (-\infty, \infty]$$

is essentially smooth, lower semicontinuous, and finite in some neighborhood of 0. Then $\{\mu_\epsilon\}$ satisfies the LDP in $(\mathcal{X}, \mathcal{B})$ with the convex, good rate function

$$\Lambda^*(x) = \sup_{\lambda \in \mathcal{W}} \{\langle \lambda, x\rangle - \Lambda(\lambda)\} . \tag{6.4.71}$$

(b) Alternatively, if for any $\lambda_1, \ldots, \lambda_d \in \mathcal{W}$, there exists a compact set $K \subset \mathbb{R}^d$ such that $\mu_\epsilon \circ p_{\lambda_1,\ldots,\lambda_d}^{-1}(K) = 1$, and moreover $\{\mu_\epsilon \circ p_{\lambda_1,\ldots,\lambda_d}^{-1}, \epsilon > 0\}$ satisfies the LDP with a convex rate function, then $\Lambda : \mathcal{W} \to \mathbb{R}$ exists, is finite everywhere, and $\{\mu_\epsilon\}$ satisfies the LDP in $(\mathcal{X}, \mathcal{B})$ with the convex, good rate function $\Lambda^*(\cdot)$ as defined in (6.4.71).

Remark: Since $\mathcal{X}$ satisfies Assumption 6.4.67, the only continuous linear functionals on $\mathcal{X}$ are of the form $x \mapsto \langle \lambda, x\rangle$, where $\lambda \in \mathcal{W}$. Consequently, $\mathcal{X}^*$ may be identified with $\mathcal{W}$, and $\Lambda^*(\cdot)$ is the Fenchel–Legendre transform of $\Lambda(\cdot)$ as defined in Section 6.4.4.

Recall that a function $f : \mathcal{X}^* \to \mathbb{R}$ is *Gateaux differentiable* if, for every $\lambda, \theta \in \mathcal{X}^*$, the function $f(\lambda + t\theta)$ is differentiable with respect to t at $t = 0$. In the next corollary, Gateaux differentiability of $\Lambda(\cdot)$ results with the LDP, dispensing with Assumption 6.4.68.

Corollary 6.4.41 *Let $\{\mu_\epsilon\}$ be an exponentially tight family of Borel probability measures on the locally convex Hausdorff topological vector space $\mathcal{E}$. Suppose $\Lambda(\cdot) = \lim_{\epsilon\to 0} \epsilon\Lambda_{\mu_\epsilon}(\cdot/\epsilon)$ is finite valued and Gateaux differentiable. Then $\{\mu_\epsilon\}$ satisfies the LDP in $\mathcal{E}$ with the convex, good rate function Λ^*.*

6.5 Sample Path LDPs

The finite dimensional LDPs considered in Section 6.3 allow computations of the tail behavior of rare events associated with various sorts of empirical means. In many problems, the interest is actually in rare events that depend on a collection of random variables, or, more generally, on a random process. Whereas some of these questions may be cast in terms of empirical measures, this is not always the most fruitful approach. Interest often lies in the probability that a *path* of a random process hits a particular set. Questions of this nature are addressed here. We start with the case of a random walk, the simplest example of all. The Brownian motion counterpart is then an easy application of exponential equivalence, and the diffusion case follows by suitable approximate contractions.

The material for this section is taken from Sections 5.1/5.2, 5.6 and 5.7 of [DeZ98] to which the reader is referred for additional details, bibliography, and proofs.

6.5.1 Sample Path Large Deviations for Random Walk and for Brownian Motion

Let $X_1, X_2, \ldots$ be a sequence of i.i.d. random vectors taking values in $\mathbb{R}^d$, with $\Lambda(\lambda) \triangleq \log E(e^{\langle \lambda, X_1\rangle}) < \infty$ for all $\lambda \in \mathbb{R}^d$. Cramér's theorem (Theorem 6.3.9) allows the analysis of the large deviations of $\frac{1}{n}\sum_{i=1}^n X_i$. Similarly, the large deviations behavior of the pair of random variables $\frac{1}{n}\sum_{i=1}^n X_i$ and $\frac{1}{n}\sum_{i=1}^{[n/2]} X_i$ can be obtained, where $[c]$ as usual denotes the integer part of c. In Section 6.5.1, the large deviations joint behavior of a family of random variables indexed by t is considered.

Define

$$Z_n(t) = \frac{1}{n}\sum_{i=1}^{[nt]} X_i, \quad 0 \le t \le 1, \tag{6.5.72}$$

and let μ_n be the law of $Z_n(\cdot)$ in $L_\infty([0,1])$. Throughout, $|x| \triangleq \sqrt{\langle x, x\rangle}$ denotes the Euclidean norm on $\mathbb{R}^d$, $\| f \|$ denotes the supremum norm on $L_\infty([0,1])$, and $\Lambda^*(x) \triangleq \sup_{\lambda\in\mathbb{R}^d}[\langle \lambda, x\rangle - \Lambda(\lambda)]$ denotes the Fenchel–Legendre transform of $\Lambda(\cdot)$.

The following theorem is the first result of this section.

Theorem 6.5.1 (Mogulskii) *The measures μ_n satisfy in $L_\infty([0,1])$ the LDP with the good rate function*

$$I(\phi) = \begin{cases} \int_0^1 \Lambda^*(\dot\phi(t))\,dt, & \text{if } \phi \in \mathcal{AC}, \phi(0) = 0 \\ \\ \infty & \text{otherwise}, \end{cases} \tag{6.5.73}$$

where $\mathcal{AC}$ denotes the space of absolutely continuous functions, i.e.,

$$\mathcal{AC} \triangleq \Big\{ \phi \in C([0,1]) : \\ \sum_{\ell=1}^k |t_\ell - s_\ell| \to 0, s_\ell < t_\ell \le s_{\ell+1} < t_{\ell+1} \Longrightarrow \sum_{\ell=1}^k |\phi(t_\ell) - \phi(s_\ell)| \to 0 \Big\}.$$

Remarks:
(a) Recall that $\phi : [0,1] \to \mathbb{R}^d$ absolutely continuous implies that ϕ is differentiable almost everywhere; in particular, that it is the integral of an $L_1([0,1])$ function.
(b) Since $\{\mu_n\}$ are supported on the space of functions continuous from the right and having left limits, of which $\mathcal{D}_I$ is a subset, the preceding LDP holds in this space when equipped with the supremum norm topology. In fact, all steps of the proof would have been the same had we been working in that space, instead of $L_\infty([0,1])$, throughout.
(c) Theorem 6.5.1 possesses extensions to stochastic processes with jumps at random times; To avoid measurability problems, one usually works in the space of functions continuous from the right and having left limits, equipped with a topology which renders the latter Polish (the Skorohod topology). Results may then be strengthened to the supremum norm topology by using exponential tightness.

The proof of Theorem 6.5.1 is based on the following three lemmas.

Lemma 6.5.2 *Let $\tilde{\mu}_n$ denote the law of $\tilde{Z}_n(\cdot)$ in $L_\infty([0,1])$, where*

$$\tilde{Z}_n(t) \stackrel{\triangle}{=} Z_n(t) + \left(t - \frac{[nt]}{n}\right) X_{[nt]+1} \tag{6.5.74}$$

is the polygonal approximation of $Z_n(t)$. Then the probability measures μ_n and $\tilde{\mu}_n$ are exponentially equivalent in $L_\infty([0,1])$.

Lemma 6.5.3 *Let $\mathcal{X}$ consist of all the maps from $[0,1]$ to $\mathbb{R}^d$ such that $t=0$ is mapped to the origin, and equip $\mathcal{X}$ with the topology of pointwise convergence on $[0,1]$. Then the probability measures $\tilde{\mu}_n$ of Lemma 6.5.2 (defined on $\mathcal{X}$ by the natural embedding) satisfy the LDP in this Hausdorff topological space with the good rate function $I(\cdot)$ of (6.5.73).*

Lemma 6.5.4 *The probability measures $\tilde{\mu}_n$ are exponentially tight in the space $C_0([0,1])$ of all continuous functions $f : [0,1] \to \mathbb{R}^d$ such that $f(0) = 0$, equipped with the supremum norm topology.*

Proof of Theorem 6.5.1: By Lemma 6.5.3, $\{\tilde{\mu}_n\}$ satisfies the LDP in $\mathcal{X}$. Note that $\mathcal{D}_I \subset C_0([0,1])$, and by (6.5.72) and (6.5.74), $\tilde{\mu}_n(C_0([0,1])) = 1$ for all n. Thus, by Lemma 6.4.3, the LDP for $\{\tilde{\mu}_n\}$ also holds in the space $C_0([0,1])$ when equipped with the relative (Hausdorff) topology induced by $\mathcal{X}$. The latter is the pointwise convergence topology, which is generated by the sets $V_{t,x,\delta} \stackrel{\triangle}{=} \{g \in C_0([0,1]) : |g(t) - x| < \delta\}$ with $t \in (0,1]$, $x \in \mathbb{R}^d$ and $\delta > 0$. Since each $V_{t,x,\delta}$ is an open set under the supremum norm, the latter topology is finer (stronger) than the pointwise convergence topology. Hence, the exponential tightness of $\{\tilde{\mu}_n\}$ as established in Lemma 6.5.4 allows, by Corollary 6.4.13, for the strengthening of the LDP to the supremum norm topology on $C_0([0,1])$. Since $C_0([0,1])$ is a closed subset of $L_\infty([0,1])$, the same LDP holds in $L_\infty([0,1])$ by again using Lemma 6.4.3, now in the opposite direction. Finally, in view of Lemma 6.5.2, the LDP of $\{\mu_n\}$ in the metric space $L_\infty([0,1])$ follows from that of $\{\tilde{\mu}_n\}$ by an application of Theorem 6.4.15. □

The projective limit approach, which is the key to Lemma 6.5.3 hinges upon the following finite dimensional result. This in turn is a consequence of Cramér's Theorem 6.3.9.

Lemma 6.5.5 *Let J denote the collection of all ordered finite subsets of $(0,1]$. For any $j = \{0 < t_1 < t_2 < \cdots < t_{|j|} \le 1\} \in J$ and any $f : [0,1] \to \mathbb{R}^d$, let $p_j(f)$ denote the vector $(f(t_1), f(t_2), \ldots, f(t_{|j|})) \in (\mathbb{R}^d)^{|j|}$. Then the sequence of laws $\{\mu_n \circ p_j^{-1}\}$ satisfies the LDP in $(\mathbb{R}^d)^{|j|}$ with the good rate function*

$$I_j(\mathbf{z}) = \sum_{\ell=1}^{|j|} (t_\ell - t_{\ell-1}) \Lambda^* \left(\frac{z_\ell - z_{\ell-1}}{t_\ell - t_{\ell-1}} \right), \tag{6.5.75}$$

where $\mathbf{z} = (z_1, \ldots, z_{|j|})$ *and* $t_0 = 0,\ z_0 = 0$.

We next turn to the diffusion counterpart of Theorem 6.5.1. Let $w_t,\ t \in [0,1]$ denote a standard Brownian motion in $\mathbb{R}^d$. Consider the process

$$w_\epsilon(t) = \sqrt{\epsilon} w_t \,,$$

and let ν_ϵ be the probability measure induced by $w_\epsilon(\cdot)$ on $C_0([0,1])$, the space of all continuous functions $\phi : [0,1] \to \mathbb{R}^d$ such that $\phi(0) = 0$, equipped with the supremum norm topology. Note that $\| w_\epsilon \| \underset{\epsilon \to 0}{\longrightarrow} 0$ in probability (actually, almost surely) and exponentially fast in $1/\epsilon$ as implied by the following useful (though elementary) consequence of the reflection principle.

Lemma 6.5.6 *For any integer d and any* $\tau, \epsilon, \delta > 0$,

$$Prob\left(\sup_{0 \le t \le \tau} |w_\epsilon(t)| \ge \delta \right) \le 4de^{-\delta^2/2d\tau\epsilon} \quad . \tag{6.5.76}$$

The LDP for $w_\epsilon(\cdot)$ is stated in the following theorem. Let $H_1 \triangleq \{\int_0^t f(s)ds : f \in L_2([0,1])\}$ denote the space of all absolutely continuous functions with square integrable derivative equipped with the norm $\|g\|_{H_1} = [\int_0^1 |\dot{g}(t)|^2\, dt]^{\frac{1}{2}}$.

Theorem 6.5.7 (Schilder) $\{\nu_\epsilon\}$ *satisfies, in* $C_0([0,1])$, *an LDP with good rate function*

$$I_w(\phi) = \begin{cases} \frac{1}{2} \int_0^1 |\dot{\phi}(t)|^2\, dt, & \phi \in H_1 \\ \infty & \text{otherwise}. \end{cases}$$

Proof. Observe that the process

$$\hat{w}_\epsilon(t) \triangleq w_\epsilon\left(\epsilon\left[\frac{t}{\epsilon}\right]\right)$$

is for $\epsilon_n = \frac{1}{n}$ merely the process $Z_n(\cdot)$ of (6.5.72), for the particular choice of X_i, which are standard Normal random variables in $\mathbb{R}^d$ (namely, of zero mean and of the identity covariance matrix). Combining Theorem 6.5.1 with exponential equivalence leads first to the LDP for $\hat{w}_\epsilon(\cdot)$, and then using Lemma 6.5.6 to the LDP for $w_\epsilon(\cdot)$. □

6.5.2 The Freidlin–Wentzell Theory

The results of Section 6.5.1 are extended here to the case of strong solutions of stochastic differential equations. Note that these, in general, do not possess independent increments. However, some underlying independence exists in the process via the Brownian motion, which generates the diffusion. This is exploited in Section 6.5.2, where large deviations principles are derived by applying various contraction principles.

First consider the following relatively simple situation. Let $\{x_t^\epsilon\}$ be the diffusion process that is the unique solution of the stochastic differential equation

$$dx_t^\epsilon = b(x_t^\epsilon)dt + \sqrt{\epsilon}dw_t \qquad 0 \le t \le 1, \quad x_0^\epsilon = 0\,, \tag{6.5.77}$$

where $b : \mathbb{R} \to \mathbb{R}$ is a uniformly Lipschitz continuous function (namely, $|b(x) - b(y)| \le B|x - y|$). The existence and uniqueness of the strong solution $\{x_t^\epsilon\}$ of (6.5.77) is standard. Let $\tilde{\mu}_\epsilon$ denote the probability measure induced by $\{x_t^\epsilon\}$ on $C_0([0,1])$. Then $\tilde{\mu}_\epsilon = \mu_\epsilon \circ F^{-1}$, where

μ_ϵ is the measure induced by $\{\sqrt{\epsilon}w_t\}$, and the deterministic map $F : C_0([0,1]) \to C_0([0,1])$ is defined by $f = F(g)$, where f is the unique continuous solution of

$$f(t) = \int_0^t b(f(s))ds + g(t)\,, \quad t \in [0,1]\,. \tag{6.5.78}$$

The LDP associated with x_t^ϵ is therefore a direct application of the contraction principle with respect to the map F.

Theorem 6.5.8 *$\{x_t^\epsilon\}$ satisfies the LDP in $C_0([0,1])$ with the good rate function*

$$I(f) \triangleq \begin{cases} \frac{1}{2}\int_0^1 |\dot{f}(t) - b(f(t))|^2 dt & , \quad f \in H_1 \\ \infty & , \quad f \notin H_1\,. \end{cases} \tag{6.5.79}$$

Now, let $\{x_t^\epsilon\}$ be the diffusion process that is the unique solution of the stochastic differential equation

$$dx_t^\epsilon = b(x_t^\epsilon)dt + \sqrt{\epsilon}\sigma(x_t^\epsilon)dw_t, \quad 0 \le t \le T, \quad x_0^\epsilon = x\,, \tag{6.5.80}$$

where $x \in \mathbb{R}^d$ is deterministic, $b : \mathbb{R}^d \to \mathbb{R}^d$ is a uniformly Lipschitz continuous function, all the elements of the diffusion matrix σ are bounded, uniformly Lipschitz continuous functions, and $w_.$ is a standard Brownian motion in $\mathbb{R}^d$. The existence and uniqueness of the strong solution $\{x_t^\epsilon\}$ of (6.5.80) is standard.

The map defined by the process $x_.^\epsilon$ on $C([0,T])$ is measurable but need not be continuous, and thus the proof of Theorem 6.5.8 does not apply directly. Indeed, this noncontinuity is strikingly demonstrated by the fact that the solution to (6.5.80), when w_t is replaced by its polygonal approximation, differs in the limit from x^ϵ by a nonzero (Wong–Zakai) correction term. On the other hand, this correction term is of the order of ϵ, so it is not expected to influence the large deviations results. Such an argument leads to the guess that the appropriate rate function for (6.5.80) is

$$I_{x,T}(f) = \inf_{\{g \in H_1(0,T]) : f(t) = x + \int_0^t b(f(s))ds + \int_0^t \sigma(f(s))\dot{g}(s)ds\}} \frac{1}{2}\int_0^T |\dot{g}(t)|^2 dt\,, \tag{6.5.81}$$

where the infimum over an empty set is taken as $+\infty$, and $|\cdot|$ denotes both the usual Euclidean norm on $\mathbb{R}^d$ and the corresponding operator norm of matrices. The spaces H_1, and $L_2([0,T])$ for $\mathbb{R}^d$-valued functions are defined using this norm.

Theorem 6.5.9 *If all the entries of b and σ are bounded, uniformly Lipschitz continuous functions, then $\{x_t^\epsilon\}$, the solution of (6.5.80), satisfies the LDP in $C([0,T])$ with the good rate function $I_{x,T}(\cdot)$ of (6.5.81).*

Remark: For $\sigma(\cdot)$, a square matrix, and nonsingular diffusions, namely, solutions of (6.5.80) with $a(\cdot) \triangleq \sigma(\cdot)\sigma'(\cdot)$ which is uniformly positive definite, the preceding formula for the rate function simplifies considerably to

$$I_{x,T}(f) = \begin{cases} \frac{1}{2}\int_0^T (\dot{f}(t) - b(f(t))' a^{-1}(f(t))(\dot{f}(t) - b(f(t)))\, dt & , \quad f \in H_1^x \\ \infty & , \quad f \notin H_1^x\,, \end{cases}$$

where $H_1^x \triangleq \{f : f(t) = x + \int_0^t \phi(s)ds,\ \phi \in L_2([0,T])\}$.

The proof is based on approximating the process $x_.^\epsilon$ in the sense of Theorem 6.4.19 by the solution of the stochastic differential equations

$$dx_t^{\epsilon,m} = b(x^{\epsilon,m}_{\frac{[mt]}{m}})dt + \sqrt{\epsilon}\sigma(x^{\epsilon,m}_{\frac{[mt]}{m}})dw_t, \quad 0 \le t \le T, \quad x_0^{\epsilon,m} = 0\,. \tag{6.5.82}$$

Indeed, $\{x^{\epsilon,m}\}$, $m = 1, 2, \ldots$, are shown, by martingale inequalities, to be exponentially good approximations of $\{x^\epsilon\}$. This is achieved by means of the following lemma:

Lemma 6.5.10 *Let b_t, σ_t be progressively measurable processes, and let*

$$dz_t = b_t dt + \sqrt{\epsilon}\sigma_t dw_t \,, \tag{6.5.83}$$

where z_0 is deterministic. Let $\tau_1 \in [0,T]$ be a stopping time with respect to the filtration of $\{w_t, t \in [0,T]\}$. Suppose that the coefficients of the diffusion matrix σ are uniformly bounded, and for some constants M, B, ρ and any $t \in [0,\tau_1]$,

$$\begin{aligned} |\sigma_t| &\le M\left(\rho^2 + |z_t|^2\right)^{1/2} \\ |b_t| &\le B\left(\rho^2 + |z_t|^2\right)^{1/2} . \end{aligned} \tag{6.5.84}$$

Then for any $\delta > 0$ and any $\epsilon \le 1$,

$$\epsilon \log Prob\left(\sup_{t\in[0,\tau_1]} |z_t| \ge \delta\right) \le K + \log\left(\frac{\rho^2 + |z_0|^2}{\rho^2 + \delta^2}\right),$$

where $K = 2B + M^2(2+d)$.

The following theorem strengthens Theorem 6.5.9 by allowing for ϵ dependent initial conditions.

Theorem 6.5.11 *Assume the conditions of Theorem 6.5.9. Let $\{X_t^{\epsilon,y}\}$ denote the solution of (6.5.80) for the initial condition $X_0 = y$. Then:*
(a) For any closed $F \subset C([0,T])$,

$$\limsup_{\substack{\epsilon\to 0\\ y\to x}} \epsilon \log Prob(X_\cdot^{\epsilon,y} \in F) \le -\inf_{\phi\in F} I_{x,T}(\phi) . \tag{6.5.85}$$

(b) For any open $G \subset C([0,T])$,

$$\liminf_{\substack{\epsilon\to 0\\ y\to x}} \epsilon \log Prob(X_\cdot^{\epsilon,y} \in G) \ge -\inf_{\phi\in G} I_{x,T}(\phi) . \tag{6.5.86}$$

The following immediate corollary of Theorem 6.5.11 is used in Section 6.5.3.

Corollary 6.5.12 *Assume the conditions of Theorem 6.5.9. Then for any compact $K \subset \mathbb{R}^d$ and any closed $F \subset C([0,T])$,*

$$\limsup_{\epsilon\to 0} \epsilon \log \sup_{y\in K} Prob(X_\cdot^{\epsilon,y} \in F) \le -\inf_{\substack{\phi\in F\\ y\in K}} I_{y,T}(\phi) . \tag{6.5.87}$$

Similarly, for any open $G \subset C([0,T])$,

$$\liminf_{\epsilon\to 0} \epsilon \log \inf_{y\in K} Prob(X_\cdot^{\epsilon,y} \in G) \ge -\sup_{y\in K}\inf_{\phi\in G} I_{y,T}(\phi) . \tag{6.5.88}$$

6.5.3 Application: The Problem of Diffusion Exit from a Domain

Consider the system

$$dx_t^\epsilon = b(x_t^\epsilon)dt + \sqrt{\epsilon}\sigma(x_t^\epsilon)dw_t, \quad x_t^\epsilon \in \mathbb{R}^d, \quad x_0^\epsilon = x \,, \tag{6.5.89}$$

in the open, bounded domain G, where $b(\cdot)$ and $\sigma(\cdot)$ are uniformly Lipschitz continuous functions of appropriate dimensions and $w_\cdot$ is a standard Brownian motion. The following assumption prevails throughout Section 6.5.3.

Assumption (A-1) *The unique stable equilibrium point in G of the d-dimensional ordinary differential equation*

$$\dot{\phi}_t = b(\phi_t) \tag{6.5.90}$$

is at $0 \in G$, *and*

$$\phi_0 \in G \;\Rightarrow\; \forall t > 0,\; \phi_t \in G \text{ and } \lim_{t\to\infty} \phi_t = 0\,.$$

When ϵ is small, it is reasonable to guess that the system (6.5.89) tends to stay inside G. Indeed, suppose that the boundary of G is smooth enough for

$$\tau^\epsilon \stackrel{\triangle}{=} \inf\{t > 0 :\; x_t^\epsilon \in \partial G\}$$

to be a well-defined stopping time. Under mild conditions, $P(\tau^\epsilon < T) \underset{\epsilon\to 0}{\longrightarrow} 0$ for any $T < \infty$. (This fact follows for example from Theorem 6.5.13.) From an engineering point of view, (6.5.89) models a tracking loop in which some parasitic noise exists. The parasitic noise may exist because of atmospheric noise (e.g., in radar and astronomy), or because of a stochastic element in the signal model (e.g., in a phase lock loop). From that point of view, exiting the domain at ∂G is an undesirable event, for it means the loss of lock. An important question (both in the analysis of a given system and in the design of new systems) would be how probable is the loss of lock.

In many interesting systems, the time to lose lock is measured in terms of a large multiple of the natural time constant of the system. For example, in modern communication systems, where the natural time constant is a bit duration, the error probabilities are in the order of 10^{-7} or 10^{-9}. In such situations, asymptotic computations of the exit time become meaningful.

Another important consideration in designing such systems is the question of where the exit occurs on ∂G, for it may allow design of modified loops, error detectors, etc.

Throughout, E_x denotes expectations with respect to the diffusion process (6.5.89), where $x_0^\epsilon = x$. The following classical theorem characterizes such expectations, for *any* ϵ, in terms of the solutions of appropriate partial differential equations.

Theorem 6.5.13 *Assume that for any* $y \in \partial G$, *there exists a ball* $B(y)$ *such that* $\overline{G} \cap \overline{B(y)} = \{y\}$, *and for some* $\eta > 0$ *and all* $x \in G$, *the matrices* $\sigma(x)\sigma'(x) - \eta I$ *are positive definite. Then for any Hölder continuous function g (on G) and any continuous function f (on* ∂G*), the function*

$$u(x) \stackrel{\triangle}{=} E_x\left[f(x^\epsilon_{\tau^\epsilon}) + \int_0^{\tau^\epsilon} g(x_t^\epsilon)dt \right]$$

has continuous second derivatives on G, is continuous on $\overline{G}$, *and is the unique solution of the partial differential equation*

$$\begin{aligned} L^\epsilon u &= -g \quad in \quad G\,, \\ u &= f \quad on \quad \partial G\,, \end{aligned}$$

where the differential operator L^ϵ *is defined via*

$$L^\epsilon v \stackrel{\triangle}{=} \frac{\epsilon}{2} \sum_{i,j} (\sigma\sigma')_{ij}(x) \frac{\partial^2 v}{\partial x_i \partial x_j} + b(x)'\nabla v\,.$$

The following corollary, obtained by substituting $f \equiv 0$ and $g \equiv 1$ or $g \equiv 0$, is of particular interest.

Corollary 6.5.14 *Assume the conditions of Theorem 6.5.13. Let $u_1(x) = E_x(\tau^\epsilon)$. Then u_1 is the unique solution of*

$$L^\epsilon u_1 = -1, \quad in \quad G \quad ; \quad u_1 = 0\,, \quad on \quad \partial G\,. \tag{6.5.91}$$

Further, let $u_2(x) = E_x(f(x^\epsilon_{\tau^\epsilon}))$. Then for any f continuous, u_2 is the unique solution of

$$L^\epsilon u_2 = 0, \quad in \quad G \quad ; \quad u_2 = f\,, \quad on \quad \partial G\,. \tag{6.5.92}$$

In principle, Corollary 6.5.14 enables the computation of the quantities of interest for any ϵ. However, in general for $d \geq 2$, neither (6.5.91) nor (6.5.92) can be solved explicitly. Moreover, the numerical effort required in solving these equations is considerable, in particular when the solution over a range of values of ϵ is of interest. In view of that, the exit behavior analysis from an asymptotic standpoint is crucial.

Since large deviations estimates are for neighborhoods rather than for points, it is convenient to extend the definition of (6.5.89) to $\mathbb{R}^d$. From here on, it is assumed that the original domain G is smooth enough to allow for such an extension preserving the uniform Lipschitz continuity of $b(\cdot)$, $\sigma(\cdot)$.

Motivated by Theorem 6.5.9, define the cost function

$$\begin{aligned} V(y,z,t) &\triangleq \inf_{\{\phi\in C([0,t]):\phi_t=z\}} I_{y,t}(\phi) \qquad\qquad (6.5.93)\\ &= \inf_{\{u_. \in L_2([0,t]):\phi_t=z \text{ where } \phi_s=y+\int_0^s b(\phi_\theta)d\theta+\int_0^s \sigma(\phi_\theta)u_\theta d\theta\}} \frac{1}{2}\int_0^t |u_s|^2 ds\,, \end{aligned}$$

where $I_{y,t}(\cdot)$ is the good rate function of (6.5.81), which controls the LDP associated with (6.5.89). This function is also denoted as $I_y(\cdot)$, $I_t(\cdot)$ or $I(\cdot)$ if no confusion may arise. Heuristically, $V(y,z,t)$ is the cost of forcing the system (6.5.89) to be at the point z at time t when starting at y. Define

$$V(y,z) \triangleq \inf_{t>0} V(y,z,t)\ .$$

The function $V(0,z)$ is called the *quasi-potential.* The treatment to follow is guided by the heuristics that as $\epsilon \to 0$, the system (6.5.89) wanders around the stable point $x = 0$ for an exponentially long time, during which its chances of hitting any closed set $N \subset \partial G$ are determined by $\inf_{z\in N} V(0,z)$. The rationale here is that any excursion off the stable point $x = 0$ has an overwhelmingly high probability of being pulled back there, and it is not the time spent near any part of ∂G that matters but the *a priori* chance for a direct, fast exit due to a rare segment in the Brownian motion's path. Caution, however, should be exercised, as there are examples where this rationale fails.

For use below, we introduce the following basic assumptions.

Assumption (A-2) *All the trajectories of the deterministic system (6.5.90) starting at $\phi_0 \in \partial G$ converge to 0 as $t \to \infty$.*

Assumption (A-3) $\overline{V} \triangleq \inf_{z\in\partial G} V(0,z) < \infty$.

Assumption (A-4) *There exists an $M < \infty$ such that, for all $\rho > 0$ small enough and all x, y with $|x-z| + |y-z| \leq \rho$ for some $z \in \partial G \cup \{0\}$, there is a function u satisfying that $\|u\| < M$ and $\phi_{T(\rho)} = y$, where*

$$\phi_t = x + \int_0^t b(\phi_s)ds + \int_0^t \sigma(\phi_s)u_s ds$$

and $T(\rho) \to 0$ *as* $\rho \to 0$.

Assumption (A-2) prevents consideration of situations in which ∂G is the *characteristic boundary* of the domain of attraction of 0. Such boundaries arise as the separating curves of several isolated minima, and are of meaningful engineering and physical relevance. Some of the results that follow hold for characteristic boundaries. However, caution is needed in that case. Assumption (A-3) is natural, for otherwise all points on ∂G are equally unlikely on the large deviations scale. Assumption (A-4) is related to the controllability of the system (6.5.89) (where a smooth control replaces the Brownian motion). Note, however, that this is a relatively mild assumption. In particular, if the matrices $\sigma(x)\sigma'(x)$ are positive definite for $x = 0$, and uniformly positive definite on ∂G, then Assumption (A-4) is satisfied.

The following theorem, provides the precise exponential growth rate of τ^ϵ, as well as valuable estimates on the exit measure.

Theorem 6.5.15
(a) Assume (A-1), (A.3), (A-4). For all $x \in G$ *and all* $\delta > 0$,

$$\lim_{\epsilon \to 0} P_x(e^{(\overline{V}+\delta)/\epsilon} > \tau^\epsilon > e^{(\overline{V}-\delta)/\epsilon}) = 1\,. \tag{6.5.94}$$

Moreover, for all $x \in G$,

$$\lim_{\epsilon \to 0} \epsilon \log E_x(\tau^\epsilon) = \overline{V}\,. \tag{6.5.95}$$

(b) Assume (A-1)–(A-4). If $N \subset \partial G$ *is a closed set and* $\inf_{z\in N} V(0,z) > \overline{V}$, *then for any* $x \in G$,

$$\lim_{\epsilon \to 0} P_x(x^\epsilon_{\tau^\epsilon} \in N) = 0\,. \tag{6.5.96}$$

In particular, if there exists $z^* \in \partial G$ *such that* $V(0,z^*) < V(0,z)$ *for all* $z \neq z^*$, $z \in \partial G$, *then*

$$\forall \delta > 0, \forall x \in G, \ \lim_{\epsilon \to 0} P_x(|x^\epsilon_{\tau^\epsilon} - z^*| < \delta) = 1\,. \tag{6.5.97}$$

Remarks:
(a) When the quasi-potential $V(0,\cdot)$ has multiple minima on ∂G, then the question arises as to where the exit occurs. In symmetrical cases, it is easy to see that each minimum point of $V(0,\cdot)$ is equally likely. In general, by part (b) of Theorem 6.5.15, the exit occurs from a neighborhood of the set of minima of the quasi-potential. However, refinements of the underlying large deviations estimates are needed for determining the exact weight among the minima.
(b) The results of Section 6.5.3 can be, and were indeed, extended in various ways to cover general Lévy processes, dynamical systems perturbed by wide-band noise, queuing systems, partial differential equations, etc.
(c) Often, there is interest in the characteristic boundaries for which Assumption (A-2) is violated. This is the case when there are multiple stable points of the dynamical system (6.5.90), and G is just the attraction region of one of them. The exit measure analysis used for proving part (b) of the preceding theorem could in principle be incorrect. That is because the sample path that spends increasingly large times inside G, while avoiding the neighborhood of the stable point $x = 0$, could contribute a nonnegligible probability.
(d) The heuristics behind the proof of Theorem 6.5.15 are as follows: on a fixed time interval T, with T large enough, the exit from the domain is extremely unlikely (of probability roughly $p_\epsilon := e^{-\overline{V}/\epsilon}$), and if exit occurs it must follow, with overwhelming probability, the minimizing paths in (6.5.93) which end on the boundary of G at time T. If exit did

not occur, again with overwhelming probability, the path returns to a neighborhood of the origin. Since the large deviation estimates are uniform in the initial condition, and since $V(y,z)$ is continuous in both variables, the situation is well approximated by independent Bernoulli trials with probability of success p_ϵ. Thus, the number of trials before success occurs is of the order of p_ϵ^{-1}, and the time before first success is of the same (exponential) order.

6.6 LDPs for Empirical Measures

We start this section by providing the general statement of Cramér's and Sanov's theorems, as well as the outline of proof. A new ingredient makes its appearance in this outline; namely, subadditivity is exploited. We then turn to the LDP for the empirical measures of Markov processes and of mixing sequences, concluding with applications to the Gibbs conditioning principle in statistical mechanics and to hypothesis testing in statistics.

The material for this section is mostly taken from Chapter 6 and Sections 3.4 and 7.3 of [DeZ98] to which the reader is refered for details, proofs, and bibliography.

6.6.1 Cramér's Theorem in Polish Spaces

A general version of Cramér's theorem for i.i.d. random variables is presented here. Sanov's theorem is derived in Section 6.6.2 as a consequence of this general formulation. The core new idea in the derivation presented here, namely, the use of subadditivity as a tool for proving the LDP, is applicable beyond the i.i.d. case.

Let μ be a Borel probability measure on a locally convex, Hausdorff, topological real vector space $\mathcal{X}$. On the space $\mathcal{X}^*$ of continuous linear functionals on $\mathcal{X}$, define the logarithmic moment generating function

$$\Lambda(\lambda) \triangleq \log \int_{\mathcal{X}} e^{\langle \lambda, x \rangle} d\mu \,, \tag{6.6.98}$$

and let $\Lambda^*(\cdot)$ denote the Fenchel–Legendre transform of Λ.

For every integer n, suppose that $X_1, \dots, X_n$ are i.i.d. random variables on $\mathcal{X}$, each distributed according to the law μ; namely, their joint distribution μ^n is the product measure on the space $(\mathcal{X}^n, (\mathcal{B}_{\mathcal{X}})^n)$. We would like to consider the partial averages

$$\hat{S}_n^m \triangleq \frac{1}{n-m} \sum_{\ell=m+1}^{n} X_\ell \,,$$

with $\hat{S}_n \triangleq \hat{S}_n^0$ being the empirical mean. Note that $\hat{S}_n^m$ are always measurable with respect to the σ-field $\mathcal{B}_{\mathcal{X}^n}$, because the addition and scalar multiplication are continuous operations on $\mathcal{X}^n$. In general, however, $(\mathcal{B}_{\mathcal{X}})^n \subset \mathcal{B}_{\mathcal{X}^n}$ and $\hat{S}_n^m$ may be nonmeasurable with respect to the product σ-field $(\mathcal{B}_{\mathcal{X}})^n$. When $\mathcal{X}$ is separable, $\mathcal{B}_{\mathcal{X}^n} = (\mathcal{B}_{\mathcal{X}})^n$, and there is no need to further address this measurability issue. In most of the applications we have in mind, the measure μ is supported on a convex subset of $\mathcal{X}$ that is made into a Polish (and hence, separable) space in the topology induced by $\mathcal{X}$. Consequently, in this setup, for every $m, n \in \mathbb{Z}_+$, $\hat{S}_n^m$ is measurable with respect to $(\mathcal{B}_{\mathcal{X}})^n$.

Let μ_n denote the law induced by $\hat{S}_n$ on $\mathcal{X}$. In view of the preceding discussion, μ_n is a Borel measure as soon as the convex hull of the support of μ is separable. The following (technical) assumption formalizes the conditions required for our approach to Cramér's theorem.

Assumption 6.6.99 *(a) $\mathcal{X}$ is a locally convex, Hausdorff, topological real vector space. $\mathcal{E}$ is a closed, convex subset of $\mathcal{X}$ such that $\mu(\mathcal{E}) = 1$ and $\mathcal{E}$ can be made into a Polish space with respect to the topology induced by $\mathcal{X}$.*
(b) The closed convex hull of each compact $K \subset \mathcal{E}$ is compact.

The following is the extension of Cramér's theorem (Theorem 6.3.9).

Theorem 6.6.1 *Let Assumption 6.6.99 hold. Then $\{\mu_n\}$ satisfies in $\mathcal{X}$ (and $\mathcal{E}$) a weak LDP with rate function Λ^*. Moreover, for every open, convex subset $A \subset \mathcal{X}$,*

$$\lim_{n\to\infty} \frac{1}{n} \log \mu_n(A) = -\inf_{x\in A} \Lambda^*(x)\,. \tag{6.6.100}$$

Remarks:
(a) If, instead of part (b) of Assumption 6.6.99, both the exponential tightness of $\{\mu_n\}$ and the finiteness of $\Lambda(\cdot)$ are assumed, then the LDP for $\{\mu_n\}$ is a direct consequence of Corollary 6.4.41.
(b) By Mazur's theorem, part (b) of Assumption 6.6.99 follows from part (a) as soon as the metric $d(\cdot,\cdot)$ of $\mathcal{E}$ satisfies, for all $\alpha \in [0,1]$, $x_1, x_2, y_1, y_2 \in \mathcal{E}$, the convexity condition

$$d(\alpha x_1 + (1-\alpha)x_2, \alpha y_1 + (1-\alpha)y_2) \le \max\{d(x_1,y_1), d(x_2,y_2)\}\,. \tag{6.6.101}$$

This condition is motivated by the two applications we have in mind, namely, either $\mathcal{X} = \mathcal{E}$ is a separable Banach space, or $\mathcal{X} = M(\Sigma), \mathcal{E} = M_1(\Sigma)$ as in Section 6.6.2. It is straight forward to verify that (6.6.101) holds true in both cases.
(c) Observe that $\hat{S}_n^m$ are convex combinations of $\{X_\ell\}_{\ell=m}^n$, and hence with probability one belong to $\mathcal{E}$. Consider the sample space $\Omega = \mathcal{E}^{\mathbb{Z}_+}$ of semi-infinite sequences of points in $\mathcal{E}$ with the product topology inherited from the topology of $\mathcal{E}$. Since $\mathcal{E}$ is separable, the Borel σ-field on Ω is $\mathcal{B}_\Omega = (\mathcal{B}_\mathcal{E})^{\mathbb{Z}_+}$, allowing the *semi-infinite* sequence $X_1, \ldots, X_\ell, \ldots$ to be viewed as a random point in Ω, where the latter is equipped with the Borel product measure $\mu^{\mathbb{Z}_+}$, and with $\hat{S}_n$ being measurable maps from $(\Omega, \mathcal{B}_\Omega)$ to $(\mathcal{E}, \mathcal{B}_\mathcal{E})$. This viewpoint turns out to be particularly useful when dealing with Markov extensions of Theorem 6.6.1.
(d) Cramér's Theorem in $\mathbb{R}^d$ is a direct corollary of Theorem 6.6.1 for $\mathcal{X} = \mathcal{E} = \mathbb{R}^d$.

The proof of Theorem 6.6.1 combines the following key lemmas with a variant of Theorem 6.4.34

Lemma 6.6.2 *Let part (a) of Assumption 6.6.99 hold true. Then, the sequence $\{\mu_n\}$ satisfies the weak LDP in $\mathcal{X}$ with a convex rate function $I(\cdot)$.*

Lemma 6.6.3 *Let Assumption 6.6.99 hold true. Then, for every open, convex subset $A \subset \mathcal{X}$,*

$$\lim_{n\to\infty} \frac{1}{n} \log \mu_n(A) = -\inf_{x\in A} I(x)\,,$$

where $I(\cdot)$ is the convex rate function of Lemma 6.6.2.

We bring below the proof of Lemma 6.6.2, as it exhibits the use of subadditivity in large deviation proofs.

Definition 6.6.4 *A function $f : \mathbb{Z}_+ \to [0,\infty]$ is called subadditive if $f(n+m) \le f(n) + f(m)$ for all $n, m \in \mathbb{Z}_+$.*

Lemma 6.6.5 (Subadditivity) *If $f : \mathbb{Z}_+ \to [0,\infty]$ is a subadditive function such that $f(n) < \infty$ for all $n \ge N$ and some $N < \infty$, then*

$$\lim_{n\to\infty} \frac{f(n)}{n} = \inf_{n\ge N} \frac{f(n)}{n} < \infty\,.$$

The following observation is key to our application of subadditivity.

Lemma 6.6.6 *Let part (a) of Assumption 6.6.99 hold true. Then, for every convex $A \in \mathcal{B}_{\mathcal{X}}$, the function $f(n) \stackrel{\triangle}{=} -\log \mu_n(A)$ is subadditive.*

Proof. Without loss of generality, it may be assumed that $A \subset \mathcal{E}$. Now,

$$\hat{S}_{m+n} = \frac{m}{m+n}\hat{S}_m + \frac{n}{m+n}\hat{S}^m_{m+n}\,.$$

Therefore, $\hat{S}_{m+n}$ is a convex combination (with deterministic coefficients) of the independent random variables $\hat{S}_m$ and $\hat{S}^m_{m+n}$. Thus, by the convexity of A,

$$\{\omega : \hat{S}^m_{m+n}(\omega) \in A\} \cap \{\omega : \hat{S}_m(\omega) \in A\} \subset \{\omega : \hat{S}_{m+n}(\omega) \in A\}\,.$$

Since, evidently,

$$\mu^{n+m}(\{\omega : \hat{S}^m_{m+n}(\omega) \in A\}) = \mu^n(\{\omega : \hat{S}_n(\omega) \in A\})\,,$$

it follows that

$$\mu_n(A)\mu_m(A) \le \mu_{n+m}(A)\,, \tag{6.6.102}$$

or alternatively, $f(n) = -\log \mu_n(A)$ is subadditive. □

The last tool needed for the proof of Lemma 6.6.2 is the following lemma.

Lemma 6.6.7 *Let part (a) of Assumption 6.6.99 hold true. If $A \subset \mathcal{E}$ is (relatively) open and $\mu_m(A) > 0$ for some m, then there exists an $N < \infty$ such that $\mu_n(A) > 0$ for all $n \ge N$.*

Proof of Lemma 6.6.2: Fix an open, convex subset $A \subset \mathcal{X}$. Since $\mu_n(A) = \mu_n(A \cap \mathcal{E})$ for all n, either $\mu_n(A) = 0$ for all n, in which case $\mathcal{L}_A = -\lim_{n\to\infty} \frac{1}{n}\log \mu_n(A) = \infty$, or else the limit

$$\mathcal{L}_A = -\lim_{n\to\infty} \frac{1}{n} \log \mu_n(A)$$

exists by Lemmas 6.6.5, 6.6.6, and 6.6.7.

Let $\mathcal{C}^o$ denote the collection of all open, convex subsets of $\mathcal{X}$. Define

$$I(x) \stackrel{\triangle}{=} \sup\{\mathcal{L}_A : x \in A,\ A \in \mathcal{C}^o\}\,.$$

Applying Theorem 6.4.6 for the base $\mathcal{C}^o$ of the topology of $\mathcal{X}$, it follows that μ_n satisfies the weak LDP with this rate function. To prove that $I(\cdot)$ is convex, we shall apply Lemma 6.4.8. To this end, fix $A_1, A_2 \in \mathcal{C}^o$ and let $A \stackrel{\triangle}{=} (A_1 + A_2)/2$. Then since $(\hat{S}_n + \hat{S}^n_{2n})/2 = \hat{S}_{2n}$, it follows that

$$\mu_n(A_1)\mu_n(A_2) = \mu^{2n}(\{\omega : \hat{S}_n \in A_1\} \cap \{\omega : \hat{S}^n_{2n} \in A_2\}) \le \mu_{2n}(A)\,.$$

Thus, by taking n-limits, the convexity condition (6.4.38) is verified, namely,

$$\limsup_{n\to\infty} \frac{1}{n} \log \mu_n(A) \ge \limsup_{n\to\infty} \frac{1}{2n} \log \mu_{2n}(A) \ge -\frac{1}{2}(\mathcal{L}_{A_1} + \mathcal{L}_{A_2})\,.$$

With (6.4.38) established, Lemma 6.4.8 yields the convexity of I and the proof is complete. □

6.6.2 Sanov's Theorem

This section is about the large deviations of the empirical law of a sequence of i.i.d. random variables; namely, let Σ be a Polish space and let $Y_1, \ldots, Y_n$ be a sequence of independent, Σ-valued random variables, identically distributed according to $\mu \in M_1(\Sigma)$, where $M_1(\Sigma)$ denotes the space of (Borel) probability measures on Σ. With δ_y denoting the probability measure degenerate at $y \in \Sigma$, the *empirical law* of $Y_1, \ldots, Y_n$ is

$$L_n^{\mathbf{Y}} \stackrel{\triangle}{=} \frac{1}{n} \sum_{i=1}^{n} \delta_{Y_i} \in M_1(\Sigma). \tag{6.6.103}$$

Sanov's theorem about the large deviations of $L_n^{\mathbf{Y}}$ is proved in Theorem 6.3.7 for a finite set Σ. Here, the general case is considered. First, the LDP with respect to the weak topology is deduced, based on Cramér's theorem (Theorem 6.6.1). The LDP with respect to a somewhat stronger topology (the τ-topology) is then presented. The latter result may be derived by the projective limit approach of Section 6.4.5.

To set up the framework for applying the results of Section 6.6.1, let $X_i = \delta_{Y_i}$ and observe that $X_1, \ldots, X_n$ are i.i.d. random variables taking values in the real vector space $M(\Sigma)$ of finite (signed) measures on Σ. Moreover, the empirical mean of $X_1, \ldots, X_n$ is $L_n^{\mathbf{Y}}$ and belongs to $M_1(\Sigma)$, which is a convex subset of $M(\Sigma)$. Hence, our program is to equip $\mathcal{X} = M(\Sigma)$ with an appropriate topology and $M_1(\Sigma) = \mathcal{E}$ with the relative topology induced by $\mathcal{X}$, so that all the assumptions of Cramér's theorem (Theorem 6.6.1) hold and a weak LDP for $L_n^{\mathbf{Y}}$ (in $\mathcal{E}$) follows. A full LDP is then deduced by proving that the laws of $L_n^{\mathbf{Y}}$ are exponentially tight in $\mathcal{E}$, and an explicit formula for the rate function in terms of relative entropy is derived by an auxiliary argument.

To this end, let $C_b(\Sigma)$ denote the collection of bounded continuous functions $\phi : \Sigma \to \mathbb{R}$, equipped with the supremum norm, i.e., $\|\phi\| = \sup_{x \in \Sigma} |\phi(x)|$. Equip $M(\Sigma)$ with the *weak topology* generated by the sets $\{U_{\phi,x,\delta}, \phi \in C_b(\Sigma), x \in \mathbb{R}, \delta > 0\}$, where

$$U_{\phi,x,\delta} \stackrel{\triangle}{=} \{\nu \in M(\Sigma) : |\langle \phi, \nu \rangle - x| < \delta\}, \tag{6.6.104}$$

and throughout, $\langle \phi, \nu \rangle \stackrel{\triangle}{=} \int_\Sigma \phi d\nu$ for any $\phi \in C_b(\Sigma)$ and any $\nu \in M(\Sigma)$. The Borel σ-field generated by the weak topology is denoted $\mathcal{B}^w$.

Define the *relative entropy* of the probability measure ν with respect to $\mu \in M_1(\Sigma)$ as

$$H(\nu|\mu) \stackrel{\triangle}{=} \begin{cases} \int_\Sigma f \log f d\mu & \text{if } f \stackrel{\triangle}{=} \frac{d\nu}{d\mu} \text{ exists} \\ \infty & \text{otherwise}, \end{cases}$$

where $d\nu/d\mu$ stands for the Radon-Nikodym derivative of ν with respect to μ when it exists.

Remark: The function $H(\nu|\mu)$ is also referred to as *Kullback-Leibler distance* or *divergence* in the literature. It is worth noting that although $H(\nu|\mu)$ is called a distance, it is not a metric, for $H(\nu|\mu) \neq H(\mu|\nu)$. Moreover, even the symmetric sum $(H(\nu|\mu) + H(\mu|\nu))/2$ does not satisfy the triangle inequality.

We have the following alternative formula for $H(\cdot|\mu)$.

Lemma 6.6.8 *Let $\Lambda(\phi) = \log \int_\Sigma e^\phi d\mu$. Then, for any $\nu \in M_1(\Sigma)$*

$$H(\nu|\mu) = \sup_{\phi \in C_b(\Sigma)} \{\langle \phi, \nu \rangle - \Lambda(\phi)\}.$$

Theorem 6.6.9 (Sanov) *The empirical measures $L_n^{\mathbf{Y}}$ satisfy the LDP in $M_1(\Sigma)$ equipped with the weak topology, with the convex, good rate function $H(\cdot|\mu)$.*

We present below a sketch of the proof of Sanov's theorem:

Since the collection of linear functionals $\{\nu \mapsto \langle \phi, \nu \rangle : \phi \in C_b(\Sigma)\}$ is separating in $M(\Sigma)$, this topology makes $M(\Sigma)$ into a locally convex, Hausdorff topological vector space, whose topological dual is the preceding collection, hereafter identified with $C_b(\Sigma)$. Moreover, $M_1(\Sigma)$ is a closed subset of $M(\Sigma)$, and $M_1(\Sigma)$ is a Polish space when endowed with the relative topology and the Lévy metric. Note that the topology thus induced on $M_1(\Sigma)$ corresponds to the weak convergence of probability measures, and that the Lévy metric satisfies the convexity condition (6.6.101).

The preceding discussion leads to the following immediate corollary of Theorem 6.6.1.

Corollary 6.6.10 *The empirical measures $L_n^{\mathbf{Y}}$ satisfy a weak LDP in $M_1(\Sigma)$ (equipped with the weak topology and $\mathcal{B} = \mathcal{B}^w$) with the convex rate function*

$$\Lambda^*(\nu) = \sup_{\phi \in C_b(\Sigma)} \{\langle \phi, \nu \rangle - \Lambda(\phi)\}, \quad \nu \in M_1(\Sigma) \ . \tag{6.6.105}$$

The strengthening of this corollary to a full LDP with a good rate function $H(\cdot|\mu)$ is accomplished by combining Lemma 6.6.8 and

Lemma 6.6.11 *The laws of $L_n^{\mathbf{Y}}$ of (6.6.103) are exponentially tight.*

Proof. There exist compact sets $\Gamma_\ell \subset \Sigma$, $\ell = 1, 2, \ldots$ such that

$$\mu(\Gamma_\ell^c) \le e^{-2\ell^2}(e^\ell - 1) \ . \tag{6.6.106}$$

Then, for any ℓ, the set of measures

$$K^\ell = \left\{\nu : \ \nu(\Gamma_\ell) \ge 1 - \frac{1}{\ell}\right\}$$

is closed. For $L = 1, 2, \ldots$ define the compact set

$$K_L \stackrel{\triangle}{=} \bigcap_{\ell=L}^{\infty} K^\ell \subset M_1(\Sigma) \ .$$

Chebycheff's bound implies then that

$$\mathrm{Prob}(L_n^{\mathbf{Y}} \notin K^\ell) \le e^{-n\ell} \ .$$

Hence, using the union of events bound,

$$\limsup_{n \to \infty} \frac{1}{n} \log \mathrm{Prob}(L_n^{\mathbf{Y}} \in K_L^c) \le -L \ .$$

Thus, the laws of $L_n^{\mathbf{Y}}$ are exponentially tight. □

Next, we present a generalized version of Sanov's theorem, due to de Acosta [deA94], with minimal topological assumptions.

Let $(\Sigma, \mathcal{S})$ be a measurable space and let $B(\Sigma)$ be the space of bounded real-valued $\mathcal{S}$-measurable functions defined on Σ. The τ-topology on the space $M_1(\Sigma)$ of probability measures on $(\Sigma, \mathcal{S})$ is the smallest topology such that for each $f \in B(\Sigma)$, the map $f \mapsto \int f \, d\nu : M_1(\Sigma) \to \mathbb{R}$ is continuous. For $A \subset M_1(\Sigma)$, we denote by $\mathrm{cl}_\tau(A)$ (resp., $\mathrm{int}_\tau(A)$) the closure (resp., interior) of A in the τ-topology. The σ-algebra $\mathcal{B}$ on $M_1(\Sigma)$ is defined to be the smallest σ-algebra such that for each $f \in B(\Sigma)$, the map $f \mapsto \int f \, d\nu : M_1(\Sigma) \to \mathbb{R}$ is measurable. Let $Y_1, \ldots, Y_n$ denote i.i.d., Σ-valued random variables of law μ. Note that Σ is not required to be a topological space.

Theorem 6.6.12 *For every set $A \in \mathcal{B}$*

$$\limsup_{n\to\infty} \frac{1}{n} \log P(L_n^{\mathbf{Y}} \in A) \leq - \inf_{\nu \in \mathrm{cl}_\tau(A)} H(\nu|\mu),$$

$$\liminf_{n\to\infty} \frac{1}{n} \log P(L_n^{\mathbf{Y}} \in A) \geq - \inf_{\nu \in \mathrm{int}_\tau(A)} H(\nu|\mu).$$

Proof. See [deA94].

6.6.3 LDP for Empirical Measures of Markov Chains

Let Σ be a Polish space, and let $M_1(\Sigma)$ denote the space of Borel probability measures on Σ equipped with the Lévy metric, making it into a Polish space with convergence compatible with the weak convergence. Let $\pi(\sigma, \cdot)$ be a transition probability measure (also called Markov or transition kernel), i.e., for all $\sigma \in \Sigma$, $\pi(\sigma,\cdot) \in M_1(\Sigma)$ and $\sigma \mapsto \pi(\sigma, A)$ is measurable for each $A \in \mathcal{B}_\Sigma$.

Let $\Omega = \Sigma^{\mathbb{Z}_+}$ be the space of semi-infinite sequences with values in Σ, equipped with the product topology, and denote by Y_n the coordinates in the sequence, i.e., $Y_n(\omega_1, \ldots, \omega_n, \ldots) = \omega_n$. Ω is a Polish space and its Borel σ-field is precisely $(\mathcal{B}_\Sigma)^{\mathbb{Z}_+}$. Let $\mathcal{F}_n$ denote the σ-field generated by $\{Y_m,\ 1 \leq m \leq n\}$. Fixing the initial measure $P_1 \in M_1(\Sigma)$, a measure P on Ω can be uniquely constructed by the relations $P(Y_{n+1} \in \Gamma|\mathcal{F}_n) = \pi(Y_n, \Gamma)$, a.s. P for every $\Gamma \in \mathcal{B}_\Sigma$ and every $n \in \mathbb{Z}_+$. That is, let the marginals $P_n \in M_1(\Sigma^n)$ be such that for any $n \geq 1$ and any $\Gamma \in \mathcal{B}_{\Sigma^n}$,

$$\begin{aligned} P(\{\sigma \in \Sigma^{\mathbb{Z}_+} : (\sigma_1, \ldots, \sigma_n) \in \Gamma\}) &= \int_\Gamma P_1(dx_1) \prod_{i=1}^{n-1} \pi(x_i, dx_{i+1}) \\ &= \int_\Gamma P_n(dx_1, \ldots, dx_n). \end{aligned}$$

Define the (random) probability measure

$$L_n^{\mathbf{Y}} \stackrel{\triangle}{=} \frac{1}{n} \sum_{i=1}^{n} \delta_{Y_i} \in M_1(\Sigma),$$

and denote by μ_n the probability distribution of the $M_1(\Sigma)$-valued random variable $L_n^{\mathbf{Y}}$. We derive the LDP for μ_n, which, obviously, may also lead by contraction to the LDP for the empirical mean.

The following uniformity assumption, due to de Acosta [deA90] is sufficient for the LDP to hold (for any fixed initial measure P_1).

Assumption (DU) $\pi(\cdot,\cdot)$ is an irreducible Feller kernel, such that for some $\ell \geq 1$ the collection $\{\pi^\ell(\sigma,\cdot) : \sigma \in \Sigma\}$ is tight and there exists an irreducibility measure ϕ (that is, $\phi \in M_1(\Sigma)$ such that $\phi(A) > 0$ implies $\sum_{n=1}^\infty \pi^n(\sigma, A) > 0$ for all $\sigma \in \Sigma$), such that $\phi(\Gamma) = 0$ implies that for all $\sigma \in \Sigma$, $\pi^m(\sigma, \Gamma) = 0$ for some $m = m(\sigma) \geq 1$.

Theorem 6.6.13 *Assume (DU). Then $\{\mu_n\}$ satisfies a full LDP in $M_1(\Sigma)$ with the convex, good rate function*

$$I_1(\nu) = \sup_{u \in B(\Sigma), u \geq 1} \left\{ - \int_\Sigma \log\left(\frac{\pi u}{u}\right) d\nu \right\}. \tag{6.6.107}$$

Moreover,

$$I_1(\nu) = \sup_{f \in C_b(\Sigma)} \{\langle f, \nu \rangle - \Lambda(f)\} , \tag{6.6.108}$$

where for any $f \in C_b(\Sigma)$,

$$\Lambda(f) = \limsup_{n \to \infty} \frac{1}{n} \log \sup_{P_1 \in M_1(\Sigma)} E\Big(\exp(\sum_{i=1}^{n} f(Y_i))\Big) . \tag{6.6.109}$$

Proof. See [deA90].

The LDP of Theorem 6.6.13 may easily be extended to the empirical measure of k-tuples, i.e.,

$$L_{n,k}^{\mathbf{Y}} \stackrel{\triangle}{=} \frac{1}{n} \sum_{i=1}^{n} \delta_{(Y_i, Y_{i+1}, \ldots, Y_{i+k-1})} \in M_1(\Sigma^k) ,$$

where hereafter $k \geq 2$. The starting point for the derivation of the LDP for $L_{n,k}^{\mathbf{Y}}$ lies in the observation that if the sequence $\{Y_n\}$ is a Markov chain with state space Σ and transition kernel $\pi(x, dy)$, then the sequence $\{(Y_n, \ldots, Y_{n+k-1})\}$ is a Markov chain with state space Σ^k and transition kernel

$$\pi_k(x, dy) = \pi(x_k, dy_k) \prod_{i=1}^{k-1} \delta_{x_{i+1}}(y_i) ,$$

where $y = (y_1, \ldots, y_k)$, $x = (x_1, \ldots, x_k) \in \Sigma^k$. Moreover, if π satisfies Assumption (DU), then so does π_k (see [deA90] for details). The following corollary is thus obtained by applying Theorem 6.6.13 to $L_{n,k}^{\mathbf{Y}}$.

Corollary 6.6.14 *Assume π satisfies Assumption (DU). Then $L_{n,k}^{\mathbf{Y}}$ satisfies (in the weak topology of $M_1(\Sigma^k)$) the LDP with the good rate function*

$$I_k(\nu) \stackrel{\triangle}{=} \sup_{u \in B(\Sigma^k), u \geq 1} \left\{ - \int_{\Sigma^k} \log\left(\frac{\pi_k u}{u}\right) d\nu \right\} .$$

To further identify $I_k(\cdot)$, the following definitions and notations are introduced.

Definition 6.6.15 *A measure $\nu \in M_1(\Sigma^k)$ is called shift invariant if, for any $\Gamma \in \mathcal{B}_{\Sigma^{k-1}}$,*

$$\nu(\{\sigma \in \Sigma^k : (\sigma_1, \ldots, \sigma_{k-1}) \in \Gamma\}) = \nu(\{\sigma \in \Sigma^k : (\sigma_2, \ldots, \sigma_k) \in \Gamma\}) .$$

Next, for any $\mu \in M_1(\Sigma^{k-1})$, define the probability measure $\mu \otimes_k \pi \in M_1(\Sigma^k)$ by

$$\mu \otimes_k \pi(\Gamma) = \int_{\Sigma^{k-1}} \mu(dx) \int_{\Sigma} \pi(x_{k-1}, dy) 1_{\{(x,y) \in \Gamma\}} , \quad \forall \Gamma \in \mathcal{B}_{\Sigma^k} .$$

Theorem 6.6.16 *For any transition kernel π, and any $k \geq 2$,*

$$I_k(\nu) = \begin{cases} H(\nu | \nu_{k-1} \otimes_k \pi) & , \quad \nu \text{ shift invariant} \\ \infty & , \quad \text{otherwise} , \end{cases}$$

where ν_{k-1} denotes the marginal of ν on the first $(k-1)$ coordinates.

The LDP of Corollary 6.6.14 and Theorem 6.6.16 enables the deviant behavior of empirical means of fixed length sequences to be dealt with as the number of terms n in the empirical sum grows. Often, however, some information is needed on the behavior of sequences whose length is not bounded with n. It then becomes useful to consider sequences of infinite length. Formally, one could form the empirical measure

$$L_{n,\infty}^{\mathbf{Y}} \triangleq \frac{1}{n} \sum_{i=1}^{n} \delta_{T^i \mathbf{Y}},$$

where $\mathbf{Y} = (Y_1, Y_2, \dots)$ and $T^i\mathbf{Y} = (Y_{i+1}, Y_{i+2}, \dots)$, and inquire about the LDP of the random variable $L_{n,\infty}^{\mathbf{Y}}$ in the space of probability measures on $\Sigma^{\mathbb{Z}_+}$. Since such measures may be identified with probability measures on processes, this LDP is referred to as *process level* LDP.

A natural point of view is to consider the infinite sequences $\mathbf{Y}$ as limits of finite sequences, and to use a projective limit approach. Therefore, the discussion on the process level LDP begins with some topological preliminaries and an exact definition of the probability spaces involved. Since the projective limit approach necessarily involves weak topology, only the weak topologies of $M_1(\Sigma)$ and $M_1(\Sigma^{\mathbb{Z}_+})$ will be considered.

As in the beginning of this section, let Σ be a Polish space, equipped with the metric d and the Borel σ-field $\mathcal{B}_\Sigma$ associated with it, and let Σ^k denote its kth-fold product, whose topology is compatible with the metric $d_k(\sigma, \sigma') = \sum_{i=1}^k d(\sigma_i, \sigma_i')$. The sequence of spaces Σ^k with the obvious projections $p_{m,k} : \Sigma^m \to \Sigma^k$, defined by $p_{m,k}(\sigma_1, \dots, \sigma_m) = (\sigma_1, \dots, \sigma_k)$ for $k \le m$, form a projective system with projective limit that is denoted $\Sigma^{\mathbb{Z}_+}$, and canonical projections $p_k : \Sigma^{\mathbb{Z}_+} \to \Sigma^k$. Since Σ^k are separable spaces and $\Sigma^{\mathbb{Z}_+}$ is countably generated, it follows that $\Sigma^{\mathbb{Z}_+}$ is separable, and the Borel σ-field on $\Sigma^{\mathbb{Z}_+}$ is the product of the appropriate Borel σ-fields. Finally, the projective topology on $\Sigma^{\mathbb{Z}_+}$ is compatible with the metric

$$d_\infty(\sigma, \sigma') = \sum_{k=1}^{\infty} \frac{1}{2^k} \left[\frac{d_k(p_k\sigma, p_k\sigma')}{1 + d_k(p_k\sigma, p_k\sigma')} \right],$$

which makes $\Sigma^{\mathbb{Z}_+}$ into a Polish space. Consider now the spaces $M_1(\Sigma^k)$, equipped with the weak topology and the projections $p_{m,k} : M_1(\Sigma^m) \to M_1(\Sigma^k)$, $k \le m$, such that $p_{m,k}\nu$ is the marginal of $\nu \in M_1(\Sigma^m)$ with respect to its first k coordinates. The projective limit of this projective system is merely $M_1(\Sigma^{\mathbb{Z}_+})$ as stated in the following lemma.

Lemma 6.6.17 *The projective limit of $(M_1(\Sigma^k), p_{m,k})$ is homeomorphic to the space $M_1(\Sigma^{\mathbb{Z}_+})$ when the latter is equipped with the weak topology.*

Returning to the empirical process, observe that for each k, $p_k(L_{n,\infty}^{\mathbf{Y}}) = L_{n,k}^{\mathbf{Y}}$. The following is therefore an immediate consequence of Lemma 6.6.17, the Dawson–Gärtner theorem (Theorem 6.4.37), and the LDP of Corollary 6.6.14 and Theorem 6.6.16.

Corollary 6.6.18 *Assume that (DU) holds. Then the sequence $\{L_{n,\infty}^{\mathbf{Y}}\}$ satisfies the LDP in $M_1(\Sigma^{\mathbb{Z}_+})$ (equipped with the weak topology) with the good rate function*

$$I_\infty(\nu) = \begin{cases} \sup_{k \ge 2} H(p_k\nu | p_{k-1}\nu \otimes_k \pi) & , \quad \nu \text{ shift invariant} \\ \infty & , \quad \text{otherwise} \end{cases}$$

where $\nu \in M_1(\Sigma^{\mathbb{Z}_+})$ is called shift invariant if, for all $k \in \mathbb{Z}_+$, $p_k\nu$ is shift invariant in $M_1(\Sigma^k)$.

Our goal now is to derive an explicit expression for $I_\infty(\cdot)$. For $i = 0, 1$, let $\mathbb{Z}_i = \mathbb{Z} \cap (-\infty, i]$ and let $\Sigma^{\mathbb{Z}_i}$ be constructed similarly to $\Sigma^{\mathbb{Z}_+}$ via projective limits. For any $\mu \in M_1(\Sigma^{\mathbb{Z}_+})$ shift invariant, consider the measures $\mu_i^* \in M_1(\Sigma^{\mathbb{Z}_i})$ such that for every $k \geq 1$ and every $\mathcal{B}_{\Sigma^k}$ measurable set Γ,

$$\mu_i^*(\{(\ldots, \sigma_{i+1-k}, \ldots, \sigma_i) : (\sigma_{i+1-k}, \ldots, \sigma_i) \in \Gamma\}) = p_k \mu(\Gamma) .$$

Such a measure exists and is unique by the consistency condition satisfied by μ and Kolmogorov's extension theorem. Next, for any $\mu_0^* \in M_1(\Sigma^{\mathbb{Z}_0})$, define the Markov extension of it, denoted $\mu_0^* \otimes \pi \in M_1(\Sigma^{\mathbb{Z}_1})$, such that for any $\phi \in B(\Sigma^{k+1})$, $k \geq 1$,

$$\int_{\Sigma^{\mathbb{Z}_1}} \phi(\sigma_{-(k-1)}, \ldots, \sigma_0, \sigma_1) d\mu_0^* \otimes \pi$$
$$= \int_{\Sigma^{\mathbb{Z}_0}} \int_{\Sigma} \phi(\sigma_{-(k-1)}, \ldots, \sigma_0, \tau) \pi(\sigma_0, d\tau) d\mu_0^* .$$

In these notations, for any shift invariant $\nu \in M_1(\Sigma^{\mathbb{Z}_+})$,

$$H(p_k \nu | p_{k-1} \nu \otimes_k \pi) = H(\bar{p}_k \nu_1^* | \bar{p}_k (\nu_0^* \otimes \pi)) ,$$

where for any $\mu \in M_1(\Sigma^{\mathbb{Z}_1})$, and any $\Gamma \in \mathcal{B}_{\Sigma^k}$,

$$\bar{p}_k \mu(\Gamma) = \mu(\{(\sigma_{k-2}, \ldots, \sigma_0, \sigma_1) \in \Gamma\}) .$$

The characterization of $I_\infty(\cdot)$ is a direct consequence of the following classical lemma.

Lemma 6.6.19 (Pinsker) *Let Σ be Polish and $\nu, \mu \in M_1(\Sigma^{\mathbb{Z}_1})$. Then*

$$H(\bar{p}_k \nu | \bar{p}_k \mu) \nearrow H(\nu | \mu) \text{ as } k \to \infty.$$

Combining Corollary 6.6.18, Lemma 6.6.19 and the preceding discussion, the following identification of $I_\infty(\cdot)$ is obtained.

Corollary 6.6.20

$$I_\infty(\nu) = \begin{cases} H(\nu_1^* | \nu_0^* \otimes \pi) & , \quad \nu \text{ shift invariant} \\ \infty & , \quad \text{otherwise} . \end{cases}$$

6.6.4 Mixing Conditions and LDP

The goal of Section 6.6.4 is to establish the LDP for stationary processes satisfying a certain mixing condition. Bryc's theorem (Theorem 6.4.30) is first applied to establish the LDP of the empirical mean for a class of stationary processes taking values in a convex compact subset of $\mathbb{R}^d$. This result is then combined with the projective limit approach to yield the LDP for the empirical measures of a class of stationary processes taking values in Polish spaces.

Let $X_1, \ldots, X_n, \ldots$ be a stationary process taking values in a convex, compact set $K \subset \mathbb{R}^d$. Let

$$\hat{S}_n^m = \frac{1}{n-m} \sum_{i=m+1}^{n} X_i ,$$

with $\hat{S}_n = \hat{S}_n^0$ and μ_n denoting the law of $\hat{S}_n$. The following mixing assumption implies the LDP for μ_n.

Assumption 6.6.110 *For any continuous $f : K \to [0,1]$, there exist $\beta(\ell) \geq 1$, $\gamma(\ell) \geq 0$ and $\delta > 0$ such that*

$$\lim_{\ell\to\infty} \gamma(\ell) = 0 \ , \ \limsup_{\ell\to\infty} (\beta(\ell) - 1)\ell(\log \ell)^{1+\delta} < \infty \,, \tag{6.6.111}$$

and when ℓ and $n + m$ are large enough,

$$\begin{aligned} E[f(\hat{S}_n)^n f(\hat{S}_{n+m+\ell}^{n+\ell})^m] \ \geq\ & E[f(\hat{S}_n)^n]E[f(\hat{S}_m)^m] \\ & -\gamma(\ell)\left\{E[f(\hat{S}_n)^n]E[f(\hat{S}_m)^m]\right\}^{1/\beta(\ell)} \end{aligned} \tag{6.6.112}$$

Indeed, an application of subadditivity and Theorem 6.4.30 yields

Theorem 6.6.21 *Let Assumption 6.6.110 hold. Then $\{\mu_n\}$ satisfies the LDP in $\mathbb{R}^d$ with the good convex rate function $\Lambda^*(\cdot)$, which is the Fenchel–Legendre transform of*

$$\Lambda(\lambda) = \lim_{n\to\infty} \frac{1}{n} \log E[e^{n\langle\lambda, \hat{S}_n\rangle}] \,. \tag{6.6.113}$$

In particular, the limit (6.6.113) exists.

Remark: Assumption 6.6.110, and hence Theorem 6.6.21, hold when $X_1, \ldots, X_n, \ldots$ is a bounded, ψ-mixing process. Other strong mixing conditions that suffice for Theorems 6.6.21 and 6.6.23 to hold are provided in [BryD96].

The main ingredient needed for the application of (approximate) subadditivity is

Lemma 6.6.22 (Hammersley) *Assume $f : \mathbb{Z}_+ \to \mathbb{R}$ is such that for all $n, m \geq 1$,*

$$f(n+m) \leq f(n) + f(m) + \epsilon(n+m) \,, \tag{6.6.114}$$

where $\epsilon(n)$ is non-decreasing such that

$$\sum_{r=1}^{\infty} \frac{\epsilon(r)}{r(r+1)} < \infty \,. \tag{6.6.115}$$

Then $\bar{f} = \lim_{n\to\infty} [f(n)/n]$ exists.

Remark: Hammersley in [Ham62] shows that (6.6.115) is necessary for the existence of $\bar{f} < \infty$, and provides explicit upper bounds on $\bar{f} - f(m)/m$ for every $m \geq 1$.

The previous theorem coupled with Corollary 6.4.40 allow one to deduce the LDP for quite a general class of processes. Let Σ be a Polish space, and $B(\Sigma)$ the space of all bounded, measurable real-valued functions on Σ, equipped with the supremum norm.

Let $\Omega = \Sigma^{\mathbb{Z}_+}$, let P be a stationary and ergodic measure on Ω, and let $Y_1, \ldots, Y_n, \ldots$ denote its realization. Throughout, P_n denotes the nth marginal of P, i.e., the measure on Σ^n whose realization is $Y_1, \ldots, Y_n$. As in Section 6.6.2, $L_n^{\mathbf{Y}} = \frac{1}{n}\sum_{i=1}^n \delta_{Y_i} \in M_1(\Sigma)$, and μ_n denotes the probability measure induced on $M_1(\Sigma)$ by $L_n^{\mathbf{Y}}$.

For any given integers $r \geq k \geq 2$, $\ell \geq 1$, a family of functions $\{f_i\}_{i=1}^k \in B(\Sigma^r)$ is called *ℓ-separated* if there exist k disjoint intervals $\{a_i, a_i+1, \ldots, b_i\}$ with $a_i \leq b_i \in \{1, \ldots, r\}$ such that $f_i(\sigma_1, \ldots, \sigma_r)$ is actually a bounded measurable function of $\{\sigma_{a_i}, \ldots, \sigma_{b_i}\}$ and for all $i \neq j$ either $a_i - b_j \geq \ell$ or $a_j - b_i \geq \ell$.

Assumption (H-1) *There exist $\ell, \alpha < \infty$ such that, for all $k, r < \infty$, and any ℓ-separated functions $f_i \in B(\Sigma^r)$,*

$$E_P(|f_1(Y_1, \ldots, Y_r) \cdots f_k(Y_1, \ldots, Y_r)|) \leq \prod_{i=1}^k E_P(|f_i(Y_1, \ldots, Y_r)|^\alpha)^{1/\alpha} \,. \tag{6.6.116}$$

Assumption (H-2) *There exist a constant ℓ_0 and functions $\beta(\ell) \geq 1$, $\gamma(\ell) \geq 0$ such that, for all $\ell > \ell_0$, all $r < \infty$, and any two ℓ-separated functions $f, g \in B(\Sigma^r)$,*

$$|E_P(f(Y_1, \dots, Y_r))E_P(g(Y_1, \dots, Y_r)) - E_P(f(Y_1, \dots, Y_r)g(Y_1, \dots, Y_r))|$$

$$\leq \gamma(\ell)E_P\left(|f(Y_1, \dots, Y_r)|^{\beta(\ell)}\right)^{1/\beta(\ell)} E_P\left(|g(Y_1, \dots, Y_r)|^{\beta(\ell)}\right)^{1/\beta(\ell)},$$

and $\lim_{\ell\to\infty} \gamma(\ell) = 0$, $\limsup_{\ell\to\infty}(\beta(\ell) - 1)\ell(\log \ell)^{1+\delta} < \infty$ *for some* $\delta > 0$.

Remarks:
(a) Conditions of the type (H-1) and (H-2) are referred to as *hypermixing* conditions. Hypermixing is tied to analytical properties of the semigroup in Markov processes. For details, consult the excellent exposition in [DeuS89]. Note, however, that in (H-2) of the latter, a less stringent condition is put on $\beta(\ell)$, whereas in (H-1) there, $\alpha(\ell)$ converges to one.
(b) The particular case of $\beta(\ell) = 1$ in Assumption (H-2) corresponds to ψ-mixing [Bra86], with $\gamma(\ell) = \psi(\ell)$.

Assumptions (H-1) and (H-2) lead to the following LDP for $L_n^{\mathbf{Y}}$.

Theorem 6.6.23 *Let $Y_1, \dots, Y_n, \dots$ be the stationary process defined before. Assume (H-1), (H-2). Then $L_n^{\mathbf{Y}}$ satisfies in $M_1(\Sigma)$ the LDP with the convex good rate function*

$$\Lambda^*(\nu) = \sup_{f\in B(\Sigma)} \left(\langle f, \nu\rangle - \Lambda(f)\right),$$

where for any $f \in B(\Sigma)$,

$$\Lambda(f) = \lim_{n\to\infty} \frac{1}{n} \log E\left(\exp\left(\sum_{i=1}^{n} f(Y_i)\right)\right).$$

In particular, the preceding limit exists.

6.6.5 Application: The Gibbs Conditioning Principle

Let Σ be a Polish space and $Y_1, Y_2, \dots, Y_n$ a sequence of Σ-valued i.i.d. random variables, each distributed according to the law $\mu \in M_1(\Sigma)$. Let $L_n^{\mathbf{Y}} \in M_1(\Sigma)$ denote the empirical measure associated with these variables. Given a functional $\Phi : M_1(\Sigma) \to \mathbb{R}$ (the *energy* functional), we are interested in computing the law of Y_1 under the constraint $\Phi(L_n^{\mathbf{Y}}) \in D$, where D is some measurable set in $\mathbb{R}$ and $\{\Phi(L_n^{\mathbf{Y}}) \in D\}$ is of positive probability. This situation occurs naturally in statistical mechanics, where Y_i denote some attribute of independent particles (e.g., their velocity), Φ is some constraint on the ensemble of particles (e.g., an average energy per particle constraint), and one is interested in making predictions on individual particles based on the existence of the constraint. The distribution of Y_1 under the energy conditioning alluded to before is then called the *micro-canonical* distribution of the system.

For every measurable set $A \subset M_1(\Sigma)$ such that $\{L_n^{\mathbf{Y}} \in A\}$ is of positive probability, and every bounded measurable function $f : \Sigma \to \mathbb{R}$, due to the exchangeability of the Y_i-s,

$$E(f(Y_1)|L_n^{\mathbf{Y}} \in A) = E(\langle f, L_n^{\mathbf{Y}}\rangle | L_n^{\mathbf{Y}} \in A). \tag{6.6.117}$$

Thus, for $A \triangleq \{\nu : \Phi(\nu) \in D\}$, computing the conditional law of Y_1 under the conditioning $\{\Phi(L_n^{\mathbf{Y}}) \in D\} = \{L_n^{\mathbf{Y}} \in A\}$ is equivalent to the computation of the conditional expectation of $L_n^{\mathbf{Y}}$ under the same constraint. It is this last problem that is treated in the rest of this section, in a slightly more general framework.

Throughout this section, $M_1(\Sigma)$ is equipped with the τ-topology and the corresponding σ-field $\mathcal{B}$. (For the definitions see end of Section 6.6.2.)

For any $\mu \in M_1(\Sigma)$, let $\mu^n \in M_1(\Sigma^n)$ denote the induced product measure on Σ^n and let Q_n be the measure induced by μ^n in $M_1(\Sigma)$ through $L_n^{\mathbf{Y}}$. Let A_δ, $\delta > 0$ be nested measurable sets, i.e., $A_\delta \subseteq A_{\delta'}$ if $\delta < \delta'$. Let F_δ be nested *closed* sets such that $A_\delta \subseteq F_\delta$. Define $F_0 = \cap_{\delta>0} F_\delta$ and $A_0 = \cap_{\delta>0} A_\delta$ (so that $A_0 \subseteq F_0$). The following assumption prevails in this section.

Assumption (A-1) *There exists a $\nu_* \in A_0$ (not necessarily unique) satisfying*

$$H(\nu_*|\mu) = \inf_{\nu \in F_0} H(\nu|\mu) \stackrel{\triangle}{=} I_F < \infty \,,$$

and for all $\delta > 0$,

$$\lim_{n\to\infty} \nu_*^n(\{L_n^{\mathbf{Y}} \in A_\delta\}) = 1 \,. \tag{6.6.118}$$

Think of the following situation as representative: $A_\delta = \{\nu : |\Phi(\nu)| \le \delta\}$, where $\Phi : M_1(\Sigma) \to [-\infty, \infty]$ is only lower semicontinuous, and thus A_δ is neither open nor closed. (For example, the energy functional $\Phi(\nu) = \int_\Sigma (\| x \|^2 - 1)\nu(dx)$ when Σ is a separable Banach space.) The nested, closed sets F_δ are then chosen as $F_\delta = \{\nu : \Phi(\nu) \le \delta\}$ with $F_0 = \{\nu : \Phi(\nu) \le 0\}$, while $A_0 = \{\nu : \Phi(\nu) = 0\}$. We are then interested in the conditional distribution of Y_1 under a constraint of the form $\Phi(L_n^{\mathbf{Y}}) = 0$ (for example, a specified average energy). The following is a direct consequence of Theorem 6.6.12.

Theorem 6.6.24 *Assume (A-1). Then $\mathcal{M} \stackrel{\triangle}{=} \{\nu \in F_0 : H(\nu|\mu) = I_F\}$ is a nonempty, compact set. Further, for any measurable Γ with $\mathcal{M} \subset \Gamma^o$,*

$$\limsup_{\delta\to 0} \limsup_{n\to\infty} \frac{1}{n} \log \mu^n(L_n^{\mathbf{Y}} \notin \Gamma |\, L_n^{\mathbf{Y}} \in A_\delta) < 0 \,.$$

The following corollary shows that if ν_* of (A-1) is unique, then $\mu^n_{\mathbf{Y}^k|A_\delta}$, the law of $\mathbf{Y}^k = (Y_1, \ldots, Y_k)$ conditional upon the event $\{L_n^{\mathbf{Y}} \in A_\delta\}$, is approximately a product measure.

Corollary 6.6.25 *If $\mathcal{M} = \{\nu_*\}$ then $\mu^n_{\mathbf{Y}^k|A_\delta} \to (\nu_*)^k$ weakly in $M_1(\Sigma^k)$ for $n \to \infty$ followed by $\delta \to 0$.*

Proof. Assume $\mathcal{M} = \{\nu_*\}$ and fix $\phi_j \in C_b(\Sigma)$, $j = 1, \ldots, k$. By the invariance of $\mu^n_{\mathbf{Y}^n|A_\delta}$ with respect to permutations of $\{Y_1, \ldots, Y_n\}$,

$$\langle \prod_{j=1}^k \phi_j, \mu^n_{\mathbf{Y}^k|A_\delta} \rangle = \frac{(n-k)!}{n!} \sum_{i_1 \ne \cdots \ne i_k} \int_{\Sigma^n} \prod_{j=1}^k \phi_j(y_{i_j}) \mu^n_{\mathbf{Y}^n|A_\delta}(d\mathbf{y}) \,.$$

Since,

$$E(\prod_{j=1}^k \langle \phi_j, L_n^{\mathbf{Y}} \rangle \,|\, L_n^{\mathbf{Y}} \in A_\delta) = \frac{1}{n^k} \sum_{i_1, \ldots, i_k} \int_{\Sigma^n} \prod_{j=1}^k \phi_j(y_{i_j}) \mu^n_{\mathbf{Y}^n|A_\delta}(d\mathbf{y}) \,,$$

and ϕ_j are bounded functions, it follows that

$$|\langle \prod_{j=1}^k \phi_j, \mu^n_{\mathbf{Y}^k|A_\delta} \rangle - E(\prod_{j=1}^k \langle \phi_j, L_n^{\mathbf{Y}} \rangle \,|\, L_n^{\mathbf{Y}} \in A_\delta)| \le C\Big(1 - \frac{n!}{n^k(n-k)!}\Big) \underset{n\to\infty}{\longrightarrow} 0 \,.$$

For $\mathcal{M} = \{\nu_*\}$, Theorem 6.6.24 implies that for any $\eta > 0$,

$$\mu^n(|\langle \phi_j, L_n^{\mathbf{Y}} \rangle - \langle \phi_j, \nu_* \rangle| > \eta \,|\, L_n^{\mathbf{Y}} \in A_\delta) \to 0$$

as $n \to \infty$ followed by $\delta \to 0$. Since $\langle \phi_j, L_n^{\mathbf{Y}} \rangle$ are bounded,

$$E(\prod_{j=1}^k \langle \phi_j, L_n^{\mathbf{Y}} \rangle \,|\, L_n^{\mathbf{Y}} \in A_\delta) \to \langle \prod_{j=1}^k \phi_j, (\nu_*)^k \rangle ,$$

so that

$$\limsup_{\delta \to 0} \limsup_{n \to \infty} \langle \prod_{j=1}^k \phi_j, \mu^n_{\mathbf{Y}^k | A_\delta} - (\nu_*)^k \rangle = 0 .$$

Recall that $C_b(\Sigma)^k$ is convergence determining for $M_1(\Sigma^k)$, hence it follows that $\mu^n_{\mathbf{Y}^k|A_\delta} \to (\nu_*)^k$ weakly in $M_1(\Sigma^k)$. □

Having stated a general conditioning result, it is worthwhile checking Assumption (A-1) and the resulting set of measures $\mathcal{M}$ for some particular choices of the functional Φ. Two options are considered in detail in the sequel. Noninteracting particles, in which case $n^{-1}\sum_{i=1}^n U(Y_i)$ is specified, and interacting particles, in which case $n^{-2}\sum_{i,j=1}^n U(Y_i, Y_j)$ is specified.

Let $U : \Sigma \to [0, \infty)$ be a Borel measurable function. Define the functional $\Phi : M_1(\Sigma) \to [-1, \infty]$ by

$$\Phi(\nu) = \langle U, \nu \rangle - 1 ,$$

and consider the constraint

$$\{L_n^{\mathbf{Y}} \in A_\delta\} \stackrel{\triangle}{=} \{|\Phi(L_n^{\mathbf{Y}})| \le \delta\} = \{|\frac{1}{n}\sum_{i=1}^n U(Y_i) - 1| \le \delta\}.$$

Let $Z_\beta = \int_\Sigma e^{-\beta U(x)} \mu(dx)$, $\beta_\infty \stackrel{\triangle}{=} \inf\{\beta : \ Z_\beta < \infty\}$, and define the Gibbs measures γ_β, $\beta > \beta_\infty$ where

$$\frac{d\gamma_\beta}{d\mu} = \frac{e^{-\beta U(x)}}{Z_\beta} .$$

The following lemma is obtained by elementary analysis.

Lemma 6.6.26 *Assume that $\mu(\{x : U(x) > 1\}) > 0$, $\mu(\{x : U(x) < 1\}) > 0$, and either $\beta_\infty = -\infty$ or*

$$\lim_{\beta \searrow \beta_\infty} \langle U, \gamma_\beta \rangle > 1 . \qquad (6.6.119)$$

Then there exists a unique $\beta^* \in (\beta_\infty, \infty)$ such that $\langle U, \gamma_{\beta^*} \rangle = 1$.

One now checks the following:

Theorem 6.6.27 *Let U,μ and β^* be as in the preceding lemma. If either U is bounded or $\beta^* \ge 0$, then Theorem 6.6.24 applies, with $\mathcal{M}$ consisting of a unique Gibbs measure γ_{β^*}.*

In particular, Theorem 6.6.27 states that the conditional law of Y_1 converges, as $n \to \infty$, to the Gibbs measure γ_{β^*}.

We next move to the case where interaction is present, which is even more interesting from a physical point of view. To build a model of such a situation, let $M > 1$ be given, let $U : \Sigma^2 \to [0, M]$ be a continuous, symmetric, bounded function, and define $\Phi(\nu) = \langle U\nu, \nu\rangle - 1$ and $A_\delta = \{\nu : |\Phi(\nu)| \le \delta\}$ for $\delta \ge 0$. Throughout, $U\nu$ denotes the bounded, continuous function

$$U\nu(x) = \int_\Sigma U(x,y)\nu(dy) .$$

The reason for choosing U continuous is that then, the functional $\nu \mapsto \langle U\nu, \nu\rangle$ is continuous with respect to the τ-topology on $M_1(\Sigma)$.

With $Z_\beta = \int_\Sigma e^{-\beta U\gamma_\beta(x)}\mu(dx)$, let $\frac{d\gamma_\beta}{d\mu} = \frac{e^{-\beta U\gamma_\beta(x)}}{Z_\beta}$ and make the following additional assumptions.

Assumption (A-2) *For any ν_i such that $H(\nu_i|\mu) < \infty$, $i = 1, 2$,*

$$\langle U\nu_1, \nu_2\rangle \le \frac{1}{2}(\langle U\nu_1, \nu_1\rangle + \langle U\nu_2, \nu_2\rangle) .$$

Assumption (A-3) $\int_{\Sigma^2} U(x,y)\mu(dx)\mu(dy) \ge 1$.
Assumption (A-4) *There exists a probability measure ν with $H(\nu|\mu) < \infty$ and $\langle U\nu, \nu\rangle < 1$.*

Note that, unlike the noninteracting case, here even the existence of Gibbs measures needs to be proved.

Theorem 6.6.28 *Assume (A-2)–(A-4). Then Theorem 6.6.24 applies, with $\mathcal{M}$ consisting of a unique Gibbs measure γ_{β^*}, where*

$$\beta^* = \triangleq \inf\{\beta \ge 0 : \langle U\gamma_\beta, \gamma_\beta\rangle \le 1\} .$$

We now return to the general setup of Theorem 6.6.24. Our goal is to explore the structure of the conditional law $\mu^n_{\mathbf{Y}^k|A_\delta}$ (the law of $Y_1, \cdots, Y_k$ conditional on $L^{\mathbf{Y}}_n \in A_\delta$) when $k = k(n) \to_{n\to\infty} \infty$. The motivation is clear: we wish to consider the effect of Gibbs conditioning on subsets of the system whose size increases with the size of the system.

To this end we make the following simplifying assumption.
Assumption (A-5) $F_\delta = A_\delta \equiv A$ *is a closed, measurable convex set of probability measures on a compact metric space (Σ, d) such that*

$$I_F \triangleq \inf_{\nu\in A} H(\nu|\mu) = \inf_{\nu\in A^o} H(\nu|\mu) < \infty .$$

With $H(\cdot|\mu)$ strictly convex on the compact convex sets $\{\nu : H(\nu|\mu) \le \alpha\}$, there exists a unique $\nu_* \in A$ such that $H(\nu_*|\mu) = I_F$. Relying upon the convexity of A and using various properties of $H(\cdot|\cdot)$ leads to the following refinement of Corollary 6.6.25.

Theorem 6.6.29 *Assume (A-5), and further that*

$$\mu^n(L^{\mathbf{Y}}_n \in A)e^{nI_F} \ge g_n > 0 . \qquad (6.6.120)$$

Then, for any $k = k(n)$,

$$H\left(\mu^n_{\mathbf{Y}^k|A}\middle|(\nu_*)^k\right) \le \frac{1}{\lfloor \frac{n}{k(n)} \rfloor}\log(1/g_n) . \qquad (6.6.121)$$

Thus, refinements of Sanov's theorem as in (6.6.120) allow for the extension of the Gibbs conditioning principle to blocks of size $k(n)$. A particular concrete application is the following:

Corollary 6.6.30 *Let $A = \{\nu \in M_1[0,1] : \langle U, \nu \rangle \leq 1\}$ for a bounded nonnegative Borel function $U(\cdot)$, such that $\mu \circ U^{-1}$ is a non-lattice law, $\int_0^1 U(x)d\mu(x) > 1$ and $\mu(\{x : U(x) < 1\}) > 0$. Then (A-5) holds with $\nu_* = \gamma_{\beta^*}$ of Theorem 6.6.27 and for $n^{-1}k(n)\log n \to_{n\to\infty} 0$,*

$$H\left(\mu^n_{\mathbf{Y}^{k(n)}|A}\Big|(\nu_*)^{k(n)}\right) \underset{n\to\infty}{\longrightarrow} 0\,. \tag{6.6.122}$$

In particular,

$$\left\|\mu^n_{\mathbf{Y}^{k(n)}|A} - (\nu_*)^{k(n)}\right\|_{var} \underset{n\to\infty}{\longrightarrow} 0\,.$$

6.6.6 Application: The Hypothesis Testing Problem

For Σ a Polish space, let $Y_1, \ldots, Y_n$ be distributed either according to the law μ_0^n (hypothesis H_0) or according to μ_1^n (hypothesis H_1), where μ_i^n denotes the product measure of $\mu_i \in M_1(\Sigma)$.

Definition 6.6.31 *A decision test $\mathcal{S}$ is a sequence of measurable (with respect to the product σ-field) maps $\mathcal{S}^n : \Sigma^n \to \{0,1\}$, with the interpretation that when $Y_1 = y_1, \ldots, Y_n = y_n$ is observed, then H_0 is accepted (H_1 rejected) if $\mathcal{S}^n(y_1, \ldots, y_n) = 0$, while H_1 is accepted (H_0 rejected) if $\mathcal{S}^n(y_1, \ldots, y_n) = 1$.*

The performance of a decision test $\mathcal{S}$ is determined by the error probabilities

$$\alpha_n \stackrel{\triangle}{=} \text{Prob}_{\mu_0}(\mathcal{S}^n \text{ rejects } H_0), \quad \beta_n \stackrel{\triangle}{=} \text{Prob}_{\mu_1}(\mathcal{S}^n \text{ rejects } H_1).$$

The aim is to minimize β_n. If no constraint is put on α_n, one may obtain $\beta_n = 0$ using the test $\mathcal{S}^n(y_1, \ldots, y_n) \equiv 1$ at the cost of $\alpha_n = 1$. Thus, a sensible criterion for optimality, originally suggested by Neyman and Pearson, is to seek a test that minimizes β_n subject to a constraint on α_n. Suppose now that the probability measures μ_0, μ_1 are known *a priori* and that they are *equivalent measures*, so the likelihood ratios $L_{0\|1}(y) = d\mu_0/d\mu_1(y)$ and $L_{1\|0}(y) = d\mu_1/d\mu_0(y)$ exist. In order to avoid trivialities, it is further assumed that μ_0 and μ_1 are distinguishable, i.e., they differ on a set whose probability is positive.

Let $X_j \stackrel{\triangle}{=} \log L_{1\|0}(Y_j) = -\log L_{0\|1}(Y_j)$ be the observed log-likelihood ratios. These are i.i.d. real valued random variables that are nonzero with positive probability. Moreover, let

$$\overline{x}_1 \stackrel{\triangle}{=} E_{\mu_1}[X_1] = E_{\mu_0}[X_1 e^{X_1}] > E_{\mu_0}[X_1] = E_{\mu_1}[X_1 e^{-X_1}] \stackrel{\triangle}{=} \overline{x}_0\,.$$

Definition 6.6.32 *A Neyman–Pearson test is a test in which for any $n \in \mathbb{Z}_+$, the normalized observed log-likelihood ratio*

$$\hat{S}_n \stackrel{\triangle}{=} \frac{1}{n}\sum_{j=1}^n X_j$$

is compared to a threshold γ_n and H_1 is accepted (rejected) when $\hat{S}_n > \gamma_n$ (respectively, $\hat{S}_n \leq \gamma_n$).

Neyman–Pearson tests are optimal in the sense that there are neither tests with the same value of α_n and a smaller value of β_n nor tests with the same value of β_n and a smaller value of α_n.

The exponential rates of α_n and β_n for Neyman–Pearson tests with constant thresholds $\gamma \in (\overline{x}_0, \overline{x}_1)$ are thus of particular interest. These may be cast in terms of the large deviations of $\hat{S}_n$. In particular, since X_j are i.i.d. real valued random variables, the following theorem is a direct application of Theorem 6.3.9.

Theorem 6.6.33 *The Neyman–Pearson test with the constant threshold $\gamma \in (\overline{x}_0, \overline{x}_1)$ satisfies*

$$\lim_{n\to\infty} \frac{1}{n} \log \alpha_n = -\Lambda_0^*(\gamma) < 0 \tag{6.6.123}$$

and

$$\lim_{n\to\infty} \frac{1}{n} \log \beta_n = \gamma - \Lambda_0^*(\gamma) < 0 \,, \tag{6.6.124}$$

where $\Lambda_0^*(\cdot)$ is the Fenchel–Legendre transform of $\Lambda_0(\lambda) \triangleq \log E_{\mu_0}\, [e^{\lambda X_1}]$.

A corollary of the preceding theorem is Chernoff's asymptotic bound on the best achievable Bayes probability of error,

$$P_n^{(e)} \triangleq \alpha_n \text{Prob}(H_0) + \beta_n \text{Prob}(H_1) \,.$$

Corollary 6.6.34 (Chernoff's bound) *If $0 < Prob(H_0) < 1$, then*

$$\inf_{\mathcal{S}} \liminf_{n\to\infty} \{\frac{1}{n} \log P_n^{(e)}\} = -\Lambda_0^*(0) \,,$$

where the infimum is over all decision tests.

Remarks:
(a) Note that by Jensen's inequality, $\overline{x}_0 < \log E_{\mu_0}[e^{X_1}] = 0$ and $\overline{x}_1 > -\log E_{\mu_1}[e^{-X_1}] = 0$, and these inequalities are strict, since X_1 is nonzero with positive probability. Theorem 6.6.33 and Corollary 6.6.34 thus imply that the best Bayes exponential error rate is achieved by a Neyman–Pearson test with *zero threshold.*
(b) $\Lambda_0^*(0)$ is called Chernoff's information of the measures μ_0 and μ_1. **Proof.** It suffices to consider only Neyman–Pearson tests. Let α_n^* and β_n^* be the error probabilities for the zero threshold Neyman–Pearson test. For any other Neyman–Pearson test, either $\alpha_n \geq \alpha_n^*$ (when $\gamma_n \leq 0$) or $\beta_n \geq \beta_n^*$ (when $\gamma_n \geq 0$). Thus, for any test,

$$\frac{1}{n} \log P_n^{(e)} \geq \frac{1}{n} \log\left[\min\{\text{Prob}(H_0)\,,\, \text{Prob}(H_1)\}\right] + \min\{\frac{1}{n} \log \alpha_n^*\,,\, \frac{1}{n} \log \beta_n^*\} \,.$$

Hence, as $0 < \text{Prob}(H_0) < 1$,

$$\inf_{\mathcal{S}} \liminf_{n\to\infty} \frac{1}{n} \log P_n^{(e)} \geq \liminf_{n\to\infty} \min\{\frac{1}{n} \log \alpha_n^*\,,\, \frac{1}{n} \log \beta_n^*\} \,.$$

By (6.6.123) and (6.6.124),

$$\lim_{n\to\infty} \frac{1}{n} \log \alpha_n^* = \lim_{n\to\infty} \frac{1}{n} \log \beta_n^* = -\Lambda_0^*(0) \,.$$

Consequently,

$$\liminf_{n\to\infty} \frac{1}{n} \log P_n^{(e)} \geq -\Lambda_0^*(0) \,,$$

with equality for the zero threshold Neyman–Pearson test. □

Another related result is the following lemma, which determines the best exponential rate for β_n when α_n are bounded away from 1.

Lemma 6.6.35 (Stein's lemma) *Let β_n^ϵ be the infimum of β_n among all tests with $\alpha_n < \epsilon$. Then, for any $\epsilon < 1$,*

$$\lim_{n\to\infty} \frac{1}{n} \log \beta_n^\epsilon = \overline{x}_0 \,.$$

Bibliography

[As81] Geodésique et diffusions en temps petit. Astérisque, 84–85, 1981.

[Aze80] R. Azencott. Grandes déviations et applications. In P. L. Hennequin, editor, *Ecole d'Été de Probabilités de Saint-Flour VIII–1978*, Lecture Notes in Math. 774, pages 1–176. Springer-Verlag, Berlin, 1980.

[Bah71] R. R. Bahadur. *Some limit theorems in statistics*, volume 4 of *CBMS–NSF regional conference series in applied mathematics*. SIAM, Philadelphia, 1971.

[BaZ79] R. R. Bahadur and S. L. Zabell. Large deviations of the sample mean in general vector spaces. *Ann. Probab.*, 7:587–621, 1979.

[BeLa91] G. Ben Arous and R. Léandre. Décroissance exponentielle du noyau de la chaleur sur la diagonale. *Prob. Th. Rel. Fields*, 90:175–202, 1991.

[BeLd93] G. Ben Arous and M. Ledoux. Schilder's large deviations principle without topology. In *Asymptotic problems in probability theory: Wiener functionals and asymptotics (Sanda/Kyoto, 1990)*, pages 107–121. Pitman Res. Notes Math. Ser., 284, Longman Sci. Tech., Harlowi, 1993.

[Bor67] A. A. Borovkov. Boundary–value problems for random walks and large deviations in function spaces. *Th. Prob. Appl.*, 12:575–595, 1967.

[Bra86] R. C. Bradley. Basic properties of strong mixing conditions. In E. Eberlein and M. Taqqu, editors, *Dependence in Probability and Statistics*, pages 165–192. Birkhäuser, Basel, Switzerland, 1986.

[Bry90] W. Bryc. Large deviations by the asymptotic value method. In M. Pinsky, editor, *Diffusion Processes and Related Problems in Analysis*, pages 447–472. Birkhäuser, Basel, Switzerland, 1990.

[BryD96] W. Bryc and A. Dembo. Large deviations and strong mixing. *Ann. Inst. H. Poincaré Probab. Stat.*, 32:549–569, 1996.

[Buc90] J. A. Bucklew. *Large Deviations Techniques in Decision, Simulation, and Estimation.* Wiley, New York, 1990.

[CsK81] I. Csiszár and J. Körner. *Information Theory: Coding Theorems for Discrete Memoryless Systems.* Academic Press, New York, 1981.

[DaG87] D. A. Dawson and J. Gärtner. Large deviations from the McKean-Vlasov limit for weakly interacting diffusions. *Stochastics*, 20:247–308, 1987.

[deA90] A. de Acosta. Large deviations for empirical measures of Markov chains. *J. Theoretical Prob.*, 3:395–431, 1990.

[deA94] A. de Acosta. On large deviations of empirical measures in the τ-topology. Studies in applied probability. *J. Appl. Prob.*, 31A:41–47, 1994.

[DeZ95] A. Dembo and O. Zeitouni. Large deviations via parameter dependent change of measure and an application to the lower tail of Gaussian processes. In E. Bolthausen, M. Dozzi and F. Russo, editors, *Progress in Probability, volume 36*, pages 111–121. Birkhäuser, Basel, Switzerland, 1995.

[DeZ98] A. Dembo and O. Zeitouni. *Large Deviations Techniques and Applications - 2nd edition.* Springer, New York, 1998.

[DeuS89] J. D. Deuschel and D. W. Stroock. *Large Deviations.* Academic Press, Boston, 1989.

[DuE97] P. Dupuis and R. S. Ellis. *A Weak Convergence Approach to the Theory of Large Deviations.* Wiley, New York, 1997.

[DV75a] M. D. Donsker and S. R. S. Varadhan. Asymptotic evaluation of certain Markov process expectations for large time, I. *Comm. Pure Appl. Math.*, 28:1–47, 1975.

[DV75b] M. D. Donsker and S. R. S. Varadhan. Asymptotic evaluation of certain Markov process expectations for large time, II. *Comm. Pure Appl. Math.*, 28:279–301, 1975.

[DV76] M. D. Donsker and S. R. S. Varadhan. Asymptotic evaluation of certain Markov process expectations for large time, III. *Comm. Pure Appl. Math.*, 29:389–461, 1976.

[DV83] M. D. Donsker and S. R. S. Varadhan. Asymptotic evaluation of certain Markov process expectations for large time, IV. *Comm. Pure Appl. Math.*, 36:183–212, 1983.

[DV85] M. D. Donsker and S. R. S. Varadhan. Large deviations for stationary Gaussian processes. *Comm. Math. Physics*, 97:187–210, 1985.

[DV87] M. D. Donsker and S. R. S. Varadhan. Large deviations for noninteracting particle systems. *J. Stat. Physics*, 46:1195–1232, 1987.

[Ell84] R. S. Ellis. Large deviations for a general class of random vectors. *Ann. Probab.*, 12:1–12, 1984.

[Ell85] R. S. Ellis. *Entropy, Large Deviations and Statistical Mechanics.* Springer-Verlag, New York, 1985.

[FW84] M. I. Freidlin and A. D. Wentzell. *Random Perturbations of Dynamical Systems.* Springer-Verlag, New York, 1984.

[Gär77] J. Gärtner. On large deviations from the invariant measure. *Th. Prob. Appl.*, 22:24–39, 1977.

[Ham62] J. M. Hammersley. Generalization of the fundamental theorem on subadditive functions. *Math. Proc. Camb. Philos. Soc.*, 58:235–238, 1962.

[INN85] I. Iscoe, P. Ney and E. Nummelin. Large deviations of uniformly recurrent Markov additive processes. *Adv. in Appl. Math.*, 6:373–412, 1985.

[Kif90] Y. Kifer. Large deviations in dynamical systems and stochastic processes. *Trans. Amer. Math. Soc.*, 321:505–524, 1990.

[Kif92] Y. Kifer. Averaging in dynamical systems and large deviations. *Invent. Math.*, 110:337–370, 1992.

[KL99] C. Kipnis and C. Landim. Scaling limits of interacting particle systems. *Springer*, New York, 1999.

[KuS91] S. Kusuoka and D. W. Stroock. Precise asymptotics of certain Wiener functionals. *J. Funct. Anal.*, 1:1–74, 1991.

[KuS94] S. Kusuoka and D. W. Stroock. Asymptotics of certain Wiener functionals with degenerate extrema. *Comm. Pure Appl. Math.*, 47:477–501, 1994.

[Lan73] O. E. Lanford. Entropy and equilibrium states in classical statistical mechanics. In A. Lenard, editor, *Statistical Mechanics and Mathematical Problems*, Lecture Notes in Physics 20, pages 1–113. Springer-Verlag, New York, 1973.

[Mar98] F. Martinelli. Glauber Dynamics for Discrete Spin Models. In P. L. Hennequin, editor, *Ecole d'Été de Probabilités de Saint-Flour XXV–1997*, Lecture Notes in Math., 1998.

[Mog76] A. A. Mogulskii. Large deviations for trajectories of multi dimensional random walks. *Th. Prob. Appl.*, 21:300–315, 1976.

[Nag79] S. V. Nagaev. Large deviations of sums of independent random variables. *Ann. Probab.*, 7:745–789, 1979.

[NN87a] P. Ney and E. Nummelin. Markov additive processes, I. Eigenvalues properties and limit theorems. *Ann. Probab.*, 15:561–592, 1987.

[NN87b] P. Ney and E. Nummelin. Markov additive processes, II. Large deviations. *Ann. Probab.*, 15:593–609, 1987.

[Puk91] A. A. Pukhalskii. On functional principle of large deviations. In V. Sazonov and T. Shervashidze, editors, *New Trends in Probability and Statistics*, pages 198–218. VSP Moks'las, Moskva, 1991.

[Rue67] D. Ruelle. A variational formulation of equilibrium statistical mechanics and the Gibbs phase rule. *Comm. Math. Physics*, 5:324–329, 1967.

[Sal97] L. Saloff-Coste. Markov Chains. In P. L. Hennequin, editor, *Ecole d'Été de Probabilités de Saint-Flour XXIV–1996*, Lecture Notes in Math. 1665, pages 301–413. Springer-Verlag, New York, 1997.

[San57] I. N. Sanov. On the probability of large deviations of random variables. In Russian, 1957. (English translation from *Mat. Sb. (42)*) in Selected Translations in Mathematical Statistics and Probability I (1961), pp. 213–244).

[Sch66] M. Schilder. Some asymptotic formulae for Wiener integrals. *Trans. Amer. Math. Soc.*, 125:63–85, 1966.

[StZ91] D. W. Stroock and O. Zeitouni. Microcanonical distributions, Gibbs states, and the equivalence of ensembles. In R. Durrett and H. Kesten, editors, *Festschrift in honour of F. Spitzer*, pages 399–424. Birkhäuser, Basel, Switzerland, 1991.

[SW95] A. Shwartz and A. Weiss. *Large deviations for performance analysis.* Chapman and Hall, London, 1995.

[Tal95] M. Talagrand. Concentration of measure and isoperimetric inequalities in product spaces. *Publ. Mathématiques de l'I.H.E.S.*, 81:73–205, 1995.

[Tal96] M. Talagrand. New concentration inequalities in product spaces. *Invent. Math.*, 126:505–563, 1996.

[Var66] S. R. S. Varadhan. Asymptotic probabilities and differential equations. *Comm. Pure Appl. Math.*, 19:261–286, 1966.

[Var84] S. R. S. Varadhan. *Large Deviations and Applications.* SIAM, Philadelphia, 1984.

[VF70] A. D. Ventcel and M. I. Freidlin. On small random perturbations of dynamical systems. *Russian Math. Surveys*, 25:1–55, 1970.

[VF72] A. D. Ventcel and M. I. Freidlin. Some problems concerning stability under small random perturbations. *Th. Prob. Appl.*, 17:269–283, 1972.

[Zab92] S. L. Zabell. Mosco convergence and large deviations. In M. G. Hahn, R. M. Dudley and J. Kuelbs, editors, *Probability in Banach Spaces, 8*, pages 245–252. Birkhäuser, Basel, Switzerland, 1992.

Chapter 7

Stability and Stabilizing Control of Stochastic Systems

P. V. PAKSHIN
Department of Applied Mathematics
Nizhny Novgorod State Technical University at Arzamas
19, Kalinina Str.,
Arzamas, 607220, Russia

List of frequently used symbols and notations

X the system state

U the system input(control)

A the state matrix

B the input (control) matrix

K the linear feedback control matrix

W the Wiener process

Y the homogeneous Markov chain with discrete set of states

H positive definite or at least positive semidefinite solution of Lyapunov matrix equation or Riccati matrix equation

$\widehat{X}$ the expected value of the current system state

x' (A') transpose of the vector x (of the matrix A)

I_n the $n \times n$ identity (unit) matrix

I identity (unit) matrix

$|x|$ $(|A|)$ the Euclidean norm of the vector x (of the matrix A)

P the probability

$\mathcal{E}$ the expectation operator

$\mathcal{A}$ the differential generator of a Markov process

$\mathcal{L}$ the weak infinitesimal operator of a homogeneous Markov process

V the Lyapunov or Lyapunov-Bellman function

R^n n-dimensional Euclidean space

$\mathcal{I}$ time interval: $\mathcal{I} = [t_0, T], T < \infty$, or $\mathcal{I} = [t_0, \infty)$

$\mathcal{Y}$ the set of states of the Markov chain Y

$\mathcal{N}$ the finite set of states of the Markov chain Y:$\mathcal{N} = \{1, \dots, \nu\}$

$\mathcal{U}$ the set of admissible controls

$\mathcal{B}^n$ the Borel set in R^n

$\mathcal{C}$ the class of functions $f(t)$, continuous on $[0, T]$ with values in R^n

a.a.,a.s., w.p.1. almost all, almost surely, with probability one
u.h.c., (α), u.l.c. uniformly Hölder continuous (exponent α), uniformly Lipschitz continuous

Introduction

Stochastic control theory is a very important direction in modern stochastic analysis and applications. For the solution of stochastic control problems one needs to obtain systems state information. From this point of view the stochastic control systems are separated into the two classes: systems with complete state information and systems with partial state information (partially observed systems), i.e. only a function of the state, possibly corrupted by noise, is observable. Usually the control synthesis problem is formulated as an optimal control problem: obtaining such a control that minimizes an integral cost functional over the set of admissible controls. For a system with complete state information two approaches for the optimal control problem are used: dynamic programming leading to the Hamilton-Jacobi-Bellman (HJB) equation and the maximum principle. For systems with incomplete state information one needs to estimate the state first, but in the general nonlinear case the estimation and control problems are not separated. One a way to solve these problems jointly is based on the use of the Dunkan-Mortensen-Zakai (DMZ) equation, often called shortly the Zakai equation [21, 17, 70, 103]. The DMZ equation of nonlinear filtering of stochastic processes is a linear, stochastic, partial differential equation which describes in a recursive manner the evolution of the unnormalized conditional distribution of the state process,$\{x(t),\ t \geq 0\}$, given the observation $\{y(t),\ t \geq 0\}$. To solve the stochastic control problem of partially observed systems it is possible to reformulate this problem as one of complete information in which the control is a functional of an information state. It turns out that the information state satisfies a controlled version of the DMZ equation [17, 70]. For a stochastic linear dynamic system observed via a linear channel corrupted by noise the joint problem of optimal control and estimation (filtering) can be reduced to two independent problems of control and filtering. This structural property of the optimal system depends on whether or not the cost functional is quadratic, and whether or not the optimal feedback control happens to be linear in the system state or its expectation. A special result of this type for the standard linear-guadratic Gaussian (LQG) control problem is called the "separation theorem" or the "separation principle." The separation principle allows using well-known Kalman-Bucy filtering results for estimation of the systems state. As a rule the control law must guarantee stability of the stochastic system in the suitable sense. In most cases the systems with random inputs but with nonrandom operator are considered. Here it is possible to use the results of the deterministic stability theory. In the meantime often we also have a random disturbance of parameters and in general the operator of the system will be random too. To study dynamic properties of this class of systems *the stochastic stability* concept is used. The concept of stochastic stability and stabilization was introduced in pioneering works by Kats and Krasovskii [42], Bertram and Sarachik [7], Krasovskii and Lidskii [51]. The stochastic stability and stabilization theory has been well-established mainly for the Ito stochastic differential equations. A systematic exposition of this theory is presented in the well known monographs by Khasminskii [45] and by Kushner [54]. These fundamental books, addressed first and foremost to pure matematicians, contain, basically, results of a general nature and hardly reflect the applied side of the problem. This is one of the reasons why the ideas and methods of stochastic stability and stabilization theory have not been wide spread in practice. In applications the task of stochastic stability and stabilization theory is to obtain criteria and algorithms suitable for the direct implementation in the design of stochastic dynamic systems (the system with random operator). It so happens that the publications of applied nature in the area of stochastic stability and stabilization are highly scattered in periodicals. This is the second cause which impedes the development of the applied theory. In this connection the purpose of this survey paper is to present stochastic stabilizing control results for both categories of readers: theoreticiarys and practicians. This style was stimulated to a large degree by the Wonham's paper [96] and, especially, by the book by Kats [41]. We consider only the systems described by ordinary stochastic differential equations. The reader is referred to monographs by Meyn and Tweedie [65], and Pakshin [72] to study stochastic stability and stabilization problems for discrete systems; see also the papers [34, 35] and references therein. The stochastic systems with time delay are studied in books by Kolmanovskii and Myshkis [46], Kolmanovskii and Shaikhet [47], by Korenevskii [48]; see also the references therein.

7.1 Stochastic mathematical models of systems

7.1.1 Models of differential systems corrupted by noise

A wide variety of problems in the study of dynamic systems leads to stochastic differential equations of Ito type

$$dX_t = a(t, X(t))dt + b(t, X(t))dW(t),\ t \in \mathcal{I}, \tag{7.1.1}$$

where $X(t)$ is the n-dimensional state vector, $W(t)$ is the m-dimensional standard Wiener process, $a(t, x)$ is n-dimensional vector function and $b(t, x)$ is $n \times m$ matrix function, $\mathcal{I} = [t_0, T], T < \infty$, or $\mathcal{I} = [t_0, \infty)$. The equation (7.1.1) means that $X(t)$ is the stochastic process, satisfying the following stochastic integral equation

$$X(t) = X(t_0) + \int_{t_0}^{t} a((s), X(s))ds + \int_{t_0}^{t} b(s, X(s))dW(s). \tag{7.1.2}$$

The third term in the right hand side of (7.1.2) is so called *Ito stochastic integral*, see [19, 20, 38, 71]. It is supposed that both $a(t, x)$ and $b(t, x)$ are measurable functions for all $t \in \mathcal{I}, x \in R^n$ and satisfy the growth condition

$$|a(t, x)| + |b(t, x)| \le K(1 + |x|), t \in \mathcal{I},\ x \in R^n, \tag{7.1.3}$$

for some constant K, and the uniform Lipschitz condition

$$|a(t,x) - a(t,y)| + |b(t,x) - b(t,y)| \le k|x-y|,\ t \in \mathcal{I},\ x, y \in R^n, \tag{7.1.4}$$

for some constant k. If these conditions are valid and $X(t_0)$ is an arbitrary finite random variable, which is independent on the increments of the Wiener process, then the equation (7.1.2) uniquelly defines the stochastic process $X(t), t \in \mathcal{I}$ with the following properties:

1) The process $X(t)$ has continuous paths with probability one (w.p.1.).

2) If $\mathcal{E}[|X(t_0|^2] < \infty$, then

$$\mathcal{E}[\max_{t_0 \le t \le t_1} |X(t)|^2] < \infty,\ t_1 \in \mathcal{I}, \tag{7.1.5}$$

where $\mathcal{E}$ denotes the expectation operator.

3) For every t the random variable $X(t)$ is independent on the increments of the Wiener process $(W(t_1) - W(s)), t \le s < t_1$.

4) The stochastic process $X(t), t \in \mathcal{I}$ is the Markov process.

The model (7.1.1) can be motivated in the following way. Consider the ordinary differential equation

$$\frac{dX(t)}{dt} = a(t, X_t). \tag{7.1.6}$$

In many practical situations, for example in engineering, the right hand side of (7.1.6) may be corrupted by a noise process, such that

$$\frac{dX(t)}{dt} = a(t, X(t)) + b(t, X(t))V(t), \tag{7.1.7}$$

where $V(t)$ is m-dimensional Gaussian "physical" white noise, i.e., the m-dimensional vector whose components are scalar Gaussian processes with a correlation time much smaller than the time response of the considered system. Such a nonrigorous mode is often refered to as the *Langevin equation*, see [2, 56, 95, 100]. It is natural that under the noise action the distribution of $\frac{dX(t)}{dt}$ will only depend on t and $X(t)$, and the question is to construct a reasonable mathematical model of noise term $b(t, X_t)V(t)$ in this equation. It is clear that the process $V(t)$ will have (at least approximately) these properties:

(i) if $t_1 \ne t_2$ then $V(t_1)$ and $V(t_2)$ are independent;

(ii) the process $V(t)$ is stationary, i.e. the joint distribution of $\{V(t_1 + t), \dots, V(t_k + t)\}$ does not depend on t;

(iii) $\mathcal{E}[V_t] = 0$ for all t.

However, it turns out, there does not exist any reasonable stochastic process, satisfying (i) and (ii). Such a $V(t)$ cannot have continuous paths. Nevertheless it is possible to use the theory of *generalized stochastic processes*, see [33]. In this case the process V_t is represented as a generalized stochastic process called the *white noise process.* The other way is to avoid the construction of the generalized stochastic process and rather try to rewrite equation (7.1.7) in a form that suggests a replacement of V_t by a proper stochastic process. Let $t_0 < t_1 < \dots < t_s = t$ and consider a discrete version of (7.1.7):

$$X(t_{k+1}) - X(t_k) = a(t_k, X(t_k))\Delta t_k + b(t_k, X(t_k))V(t_k)\Delta t_k. \tag{7.1.8}$$

Now, we replace $V(t_k)\Delta t_k$ by $\Delta W(t_k) = W(t_{k+1}) - W(t_k)$, where $W(t)$ is some suitable stochastic process. The assumptions (i), (ii) and (iii) on V_t suggest that W_t should have *stationary independent increments* with zero mean. It turns out, that the only such process with continuous paths is the Wiener process, or in other terms *the Brownian motion process*, see [52]. Thus we obtain from (7.1.7):

$$X(t_k) = X(t_0) + \sum_{i=0}^{k-1} a(t_i, X(t_i))\Delta t_i + \sum_{i=0}^{k-1} b(t_i, X(t_i))\Delta W(t_i), \tag{7.1.9}$$

where $\Delta t_i = t_{i+1} - t_i$ Under the regularity properties (7.1.3),(7.1.4) there exists the limit of the right hand side of (7.1.9) in the mean square sense and by applying the usual integration notation we obtain (7.1.2). The stochastic process $X(t)$ defined by (7.1.2) has continuous paths. It is adopted as a convention that (7.1.7) really means that $X(t)$ is a stochastic process satisfying (7.1.2). The reader is refered to [1, 19, 20, 29, 71] for more exact formulations and detailed proofs. It is very important that there exist other interpretations of (7.1.7). Let us consider the following discrete version of (7.1.7)

$$X(t_k) = X(t_0) + \sum_{i=0}^{k-1} a(t_i, X_i^*)\Delta t_i + \sum_{i=0}^{k-1} b(t_i, X_i^*)\Delta W(t_i), \tag{7.1.10}$$

where $X_i^* = (X(t_{i+1)} + X(t_i))/2$. When $\Delta t_i \to 0$ this equation converges (by the regularity properties above) to the stochastic integral equation

$$X(t) = X(t_0) + \int_{t_0}^{t} a((s), X(s))ds + \int_{t_0}^{t} b(s, X(s)) \circ dW(s). \tag{7.1.11}$$

The last term in the right hand side of (7.1.11) is known as the *Stratonovich stochastic integral* [81]. In general this integral is different from the Ito integral and this implies that the stochastic processes defined by (7.1.2) and (7.1.11) are different too. The Stratonovich interpretation in some situations may be most appropriate. The argument that indicates it is in following [94, 95]. Choose t-continuously differentiable processes $W^{(k)}(t,\omega)$ such that for almost all (a.a.) $\omega W^{(k)}(t) \to W(t,\omega)$ as $k \to \infty$ uniformly in t in bounded intervals. For each ω let $X^{(k)}(t,\omega)$ be the solution of the corresponding deterministic differential equation

$$\frac{dX(t)}{dt} = a(t, X(t)) + b(t, X(t))\frac{dW^{(k)}(t)}{dt}.$$

Then $X^{(k)}(t,\omega)$ converges to some function $X(t,\omega)$ in the same sence: for a.a. ω we have that $X^{(k)}(t,\omega) \to X(t,\omega)$ as $k \to \infty$ uniformly in t in bounded intervals. It turns out, see [94, 82, 95] that this solution $X(t)$ coincides with the solution of (7.1.11) obtained by using the Stratonovich integral. Therefore, from this point of view it seems reasonable to use (7.1.11) and not the Ito interpretation (7.1.2) as the model for the original noise corrupted system (7.1.7). It is shown that the solution $X(t)$ of the Stratonovich equation (7.1.11) satisfies the following *modified Ito equation*

$$X(t) = X(t_0) + \int_{t_0}^{t} \hat{a}((s), X(s))ds + \int_{t_0}^{t} b(s, X(s))dW(s), \tag{7.1.12}$$

where

$$\hat{a}(t,x) = a(t,x) + \frac{1}{2}d(t,x),$$

$$d_i(t,x) = \sum_{j=1}^{m}\sum_{k=1}^{n} \frac{\partial b_{ij}}{\partial x_k} b_{kj}, \; i = 1, \dots, n.$$

For a more detailed study of the general theory of stochastic differential equations and their applications the reader is referred to [1, 2, 27, 28, 29, 71, 77]

7.1.2 Models of differential systems with random jumps

Many dynamical systems, especially the systems with a certain switching mechanism or (and) jumping disturbances, cannot be adequately represented by (7.1.1) This class of systems is described by the differential equation [41, 64, 96]

$$dX_t = a(t, X(t), Y(t))dt + b(t, X(t), Y(t))dW(t),\ t \in \mathcal{I}. \tag{7.1.13}$$

In general $Y(t)$ in (7.1.13) is the r-dimensional random vector, such that $Y(t) \in \mathcal{Y} \subset R^r$ for all $t \in \mathcal{I}$. The components of $Y(t)$ are independent Markov processes whose transition probabilities $P_l(\tau, \eta; t, \mathcal{B}^1)$ $(l = 1, \dots, r)$, $\mathcal{B}^1$ is a Borel set in R^1, having these properties:

$$\begin{aligned} P[Y_l(t+\Delta t) < \beta \big| Y_l(t+\Delta t) \neq \eta, Y_l(t) = \eta] &= q_l(t, \eta, \beta)\Delta t + o(\Delta t), \\ P[Y_l(\tau) = \eta, t < \tau \leq t + \Delta t \big| Y_l(t) = \eta] &= 1 - q_l(t, \eta)\Delta t + o(\Delta t), \end{aligned} \tag{7.1.14}$$

where $q_l(t, \eta, \beta), q_l(t, \eta)$ are given functions, so that $q_l(t, \eta, \beta) \to q_l(t, \eta)$, as $\beta \to +\infty$. By the corresponding regularity properties almost all the paths of $Y_l(t)$ are piecewise constant and right continuous functions [29]. In many cases it is supposed that $Y(t)$ is homogenious scalar Markov chain with finite set of states $\mathcal{Y} = \mathcal{N} = \{1, 2, \dots, \nu\}$ and with transition probabilities

$$\begin{aligned} P[Y(t+\Delta t) = j \big| Y(t) = i \neq j] &= q_{ij}\Delta t + o(\Delta t), \\ P[Y(\tau) = i, t < \tau \leq t + \Delta t \big| Y(t) = i] &= 1 - q_i\Delta t + o(\Delta t), \\ q_i &= \sum_{j \neq i}^{\nu} q_{ij}. \end{aligned} \tag{7.1.15}$$

Let $[\tau - h, \tau)$ be a random interval such that $Y(t) = i \in \mathcal{N}$ for all $t \in [\tau - h, \tau)$. Then the system (7.1.13) will be described by

$$dX_t = a(t, X(t), i)dt + b(t, X(t), i)dW(t),\ t \in [\tau - h, \tau), \tag{7.1.16}$$

$$X(t-h) = X_h,\ Y(t-h) = i. \tag{7.1.17}$$

for every such interval. If $\tau > t_0$ is the transition (jump) time from $Y(\tau - 0) = i$ to $Y(\tau) = j \neq i$ then in the next interval $[\tau, \tau + \theta)$, where $Y(t) = j$ the system will be described by (7.1.16) with the replacement i to j, but we cannot correct to define the initial condition X_τ without additional assumptions regarding the considered system. As a rule the following types of systems are considered, see [41]:

1) The state vector $X(t)$ is changed continuously for all jump moments of $Y(t)$. This means that if τ is a transition time of $Y(t)$, then

$$X(\tau - 0) = X(\tau). \tag{7.1.18}$$

2) The value of the state vector $X(t)$ after the jump moment of $Y(t)$ uniquelly depends on the value of this vector before the jump moment, so that if $Y(\tau - 0) = i$ and $Y(\tau) = j \neq i$ then

$$X(\tau) = \phi_{ij}(X(\tau - 0)),\ i \neq j, \tag{7.1.19}$$

where $\phi_{ij}(x)$ is the continuous n-dimensional vector function, satisfying the condition $\phi_{ij}(0) = 0$. In the particular case $\phi_{ij}(x)$ is a linear function, then there exists an $n \times n$ matrix Φ_{ij} such that

$$X(\tau) = \Phi_{ij} X(\tau - 0). \tag{7.1.20}$$

3) The conditional distribution of the state vector $X(\tau)$ after the jump moment is given:

$$P\{X(\tau) \in [z, z + dz) \Big| X(\tau - 0) = x\} = p_{ij}(\tau, z \Big| x) dz + o(dz), \tag{7.1.21}$$

where $p_{ij}(\tau, z \Big| x)$ is the conditional density of the distribution.

So, for correct mathematical description of the dynamical system with random jumps the following elements are necessary:

1) The differential equation (7.1.13) with initial conditions

$$X(t_0) = X_0, \; Y(t_0) = Y_0. \tag{7.1.22}$$

2) The probabilistic description of the Markov process $Y(t)$ in the form given by (7.1.14) or (7.1.15).

3) The conditional distribution of the state vector after the jump moment (7.1.21) or particular conditions (7.1.18)-(7.1.20).

This description uniquelly defines the Markov process $[X(t), Y(t)]$. Almost all the paths $[X(t, \omega), Y(t, \omega)]$ are continuous on the right functions. Note that the component $X(t)$ is *not* the Markov one.

7.1.3 Differential generator

Consider a scalar function $V(t, x, y)$ defined in the domain

$$x \in R^n, \; y \in \mathcal{Y}, \; t \geq t_0 \tag{7.1.23}$$

and continuously differentiable in all the variables in this domain as often as is required in the process of solution of a stated problem. Roughly speaking the differental generator is average value of the derivative of $V(t, x, y)$ along all the paths of Markov process $[X(t), Y(t)]$ defined by (7.1.13) and by additional conditions below, starting from the point (x, y) at the moment s. In this connection, in the classic work by Kats and Krasovskii [42] and also in the book by Kats [41] this operator is called *the average derivative.*

Definition 7.1.1 *The operator*

$$\mathcal{A}V(s, x, y) = \lim_{t \to s+0} \frac{1}{t - s} \{\mathcal{E}[V(t, X(t), Y(t)) \Big| X(s) = x, \; Y(s) = y] - V(s, x, y)\} \tag{7.1.24}$$

is called the differential generator (the average derivative) by virtue of the system (7.1.13) at the point (s, x, y).

The differential generator is defined by *the weak infinitesimal operator* of the Markov process $[X(t), Y(t)]$, if this process is homogeneous and function V does not depend on t, and by analogous operator in the inhomogeneous case. To explain it in more detail, let $P(s, x, y; \; t, \mathcal{B}^n, \mathcal{B}^r)$ $(s \leq t, \; x \in R^n, y \in \mathcal{Y}, \mathcal{B}^n$ and $\mathcal{B}^r$ are Borel sets of R^n and $R^r)$ be the transition probabilty of the Markov process $[X(t), Y(t)]$. This function defines the family of linear operators

$$\mathcal{T}^t V(s,x,y) = \int P(s,x,y;\ t,du,dz)V(t,u,z) = \mathcal{E}[V(t,X(t),Y(t))\big| X(s)=x,\ Y(s)=y]$$

and the differential generator $\mathcal{L}V$ at the point (s,x,y) is defined by

$$\mathcal{A}V(s,x,y) = \lim_{t\to s+0} \frac{\mathcal{T}^t V(s,x,y) - V(s,x,y)}{t-s}. \tag{7.1.25}$$

In the particular case, if the Markov process $[X(t), Y(t)]$ is homogeneous and V does not depend on t, i.e. $V = V(x,y)$ we have

$$\mathcal{A}V(x,y) = \mathcal{L}V(x,y) = \lim_{t\to s+0} \frac{\mathcal{T}^t V(x,y) - V(x,y)}{t-s}.$$

The operator $\mathcal{L}V(x,y)$ is called the weak infinitesimal operator [20, 54] of the homogeneous Markov process.

Formulae for the differential generator

Consider the system (7.1.13) in domain (7.1.23). Suppose that $Y(t)$ is a scalar Markov chain with finite state space $\mathcal{Y} = \mathcal{N} = \{1,2,\dots,\nu\}$ and with the transition probabilities given by (7.1.15). At the moment τ of transition of the Markov chain $Y(t)$ from $Y(\tau-0)=i$ to $Y(\tau)=j$ the state vector $X(t)$ have a jump from $X(\tau-0)=x$ to $X(\tau)=z$ with a conditional density of distribution $p_{ij}(\tau,z)\big|x)$ given by (7.1.21). It is supposed that this density is continuous in τ and has compact support, such that

$$h_1|x| \le |z| \le h_2|x|, 0 < h_1 < h_2, p_{ij}(\tau,z\big|0) = \delta(z).$$

These conditions do not allow the process $X(t)$ to be zero as a result of the jump. Under their validity the differential generator by virtue of the system (7.1.13) at the point (s,x,i) is given by the following formula

$$\mathcal{A}V(s,x,i) = \frac{\partial V}{\partial s} + \left[\frac{\partial V}{\partial x}\right]' a(s,x,i) + \frac{1}{2}\mathrm{tr}\left[\frac{\partial^2 V}{\partial s^2} b(s,x,i)b'(s,x,i)\right] + \sum_{j\ne i}^{\nu}[\int V(s,x,j)p_{ij}(s,z\big|x)dz - V(s,x,i)]q_{ij}, \tag{7.1.26}$$

where $'$ denotes the transpose symbol. If at the moment of the jump the vector $X(t)$ is changed by the deterministic law (7.1.19), then from the previous formula it follows that

$$\mathcal{A}V(s,x,i) = \frac{\partial V}{\partial s} + \left[\frac{\partial V}{\partial x}\right]' a(s,x,i) + \frac{1}{2}\mathrm{tr}\left[\frac{\partial^2 V}{\partial s^2} b(s,x,i)b'(s,x,i)\right] + \sum_{j\ne i}^{\nu}[V(s,\phi_{ij}(x),j) - V(s,x,i)]q_{ij}. \tag{7.1.27}$$

In the particular case, when at the moment of the jump the vector $X(t)$ is changed continuously by the formula (7.1.18) we have

$$\mathcal{A}V(s,x,i) = \frac{\partial V}{\partial s} + \left[\frac{\partial V}{\partial x}\right]' a(s,x,i) +$$

$$\frac{1}{2}\mathrm{tr}\left[\frac{\partial^2 V}{\partial s^2} b(s,x,i)b'(s,x,i)\right] + \sum_{j\neq i}^{\nu}[V(s,x,j) - V(s,x,i)]q_{ij}. \tag{7.1.28}$$

If the Markov process $[X(t), Y(t)]$ is homogeneous and V does not depend on t, i.e. $V = V(x,y)$, these formulae are valid for calculaton of the weak infinitesimal operator $\mathcal{A}V(s,x,i)$. In this case the term $\frac{\partial V(s,x,y)}{\partial s} \equiv 0$. For more details and proofs see for instance [41, 45, 54]. A very important role in the proofs of many stability and control results plays the so called Ito-Dynkin formula [20, 45]:

$$\mathcal{E}[V(t, X(t), Y(t))\Big|X(s) = x,\ Y(s) = y] = V(s,x,y) +$$

$$\mathcal{E}[\int_s^t [\mathcal{A}[V(u, X(u), Y(u))]du\Big|X(s) = x,\ Y(s) = y]. \tag{7.1.29}$$

This formula is a stochastic analogue of the well-known Newton-Leibnitz formula

$$F(t, X(t)) = F(s, X(s)) + \int_s^t dF(u, X(u)).$$

7.2 Stochastic control problem

7.2.1 Preliminaries

Consider a system described by the differential equation

$$dX(t) = a(t, X(t), U(t), Y(t))dt + b(t, X(t), Y(t))dW(t),\ t \in \mathcal{I}, \tag{7.2.1}$$

where all the notations are the same as in (7.1.13), the difference is that now function a depends on k-dimensional control vector U. Generally speaking the stochastic control problem is to obtain a stochastic process $U(t)$ from the given set $\mathcal{U} \subset R^k$ of admissible controls such that the stochastic process $X(t)$ described by (7.2.1) will have some prescribed properties. As a rule the problem is formalized in such a way that the desired properties are achieved, when the control law minimizes a functional (in the other words performance function or objective function) along the paths of the considered system. This functional may be written as follows

$$J(s,x,y,U) = \mathcal{E}[\int_s^T L(t, X(t), Y(t), U(t)dt + \Psi(X(T), Y(T))\Big|X(s) = x], \tag{7.2.2}$$

such that it should be well defined for all admissible controls $U(t) \in \mathcal{U}$. So, the original problem is reduced to the optimal control problem: to find a function $U = U^*(t,\omega)$ from the set of admissible controls such that

$$V^0(s,x,y) = \min_{U\in\mathcal{U}} J(s,x,y,U)$$

Such a control, if it exists, is called an *optimal control* and scalar $V^0(s,x)$ is called the *optimal performance*. The types of control functions that may be considered are:

1) Functions of the form $U(t,\omega) = u(t)$, i.e. not depending on ω. These controls are sometimes called *deterministic, program* or open loop controls.

2) Functions of the form $U(t,\omega) = \mathbf{u}(t, X(t,\omega), Y(t,\omega)$ for some function $\mathbf{u} : \mathcal{I}\times R^n \times \mathcal{Y} \mapsto \mathcal{U}$. In this case it is assumed that U does not depend on the starting point $\{s, x, y\}$: the value at the time t is chosen only depends on the state of the system at this time. These are called *Markov controls*, because with such U the corresponding process $\{X(t,\omega), Y(t.\omega)\}$ becomes a Markov process. Markov control is the particular form of more general case of *closed loop* or *feedback* control. In the following we will not consider this general case and we will not distinguish between Markov and feedback controls.

3) Only a partially observed state of the system possibly corrupted by noise is available for the control purpose. In this situation the stochastic control problem is linked to the *filtering problem*. In fact, if the equation (7.2.1) is linear, its right hand side is not dependent on $Y(t)$ and the performance function is integral quadratic (i.e. the functions L and Ψ in (7.2.2) are quadratic in X and U); then the stochastic control problem splits into a linear filtering problem and a correspoding esimated state feedback control problem. This fact, known as the separation principle, will be considered below in more details.

It is more natural to obtain program control using deterministic models. From this point of view, control law can be conceptually split into two parts: the program part and the feedback (stabilizing) part. The program part is more often obtained in an open-loop fashion for a more general design objective, such as maximum throughput of a manufacturing system or minimal heating along a spacecraft re-entry path. Optimal trajectories are generated assumming that the environment and initial conditions are fixed. This serves as an ideal reference but it cannot be expected that the plant will actually follow the optimized trajectory. For various reasons, including modeling errors, changes in the environment, etc., deviation from the reference can occur and should be compensated. This is achieved in a closed-loop fashion with the stabilization term: by feeding back some measure of the deviation, it is posssible to stabilize the actual trajectory around the reference, so that the desired behavior is obtained.

In this connection the primary focus in stochastic control is more on the closed-loop part of control law, assuming that a certain reference trajectory has been obtained. Here linear stochastic models play a very important role, such as linearized approximations around the desired trajectory of the original nonlinear plant state dynamics.

As a rule *the stabilizing control problem* means a feedback control with the infinite time horizon. Often this problem approximates the practical case, when the time of control is sufficiently long in comparison with the time response of the controlled system. In this case the stability properties of the system under study play an important role.

So, in this section we consider some approaches to the solution of stochastic optimal feedback control problems for both systems with complete state information and for partially observed systems. In the next sections the stochastic stability concept will be introduced and the stabilizing control problems will be considered.

7.2.2 Stochastic dynamic programming

A heuristic derivation

We consider for simplicity the case of the system (7.2.1) without jumps:

$$dX(t) = a(t, X(t), U(t))dt + b(t, X(t))dW(t), \ t \in \mathcal{I}, \tag{7.2.3}$$

Generally speaking we use only the fact that there exists a Markov process $X(t)$ with differential generator $\mathcal{A}$. The control $U(t)$ is said to be feedback control if it is a function $\mathbf{u} : \mathcal{I} \times R^n \mapsto \mathcal{U}$, such that for every $U(t) = \mathbf{u}(t, X(t)) \in \mathcal{U} \subset R^k$ and given nonrandom initial condition $X(s) = x$ there exists the unique solution $X^{\mathbf{u}}_{s,x}$ of (7.2.3) in the sense of Ito and functional

$$J(s,x,\mathbf{u}) = \mathcal{E}[\int_s^T L(t, X(t), \mathbf{u}(t, X(t))dt + \Psi(X(T))\Big|X(s) = x], \qquad (7.2.4)$$

is well defined. The optimization problem is to minimize (7.2.4) along the paths of (7.2.3) and over the set $\mathcal{U}$. We denote $\mathcal{A}_u$ the differential generator with $a = a(t, x, u)$, where $u \in R^k$ is arbitrary and $\mathcal{L}_{\mathbf{u}}$ the same differential generator with formal substitution $u = \mathbf{u}(t, x)$. Assuming the existence of an optimal solution, noted $\mathbf{u}^*$, we consider the optimal expected performance as a function of the initial data s and x and we introduce the cost function $V^0(s, x)$ as the optimal performance for the problem with initial data s and x [24]:

$$V^0(s,x) = \min_{\mathbf{u}\in\mathcal{U}} J(s,x,\mathbf{u}) \qquad (7.2.5)$$

For the optimal feedback control $\mathbf{u}^*$ we have $V^0(s,x) = J(s,x,\mathbf{u}^*)$ for all s, x. To write the equation for V^0 we fix a control $\mathbf{u}$ and use Ito-Dynkin formula (7.1.29)

$$V^0(s,x) = -\mathcal{E}[\int_s^t [\mathcal{A}_{\mathbf{u}} V^0(\tau, X(\tau))]d\tau\Big|X(s) = x] + \\ \mathcal{E}[V^0(t, X(t))\Big|X(s) = x], \; s < t \le T, \qquad (7.2.6)$$

Now assume that we use the optimal control $\mathbf{u}^*$ for $\tau > t$ and $\mathbf{u}$ for $\tau \le t$. In other words, let

$$\mathbf{u}_1(\tau, x) = \begin{cases} \mathbf{u}(\tau, x) & \text{if} \quad \tau \le t, \\ \mathbf{u}^*(\tau, x) & \text{if} \quad \tau > t. \end{cases}$$

Then using conditional expectation properties from (7.2.4)-(7.2.6) we have

$$J(s,x,\mathbf{u}_1) = \mathcal{E}[\int_s^t [L(\tau, X(\tau), U(\tau)]d\tau\Big|X(s) = x] + \\ \mathcal{E}[J(t, X(t), \mathbf{u}^*)\Big|X(s) = x], \; s < t \le T,$$

and hence

$$V^0(s,x) \le J(s,x,\mathbf{u}_1), \; V^0(t, X(t)) = J(t, X(t), \mathbf{u}^*), \\ V^0(s,x) \le \mathcal{E}[\int_s^t [L(\tau, X(\tau), U(\tau)]d\tau\Big|X(s) = x] + \\ \mathcal{E}[V^0(t, X(t))\Big|X(s) = x]. \qquad (7.2.7)$$

We have equality in (7.2.7) if $\mathbf{u} = \mathbf{u}^*$ on $[s, t]$ or in other words we can write

$$V^0(s,x) = \min_{U\in\mathcal{U}} \mathcal{E}[\int_s^t [L(\tau, X(\tau), U(\tau))]d\tau\Big|X(s) = x] + \\ \mathcal{E}[V^0(t, X(t))\Big|X(s) = x], \; s \le \tau \le t. \qquad (7.2.8)$$

The equation (7.2.8) formally expresses the well known “intuitively obvious” Bellman's optimality principle [5] for the considered class of stochastic systems.

Subtract (7.2.6) from (7.2.7) and divide by $t-s$. Then taking into account that $x = X(s)$ and moving to limit as $t \to s^+$ we obtain

$$\mathcal{A}_u V^0(s,x) + L(s,x,u) \geq 0, \tag{7.2.9}$$

where $u = \mathbf{u}(s,x)$. We have equality in (7.2.9) if $u = \mathbf{u}^*(s,x)$. So, the function V^0 satisfies the equation

$$\min_{u \in \mathcal{U}} [\mathcal{L}_u V^0(s,x) + L(s,x,u)] = 0. \tag{7.2.10}$$

The boundary condition

$$V^0(T,x) = \Psi(x) \tag{7.2.11}$$

immediately follows from the definition of the cost function (7.2.5). In stochastic control theory the equation (7.2.10) is called the *dynamic programming equation with continuous time* or *Hamilton-Jacobi-Bellman equation*. It is easy to see from (7.2.7) that

$$\mathcal{E}[V^0(t,X(t))\big|X(t_1)] \geq V^0(t_1,X(t_1)), \; s \leq t_1 \leq t < T.$$

This means that the process $V^0(t,X(t))$ is submartingale with respect to the family of σ-algebras generated by process $X(\tau)$, $\tau < t$. An interesting viewpoint is to observe that for the optimal control $\mathbf{u}^*$ and corresponding process $X^*(t)$ the optimal cost function satisfies the martingale property

$$\mathcal{E}[V^0(t,X^*(t))\big|X^*(t_1)] = V^0(t_1,X^*(t_1)), \; s \leq t_1 \leq t < T,$$

so that the process $V^0(t,X^*(t))$ is the martingale with respect to the same family of σ-algebras.The reader is refered to [21] for more detail on martingale applications to stochastic control.

An exact derivation

Here we formulate rigorous conditions for stochastic dynamic programming approach, see [96, 24]. First we define the class $\mathcal{U}_0$ of admissible controls. Let Φ be the class of functions

$$\varphi : [t_0,T] \times R^n \mapsto R^k$$

with the following properties: $\varphi(t,x)$ is piecewise continuous in t for every fixed x, satisfying the growth condition

$$|\varphi(t,x)| \leq K_g(1+|x|), \; (t,x) \in [t_0,T] \times R^n$$

and the uniform Lipschitz condition

$$|\varphi(t,x) - \varphi(t,\xi)| \leq K_L|x-\xi|, \; t \in [t_0,T], \; x,\xi \in R^n,$$

where K_g, K_L are positive constants. We can write $u \in \mathcal{U}_0$ if

$$u = \mathbf{u}(t,x) = \varphi(t,x), \; t \in [t_0,T],$$

for a certain function $\varphi \in \Phi$. Under these conditions the equation (7.2.3) with $U(t) = \varphi(t,X(t))$ is well defined as an Ito equation and has a unique solution. We also assume that the functions L and Ψ in (7.2.4) are continuous and satisfy the polynomial growth conditions:

$$|L(t,x,u)| \le C_L(1+|x|+|u|)^p,$$
$$|\Psi(t,x)| \le C_\Psi(1+|x|)^p$$

for some positive C_L and C_Ψ and integer p. These conditions guarantee that functional (7.2.4) is finite for any admissible control $\mathbf{u} \in \mathcal{U}_0$. We say that the admissible control $\mathbf{u} = \mathbf{u}^*$ is optimal if it minimizes functional (7.2.4) with $s = t_0$.

Theorem 7.2.1 *Let the function $V(s,x)$ be a solution of the dynamic programming equation*

$$\min_{u \in \mathcal{U}} [\mathcal{A}_u V(s,x) + L(s,x,u)] = 0, \quad s,x \in [t_0,T) \times R^n \tag{7.2.12}$$

with the boundary condition

$$V(T,x) = \Psi(T,x), \quad x \in R^n$$

and this function has the following properties:

(i) $V(t,x)$ is continuous in $[t_0,T] \times R^n$, has continuous first and second partial derivatives in this domain and satisfies the condition of the polynomial growth;

(ii) $V(s,x) \le J(s,x,\mathbf{u})$ for any admissible state feedback control $\mathbf{u} \in \mathcal{U}_0$ and $s,x \in [t_0,T] \times R^n$;

(iii) If $\mathbf{u}^* \in \mathcal{U}_0$ is admissible control, such that

$$\mathcal{A}_{\mathbf{u}^*} V(s,x) + L(s,x,\mathbf{u}^*) = \min_{u \in \mathcal{U}} [\mathcal{A}_u V(s,x) + L(s,x,u)], \quad s,x \in [t_0,T) \times R^n,$$

then

$$V(s,x) = J(s,x,\mathbf{u}^*),$$

i.e. control $\mathbf{u}^*$ is optimal.

In general, solving the Bellman equation (7.2.12) is very complicated. In the following section we consider a linear case, when it is possible to obtain the solution in analytic form.

Linear regulator problem

The regulation objective is to stabilize deviations around the nominal level using a feedback control action, so that the plant will stay near a nominal trajectory, determined by optimal program law. As explained above, the linear models play a very important role. Consider the system (7.2.3) in the special linear case [96], when it is described by the following equation

$$dX(t) = [A(t)X(t) + B(t)U(t)]dt + \sum_{l=1}^{m_1} A_l(t)X(t)dW_{1l}(t) + \sum_{s=1}^{m_2} B_s(t)U(t)dW_{2s}(t) + C(t)dW_3(t), \quad t_0 \le t \le T, \tag{7.2.13}$$

where W_1, W_2 and W_3 are m_1, m_2 and m_3-dimensional independent standard Wiener processes. In this case the vector $X(t)$ can be considered as a small deviation from the nominal value and the regulator task is to stabilize this vector around zero. A possible approach is to compute this regulator so as to minimize the expected value of an integral of a quadratic function of $X(t)$ and $U(t)$.

$$J(t_0, x, \mathbf{u}) = \mathcal{E}[\int_{t_0}^{T} (X'(t)M(t)X(t) + U'(t)R(t)U(t))dt + X'(T)DX(T)\Big|X(t_0) = x], \tag{7.2.14}$$

where $M(t) = M'(t)$ and $R(t) = R'(t)$ are piecewise continous in $[t_0, T]$ positive semidefinite and positive definite matrix functions, D is symmetric positive definite constant matrix. It is supposed that the control is unbounded ($\mathcal{U} = R^k$). To apply Theorem 7.2.1 we will find a solution of dynamic programming equation (7.2.12) in the form

$$V(t, x) = h(t) + x'H(t)x, \ t_0 \le t \le T, \tag{7.2.15}$$

where $h(t)$ is a scalar function, $H(t)$ is a symmetric nonnegative definite matrix function. As a result we obtain the following theorem.

Theorem 7.2.2 *Let function $V(t, x)$ be given by (7.2.15), the matrix $H(t)$ be the solution of differential equation*

$$\dot{H}(t) + A'(t)H(t) + H(t)A(t) = H(t)B(t)[\Gamma(t, H(t)) + R(t)]^{-1}B'(t)H(t) + \Delta(t, H(t)) + M(t), \ t_0 \le t \le T \tag{7.2.16}$$

with the boundary condition

$$H(T) = D,$$

where

$$\Gamma(t, H(t)) = \sum_{l=1}^{m_1} A_l'(t)H(t)A_l(t), \ \Delta(t, H(t)) = \sum_{s=1}^{m_2} B_s'(t)H(t)B_s(t),$$

and the function $h(t)$ is defined by the formula

$$h(t) = \int_{t}^{T} \mathrm{tr}[C'(\tau)H(\tau)C(\tau)]d\tau.$$

Then the optimal control is given by

$$U^*(t) = \varphi^*(t, X(t) = -K(t)X(t), \tag{7.2.17}$$

where

$$K(t) = [\Gamma(t, H(t)) + R(t)]^{-1},$$

and $V(t_0, x)$ is the minimal value of functional (7.2.14) (the optimal performance).

7.2.3 Stochastic maximum principle

There have been many efforts to extend Pontryagin's maximum principle [76] to the optimization of stochastic systems, see [53, 83, 23, 84, 4, 87, 43, 21] and references therein. Correspondingly there exist different versions of this result. We consider a special linear case [83, 84] to show the idea of approach. Assume that the system to be controlled is described by the differential equation

$$\dot{X}(t) = A(Y(t))X(t) + B(Y(t))U(t),\ t \in [t_0, T], \qquad X(t_0) = x_0,\ Y(t_0) = y_0, \tag{7.2.18}$$

where all the notations are the same as earlier. At time t the controller observes both continuous variable $X(t)$ and discrete variable $Y(t)$. Based upon this observation the controller selects a control action $U(t)$, i.e.,

$$U(t) = \mathbf{u}(t, X(t), Y(t)) \tag{7.2.19}$$

Define the set of admissible controls $\mathcal{U}$ as the set of all functions on $[t_0, T] \times R^n \times \mathcal{Y} \mapsto R^k$ such that with probability one the equation

$$\dot{X}(t) = A(Y(t))X(t) + B(Y(t))\mathbf{u}(t, X(t), Y(t)),\ t \in [0, T], \qquad X(t_0) = x_0,\ Y(t_0) = y_0 \tag{7.2.20}$$

has a unique solution, which is continuous in the pair (t_0, x_0), continuable to all of $[0, T]$, and for fixed (t, t_0) satisfies a Lipschitz condition with respect to x_0 in every bounded region of R^n. Denote the solution to (7.2.20) by $X^{\mathbf{u}}(t; t_0, x_0, y_0)$. The problem is to find that element $\mathbf{u} = \mathbf{u}^*$ of $\mathcal{U}$ which minimizes the functional

$$J[\mathbf{u}; t_0, x_0, y_0)] = \mathcal{E} \left[\int_{t_0}^{T} L[\tau, X^{\mathbf{u}}(\tau; t_0, x_0, y_0), \mathbf{u}(\tau; X^{\mathbf{u}}(\tau; t_0, x_0, y_0), y(\tau)), Y(\tau)] d\tau \Big| t_0, x_0, y_0 \right], \tag{7.2.21}$$

where $L(t, x, u, y)$ is nonnegative and continuously differentiable with respect to x and u. To formulate necessary conditions we suppose that there exists such an element $\mathbf{u}^*$ that for every $t_0 \in [0, T]$, every $x_0 \in R^n$, and every $y_0 \in \mathcal{Y}$

$$J(\mathbf{u}^*; t_0, x_0, y_0) = \min_{\mathbf{u} \in \mathcal{U}} J(\mathbf{u}; t_0, x_0, y_0). \tag{7.2.22}$$

Let $\mathbf{u} \in \mathcal{U}$. Define the *stochastic Hamiltonian* at the point (s, ξ) as

$$\mathcal{H}(s, \xi, \mathbf{u}) = \psi'(s, \xi)[A(Y(s))\xi + B(Y(s))\mathbf{u}(s, \xi, Y(s))] - L[s, \xi, \mathbf{u}(s, \xi, Y(s)), Y(s)], \tag{7.2.23}$$

where $\psi(s, \xi)$ is the so called *co-state* or *adjoint vector* satisfying the following differential equation, integrated backward in time

$$\dot{\psi}(t, X^{\mathbf{u}}(t; t_0, x_0, y_0)) = -\frac{\partial \mathcal{H}}{\partial x}(t, X^{\mathbf{u}}(t; t_0, x_0, y_0)),\ \psi(T, x) = 0 \text{ for all } x \tag{7.2.24}$$

Calculating the right hand side of (7.2.24) with $\mathbf{u} = \mathbf{u}^*$ we have

$$\dot{\psi}(t, X^{\mathbf{u}^*}(t; t_0, x_0, y_0)) = -[A(Y(t) + B(Y(t))\mathbf{u}_x^*(t, X^{\mathbf{u}^*}(t; t_0, x_0, y_0), Y(t))]'\psi(t, X^{\mathbf{u}^*}(t; t_0, x_0, y_0)) + [L_x + L_u \mathbf{u}_x^*(t, X^{\mathbf{u}^*}(t; t_0, x_0, y_0), Y(t))]',\ t \in [t_0, T],\ \psi(T, x) = 0 \text{ for all } x. \tag{7.2.25}$$

Suppose that the partial derivative $\mathbf{u}_x^*$ exists everywhere except perhaps in some Borel set in $[0, T] \times R^n$ of Lebesgue measure zero, and that $\mathbf{u}^*$ satisfies a Lipschitz condition with respect to x. Then for fixed t_0 it can be shown that a unique solution to (7.2.25) exists a.s. for almost all $x_0 \in R^n$. Under the assumptions above the following result is established.

Theorem 7.2.3 *Let* $\mathbf{u} \in \Gamma$ *be an admissible control. Then*

$$\mathcal{E}[\mathcal{H}(t, X(t), \mathbf{u}^*)\Big| t, X(t), Y(t)] \geq \mathcal{E}[\mathcal{H}(t, X(t), \mathbf{u})\Big| t, X(t), Y(t)]$$

a.s.

Let

$$L(t, x, u, y) = x'M(y)x + u'R(y)u,$$

where $M(y)$ and $R(y)$ $y \in \mathcal{Y}$ are symmetric nonnegative definite and positive definite matrices, and control is unbounded. In this case the stochastic Hamiltonian (7.2.23) has the form

$$\begin{aligned}\mathcal{H}(t, X(t), U(t)) = \psi'(t, X(t))[A(Y(t))X(t) + B(Y(t))\mathbf{u}(t, X(t), Y(t))] - \\ X'(t)M(Y(t))X(t) - \mathbf{u}'(t, X(t), Y(t))R(Y(t))\mathbf{u}(t, X(t), Y(t)),\end{aligned} \quad (7.2.26)$$

Applying Theorem 7.2.3 with Hamiltonian (7.2.26) and taking into account that there are no constraints we obtain directly

$$\mathbf{u}^* = \frac{1}{2}R^{-1}(Y(t))B'(Y(t))\mathcal{E}[\psi(t, X(t))\Big| X(t), Y(t)]$$

To determine the solution, assume that the co-state takes the form

$$\psi(t, X(t)) = -2H(t)X(t), \ t \in [t_0, T]$$

where $H(t)$ is a random symmetric matrix conditionally independent on X and differentiable everywhere. Then

$$\mathbf{u}^* = -R^{-1}(Y(t))B'(Y(t))\mathcal{E}[H(t)\Big| Y(t)]X(t). \quad (7.2.27)$$

With the notation $H_i(t) = \mathcal{E}[H(t)\Big| Y(t) = i]$ yields the optimal control as

$$\mathbf{u}^* = -R^{-1}(i)B'(i)H_i(t)X(t), \text{ when } Y(t) = i$$

The matrix $H_i(t)$ $(i \in \mathcal{N})$ satisfies the set of coupled matrix Riccati differential equations, integrated backward in time

$$\begin{aligned}\dot{H}_i(t) + A'(i)H_i(t) + H_i(t)A(i) - H_i(t)B(i)R^{-1}(i)B'(i)H_i(t) + \\ \sum_{j \neq i}^{\nu} q_{ij}(H_j(t) - H_i(t)), \ t_0 \leq t \leq T\end{aligned} \quad (7.2.28)$$

with boundary condition

$$H_i(T) = 0$$

for all $i \in \mathcal{N}$. The reader is refered to [21, 23, 53, 87] for more detail regarding to Ito differential equation (7.1.1). The various versions of the stochastic matrix principle for systems with jumps are presented in [21, 43, 83, 84, 150] and references therein.

7.2.4 Separation principle

General formulation

As it was formulated at the beginning the separation principle is usually used to convert a partially observed system to a "completely" observed system, so we can use to obtain optimal control of partially observed systems the same methods used as in case of systems with completely state information. It typically works for linear or almost linear systems. The purpose of this section is to show that the problem of optimal control for a stochastic linear dynamic system, observed via a noise linear channel, can be reduced to two independent problems of the control and filtering respectively. Under suitable conditions, solution of the latter problems are shown to exist [18, 24, 101] and references therein. Consider the system described by linear stochastic differential equations

$$dX(t) = [A(t)X(t) + b(t, U(t))]dt + C(t)dW_1(t),\ 0 \le t \le T, \quad X(0) = x_0, \tag{7.2.29}$$

$$dZ(t) = F(t)X(t)dt + G(t)dW_2(t),\ 0 \le t \le T, \quad Z(0) = 0, \tag{7.2.30}$$

where the control vector U takes values in a convex compact subset $\mathcal{U} \subset R^k$; $Z(t) \in R^n$ is channel output; W_1, W_2 are independent standard Wiener processes in R^{d_1}, R^{d_2}. The problem is to control $X(\cdot)$ in such a way as to minimize functional

$$J[U] = \mathcal{E}\left[\int_0^T L(t, X(t), U(t))dt\right]. \tag{7.2.31}$$

The control is based on the a priori distribution of X_0 and on the information provided by the channel output $Z(\cdot)$. Since the controller is not clairvoyant, $U(t)$ must be assumed to depend only on the $Z(s)$ for $0 \le s \le T$. To express this nonanticipative dependence we introduce, following [26, 101] a suitable class of control functionals. Let $\mathcal{C}$ denote the class of functions $f(t)$, continuous on $[0,T]$ with values in R^n and write for the *past of f at time* t,

$$(\pi_t f)(s) = \begin{cases} f(s) & \text{for} \quad 0 \le s \le t, \\ f(t) & \text{fo} \quad t \le s \le T \end{cases} \tag{7.2.32}$$

and $U(t) = \mathbf{u}(t, \pi_t Z)$ Clearly $\pi_t f \in \mathcal{C}$ if $f \in \mathcal{C}$. Let

$$\psi : [0, T] \times \mathcal{C} \mapsto \mathcal{U},$$

be a mapping with the properties: $\psi(t, f)$ is Hölder continuous in t for each $f \in \mathcal{C}$ and satisfies a uniform Lipschitz condition

$$|\psi(t, f) - \psi(t, g)| < c_1 \|f - g\|, \tag{7.2.33}$$

where $t \in [0, T]$, $f, g \in \mathcal{C}$ and $\|\cdot\|$ denote sup norm in $\mathcal{C}$. Let Ψ denote the class of functionals ψ. We call the control $\mathbf{u}(\cdot, \cdot)$ admissible and write $\mathbf{u} \in \mathcal{U}_a$ if

$$U(t) = \mathbf{u}(t, \pi_t Z) = \psi(t, \pi_t Z),\ 0 \le t \le T,$$

for some $\psi \in \Psi$. The problem is to find $\mathbf{u}^0 \in \mathcal{U}_a$ such that

$$J[\mathbf{u}^0] = \min\{J[\mathbf{u}] : \mathbf{u} \in \mathcal{U}_a\}.$$

The corresponding functional ψ^0 is optimal. It is shown [101] that $J[\mathbf{u}]$ is well defined. The separation theorem states that an optimal control exists in a subclass $\widehat{\mathcal{U}}_a$ of controls which depend only on the expected value of the current state given the past of Z. More precisely, let

$$\mathcal{Z}_t = \sigma\{Z(s), 0 \le s \le t\},$$

$$\widehat{X} = \mathcal{E}[X(t) \big| \mathcal{Z}_t].$$

Write $\widehat{\Psi}$ for the class of functions

$$\widehat{\psi} : [0, T] \times R^n \mapsto \mathcal{U},$$

such that

$$|\widehat{\psi}(t, \xi) - \widehat{\psi}(s, \xi)| + |\widehat{\psi}(t, \xi) - \widehat{\psi}(t, \eta)| \leq c_2(R)|t - s|^\alpha + c_3|\xi - \eta| \qquad (7.2.34)$$

in every domain $0 \leq s, t \leq T$, $|\xi| < R$, $|\eta| < R$, where c_3 and $\alpha \in (0, \frac{1}{2})$ are independent on R. We write $\mathbf{u} \in \widehat{\mathcal{U}}_a$ if

$$U(t) = \mathbf{u}(t, \pi_t Z) = \widehat{\psi}[t, \widehat{X}(t)], \; t \in [0, T],$$

for some $\widehat{\psi} \in \widehat{\Psi}$. It is shown in [101] that $\widehat{\mathcal{U}}_a \subset \mathcal{U}_a$. The following additional assumptions will be made. We write u.h.c. (α) for "uniformly Hölder continuous (exponent α)," and u.l.c. for "uniformly Lipschitz continuous," where the uniformity is to hold over the whole range of the relevant arguments, unless otherwise stated:

(A.1) The matrices A, C are u.h.c.(α) in t and F, G are continuously differentiable in $[0, T]$.

(A.2) $G(t)G'(t) \geq cI$, $t \in [0, T]$.

(A.3) $|\det[F(t)]| \geq c$, $t \in [0, T]$.

(A.4) b, b_u, b_{uu} are continuous on $[0, T] \times \mathcal{U}$ (a subscript denotes partial differentiation) and b, b_u are u.h.c.(α) in t.

(A.5) L and L_u are bounded, u.h.c.(α) in t and u.l.c. in x. L_{uu} is bounded and continuous on $[0, T] \times R^n \times \mathcal{U}$.

(A.6) $[b'(t, u)p + L(t, x, u)]_{uu} \geq c_6 I$ for all $(t, x, u, p) \in [0, T] \times R^n \times \mathcal{U} \times \{p : |p| \leq \pi\}$, where π is an a priori upper bound of the space derivative $V_\xi(t, \xi)$ of the solution $V(t, \xi)$ of Bellman's equation below.

(A.7) X_0 is a Gaussian random variable independent of the processes $W_1(t), W_2(t)$ and with positive definite covariance matrix S_0.

The foregoing restrictions are mainly technical. Assumption (A.3) would rarely be met in practice, where typically $\dim Y < \dim X$; this condition is needed to guarantee that a certain elliptic operator will be nondegenerate. A square nonsingular matrix F could be constructed artificially, if necessary, by adjoining to the channel equation (7.2.30) a suitable term of form

$$d\tilde{Z}(t) = \epsilon \tilde{F} X(t) dt + \tilde{G} d\tilde{W}_2(t).$$

If $\epsilon > 0$ is sufficiently small, then from a practical viewpoint the components $\tilde{Z}(t)$ of the observation vector contribute negligible information to the controller. However, details of such an approximation have yet to be worked out. The number π in (A.6) is an a priori bound on the space derivative of the solution of Bellman's equation. In the special, but very important case, where $b(t, u)$ is linear in u, the estimate π is not required, and (A.6) can be replaced by

$$(\text{A.6})' \quad L_{uu}(t, x, u) \geq cI, \; t, x, u \in [0, T] \times R^n \times \mathcal{U}.$$

The crucial assumptions for the separation theorem below are the following: (i) the basic equations have the form (7.2.29), (7.2.30); (ii) the (formal) perturbations $\frac{dW_i}{dt}$, $i = 1, 2$ are "white Gaussian noise"; (iii) X_0 is Gaussian and independent on the W_i; (iv) $J[U]$ is a functional additive in t.

Theorem 7.2.4 (Separation theorem). *Subject to the assumptions stated, an admissible optimal control exists in the form of*

$$U^0(t) = \widehat{\psi}^0[t, \widehat{X}(t)], \ \ t \in [0, T]$$

for some $\widehat{\psi}^0 \in \widehat{\Psi}$.

Stochastic differential equation for the expected value of the current state

Let $U(\cdot)$ be admissible and write

$$\beta(t) = b[t, U(t)] = b[t, \psi(t, \pi_t Z)]. \tag{7.2.35}$$

It is shown in [101] that the random variable $\beta(t)$ is $\mathcal{Z}_t$ measurable. Next let

$$X(t) = \tilde{X}(t) + X^*(t),$$

where $\tilde{X}(t)$ is the process determined by the stochastic differential equation

$$d\tilde{X}(t) = A(t)\tilde{X}(t) + C(t)dW_1(t), \ \tilde{X}(0) = X_0, \ t \in [0, T], \tag{7.2.36}$$

and X^* is defined by

$$\frac{dX^*(t)}{dt} = A(t)X^*(t) + \beta(t), \ X^*(0) = 0, \ t \in [0, T]. \tag{7.2.37}$$

Since $X^*(t)$ is $\mathcal{Z}_t$-measurable there follows

$$\widehat{X}(t) = \{\mathcal{E}\tilde{X}(t)\big|\mathcal{Z}_t\} + X^*(t). \tag{7.2.38}$$

Now define a process $\tilde{Z}(t)$ according to

$$\begin{gathered} d\tilde{Z}(t) = dZ(t) - F(t)X^*(t)dt = \\ F(t)\tilde{X}(t)dt + G(t)dW_2(t), \ \tilde{Z}(0) = 0, \ t \in [0, T], \end{gathered} \tag{7.2.39}$$

and let

$$\tilde{\mathcal{Z}}(t) = \sigma\{\tilde{Z}(s), \ 0 \leq s \leq t\}.$$

By (7.2.37) $X^*(t)$ is $\mathcal{Z}_t$-measurable, then by (7.2.39), $\tilde{Z}(t)$ is $\mathcal{Z}_t$-measurable. It is proved in [101] that $Z(t)$ is $\tilde{\mathcal{Z}}_t$-measurable and thus that $\mathcal{Z}_t = \tilde{\mathcal{Z}}_t$. Now we have from (7.2.38)

$$\widehat{X}(t) = \bar{X}(t) + X^*(t), \tag{7.2.40}$$

where

$$\bar{X}(t) = \mathcal{E}\{\tilde{X}(t)\big|\mathcal{Z}(t)\} = \mathcal{E}\{\tilde{X}(t)\big|\tilde{\mathcal{Z}}(t)\}.$$

To compute $\bar{X}(t)$ we note that the equations (7.2.36), (7.2.39) have the form (7.2.29), (7.2.30) with $b \equiv 0$ and well-known Kalman-Bucy filtering results can be applied [18, 24, 39, 40, 96]. Introduce the conditional covariance matrix

$$S(t) = \mathcal{E}\{[X(t) - \widehat{X}(t)][X(t) - \widehat{X}(t)]'\big|\mathcal{Z}_t\} = \mathcal{E}\{[\tilde{X}(t) - \bar{X}(t)][\tilde{X}(t) - \bar{X}(t)]'\big|\tilde{\mathcal{Z}}_t\},$$

where the second equality holds because $X(t) = \tilde{X}(t) + X^*(t)$ and $\mathcal{Z}_t = \tilde{\mathcal{Z}}_t$. By the result of Kalman-Bucy filtering, applied to (7.2.36)-(7.2.39), $S(t)$ is the unique solution of the Riccati equation

$$\frac{dS(t)}{dt} = AS + SA' + CC' - SF'(GG')^{-1}FS, \ t \in [0, T], \ S(0) = S_0. \tag{7.2.41}$$

Then $\bar{X}$ is determined by

$$d\bar{X}(t) = A\bar{X}(t)dt + SF'(GG')^{-1}(d\tilde{Z} - F\bar{X}dt),\ t \in [0,T], \tag{7.2.42}$$

with the initial condition

$$\bar{X}(0) = \mathcal{E}[\tilde{X}(0)\big|\tilde{\mathcal{Z}}_0] = \mathcal{E}[\tilde{X}(0)] = \mathcal{E}[X_0].$$

Combining (7.2.36)-(7.2.39) and (7.2.40) and (7.2.42) we obtain

$$d\widehat{X}(t) = A\widehat{X}(t)dt + \beta(t)dt + SF'(GG')^{-1}(dZ - F\widehat{X}dt),\ t \in [0,T],\ \widehat{X}(0) = \mathcal{E}[X_0]. \tag{7.2.43}$$

Equation (7.2.43) exhibits the process $\widehat{X}(t)$ as the solution of an equation "forced" by the channel output increments dZ and by the control term β. It is very important that it is possible to replace the differential $dZ - F\widehat{X}dt$ by the suitable scaled differential of a Wiener process. This can be justified by the observation that linear least square estimation is equivalent to an orthogonal projection of the estimated variable on the data, see [96, 101] for more detail. As a result we have finally

$$d\widehat{X}(t) = A\widehat{X}(t)dt + b[t, \widehat{\psi}(t, \widehat{X})]dt + SF'(GG')^{-\frac{1}{2}}d\widehat{W},\ \widehat{X}(0) = \mathcal{E}[X_0]. \tag{7.2.44}$$

Under the regularity conditions (7.2.34) and (A.4) the equation (7.2.44) determines a diffusion process on $[0,T]$. Let $\xi \in R^n$ denotes a value of $\widehat{X}$ and let $V : [0,T] \times R^n \mapsto R^1$ have continuous derivatives up to second order. The differential generator of the process $\widehat{X}$ is the operator $\widehat{\mathcal{A}}(\widehat{\psi})$ given by

$$\widehat{\mathcal{A}}(\widehat{\psi})V(t,\xi) = \frac{1}{2}\mathrm{tr}[\widehat{C}'V_{\xi\xi}(t,\xi)\widehat{C}] + [A\xi + b[t, \widehat{\psi}(t,\xi)]]'V_{\xi}(t,\xi) + V_t(t,\xi), \tag{7.2.45}$$

where $\widehat{C} = SF'(GG')^{-\frac{1}{2}}$ and $V_t,\ V_{\xi}(V_{\xi\xi})$ denotes the vector (matrix) of first (second) partial derivatives of V. It is also shown [101, 96] that the conditional distribution of $X(t)$ given $\mathcal{Z}_t$ is Gaussian one and that, if $0 \le t_1 \le t_2 \le t_3 \le T$, the increments $\widehat{W}(t_3) - \widehat{W}(t_2)$ are independent on $\mathcal{Z}_{t_1}$. The reader is refered to [21, 61, 62] for more general consideration of the discussed problems.

Optimality criterion and application to linear regulator problem

Let $\mathcal{G}(x,t,\xi)$ be the Gaussian probability density in R^n with mean ξ and covariance matrix $S(t)$:

$$\mathcal{G} = (2\pi)^{-\frac{n}{2}}[\det S(t)]^{-\frac{1}{2}}\exp[-\frac{1}{2}(x-\xi)'S^{-1}(t)(x-\xi)].$$

By the results of previous section if u is a fixed vector of $\mathcal{U}$, then

$$\widehat{L}(t,\xi,u) = \mathcal{E}[L(t,X(t),u)\big|\widehat{X}(t) = \xi] = \int_{R^n} L(t,x,u)\mathcal{G}(x,t,\xi)dx.$$

It is verified in [96, 101] that $\widehat{L}$ satisfies the conditions imposed on L in (A.5). On this assumption the following sufficient optimality conditions are established

Theorem 7.2.5 (Optimality criterion). *Suppose that there exist an element $\widehat{\psi}^0 \in \widehat{\Psi}$ and a function $V : [0,T] \times R^n \mapsto R^1$ such that*

(i) $V, V_t, V_{\xi}, V_{\xi\xi}$ are continuous and

$$|V| + |V_t| + |\xi||V_{\xi}| + |V_{\xi\xi}| \le c(1 + |\xi|^2),$$

(ii)

$$\widehat{\mathcal{A}}(\widehat{\psi}^0)V(t,\xi) + \widehat{L}[t,\xi,\widehat{\psi}^0(t,\xi)] = 0, \tag{7.2.46}$$

$$\widehat{\mathcal{A}}(u)V(t,\xi) + \widehat{L}(t,\xi,u) \geq 0 \tag{7.2.47}$$

for all $(t,\xi,u) \in [0,T] \times R^n \times U$, *and*

$$V(T,\xi) = 0, \ \xi \in R^n. \tag{7.2.48}$$

Then the control $U(t) = \widehat{\psi}^0(t, \widehat{X}(t))$ *is optimal in* $\mathcal{U}_a$.

Observe that (7.2.46), (7.2.47) are formally equivalent to Bellman's functional equation

$$\min_{u\in\mathcal{U}}[\widehat{\mathcal{L}}(u)V(t,\xi) + \widehat{L}(t,\xi,u)] = 0, \ (t,\xi) \in [0,T] \times R^n \tag{7.2.49}$$

with boundary condition (7.2.48). If Bellman's equation can be solved explicily for functions V and $\widehat{\psi}^0$, which satisfies the hypothesis of Theorem 7.2.5, then of course, many of the restrictive conditions, imposed in general discussion become irrelevant. A well known result is the mentioned above LQG problem i.e. the linear regulator problem using a linear channel output information corrupted by Gaussian noise [2, 18, 36, 96]. Consider this problem in more detail. In (7.2.29) let

$$b[t,u] = B(t)u,$$

let u range over R^k and let

$$L(t,x,u) = x'M(t)x + u'R(t)u,$$

where $M(t)$ and $R(t)$ are respectively positive semidefinite and positive definite, with $R^{-1}(t)$ bounded on $[0,T]$. In this case

$$\widehat{L}(t,\xi,u) = \xi'M(t)\xi + u'R(t)u + \mathrm{tr}[M(t)S(t)],$$

and Bellman's equation has the following form

$$V_t + \frac{1}{2}\mathrm{tr}[\widehat{C}'V_{\xi\xi}\widehat{C}] + \xi'A'V_\xi - \frac{1}{4}V_\xi' BR^{-1}B'V_\xi + \\ \xi'M\xi + \mathrm{tr}(MS) = 0. \tag{7.2.50}$$

The equation (7.2.50) has a quadratic solution

$$V(t,\xi) = \xi'H(t)\xi + h(t),$$

where $H(t)$ is the unique solution of the matrix differential Riccati equation

$$\frac{dH}{dt} + A'H + HA - HBR^{-1}B'H + M = 0, \ t \in [0,T], \ H(T) = 0 \tag{7.2.51}$$

and $h(t)$ is given by

$$\frac{dp}{dt} + \mathrm{tr}(\widehat{C}'H\widehat{C}) + \mathrm{tr}(MS), \ t \in [0,T], \ h(T) = 0.$$

The optimal control is then

$$U(t) = \widehat{\psi}(t,\widehat{X}) = -\frac{1}{2}R^{-1}(t)B'(t)V_\xi(t,\widehat{X}) = -R^{-1}(t)B'(t)H(t)\widehat{X}.$$

Here $H(t)$ and hence $U(t)$ are actually independent of the channel coefficient matrices F, G. Moreover the optimal control is the same function of $\widehat{X}$ as in the case of complete state information. For this solution of (7.2.50) to exist it is sufficient with the stated conditions on M and R that all parameter matrices will be piecewise continuous and that (A.2) holds. The reader is refered to [2, 18, 96] for more detail of the LQG problem.

7.3 Definition of stochastic stability and stochastic Lyapunov function

7.3.1 Classic stability concept

The stability theory of stochastic systems follows in general the classic Lyapunov stability concept, see [41, 45, 54]. An important application of this theory is the stochastic stabilizing control problem. The stabilizing control law should guarantee stochatic stability of the system in an appropriate sense. Consder the system (7.1.13). Suppose that the initial conditions (7.1.22) are deterministic and let $\tilde{X}(t)$ be the solution of (7.1.13), satisfying these initial conditions. Roughly speaking the solution $\tilde{X}(t)$ is stable if for bounded changes of initial conditions the solution $\tilde{X}(t)$ has bounded changes too. The process $\tilde{X}(t)$ is called an undisturbed motion with respect to given initial condition and the changes of the initial conditions are called disturbances. For more uniform mathematical definitions note that without a loss of generality we can suppose $\tilde{X}(t) \equiv 0$. In this case it is necessary that

$$a(t,0,y) \equiv 0, \; b(t,0,y) \equiv 0. \tag{7.3.1}$$

Under the conditions (7.3.1) the equation (7.1.13) has the solution $X(t) \equiv 0$. This solution is called the *trivial solution.* As a rule it is supposed that the undisturbed motion is the trivial solution of (7.1.13). The set $\mathcal{D} = \{0, \mathcal{Y}\}$ is the invariant set for the Markov process $[X(t), Y(t)]$ [45] in the sense that

$$P\{[X(t), Y(t)] \in \mathcal{D} | X(t_0) = x_0 \in \mathcal{D}, \; Y(t_0) = y_0 \in \mathcal{D}\} = 1.$$

From this point of view the stability of the trivial solution $X(t) \equiv 0$ means the stability of the corresponding invariant set of the Markov process $[X(t), Y(t)]$. We follow [41] and partially [45] in definitions of stability below.

7.3.2 Weak Lyapunov stability

Definition 7.3.1 (Weak stability in probability.) *The trivial solution $X(t) \equiv 0$ of the system (7.1.13) (the invariant set $\mathcal{D} = \{0, \mathcal{Y}\}$ of the Markov process $[X(t), Y(t)]$) is called weakly stable in probability if for any numbers $\epsilon > 0, p > 0$ sufficiently small there exists a number $\delta > 0$ such that if*

$$|x_0| \leq \delta, \; y_0 \in \mathcal{Y}, \; t_0 \geq 0 \tag{7.3.2}$$

then for every $t \geq t_0$

$$P[|X(t)| < \epsilon \Big| X(t_0) = x_0, Y(t_0) = y_0] > 1 - p. \tag{7.3.3}$$

Definition 7.3.2 (Weak asymptotic stability in probability.) *If the trivial solution $X(t) \equiv 0$ is weakly stable in probability and for any number $\gamma > 0$ and initial condition from the domain $|x_0| \leq h_0$, the following equality holds*

$$\lim_{t \to \infty} P[|X(t)| < \gamma \Big| X(t_0) = x_0, Y(t_0) = y_0] = 1, \tag{7.3.4}$$

then this solution is called weakly asymptotically stable in probability. The constant h_0 defines the domain of attraction of the trivial solution (the undisturbed motion).

7.3.3 Strong Lyapunov stability

The definitions of weak stability do not characterize the behavior of the paths of process $X(t)$. The trivial solution (the undisturbed motion) can be weakly stable, but almost all the paths can leave the domain $|X(t)| < \epsilon$ in different moments. In this connection more often the strong stability concept is used.

Definition 7.3.3 (Stability in probability.) *The trivial solution $X(t) \equiv 0$ of the system (7.1.13) is called (strongly) stable in probability if for any numbers $\epsilon > 0, p > 0$ sufficiently small there exists a number $\delta > 0$ such that from condition (7.3.2) it follows that*

$$P[\sup_{t \geq t_0} |X(t)| < \epsilon \Big| X(t_0) = x_0, Y(t_0) = y_0] > 1 - p. \tag{7.3.5}$$

This definition means that the path of $X(t)$ with the initial disturbance sufficiently small does not leave an arbitrary small neighborhood of trivial solution with probability tending to one.

Definition 7.3.4 (Asymptotic stability in probability.) *If the trivial solution $X(t) \equiv 0$ is stable in probability and for any number $\gamma > 0$ and the initial conditions from the domain $|x_0| \leq h_0$, the following equality holds*

$$\lim_{T \to \infty} P[\sup_{t \geq t_0 + T} |X(t)| < \gamma \Big| X(t_0) = x_0, Y(t_0) = y_0] = 1, \tag{7.3.6}$$

then this solution is called (strongly) asymptotically stable in probability. The constant h_0 defines the domain of attraction of the trivial solution (the undisturbed motion).

The case is interesting in many applications when the domain of attraction covers the entire state space.

Definition 7.3.5 (Asymptotic stability in probability in large.) *The trivial solution $X(t) \equiv 0$ is said to be asymptotically stable in probability in large if for any bounded domain $|x_0| \leq h_0$ and for numbers $\gamma > 0,\ p > 0,\ q > 0$ there exists a bounded domain $|x| < h_1$ and a number $T > 0$, such that*

$$P[\sup_{t \geq t_0} |X(t)| < h_1 \Big| X(t_0) = x_0, Y(t_0) = y_0] > 1 - p,$$

$$P[\sup_{t \geq t_0 + T} |X(t)| < \gamma \Big| X(t_0) = x_0, Y(t_0) = y_0] > 1 - q - p. \tag{7.3.7}$$

7.3.4 Mean square and p-stability

Sometimes it is more convenient to restrict attention to the stochastic moments of the solution. In this case we define the stochastic stability as *stability in the pth mean, p-th mean stability* or *pstability* [45] in particular, when $p = 2$, as *stability in the mean square* or *mean square stability*

Definition 7.3.6 (p-stability.) *The trivial solution $X(t) \equiv 0$ of the system (7.1.13) is called p-stable (stable in the pth mean) if for any $\epsilon > 0$ there exists $\delta > 0$ such that for any solution with the initial conditions satisfying (7.3.2), the following inequality holds*

$$\mathcal{E}[|X(t)|^p \Big| X(t_0) = x_0, Y(t_0) = y_0] < \epsilon,\ t \geq t_0. \tag{7.3.8}$$

Definition 7.3.7 (Asymptotic p-stability.) *The trivial solution $X(t) \equiv 0$ of the system (7.1.13) is called asymptotically p-stable (asymptotically stable in the pth mean) if it is p-stable and for all the solutions, with $|x_0| \leq h_0, y_0 \in \mathcal{Y}$*

$$\lim_{t\to\infty} \mathcal{E}[|X(t)|^p \Big| X(t_0) = x_0, Y(t_0) = y_0] = 0. \tag{7.3.9}$$

We say in this case that domain $|x_0| \leq h_0$ belongs to the domain of attraction of the solution $X(t) \equiv 0$.

Definition 7.3.8 (Exponential p-stability.) *The trivial solution $X(t) \equiv 0$ of the system (7.1.13) is called exponentially p-stable (exponentially stable in the pth mean) if for any $x_0 \in R^n$, $y_0 \in \mathcal{Y}$ and $t \geq t_0 \geq 0$ there exists $\alpha > 0$ and $\beta > 0$ constant such that*

$$\mathcal{E}[|X(t)|^p \Big| X(t_0) = x_0, Y(t_0) = y_0] \leq \beta |X(t_0)|^p e^{-\alpha(t-t_0)}. \tag{7.3.10}$$

All these definitions do not require p to be an integer. When only the p positive integers are considered the following definition is widely used [10, 89, 45].

Definition 7.3.9 *The pth moment of the solution of the system (7.1.13) is called asymptotically stable (with p a positive integer) if for all nonnegative integers $p_1, p_2, \dots, p_n$ such that $p_1 + p_2 + \ldots + p_n = p$ we have:*

(i) for any positive $\epsilon > 0$ there exists $\delta > 0$ such that (7.3.2) implies

$$|\mathcal{E}[X_1^{p_1} X_2^{p_2} \dots X_n^{p_n} \Big| X(t_0) = x_0, Y(t_0) = y_0]| < \epsilon,$$

(ii) for all the solutions with $|x_0| \leq h_0, y_0 \in \mathcal{Y}$

$$\lim_{t\to\infty} \mathcal{E}[X_1^{p_1} X_2^{p_2} \dots X_n^{p_n} \Big| X(t_0) = x_0, Y(t_0) = y_0] = 0,$$

where X_i denotes ith component of X.

For even integers p the properties expressed by Definitions 7.3.7 and 7.3.9 are equivalent. For odd integers p the property of Definition 7.3.7 is equivalent or stronger than the property of Definition 7.3.9.

7.3.5 Recurrence and positivity

An alternative "weak" counterpart to classic Lyapunov stability is the property that $X(t)$ will be *recurrent* or, more strongly, that $X(t)$ will be *positive* [45, 97, 98]. Roughly speaking, $X(t)$ is recurrent if for every initial state, any ball in the state space is hit eventually w.p.1; $X(t)$ is positive if, in addition, the hitting time has finite expectation. Under additional restrictions, the positivity of X is equivalent to the existence of a unique invariant probability measure μ: that is if the distribution of $X(t_0)$ is μ then so is that of $X(t)$ for all $t > 0$. So, consider the homogeneous version of the system (7.1.1).

$$dX(t) = a(X(t))dt + B(X(t))dW(t),\ t \in \mathcal{I}, \tag{7.3.11}$$

where $B(x)$ is the $n \times n$ nonsingular matrix and $W(t)$ is the n-dimensional Wiener process. The following assumptions are made with respect to (7.3.11):

(i) $X(t_0)$ is a random variable independent on the increment of the Wiener process,

(ii) for some constant k_1

$$|a(x) - a(z)| + |B(x) - B(z)| \leq k_1|x - z|, \ x, z \in R^n, \tag{7.3.12}$$

(iii) for some constant k_2

$$z'B(x)B'(x)z \geq k_2 z'z, \ x, z \in R^n. \tag{7.3.13}$$

Definition 7.3.10 *The process $X(t)$ defined by (7.3.11) is said to be recurrent if there exists a compact subset $\mathcal{K} \subset R^n$ such that for every $x \in R^n$*

$$P[X(t) \in \mathcal{K} | X(t_0) = x] = 1.$$

Definition 7.3.11 *Let $\mathcal{G}$ be a nonempty open set in R^n and let τ_G be the first time the boundary of $\mathcal{G}$ is reached. The process $X(t)$ defined by (7.3.11) is said to be positive if it is recurrent and if*

$$\mathcal{E}[\tau_G | X(t_0) = x] < \infty$$

for arbitrary $\mathcal{G} \subset R^n$ and $x \in R^n \setminus \mathcal{G}$.

7.3.6 Stochastic Lyapunov function

The stochastic Lyapunov function plays the same role in the study of stochastic stability as the Lyapunov function does for deterministic stability analysis. It turns out that, roughly speaking, the key step is to prove that a candidate positive function of the system variables possesses the *supermartingale property*, see [45, 54], but it is important that Kats and Krasovskii, Bertram and Sarachik in their pioneering works [42, 7] originally proved stability results by Lyapunov function methods without reference to martingale theory. Consider a scalar function $V(t, x, y)$ defined in the domain (7.1.23) and continuously differentiable in all the variables in this domain as often as required in the process of solution of a stability problem and $V(t, 0, y) \equiv 0$. This leads to original definitions by Kats and Krasovskii [42, 41].

Definition 7.3.12 *The function $V(t, x, y)$ is positive definite (negative definite) if*

$$\inf_{y \in \mathcal{Y}, t \geq t_0} V(t, x, y) = W(x),$$
$$\left(\sup_{y \in \mathcal{Y}, t \geq t_0} V(t, x, y) = -W(x), \right),$$

where $W(x)$ is the positive definite function in Lyapunov's sense, i.e. $W(0) = 0$ and $W(x) > 0$ if $x \neq 0$.

Definition 7.3.13 *The function $V(t, x, y)$ admits an infinitesimal lower limit if there exists a continuous function $W(x)$, $W(0) = 0$, such that*

$$|V(t, x, y)| \leq W(x).$$

This means that $V(t, x, y) \to 0$, as $x \to 0$ uniformly in $t \geq t_0$, $y \in \mathcal{Y}$.

Definition 7.3.14 *The function $V(t, x, y)$ admits an infinite upper limit in the domain given by (7.1.23) if*

$$|V(t, x, y)| \geq W(x)$$

and $W(x) \to \infty$ as $x \to \infty$ (or as $|x| \to h$ if equation (7.1.13) is defined in bounded domain given by $|x| < h$, $y \in \mathcal{Y}$, $t \geq t_0$).

Definition 7.3.15 *The function $V(t,x,y)$ is said to be positive (negative) semidefinite in the domain* (7.1.23) *if it cannot be negative (positive) in this domain.*

Generally speaking the positive definite function $V(t,x,y)$ is called the *stochastic Lyapunov function* if the value of its differential generator (average derivative) $\mathcal{A}V(t,x,y)$ along the paths of the considered system is at least negative semidefinite. More exact formulation depends on the type (definition) of stochastic stability and will be given in stability theorems below.

7.4 General stability and stabilization theorems

7.4.1 Stability in probability theorems

As in the deterministic case using the stochastic Lyapunov function we obtain in general only sufficient conditions of stability, and the main difficulty is to find a suitable Lyapunov function. The following theorems were originally presented by Kats and Krasovskii [41, 42].

Theorem 7.4.1 *Let for the system described by the equations (7.1.13), (7.1.15) with the jump conditions (7.1.21) there exist a positive definite function $V(t,x,y)$, such that $\mathcal{A}V(t,x,y)$ is a semidefinite function in the domain (7.1.23). Then the trivial solution of this system is stable in probability.*

Theorem 7.4.2 *Let for the system described by the equations (7.1.13), (7.1.15) with the jump conditions (7.1.21) there exist a positive definite function $V(t,x,y)$, which admits infinitesimal lower limit and such that $\mathcal{A}V(t,x,y)$ is negative definite in the domain (7.1.23). Then the trivial solution of this system is asymptotically stable in probability.*

Theorem 7.4.3 *If the function $V(t,x,y)$ satisfies all the conditions of Theorem 7.4.2 and has infinite lower limit, then the trivial solution of this system is asymptotically stable in probability in large.*

All the proofs of these theorems are based on the supermartingale property of $V(t,x,y)$ [41, 45, 54] and effectively use the Ito-Dynkin formula (7.1.29).

7.4.2 Recurrence and positivity theorems

Consider system (7.3.11) and fomulate theorems like the Lyapunov ones for it, which give a sufficient recurrence and positivity conditions [97, 98], see also [45, 102].

Theorem 7.4.4 *If there exists a function $V(x)$ with properties*

(i) $V(x)$ is defined for $x \in \bar{D}_V$, where $D_V = \{x : |x| > R\}$ $(0 < R < \infty$ is arbitrary);

(ii) $V(x)$ is continuous in $\bar{D}_V$ and is twice continuously differentiable in D_V;

(iii) $V(x) \geq 0$ $x \in \bar{D}_V$ and $V(x) \to +\infty$ as $|x| \to \infty$

and if along the paths of the system (7.3.11)

$$\mathcal{L}V(x) \leq 0, \; x \in D_V,$$

then the process $X(t)$ defined by (7.3.11) is recurrent.

Theorem 7.4.5 *If there exists a function $V(x)$ with the same properties (i)-(iii) as in Theorem 7.4.4 and if along the paths of the system (7.3.11)*

$$\mathcal{L}V(x) \leq -1,\ x \in D_V,$$

then the process $X(t)$ defined by (7.3.11) is positive.

Under the conditions of positivity there exists a unique probability invariant measure μ defined on the Borel sets $\mathcal{B}^n \subset R^n$: that is, if P denotes probability measure on the paths of $X(t)$ and if

$$P[X(t_0) \in \mathcal{B}^n] = \mu(\mathcal{B}^n),$$

then

$$P[X(t) \in \mathcal{B}^n] = \mu(\mathcal{B}^n),\ t \geq t_0.$$

Let $L(x) \geq 0$ be Hölder continuous on the compact subsets of R^n. The problem is to obtain a condition that

$$\mathcal{E}_\mu[L(x)] = \int_{R^n} L(x)\mu(dx)$$

will be finite. Sufficient conditions in terms of Lyapunov like functions are given by the following theorems.

Theorem 7.4.6 *Let the process $X(t)$ defined by (7.3.11) be positive. If there exists a function $V(x)$ with the same properties (i)-(iii) as in Theorem 7.4.4 and if along the paths of the system (7.3.11)*

$$\mathcal{L}V(x) \leq -L(x),\ x \in D_V,$$

then

$$\mathcal{E}_\mu[L(x)] < \infty.$$

The next theorem allows, in addition, to estimate $\mathcal{E}_\mu[L(x)]$.

Theorem 7.4.7 *Let process $X(t)$ defined by (7.3.11) be positive. If there exist a function $V(x)$ such that the properties (i)-(iii) of Theorem 7.4.4 are valid with $D_V = R^n$ and a positive constant k such that along the paths of the system (7.3.11)*

$$\mathcal{L}V(x) \leq k - L(x),\ x \in R^n,$$

then

$$\mathcal{E}_\mu[L(x)] \leq k.$$

7.4.3 *p*th mean stability theorems and their inversion

Stability in the mean square is studied in many works. It is clear that on the one hand the mean square analysis is more simple than direct calculation or estimation of some probabilistic measures. On the other hand it turns out, that exponential stability in the mean square is the sufficient condition of (strong) asymptotic stability in probability in large. Consider the system (7.1.13) with the jump condition (7.1.19) and suppose that there exists a constant $0 < h_1 \leq h_2$, such that

$$h_1|x| \leq |\phi_{ij}(x)| \leq h_2|x|,\ i,j \in \mathcal{N}. \tag{7.4.1}$$

Theorem 7.4.8 *Let for the system described by the equations (7.1.13), (7.1.15) with the jump conditions (7.1.19) there exists a positive definite function $V(t,x,y)$ such that in the domain (7.1.23)*

$$V(t,x,y) \geq c_1|x|^2$$

and $\mathcal{A}V(t,x,i)$ is negative semidefinite function ($\mathcal{A}V(t,x,i) \leq 0$), where c_1 is positive constant. Then the trivial solution of this system is stable in the mean square.

Theorem 7.4.9 *Let for the system described by the equations (7.1.13), (7.1.15) with the jump conditions (7.1.19) there exists a positive definite function $V(t,x,y)$ such that in the domain (7.1.23)*

$$c_1|x|^2 \le V(t,x,y) \le c_2|x|^2, \; \mathcal{A}V(t,x,y) \le -c_3|x|^2, \tag{7.4.2}$$

where c_1, c_2, c_3 are positive constants. Then the trivial solution of this system is exponentially stable in the mean square.

This theorem admits the following important converse.

Theorem 7.4.10 *If the trivial solution of the system, described by equations (7.1.13), (7.1.15) with jump conditions (7.1.19) is exponentially stable in the mean square, then in the domain (7.1.23) there exists a function $V(t,x,y)$ satisfying conditions (7.4.2).*

It follows that under conditions of this theorem the trivial solution of the system (7.1.13) is asymptotically stable in probability in large. So, exponential stability in the mean square implies asymptotic stability in probability in large. It turns out that in the case of exponential stability in the mean square a more strong property holds: almost all the paths of the process $X(t)$ are exponentially stable according to the following theorem.

Theorem 7.4.11 *If the trivial solution of the system (7.1.13) is exponentially stable in the mean square then there exists a constant $\beta > 0$ such that for any $x_0 \in R^n$, $y_0 \in \mathcal{Y}$, $t_0 \ge 0$ almost all the paths $[X(t)Y(t)]$ satisfy conditions*

$$|X(t)| \le C \exp^{-\beta t},$$

where a random quantity C is finite w.p.1.

The stability in the pth mean (p-stability) was studied by Nevelson and Khas'minskii [45]. Let U be some domain with closure $\tilde{U}$ in the space $E = I \times R^n$ and $U^\epsilon(0) = \{(t,x) : |x| < \epsilon\}$. We say that the function $V(t,x)$ belongs to the class $C_2^0(U)$ ($V(t,x) \in C_2^0(U)$) if it is twice continuously differentiable in x and once in t everywhere in U excepting (maybe) the set $x = 0$ and is continuous in the closed set $\tilde{U} \setminus U^\epsilon(0)$ for any $\epsilon > 0$. The main results are contained in the following theorems.

Theorem 7.4.12 *Let there exist a function $V(t,x) \in C_2^0(E))$, satisfying for some positive constants c_1, c_2, c_3 the inequalities*

$$c_1|x|^p \le V(t,x) \le c_2|x|^p, \; \mathcal{A}V(t,x) \le -c_3|x|^p. \tag{7.4.3}$$

Then the trivial solution of the system (7.1.1) is exponentially p-stable.

Theorem 7.4.13 *Let the trivial solution of the system (7.1.1) be exponentially p-stable and $a(t,x)$ and $b(t,x)$ have continuous bounded deruivatives of both first and second orders. Then there exists a function $V(t,x) \in C_2^0(E))$, satisfying the inequalities (7.4.3) and for some $c_4 > 0$ the inequalities*

$$\left|\frac{\partial V}{\partial x_i}\right| < c_4|x|^{p-1}, \; \left|\frac{\partial^2 V}{\partial x_i \partial x_j}\right| < c_4|x|^{p-2}, \; i,j = 1,\dots,n.$$

Stability theorems for linear systems

In the linear case the system (7.1.1) has the form

$$dX(t) = A(t)X(t)dt + \sum_{l=1}^{m} A_l(t)X(t)dW_l(t), \tag{7.4.4}$$

where $A(t)$ and $A_l(t)$ $(l = 1, \dots, m)$ are $n \times n$ matrices with bounded Euclidean norms.

Theorem 7.4.14 *The trivial solution $X(t) \equiv 0$ of the linear system (7.4.4) is exponentially p-stable if and only if there exists homogeneous in x pth order function $V(t,x)$, satisfying conditions*

$$c_1|x|^p \le V(t,x) \le c_2|x|^p, \ \mathcal{A}V(t,x) \le -c_3|x|^p,$$
$$\left|\frac{\partial V}{\partial x_i}\right| < c_4|x|^{p-1}, \ \left|\frac{\partial^2 V}{\partial x_i \partial x_j}\right| < c_4|x|^{p-2}, \ i,j = 1, \dots, n,$$

where $c_1, \cdots, c_4$ are some positive constants.

If p is an even number $(p = 2, 4, \dots)$ then it turns out that $V(t,x)$ is a form of order p and the following theorem is true.

Theorem 7.4.15 *For exponential p-stability of the even order of the trivial solution $X(t) \equiv 0$ of the linear system (7.4.4) it is necessary that for any and sufficient that for some positive definite form $W(t,x)$ of order p, whose coefficients are continuous and bounded functions of t, a positive definite form $V(t,x)$ of the same order has been found, such that*

$$\mathcal{A}V(t,x) = -W(t,x).$$

The system with jumping disturbances in the linear case is described by the equation

$$dX(t) = A(t, Y(t))X(t)dt + \sum_{l=1}^{m} A_l(t, Y(t))X(t)dW_l(t). \tag{7.4.5}$$

Suppose that $Y(t)$ is the homogeneous scalar Markov chain with finite set of states $\mathcal{Y} = \mathcal{N} = \{1, \dots, \nu\}$ and with transition probabilities satisfying (7.1.15). The jump condition for the vector $X(t)$ is given by (7.1.20).

Theorem 7.4.16 *Let the trivial solution $X(t) \equiv 0$ of the linear system (7.4.5), (7.1.15), (7.1.20) be exponentially stable in the mean square. Then for any positive definite quadratic form $W(t,x,y)$ of variables $x_1, \dots, x_n$ whose coefficients are continuous and bounded functions of t, $t \ge t_0, y \in \mathcal{Y}$, there exists a positive definite quadratic form $W(t,x,y)$, satisfying inequalities (7.4.2), and such that*

$$\mathcal{A}V(t,x,y) = -W(t,x,y).$$

In the stationary case the equation (7.4.5) has the form

$$dX(t) = A(Y(t))X(t)dt + \sum_{l=1}^{m} A_l(Y(t))X(t)dW_l(t). \tag{7.4.6}$$

Theorem 7.4.17 *Let the trivial solution $X(t) \equiv 0$ of the linear system (7.4.6), (7.1.15), (7.1.20) be exponentially stable in the mean square. Then for any positive definite quadratic form $W(x,y)$ of variables $x_1, \dots, x_n$ there exists a unique positive definite quadratic form $V(x,y)$ such that*

$$\mathcal{A}V(x,y) = -W(x,y). \tag{7.4.7}$$

The reader is referred to [41, 45] for the proofs of the formulated theorems.

7.4.4 Stability in the first order approximation

Consider the system with random jumps (7.1.13) and rewrite differential equation (7.1.13) in the form

$$dX(t) = [A(t, Y(t))X(t) + \alpha(t, X(t), Y(t))]dt + \sum_{l=1}^{m}[A_l(t, Y(t))X(t) + \beta_l(t, X(t), Y(t))]dW_l(t). \tag{7.4.8}$$

The jump condition of the vector X is given by (7.1.19) and one rewrites this condition in analogous form

$$X(\tau) = \Phi_{ij}X(\tau - 0) + \psi_{ij}(X(\tau - 0)), \tag{7.4.9}$$

where τ is the random moment of the jump of $Y(t)$ from ith to jth state. Here $A(t, y), A_l(t, y)$ are $n \times n$ matrices, whose components are bounded and continuous functions for all $t \geq t_0$ and $y \in \mathcal{Y}$; $\alpha(t, x, y)$ and $\beta_l(t, x, y)$, are vector functions, satisfying for all $t \geq t_0$, $x \in R^n$ and $y \in \mathcal{Y}$ the growth condition (7.1.3), the Lipschitz condition (7.1.4), and such that $\alpha(t, 0, y) \equiv 0$, $\beta_l(t, 0, y) \equiv 0$; Φ_{ij} are constant $n \times n$ matrices; ψ_{ij} are continuous functions, such that $\psi_{ij}(0) \equiv 0$, W_l are independent components of standard m-dimensional Wiener process $W(t)$. We consider together with the system (7.4.8) the linear system

$$dX(t) = A(t, Y(t))X(t)dt + \sum_{l=1}^{m} A_l(t, Y(t))X(t)dW_l(t) \tag{7.4.10}$$

with linear jump condition of vector X

$$X(\tau) = \Phi_{ij}X(\tau - 0), \tag{7.4.11}$$

We say that the system (7.4.10),(7.4.11) is the *first order approximation system*. The problem is to study when from the fact of stochastic stability of the first order approximation *linear* system (7.4.10), (7.4.11) it follows that the *nonlinear* system (7.4.8), (7.4.9) is stochastically stable too.

Theorem 7.4.18 *If the trivial solution $X(t) \equiv 0$ of the system (7.4.10) (7.4.11) is exponentially stable in the mean square and for all $t \geq t_0$, $x \in R^n$, $y \in \mathcal{Y}$ and $\gamma > 0$ sufficiently small*

$$|\alpha(t, x, y)| \leq \gamma|x|, \ |\beta_l(t, x, y)| \leq \gamma|x|, |\psi_{ij}(x)| \leq \gamma|x|, \tag{7.4.12}$$

then the solution $X(t) \equiv 0$ of the system (7.4.8), (7.4.9) is asymptotically stable in probability in large and is exponentially stable in the mean square.

When all the functions in the right hand side are slowly changed in time it is possible to use "frozen" coefficients method. For simplicity we consider the linear nonstationary system (7.4.10) with the jump condition (7.4.11) and assume that

$$\left|\frac{\partial A}{\partial t}\right| \leq \varphi(t), \left|\frac{\partial A_l}{\partial t}\right| \leq \varphi(t), \tag{7.4.13}$$

where φ is the bounded continuous function for which there exists a number $T > 0$, such that for all $t_0 > 0$ and some $\gamma > 0$

$$\frac{1}{T}\int_{t_0}^{t_0+T} \varphi(t)dt < \gamma \tag{7.4.14}$$

Consider the "frozen" linear stationary system as the first order approximation system. The motivation is that for a stationary system it is possible to obtain more effective testable stability conditions. The "frozen" system is described by

$$dX_\mu(t) = A(\mu, Y(t))X(t)dt + \sum_{l=1}^{m} A_l(\mu, Y(t))X(t)dW_l(t) \tag{7.4.15}$$

with the same jump condition (7.4.11). Assume that the first order approximation system (7.4.15), (7.4.11) is exponentially stable in the mean square uniformly in $\mu > t_0$, $y_0 \in \mathcal{Y}$. This means that for any solution X_μ of this system for all $t \geq t_0$ we have

$$\mathcal{E}[|X_\mu(t)|^2 \big| X(t_0) = x_0,\ Y(t_0) = y_0] \leq C|x_0|^2 e^{-\alpha(t-t_0)}, \tag{7.4.16}$$

where $C \geq 1$ and $\alpha > 0$ are not dependent on $\mu \leq t_0$, $y_0 \in \mathcal{Y}$.

Theorem 7.4.19 *If the first order approximation system (7.4.15), (7.4.11) satisfies the condition (7.4.16) and the function φ from (7.4.13) for some γ sufficiently small satisfies condition (7.4.14), then the trivial solution $X(t) \equiv 0$ of the system (7.4.10), (7.4.11) is exponentially stable in the mean square.*

The choice of the first order approximation system depends on the properties of the original system. The reader is refered to [41], where either a certain deterministic system or a stochastic system without jumps is used as the first order approximation system. The reader is refered to [45] for a more detailed study of stability in the first order approximation of systems, descriebed by Ito differential equation (7.1.1) without jumps.

7.4.5 Stabilization problem and fundamental theorem

Consider a system described by the differential equation (7.2.1) and suppose that $U(t) = \mathbf{u}(t, X(t), Y(t))$; as it is defined above such a particular form of the state feedback control control is called the Markov control. We say that $u = \mathbf{u}(t, x, y)$ is an admissible function if $a(t, x, \mathbf{u}(t, x), y)$ is continuously differentiable in the domain (7.1.23), $a(t, 0, 0, y) \equiv 0$ and $u(t, 0, y) \equiv 0$. Let $\mathcal{U}$ be a class of admissible controls. Then every $\mathbf{u} \in \mathcal{U}$ generates the Markov process $[X_\mathbf{u}(t)Y(t)]$ as a solution of (7.2.1) with the given initial conditions. We suppose that

$$X(t_0) = x_0 \in R^n,\ Y(t_0) = y_0 \in \mathcal{Y},\ t_0 \geq 0, \tag{7.4.17}$$

the description of $Y(t)$ is given by (7.1.14) or (7.1.15) and the jump condition of vector $X(t)$ is given by (7.1.21) or by their particular cases (7.1.18)-(7.1.20) The stabilization problem is in the following: to find an admissible control such that the trivial solution $X(t) \equiv 0$ of system (7.2.1) is stochastically stable in some suitable sense e.g. asymptotically stable in probability in large. It is obvious that the solution of this problem is nonunique and as a rule it is supposed that the stabilizing control provides some additional condition. In many cases this condition is to minumize a functional along the motions of the system. It is the optimal stabilization problem [41, 45, 51, 96, 99]. Let us formulate this problem exactly: to find an admissible control $\mathbf{u}^0(t, x, y)$ for the system (7.2.1) such that:

1) The trivial solution $X(t) \equiv 0$ with $U(t) = \mathbf{u}^0(t, X(t), Y(t))$ is asymptotically stable in probability in large (or in another suitable sense).

2) The functional

$$J_\mathbf{u}(t_0, x^0, y^0) =$$

$$\int_{t_0}^{\infty} \mathcal{E}[L(t, X(t), \mathbf{u}(t, X(t), Y(t))Y(t)) \big| X(t_0) = x_0, Y(t_0) = y_0]dt, \tag{7.4.18}$$

where $L(t, x, u, y)$ is a nonnegative function defined in the domain (7.1.23), with $\mathbf{u} = \mathbf{u}^0(t, x, y)$ converges and for all initial conditions, satisfying (7.4.17)

$$J_{\mathbf{u}^0}(t_0, x_0, y_0) = \min_{\mathbf{u} \in \mathcal{U}} J_u(t_0, x_0, y_0). \tag{7.4.19}$$

Theorem 7.4.20 *Let for system (7.2.1) there exists a scalar function $V^0(t, x, y)$ and a vector function $\mathbf{u}^0(t, x, y) \in R^k$ defined in the domain (7.1.23) such that:*

1) *The function $V^0(t, x, y)$ is positive definite in x in the domain (7.1.23) and admits both infinitesimal lower limit and infinite upper limit.*

2) *The function $L(t, x, \mathbf{u}^0(t, x, y), y)$ from the functional (7.4.18) is positive definite in x.*

3) *The differential generator (the average derivative) by virtue of the system (7.2.1) with $u = \mathbf{u}^0(t, x, y)$ satisfies the conditions*

$$\mathcal{A}_{\mathbf{u}^0} V^0(t, x, y) = -L(t, x, \mathbf{u}^0, y). \tag{7.4.20}$$

4) *The value $\mathcal{A}_u V^0(t, x, y) + W(t, x, u, y)$ is minimized by $u = \mathbf{u}^0$ i.e.*

$$\mathcal{A}_{\mathbf{u}^0} V^0(t, x, y) + L(t, x, \mathbf{u}^0, y) = \min_{u \in R^k} [\mathcal{A}_u V^0(t, x, y) + L(t, x, u, y)] = 0. \tag{7.4.21}$$

Then the function $\mathbf{u}^0(t, x, y)$ is optimal stabilizing control law and the following equality is true

$$\begin{aligned} V^0(t_0, x_0, y_0) &= \\ \int_{t_0}^{\infty} \mathcal{E}[L(t, X(t), \mathbf{u}^0(t, X(t), Y(t)), Y(t)) \big| X(t_0) = x_0, Y(t_0) = y_0] dt &= \\ \min_{\mathbf{u} \in \mathcal{U}} \int_{t_0}^{\infty} \mathcal{E}[L(t, X(t), \mathbf{u}(t, X(t), Y(t)), Y(t)) \big| X(t_0) = x_0, Y(t_0) = y_0] dt &= \\ J_{\mathbf{u}^0}(t_0, x_0, y_0), \end{aligned} \tag{7.4.22}$$

where $X(t)$ denotes the solution of (7.2.1) with the corresponding state feedback control.

It is clear that it is possible to unite the condition (7.4.20) (7.4.21) in the Bellman's functional equation

$$\min_{u \in R^k} [\mathcal{A}_u V^0(t, x, y) + L(t, x, u, y)] = 0. \tag{7.4.23}$$

The solution $V^0(t, x, y)$ of this equation is called Lyapunov-Bellman function or optimal Lyapunov function.

7.5 Instability

7.5.1 Classic stochastic instability concept

The classic stochastic instability concept is based on the generalization of the Lyapunov instability concept to the stochastic systems. Unfortunately the study of this type of instability is more complicated than the study of stability. Roughly speaking the paths of the stochastic system can leave the instability region as a result of random actions. The reader is refered to [45] for examples and more details. Consider the system (7.1.1) and denote U_r

the set $\{|x| < r\}$ in R^n. To avoid the problems above the following nondegeneracy condition will be used for this system

$$z'b(t,x)b'(t,x)z > m(x)|z|^2, \ x, z \in R^n, \tag{7.5.1}$$

where $m(x)$ is a continuous function such that $m(x) > 0$ if $x \neq 0$.

Definition 7.5.1 (Instability in probability.) *The trivial solution $X(t) \equiv 0$ of the system (7.1.1) is called instable in probability if for some numbers $\epsilon > 0, p > 0$ does not exist, a number $\delta > 0$ such that from the condition*

$$|x_0| \leq \delta, \ t_0 \geq 0$$

follows that

$$P[\sup_{t \geq t_0} |X(t)| < \epsilon \big| X(t_0) = x_0] > 1 - p.$$

Theorem 7.5.2 *Let there exists a function $V(t,x) \in C_2^0(\{t > 0\} \times U_r)$, satisfying the conditions*

$$\mathcal{A}V(t,x) \leq 0, \ x \in U_r, \ x \neq 0, \tag{7.5.2}$$

$$\liminf_{x \to 0} \inf_{t > 0} V(t,x) = \infty \tag{7.5.3}$$

and nondegeneracy condition (7.5.1) holds. Then the trivial solution $X(t) \equiv 0$ of system (7.1.1) is instable in probability.

Definition 7.5.3 (p-instability) *The trivial solution $X(t) \equiv 0$ of the system (7.1.1) is called exponentially p-unstable ($p > 0$) if for some positive C and α*

$$\mathcal{E}[|X(t)|^{-p} \big| X(t_0) = x] < C|x|^{-p} e^{-\alpha(t-t_0)}.$$

This definition is more strong because from exponential p-instability for some p it follows that the system (7.1.1) is instable in probability.

Theorem 7.5.4 *If there exists a function $V(t,x) \in C_2^0(R^n)$ satisfying the conditions*

$$c_1|x|^{-p} \leq V(t,x) \leq c_2|x|^{-p}, \tag{7.5.4}$$

$$\mathcal{A}V(t,x) \leq -c_3|x|^{-p}, \tag{7.5.5}$$

then the trivial solution $X(t) \equiv 0$ of the system (7.1.1) is exponentially p-unstable for $t \geq 0$. Moreover there exists a constant $\gamma > 0$ such that for any $t_0 \geq 0$, $X(t_0) = x \neq 0$

$$|X(t)| > C_{t_0,x} e^{\gamma t}, \ t \geq t_0$$

w.p.1. and the random variable $C_{t_0,x}$ is a.s. positive.

Theorem 7.5.5 *Let the trivial solution of the system (7.1.1) exponentially p-unstable and $a(t,x)$ and $b(t,x)$ have continuous bounded derivatives of both first and second orders. Then there exists a function $V(t,x) \in C_2^0(R^n)$, satisfying the inequalities (7.5.4), (7.5.5) and for some $c_4 > 0$ the inequalities*

$$\left|\frac{\partial V}{\partial x_i}\right| < c_4|x|^{-p-1}, \ \left|\frac{\partial^2 V}{\partial x_i \partial x_j}\right| < c_4|x|^{-p-2}, \ i,j = 1,\ldots,n. \tag{7.5.6}$$

Now we consider the linear case, when the system is described by the equation

$$dX(t) = A(t)X(t)dt + \sum_{l=1}^{m} A_l(t)X(t)dW_l(t). \tag{7.5.7}$$

It is assumed that $|A(t)|$ amd $A_l(t)$ are the bounded functions.

Theorem 7.5.6 *The trivial solution $X(t) \equiv 0$ is exponentially p-unstable if and only if there exists a uniform in x of order $-p$ function $V(t,x)$, satisfying for some positive constants $c_1 - c_4$ the conditions*

$$c_1|x|^{-p} \le V(t,x) \le c_2|x|^{-p}, \ \mathcal{A}V(t,x) \le -c_3|x|^{-p}$$
$$\left|\frac{\partial V}{\partial x_i}\right| < c_4|x|^{-p-1}, \ \left|\frac{\partial^2 V}{\partial x_i \partial x_j}\right| < c_4|x|^{-p-2}, \ i,j = 1,\dots,n.$$

Assume that in the sufficiently small neighborhood of the point $x = 0$ the parameters of the system (7.1.1) satisfy the inequality

$$|a(t,x) - Ax| + |\sum_{l=1}^{m} b_l(t,x) - A_l x| < \gamma|x|, \tag{7.5.8}$$

for some $\gamma > 0$ sufficiently small, where $b_l(t,x)$, $(l = 1,\dots,m)$ are columns of the matrix $b(t,x)$ in (7.1.1), A and A_l are constant matrices. In this case it is possible to use (7.5.7) as the first order approximation system for instability analysis of (7.1.1).

Theorem 7.5.7 *Let the coeffecients of linear system (7.5.7) be bounded functions of t, the trivial solution $X(t) \equiv 0$ of this system is exponentially p-unstable for some $p > 0$ and for $\gamma > 0$ sufficiently small, depending on $\sup_{t>0}|A_l(t)|$ and on the constants $c_1 - c_4$ from (7.5.4)-(7.5.6) only, the inequality (7.5.8) holds. Then the solution $X(t) \equiv 0$ of the system (7.1.1) is instable in probability.*

7.5.2 Nonpositivity and nonrecurrence

In this section the sufficient conditions are given for the process $X(t)$ described by (7.3.11) to be nonrecurrent or at least nonpositive [97]. We say that a domain in R^n is normal domain if it is nonempty, open and simply connected set in R^n with the smooth boundary. We introduce function $V(x)$ with the following properties.

(i) $V(x)$ is defined for $x \in \bar{D}_V$, where $D_V = \{x : |x| > R\}$ $(0 < R < \infty$ is arbitrary).

(ii) $V(x)$ is continuous in $\bar{D}_V$ and is twice continuously differentiable in D_V.

(iii) $V(x)$ is bounded above for $x \in D_V$.

(iv) There is a normal domain $\mathcal{G}$ with boundary Γ such that $D_V \supset R^n \setminus \mathcal{G}$ and $\max\{V(x) : x \in \Gamma\} < \sup\{V(x) : x \in R^n \setminus \mathcal{G}\}$.

(v) $\mathcal{L}V(x) \ge 0, \ x \in D_V$.

Theorem 7.5.8 *If there exists a function $V(x)$ with properties (i)-(v) then the process $X(t)$ defined by (7.3.11) is nonrecurrent.*

The following theorem is sometimes useful to identify processes which are recurrent, but not positive. Let $V_1(x), V_2(x)$ be a pair of functions with the properties (i), (ii), and with the additional properties:

(1) There is a sequence $\{x_n\}$ in D_V such that $|x_n| \to \infty$ and $V_1(x_n) \to \infty$.

(2) $V_2(x) > 0,\ x \in D_V$.

(3) $\limsup_{\rho\to 0} \frac{\max\{V_1(x):|x|=\rho\}}{\min\{V_2(x):|x|=\rho\}}$.

(4) $\mathcal{L}V_1(x) \geq 0,\ \mathcal{L}V_2(x) \leq +1,\ x \in D_V$.

Theorem 7.5.9 *If there exists a pair of functions with properties (i), (ii) and (1)-(4) then the process $X(t)$ defined by (7.3.11) is nonpositive.*

7.6 Stability criteria and testable conditions

7.6.1 General stability tests for linear systems

Consider the linear stationary system (7.4.6) with the jump conditions of vector $X(t)$ given by (7.1.20) and with $Y(t)$ described by (7.1.15). Let $W(x,y) = x'M(y)x$ be positive definite in the domain (7.1.23). According to Theorem 7.4.17 the system (7.4.6) is exponentially stable in the mean square if and only if there exists a unique positive definite quadratic form $V(x,y) = x'H(y)x$, satisfying equation(7.4.7). Calculating the left hand side of (7.4.7) by virtue of the system (7.4.6) we obtain the following system of coupled linear matrix equations of Sylvester type [41]:

$$A'(i)H(i) + H(i)A(i) + \sum_{l=1}^{m} A_l'(i)H(i)A_l(i) +$$

$$\sum_{j\neq i}^{\nu}(\Phi_{ij}'H(j)\Phi_{ij} - H(i))q_{ij} = -M(i),\ i \in \mathcal{N}. \tag{7.6.1}$$

The solvability conditions of (7.6.1) give the necessary and sufficient conditions of exponential stability in the mean square of the system (7.4.6) in their parameter space. A general way to obtain these conditions is in the following: form the long vector from the rows of the matrices $[H(1),\dots,H(\nu)]$ and rewrite the system (7.6.1) as a standard vector linear equation using Kronecker products. The solvability conditions of this equation are well known. For more detail consider the system (7.4.6) without jumps of vector $X(t)$, such that the condition (7.1.18) holds. The system of equations (7.6.1) in this case has the form

$$A'(i)H(i) + H(i)A(i) + \sum_{l=1}^{m} A_l'(i)H(i)A_l(i) +$$

$$\sum_{j\neq i}^{\nu}(H(j) - H(i))q_{ij} = -M(i),\ i \in \mathcal{N}. \tag{7.6.2}$$

Determine the $n^2\nu \times n^2\nu$ matrix G with the block elements

$$G_{ii} = (A(i) - \frac{1}{2}q_iI_n)' \otimes I_n + I_n \otimes (A(i) - \frac{1}{2}q_iI_n) + \sum_{l=1}^{m} A_l'(i) \otimes A_l'(i),$$

$$G_{ij} = q_{ij}I_n \otimes I_n,\ i \neq j,\ i,j \in \mathcal{N}.$$

Denote by $\mathbf{h}, \mathbf{m}$ vectors of length $n^2\nu$, constructed from the consequtively-taken rows of the matrices $H(i)$ and $M(i)$ $(i \in \mathcal{N})$ which satisfy equation (7.6.2). Then the system of matrix equations (7.6.2) can be rewritten as single vector linear algebraic equation:

$$G\mathbf{h} = -\mathbf{m}.$$

Theorem 7.6.1 *The system (7.4.6),(7.1.18) is exponentially stable in the mean square if and only if the matrix G is Hurwitz.*

This approach was used for by Kleinman [49] and Willems [88] and other authors.

7.6.2 Some particular stability criteria for linear systems

The way above is connected with very complicated calculations. In some particular cases it is possible to find more effective stability conditions. First we consider the system (7.4.6) without jumps of vector $X(t)$, and without the noise term in the right hand side. This system can be described by ordinary linear differential equation with random matrix:

$$\dot{X}(t) = A(Y(t))X(t). \tag{7.6.3}$$

The case when

$$A(i) = A + \mathbf{b}\mathbf{c}'h(i),\ i \in \mathcal{N}, \tag{7.6.4}$$

where $\mathbf{b}, \mathbf{c}$ are n-dimensional vectors and $h(i)$ is a scalar is considered in [8]. Assume that matrix Q can be reduced to diagonal form. We denote $\lambda_1, \lambda_2, \dots, \lambda_\nu$ the eigenvalues and $d_1, d_2, \dots, d_\nu$ the eigenvectors of Q and construct matrix $D = [d_1, d_2 \dots d_\nu]$. Let $W(p)$ be the matrix transfer function of the linear differential system

$$\begin{aligned} \dot{Z}(t) &= AZ(t) + Z(t)A' + \mathbf{b}\mathbf{v}'(t) + \mathbf{v}(t)\mathbf{b}', \\ \mathbf{u}(t) &= Z(t)\mathbf{c} \end{aligned} \tag{7.6.5}$$

from the vector input $\mathbf{v}$ to the vector output $\mathbf{u}$ and $\Delta(p)$ be characteristic polynomial of the matrix differential equation in (7.6.5) of $n(n+1)/2$ degree.

Theorem 7.6.2 *The trivial solution of the system (7.6.3), (7.6.4) is asymptotically stable in the mean square if and only if the polynomial*

$$\begin{aligned} &\Delta(p-\lambda_1)\dots\Delta(p-\lambda_\nu)\det[I_{n\nu} - \\ &\operatorname{diag}[W(p-\lambda_1)\dots[W(p-\lambda_\nu)]D'\operatorname{diag}[h(1)\dots h(\nu)][D']^{-1} \otimes I_n] \end{aligned}$$

be Hurwitz

An effective algoritm for obtaining the matrix transfer function $W(p)$ is also presented in [8]. Now, consider the system described by the linear stationary Ito equation

$$dX(t) = AX(t)dt + \sum_{l=1}^{m} A_l X(t) dW_l(t). \tag{7.6.6}$$

In this case the equations (7.6.1) are reduced to one matrix equation

$$A'H + HA + \sum_{l=1}^{m} A_l' H A_l = -M. \tag{7.6.7}$$

The system (7.6.6) was studied by many authors, see for instance [45, 49, 58, 59, 60, 88]. Suppose that

$$A_l = \mathbf{q}_l \mathbf{r}_l',\ l = 1, \dots, m. \tag{7.6.8}$$

Define matrix R with elements $\rho_{lj}(l, j = 1, \dots, m)$ given by the formula

$$\rho_{lj} = \mathbf{q}_j' H_l \mathbf{q}_j,\ l, j = 1, \dots, m,$$

where matrix H_l is the solution of the following Lyapunov equation

$$A'H_l + H_l A = -\mathbf{r}_l \mathbf{r}_l'\ l = 1, \dots, m. \tag{7.6.9}$$

Theorem 7.6.3 *The trivial solution $X(t) \equiv 0$ of the system (7.6.6), (7.6.8) is exponentially stable in the mean square if and only if matrix A is Hurwitz and eigenvalues of the matrix R are smaller than one in modulus.*

The reader is referred to [60] for the proof of this theorem. It can be shown that ρ_{lj} can be expressed by the formula

$$\rho_{lj} = \frac{1}{2\pi}\int_{-\infty}^{\infty} \chi_{lj}(-i\omega)\chi_{lj}(i\omega)d\omega, \tag{7.6.10}$$

$$\chi_{lj}(p) = \mathbf{r}_l'(pI - A)^{-1}\mathbf{q}_j,\ l, j = 1, \ldots, m. \tag{7.6.11}$$

The integral (7.6.10) is well known in the complex analysis and control theory [37].

In some cases, in particular, when the system is described by differential equatiion of nth order with the random coefficients

$$Z^{(n)}(t) + [c_1 + v_1(t)]Z^{(n-1)}(t) + \ldots + [c_n + v_n(t)]Z(t) = 0, \tag{7.6.12}$$

where $v_i(t)$ $(i = 1, \ldots, n)$ are correlated white noise type processes, we have $\mathbf{q}_j = \mathbf{q}$, $j = 1, \ldots, n$. It is natural that not all the coefficients can be disturbed by noise and in general

$$\mathbf{q}_j = \mathbf{q},\ j = 1, \ldots, m \leq n \tag{7.6.13}$$

(see [45, 88] for details of transformation of (7.6.12) into (7.6.6)).

Theorem 7.6.4 *The trivial solution $X(t) \equiv 0$ of the system (7.6.6),(7.6.8), (7.6.13) is exponentially stable in the mean square if and only if matrix A is Hurwitz and there exists a solution H, of Lyapunov matrix equation*

$$A'H + HA + \sum_{l=1}^{m} \mathbf{r}_l\mathbf{r}_l' = 0, \tag{7.6.14}$$

satisfying the inequality

$$\mathbf{q}'H\mathbf{q} < 1. \tag{7.6.15}$$

Taking into account that matrix A is Hurwitz rewrite (7.6.15) in the form

$$\frac{1}{2\pi}\int_{-\infty}^{\infty} \chi'(-i\omega)\chi(i\omega)d\omega < 1, \tag{7.6.16}$$

where

$$\chi(p) = [\chi_1(p)\chi_2(p)\ldots\chi_m(p)],\ \chi_l(p) = \mathbf{r}_l'(pI - A)^{-1}\mathbf{q},\ l = 1, \ldots, m.$$

The integral in the left hand side of (7.6.16) can be represented as

$$I_c = \frac{1}{2\pi}\int_{-\infty}^{\infty} \chi'(-i\omega)\chi(i\omega)d\omega = \frac{1}{2\pi}\int_{-\infty}^{\infty} [g(i\omega)/h(-i\omega)h(i\omega)]d\omega, \tag{7.6.17}$$

where $g(p) = b_{n-1}p^{2(n-1)} + b_{n-2}p^{2(n-2)} + \ldots + b_0$, $h(p) = p^n + a_{n-1}p^{n-1} + \ldots + a_1p + a_0$ is the characteristic polynomial of matrix A. According to the classic formula for computing the integral in the right hand side [37] we get

$$I_c = (-1)^{n+1}\Delta_b/2\Delta_n,$$
$$\Delta_b = \det[g_{ij}]_{i,j=1}^n,$$

$$g_{ij} = \begin{cases} b_{n-i} & \text{if} \quad j = 1, \\ a_{n+j-2i} & \text{if} \quad j > 1,\ i, j = 1, 2, \ldots, n, \end{cases}$$

and Δ_n is nth Hurwitz determinant for the polynomial $h(p)$.

Theorem 7.6.5 *The trivial solution $X(t) \equiv 0$ of the system (7.6.6), (7.6.8), (7.6.13) is exponentially stable in the mean square if and only if matrix A is Hurwitz and the inequality (7.6.16) holds.*

Theorem 7.6.6 *The trivial solution $X(t) \equiv 0$ of the system (7.6.6), (7.6.8), (7.6.13) is exponentially stable in the mean square if and only if matrix A is Hurwitz and Hurwitz determinant Δ_n satisfies the inequality*

$$\Delta_n > 2(-1)^{n+1}\Delta_b.$$

Simple sufficient stability and instability conditions for the system (7.6.6) are obtained in [79, 80].

7.6.3 Stability of the pth moments of linear systems

Consider moment stability problem. For the system (7.6.6) one can obtain, in principle, the pth moment stability conditions for an arbitrary p if one uses the special power transformation technique [10]. For this the purpose the vector $X^{[p]}$ is introduced, whose components are the forms (monomials) of degree p in $X_1, \ldots, X_n$, the components of X:

$$X^{[p]} = \text{col}[X_1^p, \alpha_1 X_1^{p-1} X_2, \alpha_2 X_1^{p-2} X_2 X_3, \ldots, X_n^p].$$

The dimension of vector $X^{[p]}$ is the number of linearly independent degree p forms in n variables and is given by

$$n_p = C_{n+p-1}^p = \frac{(n+p-1)!}{p!(n-1)!}.$$

The scale factors α_i are chosen in such a way as to validate the equality

$$|X^{[p]}|^2 = |X|^{2p}.$$

We define the $n_p \times n_p$ matrix $A_{[p]}$ in the following way: if X satisfies the ordinary linear differential equations

$$\dot{X}(t) = AX(t)$$

then $X^{[p]}$ satisfies the following linear differential equations

$$\dot{X}^{[p]} = A_{[p]} X^{[p]}(t).$$

Using the properties of this transformation [10] by virtue of (7.6.6) the differential equation for $X^{[p]}$ is easily expressed in terms of $A_{[p]}$ and $A_{l[p]}$ matrices:

$$dX^{[p]}(t) = A_p X^{[p]}(t)dt + \sum_{l=1}^{m} A_{l[p]} X^{[p]}(t) dw_l(t), \tag{7.6.18}$$

where

$$A_p = [A - \frac{1}{2}\sum_{l=1}^{m} A_l^2]_{[p]} + \frac{1}{2}\sum_{l=1}^{m}(A_{l[p]})^2.$$

Evaluating the expectation, we obtain an equation for the pth order moment:

$$\frac{d}{dt}\mathcal{E}[X^{[p]}] = A_p \mathcal{E}[X^{[p]}]. \tag{7.6.19}$$

Thus the pth moment stability conditions can be obtained by analyzing the stability of the deterministic linear system (7.6.19).

Theorem 7.6.7 *The pth moment of the solution of equation (7.6.6) is asymptotically stable for all X_0 if and only if the matrix A_p is Hurwitz.*

Note that for even p this theorem simultaneously gives p-stability conditions. Simple sufficient p-stability conditions for system (7.6.3) are obtained in [75]

The reader is refered to [3, 11, 12, 32, 89, 90, 91] for more details in study of this direction.

7.6.4 Absolute stochastic stability

Consider a stochastic system described by the Ito equation

$$dX(t) = [AX(t) + \mathbf{b}U(t)]dt + \sum_{l=1}^{m} A_l X(t) dW_l(t),$$
$$U(t) = f[Z(t), t], \; Z(t) = \mathbf{c}'X(t), \tag{7.6.20}$$

where $U(t), Z(t)$ are scalar input and output variables; $\mathbf{b}, \mathbf{c}$ are constant n-dimensional vectors; $f(z,t)$ is a nonlinear function which satisfies the conditions

$$f(0,t) \equiv 0, \; 0 \le f(z,t)z \le Kz^2, \; K > 0; \tag{7.6.21}$$

The remaining notations correspond to those adopted earlier.

Definition 7.6.8 *The system (7.6.20) is said to be absolutely stochastically stable if it is stochastically stable in the sense of one of the adopted definitions independently of the specific nonlinearity from the examined class.*

We suppose that

$$A_l = \mathbf{b}\mathbf{r}_l \; l = 1, \dots m, \tag{7.6.22}$$

where $\mathbf{r}_l$ $(l = 1, \dots m)$ are constant n-dimensional vectors. The absolute exponential stability in the mean square (absolute ESMS) of system (7.6.20), (7.6.21) was investigated by Levit [57] and Pakshin [73]. Applying Theorem 7.4.12 with $p = 2$ and with a quadratic form Lyapunov function

$$V(x) = x'Hx,$$

where $H = H'$ is constant positive definite matrix, and using S-procedure (see [3, 9]) one reduces the stability problem to finding the conditions for the solvability of the Lur'ie equations

$$A'H + HA + \alpha \sum_{l=1}^{m} \mathbf{r}_l \mathbf{r}_l' = -\mathbf{h}\mathbf{h}' - \epsilon D,$$
$$H\mathbf{b} + \frac{1}{2}\mathbf{c} = \mathbf{h}\kappa, \tag{7.6.23}$$
$$\kappa^2 = K^{-1},$$

under the supplemental constraint

$$\mathbf{b}'H\mathbf{b} < \alpha, \tag{7.6.24}$$

where $\alpha > 0$ and κ are scalars, $\mathbf{h}$ is n-dimensional vector, $D = D'$ is a positive definite matrix, and ϵ is an arbitrary small positive number. It is assumed that the matrix A

is Hurwitz, the pair $(A, \mathbf{b})$ is completely controllable and the pair $(\mathbf{c}', A)$ is completely observable, see [55] for the definitions. We denote

$$W(\lambda) = \mathbf{c}'(A - \lambda I)^{-1}\mathbf{b}, \tag{7.6.25}$$
$$\delta(\lambda) = \det(\lambda I - A) = \lambda^n + \delta_1\lambda^{-1} + \delta_2\lambda^{n-2} + \ldots + \delta_n, \tag{7.6.26}$$
$$W_l(\lambda) = \mathbf{r}_l(A - \lambda I)\mathbf{b}. \tag{7.6.27}$$

Theorem 7.6.9 *For the system (7.6.20)-(7.6.22) to be absolutely ESMS it is sufficient that the inequalities*

$$\Pi(i\omega) = K^{-1} + \mathrm{Re}W(i\omega) - \alpha\sum_{l=1}^{m}|W_l(i\omega)|^2 > 0, \tag{7.6.28}$$

$$\alpha - \frac{1}{2}\beta_1 + K^{-\frac{1}{2}}\kappa_1 - K^{-1}\delta_1 > 0, \tag{7.6.29}$$

be satisfied for all real valued ω, where β_1 is the coefficient of the $(n-1)$th power term of the numerator of the transfer function (7.6.25), κ_1 is the coefficient of the $(n-1)$th power term of the Hurwitz polynomial $\Psi(\lambda)$ with the highest power term coefficient $\kappa = K^{-\frac{1}{2}}$, which is determined uniquely from the factorization equation

$$\Psi(-i\omega)\Psi(i\omega) = \Pi(i\omega)\delta(-i\omega)\delta(i\omega).$$

Remark *The conditions of this theorem are necessary and sufficient for solvability of the system (7.6.23), (7.6.24).*

The problem of absolute stochastic stability was also studied in [3, 63, 78]. In all these papers the problem is reduced somehow to finding the conditions for the solvability of the Lur'ie equations under some supplemental constrains. On the other hand this problem can be reduced to finding the conditions for the solvability of matrix equations of a more general form than the standard Lur'ie equations. For the system (7.6.20), (7.6.21) equations of this type are:

$$\begin{aligned} A'H + HA + \sum_{l=1}^{m} A_l'HA_l &= -\mathbf{h}\mathbf{h}' - \epsilon D, \\ H\mathbf{b} + \frac{1}{2}\mathbf{c} &= \mathbf{h}\kappa, \\ \kappa^2 &= K^{-1}, \end{aligned} \tag{7.6.30}$$

This direction was developed and generalized in [14, 15, 16, 22, 86]. An algebraic approach was developed in [48].

7.6.5 Robust stability

It is very interesting to obtain conditions of stochastic stability of system (7.1.13) independently of the jump intensities. We consider here this problem for the particular linear case, when the system is described by (7.4.6). For more easy formulation of the results denote $A_l = \sigma_l F_l$; then the scalar factors σ_l $(l = 1, \ldots, m)$ will indicate the intensities or disturbances. We introduce the following definitions.

Definition 7.6.10 *The system (7.4.6) is said to be robustly stable against the jump intensities if it is asymptotically stable in the mean square independently of q_{ij} $(i, j \in \mathcal{N})$ for given noise intensities σ_l $(l = 1, \ldots, m)$.*

Definition 7.6.11 *The system (7.4.6) is said to be perfectly robustly stable against the jump intensities if it is asymptotically stable in the mean square independently of q_{ij} $(i, j \in \mathcal{N})$ for all noise intensities σ_l $(l = 1, \ldots, m)$.*

Note that in both cases the stability region in parameter space of the system will not depend on q_{ij} $(i, j \in \mathcal{N})$. According to the second definition this region is allowed to depend on noise intensities, but it is not empty for all σ_l $(l = 1, \ldots, m)$. Let us consider matrices

$$G_{ii} = A'(i) \otimes I + I \otimes A'(i) + \sum_{l=1}^{m} \sigma_l^2 A_l'(i) \otimes A_l'(i),\ i \in \mathcal{N}, \tag{7.6.31}$$

Define for some fixed $k \in \mathcal{N}$ matrices $M(i)$ according to the formula

$$M(i) = -(A'(i)H(k) + H(k)A(i) + \Delta_i(H(k)),\ i \in \mathcal{N}, \tag{7.6.32}$$

where

$$\Delta_i(H(j)) = \sum_{l=1}^{m} \sigma_l^2 A_l'(i)H(j)A_l(i).$$

Theorem 7.6.12 *Let all matrices (7.6.31) be Hurwitz and let there exist at least one index $k \in \mathcal{N}$ and a positive definite matrix $M(k) = M'(k)$ such that all matrices $M(i)$ $(i \in \mathcal{N})$ of (7.6.32) are positive definite. Then the system (7.4.6) is robustly stable against the jump intensities.*

Remark 1 *The conditions of Theorem 7.6.12 are equivalent to the existence of constant matrix $H = H'$, satisfying the following linear matrix inequalities*

$$A'(i)H + HA(i) + \Delta_i(H) < 0 \ \ i \in \mathcal{N}. \tag{7.6.33}$$

LMI theory and the LMI toolbox of MATLAB software [9] can be effectively used to solve (7.6.33).

Now we consider the perfect robust stability problem. Suppose that for all $i \in \mathcal{N}$, $A_l(i) = A_l$ $(l = 1, \ldots, N)$. Let $H_0(k)$ denote the solution of Lyapunov's equation

$$A'(k)H(k) + H(k)A(k) + M(k) = 0 \tag{7.6.34}$$

with $M(k) = M_0(k)$ where $M_0(k) \geq 0$, but $x'M_0(k)x > 0$ for all $x \notin \Omega$, $\Omega = \{x : A_l x = 0,\ l = 1, \ldots, m\}$; $H_\epsilon(k)$ denotes the solution of the equation (7.6.34) with $M(k) = M_\epsilon(k) = M_0(k) + \epsilon M_1$, $\epsilon > 0$, M_1 a positive definite matrix, and $M_\epsilon(i) = -A'(i)H_\epsilon(k) - H_\epsilon(k)A(i)$ $(i \neq k)$.

Theorem 7.6.13 *Let the matrices $A(i)$ $(i \in \mathcal{N})$ be Hurwitz and let us assume further that there exists at least one number $k \in \mathcal{N}$ such that $\Delta(H_0(k)) = 0$ and that for $\epsilon > 0$ sufficiently small we have*

$$x'M_\epsilon(i)x > 0,\ x \in \Omega,\ \ x'M_\epsilon(i)x \geq x'M_\epsilon(k)x,\ x \notin \Omega.$$

Then the system (7.4.6) is perfectly robustly stable against the jump intensities.

Remark 2 *Since $A(k)$ matrix is Hurwitz, the solution of Lyapunov's equation (7.6.34) is given by the formula*

$$H_0(k) = \int_0^\infty \exp(A'(k)t)M_0(k)\exp(A(k)t)dt.$$

Then, it is easy to see that the condition $\Delta(H_0(k)) = 0$ *is equivalent to*

$$\begin{bmatrix} M_0(k) \\ M_0(k)A(k) \\ \cdots \\ M_0(k)A^{n-1}(k) \end{bmatrix} A_l = 0,\ l = 1, \dots, m.$$

The reader is refered to [74] for the proofs and more detail. Some other approaches to robustness of stochastic systems based on deterministic ideas are presented in [6, 85].

7.7 Stabilizing control of linear system

7.7.1 General linear systems

Consider the system (7.2.1) in the linear case

$$dX(t) = [A(Y(t))X(t) + B(Y(t))U(t)]dt + \sum_{l=1}^{m} A_l(Y(t))X(t)dW_l(t), \tag{7.7.1}$$

with the initial condition (7.4.17) and with the jump condition of vector $X(t)$ given by (7.1.20). Suppose that the Lyapunov-Bellman function has the form

$$V^0(x, y) = x'H(y)x,\ H(y) = H'(y) > 0 \tag{7.7.2}$$

and

$$L(x, u, y) = x'M(y)x + u'R(y)u,\ M(y) = M'(y) \geq 0.\ R(y) = R'(y) > 0 \tag{7.7.3}$$

Applying Theorem 7.4.20 we obtain that matrix $H(y)$, $y \in \mathcal{N}$ satisfies the following system of coupled matrix quadratic equations

$$A(i)'H(i) + H(i)A(i) - H(i)B(i)R^{-1}(i)B'(i)H(i) + \sum_{l=1}^{m} A_l'(i)H(i)A(i) +$$

$$\sum_{j \neq i}^{\nu} [\Phi_{ij}' H(j)\Phi_{ij} - H(i)]q_{ij} + M(i) = 0, i \in \mathcal{N} \tag{7.7.4}$$

and control law, which stabilizes the system (7.7.1) in the sense that this system is exponentially stable in the mean square is given by

$$U(t) = -K(i)X(t), \text{ if } Y(t) = i, \tag{7.7.5}$$

where $K(i) = R^{-1}(i)B'(i)H(i)$ $(i \in \mathcal{N})$. Simultaneously this control law minimizes the functional (7.4.18) along the trajectories of the system (7.7.1) with function L given by the formula (7.7.3). It is very important to obtain the conditions of existence of stabilizing control. These conditions for the linear system (7.7.1) can be expressed as solvability conditions of matrix equations (7.7.4). The following theorem gives a sufficient condition of stabilizability in the case when at the moments of jumps of the Markov chain $Y(t)$ the vector $X(t)$ is changed continuously.

Theorem 7.7.1 *Consider the system (7.7.1) with the continuous change of vector $X(t)$, satisfying the condition (7.1.18). Assume that the pairs $(A(i), B(i))$ $(i \in \mathcal{N})$ are stabilizable, the pairs $(\sqrt{M(i)}, A(i))$ $(i \in \mathcal{N})$ are observable and the following inequality is true*

$$\max_{i\in\mathcal{N}} \inf_{K} |q_i \int_0^\infty \exp(-sq_i)\exp[s(A(i)-B(i)K)']\exp[s(A(i)-B(i)K]dt| < 1. \qquad (7.7.6)$$

Then:

(i) there exists a unique positive definite solution $H(i)$ $(i \in \mathcal{N})$ of the system of coupled matrix quadratic equations

$$A(i)'H(i) + H(i)A(i) - H(i)B(i)R^{-1}(i)B'(i)H(i) +$$
$$\sum_{l=1}^{m} A_l'(i)H(i)A(i) + \sum_{j\neq i}^{\nu}[H(j)-H(i)]q_{ij} + M(i) = 0, i \in \mathcal{N}; \qquad (7.7.7)$$

(ii) the control law (7.7.5) stabilizes the system (7.7.1) in the sense that this system is exponentially stable in the mean square;

(iii) matrices

$$A(i) - B(i)K(i) - \frac{1}{2}q_i I, \ i \in \mathcal{N}$$

are Hurwitz.

Remark 3 *Under the conditions of Theorem 7.7.1 the solution $H_i(t)$ of coupled differential equations (7.2.28) has property $H_i(t_0) \to H_i$ $i \in \mathcal{N}$, as $t_0 \to -\infty$, where H_i is the solution of (7.7.4).*

The reader is referred to [100, 96] for the proofs and more detail.

7.7.2 Linear systems with parametric noise

General stabilizability conditions

Consider the important case, when the system may be described by the Ito differential equation

$$dX(t) = [AX(t) + BU(t)]dt + \sum_{l=1}^{m_1} A_l X(t)dW_{1l}(t) + \sum_{s=1}^{m_2} B_s U(t)dW_{2s}(t), \qquad (7.7.8)$$

where W_1 and W_2 are m_1-dimensional and m_2-dimensional independent standard Wiener processes. For easy formulations of theorems take $A_l = \sigma_l F_l$, $l = 1, \ldots m_1$, $B_l = \rho_l G_l$, $l = 1, \ldots m_2$; then the scalar factors will indicate the intensities or disturbances.

Definition 7.7.2 *System (7.7.8) is said to be stabilizable in the mean square sense if there exists a matrix K such that the system*

$$dX(t) = [A - BK]X(t)dt + \sum_{l=1}^{m_1} A_l X(t)dW_{1l}(t) + \sum_{s=1}^{m_2} B_s U(t)dW_{2s}(t), \qquad (7.7.9)$$

is exponentially stable in the mean square.

The following fundamental theorem gives a necessary and sufficient condition for the mean square stabilizability of (7.7.8). It is stated in terms of the nonlinear (quadratic) matrix equation

$$A'H + HA - HB[R + \Gamma(H)]^{-1}B'H + \Delta(H) + M = 0 \tag{7.7.10}$$

in the symmetric matrix H for given symmetric R and M of dimension $n \times n, n \times n$ and $k \times k$ respectively, where

$$\Delta(H) = \sum_{l=1}^{m_1} \sigma^2 F_l' H F_l, \; \Gamma(H) = \sum_{s=1}^{m_2} \rho^2 G_s' H G_s.$$

Theorem 7.7.3 *A sufficient condition for mean square stabilizability of (7.7.8) is that there exists positive definite matrices M and R, for which (7.7.10) has a positive definite solution H. A necessary condition for mean square stabilizability of (7.7.8) is that (7.7.10) has a positive definite solution H for any given positive definite matrices M and R.*

Systems with state dependent noise only

Consider the particular case of the system (7.7.8) in which there is only state dependent noise

$$dX(t) = [AX(t) + BU(t)]dt + \sum_{l=1}^{m_1} \sigma_l F_l X(t) dW_{1l}(t). \tag{7.7.11}$$

The matrix Riccati involved in the application of Theorem 7.7.3 correspondingly becomes

$$A'H + HA - HBR^{-1}B'H + \Delta(H) + M = 0. \tag{7.7.12}$$

Consider also the algebraic matrix Riccati equation

$$A'H + HA - \frac{1}{\beta}HBB'H + M = 0 \tag{7.7.13}$$

with $\beta > 0$ and $M = M' \geq 0$. It is well known [55] that if the pair (A, B) is stabilizable and the pair $(\sqrt{M}, A)$ is observable then there exists a unique positive definite solution H^+ of (7.7.13) such that $A - \frac{1}{\beta}BB'H^+$ is a Hurwitz matrix. Moreover H^+ is monotone nonincreasing with decreasing β and

$$H_0 = \lim_{\beta \to 0} H^+$$

is well-defined for all fixed M and is positive semidefinite.

Let Ω denote the subspace of R^n spanned by the columns of the matrices F_l', $l = 1, \ldots, m_1$, i.e.,

$$\Omega = \{x \in R^n \,|\, x \perp N(F_l) \text{ for all } l\},$$

where N denotes the null space. Application of Theorem 7.7.3 to the case under consideration leads to the following criterion for stabilizability.

Theorem 7.7.4 *The system (7.7.11) is mean square stabilizable if and only if*

(i) the pair (A, B) is stabilizable,

(ii) there exists a matrix $M_* = M_*'$ with $M_* \geq 0$, but $M_* > 0$ on Ω such that

$$\Delta(H_0) \leq M_*, \text{ but } \Delta(H_0) < M_* \text{ on } \Omega.$$

Stabilizability for arbitrary state dependent noise intensities It is very interesting to have a condition on the parameter matrices A, B and $F_l,\ l = 1, \ldots m_1$ of system (7.7.11) such that for *all values* of the noise intensities σ_l there exists a stabilizing gain matrix. The following result is an immediate consequense of Theorem 7.7.4.

Theorem 7.7.5 *System (7.7.11) is mean square stabilizable for all noise intensities σ_l if pair (A, B) is stabilizable and if there exists a symmetric matrix M with $M \geq 0$, but $M > 0$ on Ω such that*

$$F_l' H_0 F_l = 0,\ l = 1, \ldots m_1.$$

Necessary conditions for (7.7.11) to be mean square stabilizable for all σ_l are that the pair (A, B) is stabilizable and that $F_l' H_0 F_l = 0,\ l = 1, \ldots m_1$ for some semidefinite matrix M.

Remark 4 *Theorem 7.7.5 gives a necessary and sufficient condition if Ω is one-dimensional.*

Assume that

$$\kappa = \dim\{\mathcal{R}(F_1) \oplus \ldots \oplus \mathcal{R}(F_{m_1})\} < \dim\{\mathcal{R}(B)\}$$

and let C be a $\kappa \times n$ matrix such that $\mathcal{R}(C) = \mathcal{R}(F_1) \oplus \ldots \oplus \mathcal{R}(F_{m_1})$, where $\mathcal{R}$ denotes range space and $\oplus$ is the direct sum symbol.

Corollary 7.7.6 *The system (7.7.11) is mean square stabilizable for all noise intensities σ_l if there exists matrix an $n \times \kappa$ matrix B_1 such that $\mathcal{R}(B_1) \subset \mathcal{R}(B)$ and such that the polynomial*

$$\frac{\det C(sI - A)^{-1} B_1}{\det(sI - A)}$$

has no zeroes with the positive real part.

Corollary 7.7.7 *The system (7.7.11) is mean square stabilizable for all noise intensities σ_l if pair (A, B) is stabilizable and if $\mathcal{R}(F_l) \subset \mathcal{R}(B)$ for all $l = 1, \ldots m_1$.*

Consider as a special case of (7.7.11) the system with a single input, a single noise term and a matrix F_1 of rank one:

$$dX(t) = [AX(t) + \mathbf{b}U(t)]dt + \sigma \mathbf{b}_1 \mathbf{c}_1' X(t) dW(t), \tag{7.7.14}$$

where $\mathbf{b}$, $\mathbf{b}_1$ and $\mathbf{c}_1$ are n-dimensional vectors, $W(t)$ is a standard scalar Wiener process, σ is a scalar which indicates the intensity of the disturbance. Then we have:

Corollary 7.7.8 *Let the pair $(\mathbf{c}_1, A)$ be detectable. Then the system (7.7.14) is mean square stabilizable for all noise intensities σ if and only if*

(i) thc pair $(A, \mathbf{b})$ is stabilizable;

(ii) the rational function

$$\frac{\mathbf{c}_1 (sI - A)^{-1} \mathbf{b}_1}{\mathbf{c}_1 sI - A)\mathbf{b}}$$

has no poles with the positive real part, after possible cancelation of common factors.

The reader is refered to [92] and [99] for the proofs and more detailed study of the state dependend noise case.

Systems with control dependent noise only

Consider another important particular case of the system (7.7.8), in which there is only control dependent noise

$$dX(t) = [AX(t) + BU(t)]dt + \sum_{s=1}^{m_2} \rho_s G_s U(t) dW_{2s}(t). \tag{7.7.15}$$

The matrix Riccati involved in the application of Theorem 7.7.3 correspondingly becomes

$$A'H + HA - HB[\Gamma(H) + R^{-1}]B'H + M = 0. \tag{7.7.16}$$

Consider also the algebraic matrix Riccati equation

$$A'H + HA - HBS^{-1}B'H + \alpha T = 0. \tag{7.7.17}$$

with $S = S' > 0$, $T = T' \geq 0$ and $\alpha > 0$. If the pair (A, B) is stabilizable then there exists a unique positive definite solution H^+ of (7.7.17) which is monotone decreasing with α and

$$H_* = \lim_{\alpha \to 0} H^+$$

is well-defined for all fixed $S, T > 0$ and at least positive semidefinite. Application of Theorem 7.7.3 to the case under consideration leads to the following criterion for stabilizability.

Theorem 7.7.9 *The system (7.7.15) is mean square stabilizable if and only if*

(i) the pair (A, B) is stabilizable,

(ii) there exists a matrix $S = S' > 0$ such that

$$\Gamma(H_*) < S.$$

For the special case that there is only a scalar control, i.e. for the system

$$dX(t) = [AX(t) + \mathbf{b}U(t)]dt + \sum_{s=1}^{m_2} \rho_s \mathbf{g}_s U(t) dW_s(t), \tag{7.7.18}$$

with $\mathbf{b}$ and $\mathbf{g}_s$ ($s = 1, \ldots m_2$) n-dimensional vectors, then one can carry the computation further. Let $H_1 = \lim_{\alpha \to 0} H$, where H is the unique positive definite solution of the algebraic Riccati equation

$$A'H + HA - H\mathbf{b}\mathbf{b}'H + \alpha T = 0. \tag{7.7.19}$$

Corollary 7.7.10 *The system (7.7.18) is mean square stabilizable if and only if the pair (A,$\mathbf{b}$) is stabilizable and*

$$\Gamma(H_1) < 1.$$

Stabilizability for arbitrary control dependent noise intensities Now we present the conditions on the parameter matrices A, B and G_s ($s = 1, \ldots m_2$) of the system (7.7.15) such that for all values of the noise intensities ρ_s there exists a stabilizing control.

Corollary 7.7.11 *The system (7.7.18) is mean square stabilizable for all noise intensities ρ_s if and only if*

(i) the pair $(A, \mathbf{b})$ is stabilizable;

(ii) the vectors $\mathbf{g}_s$ $(s = 1, \dots m_1)$ *belong to the invariant subspace of* A *spanned by its (generalized) eigenvectors corresponding to eigenvalues with nonpositive real parts.*

In the multivariable case this condition is only sufficient, but not necessary:

Corollary 7.7.12 *The system (7.7.15) is mean square stabilizable for all noise intensities* ρ_s *if*

(i) the pair (A, B) *is stabilizable;*

(ii) the columns of G_s $(s = 1, \dots m_1)$ *belong to the invariant subspace of* A *spanned by its (generalized) eigenvectors corresponding to eigenvalues with nonpositive real parts.*

The reader is refered to [92] and [31] for the proofs and more detailed study of the control dependend noise case.

Systems with state and control dependent noise

For the case in which one wants to obtain stabilizability criteria for system (7.7.8) with both state and control dependent noise present, it is necessary to study the full nonlinear matrix equation (7.7.10).

Theorem 7.7.13 *Let the pair* (A, B) *be stabilizable and*

$$\inf_K \Big| \int_0^\infty e^{t(A-BK)'}[K'\Gamma(I)K + \Delta(I)]e^{t(A-BK)}dt \Big| < 1.$$

Then the system (7.7.8) is mean square stabilizable.

Remark 5 *Under the conditions of Theorem 7.7.13 the solution* $H(t)$ *of (7.2.16) with constant matrices* A, B, M, R *has property* $H(t_0) \to H$ *as* $t_0 \to -\infty$*, where* H *is the solution of (7.7.10).*

The reader is referred to [96] and [100] for the proof. This result is very complicated for computations. In particular cases some rather explicit criteria are needed. Consider the system

$$dX(t) = [AX(t) + \mathbf{b}U(t)]dt + \sigma\mathbf{b}_1\mathbf{c}_1'dW_1 + \sum_{s=1}^{m_2} \rho_s\mathbf{g}_sU(t)dW_{2s}(t), \tag{7.7.20}$$

with $\mathbf{b}, \mathbf{b}_1, \mathbf{c}$ and $\mathbf{g}_s$ $(s = 1, \dots m_2)$ n-dimensional vectors. Together with this system consider the associated algebraic Riccati equation

$$A'H + HA - \frac{1}{\alpha}H\mathbf{b}\mathbf{b}'H + \mathbf{c}_1\mathbf{c}_1' = 0 \tag{7.7.21}$$

where $\alpha > 0$ is a parameter. If triple $(A, \mathbf{b}, \mathbf{c}_1)$ is completely controllable and observable then as it is well known there exists for each $\alpha > 0$ a unique positive definite solution $H(\alpha)$ of (7.7.21).

Theorem 7.7.14 *Let triple* $(A, \mathbf{b}, \mathbf{c}_1)$ *be completely controllable and observable. Then the system (7.7.20) is mean square stabilizable if and only if*

(i) there exists a solution $\alpha^* > 0$ *of the equation*

$$\sigma^2\mathbf{b}_1'H(\alpha^*)\mathbf{b}_1 = 1;$$

(ii) *the inequality*

$$\sum_{s=1}^{m_2} \rho_s^2 \mathbf{g}_s' H(\alpha^*) \mathbf{g}_s < \alpha^*$$

holds for this α^*.

The proof is presented in [92]. The reader is referred to [30, 44, 50, 64, 67, 69] and references therein for more detailed study of this direction.

7.7.3 Robust stabilizing control

Robust stabilization of systems with state dependent noise

In this section we present conditions on the parameter matrices A, B and F_i $(i = 1, \dots m)$ of the system (7.7.11) for which there exists a feedback gain matrix K such that the closed loop system

$$dX(t) = [A - BK]X(t)dt + \sum_{l=1}^{m_1} \sigma_l F_l X(t) dW_{1l}(t) \qquad (7.7.22)$$

is asymptotically stable in the mean square for all noise intensities σ_l $(l = 1, \dots, m)$. These conditions are different from the ones obtained earlier in Section 7.7.2, because in 7.7.2 the feedback gain matrix is allowed to be a function of σ_l $(l = 1, \dots, m)$. In this section we consider the case in which this feedback gain matrix need not be a function of the noise intensities. So, we consider the stabilizability of (7.7.11) by means of a time invariant state feedback law

$$U(t) = -KX(t). \qquad (7.7.23)$$

Definition 7.7.15 *The system (7.7.11) is said to be perfectly robustly stabilizable if there exists a feedback control (7.7.23) such that (7.7.22) is asymptotically stable in the mean square for all noise intensities* σ_l $(l = 1, \dots, m)$.

Definition 7.7.16 *The system (7.7.11) is said to be robustly stabilizable for all noise intensities (from the given domain) if there exists a feedback control (7.7.23) such that (7.7.22) is asymptotically stable in the mean square for all noise intensities satisfying*

$$\sigma_l \le s_l, \; l = 1, \dots, m.$$

The property expressed by Definition 7.7.16 is somewhat weaker than the property expressed by Definition 7.7.15 in that the feedback matrix K may depend on the bounds s_i; some entires of K may increase without bound as some of these bounds s_i tend to infinity.

Theorem 7.7.17 *The System (7.7.11) is perfectly robustly stabilizable if and only if there exists a matrix* K, *such that matrix* $A - BK$ *is Hurwitz and in a suitable basis the matrices* $\bar{A} = A - BK$ *and* F_l $(l = 1, \dots m)$ *take the block triangular form:*

$$\bar{A} = \begin{bmatrix} \bar{A}_{11} & \bar{A}_{12} & \dots & \bar{A}_{1p} \\ 0 & \bar{A}_{22} & \dots & \bar{A}_{2p} \\ \vdots & \vdots & \ddots & \vdots \\ 0 & 0 & \dots & \bar{A}_{pp} \end{bmatrix}, \quad F_l = \begin{bmatrix} 0 & F_{l12} & \dots & F_{l1p} \\ 0 & 0 & \dots & F_{l2p} \\ \vdots & \vdots & \ddots & \vdots \\ 0 & 0 & \dots & 0 \end{bmatrix}.$$

A series of formalized robustness criteria based on the geometric theory of linear multivariable systems was obtained by Willems and Willems [93]; the reader is referred to [93] for more details.

Robust stabilization of systems with random jumps

Consider the linear system (7.4.6) with the control action

$$dX(t) = [A(Y(t))X(t) + B(Y(t))U(t)]dt + \sum_{l=1}^{m} \sigma_l F_l(Y_(t))X(t)dW_l(t), \qquad (7.7.24)$$

where $U(t)$ is a k-dimensional control vector, and $B(i)$ is an $n \times k$ matrix. Assume that at the jump moments of $Y(t)$ the vector $X(t)$ is changed continuously, so that (7.1.18) is valid.

We obtain a state feedback control law in the form of (7.7.23), which guarantees robust stability of the closed loop system (7.7.24) (7.7.23) against the jump intensities.

Theorem 7.7.18 *If for some positive definite matrices $R(i)$ and $M(i)$ $i \in \mathcal{N}$ there exist the constant matrices $H > 0$ and K, satisfying the equations*

$$(A(i) - B(i)K)'H + H(A(i) - B(i)K) + \Delta_i(H) + \\ K'R(i)K + M(i) = 0, \; i \in \mathcal{N}, \qquad (7.7.25)$$

$$K = [\sum_{i=1}^{\nu} R(i)]^{-1} \sum_{i=1}^{\nu} B'(i)H, \qquad (7.7.26)$$

then the closed loop system (7.7.24), (7.7.23), (7.7.26) is robustly stable against the jump intensities.

Denote $A_\Sigma = \sum_{i=1}^{\nu} A(i)$ and analogously B_Σ R_Σ and M_Σ. The following assertion is more effective from the point of view of computation.

Corollary 7.7.19 *Let for some positive definite matrix R_Σ and positive semidefinite matrix M_Σ there exist the constant matrices $H > 0$ and K, satisfying the relations*

$$A'_\Sigma H + HA_\Sigma - HB_\Sigma R_\Sigma^{-1} B'_\Sigma H + \sum_{i=1}^{\nu} \sum_{l=1}^{m} A'_l(i) H A_l(i) + M_\Sigma = 0, \qquad (7.7.27)$$

$$K = R_\Sigma^{-1} B'_\Sigma(i) H, \qquad (7.7.28)$$

$$(A(i) - B(i)K)'H + H(A(i) - B(i)K) + \Delta_i(H) < 0, \; i \in \mathcal{N}, \qquad (7.7.29)$$

then the closed loop system (7.7.24), (7.7.23), (7.7.28) is robustly stable against the jump intensities.

The matrix quadratic equation (7.7.27) can be solved by using a consecutive approximation of Riccati equations; the inequalities (7.7.29) are well known linear matrix inequalities [9]. For the systems without the state dependent noise ($\sigma_l = 0$, $l = 1, \ldots, m$) the equation (7.7.27) is ordinary Riccati equation. The reader is referred to [74] for more details.

Bibliography

[1] L.Arnold. *Stochastic Differential Equations. Theory and Applications.* John Wiley, 1974.

[2] K.J. Aström. *Introduction to Stochastic Control Theory.* Academic Press, 1970.

[3] A.I. Barkin, A.L. Zelentsovskii and P.V. Pakshin. *Absolute Stability of Deterministic and Stochastic Control Systems.* MAI, Moscow 1992. (Russian.)

[4] Batkov A.M.et al. *Optimization methods in statistical control problem.* Mashinostroenie, Moscow 1974. (Russian.)

[5] R.Bellman. *Dynamic programming.* Princeton University Press, 1957.

[6] K. Benjelloun, E.K. Boukas, O.L.V. Costa and P.Shi. Design of robust controller for linear systems with Markovian jumping parameters. *Mathematical Problems in Engineering*, 44:269–288, 1998.

[7] J.E. Bertram and P.E. Sarachik. Stability of circuits with randomly time-varying parameters. *Trans. IRE* (CT-6):260-270, 1959.

[8] E.N. Berezina and M.V. Levit. Moment equations and stability of linear system with scalar parametric disturbance of Markov chain type. *Prikl. Matematika i Mekhanika*, 44(5):792–901, 1980. (Russian.)

[9] S. Boyd, E. El Ghaoui, E. Feron and V. Balakrishnan. *Linear matrix inequalities in control and system theory.* SIAM, 1994.

[10] R.W. Brockett. Lie algebras and Lie groups in control theory. In: *Geometric methods in systems theory.* R.W. Brockett and D.Q. Mayne, Eds., Reidel, 1973.

[11] R.W. Brockett. Lie theory and control systems defined in spheres. *SIAM J. Appl. Math.*, 25(2):213–225, 1973.

[12] R.W. Brockett. Parametrically stochastic linear differential equations. *Mathematical Programming Study*, 5:8–21, 1976.

[13] R.W. Brockett and J.C. Willems. Average value criteria for stochastic stability. In: *Lecture Notes in Mathematics*, 294:252-272. Springer Verlag, 1972.

[14] V.A. Brusin. Global stability and dichotomy of a class of nonlinear systems with stochastic parameters. *Sibirskii Math. J.*, 22:57–73, 1981. (Russian.)

[15] V.A. Brusin and V.A. Ugrinovskii. Stochastic stability of a class of nonlinear differential equations of Ito type. *Sibirskii Math. J.*, 28:381–393, 1987. (Russian.)

[16] V.A. Brusin and V.A. Ugrinovskii. Absolute stability approach to stochastic stability of infinite dimensional nonlinear systems. *Automatica*, 31:1453–1458, 1995.

[17] C.D. Charalambous and R.J. Elliott. Information states in stochastic control and filtering: A Lie algebraic theoretic approach. *IEEE Trans. Automatic Control*, 45(4):653–674, 2000.

[18] M.H.A. Davis. *Linear estimation and stochastic control.* Chapman and Hall, 1977.

[19] J.L. Doob. *Stochastic Processes.* J.Wiley, 1953.

[20] E.B. Dynkin. *Markov processes*, vol.1, 2. Springer Verlag, 1965. (Transl.)

[21] R.J. Elliott. *Stochastic calculus and applications*, Springer Verlag, 1982.

[22] R.F. Estrada. Passive stochastic feedback stability. Pt. I, II. *Int. J. Control*, 18(2):255–272, 1972.

[23] W.H. Fleming. Optimal control of partially observable diffusions. *SIAM. J. Control*, 6:194–214, 1968.

[24] W.H. Fleming and R.W. Rishel. *Deterministic and stochastic optimal control*, Springer-Verlag, 1975.

[25] W.H. Fleming and H.M. Soner. *Controlled Markov processes and viscosity solutions*, Springer- Verlag, 1992.

[26] W.H. Fleming and M. Nisio. On the existence of optimal stochastic control. *J. Math. and Mechanics*, 15:777–794, 1966.

[27] A.Friedman. *Stochastic Differential Equations and Applications*, vol.I. Academic Press, 1975.

[28] A.Friedman. *Stochastic Differential Equations and Applications*, vol.II. Academic Press, 1976.

[29] I.I. Gihman and A.V.Skorohod. *Stochastic Differential Equations.* Springer Verlag, 1974. (Transl.)

[30] U.G. Haussmann. Optimal stationary control with state and control dependent noise. *SIAM J. Control*, 9(2):184–198, 1971.

[31] U.G. Haussmann. Stability of linear systems with control dependent noise. *SIAM J. Control*, 11(2):382–394, 1973.

[32] U.G. Haussmann. On the existence of moments of stationary linear systems with multiplicative noise. *SIAM J. Control*, 12(1):99–105, 1974.

[33] T. Hida, H-H. Kuo, J. Potthoff and L. Streit. *White noise. An Infinite Dimensional Approach.* Kluwer, 1993.

[34] Y.Ji and H.J. Chizeck. Jump linear quadratic Gaussian control:steady state solution and testable conditions. *Control Theory and advanced technology*, 6(3):289–319, 1990.

[35] Y.Ji, H.J. Chizeck, X. Feng and K.A. Loparo. Stability and control of discrete time jump linear systems. *Control Theory and advanced technology*, 7(2):247–270, 1991.

[36] P.D. Joseph and J.T.Tou. On linear control theory. *AIEE Trans. on Appl and Ind. Pt.II* , 80:193–196, 1961.

[37] E.I. Jury. *Inners and Stability of Dynamic Systems.* John Wiley, 1974.

[38] K. Ito. On stochastic differential equations. *Mem. Amer. Math. Soc.*, 4:(1-51), 1951.

[39] R.E. Kalman. A new approach to linear filtering an prediction problems. *Transactions of the ASME, ser.D: J.of Basic Engr.*, 82:35–45, 1960.

[40] R.E. Kalman and R.E. Bucy. New results in linear filtering an prediction theory. *Transactions of the ASME, ser.D: J.of Basic Engr.*, 83:95–108, 1961.

[41] I.Ya. Kats. *Lyapunov function method in problems of stability and stabilization of systems with random structure.* Ekaterinburg, UGAPS, 1998. (Russian.)

[42] I.Ya. Kats and N.N. Krasovskii. On the stability of systems with random attributes. *Journal of Applied Mathematics and Mechanics*, 24:1225–1246, 1960. (Transl.)

[43] I.Ye. Kazakov and V.M. Artem'ev. *Optimization of dynamical with random structure.* Moscow, Nauka, 1990. (Russian.)

[44] Yu.F. Kazarinov. On the stabilization criterion of linear stochastic system with parametric exitation of white noise type. *Prikl. Matematika i Mekhanika*, 41(2):245–250, 1977. (Russian.)

[45] R.Z. Khasminskii. *Stochastic stability of differential equations.* Sijthoff and Noordhoff, Alphen, 1980. (Transl.)

[46] V.B. Kolmanovskii and A.D. Myshkis. *Introduction to the theory and applications of functional differential equations.* Kluwer,.1999.

[47] V.B. Kolmanovskii and L.E. Shaikhet. *Control of Systems with Aftereffect.* AMS, 1996.

[48] D.G. Korenevskii. *Stability of dynamic systems under random perturbation of the parameters. Algebraic Critreria.* Kiev, Naukova Dumka 1989. (Russian.)

[49] D.L.Kleinman.* On the stability of linear stochastic systems. *IEEE Trans. Automatic Control*, AC-14 (4):429–430, 1969.

[50] N.N. Krasovskii. Stabilization of the systems in which noise is dependent on the value of the control signal. *Engrg. Cybernetics*, 2:94–102, 1965. (Transl.)

[51] N.N. Krasovskii and E.A. Lidskii. Analytical design of controllers in systems with random attributes I, II, III. *Automation and Remote Control*, 22(9):1021–1025, (10): 1141–1146, (11):(1289–1294, 1961. (Transl.)

[52] F.B.Knight. *Essentials of Brownian motion.* American Math. Soc, 1981.

[53] H.J. Kushner. On the stochastic maximum principle: fixed time of control. *J. Math. Anal. Appl*, 11:78–92, 1965.

[54] H.J. Kushner. *Stochastic Stability and Control* Academic Press, 1967.

[55] H. Kwakernaak and R. Sivan. The maximally achievable accuracy of linear optimal regulators and linear optimal filters. *IEEE Trans. Automatic Control*, AC-17(1):79–86, 1972.

[56] P. Langevin. Sur la theorie du mouvement brownien. *Compt. Rend. Acad. Sci. Paris*, 146:530–533, 1908.

[57] M.V. Levit. Frequency-domain criterion of absolute stochastic stability for nonlinear systems of differential equations of Ito. *Uspekhi Matem. Nauk*, 27(4):215–216, 1972. (Russian.)

[58] M.V. Levit. Algebraic criterion of stochastic stability of linear system with parametric exitation of correlated white noises. *Prikl. Matematika i Mekhanika*, 36(3):546–551, 1972. (Russian.)

[59] M.V. Levit. Stability of linear multivariable stochastic systems with white noise. *Avtomatika i Telemechanika*, (10):38–50, 1977. (Russian.)

[60] M.V. Levit and V.A. Yacubovich. Algebraic criterion of stochastic stability for linear systems with parametric action of white noise type. *Prikl. Matematika i Mekhanika*, 36(1):142–148, 1971. (Russian.)

[61] R.S. Liptser and A.N.Shiryayev. *Statistic of random processes, vol.I.* Springer-Verlag, 1977.

[62] R.S. Liptser and A.N.Shiryayev. *Statistic of random processes, vol.II.* Springer-Verlag, 1978.

[63] A.K. Mahalanabis and S. Purkayastha. Frequency-domain criteria for stability of a class of nonlinear stochastic systems. *IEEE Trans. Automatic Control*, AC-18 (3):266–270,1973.

[64] M. Mariton. *Jump linear systems in automatic control.* Marcel Dekker, 1990.

[65] S.P. Meyn and R.L. Tweedie. *Markov Chains and Stochastic Stability.* Springer Verlag, 1993.

[66] G.N.Milstein. Mean square stability of linear system under the action of the Markov chain. *Prikl. Matematika i Mekhanika*, 36(3):537–545, 1972. (Russian.)

[67] G.N.Milstein. Design of stabilizing controller with incomplete state information for linear stochastic systems with multiplicative noises. *Avtomatika i Telemekhanika*, (5):98–106, 1982. (Russian.)

[68] G.N.Milstein and L.B. Ryashko. Optimal stabilization of linear stochastic systems. *Prikl. Matematika i Mekhanika*, 40(6):1034–1039, 1976. (Russian.)

[69] G.N.Milstein and L.B. Ryashko. Optimal stabilization of linear stochastic systems. *Prikl. Matematika i Mekhanika*, 40(6):1034–1039, 1976. (Russian.)

[70] R.E. Mortensen. Stochastic optimal control with noise observations. *Int. J. Control*, 4:455–464, 1966. (Russian.)

[71] B. Øksendal. *Stochastic differential equations. An Introduction with Application*, 4-th edition. Springer Verlag, 1995.

[72] P.V. Pakshin. *Discrete Systems with Random Parameter and Structure.* Nauka Fizmatlit, 1994. Russian.

[73] P.V. Pakshin. Stability of a class of nonlinear stochastic systems. *Avtomatika i Telemekhanika*, (4):27–36, 1974. (Russian.)

[74] P.V. Pakshin. Robust stability and stabilization of the family of jumping stochastic systems. *Nonlinear Analysis, Theory, Methods and Applications*, 30(5):2855–2866, 1997.

[75] Yu.I. Paraev. On the stability of linear systems with randomly varying structure. *Avtomatika i Telemekhanika*, (8):165–168, 1982. (Russian.)

[76] L.S.Pontryagin, V.G.Boltyanskii, R.V.Gamkredze and E.F. Mishchenko. *The matematical theory of optimal processes.* Interscience, 1962. Transl.

[77] V.S. Pugachev and I.N. Sinitsyn. *Stochastic differential systems* Nauka, Moscow 1985. Russian.

[78] L. Socha. Application of Yacubovich criterion for stability of nonlinear stochastic systems. *IEEE Trans. Automatic Control*, AC-25(2):350–352, 1980.

[79] T. Sasagawa. On the exponential stability and instability of linear stochastic systems. *Int. J. Control*, 33(2):363–370, 1980.

[80] T. Sasagawa. A note on exponential asymptotic properties of linear stochastic systems. *Int. J. Control*, 33(6):1155–1163, 1980.

[81] R.L. Stratonovich. A new representation for stochastic integral and equations. *SIAM J. Control*, 4:362–371, 1966. (Transl.)

[82] H.J. Sussmann. On the gap between deterministic and stochastic ordinary differential equations. *The Annals of Prob.*, 60:19–41, 1978.

[83] D.D. Sworder. On the stochastic maximum principle. *J. Math. Anal. Appl.*, 24:627–635, 1968.

[84] D.D. Sworder. Feedback control of a class of linear systems with jump parameters. *IEEE Trans. Automatic Control*, AC-14(1):9–14, 1969.

[85] V.A. Ugrinovskii. On the robustness of linear systems with randomly changed parameters. *Avtomatica i Telemekhanika*, (4):90–99,1994. (Russian.)

[86] V.A. Ugrinovskii. Stochastic analog of frequency domain theorem. *Izv. VUZov. Matematika*, (10):37–43,1987. (Russian.)

[87] V.M. Warfield. A stochastic maximum principle. *SIAM J. Control Optim.*, 14:803–826,1976.

[88] J.L. Willems. Mean square stability criteria for stochastic feedback systems. *Int.J. Systems Sci.*, 4(4):545–564,1973.

[89] J.L. Willems. Stability of higher order moments for linear stochastic systems. *Ingenieur-Archiv*, 44:123–129,1975.

[90] J.L. Willems. Stability criteria for stochastic systems with colored multiplicative noise. *Acta Mechanica*, 23:171–178,1975.

[91] J.L. Willems and D. Aeyels. An equivalence result for moment stability criteria for parametric stochastic systems and Ito equations *Int. J. Syst. Sci.*, 7(5):577–590,1976.

[92] J.L. Willems and J.C Willems. Feedback stabilizability for stochastic systems with state and control dependent noise. *Automatica*, 12:277–283,1976.

[93] J.L. Willems and J.C Willems. Robust stabilization of uncertain systems. *SIAM J. Control and Optimization*, 21(3):352–374,1983.

[94] E.Wong and M.Zakai. Riemann-Stiltjes approximation of stochastic integrals. *Z. Warscheinlichkeitstheorie verw. Geb.*, 12:87–97,1969.

[95] E.Wong and M.Zakai. On the relation between ordinary and stochastic differential equations. *Int. J. Engrg. Sci.*, 3:213–229,1965.

[96] W.M.Wonham. Random differential equations in control theory. In: *Probabilistic Methods in Applied Mathematics*, 2:131-212. A.T. Bharucha-Reid, Ed., Academic Press, 1970.

[97] W.M.Wonham. Liapunov criteria for weak stochastic stability. *J. Differential Equations*, 2:195–207, 1966.

[98] W.M.Wonham. A Liapunov method for estimation of statistical averages. *J. Differential Equations*, 2:365–377, 1966.

[99] W.M.Wonham. Optimal stationary control of a linear system with state-dependent noise. *SIAM J. Control*, 5:486–500, 1967.

[100] W.M.Wonham. On a matrix Riccati equation of stochastic control. *SIAM J. Control*, 6:681–697, 1968.

[101] W.M.Wonham. On the separation theorem of stochastic control. *SIAM J. Control*, 6:312–326, 1968.

[102] M.Zakai. A Lyapunov criterion for the existence of stationary probability distribution for systems perturbed by noise. *SIAM J. Control*, 7(3):390–397, 1969.

[103] M.Zakai. On the optimal filtering of diffusion processes. *Z. Warscheinlichkeitstheorie verw. Geb.*, 11:230–243,1969.

Chapter 8

Stochastic Differential Games and Applications

K.M. RAMACHANDRAN
Department of Mathematics
University of South Florida
Tampa, FL 33620-5700

This chapter deals with stochastic differential games in a completely competitive situation. There is considerable research in this area. We have attempted to put together some representative works on this topic. First we consider two person zero-sum stochastic differential games. In here, a solution is obtained using martingale techniques. Also, recent works using the viscosity solution method are briefly explained. Additionally, a stochastic differential game with multiple modes is presented. Next an N-person stochastic differential game problem in the relaxed control framework is analyzed using the method of occupation measures. An equilibrium solution (in the sense of Nash) is derived. Later, the powerful methods of weak convergence is adapted to study stochastic differential games where the dynamics is driven by the wideband noise process rather than the ideal white noise process. A game problem with imperfect information is also analyzed. Finally, we have mentioned some applications of stochastic differential games and explained in some detail a stochastic differential game of institutional investor speculation.

8.1 Introduction

The origins of game theory and their development could be traced to the pioneering work of Von Neumann and Morgenstern [112]. Due to the introduction of guided interceptor missiles in the 1950s, the questions of pursuit and evasion took center stage. The mathematical formulation and study of differential games was initiated by Rufus Isaacs, who was then with the Mathematics Department of the RAND Corporation, in a series of RAND Corporation memoranda that appeared in 1954, [52, 53, 54, 55]. This work and his further research were incorporated into a book [56] which inspired much further work and interest in this area. The relationship between differential games and optimal control theory and the publication of [56] at a time when interest in optimal control theory was very great served to further stimulate interest in differential games [17]. For good coverage on the connection between control theory and game theory, readers are referred to [67]. Earlier works on differential games and optimal control theory appeared almost simultaneously, independently of each other. At first, it seems natural to view a differential game as a control process where the controls are divided among various players who are willing to use them for objectives which possibly conflict with each other. However a much deeper study will reveal that the development of the two fields followed different paths. Both have the evolutionary aspect in common, but differential games have in addition a game-theoretic aspect. As a result, the techniques developed for the optimal control theory cannot be simply reused.

In the 1960s researchers started working on what have been called stochastic differential games. These games are stochastic in the sense that noise is added to the players' observations of the state of the system or to the transition equation itself. A stochastic differential game problem was solved in [50] using variational techniques where one player controlled the state and attempted to minimize the error and confuse the other player who could only make noisy measurements of the state and attempted to minimize his/her error estimate. Later in [9], a problem of pursuit-evasion is considered where the pursuer has perfect knowledge whereas the evader can only make noisy measurements of the state of the game. In [2, 94], a definition of a stochastic differential game is given. A connection between stochastic differential games and control theory is discussed in [78]. In the 1970s, rigorous discussion of existence and uniqueness results for stochastic differential games using martingale problem techniques and variational inequality techniques ensued, [15, 16, 14, 27, 24], among many others. There are many aspects of differential games such as pursuit evasion games, zero-sum games, cooperative and noncooparative games and other types of dynamic games. Dealing with all of the aspects is beyond the scope of an article of this size. For some survey papers on such diverse topics as pursuit-evasion games, viscosity solutions, discounted stochastic games, numerical methods, and others, we refer to [3], which serves as a rich source of information on these topics. In this article we will restrict ourselves to mostly strictly noncooparative stochastic differential games.

The early works on differential games are based on the dynamic programming method now known as Hamiltonian-Jacobi-Isaacs (HJI). Many authors worked on making the concept of value of a differential game precise and providing a rigorous derivation of HJI equation, which does not have a classical solution in most cases. For the HJI equations smooth solutions do not exist in general and nonsmooth solutions are highly non-unique. Some of the works in this direction include [17, 26, 24, 32, 36, 60, 94, 108, 109, 110]. In the 1980s, a new notion of generalized solutions for Hamilton-Jacobi equations, (namely, viscosity solutions), [22, 33, 71, 72, 73, 79, 99], provided a means of characterizing the value function as the unique solution of the HJI equation satisfying suitable boundary conditions. This method also provided the tools to show the convergence of the algorithms based on Dynamic Programming to the correct solution of the differential game and to establish the rate of convergence. A rigorous analysis of the viscosity solution of the Hamilton-Jacobi-Bellman-Isaacs equations in infinite dimensions is given in [105]. In the 1990s, a method based on an occupation measure approach is introduced for stochastic differential games in a relaxed control setting in which the differential game problem reduces to a static game problem on the set of occupation measures, the dynamics of the game being captured in these measures [18]. The major advantage of this method is that it enabled one to consider the dynamic game problems in much more physically appropriate wideband noise settings and use the powerful weak convergence methods, [84, 85, 88]. As a result, discrete games and differential games could be considered in a single setting.

The information structure plays an important role in stochastic differential games. All the above-referenced works assume that all the players of the game have full information of the state. This need not be the case in many applications. The interplay of information structure in the differential games is described in [37, 51, 82, 86, 72]. The stochastic differential game problems with incomplete information are not as much developed as the stochastic control problems with partial observations.

One of the earlier works on obtaining computational method for stochastic differential games is given in [43]. Following the work on numerical solutions for stochastic control [65] and many references in there, currently there are some efforts in deriving numerical schemes for stochastic differential games. For a numerical scheme for the viscosity solution of the Isaacs' equation, we refer to [10]. Also, as a result of weak convergence analysis [84, 88], it is easier to obtain numerical methods for stochastic differential games similar to that of [65] and to develop new computational methods as in [65].

In this article, first we will deal with two person zero-sum stochastic differential games for which the existence concepts will be derived using martingale methods. In this section, we will also briefly mention the viscosity solution method and a game problem with multiple modes. The N-person noncooperative stochastic differential games along with the concept of Nash equilibrium using more recent efforts with occupation measure approach is described in the next section. Recent works using the weak convergence methods for stochastic differential games will be the topic of Section 4. Some applications of stochastic differential games will be mentioned at the end and a stochastic differential game of institutional investor speculation will be explained in some detail. Some concluding remarks will be given in Section 6.

8.2 Two person zero-sum differential games

The object of this section is to present the concept of solutions and strategies as well as existence and uniqueness results for the two person zero-sum stochastic differential games. First, we will present the earlier work on stochastic differential games using martingale methods. Almost all of the material on this subsection comes from [24]. In the next subsection, we will briefly mention the recent results obtained on two person zero-sum stochastic differential games using the concept of viscosity solutions, [100]. There are various other methods used in studying stochastic differential games. In [14], two player stochastic differential games with stopping is analyzed using the method of two sided variational inequalities. Also refer to [15] and [16] for more results in this direction. A zero-sum Markov games with stopping and impulsive strategies is discussed in [104].

8.2.1 Two person zero-sum games: martingale methods

The evolution of the system is described by the stochastic differential equations

$$dx(t) = f(t,x,u_1,u_2)dt + \sigma(t,x)dB(t) \tag{8.2.1}$$

$$x(0) = x_0 \in \mathbf{R}^n, t \in [0,1] \tag{8.2.2}$$

where B is an n-dimensional Brownian motion; $u_i \in \mathcal{U}_i, i = 1,2$ are control functions.

There are two controllers, or players, I and II. Game is zero sum, player I is choosing his control to maximize the payoff and player II is choosing his control to minimize the payoff.

$\mathcal{F}_t = \sigma\{x(s) : s \leq t\}$ is the σ-algebra generated on $\mathcal{C}$, the space of continuous functions from $[0,1] \to \mathbf{R}^n$, up to time t. Assume that $f : [0,1] \times \mathcal{C} \times \mathcal{U}_1 \times \mathcal{U}_2 \to \mathbf{R}^n$ and σ, a nonsingular $n \times n$ matrix, satisfy the usual measurability and growth conditions. Given an n-dimensional Brownian motion $B(t)$ on a probability space (Ω, P), these conditions on σ ensures the stochastic equation

$$x(t) = x_0 + \int_0^t \sigma(s,x)dB(t)$$

has unique solution with sample path in $\mathcal{C}$. Let $\Im_t = \sigma\{B(s) : s \leq t\}$.

Assume that the spaces $\mathcal{U}_1$ and $\mathcal{U}_2$ are compact metric spaces and suppose that f is continuous in variables $u_1 \in \mathcal{U}_1$ and $u_2 \in \mathcal{U}_2$. The admissible feedback controls $\mathcal{A}_{1s}^t$ for the player I, over $[s,t] \subset [0,1]$, are measurable functions $u_1 : [s,t] \times \mathcal{C} \to \mathcal{U}_1$ such that for each τ, $s \leq \tau \leq t$, $u_1(\tau,.)$ is $\mathcal{F}_t$ measurable and for each $x \in \mathcal{C}$, $u_1(.,x)$ is Lebesgue measurable. The admissible feedback controls $\mathcal{A}_{2s}^t$ for the player II, over $[s,t] \subset [0,1]$, are measurable functions $u_2 : [s,t] \times \mathcal{C} \to \mathcal{U}_2$ with similar properties. Let $\mathcal{A}_i = \mathcal{A}_{i0}^1, i = 1,2$. For $u_i \in \mathcal{A}_{is}^t, i = 1,2$, write

$$f^{u_1,u_2}(\tau,x) = f(\tau,x,u_1(\tau,x),u_2(\tau,x)).$$

Then conditions on f ensure that

$$E[\exp \xi_s^t(f^{u_1,u_2}) \mid \mathcal{F}_s] = 1 a.s.P$$

where

$$\begin{aligned} \xi_s^t(f^{u_1,u_2}) &= \int_s^t \{\sigma^{-1}(\tau,x) f^{u_1,u_2}(\tau,x)\}' dB(\tau) \\ &\quad -1/2 \int_s^t \left|\sigma^{-1}(\tau,x) f^{u_1,u_2}(\tau,x)\right|^2 d\tau. \end{aligned}$$

For each $u_i \in \mathcal{A}_i$ a probability measure P_{u_1,u_2} is defined through

$$\frac{dP_{u_1,u_2}}{dP} = \exp \xi_0^1(f^{u_1,u_2}).$$

Then by Girsanov's theorem, [74], we have the following result.

Theorem 8.2.1 *Under the measure P_{u_1,u_2} the process $w^{u_1,u_2}(t)$ is a Brownian motion on Ω, where*

$$dw^{u_1,u_2}(t) = \sigma^{-1}(t,x)(dx(t) - f^{u_1,u_2}(t,x)dt).$$

Corresponding to controls $u_i \in \mathcal{A}_i, i = 1,2$ the expected total cost is

$$J(u_1,u_2) = E_{u_1,u_2}[g(x(1)) + \int_0^1 h^{u_1,u_2}(t,x)dt] \tag{8.2.3}$$

where h and g are real valued and bounded, $g(x(1))$ is $\mathcal{F}_1$ measurable and h satisfies the same conditions as the components of f. Also E_{u_1,u_2} denotes the expectation with respect to P_{u_1,u_2}. For a zero sum differential game, player I wishes to choose u_1 so that $J(u_1,u_2)$ is maximized and player II wishes to choose u_2 so that $J(u_1,u_2)$ is minimized.

Now the principle of optimality will be derived. Suppose that player II uses the control $u_2(t,x) \in \mathcal{A}_2$ through out the game. Then if player I uses the control $u_1(t,x) \in \mathcal{A}_1$, the cost incurred from time t onwards, given $\mathcal{F}_t$ is independent of the controls used up to time t and is given by

$$\psi_t^{u_1,u_2} = E_{u_1 u_2}[g(x(1)) + \int_t^1 h^{u_1,u_2}(s,x)ds \mid \mathcal{F}_t].$$

Because $L^1(\Omega)$ is a complete lattice, the suprenum

$$W_t^{u_2} = \bigvee_{u_1 \in \mathcal{A}_1} \psi_t^{u_1,u_2} \tag{8.2.4}$$

exists, and represents the best that player I can attain from t onwards, given that player II is using control u_2. Let $u_1(u_2)$ represent the response of player I to the control u_2 used by player II. Then we have

Theorem 8.2.2

(a) $u_1^*(u_2)$ *is the optimal reply to* u_2 *iff*

$$W_t^{u_2} + \int_0^t h^{u_1^*,u_2}(s)ds$$

is a martingale on $\left(\Omega, \Im_t, P_{u_1^*(u_2),u_2}\right)$.

(b) *In general, for* $u_1 \in \mathcal{A}_1$,

$$W_t^{u_2} + \int_0^t h^{u_1,u_2}(s)ds$$

is a super martingale on $(\Omega, \Im_t, P_{u_1,u_2})$.

From martingale representation results, one can see that u_1^* is the optimal reply for player I iff there is a predictable process $g_t^{u_2}$ such that

$$\int_0^1 |g_s^{u_2}|^2 \, ds < \infty a.s.$$

and

$$W_t^{u_2} + \int_0^t h^{u_1^*,u_2}(s)ds = W_0^{u_2} + \int_0^t g_s^{u_2} dw_s^{u_1^*(u_2),u_2}.$$

For any other $u_1 \in \mathcal{A}_1$ the supermartingale $W_t^{u_2} + \int_0^t h^{u_1,u_2}(s)ds$ has a unique Doob-Meyer decomposition as

$$W_0^{u_2} + M_t^{u_1,u_2} + A_t^{u_1,u_2}, \tag{8.2.5}$$

where $M_t^{u_1,u_2}$ is a martingale on $(\Omega, \Im_t, P_{u_1,u_2})$ and $A_t^{u_1,u_2}$ is a predictable decreasing process. From the representation (8.2.5),

$$\begin{aligned} W_t^{u_2} + \int_0^t h^{u_1^*,u_2}(s)ds = W_0^{u_2} &+ \int_0^t g^{u_2}\sigma^{-1}(dx_s - f_s^{u_1,u_2}ds) \\ &- \int_0^t [(g^{u_2}\sigma^{-1} f_s^{u_1^*(u_2),u_2} + h_s^{u_1^*(u_2),u_2}) \\ &- (g^{u_2}\sigma^{-1} f_s^{u_1,u_2} + h_s^{u_1^*(u_2),u_2})]ds. \end{aligned}$$

Again from Theorem 8.2.1, $dw_s^{u_1,u_2} = \sigma^{-1}(dx_s - f_s^{u_1,u_2}ds)$ is a Brownian motion on $(\Omega, \Im_t, P_{u_1,u_2})$ and hence the stochastic integral is a predictable process, so by uniqueness of the Doob-Meyer decomposition

$$M_t^{u_1,u_2} = \int_0^t g^{u_2} dw^{u_1,u_2}, \tag{8.2.6}$$

$$A_t^{u_1,u_2} = \int_0^t [(g^{u_2}\sigma^{-1} f_s^{u_1^*(u_2),u_2} + h_s^{u_1^*(u_2),u_2}) - (g^{u_2}\sigma^{-1} f_s^{u_1,u_2} + h_s^{u_1^*(u_2),u_2})]ds. \tag{8.2.7}$$

Since $A_t^{u_1,u_2}$ is decreasing one can obtain the following principle of optimality:

Theorem 8.2.3 *If $u_1^*(u_2)$ is the best reply for player I then, almost surely,*

$$g^{u_2}\sigma^{-1} f_s^{u_1^*(u_2),u_2} + h_s^{u_1^*(u_2),u_2} \geq g^{u_2}\sigma^{-1} f_s^{u_1,u_2} + h_s^{u_1^*(u_2),u_2} \tag{8.2.8}$$

That is, if the optimal reply for player I exists, it is obtained by maximizing the Hamiltonian

$$g^{u_2}\sigma^{-1} f_s^{u_1,u_2} + h_s^{u_1,u_2}. \tag{8.2.9}$$

We will establish existence of optimal control $u_1^*(u_2) \in \mathcal{A}_1$ for player I in reply to any control $u_2 \in \mathcal{A}_2$ used by player II. Now we will make the payoff (8.2.3) into a completely terminal payoff by introducing a new state variable x_{n+1} and a new Brownian motion B_{n+1} on a probability space (Ω', P'). Suppose x_{n+1} satisfies the equation

$$dx_{n+1} = h(t, x, u_1, u_2)\, dt + dB_{n+1} \tag{8.2.10}$$
$$x_{n+1}(0) = 0. \tag{8.2.11}$$

The $(n+1)$ dimensional process (x, x_{n+1}) is defined on the product space $(\Omega^+, P^+) = (\Omega \times \Omega', P \times P')$. If we write $x^+ = (x, x_{n+1})$, $f^+ = (f, h)$, $\sigma^+ = \begin{bmatrix} \sigma & 0 \\ 0 & 1 \end{bmatrix}$, and $w_{n+1} = B_{n+1}$, then $w^+ = (w, w_{n+1})$ is an $n+1$ dimensional Brownian motion on Ω^+.

Define a new probability measure P_{u_1,u_2}^+ on Ω^+ by putting

$$\frac{dP_{u_1,u_2}^+}{dP} = \exp \xi_0^1 \left(f_{u_1,u_2}^+\right)$$

Let E_{u_1,u_2}^+ denote the expectation with respect to P_{u_1,u_2}^+. Since w_{n+1} is a Brownian motion and h and g are independent of x_{n+1}, the expected payoff corresponding to the controls u_1 and u_2 is

$$E_{u_1,u_2}^+[g(x(1)) + x_{n+1}(1)] = E_{u_1,u_2}[g(x(1)) + \int_0^1 h(s, x, u_1, u_2)ds]. \tag{8.2.12}$$

Define

$$W_{u_2}^+(t) = \bigvee_{u_1 \in \mathcal{U}_1} E_{u_1,u_2}^+ \left[g(x(1)) + x_{n+1}(1) \mid \Im_t^+\right],$$

the supremum being in $L^1(\Omega^+)$. Let C^+ denote the $\mathbf{R}^{n+1}$ valued continuous function on $[0,1]$ and $\Im_t^+$ the σ-field on C^+ generated up to time t. Let $\Phi^+ = \{\phi : [0,1] \times C^+ \to \mathbf{R}^{n+1}\}$ which satisfy:

(i) for each $t \in [0,1]$, $\phi(t,.)$ is $\Im_t^+$ measurable,

(ii) for each $x \in C^+$, $\phi(., x)$ is Lebesgue measurable, and

(iii) $\left|(\sigma^+)^{-1}(t,x)\phi(t,x)\right| \leq M(1 + \|x\|_t)$ where $\|x\|_t = \sup_{0 \leq s \leq t} |x(s)|$.

Write $\mathcal{D} = \{\exp \xi_0^1(\phi) : \phi \in \Phi^+\}$. Because ϕ has linear growth $E^+ \exp \xi_0^1(\phi) = 1$ for all $\phi \in \Phi^+$, where E^+ denotes the expectation with respect to P^+. Since $\mathcal{D}$ is weakly compact, we have the following result.

Theorem 8.2.4 *There is a function $H \in \Phi^+$ such that $\left(W^+_{u_2}(t), \Im^+_t, P^*\right)$ is a martingale. Here P^* is defined on Ω^+ by*

$$\frac{dP^*}{dP^+} = \exp \xi_0^1(H). \tag{8.2.13}$$

If there is an optimal reply $u_1^*(u_2)$ for player I, take $H = f^+_{u_1^*(u_2),u_2}$.

This result states that, even if there not an optimal control, there is always a 'drift term' $H \in \Phi^+$ whose corresponding measure gives the maximum value function

$$\begin{aligned} W^+_{u_2}(t) &= \bigvee_{u_1 \in \mathcal{U}_1} E^+_{u_1,u_2}\left[g(x(1)) + x_{n+1}(1) \mid \Im^+_t\right] \\ &= E^*\left[g(x(1)) + x_{n+1}(1) \mid \Im^+_t\right] \end{aligned}$$

where E^* denotes expectation with respect to P^*.

Under P^*, using Girsanov's theorem, we are considering an $n+1$ dimensional Brownian motion w^* on (Ω^+, P^*) defined by

$$\begin{pmatrix} dw^* \\ dw^*_{n+1} \end{pmatrix} = \begin{pmatrix} \sigma^{-1} & 0 \\ 0 & 1 \end{pmatrix} \begin{pmatrix} dx - \widehat{H}dt \\ dx_{n+1} - H_{n+1}dt \end{pmatrix}.$$

where $\widehat{H}$ denotes the first n coordinates of H.

Since $h\left(t, x, u_1(t,x), u_2(t,x)\right)$ is independent of x_{n+1}, for any controls, the weak limit H_{n+1} is independent of x_{n+1}, so any control $u_1 \in U_1$:

$$\begin{aligned} &E^+_{u_1,u_2}\left[g(x(1)) + \int_0^1 h\left(s, x, u_1, u_2\right) ds + w_{n+1}(1) - w_{n+1}(t) \mid \Im^+_t\right] + x_{n+1}(t) \\ &= E_{u_1,u_2}\left[g(x(1)) + \int_t^1 h\left(s, x, u_1, u_2\right) ds + w_{n+1}(1) - w_{n+1}(t) \mid \Im_t\right] + x_{n+1}(t) \end{aligned}$$

Taking supremum to obtain $W^+_{u_2}$ we see

$$W^+_{u_2}(t) = W^{u_2}_t + \int_0^t H_{n+1}(s)ds + w^*_{n+1}(t).$$

Therefore

$$W^{u_2}_t + \int_0^t H_{n+1}(s)ds + w^*_{n+1}(t) = E^*[g(x(1)) + x_{n+1}(1) \mid \Im^+_t].$$

Taking expectation with respect to $\Im_t \subset \Im^+_t$ we have

$$W^{u_2}_t + \int_0^t H_{n+1}(s)ds = E^*\left[g(x(1)) + x_{n+1}(1) \mid \Im_t\right].$$

Hence, $W^{u_2}_t + \int_0^t H_{n+1}(s)ds$ is a martingale on $(\Omega, \Im_t, P^*)$, and so can be represented as a stochastic integral, $B^{u_2} + \int_0^t g^* dw^*$, with respect to n-dimensional Brownian motion w^* defined on $(\Omega, \Im_t, P^*)$ by

$$dw^* = \sigma^{-1}dx - \sigma^{-1}Hdt. \tag{8.2.14}$$

Here $B^{u_2} = W_0^{u_2}$ and g^* is a predictable process. Under any other control $u_1 \in U_1$, as in Theorem 2.2, $W_t^{u_2} + \int_0^t h_s^{u_1,u_2} ds$ is a supermartingale and hence

$$W_t^{u_2} + \int_0^t h_s^{u_1,u_2} ds$$
$$= B^{u_2} + \int_0^t g^* dw_s^{u_1,u_2} + \int_0^t (g^*\sigma^{-1} f_s^{u_1,u_2} + h_s^{u_1,u_2}) - (g^*\sigma^{-1}\widehat{H}_s + H_{n+1}(s))ds. \tag{8.2.15}$$

Since $w_s^{u_1,u_2}$ is a Brownian motion on (Ω, P_{u_1,u_2}) defined by

$$dw_s^{u_1,u_2} = \sigma^{-1}(dx_s - f_s^{u_1,u_2} ds),$$

the first integral on the right hand side of (8.2.15) is a stochastic integral and the second a decreasing process. Hence we have almost surely

$$g^*\sigma^{-1}\widehat{H} + H_{n+1} \geq g^*\sigma^{-1} f^{u_1,u_2} + h^{u_1,u_2} \tag{8.2.16}$$

If there is a process $u_1^*(u_2)$ such that, almost surely,

$$g^*\sigma^{-1}\widehat{H} + H_{n+1} = g^*\sigma^{-1} f^{u_1^*,u_2} + h^{u_1^*,u_2}$$

then

$$W_t^{u_2} + \int_0^t h_s^{u_1^*,u_2} ds = B^{u_2} + \int_0^t g^* dw^+_{u_1^*(u_2),u_2} \tag{8.2.17}$$

and so is a martingale. Therefore, $u_1^*(u_2)$ would be an optimal reply to u_2.

For the above process g^*, since f and h are continuous in the control variables u_1 and u_2 and the control spaces are compact, there is a measurable feedback control $u_1^*(u_2)$ such that almost surely

$$g^*.\sigma^{-1} f^{u_1^*(u_2),u_2} + h^{u_1^*(u_2),u_2} \geq g^*.\sigma^{-1} f^{u_1,u_2} + h^{u_1,u_2}. \tag{8.2.18}$$

We will now show that such a control $u_1^*(u_2)$ is an optimal reply for Player I.

Let

$$\Gamma_s(u_1, u_2) = g^*.\sigma^{-1} f_s^{u_1,u_2} + h_s^{u_1,u_2}$$

and

$$\widehat{\Gamma}_s = g^*.\sigma^{-1}\widehat{H}_s + H_{m+1}(s),$$

and let $u_1^*(u_2)$ is selected as in (8.2.18) so that $\Gamma_s(u_1^*, u_2) \geq \Gamma_s(u_1, u_2)$. Then

$$W_t^{u_2} + \int_0^t h_s^{u_1,u_2} ds = B^{u_2} + \int_0^t g^* dw^+_{u_1,u_2} + \int_0^t (\Gamma_s(u_1, u_2) - \widehat{\Gamma}_s) ds.$$

Taking expectations with respect to $\mu^+_{u_1,u_2}$ at $t = 1$:

$$E^+_{u_1,u_2}[g(x(1)) + \int_0^1 h_s^{u_1,u_2} ds] = B^{u_2} + E^+_{u_1,u_2}[\int_0^1 (\Gamma_s(u_1, u_2) - \widehat{\Gamma}_s) ds]$$
$$\leq B^{u_2} + E^+_{u_1,u_2}[\int_0^1 (\Gamma_s(u_1^*(u_2), u_2) - \widehat{\Gamma}_s) ds] \tag{8.2.19}$$

The left hand side of the inequality (8.2.19) is just $\psi_0^{u_1,u_2}$, so for any $n \in Z^+$ there is a control $u_{1n} \in U_1$ such that

$$-E^+_{u_{1n},u_2}\left[\int_0^1 \left(\Gamma_s\left(u_1^*\left(u_2\right), u_2\right) - \widehat{\Gamma}_s\right) ds\right] < 1/n.$$

Writing

$$-X = \int_0^1 (\Gamma_s(u_1^*(u_2), u_2) - \widehat{\Gamma}_s) ds$$

The inequality (8.2.16) implies X is positive almost surely, and $E^+\phi_n X \to 0$, where $\phi_n = \exp \xi_0^1(f^+_{u_{1n},u_2})$. Let $X^N = \min(N, X)$ for $N \in Z^+$, so $0 \le X^N \le X$ and $E^+\phi_N X^N \to 0$. By weak compactness of $\mathcal{D}$ there is a $\phi \in \mathcal{D}$ such that the $\phi_n$converge to ϕ weakly, so

$$\lim_{n\to\infty} E^+\phi_n X^N = E^+\phi X^N = 0.$$

Since $\phi > 0$ a.s., we have $X^N = 0$ a.s. Therefore $X = 0$ a.s., and hence

$$\Gamma_s(u_1^*(u_2), u_2) = \widehat{\Gamma}_s \text{ a.s.} \tag{8.2.20}$$

Therefore we conclude that an optimal reply $u_1^*(u_2)$ exists for player I in reply to any control $u_2 \in U_2$ used by player II.

We will now establish the existence, and obtain a characterization, of the optimal feedback control that player II should use if he chooses his control first. Assume that the player I will always play his best reply $u_1^*(u_2) \in U_1$ in response to any control $u_2 \in U_2$. Now the problem is how player II, who is trying to minimize the payoff (8.2.3), should choose a $u_2^* \in U_2$ such that

$$\inf_{u_2\in U_2} \sup_{u_1\in U_1} J(u_1, u_2) = \inf_{u_2\in U_2} J(u_1^*(u_2), u_2). \tag{8.2.21}$$

For any $u_2 \in U_2$ and $t \in [0, 1]$, if player I plays $u_1^*(u_2)$, the expected terminal payoff is

$$\psi_{u_2}(t) = E_{u_1^*(u_2),u_2}[g(x(1) + \int_0^1 h^{u_1^*(u_2),u_2} ds \mid \Im_t].$$

Since $L^1(\Omega)$ is a complete lattice the infimum

$$V_t^+ = \bigwedge_{u_2\in U_2} \psi_{u_2}(t) \tag{8.2.22}$$

exists in $L^1(\Omega)$. V_t^+ in (8.2.22) is called the {textitupper value function of the differential game, and

$$V_0^+ = \inf_{u_2\in U_2} \sup_{u_1\in U_1} J(u_1, u_2) \tag{8.2.23}$$

is the *upper value* of the game. One can obtain the following result [24].

Theorem 8.2.5

(a) *$u_2^* \in U_2$ is optimal for player II if and only if*

$$V_t^+ + \int_0^t h^{u_1^*(u_2^*),u_2^*} ds$$

is a martingale on $(\Omega, \mathcal{A}_t, P_{u_1^*(u_2^*),u_2^*})$.

(b) *In general, for $u_2 \in U_2$,*

$$V_t^+ + \int_0^t h^{u_1^*(u_2),u_2} ds$$

is a submartingale on $\left(\Omega, \mathcal{A}_t, P_{u_1^*(u_2),u_2}\right)$.

From the above martingale representation, $u_2^* \in U_2$ is optimal for player II playing first if and only if there is a predictable process g_t^* such that

$$\int_0^1 |g_s^*|^2 \, ds < \infty a.s.$$

and

$$V_t^+ + \int_0^t h^{u_1^*(u_2^*),u_2^*} ds = B^* + \int_0^t g^* dw_s^*.$$

Here the w^* is the Brownian motion given by

$$dw^* = \sigma^{-1}(dx - f^{u_1^*(u_2^*),u_2^*} ds)$$

on $\left(\Omega, P_{u_1^*(u_2^*),u_2^*}\right)$. For a general $u_2 \in U_2$ the submartingale

$$V_t^+ + \int_0^t h^{u_1^*(u_2),u_2} ds$$

has a unique Doob-Mayer decomposition $B^* + M_t^{u_2} + A_t^{u_2}$, where $M_t^{u_2}$ is a martingale on $\left(\Omega, P_{u_1^*(u_2),u_2}\right)$ and $A_t^{u_2}$ is a predictable increasing process. Also, if $u_2^* \in U_2$ is optimal for player II playing first, then almost surely

$$g^*.\sigma^{-1} f_s^{u_1^*(u_2^*),u_2^*} + h_s^{u_1^*(u_2^*),u_2^*} \leq g^*.\sigma^{-1} f_s^{u_1^*(u_2),u_2} + h_s^{u_1^*(u_2),u_2}. \tag{8.2.24}$$

Conversely, without a priori assuming there is an optimal control $u_2^* \in U_2$, one can obtain an integral representation for V_t^+, and show that the measurable strategy, obtained by minimizing a Hamiltonian $g^*.\sigma^{-1} f_s^{u_1^*(u_2),u_2} + h_s^{u_1^*(u_2),u_2}$, exists and is optimal.

Theorem 8.2.6 *There is a predictable process g^* and $u_2^* \in U_2$ is optimal if and only if u_2^* minimizes the Hamiltonian*

$$\Gamma_s(u_1^*(u_2), u_2) = g^*.\sigma^{-1} f_s^{u_1^*(u_2),u_2} + h_s^{u_1^*(u_2),u_2}, a.s.in(s,\omega). \tag{8.2.25}$$

The Isaacs condition

We have seen that

$$V_0^+ = \inf_{u_2 \in U_2} \sup_{u_1 \in U_1} J(u_1, u_2)$$

represents the best outcome that players I and II can ensure if player II chooses his feedback control first. Now we will define the *lower value* of the game,

$$V_0^- = \sup_{u_1 \in U_1} \inf_{u_2 \in U_2} J(u_1, u_2).$$

For $t \in [0,1]$, $x \in C$, $u_1 \in U_1$, $u_2 \in U_2$ and $p \in \mathbf{R}^n$ write

$$L(t, x, p; u_1, u_2) = p.\sigma^{-1}(t, x) f(t, x, u_1, u_2) + h(t, x, u_1, u_2). \tag{8.2.26}$$

The game is said to satisfy the *Isaacs condition* if, for all such t, x, p,

$$\min_{u_2 \in U_2} \max_{u_1 \in U_1} L(t, x, p; u_1, u_2) = \max_{u_1 \in U_1} \min_{u_2 \in U_2} L(t, x, p; u_1, u_2). \tag{8.2.27}$$

We say the game satisfies a *saddle-point condition* if the upper and lower values of an 'infinitesimal' game are equal, then $V_0^+ = V_0^-$. Next result states that the game has a value under Isaacs condition.

Theorem 8.2.7 *If the game satisfies the Isaacs condition then* $V_0^+ = V_0^-$.

Proof. Note that for $u_i \in U_i$, $i = 1, 2$

$$\Gamma_s(u_1, u_2) = L(s, x, g^*; u_1(t, x), u_2(t, x))$$

where g^* is the predictable process introduced earlier. Also, for any $u_2 \in U_2$ we proved that there exists a strategy $u_1^*(u_2) \in U_1$ such that

$$\Gamma_s(u_1^*(u_2), u_2) = \max_{u_1 \in U_1} \Gamma_s(u_1^*(u_2), u_2)$$

and then that there is a $u_2^* \in U_2$ such that

$$\begin{aligned}\Gamma_s(u_1^*(u_2^*), u_2^*) &= \min_{u_2 \in U_2} \Gamma_s(u_1^*(u_2), u_2) a.s. \\ &= \min_{u_2 \in U_2} \max_{u_1 \in U_1} \Gamma_s(u_1, u_2) a.s.\end{aligned}$$

We also had a representation

$$V_t^+ + \int_0^t h^{u_1^*(u_2^*), u_2^*} ds = B^* + \int_0^t g^* dw_s^* a.s.$$

Because f and u_1 are continuous in u_1 and u_2 and U_1 and U_2 are compact, for any $u_1 \in U_1$ there exists a strategy $u_2^*(u_1) \in U_2$ such that

$$\Gamma_s(u_1, u_2^*(u_1)) = \min_{u_2 \in U_2} \Gamma_s(u_1, u_2) a.s.$$

Similarly there is a $u_1^* \in U_1$ such that

$$\begin{aligned}\Gamma_s(u_1^*, u_2^*(u_1^*)) &= \max_{u_1 \in U_1} \Gamma_s(u_1, u_2^*(u_1)) a.s. \\ &= \max_{u_1 \in U_1} \min_{u_2 \in U_2} \Gamma_s(u_1, u_2) a.s.\end{aligned}$$

Since the Isaacs condition (8.2.27) holds

$$\Gamma_s(u_1^*, u_2^*(u_1^*)) = \Gamma_s(u_1^*(u_2^*), u_2^*) a.s.$$

Now for any $u_2 \in U_2$

$$\Gamma_s(u_1^*, u_2^*(u_1^*)) \leq \Gamma_s(u_1^*, u_2) a.s.$$

and for any $u_1 \in U_1$

$$\Gamma_s(u_1, u_2^*) \leq \Gamma_s(u_1^*(u_2^*), u_2^*) a.s.$$

hence

$$\Gamma_s(u_1, u_2^*) \leq \Gamma_s(u_1^*, u_2^*) \leq \Gamma_s(u_1^*, u_2) a.s.$$

Therefore

$$V_t^+ + \int_0^t h^{u_1^*, u_2^*} ds = B^* + \int_0^t g^* dw_s^{u_1^*, u_2^*} a.s.$$

where

$$dw_s^{u_1^*, u_2^*} = \sigma^{-1}(dx_s - f_s^{u_1^*, u_2^*} ds)$$

is a Brownian motion under $P_{u_1^*, u_2^*}$. For any other $u_1 \in U_1$:

$$V_t^+ + \int_0^t h^{u_1, u_2^*} ds = B^* + \int_0^t g^* dw_s^{u_1, u_2^*} + \int_0^t (\Gamma_s(u_1, u_2^*) - \Gamma_s(u_1^*, u_2^*)) ds.$$

Taking expectations at $t = 1$ with respect to P_{u_1, u_2^*}

$$E_{u_1, u_2^*}[g(x(1)) + \int_0^1 h_s^{u_1, u_2^*} ds] = J(u_1, u_2^*) \leq J^* = J(u_1^*, u_2^*).$$

Similarly one can show that

$$J(u_1^*, u_2^*) \leq J(u_1^*, u_2).$$

Therefore, if Isaacs condition is satisfied

$$\sup_{u_1 \in U_1} \inf_{u_2 \in U_2} J(u_1, u_2) = \inf_{u_2 \in U_2} \sup_{u_1 \in U_1} J(u_1, u_2) = J^*,$$

hence the upper and lower value of the differential game are equal. One can also show that if the upper and lower values are equal then

$$\max_{u_1 \in U_1} \min_{u_2 \in U_2} L(t, x, g^*; u_1, u_2) = \min_{u_2 \in U_2} \max_{u_1 \in U_1} L(t, x, g^*; u_1, u_2) \, a.s. \quad \square$$

□

In this subsection, using the martingale methods we have proved the existence of value for the game under the Isaacs condition as well as characterized the optimal strategies.

8.2.2 Two person zero-sum games and viscosity solutions

In this subsection, we present briefly some key elements of the viscosity solutions method for the theory of two person zero-sum stochastic differential games. For more details we refer to [35] and [34]. For $s \in [t, T]$, consider the dynamics

$$dx_s = f(x_s, s, u_{1s}, u_{2s}) \, ds + \sigma(x_s, s, u_{1s}, u_{2s}) \, dw_s \tag{8.2.28}$$

with initial condition

$$x_t = x\,(x \in \mathbf{R}^n), \tag{8.2.29}$$

where w is a standard m-dimensional Brownian motion. The payoff is given by

$$J_{x,t}(u_1, u_2) = E_{x,t}\left\{\int_t^T h(x_s, s, u_{1s}, u_{2s})ds + g(x_T)\right\}, \tag{8.2.30}$$

Here u_1 and u_2 are stochastic processes taking values in the given compact sets $U_1 \subset \mathbf{R}^k$ and $U_2 \subset \mathbf{R}^l$. Assume that $f : \mathbf{R}^n \times [0,T] \times U_1 \times U_2 \to \mathbf{R}^n$ is uniformly continuous and satisfies, for some constant C_1 and all t, $\widehat{t} \in [0,T]$, x, $\widehat{x} \in \mathbf{R}^n$, $u_i \in U_i$, $i = 1, 2$,

$$\begin{cases} |f(x,t,u_1,u_2)| \le C_1, \\ \left|f(x,t,u_1,u_2) - f(\widehat{x},\widehat{t},u_1,u_2)\right| \le C_1\left(|x-\widehat{x}| + |t-\widehat{t}|\right). \end{cases} \tag{8.2.31}$$

$h : \mathbf{R}^n \times [0,T] \times U_1 \times U_2 \to \mathbf{R}$ is uniformly continuous and satisfies, for some constant C_2,

$$\begin{cases} |h(x,t,u_1,u_2)| \le C_2, \\ \left|h(x,t,u_1,u_2) - h(\widehat{x},\widehat{t},u_1,u_2)\right| \le C_2\left(|x-\widehat{x}| + |t-\widehat{t}|\right). \end{cases} \tag{8.2.32}$$

and $g : \mathbf{R}^n \to \mathbf{R}^n$ satisfies

$$\begin{cases} |g(x)| \le C_3, \\ |g(x) - g(\widehat{x})| \le C_3\left(|x-\widehat{x}|\right). \end{cases} \tag{8.2.33}$$

Also the $n \times m$ matrix σ is bounded uniformly continuous and Lipschitz continuous with respect to x. On a probability space $(\Omega, \Im, P)$, set

$$U_i(t) \equiv \{u_i : [t,T] \to U_i \text{ measurable}\}, \quad i = 1, 2.$$

These are the sets of all controls for players I and II. We consider the controls that agree a.e. are the same.

Define any mapping

$$\alpha : U_2(t) \to U_1(t)$$

to be a *strategy* for I (beginning at time t) provided for each $s \in [t,T]$ and u_2, $\widehat{u}_2 \in U_2(t)$

$$if u_2 = \widehat{u}_2 \text{ a.e. in } [t,s], then \alpha[u_2] = \alpha[\widehat{u}_2] a.s.in[t,s]. \tag{8.2.34}$$

Similarly a mapping

$$\beta : U_1(t) \to U_2(t)$$

is a *strategy* for player II provided for each $s \in [t,T]$ and u_1, $\widehat{u}_1 \in U_1(t)$

$$if u_1 = \widehat{u}_1 \text{ a.e. in } [t,s], then \beta[u_1] = \beta[\widehat{u}_1] a.e.in[t,s]. \tag{8.2.35}$$

Denote by $\Gamma_i(t)$, $i = 1, 2$, the set of all strategies for players I and II, respectively, beginning at time t. At this point we note that there are some serious measurability problems that need to be addressed in the characterization of strategies for stochastic games. For a detailed account on the concept of measurability in the stochastic case and how to overcome this difficulty, we refer to [34]. Define the lower and upper values V and U by

$$V(x,t) = \inf_{\beta \in \Gamma_2(t)} \sup_{u_1 \in U_1(t)} J_{x,t}(u_1, \beta[u_1]) \tag{8.2.36}$$

and

$$U(t,x) = \sup_{\alpha\in\Gamma_1(t)} \inf_{u_2\in U_2(t)} J_{x,t}(\alpha(u_2),u_2) \tag{8.2.37}$$

The U and V satisfy the dynamic programming principle which for simplicity is stated with $h \equiv 0$. The proof of this result rests on the results about uniqueness of viscosity solutions to fully nonlinear second-order PDE as well as some appropriate discretization of the game in time but not in space and we refer the reader to [34].

Theorem 8.2.8 *Let $t,\tau \in [0,T]$ be such that $t \le \tau$. For every $x \in \mathbf{R}^n$*

$$V(x,t) = \inf_{\beta\in\Gamma_2(t)} \sup_{u_1\in U_1(t)} E_{x,t}\{V(x_\tau,\tau)\}, \tag{8.2.38}$$

and

$$U(x,t) = \sup_{\alpha\in\Gamma_1(t)} \inf_{u_2\in U_2(t)} E_{x,t}\{U(x_\tau,\tau)\}. \tag{8.2.39}$$

With this result, one can study the connections between U and V and the associated Bellman-Isaacs equations which are of the form

$$\begin{cases} y_t + H(D^2y, Dy, x, t) = 0 \text{ in } \mathbf{R}^n \times [0,T], \\ y = g \text{ on } \mathbf{R}^n \times \{T\}, \end{cases} \tag{8.2.40}$$

with

$$\begin{aligned} H(A,p,x,t) &= H^-(A,p,x,t) \\ &= \max_{u_1\in U_1} \min_{u_2\in U_2} [\frac{1}{2}tr(a(x,t,u_1,u_2)A + f(x,t,u_1u_2).p + h(x,t,u_1,u_2)] \end{aligned} \tag{8.2.41}$$

and

$$\begin{aligned} H(A,p,x,t) &= H^+(A,p,x,t) \\ &= \min_{u_2\in U_2} \max_{u_1\in U_1} [\frac{1}{2}tr(a(x,t,u_1,u_2)A + f(x,t,u_1u_2).p + h(x,t,u_1,u_2)] \end{aligned} \tag{8.2.42}$$

where

$$a = \sigma\sigma^T.$$

We will now give the definition of viscosity solution for (8.2.40) and a comparison principle.

Definition 8.2.9 *A continuous function $y : \mathbf{R}^n \times [0,T] \to \mathbf{R}$ is a viscosity solution (resp. supersolution) of (8.2.40) if*

$$y \le g \textit{ on } \mathbf{R}^n \times \{T\}, \tag{8.2.43}$$

(resp.

$$y \ge g \textit{ on } \mathbf{R}^n \times \{T\}), \tag{8.2.44}$$

and

$$\phi_t(x,t) + H(D^2\phi(x,t), D\phi(x,t), x, t) \ge 0, \tag{8.2.45}$$

(resp.

$$\phi_t(x,t) + H(D^2\phi(x,t), D\phi(x,t), x, t) \le 0), \tag{8.2.46}$$

for every smooth function ϕ and any local maximum (resp. minimum) (x,t) of $y - \phi$.

Following result is obtained in [57].

Theorem 8.2.10 *Assume that the functions f, g, h, and σ are bounded and Lipschitz continuous. If z and $\tilde{z}$ (resp. y and $\tilde{y}$) are viscosity subsolution and supersolution of* (8.2.40) *with H given by* (8.2.41) *(resp. of* (8.2.40) *with H given by* (8.2.42)) *with terminal data g and $\tilde{g}$ and if $g \leq \tilde{g}$ on $\mathbf{R}^n \times \{T\}$, then $z \leq \tilde{z}$ (resp., $y \leq \tilde{y}$) on $\mathbf{R}^n \times [0,T]$.*

Following is the main result for the zero-sum stochastic differential game problem with two players which is stated with out proof. The proof is given in [34] which is tedious and involve several approximation procedures.

Theorem 8.2.11

(i) *The lower value V is the unique viscosity solution of* (8.2.40) *with H as in* (8.2.41).

(ii) *The upper value U is the unique viscosity solution of* (8.2.40) *with H as in* (8.2.42).

For the dynamics in (8.2.28) with initial time $t = 0$, and for a discounted payoff

$$J(u_1, u_2) = E\left\{\int_0^\infty e^{-\lambda s} h\left(x(s), u_1(s), u_2(s)\right) ds\right\}, \tag{8.2.47}$$

the existence of value function is obtained by [106] using a different approach. The so-called sub- and super-optimality inequalities of dynamic programming are used in the proofs. In this approach to the existence of value functions, one starts with solutions of the upper and lower Bellman-Isaacs equations which exist by the general theory and then prove that they must satisfy certain optimality inequalities which in turn yield solutions that are equal to the value functions.

8.2.3 Stochastic differential games with multiple modes

In [28], two person stochastic differential games with multiple modes are studied. The state of the system at time t is given by a pair $(x(t), \theta(t)) \in \mathbf{R}^n \times S$, where $S = \{1, 2, \ldots, N\}$. The discrete component $\theta(t)$ describes the various modes of the system. The continuous component $x(t)$ is governed by a "controlled diffusion process" with drift vector which depends on the discrete component $\theta(t)$. Thus $x(t)$ switches from one diffusion path to another at random times as the mode $\theta(t)$ changes. The discrete component $\theta(t)$ is a "controlled Markov chain" with transition rate matrix depending on the continuous component. The evolution of the process $(x(t), \theta(t))$ is given by the following equations

$$dx(t) = b(x(t), \theta(t), u_1(t), u_2(t))dt + \sigma(x(t), \theta(t))dw(t), \tag{8.2.48}$$

$$P(\theta(t + \delta t) = j \mid \theta(t) = i, x(s), \theta(s), s \leq t) = \lambda_{ij}(x(t))\delta t + o(\delta t), i \neq j, \tag{8.2.49}$$

for $t \geq 0$, $x(0) = x \in \mathbf{R}^n$, $\theta(0) = i \in S$, where b, σ, λ are suitable functions. In a zero sum game player I is trying to maximize and player II is trying to minimize the expected payoff

$$J_{x,i}(u_1, u_2) = E_{x,i}\left[\int_0^\infty e^{-\alpha t} r(x(t), \theta(t), u_1(t), u_2(t))dt\right] \tag{8.2.50}$$

over their respective admissible strategies, where $\alpha > 0$ is the discount factor and $r : \mathbf{R}^n \times S \times U_1 \times U_2 \to \mathbf{R}$ is the payoff function and is defined by

$$r(x, i, u_1, u_2) = \int_{V_2}\int_{V_1} \bar{r}(x, i, v_1, v_2)\, u_1(dv_1)\, u_2(dv_2).$$

Here V_l, $l = 1, 2$ are compact metric spaces and $U_l = \mathcal{P}(V_l)$ the space of probability measures on V_l endowed with the topology of weak convergence and $\bar{r} : \mathbf{R}^n \times S \times V_1 \times V_2 \to \mathbf{R}$. Also let

$$\bar{b} : \mathbf{R}^n \times S \times V_1 \times V_2 \to \mathbf{R}^n$$

$$\sigma : \mathbf{R}^n \times S \to \mathbf{R}^{n\times n}$$

$$\lambda_{ij} : \mathbf{R}^n \to \mathbf{R}, 1 \le i, j \le N,\ \lambda_{ij} \ge 0,\ i \ne j,\ \sum_{j=1}^{N} \lambda_{ij} = 0.$$

The following assumption is made.

(A2.1)

(i) For each $i \in S$, $\bar{b}(., i, ., .)$, $\bar{r}(., i, ., .)$ is bounded, continuous and Lipschitz in its first argument uniformly with respect to the rest.

(ii) For each $i \in S$, $\sigma(., i)$ is bounded and Lipschitz with the least eigenvalue of $\sigma\sigma'(., i)$ uniformly bounded away from zero.

(iii) For $i, j \in S$, $\lambda_{ij}(.)$ is bounded and Lipschitz continuous.

Define

$$b_k(x, i, u_1, u_2) = \int_{V_1}\int_{V_2} \bar{b}_k(x, i, v_1, v_2)\, u_1(dv_1)\, u_2(dv_2), k = 1, \ldots, n$$

and

$$b(x, i, u_1, u_2) = [b_1(x, i, u_1, u_2), \ldots, b_n(x, i, u_1, u_2)]'.$$

If $u_l(.) = v_l(x(.), \theta(.))$ for a measurable $v_l : \mathbf{R}^n \times S \to U_l$, then $u_l(.)$ is called a Markov strategy for the lth player. Let M_l denote the set of Markov strategies for player l. A strategy $u_l(.)$ is called pure if u_l is a Dirac measure, i.e., $u_l(.) = \delta_{v_l}(.)$, where $v_l(.)$ is a V_l valued nonanticipative process. For $p \ge 1$ define

$$W^{2,p}_{loc}(\mathbf{R}^n \times S) = \{f : \mathbf{R}^n \times S \to \mathbf{R} : foreach i \in S,\ f(., i) \in W^{2,p}_{loc}(\mathbf{R}^n)\}.$$

$W^{2,p}_{loc}(\mathbf{R}^n \times S)$ is endowed with the product topology of $\left(W^{2,p}_{loc}(\mathbf{R}^n)\right)^N$. For $f \in W^{2,p}_{loc}(\mathbf{R}^n \times S)$ write

$$L^{v_1,v_2} f(x, i) = L^{v_1,v_2}_i f(x, i) + \sum_{j=1}^{N} \lambda_{ij} f(x, j) \tag{8.2.51}$$

where

$$L^{v_1,v_2}_i f(x, i) = \sum_{j=1}^{n} \bar{b}_j(x, i, v_1, v_2) \frac{\partial f(x, i)}{\partial x_j} + \frac{1}{2} \sum_{j,k=1}^{n} a_{jk}(x, i) \frac{\partial^2 f(x, i)}{\partial x_j \partial x_k}. \tag{8.2.52}$$

Here $a_{jk}(x,i) = \sum_{l=1}^{n} \sigma_{jl}(x,i)\sigma_{kl}(x,i)$. Define

$$L^{u_1,u_2} f(x,i) = \int_{V_1}\int_{V_2} L^{v_1,v_2} f(x,i) u_1(dv_1)\, u_2(dv_2). \tag{8.2.53}$$

The Isaacs equation for this problem is given by

$$\begin{aligned}\inf_{u_2\in U_2}\sup_{u_1\in U_1} &\left[L^{u_1,u_2}\phi(x,i) + r(x,i,u_1,u_2)\right] \\ &= \sup_{u_1\in U_1}\inf_{u_2\in U_2}\left[L^{u_1,u_2}\phi(x,i) + r(x,i,u_1,u_2)\right] \\ &= \alpha\phi(x,i).\end{aligned} \tag{8.2.54}$$

This is a quasilinear system of uniformly elliptic equations with weak coupling in the sense that the coupling occurs only in the zeroth order term. Now we will state the following results from [28] and for the proofs, we refer to [28].

Theorem 8.2.12 . *Under* (A2.1) *the equation* (8.2.54) *has a unique solution in* $C^2(\mathbf{R}^n \times S) \cap C_b(\mathbf{R}^n \times S)$.

Next result characterizes the optimal Markov strategies for both players.

Theorem 8.2.13 *Assume* (A2.1). *Let* $u_1^* \in M_1$ *be such that*

$$\begin{aligned}\inf_{u_2\in U_2} &\left[\sum_{j=1}^{n} b_j(x,i,u_1^*(x,i),u_2)\frac{\partial V(x,i)}{\partial x_j}\right. \\ &\left. + \sum_{j=1}^{N}\lambda_{ij}(x)V(x,j) + r(x,i,u_1^*(x,i),u_2)\right] \\ &= \sup_{u_1\in U_1}\inf_{u_2\in U_2}\left[\sum_{j=1}^{n} b_j(x,i,u_1,u_2)\frac{\partial V(x,i)}{\partial x_j}\right. \\ &\left. + \sum_{j=1}^{N}\lambda_{ij}(x)V(x,j) + r(x,i,u_1,u_2)\right]\end{aligned} \tag{8.2.55}$$

for each i and a.e. x. Then u_1^* is optimal for player I. Similarly, let $u_2^* \in M_2$ be such that

$$\begin{aligned}\sup_{u_1\in U_1} &\left[\sum_{j=1}^{n} b_j(x,i,u_1,u_2^*(x,i))\frac{\partial V(x,i)}{\partial x_j}\right. \\ &\left. + \sum_{j=1}^{N}\lambda_{ij}(x)V(x,j) + r(x,i,u_1,u_2^*(x,i))\right] \\ &= \inf_{u_2\in U_2}\sup_{u_1\in U_1}\left[\sum_{j=1}^{n} b_j(x,i,u_1,u_2)\frac{\partial V(x,i)}{\partial x_j}\right. \\ &\left. + \sum_{j=1}^{N}\lambda_{ij}(x)V(x,j) + r(x,i,u_1,u_2)\right]\end{aligned} \tag{8.2.56}$$

for each i and a.e. x. Then u_2^* is optimal for player II.

This kind of game typically occurs in a pursuit-evasion problem where an interceptor tries to destroy a specific target. Due to swift maneuvering of the evader and the corresponding response by the interceptor the trajectories keep switching rapidly.

In [43], the problem of the numerical solution of the nonlinear partial differential equation associated with the game is considered. In general, due to the nonlinearities and to the nonellipticity or nonparabolicity of these equations, the available theory is not very helpful in choosing finite difference approximations, guaranteeing the convergence of the iterative procedures, or providing an interpretation of the approximation. For a specific problem, a finite difference scheme is given in [43] so that the convergence of the iterative process is guaranteed. With the development of weak convergence theory for game problems, [84], and the numerical methods described in [65], it is possible to develop computational methods for stochastic differential games.

8.3 N-Person stochastic differential games

Now we will deal with the stochastic differential game problem where N players are simultaneously controlling the evolution of a system. The approach that we are going to use in this section is based on occupation measures as described in [18]. In this framework the game problem is viewed as a multidecision optimization problem on the set of canonically induced probability measures on the trajectory space by the joint state and action processes. Each of the payoff criteria, such as discounted on the infinite horizon, limiting average, payoff up to an exit time etc., are associated with the concept of an occupation measure so that the total payoff becomes the integral of some function with respect to this measure. Then the differential game problem reduces to a static game problem on the set of occupation measures, the dynamics of the game being captured in these measures. This set is shown to be compact and convex. A fixed point theorem for point-to-set mapping is used to show the existence of equilibrium in the sense of Nash.

Let V_i, $i = 1, 2, \ldots, N$ be compact metric spaces and $U_i = \mathcal{P}(V_i)$ be the space of probability measures on V_i with Prohorov topology. Let $V = V_1 \times V_2 \times \cdots \times V_N$ and $U = U_1 \times U_2 \times \cdots \times U_N$. Let

$$\overline{m}(.,.) = [\overline{m}_1(.,.), \ldots, \overline{m}_d(.,.)]^T : \mathbf{R}^d \times V \to \mathbf{R}$$
$$\sigma = [[\sigma_{ij}(.)]], 1 \le i, j \le d, : \mathbf{R}^d \to \mathbf{R}^{d\times d}$$

be bounded continuous maps such that $\overline{m}$ is Lipschitz in its first argument uniformly with respect to the rest and σ is Lipschitz with the least eigenvalue of $\sigma\sigma^T(.)$ uniformly bounded away from zero. Define, for $x \in \mathbf{R}^d$, $u = (u_1, \ldots, u_N) \in U$,

$$m(.,.) = [m_1(.,.), \ldots, m_d(.,.)]^T : \mathbf{R}^d \times U \to \mathbf{R}^d$$

by

$$m_i(x,u) = \int\limits_{V_N} \cdots \int\limits_{V_1} \overline{m}_i(x, y_1, \ldots, y_N) u_1(dy_1) \ldots u_N(dy_N)$$
$$\doteq \int\limits_{V} \overline{m}_i(x,y) u(dy)$$

where $y \in V$. Let $x(.)$ be an $\mathbf{R}^d$-valued process given by the following controlled stochastic differential equation of Ito type

$$dx(t) = m(x(t), u(t))dt + \sigma(x(t))dw(t), t \ge 0, \quad (8.3.57)$$
$$x(0) = x_0, \quad (8.3.58)$$

where,

(i) x_0 is a prescribed random variable,

(ii) $w(.) = [w_1(.), \ldots, w_d(.)]^T$ is a standard Wiener process independent of x_0,

(iii) $u(.) = (u_1(.), \ldots, u_N(.))$, where $u_i(.)$ is a U_i-valued process satisfying: for $t_1 \geq t_2 \geq t_3$, $w(t_1) - w(t_2)$ is independent of $u(t)$, $t \leq t_3$.

Such a process $u_i(.)$ will be called an *admissible strategy* for the ith player. If $u_i(.) = v_i(x(.))$ for a measurable $v_i : \mathbf{R}^d \to U_i$, then $u_i(.)$ is called a *Markov strategy* for the ith player. A strategy $u_i(.)$ is called pure if u_i is a Dirac measure, i.e., $u_i(.) = \delta_{y_i(.)}$ where $y_i(.)$ is a V_i-valued process. If for each $i = 1, \ldots, N$, $u_i(.) = v_i(x(.))$ for some measurable $v_i : \mathbf{R}^d \to U_i$, then (8.3.57) admits a unique strong solution which is Feller process [113]. Let A_i, M_i, $i = 1, 2, \ldots, N$ denote the set of arbitrary admissible, resp. Markov strategies for the ith player. An N-tuple of Markov strategies $v = (v_1, \ldots, v_N) \in M$ is called *stable* if the corresponding process is positive recurrent and thus has a unique invariant measure $\eta(v)$. For any $f \in W^{2,p}_{loc}(\mathbf{R}^d)$, $p \geq 2$, $x \in \mathbf{R}^d$, $u \in V$, let

$$(Lf)(x, u) = \frac{1}{2} \sum_{i,j,k=1}^{d} \sigma_{ik}(x)\sigma_{jk}(x) \frac{\partial^2 f(x)}{\partial x_i \partial x_j} + \sum_{i=1}^{d} \overline{m}_i(x, u) \frac{\partial f(x)}{\partial x_i} \tag{8.3.59}$$

and for any $v \in U$

$$(L_v f)(x) = \int_{V_N} \cdots \int_{V_1} (Lf)(x, y) v_1(x)(dy_1) \ldots v_N(x)(dy_N). \tag{8.3.60}$$

For an N-tuple $y = (y_1, \ldots, y_N)$, denote $y^{\widehat{k}} = (y_1, \ldots, y_{k-1}, y_{k+1}, \ldots, y_N)$ and $\left(y^{\widehat{k}}, \widetilde{y}^k\right) = (y_1, \ldots, y_{k-1}, \widetilde{y}_k, y_{k+1}, \ldots, y_N)$.

For each $k = 1, \ldots, N$, let $\overline{r}_k : \mathbf{R}^d \times V \to \mathbf{R}$ be bounded continuous functions. When the state is x and actions $v \in V$ are chosen by the players then the player k receives a payoff $\overline{r}_k(x, v)$. For $x \in \mathbf{R}^d$, $u \in U$, let $r_k : \mathbf{R}^d \times U \to \mathbf{R}$ be defined by

$$r_k(x, u) = \int_{V_N} \cdots \int_{V_1} \overline{r}_k(x, y_1, \ldots, y_n) u_1(dy_1) \ldots u_N(dy_N) \tag{8.3.61}$$

Each player wants to maximize his accumulated income. We will now consider two evaluation criteria: discounted payoff on the infinite horizon, and ergodic payoff.

8.3.1 Discounted payoff on the infinite horizon

Let $\lambda > 0$ be the discount factor and let $u \in A = A_1 \times \cdots \times A_N$. Let $x(.)$ be the solution of (8.3.57) corresponding to u. The discounted payoff to player k for initial condition $x \in \mathbf{R}^d$ is defined by

$$R^k_\lambda[u](x) = E_u[\int_0^\infty e^{-\lambda t} r_k(x(t), u(t)) dt \mid x(0) = x]. \tag{8.3.62}$$

For an initial law $\pi \in \mathcal{P}(\mathbf{R}^d)$ the payoff is

$$R^k_\lambda[u](\pi) = \int_{\mathbf{R}^d} R^k_\lambda[u](x) \pi(dx). \tag{8.3.63}$$

An N-tuple of strategies $u^* = (u_1^*, \dots, u_N^*) \in A_1 \times \cdots \times A_N$ is said to be a discounted equilibrium (in the sense of Nash) for initial law π if for any $k = 1, \dots, N$,

$$R_\lambda^k[u^*](\pi) \geq R_\lambda^k[u^{*\widehat{k}}, u_k](\pi) \tag{8.3.64}$$

for any $u_k \in A_k$. The existence of a discounted equilibrium will be shown later.

8.3.2 Ergodic payoff

Let $u \in A$ and let $x(.)$ be the corresponding process with initial law π. The ergodic payoff to player k is given by

$$C_k[u](\pi) = \liminf_{T \to \infty} \frac{1}{T} E_u[\int_0^T r_k(x(t), u(t))dt] \tag{8.3.65}$$

The concept of equilibrium for the ergodic criterion is defined similarly. Under a Lyapunov stability condition (assumption (A3.1) introduced later) all $v \in M$ will be stable. For such a v, (8.3.65) is equal to

$$\rho_k[v] = \int_{\mathbf{R}^d} r_k(x, v(x))\eta[v](dx) \tag{8.3.66}$$

where $\eta[v] \in \mathcal{P}(\mathbf{R}^d)$ is the invariant measure of the process $x(.)$ governed by v. It will be shown that there exists a $v^* \in M$ such that for any $k = 1, \dots, N$

$$\rho_k[v^*] \geq \rho_k[v^{*\widehat{k}}, v_k]$$

for any $v_k \in M_k$. Thus v^* will be an ergodic equilibrium. Now we will explain the concept of occupation measures.

Occupation measures

Let

$$M_k = \{v : \mathbf{R}^d \to U_k \mid v \text{ measurable}\}, \quad k = 1, 2, \dots, N.$$

For $n \geq 1$, let Λ_n be the cube of side $2n$ in $\mathbf{R}^d$ with sides parallel to the axes and center at zero. Let B_n denote the closed unit ball of $L_\infty(\Lambda_n)$ with the topology obtained by relativizing to it the weak topology of $L_2(\Lambda_n)$. Then B_n is compact and metrizable, for example by the metric

$$d_n(f, g) = \sum_{m=1}^{\infty} 2^{-m} \left| \int_{\Lambda_n} f e_m dx - \int_{\Lambda_n} g e_m dx \right|$$

where $\{e_m\}$ is an orthonormal basis of $L_2(\Lambda_n)$. Let $\{f_i\}$ be a countable dense subset of the unit ball of $C(V_k)$. Then $\{f_i\}$ separates points of U_k. For each $v \in M_k$, define $g_{v_i} : \mathbf{R}^d \to \mathbf{R}$ by

$$g_{v_i}(x) = \int_{V_k} f_i dv(x), i \geq 1,$$

and $g_{v_i n}(.)$ denote the restriction of $g_{v_i}(.)$ to Λ_n for each n. Define a pseudometric $d_k(.,.)$ on M_k by

$$d_k(v, u) = \sum_{i,n=1}^{\infty} 2^{-(n+1)} \frac{d_n(g_{v_i n}, g_{u_i n})}{[1 + d_n(g_{v_i n}, g_{u_i n})]}$$

Replacing M_k by its quotient with respect to a.e. equivalence, $d_k(.,.)$ becomes a metric. The following is from [19]

Theorem 8.3.1 *M_k is compact under the metric topology of $d_k(.,.)$. Let $f \in L_2(\mathbf{R}^d)$, $g \in C_b\left(\mathbf{R}^d \times V_k\right)$ and $v_n \to v$ in M_k. Then*

$$\int_{\mathbf{R}^d} f(x) \int_{V_k} g(x,.)dv_n dx \to \int_{\mathbf{R}^d} f(x) \int_{V_k} g(x,.)dvdx$$

Conversely, if above holds for all such f, g then $v_n \to v$ in M_k.

Endow M with the product topology of M_k. Let $v \in M$ and $x(.)$ be the process governed by v with a fixed initial law. Let $L(v)$ denote the law of $x(.)$.

Theorem 8.3.2 *The map $v \to L(v) : M \to \mathcal{P}\left(C[0,\infty); \mathbf{R}^d\right)$ is componentwise continuous, i.e., for each $k = 1, 2, \ldots, N$, if $v_k^n \to v_k^\infty$ in M_k, and $v_i \in M_i$, $i \neq k$, then $L\left(v^{\widehat{k}}, v_k^n\right) \to L\left(v^{\widehat{k}}, v_k^\infty\right)$ in $\mathcal{P}(C[0,\infty); \mathbf{R}^d)$.*

Now we will introduce occupation measures for both discounted and ergodic payoff criterion. First consider the discounted case. Let $u \in A$ and $x(.)$ be the corresponding process. The *discounted occupation measure* for initial condition $x \in \mathbf{R}^d$ denoted by $\nu_{\lambda x}[u] \in \mathcal{P}\left(\mathbf{R}^d \times V\right)$ is defined by

$$\int_{\mathbf{R}^d \times V} f d\nu_{\lambda x}[u]$$
$$= \lambda^{-1} E_u\Big[\int_0^\infty \int_{V_N} \cdots \int_{V_1} e^{-\lambda t} f(x(t), y_1, \ldots, y_N) u_1(t)(dy_1) \ldots u_N(t)(dy_N) dt \mid x_0 = x\Big] \tag{8.3.67}$$

for $f \in C_b(\mathbf{R}^d \times V)$ and for an initial law $\pi \in \mathcal{P}\left(\mathbf{R}^d\right)$, $\nu_{\lambda\pi}[u]$ is defined by

$$\int f d\nu_{\lambda\pi}[u] = \int_{\mathbf{R}^d} \pi(dx) \int_{\mathbf{R}^d \times V} f d\nu_{\lambda x}[u] \tag{8.3.68}$$

In terms of $\nu_{\lambda\pi}[u]$, (8.3.63) becomes

$$R_\lambda^k[u](\pi) = \lambda \int \bar{r} d\nu_{\lambda x}[u] \tag{8.3.69}$$

Let

$$\nu_{\lambda\pi}[A] = \{\nu_{\lambda\pi}[u] \mid u \in A\} \tag{8.3.70}$$

$\nu_{\lambda\pi}\left[M_1, A_2, \ldots, A_N\right]$, $\nu_{\lambda\pi}\left[M_1, \ldots, M_N\right]$ are defined analogously. Then from [18] we have the following result.

Lemma 8.3.3 *For any $k = 1, 2, \ldots, N$,*

$$\nu_{\lambda\pi}[M_1, \ldots, M_{k-1}, A_k, M_{k+1}, \ldots, M_N] = \nu_{\lambda\pi}[M_1, \ldots, M_N].$$

Let $v \in M$. By Krylov's inequality it can be shown that $\nu_{\lambda\pi}[v]$ is absolutely continuous with respect to the Lebesgue measure on $\mathbf{R}^d$ and hence has a density $\phi_{\lambda\pi}[v]$. For $f \in W^{2,p}_{loc}(\mathbf{R}^d)$ define

$$L_v^\lambda f(x) = (L_v f)(x) - \lambda f(x). \tag{8.3.71}$$

Then $\phi_{\lambda\pi}[v]$ is the unique solution in $L_1(\mathbf{R}^d)$ of: for every $f \in C_0^\infty(\mathbf{R}^d)$

$$\int L_v^\lambda f(x)\phi(x)dx = -\int f(x)\pi(dx) \tag{8.3.72}$$

$$\int \phi(x)dx = 1,\ \phi \geq 0. \tag{8.3.73}$$

Now from [18] we have following results.

Lemma 8.3.4 *$\nu_{\lambda\pi}[M_1,\ldots,M_N]$ is componentwise convex, i.e., for any fixed k and prescribed $v_i \in M_i$, $i \neq k$*

$$\nu_{\lambda\pi}[v^{\widehat{k}}, M_k] = \{\nu_{\lambda\pi}[v^{\widehat{k}}, v_k] : v_k \in M_k\}$$

is convex.

Lemma 8.3.5 *$\nu_{\lambda\pi}[M_1,\ldots,M_N]$ is componentwise compact, i.e., for any fixed k and prescribed $v_i \in M_i$, $i \neq k$,*

$$\nu_{\lambda\pi}[v^{\widehat{k}}, M_k] = \{\nu_{\lambda\pi}[v^{\widehat{k}}, v_k] : v_k \in M_k\}$$

is compact.

For the ergodic payoff criterion we will impose the following Lyapunov type stability condition.

(A3.1) There exists a twice continuously differentiable function $w : \mathbf{R}^d \to R_+$ such that

(i) $\lim_{\|x\|\to\infty} w(x) = \infty$ uniformly in $\|x\|$.

(ii) There exist $a > 0$, $\epsilon_0 > 0$ such that for $\|x\| > a$,

$$Lw(x,u) < -\epsilon_0 \text{ for all } u \in V$$

$$\|\nabla w\|^2 \geq (\overline{\lambda})^{-1}$$

where $\overline{\lambda}$ is the ellipticity constant of $\sigma\sigma^T$.

(iii) $w(x)$ and $\|\nabla w\|$ have polynomial growth.

For $v \in M$, let $x(.)$ be the corresponding process. Also, for $\|x\| > a$, let

$$\tau_a = \inf\{t \geq 0 \mid \|x(t)\| = a\}.$$

The following result is a consequence of Assumption (A3.1).

Lemma 8.3.6

(i) *All $v \in M$ are stable.*

(ii) *$E_v[\tau_a \mid x(0) = x] \leq w(x)/\epsilon_0$, for any v.*

(iii) $\int w(x)\eta[v](dx) < \infty$ *for any* v.

(iv) *Under any* v *and* $x \in \mathbf{R}^d$

$$\lim_{t\to\infty} \frac{1}{t} E_v[w(x(t))] = 0.$$

(v) *The set* $I = \{\eta[v] \mid v \in M\}$ *is componentwise compact in* $\mathcal{P}\left(\mathbf{R}^d\right)$.

For $v \in M$, the ergodic occupation measure, denoted by $\nu_E[v] \in \mathcal{P}(\mathbf{R}^d \times V)$ is defined as

$$\nu_E[v]\,(dx, dy_1, \ldots, dy_N) = \eta[v] \prod_{i=1}^{N} v_i(x)\,(dy_i) \tag{8.3.74}$$

Let

$$\nu_E[M] = \{\nu_E[v] \mid v \in M\}. \tag{8.3.75}$$

For $v \in M$, let $x(.)$ be the process governed by v. Then

$$\eta[v](dx) = \left(\int p(t, y, x)\eta[v](dy)\right) dx$$

where $p(.,.,.)$ is the transition density of $x(.)$ under v. Thus $\eta[v]$ itself has a density which we denote by $\phi[v](.)$. Then $\phi[v]$ is the unique solution of: for every $f \in C_0^\infty\left(\mathbf{R}^d\right)$

$$\int L_v f(x)\phi(x)dx \qquad = 0 \tag{8.3.76}$$

$$\int \phi(x)dx \quad = 1,\ \phi \geq 0. \tag{8.3.77}$$

As for the discounted case, we now have following results.

Lemma 8.3.7 $\nu_e[M]$ *is componentwise convex and compact.*

For any fixed $k \in \{1, 2, \ldots, N\}$, let $v_i \in M_i$, $i \neq k$ and $u_k \in A_k$. Let $x(.)$ be the process governed by $\left(v^{\widehat{k}}, u_k\right)$. Define $\mathcal{P}\left(\mathbf{R}^d \times V\right)$-valued empirical process ν_t as follows: For $B \subset \mathbf{R}^d$, $A_i \subset U_i$, $i = 1, \ldots, N$, Borel,

$$\nu_t\,(B \times A_1 \times \cdots \times A_N) = \frac{1}{t} \int_0^t I\{x(s) \in B\} \prod_{i=1, i\neq k}^{N} v_i(x(s))\,(A_i)\, u_k(s)\,(A_k)\, ds. \tag{8.3.78}$$

Lemma 8.3.8 *The process* $\{\nu_t\}$ *is a.s. tight and outside a set of zero probability, each limit point* ν *of* $\{\nu_t\}$ *as* $t \to \infty$ *belongs to* $\nu_E[M]$.

Existence of an equilibrium

We make the following assumption.

(A3.2) $\overline{m}$ and $\overline{r}$ are of the form

$$\overline{m}(x, u_1, \ldots, u_N) = \sum_{i=1}^{N} \overline{m}^i(x, u_i)$$

$$\overline{r}(x, u_1, \ldots, u_N) = \sum_{i=1}^{N} \overline{r}_i(x, u_i)$$

where $\overline{m}^i : \mathbf{R}^d \times V_i \to \mathbf{R}^d$ and $\overline{r}_i : \mathbf{R}^d \times V_i \to \mathbf{R}$ and they satisfy the same conditions as $\overline{m}$ and $\overline{r}$.

Let $v \in M$. Fix a $k \in \{1, 2, \ldots, N\}$ and $\pi \in \mathcal{P}(\mathbf{R}^d)$. Then by Lemma 8.3.3

$$\sup_{u_k \in A_k} R_\lambda^k[v^{\widehat{k}}, u_k](\pi) = \sup_{\overline{v}_k \in M_k} R_\lambda^k[v^{\widehat{k}}, \overline{v}_k](\pi).$$

Since M_k is compact and $\overline{r}_k$ is continuous, the supremum on the right hand side above can be replaced by maximum. Then there exists a $v_k^* \in M_k$ such that

$$\sup_{u_k \in A_k} R_\lambda^k[v^{\widehat{k}}, u_k](\pi) = \max_{\overline{v}_k \in M_k} R_\lambda^k[v^{\widehat{k}}, \overline{v}_k](\pi) = R_\lambda^k[v^{\widehat{k}}, v_k^*](\pi). \tag{8.3.79}$$

This optimal discounted response strategy for player k, v_k^* can be chosen to be independent of π. Define $\widetilde{R}_\lambda^k[v] : \mathbf{R}^d \to \mathbf{R}$ by

$$\widetilde{R}_\lambda^k[v](x) = \max_{\overline{v}_k \in M_k} R_\lambda^k[v^{\widehat{k}}, \overline{v}_k](x)$$

Then we can obtain the following result.

Lemma 8.3.9 *$\widetilde{R}_\lambda^k[v](.)$ is the unique solution in $W_{loc}^{2,p}(\mathbf{R}^d) \cap C_b(\mathbf{R}^d)$, $2 \leq p < \infty$, of*

$$\lambda\phi(x) = \sup_{\overline{v}_k} \left[L_{v^{\widehat{k}}, \overline{v}_k} \phi(x) + r(x, v^{\widehat{k}}(x), \overline{v}_k) \right] \tag{8.3.80}$$

in $\mathbf{R}^d$. A strategy $v_k^* \in M_k$ is discounted optimal response for player k given v if and only if

$$\begin{aligned} &\left[\sum_{i=1}^{d} m_i(x, v^{\widehat{k}}(x), v_k^*(x)) \frac{\partial \widetilde{R}_\lambda^k[v](x)}{\partial x_i} + r(x, v^{\widehat{k}}(x), v_k^*(x)) \right] \\ &= \sup_{v_k} \left[\sum_{i=1}^{d} m_i(x, v^{\widehat{k}}(x), v_k(x)) \frac{\partial \widetilde{R}_\lambda^k[v](x)}{\partial x_i} + r(x, v^{\widehat{k}}(x), v_k(x)) \right] \text{ a.e.} \end{aligned} \tag{8.3.81}$$

Next result from [18] gives the existence of discounted equilibrium in the set of Markov strategies.

Theorem 8.3.10 *There exists a discounted equilibrium $v^* = (v_1^*, \ldots, v_N^*) \in M$.*

Proof. Let $v \in M$ and $\overline{v}_k \in U_k$. Set

$$F_k\left(x, v^{\widehat{k}}, \overline{v}_k\right) = \sum_{i=1}^{d} m_i\left(x, v^{\widehat{k}}(x), \overline{v}_k\right) \frac{\partial \widetilde{R}_\lambda^k[v](x)}{\partial x_i} + r\left(x, v^{\widehat{k}}(x), \overline{v}_k\right) \tag{8.3.82}$$

Let

$$G_k[v] = \left\{ v_k^* \in M_k \mid F_k\left(x, v^{\widehat{k}}(x), v_k^*(x)\right) = \sup_{\overline{v}_k \in U_k} F_k\left(x, v^{\widehat{k}}, \overline{v}_k\right) \text{ a.e.} \right\}. \tag{8.3.83}$$

Then $G_k[v]$ is non-empty, convex, closed and hence compact. Set $G[v] = \prod_{k=1}^{N} G_k[v]$. Then $G[v]$ is non-empty convex and compact subset of M. Thus $v \to G[v]$ defines a point-to-set

map from M to 2^M. This map is upper semicontinuous. Hence by Fan's fixed point theorem [29], there exists a $v^* \in M$ such that $v^* \in G[v^*]$. This v^* is a discounted equilibrium. □
□

Next we will discuss the existence result for the ergodic payoff. Let $v \in M$ and fix a $k \in \{1, 2, \ldots, N\}$. Let $v_k^* \in M_k$ be such that

$$\rho_k^*[v] \doteq \rho_k[v^{\widehat{k}}, v_k^*] = \max_{\overline{v}_k \in M_k} \rho[v^{\widehat{k}}, v_k^*]$$

where $\rho_k[v]$ is defined in (8.3.66). If all but player k uses strategies $v^{\widehat{k}}$ then, by Lemma 8.3.8, player k cannot obtain a higher payoff than $\rho_k^*[v]$ by going beyond M_k a.s. This v_k^* is said to be an *ergodic optimal* response for player k given v. Consider the following

$$\rho = \sup_{\overline{v}_k} \left[L_{v^{\widehat{k}}, \overline{v}_k} \phi(x) + r\left(x, v^{\widehat{k}}(x), \overline{v}_k\right)\right] \tag{8.3.84}$$

where ρ is a scalar and $\phi : \mathbf{R}^d \to \mathbf{R}$. Then we have the following result.

Lemma 8.3.11 *The equation* (8.3.84) *has a unique solution* $(\phi_k[v], \rho_k^*[v])$ *in the class of functions* $W_{loc}^{2,p}(\mathbf{R}^d) \cap O(w(.))$, $2 \le p < \infty$, *satisfying* $\phi[v] = 0$. *A* $v_k^* \in M_k$ *is ergodic optimal response for player* k *given* v *if and only if*

$$\begin{aligned} &\left[\sum_{i=1}^d m_i\left(x, v^{\widehat{k}}(x), v_k^*(x)\right) \frac{\partial \phi_k[v](x)}{\partial x_i} + r\left(x, v^{\widehat{k}}(x), v_k^*(x)\right)\right] \\ &= \sup_{v_k} \left[\sum_{i=1}^d m_i\left(x, v^{\widehat{k}}(x), v_k(x)\right) \frac{\partial \phi_k[v](x)}{\partial x_i} + r\left(x, v^{\widehat{k}}(x), v_k(x)\right)\right] \ a.e. \end{aligned} \tag{8.3.85}$$

Following result from [18] gives the existence result for an ergodic equilibrium.

Theorem 8.3.12 *There exists an ergodic equilibrium* $v^* \in M$.

Proof. Let $v \in M$ and $\overline{v}_k \in U_k$. Set

$$J_k\left(x, v^{\widehat{k}}, \overline{v}_k\right) = \left[\sum_{i=1}^d m_i\left(x, v^{\widehat{k}}(x), \overline{v}_k(x)\right) \frac{\partial \phi_k[v](x)}{\partial x_i} + r\left(x, v^{\widehat{k}}(x), \overline{v}_k(x)\right)\right]$$

Let

$$H_k(v) = \left\{\widetilde{v}_k \in M_k \mid J_k\left(x, v^{\widehat{k}}, \widetilde{v}_k(x)\right) = \sup_{\overline{v}_k \in U_k} J_k\left(x, v^{\widehat{k}}, \overline{v}_k\right) a.e.\right\}$$

Set $H[v] = \prod_{k=1}^N H_k(v)$. Then $H(v)$ is a non-empty, convex, compact subset of M. As in the discounted case, an application of Fan's fixed point theorem yields a $v^* \in M$ such that $v^* \in H[v^*]$. This v^* is an ergodic equilibrium. □ □

In this section we have used a non-anticipative relaxed control framework to show the existence of an equilibrium for an N-person stochastc differential game. Using this approach, one could also show the existence of value and optimal strategies for a two person strictly competitive differential game that we have discussed in Section 2. Other payoff criteria could also be considered. Using the approach described here, one could obtain similar results for feedback randomized strategies.

8.4 Weak convergence methods in differential games

In this section, we will present weak convergence and martingale techniques applied to stochastic differential games. In [32], the convergence problem for a deterministic game was considered. An analogous problem for optimal stopping by two players was discussed in [21]. First we will present the weak convergence method for an N-person stochastic differential game. Weak convergence methods applied to two person stochastic differential games with complete observations could be found in [84]. Later the weak convergence method will be used for the analysis of partially observed stochastic differential games. We will begin this section by giving some weak convergence preliminaries; for more details we refer to [62].

8.4.1 Weak convergence preliminaries

Let $D^d[0,\infty)$ denote the space of R^d valued functions which are right continuous and have left-hand limits endowed with the Skorohod topology. Following [62, 68], we define the notion of 'p-lim' and an operator $\widehat{A}^\epsilon$ as follows. Let $\{\Im_t^\epsilon\}$ denote the minimal σ-algebra over which $\{x^\epsilon(s), \xi^\epsilon(s), s \le t\}$ is measurable, and let E_t^ϵ denote the expectation conditioned on $\Im_t^\epsilon$. Let $\widetilde{M}$ denote the set of real valued functions of (ω, t) that are nonzero only on a bounded t-interval. Let

$$\overline{M}^\epsilon = \left\{ f \in \widetilde{M}; \sup_t E\left|f(t)\right| < \infty and f(t) is \Im_t^\epsilon \text{ measurable} \right\}.$$

Let $f(.), f^\Delta(.) \in \overline{M}^\epsilon$, for each $\Delta > 0$. Then $f = p\text{-}\lim_\Delta f^\Delta$ if and only if

$$\sup_{t,\Delta} E\left|f^\Delta(t)\right| < \infty$$

and $\lim_{\Delta \to 0} E\left|f(t) - f^\Delta(t)\right| = 0$, for each t. $f(.)$ is said to be in the domain of $\widehat{A}^\epsilon$, i.e., $f(.) \in D\left(\widehat{A}^\epsilon\right)$, and $\widehat{A}^\epsilon f = g$ if

$$p\text{-}\lim_{\Delta \to 0} \left(\frac{E_t^\epsilon f(t+\Delta) - f(t)}{\Delta} - g(t) \right) = 0.$$

If $f(.) \in D\left(\widehat{A}^\epsilon\right)$, then

$$f(t) - \int_0^t \widehat{A}^\epsilon f(u)du \text{ is a martingale,}$$

and

$$E_t^\epsilon f(t+s) - f(t) = \int_t^{t+s} E_t^\epsilon \widehat{A}^\epsilon f(u)du, w.p.1.$$

The $\widehat{A}^\epsilon$ operator plays the role of an infinitesimal operator for a non-Markov process. In our case, it becomes a differential operator by the martingale property and the definition of p-limit. We will use the terms such as "tight," Skorohod imbedding etc. without explanation, reader can obtain these from [62]. The following result will be used to conclude that various terms will go to zero in probability.

Lemma 8.4.1 *Let $\xi(.)$ be a ϕ-mixing process with mixing rate $\phi(.)$, and let $h(.)$ be a function of ξ which is bounded and measurable on $\Im_t^\infty$. Then, there exist K_i, $i = 1, 2, 3$ such that*

$$\left|E\left(h(t+s)/\Im_0^t\right) - Eh(t+s)\right| \leq K_1\phi(s).$$

If $t < u < v$, and $Eh(s) = 0$ for all s, then,

$$\left|E\left(h(u)h(v)/\Im_\tau^t\right) - Eh(u)h(v)\right| \leq \begin{cases} K_2\phi(v-u), & u < \tau < v \\ K_3\phi(u-t), & t < \tau < u \end{cases},$$

where $\Im_\tau^t = \sigma\{\xi(s); \tau \leq s \leq t\}$.

In order to obtain the weak convergence result, the following condition needs to be verified:

$$\lim_{n\to\infty} \limsup_{\epsilon\to 0} P\left(\sup_{t\leq T} |x^\epsilon(t)| \geq n\right) = 0$$

for each $T < \infty$. Direct verification of this is very tedious. Instead, one can utilize the method of K-truncation. This is as follows. For each $K > 0$, let

$$S_K = \{x : |x| \leq K\} betheK\text{-ball},$$

Let $x^{\epsilon,K}(0) = x^\epsilon(0), x^{\epsilon,K}(t) = x^\epsilon(t)$ up until first exit from S_K, and

$$\lim_{n\to\infty} \limsup_{\epsilon\to 0} P\left(\sup_{t\leq T} \left|x^{\epsilon,K}(t)\right| \geq n\right) = 0 \text{ for each } T < \infty.$$

$x^{\epsilon,K}(t)$ is said to be the K-truncation of $x^\epsilon(.)$. Let

$$q^K(x) = \begin{cases} 1 & forx \in S_K \\ 0 & forx \in R^d - S_{K+1} \\ smooth & otherwise. \end{cases}$$

Define $a_K(x, \alpha) = a(x, \alpha)q^K(x)$ and $g_K(x, \xi) = g(x, \xi)q^K(x)$. Let $x^{\epsilon,K}(.)$ denote the solution of (8.4.94) corresponding to the use of truncated coefficients. Then $x^{\epsilon,K}(.)$ is bounded uniformly in t and $\epsilon > 0$.

For proving the main weak convergence result, Theorem 8.4.5, we will use following results from [62].

Lemma 8.4.2 *Let $\{y^\epsilon(.)\}$ be tight on $D^d[0, \infty)$. Suppose that for each $f(.) \in C_0^3$, and each $T < \infty$, there exists $f^\epsilon(.) \in D(\widehat{A}^\epsilon)$ such that*

$$p\text{-}\lim\left(f^\epsilon(.) - f\left(y^\epsilon(.)\right)\right) = 0 \tag{8.4.86}$$

and

$$p\text{-}\lim_{\epsilon}\left(\widehat{A}^\epsilon f^\epsilon(.) - \widehat{A}f\left(y^\epsilon(.)\right)\right) = 0 \tag{8.4.87}$$

Then $y^\epsilon(.) \Rightarrow y(.)$, the solution of the martingale problem for the operator $\widehat{A}$.

Lemma 8.4.3 *Let the K-truncations $\left\{y^{\epsilon,K}\right\}$ be tight for each K, and that the martingale problem for the diffusion operator A has a unique solution $y(.)$ for each initial condition. Suppose that $y^K(.)$ is a K-truncation of $y(.)$ and it solves the martingale problem for operator A^K. For each K and $f(.) \in D$, let there be $f^\epsilon(.) \in D(A^\epsilon)$ such that (8.4.86) and (8.4.87) hold with $y^{\epsilon,K}(.)$ and A^K replacing y^ϵ and A, respectively. Then $y^\epsilon(.) \Longrightarrow y(.)$.*

Now we will outline a general method one can follow to show that a sequence of solutions to a wide bandwidth noise driven ODE converge weakly to a diffusion, and identify the limit diffusion [62, 84]. Let $z^\epsilon(.)$ be defined by

$$dz^\epsilon = a\left(z^\epsilon\right)dt + \frac{1}{\epsilon}b\left(z^\epsilon\right)\xi\left(t/\epsilon^2\right)dt \tag{8.4.88}$$

where $\xi(.)$ is a second order stationary right continuous process with left hand limits and integrable correlation function $\overline{R}(.)$, and the functions $a(.)$ and $b(.)$ are continuous, $b(.)$ is continuously differentiable and (8.4.88) has a unique solution. Define $\overline{R_0} = \int\limits_{-\infty}^{\infty} E\xi(u)\xi'(0)du$ and assume that

$$E\left|\int\limits_s^t du\left[E\left(\xi(u)\xi'(s)/\xi(\iota), \iota \geq 0\right) - \overline{R}(u-s)\right]\right| \to 0 \quad \text{as} \quad t, s \to \infty.$$

Define the infinitesimal generator A and function $\overline{K} = \left(\overline{K_1}, \dots\right)$ by

$$Af(z) = f_z'(z)a(z) + \int\limits_0^\infty E\left[f_z'(z)b(z)\xi(t)\right]_z' b(z)\xi(0)dt$$

$$\equiv \sum_i f_{z_i}(z)\overline{K_i}(z) + \frac{1}{2}trace\left\{f_{z_i z_j}(z)\right\}\left\{b(z)\overline{R_0}b(z)\right\} \tag{8.4.89}$$

where $\overline{K} = \left(\overline{K}_1, \dots\right)$ are the coefficients of the first derivatives $\left(f_{z_1}, \dots\right)$ in (8.4.89). The operator A is the generator of

$$dz = \overline{K}(z)dt + b(z)\overline{R}_0^{\frac{1}{2}}dw \tag{8.4.90}$$

where $w(.)$ is the standard Wiener process. In order to obtain that $z^\epsilon(.) \Rightarrow z(.)$ of (8.4.90), by the martingale problem solution, it is enough to show that

$$p\text{-}\lim_{\epsilon}\left(\widehat{A}^\epsilon f^\epsilon(.) - Af\left(z^\epsilon(.)\right)\right) = 0. \tag{8.4.91}$$

Then by Lemma 8.4.2, $z(.)$ satisfies (8.4.90).

8.4.2 Weak convergence in N-person stochastic differential games

Problem description

As in Section 3, let the diffusion model be given in a non-anticipative relaxed control frame work. For convenience, we will redescribe some of the concepts from that section. However, in this section, the entire differential game problem is discribed in the pathwise sense, that is, there is no expected value in the payoff functionals. Let U_i, $i = 1, \dots, N$ be compact metric spaces (we can take U_i as compact subsets of R^d), and $M_i = P(U_i)$, the space of probability measures on U_i with Prohorov topology. Use the notation $m^{\widehat{k}} = (m_1, \dots, m_{k-1}, m_{k+1}, \dots, m_N)$ and $\left(m^{\widehat{k}}, m^{\widetilde{k}}\right) = (m_1, \dots, m_{k-1}, \widetilde{m}_k, m_{k+1}, \dots, m_N)$.

Let $m = (m_1, \ldots, m_N) \in M = M_1 \times \cdots \times M_N$ and $U = U_1 \times \cdots \times U_N$, $x(.) \in R^d$ be an R^d-valued process given by the following controlled stochastic differential equation

$$dx(t) = \int_U a(x(t), \alpha) m_t(d\alpha) dt + \overline{g}(x(t)) dt + \sigma(x(t)) dw(t) \qquad (8.4.92)$$
$$x(0) = x_0$$

where we use the notation $a(.,.) = (a_1(.,.), \ldots, a_N(.,.))' : R^d \times U \to R$, $\alpha = (\alpha_1, \ldots, \alpha_N)$, $\sigma = [[\sigma_{ij}]], 1 \le i, j \le d : R^d \to R^{d \times d}$, and

$$\int_U a_i(x, \alpha) m_t(d\alpha) \doteq \int_{U_N} \cdots \int_{U_1} a_i(x, \alpha_1, \ldots, \alpha_N) m_{1t}(d\alpha_1) \ldots m_{Nt}(d\alpha_N).$$

The pathwise average payoff per unit time for player k is given by

$$J_k[m] = \liminf_{T \to \infty} \frac{1}{T} \int_0^T \int r_k(x(s), \alpha) m_s(d\alpha) ds \qquad (8.4.93)$$

Let $w(.)$ in (8.4.92) be a Wiener process with respect to a filtration $\{\Im_t\}$ and let Ω_i, $i = 1, 2, \ldots, N$ be a compact set in some Euclidean space. A measure valued random variable $m_i(.)$ is an *admissible strategy* for the ith player if

$$\int_0^t \int f_i(s, \alpha_i) m_i(ds d\alpha_i)$$

is progressively measurable for each bounded continuous $f_i(.)$ and $m_i([0,t] \times \Omega_i) = t$, for $t \ge 0$. If $m_i(.)$ is admissible then there is a derivative $m_{it}(.)$ (defined for almost all t) that is non-anticipative with respect to $w(.)$ and

$$\int_0^t \int f_i(s, \alpha_i) m_i(ds d\alpha_i) = \int_0^t ds \int f_i(s, \alpha_i) m_{is}(d\alpha_i)$$

for all t with probability one (w.p.1). The results derived in this work are for the Markov strategies. We will denote by A_i the set of admissible strategies and M_{ai} the set of Markov strategies for the player i. One can introduce appropriate metric topology under which M_{ai} is compact [18].

In the relaxed control settings, one chooses at time t a probability measure m_t on the control set M rather than an element $u(t)$ in U. We call the measure m_t the relaxed control at time t. Any ordinary control can be represented as a relaxed control via the definition of the derivative $m_t(d\alpha) = \delta_{u(t)}(\alpha) d\alpha$. Hence, if m_t is an atomic measure concentrated at a single point $m(t) \in M$ for each t, then the relaxed control will be called ordinary control. We will denote the ordinary control by $u_m(t) \in M$.

An N-tuple of strategies $m^* = (m_1^*, \ldots, m_N^*) \in A_1 \times \cdots \times A_N$ is said to be *ergodic equilibrium* (in the sense of Nash) for initial law π if for $k = 1, \ldots, N$,

$$J_k[m^*](\pi) \ge J_k[m^{*\widehat{k}}, m_k](\pi)$$

for any $m_k \in A_k$. Fix a $k \in \{1, \ldots, N\}$. Let $m_k^* \in M_{ak}$ be such that

$$J_k^*[m] \doteq J_k[m^{\widehat{k}}, m_k^*] = \max_{m_k \in M_k} J[m^{\widehat{k}}, m_k].$$

If all but player k use strategies $m^{\widehat{k}}$ then player k can not get a higher payoff than $J_k^*[m]$ by going beyond M_{ak} a.s. We say that m_k^* is *ergodic optimal response* for player k given m. An N-tuple of strategies $m^\delta = (m_1^\delta, \ldots, m_N^\delta)$ is a *δ-ergodic equilibrium* for initial law π if for any $k = 1, \ldots, N$,

$$J_k[m^*](\pi) \geq \sup_{m_k \in A_k} J_k[m^{\widehat{k}}, m_k] - \delta.$$

The wide band noise system considered in this section is of the following type:

$$dx^\epsilon = \int a(x^\epsilon, \alpha)\, m_t^\epsilon(d\alpha)dt + G(x^\epsilon, \xi^\epsilon(t))\, dt + \frac{1}{\epsilon} g(x^\epsilon, \xi^\epsilon)\, dt \tag{8.4.94}$$

and pathwise average payoff per unit time for player k is given by

$$J_k[m^\epsilon] = \liminf_{T\to\infty} \frac{1}{T} \int_0^T\!\!\int r_k(x^\epsilon(s), \alpha)\, m_s^\epsilon(d\alpha)ds \tag{8.4.95}$$

An *admissible relaxed strategy* $m_k^\epsilon(.)$ for the kth player with system (8.4.94) is a measure valued random variable satisfying that $\int_0^t\!\int f(s,\alpha)m^\epsilon(dsd\alpha)$ is progressively measurable with respect to $\{\Im_t^\epsilon\}$, where $\Im_t^\epsilon$ is the minimal σ-algebra generated by $\{\xi^\epsilon(s), x^\epsilon(s), s \leq t\}$. Also $m^\epsilon([0,t] \times U) = t$ for all $t \geq 0$. Also, there is a derivative m_t^ϵ, where $m_t^\epsilon(B)$ are $\Im_t^\epsilon$ measurable for Borel B. We will use following assumptions, which are very general. For a detailed description on these types of assumptions, we refer to [62] and [65].

(A4.1) $a_i(.,.), G(.,.), g(.,), g_x(.,.)$ are continuous and are bounded by $O(1+|x|)$. $G_x(.,\xi)$ is continuous in x for each ξ and is bounded. $\xi(.)$ is bounded, right continuous, and $EG(x,\xi(t)) \to 0, Eg(x,\xi(t)) \to 0$ as $t \to \infty$, for each x.

(A4.2) $g_{xx}(.,\xi)$ is continuous for each ξ, and is bounded.

(A4.3) Let $W(x,\xi)$ denote either $\epsilon G(x,\xi), G_x(x,\xi), g(x,\xi)$ or $g_x(x,\xi)$. Then for compact Q,

$$\epsilon \sup_{x\in Q} \left| \int_{t/\epsilon^2}^{\infty} E_t^\epsilon W(x,\xi(s))ds \right| \xrightarrow{\epsilon} 0$$

in the mean square sense, uniformly in t.

(A4.4) Let g_i denote the ith component of g. There are continuous $\overline{g}_i(.), b(.) = \{b_{ij}(.)\}$ such that

$$\int_t^\infty E g_{i,x}(x,\xi(s))g(x,\xi(t))ds \to \overline{g}_i(x),$$

$$\int_t^\infty E g_i(x,\xi(s))g_j(x,\xi(t))ds \to \frac{1}{2} b_{ij}(x),$$

as $t \to \infty$, and the convergence is uniform in any bounded x-set.

NOTE: Let $b(x) = \{b_{ij}(x)\}$. For $i \neq j$, it is not necessary that $b_{ij} = b_{ji}$. In that case define $\widetilde{b}(x) = \frac{1}{2}[b(x) + b'(x)]$ as the symmetric covariance matrix, then use b for the new $\widetilde{b}$. Hence for notational simplicity, we will not distinguish between $b(x)$ and $\widetilde{b}(x)$.

(A4.5) For each compact set Q and all i, j,

(a) $\sup_{x\in Q} \epsilon^2 \left| \int_{t/\epsilon^2}^{\infty} d\tau \int_{\tau}^{\infty} ds[E_{t/\epsilon^2} g'_{i,x}(x,\xi(s))g(x,\xi(t)) - Eg'_{i,x}(x,\xi(s))g(x,\xi(t))] \right| \to 0$;

(b) $\sup_{x\in Q} \epsilon^2 \left| \int_{t/\epsilon^2}^{\infty} d\tau \int_{\tau}^{\infty} ds[E_{t/\epsilon^2} g_i(x,\xi(s))g_j(x,\xi(t)) - Eg_i(x,\xi(s))g_j(x,\xi(t))] \right| \to 0$; in the mean square sense as $\epsilon \to 0$, uniformly in t.

Define $\overline{a}(x,\alpha) = a(x,\alpha) + \overline{g}(x)$ and the operator A^m as

$$A^m f(x) = \int A^\alpha f(x) m_x(d\alpha),$$

where

$$A^\alpha f(x) = f'_x(x)\overline{a}(x,\alpha) + \frac{1}{2}\sum_{i,j} b_{ij}(x) f_{x_i x_j}(x).$$

For a fixed control α, A^α will be the operator of the process that is the weak limit of $\{x^\epsilon(.)\}$.

(A4.6) The martingale problem for operator A^m has a unique solution for each relaxed admissible Markov strategy $m_x(.)$, and each initial condition. The process is a Feller process. The solution of (8.4.94) is unique in the weak sense for each $\epsilon > 0$. Also $b(x) = \sigma(x)\sigma'(x)$ for some continuous finite dimensional matrix $\sigma(.)$.

For an admissible relaxed policy for (8.4.94) and (8.4.92), respectively, define the occupation measure valued random variables $P_T^{m,\epsilon}(.)$ and $P_T^m(.)$ by, respectively,

$$P_T^{m,\epsilon}(B\times C) = \frac{1}{T}\int_0^T I_{\{x^\epsilon(t)\in B\}} m_t^\epsilon(C)\, dt,$$

$$P_T^{m}(B\times C) = \frac{1}{T}\int_0^T I_{\{x(t)\in B\}} m_t(C)\, dt$$

where B and C are Borel subsets in $\mathbf{R}^d$ and $[0,t]\times U$, respectively.

Let $\{m^\epsilon(.)\}$ be a given sequence of admissible relaxed controls.

(A4.7) For a fixed $\delta > 0$,

$$\{x^\epsilon(t), small\ \epsilon > 0, t \in dense\ set\ in\ [0,\infty), m^\epsilon\ used\}$$

is tight.

NOTE: The assumption (A4.7) implies that the set of measure valued random variables

$$\{P_T^{m^\epsilon,\epsilon}(.), small\ \epsilon > 0, T < \infty\}$$

are tight.

(A4.8) For $\delta > 0$, there is an N-tuple of Markov strategies $m^\delta = (m_1^\delta, \ldots, m_N^\delta)$ which is a *δ-ergodic equilibrium* for initial law π for (8.4.92) and (8.4.93), and for which the martingale problem has a unique solution for each initial condition. The solution is a Feller process and there is a unique invariant measure $\mu\left(m^{\delta}\right)$.

NOTE: Existence of such an invariant measure is assured if the process is positive recurrent. Also, under the conditions of Theorem 8.4.4 below, the assumption (A4.8) will follow.

(A4.9) $r_k(.,.)$ is bounded and continuous. Also, $r\left(x, m_1, \ldots, m_N\right)=\sum\limits_{k=1}^{N} r_k\left(x, m_k\right)$ and $a\left(x, m_1, \ldots, m_N\right)=\sum\limits_{k=1}^{N} a_k\left(x, m_k\right)$.

In Section 3, under the Lyapunov type stability condition and (A4.9), we have shown the following result.

Theorem 8.4.4 *There exists an ergodic equilibrium* $m^*=\left(m_1^*, \ldots, m_N^*\right) \in M_{a1} \times \cdots \times M_{aN}$.

Weak Convergence result

The following result gives the main weak convergence and δ–optimality result for the ergodic payoff criterion.

Theorem 8.4.5 *Assume* (A4.1)–(A4.9). *Let* (8.4.94) *have a unique solution for each admissible relaxed policy and each* ϵ. *Then for* m^{δ} *of* (A4.8), *the following holds:*

$$\varliminf_{\epsilon, T} P\{J_k\left(m^{\epsilon}\right) \geq J_k\left(m^{\delta}\right)-\delta\}=1 \tag{8.4.96}$$

for any sequence of admissible relaxed policies $m^{\epsilon}(.)$.

Proof. The correct procedure of proof is to work with the truncated processes $x^{\epsilon, K}(.)$ and to use the piecing together idea of Lemma 8.4.3 to get convergence of the original $x^{\epsilon}(.)$ sequence, unless $x^{\epsilon}(.)$ is bounded on each $[0, T]$, uniformly in ϵ. For notational simplicity, we ignore this technicality. Simply suppose that $x^{\epsilon}(.)$ is bounded in the following analysis. Otherwise, one can work with K–truncation. Let $\widehat{D}$ be a measure determining set of bounded real-valued continuous functions on R^d having continuous second partial derivatives and compact support. Let $m_t^{\epsilon}(.)$ be the relaxed Markov policies of (A4.7). Whenever convenient, we write $x^{\epsilon}(t)=x$. For the test function $f(.) \in \widehat{D}$, define the perturbed test functions (the change of variable $s/\epsilon^2 \rightarrow s$ will be used through out the proofs)

$$\begin{aligned}
f_0^{\epsilon}(x, t) &= \int_t^{\infty} E_t^{\epsilon} f_x^{\prime}(x) G(x, \xi^{\epsilon}(s)) ds \\
&= \epsilon^2 \int_{t/\epsilon^2}^{\infty} E_t^{\epsilon} f_x^{\prime}(x) G(x, \xi(s)) ds
\end{aligned}$$

$$\begin{aligned}
f_1^{\epsilon}(x, t) &= \frac{1}{\epsilon} \int_t^{\infty} E_t^{\epsilon} f_x^{\prime}(x) g(x, \xi^{\epsilon}(s)) ds \\
&= \epsilon \int_{t/\epsilon^2}^{\infty} E_t^{\epsilon} f_x^{\prime}(x) g(x, \xi(s)) ds
\end{aligned}$$

$$f_2^\epsilon(x,t) = \frac{1}{\epsilon^2}\int_t^\infty ds \int_s^\infty d\tau \{E_t^\epsilon[f_x'(x)g(x,\xi^\epsilon(\tau))]_x' g(x,\xi^\epsilon(s)) - E[f_x'(x)g(x,\xi^\epsilon(\tau))]_x' g(x,\xi^\epsilon(s))$$
$$= \epsilon^2 \int_{t/\epsilon^2}^\infty ds \int_s^\infty d\tau \{E_t^\epsilon[f_x'(x)g(x,\xi(\tau))]_x' g(x,\xi(s)) - E[f_x'(x)g(x,\xi(\tau))]_x' g(x,\xi(s))$$

From (A4.1), (A4.2), (A4.3), and (A4.5), $f_i^\epsilon(.) \in D(A^\epsilon)$ for $i = 0, 1, 2$. Define the perturbed test function

$$f^\epsilon(t) = f(x^\epsilon(t)) + \sum_{i=0}^2 f_i^\epsilon(x^\epsilon(t), t).$$

The reasons for defining f_i^ϵ are to facilitate the averaging of the "noise" terms involving ξ^ϵ terms. By the definition of the operator A^ϵ and its domain $D(A^\epsilon)$, we will obtain that $f(x^\epsilon(.))$ and the $f_i^\epsilon(x^\epsilon(.), .)$ are all in $D(A^\epsilon)$, and

$$A^{m^\epsilon,\epsilon} f(x^\epsilon(t)) = f_x'(x^\epsilon(t))[\sum_{i=1}^N \int a_i(x^\epsilon(t),\alpha)m_t^\epsilon(d\alpha) + G(x^\epsilon(t),\xi^\epsilon(t)) + \frac{1}{\epsilon}g(x^\epsilon(t),\xi^\epsilon(t))]. \tag{8.4.97}$$

From this we can obtain,

$$\begin{aligned} A^{m^\epsilon,\epsilon} f_0(x^\epsilon(t)) &= -f_x'(x^\epsilon(t))G(x^\epsilon(t),\xi^\epsilon(t)) \\ &\quad + \int_t^\infty ds[E_t^\epsilon f_x'(x^\epsilon(t))\, G(x^\epsilon(t),\xi^\epsilon(s))]_x' \,\dot{x}^\epsilon(t) \\ &= -f_x'(x^\epsilon(t))G(x^\epsilon(t),\xi^\epsilon(t)) \\ &\quad + \epsilon^2 \int_{t/\epsilon^2}^\infty ds[E_t^\epsilon f_x'(x^\epsilon(t))\, G(x^\epsilon(t),\xi(s))]_x' \,\dot{x}^\epsilon(t) \end{aligned} \tag{8.4.98}$$

Note that the first term in (8.4.98) will cancel with the $f_{x'}G$ term of (8.4.97). The $p\text{-}\lim_\epsilon$ of the last term in (8.4.98) is zero.

$$\begin{aligned} A^{m^\epsilon,\epsilon} f_1(x^\epsilon(t)) &= -\frac{1}{\epsilon} f_x'(x^\epsilon(t))\, g(x^\epsilon(t),\xi^\epsilon(t)) \\ &\quad + \frac{1}{\epsilon}\int_t^\infty ds[E_t^\epsilon f_x'(x^\epsilon(t))\, g(x^\epsilon(t),\xi^\epsilon(s))]_x' \dot{x}^\epsilon(t) \\ &= -\frac{1}{\epsilon} f_x'(x^\epsilon(t))g(x^\epsilon(t),\xi^\epsilon(t)) \\ &\quad + \epsilon \int_{t/\epsilon^2}^\infty ds[E_t^\epsilon f_x'(x^\epsilon(t))\, g(x^\epsilon(t),\xi(s))]_x' \dot{x}^\epsilon(t) \end{aligned} \tag{8.4.99}$$

The first term on the right of (8.4.99) will cancel with the $\frac{f'_x g}{\epsilon}$ term in (8.4.97). The only component of the second term on the right of (8.4.99) whose $\underset{\epsilon}{p\text{-lim}}$ is not zero is

$$\frac{1}{\epsilon^2}\int_t^\infty ds\{E_t^\epsilon[f'_x(x^\epsilon(t))\,g(x^\epsilon(t),\xi^\epsilon(s))]'_x g(x^\epsilon(t),\xi^\epsilon(t)).$$

This term will cancel with the first term of (8.4.100)

$$\begin{aligned} A^{m^\epsilon,\epsilon} f_2(x^\epsilon(t)) &= -\frac{1}{\epsilon^2}\int_t^\infty ds\{E_t^\epsilon[f'_x(x^\epsilon(t))\,g(x^\epsilon(t),\xi^\epsilon(s))]'_x g(x^\epsilon(t),\xi^\epsilon(t)) \\ &\quad - E[f'_x(x^\epsilon(t))g(x,\xi^\epsilon(s))]'_x g(x,\xi^\epsilon(t))\,|_{x=x^\epsilon(t)}\} \\ &\quad + [f_2^\epsilon(x,t)]'_x \dot{x}^\epsilon\,|_{x=x^\epsilon(t)} \\ &= -\int_{t/\epsilon^2}^\infty ds\{E_t^\epsilon[f'_x(x^\epsilon(t))\,g(x^\epsilon(t),\xi(s))]'_x g(x^\epsilon(t),\xi^\epsilon(t)) \\ &\quad - E[f'_x(x^\epsilon(t))\,g(x,\xi(s))]'_x g(x,\xi^\epsilon(t))\,|_{x=x^\epsilon(t)}\} \\ &\quad + [f_2^\epsilon(x,t)]'_x \dot{x}^\epsilon\,|_{x=x^\epsilon(t)}\,. \end{aligned} \tag{8.4.100}$$

The $\underset{\epsilon}{p\text{-lim}}$ of the last term of the right side of (8.4.100) is zero.

Evaluating $A^{m^\epsilon,\epsilon} f^\epsilon(t) = A^{m^\epsilon,\epsilon}\left[f(x^\epsilon(t)) + \sum_{i=0}^{2} f_i^\epsilon(x^\epsilon(t),t)\right]$ and by deleting terms that cancel yield

$$\begin{aligned} A^{m^\epsilon,\epsilon} f^\epsilon(t) &= f'_x(x^\epsilon(t))\sum_{i=1}^{N}\int a_i(x^\epsilon(t),\alpha)m_t^\epsilon(d\alpha) \\ &\quad + \int_{t/\epsilon^2}^\infty E\left[f'_x(x^\epsilon(t))g(x,\xi(s))\right]' g(x,\xi(t/\epsilon^2))ds. \end{aligned} \tag{8.4.101}$$

As a result, we get

$$p\text{-}\lim\left(f^\epsilon(t) - f(x^\epsilon(.))\right) = 0 \tag{8.4.102}$$

$$\underset{\epsilon}{p\text{-}\lim}\left|A^{m^\epsilon,\epsilon} f(x^\epsilon(t)) - A^{m^{\epsilon,\epsilon}} f^\epsilon(t)\right| = 0. \tag{8.4.103}$$

Hence, by Lemma 8.4.2,

$$M_f^\epsilon(t) = f^\epsilon(t) - f^\epsilon(0) - \int_0^t A^{m^\epsilon} f^\epsilon(s)ds$$

is a zero mean martingale.

Let $[t]$ denote the greatest integer part of t. Write

$$\frac{M_f^\epsilon(t)}{t} = \frac{1}{t}\left[\left(M_f^\epsilon(t) - M_f^\epsilon([t])\right) + M_f^\epsilon(0)\right] + \frac{1}{t}\sum_{k=0}^{[t]-1}\left[M_f^\epsilon(k+1) - M_f^\epsilon(k)\right].$$

Using the fact that $f(.)$ is bounded and (8.4.103), and martingale property of $M_f^\epsilon(.)$, we get $E\left[\frac{M_f^\epsilon(t)}{t}\right]^2 \to 0$ as $t \to \infty$ and $\epsilon \to 0$, which in turn implies that $\frac{M_f^\epsilon(t)}{t} \xrightarrow{P} 0$ as $t \to \infty$ and $\epsilon \to 0$ in any way at all. From (8.4.103), and the fact that $\frac{M_f^\epsilon(t)}{t}$, $\frac{f^\epsilon(t)}{t}$, and $\frac{f^\epsilon(0)}{t}$ all go to zero in probability implies that as $t \to \infty$ and $\epsilon \to 0$,

$$\frac{1}{t}\int_0^t A^{m^\epsilon} f(x^\epsilon(s))ds \xrightarrow{P} 0. \tag{8.4.104}$$

By the definition of $P_T^{m^\epsilon,\epsilon}(.)$, (8.4.104) can be written as

$$\int A^\alpha f(x) P_T^{m^\epsilon,\epsilon}(dxd\alpha) \xrightarrow{P} 0 \text{ as } T \to \infty and \epsilon \to 0. \tag{8.4.105}$$

For the policy $m^\delta(.)$, choose a weakly convergent subsequence of set of random variables $\left\{P_T^{m^\delta,\epsilon}(.),\epsilon,T\right\}$, indexed by ϵ_n, T_n, with limit $\widehat{\mu}(.)$. Let this limit $\widehat{P}(.)$ be defined on some probability space $\left(\widetilde{\Omega},\widetilde{P},\widetilde{\Im}\right)$ with generic variable $\widetilde{\omega}$. Factor $\widehat{P}(.)$ as $\widehat{P}(dxd\alpha) = m_x^\delta(d\alpha)\mu(dx)$. We can suppose that $m_x(C)$ are x-measurable for each Borel C and $\widetilde{\omega}$.

Now (8.4.105) implies that for all $f(.) \in \widehat{D}$,

$$\iint A^\alpha f(x) m_x^\delta(d\alpha)\widehat{\mu}(dx) = 0 \text{ for } \widetilde{P}-almostall\widetilde{\omega}. \tag{8.4.106}$$

Since $f(.)$ is measure determining, (8.4.106) implies that almost all realizations of $\widehat{\mu}$ are invariant measures for (8.4.92) under the relaxed policies m^δ. By uniqueness of the invariant measure, we can take $\mu\left(m^\delta,.\right) = \widehat{\mu}(.)$ does not depend on the chosen subsequence ϵ_n, T_n. By the definition of $P_T^{m^\delta,\epsilon}(.)$,

$$\frac{1}{t}\iint_0^t r_k(x^\epsilon(s),\alpha)m^\delta(d\alpha)ds = \iint_0^t r_k(x^\epsilon(s),\alpha)P_T^{m^\delta,\epsilon}(d\alpha dx)$$

$$\xrightarrow{P} \iint_0^t r_k(x,\alpha)m_x^\delta(d\alpha)\widehat{\mu}(dx) = J_k(m^\delta).$$

Since $m^\delta(.)$ is a δ-equilibrium policy, by the definition of δ-equilibrium, for almost all $\widetilde{\omega}$ we have $J_k(m^\epsilon) \geq J_k\left(m^\delta\right) - \delta$. Since this is true for all the limits of the tight set $\left\{P_T^{m^\delta,\epsilon}(.);\epsilon,T\right\}$, (8.4.96) follows. □ □

It is important to note that, as a result of Theorem 8.4.5, if one needs a δ-optimal policy for the physical system, it is enough to compute the optimal policies for the diffusion model and use it for the physical system. There is no need to compute optimal policies for each ϵ.

Since relaxed control is a device with primarily a mathematical use, it is desirable to have a chattering type result for an N-person games. The following result captures the spirit of such a result for the δ-optimal strategies, which states that for any near equilibrium, relaxed strategy, there is an ordinary strategy which gives a δ-optimal value.

Corollary 8.4.6 *Let the conditions of Theorem 8.4.5 hold. Then there exists an ordinary control policy $u_m^\delta(t) \in M$ such that*

$$\underline{\lim}_{\epsilon,T} P\{J_k(m^\epsilon) \geq J_k(u_m^\delta(t)) - \delta\} = 1 \tag{8.4.107}$$

Proof. Following the reasoning of Lemma 2 [31, page 153], we conclude that corresponding to the relaxed control policy $m^{\delta/2}(t)$ of (A4.8), there exists an ordinary control policy $u_m^{\delta}(t)$ such that

$$\left|J_k(m^{\delta/2}) - J_k(u_m^{\delta}(t))\right| < \delta/2, \text{ a.s.} \tag{8.4.108}$$

Also from equation (8.4.96) (with $\delta = \delta/2$), we have

$$\underline{\lim}_{\epsilon,T} P\{J_k(m^{\epsilon}) \geq J_k(m^{\delta/2}) - \delta/2\} = 1 \tag{8.4.109}$$

Equation (8.4.107) now follows from (8.4.108) and (8.4.109), Since

$$\underline{\lim}_{\epsilon,T} P\{J_k(m^{\epsilon}) \geq J_k(m^{\delta/2}) - \delta/2\} = 1 \Leftrightarrow$$

$$\underline{\lim}_{\epsilon,T} P\{J_k(m^{\epsilon}) \geq J_k(m^{\delta/2}) - J_k(m^{\delta}(t)) + J_k(m^{\delta}(t)) - \delta/2\} = 1 \Leftrightarrow$$

$$\underline{\lim}_{\epsilon,T} P\{J(m^{\epsilon}) \geq -\delta/2 + J_k(m^{\delta}(t)) - \delta/2\}$$
$$= \underline{\lim}_{\epsilon,T} P\{J(m^{\epsilon}) \geq J_k(m^{\delta}(t)) - \delta\} = 1$$

for

$$-\delta/2 \leq J_k(m^{\delta/2}) - J_k(u_m^{\delta}(t)) \leq \delta/2$$

from (8.4.109). □ □

Pathwise discounted payoffs

In place of the ergodic payoff, now consider the pathwise discounted payoffs for the player k given by

$$R_k^{\lambda,\epsilon}(m^{\epsilon}) = \lambda \int_0^{\infty} e^{-\lambda s} \int r_k(x^{\epsilon}(s), \alpha) m_s(d\alpha) ds \tag{8.4.110}$$

Now we will state the pathwise result for discounted payoff and suggest the necessary steps needed in the proof.

Theorem 8.4.7 *Let m^{ϵ} be a sequence of δ-optimal discounted payoffs and m^{δ} be δ-equilibrium policies for* (8.4.92). *Under the conditions of Theorem* 8.4.5, *the following limits hold:*

$$R_k^{\lambda,\epsilon}(m^{\delta}) \xrightarrow{P} J_k(m^{\delta}) as \lambda \to 0, \epsilon \to 0, \tag{8.4.111}$$

$$\underline{\lim}_{\epsilon,T} P\{R_k^{\lambda,\epsilon}(m^{\epsilon}) \geq J_k(m^{\delta}) + \delta\} = 1 \tag{8.4.112}$$

Proof. The proof is essentially same as of Theorem 8.4.5. We will only explain the differences needed. Define the discounted occupation measures

$$P_{\lambda}^{m^{\epsilon},\epsilon}(B \times C) = \lambda \int_0^{\infty} e^{-\lambda t} I_{\{x^{\epsilon}(t) \in B\}} m_t(C) dt$$

$$P_{\lambda}^{m}(B \times C) = \lambda \int_0^{\infty} e^{-\lambda t} I_{\{x(t) \in B\}} m_t(C) dt$$

Then (8.4.110) can be written as

$$R_k^{\lambda,\epsilon}(m^\epsilon) = \int r_k(x(s),\alpha) P_\lambda^{m^\epsilon,\epsilon}(dx d\alpha).$$

By tightness condition (A4.7), the $\{P_\lambda^{m^\epsilon,\epsilon}(.)\}$ and $\{P^{m^\delta,\epsilon}(.)\}$ are tight. Define

$$f_\lambda^\epsilon(t) = \lambda e^{\lambda t} f^\epsilon(t).$$

This will be used in the place of $f^\epsilon(.)$ defined in Theorem 8.4.5. Then

$$A^{m^\epsilon,\epsilon} f_\lambda^\epsilon(t) = -\lambda^2 e^{\lambda t} f^\epsilon(t) + \lambda e^{\lambda t} A^{m^\epsilon,\epsilon} f^\epsilon(t).$$

Define the martingale

$$f_\lambda^\epsilon(t) - f_\lambda^\epsilon(0) - \int_0^t A^{m^\epsilon,\epsilon} f_\lambda^\epsilon(s) ds$$

$$= \lambda e^{\lambda t} f^\epsilon(t) - \lambda f^\epsilon(0) - \int_0^t \left[-\lambda^2 e^{\lambda s} f^\epsilon(s) + \lambda e^{\lambda s} A^{m^\epsilon,\epsilon} f^\epsilon(s)\right] ds.$$

As in Theorem 8.4.5,

$$\lim_{\substack{t\to\infty \\ (\lambda,\epsilon)\to 0}} \lambda \int_0^t e^{-\lambda s} A^{m^\epsilon,\epsilon} f(x^\epsilon(s)) ds = 0.$$

Thus

$$\lim_{(\lambda,\epsilon)\to 0} \iint A^\alpha f(x) P_\lambda^{m^\epsilon,\epsilon}(dx d\alpha) = 0$$

Now choose weakly convergent subsequences of the $\{P_\lambda^{m^\epsilon,\epsilon}(.)\}$ or $\{P_\lambda^{m^\delta,\epsilon}(.)\}$ and continue as in the proof of Theorem 8.4.5 to get (8.4.111) and (8.4.112). ⊔ □

Discrete parameter (stochastic) games

The discrete parameter system is given by

$$X_{n+1}^\epsilon = X_n^\epsilon + \epsilon G(X_n^\epsilon) + \epsilon \sum_{i=1}^N \int a_i(X_n^\epsilon, \alpha_i) m_{in}(d\alpha_i) + \sqrt{\epsilon} g(X_n^\epsilon, \xi_n^\epsilon) \tag{8.4.113}$$

where $\{\xi_n^\epsilon\}$ satisfies the discrete parameter version of (A2) and $m_{in}(.)$, $i = 1, \dots, N$ be the relaxed control strategies depending only on $\{X_i, \xi_{i-1}, i \le n\}$. It should be noted that, in the discrete case, strategies would not be relaxed, one need to interpret this in the asymptotic sense, i.e., the limiting strategies will be relaxed. Let E_n^ϵ denote the conditional expectation with respect to $\{X_i, \xi_{i-1}, i \le n\}$. Define $x^\epsilon(.)$ by $x^\epsilon(t) = X_n^\epsilon$ on $[n\epsilon, n\epsilon + \epsilon)$ and $m_i(.)$ by

$$m_i(B_i \times [0,t]) = \epsilon \sum_{n=0}^{[t/\epsilon]-1} m_{in}(B_i) + \epsilon(t - \epsilon[t/\epsilon]) m_{[t/\epsilon]}(B_i), i = 1, \dots, N.$$

(A4.10)

(i) For V equals either $a(.,.), g$ or g_x, and for Q compact,

$$E\sup_x \left| \sum_{n+L_1}^{L} E_n^\epsilon V(x, \xi_i^\epsilon) \right| \to 0,$$

as L, n and $L_1 \to \infty$, with $L > n + L_1$ and $L - (n + L_1) \to \infty$.

(ii) There are continuous functions $c(i, x)$ and $c_0(i, x)$ such that for each x

$$\frac{1}{L}\sum_{n=\ell}^{\ell+L} E_\ell^\epsilon g(x, \xi_{n+i}^\epsilon) g'(x, \xi_n^\epsilon) \xrightarrow{P} c(i, x)$$

$$\frac{1}{L}\sum_{n=\ell}^{\ell+L} E_\ell^\epsilon g_x' \left(x, \xi_{n+i}^\epsilon\right) g\left(x, \xi_n^\epsilon\right) \xrightarrow{P} c_0(i, x)$$

as ℓ and $L \to \infty$.

(iii) For each $T < \infty$ and compact Q,

$$\epsilon \sup_{x\in Q} \left| \sum_{j=n}^{T/\epsilon} \sum_{k=j+1}^{T/\epsilon} \left[E_n^\epsilon g_{i,x}' \left(x, \xi_k\right) g\left(x, \xi_j\right) - E g_{i,x}' \left(x, \xi_k\right) g\left(x, \xi_j\right)\right] \right| \to 0 i \leq n,$$

$$\epsilon \sup_{x\in Q} \left| \sum_{j=n}^{T/\epsilon} \sum_{k=j+1}^{T/\epsilon} \left[E_n^\epsilon g' \left(x, \xi_k\right) g\left(x, \xi_j\right) - E g' \left(x, \xi_k\right) g\left(x, \xi_j\right)\right] \right| \to 0,$$

in the mean as $\epsilon \to 0$ uniformly in $n \leq T/\epsilon$. Also, the limits hold when the bracketed terms are replaced by their x-gradient$/\sqrt{\epsilon}$.

Define

$$\widetilde{a}(x) = \sum_{1}^{\infty} c_0(i, x)$$

and

$$\widetilde{c}(x) = c(0, x) + 2\sum_{1}^{\infty} c(i, x) = \sum_{-\infty}^{\infty} c(i, x)$$

With some minor modifications in the proof of Theorem 8.4.5, we can obtain the following result (refer to [62] and [87] for convergence proofs in similar situations).

Theorem 8.4.8 *Assume* (A4.1) *to* (A4.3), (A4.6) *to* (A4.10). *Then the conclusions of Theorem* 8.4.5 *hold for model* (8.4.113).

8.4.3 Partially observed stochastic differential games and weak convergence

In practical differential games difficulties are often encountered in obtaining information about the state of the system due to time lag, high cost of obtaining data, or simply asymmetry in availability of information due to the nature of the problems in a competitive environment. Stochastic differential games with imperfect state informations are inherently very difficult to analyze. In the literature, there are various information structures considered such as both players will have the same information as in the form as a broadcasting channel, [51, 72], or the two players will have available only noise-corrupted output measurements, [90, 91]. There are various other possibilities, such as one player will have full information where as the other player will have only partial information or only a deterministic information. A fixed duration stochastic two-person nonzero-sum differential game in which one player has access to closed-loop nonanticipatory state information while the other player makes no observation is considered in [4]. A comprehensive study on partially observed stochastic differential games is still far from solved. In this subsection, we will present a linear system with quadratic cost functional and imperfect state information. Solution to the diffusion model is given and a weak convergence method is described. We will also deal with a form of nonlinearity.

The system under consideration is of the following type, where both players have the same information such as from a broadcasting channel

$$dx = [A(t)x + B(t)u - C(t)v]dt + Ddw_1(t) \tag{8.4.114}$$

with observation data

$$dy = Hxdt + Fdw_2(t) \tag{8.4.115}$$

and payoff

$$J(u,v) = E\left\{x^{'}(T)Sx(T) + \int_0^T \left[u^{'}Ru - v^{'}Qv\right]dt\right\} \tag{8.4.116}$$

In here, we are concerned with a partially observed two person zero-sum stochastic differential games driven by wide band noise. The actual physical system will be more naturally modeled by

$$\dot{x}^\epsilon = Ax^\epsilon + Bu - Cv + D\xi_1^\epsilon \tag{8.4.117}$$

with observations

$$\dot{y}^\epsilon = Hx^\epsilon + \xi_2^\epsilon \tag{8.4.118}$$

where ξ_i^ϵ, $i = 1, 2$ are wide band noise processes. Let the payoff be given in linear quadratic form

$$J^\epsilon(u^\epsilon, v^\epsilon) = E\left\{x^{\epsilon^{'}}(T)Sx^\epsilon(T) + \int_0^T \left[u^{\epsilon^{'}}Ru^\epsilon - v^{\epsilon^{'}}Qv^\epsilon\right]dt\right\} \tag{8.4.119}$$

for some $T < \infty$.

Typically, one decides upon a suitable model (8.4.114), (8.4.115), (8.4.116), obtains a good or optimal policy pair, and uses this policy to the actual physical system. In this case, the value of the determined policy for the physical system is not clear, as well as the value of the output of the filter for making estimates of functional of the physical process $x^\epsilon(.)$ which is approximated by $x(.)$. The filter output will rarely be nearly optimal for use in making such estimates, and the policies based on the filter outputs will rarely be 'nearly' optimal. Very little attention has been devoted to the case of game problems. Under quite broad conditions, we will obtain a very reasonable class of alternative filters and policies for the physical system with respect to which it is nearly optimal.

For a general filtering theory, we refer to [74]. We begin with a discussion of filtering and game problem for the ideal white noise linear model (8.4.114), (8.4.115), (8.4.116) and use the Kalman-Bucy filter for this model to obtain an optimal strategy pair for the game problem. Then we will describe the wide bandwidth analogue and give results on filtering and near optimal policies. Also we will include the study on the asymptotic in the time and bandwidth problems. Some extensions to partly nonlinear observations will also be given.

The diffusion model

Consider the linear quadratic Gaussian (LQG) games. We have

$$dx = [A(t)x + B(t)u - C(t)v]dt + Ddw_1(t) \tag{8.4.120}$$

where A,B,C,D are $n \times n, n \times m, n \times s$, and $n \times r$ matrices whose elements are continuous in $[0,T]$. $x \in R^n$ is the state vector with initial state x_0, which is normally distributed with $N(\overline{x}_0, M_0)$. Players I and II are endowed with measurements

$$dy = dy_1 = dy_2 = Hxdt + Fdw_2(t) \tag{8.4.121}$$

where F is of full rank with $p \times q, q > p$ matrix. The objective functional is defined with

$$J(u,v) = E\{x'(T)Sx(T) + \int_0^T [u'Ru - v'Qv]dt\} \tag{8.4.122}$$

where $S \geq 0, R(t) > 0, Q(t) > 0$ are $n \times n, m \times m$, and $n \times s$ symmetric matrices whose elements are continuous on [0,T]. Let $R_0 = FF'$ be positive definite (denoted by $R_0 > 0$). Note that the $-v'Qv$ term is due to the fact that v is minimizing.

The policies u and v take values in compact sets U and V, and sets Ξ_1 and Ξ_2 denote the set of U and V-valued measurable (t, ω) functions on $[0,T] \times C[0,T]$, ($C[0,T]$ is the space of real valued continuous functions on $[0,T]$ with the topology of uniform convergence) which are continuous $w.p.1.$ relative to the Wiener measure. Let Ξ_{1t} and Ξ_{2t} denote the subclass which depends only on the function values up to time t. Let $\Xi = \Xi_1 \times \Xi_2$ and $\Xi_t = \Xi_{1t} \times \Xi_{2t}$. We view functions in Ξ as the data dependent policies with values $u(y(.),t)$ and $v(y(.),t)$ at time t and data $y(.)$. Let $\overline{\Xi}$ denote the sub class of functions $(u,v) \in \Xi$ such that $(u(.,t), v(.,t)) \in \Xi_t$ for all t and with the use of policies $(u(y,.), v(y,.))$, (8.4.120) has a unique solution in the sense of distributions. These pairs $(u(y,.), v(y,.))$ are the *admissible strategies.* We say that an admissible pair $(u^*(t), v^*(t))$ is a *saddle point* for the game iff

$$J(u(t), v^*(t)) \leq J(u^*(t), v^*(t)) \leq J(u^*(t), v(t)) \tag{8.4.123}$$

where $u(t)$ and $v(t)$ any admissible control laws. We call $(u^*(t), v^*(t))$ the optimal strategic pair. Admissible strategies $\widehat{u}$ and $\widehat{v}$ are called δ-*optimal* for players I and II respectively if

$$\sup_u J(u, \widehat{v}) - \delta \leq J(u^*, v^*) \leq \inf_v J(\widehat{u}, v) + \delta. \tag{8.4.124}$$

Let $G_t = \sigma\{y(s), s \leq t\}$. Let $\widehat{x}(\tau) = E\{x(\tau)/G_\tau; u(\tau), v(\tau)\}$. For (8.4.120), (8.4.121), the classical Kalman-Bucy filter equations are

$$d\widehat{x} = (A\widehat{x} + Bu - Cv)dt + L(t)(dy - H\widehat{x}dt) \tag{8.4.125}$$

$$L(t) = P(t)H'(t)R_0^{-1}(t)$$

with $\widehat{x}_0 = \overline{x}_0$ and $P(t) = E\{(x(t) - \widehat{x}(t))(x(t) - \widehat{x}(t))'\}$ is the error covariance matrix and is the unique solution to the matrix Riccati equation:

$$\dot{P} = FP + PF' - PN(y)P + DD' \tag{8.4.126}$$

$P_0 = M_0$, where $N(y) = H'R_0^{-1}H$, and the Riccati equation

$$\dot{\Sigma} = -\Sigma A - A'\Sigma + \Sigma[BR^{-1}B' - CQ^{-1}C']\Sigma \tag{8.4.127}$$

with the boundary condition

$$\Sigma(T) = S'(T)S(T).$$

The following result can be obtained from [51] and [72].

Theorem 8.4.9 *The optimal strategy pair for the problem* (8.4.120), (8.4.121), (8.4.122) *exists. The optimal pair at time t is*

$$u^*(t) = -R^{-1}(t)B'(t)\Sigma(t)\widehat{x}(t) \tag{8.4.128}$$
$$v^*(t) = -Q^{-1}(t)C'(t)\Sigma(t)\widehat{x}(t) \tag{8.4.129}$$

Furthermore,

$$J(u^*, v^*) = \int_0^T Tr\Sigma(s)\left[DD' + (B(s)R^{-1}B'(s) - C(s)Q^{-1}C'(s))\Sigma(s)P(s)\right] ds + Tr\Sigma_0 M_0 \tag{8.4.130}$$

where P satisfies (8.4.126).

Finite time filtering and game, wide band noise case

Now consider the wide bandwidth analogue of the previous filtering and game problem. Let the system be defined by

$$\dot{x}^\epsilon = Ax^\epsilon + Bu - Cv + D\xi^\epsilon \tag{8.4.131}$$

with observations $\dot{y}^\epsilon(.)$, where

$$\dot{y}^\epsilon = Hx^\epsilon + \xi_2^\epsilon \tag{8.4.132}$$

where $\int_0^t \xi_i^\epsilon(s)ds = W_i^\epsilon(t)$, $i = 1, 2$, $W_1^\epsilon(.)$ and $W_2^\epsilon(.)$ are mutually independent. Let $W_i^\epsilon(.) \Longrightarrow W_i(.)$, standard Wiener processes. Let the corresponding objective functional be given by

$$J^\epsilon(u, v) = E\{x^{\epsilon'}(T)Sx^\epsilon(T) + \int_0^T [u'Ru - v'Qv]\, dt \tag{8.4.133}$$

In practice, with physical wide band observation noise and state process are not driven by the ideal white noise, one uses (8.4.126), (8.4.127) and the natural adjustment of (8.4.125), that is

$$\widehat{x}^\epsilon = (A\widehat{x}^\epsilon + Bu - Cv) + L(t)\left[\dot{y}^\epsilon - H\widehat{x}^\epsilon\right] \tag{8.4.134}$$

First of all we want to know in what way the triple (8.4.134), (8.4.126), (8.4.127) makes sense. In general, it is not an optimal filter for the physical observation. Instead of asking whether it is nearly optimal, we will ask, with respect to what class of alternative estimators is it nearly optimal when estimating the specific functional of $x^\epsilon(.)$? Another problem is that if one obtains a policy (optimal or not) based on the white noise driven limit model, the policy will be a function of the outputs of the filters. The value of applying this to the actual wide bandwidth noise system is not clear. If one uses the model (8.4.120), (8.4.121), (8.4.125) to get a optimal (or nearly optimal) policy pair for the value (8.4.122), and apply this to the physical system, the question then is with respect to what class of comparison policies is such a policy nearly optimal? In both the cases, weak convergence theory can provide some answers. In order to obtain weak convergence of $(x^\epsilon(.), y^\epsilon(.))$ of (8.4.117) and (8.4.118), we need to use above method outlined for equations (8.4.88) through (8.4.91).

In subsequent results, in order to avoid lengthy calculations, we will not give the weak convergence proofs. The reader can obtain necessary steps from [62] and [84].

Even when $W_2^\epsilon(.) \Longrightarrow W_2(.)$, a nondegenerate Wiener process, $y^\epsilon(.)$ might contain a great deal more information about $x^\epsilon(.)$ than $y(.)$ does about $x(.)$. We give the following example from [66] for an extreme case when $B \equiv 0$ and $C \equiv 0$. Call the corresponding process $z^\epsilon(.)$.

Example 8.4.10 *Let t_i^ϵ, $i > 0$ be a strictly increasing sequence of real numbers for each ϵ, such that $t_i^\epsilon \xrightarrow{i} \infty$ and $\sup_i \left|t_{i+1}^\epsilon - t_i^\epsilon\right| \xrightarrow{\epsilon} 0$. Define $\Delta_i^\epsilon = t_{2i+1}^\epsilon - t_{2i}^\epsilon$, and for any $t > 0$, let $\sum_{t_i^\epsilon \leq t} \Delta_i^\epsilon \xrightarrow{\epsilon} 0$. Define a new observation noise $\xi_y^\epsilon(.)$ by resetting $\xi_y^\epsilon(t) = 0$ for $t \in \left[t_{2i}^\epsilon, t_{2i+1}^\epsilon\right)$, all i. The integral of the $\xi_y^\epsilon(.)$ still converges weakly to the Wiener process $W_2(.)$. But $Hz^\epsilon(.)$ is exactly known for small ϵ.*

The following result [66], shows that we never gain information on going to the limit.

Lemma 8.4.11 *Let $(Z_n, Y_n) \Longrightarrow (Z, Y)$. Then*

$$\overline{\lim_n} E\left[Z_n - E(Z_n/Y_n)\right]^2 \leq E\left[Z - E(Z/Y)\right]^2$$

A class of estimators.

By earlier assumptions, we have $(Z^\epsilon(.), W_2^\epsilon(.)) \Longrightarrow (Z(.), W_2(.))$ as $\epsilon \to 0$. By the weak convergence and independence of $z^\epsilon(.)$ and $\xi_2^\epsilon(.)$, $w_1(.)$ is independent of $w_2(.)$. The weak limit of $y^\epsilon(.)$ is $y(.)$, the solution of (8.4.121).

Let $\aleph$ denote the class of measurable functions on $C[0,\infty)$ which are continuous w.p.1 relative to Wiener measure. Hence, they are continuous w.r.t. the measure of $y(.)$. Let $\aleph_t$ denote the sub class which depends only on the function values up to time t. For arbitrary $f(.) \in \aleph$ or in $\aleph_t$, $f\left(y^\epsilon(.)\right)$ will denote an alternative estimator of a functional of $z^\epsilon(.)$. We consider $\aleph$ and $\aleph_t$ as a class of data processors.

We now obtain a robustness result. Let (m_t^ϵ, q) be the integral of a function $q(z)$ with respect to the Gaussian distribution with mean $\widehat{z}^\epsilon(t)$ and the covariance $p(t)$. We will assume the following:

(A4.11) $\left\{(m_t^\epsilon, q)^2, q^2\left(z^\epsilon(t)\right), F^2\left(y^\epsilon(.)\right)\right\}$ is uniformly integrable.

The following theorem states that, for a small ϵ, the ersatz conditional distribution is 'nearly optimal' with respect to a specific class of alternative estimators.

Theorem 8.4.12 *Assume* (A4.11) *and that $w_2^\epsilon(.) \Longrightarrow w_2(.)$, a standard Wiener process. Then*

$$\left(\widehat{z}^\epsilon(.), z^\epsilon(.), w_2^\epsilon(.)\right) \Longrightarrow \left(\widehat{z}(.), z(.), w_2(.)\right).$$

Also,

$$\lim_\epsilon E\left[q\left(z^\epsilon(t)\right) - F\left(y^\epsilon(.)\right)\right]^2 \geq \lim_\epsilon E\left[q\left(z^\epsilon(t)\right) - (m_t^\epsilon, q)\right]^2 \tag{8.4.135}$$

Proof. The weak convergence is clear from the assumptions. Since $F(.)$ is w.p.1. continuous, we also have

$$\left(q\left(z^\epsilon(t)\right), F(y^\epsilon(.)), (m_t^\epsilon, q)\right) \Longrightarrow \left(q(z(t)), F(y(.)), (m_t, q)\right).$$

Hence

$$(m_t, q) = \int q(z) dN(\widehat{z}(t), P(t), dz)$$

and $N(\widehat{z}, P, .)$ is normal distribution with mean $\widehat{z}$ and covariance P. Hence

$$\lim_{\epsilon} E\left[q\left(z^{\epsilon}(t)\right) - F(y\epsilon(.))\right]^2 = E[q(z(t)) - F(y(.))]^2$$

and

$$\lim_{\epsilon} E\left[q\left(z^{\epsilon}(t)\right) - (m_t^{\epsilon}, q)\right]^2 = E[q(z(t)) - E[q(z(t))/y(s), s \leq t]]^2.$$

Since the conditional expectation is the optimal estimator, (8.4.135) follows. □

Now we will give the 'near optimality' result for the policies. Let $\mathcal{M}_1$ (respectively $\mathcal{M}_2$) denote the class of U (respectively, V) valued continuous functions $u(.,.)$ (respectively, $v(.,.)$) such that with the use of policy value $(u(\widehat{x}(t), t), v(\widehat{x}(t), t))$ at time t, (8.4.120), (8.4.125), has a unique (weak sense) solution. In Theorem 4.9, we have shown that there are optimal strategy pairs (u^*, v^*) and a value J^* for the system (8.4.120), (8.4.125) with payoff (8.4.122). Hence, we can assume the following.

(A4.12) Let the strategy pair $(u^*(.,.), v^*(.,.))$ be in $\mathcal{M}$ and let this strategy be unique. Assume (u^*, v^*) is admissible for $x^{\epsilon}(.), \widehat{x}^{\epsilon}(.)$ of (8.4.131), (8.4.134) for small ϵ.

Theorem 8.4.13 *Assume* (A4.11), (A4.12). *Let* $x^{\epsilon}(.)$ *and* $\widehat{x}^{\epsilon}(.)$ *denote the process and its estimate with* $(u^*(.,.), v^*(.,.))$ *used. Then*

$$\{x^{\epsilon}(.), \widehat{x}^{\epsilon}(.), u^*, v^*\} \Longrightarrow (x(.), \widehat{x}(.), u^*, v^*)$$

and the limit satisfies (8.4.120), (8.4.125). *Also,*

$$J^{\epsilon}(u^*, v^*) \rightarrow J(u^*, v^*) = J^* \tag{8.4.136}$$

In addition, let $\widehat{u}(.,.)$ *and* $\widehat{v}(.,.)$ *be a* δ*-optimal strategy pair for players* I *and* II*, respectively, with* $(x(.), \widehat{x}(.))$ *of* (8.4.120), (8.4.125). *Then*

$$\underline{\lim}_{\epsilon} \left| \sup_{u \in \mathcal{M}_{\infty}} J\left(u\left(y^{\epsilon}, .\right), \widehat{v}\left(\widehat{x}^{\epsilon}, .\right)\right) - J^{\epsilon}\left(u^*, v^*\right) \right| \leq \delta \tag{8.4.137}$$

and

$$\overline{\lim}_{\epsilon} \left| \inf_{v \in \mathcal{M}_{\in}} J\left(\widehat{u}\left(\widehat{x}^{\epsilon}, .\right), v\left(y^{\epsilon}, .\right)\right) - J^{\epsilon}(u^*, v^*) \right| \leq \delta \tag{8.4.138}$$

Proof. Weak convergence is straightforward. By the assumed uniqueness, the limit $(x(.), \widehat{x}(.), u^*, v^*)$ satisfies (8.4.120),(8.4.125). Also, by this weak convergence and the fact that $T < \infty$, by the bounded convergence,

$$\lim_{\epsilon} J^{\epsilon}(u^*, v^*) = J(u^*, v^*).$$

To show (8.4.137) and (8.4.138), repeat the procedure with admissible strategies $(u(y^{\epsilon}, .), v(y^{\epsilon}, .))$. The limit $(x(.), u(y, .)\ ,\ v(y, .))$ might depend on the chosen subsequence. For any convergent subsequence $\{\epsilon_n\}$, we obtain

$$\lim_{\epsilon = \epsilon_n \rightarrow 0} J^{\epsilon}\left(u\left(y^{\epsilon}, .\right), v\left(y^{\epsilon}, .\right)\right) = J(u, v).$$

Now by the definition of δ-optimality (8.4.124), (8.4.137) and (8.4.138) follows. □

Large time problem

When the filtering system with wide band noise operates over a very long time interval, there are two limits involved, since both $t \to \infty$ and $\epsilon \to 0$. It is then important that the results do not depend on how $t \to \infty$ and $\epsilon \to 0$. We will make the following assumptions.

(A4.13) A is stable, $[A, H]$ is observable and $[A, D]$ is controllable.

(A4.14) $\xi_i(.), i = 1, 2$ are right continuous second order stationary processes with integrable covariance function $\overline{S}(.)$. $\xi_i^\epsilon(t) = \frac{1}{\epsilon}\xi_i\left(t/\epsilon^2\right)$. Also, if $t_\epsilon \to \infty$ as $\epsilon \to 0$, then $W_2^\epsilon(t_\epsilon + .) - W_2^\epsilon(t_\epsilon) \Longrightarrow W_2(.)$.

(A4.15) If $z^\epsilon(t_\epsilon) \Longrightarrow z(0)$ (a random variable) as $\epsilon \to 0$, then $z^\epsilon(t_\epsilon + .) \Longrightarrow z(.)$ with initial condition $z(0)$. Also $\sup_{\epsilon,t} E\left|z^\epsilon(t)\right|^2 < \infty$.

(A4.16) For each $\epsilon > 0$, there is a random process $\zeta^\epsilon(.)$ such that $\{\zeta^\epsilon(t), t < \infty\}$ is tight and for each strategy pair $(u(.), v(.)) \in \mathcal{M}$.

$\{x^\epsilon(.), \widehat{x}^\epsilon(.), z^\epsilon(.), \widehat{z}^\epsilon(.), \zeta^\epsilon(.), \xi_1^\epsilon(.), \xi_2^\epsilon(.)\}$ is a right continuous homogeneous Markov-Feller process with left hand limits. We have following result for filtering from [66].

Theorem 8.4.14 *Assume* (A4.13)–(A4.15) *and let* $q(.)$ *be a bounded continuous function. Let* $F(.) \in \aleph_t$. *Define* $y^\epsilon(s) = 0$, *for* $s \le 0$. *Define* $y^\epsilon(-\infty, t, .)$ *to be the 'reversed' function with values* $y^\epsilon(-\infty, t : \tau) = y^\epsilon(t - \tau)$ *for* $0 \le \tau < \infty$. *Then, if* $t_\epsilon \to \infty$ *as* $\epsilon \to 0$,

$$\{z^\epsilon(t_\epsilon + .), \widehat{z}^\epsilon(t_\epsilon + .), W_2^\epsilon(t_\epsilon + .) - W_2^\epsilon(t_\epsilon)\} \Longrightarrow (z(.), \widehat{z}(.), W_2(.)) \tag{8.4.139}$$

where $z(.)$ *and* $\widehat{z}(.)$ *are stationary. Also*

$$\lim_{\epsilon,t} E\left[q\left(z^\epsilon(t)\right) - F\left(y^\epsilon(-\infty, t; .)\right)\right]^2 \ge \lim_{\epsilon,t} E\left[q\left(z^\epsilon(t)\right) - (m_t^\epsilon, q)\right]^2. \tag{8.4.140}$$

The limit of (m_t^ϵ, q) *is the expectation with respect to the stationary* $(\widehat{z}(.), P(0))$ *system.*

Now we will use an ergodic payoff functional of the form

$$\rho^\epsilon(u, v) = \limsup_{T\to\infty} \frac{1}{T} E\left[\int_0^T k\left(x^\epsilon(t), z^\epsilon(t), u(t), v(t)\right) dt\right] \tag{8.4.141}$$

$$\rho(u, v) = \limsup_{T\to\infty} \frac{1}{T} E\left[\int_0^T k(x(t), z(t), u(t), v(t)) dt\right] \tag{8.4.142}$$

where $k(., ., .)$ is a bounded continuous function.

Ergodic optimal strategies for players I and II are defined similarly to the finite horizon case. We will assume the following.

(A4.17) There is an optimal strategy pair $(u^*, v^*) \in \mathcal{M}$ for (8.4.114), (8.4.115), and (8.4.142) with (8.4.114), (8.4.125) has a unique invariant measure $\mu^{(u,v)}(.)$.

The assumptions are not very restrictive. For detailed discussion on these type of assumptions, we refer to [62, 65].

Theorem 8.4.15 *Assume* (A4.13)–(A4.17). *Then the conclusions of Theorem* 8.4.13 *hold for the model* (8.4.117), (8.4.118) *with payoff* (8.4.141).

Proof. For a fixed $(u,v) \in \mathcal{M}$, we define

$$P_T^\epsilon(.) = \frac{1}{T} E_x \int_0^T P\{X^\epsilon(t) \in ./X^\epsilon(0)\} dt,$$

where $X^\epsilon(.)$ is the process corresponding to $(u(\widehat{x}^\epsilon(.), \widehat{z}^\epsilon(.)), v(\widehat{x}^\epsilon(.), \widehat{z}^\epsilon(.)))$. By (A4.16), $\{P_T^\epsilon(.), T \geq 0\}$ is tight. Also

$$\rho^\epsilon(u(\widehat{x}^\epsilon(.), \widehat{z}^\epsilon(.)), v(\widehat{x}^\epsilon(.), \widehat{z}^\epsilon(.))) = \limsup_T \int r(x, z, u(\widehat{x}, \widehat{z}), v(\widehat{x}, \widehat{z})) P_T^\epsilon(dX)$$

where $X = (x, z, \widehat{x}, \widehat{z})$. Let $T_n^\epsilon \to \infty$ be a sequence such that it attains the limit $\limsup_T$, and for which $P_{T_n^\epsilon}^\epsilon(.)$ converges weakly to a measure $P^\epsilon(.)$. Again by (A4.16), $P^\epsilon(.)$ is an invariant measure for $X^\epsilon(.)$. Also, by construction of $P^\epsilon(.)$,

$$\rho^\epsilon(u(\widehat{x}^\epsilon(.), \widehat{z}^\epsilon(.)), v(\widehat{x}^\epsilon(.), \widehat{z}^\epsilon(.))) = \limsup_T \int r(x, z, u(\widehat{x}, \widehat{z}), v(\widehat{x}, \widehat{z})) P^\epsilon(dX).$$

Now by a weak convergence argument and (A4.17),

$$\begin{aligned} &\rho^\epsilon(u(\widehat{x}^\epsilon(.), \widehat{z}^\epsilon(.)), v(\widehat{x}^\epsilon(.), \widehat{z}^\epsilon(.))) \to \rho(\widehat{x}, \widehat{z}) \\ &= \int r(x, z, u(\widehat{x}, \widehat{z}), v(\widehat{x}, \widehat{z})) \mu^{(u,v)}(dx\, dz\, d\widehat{x} d\widehat{z}) \end{aligned}$$

The rest of the proof is similar (with minor modifications) to that of Theorem 8.4.12 and hence we omit it. □ □

Partly nonlinear observations

The ideas of previous subsections are useful in the case of nonlinear observations. However, we need the limit system to be linear. Consider the observations with a normalizing term $\frac{1}{\epsilon}$.

$$\dot{y}^\epsilon = h(Hx^\epsilon + \xi_2^\epsilon(t))/\epsilon \tag{8.4.143}$$

$$y^\epsilon(0) = 0,$$
$$h(x) = \operatorname{sign}(x).$$

We assume the following:

(A4.18) $\xi_2^\epsilon(t) = \frac{1}{\epsilon}\xi_2(t/\epsilon^2)$, where $\xi_2(.)$ is a component of a stationary Gauss-Markov process whose correlation function goes to zero as $t \to \infty$.

Let $v_0^2 = E(\xi_2^\epsilon(t))^2$. Then the average of (8.4.143) over the noise ξ_2^ϵ is

$$\left(\frac{2}{\pi v_0^2}\right)^{\frac{1}{2}} Hx^\epsilon(t) + \delta_\epsilon$$

where $\delta_\epsilon \to 0$ as $\epsilon \to 0$, uniformly for $x^\epsilon(t)$ in any bounded set. The limit observation system is given by

$$dy = \left(\frac{2}{\pi v_0^2}\right)^{\frac{1}{2}} Hx dt + 2\Gamma_0^{\frac{1}{2}} dw_2. \tag{8.4.144}$$

For (8.4.120), (8.4.144), the Kalman-Bucy filter equations are

$$d\widehat{x} = (A\widehat{x} + Bu - Cv)\,dt + L(t)\left(dy - \left(\frac{2}{\pi v_0^2}\right)^{\frac{1}{2}} H\widehat{x}dt\right) \tag{8.4.145}$$

$$L(t) = P(t)H'\left(\frac{2}{\pi v_0^2}\right)^{\frac{1}{2}}\frac{1}{4\Gamma_0}$$

where $P(t)$ satisfies the Riccati equations

$$\dot{P} = FP + PF'PH'HP\left(\frac{1}{\Gamma_0}\right)\left(\frac{2}{\pi v_0^2}\right) \tag{8.4.146}$$

and (8.4.127), where

$$\Gamma_0 = \frac{1}{\pi}\int_0^{\infty} \sin^{-1}\kappa(t)dt,$$

with $\kappa(t)$ being the correlation function of $\xi_2(.)$. Define

$$\dot{\widehat{x}}^{\epsilon} = (A\widehat{x}^{\epsilon} + Bu - Cv) + L(t)\left[\dot{y}^{\epsilon} - \left(\frac{2}{\pi v_0^2}\right)^{\frac{1}{2}} H\widehat{x}^{\epsilon}\right]. \tag{8.4.147}$$

Now we will give the main result of this section.

Theorem 8.4.16 *Assume* (A4.11), (A4.12), *and* (A4.18). *Then the conclusions of Theorem* 8.4.12 *and Theorem* 8.4.13 *continue to hold.*

Remark. All the analysis can be carried out for a 'soft' limiter of the form $h(x) = \text{sign}(x)$ for $|x| > c > 0$, $h(x) = x/c$ for $|x| < c$.

In here, we obtained filtering and near optimality results for linear stochastic differential games with wide band noise perturbations. It is clear from Example 8.4.10 that the limits of $\{u^{\epsilon}(y^{\epsilon},.), v(y^{\epsilon},.)\}$ would not necessarily be dependent only on the limit data y—even when $y^{\epsilon}(.) \Longrightarrow y(.)$. The case of partly nonlinear observations is also considered. Using the methods of this subsection, we can extend the results to the conditional Gaussian problem, in which, the coefficients of x^{ϵ} and ξ_2^{ϵ} in the observation equation (8.4.118) can depend on the estimate $\widehat{x}^{\epsilon}$ and on $P^{\epsilon}(.)$.

8.5 Applications

Stochastic differential game models are increasingly used in various fields. Military applications of differential games are well known. There is much research in the fields of mathematical finance and economics.

In [120], stochastic differential game techniques are applied to compare the performance of a medium-range air-to-air missile for different values of the second ignition time in a two-pulse rocket motor. The measure of performance is the probability that it will reach a lock-on-point with a favorable range of guidance and flight parameters, during a fixed time interval. A similar problem is considered in [121].

In mathematical finance for example, consider two investors (players) who have available two different, but possibly correlated, investment opportunities. This could be modeled as stochastic dynamic investment games in continuous time [20]. There is a single payoff function which depends on both investors' wealth processes. One player chooses a dynamic portfolio strategy in order to maximize this expected payoff while his opponent is simultaneously choosing a dynamic portfolio strategy so as to minimize the same quality. This leads to a stochastic differential game with controlled drift and variance. Consider games with payoffs that depend on the achievement of relative performance goals and/or shortfalls. [20] provides conditions under which a game with a general payoff function has an achievable value, and gave an explicit representation for the value and resulting equilibrium portfolio strategies in that case. It is shown that nonperfect correlation is required to rule out trivial solutions. This result allow a new interpretation of the market price of risk in a Black-Scholes world. Games with discounting are also discussed as are games of fixed duration related to utility maximization. In [6], a stochastic model of monetary policy and inflation in continuous-time has been studied. We refer to [98] for a review of:

(i) the development of the general equilibrium option pricing model by Black and Scholes, and the subsequent modifications of this model by Merton and others;

(ii) the empirical verification of these models; and

(iii) applications of these models to value other contigent claim assets such as the debt and equity of a levered firm and dual purpose mutual funds.

Economists are interested in bargaining not only because many transactions are negotiated but also because, conceptually, bargaining is precisely the opposite of the "perfect competition" among infinitely many traders, in terms of which economists often think about the markets. With the advances in game theory, attempts were made to develop theories of bargaining which would predict particular outcomes in the contract curve. John Nash initiated work on this direction. Nash's approach of analyzing bargaining with complementary models—abstract models which focus on outcomes, in the spirit of "cooperative" game theory, and more detailed strategic models, in the spirit of 'non-cooperative" game theory—has influenced much of the game theoretic applications in economics. We refer to [39], [92] and [93] for more details as well as details on some new approaches based on experimental economics. For a study on stochastic differential games in economic modeling, refer to [49]. We will now describe the idea of Nash equilibrium applied to the study of institutional investor speculation. The material described in the next subsection mainly comes from [123].

8.5.1 Stochastic equity investment model with institutional investor speculation

Recent time has witnessed mounting concern and interest in the growing power of institutional investors (fund houses of various kinds) in financial markets. The shares of corporations have been increasingly concentrated in the hands of institutional investors and these investors have become the major holders of corporate stock. Since the asset prices are mainly influenced by trading, a large volume of speculative buying and selling by institutional investors often produce a profound effect on market volatility. The asset prices might fluctuate for reasons having to do more with speculative activities than with information about true fundamental values which lead to studying investment behavior in a strategically interactive framework. Since the financial assets are traded continuously, it is reasonable to assume that the price dynamics are a continuous time stochastic process.

Let $R(s)$ be the gross revenue/earning of a firm at time $s \in [0, \infty)$ and let m be the corresponding outlay generating this return. The net return/earnings of the stock of the firm at time s is then $R(s) - m$. The value of the firm at any time t with the discount rate

r can be obtained as

$$V(t) = \int_t^\infty [R(s) - m] \exp[-r(s-t)] ds. \tag{8.5.148}$$

The value $V(t)$, normalized with respect to the total number of shares issued, reflects actually the price of the firm's stock and is denoted by $P(t)$. The future gross revenues are not known with certainty and vary over time according to the following dynamics:

$$dR(s) = k\left[\overline{R} - R(s)\right] ds + R(s)\mho dw(s), \tag{8.5.149}$$

where $w(s)$ is a Wiener process. The term $\mho$ is a scalar factor governing the magnitude of the stochastic element. Gross revenue tends to perturb around a central tendency $\overline{R}$, and k is the positive parameter gauging the rate adjustment of gross revenues toward their central tendency. Hence the net return of the firm center around $\overline{R}-m$. Also, $R(s)$ remains positive if its initial value is positive. To simplify the derivation of a closed form solution, the proportion of m to $\overline{R}$ is assumed to be equal to $k/(r+k)$.

An issue concerning institutional investors is that they are capable of initiating large block transactions. Since asset prices are influenced largely by trading, a large volume of speculative buying and selling by institutional investors often produces a significant effect on market volatility. The following model reflects the sensitivity of market price to institutional investors actions. Let there be n institutional investors in the market. In [123], it is assumed that n is less that three and the price dynamics is given as

$$\begin{aligned} dP(s) = & \left\{ -a\left[\sum_{j=1}^{n} u_j\right]^{1/3} - (k/r)\left[rP(s) - \left(\overline{R}-m\right)\right]\right\} ds \\ & + P(s)\mho dw(s), \end{aligned} \tag{8.5.150}$$

where u_j is the quantity of stock sold by institutional investor j. Negative u_j represents the quantity of stock purchased. The parameter a gauges the sensitivity of market price to the large traders action. The dynamics (8.5.150) show that institutional buying would create an upward pressure on equity price and that institutional selling would exert a downward pressure. Denoting the quantity of stock held by institutional investor i at time s by $x_i(s)$ and the discount rate by r, the ith investor seeks to maximize the payoff

$$J_i\left(u_i, P, R, x, t\right) = E_0\left[\int_0^\infty \{P(s)u_i(s) + [R(s)-m]x_i(s)\} \exp(-rs) ds\right], \tag{8.5.151}$$

subject to the stock dynamics

$$dx_i(s) = -u_i(s) ds, \tag{8.5.152}$$

earning variation (8.5.149) and price dynamics (8.5.150). The term $P(s)u_i(s)$ represents the revenue/outlay from selling/buying of stocks at time s, and the dividend yield is $[R(s) - m]x_i(s)$. Equation (8.5.152) shows that the quantity of stock held by institutional investor i varies according to their buying and selling the stock.

Now we consider the equilibrium outcome in the equity market defined by (8.5.149), (8.5.150), (8.5.151) and (8.5.152). The solution concept adopted is a feedback Nash equilibrium (FNE). The institutional investors use feedback buying and selling strategies, which at each point of time s depend on the observed values of stock price, the firm dividend, and the quantity of stock held by each institutional investor. Let $x = (x_1, \ldots, x_n)$ be the vector of stock holdings of institutional investors.

Definition 8.5.1 *A feedback buying and selling strategy of institutional investor i is a decision rule $u_i(s) = \Phi(P, R, x, s)$ such that Φ_i is uniformly Lipschitz continuous in P, R, and x at every instant s in the game horizon. The set of feasible feedback strategies for institutional investor i is denoted by A_i.*

These feedback strategies satisfy the property that investor actions are based on observed market information at each time instant. The maximized payoff of the ith institutional investor is denoted by

$$V^i(P, R, x, t) = \max_{u_i \in A_i} J_i\left(u_i, P, R, x, t\right) \tag{8.5.153}$$

By the principle of optimality, $V^i(P, R, x, t)$ must satisfy the following HJB equations:

$$\begin{aligned} V_t^i = \max_{u_i \in A_i} \Big\{ & V_{x_i}^i u_i - [P u_i + (R-m) x_i \exp(-rt) \\ & + V_P^i \left(-a \left(\sum_{j=1}^n u_j \right)^3 - (k/r)\left[rP - (\overline{R} - m)\right] \right) + V_R^i \left[k\left(\overline{R} - R\right)\right] \\ & + (1/2) V_{PP}^i \mho^2 P^2 + (1/2) V_{RR}^i \mho^2 R^2 + V_{PR}^i \mho^2 PR] \Big\}, \end{aligned} \tag{8.5.154}$$

$i = 1, \ldots, n$. Equations (8.5.154) characterizes the maximized payoffs and give conditions from which the optimal feedback strategies of the institutional investors are derived. From this, the following set of first order equations are obtained:

$$P \exp(-rt) = V_{x_i}^i + V_P^i 3a \left(\sum_{j=1}^n u_j \right)^2, \; i = 1, \ldots, n. \tag{8.5.155}$$

The left hand side term of (8.5.155) is the price (in present value) of a unit of the firm's stock. The term $V_{x_i}^i$ measures the change in maximized payoff due to marginal change in the quantity of stock held by the institutional investor i. The term V_P^i is the change in the investor i maximized payoff brought about by a marginal change in price and can be interpreted as the marginal value of maintaining price at P. The marginal effect on the stock price brought about by buying and selling is represented by the term $3a\left(\sum_1^n u_j\right)^2$. The right hand side of (8.5.155) reflects the marginal cost (gain) of selling (buying) and the left hand side shows the marginal cost (gain) of selling (buying). In an optimal situation, institutional investors would buy or sell up to the point where the marginal gain equals the marginal cost of trading the stock. Since the marginal effect of one institutional investor buying and selling on the stock price is related to the actions of other institutional investors, the optimal strategies are interrelated. The best (optimal) response/reaction functions of the institutional investor i to the actions of the competitors at time t can be expressed as

$$u_i = \left\{ \left[P \exp(-rt) - V_{x_i}^i\right] / 3a V_p^i \right\}^{1/2} - \sum_{j=1, j \neq i}^n u_j. \tag{8.5.156}$$

The derivation of institutional investor i's optimal strategy at any time is a decision making process which takes into consideration three types of factors:

(i) current observed market information $(P(t), R(t), x(t), r)$,

(ii) optimal strategies chosen by competing institutional investors, and

(iii) marginal value of holding the stock and marginal value of maintaining price at P.

The first type of factor is available at each instant of time. The second factor is derived from the premise that investors are rational and they choose their actions with full consideration of their competitor's rational behavior. The third type of factor is the result of intertemporal optimization.

Substituting u_i, $i = 1, \ldots, n$ obtained in (8.5.156) into the HJB equations (8.5.154), one gets a set of parabolic partial differential equations. Now the task is to find a set of twice differentiable functions $V^i : \mathbf{R}^3 \times [0, \infty) \to \mathbf{R}$ that is governed by this set of partial differential equations. The smooth functions yield the optimal payoffs of the institutional investors and solve the game. The optimal payoffs are obtained in [123] as

$$\begin{aligned} &V^i(P, R, x, t) \\ &= \left\{ A[P - R/(r+k)]^{4/3} + [R/(r+k)]x_i \right\} \exp(-rt), i = 1, \ldots, n, \end{aligned} \tag{8.5.157}$$

where A is a constant,

$$A = \left\{ \left[a^{-1/2}(1/2n - 1/6) \right] \div \left[r + (4/3)k - (2/9)\mho^2 \right] \right\}^{2/3}.$$

The value function $V^i(P, R, x, t)$ yields the equilibrium payoff of institutional investor i. Following [95] it is assumed that $\mho^2 \leq k$. This assumption guarantees that A is positive. From (8.5.157), one can derive two marginal valuation measures. The institutional investor i's marginal value of maintaining price at P can be derived as

$$V_P^i = (4A/3)[P - R/(r+k)]^{1/3} \exp(-rt). \tag{8.5.158}$$

The investor marginal value of holding the stock can be obtained as

$$V_{x_i}^i = [R/(r+k)] \exp(-rt). \tag{8.5.159}$$

The marginal value of stock holding is always positive. It is increasing in the current earnings and reflects the fact that higher yields raise the value of holding the stock. At the same time, it is negatively related to the discount rate and exhibits the property that the gains from investing in the stock decline as the discount rate rises. Also from (8.5.158), the investor marginal value of maintaining price at P is positive (negative) when P is greater (less) than $R/(r+k)$.

Now we can derive a feedback Nash equilibrium of the equity market with speculating investors. Substituting V_P^i in (8.5.158) and $V_{x_i}^i$ in (8.5.159) into the optimal strategies given in (8.5.156), the feedback Nash equilibrium buying and selling strategies of institutional investor i is obtained as

$$\Phi_i(P, R, x, t) = (1/n)(1/4Aa)^{1/2}[P - R/(r+k)]^{1/3}, i = 1, \ldots, n. \tag{8.5.160}$$

The set of feedback buying and selling strategies in (8.5.160) constitutes a feedback Nash equilibrium of the equity market as characterized by (8.5.149), (8.5.150), (8.5.151), and (8.5.152). These buying and selling strategies are decision rules contigent upon the current values of the price and earnings.

To examine the impact of the institutional investor speculation on stock price volatility, substitute the feedback strategies in (8.5.160) into (8.5.150) to obtain the equilibrium price dynamics

$$\begin{aligned} dP(s) = &\left\{ -a(1/4aA)^{3/2}[P(s) - R(s)/(r+k)] \right. \\ &\left. -(k/r)\left[rP(s) - r\overline{R}/(r+k) \right] \right\} ds + P(s)\mho dw(s). \end{aligned} \tag{8.5.161}$$

This along with (8.5.149) characterize the joint behavior of the stock price and earnings of the firm. In [95], for the equity market with numerous ordinary investor, the change in stock price of the firm is modeled by

$$dP(s) = -(k/r)\left[rP(s) - \left(\overline{R} - m\right)\right] ds + P(s)\mho dw(s). \tag{8.5.162}$$

A comparison between (8.5.161) and (8.5.162) shows additional movements, symbolized as the first term in the right-hand side of (8.5.161), in the price dynamics caused by institutional investors. In [123] an analysis is given to show that the prices tend to rise in spite of the fact that they have been valued above their intrinsic value and prices tend to drop although $P(s)$ is below its intrinsic value in the presence of institutional speculation. Hence one could conclude that the market is more volatile in the presence of institutional speculation. Following results are proved in [123]:

(i) The greater the discrepancy between P and $R/(r+k)$, the higher the profit of an institutional investor, and

(ii) The greater the degree of uncertainty in the market, the higher the speculative profits.

This implies that institutional investors are more attracted to markets with high uncertainty, such as emerging markets.

8.6 Conclusion

In this presentation, we have attempted to explain stochastic differential games in competitive situations. For the analysis of stochastic differential games, we have presented some probability techniques such as martingale methods and weak convergence methods, and some analytical methods such as viscosity solution techniques. We have also mentioned some applications of stochastic differential games and presented in some detail a stochastic differential game of institutional investor speculation. We have given a substantial, yet by no means exhaustive, bibliography. It needs to be noted that even though there have been some attempts made at obtaining numerical methods for stochastic differential games, it is still a wide open area that needs the attention of investigators. There are many other solution concepts in stochastic differential games. We did not make any effort in presenting or referring to the bibliography on stochastic differential games which are not completely competitive in nature. The area of stochastic differential games with imperfect informations needs much more work. Recently, there have been some works initiated in risk-sensitive stochastic differential games, [5, 105]. Another direction of interest is backward equations and stochastic differential games [44, 45].

Bibliography

[1] S.I. Aihara and A. Bagchi, Linear-quadratic stochastic differential games for distributed parameter systems: Pursuit-evasion differential games. *Comput. Math. Appl.*, 13:247–259, 1987.

[2] R. Bafico. On the definition of stochastic differential games and the existence of saddle points. it Ann. Mat. Pura Appl., 96: 41–67 1972.

[3] M. Bardi and T.E.S. Raghavan and T. Parthasarathy. *Stochastic and differential games: Theory and numerical methods.* Birkhäuser, 1999.

[4] T. Başar, Existence of unique Nash equilibrium solutions in nonzero-sum stochastic differential games. In *Differential games and control theory, II, (Proc. 2nd Conf.. Univ. Rhode Island, Kingston, RI, 1976)*, pages 201–228. Dekker, New York, 1976.

[5] T. Başar. Nash equilibrium of risk-sensitive nonlinear differential games. *J. Optim. Theory Appl.*, 100: 479–498, 1999.

[6] T. Başar. *A continuous-time model of monetary policy and inflation: a stochastic differential game*, volume 353 of *Lecture Notes Econom. and Math. Systems*, pages 3–17. Springer, 1991 (Modena, 1989).

[7] T. Başar. On the existence and uniqueness of closed-loop sampled data Nash controls in linear-quadratic stochastic differential games. In K. Iracki, K. Malonowsi, and S. Walukiewicz, editor, *Optimal Techniques*, volume 22 of *Lecture Notes in Control and Information Sciences*, pages 193–203. Springer-Verlag, 1980.

[8] T. Başar and P. Bernhard. *H_∞-Optimal control and related minimax design problems: A dynamic game approach.* Birkhäuser, 2nd edition, 1995.

[9] T. Başar and A. Haurie. Feedback equilibria in differential games with structural and modal uncertainties In J.B. Cruz, Jr., editor, *Advances in large scale systems.* volume 1, pages 163–301. JAI, Greenwich, CT, 1984.

[10] T. Başar and A. Haurie. *Advances in dynamic games and applications.* Birkhäuser, 1994.

[11] T. Başar and G.J. Olsder. *Dynamic noncooperative game theory.* Academic Press, 2nd edition, 1995.

[12] V.E. Benes. Existence of optimal strategies based on a specific information for a class of stochastic decision problems. *SIAM J. Control*, 8: 179-188, 1970.

[13] R.D. Behn and Y.C. Ho. On a class of linear stochastic differential games. *IEEE Trans. Automatic Control*, AC-13:227-240, 1968.

[14] A. Bensoussan and J.L. Lions. Stochastic differential games with stopping times. In *Differential games and control theory, II, (Proc. 2nd Conf.. Univ. Rhode Island, Kingston, RI, 1976)*, volume 30 of *Lectures Notes in Pure and Applied Mathematics*, pages 377-399, New York, 1976. Dekker.

[15] A. Bensoussan and A. Friedman. Nonlinear variational inequalities and differential games with stopping times. *J. Functional Analysis*, 16: 305-352, 1974.

[16] A. Bensoussan and A. Friedman. Nonzero-sum stochastic differential games with stopping times and free boundary problems. *Trans. Amer. Math. Soc.*, 231: 275–327, 1977.

[17] L.D. Berkovitz. Two person zero sum differential games: an overview. In J. D. Grote, editor *The theory and application of differential games, (Proc. NATO Advanced Study Inst.. Univ. Warwick, Coventry, England, 1974)*, pages 13–22, Dordrecht, 27 August-6 Septebmer 1975, Riedel.

[18] V.S. Borkar and M.K. Ghosh. Stochastic differential games: An occupation measure based approach. *Journal of Optimization Theory and Applications*, 73:359–385, 1992.

[19] V.S. Borkar. Optimal control of diffusion processes. In *Pitman Research Notes in Mathematics*, volume 203. Longman Scientific & Technical, Harlow, 1989, (Co-published in the United States with John Wiley & Sons Inc.. New York).

[20] S. Browne. Stochastic differential portfolio games. Working Paper Series PW-97-17, PaineWebber, 1997.

[21] R.J. Chitashvili and N.V. Elbakidze. *Optimal stopping by two players*, pages 10-53. Translation Series - Mathematics & Engineering. Optimization Software, New York, 1984.

[22] M.G. Crandall and P.L. Lions. Viscosity solutions of Hamilton-Jacobi equations. *Trans. of the AMS*, 277:1–42, 1983.

[23] W.B. Davenport. Signal to noise ratios in band pas limiters. *J. Appl. Phys.*, 24, 1953.

[24] R.J. Elliott. The existence of optimal strategies and saddle points in stochastic differential games. In *Differential games and applications, (Proc. Workshop, Enschede, 1977)*, volume 3 of *Lectures Notes in Control and Information Sciences*, pages 123-135. Springer, Berlin, 1977.

[25] R.J. Elliott. Stochastic differential games and alternate play. In *International Symposium, IRIA, LABORIA, Rocquencourt, 1974*, volume 107, of *Lectures Notes in Economics and Mathematical Systems*, pages 97-106. Springer-Verlag, Berlin, 1975.

[26] R.J. Elliott. Introduction to differential games II. In J. Grote and D. Reidel, editors, *Stochastic games and parabolic equations*, The Theory and Application of Differential Games, pages 34-43. Dordrecht, Holland, 1975.

[27] R.J. Elliott. The existence of value in stochastic differential games. *SIAM Journal of Control*, 14:85–94 1976.

[28] L.C. Evans and P.E. Souganidis. Differential games and representation formulas for solutions of Hamilton-Jacobi-Isaacs equations. *Ind. Univ. Math. J.* 33:773–797, 1984.

[29] K. Fan. Fixed points and minimax theorems in locally convex topological linear spaces. *Proc. Nat. Acad. Sci. U.S.A.*, 38:121–126 195.

[30] J.A. Filar and K. Vrieze. *Competitive Markov Decision Processes.* Springer-Verlag, 1997.

[31] W.H. Fleming. Generalized solutions in optimal stochastic control. *Differential games and control theory, II, (Proc. 2nd Conf., Univ. Rhode Island, Kingston, RI, 1976)* , volume 30 of *Lecture Notes in Pure and Applied Mathematics*,pages 147–165, New York, 1977. Dekker.

[32] W.H. Fleming. The convergence problem for differential games. *J. Math. Analysis and Applications*, 3:102–116, 1961.

[33] W.H. Fleming and H.M. Soner. *Controlled Markov processes and viscosity solutions.* Springer-Verlag, 1993.

[34] W.H. Fleming and P.E. Souganidis. On the existence of value functions of two-player, zero-sum stochastic differential games. *Indiana Univ. Math. J.*,38:293–314, 1989.

[35] W.H. Fleming and P.E. Souganidis. *Two player, zero sum stochastic differential games*, pages 11-164. Gauthier-Villars, 1988.

[36] A. Friedman. *Differential games.* Wiley, 1971.

[37] A. Friedman. *Stochastic differential equations and applications*, volume 2, Academic Press, 1976.

[38] A. Friedman. Stochastic differential games. *J. Differential Equations*, 11:79–108, 1972.

[39] S.D. Gaidov. On the Nash-bargaining solution in stochastic differential games. *Serdica*, 16:120–125, 1990.

[40] S.D. Gaidov. Mean-square strategies in stochastic differential games. *Problems Control Inform. Theory/Problemy Upravlen. Teor. Inform.*, 18:161–168, 1989.

[41] M.K. Ghosh and K.S. Kumar. Zero-sum stochastic differential games with reflecting diffusions. *Math. Appl. Comput.*, 16:237–246, 1997.

[42] M.K. Ghosh and S.I. Marcus. Stochastic differential games with multiple modes. *Stochastic Analysis and Applications*, 16:91–105, 1998.

[43] P. Hagedorn and H.W. Knobloch and G.J. Olsderm, editors. *Differential games and applications*, volume 31 of *Lecture Notes in Control and Information Sciences.* Springer-Verlag, Berlin, 1977.

[44] S. Hamadène. Backward-forward SDE's and stochastic differential games. *Stochastic Process. Appl.*, 77:1–15, 1998.

[45] S. Hamadène and J.P. Lipeltier. Backward equations, stochastic control and zero-sum stochastic differential games. *Stochastics Stochastics Rep.*, 54:221–231, 1995.

[46] R.P. Hàmalainen and H. Ehtamo. *Advances in Dynamic Games and Applications*, volume 1 of *annals of the ISDG.* Birkhäuser, 1994.

[47] R.P. Hàmalainen and H. Ehtamo. *Dynamic games in economic analysis*, volume 157 of *Lectures Notes in Control and Information Sciences.* Springer-Verlag, 1991.

[48] R.P. Hàmalainen and H. Ehtamo. *Differential games: Developments in modelling and computation*, volume 156 of *Lectures Notes in Control and Information Sciences.* Springer-Verlag, 1991.

[49] A. Haurie. *Stochastic differential games in economic modeling*, volume 197 of *Lecture Notes in Control and Information Sciences.* Springer, 1994.

[50] Y.C. Ho. Optimal terminal maneuver and evasion strategy. *SIAM J. Control*, 4:421–428, 1966.

[51] Y.C. Ho. On maximum principle and zero-sum stochastic differential games. *JOTA*, 13, 1974.

[52] R. Isaacs. Differential Games I. Research Memoranda RM-1391, The RAND Corporation, 1954.

[53] R. Isaacs. Differential Games II. Research Memoranda RM-1399, The RAND Corporation, 1954.

[54] R. Isaacs. Differential Games III. Research Memoranda RM-1411, The RAND Corporation, 1954.-1411.

[55] R. Isaacs. Differential Games IV. Research Memoranda RM-1486, The RAND Corporation, 1954.

[56] R. Isaacs. *Differential Games.* John Wiley and Sons, New York, 1965.

[57] H. Ishii. On uniqueness and existence of viscosity solutions for fully nonlinear second order elliptic pde. *Comm. Pure Appl. Math.*, 42:14–45 1989.

[58] S. Jørgensen and D.W.K. Yeung. Stochastic differential game model of a common p roperty fishery. *J. Optim. Theory Appl.*, 90:381–403, 1996.

[59] R.E. Kalman and R.S. Bucy. New results in linear filtering and prediction theory. *Trans. ASME, J. Basic Eng., Ser. D.*, 83:95–108, 1961.

[60] N.J. Kalton and N.N. Krasovskii and A.I. Subbotin. *Positional differential games.* Nauka, 1974, (Springer, 1988).

[61] N.V. Krylov. *Controlled Diffusion Processes.* Springer, New York, 1980.

[62] H.J. Kushner. *Approximation and weak convergence methods for random processes, with applications to stochastic systems theory.* MIT Press, 1984.

[63] H.J. Kushner. *Weak convergence methods and singularly perturbed stochastic control and filtering problems.* Birkhäuser, 1990.

[64] H.J. Kushner and S.G. Chamberlain. On stochastic differential games: sufficient conditions that a given strategy be a saddle point, and numerical procedures for the solution of the game. *J. Math. Anal. Appl.*, 26:560–575, 1969.

[65] H.J. Kushner and P.G. Dupuis. *Numerical methods for stochastic control problems in continuous time.* Springer-Verlag, 1992.

[66] H.J. Kushner and W. Runggaldier. Filtering and control for wide bandwidth noise driven systems. Report #86-8, LCDS, 1986.

[67] N.N. Krasovskii and A.I. Subbotin. *Game theoretical control problems.* Springer-Verlag, 1988.

[68] T.G. Kurtz. Semigroups of conditional shifts and approximations of Markov processes. *Annals of Probability,* 4, 1975.

[69] G. Leitmann. *Multicriteria decision making and differential games.* Plenum Press, 1976.

[70] J. Lewin. *Differential games.* Springer, 1994.

[71] P.L. Lions and P.E. Souganidis. Differential games, optimal control and directional derivatives of viscosity solutions of Bellman's and Isaacs' equations. *SIAM J. of Control and Optimization,* 23:566–583, 1985.

[72] P.L. Lions and P.E. Souganidis. Differential games, optimal control and directional derivatives of viscosity solutions of Bellman's and Isaacs' equations, II. *SIAM J. of Control and Optimization,* 24:1086–1089, 1986.

[73] P.L. Lions and P.E. Souganidis. Viscosity solutions of second-order equations, stochastic control and stochastic differential games. In *Stochastic differential systems, stochastic control theory and applications,* pages 293-309. Springer-Verlag, 1988.

[74] R.S. Lipster and A.N. Shiryaev. *Statistics of Random Processes.* Springer-Verlag, 1977.

[75] D. Lund and B. Øksendal. *Stochastic models and option values: Applications to resources, environment and investment problems.* North-Holland, 1991.

[76] R.C. Merton. Theory of finance from the perspective of continuous time. *Journal of Financial and Quantitative Analysis,* pages 659-674, 1975.

[77] H. Morimoto and M. Ohashi. On linear stochastic differential games with average cost criterions. *J. Optim. Theory Appl.,* 64:127–140, 1990.

[78] W.G. Nicholas. Stochastic differential games and control theory. Master's thesis, Virginia Polytechnic Institute and State University, Blackburg, Virginia, 1971.

[79] M. Nisio. Stochastic differential games and viscosity solutions of Isaacs equations. *Nagoya Math. J.,* 110:163–184, 1988.

[80] M. Nisio. On infinite-dimensional stochastic differential games. *Osaka J. Math.,* 35:15–33, 1998.

[81] G.J. Olsder. *New trends in dynamic games and applications.* Birkhäuser, Boston, 1995.

[82] G.J. Olsder. On observation costs and information structures in stochastic differential games. In *Differential games and applications, (Proc. Workshop, Enschede, 1977),* volume 3 of *Lecture Notes in Control and Information Sciences,* pages 172 - 185. Springer, Berlin, 1977.

[83] G.J. Olsder. *New trends in dynamic games and applications.* Birkäuser, 1995.

[84] K.M. Ramachandran. Stochastic differential games with a small parameter. *Stochastics and Stochastics Reports*, 43:73–91 1993.

[85] K.M. Ramachandran. N-Person stochastic differential games with wideband noise perturbation. *Journal of Combinatorics, & Information System Sciences*, 21(3-4):245-260, 1996.

[86] K.M. Ramachandran. Weak convergence of partially observed zero-sum stochastic differential games. *Dynamical Systems and Applications*, 4(3):329-340, 1995.

[87] K.M. Ramachandran. Discrete parameter singular control problem with state dependent noise and non-smooth dynamics. *Stochastic Analysis and Applications*, 12:261–276, 1994.

[88] K.M. Ramachandran and A.N.V. Rao. Deterministic approximation to two person stochastic game problems. *Dynamics of Continuous, Discrete and Impulsive Systems.* 1998, (To appear).

[89] K.M. Ramachandran and A.N.V. Rao.N-person stochastic differential games with wideband noise perturbations: Pathwise average cost per unit time problem. Preprint, 1999.

[90] I.B. Rhodes and D.G. Luenberger. Differential games with imperfect state information. *IEEE Trans. Automatic Control.* AC-14:29-38, 1969.

[91] I.B. Rhodes and D.G. Luenberger. Stochastic differential games with constrained state estimators. *IEEE Trans. on Automatic Control.*AC-14:476-481, 1969.

[92] A.E. Roth. *Game-Theoretic models of bargaining.* Cambridge, 1985.

[93] A.E. Roth. Bargaining experiments. In J. Kagel and A. E. Roth, editors *Hondabook of Experimental Economics*, pages 253-348. Princeton University Press, 1995.

[94] Emilio Roxin and Chris P. Tsokos. On the Definition of a Stochastic Differential Game. *Mathematical Systems Theory*, 4(1):60–64, 1970.

[95] P.A. Samuelson. Rational theory of warrant pricing. *Industrial Management Review*, 6:13–31, 1965.

[96] L.S. Shapley. Stochastic games. *Proceedings of the National Academy of Science U.S.A.*, 39:1095–1100, 1953.

[97] K. Shell. The theory of Hamiltonian dynamical systems, and an application to economics. In J.D. Grote, editor, *The Theory and Application of Differential Games*, pages 189-199. D. Reidel Publishing company, 1975.

[98] C.W. Smith, Jr.,Option Pricing: A review. *Journal of Financial Economics*, 3:3–51, 1976.

[99] P.E. Souganidis. Approximation schemes for viscosity solutions of Hamilton-Jacobi equations with applications to differential games. *Journal of Nonlinear Analysis, T.M.A.*, 9:217–257, 1985.

[100] P.E. Souganidis. Two player, zero-sum differential games and viscosity solutions. In M. Barji, T.E.S. Raghaven, and T. Parthasarathy, editors, *Stochastic and differential games: Theory and numerical methods*, pages 69-104. Birkhäuser, 1999.

[101] J.L. Speyer. A stochastic differential game with controllable statistical parameters. *IEEE Trans. Systems Sci. Cybernetics*, SSC-3:17-20, 1967.

[102] J.L. Speyer and S. Samn and R. Albanese. A stochastic differential game theory approach to human operators in adversary tracking encounters. *IEEE Trans. Systems Man Cybernet.*, 10755–762, 1980.

[103] F.K. Sun and Y.C. Ho. Role of information in the stochastic zero-sum differential game. In G. Leitmann, editor, *Multicriteria decision making and differential games.* Plenum Press, 1976 .

[104] L. Stetner. Zero-sum Markov games with stopping and impulsive strategies. *Appl. Math. Optim.*, 9:1–24, 1982.

[105] A. Swiech. Risk-sensitive control and differential games in infinite dimensions. Preprint, 1999.

[106] A. Swiech. Another approach to existence of value functions of stochastic differential games. Preprint, 1999.

[107] K. Szajowski. Markov stopping games with random priority. *Zeitschrift f ur Operations Research*, 39(1):69-84, 1993.

[108] K. Uchida. On existence of a Nash equilibrium point in N-person nonzero sum stochastic differential games. *SIAM J. Control Optim.*, 16:142–149, 1978.

[109] P.P. Varaiya. N-person stochastic differential games. In J. Grote and D. Reidel, editors, *The Theory and Application of Differential Games.* pages 97-107. Dordrecht, Holland, 1975.

[110] P.P. Varaiya. N-player stochastic differential games. *SIAM J. Control Optim.*,4:538–545, 1976.

[111] P.P. Varaiya and J. Lin. Existence of saddle points in differential games. *SIAM Jour. Control*, pages 141-157, 1969.

[112] J. von Neumann and O. Morgenstern. *Theory of games and economic behavior.* Princeton University Press, 1944.

[113] A.Ju. Veretennikov. On strong solution and explicit formulas for solutions of stochastic integral equations. *Math. USSR-Sb.*, 39:387–403, 1981.

[114] T.L. Vincent. An evolutionary game theory for differential equation models with reference to ecosystem management. In t. Başar and A. Haurie, editors, *Advances in dynamic games and applications*, pages 356-374. Birkhäuser, 1994.

[115] B. Wernerfelt. Uniqueness of Nash equilibrium for linear-convex stochastic differential games. *J. Optim. Theory Appl.*, 53:133–138, 1987.

[116] W. Willman. Formal solution of a class of stochastic differential games. *IEEE Trans. on Automatic Control*, AC-14:504-509, 1969.

[117] Y. Yavin. The numerical solution of three stochastic differential games. *Comput. Math. Appl.*, 10:207–234, 1984.

[118] Y. Yavin. Computation of Nash equilibrium pairs of a stochastic differential game. *Optimal Control Appl. Methods*, 2:443–464, 1981.

[119] Y. Yavin. Computation of suboptimal Nash strategies for a stochastic differential game under partial observation. *International Journal of Systems Science*, 13:1093–1107, 1982.

[120] Y. Yavin. Applications of stochastic differential games to the suboptimal design of pulse motors: pursuit-evasion differential games, III. *Computational Mathematics & Applications*, 26:87–95, 1993.

[121] Y. Yavin and R. de Villiers. Application of stochastic differential games to medium-range air-to-air missiles. *Journal of Optimization Theory & Applications*, 67:355–367, 1990.

[122] D. Yeung. A feedback Nash equilibrium solution for noncooperative innovations in a stochastic differential framework. *Stochastic Analysis & Applications*, 9:195–213, 1991.

[123] D.W.K. Yeung. A stochastic differential game of Institutional Investor speculation. *Journal of Optimization Theory & Applications*, 102:463–477, 1999.

[124] D.W.K. Yeung and M.T. Cheung. Capital accumulation subject to pollution control: a differential game with a feedback Nash equilibrium. In T. Başar and A. Haurie, editors, *Advances in dynamic games and applications*, pages 289–300. Birkhäuser, 1994.

Chapter 9

Stochastic Manufacturing Systems: A Hierarchical Control Approach

Q. Zhang
Department of Mathematics
University of Georgia
Athens, GA 30602

Most manufacturing systems are large, complex, and subject to uncertainty. Obtaining exact feedback policies to run these systems is nearly impossible. It is a common practice to manage such systems in a hierarchical fashion. This chapter surveys a hierarchical control approach for dealing with large-scale manufacturing systems. Various production models and system configurations are discussed. Both the discounted and long-run average cost criteria are considered.

9.1 Introduction

This chapter is concerned with decision making in manufacturing systems under uncertainty. It focuses on an important method in dealing with the optimization of large, complex systems – hierarchical control approach. The basic idea is to reduce the overall complex problem into manageable approximate problems or subproblems, to solve these problems, and to construct a solution of the original problem from the solutions of these simpler problems.

Manufacturing systems are usually large and complex, characterized by several decision subsystems. Moreover, these systems are subject to various discrete events, such as purchasing new equipment and machine failures and repairs. Management must recognize and react to these events. Because of the large size of these systems and the presence of these events, obtaining exact optimal feedback policies to run these systems is nearly impossible both theoretically and computationally.

[1]keywords: manufacturing system, hierarchical control, Markov chains
[2]90B30, 93A13, 93E20
[3]This research was supported in part by the ONR Grant N00014-96-1-0263.

The recognition of the difficulty in solving production planning problems in stochastic manufacturing systems has resulted in various attempts to obtain suboptimal or near-optimal controls. Even the research dealing with approximate solutions of the problem have without exception addressed small-sized problems. In practice, therefore, these systems are managed in a hierarchical fashion. There has been a growing interest in showing that hierarchical decision making in the context of a goal-seeking manufacturing systems leads to a near optimization of its objective.

There are several different, and not mutually exclusive, ways in which the reduction of the complexity is accomplished. These include decomposing the problem into problems of the smaller subsystems with a proper coordinating mechanism, aggregating products and subsequently disaggregating them, and replacing random processes with their averages.

It is the last method to which our approach based on singular perturbations or time scale separation is related. In this approach, different types of events taking place in the system have different frequencies of their occurrence, which define the hierarchical levels. For obtaining the decisions at each level, as suggested by Gershwin [18], quantities that vary slowly (variables that correspond to higher levels) are treated as static. Quantities that vary much faster (variables at lower levels) are modeled in a way that ignores the variations, thus, replacing fast moving variables by their averages. For example, changes in demand may occur far more slowly than breakdowns and repairs of production machines as formulated in Sethi, Taksar, and Zhang [30]. This suggests that capital expansion decisions that respond to demand are relatively longer term decisions than decisions regarding production. It is then possible to base capital expansion decisions on the average existing production capacity, and expect these decisions to be nearly optimal even though the rapid changes in machine states are ignored. Having the longer term decisions in hand, one can then solve the simpler problem of obtaining production rates. More specifically, it is shown in [30] that the two-level decisions constructed in this manner are asymptotically optimal as the rate of fluctuation in the production capacity becomes large in comparison with the rates with which other events occur.

In this chapter, we begin with a manufacturing system which consists of machines that are subject to breakdown and repair. More complex systems including multilevel systems are discussed subsequently. The objective of the system is to obtain the rate of production over time in order to meet the demands at the minimum expected discounted (or long-run average) costs of production and inventory/shortages over the infinite horizon. We assume that the rates of machine breakdown and repair are much larger than the rate of fluctuation in demand and the rate of discounting [27]. The idea of hierarchical control is to derive a limiting control problem which is simpler to solve than the given problem. This limiting problem is obtained by replacing the stochastic machine availability process by the average total capacity of machines and by appropriately modifying the objective function. From its optimal control, one constructs an asymptotically optimal control of the original, more complex, problem. The idea of hierarchical approach is closely related to that of singular perturbations. For literature on singular perturbations, we refer the reader to the papers Kokotovic [26], Phillips and Kokotovic [28]. For more recent references, see Zhang and Yin [49]. This chapter focuses on hierarchical production planning in manufacturing systems. The research in manufacturing has been an active area in the recent years. The developments can be found in, for example, Caramanis and Liberopoulos [9], Caramanis and Sharifnia [10], Fleming, Sethi, and Soner [14], Gershwin [18], Haurie and van Delft [20], Hu and Caramanis [23], Jiang and Sethi [24], Kimemia and Gershwin [25], and Sharifnia [39], among others.

This chapter consists of three parts: The first part is concerned with hierarchical control with discounted costs and the second part considers the problem with long-run average costs. The third part presents analytical solutions to three relatively simple but illustrative control problems. Such solutions are useful for constructing hierarchical control discussed in the first two parts.

Part I: Control with Discounted Costs

This part is divided into several sections. We start from a single machine – single part production system, and then move to other systems with different configurations.

9.2 Single Machine System

We begin with a simple example of production system with a single machine that produces a single part type. Let $x(t) \in R^1$ denote the surplus (state), $u(t) \in R^1$ the production rate (control), and $z \in R^1$ is the constant demand rate. They satisfy

$$\dot{x}(t) = u(t) - z, \; x(0) = x. \tag{9.2.1}$$

We consider the case when the underlying machine is subject to breakdown and repair. If the machine is operational (denoted by 1), then one can produce at the maximum unit rate; if the machine is under repair (denoted by 0), then nothing can be produced. Let $\alpha(t) \in \mathcal{M} = \{0, 1\}$ denote such machine capacity process. Then the production rate $u(t)$ must satisfy

$$0 \le u(t) \le \alpha(t).$$

Assume $\alpha(t)$ to be a two-state Markov chain generated by

$$Q = \begin{pmatrix} -\mu & \mu \\ \lambda & -\lambda \end{pmatrix}.$$

Here $\lambda > 0$ is the breakdown rate and $\mu > 0$ is the repair rate.

Given $x(0) = x$ and $\alpha(0) = \alpha$, we consider the cost function $J(x, \alpha)$ defined by

$$J(x, \alpha, u(\cdot)) = E \int_0^\infty e^{-\rho t}(h(x(t)) + c(u(t)))dt, \tag{9.2.2}$$

where $\rho > 0$ is the discount rate and $h(\cdot)$ is the cost of surplus and $c(\cdot)$ is the cost of production. The problem is to find a control $u(\cdot)$ that minimizes $J(x, \alpha, u(\cdot))$.

Let us consider a special case with $0 < z < 1$, $c(u) = 0$, and $h(x) = h_+ x^+ + h_- x^-$, where $x^+ = \max\{x, 0\}$ and $x^- = \max\{-x, 0\}$. We aim at obtaining a closed-form solution.

The corresponding Hamilton-Jacobi-Bellman (HJB) equations for this problem are as follows:

$$\begin{cases} \rho v(x,0) = & -z v_x(x,0) + h(x) + \mu(v(x,1) - v(x,0)) \\ \rho v(x,1) = & \min\limits_{0 \le u \le 1} (u - z) v_x(x,1) + h(x) + \lambda(v(x,0) - v(x,1)). \end{cases} \tag{9.2.3}$$

In order to solve these equations, we need to introduce the following matrices. Let

$$A_1 = \begin{pmatrix} -\dfrac{\rho + \mu}{z} & \dfrac{\mu}{z} \\ & \\ -\dfrac{\lambda}{1 - z} & \dfrac{\rho + \lambda}{1 - z} \end{pmatrix}, \; A_2 = \begin{pmatrix} -\dfrac{\rho + \mu}{z} & \dfrac{\mu}{z} \\ & \\ \dfrac{\lambda}{z} & -\dfrac{\rho + \lambda}{z} \end{pmatrix},$$

and

$$b_1 = \begin{pmatrix} \frac{1}{z} \\ \\ -\frac{1}{1-z} \end{pmatrix}, \quad b_2 = \begin{pmatrix} \frac{1}{z} \\ \\ \frac{1}{z} \end{pmatrix}, \quad b_3 = \begin{pmatrix} 1 \\ \\ 0 \end{pmatrix}.$$

Let $a_+ > 0$ and $a_- < 0$ denote the two eigenvalues of the matrix A_1. Akella and Kumar [2] define

$$x^* = \max\left(0, \frac{1}{a_-} \log\left[\frac{h_+}{h_+ + h_-}\left(1 + \frac{\rho z}{\lambda z - (\rho + \mu + z a_-)(1-z)}\right)\right]\right).$$

Then they prove that the value functions are given as follows:

$$xastg0 \begin{pmatrix} v(x,0) \\ v(x,1) \end{pmatrix} = \begin{cases} e^{A_1 x}\{e^{-A_1 x^*}[A_1^{-2} b_1 h_+ - A_1^{-1} b_3 h_+ \lambda^{-1}] \\ \quad -A_1^{-2} b_1 (h_+ + h_-)\} \\ \quad +A_1^{-1} b_1 h_- x + A_1^{-2} b_1 h_- & \text{if } x \le 0 \\ \\ e^{A_1(x - x^*)}[A_1^{-2} b_1 h_+ - A_1^{-1} b_3 h_+ \lambda^{-1}] \\ \quad -A_1^{-1} b_1 h_+ x - A_1^{-2} b_1 h_+ & \text{if } 0 \le x \le x^* \\ \\ e^{A_2(x - x^*)}[A_2^{-2} b_2 h_+ - A_2^{-1} b_3 h_+ \lambda^{-1}] \\ \quad -A_2^{-1} b_2 h_+ x - A_2^{-2} b_2 h_+ & \text{if } x \ge x^*. \end{cases} \tag{9.2.4}$$

It can be shown that x^* minimizes $v(x, 1)$ over $x \in R$. The optimal feedback control $u^*(x, \alpha)$ can be written as follows:

$$u^*(x,\alpha) = \begin{cases} 0 & \text{if } \alpha = 1,\ x > x^* \text{ or } \alpha = 0, \\ z & \text{if } \alpha = 1,\ x = x^*, \\ 1 & \text{if } \alpha = 1,\ x < x^*. \end{cases} \tag{9.2.5}$$

This kind of policy is referred to as hedging point policy. When the machine is up, produce at maximum rate if the surplus x is below the threshold level x^*, produce nothing if x is above x^*, and produce exactly as demand rate if $x = x^*$.

For a given system with more than two machine states, i.e., $\alpha(t) \in \mathcal{M} = \{0, 1, \ldots, m\}$ with $m > 1$. In this case, the problem is more involved. As a result, a closed-form solution will be difficult to obtain. In order to deal with the problem, one has to resort to approximation schemes. One important method is that of hierarchical control approach. It is typical for failure-prone systems that the demand rate usually fluctuates much slower than the rate of machine breakdown and repair. Therefore, it is reasonable to consider the capacity process $\alpha^\varepsilon(t)$ as a function of ε which characterizes the relative rate of its fluctuation. As ε gets smaller and smaller, the process $\alpha^\varepsilon(\cdot)$ jumps more and more rapidly in $\mathcal{M}$. We can formulate $\alpha^\varepsilon(\cdot)$ as a Markov chain with generator $Q^\varepsilon = Q/\varepsilon$ where $Q = (q_{ij})$ is an $(m+1) \times (m+1)$ matrix such that $q_{ij} \ge 0$ for $i \ne j$ and $q_{ii} = -\sum_{j \ne i} q_{ij}$, $i, j \in \mathcal{M}$. We assume Q to be irreducible and let ν denote the corresponding stationary distribution, i.e., $\nu = (\nu_0, \dots, \nu_m)$ is the only positive solution to

$$\nu Q = 0 \text{ and } \sum_{j=0}^{m} \nu_j = 1.$$

In system (9.2.1), the production rate $u(t) \geq 0$ must satisfy $p \cdot u(t) \leq \alpha^\varepsilon(t)$ for some vector $p \geq 0$, where $a \cdot b$ denotes the usual inner product of two vectors.

We consider a control $u(\cdot) = \{u(t) : t \geq 0\}$ to be *admissible* if $u(t) \geq 0$ is an $\mathcal{F}_t = \sigma\{\alpha^\varepsilon(s), s \leq t\}$ adapted measurable process and $p \cdot u(t) \leq \alpha^\varepsilon(t)$ for $t \geq 0$. We use $\mathcal{A}^\varepsilon$ to denote the set of all admissible controls.

We consider the cost function $J^\varepsilon(x, \alpha, u(\cdot))$ defined in (9.2.2). The problem is to find an admissible control $u(\cdot)$ that minimizes $J^\varepsilon(x, \alpha, u(\cdot))$.

We consider $h(x)$ and $c(u)$ to be convex functions. For all x, x', there exist positive constants C and k_g such that

$$0 \leq h(x) \leq C(1 + |x|^{k_g}) \text{ and } |h(x) - h(x')| \leq C(1 + |x|^{k_g} + |x'|^{k_g})|x - x'|.$$

Our control problem can be written as follows:

$$\mathcal{P}^\varepsilon : \begin{cases} \text{minimize} & J^\varepsilon(x, \alpha, u(\cdot)) = E \int_0^\infty e^{-\rho t}[h(x(t)) + c(u(t))]dt, \\ \text{subject to} & \dot{x}(t) = u(t) - z, \ x(0) = x, \ u(\cdot) \in \mathcal{A}^\varepsilon, \\ \text{value function} & v^\varepsilon(x, \alpha) = \inf_{u(\cdot) \in \mathcal{A}^\varepsilon} J^\varepsilon(x, \alpha, u(\cdot)). \end{cases} \tag{9.2.6}$$

The value function $v^\varepsilon(x, \alpha)$ is convex in x for each α. The value function v^ε satisfies (in the sense of viscosity solutions; see Sethi and Zhang [35]) the HJB equations

$$DP5\rho v^\varepsilon(x, \alpha) = \min_{u \geq 0, p \cdot \leq \alpha} [(u - z) \cdot v_x^\varepsilon(x, \alpha) + h(x) + c(u)] + Q^\varepsilon v^\varepsilon(x, \cdot)(\alpha), \tag{9.2.7}$$

for $\alpha \in \mathcal{M}$, where $Q^\varepsilon f(\cdot)(i) = \sum_{j \neq i} q_{ij}(f(j) - f(i))$ for a function f on $\mathcal{M}$. Clearly, these HJB equations are not easy to solve, especially when m is large. We now try to find approximation solutions instead. As in Sethi et al. [37], we consider a control problem in which the stochastic machine capacity process is averaged out. Let $\mathcal{A}^0$ denote the control space

$$\mathcal{A}^0 = \left\{U(t) = (u^0(t), u^1(t), \cdots, u^m(t)) : u^i(t) \geq 0, p \cdot u^i(t) \leq i\right\}.$$

We define the control problem $\mathcal{P}^0$ as follows:

$$\mathcal{P}^0 : \begin{cases} \text{minimize} & J^0(x, U(\cdot)) = E \int_0^\infty e^{-\rho t}\left(h(x(t)) + \sum_{i=0}^{m} \nu_i c(u^i(t))\right)dt, \\ \text{subject to} & \dot{x}(t) = \sum_{i=0}^{m} \nu_i u^i(t) - z, \ x(0) = x, \ U(\cdot) \in \mathcal{A}^0, \\ \text{value function} & v(x) = \inf_{U(\cdot) \in \mathcal{A}^0} J^0(x, U(\cdot)). \end{cases} \tag{9.2.8}$$

Sethi and Zhang [35] construct a solution of $\mathcal{P}^\varepsilon$ from a solution of $\mathcal{P}^0$ and show it to be asymptotically optimal as stated below.

Theorem 9.2.1 ([35]) (i) *There exists a constant* C *such that*

$$|v^\varepsilon(x,\alpha) - v(x)| \le C(1+|x|^{k_g})\sqrt{\epsilon}.$$

(ii) *Let* $U(\cdot) \in \mathcal{A}^0$ *denote an optimal (or* ε*-optimal) control for* $\mathcal{P}^0$. *Then*

$$u^\varepsilon(t) = \sum_{i=0}^{m} 1_{\{\alpha^\varepsilon(t)=i\}} u^i(t)$$

is asymptotically optimal, i.e.,

$$|J^\varepsilon(x,\alpha,u^\varepsilon(\cdot)) - v^\varepsilon(x,\alpha)| \le C(1+|x|^{k_g})\sqrt{\epsilon}. \tag{9.2.9}$$

(iii) *Assume in addition that* $c(u)$ *is twice differentiable with*

$$\frac{\partial^2 c(u)}{\partial u_i \partial u_j} \ge c_0 I_{n\times n} > 0,$$

h *is differentiable, and constants* C *and* $k_h > 0$ *exist such that*

$$|h(x+y) - h(x) - h_x(x)\cdot y| \le C(1+|x|^{k_h})|y|^2.$$

Then, there exists a locally Lipschitz optimal feedback control $U^*(x)$ *for* $\mathcal{P}^0$. *Let*

$$u^*(x,\alpha) = \sum_{i=0}^{m} 1_{\{\alpha=i\}} u^{i*}(x). \tag{9.2.10}$$

Then, $u^\varepsilon(t) = u^*(x(t),\alpha^\varepsilon(t))$ *is an asymptotically optimal feedback control for* $\mathcal{P}^\varepsilon$.

Remark. Gershwin [18] constructs a solution for $\mathcal{P}^\varepsilon$ by solving a secondary optimization problem and conjectures his solution to be asymptotically optimal. Sethi and Zhang [35] prove the conjecture. It should be noted, however, that the conjecture cannot be extended to include the simple two-machine example in [18] with one flexible and another inflexible machine. The presence of the inflexible machine requires aggregation of some products at the level of $\mathcal{P}^0$ and subsequent disaggregation in the construction of a solution for $\mathcal{P}^\varepsilon$.

Remark. One may also consider the generator of $\alpha^\varepsilon(\cdot)$ with more general structure such as

$$Q^\varepsilon = \frac{1}{\varepsilon}\widetilde{Q} + \widehat{Q},$$

where $\widetilde{Q}$ can be written as a canonical form including recurrent as well as transient states. In addition, the generator Q^ε can also be time-dependent. For related results on the structure of the underlying Markov chain and application to manufacturing systems, we refer to the book Yin and Zhang [49] for details.

9.3 Flowshops

In this section, we consider a production system with machines in tandem. To illustrate without undue technical difficulties, we only consider a two-machine flowshop depicted in Fig. 1:

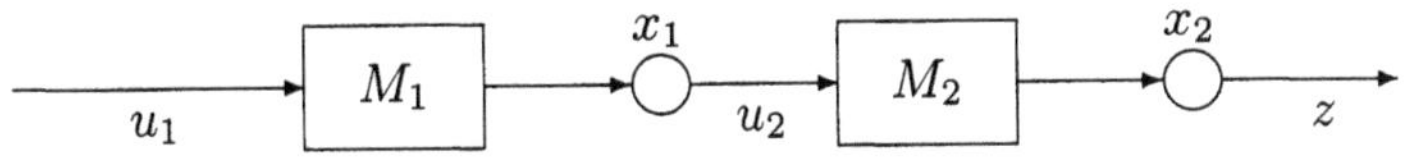

Fig. 1. A manufacturing system with 2-machines in tandem

As in Section 2, assume each machine is subject to breakdown and repair. Again, we use 1 to represent the state of machine when it is up and 0 when it is down. Let $\mathcal{M} = \{\alpha^1, \alpha^2, \alpha^3, \alpha^4\}$ denote the state space of the capacity process $\alpha^\varepsilon(\cdot)$, where $\alpha^j = (\alpha_1^j, \alpha_2^j)$ and $\alpha^1 = (0,0)$, $\alpha^2 = (0,1)$, $\alpha^3 = (1,0)$, $\alpha^4 = (1,1)$. Let $\alpha^\varepsilon(t) = (\alpha_1^\varepsilon(t), \alpha_2^\varepsilon(t))$ be a Markov chain generated by an irreducible generator $Q^\varepsilon = Q/\varepsilon$. The number of parts in the buffer between the first and the second machine is termed work-in-process and denoted as $x_1(t) \geq 0$ and the difference of the real and planned cumulative productions is called surplus at the second machine represented as $x_2(t)$. Let $S = [0, \infty) \times R^1$ denote the state constraint domain and let z denote the constant demand rate. Then, the system equations are given by

$$\begin{cases} \dot{x}_1(t) = u_1(t) - u_2(t), & x_1(0) = x_1, \\ \dot{x}_2(t) = u_2(t) - z, & x_2(0) = x_2. \end{cases}$$

A control $u(t) = (u_1(t), u_2(t))$ is *admissible* with respect to $x = (x_1, x_2) \in S$ if: (i) $u(\cdot)$ is adapted to $\sigma\{\alpha^\varepsilon(s) : 0 \leq s \leq t\}$, (ii) $0 \leq u_i(t) \leq \alpha_i(t)$, (iii) $x(t) \in S$ for all $t \geq 0$. We use $\mathcal{A}^\varepsilon(x)$ to denote the class of admissible controls. Then, our control problem $\mathcal{P}^\varepsilon$ can be written as follows:

$$\mathcal{P}^\varepsilon : \begin{cases} \text{minimize} & J^\varepsilon(x, \alpha, u(\cdot)) = E \int_0^\infty e^{-\rho t}[h(x(t)) + c(u(t))]dt, \\ \text{subject to} & \begin{cases} \dot{x}_1(t) = u_1(t) - u_2(t), & x_1(0) = x_1, \\ \dot{x}_2(t) = u_2(t) - z, & x_2(0) = x_2, \\ u(\cdot) \in \mathcal{A}^\varepsilon(x), \end{cases} \\ \text{value function} & v^\varepsilon(x, \alpha) = \inf_{u(\cdot) \in \mathcal{A}^\varepsilon(x)} J^\varepsilon(x, \alpha, u(\cdot)). \end{cases} \tag{9.3.11}$$

For $x \in S$, let $\mathcal{A}^0$ denote the set of the following deterministic measurable controls

$$U(\cdot) = (u^1(\cdot), \cdots, u^4(\cdot)) = ((u_1^1(\cdot), u_2^1(\cdot)), \cdots, (u_1^4(\cdot), u_2^4(\cdot))),$$

such that $0 \leq u_i^j(t) \leq \alpha_i^j$ for all $t \geq 0$, $i = 1, 2$ and $j = 1, \ldots, 4$.

We define the limiting problem

$$\mathcal{P}^0 : \begin{cases} \text{minimize} & J(x, U(\cdot)) = \int_0^\infty e^{-\rho t}\Big(h(x(t)) + \sum_{j=1}^4 \nu_j c(u^j(t))\Big) dt, \\ \text{subject to} & \begin{cases} \dot{x}_1(t) = \sum_{j=1}^4 \nu_j u_1^j(t) - \sum_{j=1}^4 \nu_j u_2^j(t), & x_1(0) = x_1, \\ \dot{x}_2(t) = \sum_{j=1}^4 \nu_j u_2^j(t) - z, & x_2(0) = x_2, \\ U(\cdot) \in \mathcal{A}^0 \end{cases} \\ \text{value function} & v(x) = \inf_{U(\cdot)\in\mathcal{A}^0} J(x, U(\cdot)), \end{cases}$$

where $\nu = (\nu_1, \cdots, \nu_4) > 0$ is the stationary distribution of Q^ε.

It can be shown in [33] that, for a given $\delta > 0$, there exist positive constants C and ε_0, such that for all $0 < \varepsilon \leq \epsilon_0$ and $x \in S$, we have

$$|v^\varepsilon(x, \alpha) - v(x)| = O(\varepsilon^{\frac{1}{2}-\delta}). \tag{9.3.12}$$

Next, for a given $x \in S$, we describe the flow of constructing an asymptotic optimal control $u^\varepsilon(\cdot) \in \mathcal{A}^\varepsilon(x)$ of the original problem $\mathcal{P}^\varepsilon$ beginning with any near-optimal control $\bar{U}(\cdot) \in \mathcal{A}^0$ of the limiting problem $\mathcal{P}^0$.

Let us fix an initial state $x \in S$. Let $\bar{\mathcal{U}}u(\cdot) = (\bar{u}^1(\cdot), \cdots, \bar{u}^4(\cdot)) \in \mathcal{A}^0$, where $\bar{u}^j(t) = (\bar{u}_1^j(t), \bar{u}_2^j(t))$ is an $\varepsilon^{\frac{1}{2}-\delta}$-*optimal control* for $\mathcal{P}^0$, i.e.,

$$|J(x, \bar{\mathcal{U}}u(\cdot)) - v(x)| \leq K\varepsilon^{\frac{1}{2}-\delta}.$$

Let

$$t^* = \inf\Big\{t : \int_0^t \sum_{j=1}^4 (\alpha_1^j - \bar{u}_1^j(s) + \bar{u}_2^j(s))ds \geq \varepsilon^{\frac{1}{2}-\delta}\Big\}.$$

Using t^*, we define another control process $\tilde{\mathcal{U}}u(t) = (\tilde{u}^1(\cdot), \cdots, \tilde{u}^4(\cdot))$ as follows: For $j = 1, \cdots, 4$,

$$\tilde{u}^j(t) = (\tilde{u}_1^j(t), \tilde{u}_2^j(t)) = \begin{cases} (\alpha_1^j, 0) & \text{if } t < t^*, \\ (\bar{u}_1^j(t), \bar{u}_2^j(t)) & \text{if } t \geq t^*. \end{cases} \tag{9.3.13}$$

It is easy to check that $\tilde{U}(\cdot) \in \mathcal{A}^0$. Let

$$w(t) = (w_1(t), w_2(t)) = \sum_{j=1}^4 1_{\{\alpha^\varepsilon(t)=\alpha^j\}}(\tilde{u}_1^j(t), \tilde{u}_2^j(t)) = \sum_{j=1}^4 1_{\{\alpha^\varepsilon(t)=\alpha^j\}}\tilde{u}^j(t), \tag{9.3.14}$$

and let $y(t) = (y_1(t), y_2(t))$ be the corresponding trajectory defined as

$$\begin{cases} y_1(t) = x_1 + \int_0^t (w_1(s) - w_2(s))ds \\ y_2(t) = x_2 + \int_0^t (w_2(s) - z)ds. \end{cases}$$

Note that $E|y(t) - \tilde{x}(t)|^2 \leq C(1+t^2)\varepsilon$. However, $y(t)$ may not be in S for some $t \geq 0$. To obtain an admissible control for $\mathcal{P}^\varepsilon$, we need to modify $w(t)$ so that the state trajectory stays in S. This is done as follows. Let

$$u^\varepsilon(t) = (u_1^\varepsilon(t), u_2^\varepsilon(t)) := w(t)1_{\{y_1(t)\geq 0\}}. \tag{9.3.15}$$

Then, for the control $u^\varepsilon(\cdot) \in \mathcal{A}^\varepsilon(x)$ constructed using (9.3.13)-(9.3.15) above, it is shown in Sethi et al. [36] that

$$|J^\varepsilon(x, \alpha, u^\varepsilon(\cdot)) - v^\varepsilon(x, \alpha)| = O(\varepsilon^{\frac{1}{2}-\delta}). \tag{9.3.16}$$

For optimal control and hierarchical control of general flowshops, we refer to the papers Presman et al. [29] and Sethi et al. [36]; see also Sethi and Zhang [33] for complete treatment of the subject.

9.4 Jobshops

In this section, let us discuss briefly general production systems. For more details, see Sethi and Zhang [33].

Sethi and Zhou [38] consider hierarchical production planning in a general manufacturing system consisting of a network of machines which generalizes both the parallel and the tandem machine models; see also Bai and Gershwin [4]. As in the flowshop models, the optimal control problem for the system is a state-constrained problem, since the number of parts in any buffer between any two machines must remain non-negative.

Sethi and Zhou [38] establish a graph-theoretical framework that appropriately describes and uniquely determines the system dynamics along with the state and control constraints. Within their framework, one can model a large class of manufacturing systems of interest. The concept of a "dynamic job shop" is introduced by interpreting a system with a network of machines as a directed graph along with a "placement of machines" that reflects system dynamics and the control constraints. To illustrate, let us consider the system given in Fig. 2.

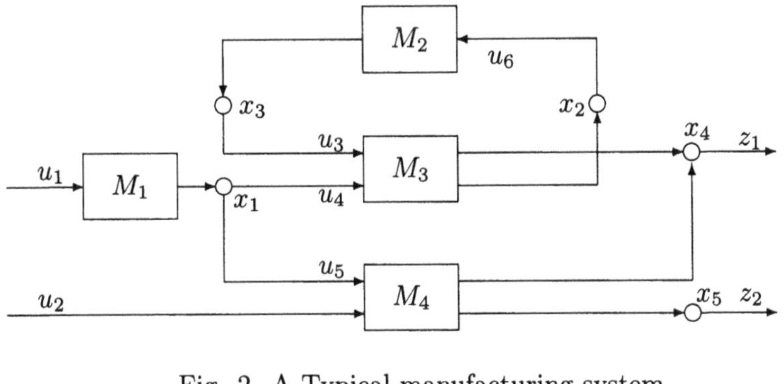

Fig. 2. A Typical manufacturing system

Here, we have four machines $M_1, \cdots, M_4$, two distinct products, and five buffers. Each machine M_i, $i = 1, 2, 3, 4$, has capacity $\alpha_i(t)$ at time t, and each product $j = 1, 2$ has demand z_j. As indicated in the figure, x_i, $i = 1, 2, \cdots, 5$, known as the state variables are associated with the buffers. More specifically, x_i denotes the inventory/backlog of part type i, $i = 1, 2, \cdots, 5$. Control variables u_i, $i = 1, 2 \cdots, 6$, represent production rates. More specifically, u_1 and u_2 are the rates at which raw parts coming from outside are converted to part types 1 and 5, respectively, and u_3, u_4, u_5 and u_6 are the rates of conversion from part types 3,1,1, and 2 to part types 4,2,4 and 3, respectively. The corresponding system dynamics is given by

$$\begin{cases} \dot{x}_1(t) = u_1(t) - u_4(t) - u_5(t), \\ \dot{x}_2(t) = u_4(t) - u_6(t), \\ \dot{x}_3(t) = u_6(t) - u_3(t), \\ \dot{x}_4(t) = u_3(t) + u_5(t) - z_1, \\ \dot{x}_5(t) = u_2(t) - z_2. \end{cases} \tag{9.4.17}$$

and the process $u(t) = (u_1(t), \cdots, u_6(t))$ must satisfy the capacity constraints

$$u_1(t) \leq \alpha_1(t), u_2(t) + u_5(t) \leq \alpha_4(t), u_3(t) + u_4(t) \leq \alpha_3(t), u_6(t) \leq \alpha_2(t). \tag{9.4.18}$$

Moreover, part types 1,2 and 3 are intermediate items to be further processed in the system. For $i = 1, 2, 3$, buffer i is between some two machines and is known as an internal buffer. Since internal buffers provide inputs to machines, a fundamental physical fact about them is that they must not have shortages. In other words, we must have

$$x_i(t) \geq 0, \; i = 1, 2, 3. \tag{9.4.19}$$

The remaining buffers 4 and 5 are called external buffers, since it is from these buffers that we must meet the demands for final products facing the system. Since we permit backlogging of demand, the inventories in the external buffers are allowed to be negative. Indeed, $x_4(t)$ and $x_5(t)$ are called surpluses with positive values meaning inventories and negative values meaning backlogs.

State constraint domain, admissible controls, the limiting problem, and the associated value functions can be defined similarly as in the last section. Sethi and Zhou [38] verify the Lipschitz continuity of the value functions and show that (9.3.12) holds. They construct controls for the original problem from an optimal control of the limiting problem in a way similar to (9.3.13)-(9.3.15). Finally, they show that the constructed controls are asymptotically optimal as in (9.3.16).

9.5 Production–Capacity Expansion Models

In practice, if a manufacturing firm faces higher demand for its product, it is natural for the firm to increase its production to meet the demand, moreover, if necessary, to increase investment in order to increase its production capacity. In this section, we consider the case when some additional production capacity can be purchased at a future time $0 \leq \tau \leq \infty$, at a cost of K.

We use the single machine model studied in Section 2. The control variable is a pair $(\tau, u(\cdot))$ of a Markov time $\tau \geq 0$ and a production process $u(\cdot)$ over time. Consider the cost function

$$J^\varepsilon(x, \alpha, \tau, u(\cdot)) = E\left(\int_0^\infty e^{-\rho t} G(x(t), u(t))dt + Ke^{-\rho\tau}\right), \tag{9.5.20}$$

where $\alpha^\varepsilon(0) = \alpha$ is the initial capacity and $\rho > 0$ is the discount rate, and $G(x, u) = h(x) + c(u)$. The problem is to find an admissible control $(\tau, u(\cdot))$ that minimizes $J^\varepsilon(x, \alpha, \tau, u(\cdot))$.

Define $\alpha_1^\varepsilon(t)$ and $\alpha_2^\varepsilon(t)$ as two Markov chains with state spaces $\mathcal{M}_1 = \{0, 1, \ldots, m_1\}$ and $\mathcal{M}_2 = \{0, 1, \ldots, m_1 + m_2\}$, respectively. Here, $\alpha_1^\varepsilon(t) \geq 0$ denotes the existing production capacity process and $\alpha_2^\varepsilon(t) \geq 0$ denotes the capacity process of the system if it were to be supplemented by the additional new capacity at time $t = 0$.

Let $\mathcal{F}_1(t) = \sigma\{\alpha_1^\varepsilon(s) : s \leq t\}$ and $\mathcal{F}_2(t) = \sigma\{\alpha_2^\varepsilon(t) : s \leq t\}$. Define further a new process $\alpha^\varepsilon(t)$ as follows: For each $\mathcal{F}_1(t)$-Markov time $\tau \geq 0$,

$$\alpha^\varepsilon(t) = \begin{cases} \alpha_1^\varepsilon(t) & \text{if } t < \tau \\ \alpha_2^\varepsilon(t - \tau) & \text{if } t \geq \tau \end{cases} \quad \text{and } \alpha^\varepsilon(\tau) = \alpha_2^\varepsilon(0) := \alpha_1^\varepsilon(\tau) + m_2. \tag{9.5.21}$$

Here m_2 denotes the maximum additional capacity resulting from investment in the new capacity.

We assume the following conditions: $\alpha_1^\varepsilon(t) \in \mathcal{M}_1$ and $\alpha_2^\varepsilon(t) \in \mathcal{M}_2$ are Markov processes with generators $\varepsilon^{-1}Q_1$ and $\varepsilon^{-1}Q_2$, respectively. Moreover, Q_1 and Q_2 are both irreducible.

We say that a control $(\tau, u(\cdot))$ is *admissible* if (1) τ is an $\mathcal{F}_1(t)$-Markov time; (2) $u(t)$ is $\mathcal{F}(t) = \sigma\{\alpha^\varepsilon(t) : s \leq t\}$ adapted and $p \cdot u(t) \leq \alpha^\varepsilon(t)$ for $t \geq 0$. We use $\mathcal{A}^\varepsilon$ to denote the set of all admissible controls $(\tau, u(\cdot))$. Then the problem is:

$$\mathcal{P}^\varepsilon : \begin{cases} \min\limits_{(\tau, u(\cdot)) \in \mathcal{A}^\varepsilon} & J^\varepsilon(x, \alpha, \tau, u(\cdot)), \\ \text{subject to} & \dot{x}(t) = u(t) - z, \; x(0) = x. \end{cases}$$

Let $v^\varepsilon(x, \alpha)$ denote the value function of the problem. We define an auxiliary value function $v_a^\varepsilon(x, \alpha')$ to be K plus the optimal cost with the capacity process $\alpha_2^\varepsilon(t)$ with the initial capacity $\alpha' \in \mathcal{M}_2$ and no future capital expansion possibilities. Then the HJB equations are as follows:

$$\begin{aligned} \min\Big\{ & \min_{u \geq 0, p \cdot \leq \alpha} [(u - z) \cdot (v^\varepsilon)_x(x, \alpha) + G(x, u)] + \varepsilon^{-1} Q_1 v^\varepsilon(x, \cdot)(\alpha) \\ & -\rho v^\varepsilon(x, \alpha), v_a^\varepsilon(x, \alpha + m_2) - v^\varepsilon(x, \alpha)\Big\} = 0, \alpha \in \mathcal{M}_1, \end{aligned} \tag{9.5.22}$$

$$\begin{aligned} & \min_{u \geq 0, p \cdot \leq \alpha} [(u - z) \cdot (v_a^\varepsilon)_x(x, \alpha) + G(x, u)] \\ & \quad + \varepsilon^{-1} Q_2 v_a^\varepsilon(x, \cdot)(\alpha) - \rho(v_a^\varepsilon(x, \alpha) - K) = 0, \alpha \in \mathcal{M}_2. \end{aligned} \tag{9.5.23}$$

Let $\nu^{(1)} = (\nu_0^{(1)}, \nu_1^{(1)}, \ldots, \nu_{m_1}^{(1)})$ and $\nu^{(2)} = (\nu_0^{(2)}, \nu_1^{(2)}, \ldots, \nu_{m_1+m_2}^{(2)})$ denote the corresponding stationary distributions of Q_1 and Q_2, respectively.

We now proceed to develop a limiting problem. We first define the control sets for the limiting problem. Let

$$\begin{aligned} \mathcal{U}_1 &= \{(u^0, \ldots, u^{m_1}) : \; u^i \geq 0, p \cdot u^i \leq i\}, \\ \mathcal{U}_2 &= \{(u^0, \ldots, u^{m_1+m_2}) : \; u^i \geq 0, p \cdot u^i \leq i\}. \end{aligned}$$

Then

$$\mathcal{U}_1 \subset R^{n \times (m_1+1)} \text{ and } \mathcal{U}_2 \subset R^{n \times (m_1+m_2+1)}.$$

We use $\mathcal{A}^0$ to denote the set of the following (*admissible controls* for the limiting problem): (1) a deterministic time σ; (2) a deterministic $\mathcal{U}u(t)$ such that for $t < \sigma$, $\mathcal{U}u(t) = (u^0(t), \ldots, u^{m_1}(t)) \in \mathcal{U}_1$ and for $t \geq \sigma$, $\mathcal{U}u(t) = (u^0(t), \ldots, u^{m_1+m_2}(t)) \in \mathcal{U}_2$.

Let

$$J(x,\sigma,\mathcal{U}u(\cdot)) = \int_0^\sigma e^{-\rho t}\sum_{i=0}^{m_1}\nu_i^{(1)}G(x(t),u^i(t))dt + \int_\sigma^\infty e^{-\rho t}\sum_{i=0}^{m_1+m_2}\nu_i^{(2)}G(x(t),u^i(t))dt + e^{-\rho\sigma}K$$

and let

$$\bar{u}(t) = \begin{cases} \sum_{i=0}^{m_1}\nu_i^{(1)}u^i(t) & \text{if } t<\sigma, \\ \sum_{i=0}^{m_1+m_2}\nu_i^{(2)}u^i(t) & \text{if } t\geq\sigma. \end{cases}$$

We can now define the following limiting optimal control problem:

$$\mathcal{P}^0: \begin{cases} \min_{(\sigma,\mathcal{U}u(\cdot))\in\mathcal{A}^0} & J(x,\sigma,\mathcal{U}u(\cdot)) \\ \text{subject to} & \dot{x}(t)=\bar{u}(t)-z,\ x(0)=x. \end{cases} \tag{9.5.24}$$

Let $(v(x), v_a(x))$ denote the value functions for $\mathcal{P}^0$. Let $(\sigma, U(\cdot)) \in \mathcal{A}^0$ denote any admissible control for the limiting problem $\mathcal{P}^0$, where

$$U(t) = \begin{cases} (u^0(t),\ldots,u^{m_1}(t))\in\mathcal{U}_1 & \text{if } t<\sigma, \\ (u^0(t),\ldots,u^{m_1+m_2}(t))\in\mathcal{U}_2 & \text{if } t\geq\sigma. \end{cases}$$

We take

$$u^\varepsilon(t) = \begin{cases} \sum_{i=0}^{m_1}1_{\{\alpha^\varepsilon(t)=i\}}u^i(t) & \text{if } t<\sigma, \\ \sum_{i=0}^{m_1+m_2}1_{\{\alpha^\varepsilon(t)=i\}}u^i(t) & \text{if } t\geq\sigma. \end{cases}$$

Then the control $(\sigma, u^\varepsilon(\cdot))$ is admissible for $\mathcal{P}^\varepsilon$.

Let

$$\mathcal{S} = \{x : v_a(x) = v(x)\}. \tag{9.5.25}$$

Then $\mathcal{S}$ defines a switching set for $\mathcal{P}^0$. Let $u^*(x)$ denote the minimizer of the HJB equation and let $x(t)$ denote the state trajectory that satisfies

$$\dot{x}(t) = u^*(t,x(t)) - z,\ x(0) = x.$$

Then the optimal purchasing time σ in $\mathcal{P}^0$ is given as follows:

$$\sigma = \inf\{t : x(t)\in\mathcal{S}\}.$$

It can be shown that $(\sigma, u^*(t,x))$ is optimal for $\mathcal{P}^0$.

Theorem 9.5.1 ([30]) (i) *There exists a constant* C *such that*

$$|v^\varepsilon(x,\alpha) - v(x)| + |v_a^\varepsilon(x,\alpha) - v_a(x)| \le C(1+|x|^{k_g})\sqrt{\varepsilon}.$$

(ii) *Let* $(\sigma, U(\cdot)) \in \mathcal{A}^0$ *be an* ε*-optimal control for the limiting problem* $\mathcal{P}^0$ *and let* $(\sigma, u^\varepsilon(\cdot)) \in \mathcal{A}^\varepsilon$ *be the control constructed above. Then,* $(\sigma, u^\varepsilon(\cdot))$ *is asymptotically optimal with error bound* $\sqrt{\varepsilon}$*, i.e.,*

$$|J^\varepsilon(x,\alpha,\sigma,u^\varepsilon(\cdot)) - v^\varepsilon(x,\alpha)| \le C(1+|x|^{k_g})\sqrt{\varepsilon}.$$

Example. Let us consider a production system having an existing (failure-prone) machine. When operational, it has a unit production capacity; when broken down, it has zero capacity, i.e, $m_1 = 1$. We assume that the demand for the firm's product is higher than the average production capacity of the existing machine. However, the firm has some initial inventory of its product to absorb the excess demand for a few initial periods. The firm may have to increase its production capacity at some future time $\tau \ge 0$. Therefore, the firm has an option to purchase a new machine, identical to the existing machine, at a given fixed cost of K in order to double its average production capacity.

The problem is to find the optimal time of purchase as well as the optimal production simultaneously, which is given as follows:

$$\begin{cases} \displaystyle\min_{(\tau,u(\cdot))\in\mathcal{A}^\varepsilon(\alpha)} \quad J^\varepsilon(x,\alpha,\tau,u(\cdot)) = E\left[\int_0^\infty e^{-\rho t}|x(t)|dt + Ke^{-\rho\tau}\right], \\ \text{subject to} \quad \dot{x}(t) = u(t) - z, \ x(0) = x. \end{cases} \tag{9.5.26}$$

We take $0 < z < 1$, $\mathcal{M}_1 = \{0,1\}$, and $\mathcal{M}_2 = \{0,1,2\}$ and assume also that

$$Q_1 = \begin{pmatrix} -1 & 1 \\ 1 & -1 \end{pmatrix} \text{ and } Q_2 = \begin{pmatrix} -1 & 1 & 0 \\ 1 & -2 & 1 \\ 0 & 1 & -1 \end{pmatrix}.$$

In this example, the stationary distributions are

$$\nu = \begin{cases} \left(\dfrac{1}{2}, \dfrac{1}{2}\right) & \text{if } t < \sigma, \\ \left(\dfrac{1}{3}, \dfrac{1}{3}, \dfrac{1}{3}\right) & \text{if } t \ge \sigma, \end{cases}$$

and the average capacities are $\bar{\alpha}_1 = 1/2$ and $\bar{\alpha}_2 = 1$.

The limiting problem $\mathcal{P}^0$ is the following:

$$\begin{cases} \displaystyle\min_{(\sigma,u(\cdot))\in\mathcal{A}^0} \quad J^0(x,\sigma,u(\cdot)) = \int_0^\infty e^{-\rho t}|x(t)|dt + Ke^{-\rho\sigma}, \\ \text{subject to} \quad \dot{x}(t) = u(t) - z, \ x(0) = x. \end{cases}$$

The value functions $v(x)$ and $v_a(x)$ can be shown to be the unique viscosity solutions to the following HJB equations:

$$\begin{cases} \min\left\{\displaystyle\min_{u\in\mathcal{U}_{\bar{\alpha}_1}} [(u-z)v_x(x) + |x| - \rho v(x)], v_a(x) - v(x)\right\} = 0, \\ \displaystyle\min_{u\in\mathcal{U}_{\bar{\alpha}_2}} [(u-z)(v_a)_x(x) + |x|] - \rho(v_a(x) - K) = 0. \end{cases}$$

We solve the HJB equations by considering the following five possible cases.

Case (i): $0 < \rho^2 K < \bar{\alpha}_2 - \bar{\alpha}_1$ and $\bar{\alpha}_1 < z$.
Define x^* and $\hat{x}$ as follows:

$$\begin{cases} x^* = \left(\dfrac{\bar{\alpha}_2 - z}{\rho}\right) \log\left(\dfrac{\bar{\alpha}_2 - \bar{\alpha}_1 - \rho^2 K}{\bar{\alpha}_2 - \bar{\alpha}_1}\right) < 0, \\ \hat{x} = \left(\dfrac{z - \bar{\alpha}_1}{\rho}\right) \log\left(2 - \left(\dfrac{\bar{\alpha}_2 - \bar{\alpha}_1 - \rho^2 K}{\bar{\alpha}_2 - \bar{\alpha}_1}\right)^{\frac{\bar{\alpha}_2 - \bar{\alpha}_1}{z - \bar{\alpha}_1}}\right) > 0. \end{cases}$$

The value functions $v_a(x)$ and $v(x)$ can be written in terms of x^* and $\hat{x}$ as follows:

$$v_a(x) = \begin{cases} \dfrac{z}{\rho^2}\left[e^{-\rho x/z} + \dfrac{\rho x}{z} - 1\right] + K & \text{if } x \geq 0, \\ \dfrac{\bar{\alpha}_2 - z}{\rho^2}\left[e^{\rho x/(\bar{\alpha}_2 - z)} - \dfrac{\rho x}{\bar{\alpha}_2 - z} - 1\right] + K & \text{if } x < 0, \end{cases}$$

$$v(x) = \begin{cases} \displaystyle\int_0^{(x-\hat{x})/z} e^{-\rho t}|x - zt|dt + e^{-\rho(x-\hat{x})/z}\int_0^{(\hat{x}-x^*)/(z-\bar{\alpha}_1)} e^{-\rho t}|\hat{x} + (\bar{\alpha}_1 - z)t|dt \\ \quad + e^{-\rho[(x-\hat{x})/z + (\hat{x}-x^*)/(z-\bar{\alpha}_1)]} v_a(x^*) & \text{if } x > \hat{x}, \\ \displaystyle\int_0^{(x-x^*)/(z-\bar{\alpha}_1)} e^{-\rho t}|x + (\bar{\alpha}_1 - z)t|dt + e^{-\rho(x-x^*)/(z-\bar{\alpha}_1)} v_a(x^*) \\ & \text{if } x^* \leq x \leq \hat{x}, \\ (\bar{\alpha}_2 - z)\rho^{-2}\left[e^{\rho x/(\bar{\alpha}_2 - z)} - \dfrac{\rho x}{\bar{\alpha}_2 - z} - 1\right] + K & \text{if } x < x^*. \end{cases}$$

Case (ii): $0 < \rho^2 K < \bar{\alpha}_2 - \bar{\alpha}_1$ and $\bar{\alpha}_1 = z$.
Let

$$x^* = \left(\frac{\bar{\alpha}_2 - z}{\rho}\right) \log\left(\frac{\bar{\alpha}_2 - \bar{\alpha}_1 - \rho^2 K}{\bar{\alpha}_2 - \bar{\alpha}_1}\right) \leq 0.$$

Then,

$$v(x) = \begin{cases} z\rho^{-2}\left[e^{-\rho x/z} - 1\right] + \dfrac{x}{\rho} & \text{if } x > 0, \\ -\dfrac{x}{\rho} & \text{if } x^* \leq x \leq 0, \\ \dfrac{\bar{\alpha}_2 - z}{\rho^2}\left[e^{\rho x/(\bar{\alpha}_2 - z)} - \dfrac{\rho x}{\bar{\alpha}_2 - z} - 1\right] + K & \text{if } x < x^*, \end{cases}$$

and $v_a(x)$ is as in Case (i).

Case (iii): $0 < \rho^2 K < \bar{\alpha}_2 - \bar{\alpha}_1$ and $\bar{\alpha}_1 > z$.
Let $x^*(\leq 0)$ denote the only value such that

$$(\bar{\alpha}_1 - z)e^{\rho x^*/(\bar{\alpha}_1 - z)} - (\bar{\alpha}_2 - z)e^{\rho x^*/(\bar{\alpha}_2 - z)} = K\rho^2 - (\bar{\alpha}_2 - \bar{\alpha}_1).$$

Then,

$$v(x) = \begin{cases} z\rho^{-2}\left[e^{-\rho x/z} + \rho x/z - 1\right] & \text{if } x \geq 0, \\ (\bar{\alpha}_1 - z)\rho^{-2}\left[e^{\rho x/(\bar{\alpha}_1 - z)} - \rho x/(\bar{\alpha}_1 - z) - 1\right] & \text{if } x^* \leq x < 0, \\ (\bar{\alpha}_2 - z)\rho^{-2}\left[e^{\rho x/(\bar{\alpha}_2 - z)} - \rho x/(\bar{\alpha}_2 - z) - 1\right] + K & \text{if } x < x^*, \end{cases}$$

and $v_a(x)$ is as in Case (i).

Case (iv): $K = 0$.

In this case $v(x) = v_a(x)$ for all x, where $v_a(x)$ is as in Case (i) with $K = 0$. This means that the optimal purchase time $\sigma = 0$.

Case (v): $\rho^2 K \geq \bar{\alpha}_2 - \bar{\alpha}_1$. In this case, the optimal $\sigma = \infty$.

The value function

$$v(x) = \begin{cases} \displaystyle\int_0^{(x-\hat{x})/z} e^{-\rho t}(x - zt)dt + e^{-\rho(x-\hat{x})/z}\int_0^{\infty} e^{-\rho t}|\hat{x} - (z-\bar{\alpha}_1)t|dt & \\ & \text{if } x \geq \hat{x}, \\ \displaystyle\int_0^{\infty} e^{-\rho t}|x - (z-\bar{\alpha}_1)t|dt & \text{if } x < \hat{x}, \end{cases}$$

where

$$\hat{x} = \begin{cases} 0 & \text{if } \bar{\alpha}_1 \geq z \\ \frac{\log 2}{\rho}(z - \bar{\alpha}_1) > 0 & \text{if } \bar{\alpha}_1 < z. \end{cases}$$

Again, $v_a(x)$ is as in Case (i).

We have now obtained the value function in each of the five cases.

In this example, the switching set is given as follows:

$$\mathcal{S} = \begin{cases} (-\infty, \infty) & \text{if } K = 0 \\ (-\infty, x^*] & \text{if } 0 < \rho^2 K < (\bar{\alpha}_2 - \bar{\alpha}_1) \\ \emptyset & \text{if } \rho^2 K \geq (\bar{\alpha}_2 - \bar{\alpha}_1). \end{cases}$$

Let $\sigma = \inf\{t : x(t) \in \mathcal{S}\}$. If $\bar{\alpha}_1 < z$, then let

$$u^*(t, x) = \begin{cases} \begin{cases} 0 & \text{if } x > \hat{x} \\ \bar{\alpha}_1 & \text{if } x \leq \hat{x} \end{cases} & \text{if } t < \sigma, \\ \begin{cases} 0 & \text{if } x > 0 \\ z & \text{if } x = 0 \\ \bar{\alpha}_2 & \text{if } x < 0 \end{cases} & \text{if } t \geq \sigma, \end{cases}$$

and if $\bar{\alpha}_1 \geq z$, let

$$u^*(t, x) = \begin{cases} \begin{cases} 0 & \text{if } x > 0 \\ z & \text{if } x = 0 \\ \bar{\alpha}_1 & \text{if } x < 0 \end{cases} & \text{if } t < \sigma, \\ \begin{cases} 0 & \text{if } x > 0 \\ z & \text{if } x = 0 \\ \bar{\alpha}_2 & \text{if } x < 0 \end{cases} & \text{if } t \geq \sigma, \end{cases}$$

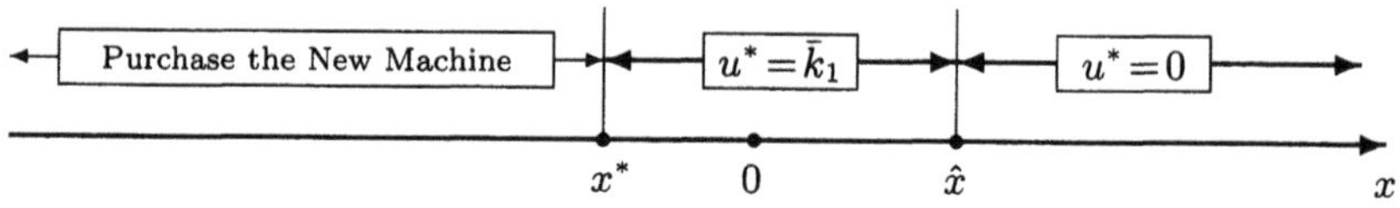

Fig. 3. Machine purchase policy and production policy for $t<\sigma$

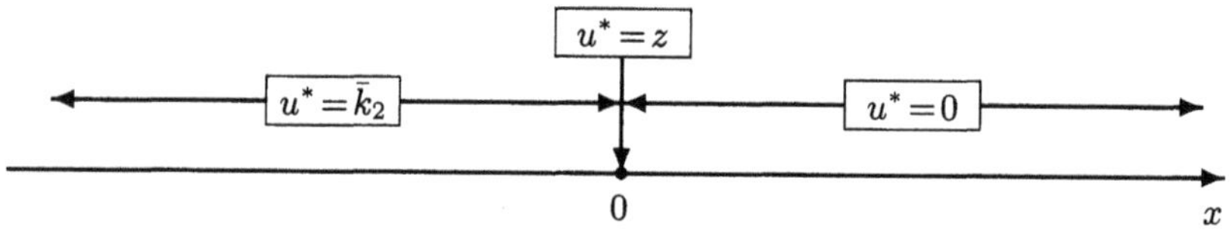

Fig. 4. Production policy for $t\geq\sigma\geq 0$

where σ is defined in (9.5.25). The optimal decision $(\sigma, u^*(t,x))$ in Case (i) is depicted in Figs. 3 and 4.

Let $(\sigma, u^\varepsilon(t,x))$ denote the scaled decision as constructed below:

$$u^\varepsilon(t,\alpha,x) = \begin{cases} \alpha u^*(t,x)/\bar{\alpha}_1 & \text{if } t < \sigma, \\ \alpha u^*(t,x)/\bar{\alpha}_2 & \text{if } t \geq \sigma. \end{cases}$$

Then we have $(\sigma, u^\varepsilon(t,\alpha,x))$ is asymptotically optimal for $\mathcal{P}^\varepsilon$.

9.6 Production-Marketing Models

In this section, we discuss the model developed in Sethi and Zhang [34], which considers the case when both capacity and demand are finite state Markov chains constructed from generators that depend on the production and promotional decisions, respectively. Due to the complexity of the manufacturing systems, traditionally, marketing decision making and other decision related areas such as productions are often treated independently. Clearly, a marketing model with addition of production is more realistic and useful from a practical point of view. In this connection, Abad [1] proposed a decentralized marketing-production planning model and solved the problem by applying Pontryagin's maximum principle. Sethi and Zhang [33] considered a marketing-production model in which the demand is assumed to be a Markov decision process. The main focus of that paper is reduction of dimensionality of the underlying problem via a hierarchical control approach;

In order to specify their marketing-production problem, let $\alpha^\varepsilon(t) \in \mathcal{M}$ as in Section 2 and $z(\delta,t) \in \{z^0, z^1, \ldots, z^d\}$, for a given δ, denote the capacity process and the demand process, respectively.

We say that a control $(u(\cdot), w(\cdot)) = \{(u(t), w(t)) : t \geq 0\}$ is *admissible* if $(u(\cdot), w(\cdot))$ is right-continuous having left-hand limit (RCLL), is $\sigma\{(\alpha^\varepsilon(s), z(\delta, s)) : s \leq t\}$ adapted, and satisfies $u(t) \geq 0$, $p \cdot \underset{\sim}{(t)} \leq \alpha^\varepsilon(t)$ and $0 \leq w(t) \leq 1$ for all $t \geq 0$. We use $\mathcal{A}^{\varepsilon,\delta}$ to denote the set of all admissible controls. Then our control problem can be written as follows:

$$
\mathcal{P}^{\varepsilon,\delta}:\begin{cases}\text{maximize} & J^{\varepsilon,\delta}(x,\alpha,z,u(\cdot),w(\cdot)) \\ & = E\int_0^\infty e^{-\rho t}G(x(t),z(\delta,t),u(t),w(t))dt, \\ \text{subject to} & \begin{cases}\dot{x}(t)=u(t)-z(\delta,t), & x(0)=x, \\ \alpha^\varepsilon(t)\sim\varepsilon^{-1}Q^m(u(t)), & \alpha^\varepsilon(0)=\alpha, \\ z(\delta,t)\sim\delta^{-1}Q^d(w(t)), & z(\delta,0)=z, \\ (u(\cdot),w(\cdot))\in\mathcal{A}^{\varepsilon,\delta}, \end{cases} \\ \text{value function} & v^{\varepsilon,\delta}(x,\alpha,z)=\inf_{(u(\cdot),w(\cdot))\in\mathcal{A}^{\varepsilon,\delta}} J^{\varepsilon,\delta}(x,\alpha,z,u(\cdot),w(\cdot)).\end{cases} \tag{9.6.27}
$$

where by $\alpha^\varepsilon(t)\sim\varepsilon^{-1}Q^m(u(t))$, we mean that the Markov process $\alpha^\varepsilon(t)$ has the generator $\varepsilon^{-1}Q^m(u(t))$.

We use $\mathcal{A}^{0,\delta}$ to denote the admissible control space

$$
\begin{aligned}\mathcal{A}^{0,\delta} = \; & \{(U(t),w(t))=(u^0(t),u^1(t),\ldots,u^m(t),w(t)):u^i(t)\ge 0, p\cdot u^i(t)\le i, \\ & 0\le w\le 1,(U(t),w(t)) \text{ is } \sigma\{z(\delta,s):\; s\le t\} \text{ adapted and RCLL}\}.\end{aligned}
$$

The limiting problem is given as follows:

$$
\mathcal{P}^{0,\delta}:\begin{cases}\text{maximize} & J^{0,\delta}(x,z,U(\cdot),w(\cdot)) \\ & = \int_0^\infty e^{-\rho t}\sum_{i=0}^m \nu_i^m(U(t))G(x(t),z(\delta,t),u^i(t),w(t))dt, \\ \text{subject to} & \begin{cases}\dot{x}(t)=\sum_{i=0}^m \nu_i^m(U(t))u^i(t)-z(\delta,t), & x(0)=x, \\ z(\delta,t)\sim\frac{1}{\delta}Q^d(w(t)), & z(\delta,0)=z, \\ (U(\cdot),w(\cdot))\in\mathcal{A}^{0,\delta}, \end{cases} \\ \text{value function} & v^{0,\delta}(x,z)=\inf_{(U(\cdot),w(\cdot))\in\mathcal{A}^{0,\delta}} J^{0,\delta}(x,z,U(\cdot),w(\cdot)).\end{cases} \tag{9.6.28}
$$

Let $(U(\cdot),w(\cdot))\in\mathcal{A}^{0,\delta}$ denote an optimal open-loop control for $\mathcal{P}^{0,\delta}$. We construct

$$
u^{\varepsilon,\delta}(t)=\sum_{i=0}^m 1_{\{\alpha^\varepsilon(t)=i\}}u^i(t) \text{ and } w^{\varepsilon,\delta}(t)=w(t).
$$

Then $(u^{\varepsilon,\delta}(t),w^{\varepsilon,\delta}(t))\in\mathcal{A}^{\varepsilon,\delta}$, and it is asymptotically optimal, i.e.,

$$
\lim_{\delta\to 0}|J^{\varepsilon,\delta}(x,\alpha,z,u^{\varepsilon,\delta}(\cdot),w^{\varepsilon,\delta}(\cdot))-v^{\varepsilon,\delta}(x,\alpha,z)|=0.
$$

Similarly, let $(U(x,z),w(x,z))\in\mathcal{A}^{0,\delta}$ denote an optimal feedback control for $\mathcal{P}^{0,\delta}$. Suppose that $(U(x,z),w(x,z))$ is locally Lipschitz for each z. Let

$$
u^{\varepsilon,\delta}(t)=\sum_{i=0}^m 1_{\{\alpha^\varepsilon(t)=i\}}u^i(x(t),\alpha^\varepsilon(t),z(\delta,t)) \text{ and } w^{\varepsilon,\delta}(t)=w(x(t),z(\delta,t)).
$$

The feedback control $(u^{\varepsilon,\delta}(\cdot), w^{\varepsilon,\delta}(\cdot))$ is asymptotically optimal for $\mathcal{P}^{\varepsilon,\delta}$, i.e.,

$$\lim_{\delta\to 0} |J^{\varepsilon,\delta}(x,\alpha,z,u^{\varepsilon,\delta}(\cdot),w^{\varepsilon,\delta}(\cdot)) - v^{\varepsilon,\delta}(x,\alpha,z)| = 0.$$

We have considered only the hierarchy that arises from a fixed δ and a small ε. In this case, promotional decisions are obtained under the assumption that the available production capacity is equal to the average capacity. Subsequently, production decisions taking into account the stochastic nature of the capacity can be constructed. Other possible hierarchies result when both δ and ε are small or when ε is fixed and δ is small. The details can be found in Sethi and Zhang [33].

9.7 Risk-Sensitive Control

In this section, we consider robust production plans with a risk sensitive cost criterion. This consideration is motivated by the following observations. First, since most manufacturing systems are large and complex, it is difficult to establish accurate mathematical models to describe these systems. Modeling errors are inevitable. Second, in practice, an optimal policy for a subdivision of a big corporation is usually not an optimal policy for the whole corporation. Optimal solutions with the usual cost criterion may not be desirable in many real problems. An alternative approach is to consider robust controls. In some manufacturing systems, it is more desirable to consider controls that are robust enough to attenuate uncertain disturbances, which include modeling errors, and therefore to achieve the system stability. Robust control design is particularly important in manufacturing systems with unfavorable disturbances. There are two kinds of system disturbances in the system under consideration: (1) unfavorable internal disturbances – usually associated with unfavorable machine capacity fluctuations; (2) unfavorable external disturbances such as fluctuations in demand.

The basic idea of the risk-sensitive control is to consider a risk sensitive cost function that penalizes heavily on costs associated with large state trajectories and controls. Related literature on risk sensitive control and robust control can be found in Whittle [47], Fleming and McEneaney [13], Basar and Bernhard [5], Barron and Jensen [6], and references therein. For details discussed in this section, see Zhang [50].

As the rate of fluctuation of the production capacity process goes to infinity, we show that the risk sensitive control problem can be approximated by a limiting problem in which the stochastic capacity process can be averaged out and replaced by its average. We also show that the value function of the limiting problem satisfies the Isaacs equation of a zero-sum, two-player differential game. Then, we use a near optimal control of the limiting problem to construct a nearly optimal control for the original risk sensitive control problem.

The system equation is given by

$$\dot{x}(t) = u(t) - z(t),\ x_0 = a \in R^n \ (a \text{ is given}).$$

Let $J^{\varepsilon,\sqrt{\varepsilon}}(u(\cdot))$ denote the risk sensitive cost function defined by

$$J^{\varepsilon,\sqrt{\varepsilon}}(u(\cdot)) = \sqrt{\varepsilon}\log E\left[\exp\left\{\frac{1}{\sqrt{\varepsilon}}\int_0^\infty e^{-\rho t}[h(x(t)) + c(u(t))]dt\right\}\right]. \tag{9.7.29}$$

The problem is to find an admissible control $u(\cdot)$ that minimizes $J^{\varepsilon,\sqrt{\varepsilon}}(u(\cdot))$.

We now specify the production constraints. For each $i \in \mathcal{M} = \{0,1,2,\ldots,m\}$, let

$$\mathcal{U}(i) = \{l = (l_1,\ldots,l_n) \geq 0 : p\cdot l \leq i\} \subset R^n. \tag{9.7.30}$$

With this definition, the production constraint at time t is $u(t) \in \mathcal{U}(\alpha^\varepsilon(t))$.

We assume the demand rate $z(t)$ is a bounded process which is independent of $\alpha^\varepsilon(t)$.

We say that a control $u(\cdot) = \{u(t) : t \geq 0\}$ is *admissible* if $u(t)$ is a $\sigma\{\alpha^\varepsilon(s), z(s) : s \leq t\}$ adapted measurable process and $u(t) \in \mathcal{U}(\alpha^\varepsilon(t))$ for all $t \geq 0$. Then our control problem can be written as follows:

$$\mathcal{P}^{\varepsilon,\sqrt{\varepsilon}} : \begin{cases} \text{minimize} & J^{\varepsilon,\sqrt{\varepsilon}}(u(\cdot)) \\ & = \sqrt{\varepsilon}\log E\left[\exp\left\{\dfrac{1}{\sqrt{\varepsilon}}\displaystyle\int_0^\infty e^{-\rho t}[h(x(t)) + c(u(t))]dt\right\}\right], \\ \text{subject to} & \dot{x}(t) = u(t) - z(t),\ x_0 = a,\ u(\cdot) \in \mathcal{A}e, \\ \text{value function} & v^{\varepsilon,\sqrt{\varepsilon}} = \displaystyle\inf_{u(\cdot)\in\mathcal{A}e} J^{\varepsilon,\sqrt{\varepsilon}}(u(\cdot)). \end{cases} \tag{9.7.31}$$

Let $\mathcal{Z}_t = \sigma\{z(s) : s \leq t\}$. We consider the following control space:

$$\mathcal{A}^0 = \{U(\cdot) = (u^0(\cdot), u^1(\cdot), \ldots, u^m(\cdot)) : u^i(t) \in \mathcal{U}(i), \text{ and } U(t) \\ \text{is a } \mathcal{Z}_t \text{ adapted measurable process}\}$$

and two control problems $\mathcal{P}^{0,\sqrt{\varepsilon}}$ and $\mathcal{P}^{0,0}$ defined as follows:

and

$$\mathcal{P}^{0,0} : \begin{cases} \text{minimize} & J^{0,0}(U(\cdot)) = \left\|\displaystyle\int_0^\infty e^{-\rho t}[h(x(t)) + \sum_{i=0}^m \nu_i c(u^i(t))]dt\right\|_\infty, \\ \text{subject to} & \dot{x}(t) = \displaystyle\sum_{i=0}^m \nu_i u^i(t) - z(t),\ x_0 = a, \\ & U(\cdot) = (u^0(\cdot), \ldots, u^m(\cdot)) \in \mathcal{A}^0, \\ \text{value function} & v^{0,0} = \displaystyle\inf_{U(\cdot)\in\mathcal{A}^0} J^{0,0}(U(\cdot)). \end{cases} \tag{9.7.32}$$

It can be seen below that, when ε is small, $\mathcal{P}^{\varepsilon,\sqrt{\varepsilon}}$ can be approximated by $\mathcal{P}^{0,\sqrt{\varepsilon}}$ and $\mathcal{P}^{0,\sqrt{\varepsilon}}$ can be approximated further by $\mathcal{P}^{0,0}$. Therefore, $\mathcal{P}^{\varepsilon,\sqrt{\varepsilon}}$ can be approximated by $\mathcal{P}^{0,0}$. Then, a near optimal control for $\mathcal{P}^{0,0}$ will be used to construct controls for $\mathcal{P}^{\varepsilon,\sqrt{\varepsilon}}$ that are nearly optimal.

Theorem 9.7.1 ([50]) *There exist constants $\varepsilon_0 > 0$ and C such that, for $0 < \varepsilon \leq \varepsilon_0$,*

$$|v^{\varepsilon,\sqrt{\varepsilon}} - v^{0,\sqrt{\varepsilon}}| \leq C\sqrt{\varepsilon}.$$

We show that $\mathcal{P}^{0,\sqrt{\varepsilon}}$ can be approximated by $\mathcal{P}^{0,0}$ and the value function of $\mathcal{P}^{0,0}$ is a viscosity solution to the Isaacs equation of a zero-sum, two-player differential game. To simplify the notation, we take $\delta = \sqrt{\varepsilon}$ and consider the following control problem $\mathcal{P}^{0,\delta}$.

$$\mathcal{P}^{0,\delta}: \begin{cases} \text{minimize} & J^{0,\delta}(U(\cdot)) \\ & = \delta \log E\left[\exp\left\{\frac{1}{\delta}\int_0^\infty e^{-\rho t}[h(x(t)) + \sum_{i=0}^m \nu_i c(u^i(t))]dt\right\}\right], \\ \text{subject to} & \dot{x}(t) = \sum_{i=0}^m \nu_i u^i(t) - z(t),\ x_0 = a, \\ & U(\cdot) = (u^0(\cdot), \dots, u^m(\cdot)) \in \mathcal{A}^0, \\ \text{value function} & v^{0,\delta} = \inf_{U(\cdot)\in\mathcal{A}^0} J^{0,\delta}(U(\cdot)). \end{cases} \tag{9.7.33}$$

Theorem 9.7.2 ([50]) *$v^{0,\delta}$ is a monotone increasing function of $\delta > 0$ and*

$$\lim_{\delta\to 0} v^{0,\delta} = v^{0,0}.$$

For each $U(\cdot) \in \mathcal{A}^0$.

$$J^{0,\delta}(U(\cdot)) \uparrow J^{0,0}(U(\cdot)) \text{ as } \delta \downarrow 0. \tag{9.7.34}$$

We write $v^{0,0}(x)$ as the value function of $\mathcal{P}^{0,0}$ with the initial value $x_0 = x$. Note that $|\xi|_\infty = \inf_{P(F)=0} \sup_{\omega\in\Omega-F} |\xi(\omega)|$ for any random variable ξ. Let $\Gamma_u = \{U = (u^0, u^1, \dots, u^m) \in R^{n\times(m+1)}$ such that $u^i \in \mathcal{U}(i)\}$ and let Γ_z denote a compact subset of R^n. We consider functions $z(t) \in \Gamma_z$ $(t \geq 0)$ that are right continuous and have left hand limits. Let $\mathcal{Z}$ denote the metric space of such functions that is equipped with the Skorohod topology $d(\cdot,\cdot)$.

We assume $z(\cdot) = z(\cdot)(\omega) \in \Gamma_z$ a.s. and for each $z^0 = z^0(\cdot) \in \mathcal{Z}$ and any $\delta_0 > 0$, $P(d(z(\omega), z^0) \leq \delta_0) > 0$.

Theorem 9.7.3 ([50]) *$v^{0,0}(x)$ is the only viscosity solution to the following Isaacs equation*

$$\begin{aligned} \rho v^{0,0}(x) = & \min_{U\in\Gamma_u} \max_{z\in\Gamma_z} \left[\left(\sum_{i=0}^m \nu_i u^i - z\right) v_x^{0,0}(x) + h(x) + \sum_{i=0}^m \nu_i c(u^i)\right] \\ = & \max_{z\in\Gamma_z} \min_{U\in\Gamma_u} \left[\left(\sum_{i=0}^m \nu_i u^i - z\right) v_x^{0,0}(x) + h(x) + \sum_{i=0}^m \nu_i c(u^i)\right]. \end{aligned} \tag{9.7.35}$$

Theorem 9.7.4 ([50]) *The following assertions hold.*

(i)

$$\lim_{\varepsilon\to 0} v^{\varepsilon,\sqrt{\varepsilon}} = v^{0,0}. \tag{9.7.36}$$

(ii) *Let $U(\cdot) = (u^0(\cdot), \dots, u^m(\cdot)) \in \mathcal{A}^0$ denote a stochastic open loop ϵ'-optimal control for $\mathcal{P}^{0,0}$, i.e.,*

$$0 \leq J^{0,0}(U(\cdot)) - v^{0,0} \leq \epsilon'.$$

Let $u^\varepsilon(t) = \sum_{i=0}^m 1_{\{\alpha^\varepsilon(t)=i\}} u^i(t)$, where 1_A denotes the indicator of a set A. Then, $u^\varepsilon(\cdot) \in \mathcal{A}^\varepsilon$ and

$$\limsup_{\varepsilon\to 0} |J^{\varepsilon,\sqrt{\varepsilon}}(u^\varepsilon(\cdot)) - v^{\varepsilon,\sqrt{\varepsilon}}| \leq \epsilon'. \tag{9.7.37}$$

(iii) *Let* $U(\cdot) = U(z(\cdot), x(\cdot)) = (u^0(z(\cdot), x(\cdot)), \ldots, u^m(z(\cdot), x(\cdot)))$ *denote a feedback* ϵ'-*optimal control for* $\mathcal{P}^{0,0}$, *i.e.*, $0 \le J^{0,0}(U(\cdot)) - v^{0,0} \le \epsilon'$. *Let*

$$u^\varepsilon(\cdot) = u^\varepsilon(\alpha^\varepsilon(\cdot), z(\cdot), x(\cdot)) = \sum_{i=0}^{m} 1_{\{\alpha^\varepsilon(\cdot)=i\}} u^i(z(\cdot), x(\cdot)).$$

Assume that $U(z,x)$ *is locally Lipschitz in* x, *i.e., for some* $k_5 > 0$,

$$|U(z,x) - U(z,x')| \le C(1 + |x|^{k_5} + |x'|^{k_5})|x - x'|.$$

Then, $u^\varepsilon(\cdot) = u^\varepsilon(\alpha^\varepsilon(\cdot), z(\cdot), x(\cdot)) \in \mathcal{A}^\varepsilon$ *and*

$$\limsup_{\varepsilon \to 0} |J^{\varepsilon,\sqrt{\varepsilon}}(u^\varepsilon(\cdot)) - v^{\varepsilon,\sqrt{\varepsilon}}| \le \epsilon'. \tag{9.7.38}$$

Part II: Control with Long-Run Average Costs

A discounted cost weights more on recent events, while a long-run average cost focuses on long term development. In this part we review results on problems with long-run average costs. In this part, we only consider single machine systems discussed in Section 2. Related literature on control with long-run average costs can be found in Bensoussan and Nagai [7], Bielecki and Kumar [8], and references therein.

9.8 Optimal Control

In this section, we consider a single product manufacturing system with stochastic production capacity and constant demand for its production over time.

For any admissible $u(\cdot)$, define

$$J(x,k,u(\cdot)) = \limsup_{T\to\infty} \frac{1}{T} E \int_0^T \big(h(x(t)) + c(u(t))\big)\, dt. \tag{9.8.39}$$

Our goal is to choose $u(\cdot) \in \mathcal{A}(k)$ so as to minimize the cost functional $J(x,k,u(\cdot))$.

We assume the the cost functions $h(\cdot)$ and $c(\cdot)$ to be smooth and convex functions. Moreover, the average capacity $\bar{\alpha} \equiv \sum_{i=0}^{m} i\nu_i > z$ and $z \notin \mathcal{M}$.

An admissible control $u(\cdot)$ is called *stable* if it satisfies the condition

$$\lim_{T\to\infty} \frac{E|x(T)|^{\kappa+1}}{T} = 0. \tag{9.8.40}$$

The HJB equation associated with the long-run average cost optimal control problem takes the following form:

$$\lambda = F(k, W_x(x,k)) + h(x) + QW(x,\cdot)(k), \tag{9.8.41}$$

where $F(k,r) = \inf_{0\le u\le k}\{(u-z)r + c(u)\}$, λ is a constant and W is a real-valued function defined on $R \times \mathcal{M}$.

Let $\mathcal{G}$ denote the family of real-valued functions $W(\cdot,\cdot)$ defined on $R \times \mathcal{M}$ such that (i) $W(\cdot,k)$ is convex; (ii) $W(\cdot,k)$ is continuously differentiable; (iii) $W(\cdot,k)$ has polynomial growth. A solution to the HJB equation (9.8.41) is a pair (λ, W) with λ a constant and $W \in \mathcal{G}$. The function W is called a *potential function* for the control problem, if λ is the minimum long-run average cost.

Theorem 9.8.1 ([31]) (i) (λ^*, V) *is a viscosity solution to the HJB equation* (9.8.41). *Moreover, the constant* λ^* *is unique.*

(ii) *The function* $V(x,k)$ *is continuously differentiable in* x, *and* (λ^*, V) *is a classical solution to the* HJB *equation. Moreover,* $V(x,k)$ *is convex in* x *and*

$$|V(x,k)| \le C(1 + |x|^{\kappa+1}).$$

Theorem 9.8.2 ([31]) *Let* (λ, W) *be a solution to the* HJB *equation* (9.8.41). *Then*

(i) *If there is a control* $u^*(\cdot) \in \mathcal{A}(k)$ *such that*

$$F\left(\alpha(t), W_x(x^*(t), \alpha(t)\right) = (u^*(t) - z)W_x(x^*(t), \alpha(t)) + c(u^*(t)) \tag{9.8.42}$$

for a.e. $t \ge 0$ *with probability* 1, *where* $x^*(\cdot)$ *is the surplus process corresponding to the control* $u^*(\cdot)$, *and*

$$\lim_{T\to\infty} \frac{W(x^*(T), \alpha(T))}{T} = 0, \tag{9.8.43}$$

then

$$\lambda = J(x, k, u^*(\cdot)).$$

(ii) For any $u(\cdot) \in \mathcal{A}(k)$, *we have* $\lambda \le J(x,k,u(\cdot))$, *i.e.,*

$$\limsup_{t\to\infty} E \int_0^t (h(x(t)) + c(u(t)))\, dt \ge \lambda.$$

(iii) *For any (stable) control policy* $u(\cdot) \in \mathcal{B}(k)$, *we have*

$$\liminf_{t\to\infty} \frac{1}{t} E \int_0^t (h(x(t)) + c(u(t)))\, dt \ge \lambda. \tag{9.8.44}$$

We know that the function $V \in \mathcal{G}$, and that it is also a solution of the HJB equation (24). The function V is sometimes referred to as the relative value function. Let us now define a control policy $u^*(\cdot,\cdot)$ via the relative function $V(\cdot,\cdot)$ as follows:

$$u^*(x,k) = \begin{cases} 0 & \text{if} \quad V_x(x,k) > -c_u(0), \\ (c_u)^{-1}(-V_x(x,k)) & \text{if} \quad -c_u(k) \le V_x(x,k) \le -c_u(0), \\ k & \text{if} \quad V_x(x,k) < -c_u(k), \end{cases} \tag{9.8.45}$$

if the function $c(\cdot)$ is strictly convex, or

$$u^*(x,k) = \begin{cases} 0 & \text{if} \quad V_x(x,k) > -c, \\ \min\{k, z\} & \text{if} \quad V_x(x,k) = -c, \\ k & \text{if} \quad V_x(x,k) < c, \end{cases} \tag{9.8.46}$$

if $c(u) = cu$. Therefore, the control policy $u^*(\cdot,\cdot)$ satisfies the condition (9.8.42).

From the convexity of the function $V(\cdot,k)$, there are x_k, y_k, $-\infty < y_k < x_k < \infty$ such that

$$U(x) = (x_k, \infty) \quad \text{and} \quad L(k) = (-\infty, y_k).$$

The control policy $u^*(\cdot,\cdot)$ can be written as

$$u^*(x,k) = \begin{cases} 0 & x > x_k, \\ (c_u)^{-1}(-V_x(x,k)) & y_k \le x \le x_k, \\ k & x < y_k. \end{cases}$$

Theorem 9.8.3 ([31]) *The control policy $u^*(\cdot,\cdot)$, defined in (9.8.45) or (9.8.46) as the case may be, is optimal.*

When $c(u) = 0$, i.e., there is no production cost in the model, the optimal control policy can be chosen to be a hedging point policy, which has the following form: There are real numbers x_k, $k = 1, \ldots, m$, such that

$$u^*(x,k) = \begin{cases} 0 & x > x_k \\ \min\{k,z\} & x = x_k \\ k & x < x_k. \end{cases}$$

9.9 Hierarchical Control

In this section, we consider a slight variation of the model studied in Section 2. With the production rate $u(t) \in R^n$, $u(t) \ge 0$, the total surplus $x(t) \in R^n$, and a constant demand rate $z \in R^n$, $z \ge 0$, the system dynamics satisfy the differential equation

$$\dot{x}(t) = -ax(t) + u(t) - z, \quad x(0) = x \in R^n,$$

where $a = (a_1, \ldots, a_n)$ is a constant vector with $a_i > 0$. The attrition rate a_i represents the deterioration rate of the inventory of the finished product type i when $x_i(t) > 0$, and it represents a rate of cancelation of backlogged orders when $x_i(t) < 0$. We assume symmetric deterioration and cancellation rates for product i only for convenience in exposition.

Let $\alpha^\varepsilon(t) \in \mathcal{M} = \{0, 1, \ldots, m\}$, $t \ge 0$, denote a Markov process generated by Q/ε.

A function $f(x,k)$ defined on $R^n \times \mathcal{M}$ is called an admissible feedback control or simply a feedback control, if (i) for any given initial surplus x and production capacity k, the equation

$$\dot{x}(t) = -ax(t) + f(x(t), \alpha^\varepsilon(t)) - z$$

has a unique solution. For any admissible $u(\cdot)$, define the expected long-run average cost

$$J^\varepsilon(u(\cdot)) = \limsup_{T\to\infty} \frac{1}{T} E \int_0^T (h(x(t)) + c(u(t)))dt.$$

The problem is to obtain $u(\cdot) \in \mathcal{A}^\varepsilon(k)$ that minimizes $J^\varepsilon(u(\cdot))$. We formally summarize our control problem as follows:

$$\mathcal{P}^\varepsilon : \begin{cases} \text{minimize } J^\varepsilon(u(\cdot)) = \limsup\limits_{T\to\infty} \dfrac{1}{T} E \displaystyle\int_0^T (h(x(t)) + c(u(t)))dt, \\ \text{subject to } \dot{x}(t) = -ax(t) + u(t) - z, \; x(0) = x, \; u(\cdot) \in \mathcal{A}^\varepsilon(k), \\ \text{minimum average cost } \lambda^\varepsilon = \inf\limits_{u(\cdot)\in\mathcal{A}^\varepsilon(k)} J^\varepsilon(u(\cdot)). \end{cases}$$

The HJB equation associated with the average-cost optimal control problem in $\mathcal{P}^\varepsilon$, as shown in Sethi et al. [31], takes the form

$$\bar{\lambda}^\varepsilon = \inf_{0\le p\cdot u\le k}\left\{\frac{\partial \bar{w}^\varepsilon(x,k)}{\partial(-ax+u-z)}+c(u)\right\}+h(x)+\frac{Q}{\varepsilon}\bar{w}^\varepsilon(x,\cdot)(k),$$

where $\bar{w}^\varepsilon(x,k)$ is the potential function of the problem $\mathcal{P}^\varepsilon$, $\frac{\partial \bar{w}^\varepsilon(x,k)}{\partial(-ax+u-z)}$ denotes the directional derivative of $\bar{w}^\varepsilon(x,k)$ along the direction $(-ax+u-z)$.

Theorem 9.9.1 ([32]) *The minimum average cost λ^ε of $\mathcal{P}^\varepsilon$ is bounded in ε, i.e., there exists a constant $M_1>0$ such that*

$$0\le \lambda^\varepsilon \le M_1 \quad \textit{for all } \varepsilon>0.$$

In the remainder of this section, we derive the limiting control problem as $\varepsilon\to 0$. As in Sethi and Zhang [34], we consider the enlarged control space

$$\begin{aligned}\mathcal{A}^0 = \{U(\cdot) = (u^0(\cdot),u^1(\cdot),...,u^m(\cdot)) : u_i^k(t)\ge 0, \forall i \text{ and } p\cdot u(t)\le k,\ t\ge 0,\\ U(\cdot) \ \text{ is a deterministic process}\}.\end{aligned}$$

Then we define the limiting control problem $\mathcal{P}^0$ as follows:

$$\mathcal{P}^0:\begin{cases}\text{minimize } J(U(\cdot)) = \limsup\limits_{T\to\infty}\dfrac{1}{T}\displaystyle\int_0^T [h(x(s))+\sum_{j=0}^m \nu_j c(u^j(s))]ds,\\ \text{subject to } \dot{x}(t) = -ax(t)+\displaystyle\sum_{j=0}^m \nu_j u^j(t)-z,\ x(0)=x,\ U(\cdot)\in\mathcal{A}^0,\\ \text{minimum average cost } \lambda = \inf\limits_{U(\cdot)\in\mathcal{A}^0} J(U(\cdot)).\end{cases}$$

The average cost optimality equation associated with the limiting control problem $\mathcal{P}^0$ is

$$\bar{\lambda} = \inf_{0\le p\cdot u\le k, k\in\mathcal{M}}\left\{\frac{\partial \bar{w}(x)}{\partial(-ax+\sum_{j=0}^m \nu_j u^j - z)}+\sum_{j=0}^m \nu_j c(u^j)\right\}+h(x), \tag{9.9.47}$$

where $\bar{w}(x)$ is a potential function of the problem $\mathcal{P}^0$ and $\frac{\partial \bar{w}(x)}{\partial(-ax+\sum_{j=0}^m \nu_j u^j - z)}$ is the directional derivative of $\bar{w}(x)$ along the direction $-ax+\sum_{j=0}^m \nu_j u^j - z$.

Theorem 9.9.2 ([32]) *There exists a constant C such that for all $\varepsilon>0$,*

$$|\lambda^\varepsilon-\lambda|\le C\varepsilon^{\frac{1}{2}}.$$

This implies in particular that $\lim_{\varepsilon\to 0}\lambda^\varepsilon=\lambda$.

We next consider feedback controls. We begin with an optimal feedback control $\bar{U}(x) = (\bar{u}^0(x),\bar{u}^1(x),...,\bar{u}^m(x))$ for the limiting control problem $\mathcal{P}^0$. This is obtained by minimizing the right-hand side of (9.9.47), i.e.,

$$\begin{aligned}&\left(\textstyle\sum_{j=0}^m \nu_j\bar{u}^j(x)-z\right)w_x(x)+\textstyle\sum_{j=0}^m \nu_j c(\bar{u}^j(x))+h(x)\\ &=\inf_{0\le u^i\le i, i\in\mathcal{M}}\left\{\left(\textstyle\sum_{j=0}^m \nu_j u^j - z\right)w_x(x)+\textstyle\sum_{j=0}^m \nu_j c(u^j)\right\}+h(x).\end{aligned}$$

We then construct the control

$$f^\varepsilon(x, \alpha^\varepsilon(t)) = \sum_{j=0}^{m} 1_{\{\alpha(\varepsilon,t)=j\}} \bar{u}^j(x), \tag{9.9.48}$$

which is clearly feasible (satisfies the control constraints) for $\mathcal{P}^\varepsilon$. Furthermore, if each $\bar{u}^j(\cdot)$ is locally Lipschitz, then the system

$$\dot{x}^\varepsilon(t) = -ax^\varepsilon(t) + f^\varepsilon(x^\varepsilon(t), \alpha^\varepsilon(t)) - z, \; x(0) = x$$

has a unique solution and therefore, $f^\varepsilon(x(t), \alpha^\varepsilon(t)),\; t \geq 0$, is also an admissible feedback control for $\mathcal{P}^\varepsilon$.

Theorem 9.9.3 ([32]) *Assume the feedback control of the limiting problem $\bar{U}(\cdot)$ is locally Lipschitz. Moreover, suppose that for each $\varepsilon \in [0, \varepsilon_0]$, the equation*

$$-ax + \sum_{j=0}^{m} \nu_j^\varepsilon \bar{u}^j(x) - z = 0$$

has a unique solution θ^ε, called the threshold, and for $x \in (\theta^\varepsilon, \infty)$,

$$-ax + \sum_{j=0}^{m} \nu_j \bar{u}^j(x) - z < 0,$$

and for $x \in (-\infty, \theta^\varepsilon)$,

$$-ax + \sum_{j=0}^{m} \nu_j \bar{u}^j(x) - z > 0.$$

Then the feedback control given in (9.9.48) is asymptotically optimal, i.e.,

$$\lim_{\varepsilon \to 0} |J^\varepsilon(u^\varepsilon(\cdot)) - \lambda| = 0,$$

where $u^\varepsilon(t) = f^\varepsilon(x(t), \alpha^\varepsilon(t))$.

9.10 Risk-Sensitive Control

In this section we consider a manufacturing system with the objective of minimizing a risk sensitive cost criterion over the infinite horizon. In risk sensitive control theory, typically an exponential-of-integral cost criterion is considered.

We use the dynamic model considered in the previous section. Let $L(x, u)$ denote a cost function of the surplus and the production. The objective of the problem is to choose $u(\cdot) \in \mathcal{A}^\varepsilon$ to minimize

$$J^\varepsilon(u(\cdot)) = \limsup_{T \to \infty} \frac{\varepsilon}{T} \log E \exp\left(\frac{1}{\varepsilon} \int_0^T L(x(t), u(t))dt \right), \tag{9.10.49}$$

where $x(\cdot)$ is the surplus process corresponding to the production process $u(\cdot)$. Let $\lambda^\varepsilon = \inf_{u(\cdot) \in \mathcal{A}^\varepsilon} J^\varepsilon(u(\cdot))$.

A motivation for choosing such an exponential cost criterion is that such criteria are sensitive to large values of the exponent which occur with small probability, for example rare sequences of unusually many machine failures resulting in shortages ($x(t) < 0$).

We assume $L(x,u) \geq 0$ is continuous, bounded, and uniformly Lipschitz in x. The associated HJB equations are as follows:

$$\begin{aligned}\frac{\lambda^\varepsilon}{\varepsilon} = \inf_{0\leq u\leq \alpha} \Bigg\{ & (-ax+u-z)\frac{w_x^\varepsilon(x,\alpha)}{\varepsilon} \\ & + \exp\left(-\frac{w^\varepsilon(x,\alpha)}{\varepsilon}\right)\frac{Q}{\varepsilon}\exp\left(\frac{w^\varepsilon(x,\cdot)}{\varepsilon}\right)(\alpha) + \frac{L(x,u)}{\varepsilon}\Bigg\},\end{aligned} \tag{9.10.50}$$

where $w^\varepsilon(x,\alpha)$ is the potential function, $w_x^\varepsilon(x,\alpha)$ denotes the partial derivative of $w^\varepsilon(x,\alpha)$ with respect to x.

Theorem 9.10.1 ([16]) *The following assertions hold.*

(i) *The HJB equation* (9.10.50) *has a viscosity solution* $(\lambda^\varepsilon, w^\varepsilon(x,\alpha))$.

(ii) *The pair* $(\lambda^\varepsilon, w^\varepsilon(x,\alpha))$ *satisfies the following conditions:*
For some constant C *independent of* $\varepsilon > 0$,
(a) $0 \leq \lambda^\varepsilon \leq C_1$ *and*
(b) $|w^\varepsilon(x,\alpha) - w^\varepsilon(\tilde{x},\alpha)| \leq C_2|x-\tilde{x}|$.

(iii) *Assume that* $w^\varepsilon(x,\alpha)$ *to be Lipschitz continuous in* x. *Then,*

$$\lambda^\varepsilon = \inf_{u(\cdot)\in\mathcal{A}^\varepsilon} J^\varepsilon(u(\cdot)).$$

This theorem implies that λ^ε in $(\lambda^\varepsilon, w^\varepsilon(x,\alpha))$ as a viscosity solution is unique.

We next give a verification theorem. In order to incorporate nondifferentiability of the value function, we consider superdifferential of the function. Let $D^+f(x)$ denote the superdifferential of a function $f(x)$, i.e.,

$$D^+f(x) = \left\{ r \in R : \limsup_{h\to 0} \frac{f(x+h) - f(x) - hr}{|h|} \leq 0 \right\}.$$

Theorem 9.10.2 ([16]) *Let* $(\lambda^\varepsilon, w^\varepsilon(x,\alpha))$ *be a viscosity solution to the HJB equation in* (9.10.50). *Assume that* $w^\varepsilon(x,\alpha)$ *to be Lipschitz continuous in* x. *Let* $\psi^\varepsilon(x,\alpha) = \exp(w^\varepsilon(x,\alpha)/\varepsilon)$. *Suppose that there are* $u^*(\cdot)$, $x^*(\cdot)$, *and* $r^*(t)$ *such that*

$$\dot{x}^*(t) = -ax^*(t) + u^*(t) - z, \; x^*(0) = x,$$

$r^*(t) \in D^+\psi_x^\varepsilon(x^*(t), \alpha^\varepsilon(t))$ *satisfying*

$$\begin{aligned}\frac{\lambda^\varepsilon}{\varepsilon}\psi^\varepsilon(x^*(t),\alpha^\varepsilon(t)) &= (-ax^*(t)+u^*(t)-z)r^*(t) \\ &\quad + \frac{L(x^*(t),u^*(t))}{\varepsilon}\psi^\varepsilon(x^*(t),\alpha^\varepsilon(t)) + \frac{Q}{\varepsilon}\psi^\varepsilon(x^*(t),\cdot)(\alpha^\varepsilon(t)),\end{aligned} \tag{9.10.51}$$

a.e. in t *and w.p.1. Then,* $\lambda^\varepsilon = J^\varepsilon(u^*(\cdot))$.

We next discuss the asymptotic property of the HJB equation (9.10.50) as $\varepsilon \to 0$. First of all, note that this HJB equation is similar to that for an ordinary long-run average cost problem except for the term involving the exponential functions. In order to get rid of such term, we make use of the logarithmic transformation in Fleming and Soner [15, p. 275]).

Let $\mathcal{V} = \{v = (v(0), \dots, v(m)) \in R^{m+1} : v(i) > 0, i = 0, 1, \dots, m\}$. Define

$$Q^v = (q_{ij}^v) \text{ such that } q_{ij}^v = q_{ij}\frac{v(j)}{v(i)} \text{ for } i \neq j \text{ and } q_{ii}^v = -\sum_{j\neq i} q_{ij}^v.$$

Then, in view of the logarithmic transformation, we have, for each $i \in \mathcal{M}$,

$$\begin{aligned}\exp\left(-\frac{w^\varepsilon(x,\alpha)}{\varepsilon}\right) Q \exp\left(\frac{w^\varepsilon(x,\cdot)}{\varepsilon}\right)(i) \\ = \sup_{v\in\mathcal{V}}\left\{\frac{Q^v}{\varepsilon} w^\varepsilon(x,\cdot)(i) - Q^v(\log v(\cdot))(i) + \frac{Qv(\cdot)(i)}{v(i)}\right\}.\end{aligned}$$

The supremum is obtained at $v(i) = \exp(-w^\varepsilon(x,i)/\varepsilon)$.

The logarithmic transformation suggests that the HJB equation is equivalent to an Isaacs equation of a two-player, zero-sum dynamic stochastic game. The Isaacs equation is given as follows:

$$\lambda^\varepsilon = \inf_{0\le u\le \alpha} \sup_{v\in\mathcal{V}}\left\{(-ax+u-z)w_x^\varepsilon(x,\alpha) + \widetilde{L}(x,u,v,\alpha) + \frac{Q^v}{\varepsilon} w^\varepsilon(x,\cdot)(\alpha)\right\} \tag{9.10.52}$$

where

$$\widetilde{L}(x,u,v,i) = L(x,u) - Q^v(\log v(\cdot))(i) + \frac{Qv(\cdot)(i)}{v(i)}, \tag{9.10.53}$$

for $i \in \mathcal{M}$.

We consider the limit of the problem as $\varepsilon \to 0$. In order to define a limiting problem, we first define control sets for the limiting problem. Let

$$\Gamma_u = \{U = (u^0, \dots, u^m);\ 0 \le u^i \le i,\ i = 0, \dots, m\}$$

and

$$\Gamma_v = \{V = (v^0, \dots, v^m);\ v^i = (v^i(0), \dots, v^i(m)) \in \mathcal{V},\ i = 0, \dots, m\}.$$

For each $V \in \Gamma_v$, let $\overline{Q}^V := (q_{ij}^V)$ such that

$$q_{ij}^{v^i} = q_{ij}^V = q_{ij}\frac{v^i(j)}{v^i(i)} \text{ for } i \neq j \text{ and } q_{ii}^V = -\sum_{j\neq i} q_{ij}^V,$$

and let $\nu^V = (\nu_0^V, \dots, \nu_m^V)$ denote the stationary distribution of $\overline{Q}^V$. The next lemma says $\overline{Q}^V$ is irreducible. Therefore, there exists a unique positive ν^V for each $V \in \Gamma_v$. Moreover, ν^V depends continuously on V. It can be shown for each $V \in \Gamma_v$, $\overline{Q}^V$ is irreducible.

Theorem 9.10.3 ([16]) *Let $\varepsilon_n \to 0$ be a sequence such that $\lambda^{\varepsilon_n} \to \lambda^0$ and $w^{\varepsilon_n}(x,\alpha) \to w^0(x,\alpha)$. Then,*

(i) $w^0(x,\alpha)$ is independent of α, i.e., $w^0(x,\alpha) = w^0(x)$;

(ii) $w^0(x)$ is Lipschitz; and

(iii) $(\lambda^0, w^0(x))$ is a viscosity solution to the following Isaacs equation:

$$\begin{aligned}\lambda^0 = \inf_{U\in\Gamma_u}\sup_{V\in\Gamma_v}\Bigg\{&\left(-ax + \sum_{i=0}^m \nu_i^V u^i - z\right) w_x^0(x) + \sum_{i=0}^m \nu_i^V L(x,u^i) \\ &+ \left(\sum_{i=0}^m \nu_i^V \frac{Qv^i(\cdot)(i)}{v^i(i)} - \sum_{i=0}^m \nu_i^V \overline{Q}^V(\log v^i(\cdot))(i)\right)\Bigg\}.\end{aligned} \tag{9.10.54}$$

Let

$$\widehat{L}(x,U,V)=\sum_{i=0}^{m}\nu_i^V L(x,u^i)+\sum_{i=0}^{m}\nu_i^V\frac{Qv^i(\cdot)(i)}{v^i(i)}-\sum_{i=0}^{m}\nu_i^V\overline{Q}^V(\log v^i(\cdot))(i).$$

Note that $\widehat{L}(x,U,V)\le ||L||$, where $||\cdot||$ is the sup norm. Moreover, since $L\ge 0$, $\widehat{L}(x,U,1)\ge 0$ where $V=1$ means $v^i(j)=1$ for all i,j. Then, the equation in (9.10.54) is an Isaacs equation associated with a two-player, zero-sum dynamic game with objective

$$J^0(U(\cdot),V(\cdot))=\limsup_{T\to\infty}\frac{1}{T}\int_0^T\widehat{L}(x(t),U(t),V(t))dt$$

subject to

$$\dot{x}(t)=-ax(t)+\sum_{i=0}^{m}\nu_i^{V(t)}u^i(t)-z,\ x(0)=x,$$

where $U(\cdot)$ and $V(\cdot)$ are Borel measurable functions and $U(t)\in\Gamma_u$ and $V(t)\in\Gamma_v$ for $t\ge 0$.

One can show that

$$\lambda^0=\inf_{U(\cdot)}\sup_{V(\cdot)}J^0(U(\cdot),V(\cdot)),$$

which implies the uniqueness of λ^0.

Finally, in order to use the solution to the limiting problem to obtain a control for the original problem, a numerical scheme has to be used to obtain an approximate solution. The advantage of the limiting problem is its dimensionality, which is much smaller that of the original problem if the number of states in $\mathcal{M}$ is large.

Let $(U^*(x),V^*(x))$ denote a solution to the upper value problem. Suggested by the ideas of hierarchical control, it is expected that the control

$$u(x,\alpha)=\sum_{j=0}^{m}1_{\{\alpha=j\}}u^{i*}(x)$$

is nearly optimal for the original problem. For more details discussed in this section, see Fleming and Zhang [16].

PART III: PROBLEMS WITH CLOSED-FORM SOLUTIONS

The main advantage of hierarchical control is to reduce the system dimensionality and the computational burden. By considering a limiting problem and using its solution, one constructs a near optimal control for the original problem. In this part, we give closed-form solutions to three problems. The solutions of these problems can be used to construct controls for the corresponding original problems.

9.11 Constant Product Demand

In this section, we consider finite horizon production planning of stochastic manufacturing systems. Note that there are some distinct differences between the finite time and the infinite horizon formulations. For an infinite horizon formulation, such as the problems studied in Section 2, the dynamics of the systems are essentially homogeneous, and therefore, the hedging point (or turnpike sets) consist of constants, which completely characterize the optimal control policies. If the system performance is evaluated over a finite time horizon the threshold levels are no longer constants, but are "time dependent threshold curves." Therefore, the problem becomes much more complicated. Naturally, one expects that the essence of the turnpike sets should still work, i.e. produce at the maximum speed if the inventory level is below the turnpike, produce nothing if the inventory level is above the turnpike, and produce exactly as the demand if the inventory reaches the turnpike. Nevertheless, the time inhomogeneous nature of the sets makes it very difficult to obtain explicit optimal solutions. In order to fulfill our goal of achieving optimality, the turnpike sets must be smooth enough and be "traceable" by the trajectory of the system.

Let $x(t) \in R^1$ denote the inventory/backlog process and $u(t) \geq 0$ denote the rate of production planning of a manufacturing system. The product demand is assumed to be a constant and denoted by z. Then,

$$\dot{x}(t) = u(t) - z, \; x(s) = x, \; 0 \leq s \leq t \leq T \tag{9.11.55}$$

where T is a finite horizon.

Let $\mathcal{M} = \{\alpha_1, \alpha_2\}$ ($\alpha_1 > \alpha_2 \geq 0$) denote the set of machine states and let $\alpha(t) \in \mathcal{M}$ denote the machine capacity process. If $\alpha(t) = \alpha_1$, it means the machine is in a good condition with capacity α_1. If $\alpha(t) = \alpha_2$, the machine (or part of the machine) breaks down with a remaining capacity α_2. We assume that $\alpha_1 > z > \alpha_2$, i.e., the demand can be satisfied if the machine is in a good condition and cannot be satisfied if the machine (or part of the machine) breaks down.

The cost function $J(s, x, \alpha, u(\cdot))$ with $\alpha(s) = \alpha \in \mathcal{M}$ is defined by

$$J(s, x, \alpha, u(\cdot)) = E \int_s^T e^{-\rho t} h(x(t)) dt, \tag{9.11.56}$$

where $\rho \geq 0$ is the discount factor. Here ρ is allowed to be zero, since we are now considering a finite horizon problem. The problem is to find a production plan $0 \leq u(t) \leq \alpha(t)$ as a function of the past $\alpha(\cdot)$ that minimizes $J(s, x, \alpha, u(\cdot))$.

We make the following assumptions on the running cost function $h(x)$ and the random process $\alpha(t)$.

(A1) $h(x)$ is a convex function such that for positive constants C_h and k_h,

$$0 \leq h(x) \leq C_h(1 + |x|^{k_h}) \text{ and } h(x) > h(0) = 0 \text{ for all } x \neq 0.$$

Moreover, there exists a constant $c_h > 0$ such that

$$\frac{h_{x^+}(x_2) - h_{x^+}(x_1)}{x_2 - x_1} \geq c_h \text{ for all } -|\alpha_2 - z|T \leq x_1 < 0 \leq x_2 \leq |\alpha_2 - z|T, \tag{9.11.57}$$

where $h_{x^+}(x)$ denotes the right-hand derivative of $h(x)$.

Note that the convexity of $h(x)$ implies that both the left-hand derivative $h_{x^-}(x)$ and the right-hand derivative $h_{x^+}(x)$ exist a.e. and $h_{x^-}(x) = h_{x^+}(x) = h_x(x)$ a.e. In this section, we use mostly the right-hand derivative $h_{x^+}(x)$ to represent the derivative $h_x(x)$.

(A2) The capacity process $\alpha(t) \in \mathcal{M}$ is a two-state Markov chain governed by

$$L_\alpha f(\cdot)(i) = \begin{cases} \lambda(f(\alpha_2) - f(\alpha_1)) & \text{if } i = \alpha_1 \\ 0 & \text{if } i = \alpha_2. \end{cases} \tag{9.11.58}$$

for any function f on $\mathcal{M}$. Here $\lambda \geq 0$ is the machine breakdown rate.

Examples of $h(x)$. A few examples of $h(x)$ that satisfy Assumption (A1) can be given as follows.

(1) $h(x) = x^2$;

(2) $h(x) = h^+ \max\{0, x\} + h^- \max\{0, -x\}$ where $h^+ > 0$ and $h^- > 0$ are constants. This cost function was employed in [2].

(3) $h(x)$ is convex and piecewise linear with $h(0) = 0$, $h_{x^-}(0) < 0$, and $h_{x^+}(0) > 0$.

Assumption (A2) is a condition on the machine capacity process $\alpha(t)$. It indicates that once the machine goes down it will never come up again. Such a situation occurs when the repairing is very expensive, or no repair facilities are available. As a result, replacement is a better alternative than repair.

Definition 9.11.1. A control $u(\cdot) = \{u(t) : t \geq 0\}$ is *admissible* if $u(t)$ is an $\mathcal{F}_t = \sigma\{\alpha(s), s \leq t\}$ adapted measurable process and $0 \leq u(t) \leq \alpha(t)$ for all $0 \leq t \leq T$. $\mathcal{A}$ will denote the set of all admissible controls in the sequel.

Let $v(s, x, \alpha)$ denote the value function of the problem, i.e.,

$$v(s, x, \alpha) = \inf_{u(\cdot) \in \mathcal{A}} J(s, x, \alpha, u(\cdot)), \text{ for } \alpha \in \mathcal{M}.$$

We can show as in [53] that the value function $v(s, x, \alpha)$ is convex in x for each $s \in [0, T]$ and $\alpha \in \mathcal{M}$. Moreover, $v(s, x, \alpha) \in C([0, T], R^1)$ is the only viscosity solution for the following dynamic programming equations.

$$\begin{cases} 0 = & -v_s(s, x, \alpha_1) + \sup\limits_{0 \leq u \leq \alpha_1} [-(u - z) v_x(s, x, \alpha_1)] \\ & - \exp(-\rho s) h(x) - \lambda(v(s, x, \alpha_2) - v(s, x, \alpha_1)), \\ 0 = & v(T, x, \alpha_1) \end{cases} \tag{9.11.59}$$

and

$$\begin{cases} 0 = & -v_s(s, x, \alpha_2) + \sup\limits_{0 \leq u \leq \alpha_2} [-(u - z) v_x(s, x, \alpha_2)] - \exp(-\rho s) h(x) \\ 0 = & v(T, x, \alpha_2). \end{cases} \tag{9.11.60}$$

In the following, we modify the turnpike definition given in [33] to incorporate the variation of the turnpike sets with the changes of time.

Definition 9.11.2. $\phi(s)$ and $\psi(s)$ are said to be the *turnpike sets* for $\alpha = \alpha_1$ and $\alpha = \alpha_2$, respectively, if for all $s \in [0, T]$,

$$v(s, \phi(s), \alpha_1) = \min_x v(s, x, \alpha_1)$$
$$\text{and} \quad v(s, \psi(s), \alpha_2) = \min_x v(s, x, \alpha_2), \text{ respectively.}$$

Lemma 9.11.3 ([53]) *Let $\psi(s)$ be defined as follows:*

$$\int_s^T e^{-\rho t} h_{x^+}(\psi(s) + (\alpha_2 - z)(t - s)) dt = 0. \tag{9.11.61}$$

Then $\psi(s)$ is continuous, uniquely determined by (9.11.61) and satisfies

(a) $0 < \psi(s) < |\alpha_2 - z|(T - s)$ *for* $s \in [0, T)$ *and* $\psi(T) = 0$;

(b) $\psi(s)$ *is monotone decreasing and absolute continuous. Moreover,*

$$\dot{\psi}(s) + z > \alpha_2, \ a.e.$$

Remark. The absolute continuity of $\psi(s)$ implies that it is differentiable almost everywhere in s. $\dot{\psi}(s) + z > \alpha_2$ says that if $x(t) \le \psi(t)$ for some t_1, then $x(t)$ will stay below $\psi(t)$ for all $t_1 \le t \le T$.

Let

$$H(s,x) = h(x) + \lambda e^{\rho s} v(s,x,\alpha_2).$$

Then,

$$J(s,x,\alpha_1,u(\cdot)) = \int_s^T e^{-\lambda(t-s)} e^{-\rho t} H(t,x(t))dt.$$

Note that $H(s,x)$ is convex in x for each s. We are to show that the turnpike set for $\alpha = \alpha_1$ is given by the minimizer of $H(s,x)$, i.e.,

$$H(s,\phi(s)) = \min_x H(s,x). \tag{9.11.62}$$

To proceed, we need to consider an important property possessed by $\phi(s)$, which is described in the following definition.

Lemma 9.11.4 ([53]) *Let $\phi(s)$ be the minimizer of $H(s,x)$. Then $\phi(s)$ is a single-valued function and satisfies:*

(a) $0 \le \phi(s) < \psi(s)$ *for* $s \in [0,T)$ *and* $\phi(T) = 0$;

(b) $\phi(s)$ *is traceable, i.e.,* $\phi(s)$ *is absolutely continuous on* $[0,T]$ *and*

$$0 \le \dot{\phi}(s) + z \le z, \quad a.e. \text{ in } s \in [0,T].$$

It is easy to see that a traceable curve is always decreasing. If a function $\gamma(s)$ is traceable, then there exists a control $0 \le u(s) = \dot{\gamma}(s) + z \le z < \alpha_1$ such that the corresponding system trajectory $x(t)$ may stay on the curve $\gamma(s)$ after it reaches $\gamma(s)$.

Let

$$\tilde{H}(s,x) = h(x) + \lambda e^{\rho s} \int_s^T e^{-\rho t} h(x + (\alpha_2 - z)(t-s))dt. \tag{9.11.63}$$

Note that $\phi(s) < \psi(s)$ and

$$v(s,x,\alpha_2) = \int_s^T e^{-\rho t} h(x + (\alpha_2 - z)(t-s))dt \text{ if } x \le \psi(s).$$

It follows that

$$\min_x H(s,x) = \min_x \tilde{H}(s,x) = h(\phi(s)) + \lambda e^{\rho s} \int_s^T e^{-\rho t} h(\phi(s) + (\alpha_2 - z)(t-s))dt. \tag{9.11.64}$$

It can also be seen that $\phi(s)$ is the only solution to (9.11.64).

Theorem 9.11.5 ([53]) *Let $\phi(s)$ and $\psi(s)$ be given as in* (9.11.64) *and* (9.11.61), *respectively. Then $\phi(s)$ and $\psi(s)$ are the turnpike sets for $\alpha = \alpha_1$ and $\alpha = \alpha_2$, respectively. Moreover, the feedback control $u^*(t) = u^*(t,x^*(t),\alpha(t))$ given below is optimal:*

$$u^*(t,x,\alpha_1) = \begin{cases} 0 & \text{if } x > \phi(t) \\ \dot{\phi}(t) + z & \text{if } x = \phi(t) \\ \alpha_1 & \text{if } x < \phi(t); \end{cases}$$
$$u^*(t,x,\alpha_2) = \begin{cases} 0 & \text{if } x > \psi(t) \\ \alpha_2 & \text{if } x \le \psi(t). \end{cases} \tag{9.11.65}$$

Moreover, it is easy to see that under the control policy $u^*(t) = u^*(t, x(t), \alpha(t))$, the ordinary differential equation

$$\dot{x}^*(t) = u^*(t, x^*(t), \alpha(t)) - z, \; x^*(s) = x$$

has a unique solution.

Next, using the control given in (9.11.65), the value function $v(s, x, \alpha_1)$ can be written as follows:

$$v(s,x,\alpha_1) = \begin{cases} \int_s^T e^{-\lambda(t-s)}[e^{-\rho t}h(x-z(t-s)) + \lambda v(t, x-z(t-s), \alpha_2)]dt \\ \qquad\qquad \text{if } x \geq z(T-s) \\ \int_s^{s_1} e^{-\lambda(t-s)}[e^{-\rho t}h(x-z(t-s)) + \lambda v(t, x-z(t-s), \alpha_2)]dt \\ \quad + \int_{s_1}^T e^{-\lambda(t-s)}[e^{-\rho t}h(\phi(t)) + \lambda v(t, \phi(t), \alpha_2)]dt \\ \qquad\qquad \text{if } \phi(s) < x < z(T-s) \\ \int_s^T e^{-\lambda(t-s)}[e^{-\rho t}h(\phi(t)) + \lambda v(t, \phi(t), \alpha_2)]dt \\ \qquad\qquad \text{if } x = \phi(s) \\ \int_s^{s_2} e^{-\lambda(t-s)}[e^{-\rho t}h(x+(\alpha_1-z)(t-s)) + \lambda v(t, x+(\alpha_1-z)(t-s), \alpha_2)]dt \\ \quad + \int_{s_2}^T e^{-\lambda(t-s)}[e^{-\rho t}h(\phi(t)) + \lambda v(t, \phi(t), \alpha_2)]dt \\ \qquad\qquad \text{if } -(\alpha_1-z)(T-s) < x < \phi(s) \\ \int_s^T e^{-\lambda(t-s)}[e^{-\rho t}h(x+(\alpha_1-z)(t-s)) + \lambda v(t, x+(\alpha_1-z)(t-s), \alpha_2)]dt \\ \qquad\qquad \text{if } x \leq -(\alpha_1-z)(T-s), \end{cases}$$

where s_1 is the first time that $x-z(t-s)$ hits $\phi(t)$ and s_2 is the first time that $x+(\alpha_1-z)(t-s)$ hits $\phi(t)$, respectively. Thus,

$$x - z(s_1 - s) = \phi(s_1) \text{ and } x + (\alpha_1 - z)(s_2 - s) = \phi(s_2), \text{ respectively.}$$

Using the control (9.11.65), we can write the value function $v(s, x, \alpha_2)$ as follows:

$$v(s,x,\alpha_2) = \begin{cases} \int_s^T e^{-\rho t}h(x - z(t-s))dt & \text{if } x \geq z(T-s) \\ \int_s^{s_0} e^{-\rho t}h(x - z(t-s))dt & \\ \quad + \int_{s_0}^T e^{-\rho t}h(x - z(s_0 - s) + (\alpha_2 - z)(t - s_0))dt & \\ & \text{if } \psi(s) < x < z(T-s) \\ \int_s^T e^{-\rho t}h(x + (\alpha_2 - z)(t-s))dt & \text{if } x \leq \psi(s) \end{cases}$$

where s_0 is the first time that $x-z(t-s)$ hits $\psi(t)$. Thus, $x - z(s_0 - s) = \psi(s_0)$.

Example 9.11.6. In this example, the cost function is given by

$$h(x) = h^+ \max\{0, x\} + h^- \max\{0, -x\}.$$

Then, (9.11.61) becomes

$$\int_s^{t_1} e^{-\rho t} h^+ dt - \int_{t_1}^{T} e^{-\rho t} h^- dt = 0,$$

where t_1 is given by $\psi(s) + (\alpha_2 - z)(t_1 - s) = 0$. This yields

$$\psi(s) = \frac{\alpha_2 - z}{\rho} \log \frac{h^+ + h^- e^{-\rho(T-s)}}{h^+ + h^-}.$$

We now identify $\phi(s)$. Recall that $0 \le \phi(s) < \psi(s) \le |\alpha_2 - z|(T - s)$ and for $x < \psi(s)$,

$$v(s, x, \alpha_2) = \int_s^T e^{-\rho t} h(x + (\alpha_2 - z)(t - s))dt.$$

Moreover, for $0 \le x \le |\alpha_2 - z|(T - s)$,

$$\begin{aligned}
& v(s, x, \alpha_2) \\
= \; & \exp(-\rho s) \int_0^{\frac{x}{|\alpha_2 - z|}} e^{-\rho t} h^+ (x + (\alpha_2 - z)t)dt - \int_{\frac{x}{|\alpha_2 - z|}}^{T-s} e^{-\rho t} h^- (x + (\alpha_2 - z)t)dt \\
= \; & \exp(-\rho s) h^+ \rho^{-1} [x + (\alpha_2 - z)\rho^{-1}(1 - e^{-\rho x/|\alpha_2 - z|})] \\
& + h^- \rho^{-1} [(x + (\alpha_2 - z)(T - s)e^{-\rho(T-s)} + |\alpha_2 - z|\rho^{-1}(e^{-\rho x/|\alpha_2 - z|} - e^{-\rho(T-s)})]
\end{aligned}$$

This together with (9.11.64) yields

$$\phi(s) = \max\left\{0, -\frac{|\alpha_2 - z|}{\rho} \log \frac{\rho h^+ + \lambda(h^+ + h^- e^{-\rho(T-s)})}{\lambda(h^+ + h^-)}\right\}. \tag{9.11.66}$$

Equivalently, $\phi(s)$ can also be written as:

$$\phi(s) = 0 \text{ for all } s \in [0, T] \text{ if } \lambda h^- - \rho h^+ \le 0;$$

otherwise,

$$\phi(s) = \begin{cases} -\dfrac{|\alpha_2 - z|}{\rho} \log \dfrac{\rho h^+ + \lambda(h^+ + h^- e^{-\rho(T-s)})}{\lambda(h^+ + h^-)} & \text{if } s \le T + \dfrac{1}{\rho} \log \dfrac{\lambda h^- - \rho h^+}{\lambda h^-} \\ 0 & \text{if } s > T + \dfrac{1}{\rho} \log \dfrac{\lambda h^- - \rho h^+}{\lambda h^-}. \end{cases}$$

As $T \to \infty$, it is easily seen that

$$\phi(s) \to \max\left\{0, -\frac{|\alpha_2 - z|}{\rho} \log \frac{(\rho h^+ + \lambda h^+)}{\lambda(h^+ + h^-)}\right\},$$

which gives the same turnpike set as in [2] provided that the repair rate vanishes.

In this example, we are able to solve (9.11.61) and (9.11.64) to obtain explicitly the turnpike sets $\phi(s)$ and $\psi(s)$. It should be noted that such explicit turnpike sets are not available for general $h(\cdot)$. However, in many applications of manufacturing systems, $h(\cdot)$ appears to be piecewise linear or a linear combination of linear functions. Then (9.11.61) and (9.11.64) are solvable.

9.12 Constant Machine Capacity

This problem is a variation of the one with constant demand. Here the system having a constant machine capacity $\alpha_0 > 0$ with a random demand rate $z(t)$, is described by the following equation

$$\dot{x}(t) = u(t) - z(t),\ x(s) = x,\ 0 \le s \le t \le T < \infty \tag{9.12.67}$$

with the production constraints $0 \le u(t) \le \alpha_0$.

Let $\mathcal{Z} = \{z_1, z_2\}$ denote the set of demand rates with $0 < z_1 < \alpha_0 < z_2$.

The corresponding cost function $J(s, x, z, u(\cdot))$ with $z(s) = z \in \mathcal{Z}$ is defined by

$$J(s, x, z, u(\cdot)) = E \int_s^T e^{-\rho t} h(x(t)) dt. \tag{9.12.68}$$

The problem is to find a production plan $0 \le u(t) \le \alpha_0$ as a function of the past $z(t)$, that minimizes $J(s, x, z, u(\cdot))$.

(A1') Let Assumption (A1) be satisfied with (9.11.57) replaced by

$$\frac{h_{x^+}(x_2) - h_{x^+}(x_1)}{x_2 - x_1} \ge c_h \text{ for all } - |\alpha_0 - z_2|T \le x_1 < 0 \le x_2 \le |\alpha_0 - z_2|T,$$

with z_2 given below.

(A2') The demand process $z(t) \in \mathcal{Z}$ is also a two state Markov chain governed by

$$L_z f(\cdot)(i) = \begin{cases} \lambda'(f(z_2) - f(z_1)) & \text{if } i = z_1 \\ 0 & \text{if } i = z_2. \end{cases} \tag{9.12.69}$$

for any function f on $\mathcal{Z}$.

Let $v(s, x, z)$ denote the value function of the problem, i.e.,

$$v(s, x, z) = \inf_{u(\cdot) \in \mathcal{A}} J(s, x, z, u(\cdot)), \text{ for } z \in \mathcal{Z}.$$

It can be shown as in [53] that the value functions $v(s, x, z)$ are convex functions in x for each $s \in [0, T]$ and $z \in \mathcal{Z}$. Moreover, $v(s, x, z) \in C([0, T], R^1)$ are the only viscosity solutions for the following dynamic programming equations.

$$\begin{cases} 0 = & -v_s(s, x, z_1) + \sup_{0 \le u \le \alpha_0} [-(u - z_1) v_x(s, x, z_1)] \\ & - \exp(-\rho s) h(x) - \lambda'(v(s, x, z_2) - v(s, x, z_1)), \\ 0 = & v(T, x, z_1) \end{cases} \tag{9.12.70}$$

and

$$\begin{cases} 0 = & -v_s(s, x, z_2) + \sup_{0 \le u \le \alpha_0} [-(u - z_2) v_x(s, x, z_2)] - \exp(-\rho s) h(x) \\ 0 = & v(T, x, z_2). \end{cases} \tag{9.12.71}$$

Definition 9.12.1. $\phi(s)$ and $\psi(s)$ are said to be *turnpike sets* if

$$v(s, \phi(s), z_1) = \min_x v(s, x, z_1)$$

$$\text{and} \quad v(s, \psi(s), z_2) = \min_x v(s, x, z_2).$$

Let

$$\hat{H}(s,x) = h(x) + \lambda \int_s^T e^{-\rho t} h(x + (\alpha_0 - z_2)(t-s))dt.$$

Then $\psi(s)$ and $\phi(s)$ are determined by:

$$\int_s^T e^{-\rho t} h_{x^+}(\psi(s) + (\alpha_0 - z_2)(t-s))dt = 0 \tag{9.12.72}$$

and

$$\hat{H}(s,\phi(s)) = \min_x \hat{H}(s,x). \tag{9.12.73}$$

Lemma 9.12.2 ([53]) *$\phi(s)$ and $\psi(s)$ are single-valued absolute continuous functions. They satisfy the following properties:*
(a) $0 < \psi(s) < |\alpha_0 - z_2|(T-s)$ *for* $s \in [0,T)$ *and* $\psi(T) = 0$;
(b) $\psi(s)$ *is monotone decreasing and* $\dot{\psi}(s) + z_2 > \alpha_0$, *a.e.;*
(c) $0 \le \phi(s) < \psi(s)$ *for* $s \in [0,T)$ *and* $\phi(T) = 0$;
(d) $\phi(s)$ *is monotone decreasing and* $\dot{\phi}(s) + z_1 \ge \alpha_0 - z_2 + z_1$, *a.e.*

Note that by (4) of the above lemma, a sufficient condition for $\phi(s)$ to be traceable (i.e., $0 \le \dot{\phi}(s) + z_1 \le \alpha_0$) is $z_1 \ge z_2 - \alpha_0$.

Theorem 9.12.3 ([53]) *Suppose that (A1'), (A2') are satisfied, and $z_1 \ge z_2 - \alpha_0$. Let $u^*(t, x(t), z)$ be defined as follows:*

$$u^*(t,x,z_1) = \begin{cases} 0 & \text{if } x > \phi(t) \\ \dot{\phi}(t) + z_1 & \text{if } x = \phi(t) \\ \alpha_0 & \text{if } x < \phi(t), \end{cases}$$
$$u^*(t,x,z_2) = \begin{cases} 0 & \text{if } x > \psi(t) \\ \alpha_0 & \text{if } x \le \psi(t). \end{cases} \tag{9.12.74}$$

Then under the control $u^*(t) = u^*(t, x(t), z(t))$, the equation

$$\dot{x}^*(t) = u^*(t, x^*(t), z(t)) - z(t), \ x_0^* = x$$

has a unique solution. Therefore, the control $u^*(t)$ is optimal.

Example 9.12.4. Consider the cost function

$$h(x) = h^+ \max\{0, x\} + h^- \max\{0, -x\}.$$

Then,

$$\psi(s) = \frac{\alpha_0 - z_2}{\rho} \log \frac{h^+ + h^- e^{-\rho(T-s)}}{h^+ + h^-}.$$

$$\phi(s) = 0 \text{ if } \lambda h^- - \rho h^+ \le 0.$$

If $\lambda h^- - \rho h^+ > 0$, then

$$\phi(s) = \begin{cases} -\dfrac{|\alpha_0 - z_2|}{\rho} \log \dfrac{\rho h^+ + \lambda(h^+ + h^- e^{-\rho(T-s)})}{\lambda(h^+ + h^-)} & \text{if } s \le T + \dfrac{1}{\rho} \log \dfrac{\lambda h^- - \rho h^+}{\lambda h^-} \\ 0 & \text{if } s > T + \dfrac{1}{\rho} \log \dfrac{\lambda h^- - \rho h^+}{\lambda h^-}. \end{cases}$$

Note that the assumption $z_1 \geq z_2 - \alpha_0$ in the above theorem is a relatively conservative one. In the previous example, this condition can be relaxed to

$$z_1 \geq |\alpha_0 - z_2| \frac{\lambda h^- - \rho h^+}{\lambda(h^+ + h^-)}. \tag{9.12.75}$$

It can be seen that (9.12.75) is also necessary for the traceability of $\phi(s)$ in this example. If (9.12.75) fails, $\phi(s)$ will no longer be the turnpike of the problem since it is not traceable on $[0, T]$. Let $\tilde{\phi}(s)$ denote the turnpike set for $\alpha = \alpha_1$ defined as $v(s, \tilde{\phi}(s), z_1) = \min_x v(s, x, z_1)$. Note that $0 \leq \tilde{\phi}(s) \leq \psi(s)$. It can be shown that $h(x) + \lambda v(s, x, z_2)$ is strictly convex on $[0, \psi(s)]$, which implies that $v(s, x, z_1)$ is also strictly convex on $[0, \psi(s)]$ (see [33] for details). Therefore, $\tilde{\phi}(s)$ is a continuous function. Intuitively, the optimal control $u^*(t, x, \alpha_1)$ should be given as in (9.12.74) with $\tilde{\phi}(s)$ in place of $\phi(s)$ provided that $\tilde{\phi}(s)$ is traceable. Let

$$s_0 = T + \frac{1}{\rho} \log \frac{\lambda h^- - \rho h^+}{\lambda h^-}.$$

Then, $0 \leq \dot{\phi}(s) + z_1 \leq \alpha_0$ for $s \geq s_0$. This implies $\tilde{\phi}(s) = \phi(s)$ for $s \geq s_0$. However, if $0 > s_0 > T$ and z_2 is large enough, the traceability property of $\tilde{\phi}(s)$ will not hold, which makes the problem very complicated; an explicit optimal solution is very difficult to obtain.

9.13 Marketing-Production with a Jump Demand

In this section, we consider a marketing-production model in which a manufacturing firm seeks to maximize its overall profit by properly choosing the rates of production and advertising over time. Similar to Section 6, the marketing decision depends on how much the advertising effort is needed. Such promotional activities create additional demand of the product.

In this section, we consider a basic building-block model. We aim at obtaining analytic solutions of various control regions involved to yield managerial insight for applications. The demand rate is modeled as a process with a jump. The problem is to choose the optimal strategy so that the overall expected profit is maximized. To exploit the intrinsic properties of the system, we examine a single-machine system in order not to involve complex notation and excessive technical details. The model considered can be thought of as a macro model from a higher level management point of view. The obtained results will enable us to develop optimal strategies for more complex jobshops by considering integrated processes as single-machine systems in computational approaches.

The demand normally changes not very frequently, its sample paths displaying piecewise constant behavior. As a result, it is reasonable to model the demand as a controlled Markov chain. Typically, the demand of a new product is nondecreasing. From a management point of view, when the demand significantly decreases, it is probably time to terminate the production of such a product and to create newer models. Therefore, a Poisson process is used quite often (see [21]) to characterize the demand process. Based on such a premise, we consider one possible Poisson-like "up jump" (one state in the increasing demand direction) in the formulation. If the demand can increase with more than one "up jumps," we may choose to deal with one jump at a time. The decision that the manager faces is over a finite horizon. Although the objective function is written as a discounted infinite horizon one, by appropriate choice of the discount factor $\rho > 0$, the underlying problem is essentially "equivalent" to a finite horizon one (i.e., the future is sufficiently discounted with an exponentially decaying rate ρ).

We obtain closed-form optimal control policies. An interesting feature of these results is that the optimal market-production policy is of the hedging point type, and the hedging point depends on the amount of the marginal revenue. If the marginal revenue is small, it is not worthwhile to take any advertisement action. Otherwise, whether or not to use advertising for promotion depends on if the inventory surplus is above or below the hedging point. The derived analytical solution yields good insight on how production planning tasks can be carried out. In addition, it also provides guidelines for further study and development of numerical methods for more complex systems involving more general random demand and random machine capacity.

For $t \geq 0$, let $x(t)$, $u(t)$, and $z(t)$ denote the inventory level, the production rate, and the demand rate, respectively. They are governed by the dynamic equation:

$$\dot{x}(t) = u(t) - z(t),\ x(0) = x. \tag{9.13.76}$$

For $t \geq 0$, suppose that $x(t) \in R = (-\infty, \infty)$, that the production system has a unit production capacity constraint, $w(t)$ denotes the marketing (or advertising) rate with $w(t) \in \{0, w_d\}$ for some $w_d > 0$. Assume the demand rate $z(t)$ is a two-state Markov chain with state space $\{z_1, z_2\}$ (some $0 < z_1 < z_2 < 1$). The generator of this Markov chain is

$$Q(w(t)) = \begin{pmatrix} -kw(t) & kw(t) \\ 0 & 0 \end{pmatrix}, \tag{9.13.77}$$

for a given constant $k > 0$. Let $h(x) = c^+x^+ + c^-x^-$ denote the inventory cost function where c^+ and c^- are positive constants, and

$$x^+ = \max\{0, x\} \text{ and } x^- = \max\{0, -x\}.$$

We treat $(x(t), z(t))$ as a pair of state variables and $(u(t), w(t))$ as a pair of control variables throughout.

Definition 9.13.1. A control $(u(\cdot), w(\cdot)) = \{u(t), w(t);\ t \geq 0\}$ is *admissible* if $u(t) \in [0, 1]$ and $w(t) \in \{0, w_d\}$ and is progressively measurable with respect to the σ-algebra generated by $z(s)$, $s \leq t$. Denote the set of all admissible controls by $\mathcal{A}$.

Our objective is to choose $(u(\cdot), w(\cdot)) \in \mathcal{A}$ to maximize the total expected profit:

$$J(x, z, u(\cdot), w(\cdot)) = E \int_0^\infty e^{-\rho t}[\pi z(t) - h(x(t)) - w(t)]dt, \tag{9.13.78}$$

where $\rho > 0$ is the discount rate and π is the revenue per unit sale.

Note that z_2 is an absorbing state. Choosing $w(t) = w_d$ means to promote the product at the cost of w_d and choosing $w(t) = 0$ means that no marketing action is taken.

Note also that the optimal marketing rate when $z(t) = z_2$ should be $w^*(x, z_2) = 0$. When $z(t)$ enters the state z_2 (i.e., $z(t) = z_2$), it will remain there. Intuitively, since the demand rate satisfies $0 < z_1 < z_2 < 1$, when the maximum of the demand rate is reached, no additional advertising is needed.

Denote the value function by

$$v(x, z) = \max_{(u(\cdot), w(\cdot)) \in \mathcal{A}} J(x, z, u(\cdot), w(\cdot)).$$

The associated Hamilton-Jacobi-Bellman (HJB) equations for the value function are given as follows:

$$\begin{cases} \rho v(x, z_1) = \displaystyle\max_{u \in [0,1], w \in \{0, w_d\}} \{(u - z_1)v_x(x, z_1) + \pi z_1 - h(x) - w \\ \qquad\qquad + kw(v(x, z_2) - v(x, z_1))\} \\ \rho v(x, z_2) = \displaystyle\max_{u \in [0,1]} \{(u - z_2)v_x(x, z_2) + \pi z_2 - + h(x)\}, \end{cases} \tag{9.13.79}$$

where f_x denotes the derivative of a function f with respect to x.

For future use, denote

$$
\begin{aligned}
C_0^+ &= \frac{c^+(z_2-z_1)}{\rho^2} + \frac{\pi(z_2-z_1)}{\rho} - \frac{1}{k},\\
C_0^- &= \frac{c^-(z_2-z_1)}{\rho^2} - \frac{\pi(z_2-z_1)}{\rho} + \frac{1}{k},\\
b &= \frac{\rho+kw_d}{1-z_1},
\end{aligned}
\tag{9.13.80}
$$

and denote by $z^+ > 0$ and $z^- < 0$ the unique solutions of the following equations:

$$
\begin{aligned}
&\frac{c^+z_1}{\rho^2}e^{-\frac{\rho z^+}{z_1}} - \frac{c^+z_2}{\rho^2}e^{-\frac{\rho z^+}{z_2}} + C_0^+ = 0,\\
&\frac{c^-(z_2-z_1)}{b(1-z_2)-\rho}\left(-e^{bz^-} + \frac{b(1-z_2)}{\rho}e^{\frac{\rho z^-}{1-z_2}}\right) - \rho C_0^- = 0, \qquad \text{if } b \neq \frac{\rho}{1-z_2},\\
&c^-(z_2-z_1)(e^{bz^-} - bz^-e^{bz^-}) - \rho^2 C_0^- = 0, \qquad \text{if } b = \frac{\rho}{1-z_2},
\end{aligned}
$$

Theorem 9.13.2. *Define the production policy, the marketing policy, and the hedging point by*

$$
u^*(x,z) = \begin{cases} 1, & \text{if } x<0,\\ z, & \text{if } x=0,\\ 0, & \text{if } x>0, \end{cases}
\tag{9.13.81}
$$

$$
\begin{cases} w^*(x,z_1) = \begin{cases} 0, & \text{if } x<z^*\\ w_d, & \text{if } x\geq z^* \end{cases}\\ w^*(x,z_2)=0, \end{cases}
\tag{9.13.82}
$$

and

$$
z^* = \begin{cases}
-\infty, & \text{if } \pi(z_2-z_1) \geq \frac{\rho}{k} + \frac{c^-(z_2-z_1)}{\rho},\\
z^-, & \text{if } \frac{\rho}{k} < \pi(z_2-z_1) < \frac{\rho}{k} + \frac{c^-(z_2-z_1)}{\rho},\\
0, & \text{if } \pi(z_2-z_1) = \frac{\rho}{k},\\
z^+, & \text{if } \frac{\rho}{k} - \frac{c^+(z_2-z_1)}{\rho} < \pi(z_2-z_1) < \frac{\rho}{k},\\
\infty, & \text{if } \pi(z_2-z_1) \leq \frac{\rho}{k} - \frac{c^+(z_2-z_1)}{\rho},
\end{cases}
\tag{9.13.83}
$$

respectively. Then the feedback control policy $(u^*(x,z), w^*(x,z))$ *given by* (9.13.81)–(9.13.83) *is optimal.*

Let $\xi = \pi(z_2-z_1)$. Then ξ can be regarded as the marginal revenue rate. The marketing policy can be clearly presented in the following table.

$\xi \leq \frac{\rho}{k} - \frac{c^+(z_2-z_1)}{\rho}$	$\frac{\rho}{k} - \frac{c^+(z_2-z_1)}{\rho} < \xi < \frac{\rho}{k}$	$\xi = \frac{\rho}{k}$	$\frac{\rho}{k} < \xi < \frac{\rho}{k} + \frac{c^-(z_2-z_1)}{\rho}$	$\xi \geq \frac{\rho}{k} + \frac{c^-(z_2-z_1)}{\rho}$
$z^* = \infty$	$z^* = z^+ > 0$	$z^* = 0$	$z^* = z^- < 0$	$z^* = -\infty$

If the marginal revenue rate ξ is small ($\le \frac{\rho}{k} - \frac{c^+(z_2 - z_1)}{\rho}$), it is not worthwhile to take marketing action. Therefore $z^* = \infty$ which implies that $w^*(x, z_1) = 0$ for all x. If ξ is not very small ($\frac{\rho}{k} - \frac{c^+(z_2 - z_1)}{\rho} < \xi < \frac{\rho}{k}$), then $z^* > 0$, i.e., take marketing action only if x is large than z^*. If ξ is "moderate," then z^* gets smaller which gives better incentive to take marketing action. Finally, if ξ is so big ($\ge \frac{\rho}{k} + \frac{c^-(z_2 - z_1)}{\rho}$), then $z^* = -\infty$, which means to take marketing action right away no matter what inventory level x is.

Note that both the optimal production and marketing are of the hedging-point type. Such control policies are very attractive from a practical point of view due to their structural simplicity. We would like to mention that in general not all optimal policies are of the hedging type; see [33, Chap. 3].

Example 9.13.3. Consider the following example. Suppose

$$\frac{\rho}{k} - \frac{c^+(z_2 - z_1)}{\rho} < \pi(z_2 - z_1) < \frac{\rho}{k} \quad \text{and } z_2 = 2z_1.$$

The explicit expression of z^* is given by

$$z^* = z^+ = -\frac{2z_1}{\rho} \log\left(1 - \sqrt{\frac{\rho^2}{c^+ z_1 k} - \frac{\rho\pi}{c^+}}\right) > 0.$$

If we take in particular

$$c^+ = c^- = 1,\ z_1 = 0.3,\ z_2 = 0.6,\ k = 1,\ \pi = 1,\ \rho = 0.6,$$

then $z^* = 1.4899$.

9.14 Concluding Remarks

To conclude, we would like to point out that there are numerous applications of hierarchical control, in addition to manufacturing, in large-scale systems including ecological systems (Hirata [22]), computing systems (Courtois [11]), intelligent vehicle highway system (Godbole and Lygeros [19]), spacecraft control systems (Siljak [40]), and target tracking (Zhang [51, 52]). Many of these systems share similar structural properties with large-scale manufacturing systems. For more general treatment of the hierarchical approach, we refer the reader to Auger [3], Singh [42], Simon [41], Smith and Sage [43], Stadtler [45], Switalski [46], and Xie [48], among others.

There have been a series of advances in large-scale manufacturing, but there is still much to be done. We refer the reader to the book [33] for more detailed discussions of results and of open problems.

Bibliography

[1] P.L. Abad, Approach to decentralized marketing-production planning, *Internat. J. Syst. Sci.* **13**, 227-235, (1982).

[2] R. Akella and P. R. Kumar, Optimal Control of Production Rate in a Failure-Prone Manufacturing System, *IEEE Trans. Auto. Contr.*, **AC-31**, 116-126, (1986).

[3] P. Auger, *Dynamics and Thermodynamics in Hierarchically Organized Systems*, Pergamon Press, Oxford, England, 1989.

[4] S. Bai and S. B. Gershwin, Scheduling Manufacturing Systems with Work-in-Process Inventory, *Proc. of the 29th IEEE Conference on Decision and Control*, Honolulu, HI, 557-564, (1990).

[5] T. Basar and P. Bernhard, H^∞ - *Optimal Control and Related Minimax Design Problems*, Birkhauser, Boston, 1991.

[6] E.N. Barron and Jensen, *Total risk aversion, stochastic optimal control, and differential games*, *Appl. Math. Optim.*, **19**, pp. 313-327, (1989).

[7] A. Bensoussan and H. Nagai, An ergodic control problem arising from the principal eigenfunction of an elliptic operator, *J. Math. Soc. Japan*, **43**, 49-65, (1991).

[8] T. Bielecki and P.R. Kumar, Optimality of Zero-Inventory Policies for Unreliable Manufacturing Systems, *Operations Research*, **36**, 532-546, (1988).

[9] M. Caramanis and G. Liberopoulos, Perturbation analysis for the design of flexible manufacturing system flow controllers, *Opns. Res.* **40**, 1107-1125, (1992).

[10] M. Caramanis and A. Sharifnia, Optimal manufacturing flow controller design, *Int. J. Flex. Manuf. Syst.* **3**, 321-336, (1991).

[11] P.J. Courtois, *Decomposability: Queueing and Computer System Applications*, Academic Press, New York, 1977.

[12] G. Feichtinger, R.F. Hartl, and S.P. Sethi, Dynamic optimal control models in advertising: Recent developments, *Management Sci.* **40**, 195-226, (1994).

[13] W. H. Fleming and W. M. McEneaney, Risk sensitive control on an infinite horizon, *SIAM J. Control Optim.*, **33**, 1881-1921, (1995).

[14] W. H. Fleming, S. P. Sethi, and H. M. Soner, An optimal stochastic production planning problem with random fluctuating demand, *SIAM J. Control Optim.*, **25**, 1494-1502, (1987).

[15] W. H. Fleming and H. M. Soner, *Controlled Markov Processes and Viscosity Solutions*, Springer-Verlag, New York, 1992.

[16] W. H. Fleming and Q. Zhang, Risk-sensitive production planning in a stochastic manufacturing system, *SIAM Journal on Control and Optimization*, **36**, 1147-1170, (1998).

[17] C. Gaimon, The price-production problem: An operations and marketing interface, in *Operations Research, Methods, Models, and Applications*, 247-266, J.E. Aronson and S. Zionts (Eds.), Quorum Books, Westport, CN, 1998.

[18] Gershwin, S. B., *Manufacturing Systems Engineering*, Prentice-Hall, Englewood Cliffs, NJ, 1993.

[19] D.N. Godbole and J. Lygeros, (preprint) Hierarchical hybrid control: A case study.

[20] A. Haurie and Ch. van Delft, Turnpike properties for a class of piecewise deterministic control systems arising in manufacturing flow control, *Annals of O. R.*, **29**, 351-373, (1991).

[21] F.S. Hillier and G. J. Lieberman, *Introduction to Operations Research*, McGraw-Hill, New York, 1989.

[22] H. Hirata, Modeling and analysis of ecological systems: the large-scale system viewpoint, *Int. J. Systems Science*, **18**, 1839–1855, (1987).

[23] J. Hu and M. Caramanis, Near optimal setup scheduling for flexible manufacturing systems, *Proc. of the Third RPI International Conference on Computer Integrated Manufacturing*, Troy, NY, May 20-22, 1992.

[24] J. Jiang and S. P. Sethi, A state aggregation approach to manufacturing systems having machines states with weak and strong interactions, *Operations Research*, **39**, 970-978, (1991).

[25] J.G. Kimemia and S. B. Gershwin, An algorithm for the computer control production in flexible manufacturing systems, *IIE Trans.*, **15**, 353-362, (1983).

[26] P. Koktovic, Application of singular perturbation techniques to control problems, *SIAM Review*, **26**, 501-550, (1984).

[27] J. Lehoczky, S. P. Sethi, H. M. Soner, and M. Taksar, An asymptotic analysis of hierarchical control of manufacturing systems under uncertainty, *Mathematics Operations Research*, **16**, 596-608, (1992).

[28] R. G. Phillips and P. Koktovic, A singular perturbation approach to modelling and control of Markov chains, *IEEE Trans. Automatic Control*, **AC-26**, 1087-1094, (1981).

[29] E. Presman, S.P. Sethi, and Q. Zhang, Optimal feedback production planning in a stochastic N-machine flowshop, *Automatica*, **31**, 1325-1332, (1995).

[30] S.P. Sethi, M. Taksar, and Q. Zhang, Capacity and production decisions in stochastic manufacturing systems: An asymptotic optimal hierarchical approach, *Prod. & Oper. Mgmt.*, **1**, 367-392, (1992).

[31] S. P. Sethi, W. Suo, M. I. Taksar and Q. Zhang. Optimal production planning in a stochastic manufacturing system with long-run average cost, *Journal of Optimization Theory and Applications*, **92**, 161-188, (1997).

[32] S. P. Sethi, H. Zhang, and Q. Zhang, Hierarchical production planning in a stochastic manufacturing system with long-run average cost, *Journal of Mathematical Analysis and Applications*, Vol. 214, pp. 151-172, (1997).

[33] S. P. Sethi and Q. Zhang, *Hierarchical Decision Making in Stochastic Manufacturing Systems*, Birkhäuser Boston, Cambridge, MA, 1994.

[34] S.P. Sethi and Q. Zhang, Multilevel hierarchical decision making in stochastic marketing-production systems, *SIAM J. Control and Optim.*, **33**, 528-553, (1995).

[35] S.P. Sethi and Q. Zhang, Hierarchical production planning in dynamic stochastic manufacturing systems: asymptotic optimality and error bounds, *J. Math. Anal. and Appl.*, **181**, 285-319, (1994).

[36] S.P. Sethi, Q. Zhang, and X. Y. Zhou, Hierarchical controls in stochastic manufacturing systems with machines in tandem, *Stochastics and Stochastics Reports*, **41**, 89-118, (1992).

[37] S.P. Sethi, Q. Zhang, and X. Y. Zhou, Hierarchical controls in stochastic manufacturing systems with convex costs, *J. Opt. Theory and Appl.*, **80**, 303-321, (1994).

[38] S. P. Sethi and X. Y. Zhou, Stochastic dynamic job shops and hierarchical production planning, *IEEE Trans. Auto. Contr.*, **39**, 2061-2076, (1994).

[39] A. Sharifnia, Production control of a manufacturing system with multiple machine states, *IEEE Trans. Auto. Contr.* **AC-33**, 620-625, (1988).

[40] D.D. Siljak, *Large-Scale Dynamic Systems*, North-Holland, New York, 1978.

[41] H.A. Simon, The architecture of complexity, *Proc. of the American Philosophical Society*, **106**, 467-482, (1962). reprinted as Chapter 7 in Simon, H.A., *The Sciences of the Artificial*, 2nd Ed., The MIT Press, Cambridge, MA (1981).

[42] M.G. Singh, *Dynamical Hierarchical Control*, Elsevier, rev., 1982.

[43] N. J. Smith and A. P. Sage, An introduction to hierarchical systems theory," *Computers and Electrical Engineering*, **1**, 55-72, (1973).

[44] A.G. Sogomonian and C.S. Tang, *A modeling framework for coordinating production and production decisions within a firm, Management Sci.* **39**, 191-203, (1993).

[45] H. Stadtler, *Hierarchische Produktionsplanung bei Losweiser Fertigung*, Physica-Verlag, Heidelberg, Germany, 1988.

[46] M. Switalski, *Hierarchische Produktionsplanung*, Physica-Verlag, Heidelberg, Germany, 1989.

[47] P. Whittle, *Risk-Sensitive Optimal Control*, Wiley, New York, 1990.

[48] X L. Xie, Hierarchical production control of a flexible manufacturing system, *Applied Stochastic Models and Data Analysis,* **7**, 343-360, (1991).

[49] G. Yin and Q. Zhang, *Continuous-Time Markov Chains and Applications: A Singular Perturbation Approach*, Springer-Verlag, New York, 1998.

[50] Q. Zhang, Risk sensitive production planning of stochastic manufacturing systems: A singular perturbation approach, *SIAM J. Control Optim.*, **33**, 498-527, (1995).

[51] Q. Zhang, Nonlinear filtering and control of a switching diffusion with small observation noise, *SIAM J. Control Optim.*, **36**, 1738-1768, (1998).

[52] Q. Zhang, Optimal filtering of discrete-time hybrid systems, *J. Optim. Theory Appl.*, **100**, 123-144, (1999).

[53] Q. Zhang and G. Yin, Turnpike sets in stochastic manufacturing systems with finite time horizon, *Stochastics Stochastic Rep.* **51**, 11-40, (1994).

[54] Q. Zhang, G. Yin and E.K. Boukas, Optimal control of a marketing-production system, preprint.

Chapter 10

Stochastic Approximation: Theory and Applications

G. YIN
Department of Mathematics
Wayne State University
Detroit, MI 48202

This chapter focuses on stochastic approximation methods and applications. It presents various forms of stochastic approximation algorithms and their variants, including the basic algorithms, the most general algorithms, projection and truncation procedures, algorithms with soft constraints, global stochastic approximation algorithms, continuous-time problems, and infinite dimensional problems. Then the asymptotic properties of stochastic approximation algorithms are examined by considering their convergence, rate of convergence, asymptotic efficiency, and large deviations. The asymptotic analysis is followed by the presentation of a wide range of applications to demonstrate the utility of stochastic approximation methods.

10.1 Introduction

Half a century has passed since stochastic approximation (SA) methods were introduced by Robbins and Monro in their pioneering work [67]. Significant progress has been made in the study of such stochastic recursive algorithms. The original motivation stems from the problem of finding roots of a continuous function $f(\cdot)$, where either the precise form of the function is not known, or it is too complicated to compute; the experimenter is able to take "noisy" measurements at desired values, however. A classical example is to find the appropriate dosage level of a drug, provided only $f(x)$+noise is available, where x is the level of dosage and $f(x)$, assumed to be an increasing function, is the probability of success (leading to the recovery of the patient) at dosage level x. The classical Kiefer–Wolfowitz (KW) algorithm introduced by Kiefer and Wolfowitz [34] concerns the minimization of a real-valued function using only noisy functional measurements. The interesting theoretical issues in the analysis of iteratively defined stochastic processes and a wide variety of applications focus on the basic paradigm of stochastic difference equations. Much of the development

[1]stochastic approximation, projection, constrained algorithm, asymptotic property, convergence, rate of convergence, asymptotic efficiency, gradient estimate.
[2]60F05, 62L20, 93E20, 93E23
[3]The research was supported in part by the National Science Foundation under grant DMS-9877090.

started from a wide range of applications in optimization, control theory, economic systems, signal processing, communication theory, learning, pattern classification, neural network, and many other related fields. Owing to its importance, stochastic approximation has had a long history and has drawn much attention in the past five decades. A number of monographs have been written; each of them has its own distinct features. To mention just a few, we cite the books of Albert and Gardner [2], Wasan [84], Tsypkin [80], Nevel'son and Khasminskii [60], Kushner and Clark [43], Benveniste, Métivier, and Priouret [8], Duflo [21], Solo and Kong [76], Chen and Zhu [12], and Kushner and Yin [53] among others.

10.1.1 Historical Development

The development of stochastic approximation methods can be naturally divided into several periods. To put things in historical perspective, the early development around 1950s and 1960s used mainly basic probabilistic tools and traditional statistical assumptions (such as independent and identically distributed noise) together with certain restrictions on functions (such as assuming $f(x)$ to be increasing for instance). The book of Wasan [84] summaries much of the early development including the with probability one (w.p.1) convergence proof for multidimensional problems of Blum, the asymptotic normality study of Sacks, and the work of Fabian among others. Nevelson and Khasminskii's book [60] treats stochastic approximation as stochastic processes and deals with martingale difference type noise processes. The work of Tsypkin [80] emphasizes the adaptation aspect of applications. As time went on, many applications arising in control and optimization forced researchers to examine the algorithm more closely and indicated that for many applications the noise encountered is correlated. In the middle 1970s, Ljung studied SA from a dynamic system point of view. His idea is: In lieu of the discrete recursion, one treats a continuous-time dynamic system given by an ordinary differential equation. Such an idea was further developed in [43]. By combining analysis and probabilistic argument, Kushner and Clark set up a framework by considering asymptotic properties of suitably scaled sequences. The work of Benveniste, Métivier, and Priouret [8] emphasized the close connection of stochastic approximation and adaptive systems. One of the distinct features is the use of the Markovian setting and the treatment of the Poisson equations. Treating recursive algorithms, the book [21] emphasizes identification, estimation, and tracking. Solo and Kong's book is concerned with the stochastic approximation type of algorithms with applications to adaptive signal processing; it exploits the idea of stochastic averaging in details. The book of Chen and Zhu [12] summarizes their work of using random varying truncation bounds and applications to parameter estimation and adaptive filtering. The work of Kushner and Yin [53] presents a comprehensive development of the modern theory of stochastic approximation, or recursive stochastic algorithms, for both constrained and unconstrained problems, with step sizes that either go to zero or are constant and small (and perhaps random).

To summarize, stochastic approximation methods have been the subject of an enormous literature, both theoretical and applied, for five decades. Due to the vast amount of literature accumulated, it is very difficult or virtually impossible to provide an exhaustive list of references on stochastic approximation. Our hope is that with the references provided at the end of this article, the reader will be able to pick out suitable references of his/her needs. Moreover, it is likewise very difficult to give an extensive account on the technical development in a survey paper of this scale. As a result, we choose the road of discussing the main ideas and leaving most of the technical details aside. Appropriate references are provided however.

10.1.2 Basic Issues

In recent years, algorithms of the stochastic approximation type have found many new applications in diverse areas. New techniques have been developed for proofs of convergence and rates of convergence. Whether or not they are called stochastic approximation algorithms, many procedures frequently used in practical systems are for the purposes of locating the roots of a function and/or for function optimization. Owing to the recent extensive development of methods such as infinitesimal perturbation analysis [32] for the estimation of the pathwise derivatives of complex discrete event systems, the possibilities for the recursive on-line optimization of many such systems that arise in communications or manufacturing have been widely recognized.

Treating stochastic approximation type recursive algorithms, the main idea is to show that asymptotically the noise effects average out so that the asymptotic behavior is determined effectively by that of a "mean" ODE. Since the algorithms are recursive and iterative, the basic issues in the study of stochastic approximation methods include convergence of the algorithms, the rates of convergence, the efficiency of the procedures, and related methods in stochastic optimization.

10.1.3 Outline of the Chapter

The rest of the chapter is arranged as follows. Section 2 presents various algorithms and their variants. Section 3 deals with convergence of stochastic approximation type algorithms, and Section 4 presents rates of convergence of the corresponding algorithms. Large deviations principle is then discussed in Section 5, and asymptotic efficiency is treated in Section 6. Several recent applications of stochastic approximation algorithms are presented in Section 7. Finally, we close this chapter with a few more remarks in Section 8.

10.2 Algorithms and Variants

This section is divided into several parts. We begin with the basic algorithm in its simplest form, and then generalize it to include various variations.

10.2.1 Basic Algorithm

We begin with the simplest algorithms known as RM algorithms aiming at finding the zeros of a nonlinear function. This is then extended to function optimization problems with the use of KW algorithm.

RM Algorithm

Let $f : \mathbb{R}^r \mapsto \mathbb{R}^r$ be a continuous function. Suppose that we want to find $f(x) = 0$, but only noisy measurements

$$y_n = f(x_n) + \xi_n$$

are available, where $\{\xi_n\}$ denotes a sequence of random noise. Note that n is a positive integer representing the number of observations up to the current moment (the current iterate). For convenience, it is often thought as a "discrete time." The basic setup of the stochastic approximation algorithms proposed by Robbins and Monro takes the form

$$x_{n+1} = x_n + a_n y_n, \tag{10.2.1}$$

where $\{a_n\}$ is a sequence of nonnegative real numbers satisfying $\sum_n a_n = \infty$ and $a_n \to 0$ as $n \to \infty$. The sequence $\{a_n\}$ is usually referred to as a sequence of step sizes or gains. The conditions on the step sizes indicate that they cannot be too small. If they are too small (i.e., $\sum_n a_n < \infty$), then the iterates produced may not ever converge to the desired value. To see this, take the noise-free case $\xi_n = 0$ and suppose that $f(\cdot)$ is a bounded function. Then

$$\sum_{j=0}^{\infty} |x_{j+1} - x_j| \le \sum_{j=0}^{\infty} a_j |f(x_j)| \le \kappa.$$

[Here and throughout the paper, $\kappa > 0$ is used as a generic constant; its value may change for different usage. Thus by our convention, $\kappa + \kappa = \kappa$ and $\kappa\kappa = \kappa$.] The above argument indicates that $\sum_j (x_{j+1} - x_j)$ converges absolutely. Nevertheless, by telescoping

$$\sum_{j=0}^{n} (x_{j+1} - x_j) = x_{n+1} - x_0.$$

Thus, $x_n \not\to x^*$, the true parameter we are approximating unless x_0 is sufficiently close to x^*.

KW Algorithm

The RM algorithm in the previous subsection concerns root findings. In 1952, Kiefer and Wolfowitz proposed another type of stochastic approximation algorithm to locate the optima of a real-valued function. Suppose that we want to minimize a function $f(x)$, but only noise observations $F(x, \zeta)$ are available. Suppose that $EF(x, \zeta) = f(x)$, but we know neither the form of $F(\cdot)$ nor that of $f(\cdot)$. To approximate/estimate the optimizer, we use the finite difference approximation to the gradient of $f(x)$. Denote the finite difference interval by $\{c_n\}$ (with $c_n \to 0$ as $n \to \infty$). Use x_n to denote the nth estimate of the minimum. Suppose that for each i and each n, we can observe

$$y_{n,i} = -\frac{F(x_n + c_n e_i, \zeta_n^+) - F(x_n - c_n, \zeta_n^-)}{2c_n},$$

where $\zeta_n^{\pm}$ are random noise. Denote $y_n = (y_{n,1}, \cdots, y_{n,r})$. Then the approximation algorithm is again given by $x_{n+1} = x_n + a_n y_n$, which is the same form as that of (10.2.1). By introducing

$$\xi_n = (\xi_{n,1}, \cdots, \xi_{n,r}),\ \beta_n = (\beta_{n,1}, \cdots, \beta_{n,r}),$$

with

$$\begin{aligned}\xi_{n,i} &= [f(x_n + c_n e_i) - F(x_n + c_n e_i, \zeta_{n,i}^+)] - [f(x_n - c_n e_i) - F(x_n - c_n e_i, \zeta_{n,i}^-)] \\ \beta_{n,i} &= f_{x_i}(x_n) - \frac{f(x_n + c_n e_i) - f(x_n - c_n e_i)}{2c_n},\end{aligned}$$

where $f_x(\cdot)$ denotes the derivative (gradient) of $f(\cdot)$ w.r.t. x. Now the above algorithm can be rewritten as

$$x_{n+1} = x_n - a_n f_x(x_n) + a_n \frac{\xi_n}{2c_n} + a_n \beta_n. \tag{10.2.2}$$

In the above, ξ_n represents the noise, and β_n denotes the bias. We have used two-sided finite difference. One-sided finite difference can also be used. However, in practice, the two-sided finite difference method appears to be more preferable since it has smaller bias. This is easily seen by taking a Taylor expansion of the finite difference quotient in β_n.

10.2.2 More General Algorithms

We first present algorithms with nonadditive noise. Then we treat stochastic approximation of the most general form, in which not only the noisy appears in nonadditive form, but also the functions involved are varying with respect to time.

Algorithms with Nonadditive Noise

In various applications such as those arising in signal processing and adaptive controls, one often needs to treat stochastic approximation algorithms with nonadditive noise of the form

$$x_{n+1} = x_n + a_n f(x_n, \xi_n). \tag{10.2.3}$$

It is clear that (10.2.3) includes both (10.2.1) and (10.2.2) as special cases. Such an algorithm arises, for example, in the use of "equalization" filters in communication channels, adaptive antenna array processing etc.

Algorithms Involving Time-varying Functions

Similar to the previous case, we treat problems with the general nonadditive noise case. In addition, $f(\cdot)$ also depends on the discrete time n. The underlying algorithm is:

$$gen - sa - vf x_{n+1} = x_n + a_n f_n(x_n, \xi_n). \tag{10.2.4}$$

To be able to track slight parameter variation, one often uses an algorithm with constant step size of the form

$$x_{n+1} = x_n + \varepsilon f_n(x_n, \xi_n), \tag{10.2.5}$$

where $\varepsilon > 0$ is a small parameter. Such constant step-size algorithms are used frequently in tracking parameter variations in a time-varying system.

Passive Stochastic Approximation

Suppose that one wants to solve the equation $f(x) = 0$ on the basis of measurements $y_n = f(x_n) + \xi_n$, where $\{\xi_n\}$ is a sequence of random noise. Unlike the traditional stochastic approximation problem, the sequence $\{x_n\}$ emerges in a random manner and is not at one's disposal. How can one solve such a problem?

In [30], Härdle and Nixdorf suggested an interesting approach and termed it as *passive stochastic approximation.* The origin of such an approach can be traced back to an early work of Révész [66]. Its essence is to combine the stochastic approximation methods with nonparametric kernel estimation procedures and to approximate the root of equation $\bar{f}(x) = 0$ by another sequence $\{z_n\}$ according to

$$z_{n+1} = z_n + \frac{a_n}{h_n} K\left(\frac{x_n - z_n}{h_n}\right) y_n, \tag{10.2.6}$$

where $K(\cdot)$ is a kernel function, a_n is the step size, and h_n represents the window width. This procedure is a generalization of the conventional Robbins-Monro methods. One of the crucial points here is the utilization of the real-valued kernel function $K(\cdot)$, which is often a concave curve. If z_n and x_n are far apart, $K((x_n - z_n)/h_n)$ will be very small. As a result, only a small proportion of the measurement y_n is added to the iteration. In Yin and Yin [101], we treated the measurements of the form $y_n = f(x_n, \xi_n)$. Considering the fact that algorithms with constant step size are capable of tracking small parameter variation and are numerically robust, we considered an algorithm with constant step size and constant window width

$$z_{n+1} = z_n + \frac{\varepsilon}{\delta} K\left(\frac{x_n - z_n}{\delta}\right) f(x_n, \xi_n), \text{ for} n \geq 0. \tag{10.2.7}$$

The asymptotic analysis of such algorithms is provided under the framework of weak convergence (see also the related work [94] for with probability one convergence); applications to chemical processes are also dealt with.

10.2.3 Projection and Truncation Algorithms

An important issue in applications of stochastic approximation concerns the boundedness of the recursive iterates. In practice, one often modifies the algorithms in one way or another. Although there are no "rules of thumb," one confines the attention of the iterates to some compact set by using physical or economical constraints from the actual problems. As argued in [53], well-defined problems in applications always have either explicit bounds or implicit bounds. "For example, instability can be caused by values of a_n that are too large or values of finite difference intervals that are too small. The path must be checked for undesirable behavior, whether or not there are hard constraints. If the algorithm appears to be unstable, then one could reduce the step size and restart at an appropriate point or even reduce the size of the constraint set. The path behavior might suggest a better algorithm." Based on such a consideration, much of the book [53] is devoted to projection or truncation algorithms.

To proceed, we use (10.2.3) to describe the projection algorithms. Both fixed-projection regions and random truncation bounds will be discussed.

Projection Algorithms as Constraints

Suppose that H is a constraint set. We demand the iterate to be in the set. To do so, write the recursive algorithm as

$$x_{n+1} = \Pi_H \left(x_n + a_n f(x_n, \xi_n)\right), \tag{10.2.8}$$

where Π_H denotes the projection onto the constraint set H. Basically, if the iterate is within the projection region, we simply keep the recursion running; if it is outside the region, we project it back. Define a "reflection" or correction term z_n as

$$a_n z_n = x_{n+1} - x_n - a_n f(x_n, \xi_n),$$

i.e., it is the vector of shortest Euclidean length needed to take $x_n + a_n f(x_n, \xi_n)$ back to the constraint set H if it is not in H. Using this notation, (10.2.8) can be rewritten as

$$x_{n+1} = x_n + a_n f(x_n, \xi_n) + a_n z_n. \tag{10.2.9}$$

What kind of constraint sets can be included? In fact, a wide range of constraints can be considered; see [53, pp. 77-79] for several choices given in (A4.3.1)–(A4.3.3). One of the widely used such sets is a hyper cube. In this case, the iterates are confined to a cube with appropriate dimensions. Based on nonlinear programing type consideration, a more general region with boundaries given by differentiable functions may also be considered. Suppose $x \in \mathbb{R}^r$. Another possible candidate is an even more general set with H being an $\mathbb{R}^{r-1}$-dimensional connected surface with continuously differentiable outer normal.

Soft Constraints

The projection or truncation algorithms discussed so far may be considered as "hard constraints." The iterates are required to be in the constraint set H at all the time (for all n). Sometimes, we may wish to relax these "hard constraints," and allow them to be violated slightly from time to time. Roughly, such constraints are "soft constraints." In various applications, one often wants to use the hard constraints and the soft constraints in a combined manner.

The following example is taken from [53, Section 5.5]. The soft constraint is taken as the sphere

$$S_0 = \{x; |x| \le R_0\}.$$

Define

$$q(x) = [\text{dist}(x, S_0)]^2 = [\inf_{y \in S_0} |x - y|]^2.$$

Then,

$$q(x) = \begin{cases} (|x| - R_0)^2 & \text{for } |x| \ge R_0 \\ 0 & \text{otherwise.} \end{cases}$$

Its gradient is

$$q_x(x) = \begin{cases} 2x(1 - R_0/|x|) & \text{for } |x| \ge R_0 \\ 0 & \text{otherwise.} \end{cases}$$

The algorithm is

$$x_{n+1} = x_n + a_n y_n - a_n K_0 q_x(x_n) \tag{10.2.10}$$

for sufficiently large positive K_0. In view of (10.2.10), the iterates are allowed to be outside the sphere, and the constraint on the sphere can be violated. However, by adding a a penalty term $K_0 q_x(\,)$, we make sure that the iterates do not wander too far from the sphere, and the violation of the constraint set is in a tolerable range.

Random Truncation Bounds

With the motivation of building a truncation region without prior knowledge of the truncation set, Chen and Zhu suggested a randomly varying truncation algorithm in 1986; see [12] and the references therein.

To proceed, we let $\{M(n)\}$ be a sequence of positive real numbers, such that $M(n) \xrightarrow{n} \infty$. Define a sequence of integer-valued random variables σ_n recursively as

$$\sigma_0 = 0 \tag{10.2.11}$$
$$\sigma_{n+1} = \sigma_n + I_{\{|x_n + a_n y_|> M(\sigma_n)\}}. \tag{10.2.12}$$

Now define the stochastic approximation algorithm with randomly varying truncations as

$$\begin{aligned} x_{n+1} &= [x_n + a_n y_n] I_{\{|x_n + a_n y_n| \le M(\sigma_n)\}} \\ &\quad + \hat{x} I_{\{|x_n + a_n y_n| > M(\sigma_n)\}}. \end{aligned} \tag{10.2.13}$$

The rationale is that at each iteration, one should check if the iterate obtained is within the randomly generated bound. If it is, do nothing; otherwise, return the iterate to a fixed point. Since σ_n is monotone increasing, either $\sigma_n \to \sigma$ a finite limit, or $\sigma_n \to \infty$. The effort is then to show that a finite σ exists. When a finite σ exists, then there exists a $\bar{n}(n)$, such that for all $n \geq \bar{n}(n)$, σ_n is sufficiently close to σ and $|x_n + a_n y_n| \leq M(\sigma)$. Therefore, for any $n \geq \bar{n}(n)$, we have $|x_n| \leq M(\sigma)$, i.e., after finitely many steps, x_n will be bounded uniformly for almost all sample points ω. Thus eventually (for large n) the algorithm becomes a standard one with bounded iterates.

10.2.4 Global Stochastic Approximation

An important task in control, optimization and related fields is to locate the global minimum of $f(\cdot) : \mathbb{R}^r \mapsto [0, \infty)$, a smooth function, which has multiple local minima. The situation of interest is: We cannot calculate the gradient of $f(\cdot)$ explicitly and only noise corrupted gradient estimates or measurements, $\nabla f(x)$+noise, are available. Consequently standard deterministic algorithms are not able to produce desirable results. One needs to rely on stochastic approximation type of algorithms. Nevertheless, a stochastic approximation algorithm of the form

$$x_{n+1} = x_n - a_n(\nabla f(x_n) + \xi_n), \tag{10.2.14}$$

may lead to the convergence to a local minimum. Let S_l denote the collection of all the minima of $f(x)$. Under broad conditions (see for example, Kushner and Clark [43] or the more up-to-date treatment of Kushner and Yin [53]), $x_n \to S_l$ w.p.1. Very often the iterates will be trapped at a local minimum and will miss the global one. To overcome the difficulties, much effort has been made to design suitable procedures for the global optimization task. In the 1980s, one such global optimization methods, simulated annealing, started attracting the attentions of researchers and practitioners. In [36], Kirkpatrick, Gelatt and Vecchi proposed a method of solution by running the Metropolis algorithm [58] while gradually lowering the temperature. Further analysis on the methods via Monte Carlo techniques are contained in Kushner [41], and Gelfand and Mitter [28] among others (see also Dippon and Fabian [18] for a different treatment). The rate of convergence is analyzed in Yin [95]. Algorithms with restarting devices are considered in Yin [96]; see also the applications to image estimation problems in [98] and the references therein. To proceed, consider

$$x_{n+1} = x_n - \frac{A}{n^\gamma}(\nabla f(x_n) + \xi_n) + \frac{B}{n^{\gamma/2}\sqrt{\ln[n^{1-\gamma} + A_0]}} W_n, \quad \text{for } 0 < \gamma < 1, \tag{10.2.15}$$

and/or

$$x_{n+1} = x_n - \frac{A}{n}(\nabla f(x_n) + \xi_n) + \frac{B}{\sqrt{n \ln\ln(n + A_0)}} W_n, \tag{10.2.16}$$

where A, A_0 and B are some positive constants. Notice that there are two noise sequences, of which $\{\xi_n\}$ is a sequence of measurement noise, and $\{W_n\}$ is a sequence of added random perturbations. Following the basic premise of the annealing scheme, the purpose of the use of $\{W_n\}$ is to give the iterates enough excitation and to force x_n jumping around so that the iterates will not be trapped at one of the local minima. This idea can also be used in conjunction with KW type algorithms. In such a case, $\nabla f(x) + \xi_n$ is replaced by its gradient estimate using only values of functions at design points.

10.2.5 Continuous-time Stochastic Approximation Algorithms

Until now, we have only mentioned stochastic approximation in discrete-time. There are continuous-time version stochastic approximation algorithms. In addition to the mathematical interest, the reason for considering the continuous version algorithm stems from the fact that the continuous-time algorithms are good approximations to discrete-time problems when the sampling speed is high. It is important to establish that no problems arise should the sampling rate become very high. This point was well taken in [59] for least squares type estimation schemes.

Consider the following stochastic approximation algorithms in continuous-time:

$$\dot{x}(t) = a(t)\left(f(x(t)) + \xi(t)\right), \tag{10.2.17}$$

where $a(t) \geq 0$ is the step size satisfying

$$a(t) \to 0, \quad \int_0^\infty a(t)dt = \infty,$$

$\xi(t)$ represents the noise. In [43], continuous-time stochastic approximation problems were treated extensively in addition to the discrete-time problems. Some of the recent work include [97] among others. Because in various applications, discrete-version of the problems is more frequently encountered, in this chapter we will mainly concentrate on discrete-time algorithms.

10.2.6 Stochastic Approximation in Function Spaces

The setting of stochastic approximation can be carried over to infinite dimensional spaces, e.g., Banach spaces and/or Hilbert spaces. In addition to the pure mathematical interest, the motivation of the study stems from the fact in various optimization problems, the solutions of the problems involve finding the root or the optimum for points not living in Euclidean spaces, but in function spaces. For example, consider a system with transfer function $K(\cdot)$, input $z(\cdot)$, sampling interval Δ, and output (at sampling time $n\Delta$)

$$y_n = \int_0^T K(s)z(n\Delta - s)ds + \psi_n,$$

where $\{\psi_n\}$ is a stationary sequence of observation noise with zero mean that is independent of $z(\cdot)$. To estimate $K(\cdot)$, one can use the following recursive algorithm

$$K_{n+1}(u) = K_n(u) - \varepsilon z(n\Delta - u)\left(\int_0^T K_n(s)z(n\Delta - s)ds - y_n\right),$$

where $\varepsilon > 0$ is a constant step size. Working with, for example, $K(\cdot) \in L_2[0,T]$, the space of square integrable functions on $[0,T]$, the problem becomes a stochastic approximation type procedure in a Hilbert space. For a detailed account on the treatment of this problem, see [47]. Stochastic approximation methods in function spaces have been studied by a host of researchers. To mention just a few, see [5, 37, 70, 82, 103] among others.

10.3 Convergence

This section is concerned with convergence of stochastic approximation algorithms. To avoid the complex technical details and to bring out the salient features of the problems, we shall consider stochastic approximation algorithms with the simplest form. As a result, much of the subsequent development focuses on algorithms with additive noise. We do not attempt to present the weakest conditions here, but rather aim to present the results in their simple form. It should be mentioned that much more general systems with time-varying functions, nonadditive noise, state dependent noise, and complex projection regions can be dealt with. We refer the reader to [53] for various detailed treatments. Here, in this paper, we concentrate on the ordinary differential equation approach (ODE), which establishes connections of the discrete iteration and the continuous-time dynamic systems.

10.3.1 ODE Methods

The ODE method combines probability ideas and analysis techniques. Instead of working with the discrete iteration directly, we take a continuous-time interpolation. To get some insight on how the method works, we first give some heuristic argument.

Consider (10.2.1). Suppose that the function $f(\cdot)$ is continuous. Also for simplicity, assume the measurement or observation error is a sequence of independent and identically distributed random variables. Choose a small $\Delta > 0$ such that

$$\sum_{j=n}^{n+m_n^\Delta-1} a_j \approx \Delta \text{ or equivalently} \tag{10.3.18}$$

$$m_n^\Delta = \max\left\{ m; \sum_{j=n}^{n+m-1} a_j < \Delta \right\}. \tag{10.3.19}$$

Then iterating on (10.2.1) yields

$$x_{n+m_n^\Delta} = x_n + \sum_{j=n}^{n+m_n^\Delta-1} a_j f(x_j) + \sum_{j=n}^{n+m_n^\Delta-1} a_j \xi_j. \tag{10.3.20}$$

For Δ small enough and for continuous $f(\cdot)$, for $n \le j \le n+m_n$, $f(x_j)$ is "close" to $f(x_n)$ by the continuity. As a result

$$\sum_{j=n}^{n+m_n^\Delta-1} a_j f(x_j) \approx \sum_{j=n}^{n+m_n^\Delta-1} a_j f(x_n) \approx \Delta f(x_n),$$

and

$$x_{n+m_n^\Delta} - x_n \approx \Delta f(x_n) + \text{ error }. \tag{10.3.21}$$

How big is the error? Let us compute its variance:

$$\text{Var (error)} = \text{Var}\left(\sum_{j=n}^{n+m_n^\Delta-1} a_j \xi \right) = O\left(\sum_{j=n}^{n+m_n^\Delta-1} a_j^2 \right) = O(\Delta a_n).$$

Therefore

$$\frac{x_{n+m_n^\Delta} - x_n}{\Delta} \approx f(x_n) + \text{ error with diminishing variance.}$$

Therefore, over small a interval, the mean change of the values of the parameter is much more important than that of the noise. The noise is averaged out in the limit, and the asymptotic behavior can be approximated by the differential equation

$$\dot{x} = f(x). \tag{10.3.22}$$

To prepare us for the study of the desired asymptotic properties, let us recall the notion of equicontinuity. Let $\{f_n\}$ denote a sequence of $\mathbb{R}^r$-valued functions on $[0,\infty)$. The set is said to be *equicontinuous* in $C^r[0,\infty)$ (the set of $\mathbb{R}^r$-valued continuous functions defined on $[0,\infty)$) if $\{f_n(0)\}$ is bounded and for each T and $\varepsilon > 0$, there is a $\delta > 0$ such that for all n

$$\sup_{|t-s|\le\delta,\ |t|\le T} |f_n(t) - f_n(s)| \le \epsilon. \qquad (10.3.23)$$

The well-known *Arzelà–Ascoli* Theorem states:

Theorem 10.3.1 *Let $\{f_n\}$ be a sequence of functions in $C^r[0,\infty)$, and let the sequence be equicontinuous. Then there is a subsequence that converges to some continuous limit, uniformly on each bounded interval.*

Remark In fact, it is more convenient to work with a sequence of functions that are piecewise constant interpolation of the iterates. However, in this case, the equicontinuity and Theorem 10.3.1 need to be modified. In [53, Chapter 4], we defined the notion of equicontinuity in the extended sense and used it to study the stochastic approximation problems. In what follows, for simplicity, we use piecewise linear interpolation and use Theorem 10.3.1 to avoid the technical details.

To formulate the problem, we take piecewise linear interpolations and work on sequences of continuous functions. To do so, define

$$t_n = \sum_{j=0}^{n-1} a_j \qquad (10.3.24)$$

$$x^0(t_n) = x_n \qquad (10.3.25)$$

$$x^0(t) = \frac{t_{n+1} - t}{a_n} x_n + \frac{t - t_n}{a_n} x_{n+1}, \quad t \in (t_n, t_{n+1}). \qquad (10.3.26)$$

That is, the interpolation interval is (t_n, t_{n+1}). Next, to bring the asymptotic behavior of the process to the foreground, define a shifted sequence by

$$x^n(t) = x^0(t + t_n).$$

Under suitable conditions, it can be shown that $\{x^n(\cdot)\}$ is uniformly bounded and equicontinuous. By Ascoli–Arzelá Theorem, we can extract a convergent subsequence $x^{n_k}(\cdot)$ such that $x^{n_k}(\cdot) \to x(\cdot)$. Then we characterize the limit $x(\cdot)$ and prove that it is nothing but the solution of the ODE (10.3.22). Why is such an ordinary differential equation important? The reason is clear. The stationary points of (10.3.22) are exactly the roots of $f(\cdot)$ that we are searching for. To proceed, we state a convergence result.

First let us recall the definition of "asymptotic rate of change is zero with probability one (w.p.1)." Denote

$$m(t) = \sup \left\{ m;\ \sum_{j=0}^{m} a_j \le t \right\}$$

and

$$M^0(t) = \sum_{j=0}^{m(t)-1} a_j \xi_j,$$

where $\{\xi_n\}$ is the noise sequence. We say the rates of change of $M^0(\cdot)$ go to zero with probability one as $t \to \infty$, if for some $T > 0$,

$$\limsup_{n\uparrow\infty} \max_{j\ge n} \max_{0\le t\le T} \left| M^0(jT + t) - M^0(jT) \right| = 0 \text{ w.p.1} \qquad (10.3.27)$$

If this holds for some positive T, then it holds for all $T > 0$. Note that the w.p.1 convergence of $\sum_j a_j \xi_j$ implies (10.3.27), but not the other way around. For example, the function $\sum_{j=0}^{m(t)-1}(1/(j+1))$ for $t > 0$ satisfies (10.3.27) but $\sum_j(1/(j+1))$ does not converge. To proceed, we state a convergence result.

Theorem 10.3.2 *Suppose the following conditions are satisfied.*

- *$f(\cdot)$ is continuous.*
- *The asymptotic rate of change of $M^0(t)$ is zero w.p.1.*
- *The iterates $\{x_n\}$ are bounded w.p.1.*
- *Denote*

$$Z = \{x \in \mathbb{R}^r;\ f(x) = 0\}.$$

There is a twice continuously differentiable function $V(\cdot)$ satisfying

$$V_x'(x)f(x) < 0 \ \ \textit{for all} \ \ x \notin Z,$$

where $V_x(\cdot)$ denotes the derivative of $V(\cdot)$.

Then

$$\lim_n d(x_n, Z) = \lim_n \inf\{|x_n - y|;\ y \in Z\} = 0 \ \textit{w.p.1.}$$

If $Z = \{x^*\}$, a singleton set, then $x_n \to x^*$ w.p.1.

Remark . The function $V(\cdot)$ used above is simply a Liapunov function for the differential equation (10.3.22). The requirement indicates that we need the stationary points of the ordinary differential equation to be asymptotically stable.

In the above, for simplicity, we have assumed the iterates $\{x_n\}$ to be bounded w.p.1. This can be realized by use of truncation algorithms mentioned previously. Even without using projections or truncations the boundedness may also be proved in certain cases and sufficient conditions guaranteeing this boundedness can be obtained. For more detailed discussion on this matter, we refer to [53, Chapters 5 and 6].

10.3.2 Weak Convergence Method

First, let us recall the definition of weak convergence. Let X_n and X be $\mathbb{R}^r$-valued random variables. We say that X_n converges weakly to X iff for any bounded and continuous function $g(\cdot)$,

$$Eg(X_n) \to Eg(X).$$

$\{X_n\}$ is said to be tight iff for each $\eta > 0$, there is a compact set K_η such that

$$P(X_n \in K_\eta) \geq 1 - \eta \ \text{ for all } \ n.$$

The definitions of weak convergence and tightness extend to random variables in a metric space. The notion of weak convergence is a substantial generalization of convergence in distribution. It implies much more than just convergence in distribution since $g(\cdot)$ can be chosen in many interesting ways. On a complete separable metric space, the notion of tightness is equivalent to sequential compactness. This is known as the Prohorov's Theorem. Due to this theorem, we are able to extract convergent subsequences once tightness is verified. Let $D^r[0,\infty)$ denote the space of $\mathbb{R}^r$-valued functions that are right continuous and have left-hand limits, endowed with the Skorohod topology. For various notations and terms in weak convergence theory such as Skorohod topology, Skorohod representation etc. and many others, we refer to [22, 40] and the references therein. To carry out the weak convergence analysis, one often uses a martingale problem formulation, which to some extent is a weak sense solution of a stochastic differential equation. Consider a stochastic differential equation

$$dx(t) = b(x(t))dt + \sigma(x(t))dw(t).$$

The differential generator for the diffusion process $x(\cdot)$ given above is

$$\mathcal{L}h(x) = h'_x(x)b(x) + \frac{1}{2}\sum_{i,j} h_{x_i x_j}(x)a_{ij}(x),$$

where

$$h_{x_i x_j}(x) = \frac{\partial^2}{\partial x_i \partial x_j}h(x) \quad \text{and} \quad a(x) = \sigma(x)\sigma'(x).$$

Define

$$M_h(t) = h(x(t)) - h(x(0)) - \int_0^t \mathcal{L}h(x(s))ds.$$

If $M_h(\cdot)$ is a martingale for each $h(\cdot) \in C_0^2$ (C^2 function with compact support), then $x(\cdot)$ is said to solve a martingale problem with operator $\mathcal{L}$. The problem of identifying the weak limit of a sequence can be recast as the characterization of a solution of an appropriate martingale problem.

In studying stochastic approximation algorithms, the techniques of weak convergence have been found to be very useful. The application of weak convergence methods usually requires first tightness be proved and then the limit process be characterized. First, when treating constant-step size algorithms, the pertinent notion of convergence is in the sense of weak convergence. Second, to deal with rate of convergence issues and/or to design stopping rules for the iterates always involve the distributional convergence of sequences of suitably scaled random processes. To study the asymptotics in such a distributional setting, weak convergence is the most useful method in our tool box.

Now let us state a result in regard to the constant-step size algorithm. Take a piecewise constant interpolation as

$$x^\varepsilon(t) = \begin{cases} x_0, & \text{when } t = 0, \\ x_n, & \text{when } t \in [n\epsilon, n\epsilon + \varepsilon). \end{cases}$$

The sample paths of the process $x^\varepsilon(\cdot)$ are in $D^r[0,\infty)$.

Theorem 10.3.3 *Consider algorithm* (10.2.1) *with the deceasing step size replaced by a constant step size* $\varepsilon > 0$. *Suppose:*

- *$f(\cdot)$ is continuous.*
- *The initial condition satisfies $x_0^\varepsilon \Rightarrow x_0$.*
- *For each x,*

$$\frac{1}{n}\sum_{j=m}^{n+m} E_m\xi_j \to 0 \ \textit{in probability} \quad \textit{as} \ n \to \infty,$$

where E_m denotes the conditional expectation with respect to the σ-algebra $\mathcal{F}_m = \sigma\{x_0, \xi_j, j < m\}$. Then $\{x^\varepsilon(\cdot)\}$ is tight in $D^r[0,\infty)$, and any weakly convergent subsequence has a limit $x(\cdot)$ which is a solution of the ordinary differential equation (10.3.22). Moreover, if $\{x_n\}$ is tight, $f(\cdot)$ has a unique stable point x^*, and if $t_\varepsilon \to \infty$ as $\varepsilon \to 0$, then $x^\varepsilon(\cdot + t_\varepsilon)$ converges weakly to x^*.

Remark Note that sufficient conditions for the tightness of $\{x_n\}$ can often be obtained with the help of the perturbed test function method that is developed by Kushner and co-workers. To prove the first result of the theorem (the convergence to the mean ODE), a truncation device [40, p. 83] or [53, Section 8.5] can be used.

The condition on the noise above is of the law of large number type. The required convergence is in the sense of weak convergence. The insertion of the conditional expectation E_m makes the condition weaker than without it (e.g., it is automatically satisfied for a sequence of i.i.d. noise with zero mean). The weak convergence to the solution of the ordinary differential equation gives us a result on t belonging to a large but still bounded interval, whereas the convergence of $x^\varepsilon(\cdot + t_\varepsilon)$ illustrates the behavior of the iterates for small ε and large n (as $\varepsilon \to 0$ and $n \to \infty$ simultaneously).

One of the effective ways of analyzing stochastic approximation algorithms with state-dependent noise is the invariant measure approach of Kushner and Shwartz [46]. Not only can we treat complex noise processes, but also we can deal with discontinuity in the underlying function. A more refined argument is in [53].

10.4 Rates of Convergence

Once the convergence of a stochastic approximation algorithm is established, the next task is to ascertain the convergence rate. To begin, the first question is: For stochastic approximation, what do we mean by "rate of convergence?" To answer the question, consider, for instance, Eq. (10.2.1). Suppose that $x_n \to x^*$ (the true parameter) w.p.1 as $n \to \infty$. To study the convergence rate, we take a suitably scaled sequence

$$u_n = (x_n - x^*)/a_n^\alpha,$$

for some $\alpha > 0$ [In case of constant-step size algorithm, this is changed to $(x_n - x^*)/\varepsilon^\alpha$.] The idea is to choose α such that u_n converges (in distribution) to a nontrivial limit. The scaling factor α together with the asymptotic covariance of the scaled sequence gives us the rate of convergence. That is, the scaling α tells us the dependence of the estimation error $x_n - x^*$ on the step size, and the asymptotic covariance is a mean of assessing "goodness" of the approximation. For general references on rate of convergence, we refer the reader to [24, 43, 44, 48, 53]. For related work on convergence rate of variants of stochastic approximation, see [55, 95].

As mentioned above, by using the definition of the rate of convergence, we are effectively dealing with convergence in the distributional sense. Since the randomness is attached, as in the investigation of convergence, the rate of convergence study is very different from any purely deterministic, root-finding and/or optimization algorithms. In lieu of examining the discrete iteration directly, we are again taking continuous-time interpolations.

10.4.1 Scaling Factor α

What are the suitable scalings for the stochastic approximation algorithms? For decreasing step size algorithms, the suitable scaling is $\sqrt{a_n}$, and for constant step size algorithms, the scaling is $\sqrt{\varepsilon}$. In both cases, the factor $\alpha = 1/2$ is used. To some extent, this is dictated by the well-known central limit theorem.

10.4.2 Tightness of the Scaled Estimation Error

To validate our claim of the scaling factors, we need to show that $\{(x_n - x^*)/\sqrt{a_n}\}$ (resp. $\{(x_n - x^*)/\sqrt{\varepsilon}\}$) is tight. Such a proof can be carried out by means of a perturbed Liapunov function approach (see [40, 53]). The approach is as follows: we examine the sequence $V(x_n)$, where $V(\cdot)$ is a Liapunov function of (10.3.22), and x_n is obtained from a stochastic approximation algorithm. In proving the desired bound of $V(x_n)$, there will be some unwanted terms showing up. To get rid of them, we introduce a perturbation to the Liapunov function. The perturbation is small in magnitude, and results in appropriate cancelation in the iterate. Then we establish the bound of $V(x_n)$ via the perturbation. For a detailed account, see [53, Chapter 10].

For simplicity, consider again the simple algorithm (10.2.1) with $a_n = \varepsilon$, constant step size. We proceed to provide sufficient conditions guaranteeing the tightness of the scaled sequence.

Theorem 10.4.1 *Suppose that the following conditions are satisfied:*

- *There is a unique asymptotically stable point x^* of the ODE (10.3.22).*
- *There is a twice continuously differentiable Liapunov function $V(\cdot)$ such that*
  - $V(x) \to \infty$ *as* $|x| \to \infty$, *and* $V_{xx}(x)$ *is bounded for each* x.
  - $|V_x(x)| \le \kappa(1 + V^{1/2}(x))$.
  - $|f(x)|^2 \le \kappa(1 + V(x))$ *for each* x.
  - $V_x'(x)f(x) \le -\lambda V(x)$ *for some* $\lambda > 0$ *and each* $x \ne x^*$.
- *The noise $\{\xi_n\}$ is a sequence of stationary random variables satisfying $E\xi_n = 0$ and $E|\xi|^2 < \infty$ such that*

$$\left|\sum_{j=m}^{\infty} E_m \xi_j\right| \le \kappa \tag{10.4.28}$$

$$\left|\sum_{j=m}^{\infty} E\xi_m' \xi_j\right| \le \kappa. \tag{10.4.29}$$

Then there is an N_ε such that for all $n \ge N_\varepsilon$, $EV(x_n) = O(\varepsilon)$. If in addition,

$$V(x) = (x - x^*)'Q(x - x^*) + o(|x - x^*|^2),$$

(i.e., $V(x)$ is locally quadratic), then $\{(x_n - x^*)/\sqrt{\varepsilon};\ n \ge N_\varepsilon\}$ is tight.

We will not provide the proof. However, we will discuss the idea of perturbed Liapunov function briefly. To begin, it can be seen that by using the assumption on $f(\cdot)$ and $V(\cdot)$,

$$\begin{aligned}
&E_n V(x_{n+1}) - V(x_n) \\
&= \varepsilon V_x'(x_n) f(x_n) + E_n \varepsilon V_x'(x_n)\xi_n + E_n \epsilon^2 \frac{1}{2} y_n' V_{xx}(x_n^+) y_n \\
&\le \varepsilon V_x'(x_n) f(x_n) + \varepsilon V_x'(x_n) E_n \xi_n + O(\varepsilon^2)(1 + V(x_n) + E_n|\xi_n|^2) \\
&\le -\lambda V(x_n) + \varepsilon V_x'(x_n) E_n \xi_n + O(\varepsilon^2)(1 + V(x_n) + E_n|\xi_n|^2),
\end{aligned} \tag{10.4.30}$$

where x_n^+ is on the line segment joining x_n and x_{n+1}. The second line in (10.4.30) follows from the growth condition on $f\cdot$. Define a perturbation

$$V_n^\varepsilon = \varepsilon \sum_{j=n}^{\infty} E_n V_x'(x_n)\xi_j$$

Note that

$$|V_n^\varepsilon| = O(\varepsilon)(1+V(x_n)). \tag{10.4.31}$$

Define

$$\widetilde{V}_n = V(x_n) + V_n^\varepsilon.$$

We then proceed to calculate $E_n\widetilde{V}_{n+1} - \widetilde{V}_n$. The defined perturbation will allow us to cancel the noise term in (10.4.30). Iterating on the recursion, taking expectation, and using the order of magnitude estimate (10.4.31), we can then obtain

$$\begin{aligned} E\widetilde{V}_{n+1} &\le (1-\varepsilon\lambda_0)E\widetilde{V}_n + O(\varepsilon^2) && (10.4.32)\\ &\le (1-\varepsilon\lambda_0)^n E\widetilde{V}_0 + \sum_{j=0}^{n}(1-\varepsilon\lambda_0)^{n-j}O(\varepsilon^2) && (10.4.33)\\ &= O(\varepsilon), && (10.4.34)\end{aligned}$$

where $0 < \lambda_0 < \lambda$. Now using (10.4.31) again, we also have $EV(x_{n+1}) = O(\varepsilon)$. The desired estimate follows.

10.4.3 Local Analysis

To obtain further results on rate of convergence, we linearize $f(x)$ about x^*, and carry out local analysis. Let us consider (10.2.1) with constant step size $\varepsilon > 0$. Taking Taylor expansion about x^* leads to

$$x_{n+1} = x_n + \varepsilon f_x(x^*)(x_n - x^*) + \varepsilon\xi_n + \frac{1}{2}\varepsilon(x_n - x^*)' f_{xx}(x_n^*)(x_n - x^*),$$

where x_n^* is on the line segment joining x_n and x^*. Define $u_n = (x_n - x^*)/\sqrt{\varepsilon}$. Using this in the above equation yields

$$u_{n+1} = u_n + \varepsilon f_x(x^*)u_n + \sqrt{\varepsilon}\xi_n + O(\varepsilon^{3/2}|u_n|^2|f_{xx}(x_n^*)|). \tag{10.4.35}$$

If $f_{xx}(\cdot)$ is bounded uniformly, and the conditions of Theorem 10.4.1 are satisfied, then the expectation of the norm of the last term in (10.4.35) is of the order $O(\varepsilon^2)$ and is thus negligible. Iterating on (10.4.35) gives us

$$u_{n+1} = u_{N_\varepsilon} + \varepsilon \sum_{j=N_\varepsilon}^{n} f_x(x^*)u_j + \sqrt{\varepsilon}\sum_{j=N_\varepsilon}^{n} \xi_j + o(1), \tag{10.4.36}$$

where $o(1) \to 0$ in probability. To proceed, define a piecewise constant interpolation $u^\varepsilon(\cdot)$ as $u^\varepsilon(t) = u_n$ for $t \in [\varepsilon(n - N_\varepsilon), \varepsilon(n - N_\varepsilon + 1))$. In view of (10.4.36),

$$u^\varepsilon(t+s) = u^\varepsilon(t) + \varepsilon \sum_{j=t/\varepsilon}^{(t+s)/\varepsilon} f_x(x^*)u_j + \sqrt{\varepsilon}\sum_{j=t/\varepsilon}^{(t+s)/\varepsilon} \xi_j + o(1),$$

where $o(1) \to 0$ in probability uniformly in t. Then the following theorem can be established.

Theorem 10.4.2 *Assume the following conditions are satisfied.*

- *All the conditions of Theorem* 10.4.1 *hold.*
- $f_{xx}(\cdot)$ *is bounded uniformly.*
- *The process*

$$\sqrt{\varepsilon}\sum_{j=0}^{t/\varepsilon}\xi_j \text{ converges weakly to } w(t) \text{ a Brownian motion with covariance } Rt,$$

where R is symmetric and positive definite.

Then $u^{\varepsilon}\cdot$ is tight in $D^r[0,\infty)$, and any weakly convergent subsequence has a limit $u(\cdot)$ that is a solution of

$$du = f_x(x^*)udt + R^{1/2}d\widetilde{w} \tag{10.4.37}$$

where $R^{1/2}$ is the "square root" of R (i.e., $R = R^{1/2}(R^{1/2})'$), and $\widetilde{w}(\cdot)$ is a standard Brownian motion.

Note that in the above, we used the notation t/ε. Eq. (10.4.37) has a unique solution for each initial condition. This is understood to be its integral part. One of the main assumptions is the weak convergence of a scaled sequence of the noise to a Brownian motion. This is not a restriction at all. If $\{\xi_n\}$ is a sequence of independent and identically distributed random variables with zero mean and second moments (or a sequence of martingale difference noise), then this assumption is just the well-known Donsker's functional central limit theorem. Suppose that $\{\xi_n\}$ is a sequence of stationary φ-mixing noise with $E|\xi_n|^{2+\delta} < \infty$ for some $\delta > 0$. Denote $p = (2+\delta)/(1+\delta)$, use the mixing measure $\varphi_p(\cdot)$ defined in (3.1) on [22, p. 350], and suppose

$$\sum_j \varphi_p^{\delta/(1+\delta)}(j) < \infty.$$

Then by Theorem 7.3.1 in [22], $\sqrt{\varepsilon}\sum_{j=N_\varepsilon}^{t/\varepsilon}\xi_j$ converges weakly to a Brownian motion $w(\cdot)$ with covariance Rt, where

$$R = E\xi_0\xi_o' + \sum_{j=1}^{\infty}E\xi_0\xi_j' + \sum_{j=1}^{\infty}E\xi_j\xi_0'.$$

Theorem 10.4.2 concentrates on constant step size algorithms. There is also a decreasing step size counterpart. In the decreasing step-size case, replace ε and $\sqrt{\varepsilon}$ by a_n and $\sqrt{a_n}$, respectively. Then we can show that $(x_n - x^*)/\sqrt{a_n}$ is tight for $n \geq N$. Define $t_n = \sum_{j=0}^{n-1}a_j$ and take a piecewise constant interpolation $u^n(\cdot)$ with the interpolation interval $[t_n - t_N, t_{n+1} - t_N)$. We then proceed as in the previous case. The traditional central limit result can be obtained from Theorem 10.4.2. For instance, for the decreasing step-size case, suppose $f_x(x^*)$ is a stable matrix (i.e., all of its eigenvalues have negative real parts). Then the stationary covariance of the diffusion given in (10.4.37) is

$$\overline{R} = \int_0^\infty \exp(f_x(x^*)t)R[\exp(f_x(x^*)t)]'dt.$$

Alternatively, it is a solution of the algebraic Liapunov equation

$$\overline{R}f_x(x^*) + f_x'(x^*)\overline{R} = -R.$$

Consequently, we have $(x_n - x^*)/\sqrt{a_n}$ converges in distribution to a normal random variable with covariance $\overline{R}$ as $n \to \infty$.

10.4.4 Random Directions

One of the most important matters is to improve the performance of stochastic approximation algorithms. Let us consider KW algorithms. Each step of the KW algorithm uses $2r$ observations (if two-sided finite difference is used). Thus $2r$ steps are needed to get a derivative estimate. An alternative method is to update only one direction at each iteration using a finite difference estimate and to choose the direction randomly at each step. This results in using only two observations at each step. Such an idea appeared in Kushner and Clark [43] together with the associated convergence and rate of convergence results. At that time, it was noted that if the random directions are chosen at the unit sphere, then there is a little advantage as compared to the KW method. The recent work of Spall [77] indicated that if the random directions are chosen on the unit cube, then the performance is better than the KW algorithms. [Note that the length of the random direction vectors chosen in [77] is $\sqrt{r}$.] Further discussion is in [53, Chapter 10]. It is demonstrated in [53] that the crucial point is the choice of the random directions vector. In fact, if the random directions are chosen to be on the sphere of radius $\sqrt{r}$ then the random directions methods can be advantageous. Such an approach is particularly efficient for large dimensional problems. However, to apply the methods, care must be taken; see the discussion in [53, Chapter 10].

Introduce a sequence of random directions vectors by $\{d_n\}$. Then the random directions KW algorithm can be written as

$$x_{n+1} = x_n + a_n d_n \frac{y_n^- - y_n^+}{2c_n}, \tag{10.4.38}$$

where

$$y_n^\pm = (y_{n,1}^\pm, \dots, y_{n,r}^\pm)', \quad y_{n,i}^\pm = F(x_n \pm c_n e_i, \zeta_n^\pm),$$

and $F(\cdot)$ and $\{c_n\}$ were defined as in Section 2.2. For convergence and rate of convergence analysis results on random directions stochastic approximation algorithms, see [53, Chapter 10] and the references therein. It is conceivable such random directions methods will be very useful for a wide range of applications, especially for large-scale optimization tasks.

10.4.5 Stopping Rules

One topic that has not been discussed thus far is the design of stopping rules. In various applications, one needs to terminate the calculations if the desired precision is reached. To develop good stopping rules is an important matter. In regard to the work along this line, we mention the papers [75, 78, 61, 91]. In these references, stopping rules were proposed based on the construction of various confidence intervals. Roughly speaking, the procedure is as follows. Choose α, such that $0 < \alpha < 1$ and $1 - \alpha$ is the desired confidence coefficient. Given $\varepsilon > 0$, let $\nu_\varepsilon = \nu(\varepsilon, \alpha)$ denote the stopping rule, and $\text{Ellip}_{\nu_\varepsilon}$ be the ellipsoidal region about the true parameter x^*, and $V(\text{Ellip}_{\nu_\varepsilon})$ be the volume of the ellipsoid. As the volume shrinks, i.e., $\varepsilon \to 0$,

$$P\{x^* \in \text{Ellip}_{\nu_\varepsilon} \text{ and } V(\text{Ellip}_{\nu_\varepsilon}) \le \varepsilon^r\} \to 1 - \alpha.$$

To analyze such problems, a main task is to treat a stopped process of a suitably scaled estimation errors. In [91], this was done by means of weak convergence method and martingale averaging.

10.5 Large Deviations

This section is concerned with the large deviations approach to stochastic approximation methods. We first give the motivation of the study on the large deviations approach, and then present certain results for stochastic approximation problems.

10.5.1 Motivation

To give motivation, we first recall the notion of large deviations. Let us begin with a simple example. Let $\{\zeta_n\}$ be a sequence of i.i.d. random variables with $E\zeta_n = 0$ and $E\zeta_n^2 = \sigma^2$. For simplicity, assume the underlying distribution is Gaussian [The Gaussian assumption allows us to get an explicit representation of the logarithm of the moment generating function.]

Define S_n to be the sequence of partial sums, i.e., $S_n = \sum_{j=1}^{n} \zeta_j$. Suppose for $a > 0$, we are interested in the probability of the event $\Xi_a = \{S_n/n > a\}$. The well-known law of large numbers indicates

$$\frac{1}{n} S_n \to 0$$

either w.p.1 or in probability depending on if the strong law or the weak law is used. The central limit theorem implies

$$P\left(\frac{S_n}{n} > a\right) = P\left(\frac{S_n}{\sqrt{n}} > a\sqrt{n}\right) \to 0.$$

So Ξ_a is a rare event. Nevertheless, neither the law of large numbers nor the central limit theorem tells us how rare the event is and how small the associated probability is.

To undertake the study, we need to have detailed description beyond the normal deviation range. The large deviations approach is very useful in this regard. Use the Cramèr transformation or Legendre transformation

$$H(t) = \log E \exp(t\zeta) = \log \text{ moment generating function of } \zeta \qquad (10.5.39)$$

$$L(a) = \inf_t [H(t) - ta] = -\frac{a^2}{2\sigma^2}. \qquad (10.5.40)$$

Note that in the last line above, we used the fact that ζ is normally distributed. The Chernoff's bound reads

$$\lim_n \frac{1}{n} \log P\left(\frac{S_n}{n} > a\right) = L(a).$$

It certainly gives the description on the probability of the event Ξ_a. What has been obtained is $P((1/n)S_n > a)$ is "exponentially equivalent to" $\exp(L(a)n) = \exp(-a^2/(2\sigma^2))$.

10.5.2 Large Deviations for Stochastic Approximation

Now, let us consider (10.2.1) with $a_n = 1/n^\gamma$ (for some $0 < \gamma < 1$), for simplicity. Suppose that O is a bounded open set that is in the domain of attraction ($DA(0)$) of the ODE (10.3.22). Define the first exit time of the trajectories of $x^n(\cdot)$ from O as

$$\tau_O^n = \min\{t;\ x^n(t) \notin O\}.$$

Then under suitable conditions, for some λ_n,

$$\lim_n \lambda_n \log P(\tau_O^n \le T) = -v,$$

for some $v > 0$. What is λ_n? It turns out λ_n is precisely $\lambda_n = 1/n^\gamma$. This indicates that the probability of the trajectories exit from the bounded domain O is exponentially small, and

$$P(\tau_G^n \le T) \sim \exp(-vn^\gamma).$$

There is an analogue result for $a_n = 1/n$.

Use P_x^n to denote the probability under the condition that $x^n(0) = x_n = x$. Suppose $B_x \subset C[0,T]$, the space of continuous functions defined on $[0,T]$ with initial value x. Then

$$\begin{aligned} -\inf_{\phi\in B_x^0} S(T,\phi) &\le \liminf_n a_n \log P_x^n\{x^n(\cdot)\in B_x\} \\ &\le \limsup_n a_n \log P_x^n\{x^n(\cdot)\in B_x\} \\ &\le -\inf_{\phi\in \overline{B}_x} S(T,\phi), \end{aligned}$$

where $\overline{B}_x$ is the closure of B_x, and

$$S(T,\phi) = \int_0^T L(\dot{\phi}(s),\phi(s),s)ds$$

if $\phi(\cdot)$ is absolutely continuous and takes the value ∞ otherwise. The function $S(T,\phi)$ is the usual action functional of the theory of large deviations. In the above, $L(\cdot)$ plays the role of a cost function–penalty for the path to depart from the mean trajectory. For various developments of the large deviations approach to stochastic approximation, we refer to the paper of Dupuis and Kushner [20] and the references therein.

10.6 Asymptotic Efficiency

It has been a longtime effort to improve the performance of stochastic approximation type algorithms. In view of the discussion in the section on rate of convergence, what one wants is to have the largest α and smallest covariance possible. The exploration on the improvement of efficiency can be traced back to Chung [17].

To obtain asymptotically more efficient algorithms, one considers the following type of algorithms (see [17, 81, 54, 85] and the references therein)

$$x_{n+1} = x_n + \frac{\Gamma}{n}\left(f(x_n) + \xi_n\right), \tag{10.6.41}$$

where Γ is a matrix to be determined. Suppose

$$f(x) = H(x - x^*) + g(x),$$

where $g(x) = O(|x - x^*|^2)$ and H is a stable matrix having all of its eigenvalues living in the left half of the complex plan. Under suitable conditions (for example, those in the next section), it can be shown that for $\{x_n\}$ given by (10.6.41)

$$\sqrt{n}(x_n - x^*) \to N(0,\Sigma) \text{ in distribution as } n\to\infty,$$

where $N(0,\Sigma)$ denotes a normal distribution with 0 mean and covariance Σ, and $\Sigma = \Sigma(\Gamma)$. It can be shown that $\Sigma(\Gamma)$ satisfies the following Liapunov equation

$$(I/2 + \Gamma H)\Sigma + \Sigma(I/2 + \Gamma H)' = -\Gamma S_0 \Gamma',$$

where S_0 is the error covariance matrix. By means of algebraic comparisons, it can be shown (see Wei [85, Theorem 1]) that by choosing $\Gamma = -H^{-1}$, the optimal covariance matrix Σ^* can be obtained and is given by:

$$\Sigma^* = H^{-1}S_0(H^{-1})'.$$

Since H^{-1} is very unlikely to be known, various approximation procedures have been sought. Instead of using (10.6.41), the following algorithm is employed

$$x_{n+1} = x_n + \frac{\Gamma_n}{n}\left(f(x_n) + \xi_n\right),$$

where $\{\Gamma_n\}$ is a sequence of estimates of Γ. For multidimensional problems, no matter what kind of procedures are taken for estimating Γ (i.e., estimating every entry of Γ), the computation task involved is very intensive. As a consequence, the results in adaptive stochastic approximation are largely of theoretical nature and have not been used widely in various applications. Instead, the standard stochastic approximation algorithms have been employed extensively in a wide range of problems. In the rest of this section, we discuss two classes of algorithms. The first one uses iterate averaging, and the second one uses averaging in both the iterates and the observations. These algorithms give us asymptotic optimality without the sacrifice of using complex estimation schemes.

10.6.1 Iterate Averaging

In the late 80's Polyak [63] and Ruppert [69] independently proposed and analyzed a very interesting model for recursive algorithms of stochastic approximation type. The main idea of their approach is the use of averaging of iterates obtained from a classical stochastic approximation algorithm with slowly varying gains. Consider the following algorithm:

$$\begin{aligned} x_{n+1} &= x_n + \frac{1}{n^\gamma}\left(f(x_n) + \xi_n\right) \\ \bar{x}_{n+1} &= \bar{x}_n - \frac{1}{n+1}\bar{x}_n + \frac{1}{n+1}x_{n+1}, \end{aligned} \tag{10.6.42}$$

where $1/2 < \gamma < 1$. They concluded that such algorithms are asymptotically optimal in that they have the best scaling α and the smallest variance.

Uncorrelated noise processes were treated in [63]. Extension to φ-mixing type processes was carried out in [93], and further generalization is in [49]. We can first prove the convergence of the algorithm via the ODE approach. Then as in the classical SA problem, it can be shown that $EV(x_n) = O(n^\gamma)$. Define

$$A_{nj} = \begin{cases} \prod_{k=j+1}^{n}\left(I + H/k^\gamma\right), & n \geq j+1; \\ I; & n = j. \end{cases}$$

Choose $\nu = \nu(n)$ such that as $n \to \infty$, $\nu(n) \to \infty$ but $\nu(n)/\sqrt{n} \to 0$. Then

$$x_{n+1} = A_{n,\nu-1}x_\nu + \sum_{j=\nu}^{n}\frac{1}{j^\gamma}A_{nj}g(x_j) + \sum_{j=\nu}^{n}\frac{1}{j^\gamma}A_{nj}\xi_j,$$

and

$$\begin{aligned} \sqrt{n+1}\,\bar{x}_{n+1} &= \frac{1}{\sqrt{n+1}}\sum_{k=1}^{\nu-1}x_k + \frac{1}{\sqrt{n+1}}\sum_{k=\nu}^{n}A_{k,\nu-1}x_\nu \\ &+ \frac{1}{\sqrt{n+1}}\sum_{k=\nu}^{n}\sum_{j=\nu}^{k}\frac{1}{j^\gamma}A_{kj}g(x_j) + \frac{1}{\sqrt{n+1}}\sum_{k=\nu}^{n}\sum_{j=\nu}^{k}\frac{1}{j^\gamma}A_{kj}\xi_j. \end{aligned}$$

We then show under appropriate conditions

$$\frac{1}{\sqrt{n+1}}\sum_{k=1}^{\nu-1} x_k \xrightarrow{n} 0 \text{ w.p.1}$$
$$\frac{1}{\sqrt{n+1}}\sum_{k=\nu}^{n} A_{k,\nu-1}x_\nu \xrightarrow{n} 0 \text{ in probability}$$
$$\frac{1}{\sqrt{n+1}}\sum_{k=\nu}^{n}\sum_{j=\nu}^{k} \frac{1}{j^\gamma} A_{kj}g(x_j) \xrightarrow{n} 0 \text{ in probability.}$$

In addition,

$$\frac{1}{\sqrt{n+1}}\sum_{k=\nu}^{n}\sum_{j=\nu}^{k} \frac{1}{j^\gamma} A_{kj}\xi_j = -\frac{H^{-1}}{\sqrt{n}}\sum_{j=1}^{n} \xi_j + o(1),$$

where $o(1) \xrightarrow{n} 0$ in probability.

Next, define

$$B_n(t) = \frac{1}{\sqrt{n}}\sum_{j=1}^{[nt]} \xi_j \tag{10.6.43}$$

$$\tilde{B}_n(t) = -\frac{1}{\sqrt{n}}\sum_{j=1}^{[nt]} H^{-1}\xi_j, \tag{10.6.44}$$

where $[z]$ denotes the largest integral part of z. Under suitable conditions, we can show $B_n(\cdot)$ converges weakly to a Brownian motion $B(\cdot)$ with covariance matrix S_0, where S_0 is given by

$$S_0 = E(\xi_1\xi_1') + \sum_{k=2}^{\infty} E(\xi_1\xi_k') + \sum_{k=2}^{\infty} E(\xi_k\xi_1').$$

In addition, by the Slutsky theorem, $\tilde{B}_n(\cdot)$ converges weakly to a Brownian motion $\tilde{B}(\cdot)$ with covariance matrix

$$S = H^{-1}S_0(H^{-1})'.$$

As a result, the desired asymptotic optimality is established. For further approaches, see [49] (see also [53, Chapter 11]) among others.

10.6.2 Smoothed Algorithms

With the motivation of improving the transient performance, we study another class of stochastic approximation/optimization algorithm. The essence of this algorithm is the utilization of averaging in both iterates (or states according to systems theory terminology, or design points according to statistical terms) x's and noisy observations. It will be shown that the algorithm also possesses asymptotic optimality. The origin of the algorithm can be traced back to Bather [3]. In that reference, a scalar problem was considered, applications of stochastic approximation to the sequential estimation of LD_{50} were dealt with and some heuristic arguments were presented. Additional discussion of a scalar linear problem with i.i.d. random noise was provided in [71] by decomposing the underlying difference equations into deterministic and random parts, and deriving a representation formula.

For $1/2 < \gamma < 1$, consider the following algorithm: Choose an initial value x_1, and let $\{x_n\}$ be given by

$$x_{n+1} = \frac{1}{n}\sum_{j=1}^{n} x_j + \frac{1}{n^\gamma}\sum_{j=1}^{n} y_j, \quad n \geq 1. \tag{10.6.45}$$

Define

$$\bar{y}_n = \frac{1}{n}\sum_{j=1}^{n} y_j = \frac{1}{n}\sum_{j=1}^{n}(f(x_j) + \xi_j), \quad \bar{x}_n = \frac{1}{n}\sum_{j=1}^{n} x_j.$$

Equation (10.6.45) then can be rewritten as

$$x_{n+1} = \bar{x}_n + n\frac{1}{n^\gamma}\bar{y}_n.$$

The essential feature of this algorithm is the use of averaging in both iterates and observations. By means of averaging, the fluctuation is smoothed out. The idea is as follows. By using larger step size, the iterates are forced to get to a neighborhood of the true parameter x^* faster, and by taking averages of both iterates and observations, rough iterates are smoothed out and modified. We shall refer to this algorithm as the smoothed stochastic approximation algorithm. Rigorous proofs and justifications for multivariate cases were provided in Yin and Yin [100]. Multidimensional systems are treated and much more general noise is considered there, which is indeed needed for many applications arising in systems theory, control and optimization problems.

Let us assume:

A. There exists a unique x^* such that $f(x^*) = 0$. The function $f(\cdot)$ is Lipschitz continuous and satisfies the following conditions:

$$|f(x)|^2 \leq \kappa(1 + |x|^2) \text{ for some } \kappa > 0, \tag{10.6.46}$$
$$f(x) = H(x - x^*) + g(x), \tag{10.6.47}$$

where $|g(x)| = O(|x - x^*|^2)$, H is a stable matrix such that all of its eigenvalues have negative real parts.

B. There is a twice continuously differentiable Liapunov function $V(\cdot) : \mathbb{R}^r \to \mathbb{R}$ such that, $V(x) \geq 0$, $|V_x(x)| \leq \kappa(1 + V^{1/2}(x))$, $|V_{xx}(\cdot)|$ is bounded, $V(x) \to \infty$ as $|x| \to \infty$, and for some $\lambda > 0$ and all $x \neq x^*$, $V_x'(x)f(x) < -\lambda V(x)$.

C. $\{\xi_n\}$ is a stationary sequence satisfying:

(1) $E\xi_n = 0$, $E|\xi_n|^{2+\delta} < \infty$ for some $\delta > 0$.

(2) $\sum_{j=1}^{n} \frac{1}{j^\gamma}\xi_j$ converges w.p.1.

(3) Define $R_k = E\xi_1\xi_{k+1}'$. Suppose that $\sum_k |R_k| < \infty$.

(4) Define $r(i,j) = E_i\xi_{i+j}$ where E_i denotes the conditional expectation with respect to the σ-algebra $\mathcal{F}_i = \sigma\{\xi_j; j \leq i\}$. For each i, the following condition is satisfied:

$$\sum_j E^{1/2}|r(i,j)|^2 < \infty.$$

Without loss of generality, assume the true parameter is $x^* = 0$ henceforth. Rewrite the algorithm for x_n as

$$x_{n+1} = x_n + \frac{1}{n^\gamma}f(x_n) + \frac{1}{n^\gamma}\xi_n + \frac{1}{n^\gamma}\pi_n$$

such that $\pi_n \to 0$ as $n \to \infty$ w.p.1, $\sup_n E|\pi_n|^2 < \infty$ and $E|\pi_n|^2 \to 0$ as $n \to \infty$, where $\{\pi_n\}$ is the "left over" term defined in an obvious manner. Then we can show $\{x_n\}$ and $\{\bar{x}_n\}$ are both bounded with probability one (w.p.1). Furthermore, $x_n \to 0$ and $\bar{x}_n \to 0$ w.p.1 as $n \to \infty$.

Next, define

$$B_n(t) = \frac{1}{\sqrt{n}} \sum_{j=1}^{[nt]} \xi_j, \quad t \in [0,1]$$

where $[z]$ denotes the largest integral part of z. Then $B_n(\cdot)$ converges weakly to $B(\cdot)$, a Brownian motion with covariance $\Sigma_0 t$, where

$$\Sigma_0 = R_0 + \sum_{k=1}^{\infty} R_k + \sum_{k=1}^{\infty} R_k'. \tag{10.6.48}$$

Let

$$\tilde{B}_n(t) = \frac{[nt]}{\sqrt{n}} \bar{x}_{[nt]+1},$$

be a scaled sequence of the iterates. Then it can shown that for $t \in [0,1]$,

$$\tilde{B}_n(t) = -\frac{H^{-1}}{\sqrt{n}} \sum_{k=1}^{[nt]} \xi_j + o(1) = -H^{-1} B_n(t) + o(1),$$

where $o(1) \to 0$ (as $n \to \infty$) in probability uniformly in t. Finally, we arrive at:

Theorem 10.6.1 *Under assumptions* A–C, $\tilde{B}_n(\cdot)$ *converges weakly to a Brownian motion* $\tilde{B}(\cdot)$ *with covariance* $\Sigma^* t$ *where* $\Sigma^* = H^{-1}\Sigma_0(H^{-1})'$ *and* Σ_0 *is given by* (10.6.48).

10.6.3 Some Numerical Data

We consider a couple of simple examples in this section. These examples are for illustrative purposes and are taken from [100]. Two-dimensional systems, both linear and nonlinear, are considered. Autoregressive moving average (ARMA) noise processes are used throughout.

Example 10.6.2 We are interested in maximizing a real-valued function

$$f(x_1, x_2) = -0.605x_1^2 - 0.78x_1 - 1.665x_2^2 + 2.92x_2.$$

The gradient of this function is given by

$$\nabla f(x_1, x_2) = \begin{pmatrix} -1.21x_1 - 0.78 \\ -3.33x_2 + 2.92 \end{pmatrix}$$

Suppose that observations $\nabla f(x_1, x_2)+$ noise can be obtained, and the noise is an ARMA (1,1) process given by:

$$\xi_n = \begin{pmatrix} 0.36\xi_{n-1,1} + w_{n,1} + 0.15w_{n-1,1} \\ 0.48\xi_{n-1,2} + w_{n,2} + 0.23w_{n-1,2} \end{pmatrix}$$

where $\{w_n\}$ is a sequence of zero mean "white" noise.

The performance of the algorithms is measured by the trace of the second sample moments, henceforth referred to as trace (SSM). Since it has been proven that the iterate averaging algorithm (Algorithm A) and the smoothed algorithm (Algorithm S) are asymptotically optimal, comparisons are made through performance of these algorithms. A summary of results for the iteration with final values at $n = 1000$ is given in Table 1.

Note that the true value of the vector is $\theta = (-0.6446, 0.8769)'$. Our approximations are $\overline{x}_{1000} = (-0.6181, 0.8612)$ for Algorithm A, and $\overline{x}_{1000} = (-0.6249, 0.8646)$ for Algorithm S, respectively. It appears that although initial conditions do not affect the approximating sequences too much, they do have an impact on the results of the second sample moments. The table above was constructed from initial condition $x_1 = (6,6)'$. If the initial condition is changed to $x_1 = (3,3)'$, the traces of the second moments are reduced to 0.027392, and 0.027804 for Algorithms A, and Algorithm S, respectively.

Algorithm	Trace of SSM
Iterate Averaging	0.100133
Smoothed Algorithm	0.095412

TABLE 1. Comparison of algorithms (linear case)

Example 10.6.3 Consider the problem of finding zeros of a nonlinear function when only noise corrupted observations are available. The function is of the form:

$$f(x) = \begin{pmatrix} -(0.3x_1 - 0.75)^3 - 8 \\ \\ -(0.8x_2 + 0.60)^3 - 1 \end{pmatrix}.$$

$f(x)$+noise is observed with the same noise as before.

Algorithm	Trace of SSM
Iterate Averaging	0.065153
Smoothed Algorithm	0.061951

TABLE 2. Comparison of algorithms (nonlinear case)

Similar comparisons are made. Summary of computation results is provided in Table 2. In fact, the function grows faster than linear so the condition in the theorem is violated. To overcome the difficulties, a projection algorithm with projection region $[-10, 5] \times [-10, 5]$ was used. From the tables, it is easily seen that the performance of the Algorithm A and Algorithm S are comparable. To some extent, the algorithms with averaging stabilize the recursive computation.

10.7 Applications

The development of stochastic approximation methods has been closely related to a wide range of applications in stochastic control, identification and adaptive control, estimation and detection, signal processing, Monte Carlo optimization, management sciences, and many other related fields. In this section, we present a number of applications of stochastic approximation methods in various areas. These are only a handful of examples from diverse fields. It can be seen that many control and optimization tasks can be recast into a form that results in the use of stochastic approximation procedures.

10.7.1 Adaptive Filtering

Adaptive filtering algorithms have been used quite frequently in various applications such as estimation, adaptive control, signal processing and related fields. The underlying problem can be stated as follows. Let $x_n, y_n \in \mathbb{R}^r$, $\psi_n \in \mathbb{R}$, where $\{y_n\}$ and $\{\psi_n\}$ are sequences of measured input and reference signals, respectively, and $\{x_n\}$ is a sequence of system parameters. We adjust the system parameter x_n so that the weighted output $x'_n y_n$ best matches the reference signal ψ_n in the mean square sense, i.e., $E|x'_n y_n - \psi_n|^2$ is minimized. The calculations are done without knowing the statistics of y and ϕ. On the basis of a stationary sequence of observations $\{(y_n, \psi_n)\}$, assume

$$Ey_n y'_n = R > 0, \ E\psi_n y_n = q,$$

where R is a symmetric positive definite matrix. It is easily seen that x^*, the minimizer of $E|x'_n y_n - \psi_n|^2$, is the unique solution of the Wiener-Hopf equation $Rx^* = q$. Many algorithms for adaptive filtering, adaptive array processing, adaptive antenna systems, adaptive equalization, adaptive noise cancelation, pattern recognition, and learning etc. have been or can be recast into the same form, with only signal, training sequence and/or reference signals varying from applications to applications. The algorithm is of the form

$$x_{n+1} = x_n + a_n y_n(\psi_n - y'_n x_n), \tag{10.7.49}$$

where $\{a_n\}$ is a sequence of step sizes. The step size can be either decreasing or a constant. For the asymptotic study of such algorithms, we refer the reader to [8, 39, 43, 53, 76, 86, 90] among others.

10.7.2 Adaptive Beam Forming

Adaptive beam forming algorithms can be viewed as adaptive filters with constraints. Suppose

$$X_k \in \mathbb{R}^{r\times m}, Y_k \in \mathbb{R}^{r\times l}, \psi_k \in \mathbb{R}^{m\times l}.$$

The problem is concerned with the determination of the azimuth of a target by using a matrix composed of sensors. The outputs of sensors Y_k are weighted by a matrix X, so that $X'Y_k$ become the best approximation of the target in the mean square sense subject to the constraint

$$X'C = \Phi, \quad C \in \mathbb{R}^{r\times l}, \quad \Phi \in \mathbb{R}^{m\times l}. \tag{10.7.50}$$

The motivation for choosing this constraint comes from an application to the adaptive beam formers for tracking systems.

We wish to construct a recursive procedure which converges to X^*, the minimizer of

$$E(X'Y_k - \psi_k)(X'Y_k - \psi_k)'$$

subject to (10.7.50). It is clear that a necessary and sufficient condition for (10.7.50) to hold is $\Phi C^\dagger C = \Phi$, where $z^\dagger$ denotes the pseudo-inverse of z. By using Gauss-Markov estimations, it can be shown that the minimizer X^* not depending on k, is given by

$$X^* = C^{\dagger,\prime}\Phi' + (PAP)^\dagger(Q - AC^{\dagger,\prime}\Phi')$$

with

$$A = EY_kY_k', \; Q = EY_k\psi_k', \;\text{ and }\; P = I - CC^{\dagger}.$$

Although the above equation gives us a closed-form solution, it is evidently not informative. Since A, Q are unknown, to obtain X^* directly is impossible. Even if A, Q can be estimated sequentially, it is rather time consuming to compute the pseudo-inverse for large dimensional systems at each iteration. Therefore, we shall approximate X^* by a matrix X_k at each time k, such that X_k can be corrected based on the measurement Y_k. This leads to the following algorithm:

$$X_{k+1} = C^{\dagger,\prime}\Phi' + P[X_k + a_k(Y_k\psi_k' - Y_kY_k'X_k)], \tag{10.7.51}$$

$$X_0 = C^{\dagger,\prime}\Phi'. \tag{10.7.52}$$

For the asymptotic study of such algorithms, see [90] and the references therein.

10.7.3 System Identification and Adaptive Control

Stochastic Gradient Algorithms

Let

$$\begin{cases} A(q^-1) = 1 + a_1q^{-1} + \cdots + a_nq^{-n}, \\ B(q^{-1}) = b_0 + b_1q^{-1} + \cdots + b_mq^{-m}, \quad b_0 \neq 0 \\ C(q^{-1}) = 1 + c_1q^{-1} + \cdots + c_lq^{-l}, \end{cases}$$

where q^{-1} denotes the unit delay operator. Consider a single input single output ARMA (autoregressive moving average) system given by

$$A(q^{-1})y_k = q^{-d}B(q^{-1})u(t) + C(q^{-1})w_k, \; k \geq 1,$$

where u_k and y_k are input and output, respectively, w_k denotes the random noise, and

$$(a_1, \cdots, a_n, b_0, \cdots, b_m, c_1, \cdots, c_l) = \theta^*$$

is an unknown parameter. Our objective is to find an algorithm converging to θ^* and design a feedback control law to make both $\{u_k\}$ and $\{y_k\}$ be sample mean bounded and

$$\lim_N \frac{1}{N} \sum_{j=1}^{N} E|y_j - y_j^*|^2$$

be minimized, where $\{y_k^*\}$ is a sequence of bounded reference signals. This problem was first considered by Goodwin, Ramadge and Caines [29]. An algorithm of stochastic approximation type was constructed. The algorithm reads:

$$\theta_{k+1} = \theta_k + \frac{1}{r_k}\varphi_k(y_{k+1} - \varphi_k^T\theta_k)$$

$$r_k = r_{k-1} + \varphi_k^T\varphi_k,$$

$$\varphi_k^T\theta_k = y_{k+1}^*,$$

$$\varphi_k^T = (y_k, \cdots, y_{k-n}, u_{k-1}, \cdots, u_{k-m}, -y_{k-1}^*, \cdots, -y_{k-l}^*).$$

Asymptotic properties were obtained through the applications of martingale convergence theorem. Since early 80s, this problem has attracted much attention. Identification and adaptive controls under the influence of random noise have been studied by many people. For an extensive account on the problem and recent literature citations, see Ljung [57], Chen and Guo [11] and the references therein.

Least Squares Algorithms: Stopping Rules

Suppose that $\{u_n\}$, $\{y_n\}$, and $\{\xi_n\}$ are sequences of scalar input, output, and random disturbance, respectively. Consider a single input, single output linear system given by

$$y_n = \phi'_{n-1}\theta + x_n, \tag{10.7.53}$$

where

$$\theta' = (a, b) = (a_1, \cdots, a_p, b_1, \cdots, b_q),$$

$$\phi'_n = (y_n, y_{n-1}, \cdots, y_{n-p+1}, u_n, \cdots, u_{n-q+1}) \quad \phi_{-1} \text{ arbitrary},$$

$a_i, b_j \in \mathbb{R}$, and p, q are known positive integers representing the order of the system. It is well known that the least squares estimate of θ is given by

$$\theta_{n+1} = (\sum_{k=0}^{n} \phi_k \phi'_k)^{-1} \sum_{k=0}^{n} \phi_k y_{k+1}.$$

An on-line identification procedure of the system (10.7.53) is given by

$$\begin{aligned}
\theta_{n+1} &= \theta_n + e_n P_n \phi_n (y_{n+1} - \phi'_n \theta_n) = \theta_n + P_{n+1}\phi_n(y_{n+1} - \phi'_n\theta_n) \\
P_{n+1} &= P_n - e_n P_n \phi_n \phi'_n P_n \\
P_{n+1} &= (\sum_{k=0}^{n} \phi_k \phi'_k)^{-1} \\
e_n &= (1 + \phi'_n P_n \phi_n)^{-1}.
\end{aligned} \tag{10.7.54}$$

Due to its recursive nature, in implementing the least squares algorithm, more often than not one would like to be able to stop the procedure if a certain degree of accuracy is reached. Therefore, to design feasible stopping criteria becomes an important matter.

To proceed, let $B(\cdot, \cdot)$ denote a bilinear form, such that $B(x, A) = x'Ax$, (where x is a vector and A is a symmetric positive definite matrix with compatible dimensions). Let

$$\text{Ellip} = \{x; B(x - \mu, \Sigma^{-1}) \leq c, c > 0\},$$

where Σ is symmetric positive definite matrix. The volume of this ellipsoid is given by

$$V(\text{Ellip}) = \frac{\pi^{\frac{p+q}{2}} c^{\frac{p+q}{2}} (\det \Sigma)^{\frac{1}{2}}}{\Gamma(\frac{p+q}{2} + 1)}.$$

If $(\sum_{k=0}^{n} \phi_k \phi'_k)^{\frac{1}{2}}(\theta_n - \theta)$ is asymptotically normal, in the sense of

$$(\sum_{k=0}^{n} \phi_k \phi'_k)^{\frac{1}{2}}(\theta_n - \theta) \to N(0, \sigma^2 I) \text{ in distribution},$$

then we can define an ellipsoidal confidence region for θ as

$$\text{Ellip}_n = \left(\theta; B\left((\theta_n - \theta), \hat{\sigma}_n^{-2}(\sum_{k=0}^{n} \phi_k \phi'_k)\right) \leq c\right)$$

where $\hat{\sigma}_n^2 \to \sigma^2$ (the variance of the noise) w.p.1. Consequently,

$$B\left((\theta_n-\theta), \hat{\sigma}_n^{-2}(\sum_{k=0}^{n} \phi_k\phi_k')\right) - B\left((\theta_n-\theta), \sigma^{-2}(\sum_{k=0}^{n} \phi_k\phi_k')\right) \to 0 \text{ in probability.}$$

If we choose $c = c_\alpha$, such that $P(\chi^2_{p+q} \geq c_\alpha) = \alpha$, where χ^2_{p+q} denotes the Chi-square distribution with $p+q$ degree of freedom, then

$$P(\theta \in \text{Ellip}_n) = P\left(B((\theta_n-\theta), \hat{\sigma}_n^{-2}(\sum_{k=0}^{n} \phi_k\phi_k')) \leq c_\alpha\right) \to P(\chi^2_{p+q} \leq c_\alpha) = 1-\alpha.$$

Therefore, Ellip_n is a confidence ellipsoid having limit confidence coefficient $1-\alpha$. Note that

$$V(\text{Ellip}_n) = \frac{\hat{\sigma}_n^{\frac{p+q}{2}} \pi^{\frac{p+q}{2}} c_\alpha^{\frac{p+q}{2}} [\det(\sum_{k=0}^{n} \phi_k\phi_k')^{-1}]^{\frac{1}{2}}}{\Gamma(\frac{p+q}{2}+1)}.$$

Suppose that there exists a non-random matrix T_n such that T_nT_n' is symmetric positive definite, and $T_n^{-1}(\sum_{k=0}^{n}\phi_k\phi_k')^{\frac{1}{2}} \to I$ in probability. For any $\varepsilon > 0$, define

$$\tilde{V}_n = \frac{\sigma^{\frac{p+q}{2}} \pi^{\frac{p+q}{2}} c_\alpha^{\frac{p+q}{2}} [\det(T_n^{-1}(T_n')^{-1})]^{\frac{1}{2}}}{\Gamma(\frac{p+q}{2}+1)} \tag{10.7.55}$$

$$m_\varepsilon = \inf\{n; \tilde{V}_n \leq \varepsilon^{p+q}\}. \tag{10.7.56}$$

The stopping rule is given by

$$\tau_\varepsilon = \inf\{n; V(\text{Ellip}_n) \leq \varepsilon^{p+q}\}. \tag{10.7.57}$$

The design of the stopping rule is based on the following fact:

$$\frac{\tau_\varepsilon}{m_\varepsilon} \to 1 \text{ in probability as } \varepsilon \to 0, \text{ and} \tag{10.7.58}$$

$$\lim_{\varepsilon\to 0} P\{\theta; \theta \in \text{Ellip}_{\tau_\varepsilon} \text{ and } V(\text{Ellip}_{\tau_\varepsilon}) \leq \varepsilon^{p+q}\} = 1-\alpha. \tag{10.7.59}$$

To establish the second assertion above, we study the asymptotic properties of a stopped process $(\sum_{k=0}^{\tau_\varepsilon-1}\phi_k\phi_k')^{-1/2}\sum_{k=0}^{\tau_\varepsilon-1}\phi_k x_{k+1}$, which in turn can be treated by considering $M_{\tau_\varepsilon}(t) = T_{\tau_\varepsilon}^{-1}\sum_{k=0}^{[\tau_\varepsilon t]}\phi_k x_{k+1}$; see Yin [89] for more details.

10.7.4 Adaptive Step-size Tracking Algorithms

Similar to adaptive filtering, many problems occur in communication theory, adaptive equalizers, time-varying channels, adaptive noise cancellation or signal enhancement systems, adaptive quantizers, and other applications, one must unavoidably deal with time-varying signals and/or parameters. Thus on-line tracking algorithms are very important to handle such applications.

Suppose that the observation at time n is given by

$$y_n = \phi_n'\overline{\theta}_n + \nu_n,$$

and $\overline{\theta}_n$ is the value of the slowly time-varying physical parameter at n. The values of $y_i, \phi_i, i \leq n$ are available at time n. To track the variation of the parameter, we use

$$\theta_{n+1}^\varepsilon = \theta_n^\varepsilon + \varepsilon\phi_n(y_n - \phi_n'\theta_n^\varepsilon),$$

where θ_n^ε is the estimate of $\overline{\theta}_n$. It is clear that the choice of the step size ε above is of foremost importance. Thus, one really has two estimation problems to contend with. One is the estimation of $\overline{\theta}_n$ and the other is the estimate of the optimal choice of the step size ε.

An "adaptive" approach was suggested in Benveniste, Metivier, and Priouret [8, p. 160], and explored in Brossier [9] with extensive simulation study. The rigorous approach of the asymptotic analysis is in Kushner and Yang [50]. The algorithms suggested is of the following form. Use ε_n to denote the estimate of the optimal step size at n. Then

$$\theta_{n+1} = \theta_n + \varepsilon_n \phi_n \left[y_n - \phi_n' \theta_n \right]. \tag{10.7.60}$$

Define $e_n(\varepsilon) = y_n - \phi_n' \theta_n^\varepsilon$. Find ε that minimizes the stationary value of

$$E[y_n - \phi_n' \theta_n^\varepsilon]^2/2 = E e_n^2(\varepsilon)/2. \tag{10.7.61}$$

Use V_n^ε to denote the "derivative" $(\partial/\partial\varepsilon)\theta_n^\varepsilon$ (in the mean squares sense). The stochastic gradient w.r.t. (10.7.61) at n is $-e_n(\varepsilon)\phi_n' V_n^\varepsilon$. Choose a constant step size $\delta > 0$. Then the algorithm is given by

$$\theta_{n+1} = \theta_n + \varepsilon_n \phi_n e_n, \tag{10.7.62}$$

$$\varepsilon_{n+1} = \Pi_{[\varepsilon_-, \varepsilon_+]} \left[\varepsilon_n + \delta e_n \phi_n' V_n \right], \tag{10.7.63}$$

$$V_{n+1} = V_n - \varepsilon_n \phi_n \phi_n' V_n + \phi_n [y_n - \phi_n' \theta_n], \quad V_0 = 0, \tag{10.7.64}$$

where $\Pi_{[\varepsilon_-, \varepsilon_+]}$ denotes the projection onto the interval $[\varepsilon_-, \varepsilon_+]$ for some $0 < \delta \ll \varepsilon_- \le \varepsilon_+$. It has been shown by Krishnamurthy and Yin [38] that similar algorithms can also be applied to code-aided suppression of multiple access interference (MAI) and narrow-band interference (NBI) in DS/CDMA systems.

10.7.5 Approximation of Threshold Control Policies

The concept of hedging policy was developed by Kimemia and Gershwin in their pioneer work [35]. They showed that for a manufacturing system with unreliable machines, the optimal control that minimizes both WIP (work-in-process) and backlog is a feedback control that is determined by the current system state, e.g., machine states and inventory levels, which is characterized by some threshold values (termed hedging point in their paper). If the inventory level of certain part type is lower than its corresponding threshold, the optimal control policy is to produce at a full speed in order to reach the threshold. If the inventory level is higher than its threshold, the production of this part type should stop. The one-machine one-part-type problem was completely solved by Akella and Kumar [1] for discounted cost function under the assumption that the machine up and down times form a finite state Markov chain. They took a dynamic programming approach, and obtained the closed form solution characterized by a single threshold value represented by the solution of the corresponding Hamilton-Jacobi-Bellman equation. The problem with an average cost per unit time was dealt with in Bielecki and Kumar [6]. For further work on stochastic control based production planning problems, see Sethi and Zhang [74]. Since the hedging policies are easily implementable, they are widely used in practice. Surplus control and Kanban system are some noted representatives. To implement such a model, a threshold (the surplus level or total number of Kanbans) is set for each production stage. Although the optimality of the threshold policy has substantially eased the passage towards the optimal control, the derivation of the optimal threshold values remain to be difficult for most problems.

In [88], we devoted our attention to threshold control type policies. In lieu of solving an optimal control problem, we turned the problem around and treated an optimization problem. That is, we focused our attention on the class of threshold type controls, and aimed at obtaining the optimal threshold values.

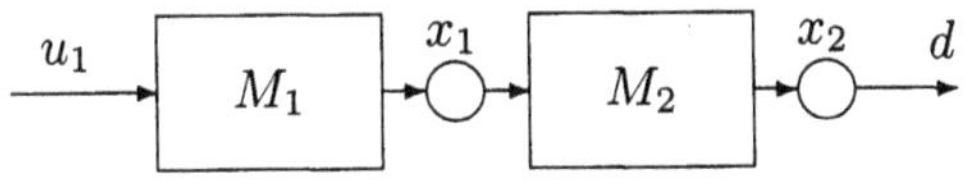

Figure 10.1: A Two-machine System

Consider a tandem two-machine system producing a single product; see Figure 10.1. The two machines are unreliable, each having two states, up and down. The up and down times are sequences of random variables. Denote the inventory levels of the machines by $x_i(t)$ and the machine capacity (a random process) by $\alpha(t)$. The production rates of the two machines $u_i(t), i = 1, 2$, and the demand rate of this product is $d(t)$.

To get the gradient estimates of the objective function, we used the methods of infinitesimal perturbation analysis developed in [32]. Define a combined process $\xi(t) = (x(t), \alpha(t))$; denote the optimal threshold by θ^*. Our task is to construct a sequence of estimates of θ^*. Consider the following stochastic optimization algorithm:

$$\theta_{n+1} = \theta_n + \varepsilon \frac{1}{T_\varepsilon} \int_{nT_\varepsilon}^{nT_\varepsilon + T_\varepsilon} b(\theta_n, \xi(t))dt, \tag{10.7.65}$$

if a continuous-time model is used and/or

$$\theta_{n+1} = \theta_n + \varepsilon \frac{1}{T_\varepsilon} \sum_{j=nT_\varepsilon}^{nT_\varepsilon + T_\varepsilon - 1} b(\theta_n, \xi_j), \tag{10.7.66}$$

if a discrete-time model is used, where

$$\frac{1}{T_\varepsilon} \int_{nT_\varepsilon}^{nT_\varepsilon + T_\varepsilon} b(\theta_n, \xi(t))dt$$

and/or

$$\frac{1}{T_\varepsilon} \sum_{j=nT_\varepsilon}^{nT_\varepsilon + T_\varepsilon - 1} b(\theta_n, \xi_j)$$

are gradient estimators and $b(\cdot)$ is an appropriate function. In (10.7.66), T_ε is understood to be an integer. Different forms of the gradient estimators are available; see for example [55] and the references mentioned there. In Yan, Yin, and Lou [88], we applied the method of infinitesimal perturbation analysis. It appears that another systematic approach is the use of the finite difference approach and its variant with the use of the random direction methods.

10.7.6 GI/G/1 Queue

Consider the optimization problem of the performance of a single server queue. Customers arrive in accordance with a renewal process. The service time distribution is controlled by a real-valued parameter θ, which is chosen to minimize the sum of the average waiting time per customer and a cost associated with the use of θ. The cost function $J(\theta)$ is given by

$$J(\theta) = \lim_N \frac{1}{N}\sum_{i=1}^{N} EX_i(\theta) + C(\theta) \equiv \widehat{J}(\theta) + C(\theta),$$

where $X_i(\theta)$ is the time that the ith customer spends in the system, and $C(\theta)$ is a known bounded real-valued (deterministic) function with a continuous and bounded gradient. The parameter values are confined to a finite interval $[a, b]$. Our interest is to minimize $J(\theta)$ over the finite interval $[a, b]$. Generally, the values of $\widehat{J}(\theta)$ are very hard to compute. A viable alternative is to use stochastic approximation methods. We can observe the queue over a longtime period, and incorporate the observed data (that are the arrival and departure times and the service time for each customer) in the estimation procedure. The observed data will then be used to obtain the gradient estimate of the cost function with respect to θ at the current value of θ, yielding a stochastic approximation algorithm.

This problem, which has attracted much attention (see [16, 48] among others), is typical for many applications in queueing networks and manufacturing systems and networks.

$$\theta_{n+1}^{\varepsilon} = \Pi_{[a,b]}\left[\theta_n^{\varepsilon} + \varepsilon\widehat{Y}_n^{\varepsilon} - \varepsilon C_\theta(\theta_n^{\varepsilon})\right],$$

where $\widehat{Y}_n^{\varepsilon}$ denotes the gradient estimate of the cost $\widehat{J}(\theta)$. Much of the recent interest on this problem lies on the use of the infinitesimal perturbation analysis method [32] to find $\widehat{Y}_n^{\varepsilon}$.

10.7.7 Distributed Algorithms for Supervised Learning

In supervised learning and pattern recognition problems, the learning systems' environment presents it with a sequence of vectors (patterns), together with a class label for each vector that indicates how the vector ought to be classified. A sequence of patterns together with the class lables are normally referred to as a "training sequence." The environment is often termed a "supervisor." The learning system adaptively adjusts its decision in order to minimize the probability of misclassification. Suppose for simplicity there are only two classes C_1 and C_2 to be selected. At each step, the environment determines a training sequence by first selecting C_1 or C_2 in accordance with the *a priori* probability $P\{C_i\}$, and then choosing a pattern vector according to the conditional probability $P\{y|C_i\}$. Associated with each y, there is a class label z, such that if class C_1 is selected, then $z = 1$; and if class C_2 is selected, then $z = 2$. Now, the training sequence consists of a sequence of pairs of the form $\{(y_n, z_n)\}$. Consequently, the decision rule which minimizes the probability of misclassification can be found by means of Bayesian *a posteriori* probabilities:

$$y \in C_1, \quad \text{if } P(C_1|y) - P(C_2|y) > 0; \tag{10.7.67}$$
$$y \in C_2, \quad \text{if } P(C_1|y) - P(C_2|y) < 0. \tag{10.7.68}$$

The decision rule depends on $P\{C_i|y\}$, $i = 1, 2$, which are not available. To circumvent this difficulty, the following alternative approach is devised [19]. Find a vector-valued parameter θ, such that $\theta' y$ approximates $P(C_1|y) - P(C_2|y)$ as well as possible in the mean square sense, and use the decision rule

$$y \in C_1, \text{ if } \theta' y > 0 \text{ and } y \in C_2, \text{ if } \theta' y < 0. \tag{10.7.69}$$

It can be shown that the objective will be achieved if the functional $J(\theta)$ given by $J(\theta) = E(\theta' y - z)^2$ is minimized. The functional $J(\theta)$ is usually unavailable, but at each n, y_n and

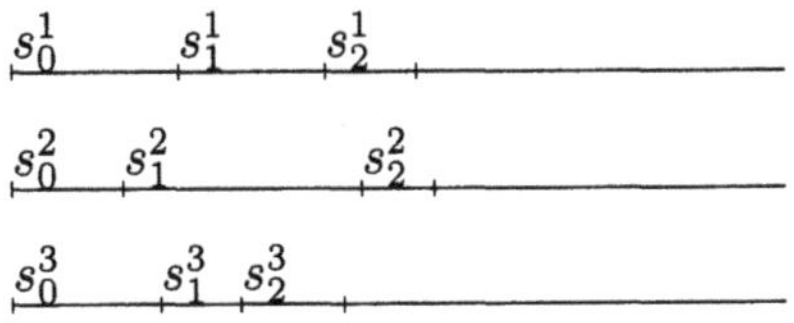

Figure 10.2: Asynchronous Random Computation Times

z_n can be observed. A sequence of approximations $\{\theta_n\}$ can be constructed, which has the same form as the adaptive filtering algorithms. If the dimensionality of the learning task is large, it makes sense to reduce its complexity by using parallel processors.

Let

$$y = (y^1, \cdots, y^r)' \in \mathbb{R}^{rp} \quad \text{and} \tag{10.7.70}$$
$$\theta = (\theta^1, \cdots, \theta^r)' \in \mathbb{R}^{rp}, \tag{10.7.71}$$

where $y^i \in \mathbb{R}^p$ and $\theta^i \in \mathbb{R}^p$. Suppose that rp is a very large number. To carry out computations for such problems in digital computers, large memory storage is needed. In our approach, we utilize r processors. The vector θ is decomposed into r blocks first. Each block consisting of p components of the vector θ, is handled by one of the parallel processors. Then, a learning algorithm is implemented in this block. These parallel processors compute and communicate with each other asynchronously and at random times. Let each processor have its own clock and the iteration on each block is carried out at renewal type of random times. To be more specific, for each $i = 1, 2, \ldots r$, let t_n^i be the nth iteration time for processor i, i.e., processor i takes t_n^i units of time to complete its nth iteration. Define s_n^i by

$$s_0^i = 0, \quad \text{and} \quad s_n^i = \sum_{j=1}^{n} t_j^i.$$

For each i, the sequence $\{t_n^i\}$ is an interarrival time and $\{s_n^i\}$ is the corresponding "renewal" time. Figure 10.2 provides an illustration of the random computation times of a simplified model with three processors.

Let $\theta_0 = (\theta_0^1, \ldots, \theta_0^r)'$ be the initial condition. For small $\varepsilon > 0$, and each $i = 1 \le r$, the distributed learning algorithm is given by:

$$\theta^i_{s^i_{n+1}} = \theta^i_{s^i_n} + \varepsilon y^i_{s^i_n}(z_{s^i_n} - \sum_{j=1}^{r} y^j_{s^i_n} \theta^j_{s^i_n}). \tag{10.7.72}$$

For simplicity, the estimated values $\theta_{s_n^i}$ are communicated to all the processors as soon as they are available. In fact, the results still hold if bounded delays incur in data transmissions. For each i, let

$$N^i(n) = \sup\{j; s_j^i \le n\}, \; \Delta_n^i = n - s^i_{N^i(n)}. \tag{10.7.73}$$

The sequence $\{N^i(n)\}$ enumerates the number of computations (iterations) up to time n for processor i, and Δ_n^i represents the time elapsed since the last iteration. Since each process has its own running clock and takes a random time to complete each iteration. The usual notion of "time," i.e., the iteration number can no longer be used as a common indicator for all the processors. To proceed, we define

$$\theta_n^i = \theta_{s_k^i}^i,\ y_n^i = y_{s_k^i}^i,\ z_{n,i} = z_{s_k^i} \text{ for } n \in [s_k^i, s_{k+1}^i).$$

The algorithm can then be written as

$$\theta_{n+1}^i = \theta_n^i + \varepsilon y_n^i (z_{n,i} - \sum_{j=1}^r y_n^j \theta_n^j) I_{\{\Delta_{n+1}^i = 0\}}.$$

At time $n+1$, if no computation takes place, then the iterate θ_{n+1}^i is equal to θ_n^i, otherwise, we incorporate the changes in the update. In the above, the dependence of i is emphasized for z_n. This is due to the fact that $z_{s_k^i}$ will generally be different from $z_{s_k^j}$ for $i \neq j$. Such algorithms can analyzed by the methods of weak convergence as in Kushner and Yin [51, 52].

10.7.8 A Heat Exchanger

Owing to the increasing needs for safe and optimal operation, parameter estimation, learning, and fault detection have received growing attention in process industries. To monitor the process performance and to effectively control it require the knowledge of certain system variables. Because of the presence of random disturbance and gross errors in the process data, however, the measurements often contain some degree of error. Due to possible instrument failure and the presence of process and measurement noise and due to technical difficulty and cost consideration, information of some of the states/properties of the system has to be deduced from certain estimation techniques.

Consider a countercurrent shell-tube lube-oil heat exchanger, a piece of equipment needed in many industrial processes. The underlying process is represented in Figure 3. The manipulated variable is the flow rate of cooling water on the shell side; and the controlled variable (output) is the lube oil temperature exiting the exchanger on the tube side. With feedback control, the oil temperature is measured and the measurement is used to adjust the cooling water flow rate. Since we are interested in estimation and learning, we have simplified the process to consider an open-loop system only. To simulate disturbance to this nonlinear process, the inlet oil is separated into two parts, a hot stream having constant flow rate and a warm stream having variable flow rate. We want to estimate the input variable, the inlet warm oil flow rate, using the noise-corrupted output y_n (exiting oil temperature) and noisy input x_n. A sequence of estimates $\{\theta_n\}$ is obtained, in which the nominal value x^* is the normal operating condition and is known in practice.

To describe the algorithms, let $\{x_n\}$ be a sequence of $\mathbb{R}^r$-valued random variables that represent the measured states or inputs, $\{y_n\}$ a sequence of $\mathbb{R}^r$-valued random variables that are measured outputs, with

$$y_n = f(x_n, \xi_n), \tag{10.7.74}$$

where $f(\cdot,\cdot)\colon \mathbb{R}^r \times \mathbb{R}^r \mapsto \mathbb{R}^r$, $\{\xi_n\}$ is a stationary sequence of $\mathbb{R}^r$-valued random disturbances. The learning/estimation task of interest can be formulated as a root searching problem for a nonlinear function $\overline{f}(\cdot)$, i.e.,

$$\text{find the solutions of } \overline{f}(x) = Ef(x, \xi_n) = 0.$$

One may wish to resort to the classical stochastic approximation method for solution. Unfortunately, such a procedure is not applicable here since $\{x_n\}$ is a sequence of random variables not depending on our choice.

On the premise that in many chemical engineering applications, nominal values of certain states are often available and therefore can be used as references for comparison, we herein propose and examine an algorithm of the form

$$\theta_{n+1} = \theta_n + \frac{\varepsilon}{\delta} K_0(\theta_n - x^*) K\left(\frac{x_n - \theta_n}{\delta}\right) y_n, \tag{10.7.75}$$

where

$$K(x) = \begin{cases} 0.75(1 - |x|^2), & |x| \le 1; \\ 0, & |x| > 1, \end{cases} \tag{10.7.76}$$

$$K_0(x) = \begin{cases} R_0^2 - |x|^2 + c_0, & \text{if } |x| \le R_0, \\ c_0, & \text{if } |x| \ge R_0, \end{cases} \tag{10.7.77}$$

with $R_0 > 0$ and $c_0 > 0$ chosen based on the knowledge of the physical system, x^* is the nominal or reference value of the state of interest, ε is the step size and δ is the window width. Under suitable conditions, it can be shown that the limit ordinary differential equation is of the form

$$\dot{\theta} = K_0(\theta - x^*)\pi(\theta)\overline{f}(\theta),$$

where (roughly speaking) $\overline{f}$ is the average of the observation and $\pi(\cdot)$ is the limit of the conditional distribution of x given the past data $\mathcal{F}_n$. The detailed analysis of this algorithm is in [102].

In fact, the proposed algorithm is an estimation procedure. However, its actual performance shows that it also has the capability of tracking slightly time-varying parameters. This is highly desirable in real applications since in many industrial processes the operating conditions deviate from their nominal values slightly but frequently.

Alternatively, one may use a soft constraint algorithm. To be more specific, let us use a spherical soft constraint $S = \{\theta; |\theta| \le \rho_0\}$. Define

$$d(\theta) = \begin{cases} (|\theta| - \rho_0)^2, & \text{if } |\theta| > \rho_0, \\ 0, & \text{if } |\theta| \le \rho_0. \end{cases}$$

The algorithm becomes

$$\theta_{n+1} = \theta_n + \frac{\varepsilon}{\delta} K\left(\frac{x_n - \theta_n}{\delta}\right) y_n - \varepsilon \rho_0 d_\theta(\theta_n - x^*),$$

where

$$d_\theta(\theta) = \begin{cases} 2\theta\left(1 - \dfrac{\rho_0}{|\theta|}\right), & |\theta| > \rho_0, \\ 0, & \text{otherwise.} \end{cases}$$

The asymptotic properties of the algorithm then can be obtained via the use of the associated limit ordinary differential equation

$$\dot{\theta} = \pi(\theta)\overline{f}(\theta) - \rho_0 d_\theta(\theta - x^*).$$

It is clear that the soft constraints prevent the iterates to be far away from the nominal value. Convergence and rate of convergence can be obtained via weak convergence methods.

10.7.9 Evolutionary Algorithms

Based upon collective processes with a population of individuals, which are search points for a given problem, the evolutionary algorithms carry out desired computing tasks by use of randomized selection, mutation and recombination. These algorithms have been applied to many problems in parameter optimization and related fields with great success. Significant progress has been made in the study of evolutionary algorithms for almost thirty years.

The evolution strategies were first introduced by Rechenberg and Schwefel in the mid-60s [65, 72]. At that time, applications in hydrodynamics such as optimizing the shape of a bent pipe and a flashing nozzle were dealt with. Different versions of the strategy were simulated [72]. The research in this subject has become a rapidly growing one ever since. Nowadays, the (μ, λ) evolution strategies, introduced in [73] are commonly used in evolution strategy research.

We consider a problem with $(1, \lambda)$ strategy. Our objective is to minimize a function $f : \mathbb{R}^d \mapsto \mathbb{R}$. The plan is to employ the $(1, \lambda)$ evolution strategy, for $\lambda \geq 2$. Loosely, the strategy can be described as follows. In each generation, one parent produces λ offspring. Among the offspring, choose the best one with respect to the evaluation of the objective function to form the next estimate.

To be more specific, generate sequences of random vectors $\{z_n(i)\}$, for $1 \leq i \leq \lambda$ that are independent and identically distributed (i.i.d.) Gaussian random variables with mean zero and covariance $\sigma^2 I_d$, where I_d denotes the $d \times d$ identity matrix such that for each n, $z_n(1), \ldots, z_n(\lambda)$ are independent. To carry out the minimization task, choose an initial estimate $x_0 \in \mathbb{R}^d$. At iteration n, add the random vector $z_n(i)$ to the current content, i.e., $x_n + z_n(i)$, for $i = 1, \ldots, \lambda$. We evaluate the corresponding values $f(x_n + z_n(i))$. Next, choose the smallest among the λ values of $f(\cdot)$. That is,

$$\begin{aligned} f(x_n + z_n(j)) &= \min_{y \in \Lambda_n} f(x_n + y), \text{ where} \\ \Lambda_n &= \{z_n(i), i = 1, \ldots, \lambda\}. \end{aligned} \tag{10.7.78}$$

Then assign $x_n + z_n(j)$ to x_{n+1}. In short

$$x_{n+1} = \text{argmin}\{f(x_n + z_n(1)), \ldots, f(x_n + z_n(\lambda))\}. \tag{10.7.79}$$

Our task now is to convert (10.7.79) to a recursive algorithm of stochastic approximation type so that the techniques in analyzing stochastic approximation type algorithms can be applied.

It is well known that the standard deviation σ is a scale factor in the problem. Since $z_n(i)$ are i.i.d. random vectors and $z_n(i) \sim N(0, \sigma I_d)$, we can rescale the sequence $z_n(i)$ or equivalently, define another sequence $\{\tilde{z}_n(i)\}$ by setting $z_n(i) = \sigma \tilde{z}_n(i)$ such that $\tilde{z}_n(i) \sim N(0, I_d)$. That is, $\tilde{z}_n(i)$ follows the standard normal distribution. Now (10.7.79) can be rewritten as

$$x_{n+1} = x_n + \sigma \sum_{i=1}^{\lambda} \tilde{z}_n(i) I_{\{f(x_n + z_n(i)) = \min_{y \in \Lambda_n} f(x_n + y)\}}, \tag{10.7.80}$$

where I is an indicator function. In evolution strategy, one often chooses σ so that it is proportional to $(1/d)H(f_x(x_n))$, where $f_x(\cdot)$ denotes the gradient of $f(\cdot)$, d is the dimension of the problem and $H(\cdot) : \mathbb{R}^d \mapsto [0, \infty)$ is an appropriate real-valued function such that $H(0) = 0$ and the only root of $H(\cdot)$ is 0. With ε denoting the proportional constant multiplied by $1/d$, the recursive formula can be written as

$$x_{n+1} = x_n + \varepsilon H(f_x(x_n)) \sum_{i=1}^{\lambda} z_n(i) I_{\{f(x_n+z_n(i))=\min_{y\in\Lambda_n} f(x_n+y)\}}. \tag{10.7.81}$$

Eq. (10.7.81) in fact, is a constant-step-size stochastic approximation algorithm with step size ε. Since normally the problems we consider are large dimensional ones, ε is relatively small. Our interest lies in obtaining convergence and rate of convergence results for the limit as $\varepsilon \to 0$. We wish to emphasize that in the actual computation, we neither change the evolution algorithm nor modify it in any way. The equivalent expression (10.7.81) is simply a convenient form that allows us to analyze the algorithm by using methods of stochastic approximation. For a detailed account on the development via stochastic approximation approach, see the recent work of Yin, Rudolph, and Schwefel [99].

10.7.10 Digital Diffusion Machines

In a recent work [87], Wong suggested a diffusion-network model, which is based on modifications of the Langevin algorithm and the Hopfield network. The motivation stems from the applications in image segmentation problems and many other optimization and estimation problems. The underlying problem can be stated as follows. Let $\mathcal{E} : [0,1]^r \mapsto \mathbb{R}$ be an "energy" function defined on the hypercube $[0,1]^r = [0,1] \times \cdots \times [0,1]$. Find the global minimizer of $\mathcal{E}(\cdot)$ by use of a neural network. Suppose that for all $t \geq 0$, $v_\alpha(t) \in [0,1]$ are the state at node α at time t and $v = (v_1, \ldots, v_r)^\tau \in [0,1]^r$ is an r-dimensional column vector (z^τ denotes the transpose of z in this section only). By injecting noise into a Hopfield network, the dynamics of the αth node are given by

$$\begin{aligned} v_\alpha(t) &= g(u_\alpha(t)) \\ du_\alpha(t) &= -\frac{\partial}{\partial v_\alpha}\mathcal{E}(v(t))dt + \widetilde{a}_\alpha(u(t))dw_\alpha(t), \end{aligned} \tag{10.7.82}$$

where for $\alpha \leq r$, $\{w_\alpha(\cdot)\}$ are independent (standard and real-valued) Brownian motions, and $\widetilde{a}_\alpha(\cdot)$ and $g(\cdot)$ are appropriate functions. It is shown in [87], by choosing $\widetilde{a}_\alpha(\cdot)$ to be $\widetilde{a}_\alpha(u(t)) = [(2T)/g'(u_\alpha(t))]^{1/2}$ (where g' denotes the derivative of g in this section only), $v(\cdot)$ is a stationary Markov process with stationary density

$$p_\infty(v) = (1/Z)\exp(-(1/T)\mathcal{E}(v)),$$

where Z is an appropriate normalizing factor so that $\int p_\infty(v)dv = 1$. Furthermore, by selecting $f(x) = g'(g^{-1}(x))$ (for each $x \in \mathbb{R}$), for each $\alpha \leq r$,

$$dv_\alpha(t) = -f(v_\alpha(t))\frac{\partial}{\partial v_\alpha}\mathcal{E}(v(t))dt + Tf'(v_\alpha(t))dt + \sqrt{2Tf(v_\alpha(t))}dw_\alpha(t), \tag{10.7.83}$$

where T goes to zero sufficiently slowly. In view of the equation above, it is worth noting that $\sqrt{2Tf(v_\alpha(t))}$ depends only on the αth node. Therefore, the noise of the system under consideration is "de-coupled" among different processors. This is an important feature that allows us to use parallel processing method efficiently and simplifies many computational tasks significantly. To take advantages of Wong's diffusion network and to overcome the difficulties of the analog implementation, in [14], Cai, Kelly, and Gong proposed a digital version of the network. The basic idea lies in the discretization of the stochastic differential equations. A number of numerical experiments are conducted for image segmentation problems. The results are rather encouraging. The heart of the approach is an approximation of the diffusion machine by a digital diffusion network [98]; much of the theoretical justification is to prove the convergence of the digitized system to that of the continuous counterpart.

To proceed, we present a recursive algorithm. The idea is to partially reset the gain sequence once a while. For each $\iota \geq 0$, and each $\alpha \leq r$,

$$\begin{aligned} v_{\alpha,\iota n+k+1} &= v_{\alpha,\iota n+k} - a_{\iota n+k} f(v_{\alpha,\iota n+k}) \frac{\partial}{\partial v_\alpha} \mathcal{E}(v_{\iota n+k}) \\ &\quad + c_{\iota n+k} f'(v_{\alpha,\iota n+k}) + b_{\iota n+k} \sqrt{f(v_{\alpha,\iota n+k})} W_{\alpha,\iota n+k}, \end{aligned} \tag{10.7.84}$$

where for some $A_0 > 1$ and some $1/2 < \gamma < 1$, the step-size sequences are given by

$$a_{\iota n+k} = 1/(\iota n+k)^\gamma, \; b_{\iota n+k} = \sqrt{2a_{\iota n+k}/\widetilde{a}_{\iota n+k}}, \; c_{\iota n+k} = a_{\iota n+k}/\widetilde{a}_{\iota n+k},$$

with

$$a_{\iota n+k} = \ln((\iota n+k)^{1-\gamma} - (\iota n)^{1-\gamma} + A_0).$$

The main task then is to prove the convergence of the digitized version of the algorithm to its continuous-time counterpart. The main tool is the method of weak convergence.

10.8 Further Remarks

This chapter delineates the methods of stochastic approximation. In addition to giving certain asymptotic results, our effort has been devoted to describe to where the methods can be applied. A diverse range of applications are given. We choose to ignore most of the technical details. In addition, the results are often mentioned in the simplest setting so as to make the main ideas clear. It should be emphasized that much of the development can be put in far more general settings to incorporate various applications.

10.8.1 Convergence

For the convergence of stochastic approximation algorithms, this chapter mainly concerns the ODE approach. There are other methods of proof available in the literature. Typically, one establishes the boundedness of the iterates first and then proves the desired convergence.

Chain Recurrence

In the recent study, there is an interesting approach that explores much of the connection of the discrete iteration with that of the continuous dynamic systems. In [4], Benaim develops the ideas of chain recurrence. Without needed information, sometimes the best one can do is to prove that x_n or the interpolated and shifted process $x^n(\cdot)$ converges w.p.1 to an invariant (or limit set) of the ODE (10.3.22). Sometimes, these limit sets turn out to be rather large. For example, consider $\dot{x} = x(1-x)$, and the set we are interested in is $[0,1]$. Then the entire interval $[0,1]$ is an invariant set for the ODE. The idea of chain recurrence can simplify the analysis. As in this example, the only chain recurrent points are 0 and 1. For further discussion on this matter, we refer the reader to [4] and [53, Chapters 5 and 6].

Differential Inclusion

Suppose that we wish to carry out an optimization task. The function under consideration is convex and continuous, but is not every where differentiable. Then the gradient of $f(\cdot)$ will be replaced by the subgradient of $f(\cdot)$. Now in lieu of (10.2.2), we have

$$x_{n+1} = x_n - a_n \gamma_n + a_n \frac{\xi_n}{2c_n},$$

where the γ_n is defined by

$$\gamma_n = (\gamma_{n,1}, \cdots, \gamma_{n,r})' \text{ with} \tag{10.8.85}$$

$$\gamma_{n,i} = \frac{f(x_n + c_n e_i) - f(x_n - c_n e_i)}{2c_n}, \text{ and} \tag{10.8.86}$$

$$\xi_n = (\xi_{n,1}, \cdots, \xi_{n,r}) \text{ with} \tag{10.8.87}$$

$$\xi_{n,i} = [f(x_n + c_n e_i) - F(x_n + c_n e_i, \zeta^+_{n,i})] - [f(x_n - c_n e_i) - F(x_n - c_n e_i, \zeta^-_{n,i})] \tag{10.8.88}$$

Note that γ_n is a subgradient. Carrying out the analysis similar to the ideas presented previously, we will get a limit result. The mean differential equation is replaced by a differential inclusion

$$\dot{x} \in -SG(x),$$

however. In the above, $SG(x)$ denotes the set of subgradients at x (see [53]).

10.8.2 Rate of Convergence

The rate of convergence issue can be addressed in conjunction with the computational budget and the noise and bias effect. One possible road along this line is the development given in L'Ecuyer and Yin [55]. Assuming that a gradient estimator is available and that both the bias and the variance of the noise of the estimator are functions of the budget devoted to its computation, the gradient estimator is used in conjunction with a stochastic approximation algorithm. Detailed analysis allows us to figure out how to allocate the total available computational budget to the successive iterations. The convergence rate is given first as a function of the number of iterations, and then as a function of the total computational effort.

Treating projection or constrained stochastic approximation algorithms, the rate of convergence is often obtained by assuming the optimizer is in the interior of the projection region. The problem of handling the rate of convergence when the optimizer is on the boundary is very difficult. One approach is to use large deviation [20]. Recently, an interesting approach was provided in Buche and Kushner [10]. The rationale is to use a reflected diffusion and consider the corresponding Skorohod problem. The authors develop the techniques and show that the associated stationary Gaussian diffusion is replaced by an appropriate stationary reflected linear diffusion.

10.8.3 Law of Iterated Logarithms

In the study of convergence rate, we have chosen the approach of weak convergence. It should be mentioned that there are also almost sure (or w.p.1) convergence rate results. One of the noted representatives is the law of the iterated logarithm. Consider (10.2.1). For simplicity, assume $r = 1$. Suppose that the noise variance is σ^2,

$$\lim_{n\to\infty} n a_n = A, \quad \text{and} \quad a = f_x(x^*)A > 1/2.$$

Then under suitable conditions, it was proved in Gaposhkin and Krasulina [26] that w.p.1,

$$\lim_{n\to\infty} \left(\frac{n}{2\log\log n}\right)^{1/2} (x_n - x^*) = A\sigma(2a-1)^{-1/2}.$$

The almost sure convergence rate has been investigated further by Heunis [31], in which he developed interesting functional laws of iterated logarithms.

10.8.4 Robustness

One of the questions not studied in detail in this chapter is robustness for stochastic approximation problems. Roughly speaking, robustness refers to the allowable tolerance and errors. In applications, one may know little about the actual dynamics or even about the statistics of the driving noise at large parameter values. It may be undesirable for single observations to have large effects on the iterates. Taking such a view point into consideration, the following algorithm was considered in Polyak and Tsypkin [64]. Let $\psi_i(\cdot), i \le r$, be bounded real-valued functions on the real line, and define $\psi(x) = (\psi_1(x^1), \dots, \psi_r(x^r))$. Let $\psi_i(\cdot)$ be monotonically nondecreasing and satisfy $\psi_i(0) = 0, \psi_i(u) = -\psi_i(-u)$ and $\psi_i(u)/u \to 0$ as $u \to \infty$. One commonly used function is $\psi_i(u) = \min\{u, K_i\}$ for $u \ge 0$, where K_i is a given constant. The algorithm of interest takes the form $x_{n+1} = x_n + a_n\psi(y_n)$, where $\{y_n\}$ is the sequence of noisy observations as obtained in (10.2.1). In the aforementioned paper, Polyak and Tsypkin examined the optimal choice of the function $\psi(\cdot)$ through minimax formulation.

In a related work, Chen, Guo, and Gao [13] studied the problem of robustness for stochastic approximations from another angle. It is a common practice to use the Liapunov function in the analysis of stochastic recursive algorithms. The following questions are particularly interesting. What kind of measurement errors can be tolerated? What kind of deviations can be allowed for the corresponding Liapunov function? It seems that the analysis of robustness plays an important role in organizing information about the behavior of the algorithms to a manageable form. The problem with the regression function evaluated at the true parameter being nonzero was considered and some simultaneous robustness analysis was given. In a broad informal sense, such a robustness analysis gives an account on the allowable tolerance and relates deviations from idealized assumptions. There are also related works in obtaining necessary and sufficient conditions on the measurement noise etc. and effort in exploring various equivalences in regard to the noise [83].

10.8.5 Parallel Stochastic Approximation

Due to rapid technological progress parallel processing methods have attracted much attention lately. Recursive algorithms of the stochastic approximation type, with distributed processors and asynchronous communications was first proposed and analyzed in Tsitsiklis, Bertsekas, and Athans [79]. Some asymptotic results were obtained and various potential applications in stochastic control and system identifications were discussed. Such decentralized algorithms have been attracting growing interest. The aforementioned model was studied further in Kushner and Yin [51]. Utilizing the weak convergence and martingale averaging techniques, convergence properties as well as rate of convergence were established under weaker conditions. Moreover, state dependent noise was treated, communication through noisy channels was dealt with and projection procedures were considered. Later, another class of parallel S.A. algorithms was suggested in Kushner and Yin [52]. Such algorithms utilize parallel processing and distributed computations in a natural way. Instead of using a single processor as in the classical setting, a collection of processors is used. Each processor operates on only part of the system vector. These processors compute and communicate with each other interactively and at generalized renewal times. Some interesting asymptotic theorems were obtained. Further work in this area can be found in the survey paper of Yin [93] (see also Kushner and Vázquez-Abad [48] and [53, Chapter 12]). Since the algorithms using parallel processors all have rather complex forms and are quite technical, we decide not to include the details in this chapter. However, appropriate references are provided.

10.8.6 Open Questions

Although stochastic approximation has been around for about 50 years, there are still many questions that need to be addressed. One of the difficult problems concerns the so-called singularly perturbed stochastic approximation. It is motivated by the ideas of singular perturbations for stochastic systems. The underlying system displays two-time behavior. Some related references for stochastic systems can be found in Kushner [42], and Yin and Zhang [105] among others. Due to the interface of the discrete time and continuous time, the asymptotic analysis is rather complex. The limit of the step size (assumed to be small, i.e., $a_n \to 0$ as $n \to \infty$ for decreasing step size or $\varepsilon \to 0$ for constant step size) and the (singular perturbation) small parameter are not interchangeable, which makes the analysis very difficult.

Another difficult task is the design of efficient global stochastic approximation algorithms. Although the simulated annealing type procedures give us the desired convergence to the global optima, the convergence rate is very slow [95]. The expected time of getting to the global optima is very long. A related question deals with the optimization of a real-valued function that is very flat near the optimum. It is clear that there are increasing demands and pressing needs to design more feasible algorithms for such optimization tasks.

10.8.7 Conclusion

As a rapidly expanding and growing discipline, stochastic approximation involves a wide spectrum of techniques that go far beyond the traditional approaches. It has given impetus, not only to the applications of applied probability and stochastic processes, but also to other areas of science and engineering. Applications of stochastic methods are growing at an increasing rate. To inherit the past and to usher in the future, we perceive unprecedented challenges and opportunities for the development of stochastic approximation methods and applications in the new millennium.

Bibliography

[1] R. Akella and P.R. Kumar, Optimal control of production rate in a failure-prone manufacturing system, *IEEE Trans. Automat. Control*, **AC-31** (1986), 116-126.

[2] A.E. Albert and L.A. Gardner, *Stochastic Approximation and Nonlinear Regression*, MIT Press, Cambridge, MA, 1967.

[3] J.A. Bather, Stochastic approximation: A generalization of the Robbins-Monro procedure, in *Prod. 4th Prague Symposium Asymptotic Statist.*, P. Mandl and M. Hušková, eds., 13-27, 1989.

[4] M. Benaim, A dynamical systems approach to stochastic approximation, *SIAM J. Control Optim.* **34** (1996), 437-472.

[5] E. Berger, Asymptotic behavior of a class of stochastic approximation procedures, *Probab. Theory Related Fields* **71** (1986), 517-552.

[6] T.R. Bielecki and P.R. Kumar, Optimality of zero-inventory policies for unreliable manufacturing systems, *Oper. Res.* **36** (1988), 532-541.

[7] P. Billingsley, *Convergence of Probability Measures*, J. Wiley & Sons, New York, 1968.

[8] A. Benveniste, M. Métivier and P. Priouret, *Adaptive Algorithms and Stochastic Approximation*, Springer-Verlag, Berlin, 1990.

[9] J.M. Brossier. *Egalization Adaptive et Estimation de Phase: Application aux Communications Sous-Marines*, Ph.D. thesis, Institut National Polytechnique de Grenoble, 1992.

[10] R. Buche and H.J. Kushner, Stochastic approximation: Rate of convergence for constrained problems, and applications to Lagrangian algorithms, preprint, 1999.

[11] H.F. Chen and L. Guo, *Identification and Stochastic Adaptive Control*, Birkhäser, Boston, 1991.

[12] H.F. Chen and Y.M. Zhu, *Stochastic Approximation*, Shanghai Sci. & Tech. Publisher, Shanghai, 1996.

[13] H.F. Chen, L. Guo, and A.J. Gao, Convergence and robustness of the Robbins-Monro algorithm truncated at randomly varying bounds, *Stochastic Process. Appl.*, **27** (1988), 217-231.

[14] X. Cai, P. Kelly, and W.B. Gong, Digital diffusion network for image segmentation, *Proc. IEEE Internat. Conf. Image Processing*, 1995.

[15] T.S. Chiang, C.R. Hwang, and S.J. Sheu, Diffusion for global optimization in $\mathbb{R}^n$, *SIAM J. Control Optim.* **25** (1987), 737-752.

[16] E.K.P. Chong and P.J. Ramadge, Optimization of queues using an IPA based stochastic algorithm with general update times, *SIAM J. Control Optim.* **31** (1993), 698-732.

[17] K.L. Chung, On a stochastic approximation method, *Ann. Math. Statist.* **25** (1954), 463-483.

[18] J. Dippon and V. Fabian, Stochastic approximation of global minimum points, *J. Statist. Plann. Inference*, **41** (1994), 327-347.

[19] R.O. Duda and P.E. Hart, *Pattern Classification and Scene Analysis*, Wiley, New York, 1973.

[20] P. Dupuis and H.J. Kushner, Stochastic approximation and large deviations: upper bounds and w.p.1 convergence, *SIAM J. Control Optim.* **27** (1989), 1108-1135.

[21] M. Duflo, *Random Iterative Models*, Springer-Verlag, New York, 1997.

[22] S.N. Ethier and T.G. Kurtz, *Markov Processes: Characterization and Convergence*, J. Wiley, New York, 1986.

[23] Yu. Ermoliev, Stochastic quasigradient Methods and their applications to system optimization, *Stochastics* **9** (1983), 1-36.

[24] V. Fabian, On asymptotic normality in stochastic approximation, *Ann. Math. Statist.* **39** (1968), 1327-1332.

[25] L. Gerencsér, On a class of mixing processes, *Stochastics* **26** (1989), 165-191.

[26] V.F. Gaposhkin and T.P. Krasulina, On the law of the iterated logarithm in stochastic approximation processes, *Theory Probab. Appl.* **20** (1975), 844-850.

[27] S. Geman and C.R. Hwang, Diffusions for global optimization, *SIAM J. Control Optim.* **24** (1986), 1031-1043.

[28] S.B. Gelfand and S.K. Mitter, Recursive stochastic algorithms for global optimization in $\mathbb{R}^d$, *SIAM J. Control Optim.* **29** (1991), 999-1018.

[29] G. Goodwin, P. Ramadge, and P. Caines, Discrete time stochastic adaptive control, *SIAM J. Control Optim.* **19** (1981), 829-853.

[30] W.K. Härdle and R. Nixdorf, Nonparametric sequential estimation of zeros and extrema of regression functions, *IEEE Trans. Inform. Theory* **IT-33** (1987), 367-372.

[31] A.J. Heunis, Asymptotic properties of prediction error estimations in approximate system identification, *Stochastics* **24** (1988), 1-43.

[32] Y.-C. Ho and X.-R. Cao, *Perturbation Analysis of Discrete Event Dynamical Systems*, Kluwer, Boston, 1991.

[33] G. Kersting, Almost sure approximation of the Robbins-Monro process by sums of independent random variables, *Ann. Probab.* **5** (1977), 954-965.

[34] J. Kiefer and J. Wolfowitz, Stochastic estimation of the maximum of a regression function, *Ann. Math. Statist.* **23** (1952), 462-466.

[35] J.G. Kimemia and S. Gershwin, An algorithm for the computer control of production in flexible manufacturing systems, *IIE Trans.*, **15** (1983), 353-362.

[36] S. Kirkpatrick, C.D. Gelatt, and M.P. Vecchi, Optimization by simulated annealing, *Science* **220** (1983), 671-680.

[37] J. Komlós and P. Révéz, On the rate of convergence of the Robbins-Monro method, *Z. Wahrsch. verb. Gebiete.* **25** (1972), 39-47.

[38] V. Krishnamurthy and G. Yin, Adaptive step size algorithms for blind interference suppression in DS/CDMA systems, preprint, 1999.

[39] P.R. Kumar and P.P. Varaiya, *Stochastic Systems: Estimation, Identification and Adaptive Control*, Prentice-Hall, Englewood Cliffts, NJ, 1986.

[40] H.J. Kushner, *Approximation and Weak Convergence Methods for Random Processes, with applications to Stochastic Systems Theory*, MIT Press, Cambridge, MA, 1984.

[41] H.J. Kushner, Asymptotic global behavior for stochastic approximation and diffusions with slowly decreasing noise effects: global minimization via Monte Carlo, *SIAM J. Appl. Math.* **47** (1987), 169-185.

[42] H.J. Kushner, *Weak Convergence Methods and Singularly Perturbed Stochastic Control and Filtering Problems*, Birkhäuser, Boston, 1990.

[43] H.J. Kushner and D.S. Clark, *Stochastic Approximation Methods for Constrained and Unconstrained Systems*, Springer-Verlag, 1978.

[44] H.J. Kushner and H. Huang, Rates of convergence for stochastic approximation type of algorithms, *SIAM J. Control Optim.* **17** (1979), 607-617.

[45] H.J. Kushner and H. Huang, Asymptotic properties of stochastic approximations with constant coefficients, *SIAM J. Control Optim.* **19** (1981), 86-105.

[46] H.J. Kushner and A. Shwartz, An invariant measure approach to the convergence of stochastic approximations with state-dependent noise, *SIAM J. Control Optim.* **22** (1984), 13-27.

[47] H.J. Kushner and A. Shwartz, Stochastic approximation and optimization of linear continuous parameter systems, *SIAM J. Optim.* **23** (1985), 774-793.

[48] H.J. Kushner and F.J. Vázquez-Abad, Stochastic approximation algorithms for systems over an infinite horizon, *SIAM J. Control Optim.* **34** (1996), 712-756.

[49] H.J. Kushner and J. Yang, Stochastic approximation with averaging of the iterates: Optimal asymptotic rate of convergence for general processes, *SIAM J. Control Optim.* **31** (1993), 1045-1062.

[50] H.J. Kushner and J. Yang. Analysis of adaptive step-size sa algorithms for parameter tracking, *IEEE Trans. Automat. Control*, **40** (1995), 1403–1410.

[51] H.J. Kushner and G. Yin, Asymptotic properties of distributed and communicating stochastic approximation algorithms, *SIAM J. Control Optim.* **25** (1987), 1266-1290.

[52] H.J. Kushner and G. Yin, Stochastic approximation algorithms for parallel and distributed processing, *Stochastics*, **22** (1987), 219-250.

[53] H.J. Kushner and G. Yin, *Stochastic Approximation Algorithms and Applications*, Springer-Verlag, New York, 1997.

[54] T.L. Lai and H. Robbins, Consistency and asymptotic efficiency of slope estimates in stochastic approximation schemes, *Z. Wahr.* **56** (1981), 329-360.

[55] P. L'Ecuyer and G. Yin, Budget-dependent convergence rate of stochastic approximation, *SIAM J. Optim.* **8** (1998), 217-247.

[56] L. Ljung, Analysis of recursive stochastic algorithms, *IEEE Trans. Automat. Control* **AC-22** (1977), 551-575.

[57] L. Ljung, *System Identification: Theory for the User*, Prentice-Hall, NJ, 1987.

[58] M. Metropolis, A.W. Rosenbluth, M.N. Rosenbluth, A.H. Teller, and E. Teller, Equations of state calculations by fast computing machines, *J. Chem. Phys.* **21** (1953), 1087-1091.

[59] J.B. Moore, Convergence of continuous time stochastic ELS parameter estimation, *Stochastic. Process. Appl.* **27** (1988), 195-215.

[60] M.B. Nevel'son and R.Z. Khasminskii, *Stochastic Approximation and Recursive Estimation*, Translation of Math. Monographs, v47, AMS, Providence, 1976.

[61] G.Ph. Pflug, Stepsize rules, stopping times and their implementation in stochastic quasigradient algorithms, in *Numerical Techniques for Stochastic Optimization*, Springer-Verlag, Berlin, 1998, 353-372.

[62] G.Ch. Pflug, *Optimization of Stochastic Models*, Kluwer, Boston, MA, 1996.

[63] B.T. Polyak, New method of stochastic approximation type, *Automat. Remote Control* **51** (1990), 937-946.

[64] B.T. Polyak and Ya.Z. Tsypkin, Optimal pseudogradient adaptation procedures, *Automat. Remote Control*, **41** (1981), 1101-1110.

[65] I. Rechenberg, Cybernetic solution path of an experimental problem, Royal Aircraft Establishment, Library translation No. 1122, Farnborough, Hants., UK, 1965.

[66] P. Révész, How to apply the method of stochastic approximation in the non-parametric estimation of regression function, *Matem. Operations Stat. Ser. Statistics* **8** (1977), 119-126.

[67] H. Robbins and S. Monro, A stochastic approximation method, *Ann. Math. Statist.* **22** (1951), 400-407.

[68] D. Ruppert, A Newton-Raphson version of the multivariate Robbins-Monro Procedure, *Ann. Statist.* **13** (1985), 236-245.

[69] D. Ruppert, Efficient estimations from a slowly convergent Robbins-Monro process, Technical Report, No. 781, School of Oper. Res. & Industrial Eng., Cornell Univ., 1988. [see also the chapter Stochastic approximation in *Handbook in Sequential Analysis*, B.K. Ghosh and P.K. Sen Eds., 503-529, Marcel Dekker, New York, 1991.]

[70] G.I. Salov, Stochastic approximation theorem in a Hilbert space and its application, *Theory Probab. Appl.* **24** (1979), 413-419.

[71] R. Schwabe, Stability results for smoothed stochastic approximation procedures, *Z. angew. Math. Mech.* **73** (1993), 639-644.

[72] H.-P. Schwefel, *Kybernetische Evolution als Strategie der experimentellen Forschung in der Strömungstechnik*, Diploma thesis, Technical University of Berlin, 1965.

[73] H.-P. Schwefel, *Evolution and Optimum Seeking*, Wiley, New York, 1994.

[74] S. P. Sethi and Q. Zhang, *Hierarchical Decision Making in Stochastic Manufacturing Systems*, Birkhäuser, Boston, 1994.

[75] R. Sielken, Stopping Times for Stochastic Approximation Procedures, *Z. Wahrsch. verw. Gebiete*, **26** (1973), 67-75.

[76] V. Solo and X. Kong, *Adaptive Signal Processing Algorithms*, Prentice-Hall, Englewood Cliffs, NJ, 1995.

[77] J.C. Spall, Multivariate stochastic approximation using a simultaneous perturbation gradient approximation, *IEEE Trans. Automat. Control* **AC-37** (1992), 331-341.

[78] D.F. Stroup and H.I. Braun, On a new stopping rule for stochastic approximation, *Z. Wahrsch. verw. Gebiete*, **60** (1982), 535-554.

[79] J.N. Tsitsiklis, D.P. Bertsekas, and M. Athans, Distributed asynchronous deterministic and stochastic gradient optimization algorithms, *IEEE Trans. Automat. Control* **AC-31** (1986), 803-812.

[80] Ya.Z. Tsypkin, *Adaptation and Learning in Automatic Systems*, Academic Press, New York, 1971.

[81] J.H. Venter, An extension of the Robbins-Monro procedure, *Ann. Math. Statist.* **38** (1967), 181-190.

[82] H. Walk, An invariant principle for the Robbins Monro process in a Hilbert space. *Z. Wahrsch. verw. Gebiete* **62** (1977), 135-150.

[83] I.J. Wang, E. Chong, and S.R. Kulkarni, Equivalent necessary and sufficient conditions on noise sequences for stochastic approximation algorithms, *Adv. Appl. Probab.* **28** (1996), 784-801.

[84] M.T. Wasan, *Stochastic Approximation*, Cambridge Press, London, 1969.

[85] C.Z. Wei, Multivariate adaptive stochastic approximation, *Ann. Statist.* **15** (1987), 1115-1130.

[86] B. Widrow and S.D. Stearns, *Adaptive Signal Processing*, Prentice-Hall, Englewood, Cliffs, NJ, 1985.

[87] E. Wong, Stochastic neural networks, *Algorithmica* **6** (1991), 466-478.

[88] H.M. Yan, G. Yin, and S.X.C. Lou, Using stochastic optimization to determine threshold levels for control of unreliable manufacturing systems, *J. Optim. Theory Appl.* **83** (1994), 511-539.

[89] G. Yin, A stopping rule for least squares identification, *IEEE Trans. Automatic Control*, **34** (1988), 659-662.

[90] G. Yin, Asymptotic properties of an adaptive beam former algorithm, *IEEE Trans. Inform. Theory*, **IT-35** (1989), 859-867.

[91] G. Yin, A stopping rule for the Robbins-Monro method, *J. Optim. Theory. Appl.* **67** (1990), 151-173.

[92] G. Yin, On extensions of Polyak's averaging approach to stochastic approximation, *Stochastics* **36** (1991), 245-264.

[93] G. Yin, Recent progress in parallel stochastic approximations, *Topics in Stochastic Systems: Modelling, Estimation and Adaptive Control*, 159-184, (L. Gerencsér and P.E. Caines Eds.), Springer-Verlag, 1991.

[94] G. Yin, Convergence and error bounds for passive stochastic algorithms using vanishing step size, *J. Math. Anal. Appl.*, **200** (1996), 474-497.

[95] G. Yin, Rates of convergence for a class of global stochastic optimization algorithms, *SIAM J. Optim.*, **10** (1999), 99-120.

[96] G. Yin, Convergence of a global stochastic optimization algorithm with partial step size restarting, to appear in *Advances Appl. Probab.*

[97] G. Yin and I. Gupta, On a continuous time stochastic approximation problem, *Acta Appl. Math.*, **33** (1993), 3-20.

[98] G. Yin, P.A. Kelly, and M.H. Dowell, Approximation of an analog diffusion network with applications to image estimation, to appear in *J. Optim. Theory Appl.*

[99] G. Yin, G. Rudolph and H.-P. Schwefel, Analyzing $(1, \lambda)$ evolution strategy via stochastic approximation methods, *Evolutionary Comp.*, **3** (1996), 473-489.

[100] G. Yin and K. Yin, Asymptotically optimal rate of convergence of smoothed stochastic recursive algorithms, *Stochastics Stochastic Rep.*, **47** (1994), 21-46.

[101] G. Yin and K. Yin, Passive stochastic approximation with constant step size and window width, *IEEE Trans. Automat. Control* **AC-41** (1996), 90-106.

[102] G. Yin, K. Yin, B. Liu, and E.K. Boukas, A class of learning/estimation algorithms using nominal values: asymptotic analysis and applications, to appear in *J. Optim. Theory. Appl.*, 1999.

[103] G. Yin and Y.M. Zhu, On H-valued Robbins-Monro processes, *J. Multi. Anal.*, **34** (1990), 116-140.

[104] G. Yin and Y.M. Zhu, Averaging procedures in adaptive filtering: an efficient approach, *IEEE Trans. Automat. Control* **AC-37** (1992), 466-475.

[105] G. Yin and Q. Zhang, *Continuous-Time Markov Chains and Applications: A Singular Perturbation Approach*, Springer-Verlag, New York, 1998.

Chapter 11

Optimization by Stochastic Methods

FRANKLIN MENDIVIL, R. SHONKWILER, AND M.C. SPRUILL
Georgia Institute of Technology
Atlanta, GA 30332

11.1 Nature of the problem

11.1.1 Introduction

This chapter is about searching for the extremal values of an objective f defined on a domain Ω, possibly a large finite set, and equally important, for where these values occur. The methods used for this problem can be analyzed as finite Markov Chains which are either homogeneous or non-homogeneous, or as renewal processes. By an optimal value we mean globally optimal, for example f_* is the minimal value and $x_* \in \Omega$ a *minimizer* if

$$f(x_*) = f_*, \quad \text{and} \quad f_* \le f(x), \quad x \in \Omega.$$

Although we strive for the optimal value, this enterprise brings forth methods which rapidly find acceptably good values. Moreover, often knowing whether a value is the optimum or not cannot be answered with certainty. More generally, one might establish a *goal* for the search. It could of course be finding a global optimizer or it might be finding an x for which $f(x)$ is within a certain fraction of the optimum or it could be based on other criteria.

In this chapter we discuss stochastic methods to treat this problem. We assume the objective function f is deterministic and returns the same value for $f(x)$ every time. Thus this chapter is not about *stochastic optimization*, even though methods discussed here can be used with probabilistic objectives. Nevertheless we assume deterministic function evaluations throughout.

Difficult optimization problems arise all the time in such fields as science, engineering, business, industry, mathematics and computer science. Specialized methods such as gradient descent methods, linear and quadratic programming and others apply very well to certain well behaved problems. For a great many problems these specialized methods will not work and more robust techniques are called for.

By a difficult problem we mean, for example, one for which there is no natural topology, or for which there are a large number of local optima (multi-modal), or for which the solution space has high cardinality. The class of NP-complete problems of Computer Science, such as the Traveling Salesman Problem are examples of such problems.

One aspect of the search problem is knowing when the optimum value has been reached; at that point the search may stop. More generally, one may wish to stop the search under a variety of circumstances such as when a fixed time has expired, when a sufficiently good value has been found, or when the incremental cost of one more iteration becomes too great. This aspect of the problem is known as the *stopping time problem* and is beyond the scope of the present chapter.

Instead, throughout we assume that either the optimal value can be recognized if discovered or that one will settle for the best value found over the course of the search. Thus the second aspect of the search problem is knowing how to conduct the search as well as possible and how to analyze the search process itself dealing with such questions as how good is the best value obtained so far, how fast does the method converge, or is the method sure to find the optimum in finite time.

Some strengths of stochastic search methods are that: they are often effective, they are robust, they are easy to implement requiring minimal programming, and they are simply and effectively parallelized. Some weakness are that: they are computationally intensive and they engender probabilistic convergence assertions.

Heuristics are used extensively in global optimization. Arguments for introducing heuristics are presented in [75]; we quote from their paper:

> The need for good heuristics in both academia and business will continue increasingly fast. When confronted with real world problems, a researcher in academia experiences at least once the painful disappointment of seeing his product, a theoretically sound and mathematically 'respectable' procedure not used by its ultimate user. This has encouraged researchers to develop new improved heuristics and rigorously evaluate their performance, thus spreading further their usage in practice, where heuristics have been advocated for a long time.

In specialized applications a heuristic can embody insight or particular information about the problem. Heuristics can often be invoked to modify a given solution into a better solution, thereby playing the role of an improvement operator. And on a grand scale, heuristics derived from natural phenomena have given rise to entire search strategies.

11.1.2 No Free Lunch

There has long been evidence that a truly universal optimization algorithm is not really possible; you cannot have an algorithm that will perform equally well on all possible problems. However, there seemed to be little or no attention paid to this in the optimization literature until the seminal work of Wolpert and McReady [74]. The idea introduced in this paper is the so-called "No Free Lunch" idea. Simply put, this idea states that if you are interested in the average performance of an algorithm, averaged over the set of all possible objective functions, then any two algorithms have the same average performance. Thus, there is no way to distinguish between them. NFL type results point out the clear need to carefully match an algorithm type to problem type, since there is no universally effective optimization algorithm.

We now discuss the NFL Theorem as presented in [74]. Let X and Y be finite sets, X will be the domain space and Y the range space. Let $d_m = \{(x_i, y_i)\}_{i=0}^{m}$ be the domain-range pairs seen by the algorithm up until time m. Then an algorithm is a function a from the set of all such histories to $X \setminus \{$ x's in history $\}$. Notice that this means that we assume that the algorithm does not revisit any previously seen domain points. Let $\vec{c}$ be a histogram of Y values seen in some history d_m. From $\vec{c}$, one can derive various measures of the "performance" of the algorithm, for example the minimum value seen so far. Finally, let $P(\vec{c} \mid f, m, a)$ denotes the conditional probability that histogram $\vec{c}$ will be seen after m iterations of algorithm a on the function f.

Theorem 11.1.1 (NFL) *For any pair of algorithms a_1 and a_2,*

$$\sum_f P(\vec{c} \mid f, m, a_1) = \sum_f P(\vec{c} \mid f, m, a_2).$$

One way to understand this theorem is to consider optimization against an adversary which randomly generates the objective function as the algorithm proceeds (see [14]). Clearly the next objective function value seen by the algorithm is an independent sequence so the expected histogram generated by two different histories (algorithms) are equal. Thus, if we restrict ourselves to algorithms which do not revisit states, all algorithms, on average, perform as well as systematically stepping through the domain space in some pre-defined order.

In many instances it is clearly impractical to ensure that an algorithm only visits new states. Many of the algorithms we discuss in this chapter allow the possibility of visiting the same state multiple times. In fact, one of the main issues in the area is the problem of how to deal with long runs of repeating the same state. How important is the condition of no-retrace to the conclusion of the NFL Theorem?

In this chapter we will use, among other measures, the expected time to find the optimum value as a measure of the performance of an algorithm. Using such measures of performance, does the No Free Lunch Theorem hold for stochastic algorithms? The answer is no, not exactly in the stated form. As an example, suppose we have two algorithms driven by the following Markov transition matrices both on the state space $\{a, b, c\}$,

$$A = \begin{pmatrix} 0 & 1 & 0 \\ 0 & 0 & 1 \\ 1 & 0 & 0 \end{pmatrix} \qquad B = \begin{pmatrix} 1/3 & 1/3 & 1/3 \\ 1/3 & 1/3 & 1/3 \\ 1/3 & 1/3 & 1/3 \end{pmatrix}.$$

Then the expected time to reach the goal, averaged over all possible functions, is 2 for algorithm A and 3 for algorithm B. The first matrix drives the algorithm cyclicly through the state space while the second matrix generates a sequence of independent, uniformly random samples from the state space. Notice that the first algorithm does not repeat states while the second will with high probability. As a third example, the algorithm driven by the Markov transition matrix

$$D = \begin{pmatrix} 0 & 1/2 & 1/2 \\ 1/2 & 1/3 & 1/6 \\ 1/2 & 1/6 & 1/3 \end{pmatrix}$$

has average expected hitting time of $11/5$. Thus, it clearly is possible to do better than purely random search by using a stochastic algorithm.

For continuous state spaces, there is an interesting related result sometimes called the "indentation argument" [20]. This principle states that knowledge of only finitely many values of f or its derivatives and the fact that f has k continuous derivatives on some region $\Omega \subset \mathbb{R}^n$, is not sufficient to determine a lower bound on $\inf f(\Omega)$. The reason for this is that it is always possible to modify f on an arbitrarily small set, away from where we have information, in such a way that we decrease $\inf f(\Omega)$ by an arbitrary amount. This modification will have no measurable effect on the rest of the function in the sense that neither the value of the function nor the value of any of the derivatives of the function will change outside this small region of change. Thus, again, to get any advantage one needs to make assumptions on f, in this case global assumptions such as a global bound on a derivative.

11.1.3 The Permanent Problem

Consider the problem of optimizing the permanent of 0/1 matrices. The permanent of an $n \times n$ matrix M is defined to be

$$\text{perm}(M) = \sum_{\sigma} \prod_{i=1}^{n} m_{i,\sigma(i)},$$

where the sum extends over all permutations σ of the first n integers. The permanent is similar to the determinant except without the alternating signs. We will only allow the matrix elements to be 0 or 1. For a given matrix size n and number d of 1's, $0 < d < n^2$, the problem is to find the matrix having maximum permanent. We refer to this as the $n : d$ permanent problem.

Two advantages of this problem are its simplicity and scalability. The problem is completely determined by the two interger values, n and d, the matrix size and its density of 1's. As n grows the problem becomes harder in two ways. The number of operations required to calculate a permanent grows as $n!$. But in addition, the number of possible permutations, σ, also grows as $n!$. Likewise, the calculation difficulty also grows with d. This is a consequence of the fact that as soon as a zero is encountered in the permanent calculation, no further terms need be considered, the permanent is immediately 0.

Another advantage of the permanent problem is that there exists a body of literature describing simulated annealing solutions and with various cooling schedules, see [57].

To illustrate our in-depth methods: simulated annealing, restarted simulated annealing and evolutionary computation, we apply each to the 14:40 permanent problem. Details of each implementation will be given in the appropriate section. With $n = 14$, the number of matrix elements is 196; this gives the number of possible arrangements of the 1's, i.e. the size of the search space for this problem, to be $\binom{196}{40}$ or about 8.3×10^{41} (even though for a small proportion of the total, many of these arrangements give the maximum permanent). As mentioned above, the difficulty in calculating a permanent grows rapidly with n, and also with d. Selecting the aforementioned values for n and d, resulted in a permanent calculation sufficiently fast that millions could be tried in a reasonable amount of time.

The accompanying figure shows the results. The performance varied greatly; evidently this is an easy problem for an evolutionary computation.

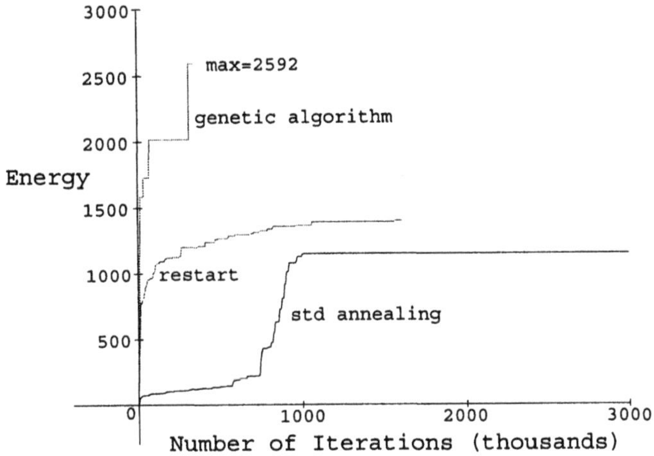

Figure 1. Best vs number of function evaluations for the three algorithms

11.2 A Brief Survey of Some Methods for Global Optimization

In subsequent sections we give in-depth discussions of simulated annealing, restart algorithms, and evolutionary computation. In this section we give brief descriptions of other representative methods.

A great many methods have been proposed for global optimization. These include grid-like subdivision methods, exhaustion, branch and bound, random search, and methods inspired by the natural world. The latter include simulated annealing and evolutionary computation. The methods given here by no means cover the field but rather are intended to be a sample of those available.

When the domain is a subset of Euclidean space, obviously its cardinality is infinite and there is no possibility of examining every point as would be possible (however impractical) when Ω is finite. On the other hand, objectives defined on Euclidean space usually have some degree of smoothness, for example a Lipschitz condition, which works in place of finiteness. An important illustration of this is in conjunction with searching for local optima; differentiable objectives can utilize gradients both to greatly improve search efficiency and to recognize attainment.

One classification of search methods is that proposed by [3], as follows:

Deterministic methods

- Covering methods
- Trajectory, tunneling methods

Probabilistic methods

- Methods based on random sampling
- Random search methods
- Methods based on a stochastic model of the objective

Although this chapter is about stochastic search methods, we include a discussion of some deterministic methods for comparison.

11.2.1 Covering Methods

An advantage of covering methods is that both aspects of the optimization problem are solved: finding an optimum and knowing that it is the optimum. In the case of discrete solution spaces, covering includes exhaustion and branch and bound methods.

Exhaustion entails computing the objective value on each and every point of the domain, which is often not feasible. The points of Ω are totally ordered in some way, $x_1, x_2, \ldots, x_N$, the objective values are computed, $y_1 = f(x_1), y_2 = f(x_2), \ldots, y_N = f(x_N)$ and compared,

$$f_* = \min_i y_i.$$

We note that the expected number of iterations, E, required to find the optimum is

$$E = \frac{N+1}{2}$$

assuming the ordering of the domain is uncorrelated with the objective. Therefore the average expected hitting time over all possible functions on Ω is $(N+1)/2$. The No Free Lunch conjecture is that this value is the best possible no matter what the search strategy.

A comparable algorithm for Euclidean spaces is the following [62].

1) Evaluate f at n equispaced points $x_1, \ldots,\ x_n$ throughout Ω and define $y_i = f(x_i)$, $i = 1, \ldots, n$.

2) Estimate f_* by $m_n = \min\{y_1, \ldots, y_n\}$.

Under the conditions that f satisfies a Lipschitz condition,

$$|f(x_1) - f(x_2)| < L\|x_1 - x_2\|, \qquad x_1, x_2 \in \Omega,$$

for some fixed $L > 0$, then the following can be proved. Given $\epsilon > 0$, define the goal to be the region

$$G_\epsilon = \{y : |y - f_*| < \epsilon\}.$$

Theorem 11.2.1 *For $i = 1, \ldots, n$, let V_i be the sphere $\|x - x_i\| \le r_i$, where*

$$r_i = \frac{|f(x_i) - m_n + \epsilon|}{L}.$$

If $\bigcup_i V_i$ covers Ω, then $m_n = \min\{y_1, \ldots, y_n\}$ belongs to G_ϵ.

Proof. Let $x_* \in V_i$, then $|x_* - x_i| < r_i$ so

$$f(x_i) - f(x_*) \le Lr_i = f(x_i) - m_n + \epsilon$$

and thus

$$f(x_*) \ge m_n - \epsilon,$$

so the goal is attained.

Prior to computing $f(x_i)$ it is impossible to determine whether $\bigcup_i V_i$ will cover all of Ω. Thus, with the x_i fixed, it is necessary to increase ϵ until this condition is satisfied. If this increases ϵ to be larger than the allowable error tolerance, the only solution is to choose more x_i's and try again.

If it is possible to obtain some additional information, then it may only be necessary to obtain more samples in parts of the space, rather than uniformly over all of Ω. This is the motivation behind another type of method in this general category, the class of subdivision algorithms. Subdivision methods are usually applied to a region $\Omega \subset \mathbf{R}^n$. The domain is covered with a coarse regular pattern, e.g. a grid, and the function values on the nodes are compared. Promising areas of the space are refined by laying down a finer grid in these areas. Finding these promising areas usually depends on some additional knowledge of the function, such as a Lipschitz condition. If the function is assumed to be Lipschitz then the values at the corners of a grid will yield estimates on the possible values inside the grid blocks, and thus identify the promising areas.

Notice that care must be taken in the estimation of the Lipschitz factor of the objective function since a poor estimate of this factor will adversely affect the performance of the algorithm.

11.2.2 Branch and bound

Branch and bound methods are related to the subdivision methods discussed above. The basic idea is to partition the domain (Branch) into various regions and obtain estimates (Bounds) on the minimum function value over these regions. Depending on the quality of the bounds, you can then eliminate some of the regions from further consideration, thus narrowing the search. In this section, we describe the general framework of a branch and bound method, leaving the details to the references [39].

The three primary operations in the branch and bound algorithm are Bounding, Selection and Refining.

Suppose our problem domain Ω is a subset of a larger set X. For example, Ω might be all the points in X that satisfy some set of constraints. At each stage of the algorithm, we have a partition $\mathcal{M}_k$ of a subset of X which contains Ω and for each element of the partition $M \in \mathcal{M}_k$, bounds $\alpha(M)$ and $\beta(M)$ such that $\beta(M) \leq \inf f(M \cap \Omega) \leq \alpha(M)$. These "local" bounds give us overall bounds, $\alpha_k = \min \alpha(M)$ and $\beta_k = \min \beta(M)$ which yield the bounds $\beta_k \leq \inf f(\Omega) \leq \alpha_k$.

To initialize the algorithm, start with the partition which consists solely of the one set X. To obtain the lower bound β_0, we use our bounding operation to compute a $\beta(X)$ such that $\beta(X) \leq \inf f(\Omega)$. To do this, we only need an underestimate of the minimum of f over X and we can choose X to make this estimate easier. For example, we could choose X to be a convex polytope and find some convex $\phi \leq f$ so that $\beta(X) = \inf \phi(X)$.

Finding the bound α involves taking an "inner" approximation $S_X \subset \Omega$ over which we can find the minimum of f. Thus, $\alpha_0 = \inf f(S_X)$. Let x_0 be the point at which this minimum is achieved.

Having initialized the algorithm, the next steps involve updating the partition $\mathcal{M}_k$, the bounds α_k, β_k , the best feasible point seen so far, x_k, and the "inner" approximation $S_{\mathcal{M}_k}$.

The first step is to remove any $M \in \mathcal{M}_k$ which are either infeasible (so that $M \cap \Omega = \emptyset$) or which we know cannot possibly contain the global minimum. This step is very important to the efficiency of the algorithm, since the more regions we can eliminate early in the algorithm, the better will be our estimate of the location of the global minimum. Note that deciding if $M \cap \Omega = \emptyset$ could be difficult, depending on the structure of the problem. However, clearly if $\beta(M) \geq \alpha_k$, we know that M does not contain the global minimum and so can be safely removed.

The next step is the selection step, where we choose which elements of the partition $\mathcal{M}_k$ are subdivided. The choice of selection rule will also be problem dependent. Examples of natural and simple selection rules are to select the "oldest" region(s) or the "largest" region(s). Both of these selection rules have the desirable property that eventually all remaining regions will be selected for subdivision.

The refining operation depends very much on the choice of sets for the partition. If the elements of the partition are convex polytops or simplices, then a natural choice of refinement is to do a simplicial refinement, where each simplex is subdivided into several smaller simplices. After we refine the regions which were selected, we again remove any of these new subregions which are either infeasible or which do not contain the minimum.

Now we must update the bounds α and β and the "inner" approximating set S. Let $\mathcal{M}_{k+1}$ be the partition consisting of the remaining sets (note that we know that $\Omega \subset \bigcup\{M : M \in \mathcal{M}_{k+1}\}$). For each $M \in \mathcal{M}_{k+1}$, we find a set $S_M \subset \Omega \cap M$. Furthermore, we use the bounding operation to find a number $\beta(M)$ so that $\beta(M) \leq \inf f(M \cap \Omega)$ if M is known to be feasible or $\beta(M) \leq \inf f(M)$ if M is uncertain. In order for our bounds to be useful and to have a chance of converging, we must choose S_M and $\beta(M)$ in such a way that if $M' \in \mathcal{M}_k$ with $M \subset M'$ (or if M is part of a refinement of M') then $S_M \supset M \cap S_{M'}$ (our "inner" approximation is growing) and $\beta(M) \geq \beta(M')$ (our lower bound is increasing). The bound $\alpha(M)$ is defined as $\alpha(M) = \inf f(S_M)$.

The overall bounds α_{k+1} and β_{k+1} are defined to be the minimum of the $\alpha(M)$'s and $\beta(M)$'s, respectively, for M in the current partition, $\mathcal{M}_{k+1}$. We update x_{k+1}, the best feasible solution seen so far, as the place where $f(x) = \alpha_{k+1}$.

Now if $\alpha_{k+1} - \beta_{k+1}$ is smaller than some error tolerance, then the algorithm has found the minimum to within this tolerance and x_{k+1} is taken to be the location of the minimum.

If the state space Ω is finite, then the 'branch and bound' is likewise exhausted except that the search proceeds in such a way that, with the points of the domain organized in a tree graph, if the function value at some node of the graph is sufficiently worse than the running best, then the rest of the branch below that point need not be searched. Clearly the difficulty is setting up the algorithm in such a way as to obtain the estimates of α and β over a region in Ω.

11.2.3 Iterative Improvement

An *iterative improvement*, or *greedy*, algorithm is one in which the successive approximations are monotonically decreasing. It is a deterministic process: if run twice starting from the same initial point, the same sequence of steps will occur. Given the current state, greedy algorithms function by taking the best point in the neighborhood of the current state, including the current state. The *neighborhood* of a state is defined to be the set of all the states that could potentially be reached in one step of the algorithm starting from the current state. If the state space is thought of as a graph, the neighbors of a state σ are all those states that are connected to σ by an edge. Eventually the greedy algorithm reaches a local minimum relative to its neighborhood system and no additional improvement is possible. When this occurs the algorithm must stop.

By its nature, an iterative improvement algorithm partitions the solution space into basins. A *basin* being all those points leading to the same local minimum. Graph theoretically, an iterative improvement algorithm can be represented as a forest of rooted trees. Each tree corresponds to a basin, with the root of the tree the local minimum of the basin.

For problems having a differentiable objective function, the gradient is generally used to compute the downhill steps required for improvement. In discrete problems, one has a "candidate neighborhood system," that is, each point has a neighborhood of candidates. Candidates are examined until one is found which most improves the objective value and that one becomes the next iteration point. Various heuristics are used for assigning a candidate neighborhood; this is the primary concern in designing a greedy algorithm for a specific problem. For example, when the domain is a Cartesian space of some sort, neighbors can be the one-coordinate perturbations of the present point.

By their very nature, greedy algorithms are good at finding local minima of the objective function. In order to find the global minimum, it is necessary to find the *goal basin*, that is the basin containing the global minimum. Clearly once this basin has been located, the algorithm will deterministically descend into the basin to find the global minimum.

Greedy algorithms generally rely on some knowledge of the problem in order to define natural and reasonable candidate neighborhood systems. Since only downhill steps are taken, it is desirable to have a heuristic which generates downhill steps with high probability.

11.2.4 Trajectory/tunneling Methods

Trajectory methods depend on the function f being defined on a smooth subset of $\mathbf{R}^n$. Given this setting, the method then constructs a set of finitely many curves in such a way that it is known that the solutions lie on one or more of these curves. An example might be to find the critical points of a function, and these points lie on the curves defined by setting all but one of the partial derivatives to be zero. Given these curves, we must find a starting point on an appropriate curve and then trace out the curve. Tracing out the curve often involves setting up a system of differential equations which define the curve and numerically solving this set of equations (thus, the curve is the trajectory of a solution to a differential equation). A physical analogy would be, thinking of the function f as a surface, to roll a marble over this surface in order to find the valleys.

Another set of methods closely related to trajectory methods are homotopy methods. In this method you choose a related, but easier, function g to minimize. You compute the minimizers of g. Then you find a homotopy between the function g and the function f. A homotopy between g and f is a continuous function $H : [0,1]\times\Omega \to \mathbf{R}$ so that $H(0,x) = g(x)$ while $H(1,x) = f(x)$; it is like a continuous "path" from g to f. The idea is that we follow the minimizers of $H(t,x)$ for each t. If we can do this all the way up to $t = 1$, then we have found the minimizers of $H(1,\cdot) = f$.

Clearly the major task in using a homotopy method is choosing the simpler function g and, most importantly, the homotopy H. It is necessary to be able to find the minima for $H(t,\cdot)$ for each t. The algorithm usually proceeds by finding the minima for $g = H(0,\cdot)$. Let us increment t by some fixed, but small amount. Then $H(t,\cdot)$ is very close to $H(0,\cdot)$ so the minima will be very similar. Thus, the minima for g are good starting points to use in an algorithm to find the minima for $H(t,\cdot)$. Continuing this way, we eventually arrive at $t = 1$ and, hopefully, the solution to the original minimization problem.

Tunneling methods involve finding a local minimum and then "tunneling" through the surrounding "hill" to find a point in the basin of another local minimum. For simplicity we describe the algorithm for a one-dimensional problem. Let f be defined on an interval $[a,b]$. Given the local minimum x_1, next find a new point z_1 by minimizing the "tunneling" function

$$T_\alpha(x) = \frac{f(x) - f(x_1)}{[(x - x_1)^2]^\alpha}.$$

This minimization is started with a point to the right of x_1. If $f(z_1) < f(x_1)$, then z_1 belongs to the basin of a local minimum with lower function value than x_1, and thus a new local search can begin to obtain the local minimum x_2. On the other hand, if $f(z_1) > f(x_1)$, then increase α and either find such a point or obtain the end point b. In this case, we know that no such local minimum exists to the right of x_1, so try points to the left of x_1.

In the multidimensional case, the tunneling function is changed to one of the form [69]

$$T_\alpha(x) = \frac{f(x) - \hat{f}_*}{\left\{\prod_{i=1}^K [(x - x_i)^T (x - x_i)]^{\alpha_i}\right\} [(x - x_m)^T (x - x_m)]^{\alpha_0}}$$

where the x_i are all the local minima with best minimum value f_* found during the previous iterations.

11.2.5 Tabu search

Tabu search is a modification of iterative improvement to deal with the problem of premature fixation in local minima. Allowing an algorithm to take "uphill" steps helps avoid this entrapment. However, this makes it possible for the algorithm to loop between several states and waste computational time. Thus, some mechanism is needed to discourage this. A Tabu search implements this by having the neighborhood system dynamically change as the search progresses. One possibility is for a tabu search to maintain a list of recently visited states and use these as "tabu" states not to revisit. This results in the actual neighbors of a state being the potential neighbors minus these tabu points.

Tabu search works by first generating a neighborhood of the current state. Then the best non-tabu element of this neighborhood is taken to be the next state. In certain situations, it may be desirable to accept a tabu state, for example if a proposed tabu state is much better than any previously seen. To allow this possibility, tabu search may includes an *aspiration level condition*, by which is meant some criteria to judge whether to allow a tabu transition.

Another possible refinement of a basic tabu search is to incorporate some type of learning into the generation of the local neighborhood. For example, a problem-dependent heuristic could favor states that look like recently seen good states.

Like all search algorithms, tabu search performs better the more problem specific information is encoded into the procedure. This is especially important in determining how to dynamically generate the local neighborhood.

11.2.6 Random Search

The simplest probabilistic method is *random search* which consists of selecting points in Ω uniformly at random for n such points. The function values are computed and the best such value encountered is reported. Suppose a goal has been established and the probability of hitting the goal is θ_0 under uniform selection over Ω. Then the probability the goal has not been found after n trials is $(1-\theta_0)^n$. Hence the probability of success is

$$S = 1 - (1-\theta_0)^n.$$

Solving for n

$$n = \frac{\log(1-S)}{\log(1-\theta_0)}.$$

The following table illustrates this equation.

Iterations needed for 90 or 99 % success using random search

	Probability of success per iteration, θ_0					
	1/20	1/50	1/100	1/1,000	1/10,000	1/100,000
90%	45	114	230	2,302	23,025	230,258
99%	90	228	459	4,603	46,050	460,515

One interpretation of the table is that in searching for 1 point from among 100,000, to attain 90% chance of success requires about 230,000 iterations or over two times as many points as in the space, clearly undesirable. Another interpretation of the same information is that in searching for 10 points from among 1,000,000, to attain 90% chance of success one needs about 230,000 iterations, about one fourth of the points, which is much better.

For the permanent problem described above, $\theta_0 \leq (14!)^2/\binom{196}{40} \approx 9.2 \times 10^{-19}$. Thus we need about 5×10^{20} iterations to be 99% sure that we have found the goal.

11.2.7 Multistart

Although we study restart methods in-depth in a subsequent section, at this point we will mention multistart (see [66]) which is a specialized "batch oriented" restart method. Multistart combines random search and iterated improvement (greedy algorithms) in a natural way. The method begins by choosing some number of random points x_i uniformly in Ω. From each of these points a local search is performed, yielding the local minima y_i. From here, we can either terminate the algorithm, taking the best of the local minima along with its corresponding minimizer as the output, or we can choose to sample some more points and perform further local searches starting from these new points.

As already mentioned, one aspect of optimal search is deciding when to stop. For multistart, some simple stopping rules have been derived (see [8]) which are based on a Bayesian estimation of both the total number of local minima (and, thus, an estimate on the percentage of these already visited) and the percentage of Ω that has been covered by the basins of these local minima. These estimates are given by

$$\frac{w(s-1)}{s-w-2} \quad \text{and} \quad \frac{(s-w-1)(s+w)}{s(s-1)}$$

as the estimates of the number of local minima and percentage of Ω covered, respectively. In these formula, w is the number of distinct minima found and s is the number of local searches performed (the number of x_i sampled randomly).

11.3 Markov Chain and Renewal Theory Considerations

Generally, optimization methods are iterative and successively approximate the extremum although the progress is not monotonic. For selecting the next solution approximation, most search algorithms use the present point or, in some cases, short histories of points or even populations of points in Ω. As a result, these algorithms are described by finite Markov chains over Ω or copies of Ω.

Markov Chain analysis can focus attention on important factors in conducting such a search, such as irreducibility, first passage times, mixing rates and others, and can provide the tools for making predictions about the search such as convergence rates and expected run times.

Associated with every Markov Chain is its *directed weighted connection graph* whose vertices are the states of the chain and whose edges are the possible transitions weighted by the associated, positive, transition probabilities. The graph defines a *topology* on the state space in terms of neighborhood systems in that the possible transitions from a given state are to its *neighbors*. By ordering the states of the chain in some fashion, $\{x_1, x_2, \ldots, x_N\}$, an equivalent representation is by means of the *transition matrix* $P(t)$,

$$P(t) = (p_{ij}(t))$$

in which $p_{ij}(t)$ is the probability of a transition from state x_i to state x_j on iteration t. If P is constant with t, the chain is *homogeneous*, otherwise *inhomogeneous*. We first consider homogeneous chains.

Retention and Acceleration

Let α_t, the *state vector*, denote the probability distribution of the chain X_t on iteration t; α_0 denotes the starting distribution. If the starting solution is chosen equally likely, α_0 will be the row vector all of whose components are $1/N$. The successive states of the algorithm are given by the matrix product

$$\alpha_t = \alpha_{t-1}P$$

and hence

$$\alpha_t = \alpha_0 P^t.$$

Now let a subset of states be designated as *goal* states, G. It is well-known that the expected hitting time E to this subset can be calculated as follows. Let $\hat{P}$ denote the matrix which results from P when the rows and columns corresponding to the goal are deleted, and let $\hat{\alpha}_t$ denoted the vector that remains after deleting the same components from α_t. Then the expected hitting time is given by

$$E = \hat{\alpha}_0(I - \hat{P})^{-1}\mathbb{1}$$

where $\mathbb{1}$ is the column vector of 1's.

This equation may be re-written as the Neumann series

$$E = \hat{\alpha}_0(I + \hat{P} + \hat{P}^2 + \hat{P}^3 + \ldots)\mathbb{1},$$

the terms of which have an important interpretation. The sum $\hat{\alpha}_t\mathbb{1}$ is exactly the probability that the process will still be "retained" in the non-goal states on the tth epoch. Since $\hat{\alpha}_0\hat{P}^t = \hat{\alpha}_t$, the term

$$chd(t) = \hat{\alpha}_0\hat{P}^t\mathbb{1}$$

calculates this retention probability. We call the probabilities $chd(\cdot)$ of not yet seeing the goal by the tth epoch the *tail probabilities* (not to be confused with measure-theoretic notions of the same name) or the *complementary hitting distribution*,

$$chd(t) = \Pr(\text{hitting time} > t), \quad t = 0, 1, \ldots.$$

In terms of $chd(\cdot)$,

$$E = \sum_0^\infty chd(t).$$

If now the sub-chain consisting of the non-goal states is irreducible and aperiodic, virtually always satisfied by these search algorithms, then by the Perron-Frobenius theorem,

$$\hat{P}^t \to \lambda^t \chi\omega \qquad \text{as} \quad t \to \infty$$

where χ is the right and ω the left eigenvectors for the principle eigenvalue λ of $\hat{P}$. The eigenvectors may be normalized so that $\omega\mathbb{1} = 1$ and $\omega\chi = 1$. Therefore asymptotically,

$$chd(t) \to \frac{1}{s}\lambda^t \qquad t \to \infty$$

where $1/s = \hat{\alpha}_0\chi$.

The left eigenvector ω has the following interpretation. Over the course of many iterations, the part of the process which remains in the non-goal sub-chain asymptotically tends to the distribution ω. The equation $\omega\hat{P} = \lambda\omega$ shows that λ is the probability that on one iteration, the process remains in the non-goal states.

The right eigenvector χ likewise has an interpretation. Since the limiting matrix is the outer product $\chi\omega$, χ is the vector of row sums of this limiting matrix. Now given any distribution vector $\hat{\alpha}$, its retention under one iteration is $\hat{\alpha}\chi\omega\mathbb{1} = \hat{\alpha}\chi$. Thus χ is the vector of relative retention values. To quickly pass from non-goal to goal states, $\hat{\alpha}$ should favor the components of χ which are smallest. Moreover, the dot product $\hat{\alpha}\chi$ is the expected retention under the distribution $\hat{\alpha}$ relative to retention, λ, under the limiting distribution.

If it is assumed that the goal can be recognized and the search stopped when attaining the goal, then we have the following theorem.

Theorem 11.3.1 *The convergence rate of a homogeneous Markov Chain search is geometric, i.e.,*

$$\lambda^{-t} Pr(X_t \notin G) \to \frac{1}{s} \quad \text{as } t \to \infty,$$

provided that the sub-chain of non-goal states is irreducible and aperiodic.

On the other hand, if goal states are not always recognized, then we may save the *best* state observed over the course of a run, the *best-so-far* random variable, see [63]. We define this to be the random variable over the chain which is the first to attain the current extreme value,

$$B_t = X_r, \quad f(X_r) \le f(X_k) \ \ 1 \le k \le t, \quad r \le k \text{ if } f(X_r) = f(X_k).$$

Now if the goal is defined in terms of objective values, then the theorem takes the following form showing that convergence is almost sure (cf. Hajek's Theorem 11.4.1).

Theorem 11.3.2 *The convergence rate of the best observation is geometric,*

$$\lambda^{-t} Pr(B_t \notin G) \to \frac{1}{s} \quad \text{as } t \to \infty,$$

provided that the sub-chain of non-goal states is irreducible and aperiodic.

Making the asymptotic substitutions for $chd(\cdot)$ in the expression for the expected hitting time, E becomes

$$\begin{aligned} E &\approx \frac{1}{s}(1 + \lambda + \lambda^2 + \dots) \\ &= \frac{1}{s}\frac{1}{1-\lambda} \end{aligned}$$

where the infinite series has been summed. We therefore arrive at the result that two scalar parameters govern the convergence of the process, *retention* λ and *acceleration* s. In most applications λ is just slightly less than 1 and s is just slightly more than 1.

In cases where repeated runs are possible, retention and acceleration can be estimated from an empirical graph of the complementary hitting distribution. Plotting $\log(chd)$ vs t gives, asymptotically, a straight line whose slope is λ and whose intercept is $-\log s$.

It is also possible to estimate retention and acceleration during a single run dynamically for the restarted iterative improvement algorithm. We discuss this further below.

The tail probabilities may also be used to calculate the *median* hitting time M. Since M is the time t such that it is just as likely to take more than t iterations as less than t, we solve for M such that $chd(M) = .5$. Under the asymptotic approximation for $chd(\cdot)$, this becomes

$$\frac{1}{s}\lambda^{M-1} = chd(M) = .5$$

from which

$$M = 1 + \frac{\log(s/2)}{\log(\lambda)}.$$

11.3.1 IIP parallel search

A major virtue of Monte Carlo methods is the ease of implementation and efficacy of parallel processing. The simplest and most universally applicable technique is parallelization by identical, independent processes, (IIP) parallel. When used for global optimization, this technique is also highly effective. (IIP) parallel is closely related to a parallelization technique for multistart in [9]. What we are calling (IIP) parallelization is referred to as *simultaneous independent search* (SIS) in [4].

One measure of the power of (IIP) parallel is seen in its likelihood of finding the goal. Suppose that a given method has q probability of success. Then running m instances of the algorithm increases the probability of success as given by

$$1-(1-q)^m.$$

This function is shown in Figure 2. For example, if the probability of finding a suitable objective value is only $q = 0.001$, then running it 400 times increases the likelihood of success to over 20%, and if 2,000 runs are done, the chances exceed 80%.

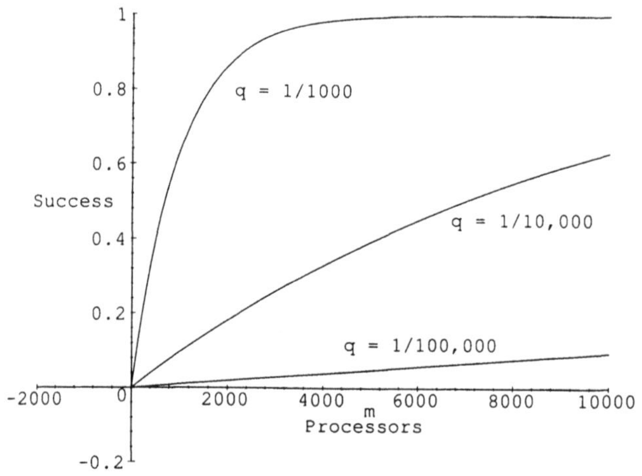

Figure 2. Probability of success vs number of parallel runs

For the purposes of this figure, the runs need not be conducted in parallel. But when they are, another benefit ensues – the possibility of superlinear speedup. By independence, the joint expected hitting time $E(m)$, meaning the expected hitting time of the first to hit, of the parallel processes is given by

$$\begin{aligned} E(m) &= \hat{\alpha}_0(I-\hat{P}^m)^{-1}\mathbb{1} \\ &= \hat{\alpha}_0(I+\hat{P}^m+(\hat{P}^m)^2+(\hat{P}^m)^3+\dots)\mathbb{1}. \\ &\approx \frac{1}{s^m}\frac{1}{1-\lambda^m}. \end{aligned}$$

If we define *speedup* $SU(m)$ to be relative to the single-processor running time, we find

$$\begin{aligned} SU(m) &= \frac{E(1)}{E(m)} \\ &\approx s^{m-1}\frac{1-\lambda^m}{1-\lambda} \\ &\approx s^{m-1}m \end{aligned}$$

where the last member follows for λ near 1.

For s and λ near one, the speed up curve will show the usual drop off with increasing m. But if s is on the order of 1.01 or bigger, then speedup will be superlinear for up to several processors. See Figure 3.

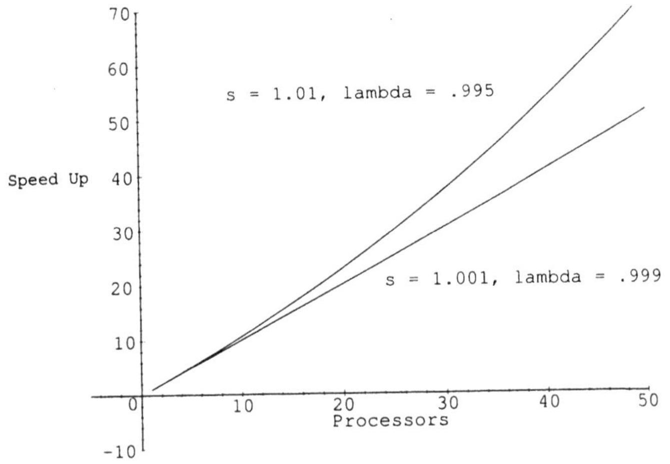

Figure 3. Speed up vs number of processors

These results show that IIP parallel is an effective technique when $s > 1$ accelerating convergence superlinearly. See reference [67].

The convergence rate for (IIP) is worked out for simulated annealing in [4]. Suppose for a given problem, N iterations in total are available and assume m parallel processes will each conduct n iterations, $mn < N$. The m processes are assumed to use the same search algorithm but otherwise are independent. Let Y_N denote the best overall ending configuration among the parallel runs, $B_{i,n}$, $1 \le i \le m$, i.e.

$$Y_N = B_{1,n} \wedge B_{2,n} \wedge \cdots \wedge B_{m,n}.$$

Then Y_N satisfies

$$\Pr(Y_N \notin G) \le \left(\frac{mK}{N}\right)^{\alpha m}$$

for some $K > 0$ and $\alpha > 0$.

A modification of (IIP) parallel, termed *periodically interacting simultaneous search*, is also treated in [4]. The method as defined there is cast in terms of simulated annealing but can be adapted to any search algorithm. Here, m processors independently undergo $s - 1$ iterations resulting in configurations $X_{s-1,1}, \ldots, X_{s-1,m}$. The next state for the kth processor, $X_{s,k}$ will be the best of the first k of these, i.e.

$$X_{s,k} = X_{s-1,1} \wedge \cdots \wedge X_{s-1,k}, \qquad 1 \le k \le m.$$

11.3.2 Restarted Improvement Algorithms

We envision a process combining a deterministic downhill operator g, acting on points of the solution space, and a uniform random selection operator U. The process starts with an invocation of U resulting in a randomly selected *starting* point. This is followed by repeated invocations of g until a local minimum is reached. Then the process is restarted with another invocation of U and so on.

As above, this process enforces a topology on the domain which is a forest of trees. The domain is partitioned into *basins* B_i, $i = 0, 1, \ldots$ as determined by the equivalence relation $x \equiv y$ if and only if $g^k(x) = g^j(y)$ for some k, j. The *settling point* or *local minimum* b of basin B is $\lim_{k\to\infty} g^k(x)$ where x is any point of B. By the *depth* of a tree we mean its maximum path length.

The transition matrix for such a process assumes the following form

$$P = \begin{bmatrix} B_0 & 0 & \ldots & 0 \\ Q & B_1 & \ldots & Q \\ \vdots & \vdots & \ddots & \vdots \\ Q & Q & \ldots & B_n \end{bmatrix},$$

where, to conserve notation, we also use B_i to denote the matrix corresponding to basin B_i. We index the points starting with the goal basin. Within a basin we index points with increasing path length from the basin bottom. Then each sub-matrix B_i has the form

$$B_i = \begin{bmatrix} p & p & p & \ldots & p \\ 1 & 0 & 0 & \ldots & 0 \\ 0 & 1 & 0 & \ldots & 0 \\ \vdots & \vdots & \vdots & \ddots & \vdots \\ 0 & 0 & \ldots & 1 & 0 \end{bmatrix},$$

where $p = 1/N$ corresponds to uniform restarting. The 1's in this matrix are in the lower triangle but not necessarily on the sub-diagonal. The blocks designated by Q are generic for the form

$$Q = \begin{bmatrix} p & p & \ldots & p \\ 0 & 0 & \ldots & 0 \\ \vdots & \vdots & \ddots & \vdots \\ 0 & 0 & \ldots & 0 \end{bmatrix}.$$

Let E denote the expected hitting time to the basin B_0 containing a minimizer, the *goal basin*. Let T_i be the expected time to reach the settling point of basin B_i. Let $|B_i|$ denote the number of points in basin B_i and θ_i the ratio $|B_i|/N$ where $N = \sum |B_i|$, i.e. θ_i is the probability of landing in basin B_i on a restart. Then by decomposition of events

$$E = \theta_0 + (1 + T_1 + E)\theta_1 + \cdots + (1 + T_n + E)\theta_n \tag{11.3.1}$$

or

$$E = \frac{1}{\theta_0}\left(1 + \sum_{i=1}^{n} T_i\theta_i\right). \tag{11.3.2}$$

As above E is also asymptotically given by

$$E = \frac{1}{s}\frac{1}{1-\lambda}.$$

Because of the special structure of P in this case, both retention and acceleration can be calculated directly.

Solving for λ and s, the Fundamental Polynomial

In the forest of trees model, it is clear that all states which are a given number of steps from a settling point are equivalent as far as the algorithm is concerned. Let $r_j(i)$ be the number of vertices j steps from the local minimizer of basin i and let $r_j = \sum_{i=1}^{n} r_j(i)$ denote the total number of vertices which are j steps from a local minimizer. In particular, $r_0 = n$ is the number of local minimizers.

Therefore the given forest of trees model in which each vertex counts 1 is equivalent to a single, linear tree in which each vertex counts equal to the number of vertices in the original forest which are at that distance from a settling point. Under the equivalency, the $\hat{P}$ matrix becomes

$$\hat{P} = \begin{bmatrix} p_0 & p_1 & p_2 & \dots & p_{n-1} & p_n \\ 1 & 0 & 0 & \dots & 0 & 0 \\ 0 & 1 & 0 & \dots & 0 & 0 \\ 0 & 0 & 1 & \dots & 0 & 0 \\ \vdots & \vdots & \vdots & \ddots & \vdots & \vdots \\ 0 & 0 & 0 & \dots & 1 & 0 \end{bmatrix}. \tag{11.3.3}$$

In this, $p_i = r_i/N$ where, as above, N is the cardinality of the domain. It is easy to calculate the characteristic polynomial of this matrix directly; expand $\det(\hat{P} - \lambda I)$ by minors along the first row,

$$-\lambda^{n+1} + p_0\lambda^n + p_1\lambda^{n-1} + \cdots + p_{n-1}\lambda + p_n.$$

Upon setting $\eta = 1/\lambda$ we get a polynomial we will refer to as the *fundamental polynomial*

$$f(\eta) = p_0\eta + p_1\eta^2 + \cdots + p_{n-1}\eta^n + p_n\eta^{n+1} - 1. \tag{11.3.4}$$

Notice that the degree of the fundamental polynomial is equal to the depth of the deepest basin.

As above, letting θ_0 be the probability of landing in the goal basin, then

$$\theta_0 + p_0 + p_1 + \cdots + p_n = 1.$$

From this we see that $f(1) = -\theta_0$ and

The derivative $f'(\eta)$ is easily seen to be positive for $\eta \geq 0$ and hence the fundamental polynomial will have a unique greater than 1 root. Denote it by η; it is the reciprocal of the Perron-Frobenius eigenvalue λ.

To calculate the acceleration s, we first find the left and right eigenvectors. The right Perron-Frobenius eigenvector, χ, of $\hat{P}$ is easily calculated. From (11.3.3) we get the recursion equations

$$\chi_k = \lambda\chi_{k+1} \qquad k = 0, \ldots, n-1.$$

And so each is given in terms of χ_0,

$$\chi_k = \eta^k\chi_0, \qquad k = 1, \ldots, n.$$

Similarly, we get recursion equations for the components of ω in terms of ω_0,

$$\omega_k = \omega_0(\eta p_k + \eta^2 p_{k+1} + \cdots + \eta^{n+1-k}p_n).$$

Recalling the normalizing conditions $\sum \omega_i = 1$, it follows that

$$\omega_0 = \frac{\eta - 1}{\eta\theta_0}.$$

And under the normalization, $\sum \omega_i\chi_i = 1$, it follows that

$$\chi_0 = \frac{1}{\omega_0\eta f'(\eta)} = \frac{\theta_0}{(\eta - 1)f'(\eta)}.$$

But $s = 1/(\chi \cdot \hat{\alpha}_0)$ where $\hat{\alpha}_0$ is the non-goal partition vector of the starting distribution,

$$\hat{\alpha}_0 = \begin{pmatrix} p_0 & p_1 & \ldots & p_n \end{pmatrix}.$$

Substituting from above, we get

$$s = \frac{\eta(\eta - 1)f'(\eta)}{\theta_0}.$$

Run time estimation of retention, acceleration and hitting time

Returning to the fundamental polynomial, we notice that its coefficients are the various probabilities for restarting a given distance from a local minimum. Thus the linear coefficient is the probability of restarting on a local minimum, the quadratic coefficient is the probability of restarting one iteration from a local minimum and so on.

As a result, it is possible to estimate the fundamental polynomial during a run by keeping track of the number of iterations spent in the downhill processes. Using the estimate of the fundamental polynomial, estimates of retention and acceleration and hence also expected hitting time can be affected. As a run proceeds, the coefficient estimates converge to their right values and so does the estimate of E.

11.3.3 Renewal Techniques in Restarting

One can restart in a more general way using other criteria. For example, one could restart if the vector process $V_n = (X_n, \cdots X_{n+r})$ of $r+1$ states with values in $\Omega^{r+1} = \Omega \times \Omega \times \cdots \times \Omega$ lies in a subset D of Ω^{r+1}.

We assume that the goal set G is a non-empty subset of the finite set Ω. Fix $r \geq 1$, let D be a subset of Ω^{r+1} and define for subsets A of Ω the sets

$$D_A = \{(x_1, x_2, \ldots, x_{r+1}) \in D : x_1 \in A\}.$$

Denote by E the (non-empty) set of x in $\Omega \backslash G$ which for some fixed $t > 0$ satisfies

$$P\big[V_n \in D \mid X_n = x\big] \geq t$$

for all n, and let $U = G \cup E$. Introduce the following two conditions, where $T_E = \min\{n : X_n \in E\}$.

(A1) $1 > P[T_E < T_G] > 0$.

(A2) There is a finite $K \geq 1$ and a number $\phi \in (0,1)$ such that uniformly for $x \in \Omega\backslash G$, and all n,

$$P[T_U > m+n \mid X_n = x] \leq K\phi^m,$$

where the probability on the left hand side is that the first epoch after n at which the X process lies in U is greater than $m+n$.

Restarting when a sequence of states lies in a subset D of Ω^{r+1} defines a new process on the original search process and under the conditions (A) the tail probabilities for the r-process V_n satisfy a renewal equation which yields their geometric convergence to zero.

If the goal is encountered then the next r are taken as identical (and our interest in the process is terminated) and otherwise the first hitting time τ_U is defined by

$$\tau_U = \min\{n \geq 1 : V_n \in D_U\}.$$

Writing

$$u_n = P[\tau_G > n],$$

one has upon decomposition of the event $\{\tau_G > n\}$ as

$$\begin{aligned}\{\tau_G > n\} = \{\tau_U > n\} \quad &\cup \quad (\{\tau_G > n\} \cap \{\tau_U = 1\}) \\ &\cup \quad \cdots \cup (\{\tau_G > n\} \cap \{\tau_U = n\})\end{aligned}$$

that

$$u_n = b_n + \sum_{j=1}^{n} P[\{\tau_G > n\} \cap \{\tau_U = j\}] = b_n + \sum_{j=1}^{n} u_{n-j} f_j,$$

where $f_n = P[\tau_G > n,\, \tau_U = n]$ and $b_n = P[\tau_U > n]$. Therefore, the tail probabilities u_n for the r-process hitting times satisfy a renewal equation.

Theorem 11.3.3 *Under the conditions* (A1) *and* (A2), $E[\tau_U] < \infty$,

$$E[\tau_G] = \frac{E[\tau_U]}{1 - P[\tau_E < \tau_G]} < \infty,$$

and there is a $\gamma \in (0,1)$ *and a finite constant* c *such that* $\gamma^{-n} u_n \to c$ *as* $n \to \infty$.

Define $\Psi_b(z) = \sum_{n \geq 0} b_n z^n$.

Corollary 11.3.4 *If* (A1), *and* (A2) *hold, if* f_n *is not periodic and there is a real solution* $\theta > 1$ *to* $\Psi_f(\theta) = 1$ *satisfying* $\Psi_b(\theta) < \infty$ *then there is a* $\rho \in (0,1)$ *and a finite positive constant* c *such that* $\rho^{-n} u_n \to c$ *as* $n \to \infty$.

By restarting, the expected time to goal of a search process can be transformed from infinite to finite. Multistart, (see [66]) where under no restarting the hitting time is infinite with positive probability, is an obvious example which has already been discussed. Under simple conditions like restarting according to a distribution which places positive mass on each state, multistart trivially satisfies the conditions (A1) and (A2) with $t = 1$. Furthermore the conditions of Corollary 11.3.4 hold and it provides an interesting formula for the Perron-Frobenius eigenvalue as the reciprocal of the root of a low degree polynomial (see [40]).

11.4 Simulated Annealing

11.4.1 Introduction

Simulated annealing (SA) is a stochastic method for function optimization that attempts to mimic the process of thermal annealing of solids. From an initial condition, a chain of states in Ω are generated that, hopefully, converge to the global minimum of the objective function f, referred to here as the *energy* E. This sequence of states dances around the state space with the amount of movement controlled by a "temperature" parameter T. The temperature of the system is lowered until the process is crystalized into the global minimum.

Simulated Annealing has its roots in the algorithm announced by [51]. This algorithm used a Monte Carlo method to simulate the evolution of a solid to thermal annealing for a fixed temperature. The current state of the solid (as represented by the state of some particle) was randomly perturbed by some small amount. If this perturbation resulted in a decrease in the (thermal) energy, then the new state was accepted. If the energy increased, then the new state was accepted with probability equal to $exp(-\Delta E/kT)$ where ΔE is the energy difference, k is Boltzmann's constant and T is the temperature. Following this rule for evolution (called the *Metropolis acceptance rule*), the probability density for the random variable of state of the system X_t, converges to the Boltzmann distribution

$$P(X_t = E) \approx \frac{1}{Z} e^{\frac{-E}{kT}} \tag{11.4.5}$$

where $Z(T)$ is a normalizing constant (called the *partition function*).

The basic Simulated Annealing algorithm can be thought of as a sequence of runs of versions of the Metropolis algorithm, but with decreasing temperature. As stated above, we use the objective function, the function to be minimized, as the energy for the system. For each temperature, the system is allowed to equilibrate before reducing the temperature. In this way, a sequence of states is obtained which is distributed according to the various Boltzmann distributions for the decreasing temperatures. However, as the temperature approaches zero, the Boltzmann distribution converges to a distribution which is completely supported on the set of global minima of the energy function, see [62]). Thus, by careful control of the temperature and by allowing the system to come to equilibrium at each temperature, the process finds the global minima of the energy function.

Since the Metropolis acceptance scheme only uses energy differences, an arbitrary constant can be added to the objective function (the energy function) and obtain the same results. Thus we can assume that the objective function is non-negative so can be thought of as an energy. However, in an implementation this constant obviously does not need to be added.

The basic (SA) algorithm is described as follows. Let Ω be the state or configuration space and f be the objective function. For each $x \in \Omega$, we have a set of "neighbors", $N(x)$, for x, the set of all possible perturbations from x. Let Q designate the proposal matrix, that is, $Q(x, y)$ is the probability that y is the result of the perturbation given that the current state is x. Thus, $N(x) = \{y : Q(x, y) \neq 0\}$. We assume that the matrix Q is irreducible so that it is possible to move from any state in Ω to any other state in Ω.

1) Initialize the state x_0 and T_0.

2) Choose a $x' \in N(x_n)$ according to the proposal scheme given by the matrix Q.

3) If $f(x') < f(x_n)$, then set $x_{n+1} = x'$.

4) If $f(x') \geq f(x_n)$, then with probability $e^{\Delta f/(kT_n)}$ set $x_{n+1} = x'$ else let $x_{n+1} = x_n$.

5) Decrease T_n to T_{n+1}.

6) If not finished, go to step 2.

Theoretical setting: Markov Chains

Generally, Simulated Annealing is analyzed in the context of Markov Chain theory. Given the state space Ω and energy function f, we denote the acceptance matrix (that generated by the Metropolis acceptance scheme) by A where

$$A(i,j) = \max\left\{1, e^{-(f(j)-f(i))/Tk}\right\}$$

where T is the temperature and k is Boltzmann's constant. Using the proposal matrix Q along with this acceptance matrix, we get the transition kernel for the chain to be

$$q(i,j) = Q(i,j)A(i,j) \quad i \neq j \qquad \text{and} \qquad 1 - \sum_j Q(i,j)A(i,j) \quad i = j.$$

If the proposal matrix Q is symmetric and irreducible and the temperature T is fixed, it is easy to show that the invariant distribution for this chain is the Boltzmann distribution (11.4.5) for the temperature T. This is the context of the original algorithm in [51].

The situation for changing (decreasing) temperatures is more difficult.

Cooling schedules

Clearly the choice of cooling schedule is critical in the performance of a simulated annealing algorithm. The decreasing temperature tends to "force" the current state towards minima, moving only downhill. However, decreasing the temperature too quickly could result in the state getting trapped in a local (nonglobal) minimum while decreasing the temperature too slowly seems to waste computational effort. A fundamental result by Hajek (see below) gives a general guideline for the cooling schedule.

Despite these theoretical results, practitioners often use other cooling schedules that decay to zero faster than the inverse log cooling schedule from Hajek's result. This is done in an attempt to speed up the algorithm.

Cooling schedules can be divided up into *fixed schedules*, or those that are preset before a run of the algorithm, and *dynamic* or *adaptive schedules*, or those that are changed during the run of the algorithm.

Common fixed cooling schedules are the inverse log cooling schedule, *inverse linear* where $T = 1/(a + bt)$ for suitable a, b, *geometric* where $T = ar^t$ for some a and $0 \leq r < 1$.

Dynamic cooling schedules are usually derived using considerations from statistical physics. One such cooling schedule is the *minimum entropy production* schedule from [2]. This schedule slows down the annealing when the internal relaxation time or where large amounts of "heat" have to be transfered out of the system (*i.e.* when we need to make sure that the system doesn't get stuck in a local minimum or meta-stable state). A disadvantage is the extra work necessary to estimate the parameters necessary for the dynamic schedule. In any particular problem, what is important is the trade-off between the extra efficiency of a dynamic schedule versus the extra work necessary to calculate the dynamic schedule.

The problem of pre-mature convergence

A common problem that plagues stochastic methods for optimization is that of *pre-mature convergence*, or "getting stuck." This is the purpose of decreasing the temperature extremely slowly. In computer runs of a Simulated Annealing algorithm it is common to see long sequences of states where there is no improvement in the solution. This is often due to the system remaining in the same state for many iterations. In fact, as the temperature decreases, it becomes more likely for this to happen and these runs of fixed states tend to get longer. Thus, several methods have evolved in order to deal with this problem.

One clear solution is to try to bypass these runs directly. Since (SA) is a Markov Chain, the time spent in chain of repeated states is completely wasted. If one could "by-pass" these states, moving directly to the next, different state, this effort could be recovered. This is one feature of the "sophisticated simulating annealing" algorithm proposed by Fox in [23].

Another, simpler, method to deal with this problem is to restart the process. We take up this idea in the next section.

11.4.2 Simulated annealing applied to the permanent problem

As an illustration of these basic ideas, we give an example of the algorithm applied to the 14:40 permanent problem.

In all the experiments reported on below, our neighborhood system was defined by allowing any 1 appearing in the matrix to swap positions above or below, left or the right with an adjacent 0. (Of course swapping with a 1 would not yield a different matrix.) In this, we allowed wrapping, that is, a 1 on the bottom row could swap positions with a 0 on the top row; similarly the first and last columns can swap values. In this way, each solution, or arrangement of d 1's, has $4d$ neighbors.

The "energy" of the annealing, to be minimized, is taken as the negative of the permanent itself so that minimizing energy, maximizes the permanent.

As for cooling schedules, we tested: geometric, inverse log, and inverse linear. In all cases we found the "phase change" temperature to be about $T = 1$. Thus we arranged for all cooling schedules to bracket this value. In order to make the comparison fair, we further arranged that each run would consist of the same number of iterations: 3 million. This meant that the starting and ending temperatures varied greatly among the different schedules.

Geometric cooling means

$$T = ab^t$$

with a and b chosen so that T ranged from 19 down to .1. Inverse log cooling is the theoretically prescribed cooling,

$$T = a/\ln(1+t).$$

With $a = 8.705194644$, temperature ranged from 12.5 down to .58. Inverse linear cooling means

$$T = a/(1+bt)$$

The parameters a and b were chosen so that temperature ranged from 19 down to .4.

Ten runs were made with each schedule. Geometric cooling worked consistently best and we only show those results.

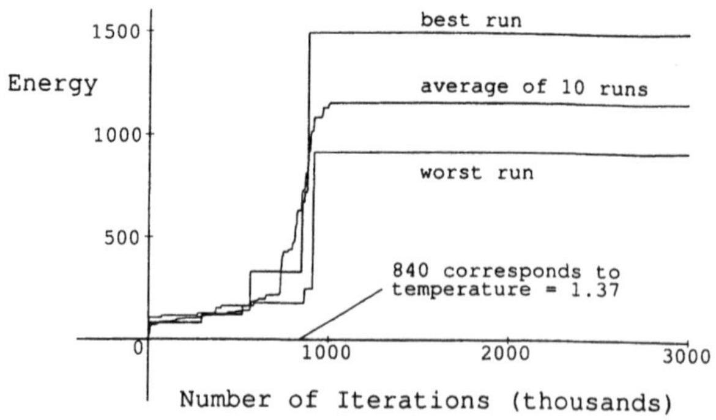

Simulated Annealing results for the permanent 14:40 problem

11.4.3 Convergence Properties of Simulated Annealing and Related Algorithms

While one obvious goal in a minimization problem is the rapid identification of some or all x which minimize f, this goal is often not attainable, and in that case other criteria, such as the rate of increase of the quality of the best solution to date, could be applied in judging an algorithm's performance. Generally the subject is difficult and relatively young so that comparatively few rigorous results are available. The methods studied and their properties are dependent upon the underlying assumptions on f and Ω. Discrete time algorithms designed for finding minima of smooth functions on subsets of $\mathbb{R}^n$, of continuous time algorithms for finding minima of arbitrary functions defined on a finite set, and all of the obvious variations have been studied. One of the most thoroughly studied techniques is simulated annealing. Let X_n be a Markov chain whose state space is Ω and whose transitions are defined by

$$P[X_{n+1} = j \mid X_n = i] = q(i,j)\exp\{-(f(j) - f(i))^+/T_n\}, \tag{11.4.6}$$

where $i, j \in \Omega$, n indicates the epoch of time, and $T_n \downarrow 0$ as $n \to \infty$. The mathematical model of the process is a time-inhomogeneous Markov chain. There is a voluminous and growing literature on Markov chains. Time-homogeneous chains are especially well understood (see, for example, [22], [44]. The chains which arise in (SA) are time inhomogeneous and far less is known about them. Cruz and Dorea [15] employ results from the theory of nonhomogeneous Markov chains (see [42]) and are able to reprove some results of Hajek.

The majority of the theoretical work on (SA) to date has been directed at the question of convergence of the probabilities $P[X_n \in G \mid X_0]$ for interesting subsets G of Ω. For example, if

$$G = \{x \in \Omega : f(x) = \min_{y \in \Omega} f(y)\}$$

then under what conditions on the algorithm, which involves a choice of the transition function q and of the cooling schedule T_n, does one have $\lim_{n\to\infty} P[X_n \in G \mid X_0] = 1$? The compilation of results below is not comprehensive. More results and sometimes in somewhat greater generality are available from the original sources.

- The energy landscape is (Ω, f, q), where Ω is a finite set, f is a function whose minimum value on Ω is sought, and q is a fixed irreducible Markov transition kernel defined on $\Omega \times \Omega$.

- For real numbers a, the level sets are $\Omega(a) = \{i \in \Omega : f(i) \le a\}$.

- The restriction of q to a subset G of Ω is $q|_A(i,j) = q(i,j)$ if i and j are in G and 0 otherwise.

- The boundary of a subset G of Ω is $B(G) = \{j \in \Omega \backslash G : \max_{i \in G} q(i,j) > 0\}$.

- For real a and under the assumption that q is symmetric, the relation $\leftrightarrow_a$ on $\Omega \times \Omega$ defined by

  $$i \leftrightarrow_a j \text{ if } \{\sup_n q^n_{|\Omega(a)}(i,j) > 0 \text{ or } i = j\}$$

  is an equivalence relation and i and j are said to communicate at level a.

- Weak reversibility holds if for any real a, $\sup_n q^n_{|\Omega(a)}(i,j) > 0$ entails $\sup_n q^n_{|\Omega(a)}(j,i) > 0$. This property also entails $\leftrightarrow_a$ being an equivalence relation on $\Omega \times \Omega$.

- The components C of $\Omega/\leftrightarrow_a$ are called cycles. As a ranges over all positive numbers the union of the cycles so obtained is what Hajek calls the collection of cups.

- The depth of a cycle C is

  $$H(C) = \max_{i \in C} \min_{j \in B(C)} (f(j) - f(i))^+.$$

- $f(C) = \min_{i \in C} f(i)$.

- $D(C) = H(C)/f(C)$

- For real $t > 0$, $D_t = \max\{D(C) : C \text{ is a cycle of } \Omega,\ f(C) \geq t + \min_\Omega f\}$.

- A state $i \in \Omega$ is a local minimum of f on Ω if no state j with $f(j) < f(i)$ communicates with i at level $f(i)$.

- Hajek's depth $d(x)$ of state x is ∞ if it is a global minimum. Otherwise it is the smallest number b such that some state y with $f(y) < f(x)$ can be reached at height $f(x)+b$ from x (it is $a - f(x)$ for the smallest a such that for some y with $f(y) < f(x)$, $x \leftrightarrow_a y$). If x is a local minimum of f then $d(x) = H(C)$, where x is at the bottom of some cup C.

- The bottom of a cup C (a cycle) is the set of $x \in C$ such that $f(x) = f(C)$. The *depth* of such a state is $H(C)$.

To illustrate these ideas, consider the following connection graph shown along with the energy for each state.

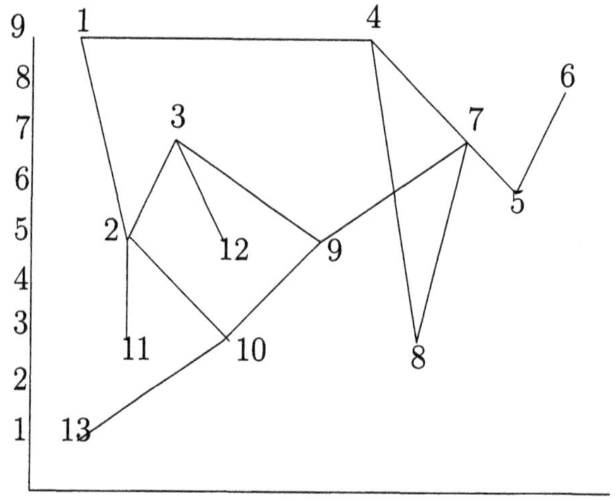

Then we have the following relationships.

- $\Omega(1) = \{13\} = \Omega(2)$

- $\Omega(3) = \{10, 13\}, \{11\}, \{8\}$

- $\Omega(5) = \{2, 9, 10, 11, 13\}, \{8\}, \{12\}$

- $\Omega(6) = \{2, 9, 10, 11, 13\}, \{12\}, \{8\}, \{5\}$

etc.

- $\Omega(9) = \Omega$
- $H(\{2, 9, 10, 11, 13\}) = 7 - 1$
- $f(\{2, 9, 10, 11, 13\}) = 1.$

I. Finite Ω Using continuous time arguments Hajek [34] proved the following about the discrete time minimization on the finite set Ω.

Theorem 11.4.1 (Hajek) *Assume that (Ω, f, q) is irreducible and satisfies weak reversibility. If $T_n \downarrow 0$ then*
(i) *For any state j that is not a local minimum of f,* $\lim_{n\to\infty} P[X_n = j] = 0$.
(ii) *Suppose that the set of states B is the bottom of a cup C and the states in B are local minima of depth $H(C)$. Then* $\lim_{n\to\infty} P[X_n \in B] = 0$ *if and only if*

$$\sum_{n\geq 1} \exp\{-H(C)/T_n\} = +\infty.$$

(iii) *Let d^* denote the maximum of all depths of local, non-global minima. If G is the set of global minima then*

$$\lim_{n\to\infty} P[X_n \in G] = 1 \tag{11.4.7}$$

if and only if

$$\sum_{n\geq 1} \exp\{-d^*/T_n\} = +\infty.$$

Note: As Hajek points out, if $T_n = c/\ln(1+n)$ then (11.4.7) holds if and only if $c \geq d^*$.
In theory one must choose $c \geq d^*$ to be assured that the algorithm converges. Hajek gives a matching problem example in which one can show that $d^* \leq 1$ for his choice of q. In any problem $c = \max f - \min f$ will work but this incurs a penalty in the convergence rate as can be seen from the next theorem of Chiang and Chow [12]. Fox [23] and Morey, et al [55] treat the problem of choosing c in more general situations.
Around the same time as Hajek's work, the rate of convergence of (SA) with logarithmic cooling was established under slightly stronger assumptions. To state the results, let $\lambda(t) = e^{-1/T(t)}$ and let $d = \max_{i\in\Omega} d(i)$, where, with $h(i,j) = \min h$ such that j can be reached at height $f(i) + h$ from i, $d(i) = \min\{h(i,j) : f(j) < f(i)\}$ if i is not a global minimum, $d(i) = \max\{h(i,j) : j \text{ is a global minimum}\}$ if i is a global minimum. Then d^* is the maximum of $d(i)$ over states which are not global minima and the following is true, assuming, WLOG, that $\min_{i\in\Omega} f(i) = 0$.

Theorem 11.4.2 ([12]) *Under irreducibility and weak reversibility and if $\int_0^\infty \lambda^d(t)dt = \infty$ and $\lambda'(t)/\lambda(t) = o(\lambda^{d^*}(t))$ as $t \to \infty$ then there exist positive constants β_i, independent of the initial distribution, such that*

$$\lim_{t\to\infty} P[X_t = i]/\lambda^{f(i)}(t) = \beta_i.$$

With the logarithmic cooling schedule $c/\ln(t)$ one has $\lambda(t) = t^{-1/c}$ and for $c \geq d$ and $c > d^*$ the theorem is true, while if $c \geq d^*$ then $\lim_{n\to\infty} P[X_n \in G] = 1$.

It can be seen from this theorem that the rate of convergence of the probabilities can be quite slow, an observation confirmed in practice, even if d^* is available.

Rates of convergence of simulated annealing algorithms have also been studied from a different perspective using Sobolev inequalities (see [38]). Holley and Strook treat continuous time irreducible, reversible processes on a finite state space Ω and the size of the Radon-Nikodym derivative f_t of the distribution of the annealing process $X(t)$ at time t with respect to stationary (Gibbs) distribution at time t is established. It follows from their work, for example, that when the cooling schedule is

$$T(t) = \frac{m}{\log(1+t)}$$

for $t \geq 0$ the L^2 norm of $f_t - 1$, with respect to the Gibbs measure, satisfies

$$\|f_t - 1\|^2 \leq \frac{4(1+4MmA)}{1-(1+t)^{-(1/2A+2mM)}} = C$$

where $A > 0$ is a constant and m and M are geometric quantities: $m = \max_{x,y\in\Omega}\{H(x,y) - f(x) - f(y)\}$, $H(x,y)$ is the minimum elevation of paths connecting x and y, and $M = \max f - \min f$. This inequality shows, for example, that

$$P[f(X(t) \geq \min f + d]^2 \leq (1+C)Q_t[f \geq \min f + d]$$

where Q_t is the equilibrium measure at temperature $T(t)$. In contrast to most studies, Holley and Stroock's analysis applies to the dynamic situation in which T is changing. For example, Ingrassia [41] investigated the spectral gaps of the discrete time processes $X_T(t)$ on the finite set Ω whose transitions are given in (11.4.6) for T_n constant and equal T. He derived bounds, also in terms of geometric quantities, on the magnitudes of the second largest and smallest eigenvalues (see also [19]) for irreducible reversible aperiodic chains and showed, for example, that for the Metropolis algorithm when T is small, the gap is $1 - \lambda_2$, where λ_2 is the second largest eigenvalue of the transition matrix.

In an effort to speed the progress of (SA) with (inverse) logarithmic cooling, many researchers tried alternative schedules which decrease to 0 more quickly, such as exponential schedules, even though, as proven by Hajek, the convergence (11.4.7) no longer holds. One such alternative is the triangular cooling schedules of Catoni. In [11], using large deviation estimates, still more details are provided on properties of the convergence of (SA). These results imply those of Hajek and, corroborating empirical observation also indicated the slow decrease of the probability that $f(X_n)$ exceeds the minimum by t or more.

Theorem 11.4.3 (Catoni) *1. For any energy landscape (Ω, f, q) there is a constant K such that for any schedule $T_n \downarrow 0$ and $t > 0$*

$$\sup_{i\in\Omega} P[f(X_n) \geq t + \min_{\Omega} f \mid X_0 = i] \geq K\frac{1}{n^{p(t)'}}$$

where $p(t) = 1/D_t$.

Catoni suggested that since computing time N is finite one should tailor the cooling schedule to this finite horizon problem. He termed these triangular cooling schedules and proved the following.

Theorem 11.4.4 (Catoni) *2. For any state space Ω and communication kernel q there exist positive constants B and K such that for any positive constant A, for any initial* *distribution p, for any positive δ and ϵ for any energy f, for any triangular schedule T_n^N, $1 \leq n \leq N$, such that*

$$\frac{1}{T_n^N} = Ae^{n\xi}$$

with

$$\xi \leq B \frac{\epsilon^{D_\delta}}{\ln(e^{-1})}$$

and

$$N \geq \frac{1}{\xi} \left[\ln_2(\epsilon^{-1}) - \ln(\delta A)\right]$$

the corresponding annealing algorithm X_n^N satisfies

$$P\left[f\left(X_N^N\right) \geq \delta + \min_\Omega f\right] \leq K\epsilon \exp\{AH(\Omega\backslash\Omega(\delta))/D_\delta\}.$$

Corollary 11.4.5 *If $d \leq D_\delta$ and $h \geq H(\Omega\backslash\Omega(\delta))$ then*

$$T_n^N = \frac{d}{h}\left(\frac{h\ln(N)}{d^2\delta}\right)^{n/N}$$

is "logarithmically almost optimal" in the sense that

$$\lim_{N\to\infty} \frac{\ln P[f(X_N^N) \geq \delta + \min_\Omega f]}{\ln(N)} = -1/D_\delta.$$

II. Continua. In addition to the work on minimizing a function f on a finite set Ω by stochastic methods, there is a large body of detailed work on minimizing a smooth function f defined on some subset Ω of $\mathbb{R}^k$.

Using large deviation results, Kushner [49] studies processes defined on $\Omega = \mathbb{R}^k$ by

$$X_{n+1} = X_n + \gamma_n b(X_n, \eta_n) + \gamma_n \sigma(X_n)\xi_n, \tag{11.4.8}$$

where η_n are random variables, ξ_n are i.i.d. Gaussian random variables,

$$\gamma_n = A/\log(n + C),$$

and there are other restrictions. Taking $E[b(x, \eta_{n+1})] = \bar{b}(x) = -B_x(x)$ for a continuously differentiable function B yields a method for locating the minimum of B. Among the properties he studies are the escape times from neighborhoods G of compact stable invariant sets K of $\dot{x} = \bar{b}(x)$. Under conditions, he shows that for A sufficiently large and x in G, after long times,

$$\lim_{m\to\infty} \frac{\log E_x[\tau^m]}{\log(m + C)} = S_G(K)/A,$$

where $x \in G$ and $S_G(K)$ is a constant related to the minimum value of an "action functional" connecting x to the boundary of G.

Another model has the candidate point X_{n+1} at epoch $n+1$ related to the candidate X_n at epoch n by

$$X_{n+1} = X_n + \gamma_n[h(X_n) + \eta_{n+1}] + \sigma_n \xi_{n+1}, \tag{11.4.9}$$

where γ_n and σ_n are sequences under control of the user of the algorithm, η_{n+1} is a random observation error (this models the error in the determination of the precise value of $\nabla h(X_n)$), and ξ_{n+1} is a random sequence. In case of minimizing a function f one can take $h(x) =$

$-\nabla f(x)$ and ξ_{n+1} is added to keep the algorithm from becoming trapped in local minima. This is a discrete time algorithm inspired by the continuous time versions first suggested in [27]. Pelletier [59] calls this a weakly disturbed algorithm if γ_n and σ_n are chosen such that $v(n) = \gamma_n \sigma_n^{-2}$ is increasing and $v(n)/\ln(n) \to \infty$ and a strongly disturbed algorithm if $v(n)$ is increasing and $v(n)/\ln(n)$ is suitably bounded. The latter case corresponds to simulated annealing as follows.

Consider a Markov chain Y_n defined on Ω with transition probability

$$P[Y_{n+1} \in A \mid Y_n = x] = \int_A s_n(x, y) dF_{xn}(y) + r_n(x) I_A(x),$$

where

$$\begin{aligned} s_n(x) &= \max\{1, a_n^{\gamma}|x|\}, \qquad \gamma > 0, \\ a_n &= A/n, \qquad b_n = \frac{\sqrt{B}}{\sqrt{n \log \log n}}, \\ s_n(x, y) &= \exp\left\{-\frac{2a_n}{b_n^2} \frac{[f(y) - f(x)]^+}{\sigma_n^2(x)}\right\}, \\ r_n(x) &= 1 - \int s_n(x, y) dF_{xn}(y), \end{aligned}$$

and F_{xn} the cdf of a $N(x, b_n^2 \sigma_n^2(x) I)$ random variable. The resulting process is an analog of the usual simulated annealing process. Furthermore, by an appropriate choice of η_{n+1}, the process in (11.4.9) represents such a process with ξ_{n+1} Gaussian.

Among the conditions for the truth of the next theorem is that the measures π^{ϵ} on Ω defined by

$$\pi^{\epsilon}(A) = \frac{1}{Z_{\epsilon}} \int_A \exp\left\{-\frac{2f(x)}{\epsilon^2}\right\} dx,$$

where $Z_{\epsilon} < \infty$, satisfy $\pi^{\epsilon} \Rightarrow \pi$.

Theorem 11.4.6 [25] *Under (several) conditions, for any bounded continuous function on* R^d

$$\lim_{n \to \infty} E_{0x}[f(Y_n)] = \pi(f).$$

Theorem 11.4.7 [59] *Under many conditions, including that $v(n)$ is increasing and $v(n)/\ln(n)$ is suitably bounded, if the function $g(a) = \int e^{-af(x)} dx$ is regularly varying at infinity with exponent $-\eta, \eta \geq 0$, then*

(i)
$$4v(n)\left[f(X_n) - \min_{y \in \Omega} f(y)\right] \Rightarrow \chi^2_{2\eta}.$$

Furthermore,

(ii) *for any real function f increasing to infinity*

$$\lim_{n \to \infty} \frac{\ln\left(P\left[f(Z_n) - \min_{y \in \Omega} f(y) \geq \frac{r f(v(n)) \ln(v(n))}{v(n)}\right]\right)}{f(v(n)) ln(v(n))} = -2r.$$

Item (i) shows that the rate of weak convergence of simulated annealing cannot be better than $c/\ln(n)$.

First hitting times

The results about (SA) cited above relate to the asymptotic distribution of the search process X_n, a non-homogeneous Markov chain. The analysis of some stochastic algorithms provided by Shonkwiler and Van Vleck [67] takes a different approach in measuring the performance of stochastic algorithms. Since it is easy to keep track of the best an algorithm has done to date, it makes sense to ask about the first hitting time of a goal state as a function of the number of epochs n it has been running.

In the case of homogeneous Markov chains the relationships between first hitting times and rate of geometric ergodicity has been investigated rather more thoroughly than in the case of non-homogeneous ones (see [70] and [45]). For (SA), geometric convergence to zero of the probability $P[T_G > n]$ that the goal has not been encountered by the algorithm through epoch n, can not hold in general. In the next section we give an example of a simple (SA) for which the expected first hitting time is infinite, thereby showing that for any $\epsilon > 0$ there is a simulated annealing problem for which one cannot have eventually

$$P[T_G > n] \leq 1/n^{1+\epsilon}.$$

11.5 Restarted Algorithms

11.5.1 Introduction

A problem faced by all global minimizing algorithms is dealing with entrapment in local minima. Evidence that stochastic algorithms can spend excessive time in states other than the goal comes most frequently and easily from simulations. For example, in simulated annealing an "optimal" cooling schedule (see [34]) for simulated annealing (SA) guarantees that the probability the search process is in the goal state tends to 1 as the number of epochs n tends to infinity; the expected time taken by (SA) to hit the goal can however be infinite as seen in the following simple "Sandia Mountain" example.

Example 11.5.1 Let $\Omega = \{0, 1, 2\}$ with $f(0) = -1$, $f(1) = 1$, and $f(2) = 0$. Potential moves will be generated by a random walk and, at the end point, with equal chance of staying put as moving. This provides for a symmetric move generation matrix. Using the usual Metropolis acceptance criteria, $a_{ij} = e^{-\max\{0, f(j)-f(i)\}/T}$, the transition matrix is given by

$$P = \begin{bmatrix} 1 - \frac{1}{2}e^{-2/T} & \frac{1}{2}e^{-2/T} & 0 \\ 1/2 & 0 & 1/2 \\ 0 & \frac{1}{2}e^{-1/T} & 1 - \frac{1}{2}e^{-1/T} \end{bmatrix}.$$

From annealing theory the temperature T should vary with iteration count t according to the equation

$$T = \frac{C}{\ell n(t+1)}.$$

where C is the depth of the deepest local non-global minimum. Here $C = 1$. Eliminating T gives the transition probabilities directly in terms of t, thus

$$p_{21}(t) = \frac{1}{2}e^{-\ell n(t+1)} = \frac{1/2}{t+1}, \qquad t = 1, 2, \ldots,$$

and

$$p_{22}(t) = 1 - p_{21}(t) = 1 - \frac{1/2}{t+1}, \qquad t = 1, 2, \ldots.$$

The expected hitting time determination involves calculating all the possible ways leading to state $x = 0$ in t iterations starting from a given state. Here however we estimate these probabilities. Hitting the goal at time k includes the possibility of remaining for $t = 1, 2, \ldots, k-2$ in state $x = 2$, then moving in two consecutive iterations to states $x = 1$ and $x = 0$. Therefore, the probability of hitting at time k is at least as large as

$$\begin{aligned} h_k &= (1 - \frac{1/2}{2})(1 - \frac{1/2}{3}) \cdots (1 - \frac{1/2}{k-1}) \frac{1}{2} \frac{1}{k} \frac{1}{2} \\ &> (1 - \frac{1}{2})(1 - \frac{1}{3}) \cdots (1 - \frac{1}{k-1}) \frac{1}{k} \frac{1}{4} \\ &= \frac{1}{4k(k-1)}, \qquad k = 2, 3, \ldots . \end{aligned}$$

It follows that the expected hitting time from state 2 is at least as large as

$$\sum_{k=2}^{\infty} k h_k = \frac{1}{4} \sum_{k=2}^{\infty} \frac{1}{k-1} = \infty.$$

∎

A simple mechanism for avoiding entrapment is *restarting*. This means terminating the present search strategy and using the initialization procedure on the next iteration instead, usually random selection.

11.5.2 The Permanent Problem using restarted simulated annealing

As in the simulated annealing application, our neighborhood system for restarted simulated annealing is defined by allowing any 1 appearing in the matrix to move one position up or down or to the left or to the right with wrap.

The "energy" of the annealing, to be minimized, was taken as the negative of the permanent itself and the cooling schedules we tested were geometric, inverse log, and inverse linear. For the restart runs, temperature ranged from on the order of 6 down to the order of 0.2. For each different cooling schedule, we tried several temperature ranges until we found one that seemed to work well. Thus, we compared the "best" runs for each cooling schedule. We took the restart repeat count, $r + 1$, to be 200.

The results displayed in the figures are the averages of 10 runs. The restart algorithm made both very rapid progress at the beginning of the runs and continued to make progress even up to the time the runs were halted. All the annealing runs with restart consistently achieved permanent values on the order of 1500. The restarting step was effective in allowing the algorithm to escape from local minima even at temperatures below the critical temperature (of approximately 1) where the phase transition occurs.

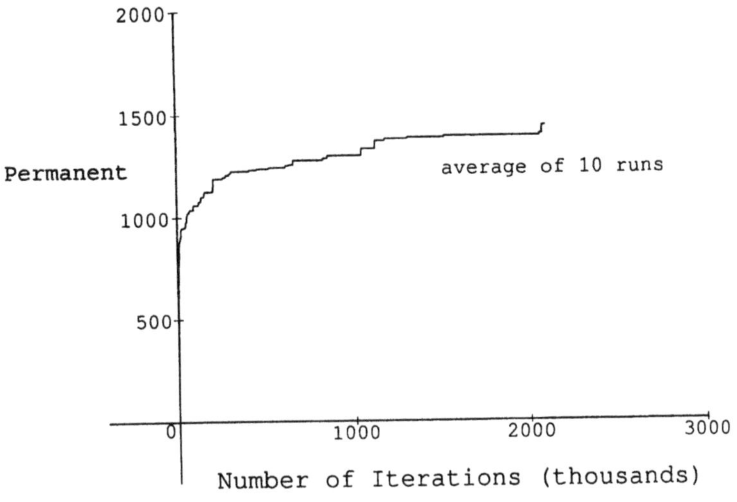

Figure 4. Restarted Simulated Annealing results for the permanent 14:40 problem

11.5.3 Restarted Simulated Annealing

The undesirably slow convergence of (SA) has motivated research such as that in Kolonko [48] and Bélisle [7] on the random adjustment of the cooling schedule, on non-random adjustments as reported, for example, in van Laarhoven and Aarts [71] or Nourani and Andresen [57], and the thorough theoretical treatment of simulating direct self-loop sequences in Fox [23] and the truncated version in Fox and Heine [24].

Although geometric decrease of the probability of not seeing the goal by epoch n does not generally hold for (SA), Mendivil, Shonkwiler and Spruill [50] have shown that it does for (SA). The algorithm is restarted whenever $f(X_{n+r}) = \cdots = f(X_n)$. Let

$$\alpha = \min_{x,y\in E}\{f(y) - f(x) : q(x,y) > 0 \text{ and } f(y) - f(x) > 0\}$$

Theorem 11.5.2 *Under the standard transition assumptions above and if there is $\beta > 1$ such that*

$$\sum_{n\geq 1} \beta^n e^{-\alpha/T(n)} < \infty \tag{11.5.10}$$

then restarting (SA) *by a distribution which places positive probability on each point in Ω for r, $1 \leq r < \infty$, sufficiently large there is a $\gamma \in (0,1)$, and a finite constant c such that $\gamma^{-n}P[\tau_G > n] \to c$ as $n \to \infty$.*

Corollary 11.5.3 *The* (RSA) *algorithm which uses the cooling schedule of $c(n) = 1/n$ will, for sufficiently large r, have tail probabilities which converge to 0 at least geometrically fast in n.*

The conditions of the Theorem are not necessary for the geometric convergence of the tail probabilities. In the following example, the geometric rate of decrease of the tail probabilities holds for a restarted simulated annealing which uses the usual logarithmic cooling schedule. This example is one for which the (SA) satisfies the conditions of Hajek's theorem but which, without restarting, has an infinite expected hitting time of the goal (see the previous section).

Example 11.5.4 The Sandia Mountain example of the previous section, as an illustration that independent identical parallel processing (IIP) can make the expected time to hit the goal go from infinite to finite, is presented here from the perspective of restarting when a state is repeated; by showing that the conditions of the Theorem are met, it can be shown that the expected time to goal can be made finite simply by restarting on the diagonal. Under restarting the tail probabilities do converge to zero geometrically quickly even using the logarithmic schedule.

Obviously, geometric or faster decrease to 0 of the tail probabilities $P[\tau_G > n]$ under (RSA) or otherwise entails a finite expected hitting time of the goal states G, but by itself geometric decrease of the tail probabilities is not a strong recommendation. Under the assumption that both processes use the same generation matrix with (SA) using a logarithmic schedule $T_n = c/\ln(n+1)$, $c \geq d^*$, and (RSA) using a linear schedule $T_n = 1/n$, assume a common position of the two algorithms at epoch n. At any instant of time at which the two processes happen to reside at the same location the cooling schedule of one, which is logarithmic, should be compared with that of the other, which is linear in the (random) age of the process, for this will indicate the relative tendencies of going downhill. If r is small then the clock will likely have been reset for (RSA), but if r is large then very likely the r-process will not have restarted at all and the epoch number will also be the current age. It is the latter instance which is of interest since (RSA) is assumed to have r "large." At a location which is not a local minimum the (SA) process will have, as the epochs tick away, an ever increasing tendency in comparison with (RSA) to proceed in uphill directions. Thus (RSA) should proceed more rapidly downhill than (SA) at points which are not local minima.

What happens when (SA) and (RSA) are at a local minimum at the same epoch? Very likely the (RSA) will be out of this "cup" (see [34]) in r steps whereas the (SA) will take some time. Since the goal cannot be reached until the process gets out of the cup this is a crucial quantity in determining the relative performance of the two methods when there are prominent or numerous local minima. The (RSA) will have an immediate chance of finding the cup containing the goal whereas, depending upon the proximity of the present cup to the one containing the goal, (SA) may be forced to negotiate many more cups.

It follows from Fox and Heine [24] and Fox [23] that the enriched neighborhood version of QUICKER-j has tail probabilities converging geometrically quickly to 0. In contrast with QUICKER, (RSA) requires the computation of only small prescribed numbers of function values in small neighborhoods.

11.5.4 Numerical comparisons

Some numerical results are presented comparing the performance of various forms of (SA) to (RSA). The comparisons were carried out for three types of problems, minimization of a univariate function, minimization of tour length for some TSP's, and finding the maximum value of the permanent of a matrix. In each case, parameters enter which have some influence on the performance of the method as we have seen. In (RSA) it seems desirable to proceed as quickly as possible to points where the function has a local minimum and then, if necessary, to restart. Rushing to restart is undesirable however, for local information about the function is indispensable in charting a course to a local minimum; by prematurely restarting, this information is lost. Therefore one should take care to stay sufficiently long in a location to examine a large enough collection of "directions" from the current point to ensure that paths to lower values are discovered. For functions on the line there are only two directions so one would expect to require very few duplications before the decision to restart is made. Were the selection of new directions deterministic, clearly at most two would be required, but the algorithm chooses these stochastically. In contrast, for a TSP on a reasonable number of cities, if the neighborhood system arises from a 2-change (see [1]) then one should presumably wait for a fairly large number of duplications to make sure enough "directions" have been examined. As a rough guide we note that in (SA) as long as the state has not changed, the generation matrix yields a sequence of iid "directions." Assuming the proportion of directions downhill is p and uniform probability spread over those directions by the generation matrix, the probability the generation matrix has not yielded a downhill after m generations is simply $(1-p)^m$. To make this quantity small, say less than β, m should be approximately $\ln(\beta)/\ln(1-p)$. On the line the most interesting places are where one direction is up and the other down so $p \sim 1/2$ seems reasonable. Furthermore, the consequences of restarting are minimal so a large a, say 1/2, also seems reasonable. Thus one should take r around 1. In a TSP with 100 cities restarting can be costly since the considerable time it takes to get downhill will likely be wasted upon restarting. Thus we take a small, say .01. It is not clear what p should be. Presumably the "surface" represented by the tour lengths could be rather rough so we'll take $p = .05$ to ensure a thorough although perhaps too lengthy examination of directions. This translates to run lengths of $r \sim 100$ and an examination of a fairly small proportion of the 4851 "directions" available under 2-change.

Example 11.5.5 For a randomly generated function the median number of epochs required to find the global minimum by (SA) under optimal cooling, with the stipulation that the search was terminated at 221 epochs if the minimum had not yet been found, was 221. For (RSA) with $r = 1$ the median number of epochs required to find the global minimum of the function was 21.

Example 11.5.6 In this example an optimal 100 city tour was sought using 2-change as the neighborhood system with equally likely probabilities for the generation matrix. The locations were scaled from a TSP instance known as kroA100 taken from a data base located on the Web at *http://softlib.rice.edu/softlib/catalog/tsplib.html.* Each of (SA) and (RSA) was run for 1000 epochs. The median best tour length found by (SA) was of 35.89 with a minimum of 34.06. For (RSA) the median best tour length found in 1000 epochs was 14.652 with a minimum of 13.481.

Example 11.5.7 A 24-city TSP instance known as gr24 obtained from the same data base as kroA above was analyzed again using 2-change for the neighborhood system and equally likely choices for the generation matrix. Each of (SA) and (RSA) was run for 500 epochs. The best tour lengths found by (SA) had a median of 2350.5 and a minimum of 1943. (RSA) with $r + 1 = 24$ had a median best tour length after 500 epochs of 1632.5 with a minimum of 1428. The optimal length is 1272. A similar result on 24 cities was obtained by running the two for 1000 epochs. Under (SA) the median was 2202 with a minimum of 1852 while for (RSA) the median best tour length was 1554.5 and minimum 1398.

Example 11.5.8 Performance of (SA) with depth 40 and (RSA) with $r = 100$ was compared on a randomly generated 100-city TSP. Median best tour length after 1000 epochs for (SA) was 43.14 and the minimum was 40.457. For (RSA) the median best was 19.177 and the minimum best was 17.983.

An alternative, more careful, analysis of the size of r is provided by closer examination of the proof of Theorem 11.5.2. Under the cooling schedule $c(n) = 1/n$ with an equally likely generation matrix the choice

$$r \geq \frac{-e^{-a}}{(1-e^{-a})^2} \frac{1}{\ln(1-p)}$$

will guarantee the conclusion of the theorem under its other hypotheses, where $p = 1 - g$ is the worst case, smallest probability of a downhill from among the points in $\Omega \backslash U$. However, this may not help in the determination of a "good" r since one would expect these quantities to be unknown.

11.6 Evolutionary Computations

11.6.1 Introduction

Evolutionary computations, including Genetic Algorithms ([37]) and Evolutionary Strategies ([6]), are optimization methods based on the paradigm of biological evolution by natural selection. As in natural selection, the essential ingredients of these methods are recombination, mutation, and selective reproduction working on a population of potential solutions. Fitness for a solution is directly related to the objective function being optimized and is greater for solutions closer to its global maximum (or minimum). The expectation is that by repeated application of the genetic and selection operations, the population will tend toward increased fitness.

An evolutionary computation is a Markov Chain X_t on *populations* over Ω under the action of three stochastic operators, *mutation*, *recombination*, and *selection* defined on Ω. Although implementation details may vary, mutation is a unary operator, recombination or *cross-over* is a binary operator and selection is a multi-argument operator. An evolutionary computation is always irreducible and aperiodic and so converges to a stationary distribution. While the existence of a stationary distribution is not of great importance, indeed these chains are never run long enough for the stationary distribution to become established, irreducibility is. Rather it is the swiftness with which the chain finds optimal or near optimal values that is paramount. Thus first passage and hitting times are of central importance. Although only general results of this nature are available at this time, see the first section of this chapter, theoretical progress is being made. We will present recent developments at the end of this section. Consequently practical implementations of evolutionary computations appeal to heuristics and experimental evidence.

The implementation of an evolutionary computation begins with the computer representation, or encoding, of the points x of the solution space Ω. Frequently this takes the form of fixed length binary strings which are called *chromosomes.* A natural mutation of such a string is to reverse, or *flip*, one or more of its bits randomly selected. Likewise, a natural recombination, of two bit strings, called *parents*, is to construct a new binary string from the bits of the parents in some random way. The most widely used technique for this is *one-point cross-over* in which the initial sequence of k bits of one parent is concatenated with the bits beyond the kth position of the second parent to produce an *offspring.* Here k is randomly chosen. Of course, a fitness evaluation must be done for each new chromosome produced.

Finally, the chromosomes selected to constitute the population in the next generation might, for example, be chosen by lottery with the probability of selection weighted according to the chromosome's fitness. This widely used method is termed *roulette wheel selection.*

These genetic operators would be tied together in a computer program as shown, for example, in Figure 11.1.

While the aforementioned typifies a standard genetic algorithm, many variants are found in the literature, some differing markedly from this norm. We will present some of these variations below. As we have discussed before, no one algorithm is right for all problems. It

Figure 11.1: A top level view of an evolutionary computation

```
initialize a population of chromosomes
repeat
    create new chromosomes from the present set by mutation and recombination
    select members of the expanded population to recover its original size
until a stop criteria is met
report the observed best
```

is often good to embed specialized knowledge about the particular problem into the evolutionary computation's components; for example, the chromosomes in a Traveling Salesman Problem evolutionary computation are universally taken to be the permutation vector of the cities. This is especially so when one has some insights about the particular problem being attempted. As mentioned above, the one design point that must be adhered to is assuring irreducibility.

Simulated annealing and evolutionary computations have several points of commonality. Both require an encoding of solutions and both proceed iteratively. Both propose new candidate solutions, evaluate them, and select a subset for the next iteration. One can think of a simulated anneal, in terms of an evolutionary program, as having a population size of one (although it could be larger). The proposal operation of an anneal could be taken as the mutation operation of this evolutionary computation. The acceptance algorithm of an anneal works as its selection operation.

The differences between the two are that evolutionary computations incorporate a second proposal operator (recombination), one requiring two arguments, and a greater than one population size to go with it. Although the selection operator of an evolutionary computation is not usually Metropolis acceptance, it could be. Boltzmann modified tournament selection chooses two structures from the present population by roulette wheel; with equal likelihood, one is designated as current and the other as candidate. Metropolis acceptance is then used to select one for the next generation. This is repeated, with replacement, until the next generation is selected.

On the other hand, simulated annealing incorporates time varying transition probabilities, although evolutionary computations can do so as well.

It is therefore feasible to take a step-wise approach in constructing an evolutionary computation. The first step is to write a multi-population mutation only algorithm. If Metropolis acceptance is used as the survival arbiter, then the algorithm is effectively a simulated anneal. Adding a binary operator on solutions and a multi-argument selection operation converts it to an evolutionary computation.

Being that evolutionary computations typically do not vary event probabilities over the course of a run, there arises a fundamental difference with simulated annealing. Theoretically, an annealing will not only find a global minimizer over its run, it will also identify it as such, since, asymptotically, the chain will be in such a state. However an evolutionary computation might well find an optimizing structure and then lose it. Theoretically this must happen since the process is irreducible and must visit all states recurrently. Therefore it is important to save the best-so-far value discovered by the algorithm, and the corresponding structure, for this will be part of the exit output, see [63].

As a random variable, the best-so-far value observed up to time t, $B(t)$, will satisfy the predicted asymptotic convergence rate for the process. In particular, as $t \to \infty$, $B(t)$ tends to the global maximum, as pointed out above. Therefore, just as in simulated annealing, globally optimizing states may be identified asymptotically.

Universal GA solvers The implementation of an evolutionary computation is an abstraction in that it operates on computer structures and utilizes an imposed definition of fitness. As a result it is possible to write a universal evolutionary programming based optimizer. The external part consists of interpreting the meaning of the genetic structures and defines the fitness function. The evolutionary computation acts like an engine, generating and testing candidate solutions. Two of these are GENESIS and GENOCOP. A comprehensive list can be found at *http://www.aic.nrl.navy.mil/galist/src.*

11.6.2 A GA for the permanent problem

We illustrate genetic algorithms by solving the 14:40 permanent problem described above. We will be using algorithm RS, see below. We give here details of this particular application, otherwise refer to the general setup below.

The points or states of the solution space are 0/1 matrices of size 14×14 having exactly 40 ones. Conceptionally, this will be our computer structure; however, to facilitate working with such a matrix, we will utilize two alternative encodings. First, we store each matrix in terms of its row structure: the number of 1's in each row and the positions of the 1's in each row. This representation allows for short cuts in the permanent calculation and greatly improves the speed of that part of the algorithm. Also, by unstacking the matrix row by row we obtain a 0/1 string structure of length $14^2 = 196$ with exactly 40 ones. This representation will be convenient for the recombination, or binary, operator. Instead of maintaining both configurations, we keep only the row by row form and calculate the binary string form from it. This, and the inverse computation, can be done quickly.

As a unary or mutation operation we take the same one used in the simulated annealing application. Namely, we randomly select a 1 in the 14×14 matrix, then randomly choose one of the four directions North, East, South or West, and exchange values with the neighbor entry in that direction. We allow wrap around, thus the East direction from the 14th column is the 1st column and the South direction from the 14th row is the 1st row. The row by row storage format of the matrix makes it easy to select a 1 at random.
The actual implementation checks to see if the value swapped is a 0 before proceeding, for otherwise it will be a wasted effort.

Next we must invent a binary or recombination operation. Let A be a 14×14 solution matrix with its 196 elements written out as one long array and B a second one likewise unstacked. At random, select a position $1 \leq k \leq 196$ in the array. Starting at position k, move along the two arrays, with wrap, comparing their elements the until the first time they differ, either A has a 1 where B has a 0 or vice-versa. Swap these two values. Moving along from that point, with wrap, continue comparing values until the first subsequent point where the two differ in the reverse way. Swap these two values. The modified A matrix is the output of the operation. Effectively this operation interchanges a 0 and a 1 in A, using B as a template, generally over a longer distance in the matrix than adjacent elements.

We take the population size to be 16. In the repeat or generation loop, we do 8 recombination operations and 8 mutation operations. Thus, after these are performed, the population size has grown to 32 and needs to be reduced back to 16. Algorithm RS selects out those for removal, one by one, according to a geometric distribution based on fitness rank.

The curve labeled "genetic algorithm" in Figure 1 (page 629) shows the results of several runs. The GA did very well on the problem, obtaining a maximum value of 2592, the best of all the methods tried.

11.6.3 Some specific Algorithms

Algorithm JH

Assume structures $x \in \Omega$ are bit strings.

Uniformly at random, with replacement, select an initial population
$\mathcal{P}(0) = \{x_1^{(0)}, \ldots, x_z^{(0)}\}$ of size z from Ω.
evalute their fitnesses $\phi(x_k^{(0)})$, $k = 1, \ldots, z$.

loop $t = 0, 1, \ldots$ **until** exit criteria met
  roulette-wheel select $x_\alpha \in \mathcal{P}(t)$
  do (with probability p_c) a recombination:
    roulette-wheel select $x_\beta \in \mathcal{P}(t)$
    perform a crossover of x_α and x_β
    select one of the resulting structures equally likely and
      designate it as y
  end do
  else // recombination not selected
    $y \leftarrow x_\alpha$
  end else
  do (with probability p_m) a mutation:
    select uniformly at random a component of y and perturb it.
  end do
  evaluate the fitness of y
  update best
  do a replacement:
    with uniform probability select $i \in \{1, 2, \ldots, z\}$ and
      replace $x_i^{(t)}$ by y to produce $\mathcal{P}(t+1)$
  end do
end loop

Algorithm DG

Assume structures $x \in \Omega$ are bit strings and population size z is an even number.

Uniformly at random, with replacement, select an initial population
$\mathcal{P}' = \{x_1^{(0)}, \ldots, x_z^{(0)}\}$ of size z from Ω.
evalute the fitnesses $\phi(x_k^{(0)})$, $k = 1, \ldots, z$.

loop $t = 0, 1, \ldots$ **until** exit criteria met
  $\mathcal{P}(t) \leftarrow \mathcal{P}'$
  $\mathcal{P}' \leftarrow$ `null`
  loop $j = 1$ to z, increment by 2
    roulette-wheel select $x_\alpha \in \mathcal{P}(t)$
    roulette-wheel select $x_\beta \in \mathcal{P}(t)$
    do (with probability p_c) a recombination:
      perform a crossover on x_α and x_β, keep both offspring x_γ and x_δ
      loop on i ranging over the components of x_γ
        with probability p_f perturb component i.
      end loop
      loop on i ranging over the components of x_δ
        with probability p_f perturb component i.

```
            end loop
            evaluate the fitnesses φ(x_γ) and φ(x_δ)
            update best
         end do
         else // recombination not selected
            x_γ ← x_α
            x_δ ← x_β
         end else
         add x_γ and x_δ to P'
      end loop
   end loop
```

Algorithm RS

The chromosomes of the population are always kept in rank order by fitness. As new chromosomes are created, they are merged into the population in their proper place according to rank. Population size is fairly small, on the order of 12 to 16.

Uniformly at random, with replacement, select an initial population
$\mathcal{P}(0) = \{x_1^{(0)}, \ldots, x_{z_0}^{(0)}\}$ of size z_0 from Ω.
Evalute and rank order $\mathcal{P}(0)$ by fitness $\phi(x_k^{(0)})$, $k = 1, \ldots, z_0$.

```
loop t = 0, 1, ... until exit criteria met
   z ← z_0
   loop j = 1 to z_0/8  // do mutations
      select i ∈ {1, 2, ..., z} uniformly at random
      z ← z + 1
      select uniformly at random a component of x_i and perturb it.
      designate the resultant structure x_z, evaluate and merge
        it into P(t)
   end loop
   loop j = 1 to z_0/2  // do recombinations
      let x_α be the structure in P(t) with rank j
      select β ∈ {1, 2, ..., z} uniformly at random
      z ← z + 1
      perform a crossover of x_α and x_β
      designate the resultant structure x_z, evaluate and merge
        it into P(t)
   end loop
   update best // check the rank 0 structure
   loop while z > z_0
      select a structure from P(t) geometrically at random
        and discard it
      z ← z − 1
   end loop
end loop
```

11.6.4 GA principles, schemata, multi-armed bandit, implicit parallelism

As previously mentioned, evolutionary computations draw their motivation and guidance from the mechanics of biological evolution. But this can lead to many complicating mechanisms such as chromosomal inversion, multiple alleles, diploidy, dominance, genotype, overlapping generations just to name a few. Even a simple evolutionary computation involves many implementation parameters. Some obvious ones are population size, number of mutations per iteration, number of recombinations per iteration, number of chromosomes to replace per iteration, number of mutations per chromosome and others. A more fundamental "parameter" is how the objective function is mapped to chromosome fitness. It may be desirable to exaggerate differences in fitness for example. Equally fundamental are the details of the three main operators, for example, the details of choosing mates. With regard to mutation, it may be desirable to only flip bits with a certain probability, what should that probability be? Finally, for how many iterations should the algorithm be run?

Thus many detailed questions arise which cannot be answered mathematically. For even simple GA's, discovering provable statements about, for example, optimal parameter determinations such as population size has been intractably difficult. To shed light on these issues and provide direction, guidance has come in the form of the Schema principle and building block hypothesis ([37]), and experimental experience.

Nevertheless, results derived from these principles and hypotheses and experiments are at best guidelines only. Having followed the guidelines, any given problem with it own unique objective and domain might not conform to the guidelines and at the very least will require a certain degree of tuning [36].

Schema Principle The Schema principle is best explained in terms of a binary coding. A string in the search space, e.g. $(1,0,0,\dots,1)$, is a vertex of the Hilbert cube in n-dimensional space. A schema is an affine coordinate subspace of n-space intersected with the cube. For example the schema in 3-space signified by $(1,0,*)$ is the set $\{(1,0,0),(1,0,1)\}$. A schema can be specified by an n-tuple of the three symbols, 0, 1, and $*$, called a *schemata* in which $*$ is the "don't care" symbol matching either a 0 or a 1. The order $o(H)$ of a schema is the number of 0's and 1's in its schemata and is thus the number of fixed positions. In terms of affine subspaces it is the co-dimension of the affine subspace. The *length* $\delta(H)$ of a schema is difference $\ell - f$ where ℓ is the position of the last fixed bit and f is the position of the first.

Any given string is a member of 2^n different schema because in each position there could be the given symbol, 0 or 1, or the don't care symbol. In a population of size z there are up to $z2^n$ schema represented (actually less because of overlap, the 0 order schema belongs to every one of them).

Suppose roulette wheel selection is used for the next generation, as in algorithm DG (661). Let $m_t(H)$ denote the number of representatives of schema H in generation t. If $\phi(H,t)$ is the average fitness of H in generation t and $\bar{\phi}(t)$ is the average fitness of the population in generation t, then the expectation $E(m_{t+1}(H))$ is given by $m_t(H)\frac{\phi(H,t)}{\bar{\phi}(t)}$.

Thus the representation of schema H increases (or decreases) as the factor $\frac{\phi(H,t)}{\bar{\phi}(t)}$. And the population thereby increases in fitness from generation to generation.

Since this is going on for each schema represented in the population (and since the present population represents the result of many previous ones, referring in particular those schema that have vanished) the processing from generation to generation represents an implicit parallel processing of schema. If population size is z, the number of short length schema processed is $O(z^3)$, see [31, p. 40].

Now consider the one-point crossover operation. The probability it will disrupt a given schema H is $\frac{\delta(H)}{n-1}$, i.e. small if $\delta(H)$ is small. Hence short schema tend to survive from generation to generation and, combining with above, if above average, increase in representation.

The probability that a one position mutation will disrupt a schema H of order $o(H)$ is $\frac{o(H)}{n}$, i.e. small for low order schema. Schema theory, short, low order, above averagely fit schema increase in representation in the population.

In the field of statistical decision theory the 2-armed bandit problem is considered. Each play of a game has two choices. Having made choice i, $i = 1, 2$, the player receives a payoff randomly drawn from a distribution with fixed mean μ_i and variance σ_i^2. Assuming these parameters are not known in advance and can only be estimated through repeated play, in T plays of the game, how many times should the ith choice be made? The answer is, the number to allocate to the worse performer up to the present time T increases essentially logarithmically in T. (And hence, T minus $O(\text{logarithmic}(T))$ in the better performer.) Put differently, the number of trials to allocate to the better performer is an exponential function of the number to allocate to the poorer one.

It can be shown that this is the same rate at which roulette wheel selection allocates fitness processing to schema [37].

On the basis of the Schema principle and experimental experience, we discuss some of the main parameters of an evolutionary computation.

Encoding The first task in constructing an evolution based search algorithm is to define a mapping, or *encoding*, between the states of the solution space Ω and computer structures. Usually there is a natural computer formulation of the states, and if so, adopting it is good practice. We shall discuss some typical situations.

Very often Ω is a Cartesian product space,

$$\Omega = \Omega_1 \times \Omega_2 \times \cdots \times \Omega_n,$$

so that $x \in \Omega$ is an n-tuple. If each component set Ω_i is finite, $\text{card}(\Omega_i) = \omega_i$ $i = 1, \ldots, n$, then a natural encoding is

$$x \longleftrightarrow (\xi_1, \xi_2, \ldots, \xi_n) \in \mathbb{Z}_{\omega_1} \times \mathbb{Z}_{\omega_2} \times \cdots \times \mathbb{Z}_{\omega_n}$$

where $\mathbb{Z}_k = \{1, 2, \ldots, k\}$. This is referred to as an *integer coding* of the solution space. In the special case that $\omega_i = 2$, for all i, then it is a *binary coding* and the components are taken as the bits 0 and 1 (instead of 1 and 2).

In the case that the component sets are intervals $[a_i, b_i]$ of the real line $\mathbb{R}$, then Ω is a subset of Euclidean space and possesses a natural topology. An encoding that takes advantage of the topology is putting $x \longleftrightarrow (\xi_1, \xi_2, \ldots, \xi_n)$ where each component $\xi_i \in [a_i, b_i]$. We show below that there are natural mutation and recombination operations of such structures. This would be a *continuous coding* of the solution space.

Alternatively, one can represent each continuous variable ξ_i, suitably scaled, as a binary string. String length will be chosen to achieve a desired level of precision for the representation. Finally the stings for each component are concatenated thus giving a binary coding for the solution space.

Not all problems have states which derive from Cartesian products. The most famous example of this is the Traveling Salesman Problem in which the solution space consists of *tours* or, mathematically, permutations of the set of cities. On the grounds that natural representations are best, the structures for this problem are typically taken as permutations of $\mathbb{Z}_{n-1}$ where n is the number of cities. In this case, the operations of mutation and recombination must be constructed so as to satisfy *closure*, that is, their resultants must remain within the set of defined structures.

Fitness If the problem at hand is one of maximization, then the simplest thing to take for fitness ϕ is the objective function f itself. However, considering that for most selection methods it is either convenient or even necessary to have a non-negative fitness function, the objective will have to be modified if it is not non-negative. If the problem is one of minimization, then again some modification of the objective will be necessary. More importantly, the choice of fitness function has been shown to have a effect on the performance of an evolutionary computation, [33]. So a prominent aspect of an implementation is choosing a mapping between the objective and the fitness function. Despite the performance effects, asymptotically, the choice of fitness function is immaterial in the sense that any two fitness functions maintaining the rank order of solutions leads to the same limit stationary distribution, see [65].

Arbitrary mappings can be described in terms of a composition $\phi(x) = \tau(f(x))$ with some mapping function τ. If the problem is one of minimization then τ will have to be inverting, for example $\tau(f) = 1/f$ if $f \neq 0$, or $\tau(f) = C - f$ for some large constant $C > \max f$, or possibly $\tau(f) = e^{-f}$. Special mention should be made of the mapping $\tau(f) = e^{\pm f}$ (e^{+f} for maximizing objectives and e^{-f} for minimizing ones). This mapping is always positive and needs no a priori knowledge about the objective.

It is easy to see how fitness can affect, for example, roulette wheel selection. If the values of f vary over only a very small range, then roulette wheel selection is not very different from uniform selection. This could be fixed with a linear mapping function, thus τ could be a simple shift or scaling, [31].

Dynamic scaling has been suggested with τ of the form

$$\tau(f(x)) = af(x) + b(t)$$

where $b(t)$ could be

$$b(t) = -\min\{f(x) : x \in \text{ population at generation } t\}.$$

This maintains strong selective pressure throughout the run.

If the selection method is based on rank order, as in algorithm RS (662), then there is no mapping issue, see [73].

Selection Distributions The uniform distribution is the simplest probability distribution and one of the most widely used. It places equal probabilistic weight on the totality of possible choices. Since computer random number generators are themselves uniform generators, this is also the easiest distribution to implement. For example, when structures are binary coded, the uniform distribution for selecting crossover points is the universal choice. If the string length is L, then

$$k = 1 + (\texttt{int})((L-1) * \text{unif}())$$

selects crossover point between the kth and $k+1$st bits from among the choices $\{1, 2, \ldots, L-1\}$ equally likely. In this, unif() is the computer function that returns a uniform floating point random number in the semi-open interval $[0, 1)$ and `int` is the greatest integer function.

One of the most important selection probability distribution is the fitness weighted lottery or roulette wheel selection. Let F denote the sum of fitnesses of the present population

$$F = \sum_{i=1}^{z} \phi(x_i).$$

Under roulette wheel selection, member x_α of the population is chosen with probability

$$p_\alpha = \phi(x_\alpha)/F.$$

Roulette wheel selection is used to impart a reproductive advantage for the better fit structures. This can be implemented either in the choice of recombination pairs or in the selection of survivors for the next generation.

The *geometric selection* is one that weights a rank ordered set, say $\{1, 2, \ldots\}$, so that the probability α is selected is

$$p_\alpha = (1 - \ell)^{\alpha - 1}\ell, \qquad 0 < \ell < 1.$$

Thus 1 has the greatest chance of being selected and the chance that another choice is selected decreases geometrically. For a finite set, $\{1, 2, \ldots, n\}$, the distribution is modified by adding the residue,

$$(1 - \ell)^n \ell + (1 - \ell)^{n+1}\ell + \cdots = (1 - \ell)^n,$$

to every choice. This distribution can be implemented as follows:

```
k = 1
loop
   if( unif() < ℓ ) return k
   k ← k + 1
   if( k > n )
      k = 1 + (int)(n * unif())
      return k
   end if
end loop
```

Population Size Population sizes reported in the literature cover a wide range from 10, to 1000 or more; most use population sizes greater than 30. When comparing results between different algorithms, it is important that the total number of function evaluations be compared and not the number of generations. The cost of a run, in terms of time and resources, is proportional to the number of function evaluations. The question then is how to optimally allocate the number of function evaluations between more per generation or more generations.

The connection with population size is that larger populations entail more function evaluations per generation. In order to appreciably modify the population from generation to generation, the number of genetic operations, and correspondingly function evaluations, per generation will also have to be large. Thus fewer generations can be run.

Population size relates to a trade-off between exploration and exploitation of the fitness surface. In large populations, there will be larger numbers of similar chromosomes, thereby ensuring greater exploitation of the local topology. Radical, poorly performing chromosomes are less likely to survive owing to their small share of the roulette-wheel. In small populations, there is a much greater chance that even dominant performers will fail to reproduce, thereby opening the way for radically different solutions to compete. Here the fitness surface is more widely explored.

Based on the criteria of maximizing the number of new schemata per individual, some reports favor population sizes for binary coded strings that vary exponentially with string length L [29], [30]. Population size 100 is used for a 30 bit problem in [31]. However most experimentalists choose population size between L and $2L$.

Again under the assumption of integer coding with q symbols, based on the criteria that every possible point in the search space should be reachable from the initial population, the calculation of the probability P that a population size Z contains at least one representative of each symbol at each place of a string of length L is [60]

$$P = \left[\frac{q!S(Z, q)}{q^Z}\right]^L.$$

In this $S(Z, q)$ is the Stirling number of the second kind (cf. [43]). Thus, for binary coding, and with $P = 99\%$, population size should be on the order of 10 for string lengths L from 20 to 200.

In Pareto optimization, that is optimization on multiple objectives simultaneously, large population sizes are necessary in order that the competing objectives be adequately represented.

Emphasis on Recombination or Mutation As previously mentioned, recombination is one of the biggest differences between evolutionary computation and simulated annealing. Moreover, the original evolutionary strategies algorithms did not use a recombination operation. As we have seen, it is the mutation operator which makes an evolutionary computation irreducible, thereby enabling its theoretical property of asymptotic convergence to the global optimum. Thus it would be possible to fashion an evolutionary computation without using recombination at all as we have pointed out by noting the possibility of a step-wise approach to writing an evolutionary computation. But this would be a hamstrung evolutionary computation indeed.

By contrast, recombination is very heavily emphasized in genetic algorithms while mutation is not. The theories about the success of genetic algorithms in terms of schemata, multi-armed bandit, and implicit parallelism all derive from the crossover operator. The upshot is that mutation rates in genetic algorithms are very small, mainly being used to avoid premature convergence. Some example recombination and mutation rates are:

source	p_{mutation}	$p_{\text{crossover}}$
[18]	0.001	0.6
[32]	0.005-0.01	0.75-0.95
[64]	0.01	0.95

Since the mutation rates above apply to each bit of an L bit string, the probability of one or more mutations occurring is

$$1 - (1 - p_m)^L$$

or about 14% for a 30 bit string when $p_m = 0.005$.

Using the criteria that the mutation should maximize the probability that a mutant is more fit than its progenitor, it has been derived that $p_{\text{mutation}} \approx 1/L$ where L is bit string length [5].

11.6.5 A genetic algorithm for constrained optimization problems

Many optimization problems, especially those in engineering design, are highly constrained, and often non-linear, resulting in a complex search space with regions of feasibility and infeasibility, see [16]. For such problems, it is necessary to find global optima not violating any constraint.

We direct the reader to the excellent work by Michalewicz and Schoenauer for a review of the literature [54].

Most approaches for handling constraints can be classified into two broad categories:

- those that exclude infeasible solutions,
- those that penalize infeasible solutions.

In turn, excluding infeasible solutions can be arranged by

- discarding them as they arise,
- the use of specialized operators that maintain feasibility,
- repairing infeasible solutions.

Discarding infeasible solutions as they arise impacts the efficiency of the algorithm. If infeasible solutions arise too frequently, then the algorithm may spend significant amounts of time looking for those few solutions that do not violate constraints. The probability that the genetic operators generate feasible offspring when applied to feasible parents is an important issue. It will take some time to find the region, but also, once found, the probability of staying within it is important.

The use of specialized operators that maintain feasibility is the most effective method for constrained problems when applicable. This approach is possible, for example, in the case of linear constraints.

When feasibility maintaining operators cannot be constructed, it still may be possible to repair or transform infeasible solutions into feasible ones. This idea works well in the case of linear equality constraints. By way of illustration, in the example above, another constraint is that the parameters p_i must be non-negative and sum to 1. After carrying out a mutation or crossover involving the p_i, the new values may be "repaired" by renormalization.

The use of specialized operators and the use of repair operators are related methods for maintaining feasibility among solutions. However, for many constrained problems, it is too hard, too costly, or even impossible to maintain feasibility.

The most prevalent technique for coping with infeasible solutions is to penalize a population member for constraint violation. In this way, penalty functions artificially create an unconstrained optimization problem. Traditionally, the weighting of a penalty for a particular problem constraint is based on judgment. Often, the algorithm must be *tuned*, that is, rerun several times before a weighting of the combination of constraint violations is found that eliminates infeasible solutions and retains feasible solutions. If the penalty is too harsh, then the few solutions found that do not violate constraints, quickly dominate the mating pool and yield suboptimal solutions. A penalty that is too lenient can allow infeasible solutions to flourish as they can have higher fitness values than feasible solutions [61].

Penalty approaches might be classified as

- static
- dynamic
- specialized.

Dynamic penalties vary with the degree of constraint violation and with either the history or the run time of the algorithm. We return to this subject below. Static penalties only vary with the degree of constraint violation. Many ideas have been promulgated for assessing the degree of constraint violation. For example, Richardson, et al. [61] tried several approaches for assigning penalty functions using the derivative of the objective function to give an indication of how far an infeasible solution is from the constraint boundary. There are also many direct methods for quantifying such distances. Generally, the degree of penalty should increase as a function of the distance from the feasible set in some norm.

To deal with the problem of having to do extensive tuning in order to find the most effective penalty level, methods have been proposed in which the relative weight allocated to the penalty varies with the progress of the algorithm. There are two types, those for which the level of penalty depends only on the run time t of the algorithm and those that allow other measures of progress, such as recent stagnation, to affect the level. This is a major distinction because the former are instances of inhomogeneous Markov chains for which there is mathematical theory, however difficult to apply, while the latter may only be analyzed experimentally. Most variable penalty approaches proposed take the fitness to be modified additively by the penalty, that is

$$\varphi = f + wM \tag{11.6.11}$$

where φ is the algorithm fitness, f is the objective to be maximized, M is a measure of the extent of constraint violation and w is a variable weight; the product wM is the penalty. But also fitness may be taken as multiplicatively modified by the effect of the penalty,

$$\varphi = \alpha f \tag{11.6.12}$$

where an "attenuation" α, depends on M and t.

Michalewicz and Attia [52] use a fitness function of the form

$$\varphi(x, r) = f(x) + \frac{1}{2r}M$$

where M is quadratic in the constraint violation and $1/(2r)$ is the varying penalty weight. The parameter r, referred to as "temperature," tends to 0 according to a "cooling schedule" $g(r,t)$. (Cooling as used here is not in the same sense as used in simulated annealing. In particular, the weight function tends to infinity as the temperature tends to 0.) In this, t counts epochs, that is, an entire genetic algorithm run conducted at a fixed temperature. A complete run of this penalty function algorithm consists of several such epochs. The initial temperature r_0 is a parameter of the problem. The cooling schedule is allowed to depend on the problem, in one case $g(r,t) = 10^{-1}g(r, t-1)$ recursively; hence

$$r_t = 10^{-t}r_0$$

giving geometric decrease in r and therefore a geometric increase in weight.

An advantage of the multiplicative form in which the penalty is applied is that it makes the method closely related to simulated annealing (but unlike annealing, there is no acceptance phase). As a consequence, this method can be proved to converge to a globally optimal feasible solution by an adaptation of a generalization of Hajek's Theorem by Catoni [10] (generalized annealing). The penalty function makes use of a single problem dependent parameter, the starting temperature T_0. As in simulated annealing, this parameter is not critical; theoretically any positive value will do. In practical terms however, T_0 should exceed the "phase transition" temperature.

Let f denote the objective function, to be maximized here, which we will assume is non-negative valued throughout its domain Ω. We will take the fitness function φ of the GA to be the product of f and an *attenuation factor* $\alpha(\cdot,\cdot)$ which depends on two parameters, M and T,

$$\varphi(x) = \alpha(M,T)f(x). \tag{11.6.13}$$

The first, $M \geq 0$, measures the extent of constraint violation in some metric, e.g. ℓ_2, and is zero in the absence of any violation. The second parameter, referred to as *temperature* $T > 0$, is a function of the running time of the algorithm; T tends to 0 (or small values) as execution proceeds. When the GA begins, we want the penalty for constraint violation to be small, or, in terms of attenuation, we want $\alpha \approx 1$, in order that the algorithm be able to utilize infeasible states as needed to find a global maximum. But toward the end of execution we want α to be zero or nearly zero since infeasible solutions are unacceptable. A function which has these properties is

$$\alpha(M,T) = e^{-M/T}. \tag{11.6.14}$$

If no constraint is violated, independent of the value of T, then $\alpha = 1$ and fitness is the unattenuated objective value. On the other hand, when T is large (relative to a non-zero M) then $\alpha \approx 1$. But as $T \to 0$, then $\alpha \to 0$ as well and hence, by equation (11.6.12), fitness tends to zero too. Thus infeasible solutions should be excluded from the GA populations at the end of a run. (In practice, a run is terminated before $T = 0$ and infeasibles may remain in the population. As the final output of the algorithm, only feasible solutions might be posted; however it may also be of value to examine good infeasible ones as well.)

A variable fitness genetic algorithm The variable fitness feature may be added to any evolutionary computation. Because fitness is a function of run time t, the fitnesses of the current population must be recalculated every time the temperature is updated. This does not entail a new objective calculation however, only an attenuation factor modification. If the original temperature is T_0 and the new one is T_1, then the adjustment factor is

$$\frac{e^{-M/T_1}}{e^{-M/T_0}} = (e^{\frac{1}{T_0} - \frac{1}{T_1}})^M.$$

This fitness modification might also impact the running best solution so care must be taken to modify that as well.

11.6.6 Markov Chain Analysis Particular to Genetic Algorithms

In this section we assume Ω is the set of all binary strings of length L. Thus

$$\Omega = \mathbb{Z}_2 \times \mathbb{Z}_2 \times \cdots \times \mathbb{Z}_2$$

for L copies. At the same time a binary string i can be identified with its base two integer representation,

$$i \longleftrightarrow i_{L-1}2^{L-1} + \cdots + i_1 2 + i_0.$$

Thus $i \in \mathbb{Z}_{N-1}$ where $N = 2^L$. Letting $\oplus$ denote bitwise EXCLUSIVE OR on Ω, the pair $(\mathbb{Z}_2 \times \cdots \times \mathbb{Z}_2, \oplus)$ is a group. Additionally it will be convenient to let $i \otimes j$ be the bitwise LOGICAL AND of i and j. Following Vose and Liepins [72], we will consider an infinite population genetic algorithm. In this way the Markov Chain is replaced by a discrete dynamical system caricature. Subsequently this development was extended to finite population size [58], but more recent work, e.g. [47], has been along the lines of the infinite population model. The infinite population assumption implies that on each iteration, the outcome will be the expected outcome of the finite population chain.

Very recently a different genetic algorithm model has been analyzed by Schmitt, Nehaniv, and Fujii, [65]. In this model, populations of size z are treated as z-tuples rather than the usual multi-sets – sets with multiplicity but no order on their elements. Otherwise no special assumptions are made. The finite dimensional linear space on which their genetic operators act is the free vector space of these populations. Results about populations as multi-sets can be recovered through projection into the quotient space over the kernel of permutations on these populations. As the authors point out, position in the z-tuple may be used to mimic spacial effect, for example, on such populations.

Returning to the dynamical systems model, the state of the system will be described by a vector $x^t \in \mathbb{R}^N$ whose ith component is equal to the proportion of i in the tth generation. Further, let $r_{i,j}(k)$ be the probability that bit vector k results from the recombination process based on parents i and j. It can be shown that if recombination is a combination of mutation and crossover, then

$$r_{i,j}(k \oplus \ell) = r_{i \oplus k, j \oplus k}(\ell).$$

Next let F be the nonnegative diagonal matrix with i,ith entry $f(i)$ and let $M = (m_{i,j})$ be the matrix whose terms $m_{i,j} = r_{i,j}(0)$. Define permutations σ_j on $\mathbb{R}^N$ by

$$\sigma_j(x_0, \ldots, x_{N-1})^T = (x_{j\oplus 0}, \ldots, x_{j\oplus(N-1)})^T$$

where T denotes transpose. Define the operator $\mathcal{M}$ by

$$\mathcal{M}(x) = \left((\sigma_0 x)^T M \sigma_0 x, \ldots, (\sigma_{N-1}x)^T M \sigma_{N-1} x\right)^T.$$

Let $\equiv$ be the equivalence relation on $\mathbb{R}^N$ defined by $x \equiv y$ if and only if there exists a $\lambda > 0$ such that $x = \lambda y$.

Theorem 11.6.1 *Over one iteration, (the expectation of) x^{t+1} is given by $x^{t+1} = F\mathcal{M}(x^t)$.*

Thus the expected behavior of the genetic algorithm is described by two matrices: fitness and selection behavior is contained in F and mixing behavior is contained in M.

Theorem 11.6.2 *The matrix M is nonnegative and symmetric, and for all i, j satisfies $1 = \sum_k m_{i\oplus k, j\oplus k}$.*

Next let $W = (w_{i,j})$ be the Walsh matrix defined by

$$w_{i,j} = \prod_{k=1}^{\ell} r_{k(\lfloor i2^{1-k}\rfloor \bmod 2)}(j)$$

where $r_k(t)$ is the Rademacher function

$$r_i(t) = 1 - 2\left(\lfloor \frac{t2^i}{N} \rfloor \bmod 2\right),$$

see [35]. The Walsh matrix is symmetric and orthogonal and satisfies

$$w_{i\oplus j,k} = w_{i,k} w_{j,k}.$$

Theorem 11.6.3 *The matrix WM_*W is lower triangular, where M_*, the twist of M is defined by*

$$m_{*i,j} = m_{i\oplus j,i}.$$

At this point we can regard the composition $\mathcal{G} = F\mathcal{M}$ as a dynamical system on the unit sphere $\mathcal{S}$ in the positive orthant of $\mathbb{R}^N$ since, except for the origin, each equivalence class of $\equiv$ has a unique member in $\mathcal{S}$. Regarding F as a map on $\mathcal{S}$, its fixed points are the eigenvectors of F which are the standard unit basis vectors $u_0, \ldots, u_{N-1}$.

Theorem 11.6.4 *The basin of attraction of the fixed point u_j of F is given by the intersection of $\mathcal{S}$ with the (solid) ellipsoid*

$$\sum_i \left(s_i \frac{f(i)}{f(j)}\right)^2 < 1.$$

The following holds for fixed points x of $\mathcal{M}$.

Theorem 11.6.5 *Let x be a fixed point of $\mathcal{M}$, then x is asymptotically stable whenever the second largest eigenvalue of M_* is less than $1/2$.*

In order to determine the second largest eigenvalue of M_*, the terms $m_{i,j}$ must be calculated. Let the genetic algorithm perform a one-point crossover every generation with probability χ and component by component bit flip with probability μ (as in algorithm DG (661)). Then it can be shown, [72], that

$$m_{i,j} = \frac{(1-\mu)^\ell}{2}\left[\eta^{|i|}\left(1-\chi+\frac{\chi}{\ell-1}\sum_{k=1}^{t-1}\eta^{-\Delta_{i,j,k}}\right) + \eta^{|j|}\left(1-\chi+\frac{\chi}{\ell-1}\sum_{k=1}^{t-1}\eta^{\Delta_{i,j,k}}\right)\right], \tag{11.6.15}$$

where $\eta = \mu/(1-\mu)$, integers are to be regarded as bit vectors when occurring in $|\cdot|$, division by zero at $\mu = 0$ and $\mu = 1$ is to be removed by continuity, and

$$\Delta_{i,j,k} = |(2^k - 1) \otimes i| - |(2^k - 1) \otimes j|.$$

The following was proved in [47].

Theorem 11.6.6 *The spectrum of M_* is*

$$(1-2\mu)^{|i|}(1-\chi wid(i)/(L-1))/2, \quad i = 0, \ldots, N-1.$$

where $wid(i)$ is the difference between the position of the highest non-zero bit and the lowest non-zero bit of i for $i > 0$ and 0 otherwise.

In particular

Corollary 11.6.7 *If $0 < \mu < 1/2$ then the second largest eigenvalue of M_* is $0.5 - \mu$.*

In addition there is a simulated annealing like result for genetic algorithms. We follow Davis and Principe [17] and Suzuki [68]. In this it is assumed that the points in Ω are sorted by decreasing fitness,

$$f(i_0) \geq f(i_1) \geq \cdots \geq f(i_{N-1}).$$

Theorem 11.6.8 *The stationary distribution $q^{(\mu)}(s)$ for mutation probability μ converges to the best population as $\mu \to 0$ and $\chi \to 0$ and the fitness ratio converges to 0,*

$$F = \max_{1\leq j\leq N-1, f(i_j)\neq f(i_{j+1}} \frac{f(i_j)}{f(i_{j+1})} \to 0.$$

That is

$$\sum \lim_{F\to 0}\lim_{\chi\to 0}\left[\lim_{\mu\to 0+} q^{(\mu)}(x)\right] \to 1,$$

where the sum is over those populations all of whose members are identical and which evaluate to the maximum fitness.

Bibliography

[1] Aarts, E. and Korst, J. (1989), *Simulated Annealing and the Boltzmann Machines.* Wiley, Chichester.

[2] Andresen, Bjarne (1996), Finite-time thermodynamics and simulated annealing. *Entropy and Entropy Generation*, ed. J. S. Shinov (Dordrecht Kluwer), 111–127.

[3] Archetti F. and Schoen F. (1984), A survey on the global optimization problem: general theory and computational approaches, *Annals of Operations Research* **1**, (1) 87-110

[4] Azencott, R. (1992), *Simulated Annealing, Parallelization Techniques*, John Wiley and Sons, New York.

[5] Bäck, Thomas (1993), Optimal Mutation Rates in Genetic Search, *Proc. of the Fifth International Conference on Genetic Algorithms*, Morgan Kaufman, San Mateo, CA, 2-8.

[6] Bäck, Thomas, Hoffmeister, Frank and Schwefel, Hans-Paul (1991), A Survey of Evolution Strategies, *Proc. of the Fourth International Conference on Genetic Algorithms*, Morgan Kaufman, San Mateo, CA, 2-9.

[7] Bélisle, Claude (1992), Convergence theorems for a class of simulated annealing algorithms on R^d. *J. Appl. Prob.* **29**, 885–895.

[8] Boender, G. and Rinnooy Kan, A. (1987), Bayesian Stopping Rules for Multistart Optimization Methods, *Math. Programming*, **37**, pp 59–80.

[9] Byrd, Richard H., Dert, Cornelius L., Rinnooy Kan, Alexander H.G., and Schnabel, Robert B. (1990), Concurrent Stochastic Methods for Global Optimization, *Math. Programming,* **46**, 1-29.

[10] Catoni, O. (1991), Sharp Large Deviations Estimates for Simulated Annealing Algorithms, *Ann. Inst. Henri Poincaré, Probabilités et Statistiques,* **27**, 3 291-383.

[11] Catoni, O. (1992), Rough large deviation estimates for simulated annaling - application to exponential schedules, *Ann. of Prob.*, 1109-1146.

[12] Chiang, T. and Chow, Y. (1988), On the convergence rate of annealing processes. *SIAM J. Control and Optimization* **26**, 1455–1470.

[13] Chung, K. (1967), Markov Chains with Stationary Transition Probabilities, Springer, Berlin.

[14] Culberson, Joseph (1998), On the futility of blind search: An algorithmic view of 'No Free Lunch,' *Evolutionary Computation Journal*, **6** 2, 109 - 128.

[15] Cruz, J. R and Dorea, C. C. (1998), Simple conditions for the convergence of simulated annealing type algorithms. *J. Appl. Probab.* **35**, no. 4, 885–892.

[16] Dasgupta, D. and Michalewicz, Z. (Eds.) (1997), *Evolutionary Algorithms in Engineering Applictions,* Springer, New York.

[17] Davis, Thomas E. and Principe, Jose C. (1991), A Simulated Annealing Like Convergence Theory for the Simple Genetic Algorithm, *Proc. of the Fourth International Conference on Genetic Algorithms*, Morgan Kaufman, San Mateo, CA, 174-181.

[18] DeJong, K. A. (1975), *An analysis of the behaviour of a class of genetic adaptive systems.* Ph.D. thesis, University of Michigan, Diss. Abstr. Int. 36(10), 5140B, University Microfilms No. 76-9381.

[19] Diaconis Persi and Stroock Daniel (1991), Geometric Bounds for eigenvalues of Markov chains, *Ann. Appl. Prob.*, vol. 1 No. 1 36-61.

[20] Diener, Immo (1995), Trajectory Methods in Global Optimization, *Handbook of Global Optimization,* eds. Reiner Horst and Panos Pardalos, Kluwer Academic, Dordrecht, 649-668.

[21] Dunham, B., Fridshal, D., Fridshal, R., North, J. H (1959), Design by Natural Selection, *Proceedings of an International Symposium on the Theory of Switching*, 192-200, Harvard U. Press, 1959 and IBM Journal of Research and Development 3, (1959) 46–53, and IBM Journal 3, 282–287.

[22] Feller, W. (1968), *An Introduction to Probability Theory*, Wiley, New York.

[23] Fox, B. (1995), Faster Simulated Annealing, *Siam J. Op.* **5** (3) 488-505.

[24] Fox, B. and Heine, G. (1995), Probabilistic search with overrides, *Annals of Applied Probability* **5**, 1087–1094.

[25] Gelfand, Saul B. and Mitter, Sanjoy K. (1993), Metropolis-type annealing algorithms for global optimization, *SIAM J. Control Optim.* vol31, 111-131.

[26] Geman, S. and Geman, D. (1984), Stochastic relaxation, Gibbs distributions, and Bayesian restoration of images, *IEEE Trans,* **PAMI-6**, no. 6, 721-741.

[27] Geman, Stuart and Hwang, Chii-Ruey (1986), Diffusions for Global Optimization, *SIAM J. Control Optim.*, Vol. 24, 1031-1043.

[28] Gidas, B. (1985), Nonstationary Markov chains and convergence of the annealing algorithm, *J. Stat. Phy.*, **39** 73-131.

[29] Goldberg, D. E. (1985), Optimal initial population size for binary-coded genetic algorithms, *TCGA Report 85001*, University of Alabama, Tuscaloosa.

[30] Goldberg, D. E. (1989), Sizing Populations for Serial and Parallel Genetic Algorithms, *Proc. of the Third International Conference on Genetic Algorithms*, Morgan Kaufman, San Mateo, CA, 70-79.

[31] Goldberg, D.E. (1989), *Genetic Algorithms in Search, Optimization, and Machine Learning*, Addison-Wesley, Reading, Mass.

[32] Grefenstette, J.J. (1986), Optimization of control parameters for genetic algorithms, *IEEE Transactions on Systems, Man and Cybernetics SMC-16(1)*, 122-128.

[33] Grefenstette, J. J. and Baker, J. E. (1989), How Genetic Algorithms work: A Critical Look at Implicit Parallelism, *Proc. of the Third International Conference on Genetic Algorithms*, Morgan Kaufman, San Mateo, CA, 20-27.

[34] Hajek, B. (1988), Cooling schedules for optimal annealing. *Math. Operat. Res.* **13**, No. 2, 311–329.

[35] Harmuth, H.F. (1970), *Transmission of Information by Orthogonal Functions*, Springer-Verlag, New York.

[36] Hart, W. E. and Belew, R. K. (1991), Optimizing an Arbitrary Function is Hard for a Genetic Algorithm, *Proc. of the Fourth International Conference on Genetic Algorithms*, Morgan Kaufman, San Mateo, CA, 190-195.

[37] Holland, J. (1975), Adaptation in Natural and Artificial Systems, Univ. of Michigan Press, Ann Arbor, MI.

[38] Holley, R. and Stroock, D. (1988), Simulated annealing via Sobolev inequalities, *Communications in Mathematical Physics*, Vol 115, 553-569.

[39] Horst, Reiner and Hoang Tuy (1993), *Global optimization: deterministic approaches*, Springer-Verlag, New York.

[40] Hu, X., Shonkwiler, R., and Spruill, M. (1997), Randomized restarts, reprint available.

[41] Ingrassia, Salvatore (1994), On the rate of convergence of the Metropolis algorithm and Gibbs sampler by geometric bounds, *Ann. Appl. Prob.*, vol. 4 347-389.

[42] Isaacson D. and R. Madsen (1976), Markov Chains Theory and Applications, Krieger Pub. Co., Malabar, FL.

[43] Jackson, B. W. and Thoro, Dmitri (1990), *Applied Combinatorics with Problem Solving*, Addison-Weseley, New York.

[44] Jerrum, M. and Sinclair, A. (1989), Approximating the permanent, *Siam J. Comput.*, **18** 1149-1178.

[45] Kendall, D.G. (1960), Geometric ergodicity and the theory of queues. *Mathematical Methods in the Social Sciences*, Arrow, Karlin, and Suppes, eds., Stanford.

[46] Kirkpatrick, S., Gelatt, C., Vecchi, M. (1983), Optimization by simulated annealing, *Science* **220**, 671-680.

[47] Koehler, Gary J. (1994), A proof of the Vose–Liepins Conjecture, *Ann. Math. and AI*, **10**, 409-422.

[48] Kolonko, M., (1995), A piecewise Markovian model for simulated annealing with stochastic cooling schedules, *J. Appl. Prob.* **32**, 649–658.

[49] Kushner H. (1987), Asymptotic global behavior for stochastic approximation and diffusions with slowly decreasing noise effects: global minimization via Monte Carlo, *SIAM J. Appl. Math*, Vol 47 169-185.

[50] Mendivil, F., Shonkwiler, R., and Spruill, C. (1999), Restarting Search Algorithms with Applications to Simulated Annealing, preprint.

[51] Metropolis, N., Rosenbluth, A., Rosenbluth, M., Teller, A., and Teller, E. (1953), Equations of State Calculations by Fast Computing Machines, *J. of Chem. Phy.*, **21**, 1087-1091.

[52] Michalewicz, Z. and Attia, N. (1994), Evolutionary Optimization of Constrained Problems, *Proceedings of the Third Annual Conference on Evolutionary Programming*, World Scientific, River Edge, N.J., 98–108.

[53] Michalewicz, Z. and Janikow, C. Z. (1991), Handling Constraints in Genetic Algorithms, *Proceedings of the Fourth International Conference on Genetic Algorithms*, Morgan Kaufmann Publishers, Inc., San Mateo, California, 151-157.

[54] Michalewicz, Z. and Schoenauer, M. (1996), Evolutionary Algorithms for Constrained Parameter Optimization Problems, *Evolutionary Computation*, **4**, 1 1–32.

[55] Morey, C., Scales, J., Van Vleck, E. (1998), A feedback algorithm for determining search parameters for Monte Carlo optimization, *J. Comput Phys*, **146** (1) 263-281.

[56] Mockus, Jonas (1989), *Bayesian Approach to Global Optimization*, Kluwer Academic Publishers, London.

[57] Nourani, Yaghout and Andresen, Bjarne (1998), A comparison of simulated annealing cooling strategies. *J. Phys. A: Math. Gen.* **31**, 8373–8385.

[58] Nix, A. and Vose, M.D. (1992), Modeling Genetic Algorithms with Markov Chains, *Ann. Math. and AI*, **5**, 79-88.

[59] Pelletier, M. (1998), Weak convergence rates for stochastic approximation with application to multiple targets and simulated annealing, *Ann. Appl. Prob.*, Vol. 8, No1, 10-44.

[60] Reeves, C. R. (1993), Using Genetic Algorithms with Small Populations, *Proc. of the Fifth International Conference on Genetic Algorithms*, Morgan Kaufman, San Mateo, CA, 92-99.

[61] Richardson, J., Palmer, M., Liepins, G., Hilliard, M. (1989), Some Guidelines for Genetic Algorithms and Penalty Functions, *Proceedings of the Third International Conference on Genetic Algorithms*, Morgan Kaufman, San Mateo, CA, 191-197.

[62] Rubinstein, Reuven Y. (1981), *Simulation and the Monte Carlo Method*, John Wiley & Sons, New York.

[63] Rudolph, G. (1994), Convergence analysis of canonical genetic algorithms, *IEEE Trans. on Neural Networks*, **5**, 96-101.

[64] Schaffer, J. D., Caruana, R. A., Eshelman, L. J., and Das, R. (1989), A study of control parameters affecting online performance of genetic algorithms for function optimization. *Proc. of the Third International Conference on Genetic Algorithms*, Morgan Kaufman, San Mateo, CA, 51-60.

[65] Schmitt, Lothar M., Nehaniv, Chrystopher L., and Fujii, Robert H. (1998), Linear analysis of genetic algorithms, *Theoretical Computer Science*, **200**, 101-134.

[66] Schoen, F. (1991), Stochastic Techniques for Global Optimization: A Survey of Recent Advances. *Journal of Global Optimization* **1**, 207–228.

[67] Shonkwiler, R. and Van Vleck, E. (1994), Parallel Speed-up of Monte Carlo Methods for Global Optimization, *J. of Complexity* **10**, 64-95.

[68] Suzuki, Joe (1997), A Further Result on the Markov Chain Model, *Foundations of Genetic Algorithms - 4*, Ed. Belew, R.K. and Vose, M.D., Morgan Kaufmann, San Francisco.

[69] Törn, Aimo and Zilinskas, Antanas (1989), *Global Optimization*, Lecture Notes in Computer Science 350, Springer-Verlag, New York.

[70] Vere-Jones, D. (1962), Geometric ergodicity in denumerable Markov chains, *Quarterly Journal of Mathematics Oxford*, Series 13, 7-28.

[71] van Laarhoven, P. and Aarts, E. (1987), Simulated Annealing: Theory and Applications, D. Reidel, Boston.

[72] Vose, Michael D. and Liepins, Gunar E. (1991), Punctuated Equilibria in Genetic Search, *Complex Systems*, **5** 31-44.

[73] Whitley, Darrell (1989), The `GENITOR` Algorithm and Selective Pressure: Why Rank-Based Allocation of Reproductive Trials is Best, *Proc. of the Third International Conference on Genetic Algorithms*, Morgan Kaufman, San Mateo, CA, 116-121.

[74] Wolpert, David and MacReady, William (1995), No Free Lunch Theorems for Search, Technical Report SFI-TR-05-010.

[75] Zanakis, S.H. and Evans, J.R. (1981), *Heuristic optimization: why, when, and how to use it*, Interfaces, **11**, 84-91.

Chapter 12

Stochastic Control Methods in Asset Pricing

THALEIA ZARIPHOPOULOU
Departments of Mathematics and MSIS
The University of Texas at Austin
Austin, TX 78712–1082

12.1 Introduction

The purpose of this paper is to offer a concise exposition of stochastic optimization methods used in mathematical finance models. These models arise in optimal *portfolio management* and in the areas of indexderivatives*derivatives* and *equilibrium asset pricing*. The main objective is to construct optimal investment policies and consumption plans, to determine equilibrium prices of primary assets and to specify prices of derivative securities and hedging strategies. As this chapter will show, the majority of the above valuation models give rise to stochastic optimization problems in which the criterion is either to maximize the expected utility, coming from wealth or consumption streams, or to minimize the expected loss, coming from a derivative position given a certain liability. The state controlled processes, modeling the current state of the valuation system are taken to be Markov diffusions with complete information of the state. The control processes represent investment policies, consumption plans, or hedging strategies. The optimal solution, or as it is otherwise known, the *value function*, gives either the maximal expected utility or the minimal expected cost. Under general conditions that are related to the Markovian structure of the underlying models, a general principle of optimality, known as the Dynamic Programing Principle holds. This result, together with stochastic calculus, yields that the value function solves the so-called Hamilton-Jacobi-Bellman (HJB) equation. In the case that the controlled processes are diffusions, the HJB equation turns out to be a second order fully non-linear

[0]The author would like to acknowledge partial support from a Romnes Fellowship, the Graduate School of the University of Wisconsin, Madison and the National Science Foundation (NSF Grant DMS-9971415)

equation of elliptic or parabolic type. If the controlled state processes do not degenerate, the value function turns out to be smooth and therefore it satisfies the HJB equation in the classical (strong) sense. Then one may also use classical verification results to determine the optimal control processes; in fact, it turns out that applying the first order conditions in HJB, yields the optimal policies in the so-called feedback form, in the sense that the optimal processes turn out to be explicit functions of the current state of the system and time.

In a number of interesting applications, the value function is not necessarily smooth and therefore it might not satisfy the HJB equation in the strong sense. Such situations arise in pricing models in *imperfect markets* in which the frictions are associated to trading constraints, transaction costs, stochastic volatility and incomplete information. These imperfections result in various degeneracies which may cause the solution to lose its regularity. Therefore, the notion of solution to the HJB equation must be relaxed and this is indeed done via the viscosity theory. Under reasonable assumptions on the state dynamics and the payoff/cost functionals, it turns out that the value function solves the HJB equation in the viscosity sense and, as a matter of fact, it is also unique. This characterization enables us to get useful results both from the analytic as well as the numerical point of view. Indeed, the general comparison results for viscosity solutions of the HJB equation have been successfully used to obtain analytic bounds on derivative prices as well as bounds on the hedging probabilities in markets with frictions; these are situations where the classical Black and Scholes approach breaks down. In other applications, for example in portfolio management models with stochastic labor income or with transaction costs, closed form solutions are not available and numerical approximations for the optimal strategies are highly desirable. Viscosity solutions have excellent stability properties which, together with the relevant uniqueness results, are used to establish convergence of a wide class of numerical schemes; the latter need to have some fundamental properties, namely to be monotone, consistent and stable with these properties arising naturally in the stochastic optimization problems at hand. Because of the important role that viscosity solutions play in the study of dynamic valuation models, a central part of this paper is dedicated to them.

An alternative valuable approach to study optimal portfolio management and derivative pricing models is based on *martingale methods*. This powerful methodology is widely used in a variety of asset pricing problems and yields rich results under rather general assumptions on the market coefficients. In subsequent sections, we provide a long list of references in which this approach is used.

The chapter is organized as follows: in Section 12.2, we present some fundamental background results on the HJB equation and its classical and viscosity solutions. Sections 12.3 and 12.4 are dedicated to stochastic optimization models of expected utility in complete markets and also in markets with frictions. In Section 12.5, we discuss models of derivative pricing which can be formulated as models of expected utility, especially in the case of incomplete markets for which the classical derivative valuation theories fail to apply.

12.2 The Hamilton-Jacobi-Bellman (HJB) equation

In this section we provide a general description of stochastic control methods for diffusion processes, we derive the relevant HJB equation and we discuss its classical and weak solutions. The overall description is rather formal since it is not intended to give the most general assumptions or to provide extensive proofs of rigorous results. We refer the technically oriented reader to the book of Fleming and Soner (1993, Chapters V and VIII) as well as to the landmark papers by Lions (1983).

We denote by X_t the *state* of our controlled valuation system and α_t the *control process.* Typically, X_t represents the state wealth process, the value of the hedging portfolio or the derivative price process. The control α_t represents an investment strategy, a consumption plan or a hedging component.

Investors have preferences reflecting their attitude towards the risk associated with the stochastic market returns. These preferences are modelled through a *utility function*, $v : \mathcal{R}^+ \rightarrow \mathcal{R}$ which is typically increasing, concave and a smooth function of the wealth or the consumption stream. An important index is the so-called *absolute*, resp. *relative, risk aversion* coefficient defined by $A(x) = -\frac{U''(z)}{U'(z)}$, resp. $R(z) = -\frac{zU''(z)}{U'(z)}$.

Trading takes place continuously in time between the available market accounts. The prices of the underlying assets, otherwise known as primitives, are determined via classical equilibrium conditions and they are assumed to be known in all models we are analyzing. A widely accepted modeling assumption is that asset prices can be modelled as Markov diffusion processes. Under this fundamental assumption of diffusion structure, a considerable volume of work has been produced in analytically defining, estimating and callibrating the asset price diffusion coefficients. Because the prices are taken to be diffusion processes, it follows that – in the absence of market frictions, like for example transaction costs – the state process X_t becomes a controlled diffusion as well. To establish some notation, we assume that the state equation can be written as

$$dX_t = r(X_t)dt + \mu(X_t, \alpha_t)dt + \sigma(X_t, \alpha_t)dW_t, \tag{12.2.1}$$

with W_t being a standard Brownian motion defined on a probability space $(\Omega, \mathcal{F}, P)$. We denote by $\mathcal{F}_t = \sigma(W_s; 0 \leq s \leq t)$ the complete filtration generated by the Brownian motion.

The coefficients r, μ and σ reflect the stochastic returns of the various assets available for trading. In the next section, we will present concrete examples and the precise role of the market returns will be explicitly stated. Note that in most models, one needs to introduce additional state variables and the problems become high dimensional. At this point, we do not address the general cases but we only use (12.2.1) to demonstrate the Dynamic Programming method.

The investors rebalance their portfolios and consume, either in a finite or an infinite trading horizon. In the former case, the utility payoff is given by (with a slight abuse of notation)

$$J(x, t; T, \alpha) = E\left[\int_t^T U_1(\alpha_s)ds + U_2(X_T)/X_t = x\right]. \tag{12.2.2}$$

The expectation is taken with respect to the probability measure P. The functions U_i, $i = 1, 2$ are the utility functions coming, respectively, from intermediate consumption and terminal wealth.

In the case of an infinite trading horizon, the payoff is of the form

$$J(x; \alpha) = E\Big[\int_0^{+\infty} e^{-\beta t} U(\alpha_t)dt/X_0 = x\Big], \tag{12.2.3}$$

with U being the utility from the intermediate consumption stream.

The *value function* is defined as

$$u(x, t) = \sup_{\mathcal{A}} J(x, t; T, \alpha) \tag{12.2.4}$$

or, as

$$u(x) = \sup_{\mathcal{A}} J(x; \alpha), \tag{12.2.5}$$

with $\mathcal{A}$ denoting the set of *admissible policies*. Typically, the admissible policies must satisfy certain integrability and measurability conditions; the latter constraint comes from the fact that the investor, who is actually playing the role of the system controller, does not have access to future information. Additionally, the admissible policies must satisfy various contraints that are associated with the specific economic model, for example, limited borrowing and/or shortselling, bankruptcy constraints and limitation to borrow against future labor income.

In the sequel, we concentrate on the case of finite horizon and we state the results for the infinite horizon models afterwards. A key role for the solution u is played by the *Dynamic Programing Principle* which yields that the value function satisfies

$$u(x,t) = \sup_{\mathcal{A}} E\Big[\int_t^{\tau} U_1(\alpha_s)ds + u(X_\tau,\tau)/X_t = x\Big]. \tag{12.2.6}$$

The random time τ is a positive $\mathcal{F}$-measurable random variable. Under certain technical conditions, one can show that it suffices to define the above supremum over the set of policies α_s that are feedback functions of the current state of the system.

Using the Dynamic Programming Principle and stochastic analysis, one can derive formally the *Hamilton-Jacobi-Bellman* equation

$$u_t + \max_{\alpha}\big[\frac{1}{2}\sigma^2(x,\alpha)u_{xx} + \mu(x,\alpha)u_x + U_1(\alpha)\big] + r(x)u_x = 0 \tag{12.2.7}$$

with terminal data

$$u(x,T) = U_2(x). \tag{12.2.8}$$

Note that no boundary conditions are given for (12.2.7). In fact, because the state X_s represents the current wealth, it has to satisfy certain constraints related to bankruptcy limitations and, more generally, to arbitrage conditions. Typically, the presence of these constraints results in lack of explicit boundary data which can be retrieved only after the value function is determined and one passes to the limit at the boundary. As discussed below, it turns out that the correct class of solutions to consider are the *constrained viscosity solutions* and it is in this class that state wealth constraints may be suitably addressed.

If it can be shown that the HJB equation admits a smooth solution, then one can argue that it coincides with the value function. Moreover, one can construct optimal control policies by applying first order conditions to the HJB equation. This result is known as the *Verification Theorem* and it is stated below without a proof. We refer the technically oriented reader to Theorem of Fleming and Soner (1993). To simplify the presentation, we assume that the utility functions U_1, U_2 are non-negative.

Theorem 12.2.1 *[(Verification Theorem)]: Let V be a classical solution of (12.2.7), (12.2.8) for $x \geq 0$, satisfying for some γ, M bounded*

$$V(x,t) \leq M(T-t)(1+x^{\gamma}), \tag{12.2.9}$$

for each $T > 0$. Then $V(x,t) \geq J(x,t;T,\alpha)$ for $\alpha \in \mathcal{A}$. Moreover, let

$$\alpha^*(x,t) = \operatorname{argmax}_{\alpha}\big[\frac{1}{2}\sigma^2(x,\alpha)V_{xx}(x,t) + \mu(x,\alpha)V_x(x,t) + U_1(\alpha)\big]. \tag{12.2.10}$$

Then the policy $\widetilde{\alpha}_s = \alpha^*(X_s^*, s)$, with X_s^* being the solution of (12.2.1) with $\widetilde{\alpha}_s$ used, is optimal and

$$V(x,t) = u(x,t) = J(x,t;\widetilde{\alpha},T). \tag{12.2.11}$$

We remark that the above version of the Verification Theorem is somehow incomplete in the sense that one needs to specify rigorously the correct probability system that would support the (optimal) control policies. We choose not to be very specific at this point since these technical issues are beyond the interests of the audience; rather, we refer to the discussion by Fleming and Soner (1993, Chapter IV).

One can derive similar results for the case of expected utility maximization problems in an infinite horizon setting. If the payoff to be maximized is (12.2.3), instead of (12.2.2), then one can derive the stationary analogue of the HJB equation (12.2.7), namely

$$\beta u = \max_{\alpha} \left[\frac{1}{2}\sigma^2(x,\alpha)u_{xx} + \mu(x,\alpha)u_x + U(\alpha)\right] + r(x)u_x. \tag{12.2.12}$$

The above equation is a fully nonlinear elliptic equation and if it has a smooth solution then similar verification results, such as the ones in Theorem 12.2.1, can be proved (see Fleming and Soner (1993), Theorem 12.3.10).

A key ingredient for the existence of classical solutions of the HJB equation and their identification with the value function, is that the underlying controlled state process X_t does not degenerate. In the context of the properties of the HJB equation, it means that the latter preserves its uniform ellipticity, i.e. $\sigma^2(x,\alpha) \geq \epsilon x^2$, $\forall \alpha$, $\forall x \neq 0$ with ϵ being a positive constant. In the majority of expected utility maximization models arising in asset pricing, this condition might be violated. The main reason is that the coefficient of the second order derivative $\sigma^2(x,\alpha)$ involves the amount invested in risky assets which is not in general bounded away from zero. In fact, this situation arises very often in models of incomplete markets, such as, for example, models with trading constraints, stochastic labor income, stochastic volatility and, more generally, with non-traded assets (see Example 4.b). Therefore, the value function might not be smooth and one needs to relax the notion of solutions to the HJB equation. As it was mentioned earlier, a rich class of weak solutions to the HJB equation are the so-called *viscosity solutions.* These solutions were introduced by Crandall and Lions (1983) for first order non-linear partial differential equations and by Lions (1983) for the second order case. For a general overview of the theory we refer to the *User's Guide* by Crandall, Ishii and Lions (1992) and to the book of Fleming and Soner (1993). The strength of this theory lies in the fact that it provides rigorous characterization of the value function as the unique solution to the HJB equation. This uniqueness result plays an instrumental role in pricing derivative securities in markets with frictions (see, for example, Section 12.5). Moreover, the strong stability properties of viscosity solutions provide excellent convergence results for a large class of numerical schemes for the value function and the optimal policies. Numerical results are highly desirable in a wide range of practical applications, because closed form solutions of the HJB equation are not in general available (see, Barles et al (1995), Barles and Souganidis (1991), Tourin and Zariphopoulou (1994)).

In stochastic optimization problems arising in optimal investment and consumption models, viscosity solutions were first employed by Zariphopoulou (1989) for the Merton problem with trading constraints (see also Zariphopoulou (1994)), and for a similar model, but with transaction costs and Markov chain parameters, by Zariphopoulou (1992). Subsequently, this class of solutions was used by Fleming and Zariphopoulou (1991), Duffie and Zariphopoulou (1993), Davis, Panas and Zariphopoulou (1993), Shreve and Soner (1994). Being also employed in a variety of asset valuation models with market imperfections by other authors (see among others, Alvarez and Tourin (1996), Barles and Soner (1998)), viscosity solutions gradually become a standard tool in the study of stochastic control problems arising in models of Mathematical Finance. Because of their important role, a considerable part of this chapter is strongly oriented towards this theory and the outlaid results follow closely the unified theme of viscosity solutions of the relevant HJB equations.

Due to the specific nature of the stochastic optimization models in asset pricing, state and control constraints are present rather frequently. As we demonstrate in the next sections, these constraints arise because of exogeneously imposed trade limitations such as prohibition of shortselling, limited borrowing, leverage and non-bankruptcy. To accommodate this feature, which as we shall see results in a lack of explicit boundary data, one needs to work with a special class of viscosity solutions, namely the *constrained viscosity solutions.* This class of solutions was introduced by Soner (1986) and Capuzzo-Dolcetta and Lions (1990) for first-order equations (see also Ishii and Lions (1990)).

Because the majority of the models we review herein are of finite horizon and two-dimensional, we present the definition of constrained viscosity solutions for the same class of problems. To this end, we consider a nonlinear second order partial differential equation of the form

$$F(X, V, DV, D^2V) = 0 \quad \text{in } D \times \ [0, T] \tag{12.2.13}$$

where D is an open subset of $\mathcal{R}^2$, DV and D^2V denote the gradient vector and the second derivative matrix of V, and the function F is continuous in all its arguments and *degenerate elliptic*, meaning that

$$F(X, p, q, A + B) \leq F(X, p, q, A) \ \text{if} B \geq 0. \tag{12.2.14}$$

Definition 12.2.2 *A continuous function $V : \overline{D} \times [0, T] \to R$ is a constrained viscosity solution of (12.2.13) if the following two conditions hold: i) V is a viscosity subsolution of (12.2.13) on $\overline{D} \times \ [0, T]$; that is, if for any $\phi \in C^{2,1}(\overline{D} \times [0, T])$ and any local maximum point $X_0 \in \overline{D} \times [0, T]$ of $V - \phi$,*

$$F(X_0, V(X_0), D\phi(X_0), D^2\phi(X_0)) \leq 0, \tag{12.2.15}$$

ii) V is a viscosity supersolution of (12.2.13) in $D \times [0, T]$; that is, if for any $\phi \in C^{2,1}(\overline{D} \times [0, T])$ and any local minimum point $X_0 \in D \times [0, T]$ of $V - \phi$,

$$F(X_0, V(X_0), D\phi(X_0), D^2\phi(X_0)) \geq 0. \tag{12.2.16}$$

12.3 Models of Optimal Investment and Consumption I.

In his seminal papers, Merton (1969), (1971) introduced an optimal portfolio management model of a single agent in a stochastic setting. Trading takes place between a riskless security (e.g. a bond) and one or more stocks whose prices are modeled as *diffusion processes.* For each stock price, the mean rate of return and volatility are assumed to be constant and known. The investor, endowed with some initial wealth, trades dynamically between the available securities and consumes part of his wealth continuously in time. He is assumed to be a "small investor" in the sense that his actions do not influence the equilibrium prices of the underlying assets. His objective is to maximize the expected utility function which models his individual preferences as well as his attitude towards the risk associated with the market uncertainty.

Merton studied, among others, the special case of power utility functions, known as *Constant Relevant Risk Aversion (CRRA) utilities* and produced closed-form solutions to the optimization problem of the single agent. An important consequence of these results is that all the risky securities can be replaced by a *mutual fund* with characteristics independent of the individual preferences. This feature facilitated the analysis of dynamic market equilibria which was developed by Merton (1973) and subsequently further generalized by others (Araujo and Montiero (1989), Dana and Pontier (1992), Duffie (1986), Duffie and Huang (1985), Huang (1987), Karatzas, Lakner, Lehoczky and Shreve (1990), (1991), and Mas-Colell ((1985), (1986)).

We start this section with the celebrated Merton model of optimal portfolio management in a finite horizon setting.

To this end, we consider a market with two securities, a bond whose price solves

$$dB_t = rB_t dt, \tag{12.3.17}$$

with $B_0 = B > 0$ and a stock whose price process satisfies the linear stochastic differential equation

$$dS_t = \mu S_t dt + \sigma S_t dW_t, \tag{12.3.18}$$

with $S_0 = S > 0$. The market parameters μ and σ are, respectively, the *mean rate of return* and the *volatility*; it is assumed that $\mu > r > 0$ and $\sigma > 0$. The process W_t is a standard Brownian motion defined on a probability space $(\Omega, \mathcal{F}, P)$.

The wealth process satisfies $X_s = \pi_s^0 + \pi_s$ with the amounts π_s^0 and π_s representing the current holdings in the bond and the stock accounts. The state wealth equation (12.2.1) reduces to

$$dX_s = rX_s ds + (\mu - r)\pi_s ds + \sigma \pi_s dW_s. \tag{12.3.19}$$

The wealth process must satisfy the state constraint

$$X_s \geq 0 \quad \text{a.e.} \quad t \leq s \leq T. \tag{12.3.20}$$

The control π_s, $t \leq s \leq T$ is admissible if it is $\mathcal{F}_s$-progressively measurable – with $\mathcal{F}_s = \sigma(W_u; t \leq u \leq s)$ – it satisfies $E\int_t^T \pi_s^2 ds < +\infty$ and, it is such that the state constraint (12.3.20) is satisfied. We denote the set of admissible policies by $\mathcal{A}$.

The value function is

$$u(x,t) = \sup_{\mathcal{A}} E\Big[\frac{1}{\gamma} X_T^\gamma / X_t = x\Big]. \tag{12.3.21}$$

The Dynamic Programming Principle yields that for every stopping time τ,

$$u(x,t) = \sup_{\mathcal{A}} E\big[u(X_\tau, \tau)/X_t = x\big]. \tag{12.3.22}$$

Using stochastic analysis and under appropriate regularity and growth conditions on the value function, we get that u solves the associated HJB equation

$$\begin{cases} u_t + \max_\pi \left[\frac{1}{2}\sigma^2\pi^2 u_{xx}(u - r)\pi u_x\right] + ru_x = 0 \\ \\ u(x,T) = \frac{1}{\gamma}x^\gamma, x \geq 0 \\ \\ u(0,t) = 0, t \in [0,T), \end{cases} \tag{12.3.23}$$

for $x \geq 0$ and $t \in [0, T)$.

Remark *The above boundary condition is not in general prespecified due to the presence of the state constraint (12.3.20). As it was mentioned in the previous section, the correct way to deal with this issue is to use that the value function is the unique constrained solution of (12.3.23) and then pass to the limit as $x \to 0$. For the case at hand though, one can derive (1.3.50) easily by observing that (12.3.20) dictates that the only admissible policy at $x = 0$ is to invest nothing in the stock account, i.e. $\pi_s = 0$, $\forall t \leq s \leq T$.*

The homogeneity of the utility function and the linearity of the state dynamics with respect to both the wealth and the control portfolio process, suggest that the value function must be of the form

$$u(x,t) = \frac{x^\gamma}{\gamma} f(t) \tag{12.3.24}$$

with $f(T) = 1$. Using the above in (12.3.23) and after some cancelations, one gets that f must satisfy the first order equation

$$f'(t) + \lambda f(t) = 0,$$

with

$$f(T) = 1,$$

where

$$\lambda = r\gamma + \frac{(\mu - r)^2 \gamma}{2(1-\gamma)\sigma^2}. \tag{12.3.25}$$

Therefore, one expects the value function to be given by

$$u(x,t) = \frac{x^\gamma}{\gamma} e^{\lambda(T-t)}. \tag{12.3.26}$$

Once the value function is determined, the optimal policy may be obtained in the so-called feedback form as follows: first, we observe that the maximum of the quadratic term appearing in (12.3.23) is achieved at the point

$$\pi^*(x,t) = -\frac{\mu - r}{\sigma^2} \frac{u_x(x,t)}{u_{xx}(x,t)}$$

or, otherwise, at

$$\pi^*(x,t) = \frac{\mu - r}{\sigma^2(1-\gamma)} x,$$

where we used (12.3.26). Next, classical verification results yield that the candidate smooth solution, given in (12.3.26), is indeed the value function and that, moreover, the policy

$$\pi_s^* = \frac{\mu - r}{\sigma^2(1-\gamma)} X_s^* \tag{12.3.27}$$

is the *optimal investment strategy*.

In other words,

$$u(x,t) = E\Big[\frac{(X_T^*)^\gamma}{\gamma} \Big/ X_t^* = x\Big],$$

where X_s^* solves

$$dX_s^* = \Big(r + \frac{(\mu - r)^2}{\sigma^2(1-\gamma)}\Big) X_s^* ds + \frac{\mu - r}{\sigma(1-\gamma)} X_s^* dW_s.$$

The solution of the optimal state wealth equation is, for $X_t = x$,

$$X_s^* = x \exp\Big[\Big(r + \frac{(\mu-r)^2}{\sigma^2(1-\gamma)} - \frac{(\mu-r)^2}{2\sigma^2(1-\gamma)^2}\Big)(s-t) + \frac{\mu-r}{\sigma(1-\gamma)}W_{s-t}\Big].$$

The Merton optimal strategy dictates that it is optimal to keep a *fixed proportion*, namely $\frac{\mu-r}{\sigma^2(1-\gamma)}$, of the current total wealth invested in the stock account. We will refer to this proportionality constant as the *Merton ratio.*

Remark *It is important to observe that the Merton model uses heavily the assumption that the stock price remains strictly positive even though the stock price does not appear explicitly. One could easily verify this constraint by looking at the actual derivation of the state wealth equation (12.3.19); we refer the reader to Merton (1969) or Karatzas et al (1987). Given that the stock price is modeled as a log-normal process, it becomes zero only if it starts at the state* 0. *In this case, the Merton model degenerates to a deterministic model with no (stochastic) optimization features. In fact, one could show that no investment takes place in the stock account and that the wealth process satisfies the deterministic equation* $dX_s = rX_s ds$ *for* $t \le s \le T$. *In this case, the value function turns out to be* $\widetilde{u}(x,t) = \frac{x^\gamma}{\gamma}e^{r\gamma(T-t)}$. *We can view this degenerate case as the limiting case of (12.3.23) as* $\mu \to r$ *or as* $\sigma \to +\infty$. *Indeed, if* $\mu = r$ *or* $\sigma = +\infty$, *the solution of the HJB equation degenerates to* $\widetilde{u}(x,t)$ *and the optimal policy, given in (12.3.27), becomes zero.*

We continue with various generalizations of the Merton model. Because the scope of this review paper is to provide a vast exposition of the literature, we chose not to present complete proofs but rather to cite the references where rigorous results can be found.

12.3.1 Merton models with intermediate consumption

We look at the case that trading takes place in an infinite horizon and intermediate consumption is allowed, say at a (nonnegative) rate C_t. Working similarly as in Merton (1973), one can show that the wealth equation becomes

$$dX_t = rX_t dt - C_t dt + (\mu - r)\pi_t dt + \sigma \pi_t dW_t, \tag{12.3.28}$$

with X_t satisfying the same state constraint (12.3.20) as before. Utility comes only from intermediate consumption and the value function is defined as the maximal expected discounted utility, namely

$$V(x) = \sup_{\mathcal{A}} E\Big[\int_0^{+\infty} e^{-\beta t}U(C_t)dt / X_0 = x\Big]. \tag{12.3.29}$$

The set of admissible policies $\mathcal{A}$ consists of policies (π_t, C_t), $t \ge 0$ which are $\mathcal{F}_t$-measurable – with $\mathcal{F}_t = \sigma(W_s : 0 \le s \le t)$ – satisfy the integrability conditions $E\int_0^T \pi_s^2 ds < +\infty$, $E\int_0^T C_s ds < +\infty$, $\forall T > 0$ and the state constraint (12.3.20).

Merton solved the above problems for the class of Constant Relative Risk Aversion (CRRA) utilities given by

$$U(c) = \frac{1}{\gamma}c^\gamma \qquad \gamma < 1 \ \ (\gamma \neq 0) \tag{12.3.30}$$

$$U(c) = \log c \qquad \gamma\text{“} = \text{”}0. \tag{12.3.31}$$

Below, we present explicit results for the case $\gamma \neq 0$. The discount factor β is assumed to satisfy the growth condition

$$\beta > \left[\frac{(\mu - r)^2}{2\sigma^2(1-\gamma)} + r\right]\gamma. \tag{12.3.32}$$

The HJB equation becomes

$$\beta V = \max_{\pi}\left[\frac{1}{2}\sigma^2\pi^2 V_{xx} + (\mu - r)\pi V_x\right] + F(V_x) + rxV_x \tag{12.3.33}$$

with

$$F(V_x) = \max_{c \geq 0}\left[-cV_x + c^{\frac{\gamma}{\gamma}}\right] = \frac{1-\gamma}{\gamma}(V_x)^{\frac{\gamma}{\gamma-1}}. \tag{12.3.34}$$

Using that the utility function is homogeneous of degree γ and that the state equation is linear with respect to the controls and the state, one gets that the value function is also homogeneous of the same degree. In fact, one can verify that for

$$K = \left[\frac{\gamma}{1-\gamma}\gamma^{\gamma/1-\gamma}\left(\beta - \frac{(\mu - r)^2\gamma}{2\sigma^2(1-\gamma)} + r\gamma\right)\right]^{\gamma-1} \tag{12.3.35}$$

determined by direct substitution in (12.3.33),

$$V(x) = Kx^{\gamma}. \tag{12.3.36}$$

Moreover, the optimal control policies π_t^* and C_t^* are given in the feedback form $\pi_t^* = \widetilde{\pi}(X_t^*)$, $C_t^* = \widetilde{C}(X_t^*)$ where

$$\widetilde{\pi}(x) = \frac{\mu - r}{\sigma^2(1-\gamma)}x \quad \text{and} \quad \widetilde{C(x)} = (\gamma K)^{\frac{1}{\gamma-1}}x, \tag{12.3.37}$$

where X_t^* is the optimal wealth trajectory, given by (12.3.28), with π_t^* and C_t^* being used.
It is worth remarking that the optimal investment and consumption rules turn out to be linear in wealth, as it was the case in the previous example.

Remark *An interesting class of models arises when the state constraint (12.3.20) is removed and bankruptcy is allowed. In this case, the value function is defined by*

$$V(x) = \sup_{\mathcal{A}} E\left[\int_0^{\tau} e^{-\beta t}U(C_t)dt + e^{-\beta\tau}P/X_0 = x\right],$$

where $\tau = \inf\{t \geq 0 : X_t = 0\}$ *is the time of bankruptcy and* P *is the value surrendered if this event occurs. A complete study of such bankruptcy models can be found in Karatzas et al (1987) as well as in the book of Sethi (1997).*

12.3.2 Merton models with non-linear stock dynamics

In the previous models, a crucial simplification was that the underlying stock price is modeled as a diffusion process with linear coefficients. This assumption enabled us to solve the optimal investment/consumption problems by introducing a single state variable, the current wealth. Even though models with lognormally distributed stock prices are frequently used, mainly because of their tractability, a rather interesting class of models are the ones with non-linear stock dynamics. Special cases are, among others, the cases of mean-reverting stock prices as well as the ones with the volatility term being an explicit function of the current stock price; a widely used model of state dependent volatility is the so-called Constant Elasticity of Variance (CEV) model (see Cox (1996)).

Models with non-linear stock dynamics were studied by Merton (1971) for the case of logarithmic utilities. Moreover, martingale techniques have been successfuly used by several authors to analyze models with stock prices solving (12.3.18) with μ and σ being replaced by $\mathcal{F}_t$-measurable processes (see for a complete overview the monograph of Karatzas (1997)). The methodology involved relies heavily on martingale representation results; the solution is provided in terms of expectations under the "correct" measure of the appropriate payoffs and the optimal processes via martingale representation theorems.

In our effort to demonstrate how one can use information directly from the HJB equation to specify the value function and the optimal policies, we present a different approach in solving the Merton problem with non-linear stock dynamics. To this end, we assume that there are two securities available, a bond whose price is given by (12.3.17) and a stock whose price solves

$$dS_s = \mu(S_s)S_s ds + \sigma(S_s)S_s dW_s, \tag{12.3.38}$$

with $S_t = S \geq 0$ and $0 \leq t < s < T$. The process W_s is a Brownian motion defined on a probability space $(\Omega, \mathcal{F}, P)$. The coefficients μ and σ are functions of the current stock price and they are assumed to satisfy all the required regularity assumptions in order to guarantee that a unique solution to (12.3.38) exists.

The investor rebalances his portfolio dynamically by choosing at any time s, for $s \in [t, T]$ and $0 \leq t \leq T$, the amounts π_s^0 and π_s to be invested respectively in the bond and the stock accounts. His total wealth satisfies the budget constraint $X_s = \pi_s^0 + \pi_s$ and the stochastic differential equation

$$\begin{cases} dX_s = rX_s ds + (\mu(S_s) - r)\pi_s ds + \sigma(S_s)\pi_s dW_s, \\ \\ X_t = x \geq 0 \qquad 0 \leq t \leq s \leq T. \end{cases} \tag{12.3.39}$$

The above state equation follows from the budget constraint and the dynamics in (12.3.38). The wealth process must also satisfy the standard non-negativity state constraint (12.3.20).

Remark *We assume that the coefficients μ and σ do not depend explicitly on time. This is assumed only to ease the presentation since the time-dependent case follows easily from the autonomous one.*

The control process π_s is said to be admissible if it is $\mathcal{F}_s$-progressively measurable, where $\mathcal{F}_s = \sigma(W_u; t \leq u \leq s)$, satisfies the integrability condition $E\int_t^T \sigma(S_s)^2 \pi_s^2 ds < +\infty$ and, is such that the state constraint (12.3.20) is satisfied. We denote by $\mathcal{A}$ the set of admissible policies.

The investor's objective is to maximize his expected utility payoff

$$J(x,S,t;\pi) = E\big[U(X_T)/X_t = x,\ S_t = S\big], \tag{12.3.40}$$

with X_s, S_s given respectively in (12.3.39) and (12.3.38).

The value function is

$$u(x,S,t) = \sup_{\mathcal{A}} J(x,S,t;\pi) \tag{12.3.41}$$

with the utility function $U : [0,+\infty) \to [0,+\infty)$ being of the form

$$U(x) = \frac{1}{\gamma}x^{\gamma}, \tag{12.3.42}$$

with $\gamma \in (0,1)$.

The special form of the above utilities together with the linearity of the wealth dynamics with respect to the state and control processes (see (12.3.39)), suggest that the value function may be written in a "separable" form. In other words, the value function may be written as $u(x,S,t) = \dfrac{x^{\gamma}}{\gamma}\ V(S,t)$. The component V is in general unknown except for some very special cases of the risk aversion parameter $1-\gamma$ and the components of the state dynamics (see Merton (1971)). As a matter of fact, V solves a nonlinear equation for which no closed form solutions are available in general.

In Zariphopoulou (1999), it is shown that under a simple power transformation, the factor V can be expressed in terms of the solution of a linear parabolic equation. This representation provides closed form solutions for the value function and the optimal policies which can in turn be used effectively in a more general class of valuation problems with stochastic components. Without stating at this point the necessary technical assumptions and the regularity properties of the solutions, we outline the main results below.

Proposition 12.3.1 *i) The value function u is given by*

$$u(x,S,t) = \frac{x^{\gamma}}{\gamma}v(S,t)^{1-\gamma}$$

where $v : R^+ \times [0,T] \to R^+$ solves the linear parabolic equation

$$\begin{cases} v_t + \dfrac{1}{2}\sigma^2(S)S^2 v_{SS} + \Big[\mu(S)S + \dfrac{\gamma(\mu(S)-r)S}{(1-\gamma)}\Big]v_S \\[2ex] +\dfrac{\gamma}{1-\gamma}\Big[r + \dfrac{(\mu(S)-r)^2}{2\sigma^2(S)(1-\gamma)}\Big]v = 0 \\[2ex] v(S,T) = 1 \quad \text{and} \quad v(0,t) = e^{\frac{r\gamma}{1-\gamma}(T-t)}, \ 0 \le t \le T. \end{cases}$$

ii) The optimal investment policy Π_s^* is given in the feedback form $\Pi_s^* = \pi^*(X_s^*, S_s, s)$ where the function $\pi^* : \mathcal{R}^+ \times \mathcal{R}^+ \times [0,T] \to R$ is defined by

$$\pi^*(x,S,t) = \Big[\frac{v_S(S,t)}{v(S,t)} + \frac{1}{1-\gamma}\frac{\mu(S)-r}{\sigma^2(S)}\Big]x.$$

12.3.3 Merton models with trading constraints

In a variety of applications, trading between the available securities may be restricted. For example, the amounts we allow to invest might be bounded from above or from below by given functions of the current wealth or, in simpler cases, by prespecified constants. The latter case arises when borrowing or shortselling is limited, if allowed at all.

Using Dynamic Programming methods and elements from the theory of viscosity solutions, Zariphopoulou ((1989), (1992)) analyzed the Merton problem when borrowing is limited and shortselling is not allowed, i.e. the allowed investment strategies π_s must satisfy $0 \leq \pi_s \leq X_s$, $0 \leq t \leq s \leq T$ a.e. Generally speaking, such kind of constraints might result in lack of smoothness of the value function and explicit solutions are not in general available. Other models with alternative trading constraints and in which the analysis relies heavily on the HJB equation, were studied by Grossman and Laroque (1989), Grossman and Vila (1992), Fleming and Zariphopoulou (1991), Fitzpatrick and Fleming (1991) and more recently by Munk (1999) in the context of derivative pricing with portfolio constraints.

Besides using the HJB equation directly, martingale methods have been successfully used, together with convex duality arguments, to produce general representation results for the value function and optimal policies for a wide range of trading constraints (see, for example, He and Pearson (1991), Cvitanić and Karatzas (1992), (1993), (1993a) and for a general overview, the monograph of Karatzas (1997)).

Below, we present a representative optimal investment and consumption model in which the constraints are of the so-called "*leverage type*,"

$$\pi_s \leq k(X_s + L), \tag{12.3.43}$$

with k, L given positive constants. Such models were studied by Grossman and Laroque (1989) and Vila and Zariphopoulou (1997); the choice of the leverage ceiling $k(X_s + L)$ is made only to simplify the presentation, since smoothness results may be readily obtained for $\pi_s \leq f(X_s)$ with $f : \mathcal{R}^+ \to \mathcal{R}$ being a smooth function of the state wealth.

To this end, we assume that trading takes place in an infinite horizon, intermediate consumption is allowed and the price S_s of the available stock solves (12.3.18). The wealth process solves (12.3.28) and the value function is

$$V(x) = \sup_{\mathcal{A}} E \int_0^{+\infty} e^{-\beta t} U(C_t) dt. \tag{12.3.44}$$

The set of admissible strategies $\mathcal{A}$ consists of $\mathcal{F}_s$-progressively measurable pairs (π_s, C_s) which satisfy the standard integrability conditions and the leverage constraint (12.3.43). The HJB equation becomes

$$\beta V = \max_{\pi \leq k(x+L)} \left[\frac{1}{2}\sigma^2\pi^2 V'' + (\mu - r)\pi V'\right] + \max_{c \geq 0}[-cV' + U(c)] + rxV_x, \; x \geq 0. \tag{12.3.45}$$

Vila and Zariphopoulou (1997) established that the above equation has a $C^2(0, +\infty)$ solution, with $V(0) = \frac{U(0)}{\beta}$ which coincides with the value function. Using the regularity of the value function, the first order conditions in (12.3.45) and classical verification results, they determined the optimal policies, π_t^* and C_t^* in the feedback form $C_t^* = (U')^{-1}(V'(X_t^*))$ and

$$\pi_t^* = \min\left\{ -\frac{\mu - r}{\sigma^2}\frac{V'(X_t^*)}{V''(X_t^*)},\; k(X_t^* + L)\right\},$$

with X_t^* being the optimal wealth process. Observe that because of the leverage constraint, the HJB equation changes form since the first maximum term in (12.3.45),

$$I(x; k, L) = \max_{\pi \leq k(x+L)} \left[\frac{1}{2}\sigma^2\pi^2 V''(x) + (\mu - r)\pi V'(x)\right],$$

satisfies

$$I(x;k,L) = \begin{cases} -\dfrac{(\mu-r)^2}{2\sigma^2}\dfrac{[V'(x)]^2}{V''(x)} & \text{if } -\dfrac{\mu-r}{\sigma^2}\dfrac{V'(x)}{V''(x)} \le k(x+L) \\ \frac{1}{2}\sigma^2 k^2 (x+L)^2 V''(x) + (\mu-r)k(x+L)V'(x) & \\ \quad \text{if } -\dfrac{\mu-r}{\sigma^2}\dfrac{V'(x)}{V''(x)} > k(x+L). & \end{cases} \tag{12.3.46}$$

This situation will be revisited in the next model when trading takes place in an inhomogeneous financial medium (see Section 12.3.4).

In the case of power utility functions with risk aversion coefficient $1-\gamma$, one may use the particular structure of (12.3.46) and (1.3.73) to analyze the nature of the optimal policies. For a variety of practical applications, an interesting question is how different is the optimal feedback rule, $\widetilde{\pi}(x) = \min\left[-\frac{\mu-r}{\sigma^2}\frac{V'(x)}{V''(x)},\ k(x+L)\right]$ from the so-called *myopic strategy*, $\pi^{\text{myopic}}(x) = \min\left[\frac{\mu-r}{\sigma^2(1-\gamma)}x,\ k(x+L)\right]$.

The proofs of the following results may be found in Section 12.4 of Vila and Zariphopoulou (1997).

Proposition 12.3.2 *The optimal strategy $\widetilde{\pi}(x)$ is at most equal to the myopic investment policy $\pi^{\text{myopic}}(x)$ and, strictly less than it for small wealth values. It coincides with π^{myopic} only if $k=1$ and $L=0$.*

We denote by $\mathcal{U}$ (resp. $\mathcal{B}$) the domains in which the trading constraint is not binding (resp. binding),

$$\mathcal{U} = \left\{x>0: -\frac{\mu-r}{\sigma^2}\frac{V'(x)}{V''(x)} < k(x+L)\right\},$$

$$\mathcal{B} = \left\{x>0: -\frac{\mu-r}{\sigma^2}\frac{V'(x)}{V''(x)} \ge k(x+L)\right\}.$$

Proposition 12.3.3 *If the discount factor β satisfies $\beta > r - \frac{(\mu-r)^2}{2\sigma^2} + \frac{k\mu}{2}$ then there exists a threshold level x^* such that $\mathcal{U} = [0, x^*)$ and $\mathcal{B} = [x^*, +\infty)$. Moreover, the optimal investment strategy $\widetilde{\pi}(x)$ is always greater than kx.*

Next, we look at the value functions V^0 and V^∞ which correspond to the optimization problem (12.3.44) but with $L=0$ and $L=\infty$ respectively. The latter case corresponds to unlimited borrowing and this is the original Merton problem; as we have seen earlier, for $\beta > r\gamma + \frac{(\mu-r)^2\gamma}{\sigma^2(1-\gamma)}$, an optimal solution exists, denoted herein by V^∞, and the optimal consumption rule $C^\infty(x) = Kx$ with K given in the previous section. If $k > \frac{\mu-r}{(1-\gamma)\sigma^2}$, the borrowing constraint (12.3.43) is not binding and the solution to the original problem (12.3.44) coincides with the solution V^∞ of the unconstrained one. If $k < \frac{\mu-r}{(1-\gamma)\sigma^2}$, then the borrowing constraint is indeed binding, as Propositions 12.3.2 and 12.3.3 indicate. The following result describes the relation between the optimal consumption rules C^∞, C^0 and C^* which correspond respectively to the problems with $L=\infty$, $L=0$ and $L<\infty$.

Proposition 12.3.4 *i) If* $0 < \gamma < 1$, $C^*(x)$ *satisfies*

$$\max\left\{C^\infty(x), C^0(x)\left(\frac{x}{x+L}\right)^{\frac{\gamma}{1-\gamma}}\right\} \le C^*(x) \le C^0(x)\left(\frac{x+L}{x}\right)^{\frac{1}{1-\gamma}},$$

ii) If $\gamma < 0$, $C^*(x)$ *satisfies*

$$C^0(x) \le C^*(x) \le \min\left\{C^0(x+L), C^\infty(x)\left(\frac{x+L}{x}\right)^{\frac{1}{1-\gamma}}\right\},$$

iii) $C^*(x)$ *satisfies* $\lim_{x\to\infty} \frac{C^*(x)}{x} = K$,

iv) For $x \in [0, x^*)$, $C^*(x) \ge C^\infty(x)$.

12.3.4 Merton models with non-homogeneous investment opportunities

In a number of real world situations, the investment opportunities become broader with higher wealth levels. In fact, when less than $10,000 is available for investment, we can usually invest only in banking accounts and mutual fund shares; of course, one can still invest in individual stocks, but in such a case it is hard to have a well-diversified portfolio. Mutual fund shares will provide all necessary investment tools for both the rich and the poor, if a form of the mutual fund separation theorem is valid and the real world provides all the necessary funds. But it is doubtful that all the risky or riskless investments in the global economy are covered by the existing array of mutual funds. For instance, many limited partnerships are not covered by public mutual funds. Furthermore, ordinary mutual funds are usually prohibited from using modern investment techniques involving options, futures, and other derivative securities. When high wealth levels are available, we are not constrained by the opportunities offered by the mutual funds; we can invest in limited partnerships, hedge funds (these funds are known to be agressive in employing modern investment techniques), individual stocks as well as banking accounts and mutual funds. There is also an explicit law which prohibits small investors from trading some securities and rule 144A stipulates that unregistered securities can be traded only by qualified institutional investors. Therefore, even for institutional investors, the investment opportunity gets better, when they get richer.

Next, we present an investment problem assuming that there exists a critical wealth level such that once an investor's wealth level exceeds it, the investment opportunity improves. We present these optimal consumption and investment rules in closed form for the case of CRRA utilities. One interesting feature of these optimal rules is that investors' consumption is much lower when there exists such a transition than in its absence, and investors generally take more risk when their wealth is below the critical level and become more risk averse once their wealth exceeds it. Namely, if investors expect that they will have a better investment opportunity when their wealth increases, they tend to increase both savings and the expected return on investments by raising the risk of the investment positions they take. Also, once their wealth crosses the critical level, they tend to reduce risk in investments, being afraid of losing the better investment opportunity they enjoy. Therefore, the optimal rules somehow provide a theoretical justification to the casually observed fact that entrepreneurs in fast-growing economies tend to take more risk than their counterparts in stabilized economies. It also gives a justification for wealthy investors to use portfolio insurance strategies.

The mathematical problem arising from the investment problem is itself rather interesting. Namely, the HJB equation for wealth levels above the critical level takes a different form than it does under the critical level.

We note that this change results in a "discontinuity" of the HJB equation across the interphase point. This situation is rather different from the one in the previous section, in which the HJB equation changed its form but in a continuous way. It is worth mentioning that discontinuous HJB arise frequently in a number of stochastic optimization models of expected utility even though the associated control problems have not been analyzed in full rigor. In general, it is not clear under what conditions one can show that the discontinuous HJB equation has a unique viscosity solution and therefore, to identify it with the value function (see Kutev and Lions (1992), Koo and Zariphopoulou (1996)).

We start with the description of the investment model. To achieve generality, we assume that there are more than one risky security at all trading times. To this end, we consider a market in which there is one riskless asset and $M+N$ risky assets. We assume that the risk-free rate is a constant r and that the price $S_j(t)$ of the j-th liquid risky asset follows a geometric Brownian motion

$$\frac{dS_j(t)}{S_j(t)} = \mu_j dt + \sum_{k=1}^{M+N} \sigma_{jk} dW_k(t), \tag{12.3.47}$$

where $(W_1(t),\ldots,W_{M+N}(t))$ is a standard Brownian motion defined on the underlying probability space $(\Omega,\mathcal{F},P)$. The market parameters, μ_j and σ_{jk}, which represent the *mean rates of return* and the *volatility matrix* coefficients of the risky securities, for $j,k=1,\ldots,M+N$, are taken to be given constants. We assume that the matrix $\Sigma = (\sigma_{ij})_{i=1,j=1}^{i=M+N,j=M+N}$ is non-singular, i.e., there is no redundant asset among the $M+N$ risky assets.

The critical wealth level is denoted by x_0. The investor's wealth process evolves according to the equation

$$\begin{cases} dX_t = [rX_t + (\tilde{\mu} - r\mathbf{1}_{M+N}, \tilde{\pi}_t) - C_t]dt + (\tilde{\pi}_t, \sigma_{ij} dW_t^*), & (t \geq 0) \\ X_0 = x, \qquad x \geq 0, \end{cases} \tag{12.3.48}$$

where $*$ denotes the transpose of a matrix,

$$\begin{cases} \pi_t = (\pi_{1,t},\ldots,\pi_{M+N,t})^*, \quad \mu = (\mu_1,\mu_2,\ldots,\mu_{M+N})^*, \\ W_t = (W_1(t),\ldots,W_{M+N}(t))^*, \quad \mathbf{1}_{M+N} = (1,1,\ldots,1)^*, \end{cases} \tag{12.3.49}$$

and by the restriction on the investment opportunity

$$\pi_{M+1,t} = \ldots . \pi_{M+N,t} = 0, \qquad \text{if} \quad X_t < x_0. \tag{12.3.50}$$

The control processes are the consumption rate C and the vector π of dollar amounts invested in the risky assets. To state their properties, we introduce the sets

$$\mathcal{L}_+ = \left\{ l \in L : \; l_t \geq 0 \text{ a.s.and } E\int_0^t l_s \, ds < +\infty \text{ for} t \geq 0) \right\}$$

and

$$\mathcal{M} = \left\{ l \in L^{M+N} : \; E\int_0^t l_s \cdot l_s \, ds < +\infty \text{ for} t \geq 0 \right\},$$

where L (resp. L^{M+N}) is the space of $\mathcal{F}_t$-progressively measurable processes (resp. vector processes in $M+N$) with $\mathcal{F}_t$ being the augmentation under P of $\sigma\{(W_{1,s},\ldots,W_{N+M,s}) : 0 \le s \le t\}$.

The set $\mathcal{A}$ of admissible controls consists of pairs (C,π) in $\mathcal{L}_+ \times \mathcal{M}$ such that $X_t \ge 0$ a.e. $(t \ge 0)$ and the trading restriction (12.3.50) is satisfied.

The investor's objective function is given by

$$\mathcal{J}(C) = E\Big[\int_0^{+\infty} e^{-\beta t} U(C_t)\,dt \Big/ X_0 = x\Big],$$

where $\beta > 0$ is the discount factor and $U : [0,+\infty) \to [0,+\infty)$ is a strictly increasing, concave, twice continuously differentiable function with $U(0) = 0$.

The value function $V : R^+ \to R$ of the investor is given by

$$V(x) = \sup_{(C,\pi)\in\mathcal{A}} \mathcal{J}(C). \tag{12.3.51}$$

We will use the following notation:

$$\kappa_1 = (\tilde{\mu} - r\mathbf{1}_M)(\Sigma_1\Sigma_1^*)^{-1}(\tilde{\mu} - \mathbf{1}_M), \quad \kappa_2 = (\mu - \mathbf{1}_{M+N})(\Sigma\Sigma^*)^{-1}(\mu - \mathbf{1}_{M+N}) \tag{12.3.52}$$

where $\tilde{\mu} = (\mu_1,\ldots,\mu_M)^*$, $\Sigma_1 = (\sigma_{i,j})_{i=1,j=1}^{i=M,j=M}$, and $\mathbf{1}_M \in \mathcal{R}^M$ is a vector whose components are equal to 1.

We will make the following assumption:

Assumption 12.3.53 $\kappa_2 > \kappa_1$.

Assumption 12.3.53 says that the investment opportunity facing the investor is *better* when $X_t \ge x_0$ than when $X_t < x_0$.

The HJB equation takes the following form:

$$\begin{aligned}\beta J(x) = &\max_{c\ge 0,\tilde{\pi}\in R^M} [\tilde{\pi}^*(\tilde{\mu} - r\mathbf{1}_M)J'(x) + (rx - C)J'(x)\\ &+ \frac{1}{2}\tilde{\pi}^*\Sigma_1\Sigma_1^*\tilde{\pi}J''(x) + U(C)], \quad \text{if } x < x_0\\ \beta J(x) = &\max_{c\ge 0,\pi}[\pi^*(\mu - r\mathbf{1}_{M+N})J'(x) + (rx - C)J'(x)\\ &+ \frac{1}{2}\pi^*\Sigma\Sigma^*\pi J''(x) + U(C)] \quad \text{if } x \ge x_0.\end{aligned} \tag{12.3.54}$$

Proposition 12.3.5 *The value function is the unique viscosity solution to the HJB equation.*

For a proof see Koo and Zariphopoulou (1996).

We will now proceed to get a closed form solution for a CRRA class utility function, i.e.,

$$U(C) = \begin{cases} C^{\frac{1-\gamma}{1-\gamma}} & if \gamma \ne 1,\\ \log C & if \gamma = 1,\end{cases} \tag{12.3.55}$$

where $\gamma > 0$ is the coefficient of relative risk aversion.

Definition 12.3.6 : *For $i = 1, 2$, define*

$$K_i = r + \frac{\beta - r}{\gamma} + \frac{1}{2}\frac{\gamma - 1}{\gamma^2}\kappa_i, \quad \text{if} \gamma \neq 1,$$

and

$$K_i = \frac{r - \beta + \kappa_i}{\beta^2}, \quad \text{if} \gamma = 1.$$

Also let $\lambda_{i,+}$ and $\lambda_{i,-}$ be the roots of the quadratic

$$\kappa_i \lambda^2 + (r - \beta - \kappa_i)\lambda - r = 0,$$

and

$$\rho_{i,+} = 2 + \lambda_{i,+}, \quad \rho_{i,-} = 1 + \lambda_{i,-}.$$

These definitions are similar to those given in Karatzas et al (1987).
It can be easily shown that for $i = 1, 2$

$$\lambda_{i,+} > 0, \qquad \lambda_{i,-} < -1$$
$$(12.3.56)$$
$$\rho_{i,+} > 0, \qquad \rho_{i,-} < 0.$$

We will also use the following simplified notation:

$$\rho_+ = \rho_{2,+}, \quad \lambda_+ = \lambda_{2,+}, \quad \rho_- = \rho_{1,-}, \quad \lambda_- = \lambda_{1,-}. \tag{12.3.57}$$

We now define functions which will be used to express the value function in closed form.

Definition 12.3.7 : *Let us assume that U is given by (3.40). For $\gamma \neq 1$, we define $C_0 > 0$, $J_1 : [0, C_0] \to R$, $X_1 : [0, C_0] \to R^+$, $J_2 : [C_0, \infty) \to R$, and $X_1 : [C_0, \infty) \to R^+$ by*

$$C_0 = \frac{\left(\frac{\lambda_-}{\rho_-} - \frac{\lambda_+}{\rho_+}\right) x_0}{\left[\frac{\lambda_-}{\rho_- K_1} - \frac{\lambda_+}{\rho_+ K_2} + \frac{1}{(1-\gamma)K_2} - \frac{1}{(1-\gamma)K_1}\right]},$$

$$J_1(C) = \frac{\lambda_- B_1}{\rho_-} C^{-\gamma\rho_-} + \frac{1}{(1-\gamma)K_1} C^{1-\gamma},$$

$$X_1(C) = B_1 C^{-\gamma\lambda_-} + \frac{1}{K_1} C,$$

$$J_2(C) = \frac{\lambda_+ B_2}{\rho_+} C^{-\gamma\rho_+} + \frac{1}{(1-\gamma)K_2} C^{1-\gamma},$$

$$X_2(C) = B_2 C^{-\gamma\lambda_+} + \frac{1}{K_2} C,$$

where

$$B_1 = x_0 C_0^{\gamma\lambda_-} - \frac{1}{K_1} C_0^{1+\gamma\lambda_-}$$

and

$$B_2 = x_0 C_0^{\gamma\lambda_+} - \frac{1}{K_1} C_0^{1+\gamma\lambda_+}.$$

For $\gamma = 1$, we define

$$C_0 = \frac{1}{\frac{1}{\beta} + \frac{K_2 - K_1}{\left(\frac{\lambda_-}{\rho_-} - \frac{\lambda_+}{\rho_+}\right)}} x_0,$$

$$J_1(C) = \frac{\lambda_- \widetilde{B}_1}{\rho_-} C^{-\rho_-} + \frac{1}{\beta} \log C + K_1 X_1(C) = \widetilde{B}_1 C^{-\lambda_-} + \frac{1}{\beta} C,$$

$$J_2(C) = \frac{\lambda_+ \widetilde{B}_2}{\rho_+} C^{-\rho_+} + \frac{1}{\beta} \log C + K_2 X_2(C) = \widetilde{B}_2 C^{-\lambda_+} + \frac{1}{\beta} C,$$

where

$$\widetilde{B}_1 = x_0 C_0^{\lambda_-} - \frac{1}{\beta} C_0^{1+\lambda_-} \quad \textit{and} \quad \widetilde{B}_2 = x_0 C_0^{\lambda_+} - \frac{1}{\beta} C_0^{1+\lambda_+}.$$

We will make the following assumption:

Assumption 12.3.58 : *$\kappa_i > 0$ for $i = 1, 2$.*

The above assumption guarantees that the investment problems without a change in the investment opportunity set and the ones in which the investment opportunity consists of the riskless asset and the first M risky assets, or the riskless asset and the $M + N$ risky assets are all well-defined.

Proposition 12.3.8 : *i) $C_0 > 0$,*
(ii) $C_0 < K_1 x_0$ if $\gamma \neq 1$ and $C_0 < \beta x_0$ if $\gamma = 1$,
(iii) $C_0 < K_2 x_0$ if $1 - \frac{\rho_-}{\lambda_-} < \gamma$ and $C_0 \geq K_2 x_0$ if $1 - \frac{\rho_-}{\lambda_-} \geq \gamma$.

Proposition 12.3.9 : *Suppose that either $\gamma \neq 1$ and*

$$\gamma \leq \frac{\left(\frac{\lambda_-}{\rho_-} - \frac{\lambda_+}{\rho_+}\right)}{\left(\frac{1}{K_1} - \frac{1}{K_2}\right)\left(\frac{1}{1-\gamma} - \frac{\lambda_-}{\rho_-}\right) K_2 \lambda_+}$$

or $\gamma = 1$ and

$$\frac{\beta(K_2 - K_1)}{\frac{\lambda_-}{\rho_-} - \frac{\lambda_+}{\rho_+}} \leq \frac{1}{\lambda_+}.$$

Then X_1 is a strictly increasing function mapping $[0, C_0]$ onto $[0, x_0]$ and X_2 is a strictly increasing function mapping $[C_0, +\infty)$ onto $[x_0, +\infty)$.

Under the previous assumptions, the function $X : R^+ \to R^+$ defined by $X(C) = X_1(C)$ for $C \leq C_0$ and $X(C) = X_2(C)$ for $C \geq C_0$, is well-defined, strictly increasing and maps $[0, +\infty)$ onto itself. We denote its inverse by X^{-1}.

We now state the main result:

Theorem 12.3.10 *Suppose that U is given in (3.40) and that, either $\gamma \neq 1$ and*

$$\gamma \leq \frac{\left(\frac{\lambda_-}{\rho_-} - \frac{\lambda_+}{\rho_+}\right)}{\left(\frac{1}{K_1} - \frac{1}{K_2}\right)\left(\frac{1}{1-\gamma} - \frac{\lambda_-}{\rho_-}\right) K_2 \lambda_+},$$

or, $\gamma = 1$ *and*

$$\frac{\beta(K_2 - K_1)}{\frac{\lambda_-}{\rho_-} - \frac{\lambda_+}{\rho_+}} \leq \frac{1}{\lambda_+}.$$

Let $J : R^+ \to R$ *be defined by*

$$J(x) = \begin{cases} J_1(X^{-1}(x)) & if x \leq x_0 \\ J_2(X^{-1}(x)) & if x \geq x_0. \end{cases}$$

Then, the following statements are true:

(i) J *coincides the value function* V.

(ii) V *belongs to* $C^2[0, x_0) \cap C^2(x_0, +\infty)$.

(iii) The optimal rule of consumption is given by

$$C_t = V'(X_t)^{-\frac{1}{\gamma}}.$$

(iv) The optimal rule of investment is given by $\pi_t^* = \widetilde{\pi}_t(X_t^*)$ *where*

$$\widetilde{\pi}(x) = \frac{V'(x)}{V''(x)} \kappa_1 (\Sigma_1 \Sigma_1^*)^{-1} (\widetilde{\mu} - r\mathbf{1}_m)$$

if $x < x_0$

$$\widetilde{\pi}(x) = -\frac{V'(x)}{V''(x)} \kappa_2 (\Sigma\Sigma^*)^{-1} (\mu - r\mathbf{1}_{M+N})$$

if $x > x_0$ *and*

$$\widetilde{\pi}(x_0) = \lim_{x \to x_0^+} -\frac{V'(x)}{V''(x)} \kappa_2 (\Sigma\Sigma^*)^{-1} (\mu - r\mathbf{1}_{M+N}),$$

if $x = x_0$,

(v) $\lim_{x \to +\infty} \left[V(x) - \frac{K_2}{1-\gamma} x^{1-\gamma} \right] = 0.$

We conclude with some properties of the optimal rules.

Proposition 12.3.11 : *Suppose that assumptions of Theorem 12.3.10 are valid.*

(i) There exists a neighborhood $N(x_0)$ *of* x_0 *such that the optimal consumption when* $X_t^* \in N(x_0)$ *is strictly smaller than* $K_1 X_t^*$.

(ii) Suppose that $\gamma > 1 - \frac{\rho_-}{\lambda_-}$. *Then, there exists a neighborhood* $N'(x_0)$ *of* x_0 *such that the optimal consumption is strictly smaller than* $K_2 X_t^*$ *when* $X_t^* \in N'(x_0)$.

(iii) Suppose that $\gamma < 1 - \frac{\rho_-}{\lambda_-}$. *Then, there exists a neighborhood* $N'(x_0)$ *of* x_0 *such that the optimal consumption is strictly greater than* $K_2 X_t^*$ *when* $X_t^* \in N'(x_0)$.

Observe that $K_1 X_t^*$ is equal to what the investor would consume if the investment opportunity consisted of the riskless asset and the first M-risky assets and did not change across x_0, and $K_2 X_t^*$ is equal to what the investor would consume if all the assets are available for investment regardless of the investor's wealth level. Therefore, the result says that the investor consumes *less* than what he or she would consume if there were only M-risky assets. Intuition tells us that if the investor anticipates a better investment opportunity when he or she gets richer, then there is more incentive to save. The following Proposition fits into this intuition.

Proposition 12.3.12 *Suppose that assumptions of Theorem 12.3.10 are valid. Then, for* $0 < x < x_0$

$$\begin{cases} -\dfrac{xV''(x)}{V'(x)} < \gamma & \text{if} \gamma > -\frac{1}{\lambda_-} \\[2ex] -\dfrac{xV''(x)}{V'(x)} > \gamma & \text{if} \gamma < -\frac{1}{\lambda_-} \end{cases}$$

and for $x > x_0$,

$$\begin{cases} -\dfrac{xV''(x)}{V'(x)} > \gamma & \text{if} \gamma \geq 1 - \frac{\rho_-}{\lambda_-} \\[2ex] -\dfrac{xV''(x)}{V'(x)} < \gamma & \text{if} \gamma < 1 - \frac{\rho_-}{\lambda_-}. \end{cases}$$

The proposition says that for the case $\gamma \geq 1$ (resp. $\gamma < 1$), the coefficient of risk aversion implied by the value function is greater (resp. smaller) than γ for wealth levels less (resp. greater) than x_0. Empirical studies give a favorable evidence that $\gamma > 1$. Therefore, the above result states that when investors anticipate an improvement in the investment opportunity, they tend to take more risk and thereby increase the expected return on investments. It also says that once their wealth crosses the critical level, they tend to reduce their risk-taking, being afraid of losing the better investment opportunity.

12.3.5 Models of Optimal Portfolio Management with General Utilities

In a variety of applications, the investors do not have preferences of constant relative risk aversion or, in other words, their utility functions are not of power form. In this case, the homogeneity of the value function is lost and explicit solutions are not in general available. Martingale methods have been successfully used to produce the value function and the optimal investment plans under a fairly general set of assumptions. For the special case of constant coefficients, one can also produce closed form solutions for the quantities of interest, working directly with the HJB equation. This approach was developed in Karatzas et al. (1987) for stationary models of optimal investment and consumption and it was later applied to similar models but in a finite horizon setting. In order to simplify the presentation and to show the main ingredients of the method, we present below the time dependent case but with no intermediate consumption; for more general settings, we refer the reader to Karatzas et al. (1987) and for an overview to the monograph of Karatzas (1997).

To this end, we recall the underlying Merton model with terminal utility $U : [0, +\infty) \to [0, +\infty)$ which is assumed to be increasing, concave, of class $C^2(0, +\infty)$ with $U(0) = 0$. The underlying securities, the bond and the stock solve the original price equations (12.3.17) and

(12.3.18); the wealth process satisfies (12.3.19) and (12.3.20) as well. The value function is defined as

$$u(x,t) = \sup_{\mathcal{A}} E\big[U(X_T)/X_t = x\big], \tag{12.3.59}$$

with $\mathcal{A}$ being the set of admissible policies defined as in Section 12.3.1.

Under the assumption of general utility functions, it is not known a priori that the value function is smooth and one needs to work with the viscosity solutions.

Theorem 12.3.13 *The value function is the unique viscosity solution of*

$$\begin{cases} u_t + \max_{\pi}\left[\frac{1}{2}\sigma^2\pi^2 u_{xx} + (\mu - r)\pi u_x\right] + rxu_x = 0, \\ \\ u(x,T) = U(x), \end{cases} \tag{12.3.60}$$

on $\overline{D} = [0,+\infty) \times [0,T]$, *in the class of concave solution that are nondecreasing in the spatial argument*

(For a proof see Zariphopoulou (1989) and Fleming and Soner (1993)).

Next, we apply formally the first order conditions in (12.3.60) which yield

$$u_t - \frac{(\mu - r)^2}{2\sigma^2}\frac{u_x^2}{u_{xx}} + rxu_x = 0. \tag{12.3.61}$$

The following transformation was used by Karatzas, Lehozcky and Shreve (1987) which transforms (12.3.60) to a *linear* partial differential equation. To this end, we parametrize the wealth variable in terms of a function $f : [0,+\infty) \times [0,T] \to [0,+\infty)$ such that

$$u_x(f(y,t),t) = y. \tag{12.3.62}$$

For conditions on the existence of such a function, see Karatzas, Lehozcky, Sethi and Shreve (1987). Successive differentiations of the above and use of (12.3.61) yield that f solves the linear parabolic problem

$$\begin{cases} f_t + \frac{1}{2}\frac{(\mu - r)^2}{\sigma^2}y^2 f_{yy} + \left(\frac{(\mu - r)^2}{\sigma^2} - r\right)yf_y = rf, \\ \\ f(y,T) = (U')^{-1}(y). \end{cases} \tag{12.3.63}$$

Clearly, it is straightforward to solve the above linear equation, which has a unique solution f; under certain natural regularity properties of the utility function, one can also show that f is smooth. As a matter of fact, the solution f can be represented via the Feynman-Kac formula as

$$f(y,t) = \widetilde{E}[U^{-1}(Y_T)/Y_t = y],$$

where the process Y_s, $t \le s \le T$, solves the stochastic differential equation

$$dY_s = \left[\frac{(\mu - r)^2}{\sigma^2} - r\right]Y_s ds + \frac{\mu - r}{\sigma}Y_s d\widetilde{W}_s,$$

with $\widetilde{W}_s$ being a Brownian motion on a probability space $(\Omega, \mathcal{G}, Q)$ and $\widetilde{E}$ is the expectation under Q.

Also, we observe that the optimal feedback policy function, say $\pi^*(x,t)$ is given by $\pi^*(x,t) = -\frac{\mu-r}{\sigma^2}\frac{u_x(x,t)}{u_{xx}(x,t)}$ or, in view of the parametrization $x = f(y,t)$ and the transformation (12.3.62), as

$$\pi^*(f(y,t),t) = -\frac{\mu-r}{\sigma^2} y f_y(y,t). \tag{12.3.64}$$

Once the solution f is determined, one can "invert" the obtained formulae and recover the value function and the optimal policies (see Karatzas (1997)).

The main ingredient of the above approach is essentially the use of the *convex dual* of the value function $\widetilde{u(x,t)} = \sup_{x>0}[u(x,t) - xy]$. Because of the special structure of the involved non-linear terms in the HJB equation, it turns out that $\widetilde{u}$ can be specified by solving a linear parabolic problem. This reduction – going from the non-linear HJB equation to the reduced linear parabolic problem – is a key component of this approach. Once $\widetilde{u}$ is found, one can recover the value function via $u(x,t) = \inf_{y>0}[\widetilde{u}(y,t) + xy]$ and subsequently the optimal policies. It is worth mentioning that in a variety of applications, useful properties of the control policies can be proved by using directly the convex dual instead of recovering the value function first and then obtain the optimal policies through the first order conditions (see, for example, Karatzas (1997)).

The above method differs in many ways from the ones we discuss herein which are based almost entirely on arguments from the theory of non-linear partial differential equations. In order to demonstrate a valuable strong alternative to the latter method, we present below an application of the methods that use heavily elements from the martingale theory and convex duality. The model we analyze is similar to one defined in (12.3.59) but more general, in the sense that intermediate consumption is also allowed.

In the exposition below, we do not include all the technical assumptions needed but we refer the reader to Section 2.4 in Karatzas (1997).

To this end, we assume that the investor can consume at intermediate times and that his expected payoff is given by

$$J(x,t;\pi,C) = E\Big[\int_t^T e^{-\beta(s-t)}U_1(C_s)ds + e^{-\beta(T-t)}U_2(X_T)/X_t = x\Big]. \tag{12.3.65}$$

The utility functions U_i, $i = 1,2$ satisfy the technical assumptions

$$\begin{cases} U_i : [0,+\infty) \to \mathcal{R},\ U_i \in C^3(0,+\infty), \\ \\ U_i(0^+) > -\infty,\ \lim\limits_{x\to\infty}\dfrac{(U_i'(x))^\alpha}{U_i''(x)} = 0,\ \lim\limits_{x\to 0}\dfrac{(U_i'(x))^2}{U_i''(x)} \text{ exists,} \end{cases}$$

for some $\alpha > 2$.

We denote, for $i = 1,2$, by $I_i(z) = (U_i')^{-1}(z)$ and by $\widetilde{U}_i(y)$ the convex-duals $\widetilde{U}_i(y) = \max_{x>0}[\widetilde{U}_i(x) - xy]$. Observe that $\widetilde{U}_i(y) = U_i(I_i(y)) - yI_i(y)$.

It is well known that the process $Z_0(t) = \exp\Big(-\frac{\mu-r}{\sigma}W_t - \frac{1}{2}\frac{(\mu-r)^2}{\sigma^2}t\Big)$, $0 \le t \le T$ is an exponential martingale where W_t is the Brownian motion driving the stock price (12.3.18). One needs to introduce the processes $Z(t,s) = \frac{Z_0(s)}{Z_0(t)}$ and

$H(t,s) = \frac{H_0(s)}{H_0(t)} = e^{-r(s-t)}Z(t,s)$ and a new (state) diffusion process Y_s, $t \leq s \leq T$ with

$$\begin{cases} dY_s = (\beta - r)Y_s ds - \frac{\mu - r}{\sigma} Y_s dW_s, \\ \\ Y_t = y > 0. \end{cases}$$

Then, $Y_s = ye^{\beta(s-t)}H(t,s)$.

Next, we represent the wealth variable via

$$\mathcal{X}(y,t) = E\Big[\int_t^T e^{-r(s-t)}Z(t,s)I_1(ye^{(\beta-r)(s-t)}Z(t,s))ds + e^{-r(T-t)}I_2(ye^{(\beta-r)(T-t)})\Big],$$

and we also define

$$G(y,t) = E\Big[\int_t^T e^{-\beta(s-t)}U_1(I_1(Y_s))ds + e^{-\beta(T-t)}U_2(I_2(Y_T))/Y_t = y\Big]$$

and

$$S(y,t) = E\Big[\int_t^T e^{-\beta(s-t)}Y_s I_1(Y_s)ds + e^{-\beta(T-t)}Y_T I_2(Y_T)\Big].$$

It can be shown that $\mathcal{X}(y,t) = yS(y,t)$ and that its inverse $\mathcal{Y}(\cdot,t)$ is well defined. The value function is then given by

$$u(x,t) = G(\mathcal{Y}(x,t),t)$$

together with its convex dual as

$$\widetilde{u}(y,t) = \sup_{x>0}[u(x,t) - xy] = G(y,t) - S(y,t) =$$

$$?? = E\Big[\int_t^T e^{-\beta(s-t)}\widetilde{U}_1(Y_s)ds + e^{-\beta(T-t)}\widetilde{U}_2(Y_T)/W_t = 0\Big]. \qquad (12.3.66)$$

It turns out that the functions G and S solve the *linear parabolic terminal time problems*

$$\begin{cases} G_t + \mathcal{L}G + U_1(I_1(y)) = 0; \quad (y,t) \in (0,+\infty) \times [0,T], \\ \\ G(y,T) = U_2(I_2(y)); \ y > 0, \end{cases}$$

and

$$\begin{cases} S_t + \mathcal{L}S + yI_1(y) = 0; \ (y,t) \in (0,+\infty) \times [0,T), \\ \\ S(y,T) = yI_2(y); \ y > 0; \end{cases}$$

where the generator $\mathcal{L}$ is given by

$$\mathcal{L}f = \frac{1}{2}\frac{(\mu - r)^2}{\sigma^2}y^2\frac{\partial^2 f}{\partial y^2} + (\beta - r)y\frac{\partial f}{\partial y} - \beta f,$$

(for details regarding boundary and growth conditions see Karatzas (1997)),

From the properties of $\widetilde{u}$ one can easily show that the latter may be determined as the solution of the linear parabolic problem

$$\begin{cases} \widetilde{u}_t + \mathcal{L}\widetilde{u} + \widetilde{U}_1(y) = 0, \ \ 0 \leq t < T \\ \\ \widetilde{u}(y,T) = \widetilde{U}_2(y). \end{cases} \qquad (12.3.67)$$

This in turn yields the value function that can be determined via the inverse dual transformation $u(x,t) = \inf_{y>0}[\widetilde{u}(y,t) + xy]$ or through the representation $u(x,t) = G(\mathcal{Y}(x,t),t)$.

Generally speaking, this approach is based on convex duality arguments and martingale theory and it has been successfully applied to a number of stochastic optimization models arising in asset and derivative pricing. It has been applied to equilibrium models (see, among others, Karatzas et al (1990), (1991)), to models of expected utility with trading constraints or other market frictions, like, for example, transaction costs (see, for example, Jouini and Kallal (1995), Cvitanić, Pham and Touzi (1997)). In the bibliography we provide additional references that use this alternative approach.

12.3.6 Optimal goal problems

Besides maximizing the individual's expected utility of terminal wealth, or the expected payoff from intermediate consumption, one might desire to maximize the probability that the state wealth reaches a prespecified level by some terminal time T. Optimization problems of achieving a financial goal arise often in *capital risk management.* Variations and extensions of the basic problem, which we present below, are directly related to the maximal probabilities of (super) hedging a derivative security.

We consider the state wealth equation (12.3.19) and we assume that $r = 0$ and $\sigma = 1$, i.e. the wealth process solves

$$dX_s = \mu\pi_s ds + \pi_s dW_s, \qquad t \leq s \leq T. \tag{12.3.68}$$

Our admissible policies π_s are taken to be $\mathcal{F}_s$-measurable, satisfying almost surely, and for $t \leq s \leq T$, the integrability condition $\int_t^T \pi_s^2 ds < +\infty$ and the state constraint

$$0 \leq X_s \leq 1. \tag{12.3.69}$$

We denote the set of admissible policies by $\mathcal{A}$.

The objective is to avoid absorption at the origin and at the same time to maximize the probability of reaching the financial goal $\overline{x} = 1$ by the expiration time T. In other words, our value function is given by

$$u(x,t) = \sup_{\mathcal{A}} P[X_T = 1/X_t = x] = \sup_{\mathcal{A}} E\big[\mathbf{1}_{\{X_T=1\}}/X_t = x\big]. \tag{12.3.70}$$

This problem was solved by Kulldorff (1993) in discrete time and subsequently by Heath (1993) in a continuous time setting; it was later revisited by Karatzas (1997). The analysis below follows closely the arguments used by Heath (1993). To simplify the presentation, we take the original time to be zero, $t = 0$ and we bring back the time dependence later.

First, one recalls that by Girsanov's theorem, the process $\widetilde{W}_s = W_s + \mu s$ is a Brownian motion under Q with the latter being a measure absolutely continuous to P, with density $Z_T = \exp\left\{ -\mu W_T - \mu^2 \frac{T}{2}\right\}$. In terms of $\widetilde{W}_s$, (12.3.68) becomes $X_t = x + \int_0^t \pi_s d\widetilde{W}_s$ with $x \in [0,1]$ being the initial condition for X_t. Thus, X_t is a local martingale under Q; as a matter of fact, it is actually a martingale because it is bounded.

Therefore, for the set $A_1 = \{\omega : X_T(\omega) = 1\}$, we have that $Q(A_1) \leq x$, which in turn yields that

$$P(A_1) \leq \sup\{P(A_2) : Q(A_2) \leq x\}. \tag{12.3.71}$$

Thus, the original problem is reduced to computing the above supremum.

Applying directly the Neyman-Pearson lemma yields that this supremum may be computed by specifying a unique number λ such that

$$Q(A_3) = x,$$

with

$$A_3 = \Big\{\omega : \frac{dP(\omega)}{dQ(\omega)} \geq \lambda\Big\}.$$

In fact, if such a unique number λ exists then $P(A_3)$ provides the solution.

Because the density $\frac{dP}{dQ} = \exp\left\{\mu\widetilde{W}_T - \frac{\mu^2}{2}T\right\}$, one needs to find λ such that

$$Q\Big\{\frac{dP}{dQ} \geq \lambda\Big\} = Q\Big\{\exp\left\{\mu\widetilde{W}_T - \frac{\mu^2}{2}T\right\}\Big\} = x.$$

It follows easily that such a λ is uniquely given by

$$\ell n\, \lambda = -\mu\sqrt{T}\,\Phi^{-1}(x) - \frac{\mu^2}{2}T,$$

where Φ is the cummulative normal distribution. Clearly, an upper bound on $P(A_1)$ is then given by

$$P\Big\{\mu W_T + \frac{\mu^2}{2}T \geq -\mu T\Phi^{-1}(x) - \frac{\mu^2}{2}T\Big\} = \Phi(\Phi^{-1}(x) + \mu\sqrt{T}),$$

and if one shows that this bound can be achieved, then this would provide the optimal solution, i.e. the maximal probability of reaching the financial target 1 by the end of the trading horizon.

In other words, a candidate for the value function starting at time t, $0 \leq t < T$, at the point $x \in (0,1)$, is given by

$$v(x,t) = \Phi(\Phi^{-1}(x) + \mu\sqrt{T-t}).$$

Next, we look at the HJB equation associated with the stochastic optimization problem (12.3.70). Observe that (12.3.70) can be viewed as a Merton problem when the interest rate $r = 0$, $\sigma = 1$, the utility from terminal wealth is given by the step function

$$U(x) = \begin{cases} 0, & if 0 \leq x < 1 \\ 1 & if x = 1, \end{cases}$$

and the wealth state process must satisfy the state constraint $0 \leq X_s \leq 1$, $t \leq s \leq T$. The standard Merton problem was presented at the beginning of this chapter; working, at least formally, along the same lines, we can derive the associated HJB equation

$$\begin{cases} u_t + \max_{\pi}\left[\frac{1}{2}\pi^2 u_{xx} + \mu\pi u_x\right] = 0, \\ u(x,T) = \begin{cases} 0, & 0 \leq x < 1 \\ 1, & x = 1. \end{cases} \end{cases} \tag{12.3.72}$$

We remark that the above equation cannot be handled directly with the analysis developed for the traditional Merton problem, due to the special form of U and the state constraint. One rigorous way to proceed is first to establish that u is the unique constrained viscosity solution of (12.3.72), (12.3.73) on $[0,1] \times [0,T]$ and in turn to verify that the candidate solution v is such a solution as well. Then, one can conclude that $v \equiv u$, i.e.

$$u(x,t) = \Phi(\Phi^{-1}(x) + \mu\sqrt{T-t}). \tag{12.3.73}$$

This verification result has not been estabished using viscosity arguments but it was proved by Heath (1993) using elements from martingale theory. It is worth observing that the value function does not achieve the terminal data $U(x)$ continuously in time since $\lim_{t \uparrow T} u(x,t) = x \neq U(x)$.

The optimal portfolio process π_s^* is established via the first order conditions in (12.3.72); they yield that the maximum is achieved at $\widetilde{\pi}(x,t) = -\mu \frac{u_x(x,t)}{u_{xx}(x,t)}$ which, in view of (12.3.73), implies

$$\widetilde{\pi}(x,t) = \frac{1}{\sqrt{T-t}} \varphi(\Phi^{-1}(x))$$

with φ being the normal density. Therefore

$$\pi_s^* = \frac{1}{\sqrt{T-t}} \varphi(\Phi^{-1}(X_s^*)),$$

with X_s^*, $t \leq s \leq T$, being the optimal wealth with the above π_s^* being used.

For other properties of the optimal policy and the optimal solution, we refer the reader to Heath (1993) and Karatzas (1997).

12.3.7 Alternative models of expected utility

Stochastic optimization models of expected utility have played a fundamental role not only in optimal portfolio management as it was discussed in detail earlier, but also in *equilibrium asset pricing* and in *derivative valuation*. The use of utility maximization in derivative pricing is discussed in subsequent chapters in the context of pricing methods in the presence of market frictions.

In asset equilibrium, the prices of the underlying securities are not known a priori, but they are determined via fundamental "supply and demand" clearing market conditions. The basic setup consists of a finite number of individuals, say M and a given number of securities a riskless bond and N risky stocks. Each agent is endowed with a utility function but all agents have the same beliefs for the asset returns. The ith agent starts with ϵ_i initial endowment and solves his expected utility optimization problem in order to determine his optimal policies, the consumption rate $C_t^{i,*}$ and the portfolio process $\pi_t^{i,j,*}$, with $1 \leq i \leq M$, $1 \leq j \leq N$.

The equilibrium prices are determined in terms of the above prolicies via the so-called market clearing conditions:

$$\sum_{i=1}^{M} C_t^{i,*} = \sum_{i=1}^{M} \epsilon^i, \quad \sum_{i=1}^{M} \pi_t^{i,j,*} = 0, \quad \sum_{i=1}^{M} X_t^{i,*} = 0,$$

for $1 \leq i \leq M$, $1 \leq j \leq N$.

In the absence of market frictions, fundamental "*aggregation*" properties hold and the entire analysis can be carried out via the so-called representative agent whose utility function is an appropriately weighted average of the individual utilities. The underlying individual optimization problems are then reduced to the basic Merton model for a single investor, the representative one, who starts with initial wealth given by the aggregate endowment

$\epsilon = \sum_{i=1}^{M} \epsilon^i$. The majority of the involved stochastic control models were studied via martingale techniques; we refer the reader to the monograph of Karatzas (1997) for an extensive review of the theory.

In all previous models, one of the fundamental assumptions was the one of time additive utilities. This assumption facilitates considerably their analysis but it does not explain certain empirical results on consumers's optimal policies. Alternative kinds of utility functions have been proposed by various authors include, among others, the utilities of the *stochastic differential utilities*, otherwise known as *recursive utilities.* This type of utility was studied in a continuous time framework by Duffie and Epstein (1992) (see, also, Schroeder and Skiadas (1999)). They incorporate a more refined structure of the aggregated information acquired through time. The associated stochastic optimization problems are mainly analyzed with techniques from the backwards and forward stochastic differential equations (see among others, Duffie and Lions (1992), El Karoui, Peng and Quenez (1997)).

Utilities with *habit formation* were introduced by Constantinides (1990) and they model how investors' satisfaction drops, according to a given decay rate, as they "get used" to certain consumption levels. The relevant expected utility models have an additional consumption state variable which decays in accordance to the individuals' habit formation. An additional consumption state variable is also needed for the model of Hindy and Huang (1993) who allow for local substitution. This feature allows for discontinuous consumption processes and, typically, the relevant models give rise to *singular stochastic control* problems. Often, the associated HJB equation contains differential and integral terms and its analysis becomes rather challenging. A rigorous treatment of this class of HJB equations can be found in Alvarez (1994) and in Alvarez and Tourin (1996).

Alternative criteria to the ones based on utility payoffs from terminal wealth or/and intermediate consumption, involve payoffs with "long-term" characteristics. Such criteria arise in certain macroeconomic growth models and give rise to stochastic optimization problems with *ergodic cost criteria.* To gain some intuition, we observe that at least heuristically, the value function of the finite horizon utility maximization problem $u(x,t)$ (see (12.2.7)) is expected to satisfy

$$u(x,T) \sim \lambda T + W(x) \quad \text{as} T \to \infty. \tag{12.3.74}$$

The coefficient λ does not depend on the initial condition x and together with W must satisfy

$$\lambda = \max_{\alpha} \left[\frac{1}{2}\sigma^2(x,\alpha)W_{xx} + \mu(x,\alpha)W_x + U_1(\alpha) \right] + r(x)W_x.$$

This HJB equation corresponds to an average cost per unit time stochastic optimization problem

$$J(x;u) = \limsup_{T\to\infty} \frac{1}{T} E \int_0^T L(X_t, u_t)dt,$$

with X_t solving (12.2.1). Therefore, the time growth coefficient in (12.3.74) coincides with the maximum average cost per unit time J (see, Bensoussan and Frehse (1992), Bensoussan et al (1998)). An interesting connection of λ with the dominant eigenvalue of certain operators can be found in Fleming and Sheu (1997) who based their analysis on logarithmic transformations of solutions to linear parabolic equations. This work also brings out the interesting connection between ergodic control and infinite time horizon *risk sensitive control.*

Models of risk sensitive control in the area of utility maximization have been proposed by Bielecki and Pliska (1999) and Fleming and Sheu (1999); see also Platen and Rebolledo (1996) and McEneaney (1997).

12.4 Models of optimal investment and consumption II

In this section, we discuss models of expected utility in financial markets with frictions. We concentrate on two kinds of such frictions, namely *transaction costs* and *stochastic labor income*. Both classes of models are rather representative in asset pricing and derivative valuation in incomplete markets. Their associated stochastic optimization problems are rather difficult to solve due to certain degeneracies inherited by the unhedgeable risks that the market frictions generate. We dedicate most of this section to the study of these models for two reasons. Firstly, the mathematical methods involved are representative of the ones used in models of mathematical finance in imperfect markets and secondly, these models will be revisited subsequently in the context of derivative pricing via utility maximization methods. At the end of the section, we provide a brief overview of other models of incomplete markets.

12.4.1 Optimal investment/consumption models with transaction costs

A crucial simplification in Merton's work is the absence of *transaction costs* on the various trades. The first to incorporate proportional transaction costs in Merton's model were Magill and Constantinides (1976) in an effort to understand how these costs affect trading policies and also to explore if the equivalence between multiple stocks and mutual funds is still preserved. Magill and Constantinides believed that transaction costs have an important impact on the trading activity of the investor; in fact, they argued that the individual must completely refrain from trading at portfolio states which are highly penalized by the transaction costs. These policies differ substantially from the ones recovered by Merton – for the same class of utility functions. Indeed, Merton's policies call for a continuous in time rebalancing of the security holdings so that a constant fraction of the current wealth remains always invested in the stock account(s). This wealth independent fraction is known as the *Merton ratio* and it depends on market parameters and the risk aversion coefficient. Thus, in the absence of transaction costs, the optimal investment process turns out to be a diffusion process with values proportional to the ones of the current wealth process. In the presence of transaction costs, Magill and Constantinides brought out an important insight about the different nature of optimal investment policies, the one of *singular trading policies*. Under these policies, lump-sum transactions take place which amount to instantaneously altering the portfolio holdings in the bond and the stock account(s). Even though Magill and Constantinides did not provide a *singular stochastic control* formulation of the underlying model, they paved the way to the correct formulation of the valuation models with transaction costs (see, also, Constantinides (1979), (1986)).

Taksar, Klass and Assaf (1988) were the first to formulate a transaction cost model as a singular stochastic control problem in the context of maximizing the long term expected rate of wealth. Subsequently, Davis and Norman (1990) provided a rigorous mathematical formulation and extensive analysis of the Merton problem in the presence of proportional costs for CRRA utilities. Their paper is considered a landmark in the literature on transaction costs and contains useful insights and fundamental results, both theoretic and numerical, for the value function and optimal investment policies. Even though these results depend heavily on the homotheticity properties of the value function, inherited by the power form of the CRRA utilities, the model of Davis and Norman is viewed as the model for benchmark transaction costs; it is presented and analyzed in detail in the next section.

Departing from the special class of CRRA utilities, Zariphopoulou ((1989), (1992)) was the first to study optimal portfolio management models with proportional transaction costs for general individual preferences. In (1989), Zariphopoulou introduced a simple investment model with two securities, a riskless bond rate and a risky security whose rate of return is modeled as a continuous-time Markov chain, and she provided characterization results for the maximal utilities.

For the case of price processes modeled as diffusion processes and CRRA preferences, a considerable body of work has been produced with modifications and extensions of the Davis and Norman (DN) model. Shreve and Soner (1994) revisited the DN model and provided additional existence and regularity results for optimal policies and the value function for a wide range of market parameters. A similar model, but in the case of a finite trading horizon, was studied by Akian, Menaldi and Sulem in (1992) who allowed for more than one risky securities and provided some regularity results for the value function. Finally, the ergodic analogue of Akian, Menaldi and Sulem was subsequently analyzed by Akian, Sulem and Taksar in (1996).

As it will be apparent from the discussion in the next sections, the stochastic optimization problems with transaction costs do not have in general closed form solutions. Thus, it is highly desirable — mainly for the practical applications — to provide numerical results for their value function and the optimal investment policies and consumption plans. Such results were first provided by Davis and Norman (1990) and later by Tourin and Zariphopoulou (1994) for general utility functions. Other numerical schemes have been proposed by Akian, Menaldi and Sulem (1996) for a model of portfolio selection with more than one risky asset and by Sulem (1997) for a mixed portfolio problem with transaction costs. Pichler (1996) developed a different class of schemes for the DN model and he also studied the probability distributions of the relevant expected gains.

As we have alredy seen in previous sections, the central object of study are the value function and the optimal investment and consummation policies. The value function is expected to satisfy the HJB equation but certain degeneracies might result in lack of sufficient regularity. Therefore, one needs to work with the weak (viscosity) solutions and this is the class of solutions we will be working with.

We continue with the description of the benchmark optimal investment/consumption model of Davis and Norman incorporating general utilities in the payoff functional. This is a model of a single agent, or a small investor as it is otherwise known, in the sense that his actions cannot influence the prices of the underlying securities.

We consider an economy with two securities, a *bond* with price B_t and a *stock* with price S_t at date $t \geq 0$. Prices are denominated in units of a consumption good, say dollars.

The bond pays no coupons, is default free and has price dynamics as in (12.3.17). The stock price is the diffusion process given by (12.3.18) where μ is the *mean rate of return* and σ is the *volatility*; μ and σ are constants such that $\mu > r$ and $\sigma \neq 0$.

The investor holds x_t dollars of the bond and y_t dollars of the stock at date t. We consider a pair of right-continuous with left limits (CADLAG), non-decreasing processes (L_t, M_t) such that L_t represents the cumulative dollar amount transferred into the stock account and M_t the cumulative dollar amount transferred out of the stock account. By convention, $L_0 = M_0 = 0$. The stock account process is

$$y_t = y + \int_0^t \mu y_\tau d\tau + \int_0^t \sigma y_\tau dW_\tau + L_t - M_t, \tag{12.4.75}$$

with $y_0 = y$.

Transfers between the stock and the bond accounts incur *proportional transaction costs.* In particular, the cumulative transfer L_t into the stock account reduces the bond account by βL_t and the cumulative transfer M_t out of the stock account increases the bond account by αM_t, where $0 < \alpha < 1 < \beta$.

The investor consumes at the rate c_t dollars out of the bond account. There are no transaction costs in transfers from the bond account into the consumption good.

The bond account process is

$$x_t = x + \int_0^t \{rx_\tau - c_\tau\}d\tau - \beta L_t + \alpha M_t, \tag{12.4.76}$$

with $x_0 = x$. The integral represents the accumulation of interest and the drain due to consumption. The last two terms represent the cumulative transfers between the stock and bond accounts, net of transaction costs.

A policy is a $\mathcal{F}_t$-progressively measurable triple (c_t, L_t, M_t). We restrict our attention to the set of admissible policies $\mathcal{A}$ such that

$$\begin{cases} c_t \geq 0 \text{ and } E\int_0^t c_\tau d\tau < \infty \text{ a.s. for } t \geq 0, \\ \text{and} \\ w_t = x_t + \binom{\alpha}{\beta} y_t \geq 0 \text{ a.s. for } t \geq 0, \end{cases} \tag{12.4.77}$$

where we adopt the notation

$$\begin{pmatrix} \alpha \\ \beta \end{pmatrix} z = \begin{cases} \alpha z & \text{if } z \geq 0 \\ \\ \beta z & \text{if } z < 0. \end{cases} \tag{12.4.78}$$

We refer to w_t as the *net worth*. It represents the investor's bond holdings, if the investor were to transfer the holdings from the stock account into the bond account, incurring in the process the transaction costs.

The investor's payoff is

$$E\left[\int_0^{+\infty} e^{-\rho t} U(c_t) dt\right],$$

over the consumption stream $\{c_t, t \geq 0\}$, where ρ is the *subjective discount rate* and the *utility function* $[0, +\infty) \to [0, +\infty)$ is assumed to have the following properties:

i) $U \in C([0, +\infty)) \cap C^1((0, +\infty))$ is increasing and concave.

ii) $U(c) \leq K(1 + c)^\gamma$, $\forall c \geq 0$, for some positive constants K and γ, with $0 < \gamma < 1$.

Given the initial endowment (x, y) in $\overline{D} = \left\{(x, y) \in \mathcal{R} \times \mathcal{R} : x + \begin{pmatrix} \alpha \\ \beta \end{pmatrix} y \geq 0\right\}$, we define the *value function* V as

$$V(x, y) = \sup_{\mathcal{A}} E\left[\int_0^{+\infty} e^{-\rho t} U(c_t) dt \mid x_0 = x, y_0 = y\right]. \tag{12.4.79}$$

To guarantee that the value function is well defined we either assume, as in Davis and Norman (1990) that

$$\rho > r\gamma + \frac{\gamma(\mu - r)^2}{2\sigma^2(1 - \gamma)}, \tag{12.4.80}$$

or, assume, as in Shreve and Soner (1994), that

$$\rho > r\gamma + \gamma^2(\mu - r)^2/2\sigma^2(1-\gamma)^2. \tag{12.4.81}$$

Either set of conditions (12.4.80) and (12.4.81), yield that the value function which corresponds to $\alpha = \beta = 1$ and $U(c) = K(1+c)^\gamma$ is finite and, therefore, all functions with $0 \le \alpha < 1, \beta \ge 1$ are finite.

We continue with some basic properties of the value function. (For their proofs and other basic properties, see Shreve and Soner (1994) or Tourin and Zariphopoulou (1994).)

Proposition 12.4.1 *i) The value function V is jointly concave in x and y, strictly increasing in x and increasing in y.*

ii) The value function V is continuous on $\overline{D}$.

We continue with a formal discussion on the derivation of the associated HJB equation.

First, we consider a random time τ and we assume that the optimal strategy of the investor is to refrain from trading and to consume at a rate say $\tilde{c}_t$, for $0 \le t \le \tau$. The Dynamic Programming Principle yields

$$V(x,y) = \sup_{\mathcal{A}_{(x,y)}} E\Big[\int_0^\tau e^{-\rho t} U(\tilde{c}_t)dt + e^{-\rho\tau} V(x_\tau, y_\tau)\Big/ x_0 = x,\ y_0 = y\Big].$$

and, in turn, that V satisfies at the point (x,y)

$$\rho V = \frac{1}{2}\sigma^2 y^2 V_{yy} + \mu y V_y + r x V_x + \max_{c\ge 0}[-cV_x + U(c)]. \tag{12.4.82}$$

Because the above policy is in general suboptimal, (4.9) holds as inequality, i.e. for all points $(x,y) \in \overline{D}$,

$$\rho V \ge \frac{1}{2}\sigma^2 y^2 V_{yy} + \mu y V_y + r x V_x + \max_{c\ge 0}[-cV_x + U(c)]. \tag{12.4.83}$$

Next, assume that at the point $(x,y) \in \overline{D}$, it is optimal to make an instantaneous transaction corresponding to the purchase of bond shares. In other words, let us assume that the investor rebalances his portfolio from (x,y) to $(x+\alpha\delta, y-\delta)$, incurring the appropriate transaction costs. Then the optimality of this decision implies

$$V(x,y) = V(x+\alpha\delta, y-\delta), \tag{12.4.84}$$

which in turn yields

$$-\alpha V_x(x,y) + V_y(x,y) = 0. \tag{12.4.85}$$

Because such a policy is in general suboptimal, (12.4.84) holds as inequality and (12.4.85) becomes

$$-\alpha V_x + V_y \ge 0, \tag{12.4.86}$$

for $(x,y) \in \overline{D}$.

Finally, let us assume that for the portfolio position $(x,y) \in \overline{D}$, it is optimal to rebalance it to the new position $(x-\beta\delta, y+\delta)$. Then optimality implies

$$V(x,y) = V(x-\beta\delta, y+\delta), \tag{12.4.87}$$

which in turn yields at the point (x, y),

$$\beta V_x(x, y) - V_y(x, y) = 0. \tag{12.4.88}$$

Like the other policies we considered, the last one is in general suboptimal which implies that (4.14) holds as inequality. In differential form this implies

$$\beta V_x - V_y \geq 0, \tag{12.4.89}$$

for all points $(x, y) \in \overline{D}$.

Combining (12.4.83), (12.4.86) and (12.4.89), we obtain the HJB equation (12.4.90), associated with (12.4.79). As a matter of fact, the HJB equation turns out to be a *Variational Inequality with gradient constraints.*

The following result was proved by Tourin and Zariphopoulou (1994) and by Shreve and Soner (1994) for the case of CRRA utilities.

Theorem 12.4.2 *The value function V is a constrained viscosity solution on $\overline{D}$ of the Hamilton-Jacobi-Bellman equation*

$$\min\left[\rho V - \frac{1}{2}\sigma^2 y^2 V_{yy} - \mu y V_y - rxV_x - \max_{c \geq 0}(-cV_x + U(c)),\right.$$

$$\left.\beta V_x - V_y, -\alpha V_x + V_y\right] = 0. \tag{12.4.90}$$

Next, we state a comparison result for constrained viscosity solutions of (12.4.79) which appears in Tourin and Zariphopoulou (1994). This result has been used to obtain convergence of the numerical schemes employed for the value function and the optimal policies and also to derive bounds on derivative prices.

Theorem 12.4.3 *Let u be an upper semi-continuous viscosity subsolution of (12.4.90) on $\overline{D}$ with sublinear growth and v be a bounded from below uniformly continuous viscosity supersolution of (12.4.90) in D. Then, $u \leq v$ on $\overline{D}$.*

We now concentrate on the special class of Constant Relative Risk Aversion (CRRA) utility functions

$$\begin{cases} U(c) = \frac{1}{\gamma}c^\gamma & for \gamma < 1, \gamma \neq 0 \\ U(c) = \log c & for \gamma = 0. \end{cases} \tag{12.4.91}$$

As we have seen in Merton's model and its variations, because the utility function is homogeneous and the state dynamics linear in the state and control variables, the value function inherits the same homogeneity. This in turn can be used effectively to produce closed form solutions for the HJB equation and explicit feedback formulae for the optimal policies.

In models with proportional transaction costs, the homotheticity properties are primarily used to reduce the dimensionality of the relevant optimization problem. This is the central feature in the benchmark work of Davis and Norman who reduced the dimensionality of (12.4.79) via the transformation

$$V(x, y) = y^\gamma F\left(\frac{x}{y}\right). \tag{12.4.92}$$

The function F solves the one-dimensional Variational Inequality

$$\min\left\{\begin{array}{c}\tilde{\rho}F - \frac{1}{2}\sigma^2 z^2 F'' - \tilde{\mu} z F' - \frac{1-\gamma}{\gamma}(F')^{\frac{\gamma}{\gamma-1}}, \\ \\ (\beta\gamma + z)F' - \gamma F,\ -(\alpha\gamma + z)F' + \gamma F\end{array}\right\} = 0$$

for $z = \dfrac{x}{y}$ and $\tilde{\rho} = \rho - \mu\gamma + \frac{1}{2}\sigma^2\gamma(1-\gamma)$ and $\tilde{\mu} = \rho - r - \sigma^2(1-\gamma)$; the non-linear term $\dfrac{1-\gamma}{\gamma}(F')^{\frac{\gamma}{\gamma-1}}$ comes from the reduced form of $\max\limits_{c\geq 0}\left\{-cV_x + \dfrac{c^\gamma}{\gamma}\right\}$ using that $V_x(x,y) = y^{\gamma-1}F'\left(\dfrac{x}{y}\right)$.

Davis and Norman analyzed the above equation and under certain assumptions on the market coefficients, they constructed a solution ψ satisfying, for some positive constants A and B, and points z_1 and z_2

$$\begin{cases}\psi(z) = A(\alpha\gamma + z)^\gamma, & z \leq z_1 \\ \tilde{\rho}\psi = \dfrac{1}{2}\sigma^2 z^2 \psi'' + \tilde{\mu} z \psi' + \dfrac{1-\gamma}{\gamma}(\psi')^{\frac{\gamma}{\gamma-1}}, & z_1 < z < z_2 \\ \psi(z) = B(\beta\gamma + z)^\gamma & z \geq z_2.\end{cases} \tag{12.4.93}$$

The function ψ was constructed as the solution of a two point boundary problem of second order with endpoints z_1 and z_2. These endpoints were specified by the so-called "principle of smooth fit" which is used to produce a smooth solution of (12.4.93).

The set of equations above indicates that when the ratio of account holdings $\dfrac{x}{y}$ is between the threshold levels z_1 and z_2, then it is optimal not to rebalance the portfolio but only to consume. In other words, the individual must refrain from trading in the region $\mathcal{NT} = \left\{(x,y) \in D : z_1 \leq \dfrac{y}{x} \leq z_2\right\}$.

If the holdings ratio, say $\dfrac{y_0}{x_0}$, is below z_1 then it is optimal to instantaneously rebalance the portfolio components by moving from the original point to the point $(\overline{y}, \overline{x})$ with $\overline{y} = z_1\overline{x}$ with $\overline{x} = \dfrac{x_0 + \beta y_0}{1 + \beta z_1}$. This corresponds to a transaction of *buying shares of stock* and this is the optimal policy that one should apply to all points $(x,y) \in \overline{D}$ with $\dfrac{y}{x} < z_1$. Similarly, if the holdings ratio $\dfrac{y_0}{x_0}$ is above z_2, then it is optimal to instantaneously rebalance the portfolio components by moving to the point $\tilde{y} = z_2\tilde{x}$ with $\tilde{x} = \dfrac{x_0 + \alpha y_0}{1 + \alpha z_2}$. This corresponds to a transaction of *selling stock shares* and this is the optimal policy for all point $(x,y) \in \overline{D}$ such that $\dfrac{y}{x} > z_2$.

The above analysis shows that the state space $\overline{D} = \{(x,y) : x + \begin{pmatrix}\alpha \\ \beta\end{pmatrix} y \geq 0\}$ depletes into three regions: the so-called *sell* ($\mathcal{S}$) and *buy* ($\mathcal{B}$) regions (sales and purchases of stock shares occur instantaneously) and the *no trading* ($\mathcal{NT}$) region (no trading takes place but only consumption from the bond account holdings). The $\mathcal{NT}$ region lies in between the $\mathcal{B}$ and the S region and the common boundaries are straight lines emanating from the origin; the latter properties are dictated by the homotheticity of the value function. Davis and Norman showed that the existence of a smooth solution of (12.4.93) provides a sufficient condition for the optimality of a policy, say (c_t^*, L_t^*, M_t^*) such that the associated state process (x_t^*, y_t^*) is a *reflecting diffusion* in the $\mathcal{NT}$ region and L_t^* and M_t^* are given from the relevant *local times* at the lower and upper boundaries, respectively.

As it was mentioned earlier, the work of Davis and Norman is a landmark in the area of transaction costs. A number of key ideas and insights were gained from their work which influenced a number of papers in the area. In particular, Shreve and Soner (1994) studied the same model and extended the DN results in several directions. Below, we present the main parts of the analysis of Shreve and Soner (1994) together with some relevant results of Davis and Norman. We choose to proceed this way mainly because Shreve and Soner used viscosity methods and therefore, we are able to continue our exposition, in a unified manner, following the previous chapter.

First, using convex analysis arguments, Shreve and Soner (1994) proved the following result.

Theorem 12.4.4 *For $\gamma < 1, \gamma \neq 0$, there exist constants $A > 0, B > 0$ such that*

$$\begin{cases} V(x,y) = \dfrac{1}{\gamma} A^{\gamma-1}(x+\alpha y)^{\gamma} & \text{for}(\mathrm{x},\mathrm{y}) \in \mathcal{S}, \\[2ex] V(x,y) = \dfrac{1}{\gamma} B^{\gamma-1}(x+\beta y)^{\gamma} & \text{for}(\mathrm{x},\mathrm{y}) \in \mathcal{B}. \end{cases}$$

For $\gamma = 0$, there exist constants A, B such that

$$\begin{cases} V(x,y) = \dfrac{1}{\rho} \log(x+\alpha y) + A & \text{for}(\mathrm{x},\mathrm{y}) \in \mathcal{S}, \\[2ex] V(x,y) = \dfrac{1}{\rho} \log(x+\beta y) + B & \text{for}(\mathrm{x},\mathrm{y}) \in \mathcal{B}. \end{cases}$$

To explore the regularity of the value function in the $(\mathcal{NT})$ region, Shreve and Soner employed, as in (DN), its homotheticity properties. They used a different scaling transformation, namely

$$V(x,y) = \begin{cases} (x+y)^{\gamma} u(\dfrac{y}{x+y}) & \text{for} \gamma < 1, \gamma \neq 0 \\[2ex] \dfrac{1}{\rho} \log(x+y) + u(\dfrac{y}{x+y}) & \text{for} \gamma = 0. \end{cases} \tag{12.4.94}$$

They subsequently studied the regularity properties of the above function $u(z)$ where the variable z is given by $z = \dfrac{y}{x+y}$. Using that (x,y) have the property $x + \begin{pmatrix} \alpha \\ \beta \end{pmatrix} y \geq 0$, (see (3.4.4)), one gets that $z \in J = \left[-\dfrac{1}{\beta - 1}, \dfrac{1}{1-\alpha} \right]$.

Below, we adopt the notation of Shreve and Soner (1994) and we state their main results.

To this end, we introduce the quantities,

$$\begin{aligned} d_1(z) &= r + (\mu - r)z - \frac{1}{2}\sigma^2(1-\gamma)z^2 \\ d_2(z) &= (\mu - r)z(1-z) - \sigma^2(1-\gamma)z^2(1-z) \\ d_3(z) &= \frac{1}{2}\sigma^2 z^2(1-z)^2 \\ d_4(z) &= \frac{1}{1-\alpha}(1-(1-\alpha)z) \\ d_5(z) &= \frac{\beta}{\beta-1}(1-\frac{\beta-1}{\beta}(1-z)). \end{aligned}$$

Direct computations in (12.4.94) show that the value function V is a smooth solution of the HJB equation (12.4.90) if and only if u is a classical solution of the second-order ordinary differential equation

$$\min\left\{\rho u - d_1(z)\gamma u - d_2(z)u' - d_3(z)u'' - \tilde{U}_\gamma(-zu' + \gamma u), \right. \tag{12.4.95}$$

$$\left. \gamma u + d_1(z)u', \gamma u - d_5(z)u'\right\} = 0 \quad \text{for} \quad \gamma \neq 0, \gamma < 1,$$

or

$$\min\left\{\rho u - \frac{1}{\rho}d_1(z) - d_2(z)u' - d_3(z)u'' - \tilde{U}_0(-zu' + \frac{1}{\rho}), \right. \tag{12.4.96}$$

$$\left. \frac{1}{\rho} + d_1(z)u', \frac{1}{\rho} - d_5(z)u'\right\} = 0 \quad \text{for} \quad \gamma = 0,$$

where

$$\tilde{U}_\gamma(\tilde{c}) = \sup_{c>0}\{-c\tilde{c} + U(c)\} = \begin{cases} \frac{1-\gamma}{\gamma}\tilde{c}^{\frac{\gamma}{\gamma-1}} & \text{for} \gamma < 1, \gamma \neq 0 \\ -1 - \log\tilde{c} & \text{for} \gamma = 0, \end{cases}$$

with U defined in (12.4.91), (??).

Using arguments from the theory of viscosity solutions, the following result was established (see Shreve and Soner (1994)).

Theorem 12.4.5 *The function u is C^1 on $J\backslash\{0\}$. If u is not also C^1 at $\{0\}$, then for every $x > 0$*

$$V(x,0) = \begin{cases} \frac{1}{\gamma}M^{\gamma-1}x^\gamma & \text{for} \gamma \neq 0 \\ \frac{1}{\rho}\log x + \frac{1}{\rho}\log\rho + \frac{r-\rho}{\rho^2} & \text{for} \gamma = 0, \end{cases}$$

where $M = \dfrac{\rho - r\gamma}{1-\gamma} > 0$. Furthermore, even if u is not C^1 at $\{0\}$, its one-sided derivatives exist and are limits of its derivatives from the appropriate sides at 0.

Subsequently, Shreve and Soner (1994) argued that $\mathcal{NT} \neq \phi$ and therefore there exist two numbers, say θ_1 and θ_2, such that

$$\mathcal{NT} = \left\{(x,y) \in \overline{D} : \theta_1 < \frac{y}{x+y} < \theta_2\right\} \tag{12.4.97}$$

Using elements from viscosity theory, they also established the following regularity result.

Theorem 12.4.6 *The function u is C^2 on $(\theta_1, \theta_2)\backslash\{0,1\}$ and, on this set satisfies, in the classical sense, the equations*

$$\begin{cases} \rho u(z) - d_1(z)\gamma u(z) - d_2(z)u'(z) - d_3(z)u''(z) - \\ \quad -\tilde{U}_\gamma(-zu'(z) + \gamma u(z)) = 0 & \text{for} \gamma \neq 0, \gamma < 1 \\ \rho u(z) - \frac{1}{\rho}d_1(z) - d_2(z)u'(z) - d_3(z)u''(z) \\ \quad -\tilde{U}_0\left(-zu'(z) + \frac{1}{\rho}\right) = 0 & \text{for} \gamma = 0. \end{cases} \tag{12.4.98}$$

Therefore, V is C^2 in the set $\mathcal{NT}\backslash\{(x,y) : x = 0$ or $y = 0\}$ and satisfies, in the classical sense, the equation

$$\rho V = \frac{1}{2}\sigma^2 y^2 V_{yy} + \mu y V_y + rxV_x + \max_{c>0}\{-cV_x + U(c)\}$$

with U as in (4.18), (4.19). Moreover, the regions $\mathcal{S} \neq \emptyset$ and $\mathcal{B}$ contain the cone $G = \{(x,y) \in \overline{D}; y < 0\}$.

The following theorem provides a *verification result* for the optimal policies (see Shreve and Soner (1994)).

Theorem 12.4.7 *The quantities θ_1 and θ_2 satisfy*

$$0 \leq \theta_1 < \theta_2 < \frac{1}{1-\alpha}.$$

Furthermore, if $(x,y) \in \overline{D}$, then there is a triple $(c, L, M) \in \mathcal{A}$ such that with the processes x_t and y_t, as defined in (4.1) and (4.2), the following conditions hold almost surely:

i) *If $(x,y) \notin \mathcal{NT}$, then $(x_0, y_0) \in \partial\mathcal{NT}$,*

ii) *the processes $(x_t, y_t) \in \mathcal{NT}$, $\forall t \geq 0$,*

iii) $L_t^* = \int_0^t \mathbf{1}_{\left\{\frac{y_s}{x_s + y_s} = \theta_1\right\}} dL_s$, $\forall t \geq 0$,

iv) $M_t^* = \int_0^t \mathbf{1}_{\left\{\frac{y_s}{x_s + y_s} = \theta_2\right\}} dM_s$, $\forall t \geq 0$,

v) $c_t^* = [V_x(x_s, y_s)]^{\frac{1}{\gamma - 1}}$, $\forall t \geq 0$.

The triple is optimal, i.e.

$$V(x,y) = E\int_0^{+\infty} e^{-\rho t}U(c_t^*)dt.$$

Next, we consider the two boundaries of the $\mathcal{NT}$ wedge region

$$\partial_1\mathcal{NT} = \{(x,y) \in D : y \geq 0, -\theta_1 x - (\theta_1 - 1)y = 0\}$$

and

$$\partial_2\mathcal{NT} = \{(x,y) \in D : y > 0, \theta_2 x + (\theta_2 - 1)y = 0\},$$

we define the reflection direction index

$$\widetilde{\gamma}(x,y) = \begin{cases} (-\beta, 1) & \text{if}(x,y) \in \partial_1 \mathcal{NT} \\ \\ (\alpha, -1) & \text{if}(x,y) \in \partial_2 \mathcal{NT} \end{cases}$$

and we let $\widetilde{\gamma}(x,y) = (\widetilde{\gamma}_1(x,y), \widetilde{\gamma}_2(x,y))$.

The above theorem states that for any pair of portfolio positions $(x,y) \in \overline{\mathcal{NT}}$, there is a solution to the *Skorohod problem:*

Skorohod problem: Find continuous processes x_t, y_t, k_t such that $x_0 = x$, $y_0 = y$, $k_0 = 0$, k is nondecreasing and the following assertions hold

i) $(x_t, y_t) \in \overline{\mathcal{NT}} \ \ \forall t \geq 0$

ii) $dx_t = [rx_t - (V_x(x_t, y_t))^{\frac{1}{\gamma-1}}]dt + \widetilde{\gamma}_1(x_t, y_t)dk_t$

iii) $dy_t = \mu y_t dt + \sigma y_t dW_t + \widetilde{\gamma}_2(x_t, y_t)dk_t$

iv) $k_t = \int_0^t \mathbf{1}_{\{(x_t, y_t) \in \partial \mathcal{NT}\}} dk_t$.

We can easily identify the above conditions with the ones presented earlier putting

$$\begin{cases} L_t = \int_0^t \mathbf{1}_{\{(x_t, y_t) \in \partial_1 \mathcal{NT}\}} dk_t \\ \\ M_t = \int_0^t \mathbf{1}_{\{(x_t, y_t) \in \partial_2 \mathcal{NT}\}} dk_t, \end{cases}$$

where

$$\partial_i \mathcal{NT} = \{(x,y) \in \overline{\mathcal{NT}} : \frac{y}{x+y} = \theta_i\}, \quad i = 1, 2.$$

Shreve and Soner (1994) used control theory arguments to establish additional regularity results for the value function across the interfaces $\theta_i \mathcal{NT}$ defined above. Finally, they produce various results for the location of the optimal exercise boundaries; these results are stated below in terms of the slopes θ_1 and θ_2.

Theorem 12.4.8 *The partial derivative V_{yy} is continuous across $\partial_2 \mathcal{NT}$, and if $\theta_2 \neq 1$, then V is C^2 across $\partial_2 \mathcal{NT}$. If $\theta_1 \neq 0$, then V_{yy} is continuous across $\partial_1 \mathcal{NT}$, and if $\theta_1 \neq 0$ and $\theta_1 \neq 1$, then V is C^2 across $\partial_1 \mathcal{NT}$.*

Proposition 12.4.9 *i) The value function V is C^2 in $S \backslash \{(x,y) : x = 0$ or $y = 0\}$.*

ii) The functions V, V_x and V_y are continuous in $D \backslash \{(x,0) : x > 0\}$.

The next propositions provide information about the location of the exercise boundaries and closed-form solutions for the value function.

Proposition 12.4.10 *If $\gamma < 1$, $\gamma \neq 0$ then there is a positive constant A with $\frac{A}{\gamma} < \frac{(\rho - r\gamma)}{\gamma(1-\gamma)}$ such that, for $(x,y) \in S$,*

$$V(x,y) = \frac{1}{\gamma} A^{\gamma-1} (x + \alpha y)^{\gamma}.$$

If $\gamma = 0$, then there is a constant $\widehat{A}$ with $\widehat{A} > (\log \rho)/\rho + (r-\rho)/\rho^2$ such that, for $(x, y) \in \mathcal{S}$,

$$V(x, y) = \frac{1}{\rho} \log(x + \alpha y) + \widehat{A}.$$

Proposition 12.4.11 *P4.3 For all $\gamma < 1$, the slope θ_2 satisfies*

$$\theta_2 < \frac{2(\mu - r)}{\alpha\sigma^2(1-\gamma) + 2(1-\alpha)(\mu - r)}.$$

Moreover, if $\sigma^2(1-\gamma) \neq \mu - r$ and the quantity $M(\gamma) = \dfrac{\rho - r\gamma}{1-\gamma} - \dfrac{\gamma(\mu-r)^2}{2\sigma^2(1-\gamma)^2} > 0$, then

$$\theta_2 > \frac{\mu - r}{\alpha\sigma^2(1-\gamma) + (1-\alpha)(\mu - r)}.$$

If $\sigma^2(1-\gamma) = \mu - r$ and $M(\gamma) > 0$, then $\theta_2 = 1$. Finally if $M(\gamma) > 0$ and $\sigma^2(1-\gamma) > \dfrac{\beta - 1}{\beta}(\mu - r)$ then the slope of the low exercise boundary satisfies

$$\theta_1 < \frac{\mu - r}{\beta\sigma^2(1-\gamma) - (\beta - 1)(\mu - r)}.$$

The slope $\theta_1 > 0$, i.e. the positive x-axis belongs to the $\mathcal{B}$.

Besides the above results, Shreve and Soner (1994) provided conditions for the value function to be well defined; these conditions are considerably more general than the ones of Davis and Norman.

Departing from the benchmark optimal investment-consumption model of Davis and Norman, other valuation models with transaction costs have been introduced and analyzed with alternative analytical methods. The main incentive to develop such models comes from the absence of closed form expressions for the optimal investment strategies in the DN model, a feature highly desirable for practical applications. In fact, as the outlaid analysis indicates, in order to analyze the transaction costs portfolio problems one needs to solve a free boundary problem. The free boundaries define the no-transaction ($\mathcal{NT}$) region in which trading is prohibited due to the high penalties from the transaction costs. The precise characterization and the accurate computation of these interphases are imperative for the practical importance of the model. In addition, more realistic models incorporate more than one risky securities as these models look at larger portfolios or at "books" of options. In this case, the regions of trading idleness have a rather complex structure. To solve such problems is a formidable task both from the theoretic as well as the numerical point of view.

To overcome these difficulties, alternative models were introduced which, from one hand, can be analyzed more effectively and, from the other hand, produce optimal trading strategies which do not deviate considerably from their theoretical counterpart. The first models in this direction are the models of Morton and Pliska (1995) and Pliska and Selby (1995); Schroder (1993) independently obtained similar results for some special cases. The key features in the approach of Morton, Pliska and Selby are the possibility of investing in more than one stock and the fact that the transaction costs are proportional to a fixed fraction of the (dynamic) portfolio value.

In the models of Morton, Pliska and Selby, the risky securities are modeled as correlated geometric Brownian motions and there is no intermediate consumption drainage. The optimality criterion differs from the DN payoff in that the aim is to maximize the long-run expected growth rate of the total portfolio value, $\liminf_{T\to\infty} E\frac{(\ell n\ V_T)}{T}$. The transaction costs are considered fixed in the sense that, each time funds are shifted between two or more securities (riskless or risky ones), a penalty is imposed equal to the fraction $(1-\epsilon)$ times the current value of the entire portfolio. Aside from the current portfolio value, the penalty is independent of the prices and the positions in the individual securities.

In Morton and Pliska (1995) the trading strategy which maximizes the relevant criterion is fully determined by an M-dimensional vector, say $\vec{b}$ and an optimal stopping time τ. The dimension of $\vec{b}$ coincides with the number of risky securities (assuming that their variance-covariance matrix is of full rank) and its components are strictly positive and sum to less than one. The optimal solution $\vec{b}^*$ and the stopping time τ are found by solving a free boundary problem which can be reduced to a linear complimentarity problem. In the case of only one risky security, the free boundary consists of only two points and the optimal solution can be achieved easily. The numerical method becomes much more complex when there are two or more risky securities because, in this case, the optimal exercise boundary consists of infinitely many points. Pliska and Selby (1995) addressed this issue by employing a novel transformation to the original free boundary problem of Morton and Pliska (1995), for the case of two risky securities. This transformation makes the problem considerably easier to solve but still does not address the issue of more than two risky securities, both from the analytic as well as the numerical point of view.

Atkinson and Wilmott (1995) studied the multi-dimensional case under the assumption that the *transaction costs are small* — but still of realistic size. They used asymptotic methods and a local analysis in the original Merton problem which is free of transaction costs. This asymptotic approach showed that the continuation region resembles an ellipsoid which actually resembles the region obtained by Morton and Pliska (1995). This property holds in a certain part of the state space, related to the local behavior of the Merton solution but the approximation breaks down in the other parts. This key difficulty was successfully addressed in Atkinson, Pliska and Wilmott (1999) who studied the non-constant coefficient version of the Atkinson and Wilmott model. By handling the non-constant coefficient case, Atkinson, Pliska and Wilmott succeeded in bypassing the difficulties in the asymptotic analysis of Atkinson and Wilmott. Asymptotic results for small transaction costs for the (DN) model in the case of many risky assets was performed by Atkinson and Al–Ali (1995).

Optimal investment models in which the stock price structure is similar to the Davis and Norman but with alternative optimization criteria have been analyzed by a number of authors. Portfolio models with *finite trading horizon* and utilities depending on the terminal wealth were studied, analytically and/or numerically, by Fleming et al (1989), Akian, Menaldi and Sulem ((1992), (1996)), Akian, Séquier and Sulem (1996), Sulem (1997) and more recently by Tiu and Zariphopoulou (1999). Other models with "long-run" type criteria have been examined first by Taksar, Klass and Assaf (1988) and subsequently by Dumas and Luciano (1991), Fleming et al (1989), Sulem (1997), Akian, Sulem and Taksar (1996).

So far, in all the above models and the ones discussed in the previous section, the common avenue of obtaining information about the value function and the optimal investment and consumption plans is via the HJB equation. An alternative and rather powerful approach is the one that uses results from *martingale theory.* The majority of the results obtained through this approach are found in models of derivative pricing with transaction costs and they are discussed in the second chapter. In the context of portfolio optimization, this methodology, together with tools from convex and functional analysis and duality theory, was employed by Cvitanić and Karatzas (1996) (we also refer the reader to the monograph of Karatzas (1997)).

Transaction costs have also been incorporated in other kinds of asset pricing models.

In many of these models, financial trades are charged by "adjustment" or "shipping" costs which can have more complex structure. The economic considerations are different and the mathematical analysis is overall less rigorous as one moves to more applied areas of finance. Optimal consumption models of durable goods have been examined by Grossman and Laroque (1989) and Eberly (1999). Other capital asset pricing models with transactions costs for divident policies, stock returns, term structure, exchange rates and asset demands are listed in the references.

In summary, transaction costs result in *irreversible losses* which in most cases cannot be valuated with the classical existing theories. There are still may challenging questions in equilibrium asset pricing theory which do not have a satisfactory answer. The difficulties come not only from the lack of a coherent modeling structure but also from the absence of good analytic and numerical techniques needed to attack the related stochastic optimization models.

12.4.2 Optimal investment/consumption models with stochastic labor income

A very important extension of Merton's model is when the individual investor is endowed with a stream of stochastic income that cannot be replicated by trading the available securities. In other words, markets are incomplete in an essential way. In the case of general time-additive utilities, this model was analyzed by Duffie and Zariphopoulou (1993) who studied the solutions of the HJB equation (see Theorem 12.4.13 below). Considerable simplification is obtained by assuming that the utility function is of the CRRA type; this case was studied by Duffie et al (1997) and Koo (1991) using pde techniques. A considerable volume of work on this subject was also produced via martingale methods and the duality approach, carried out by Cuoco (1997), He and Pearson (1991), Karatzas et al (1991); related literature also includes Duffie and Richardson (1991), El Karoui and Jeanblanc-Piqué (1991), He and Pagés (1993) and Swensson and Werner (1990).

We continue with the description of the underlying financial model and the main results of the associated stochastic optimization problem. The fundamental assumption is that individual preferences are modeled via a power function of exponent $\gamma \in (0,1)$. The majority of the results presented below are from Duffie et al (1997).

On a given probability space is a standard Brownian motion $W = (W^1, W^2)$ in $\mathcal{R}^2$. The standard augmented filtration $\{\mathcal{F}_t : t \geq 0\}$ generated by W is fixed. Riskless borrowing or lending is possible at a constant continuously compounding interest rate r. A given investor receives income at the rate Y_t, where

$$\begin{cases} dY_t = bY_t dt + aY_t dW_t^1, & (t \geq 0), \\ Y_0 = y, & (y > 0), \end{cases} \tag{12.4.99}$$

where b and a are positive constants and y is the initial level of income.

A traded security has a price process S_t given in (12.3.18) and the Brownian motion W is correlated to W^1 with correlation coefficient $\rho \in (-1,1)$; for this we can take $W_t = \rho W_t^1 + \sqrt{1-\rho^2}W_t^2$.

A consumption process is an element of the space $\mathcal{L}_+$ consisting of any non-negative $\{\mathcal{F}_t\}$-progressively measurable process C such that $E\big(\int_0^T C_t dt\big) < \infty$ for any $T > 0$. The agent's payoff function $\mathcal{J} : \mathcal{L}_+ \to \mathcal{R}^+$ from consumption is given by

$$\mathcal{J}(C) = E\Big(\int_0^{+\infty} e^{-\beta t} C_t^\gamma dt\Big), \tag{12.4.100}$$

for some risk aversion measure $1-\gamma \in (0,1)$ and discount factor $\beta > r$.

It is assumed throughout that $\beta > r$, that $|\rho| \neq 1$, and that the volatility coefficient σ is strictly positive. Cases in which $\beta \leq r$, $\sigma = 0$, or $|\rho| = 1$ are not discussed and, in general, they may lead to a different characterization of optimal policies than the one obtained herein.

The agent's *wealth process* X evolves according to the equation

$$\begin{cases} dX_t = [rX_t + (\mu - r)\Pi_t - C_t + Y_t]dt + \sigma \Pi_t dB_t, & t \geq 0), \\ X_0 = x, & (x \geq 0), \end{cases} \tag{12.4.101}$$

where x is the initial wealth endowment, and the control processes C and Π represent the consumption rate C_t and investment Π_t in the risky asset, with the remainder of wealth held in riskless borrowing or lending.

The controls C and Π are drawn, respectively, from the spaces $\mathcal{C} = \{C \in \mathcal{L}_+ : \mathcal{J}(C) < \infty\}$ and $\Phi = \{\ell : \ell \text{ is } \mathcal{F}_t\text{-progressively measurable and } E\int_0^t \ell_s^2 ds < \infty \text{ a.s. } (t \geq 0)\}$. The set $\mathcal{A}(x,y)$ of *admissible controls* consists of pairs (C,Π) in $\mathcal{C} \times \Phi$ such that $X_t \geq 0$ a.s., $(t \geq 0)$, where X_t is given by the state equation (4.27) using the controls (C,Π).

The agent's *value function* v is given by

$$v(x,y) = \sup_{(C,\Pi)\in\mathcal{A}(x,y)} \mathcal{J}(C). \tag{12.4.102}$$

Assuming, formally for the moment, that the value function v is finite-valued and twice continuously differentiable in $D \equiv (0,\infty) \times (0,\infty)$, it is natural to conjecture that v solves the HJB equation

$$\beta v = \max_{\pi} G^v(\pi) + \frac{1}{2}a^2y^2v_{yy} + \max_{c\geq 0} H^v(c) + (rx+y)v_x + byv_y, \tag{12.4.103}$$

for $(x,y) \in D$, where subscripts indicate the obvious partial derivatives and

$$\begin{cases} G^v(\pi) = \frac{1}{2}\sigma^2\pi^2 v_{xx} + \rho\pi y\sigma\tilde{\sigma}v_{xy} + (\mu - r)\pi v_{xy}, \\ H^v(c) = -cv_x + c^\gamma. \end{cases}$$

It can be shown directly from (12.4.101) and (12.4.102) that if v is finite-valued, then it is concave and is homogeneous with degree γ; that is, for any (x,y) and a positive constant k we have $v(kx,ky) = k^\gamma v(x,y)$. It therefore makes sense to define $u : [0,+\infty) \to [0,+\infty)$ by $u(z) = v(z,1)$, so that knowledge of u recovers v from the fact that $v(x,y) = y^\gamma u(x/y)$ for $y > 0$. The same idea is used, for example, in Davis and Norman (1990). This does not recover $v(x,0)$, which is known nevertheless to be the Merton's original solution without stochastic income.

If v satisfies (12.4.103) then, for $x > 0$, u solves

$$\widehat{\beta}u = \frac{1}{2}a^2z^2u'' + \max_{\pi}\left[(\frac{1}{2}\sigma^2\pi^2 - \rho\sigma a\pi z)u'' + k_1\pi u'\right] + k_2zu' + F(u'), \tag{12.4.104}$$

where

$$\begin{cases} \widehat{\beta} = \beta - b\gamma + \frac{1}{2}a^2\gamma(1-\gamma), \\ k_1 = \mu - r - (1-\gamma)\rho\sigma a, \\ k_2 = a^2(1-\gamma) + r - b, \end{cases} \tag{12.4.105}$$

and $F : [0, +\infty) \to [0, +\infty)$ is given by

$$F(p) = \max_{c \geq -1}[-cp + (1+c)^{\gamma}]. \tag{12.4.106}$$

After performing the (formal) maximization in (4.30), assuming that u is smooth and strictly concave, we get

$$\widehat{\beta}u = \frac{1}{2}a^2 z^2(1-\rho^2)u'' - \frac{k_1^2}{2\sigma^2}\frac{(u')^2}{u''} + kzu' + F(u'), \qquad (z > 0), \tag{12.4.107}$$

where

$$k = \frac{\rho k_1 a}{\sigma} + k_2. \tag{12.4.108}$$

In Duffie et al. (1997), it is shown that u can be characterized as the value function of a so-called *'dual'* investment-consumption problem. That is,

$$u(z) = \sup_{(C,\Pi)\in\mathcal{A}(z)} E\Big[\int_0^{+\infty} e^{-\widehat{\beta}t}(1+C_t)^{\gamma}dt\Big]. \tag{12.4.109}$$

where the set $\mathcal{A}(z)$ of admissible policies is defined below. To this end, we consider an "artificial" consumption-investment problem of an agent whose current wealth Z_t evolves, using a consumption process C_t and risky investment process Π_t, according to the equation

$$\begin{cases} dZ_t = [kZ_t + k_1\Pi_t - C_t]dt + \sigma\Pi_t dW_t^1 + aZ_t\sqrt{1-\rho^2}dW_t^2 \quad (t \geq 0), \\ Z_0 = z \quad (z \geq 0), \end{cases} \tag{12.4.110}$$

where z is the initial endowment and k_1 and k are given, respectively, by (12.4.105) and (12.4.108). The set $\mathcal{L}$ of consumption processes consists of any progressively measurable process C such that $C_t \geq -1$ almost surely for all t, with $E\int_0^t C_s ds < \infty$ for all t. A control pair (C, Π) for (12.4.110) consists of a consumption process C in $\mathcal{L}$ and a risky investment process $\Pi \in \Phi$ and it is admissible if $Z_t \geq 0$ a.s., $(t \geq 0)$, where Z_t is given by (12.4.110).

Observe that from one hand, the agent is forced to invest a fixed multiple of wealth in a risky asset with expected return k and 'volatility' $a\sqrt{(1-\rho^2)}$. On the other hand, he chooses the amount Π invested in another risky asset with mean return k_1 and volatility σ.

The agent's utility is given by

$$\widehat{\mathcal{J}}(C) = E\Big[\int_0^{+\infty} e^{-\widehat{\beta}t}(1+C_t)^{\gamma}dt\Big].$$

The value function $w : [0, +\infty) \to [0, +\infty)$ is defined by

$$w(z) = \sup_{(C,\Pi)\in\mathcal{A}(z)} \widehat{\mathcal{J}}(C). \tag{12.4.111}$$

The HJB equation associated with this stochastic control problem is

$$\widehat{\beta}w = \frac{1}{2}\sigma^2(1-\rho^2)z^2 w_{zz} + \max_{\pi}\Big[\frac{1}{2}\bar{\sigma}^2\pi^2 w_{zz} + k_1\pi w_z\Big]$$
$$+ \max_{c\geq -1}[-cw_z + (1+c)^{\gamma}] + kzw_z \quad (z > 0). \tag{12.4.112}$$

We observe that (12.4.111) reduces (at least formally) to (12.4.107) for smooth concave solutions.

We call problems (12.4.111) and (12.4.109) 'dual' to each other because one hedges an income stream and the other hedges an investment, and because of the relationship between their value functions: the reduced value function u of problem (12.4.109) for CRRA utility reduces to the value function w of (12.4.111) for non-CRRA utility. Conversely, it can be shown that problem (12.4.111), after substituting the CRRA utility function $\mathcal{J}$ for $\widehat{\mathcal{J}}$ and substituting the correlated Brownian motion B for W^2 in (12.4.110), has a value function equivalent to that of problem (12.4.109), after making the opposite substitutions. Thus either of these dual problems can be reduced to a version of the other with a single-state variable.

The following results can be found in Duffie et al (1997).

Theorem 12.4.12 *i) Suppose that u is an upper-semicontinuous concave viscosity subsolution of the HJB equation (12.4.112) on $[0,+\infty)$ and $u(z) \le c_0(1+z^\gamma)$ for some $c_0 > 0$; also suppose that v is bounded from below, uniformly continuous on $[0,+\infty)$ and locally Lipschitz in $(0,+\infty)$, and a viscosity supersolution of (4.38) in $(0,+\infty)$. Then $u \le v$ on $[0,+\infty)$.*

ii) The value function v is the unique constrained viscosity solution of the HJB equation (12.4.103) on $\bar{D}$ in the class of concave functions.

The next result provides a characterization of the value function w of the reduced dual problem.

Theorem 12.4.13 *i) The value function w is concave, increasing, and continuous on $[0,\infty)$.*

ii) The value function w is the unique $C[0,+\infty) \cap C^2(0,+\infty)$ solution of (12.4.112) in the class of concave functions.

iii) The value function w coincides with the function u.

It turns out that this characterization of u is crucial for proving regularity results for the value function v as well as for obtaining feedback forms for the optimal policies. By a "feedback policy,' we mean, as usual, a pair (g,h) of measurable real-valued functions on $[0,\infty) \times [0,\infty)$ defining, with current wealth x and income rate y, the *risky investment* $h(x,y)$ and *consumption rate* $g(x,y)$. Such a feedback policy (g,h) determines the stochastic differential equation for wealth given by

$$\begin{cases} dX_t = [rX_t + (\mu - r)h(X_t,Y_t) - g(X_t,Y_t) + Y_t]dt + \sigma h(X_t,Y_t)dW_t, (t \ge 0), \\ X_0 = x, \qquad (x > 0). \end{cases} \tag{12.4.113}$$

If there is a non-negative solution X to (12.4.101) and if the policy (C,Π) defined by

$$C_t = g(X_t,Y_t), \quad H_t = h(X_t,Y_t),$$

are in $\mathcal{C}$ and Φ, respectively, then (C,Π) is an admissible policy by definition of $\mathcal{A}(x,y)$.

Before stating the main conclusions, we recall that for the case $y = 0$ (implying $Y_t = 0$, $t \ge 0$), the value function v is given from Merton's (1971) work. In fact if the constant

$$K = \frac{\beta - r\gamma}{1-\gamma} - \frac{\gamma(\mu - r)^2}{2(1-\gamma)^2\sigma^2} \tag{12.4.114}$$

is strictly positive, we have

$$v(x,0) = K^{\gamma-1}x^{\gamma},$$

with optimal policies given in feedback form by

$$g(x,0) = Kx, \qquad h(x,0) = \frac{x(\mu - r)}{\sigma^2(1-\gamma)}$$

with K given in (12.4.114).

For $x > 0$ and $y > 0$, the feedback policy functions g and h defined by the first-order optimality conditions for (12.4.103), in light of the homogeneity property $v(x,y) = y^y u(x/y)$, are given by

$$g(x,y) = y\Big(\frac{u'(x/y)}{y}\Big)^{1/(\gamma-1)}, \tag{12.4.115}$$

$$h(x,y) = \frac{\rho a}{\sigma}x - y\frac{k_1}{\sigma^2}\frac{u'(x/y)}{u''(x/y)}. \tag{12.4.116}$$

The following theorem provides the verification result for the value function and the optimal policies. Its proof is in Theorem 1 by Duffie et al (1997).

Theorem 12.4.14 *Suppose $\widehat{\beta}, K$, and $r - \mu$ are all strictly positive.*

i) There is a unique $C^1([0,+\infty)) \cap C^2((0,+\infty))$ solution u of the ordinary differential equation (12.4.107) in the class of concave functions.

ii) The value function v is given by

$$v(x,0) = K^{\gamma-1}x^{\gamma},$$

$$v(x,y) = y^{\gamma}u\Big(\frac{x}{y}\Big), \qquad y > 0. \tag{12.4.117}$$

iii) There is a unique solution X_t of (12.4.101) satisfying the budget feasibility constraint $X_t \geq 0$, and an optimal policy (C^*, Π^*) is given by $C_t^* = g(X_t, Y_t)$ and $\Pi_t^* = h(X_t, Y_t)$ where g and h are given by (12.4.115)–(12.4.76), with $h(0,y) = 0$ for all y and $g(0,y) = \alpha y$ for all y, where $\alpha = (u'(0)/\gamma)^{1/(\gamma-1)}$.

iv) If $k_1 \neq 0$, starting from strictly positive wealth $(x > 0)$, the optimal wealth process, almost surely, will never hit zero, and starting from zero, almost surely, the optimal wealth process will instantaneously become strictly positive. The same conclusion holds if $k_1 = 0$ and $u'(0) > \gamma$.

12.5 Expected utility methods in derivative pricing

The area of derivative securities has been one of the fastest growing areas of finance as well as one of the most active areas of research on stochastic analysis, stochastic control and computations. Derivatives are financial instruments whose values depend on the price levels of the so-called primitive securities, like stocks. The fundamental problem of derivative valuation is in determining the derivative's fair value and in specifying the hedging policy which eliminates the risk inherent to the contract. Derivative contracts had always existed in financial environments but it was after the seminal work of Fisher Black and Myron Scholes (1973) (in collaboration with Robert Merton) that this area blossomed and started expanding rapidly. The Black and Scholes valuation approach brought to modern finance the powerful methodologies of martingale theory and stochastic calculus. Today, numerous different kinds of derivative instruments are traded all around the world and various new contracts are being created every day. The valuation of these contracts gives rise to a number of challenging problems in the areas of stochastic analysis, martingale theory, stochastic control and partial differential equations.

Despite the ever growing activity in derivatives' markets, very few questions have been successfully addressed to date when derivatives are produced, traded and hedged in markets with frictions. The most important kind of frictions comes from the stochastic nature of the volatility of the primitive stock security. Most of the research on derivatives with frictions is concentrated on the case of stochastic volatility and several methodologies have been proposed at different levels of sophistication. The majority of the theoretical results were obtained via martingale theory and convex duality arguments (see, for example, Cvitanić and Karatzas (1996), Cvitanić, Pham and Touzi (1997)) without fully involving any expected utility formulation.

Methods based on expected utility first started, and since then have been developed primarily for pricing derivative securities in markets with transaction costs. Following our unified theme to concentrate mainly on expected utility models of asset valuation, we provide below an overview of such models used in pricing with transaction costs. We also choose to proceed this way because, as it will be demonstrated below, the existing theories give rise to challenging singular stochastic control problems whose analysis is interesting in its own right.

The fundamental difficulty for pricing derivatives in the presence of transaction costs lies in the fact that the Black and Scholes approach breaks down completely. In fact, in a frictionless market, Black and Scholes (1973) and Merton (1973a) relied on an ingenious no-arbitrage argument to price an option on a stock when the interest rate is constant and the stock price follows a geometric brownian motion. They presented a self-financing, dynamic trading policy between the bond and stock accounts which replicates the payoff of the option. They then argued that absence of arbitrage dictates that the option price is equal to the cost of setting up the replicating portfolio. The appeal of the argument lies in its reliance on the absence of arbitrage alone and is independent of other aspects of the equilibrium, such as a particular asset pricing model. The precise derivation arguments of Black and Scholes are discussed in the next section.

The Achilles' heel of the argument is that the frictionless market assumption must be taken literally. The dynamic replication policy incurs an *infinite volume* of transactions over any finite trading interval, given the fact that the brownian motion which drives the stock price has infinite variation. In a market with proportional transaction costs, the dynamic replication policy incurs infinite transaction costs over any finite trading interval and cannot be self-financing, no matter how small the finite transaction costs rate is.

Merton (1990) maintained the goal of a dynamic trading policy as that of replicating the option payoff and modeled the path of the stock price as a two-period binomial process. The initial cost of the replication policy is finite and serves as an upper bound to the write price of a call which is arbitrage-free. Shen (1990) and Boyle and Vorst (1992) extended Merton's model to a multiperiod binomial process for the stock price and provided numerical solutions to the initial cost of the replicating portfolio. As the number of periods increases within the given lifetime of the call option, the initial cost of the replicating portfolio tends to infinity.

Bensaid et al (1992) and Edirisinghe, Naik and Uppal (1993) noted that a tighter upper bound on the write price of a call option is obtained by replacing the goal of replicating the payoff of the option with the goal of *dominating the payoff*. For example, the payoff of one share of stock dominates the payoff of a call option and, therefore, the cost of initially buying one share provides an upper bound to the cost of a minimum-cost dominating policy as the number of periods increases within the given lifetime of the option. Davis and Clark (1994) conjectured and Soner, Shreve and Cvitanić (1995) proved, that the cost of initially buying one share of stock is indeed the cost of the cheapest dominating policy in the presence of finite proportional transaction costs. Their result on feasible *super-replicating strategies* was subsequently generalized by Levental and Skorohod (1997).

Leland (1985) initiated a novel approach by introducing a class of *imperfectly replicating policies* in the presence of proportional transaction costs. He calculated the total cost, including transaction costs, of an imperfectly replicating policy and the "tracking error," that is the standard deviation of the difference between the payoff of the option and the payoff of the imperfectly replicating policy. Imperfectly replicating policies were further studied by Figlewski (1989), Flesaker and Hughston (1994), Henrotte (1993), Hoggard, Whalley and Wilmott (1994) and Toft (1996). Avellaneda and Parás (1994) extended the notion of imperfectly replicating policies to that of imperfectly dominating policies.

An alternative approach, initiated by Hodges and Neuberger (1989) and developed further by Davis, Panas and Zariphopoulou (1993), is the so-called *utility maximization method.* The fundamental ideas for this method stem from the economic principles of *stochastic dominance* (see, for relevant results, Perrakis and Ryan (1984), Levy (1978) and Ritchken (1985)). In this approach, the price of the derivative is determined by comparing the value functions of an investor with and without the opportunity to trade the available derivative. The individual preferences are modeled via an exponential utility and the derivative is a European call. By considering the utility functionals (with and without the derivative), this methodology incorporates the individual's attitude towards the risk which cannot be eliminated, in contradistinction to the case of no transaction costs. The above results were considerably generalized by Constantinides and Zariphopoulou (1999) who applied utility methods to establish price bounds for all types of European claims and for general preferences.

Besides the claims of European-type, the valuation of American options was examined by Davis and Zariphopoulou (1995) for the class of exponential utilities. More recently, Constantinides and Zariphopoulou (1999) extended their results to the cases of American-type and path-dependent claims, written on many stocks and for CRRA utilities. Other path-dependent claims were priced by Dewynne, Whalley and Wilmott (1994) using ideas from the Leland's valuation approach.

Finally, a considerable volume of work has been produced under the assumption that the *transaction costs are "small."* This assumption is not far from reality for a sizeable class of models and produces adequate results for the prices and, in particular for the hedging strategies. To most extent, the relevant analysis imitates the Black and Scholes methodology together with various elements from the above methods. We mention among others, the work of Whalley and Wilmott (1997), Barles and Soner (1998) and Albanese and Tompaidis (1998).

We continue by presenting the classical Black and Scholes pricing formula first and then the various valuation methodologies, namely the super-replication approach, utility maximization theory and the method of imperfectly replicating strategies.

12.5.1 The Black and Scholes valuation formula

In their seminal paper, Black and Scholes (1973) developed a theory for the valuation of derivative securities in frictionless markets. They considered the problem of determining the value of a *European call* which is written on an underlying stock whose price S_s follows the diffusion process, as described in (12.2.2). The market is also endowed with a riskless security whose price is given by (12.2.1). The European claim is written at the time say

$t > 0$ and expires at maturity time T. Its payoff, at expiration, is given by $(S_T - K)^+$ where K is the (prespecified) *exercise price.*

The valuation problem amounts to specifying the *fair value* of the security at its birth time t.

Black and Scholes had the novel idea of constructing a dynamic portfolio whose value coincides with the terminal payoff, $(S_T - K)^+$, of the call. Then they argued that the amount needed to set up this hedging portfolio, at time t, yields the correct price of the European call. Moreover, the components of this portfolio, across time, give the *perfectly replicating (hedging) strategies* which reproduce the value of the security.

Black and Scholes postulated that the call price is a smooth function of the current stock price and time. Therefore, there exists a smooth function $C : [0, +\infty) \times [0, T) \to [0, +\infty)$ such that the call price process h_s, $t \le s \le T$ can be represented as $h_s = C(S_s, s)$ with S_s being given in (12.2.2).

Applying Itô's formula to h_s yields

$$dh_s = \Big[\frac{\partial C(S_s, s)}{\partial t} + \frac{1}{2}\sigma^2 S_s^2 \frac{\partial^2 C(S_s, s)}{\partial S^2}$$

$$+\mu S_s \frac{\partial C(S_s, s)}{\partial S}\Big] ds + \sigma S_s \frac{\partial C(S_s, s)}{\partial S} dW_s. \qquad (12.5.118)$$

Next, we assume that the riskless interest rate is $r > 0$ and that the components of the replicating portfolio are β_s and δ_s. In other words, at any time s, we would have to purchase β_s bonds and δ_s shares of the underlying stock. According to the perfect replication idea of Black and Scholes, the following equalities must hold

$$\beta_s B_s + \delta_s S_s = h_s \qquad \text{a.e. } t \le s < T, \qquad (12.5.119)$$

$$\beta_T B_T + \delta_T S_T = (S_T - K)^+. \qquad (12.5.120)$$

Taking into account the price equations (12.3.17) and (12.3.18), (12.5.119) yields

$$dh_s = (\mu \delta_s S_s + r\beta_s B_s) ds + \sigma \delta_s S_s dW_s,$$

or, equivalently,

$$dh_s = [(\mu + r)\delta_s S_s - r h_s] ds + \sigma \delta_s S_s dW_s. \qquad (12.5.121)$$

We recall that the processes β_s and δ_s satisfy certain "self-financing" assumptions which in turn justify the above differential forms. Equating formally the coefficients in (12.5.118) and (12.5.121) yields

$$\delta_s = \frac{\partial C(S_s, s)}{\partial S} \qquad (12.5.122)$$

and

$$\beta_s = \frac{h_s - \delta_s S_s}{B_s},$$

as long as the following condition holds a.e.

$$r h_s = \frac{\partial C(S_s, s)}{\partial t} + \frac{1}{2}\sigma^2 S_s^2 \frac{\partial^2 C(S_s, s)}{\partial S} + r S_s \frac{\partial C(S_s, s)}{\partial S}. \qquad (12.5.123)$$

Therefore, in order to specify the components (β_s, δ_s) of the replicating portfolio, it suffices for $C = C(S,t)$ to solve the second order nonlinear partial differential equation

$$rC = \frac{\partial C}{\partial t} + \frac{1}{2}\sigma^2 S^2 \frac{\partial^2 C}{\partial S^2} + rS\frac{\partial C}{\partial S} \tag{12.5.124}$$

together with the boundary and terminal conditions, for $0 \le t \le T$, $S \ge 0$,

$$C(0,t) = 0 \text{ and } C(S,T) = (S-K)^+. \tag{12.5.125}$$

The solution of (12.5.124) and (12.5.125) is given by

$$C(S,t) = S\mathcal{N}(d_1) - e^{-r(T-t)}K\mathcal{N}(d_2)$$

where $\mathcal{N}$ is the cumulative standard normal distribution and the quantities d_1 and d_2 are defined as

$$d_1 = \frac{\ell n(S/K) + \left(r + \frac{\sigma^2}{2}\right)(T-t)}{\sigma\sqrt{T-t}}$$

and

$$d_2 = \frac{\ell n(S/K) + \left(r - \frac{\sigma^2}{2}\right)(T-t)}{\sigma\sqrt{T-t}} = d_1 - \sigma\sqrt{T-t}.$$

Equation (12.5.124) is the celebrated *Black and Scholes* equation for European type claims written on a stock with constant volatility and when the riskless interest rate is $r > 0$. The first partial derivative of the call price, $\dfrac{\partial C(S,t)}{\partial S}$ is known as the *delta* of the option and it provides the needed number of stock shares in the replicating portfolio. The important consequence of the diffusion nature of the stock price is that both components of the hedging portfolio turn out to be *diffusion processes*, given by

$$\beta_s = \frac{h_s - S_s\dfrac{\partial C(S_s,s)}{\partial S}}{B_s} \text{ and } \delta_s = \frac{\partial C(S_s,s)}{\partial S}. \tag{12.5.126}$$

Therefore, the Black and Scholes valuation analysis dictates that rebalancing of the hedging portfolio must take place *infinitely often.* It is for this reason that in the presence of transaction costs, these replicating strategies *are not feasible.* Continuous rebalancing would produce an infinite volume of transactions no matter how small the transaction costs are.

12.5.2 Super-replicating strategies

The strategies of Black and Scholes demonstrate that both components of the replicating portfolio, see (12.5.126), are diffusion processes as it follows from the diffusion nature of the underlying stock price. Clearly, these hedging strategies will immediately produce an infinite volume of transactions no matter how small the transaction costs are. Therefore, a *perfectly replicating portfolio no longer exists!*

Abandoning the idea of exact replication, one might look for a portfolio strategy which results, at expiration time T, in portfolio value at least as great as the value of the European call. Such strategies are known as *super-replicating* strategies.

Bensaid, Lesne, Pagès and Scheinkman (1992) uncovered the intriguing idea that super-replication may be feasible, in the sense that the cost of the super-replicating portfolio is actually finite. This cost then may provide a sensible bound on the price of the option. Bensaid et al. (1992) constructed super-replicating policies in a discrete-time framework. (See also Ediriskinghe, Naik and Uppal (1993).)

Unfortunately, in the (limiting) case of continuous time, the super-replication approach cannot form the basis of a viable valuation theory. In fact, Davis and Clark (1994) conjectured that the *minimal* cost of the super-replication of a European call is the *value of one share* of the underlying stock. Therefore, even though super-replication techniques might provide finite values, their minimal value yields a trivial bound, the value of one stock of share, which is of little economic interest.

Using convex analysis arguments, Soner, Shreve and Cvitanić (1995) established the conjecture of Davis and Clark. Below we state their result by adopting the notation used in the previous Chapter. The bond and stock account processes, x_s and y_s, are given by the state equations (12.4.76) and (12.4.75). The European call has exercise time T, strike price K and it is written on the underlying stock whose price is given in (12.2.2).

Theorem 12.5.1 *Consider the payoff $(S_T - K)^+$ of a European call written of a stock with price S_s, $t \leq s \leq T$, as in (12.2.2). Then in order to have at $t = T$,*

$$x_T + \binom{\alpha}{\beta} y_T \geq (S_T - K)^+ \quad a.e. \tag{12.5.127}$$

the following constraint must hold for <u>*all*</u> $t \leq s < T$

$$x_s + \binom{\alpha}{\beta}\left(y_s - \frac{S_s}{\alpha}\right) \geq 0 \quad a.e. \tag{12.5.128}$$

The above result on *trivial super-replicating strategies* was later established by Levental and Skorohod (1997) in a general framework. Levental and Skorohod assumed that the underlying stock price is a continuous semimartingale and under mild non-degenerate and stability properties, they carried out the analysis for European and American claims. Their method is based on considering a discrete-time version of the underlying model which is free of transaction costs. We refer the reader to their paper for general super-replication results; below, we state a variation of one of their propositions (see Section 12.5 on European options in Levental and Skorohod (1997) adopting the existing notation. This is done only in order to be able to refer to their results from subsequent sections and to preserve the continuity of the exposition.

Proposition 12.5.2 : *Consider a European claim with payoff $g(S_T)$ at expiration time T, where g is increasing, convex and $g(0) = 0$ with $\lim_{S\to\infty} \frac{g(S)}{S} = \ell > 0$. The stock price S_s, $t \leq s \leq T$ is the diffusion process described in (12.2.2). Then in order to have*

$$x_T + \binom{\alpha}{\beta} y_T \geq g(S_T) \quad \text{a.e.} \tag{12.5.129}$$

the following constraint must hold for <u>*all*</u> $t \leq s < T$

$$x_S + \binom{\alpha}{\beta}\left(y_s - \ell\frac{S_s}{\alpha}\right) \geq 0 \quad \text{a.e.} \tag{12.5.130}$$

Even though super-replicating strategies produce price bounds of little economic interest, these results are of fundamental importance in utility maximization theory. In fact, the *state constraints* (12.5.128) and (12.5.130) essentially characterize the set of *feasible strategies* that the writer, or the buyer of the derivative may use in their valuation strategies. As will be demonstrated in the sequel, even under stringent constraints (12.5.128), (12.5.130), the presence of risk aversion — through utility functionals — allows us to derive non-trivial bounds for the so-called reservation derivative prices.

Finally, from the previous results an interesting question arises, namely, how can the constraints (12.5.128) and (12.5.130) be relaxed, if at all, when the super-replication requirements (12.5.127) or (12.5.129) are allowed to hold with probability $1-\epsilon$, instead of almost surely. This problem is interesting especially from the practical point of view where some "slippage" might be tolerated. Numerical results for the case of European calls can be found in Tourin and Zariphopoulou (1998).

12.5.3 The utility maximization theory

The Black and Scholes valuation method produces derivative prices which are *independent* of the individual portfolio holdings as well as of the individual attitude towards risk. Clearly, these universal properties stem from the ability to exactly replicate the payoff of the security, in the absence of market frictions. As it was discussed earlier, this possibility disappears in the presence of transaction costs and thus, these universal features might not be preserved. Indeed, the utility maximization approach brings in the individual attitude towards the derivative-inherent risk, which cannot be eliminated any more.

Even though one of the main ingredients of the Black and Scholes price cannot be retrieved, this method relies on the fundamental economic principles of *stochastic dominance* which still provide adequate viable valuation conclusions. The strengths of this approach are that it can be applied to a large class of derivatives, departing from the European ones; for this class very little is known through the other existing pricing methods with transaction costs. Moreover, the derivative prices are determined via two utility maximization models which give rise to two (singular) stochastic control problems. The powerful theory of viscosity solutions facilitates considerably the analysis by providing essential comparison results for the utilities of the buyer, the writer and the one of the plain investor.

Hodges and Neuberger (1989) were the first to apply the utility approach to price European calls when the agents are endowed with exponential utilities. Their results were further developed by Davis, Panas and Zariphopoulou (1993) for the same class of options, and by Davis and Zariphopoulou (1995) for American options. Note that the exponential utilities of wealth, say $U(z)=1-e^{-\gamma z}$ have the property of constant (wealth independent) Absolute Risk Aversion, i.e. $-\dfrac{U''(z)}{U'(z)}=\gamma$.

Before we present the main results, we discuss the simple case of a one-period model, where the end of the period coincides with the expiration date of the option. The purpose of this is two-fold; first, the exposition brings out the fundamental economic ideas of stochastic dominance and secondly, the analysis demonstrates that the stochastic dominance argument breaks when *intermediate trading* is allowed as it is the case in dynamic models. The seemingly innocuous generalization of the model to allow for intermediate trading activities makes the valuation problem far more difficult.

The following arguments come from a modification of the stochastic dominance arguments of Perrakis and Ryan (1984), Levy (1978) and Ritchken (1985) to account for proportional transaction costs (see also Reisman (1998)).

i) Bounds on European options via Stochastic Dominance: single-period models

We consider an economy with two securities, a riskless bond and a risky stock. We denote by B and S the bond and the stock prices, respectively, at the beginning of the (single) period and by B_T and S_T the prices at the end of the period which is assumed to have length T.

Trading in the bond and the stock accounts occurs only at the beginning and end of the period and is subject to transaction costs. As in the continuous time model, β dollars of the bond may be converted into one dollar of the stock and, one dollar of the stock may be converted into α dollars of the bond; the constants α and β satisfy $0 < \alpha < 1 < \beta$.

The important simplifying assumption is that no trading may occur at intermediate times. This assumption is relaxed later and the implications are fully explored therein.

The investor's pre-trade endowment consists of x_0 dollars in the bond account and y_0 dollars in the stock account. The investor trades at the beginning of the period incurring transaction costs and attains a post-trade endowment of x dollars in the bond account and y dollars in the stock account.

We assume that $y > \frac{S}{\alpha}$, that is the investor invests in at least $\frac{1}{\alpha}$ shares of the stock. At the end of the period, the investor converts the stock account into the bond account and consumes

$$c(S_T) = xR_F + y\frac{S_T}{S}$$

where $R_F = \frac{B_T}{B}$.

We assume that the investor's expected utility is the expectation of $u(c(S_T))$, where $u : \mathcal{R} \to \mathcal{R}$ is increasing and concave. In the absence of the opportunity to invest in an option, the investor chooses (x, y) to maximize the expected utility.

Given (x, y), we now present the investor with the opportunity to write one cash-settled, European-style call option with expiration at the end of the period and strike price K. Let C denote the post-transaction-cost price at which the investor may write the call: if the investor writes the call, the bond account increases by C dollars at the beginning of the period and decreases by $[S_T - K]^+$ dollars at the end of the period.

To provide an upper bound to the reservation write price of a call, we adopt the stochastic dominance arguments of Perrakis and Ryan (1984), Levy (1978) and Ritchken (1985), modified to account for transaction costs.

Consider the zero-net-cost portfolio which consists of a short position in one call and a long position in $\frac{C}{\beta S}$ shares of stock. The net payoff in the bond account at the end of the period is $z(S_T)$ where

$$z(S_T) = \frac{\alpha C S_T}{\beta S} - [S_T - K]^+.$$

Note that $z(S_T) \gtrless 0$ as $S_T \lessgtr \hat{S}$ where $\hat{S}$ is defined by

$$\frac{\alpha C \hat{S}}{\beta S} - S_T + K = 0.$$

The investor has post-trade endowment (x, y) and contemplates whether to write the call. If the investor writes the call and invests the proceeds in the stock, the expected utility is

$$E[u(c(S_T) + z(S_T))] \geq E[u(c(S_T))] + E[z(S_T)u'(c(S_T) + z(S_T))]$$

(by the concavity of u)

$$\geq E[u(c(S_T))] + E[z(S_T)u'(c(\hat{S}) + z(\hat{S}))]$$

(since $z(S_T) \gtrless 0$ and $u'(c(s_T) + z(S_T)) \gtrless u'(c(\hat{S}) + z(\hat{S}))$ as $S_T \lessgtr \hat{S}$)

$$\geq E[u(c(S_T))] + u'(c(\hat{S}) + z(\hat{S}))E[z(S_T)]$$

and exceeds the expected utility from refraining to write the call, unless $E[z(S_T)] < 0$, i.e.

$$(\alpha/\beta)CE[S_T/S] - E[[S_T - K]^+] < 0.$$

Therefore,

$$C < \beta E[[S_T - K]^+]/\alpha E[S_T/S_0] \equiv \bar{C}_1$$

and $\bar{C}_1$ is an upper bound to the reservation write price of a call option.

We consider next a different zero-net-cost portfolio which consists of a short position in one call and a long position in C dollars in the bond. Proceeding as before, we conclude that the expected utility in writing the call exceeds the expected utility in not writing the call, unless

$$C < E[[S_T - K]^+]/R_F \equiv \bar{C}_2.$$

Combining the above equations we conclude that $\bar{C}$ is an upper bound to the reservation write price of a call option, where

$$\bar{C} = E[[S_T - K]^+] \min \left[R_F^{-1}, \frac{\beta/\alpha}{E[S_T/S]} \right].$$

To derive a lower bound to the reservation purchase price of a call option, let C denote the post-transaction-cost price at which the investor may purchase the call. Consider the zero-net-cost portfolio which consists of

(a) a long position in one call;

(b) a short position in $1/\beta$ shares of stock; and

(c) investment of $\alpha S_T/\beta - C$ dollars in the bond account.

Denote by $z(S_T)$ the net payoff in the bond account at the end of the period, where

$$z(S_T) = [S_T - K]^+ - S_T + \left\{ \alpha \frac{S}{\beta} - C \right\} R_F.$$

Repeating the earlier argument, we conclude that the expected utility in purchasing the call exceeds the expected utility in refraining from purchasing the call, unless $E[z(S_T)] < 0$, which yields $\underline{C}$ as a lower bound to the reservation purchase price of a call, where

$$\underline{C} = \frac{E[[S_T - K]^+]}{R_F} - \frac{E[S_T]}{R_F} + \frac{\alpha S}{\beta}.$$

It is easily shown that $\underline{C} \leq \bar{C}$. In equilibrium, transaction prices of a call option must lie in the region $[\underline{C}, \bar{C}]$. For, if a transaction occurs at a price $C < \underline{C}$, then the writer is acting suboptimally as the writer could have found a willing buyer of the call at a price as high as $\underline{C}$. Likewise, if a transaction occurs at a price $C > \bar{C}$, then the buyer of the call is acting suboptimally as the buyer could have found a willing writer of the call at a price as low as $\bar{C}$.

The stochastic dominance bounds are appealing in that they apply for any increasing and concave utility function. It turns out, however, that the derivation of these bounds breaks down when intermediate trading is permitted in the open interval $(0, T)$.

Let us reconsider the stochastic dominance argument for the reservation write price of a call. The plausible assumption was made that the investor's endowment satisfies the condition $y > \frac{S}{\alpha}$. Without intermediate trading, the consumption at the end of the period is $c(S_T)$ and has two crucial properties:

(1) it is monotone increasing in S_T with slope greater than one; and

(2) given S_T, $c(S_T)$ is independent of the stock price path ω_T over $(0, T)$.

The first property is crucial in the proof in that it implies that $c(S_T)+z(S_T)$ is increasing in S_T and therefore $u'(c(S_T)+z(S_T))$ is decreasing in S_T. The second property is crucial in the step which allowed us to take $u'(c(S_T)+z(S_T))$ outside the expectation: if c is a function of the price path ω_T, $\left[u'(c(\omega_T) + z(S_T)) \mid S_T\right]$ is a random variable and cannot be taken outside the expectation. Another problem is that, in the presence of intermediate trading, $c(\omega_T)+z(S_T)$ is not even bounded from below and the expected utility is undefined for utility functions which are only defined for consumption bounded from below. Similar problems arise in attempting to generalize the stochastic dominance argument in the derivation of a lower bound to the reservation purchase price when intermediate trading is allowed.

ii) Bounds on prices of European-type claims via utility maximization: continuous-time models

The utility maximization approach looks at the value functions of the investor with and without the opportunity to trade (write or buy) the derivative security.

If the investor chooses not to trade the available claim, his value function is given by $V(x, y)$, as it is defined in (12.4.79). Suppose now that a third asset is introduced, a cash settled European-style contingent claim with expiration at date T and payoff $g(S_T)$ at expiration. If the investor writes the claim at date t with $0 \le t \le T$, the bond account is credited with an amount, say C dollars, which represents the price of the claim, and is debited $g(S_T)$ dollars at the expiration date T. To keep the problem tractable we assume that the investor may not trade the claim in the open interval $(0, T)$.

Let x_t and y_t be the initial endowment at time t *after* the bond account has been credited with the proceeds from writing the claim. Once the claim is written, the writer's objective is to maximize his expected utility from consumption, as in case *(i)* with the extra obligation to surrender to the buyer $g(S_T)$ dollars at time T. Therefore the utility payoff of the writer is

$$E\left[\int_t^T e^{-\rho(s-t)}U(c_s)ds + e^{-\rho(T-t)}V(x_T - g(S_T), y_T) \mid x_t = x, y_t = y, S_t = S\right]$$

where V is defined in (12.4.79) and S_s is given by (12.3.18).

The *value function of the writer* is

$$J(x, y, S, t) = \sup_{\mathcal{A}_1} E\left[\int_t^T e^{-\rho(s-t)}U(c_s)ds \right. \tag{12.5.131}$$

$$\left. +e^{-\rho(T-t)}V(x_T - g(S_T), y_T) \mid x_t = x, y_t = y, S_t = S\right]$$

where $\mathcal{A}_1$ is the set of admissible policies defined below.

It is assumed that the payoff function g satisfies the following assumptions

$$\begin{cases} g : [0, +\infty) \to [0, +\infty) \quad \textit{is convex} \\ g(0) = 0 \\ \lim_{S \to \infty} \dfrac{g(S)}{S} = 1. \end{cases} \tag{12.5.132}$$

The previous exposition on feasible super-replication strategies suggests that the set of the writer's admissible policies must be determined as follows. From (5.14) we have that the writer's terminal (liquidated) wealth must be nonnegative; in other words, the terminal constraint

$$x_T + \binom{\alpha}{\beta} y_T \geq g(S_T) \quad \text{a.e.} \tag{12.5.133}$$

must be fulfilled. The payoff function g satisfies the assumptions of Levental and Skorohod (1997). Therefore, Proposition 12.5.2 yields that at *all* previous times $t \leq s < T$, the state wealth of the writer must satisfy the stringent constraint

$$x_s + \binom{\alpha}{\beta} \left(y_s - \frac{S_s}{\alpha} \right) \geq 0 \quad \text{a.e.}. \tag{12.5.134}$$

So, we define the set $\mathcal{A}_1$ of admissible policies of the investor who has written a contingent claim, as the set of $\mathcal{F}_t$-progressively measurable processes (c_t, L_t, M_t), with L_t and M_t being CADLAG which also satisfy the conditions

$$\begin{cases} \mathrm{c_s \geq 0 \text{ and } E} \displaystyle\int_{\mathrm{t}}^{\mathrm{s}} \mathrm{c_\tau d\tau < \infty \text{ a.s. for } t \leq s \leq T} \\ \text{and} \\ w_s = x_s + \binom{\alpha}{\beta} \left(y_s - \dfrac{S_s}{\alpha} \right) \geq 0 \ \text{ a.s. for} t \leq s < T, \end{cases} \tag{12.5.135}$$

and we define the set of admissible policies $\{c_t, L_t, M_t; T < t\}$ of the investor who has written a claim by $\mathcal{A}$, as given in (12.4.77) and (3.4.4). Note that for $s > T$, the option has expired and settled and the investor's problem is indistinguishable from that of an investor who has not written the claim. Thus it is natural to define the set of admissible policies for $s > T$ as $\mathcal{A}$.

The set $\mathcal{A}_1$ is a subset of $\mathcal{A}$ for $t \leq s \leq T$ in the sense that the second restriction ensures that the investor will have nonnegative net worth upon closing up the short position in the call option and, therefore, that it is feasible to write a call option in the first place. The results of Soner, Shreve and Cvitanić (1995), et al (for $g(S) = (S - K)^+$) and Levental and Skorohod (1997) (for general g) state that the set of policies in $\mathcal{A}_1$ is not overly restrictive given the goal of ensuring that it is feasible to write the claim option.

The value function $J(x, y, S, t)$ is given by (12.5.131) and is defined for $(x, y, S) \in \overline{D}_1$ where

$$\overline{D}_1 = \left\{ (x, y, S) : x + \binom{\alpha}{\beta} \left(y - \frac{S}{\alpha} \right) \geq 0, \ S \geq 0 \right\}.$$

Consider now the writer with endowment $(x, y) \in \overline{D}$ at time t before writing the claim. If the writer chooses to write the claim at price C, the endowment becomes $(x + C, y)$ and by Theorem 12.5.5 and Proposition 12.5.2, the price C must be such that $(x+C, y, S) \in \overline{D}_1$. In the case of *zero-transaction costs*, the function $C = C(S,t)$ is determined as the price that makes the writer *indifferent* between writing the claim or refraining from writing it, i.e.

$$V(x,y) = J(x + C(S,t), y, S, t).$$

In the special case $g(S) = (S - K)^+$, one can show that $C(S,t)$ is the Black and Scholes price which is of course independent of the current portfolio holdings (x, y) and the utility function. Moreover, because of the absence of transaction costs, perfect replication is possible and the constraint (12.5.134) is not binding.

In the case of *non-zero transaction costs*, the above equality *is not feasible for all* $(x, y, S) \in \overline{D}_1$ if C is allowed to depend *only* on (S, t); this fact motivates the following definitions.

Definition 12.5.3 *The reservation write price $C(x, y, S, t)$, for initial endowment (x, y), is defined as the minimum value at which the investor is willing to write the claim. Therefore, C satisfies for $(x + C(x, y, S, t), y, S) \in \overline{D}_1$*

$$V(x,y) = J(x + C(x,y,S,t), y, S, t). \tag{12.5.136}$$

Definition 12.5.4 : *The write price $\overline{C}(S,t)$ is defined as the maximum of reservation write prices across all admissible states (x, y, S). Therefore, $\overline{C}$ satisfies for all $(x+\overline{C}(S,t), y, S) \in \overline{D}_1$,*

$$V(x,y) \leq J(x + \overline{C}(S,t), y, S, t). \tag{12.5.137}$$

The above inequality guarantees that the *writer will be willing to write the option at any price higher than* $\overline{C}(S,t)$, independently of his current portfolio position.

The case of exponential utilities was first examined by Hodges and Neuberger (1989) and subsequently by Davis, Panas and Zariphopoulou (1993). Constantinides and Zariphopoulou (1999) generalized all previous results on the subject for general individual preferences and they derived an upper bound $h = h(S,t)$ for the write price which satisfies (12.5.137) on $\overline{D}_1$. The main steps for the construction and characterization of the upper bound are presented below. The proof of their main result can be found in Constantinides and Zariphopoulou (1999).

Theorem 12.5.5 *The value function is a constrained viscosity solution on $\overline{D}_1 \times [0, T)$ of the Variational Inequality*

$$\min\left[\mathcal{L}J - \overline{\mathcal{L}}J, \beta\frac{\partial J}{\partial x} - \frac{\partial J}{\partial y}, -\alpha\frac{\partial J}{\partial x} + \frac{\partial J}{\partial y}\right] = 0, \tag{12.5.138}$$

with

$$J(x, y, S, T) = V(x - g(S), y), \tag{12.5.139}$$

where the operators $\mathcal{L}$ and $\overline{\mathcal{L}}$ are given by

$$\begin{cases} \mathcal{L}J = \rho J - \frac{1}{2}\sigma^2 y^2 \frac{\partial^2 J}{\partial y^2} - \mu y \frac{\partial J}{\partial y} - rx\frac{\partial J}{\partial x} - \max_{c\geq 0}\Big(-c\frac{\partial J}{\partial x} + U(c)\Big) \\ \overline{\mathcal{L}}J = \frac{\partial J}{\partial t} + \frac{1}{2}\sigma^2 S^2 \frac{\partial^2 J}{\partial S^2} + \sigma^2 yS\frac{\partial^2 J}{\partial y \partial S} + \mu S \frac{\partial J}{\partial S}. \end{cases} \tag{12.5.140}$$

Moreover, J is the unique constrained viscosity solution of (5.21) in the class of uniformly continuous and concave functions, with respect to the state variables (x, y, S).

The underlying idea for the derivation of *analytic bounds* for the write price of a European-type claim, is to construct suitable subsolutions of the HJB equations (12.4.90) and (12.5.138) in order to use a comparison result to establish (12.5.136). The main difficulty stems from the fact that the value functions V and J are defined on different domains and that there are no explicit or closed-form solutions for the two associated free-boundary problems (12.4.79) and (12.5.131).

We start with a formal discussion in order to motivate the construction of the analytic bound. To ease the presentation, we recall that the value functions V and J solve, respectively,

$$\min\left\{\mathcal{L}V, \beta\frac{\partial V}{\partial x} - \frac{\partial V}{\partial y}, -\alpha\frac{\partial V}{\partial x} + \frac{\partial V}{\partial y}\right\} = 0$$

in

$$\overline{D} = \left\{(x, y) : x + \left(\frac{\alpha}{\beta}\right) y \geq 0\right\}$$

and

$$\min\left\{\mathcal{L}J - \overline{\mathcal{L}}J, \beta\frac{\partial J}{\partial x} - \frac{\partial J}{\partial y}, -\alpha\frac{\partial J}{\partial x} + \frac{\partial J}{\partial y}\right\} = 0$$

in

$$\overline{D}_1 = \left\{(x, y, S) : x + \left(\frac{\alpha}{\beta}\right)\left(y - \frac{S}{\alpha}\right) \geq 0\right\}$$

with the differential operators $\mathcal{L}$ and $\overline{\mathcal{L}}$ given in (12.5.140).

The goal is to construct a function $h = h(S, t)$, *independent* of (x, y), such that, for $(x + h, y, S) \in \overline{D}_1$

$$V(x, y) \leq J(x + h(S, t), y, S, t) \tag{12.5.141}$$

Using the suboptimality inequality

$$J(x + h(S, t), y, S, t) \geq J\Big(x, y + \frac{h(S, t)}{\beta}, S, t\Big)$$

and a simple transformation, we observe that (12.5.141) follows if we find an h such that

$$V\Big(x, y - \frac{h(S, t)}{\beta}\Big) \leq J(x, y, S, t) \tag{12.5.142}$$

for $(x, y, S) \in \overline{D}_1$.

The basic idea of Constantinides and Zariphopoulou (1999) for the choice of the candidate bound is first to find a price that satisfies (12.5.142) in the case that $(x, y, S) \in \partial D_1$, i.e. when the writer holds the *minimal allowed position* which amounts to the value of one stock share, taking into account the transaction costs. We then need to show that this price works for all wealth levels greater than the minimal one. The results stated below were established by Constantinides and Zariphopoulou (1999).

To this end, we start with the following lemma which gives us information about the value function J on $\partial D_1 = \left\{(x, y, S) : x + \left(\frac{\alpha}{\beta}\right)\left(y - \frac{S}{\alpha}\right) = 0\right\}$.

Lemma 12.5.6 *For $(x, y, S) \in \partial D_1$, the value function J is given by*

$$J(x, y, S, t) = E\left[e^{-\rho(T-t)}V\left(-g(S_T), \frac{S_T}{\alpha}\right) \mid S_t = S\right]. \tag{12.5.143}$$

The proof follows directly from the fact that the only admissible policy for the boundary points (x, y, S) is to move instantaneously at time t, to the point $\left(0, \frac{S}{\alpha}, S\right)$ and remain there until time T.

The next result gives the main ingredient for the construction of the candidate solution. Its proof can be found in Constantinides and Zariphopoulou (1999).

Lemma 12.5.7 *If $h_{\hat{\rho}} = h_{\hat{\rho}}(S, t)$ is such that*

$$V\left(\frac{\beta}{\alpha}S, -\frac{h_{\hat{\rho}}}{\beta}\right) = E\left[e^{-\hat{\rho}(T-t)}V(S_T - g(S_T), 0) \mid S_t = S\right] \tag{12.5.144}$$

with $0 \le h_{\hat{\rho}} \le \frac{\beta}{\alpha}S$ and $\hat{\rho} \ge \rho$, then (12.5.142) holds for $(x, y, S) \in \partial D_1$.

Next, we observe that if the $\mathcal{NT}$ region is a proper subset of the first quadrant, then the points $\left(\frac{\beta}{\alpha}S, 0\right)$ and $(S - g(S), 0)$ belong to the $\mathcal{B}$ region.

In the $\mathcal{B}$ region, the value function V satisfies $\beta V_x = V_y$, which implies that there exists a function, denoted as G such that V can be *expressed in terms of* G as

$$V(x_0, y_0) = G(x_0 + \beta y_0) \quad \text{for} \quad (x_0, y_0) \in \mathcal{B}.$$

Therefore,

$$V\left(\frac{\beta}{\alpha}S, -\frac{h(S,t)}{\beta}\right) = G\left(\frac{\beta}{\alpha}S - h(S,t)\right) \tag{12.5.145}$$

and

$$V(S_T - g(S_T), 0) = G(S_T - g(S_T)). \tag{12.5.146}$$

Combining the above equalities and (12.5.144) yields

$$G\left(\frac{\beta}{\alpha}S - h(S,t)\right) = E\left[e^{-\hat{\rho}(T-t)}G(S_T - g(S_T)) \mid S_t = S\right].$$

It follows easily from the monotonicity properties of the value function V (see, for example, Tourin and Zariphopoulou (1994)) that G is strictly increasing and therefore invertible. This in turn yields that the function h is well defined and given by

$$h(S,t) = \frac{\beta}{\alpha}S - G^{-1}\left(E\left[e^{-\hat{\rho}(T-t)}G(S_T - g(S_T)) \mid S_t = S\right]\right). \tag{12.5.147}$$

It will turn out that the above function is a candidate upper bound for the write price.

The next result is the key step in establishing the validity of $h(S,t)$ being a reservation price bound.

Proposition 12.5.8 *Assume that the $\mathcal{NT}$ region for the utility maximization problem satisfies $\mathcal{NT} \subseteq \{(x,y) : Ax \le y \le Bx, x \ge 0$ with $B > A > 0\}$. Also, assume that the utility function U satisfies $\lambda_1 \frac{c^\gamma}{\gamma} \le U(c) \le \lambda_2 \frac{c^\gamma}{\gamma}$ for some positive constants λ_1 and λ_2. Let*

$$m = \left[\left(\frac{\lambda_2\left(\beta + \frac{1}{A}\right)}{\lambda_1\left(\alpha + \frac{1}{B}\right)^\gamma}\right)^{\frac{1}{1-\gamma}} \frac{\rho - r\gamma}{\alpha\gamma} + \left(\mu + \frac{r}{\alpha A}\right) - \frac{\rho}{2}\left(1 + \frac{1}{\beta B}\right)\right]\sigma^{-2}$$

and consider the discount factor $\hat{\rho}$ in (12.5.147) given by

$$\hat{\rho} = \max\left[\rho, \mu + \frac{m\sigma^2(1-\beta A)}{2\beta A}\right]. \tag{12.5.148}$$

Then, for the candidate price h, defined in (12.5.147) with $\hat{\rho}$ as above, the function $F : D_1 \times [0,T] \to [0,+\infty)$ given by

$$F(x,y,S,t) = V\left(x, y - \frac{h(S,t)}{\beta}\right)$$

is a viscosity subsolution of the HJB equation (12.4.90).

The next theorem establishes that the candidate $h(S,t)$ is indeed a price bound.

Theorem 12.5.9 *Let h be given by*

$$h(S,t) = \frac{\beta}{\alpha}S - G^{-1}\left(E\left[e^{-\hat{\rho}(T-t)}G(S_T - g(S_T))/S_T = S\right]\right)$$

where $\hat{\rho}$ is defined in (5.31). Then the function $h(S,t)$ is an upper bound to the reservation write price.

From the above results, one can see that the "trivial" super-replicating price bound — of one stock share — is substantially improved once one employs the utility maximization method. As a matter of fact, the latter method relies on the risk aversion attitude of the investors as opposed to the super-replicating approach which is based on risk-neutrality. The weak point of the utility method is that little information is available for the hedging strategies and this can be actually retrieved only through the optimal investment strategies for the utility maximization problems (12.4.79) and (12.5.131). On the other hand, the utility method can be easily extended to other kinds of derivatives like American options, path-dependent and exotics written, actually, one more than one stock. For these kinds of derivatives very little is known through the other valuation methods for markets with transaction costs (for a complete study of these cases, we refer the reader to Constantinides and Zariphopoulou (1999)). Moreover, even though the utility maximization method departs from the fundamental and classical risk-neutral valuation theory, it could still contribute in a number of custom-made derivatives or real options, and also serve as the basis line for developing improved methods based on *general risk functionals.* Finally, the utility maximization approach can be easily applied to valuation problems with other kinds of frictions, like stochastic volatility (see, for example, Mazaheri (1998)). An interesting application of the utility method which relates small transaction costs and modified volatility can be found in Barles and Soner (1998).

12.5.4 Imperfect hedging strategies

An appealing alternative approach for the valuation of derivatives in the presence of transaction costs, is to *relax* the requirement of continuous rebalancing by allowing the adjustment of the "hedging" portfolio to take place at *discrete times.* Clearly, a correct valuation procedure based on discrete hedging is highly desirable for practical applications since continuous rebalancing is practically impossible. Generally speaking, even in the absence of transaction costs, discrete in time rebalancing does not lead to perfect hedging but nevertheless, *imperfect hedging strategies* have become a standard vehicle in valuating derivatives in practice.

This approach departs from the expected utility methodology which, as we saw previously, depends heavily on intermediate dynamic trading. Nevertheless, we choose to present the main ideas of this alternative method for the sake of completeness and also, because it is currently the most frequently used vehicle to valuate hedging strategies. A hybrid theory based on the economic principles of expected utility theory and the techniques of the imperfect hedging approach would be highly desirable.

Two important papers on discrete hedging without transaction costs were produced by Boyle and Emanuel (1980) and Wilmott (1994). In both papers, rebalancing takes place at fixed time intervals. Boyle and Emanuel (1980) provided a thorough study on *the hedging error* which is defined as the *discrepancy* between the discrete hedging strategy and the continuous in time strategy dictated by the Black and Scholes formula. They established that rehedging in fixed time intervals produces a hedging error which is proportional to the gamma of the option and chi-squared distributed. Wilmott (1994) used asymptotic expansions and found improved hedging strategies which are also related to an adjusted option value. One of the underlying ideas was to use the number of shares which minimize the variance of the hedging portfolio over the next time step. By equating the expected value on the hedged portfolio with the riskless interest rate, Wilmott (1994) found that the option should be priced at a modified constant volatility. The latter depends on the rehedging time-interval as well as the mean rate of return of the stock price.

The phenomenon of getting an *enhanced volatility* when discrete hedging takes place is rather common in derivative pricing, especially when discrete hedging is used to accomodate the effects of transaction costs. The groundwork on this subject was originated by Leland (1985) and his results are considered a benchmark in the area of imperfect hedging.

Imitating the Black and Scholes analysis and proceeding in a rather ad hoc way, Leland (1985) produced a valuation formula for European options in the presence of proportional transaction costs. He showed that the equation satisfied by the new option price resembles the Black and Scholes one (12.5.124) but with increased volatility (see equation (12.5.155) below). The enhanced volatility explodes, and so does the derivative price as the size of the hedging time intervals goes to zero.

Below, we continue with the construction of Leland's price for a European option; in order to be consistent with his calculations, we assume that the transaction costs are symmetric which corresponds, in our notation to $\alpha = \frac{1}{\beta} = 1 - k$; to simplify the exposition we also assume that the interest rate is zero. Following the Black and Scholes analysis, Leland postulated that the price of the call, say C_t, at time t, can be represented as a convex function of the stock price S_t and time, i.e. $C_t = h(S_t, t)$ with $h : [0, +\infty) \times [0, T] \to [0, +\infty)$.

Let us denote the increments of the underlying stock price as $\Delta S_s \triangleq S_{s+\Delta s} - S_s$. From equation (12.2.1) we have, for $t \leq s \leq T$,

$$\Delta S_s \simeq S_s(\mu \Delta s + \sigma \Delta W_s). \tag{12.5.149}$$

Proceeding formally, we suppose — as in the Black and Scholes case — that there exists a replicating strategy, say δ_s, with δ_s denoting the number of stocks needed at time s. Then the price of the option will change according to

$$\begin{cases} \Delta h_s & \simeq \delta_s \Delta S_s - k S_s \\ & \simeq S_s(\mu \delta_s \Delta s - k|\Delta \delta_s|) + \sigma S_s \delta_s \Delta W_s. \end{cases} \tag{12.5.150}$$

Assuming that all the necessary derivatives exist, Itô's formula yields

$$\Delta h_s \simeq \Big(\frac{\partial h}{\partial t}(S_s, s) + \mu S_s \frac{\partial h}{\partial S}(S_s, s) + \tag{12.5.151}$$

$$+\frac{1}{2}\sigma^2 S_s^2 \frac{\partial^2 h}{\partial S^2}(S_s, s)\Big)\Delta s + \sigma S_s \frac{\partial h}{\partial S}(S_s, s)\Delta W_s.$$

Equating the coefficients in (5.33) and (5.34) gives,

$$\begin{cases} \sigma S_s \delta_s = \sigma S_s \frac{\partial h}{\partial S}(S_s, s) \\ S_s\Big(\mu \delta_s - k\frac{|\Delta \delta_s|}{\Delta s}\Big) = \frac{\partial h}{\partial t}(S_s, s) + \mu S_s \frac{\partial h}{\partial S}(S_s, s) \\ \quad +\frac{1}{2}\sigma^2 S_s^2 \frac{\partial^2 h}{\partial S^2}(S_s, s). \end{cases} \tag{12.5.152}$$

The first equation above implies $\delta_s = \frac{\partial h}{\partial S}(S_s, s)$ which in turn yields

$$|\Delta \delta_s| \simeq \Big|\sigma S_s \frac{\partial^2 h}{\partial S^2}(S_s, s)\Delta W_s\Big| + m(s) \tag{12.5.153}$$

where $m(s)$ includes terms of order s or higher. Leland based his derivation of the assumption that

$$|\Delta W_s| \simeq \sqrt{\frac{2}{\pi}}\,\sqrt{\Delta s} \tag{12.5.154}$$

without really justifying his choice. Nevertheless, using this approximation together with (12.5.152) and (12.5.153) yields that the option price function $h(S, t)$ must solve

$$\frac{\partial h}{\partial t} + \frac{1}{2}\sigma^2 S^2 \Big(1 + \frac{2k}{\sigma\sqrt{\Delta t}}\sqrt{\frac{2}{\pi}}\Big)\frac{\partial^2 h}{\partial S^2} = 0. \tag{12.5.155}$$

Leland's *enhanced volatility* is then given by

$$\widetilde{\sigma} = \sigma\Big(1 + \frac{2k}{\sigma\sqrt{\Delta t}}\sqrt{\frac{2}{\pi}}\Big)^{1/2}. \tag{12.5.156}$$

The above analysis was carried out for the case of call options. Similar arguments can lead for the case of put positions to a pricing equation of the same form as (5.38) but with different enhanced volatility, namely

$$\underline{\sigma} = \sigma\Big(1 - \frac{2k}{\sigma\sqrt{\Delta t}}\sqrt{\frac{2}{\pi}}\Big)^{1/2}.$$

Therefore, in Leland's approach, "short and long" positions have different values.

As it was mentioned earlier, even though Leland did not justify his choice of approximation in (12.5.154), his formula became rather popular in practical applications mainly because it relies on discrete in time rebalancing and it also requires an implementation similar to the Black and Scholes one. Moreover, in contradistinction to the utility maximization method, Leland's approach is able to produce a specific trading strategy, albeit imperfect.

A number of researchers modified or extended Leland's work by choosing different approximations and encountering modified errors. Boyle and Vorst (1992) applied Leland's techniques to a binomial valuation model and maintained the obligation to rehedge at constant time steps. They obtained a perfectly hedging strategy and an associated option price. Additionally, they examined the behavior of the option price as the time step $\Delta t \downarrow 0$ assuming that the proportional transaction costs, λ and μ, decrease to zero at a $\sqrt{\Delta t}$ rate. With these limiting assumptions, Boyle and Vorst found that the limiting price equation preserves the Black and Scholes and Leland structure but with a different enhanced volatility, namely

$$\hat{\sigma} = \sigma\Big(1 + \frac{2k}{\sigma\sqrt{\Delta t}}\Big)^{1/2}. \tag{12.5.157}$$

The above volatility can be obtained directly from Leland's arguments provided one chooses the approximation $|\Delta W_s| \simeq \sqrt{\Delta s}$, instead of (12.5.154). Whalley and Wilmott (1996) provide a nice discussion on the similarities and differences between the various approximations and how they affect the long and short positions, attributing them mostly to the asymmetries inherent from the transaction costs.

A different valuation model, for arbitrary option payoffs, was introduced by Hoggard, Whalley and Wilmott (1994) who used the same idea of rehedging at fixed time intervals Δs but they imposed a *generalized shares costs structure*. In fact, they assume that their cost structure is of the form $k_1 + k_2\delta_s + k_3\delta_s S_s$, i.e. there is a component of fixed costs, k_1, a second component of cost $k_2\delta_s$ proportional to the number of shares rehedged and a third one, $k_2\delta_s S_s$, proportional to the current traded value. Working along the basic Leland valuation analysis, Hoggard, Whalley and Wilmott derived the following option price equation (stated for non-zero interest rates)

$$\frac{\partial h}{\partial t} + \frac{1}{2}\sigma^2 S^2 \frac{\partial^2 h}{\partial S^2} + rS\frac{\partial h}{\partial S} = rh + \frac{k_1}{\Delta t} + \Big[\sqrt{\frac{2}{\pi\Delta t}}\sigma S(k_2 + k_3 S)\Big]\,\Big|\frac{\partial^2 h}{\partial S^2}\Big|.$$

As with Leland's analysis, the above equation yields different values for short and long positions. Moreover its solution may attain negative values, a feature not desirable for an option valuation model. Hoggard, Whalley and Wilmott argued that this issue stems mainly from the ad hoc obligation to rehedge at every time step and it can be corrected by *regulating the rehedging process* taking into account the current option values. This modification calls for dynamically ceasing the rebalancing as soon as the call price goes to levels that any further rehedging would lead to negative values. This approach gives rise to a free-boundary valuation problem with similar characteristics to an American put.

Departing from the obligation to rehedge at fixed time-intervals, Whalley and Wilmott (1994) developed a model in which rebalancing takes place whenever the current position deviates considerably from the position of perfect hedging. To fix the notation, we denote by $C(S,t)$ the Black and Scholes price, with perfect hedging and by $h(S,t)$ the price under imperfect hedging. Recall that the perfect hedging position at time s, is given by the delta position $d_s(S_s, s) = \frac{\partial C}{\partial S}(S_s, s)$. In an effort to control the big losses from frequent rehedging in the presence of transaction costs, one might decide to hold $-\widetilde{d}_s(S_s, s)$ shares of the underlying without considering the extra cost of selling (or buying) for rehedging. If the variance of this position is used to measure the inherent risk exposure, one gets a risk exposure of size $\sigma^2 S_s^2(\widetilde{d}_s(S_s, s) - d_s(S_s, s))^2 \Delta s$. Since choosing $\widetilde{d}_s = d_s$ is not feasible, Whalley and Wilmott (1994) introduced an *index of tolerance* by considering the maximum expected risk in the portfolio, say H_0 and by requiring the constraint

$$|\widetilde{d}_s(S_s, s) - d_s(S_s, s)| \leq \frac{H_0}{\sigma S_s}$$

to hold at all times. Therefore, any time the above condition is violated, the position should be rebalanced.

Avellaneda and Parás (1994) studied the ill-posedness of the replication strategies by Hoggard, Whalley and Wilmott (1994) and proposed an explanation in the case of large transaction costs which intensifies the difference between the (asymmetric) short and long positions. They argued that the writer of the derivative is always obliged to rehedge dynamically his market exposure independently of the effects from the transaction costs. The buyer does not face the same stringency as all he risks is the initial premium and, after all, "hedging is done primarily to offset time-decay." Large transaction costs alter irreversibly the adjusted delta strategies and the value of the positions become eroded. Avellaneda and Parás (1994) proposed a new scheme for the valuation of the derivatives which is based on solving an obstacle problem for the Leland partial differential equation (5.38) with enhanced volatility. The obstacle problem arises from optimal stopping rules dictating when the rehedging must temporarily stop.

In a more recent paper, Avellaneda and Parás (1997), considered the issue of minimizing the total cost of the hedging strategies of option portfolios. They followed the discrete in time approach by Bensaid et al (1992) and they examined the limit of the positions as the number of trading periods becomes large. Generally speaking, Avellandea and Parás showed that, in the limit, the cost function satisfies a non-linear, diffusion equation. In particular, if the rehedging interval Δs, the volatility σ and the "roundtrip" transaction costs k satisfy $m = \frac{k}{\sigma \Delta s} < 1$, then the cost function converges to the solution of a non-linear Black and Scholes type equation. The volatility parameter of the latter depends on the local convexity of the cost function and it is adjusted either to $\sigma\sqrt{1+m}$ or to $\sigma\sqrt{1-m}$. If $m < 1$, the optimal hedging strategy *replicates* the final payoff via delta-hedging directly from the solution of the nonlinear Black and Scholes type equation. On the other hand, if $m \geq 1$ the hedging strategy of minimal cost is path-dependent and *super-replicates* the final payoff.

Henrotte (1993) also used ideas from Leland's approach and extended the concept of diffusion limits of replicating positions to hedging policies based on changes in the stock price. In (1993), Henrotte considers the asymptotic replication error and compres the performance of hedging strategies based on rebalancing at equal time steps, to strategies depending on prespecified changes of the underlying stock price. Grannan and Swindle (1999) extended the use of limiting hedging strategies by optimizing over different classes of strategies, for example, strategies which allow for varying time intervals. They also explored the induced *replication errors* and compared them to the ones of the standard approach based on constant in time intervals. The work of Grannan and Swindle (1994) was subsequently generalized by Ahn et al (1998) who considered rather general hedging strategies which include all other existing ones, for example "time-interval" strategies, "price change" strategies, renewal policies and delta-strategies based on local deviations.

As it was mentioned at the beginning of the section, Leland's analysis was not mathematically rigorous as some rather ad hoc assumptions were used. Some of his limiting results and conjectures were later revisited and corrected by Lott (1993) and more recently by Kabanov and Safarian (1997).

12.5.5 Other models of derivative pricing with transaction costs

Various other valuation techniques have been developed besides the ones mentioned in the previous sections.

Martingale theory, convex analysis and duality results have been used by a number of authors to obtain derivative prices and to construct appropriate strategies. A general approach to characterize arbitrage-free models with transaction costs was developed in Jouini and Kallal (1995). Along the lines of the super-replication method, martingale techniques were used by Cvitanić and Karatzas (1996) and by Cvitanić, Pham and Touzi (1997) for continuous time models and by Koehl, Pham and Touzi (1996) for the discrete case; see also Kusuoka (1995) for some convergence results.

A different method which relies on insights from both the utility maximization as well as the Leland's approach, uses as optimality criterion the *minimization of the "local risk."* It is based on a *local quadratic loss criterion* which was first introduced by Schweizer (1988). This method has been extensively used in the frictionless case by a number of authors, but for the case of transaction costs, it was first employed by Mercurio and Vorst (1997) for some special cases (see also Mercurio (1997)). Recently, Lamberton, Pham and Schweizer (1998) provided rigorous results for the existence of locally risk-minimizing strategies in the class of square-integrable contingent claims. The strength of this new approach is that, besides its mathematical tractability, it produces hedging strategies whose initial costs are *much lower* than those produced by the super-replicating strategies and whose replicating errors are relatively small.

In a different direction, various authors considered the derivative valuation problem assuming that the transaction costs are finite but *arbitrarily small.* The majority of these models use key insights from the utility maximization approach, see for example Barles and Soner (1998), Whalley and Wilmott (1997), Albanese and Tompaidis (1998). The model of Whalley and Wilmott was successfully tested against others in the Monte Carlo simulations of Mohamed (1994). In an arbitrary transaction, cost structure was allowed by Whalley and Wilmott (1997) when the costs are either proportional or fixed. Whalley and Wilmott produced a simple expression for the "hedging bandwidth" around the Black and Scholes delta strategies and argued that in this region rehedging is not optimal. They used asymptotic analysis to specify explicit points of optimal rehedging in the case of proportional, fixed and mixed transaction costs.

The accurate valuation of derivatives in the presence of transaction costs has become more and more desirable as new derivatives are being created every day and custom-made instruments have a rising demand. Some kinds of path-dependent derivatives, including Asians and look-backs have been examined by Dewynne, Whalley and Wilmott (1994); their valuation method is based on Leland's approach of imperfect hedging and the mathematical analysis is mostly relying on the associated non-linear Black and Scholes type equations. Using utility maximization methods, Constantinides and Zariphopoulou (2000) recently priced various kinds of exotic options as well as American instruments written on more than one security for investors with CRRA utilities.

Bibliography

[1.] H. Ahn, M. Dayal, E. Grannan and G. Swindle, *Option replication with transaction costs: general diffusion limits*, Annals of Applied Probability **8(3)** (1998), 676–707.

[2] M. Akian, J. L. Menaldi and A. Sulem, *Multi-asset portfolio selection problem with transaction costs. Probabilités numériques*, Mathematics and Computers in Simulation, **38**, (1992) 163–172.

[3] M. Akian, J. L. Menaldi and A. Sulem, *On an investment-consumption model with transaction costs*, SIAM Journal on Control and Optimization, **34** (1996), 329–364.

[4] M. Akian, P. Séquier and A. Sulem, *A finite horizon multidimensional portfolio selection problem with singular transactions*, Proceedings CDC, New Orleans **3** (1996), 2193–2198

[5] M. Akian, A. Sulem and M. Taksar, *Dynamic optimisation of a long-term growth rate for a mixed portfolio with transaction costs*, preprint (1996).

[6] C. Albanese and S. Tompaidis, *Small transaction costs asymptotics for the Black and Scholes models*, preprint (1998).

[7] O. Alvarez, *A singular stochastic control problem in an unbounded domain*, Communications in Partial Differential Equations, **19** (1994), 2075–2089.

[8] O. Alvarez and A. Tourin, *Viscosity solutions of nonlinear integro-differential equations*, Ann. Inst. Henri Poincaré, **13(3)** (1996), 203–317.

[9] A. Araujo and P. K. Monticro, *Equilibrium without uniform conditions*, Journal of Economic Theory, **48** (1989), 416–427.

[10] C. Atkinson and B. Al–Ali, *On an investment-consumption model with transaction costs: an asymptotic analysis*, preprint (1995).

[11] C. Atkinson, S. Pliska and P. Wilmott, *Portfolio management with transaction costs*, Proceedings of the Royal Society of London, A, to appear (1999).

[12] C. Atkinson and P. Wilmott, *Portfolio management with transaction costs: an asymptotic analysis of the Morton and Pliska model*, Mathematical Finance **5** (1995), 357–367.

[13] M. Avellaneda and A. Parás, *Optimal hedging portfolios for derivative securities in the presence of large transaction costs*, Applied Mathematical Finance **1** (1994), 165–193.

[14] M. Avellaneda and A. Parás, *Hedging financial derivatives in the presence of transaction costs: dynamic programming, nonlinear volatility and free boundary problems*, preprint (1997).

[15] G. Barles, J. Burdeau, M. Romano and N. Samsoen, *Critical stock price near expiration*, Mathematical Finance, **5(2)** (1995), 77–95.

[16] G. Barles, C. Daher and M. Romano, *Convergence of numerical schemes for parabolic equations arising in finance theory*, Report, Caisse Autonome de Refinancement, (1991).

[17] G. Barles and H. M. Soner, *Option pricing with transaction costs and a nonlinear Black and Scholes equation*, Finance and Stochastics, **2** (1998), 369–397.

[18] G. Barles and P. E. Souganidis, *Convergence of approximation schemes for fully nonlinear second order equations*, Journal of Asymptotic Analysis, **4** (1991), 271–283.

[19] A. Bensoussan and J. Frehse, *On Bellman equations of ergodic control in R^n*, J. Reine. Angew. Math., **429** (1992), 125–160.

[20] A. Bensoussan, J. Frehse and H. Nagai, *Some results on risk-sensitive control with full observation*, Journal of Applied Mathematics and Optimization, **1** (1998), 1–41.

[21] B. Bensaid, J. Lesne, H. Pagès and J. Scheinkman, *Derivative asset pricing with transaction costs*, Mathematical Finance, **2** (1992), 63–86.

[22] T. Bielecki and S. Pliska, *Risk sensitive asset management with transaction costs*, Finance and Stochastics, **4(1)** (1999).

[23] F. Black and M. Scholes, *The pricing of options and corporate liabilities*, Journal of Political Economy, **81** (1973), 637–654.

[24] P. P. Boyle and D. Emanuel, *Discretely adjusted option hedges*, Journal of Financial Economics, **8** (1980), 259–282.

[25] P. Boyle and T. Vorst, *Option replication in discrete time with transaction costs*, Journal of Finance, **47** (1992), 271–293.

[26] I. Capuzzo-Dolcetta and P.-L. Lions, *Hamilton-Jacobi equations with state constraints*, Transactions of the American Mathematical Society, **318** (1990), 543–583.

[27] G. M. Constantinides, *Multiperiod consumption and investment behavior with convex transactions costs*, Management Science, **25** (1979), 1127–1137.

[28] G. M. Constantinides, *Capital market equilibrium with transaction costs*, Journal of Political Economy, **94** (1986), 842–862.

[29] G. M. Constantinides, *Habit formation: A resolution of equity premium puzzle*, Journal of Political Economy, **98** (1990), 519–543.

[30] G. M. Constantinides and T. Zariphopoulou, *Bounds on prices of contingent claims in an intertemporal economy with proportional transaction costs and general preferences*, Finance and Stochastics, **3(3)** (1999), 345–369.

[31] G. M. Constantinides and T. Zariphopoulou, *Price bounds on derivative prices in an intertemporal setting with proportional costs and multiple securities*, submitted for publication (2000).

[32] J. Cox, *The constant elasticity of variance option pricing model*, Journal of Portfolio Management, special issue, Fisher Black Memorial, **23** (1996), 15–17.

[33] M. G. Crandall, H. Ishii and P.-L. Lions, *User's guide to viscosity solutions of second order partial differential equations*, Bulletin of the American Mathematical Society, **27** (1992), 1–67.

[34] M. G. Crandall and P.-L. Lions, *Viscosity solutions of Hamilton-Jacobi equations*, Transactions of the American Mathematical Society, **277** (1983), 1–42.

[35] D. Cuoco, *Optimal consumption and equilibrium prices with portfolio constraints and stochastic income*, Journal of Economic Theory, **72(1)** (1997), 33–73.

[36] J. Cvitanić and I. Karatzas, *Convex duality in convex portfolio optimization*, Annals of Applied Probability, **2** (1992), 767–818.

[37] J. Cvitanić and I. Karatzas, *Hedging contingent claims with constrained portfolios*, Annals of Applied Probability, **3** (1993), 652–681.

[38] J. Cvitanić and I. Karatzas, *On portfolio optimization under "drawdown" constraints*, IMA Journal of Applied Mathematics, **65** (1993a), 35–45.

[39] J. Cvitanić and I. Karatzas, *Hedging and portfolio optimization under transaction costs: a martingale approach*, Mathematical Finance, **6** (1996), 133–165.

[40] J. Cvitanić, H. Pham and N. Touzi, *A closed-form solution to the problem of super-replication under transaction costs*, preprint (1997).

[41] R.-A. Dana and M. Pontier, *On existence of an Arrow-Radner equilibrium in the case of complete markets. Two remarks*, Mathematics of Operations Research, **17** (1992), 148–163.

[42] M. H. A. Davis and J. M. C. Clark, *A note on super-replicating strategies*, Philosophical Transactions of the Royal Society of London A, (1994), 485–494.

[43] M. H. A. Davis and A. R. Norman , *Portfolio selection with transaction costs*, Mathematics of Operations Research, **15** (1990), 676–713,

[44] M. H. A. Davis and V. Panas, *The writing price of a European contingent claim under proportional transaction costs*, Mathematics of Computation, **13** (1994), 115–157.

[45] M. H. A. Davis, V. Panas and T. Zariphopoulou, *European option pricing with transaction costs*, SIAM Journal on Control and Optimization, **31** (1993), 470–493.

[46] M. H. A. Davis and T. Zariphopoulou, *American options and transaction fees*, Mathematical Finance, Springer-Verlag (1995).

[47] J. N. Dewynne, A. E. Whalley and P. Wilmott, *Path-dependent options and transactions costs*, Philosophical Transaction of the Royal Society of London A, **347** (1994), 517–529.

[48] D. Duffie, *Stochastic equilibria: existence, spanning number, and the "no expected gain from trade" hypothesis*, Econometrica, **54** (1986), 1161–1183.

[49] D. Duffie and L. Epstein, *Stochastic differential utility*, Econometrica, **60(2)** (1992), 353–394.

[50] D. Duffie, W. Fleming, H. M. Soner and T. Zariphopoulou, *Hedging in incomplete markets with HARA utility*, Journal of Economic Dynamics and Control, **21** (1997), 753–782.

[51] D. Duffie and C. F. Huang, *Implementing Arrow-Debreu equilibria by continuous trading of few long-lived securities*, Econometrica, **53** (1985), 1337–1356.

[52] D. Duffie and P.-L. Lions, *PDE solutions of stochastic differential utility*, Journal of Mathematical Economics, **21(6)** (1992), 577–606.

[53] D. Duffie and H. R. Richardson, *Mean-variance hedging in continuous time*, Annals of Applied Probability, **1** (1991), 1–15.

[54] D. Duffie and T. Zariphopoulou, *Optimal investment with undiversifiable income risk*, Mathematical Finance, **3** (1993), 135–148.

[55] B. Dumas and E. Luciano, *An exact solution to a dynamic portfolio choise problem under transaction costs*, Journal of Finance, **46** (1991), 577–595.

[56] J. Eberly, *Optimal consumption under uncertainty with durability and transaction costs*, Journal of Economic Dynamics and Control, to appear (1999).

[57] C. Edirisinghe, V. Naik and R. Uppal, *Optimal replication of options with transaction costs and trading restrictions*, Journal of Finance, **28** (1993), 117–138.

[58] N. El-Karoui, S. Peng and M. C. Quenez, *Backward stochastic differential equations in finance*, Mathematical Finance, **7(1)** (1997), 1–71.

[59] N. El Karoui and M. Jeanblanc-Piqué, *Martingale measures and partially observed diffusions*, Stochastic Analysis and Applications, **9(2)** (1991), 147–176.

[60] N. El Karoui and M. Jeanblanc-Piqué, *Optimization of consumption with labor income*, Finance and Stochastics, **2(4)** (1998), 409–440.

[61] S. Figlewski, *Options arbitrage in imperfect markets*, Journal of Finance, **44** (1989), 1289–1311.

[62] B. G. Fitzpatrick and W. H. Fleming, *Numerical methods for an optimal investment/consumption model*, Mathematics of Operations Research, **16** (1991), 823–841.

[63] W. H. Fleming, S. Grossman, J. L. Vila, J. L. and T. Zariphopoulou, *Optimal portfolio rebalancing with transaction costs*, preprint (1989).

[64] W.H. Fleming and S.-J. Sheu, *Asymptotics for the principal eigenvalue and eigenfunction of a nearly first-order operator with a large potential*, Annals of Probability, **25** (1997), 1953–1994.

[65] W. H. Fleming and S.-J. Sheu, *Optimal long term growth rate of expected utility of wealth*, Annals of Applied Probability, **9** (1999), 871–903.

[66] W. H. Fleming and H. M. Soner, Controlled Markov Processes and Viscosity Solutions, Springer Verlag, New York, (1993).

[67] W. H. Fleming and T. Zariphopoulou, *An optimal investment/ consumption model with borrowing*, Mathematics of Operations Research, **16** (1991), 802–822.

[68] B. Flesaker and L. P. Hughston, *Contingent claim replication in continuous time with transaction costs*, preprint (1994).

[69] E. R. Grannan and G. H. Swindle, *Minimizing transaction costs of option hedging strategies*, preprint, (1994).

[70] S. Grossman and G. Laroque, *Asset pricing and optimal portfolio choice in the presence of illiquid durable consumption goods*, Econometrica, **58(1)** (1989), 25–51.

[71] S. Grossman and J.-L. Vila, *Optimal dynamic trading strategies with leverage constraints*, Journal of Quantitative Financial Analysis, **27(2)** (1992), 151–168.

[72] H. He and H. Pagès, *Labor income, borrowing constraints, and equilibrium asset prices: A duality approach*, Economic Theory, **3** (1993), 663–696.

[73] H. He and N. D. Pearson, *Consumption and portfolios with incomplete markets and short-sale constraints: the finite dimensional case*, Mathematical Finance, **1(3)** (1991), 1–10.

[74] D. Heath, *A continuous-time version of Kulldorff's result*, preprint (1993).

[75] P. Henrotte, *Transactions costs and duplication strategies*, Graduate School of Business, Stanford University, preprint (1993).

[76] A. Hindy and C.F. Huang, *Optimal consumption and portfolio rules with duality and local substitution*, Econometrica, **61(1)** (1993), 85–121.

[77] S. D. Hodges and A. Neuberger, *Optimal replication of contingent claims under transactions costs*, The Review of Futures Markets, **8(2)** (1989), 222-239.

[78] T. Hoggard, E. Whalley and P. Wilmott, *Hedging option portfolios in the presence of transaction costs*, Advances in Futures and Options Research, **7** (1994), 21–35.

[79] C. F. Huang, *An intertemporal general equilibrium asset pricing model: the case of diffusion information*, Econometrica, **55** (1987), 117–142.

[80] H. Ishii and P.-L. Lions, *Viscosity solutions of fully nonlinear second-order elliptic partial differential equations*, Journal of Differential Equations, **83** (1990), 26–78.

[81] E. Jouini and H. Kallal, *Martingales and arbitrage in securities markets with transaction costs*, Journal of Economic Theory, **66** (1995), 178–197.

[82] Y. M. Kabanov and M. M. Safarian, *On Leland's strategy of option pricing with transactions costs*, Finance and Stochastics, **1** (1997), 239–250.

[83] I. Karatzas, *Lectures on the Mathematics of Finance*, CRM Monograph Series, AMS, (1997).

[84] I. Karatzas, *Adaptive control of a diffusion to a goal, and an associated parabolic Monge-Ampere-type equation*, Asian Journal of Mathematics, **1** (1997), 324–341.

[85] I. Karatzas, J. P. Lehoczky, S. E. Shreve, *Optimal portfolio and consumption decisions for a "small investor" on a finite horizon*, SIAM Journal on Control and Optimization, **25(6)** (1987), 1557–1586.

[86] I. Karatzas, P. Lakner, J. P. Lehoczky and S. E. Shreve, *Existence and uniqueness of multi-agent equilibrium in a stochastic, dynamic consumption/investment model*, Mathematics of Operations Research, **125** (1990), 80–128.

[87] I. Karatzas, P. Lakner, J. P. Lehoczky and S. E. Shreve, *Equilbrium models with singular asset prices*, Mathematical Finance, **15** (1991), 11–29.

[88] I. Karatzas, J. P. Lehoczky, S. E. Shreve and G. L. Xu, *Martingale and duality methods for utility maximization in an incomplete market*, SIAM Journal on Control and Optimization, **29** (1991), 702–730.

[89] I. Karatzas, J. Lehoczky, S. Sethi and S. Shreve, *Explicit solution of a general consumption/investment problem*, Mathematics of Operations Research, **11** (1987), 261–294.

[90] P.-F. Koehl, H. Pham and N. Touzi, *Option pricing under transaction costs: a martingale approach*, preprint, CREST, Paris (1996).

[91] H.-K. Koo, *Consumption and portfolio selection with labor income II: The life cycle-permanent income hypothesis*, Working Paper, Department of Finance, Washington University, (1991).

[92] H.-K. Koo and T. Zariphopoulou, *Optimal consumption and investment when opportunitie are better for the rich than for the poor*, Proceedings of International Conference in Finance, AFFI, Geneva, Switzerland (1996).

[93] M. Kulldorff, *Optimal control of favorable games with a time limit*, SIAM Journal on Control and Optimization, **31(1)** (1993), 52–69.

[94] N. Kutev and P.-L. Lions, *Nonlinear second order elliptic equations with jump discontinuous coefficients. I. Quasilinear equations*, Differential Integral Equations, **5(6)** (1992), 1201-1217.

[95] S. Kusuoka, *Limit theorem on option replication cost with transaction costs*, Annals of Applied Probability, **5** (1995), 198–221.

[96] D. Lamberton, H. Pham and M. Schweizer, *Local risk-minimization under transaction costs*, Mathematics of Operations Research, **23** (1998), 585–612.

[97] H. E. Leland, *Option pricing and replication with transaction costs*, Journal of Finance, **40** (1985), 1283–1301.

[98] S. Levental and A. Skorohod, *On the possibility of hedging options in the presence of transaction costs*, Annals of Applied Probability, **7** (1997), 410–443.

[99] H. Levy, *Equilibrium in an imperfect market: A constraint on the number of securities in the portfolio*, American Economic Review, **68** (1978), 643–658.

[100] P.-L. Lions, *Optimal control of diffusion processes and Hamilton-Jacobi-Bellman equations 1: The dynamic programming principle and applications; 2: Viscosity solutions and uniqueness*, Communications in Partial Differential Equations, **8** (1983), 1101–1174; 1229–1276.

[101] K. Lott, *Ein Verfahren zur Replikation von Optionen unter Transaktionskosten in stetiger Zeit*, Ph.D. Thesis, Universitat der Bundeswehr, Müchen (1993).

[102] M. J. P. Magill and G. Constantinides, *Portfolio selection with transaction costs*, Journal of Economic Theory, **13** (1976), 245–263.

[103] A. Mas-Colell, *The theory of general economic equilibrium: A differentiable approach*, Econometric Society Monograph, Cambridge University Press, (1985).

[104] A. Mas-Colell, *The price equilibrium existence problem in topological vector lattices*, Econometrica, **54** (1986), 1039–1053.

[105] M. Mazaheri, *Derivative pricing with stochastic volatility via a utility method*, preprint (1998).

[106] W. M. McEneaney, *A robust control framework for option pricing*, Mathematics of Operations Research, **22 (1)** (1997), 202–221.

[107] F. Mercurio, *Option pricing and hedging in discrete time with transaction costs*, Mathematics of derivative securities, Newton Institute, Cambridge University Press, Cambridge (1997).

[108] F. Mercurio and T. C. F. Vorst, *Option pricing and hedging in discrete time with transaction costs and incomplete markets*, M. A. H. Dempster and S. R. Pliska, eds, Mathematics of Derivative Securities, Cambridge University Press (1997), 190–215.

[109] R. C. Merton, *Lifetime portfolio selection under uncertainty: the continuous-time case*, Journal of Economic Theory, **3** (1969), 247–257.

[110] R. C. Merton, *Optimum consumption and portfolio rules in a continuous-time model*, Journal of Economic Theory, **3** (1971), 373–413.

[111] R. C. Merton, *An intertemporal capital asset pricing model*, Econometrica, **41** (1973), 867–887.

[112] R. C. Merton, *Theory of rational option pricing*, Bell Journal of Economics and Management Science, **4** (1973a), 141–183.

[113] R. C. Merton *Continuous Time Finance*, Basil Blackwell, Oxford, UK (1990).

[114] B. Mohamed, *Simulations of transaction costs and optimal rehedging*, preprint (1994).

[115] A. Morton and S. Pliska, *Optimal portfolio management with fixed transaction costs*, Mathematical Finance, **5(4)** (1995), 337–356.

[116] K. Munk, *The valuation of contingent claims under portfolio constraints: reservation buying and selling prices*, preprint (1999).

[117] S. Perrakis and P. J. Ryan, *Option pricing bounds in discrete time*, Journal of Finance, **39** (1984), 519–525.

[118] A. Pichler, *On transaction costs and HJB equations*, preprint (1996).

[119] E. Platen and R. Rebolledo, *Pricing via anticipative stochastic calculus*, Advances in Applied Probability, **26(4)** (1994), 1006–1021.

[120] E. Platen and R. Rebolledo, *Principles for modelling financial markets*, Journal of Applied Probability, **33(3)** (1996), 601–613.

[121] S. Pliska and M. Selby, *On a free boundary problem that arises in portfolio management*, Mathematical Models in Finance, Edited by D. Howison, F. P. Kelly and P. Wilmott, Chapman and Hall, The Royal Society, (1995), 555–561.

[122] H. Reisman, *Black and Scholes pricing and markets with transaction costs: an example*, Technion-Israel Institute of Technology, Haifa, preprint (1998).

[123] P. H. Ritchken, *On option pricing bounds*, Journal of Finance, **40** (1985), 1219–1233.

[124] M. Schroder, *Optimal portfolio selection with fixed transaction costs*, preprint (1993).

[125] M. Schroeder and C. Skiadas, *Optimal consumption and portfolio selection with stochastic differential utility*, Journal of Economic Theory, **89(1)** (1999), 68–126.

[126] M. Schweizer, *Hedging of options in a general semimartingale model*, Dissertation ETH Zürich, **8615** (1988).

[127] S. Sethi, *Optimal Consumption and Investment with Bankruptcy*, Kluwer Academic Publishers, Norwell, MA (1997).

[128] Q. Shen, *Bid-ask prices for call options with transaction costs*, Working paper, University of Pennsylvania (1990).

[129] S. E. Shreve and H. M. Soner, *Optimal investment and consumption with transaction costs*, Annals of Applied Probability, **4(3)** (1994), 206–236.

[130] H. M. Soner, *Optimal control with state space constraints*, SIAM Journal on Control and Optimization, **24** (1986), 552–562, 1110–1122.

[131] H. M. Soner, S. Shreve and J. Cvitanić, *There is no nontrivial hedging portfolio for option pricing with transaction costs*, Annals of Applied Probability, **5(2)** (1995), 327–355.

[132] A. Sulem, *Dynamic optimization for a mixed portfolio with transaction costs*, Numerical methods in Finance, Newton Institute, Cambridge University Press, Cambridge, (1997).

[133] L. E. O. Swensson and I. Werner, *Non-traded assets in incomplete markets*, European Economic Review, **37** (1990), 1149–1168.

[134] M. Taksar, M. J. Klass and D. Assaf, *A diffusion model for optimal portfolio selection in the presence of brokerage fees*, Mathematics of Operations Research, **13** (1988), 277–294.

[135] C. Tiu and T. Zariphopoulou, *On level curves of value functions in optimization models of expected utility*, Mathematical Finance, in press.

[136] K. B. Toft, *On the mean-variance tradeoff in option replication with transactions costs*, Journal of Financial and Quantitative Analysis, **31** (1996), 233–263.

[137] A. Tourin and T. Zariphopoulou, *Numerical schemes for investment models with singular transactions*, Computational Economics, **7** (1994), 287–307.

[138] A. Tourin and T. Zariphopoulou, *Portfolio selection with transactions costs*, Progress in Probability, **36** (1995), 385–391.

[139] A. Tourin and T. Zariphopoulou, *Viscosity solutions and numerical schemes for investment/ consumption models with transaction costs*, Numerical methods in Finance, Newton Institute, Cambridge University Press, Cambridge, (1997) 245–269.

[140] A. Tourin and T. Zariphopoulou, *Super-replicating strategies with probability less than one in the presence of transaction costs*, preprint (1998).

[141] J. L. Vila and T. Zariphopoulou, *Optimal consumption and portfolio choice with borrowing constraints*, Journal of Economic Theory, **7** (1997), 402–431.

[142] A. E. Whalley and P. Wilmott, *Hedge with an edge*, Risk Magazine, (October 1994).

[143] A. E. Whalley and P. Wilmott, *A review of key results in the modeling of discrete hedging and transaction costs*, Frontiers in Derivatives, Eds. Konishi and Dattatreya, (1996).

[144] A. E. Whalley and P. Wilmott, *Optimal hedging of options with small but arbitrary transaction cost structure*, preprint (1997).

[145] A. E. Whalley and P. Wilmott, *An asymptotic analysis of the Davis, Panas and Zariphopoulou model for option pricing with transaction costs*, Mathematical Finance, **7** (1997), 307–324.

[146] P. Wilmott, *Discrete charms*, Risk Magazine, (March 1994).

[147] T. Zariphopoulou, *Investment-consumption models with constraints*, Ph.D. Thesis, Brown University, (1989).

[148] T. Zariphopoulou, *Investment/consumption model with transaction costs and Markov-chains parameters*, SIAM Journal on Control and Optimization, **30** (1992), 613–636.

[149] T. Zariphopoulou, *Investment and consumption models with constraints*, SIAM Journal on Control and Optimization, **32** (1994), 59–84.

[150] T. Zariphopoulou, *Optimal investment and consumption models with nonlinear stock dynamics*, Mathematical Methods of Operations Research, **50** (1999), 271–296.

Index

Milton Keynes UK
Ingram Content Group UK Ltd.
UKHW051859071024
449327UK00025B/2023